SPACEFLIGHT AND ROCKETRY

A CHRONOLOGY

SPACEFLIGHT AND ROCKETRY

A CHRONOLOGY

David Baker

Facts On File, Inc.

AN INFOBASE HOLDINGS COMPANY

Spaceflight and Rocketry: A Chronology

Copyright © 1996 by David Baker

Facts On File, Inc.
11 Penn Plaza
New York NY 10001

Library of Congress Cataloging-in-Publication Data
Baker, David, 1944–
    Spaceflight and rocketry : a chronology / David Baker.
      p. cm.
    Includes index.
    ISBN 0-8160-1853-7
    1. Astronautics—History—Chronology. 2. Rocketry—History—
Chronology. I. Title.
TL788.5.B25 1996
629.4′09—dc20
                                  96-41843
                                    CIP

Text design by Ron Monteleone
Jacket design by Alice Soloway

Printed in the United States of America

VB TT 10 9 8 7 6 5 4 3 2 1

This book is printed on acid-free paper.

In memory of Paul,
who would have been proud to see what was accomplished.

CONTENTS

ACKNOWLEDGMENTS

I am indebted to so many people in so many countries for the primary material, collected over 30 years, that has gone into this chronology. To single out any one would be unfair to the rest. They know who they are; and to each and every one of you, my heartfelt gratitude. This book would not have happened without my publisher, who held firm when the ship hit storms and refused to turn back. Especially, my thanks to Jamie Warren for giving me the chance, to Lincoln Paine, whose hand was always on the tiller through fair weather and foul, and to Andrew Libby and Kerstin Moeller, who successfully navigated the shallows where rocks were successfully avoided. Finally, on a very personal note, my heartfelt appreciation to those closest to me who stood through thick and thin. Thank you Jane, and Tom, Alex and Rachael. Any errors found in this book are mine alone and not the responsibility of these wonderful people who made it all come together.

INTRODUCTION

As with all forms of technology, rocketry has been applied to both civil and military uses. It is impossible to separate one from the other, and no attempt has been made to do so in this book. What began as an interesting experiment in reactive flight matured quickly into rockets for warfare and fireworks for exciting public displays and offshore rescue duties. Beyond that, with a more scientific and purposeful approach, by the mid-nineteenth century the powder rocket had grown into a useful military weapon. But realization of the rocket's potential would await the invention of liquid propellant motors during the 1920s and 1930s.

It was only with the introduction of powerful motors burning liquid and solid propellants that long-range missiles became reality during the 1940s and 1950s. Towards the end of World War II, the German V-2 ushered in the age of the ballistic missile and potential access to space. From 1945, with the Cold War between East and West fanning the fires of ideological competition, the ballistic missile grew into an efficient delivery system for the first generations of atomic and thermonuclear weapons.

Rockets still had a role in the peaceful expansion of knowledge, as sounding probes and instrumented capsules shot high into the atmosphere. But it was the logical extension of the sounding rocket into a permanently orbiting satellite equipped with scientific instruments that sparked the space race. Had the United States been the first to put a satellite in orbit the race may never have begun. But it was not and, on October 4, 1957, a beeping ball from the Urals put fear into the hearts and minds of free men everywhere.

Neither President Dwight D. Eisenhower nor Premier Nikita Khrushchev thought much about the value of satellites until they witnessed their impact on an unsuspecting public; only when hurt pride sent U.S. politicians scurrying for an anvil upon which to hammer Soviet communism did the challenge become a contest. Warning rhetorically of a missile gap that did not exist in reality, John F. Kennedy promised revitalization of the nation's defenses, and when he became president in January 1961, he kept that promise to the letter.

Within six months, Kennedy had set in motion an arms race that would last three decades, and mobilized the nascent space program into a giant assault on Soviet ambitions. That they could lead the march into space had long been a dream of Russian scientists and engineers, epitomized in writings by the "father" of their space program, the nineteenth-century mathematics teacher Konstantin E. Tsiolkovsky. That they would be challenged in the pursuit of that goal was the implied right of lawmakers who set up the National Aeronautics and Space Administration (NASA) for America to be "preeminent in space."

In all its more important aspects, the race with the Russians was a phenomenon of the 1960s alone, a decade during which the moon landing goal was achieved and, in public perception at least, capitalism was seen to beat communism. At the start of the decade NASA was little more than a research and development organization built on the back of the government's National Advisory Committee for Aeronautics (NACA). By the end of the decade it had reached its peak and was already in decline. In spite of having achieved the goal Kennedy had set for it, legislators saw no enduring need for NASA to administer even bolder challenges.

Engineers and scientists quickly realized, to their genuine astonishment, that what Congress had approved in 1961 had been a one-shot boost for U.S. prestige and not a commitment to a succession of new and bolder initiatives in space. Pressure on the budget from Lyndon Johnson's Great Society programs and the war in Vietnam combined to destroy the vision of a space-based nation marching toward the cosmos. Disenchanted with technology, at odds with their country's role in Southeast Asia, the American public grew tired of spectacular acts in the weightlessness of space.

To the majority of Americans, the space program became increasingly irrelevant in a world of fuel crises and civil unrest at home and abroad. Conducted by a small group of privileged participants, the nation was made spectator to events far removed from its daily life. Yet because of this, past achievements that had accumulated year by year became a symbol of what America could achieve, given the will. NASA itself became almost a national monument to superhuman achievement, epitomizing the can-do spirit of an era writ large in retrospect, in which motives and intentions gave way to pride and nostalgia.

Paradoxically, preoccupation with environmental threats stemmed, in part, from perception of the Earth as a fragile sphere in "the vast emptiness of space" and by the discovery from satellite data that the ozone layer was being eroded by artificial pollutants. This, and the stunning quality of images showing damage done by human activity around the globe, encouraged pressure groups in a campaign against unrestricted abuse of the environment. Instead of being its salvation, many in society saw technology, epitomized in the space program, as destructive of their world and their future. And the space program responded accordingly. The 1970s were to be a decade of diversification—into remote-sensing satellites for taking stock of Earth's natural resources, into a broader exploration of the planets in the solar system to learn more of our planet's evolution and into the first commercial space endeavors with communication satellites.

By the 1980s application, rather than exploration, had become a cornerstone of the space program, and a private industry had evolved, with investors giving more to space programs than the U.S. taxpayer was willing to give to NASA. In the uncertain politics of that decade, Reaganomics and the fear of being made hostage to unrestrained Soviet power justified the commitment to weapons modernization and new goals: a space-based screen against ballistic missiles, called the Strategic Defense Initiative, and a space station called *Free-*

dom. But the SDI program was never finished and *Freedom* never flew.

What happened at the end of the 1980s and the beginning of the decade that followed brought about the collapse of Soviet communism and the totalitarian empire of beleaguered client states held rigid through fear. Ordinary people from the Baltic to the Black Sea, from Central Europe to the Sea of Japan, stood up and seized democracy for themselves, leaving their oppressors impotent and without power. There was no longer a place for the tyranny they had administered. That, in turn, collapsed the mechanism of competition between the superpowers. Space, for so long the virility symbol of science, engineering and technology, was no longer a prerequisite for international power.

The 1990s have become the decade of partnership, with plans for an international space station involving Russian, American, European and Japanese participation. Gone are the reasons that first led NASA to want a space station, for there is now no Cold War opponent whose achievements must be matched. And there is no longer profit for politicians in government-run space ventures. Instead, the burgeoning direct-broadcast and satellite communications industry alone brings greater annual revenue than NASA spends each year on operations, research and development.

The big programs of the 1970s (the Space Shuttle, moon missions, nuclear rocket stages and plans for manned flight to Mars) are now seen as sideshows. Even the reusable Shuttle is no longer believed to be a key element in the orbital assembly of international research modules planned for the turn of the century. Instead, it will be used because it is there, not because it is an essential system, and most space station hardware will probably fly on expendable rockets which are more efficient and simpler to operate. For their part, the Russians have already mothballed their own shuttle and the Europeans have cancelled a space plane called Hermes.

In 1957 it would have been inconceivable to imagine that, having begun the race into space, human beings would turn their back on the exploration of other worlds to reunite themselves with Earth. But this is what has happened, and only a token commitment to low-cost planetary exploration remains funded in the world's space programs. It is most unlikely that humans will walk the red dust of Mars during the lifetime of anyone living today. Priorities have changed and society has moved on from the driving optimism of the 1960s. But grand visions and ambitious goals are not the sole expression of a space program.

If there is a single legacy that endures and will grow, it is in the practical uses of space vehicles: weather satellites that save lives through timely warning of impending storms; the satellites that have brought a revolution in communication, between people and among people, by shrinking the world to a global village and slashing the cost of telephone calls; television and broadcast satellites that fertilize cross-cultural links with images, news and entertainment; remote-sensing satellites that compile global inventories of natural resources, record abuse of the environment and monitor pollution levels; navigation satellites that pilot lost ships to safe harbor; search-and-rescue satellites that relay distress calls from remote areas; data relay satellites that can transmit personal medical data to a physician in a foreign country when a patient falls ill on the far side of the globe, and so on.

And then there is "spin-off," that anomalous catalog of designs and inventions developed for space exploration but with practical applications here on Earth: the lives saved by cleaner operating theaters made possible by improvements in spacecraft sterilization and clean-room facilities; better and safer circuit breakers for houses; lighter and more effective thermal insulation materials; protective gear for explorers in extreme climates; miniaturized electronics for better and more reliable heart pacemakers; new materials for lighter and stronger structures to make safer and more fuel-efficient cars, airplanes and appliances; better artificial limbs for amputees, a technology derived from composite materials; or microfibers for muscle connections.

In the broader fields of human knowledge, from our own cosmic backyard to the deepest part of the universe, satellites and spacecraft have revealed the very fabric of our celestial home. It was America's first satellite, *Explorer I*, that discovered radiation around Earth trapped in belts that got their name from the scientist who first found them—James A. Van Allen. From that discovery came a long line of science satellites to explore the environment of the solar system and map radiation from the sun, the so-called "solar wind." And it was from that deep-space survey that the effect of the sun on Earth's atmosphere could be measured, forming a basis for understanding our planet.

It is to the astronomical sciences that space engineering has brought some of the greatest strides in human knowledge. Optical and nonoptical observatories seeking to map the universe in all portions of the energy spectrum have for three decades brought discoveries to shrink forever our limited view of the heavens obtained from Earth-based observatories. Quasars, pulsars, black holes, white holes, objects seen but not yet named, have all been discovered or explored by space-based observatories.

Since the early 1950s, when the only funded space programs were spy satellites developed in great secrecy, military space projects of the United States and the Soviet Union (and later Russia) have taken a major share of resources. Despite accusations to the contrary, no spacefaring power has yet placed weapons of destruction in space. Military satellites have revolutionized the applications of modern technology in war and made impressive strides in communication and navigation. The ability to target enemy forces exactly and precisely, coupled with the use of very accurate weapon systems, has led to a unique ability to minimize collateral damage and reduce casualties.

If it were not for reconnaissance and surveillance satellites, detailed inventories of global arms distribution would be almost impossible to compile with accuracy. Arms control agreements would be unworkable and the world would be a more unstable place than it is at present. Moreover, open channels between the superpowers ensure consultation at the highest level, leaving no room for misunderstanding the intentions of a potential aggressor.

Conversely, the proliferation of nuclear weapons technology among developing countries with unstable political regimes brings new threats of sudden attack and unpredictable conflict. Moreover, ballistic missiles have found their way into the armies of unscrupulous power barons seeking delivery systems for chemical, biological and nuclear weapons. In short, a bipolar world held hostage to East-West tension and the threat of a devastating nuclear war has been exchanged for a less stable, multipolar world where political alliances are fragmented and in flux.

And what of the future? Prophets got it wrong when, in the 1930s, they said it would take a hundred years to put men on the moon, and they were equally wide of the mark in the 1960s when proponents said people would be living on Mars during the 1980s. For 30 years there were two great spacefaring competitors, the Soviet Union and the United States. Now, that elite duo has been joined by the European group of nations and their separate national space programs, as well as by Japan, China, India and a host of aspiring countries with nascent satellite and rocket manufacturing organizations. It is unwise to be dogmatic when projecting future events and their timing. But the far reach is predictable.

The political imperative to mobilize armadas of spaceships and celestial cargo vessels bound for other worlds of the solar system is no longer present. For several decades into the future, broad-based exploitation of near-Earth space for giant TV, broadcast and communication satellites will predominate along with improved weather and environmental monitoring platforms. Long-term research into new vaccines, exotic materials and uniquely pure semiconductor crystals will be performed aboard the international space station, which the international partners expect to use for 20 years or more.

Beyond that, at some point in the future, expeditions will leave Earth to colonize other worlds as assuredly as the first satellite was bound to follow the development of liquid propellant rockets. When that time will come is uncertain at present, but when it does the select band chosen to represent humanity will "stand on the shoulders of giants." In a very real sense they will begin again the grand adventure that for one golden moment in 1969 seemed so near when two American astronauts set foot on the moon. Only then will that pinnacle in human achievement be fully understood for what it truly was: a landmark in human evolution, as intelligent beings walked on another world for the first time.

FOR THE READER

Over more than 36 years of the space age, through December 31, 1993, world spacefaring powers had launched 4,614 satellites and spacecraft of which more than 2,200 were still in space. Of the total, the Soviet Union and Russia had been responsible for 2,947 and the United States for 1,303. The remaining 364 had been built or purchased, mostly from manufacturers in the United States, by 31 countries or international organizations.

Many books have been written about the space program and chronologies have recorded progress in developing satellites and spacecraft for scientific research and military applications. The aim of this book is to provide a comprehensive history of rockets, missiles and space vehicles in a way that highlights incremental steps in development and ensuing stages in operations. In adapting this chronological format, it has been necessary to take liberties.

Some remote, obscure or cancelled projects appear to get disproportionate attention while other better known space ventures get very little. This is because the intention has been to compile a record of rocket and space activities that have contributed measurably to the general advancement in technology or operational capabilities. It is not simply a compilation of events and does not purport to record every launch.

For instance, the NERVA (nuclear engine for rocket vehicle application) nuclear rocket project absorbed more than $1.5 billion in 1960s research and development money but was cancelled by the Nixon administration before it achieved flight status. Nevertheless, the investment provided a wealth of test data which is now being exploited for potential nuclear rocket projects applicable to the needs of the next century.

For this reason NERVA is a very important benchmark in the development of a unique capability, and it gets significant coverage. Conversely, highly publicized Shuttle missions get less word allocation per flight than individual moon landings because each moon mission made a proportionately greater contribution to spaceflight operations and scientific knowledge.

All chronologies are subjective and reflect the opinions of the author. Wherever possible, the political and policy infrastructure attending selected decisions have been included with technical and engineering coverage. The author has taken the view that policy and politics have been as important to the ebb and flow of space achievement as the technical or engineering capabilities of a particular group. Because of that, the index will allow the reader to track political and programmatic developments as well as the operational story.

In the last several years much mystery surrounding Soviet space programs has evaporated in the wake of changes to leadership in the former Soviet Union. Much remains unknown, however, and piecing together the details of many Soviet accomplishments, and failures, is an ongoing business that will keep Russophiles occupied for years. The reader is asked to accept the inevitability that some information available when these words are read was still unknown when they were written. Nevertheless, an effort has been made to include as much as possible about programs previously veiled in secrecy, both here in the United States and in the former Soviet Union. In some cases—the details of early military reconnaissance satellites provide an example—information has been declassified especially for this chronology. For all that, however, there is a lot of information that cannot be declassified for publication because it relates to programs vital to current national security.

Some information has been included that does not directly affect the main flow of technological development but adds dimension or perspective to the primary event with which it is associated. Of such a category is the skein of events associated with the development of nuclear warheads. Rarely is their story chronicled, but to ignore them would be to repeat the mistake earlier books have made in not describing the reason why long-range ballistic missiles have been built: to deliver thermonuclear warheads across great distances with extreme accuracy.

Where several satellites in a common series—such as Intelsat communication satellites—have been launched over a period of several years, the first in the series is detailed and that entry provides reference dates for successive flights. Where large numbers of satellites have been launched in a common series, only the annual quantities are recorded to provide a general impression of the intensity of effort over a given period.

References to events where only the day and the month are given always apply to the year in which the host event is recorded, thus saving space and unnecessary repetition of numbers. If an event occurs in successive years, the full date is recorded. Entries that contain projected schedules or proposed launch dates always refer to expectations, and readers are advised to follow the actual sequence of events by accessing successive references to the same project in the index.

Title, rank and position have been applied as they were at the date in question. In general, deaths are recorded only for astronauts and for administrators or politicians for whom institutes or space facilities have been named.

Toward the end of the chronology some attempt has been made to advise readers of scheduled events coming up in the mid-1990s, but these can, at best, express plans and projections only as they were in 1994 and not the reality of changes and updates as time progresses. Even a cursory glance at the difference in past years between planned events and actual accomplishments will reveal how difficult it is to predict the future of any space program.

The author would like the reader to regard this book as a starting point for further research, and to that end there are a number of respected organizations to which he or she should refer for additional information or research opportunities. In the United States, the Smithsonian Institution in Washington,

D.C., presides over many national archives and records associated with the early days of rocketry. NASA History Offices at NASA headquarters in Washington, D.C., or at the field centers throughout the nation will provide a wealth of material for bona fide research. In the United Kingdom, the British Interplanetary Society provides publications and limited library facilities for books and documents.

In addition, there are many popular magazines available which provide useful information on space activities, including *Scientific American, Science Digest, Astronomy* and *Popular Science,* to name but a few. More specialized journals of space history are *Spaceflight,* from the British Interplanetary Society at 27/29 South Lambeth Road, London SW8 1SZ, England, and *Quest*, which can be purchased from CSPACE, P. O. Box 9331, Grand Rapids, Mich. 49509–0331, USA.

Spaceflight and Rocketry:
A Chronology

360 B.C. The earliest recorded application of the principle of action and reaction was embodied in a toy described by Aulus Gellius in his *Noctes atticae* as the ancient pigeon of Archytas. Suspended by a cord, a hollow model pigeon containing water was made to move by the action of steam generated by a small fire underneath the model and vented from a hole at the back. Later, around A.D. 62, Hero, a Greek living in Alexandria, Egypt, described in his book *Pneumatica* a primitive steam turbine in the form of a reaction wheel—a hollow globe pivoted on two free-rolling joints and made to rotate by the action of steam passing through two vent pipes bent at right angles and exhausting in opposing directions.

325 B.C. Retreating across the Punjab, in what is now northwestern India, Alexander the Great encountered "storms of lightning and thunderbolts hurled upon them from above," according to the Greek Sophist Flavius Philostratus writing 500 years later. Opinion differs on the interpretation of this statement; some historians hold the view that he encountered powder rockets. Alexander is more likely to have been deterred by less exotic means; the first confirmed use of rockets in India did not occur until A.D. 1399.

A.D. 160 In the first known fictional account of a journey through space, the Greek satirist Lucian of Samosata (in Syria) wrote a book titled *True History* which is in essence a parody of Homer's *Odyssey*. In the story a ship is lifted by a great wind and carried to the moon, where its occupants find people and animals. The story was written when Greek philosophers were coming to believe that, like the Earth, the moon was a solid body. This had an important effect on accepting the possibility of travel to at least one other world. The *True History* was republished in the 17th century when a renaissance in scientific thinking was about to oust classical teaching.

200 In his *Mémoire concernant l'histoire, les sciences, les arts, etc. des Chinois* (Memoirs on the History, Science, Arts, etc. of the Chinese), the French missionary Father Joseph Marie Amiot recorded that "firearms" were used in China around this time. Contemporary Chinese descriptions explain the operation of "fire arrows," with a length of about 4 in., which when lit will "go against the wind." There is no clear description of a self-propelled rocket, and many of these pyrotechnic devices may have been lobbed by the action of a catapult or adapted form of trebuchet, a machine for hurling stones, rockets or flaming baskets of fire.

682 The Chinese writer Sun Saumiso died, leaving evidence in his book on medicinal treatments that exploding powder was used in his time. He describes a combination of saltpeter and sulfur with acacia seeds added and then lit. Other writings confirm that knowledge about the mixing of black powder may have been well established in this period. Considering the secretive manner in which such formulas were handed down just a few centuries later, reluctance to record the chemistry openly is to be expected.

969 In China, Yo I-fang presented a "fire arrow" to Emperor T'ai Tsung and received a gift of silk in return. T'ai Tsung's brother had founded the Sung Dynasty (960–1279), and it was in this period that Chinese chemists developed a range of pyrotechnic devices for military purposes. The exact date of the rocket's origin is unknown, although the use of "fire arrows" at the defense of the city of Tzu T'ung is reported to have turned back a besieging force of 100,000 men. In 1002 it is reported that Emperor Chen Tsung, T'ai Tsung's successor, employed flaming projectiles with a range of 1,000 ft.

1045 The first recorded description of the preparation of gunpowder and the making of "flaming arrows" appeared with the publication in China of the *Wu-ching Tsung-yao* (Complete Compendium of Military Classics). In it Tseng King-Liang explains how a wide range of weapons works, and he details the preparation of explosives which, he asserts, have been in use for a long time. The work contains details of a mixture containing sulphur, saltpeter, charcoal, pitch and dried varnish. It is unclear whether the author refers to exploding devices or to the use of black powder in primitive fireworks and rockets.

1232 In one of the earliest recorded uses of rockets, the Chinese defended the besieged town of Kai-fung fu by sending

In A.D. 62 the Greek writer Hero described a steam-driven reaction device which demonstrated the principle upon which all rocket devices operate: To every action there is an equal and opposite reaction.

out "arrows of flaming fire" against the Mongols. From subsequent reports it appears that the device had a charge of black powder that on impact spread fire over an area 2,000 ft. across. There is also mention of the Chinese having made bombs that exploded on impact.

1241   The first use of gunpowder outside China came at the battle of Sejó preceding an attack on Budapest. A variety of war machines were employed here, including a device for creating foul smoke. There is the possibility that rockets were used here too, although the reference is vague and may refer only to the use of gunpowder for explosive purposes, not to the use of black powder in tubes. Many of the pyrotechnic devices so terrified the recipients that their initial reactions were clouded by fright, and clear and concise descriptions of specific weapons used against them are difficult to find.

1248   French soldiers encounter rockets for the first time during an attempted occupation of the eastern branches of the Nile River delta in Egypt. Reported in a book written in about 1268 by the French historian Sire de Joinville, the incident occurred while the French were attempting to take Damietta. The Arabs apparently hurled a flat object described as an "egg" that, having landed on the French side of the river, came hurtling across the ground toward them spitting fire through tiny openings. Propelled by three rockets, two of which were aligned to give the device stability in flight, the "egg" contained black powder and several fuses set to ignite separate chambers within as it went.

At about this date the English monk Roger Bacon (1214–ca. 1294) wrote a detailed account of the preparation of black powder. In veiled terms and with a sense of secrecy, Bacon wrote down the formulas and chemical composition in rough Latin so that only those with some knowledge of the subject could extract its meaning. In this way he hoped, naively, to control the proliferation of detailed knowledge about powder and rockets. Bacon had experimented with the chemical mixtures and modified the Chinese formulas to improve performance, setting a trend that would continue in England and continental Europe. A German contemporary of Bacon, Albertus Magnus (1193 or 1206–1280), wrote a more open account of how to make black powder and rockets.

1258 February 15   Mongol rockets containing black powder were used against Arabs at the capture of Baghdad, according

Mongol hordes attacked a small town along the Silk Route using rockets to terrorize the inhabitants and frighten the horses, as depicted in this artist's rendition.

to the writer Rashid al-Din. The rockets are described as having had a lance, or stabilizing stick, with a wick on the opposite side. This marks the introduction of rockets to southwest Asia for war fighting purposes, and Arab works only a little later than this describe the detailed preparation of black powder and rocket tubes with accuracy. Believed to have been written by the noted Syrian historian al-Hasan al-Rammah Nedjm al-din al-Ahdab between 1285 and 1295, the *Kitab al-Furisiya Wal Munasab al-Harbiya* (Treatise on Horsemanship and War Stratagems) provides a detailed account of the preparation of "materials for the sending of fire."

1270   About this year, the Greek writer Marcus Graecus prepared his *Liber ignium ad comburendos hostes* (Book of Fires and Burning the Enemy). It provides a highly detailed account of the different techniques used to prepare black powder and manufacture rockets in several countries and in different ages. Graecus also discusses the military uses of rockets and enters into a discourse on how different case designs affect the way the rocket will fly to its target and how chemical mixtures influence the duration and intensity of the burn and thereby the effectiveness of the weapon.

1271   The Chinese used flaming projectiles during the battle of Siang-yang-fu, and there is little doubt that these were rockets packed with black powder. The composition of the powder probably used in these tubes was 61% saltpeter, 18.3% charcoal, 18.3% sulphur and 2.4% a mixture referred to in Chinese writings as *Mi-to-sing*. Flaming projectiles, or rockets, were also used during a battle between the Sung and Yuan forces in 1274 and 1275, while in 1274 the Japanese are said to have encountered fire arrows launched from Mongolian ships at the battle of Tsu Shima. They were again used in the Chinese invasion of Japan in 1281. From this experience the Japanese began to develop their own rockets.

1324   In the first recorded instance of rockets having reached the westernmost parts of continental Europe, the Arabs used flaming projectiles in an attack on the small town of Huascar about 70 mi. northeast of Grenada, Spain. An eyewitness account of the battle indicates that the "Armies of Ishmael" used rocket bombs against the fortress tower. But he may have been describing an explosive projectile hurled by a cannon or a catapult. It is difficult to distinguish between Greek fire and rockets because similar descriptions were used for each.

1398   Possibly the first use of rockets on the Indian subcontinent took place when the marauding Mongol warrior Tamerlane attacked the city of Delhi. In one of the most violent, bloody and senseless military campaigns ever fought, Tamerlane is said to have used pyrotechnic devices to attack the besieged defenders. Accounts vary on the type of device used, and some contemporary writers refer only to small fire pots and grenades which may have been hurled by archers.

1420   An Italian military engineer, Joanes de Fontana, completed a sketchbook of ambitious machines for warfare. *Bellicorum instrumentorum liber* (Book of the Instruments of War) describes a variety of rocket-propelled devices. Some he disguises as floating fish, flying pigeons and running rabbits, while others, such as spiked battering rams, are clearly designed for frontal assault. Illustrations amplify the text and show designs that may only in part have been a figment of Fontana's imagination. Other contemporary works support Fontana's claim that rocket-propelled battering rams on rollers not only existed but were used in war. There is also supporting evidence for his rocket-propelled torpedoes, which were probably floating vessels directed at enemy ships by the reactive thrust of combustible black powder.

1500 In the first recorded attempt at manned flight by rocket power, the Chinese Wan-Hu attached 47 black powder rockets to a chair suspended beneath two large kites. Seated in the chair, Wan-Hu lit all 47 rockets simultaneously and promptly disappeared in the ensuing explosion. It is not clear if he became airborne in this attempt, nor is his exact objective known.

1529 Conrad Haas was appointed chief of the artillery arsenal in Sibiu, central Romania, and began to write manuscript notes about space travel and the use of rockets. Haas describes how he conducted experiments with powder rockets and put together two- and three-stage projectiles which were tested successfully. He also describes a missile capable of lobbing a powder keg at enemy fortifications. Haas recruited a small team of assistants with whom he carried out experiments in the composition of powder and the design of rocket cases. He is unique among the early experimenters in advocating the predominantly peaceful use of rockets.

1557 Leonhard Fronsperger, chief armorer for the city of Frankfurt am Main, published a book concerning contemporary developments in artillery and firearms. In only one brief paragraph does he mention rockets, but then to describe them as fireworks for peaceful displays of pyrotechnic wonder. The text does, however, contain a reference to them as "propulsion to other fireworks," which suggests a sophisticated technique of using powder rockets to send up other pyrotechnic devices which only activate at height. A great deal of secrecy surrounded the mixing of black powder and the making of fireworks, but several chemical formulas and manufacturing principles introduced by the end of the sixteenth century remain in practice today.

1566 January 5 The first accurately documented use of rockets in war took place during the battle of Talikota, India, during which the Vijayanagar army fired small devices, albeit to little effect. By this date Indian rockets may have been in use for several decades, but in the following decades the Indian war rocket was employed with increasing effect. In

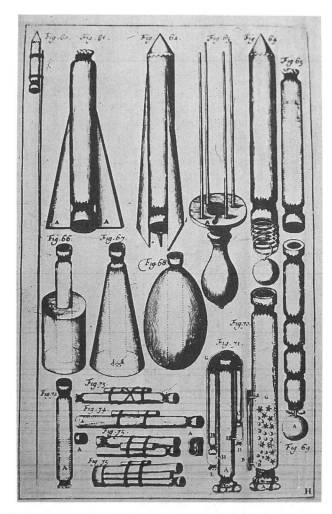

Various pyrotechnic and reaction-propelled devices are described in this early 16th-century manuscript which provides information on powders and projectiles.

Appealing for "more peace, not war" in the use of powder rockets, the 16th-century Romanian engineer Conrad Haas visualized manned flight by depicting a small space station atop a large rocket.

1589, for example, the Hindu Raja of Aununtgeer had 10,000 rocket troops. It is not known just how useful they were as weapons; as devices for bringing terror they were successful: On May 29, 1658, Dara Shukoh was struck by a rocket and his army defeated in the resulting confusion. It was through the extensive use of Indian rockets that the British would become acquainted with the weapon.

1569 In a book by a Polish writer named Martin Bielski, detailed descriptions are given of powder rocket production and the use of projectiles in war. This is the first unambiguous account of rocketry in Poland, although some historians interpret an account by the Polish monk Seweryn in 1380 as referring to rockets. Considering the diffusion of Chinese and Mongolian pyrotechnics to eastern Europe in the 13th century, it would not be surprising to find such references in 1390. But at best, Seweryn was writing about the use of rockets by invading Mongols. What Bielski writes about is the first indigenous Polish industry for manufacturing rockets.

1592 One of the first positively accurate accounts of rockets used in Spain is recorded in the *Platica Manual de Artilleria*

written by the military engineer Luis Collado. The use of rockets by the troops of King Carlos V against cities and cavalry is described. Prior to this date the use of rockets is uncertain and difficult to distinguish from pyrotechnic devices hurled by hand or a mechanical device. It appears the use of black powder rockets died out in the 17th century, only to be reintroduced during the early nineteenth century.

1634 Four years after his death a book written by the German astronomer Johannes Kepler was published for the first time. A fictional account of a trip to the moon, *Dream* benefited from Kepler's knowledge of celestial mechanics and the fact that the moon was a solid body which, because it is locked in synchronous rotation, constantly presents the same face to Earth. Kepler had satisfactorily explained how the solar system works, and this provided the foundation on which future astronauticians could develop their theories about space travel.

1638 Dutch whaling master Cornelis Pietersz IJs completed a series of trials using rocket-boosted harpoons for whaling. The tests took place in the the Arctic Sea off Spitzbergen. The rockets used by IJs were manufactured by Abraham Speeck, who made his living in Amsterdam as a painter, firework maker and general inventor. The rocket-boosted harpoon was not well suited to a small wave-tossed boat. The equipment was heavy and cumbersome and the mechanical process of operating the rocket was difficult to coordinate at sea. Parallel developments with the gun made the latter more suited to whaling, and little more was heard about this application of the rocket until the 1820s.

One of the more popular science fiction stories of the period, *The Man in the Moon: Or a Discourse of a Voyage Thither,* was published by an author who called himself Domingo Gonsales, a pseudonym for Francis Godwin, bishop of Hereford, England. In this story Domingo Gonsales is supposedly transported from the Earth to the moon in a boat pulled by birds. Assuming the moon to be 50,000 mi. distant, he arrives in eleven days, finding it much like the Earth with "mighty seas" and tall people.

1650 Casimir Siemienowicz, head of the Polish Royal Artillery under King Wladislaus IV, published *Artis magnae artileriae,* which describes the preparation and operation of rockets for military purposes. The work also gives an account of multistage rockets and provides what is probably the first technical explanation of a three-stage projectile in which each stage falls away at the end of its burn phase. For 100 years after it first appeared in Polish, it remained a primary text for artillery troops across Europe. First published at Amsterdam, it was published in French in 1651, in German in 1676 and in Dutch and English in 1729.

1657 Thirty years before Isaac Newton defined the three laws of motion, the French satirical writer Savinien Cyrano de Bergerac's book *Les Etats et empires de la lune* (The States of the Empires of the Moon) was published. For the first time it proposes the use of reactive thrust to transport people to the lunar surface. In his science fiction tale, Bergerac imagined the moon to be very like Earth. But his character was not content to visit the moon alone, and journeyed to other planets as well.

1753 In one of the first encounters with Indian rocket troops, a British Army unit under the command of Major Stringer Lawrence came under attack on the plain 3 mi. north of Facquire's Tope, India. Assaulted by a motley collection of archers, swordsmen and rocketeers, the British were taken by surprise but rallied quickly. The first British officer killed by a rocket, Lt. Col. Cay, was struck down on December 31, 1778. Generally, the rocket was merely a nuisance, but it could prove effective near horses or on ammunition stores.

1789 Innes Munroe's *Narrative of the Military Operations on the Coromandel Coast* gave Englishmen the first description of Indian rockets being used against British soldiers in India. Reports quoted sizes of between 6 lb. and 12 lb., with projectiles guided by a bamboo pole up to 10 ft. in length. As many as 5,000 rocketeers had been attached to the Indian army of Prince Hydar Ali of Mysore. Hydar Ali's son Tippoo Sahib was in charge of the rocket corps and wreaked much havoc among cavalry and infantrymen of the British Army between 1792 and 1799.

1791 A. M. Ducarne-Blangy of Picardy, France, fired rockets in what he claimed was the first attempt at producing a lifesaving device capable of propelling a rescue line to ships wrecked close inshore. There is doubt that he actually demonstrated a working lifesaver, but he indisputably fired projectiles during 1799 under the observation of a special government committee which included Admirals Rossiy and de Missiessy. The rocket he used had a diameter of 1.25 in. and carried a wire 0.13 in. in diameter attached to a stabilizer stick. The device had a range of about 540 ft. In other tests he demonstrated a rocket 1.9 in. in diameter carrying a silk line to a distance of 580 ft. The authorities were unimpressed, and he tried again in 1805 to get his device adopted, without success.

1792 April 22 Twelve days before the battle of Seringapatam, India, Tippoo Sahib, Sultan of Mysore, dispatched a rocket unit under the command of Cummer-dien Khan to the rear of the encampment where British soldiers waited to attack. With a range of approximately 3,000 ft., the Indian rocketeers fired a large number of projectiles into the camp, causing confusion and death. In one incident a single rocket killed three men and wounded four more. An eyewitness account claimed, "The rockets and musketry from 20,000 of the enemy were incessant. No hail could be thicker. Every illumination of blue lights was accompanied by a shower of rockets, . . . causing death, wounds and dreadful lacerations from the long bamboos of twenty or thirty feet."

1799 May 20 Following the fall of Seringapatam, India, a large cache of rockets found by the British were discovered to carry long swords in lieu of the more familiar long bamboo cane. These stabilized the missiles in flight and literally cut through the men and horses they were fired at.

1804 In his *Concise Account on the Origin and Progress of the Rocket System,* Col. William Congreve (1772–1828) of the Royal Laboratory of Woolwich Arsenal, London, described the principles of rocketry as they had been applied to weapons used by the Indians in Mysore on April 22, 1792. Congreve promptly set about the design of a rocket with a range of 6,000 ft. and produced a 32-lb. rocket with a length of 3 ft. 6 in., a diameter of 4 in. and stability in flight provided by a stick 15 ft. long and 1.5 in. in diameter. He experimented with a 42-lb. rocket, but Woolwich Arsenal set about the large-scale manufacture of his 32-lb. design. By August 1806 they had produced 13,109 rockets of this type for use with the Royal Artillery.

This stylistic representation of the application of war rockets during the 19th-century shows a naval attack by British forces.

1806 October 8 In the first large-scale use of rockets by the British, 18 Royal Navy boats entered the bay at Boulogne and in one 30-min. period fired more than 2,000 rockets at the French on shore. So complete was the shock and surprise that not one shot was returned, the French becoming preoccupied with putting out several fires that started just 10 min. after the assault began. The British used the 32-lb. Congreve rocket. The following year the British launched a massive rocket attack on Copenhagen in which 25,000 rockets were fired, setting fire to dwellings and warehouses.

1807 December 29 After watching helplessly as the frigate *Anson* broke up on the southwest coast of England with the loss of 100 lives, Henry Trengrouse of Helston, Cornwall, vowed to develop a device for inshore rescue and spent the next 10 years and £3,500 of his own money producing a practical lifesaving rocket. It was demonstrated at Woolwich Arsenal, London, on April 28, 1818, with two rockets and lines, the largest with a range of 760 ft. The admiralty passed over the Trengrouse rocket in favor of a design by Congreve, and few details of the former survive.

1808 The development of rockets in what was to become Europe's leading center for pyrotechnic devices began in Austria with the production of steel-cased war projectiles. Twenty-four of them were manufactured under the orders of chief fireworks master Anton Mager. Nothing further came of this until the British used their Congreve rockets during the Battle of Leipzig in 1813. This single event demonstrated the rocket to good effect and stimulated interest across Europe. In 1814 an Austrian army officer, Vincenz von Augustin, was sent to London, and while visiting the Royal Arsenal at Woolwich saw Congreve rockets for himself. He returned to Vienna in March 1815 and was given permission to build a rocket factory.

1810 In the first recorded use of rockets in Spain, William Congreve attacked the fortress of Cadiz during the Peninsular War. The results were not impressive, but Napoleon ordered a commission to study the use of rockets in war, and Congreve's rockets were used as a model for copies manufactured by the French in Seville. These were of poor quality and displayed a range of only 6,400 ft. In 1812 Lord Wellington used Congreve rockets in the liberation of Badajoz.

1811 April 4 The King of Sweden ordered the Royal Academy of Military Sciences to appoint members to a formal commission, the Brandraketcommitten "to investigate the proper composition of war rockets and to make experiments." Rockets used by the British at Copenhagen in 1807 had been brought to Sweden by the chemist Baron Jöns Jakob Berzelius. Experiments with Congreve-type rockets began in June and were followed by tests in 1812 and 1813. The rockets were 4.6-in. in diameter and demonstrated a range of 1 mi. Several improvements were made to the propellant and the rockets were modified. They had a length of 3 ft. 0.8 in., a diameter of 4.25 in. and a weight of 45 lb., or 54 lb. when attached to a 19 ft. 6 in. guide stick. The improved rockets had a range of 1.5–2 mi.

1813 October 16 The British Rocket Brigade attached to the bodyguard of the Swedish Crown Prince Carl Johann participated with Swedish forces in attacks during the three-day Battle of Leipzig. Several British officers displayed acts of bravery that brought Swedish decorations. The Rocket Brigade remained with the Crown Prince until January 18, 1814, and helped the Swedish in two more sieges.

1814 August 24 During the battle of Bladensburg, Md., the British 85th Light Infantry Regiment routed the rifle battalion of U.S. Attorney General William Pinkney with a rocket launched from undergrowth and brush along the banks of a stream. As the projectiles fell among the riflemen they scattered in panic. As the British commander George R. Gleig later noted, "Never did men with arms in their hands make better use of their legs." The rockets used in this attack were the standard 32-lb. projectiles designed by Congreve and manufactured at Woolwich Arsenal, London.

1814 September 13–14 During the night the British fleet bombarded Fort McHenry in Baltimore, Md., with Congreve rockets but were unable to gain ground. When the Washington lawyer Francis Scott Key saw the flag still flying over Fort McHenry the following morning, he was inspired to write "The Star Spangled Banner" with its reference to the "rockets' red glare." Set to a popular British drinking song, it became the national anthem of the United States.

1815 March The Austrian army officer Vincenz von Augustin returned to Vienna from London having witnessed Congreve rockets at the Woolwich Arsenal. With royal approval, von Augustin set up a Kriegsraketenanstalt (War Rocket Facility) about 20 mi. south of Vienna. Before visiting London he had also traveled to Denmark, where a secret rocket-making facility had been built on the secluded island of Hjelm in the Kattegat. Working to avoid what he saw as shortcomings with other rocket designs, von Augustin produced smaller, simpler and less cumbersome projectiles.

1815 May 31 The Austrian rocketmaker Vincenz von Augustin reported that he had prepared a stock of 2,400 rockets and trained a crew of 46 officers and men to use them in battle. At this time Napoleon, having returned from exile on the Mediterranean island of Elba, was massing an army of more than 700,000 men, and the Austrians were faced with the prospect of renewed conflict with the French. Having seen the way British rockets terrorized the enemy at the Battle of Nations in 1813, the Austrians ordered von Augustin to mobilize a *Raketenbatterie* armed with 0.2-in., 0.3-in. and 0.4-in. projectiles. It participated in the siege of Huningue, one of the last battles of the Napoleonic wars.

1820   Several German and Austrian journals published articles examining the work of Austrian army officer Vincenz von Augustin in building up a major rocket factory. Reports concluded that von Augustin's projectiles were superior to those of other European countries, including Britain, where the modern war rocket originated. In 1821 it was reported in these same journals that von Augustin had developed signal rockets capable of being seen at great distance and used for communicating with dispersed army units. The first major role in battle for von Augustin's *Raketeurs* came in 1821, when they proved highly effective in quelling an insurrection in Naples. The war with Italy proved a rich opportunity for the new weapons technology, and rocketeers aboard the frigate *Medea* off Trieste exchanged fire with artillery batteries in the Lombardy-Venetia region in 1823.

1821   William Congreve and Lt. James Nisbet Colquhon from the Royal Artillery received a patent for their "Application of Rockets to the Destruction and Capture of Whales." Not since 1638 had a serious attempt been made to develop a whaling rocket. The effectiveness and relative ease of handling afforded by the Congreve rocket once again opened the possibility of using these devices for firing harpoons. In several tests they proved their worth, and Sir William Scoresby Sr., wrote to Congreve following a whaling expedition with Congreve rockets involving Lt. Colquhon and two rocketeers from Woolwich Arsenal, saying how effective they were against whales.

1824 May 15   James Perkins of London, England, was granted a patent for his steam rocket design. It consisted of a metal container partly filled with water and a lead-tin alloy bung in a hole at the bottom. The whole container was set upon a brazier so that the water heated to steam. When it reached critical pressure the bung melted, releasing the pressure inside the container, which shot up into the air as a result. Perkins was unable to apply any purpose to his invention, and it seems to have languished without use. However, in 1841, Englishman Charles Golightly was granted a patent for a flying machine powered by steam rocket. It was never built.

1825 October 9   The British rocket designer William Congreve completed six sets of lifesaving rockets he had been asked by the Lords of the Admiralty to design. The basic rocket was his 31.7-lb. model specially modified to carry a grapple at one end of a line, rather than the conventional explosive or incendiary head. The total weight of this contraption was 59.5 lb. In operation, the rocket would be fired from the ship to throw the grapple to the beach so that a life buoy could be attached to the line and used to get people ashore. Tests showed that Congreve had done nothing to stabilize the gyrating grapple as it was hurled through the air, and he withdrew from the project.

1826 January 18   Following a disappointing series of tests with a lifesaving rocket designed by William Congreve, John Dennett demonstrated his own device to Royal Navy officers on the south coast of the Isle of Wight off the coast of southern England. A native of the isle, Dennett took the 11.9-lb. Congreve rocket and carried out improvements which afforded a ship-to-shore range of 985 ft. Success was assured and Dennett soon provided his own rocket design. The Royal

Navy set up three lifesaving stations on the Isle of Wight and officially adopted a 17.6-lb. two-stage rocket with a range of more than 1,300 ft.

1831 April 14   The German engineer Martin Westermaijer offered the Swedish ambassador in Berlin his services as a pyrotechnician, claiming he could brief the Swedish government on improved solid propellant rockets. From his experiences in an Austrian rocket factory and as adviser to the Prussian General von Braun, he said he wanted to carry out demonstration tests in Sweden and said he would pass on detailed information from other European factories. The offer was accepted and on August 18, 1831, he carried out 16 tests, of which only three failed.

1831   The Austrian rocket troops were reorganized to integrate them more effectively into existing artillery units. A major program of expansion was undertaken, and new companies and support groups were added. A large capacity for quickly making rockets and supplementing existing units was built up so that the Austrian army could respond quickly to threats from foreign powers. By 1851 the organization was converted into the Royal Imperial Rocket Corps with an establishment of 4,000 rocketeers in 18 rocket batteries. This represented the peak of the Austrian rocket forces. Most other countries had fewer than 200 men assigned to rocket batteries.

1832 December 28   Acting on experimental tests and information supplied by the German engineer Martin Westermaijer, the Swedish government created the Royal Rocket Corps under the command of Lt. J. W. Westerling. Stationed at Marieberg outside Stockholm, the corps had eight launchers and used rockets based on Austrian designs. They were basically of two types: a 2-in. rocket with a length of 15.5 in., weighing 7.5 lb. and with a guide stick of 7 ft. 6 in., and a signal rocket with a length of 8.65 in., a diameter of 1.1 in., a weight of 0.4 lb. and a 3-ft. guide stick. Range was up to 3,000 ft. for the 2-in. rocket, and a larger 2.5-in. version was available. The propellant consisted of 65% saltpeter, 21.4% sulphur, 11.2% aldercarbon and 2% alcohol.

1835   The 38-year-old English inventor William Hale moved to premises near the Royal Arsenal, Woolwich, London, and while there was so impressed by the Congreve rocket that he promptly set about modifying it. Hale made improvements by eliminating the guide stick and imparting a spin to the projectile. This gave it stability during the thrust phase and did away with the one feature of the Congreve rocket that made it prone to wind effects in flight. Hale's rockets were patented on January 11, 1844, and sold to many foreign countries. But the British government proved reluctant about paying Hale a royalty, and Hale was not paid for 26 years.

Having heard favorable reports from the Marquis of Villuma about the use of rockets in Spain, Queen Isabella II's advisers sent Lt. Col. Nuñez Arenas to England charged with purchasing a large quantity of Congreve rockets. England had gone to the aid of government forces fighting against a revolutionary order that attempted to establish Don Carlos on the throne instead of the infant Isabella, who had been appointed by King Ferdinand VII before his death. Under the agreement secured by Arenas, 5,000 Congreve rockets were to have been shipped to Navarre, but only a small number arrived. The Spanish were, nevertheless, pleased with the rockets, which they said were better than those the British had brought.

1836 The use of rockets for lifesaving purposes gathered momentum in Britain during the 19th century, and Yorkshireman Alexander G. Carte offered his own purpose-built device for sale at £18 8s. Containing rockets, launchers, lines and attachments, the complete set weighed 120 lb. By 1851 Carte lifesaving rockets were installed at 29 lifesaving stations in seven English counties and at several stations in Denmark. The Carte launch package was adapted for ship or shore use, but inadequate range imposed strict limits on its usefulness.

1840 September 8 Austrian rocket batteries operated in force against Beirut (then a part of Syria but now the capital of Lebanon), causing widespread damage, localized fires and general chaos. The Austrian rocketeers had been placed aboard ships operating alongside a British naval contingent, which was also engaged in firing rockets on the coastal region. The Austrian rockets designed and manufactured at the Kriegsraketenanstalt near Vienna had been used effectively throughout the army, and the absence of recoil when fired rendered them suitable for use on ships, particularly small vessels unable to carry cannon.

1846 March 13 A decree by the Swedish government authorized each artillery regiment to be equipped with two rocket launchers. The Royal Rocket Corps changed its name to the Firework Corps and was given responsibility for all rocket manufacture as well as any pyrotechnical devices required by the army or the government. Some amount of experimentation continued and formed the basis for William Unge's work in 1896, but the launchers were withdrawn from the artillery regiments during the 1860s.

1846 November 19 A U.S. expeditionary force to Veracruz and from there to Mexico City was authorized, with Maj. Gen. Winfield Scott in charge of a contingent of infantrymen that included rocketeers. The rocket battery of 150 men was under the command of 1st Lt. Jesse Lee Reno, who had at his disposal fifty 2.25-in. Hale rockets when they set sail from Fort Monroe aboard the bark *Saint Cloud* on February 1, 1847. The rockets were used first against the fortifications of Veracruz on March 24, 1847, and then in battles around Mexico City and against fortifications at Chapultepec on September 12–13. The rocket battery was disbanded in 1848.

1848 August Hungarian Prime Minister Lajos Batthyány asked British Foreign Secretary Lord Palmerston to send an army officer who understood rockets to Hungary. The request was ignored, and Hungary turned to its own chemists for military rockets similar to those which had given William Congreve fame across the continent of Europe. William Hale was less reticent and helped the Hungarians in their war of independence. A noncommissioned officer from the Austrian rocket corps, Sándor Mózer, carried out tests and set up a small factory in Budapest. When the city was overrun in December 1848, the work was moved to Nagyvárad (now Oradea, Romania). From this stemmed a small Hungarian rocket force which, as elsewhere, went into decline during the 1860s.

1850 May 27 A specially designed multistage Hale rocket test-fired at Shoeburyness, England, was reported upon by Lt. J. S. Aubyn of the British Army. Comprising seven 10-lb. rockets bound together with iron hoops, the 88-lb. compound projectile was observed to have reached a distance of almost three mi. The following month, further tests of the compound rocket were made and distances of up to five miles recorded.

Heavy, cumbersome and difficult to operate, the device was never used operationally.

1857 September 17 In the small town of Izhevskoye outside Moscow, Konstantin Eduardovich Tsiolkovsky was born. His prophetic writings about space travel and rocketry were to influence several generations of theorists and experimenters not only in Russia but all over the world. With a passionate interest in astronomy he was sent by his parents at the age of 16 to study in Moscow, returning three years later to become a teacher of physics and mathematics at a school in Borovsk, 55 mi. from Moscow. It was there that Tsiolkovsky began a serious study of the principles of reactive flight and space travel before moving to Kaulga at the age of 25.

1858 September 2 American whaling captain Thomas Welcome Roys took out a British patent for a whaling rocket he had designed while on a visit to Portugal. It bore a strong resemblance to the Congreve rocket, and on January 22, 1861, he took out a U.S. patent on the same device. Patents were also brought in several European countries. With New York firework maker Gustavus Adolphus Lilliendahl, Roys went to sea on trials in 1865–1866. Roys withdrew from the venture, but at a whaling station owned by the two men their rockets caught the imagination of the Danish naval Lt.-Capt. O. C. Hammer. It was he who took the rockets to Denmark and established the annual rocket spectacle at Copenhagen's Tivoli Gardens.

1858 The English engineer William Hale sold the secret of his rocket designs to the Austrians for £2,000. After an exhaustive series of trials with Hale rockets the Austrians developed their own variants which emerged as light mountain weapons. Gas guns weighing only 3 lb. each were produced for policing duties in the difficult geographical terrain in the regions that now comprise Bosnia, Serbia and Croatia. The rocket regiment was renamed Raketeur- und Gebirgsartillerie-regiment (Rocketeer and Mountain Artillery Regiment). In 1860 however, the rocket forces went into decline, emerging only briefly for the wars of 1866. It was the end of a glorious period of Austrian rocketry that had begun in 1815.

1862 July 3 The first rockets fired by Confederate troops in the American Civil War were launched by men under the command of Jeb Stuart. They caused damage and injuries among the Union troops of Col. James T. Kirk of the 10th Pennsylvania Reserves. The first Union group to be given rockets was the New York Rocket Battalion, its 260 men under the command of a British officer, Maj. Thomas W. Lion. The size of their rockets ranged from 12 in. to 20 in. long and from 2 in. to 3 in. in diameter. Maximum range was 3 mi. The first rockets fired by Union troops were set off in South Carolina during 1864. Confederate rocket batteries were set up in 1863 and 1864, supplied with projectiles manufactured at a factory in Galveston, Tex. This production later moved to Houston.

1865 March 15 The lifesaving rocket designed by Edward M. Boxer, superintendent of the Royal Arsenal, London, became the standard lifesaving device for British coastal stations 20 years after it had first been adopted. It replaced the Dennett rocket, which had been in use for almost 40 years, and was a tandem-step design: the first stage, upon burning out, ignited the second stage in front without ejecting the empty case of the first stage as would occur with a multistage rocket. The 11.9-lb. rocket was expensive, which militated against its export, despite a license from the Board of Trade. The Boxer rocket remained in service for 80 years and is credited with saving 9,407 lives.

1865   Three great works of science fiction were published: *Journey to the Moon* by Alexandre Dumas, *From the Earth to the Moon* by Jules Verne and *Voyage to Venus* by Achille Eyraud. Verne made the major contribution by describing weightlessness. In his fictional story, three men left Earth from a place in Florida to travel to the moon in a spacecraft called *Columbiad*. Fired from a giant cannon, which in reality would have crushed them in the acceleration, the three men used fireworks to adjust their course around the far side of the moon. The prophetic parallel with the first moon landing would lead the crew of Apollo 11 to select the name *Columbia* for their command module.

1869   The Rev. Edward Everett Hale began serialization of a science fiction story in the journal *Atlantic Monthly*. Called "The Brick Moon," it concerned an artificial satellite of the Earth placed in orbit by rapidly rotating water wheels. The satellite was used to determine from any point on Earth the lines of longitude essential for navigation. In this application Hale was prophetically accurate: Navigation satellites provide a similar function today. Hale was originally inspired by a conversation 30 years earlier with his brother Nathan at Harvard University.

1872   An engineer in the Royal Navy, a Mr. Quick successfully persuaded the authorities to test a new idea of his for a rocket-powered torpedo. Four 24-lb. Hale rockets were strapped together and a charge of cotton was placed towards the front, immediately behind a pointed nose section. To test the device the Navy took a 10-in. gun barrel and laid it on the beach at Shoeburyness, England. The rocket-torpedo was fired at high tide, but on emerging from the submerged barrel, the assembly burst apart and one rocket went up on a near vertical trajectory while the others went off in separate, but uncontrolled, directions. It has been said that the Duke of Cambridge, a spectator, "swore most foully" at the apparent stupidity of the idea.

1878   The New York firework maker Gustavus Adolphus Lilliendahl went into a joint venture for the manufacture of whaling rockets with the San Francisco firm of Fletcher and Suits. Taking Lilliendahl's basic whaling rocket, they modified it into a projectile 6 ft. 11 in. long with a diameter of 3 in. and a weight of 31 lb. It had a range of 165–195 ft. Put to work in the Arctic Ocean, the $50 rockets proved highly successful and caught the attention of Capt. Bernard Cogan, who had several put on his whaling boat. It was Cogan who developed the breech-loading bomb gun which became a strong competitor of the whaling rocket.

1882 October 5   Generally recognized as the "father of modern rocketry," Robert Hutchings Goddard was born in Worcester, Mass. While just a boy he was attracted to physics and mathematics and absorbed the writings of such science fiction authors Jules Verne and Herbert George Wells. At the age of 19 he wrote an article entitled "The Navigation of Space," which was not accepted for publication. Struck by the potential of liquid propellants, Goddard began in 1909 to make detailed calculations about rocket motors while studying at Worcester's Clark University. In 1911 he was awarded a Ph.D. and subsequently became professor of physics at Clark. In 1914 he was granted patents for combustion chambers, propellant delivery systems and multistage rockets.

1883   The Russian theoretician Konstantin E. Tsiolkovsky prepared a manuscript entitled "Free Space," in which he explored the possibility of using reactive force to navigate through space. He analyzed mechanical motion in space and drew up a schematic for a spacecraft, proposing the use of a gyroscopic device for stabilizing the vehicle. Tsiolkovsky had recently been admitted to the Society of Physics and Chemistry in St. Petersburg after having made contact with the famous Russian chemist Dmitri Ivanovich Mendeleyev.

1891   The first semiscientific discussion of a rocket-propelled spacecraft was made by the German Hans Ganswindt who tackled problems of propulsion not by the reactive force of expelled gases but by the ejection of steel cartridges containing dynamite. He conjectured that the explosion of one charge would transfer kinetic energy to the wall of the spacecraft and drive it in the desired direction. Multiple explosions would, he surmised, provide sufficient velocity to achieve orbit and escape speed. Ganswindt also explored the use of a rotating space station to counteract weightlessness and create artificial gravity.

1894 June 25   Hermann Julius Oberth was born in Hermannstadt, then a part of Transylvania in the Austro-Hungarian Empire and now Hungary. Oberth's writings were influenced by the writings of Jules Verne and would inspire many rocket pioneers, including the German rocket designer Wernher von Braun. The son of a doctor, Oberth entered medical school at the age of 19, and in 1914 he joined the Austro-Hungarian army and spent most of the First World War (1914–1918) in the Medical Corps. In 1919 he entered Heidelberg University and wrote a thesis about space travel which was published as a book in 1923.

1895   The Russian school teacher and prophet of space travel Konstantin Tsiolkovsky published his first science fiction book. *Musings on Earth and Heaven and Effects of Universal Gravity* proposed rotating space stations for artificial gravity, orbiting habitats constructed around nearby planets and the colonization of space as a means of relieving overpopulation on Earth. Although a work of fiction, the book prophetically described many of the steps future Russian, British and American theorists would propose.

Although he would wait 32 years before publicly announcing his claim, a South American engineer by the name of Pedro E. Paulet asserted that he carried out successful trials with a liquid propellant rocket motor in Peru during this year. Paulet made the claim in 1927 in a short letter to a Lima newspaper, stating that he had operated a conical motor 3.9 in. in diameter and powered by nitrogen peroxide and gasoline, producing a thrust of 198 lb. To this day no witnesses have testified to the validity of his claim, and despite its receiving credit in many published works covering the history of rocketry, there is not one shred of proof.

1896 September 12   Swedish Army officer Wilhelm Theodore Unge launched the first rocket wholly designed and built in Sweden. His solid propellant rockets proved inaccurate, but Unge turned to the use of rockets for lifesaving and sold some to England, India, Australia and Greece. With a useable range of 900 ft., each rocket package came complete with 1,200 ft. of line, a 10-in. rocket and a special transportation box, the total weighing 230 lb. Unge died in 1915 at the age of 70.

1901   Marking the end of the romantic era in space fantasy writing, Herbert George Wells published *First Men in the*

Moon, in which he envisaged an antigravity substance called cavorite which, when plastered over the exterior surface of a sphere, causes it to ascend to the moon. Up to this time it had been acceptable to use outlandish concepts for achieving basic objectives in space travel, but an increasing awareness on the part of the general public about science and the knowledge that flight through a vacuum would require rocket power changed the face of science fiction. From this period on, the outlandish concepts would be removed far from Earth, where the improbable could become the possible.

1903 Konstantin Tsiolkovsky published *Investigation of Space with Reactive Devices,* his first serious treatise on the technical problems associated with rocket propulsion, in which he explains his thoughts about rockets and the way the most efficient motors should be built. He concluded that the use of cryogenic liquid hydrogen and liquid oxygen would provide the best reactive force from the exhaust velocity and achieve a higher specific impulse, or measure of efficiency, than could otherwise be achieved by hydrocarbon fuels. Tsiolkovsky would do no practical experiments, but the clarity of his theoretical perception did much to stimulate interest. His book was republished in 1911, 1912 and 1914.

1904 The German engineer Alfred Maul carried out the first flight of his photo-sounding rocket. Propelled by black powder, the rocket, 3 ft. 3 in. long and 4 in. in diameter, was attached to a 13 ft. 1 in. stabilizing stick and carried a camera with a film plate having a format of 1.6 in. × 1.6 in. and a parachute for recovery. This rocket would reach a height of only 1,000 ft., but Maul worked on bigger rockets and by 1914 had produced a device that would ascend to an altitude of 2,600 ft. Maul produced a mobile unit weighing 1,000 lb. in a compact stowage unit on a two-wheeled cart. When erected, the launcher had an open truss structure and stood 7 ft. tall. The rocket was electrically ignited and worked well. The advent of the airplane made the Maul photo-sounding rocket obsolete.

1907 June 23 A pioneer of Soviet rocketry, Fredrikh Arturovich Tsander noted in personal jottings the idea of interplanetary travel. Throughout the year he developed ideas

Built by Konstantin Tsiolkovsky, this model of a hypothetical manned space vehicle embodied principles of propulsion that were sound in theory but decades ahead of their time.

about gyrostabilization, attitude control in weightlessness, the quantities of oxygen required by crew members and the amount of energy needed to reach the nearest stars. In 1909 Tsander discussed the idea of using solid fuel as the outer structure of a spacecraft so that it could be consumed en route to its destination; this is not as ridiculous as it sounds but is merely a clumsy way of extrapolating the concept of multi-stage rockets. In 1911 Tsander performed detailed calculations on liquid rocket motors.

1916 May 22 Designed by the French army Lt. Y. P. G. Le Prieur, powder rockets carried on the wing struts of a Nieuport 11 were used in combat for the first time during World War I. The French air unit Escadrille N.65 was equipped with the rockets, which were fired electrically from the cockpit. Each rocket was about 1 ft. 6 in. long, contained 7 oz. of black powder and was attached to a 6 ft. 2 in. guide stick. A good idea in principle, the rocket proved inaccurate because its path was perturbed by wind. Some successes were obtained however, primarily against observation balloons, their intended target. New and more effective phosphor ammunition brought about the demise of the Le Prieur air-to-air rocket.

1916 September 27 Robert Goddard wrote to the Smithsonian Institution asking for financial support in conducting experiments leading toward the launch of a rocket with atmospheric measuring equipment. Goddard envisaged a rocket carrying 1 lb. of instruments launched to a height of 200 mi. In subsequent communications Goddard requested the sum of $5,000 for this work. On January 5, 1917, the Smithsonian wrote Goddard approving release of this sum. When the United States went to war against Germany on April 6, 1917, Goddard informed the Smithsonian and the U.S. Army that his rockets could be used as weapons.

1918 January 22 Reacting to America's entry into the war with Germany, Robert Goddard and Smithsonian Under Secretary Charles D. Walcott, sent a letter to S. W. Stratton of the U.S. Bureau of Standards proposing the development of long-range missiles. Goddard calculated that he could build a missile weighing 32 lb., including a warhead of 3 lb. mass and a range of 120 mi. He also offered a smaller rocket weighing 9 lb. but carrying the same warhead over a range of 7 mi. Goddard was given money to work on these at Clark University in Worcester, Mass., but in June 1918 he was sent to Pasadena, Calif., to continue the work there away from prying eyes.

1918 November 6 Robert Goddard began two days of tests at the U.S. Army Aberdeen Proving Ground, Md., to demonstrate solid propellant rockets that could be fired by the infantry from special tubes. Powder propellants had been developed by the Hercules Powder Co. to Goddard's specifications. The rockets were in several sizes and weighed 5 lb., 7.5 lb. and 50 lb., fired from 2 in., 3 in. and 66 in. tubes, respectively. When the cease-fire in Europe began on November 11, 1918, Goddard's work stopped too and he returned to his professorial post at Clark University.

1919 April 7 Responding to outlandish claims made by the popular press about work he had conducted on the development of military rockets, Robert Goddard wrote to the Smithsonian suggesting that he prepare a paper to set the facts straight. The Smithsonian agreed and Goddard prepared a conservative description of potential applications for rocket propulsion. Hoping to sober up the wild speculation that surrounded all talk of experiments in rocketry, his 69-page paper, *A Method of Reaching Extreme Altitude*, suggested that rockets capable of reaching the moon were a practical possibility. Just 1,750 copies were printed.

1919 May 3 A Russian chemical engineer named Nikolai I. Tikhomirov wrote to Vladimir Ilyich Lenin urging that he be granted sufficient government funds to begin serious research into the technical development of solid and liquid propellant rockets. Born in 1860, Tikhomirov had developed several proposals which had been unsuccessful in attracting official support in the Czarist regime. Believing that the overthrow of the monarchy and the socialist revolution would grant new opportunities, he seized the chance to bring the subject before the Revolutionary Military Council, to whom Lenin passed the request. Two years later, on March 1, 1921, Tikhomirov got his wish.

1921 March 1 The Russian rocket theorist Nikolai I. Tikhomirov and his assistant V. A. Artemiev began work at no. 3 *Tikhvinski Ulitsa* in Moscow on the theoretical development of rocketry. S. S. Kamenev, commander in chief of the armed forces, had been ordered by the Revolutionary Military Council to provide facilities for Tikhomirov to begin work on rockets. Funds and practical support came irregularly and without the consistency essential for carrying out all the many concepts dreamed up by the two men. In 1922 O. G. Fillipov and S. A. Serikov of the State Scientific and Technical Institute joined Tikhomirov and Artemiev. Between 1923 and 1925 a powder rocket was tested several times in Leningrad, where the group moved in 1925.

1921 September 21 Konstantin Tsiolkovsky began to revise his 1903 book, *Investigation of Outer Space with Reactive Devices*. In unpublished notes entitled "Extension of Man into Outer Space," he discussed different forms of propulsion that could be used in space. These notes were kept with his writings

Eight years before his death in 1935, Konstantin Tsiolkovsky was visited by a Soviet writer — part of an ongoing campaign by the Communists to use the works of this rocket proponent to extol the virtues of revolutionary thought.

about the universe, while the manuscripts about rocket motors were kept with the files on reactive flight. Tsiolkovsky clearly thought through the logical steps prerequisite to an Earth-space transportation system. Tsiolkovsky died September 19, 1935, largely unknown except to the ever expanding band of space-travel proponents around the world. Only with the advent of the space age was he given public recognition. Of all his writings, the most poignant must surely be his statement that "The Earth is the cradle of mankind, but mankind cannot live forever in the cradle."

1921 September Robert Goddard switched his experiments in rocketry from solid to liquid propellant, believing that the higher specific impulse obtained from the latter would allow scientific instruments to be launched to the upper atmosphere. Powder rockets, he believed, had insufficient energy to reach extreme altitude. This was the turning point in his theoretical and practical research that set Goddard on the road to firing the world's first free-flying liquid propellant rocket on March 16, 1926.

1923 May In the first of a series of articles beginning with this issue of *Die Rakete* (The Rocket), published in Austria, the Austrian Guido von Pirquet began a discussion of interplanetary flight trajectories. The articles, titled "Fahrtrouten" (Routes for Space Travel), presented optimum flight paths, energy requirements and journey times for specific journeys. Pirquet explored the opportunities afforded by the different gravity environments of the Earth, a space station in Earth orbit and the moon. He was struck by the fact that it takes more energy to propel a spacecraft from the surface of the Earth to low Earth orbit than it does to send it from that orbit to the nearer planets.

1923 Hermann Oberth's book *Die Rakete zu den Planeten-räumen* (*The Rocket into Interplanetary Space*) was published in Munich by the German firm of R. Oldenborg. The text of a Ph.D. thesis rejected by Heidelberg University, Oberth's book represented the first serious attempt to describe the scientific and technical problems associated with spaceflight. Oberth subsidized some of the cost to have it published as a book. Renamed *The Road to Space Travel*, a second edition published in 1929 won for Oberth a 100,000 franc literary prize. Oberth used the money to fund his theoretical research into liquid rocket propulsion.

1924 April In Russia a proponent of spaceflight and rocketry, G. M. Kramarov, headed an Interplanetary Travel section of the Military Research Society of the Zhukovsky Air Force Academy in what was the world's first society for proponents of space travel. Several months later this became the Society for the Study of Interplanetary Travel. With a membership of over 200, it attracted many theorists and experimenters, among whom were Konstantin Tsiolkovsky, V. P. Vetchinkin and Fredrikh A. Tsander. Several space study groups were springing up across Russia and the socialist states under Moscow's control; a group led by D. A. Grave was set up in Kiev in 1925.

1924 The Soviets established two space and rocket study groups. One, the All-Union Society for the Study of Interplanetary Flight (OIMS) concentrated on theoretical problems associated with space travel, celestial navigation and what today would be classified as astrodynamics. The other, the Central Bureau for the Study of the Problems of Rockets (TsBIRP), would concentrate on the technical problems of space transportation. These two groups spurred broad enthusiasm across the Soviet Union, and all major cities soon had their own interest groups with regular meetings and talks.

1925 June 19 A local space study group in Kiev held an exhibition to popularize spaceflight and rocketry. Lectures were held and publicity material was gathered from many sources and handed out. The Russian public accepted this as a natural extension of many scientific developments of which most were aware but of which few had any experience. Russia had not experienced such enthusiasm for technical innovation since the days of Peter the Great. Russia was unique in this; few countries accepted spaceflight as a serious topic for intellectual debate.

1925 December 6 In his laboratory at Clark University, Robert Goddard successfully tested his liquid propellant rocket being prepared for independent flight out-of-doors. In a 24-sec. test the rocket lifted itself under the 9-lb. thrust of its motor, pressure-fed with liquid oxygen and gasoline. This was the first of a preparatory series of tests to make fine-tuned adjustments to the design of the world's first free-flying liquid propellant rocket. Other static tests in the laboratory were conducted on December 30, 1925, and on January 3 and January 20, 1926.

1925 A German engineer and architect, Walter Hohmann, published *The Attainability of Celestial Bodies,* in which he discussed flight into space and the circumnavigation of the planets. In a mathematical treatise, Hohmann defined the laws that govern motion as applied to transfer ellipses between bodies in the solar system. When measured by the amount of energy they require, Hohmann's interplanetary transfer ellipses are the most economical that are achievable with rockets, although they are not necessarily the quickest. He showed that the trajectory requiring the least amount of energy is one in which a spacecraft departs one planet in a direction tangential to its orbit about the sun and reaches the other planet in a similar manner.

1926 March 16 The first flight of a liquid propellant rocket took place at Auburn, Mass., under the direction of Robert Goddard. In a flight lasting 2.5 sec., the rocket flew a distance of 184 ft. and reached a height of 40 ft. at an average speed of 60 MPH. The device took the unfamiliar form of placing the combustion chamber and nozzle at the top of a frame comprising two vertical tubes, one carrying liquid oxygen and the other carrying gasoline from two separate tanks at the bottom placed one above the other. Heated by an alcohol burner, vaporized liquid oxygen was used to pressure-feed fuel and oxidizer to the delivery lines supporting the combustion chamber. Mrs. Goddard began making a film of the test with a small cine camera, but the film ran out before the actual flight and it was never recorded.

1926 April 3 Robert Goddard successfully directed the second flight of a liquid propellant rocket from Auburn, Mass., The same device he had used for the flight on March 16, 1926, was used again with some modifications. It flew a distance of 50 ft. and remained in the air for 4.2 sec. Two more attempts were made on April 13 and 22, but on each occasion the combustion chamber burned through before flight. Early in May Goddard decided to build a rocket twenty times the size of his first device. With a thrust of 200 lb., it had a length of just over 9 ft. and weighed 156 lb. This rocket was

never flown, but on August 31, 1927, it did achieve its design thrust before the case burned through.

1926 May 4 Robert Goddard completed design changes to the rocket he had flown successfully on March 16, 1926. For the first time he placed the combustion chamber and nozzle at the base and propellant tanks in tandem above the igniter, thus establishing the conventional layout familiar in future rocket designs. This arrangement eliminated the lengthy tubing for propellant feed which doubled as support for the combustion chamber. This configuration proved unsuccessful when applied to the original design, and Goddard turned to a completely redesigned rocket.

1927 April Responding to popular enthusiasm for space travel and planetary exploration, the Russian Association of Inventors held an exhibition of spaceships, rocket concepts and associated technologies. It publicized the works of Konstantin E. Tsiolkovsky, Hermann J. Oberth, Max Valier and Robert Esnault-Pelterie and elevated to respectability the many ideas and theories concerning future opportunities for cosmonautics. In this pre-Stalinist period there was tolerance for broad philosophical discourse on the future role of the state in high technology. Under Stalin such discussion was stamped out.

1927 June 8 French aviation pioneer and soon-to-be champion of space travel, Robert Esnault-Pelterie, gave a lecture at the Sorbonne, Paris, on the use of rockets to explore the upper atmosphere. His lecture was laced with technical detail revealing a solid grasp of fundamental problems, including the importance of escape velocity, gas expansion ratios and convergent-divergent nozzles. Esnault-Pelterie began a series of tests on liquid propellants, which he preferred to solids, and switched from liquid oxygen to tetranitromethane. It was while experimenting with this liquid on October 9, 1931, that he lost four fingers of his left hand in an explosion.

1927 July 5 In the back room of a restaurant in Breslau, Germany, the Verein für Raumschiffahrt (VfR), or Society for Space Travel, was formed with ten founding attendees. They had wanted Max Valier as president, but he was too busy with public engagements so Johannes Winkler accepted the position and agreed to edit their magazine *Die Rakete (The Rocket)*. The objectives of the VfR were to publicize rocketry and space travel and to conduct practical experiments. By September 1929 membership had grown to 870 and shortly thereafter to more than 1,000. Hermann Oberth, Walter Hohmann, Friedrich Wilhelm Sander and Franz von Hoefft were all appointed directors. Other notable members included Max Valier, Willy Ley, Robert Esnault-Pelterie, Nikolai Rynin and Jakov Perelman. Those who would play an important role in the development of German rocketry included Wernher von Braun, Klaus Riedel and Rolf Engel.

1927 September 3 Robert Goddard began work on a new liquid propellant rocket four times the size of his first device, which had flown for the last time on April 3, 1926. Rejecting a much bigger rocket which he never flew, Goddard built a 40-lb. thrust motor, and for more than a year worked on improving the fuel delivery and injection system for this design. Like its predecessors, this rocket had an open structure to minimize weight, with propellant tanks, feed lines and associated plumbing exposed to the elements. It was also designed to be launched from a supporting frame.

1927 December 26   The word *astronautique*, from the Greek words meaning to sail the stars, was coined by the French writer J. H. Rosny Sr., president of the Académie Goncourt. Robert Esnault-Pelterie adopted the word and brought it into general scientific use, the specific definition being the science and construction of space vehicles or the problems of navigation through space. Astronautics was adopted in a generalized meaning as the study of space travel in all its many forms. From this was derived the term *astronaut,* meaning a person who trains for, or engages in, space travel. Russian terminology uses the analogous terms *cosmonautics* and *cosmonaut,* from the Greek meaning to sail the cosmos, or universe.

1928 January   German rocket proponent and champion of space travel Max Valier began a series of tests to develop high-performance powder rockets at a rocket making factory in Wesermünde near Bremen. Owned by Friedrich Wilhelm Sander, the factory put out signaling and lifesaving rockets for the German navy, and Valier convinced car magnate Fritz von Opel to support experiments with powder rockets for racing cars. When Valier and Sander took the rockets to the Opel racing track at Rüsselsheim, they were transported by road because the German railway refused to transport them.

1928 February 1   Robert Esnault-Pelterie and André-Louis Hirsch founded the REP-Hirsch International Astronautics Prize, which included an award of 5,000 francs given annually until 1936, for the best original theoretical or experimental work. The first prize was awarded to Hermann Oberth, who used the money to publish his book *The Rocket into Interplanetary Space.* In 1930 and 1933 the prize went to French engineer Pierre Montagne for work on gaseous mixtures, and in 1934 it went to French engineer Louis Lamblanc for a treatise on powder rockets. In 1936 it was awarded jointly to the American Rocket Society and Alfred Africano for their design of a high-altitude rocket. Pelterie became a prolific publicist for space travel and wrote many books.

1928 February 9   In a lecture to the German-Austrian Engineering Society in Vienna, the Austrian Dr. Franz von Hoefft proposed a staging concept for winged space planes. In this proposal a rocket-powered winged space plane weighing 66,000 lb. would take off from and land on water like a hydroplane and be capable of carrying astronauts into orbit. It could also be used as the upper stage of a much larger manned shuttle. Von Hoefft also a proposed a two-stage winged space plane weighing 6,600 lb. which would propel itself by rocket to the moon. Looping around the far side of the moon, the upper stage would take photographs before returning to Earth like a glider; the film would be recovered by parachute. Another use he envisaged was that of a rocket mail service.

1928 March 3   The first launch of a smokeless-powder rocket in Russia took place when technicians from the Leningrad Gas Dynamics Laboratory fired a composite device a distance of 4,260 ft. The projectile had a diameter of 76 mm and consisted of three solid composite charges. The rocket was hurled into the air for the first 1,000 ft. of its trajectory by a mortar. The Gas Dynamics Laboratory had based much of its research on smokeless powder first supplied to the technicians by the great Russian chemist Dmitri Ivanovich Mendeleyev.

1928 March 15   Racing driver Kurt C. Volkhart became the first person propelled by rocket power when he was at the wheel of a makeshift car at Fritz von Opel's race track in Rüsslesheim. Put together by Max Valier and Friedrich Wilhelm Sander, two solid powder rockets, one with a short burn thrust of 175 lb. and another with a longer burn at 40 lb. of thrust, were attached to a wooden frame on the chassis of a spare Opel car. The run lasted only 35 sec. and the car moved just under 400 ft., but a second run was made using rockets of 175-lb. and 485-lb. thrust ignited on a rolling start with the car in neutral. This time it accelerated at 0.5 g and went from 18 MPH to 46 MPH in 1.5 sec.

1928 March 24   A young Austrian engineer named Eugene Sänger applied for membership to the German Wissenschaftliche Gesellschaft für Höhenforschung (Scientific Society for High-Altitude Research). Inspired by the writings of Hermann Oberth, Sänger would nevertheless disagree with Oberth's preference for ballistic rockets. Instead, he joined followers of the von Hoefft school who proposed manned space planes as the best way to achieve space travel. Under the Nazis, Sänger received funds for development of a rocket-powered winged bomber. Priorities were given to ballistic rockets like the A-4/V-2, and the "antipodal bomber" was never built.

1928 April 11   Kurt C. Volkhart was at the controls for the first test run of Opel-Rak 1, the chassis of an Opel sports car adapted to carry up to twelve 40-lb. thrust solid-powder rockets. For the first run, six rockets were fitted; although only five fired, a speed of 43 MPH was achieved in 6 sec. Then eight rockets were attached, but at 50 MPH the pack exploded, bringing the car to a halt without injury to the driver. During the first official public run on April 12 only five of the twelve rockets ignited, but the car reached more than 70 MPH.

1928 May 23   German car maker Fritz von Opel was at the wheel of Opel-Rak 2, a rocket-propelled vehicle developed by Max Valier with Sander rockets. Powered by 24 powder rockets, the car reached a speed of almost 125 MPH on the Avus Speedway near Berlin and was a complete success. None of the rockets exploded or failed to start, as some of them had on prior tests with Opel-Rak 1 and its test-bed predecessor. The car was an improvement over the Opel-Rak 1, and with its long, slender profile looked every bit the futuristic vehicle von Opel had envisaged.

1928 June 11   German test pilot Fritz Stamer became the first person to fly an airplane powered by rocket motors. Funded by industrialist Fritz von Opel and directed by rocket pioneer Max Valier, the event took place on the Wasserkuppe mountain in Germany. Named *Ente* (Duck), the aircraft was designed by Alexander M. Lippisch and took the form of a powered canard sailplane. It was built by the sailplane company Rhön-Rossitten Gesellschaft. Propelled by two blackpowder rockets, the *Ente* flew for 35 sec., but on the second flight this day it flew a distance of 4,000 ft. in 70 sec.

1928 June 23   In an attempt to break the existing land speed record of 207 MPH, Fritz von Opel tested a rocket-powered rail car called Opel-Rak 3 he had built following rocket tests with a car. Fitted with 10 powder rockets producing a combined thrust of just over 6,000 lb., the pilotless Rak 3 was propelled to a speed of 175 MPH on a government-owned track between Burgwedel and Celle. In a second run, this time with 30 rockets producing a thrust of 21,500 lb. thrust, the car left the rails within seconds and was wrecked. An attempted run of Opel-Rak 4 on August 4 powered by 30 rockets ended when the rail-mounted car was destroyed.

1928 June The Tikhomirov laboratory in Leningrad was reorganized as the Gas Dynamics Laboratory, responsible to the Military Research Council. This year G. E. Langemak joined the GDL and did much to advance the work of the laboratory on powder composite charges in chambers with nozzles. In 1929 B. S. Petropavlovsky replaced Tikhomirov as head of the GDL and focused development on the basic 24-mm solid propellant charge used in the flight of March 3, 1928. From this engineers developed scaled-up rocket chambers of 68 mm, 82 mm and 132 mm caliber. The latter two became the Katyusha projectiles designated RS-82 and RS-132 and used throughout World War II and after.

1928 September 30 Piloting an aircraft called Rak.1, the German industrialist Fritz von Opel covered a distance of 5,000 ft. at an average speed of 95 MPH in the first flight of a manned aircraft specifically designed for rocket power. Financed by von Opel, designed by an engineer named Hatry and developed by Max Valier, Rak.1 was a glider of monoplane construction built to carry sixteen 50-lb. thrust blackpowder rockets produced by Friedrich Wilhelm Sander at a factory near Bremen, Germany. The flight, which took place at Rebstock, near Frankfurt, was the only one attempted with the Rak.1.

1928 December 26 The third flight of a liquid propellant rocket took place at Auburn, Mass., when Robert Goddard successfully fired the 40-lb. thrust device he had been working on since September 3, 1927. As it ascended from its launch frame, the rocket tilted over and passed across the observation hut from which Goddard was watching the flight. With a velocity of more than 60 MPH, the rocket flew 204.5 ft. in this its first flight.

1928 The Russian spaceflight proponent Nikolai A. Rynin published the first of a multivolume history of spaceflight, rocketry and the associated writings of great theorists. Titled *Mezhplanetyne Soobshcheniya* (literally, *Interplanetary Communications,* or *Interplanetary Travel*), the work ran to nine volumes and was completed by 1932. The first two volumes covered the history of science fiction up to Jules Verne and Herbert George Wells. The third volume covered the problems associated with the search for extraterrestrial life, and the next three books dealt with the technical developments in rocketry. Volumes seven and eight covered the writings of the great theorists, and the ninth volume described the moon and the planets.

Writing under the pseudonym of Captain Potocnik of the Austrian Imperial Army, Hermann Noordung published *Das Problem der Befahrung des Weltraums* (*The Problem of Traveling through Space*), in which he proposed the design of a doughnut-shaped space station in what was the first serious attempt to consider design problems and seek solutions for such an enterprise. Noordung attached power generation equipment at one end of a central hub and made provision for astronomical equipment. He conceived of the ring-shape so that the station could rotate and, through inverse centripetal force, induce a sensation of gravity. Noordung also argued the advantage of a space station in geosynchronous orbit, that is, one in which it seems to remain stationary above a fixed spot on the equator by orbiting the Earth at the same rate the Earth revolves on its polar axis.

1929 February 3 Max Valier's wife, Hedwig, became the first woman propelled by a rocket motor when she rode a specially designed sled powered by six rockets at a winter sports festival on the Eibsee, Germany. Ignited in pairs, the rockets accelerated the sled to a speed of about 25 MPH in a distance of 350 ft. Hedwig Valier said later that she had persuaded Max to give her his seat on the rocket-powered bobsled. Valier had experimented with the rocket sled concept after leaving von Opel to develop his own rocket-propelled racing car and rail car. A few more tests were conducted, but Valier soon turned to the development of liquid propellant motors.

1929 April 10 Friedrich Wilhelm Sander fired a composite rocket into the air in an event that almost qualifies as the first flight of a liquid propellant motor outside the United States. The motor weighed 15 lb. and had a diameter of 8.3 in. and a length of 2 ft. 5 in. Thrust was just over 100 lb., and the motor burned for 132 sec. using both liquid and solid substances in a pyrotechnic reaction. Records are sparse about the event and claims that it reached a height of more than 6,000 ft. are dubious.

1929 May 15 Valentin P. Glushko began work at the Leningrad Gas Dynamics Laboratory with responsibility for developing liquid propellant rocket motors. Glushko is one of the greatest Russian rocket engineers of all time and his contributions to the development of liquid propellant technologies are outstanding. Nioklai Tikhomirov and M. V. Shuleikin had evaluated Glushko's electric rocket proposal and invited him into the GDL after he wrote them with a proposal for developing electric (ion) propulsion systems. Tsiolkovsky considered Glushko one of the most talented exponents of reactive flight.

1929 July 17 Directed by Robert Goddard, the fourth flight of a liquid propellant rocket and the last from Auburn, Mass., landed 171 ft. from the launch tower. It was the second flight of the 40-lb. thrust rocket first tested December 26, 1928. The rocket started to rise 13 sec. after ignition and was in free flight for 5.5 sec. This was also the first flight of an instrumented rocket, and it carried an aneroid barometer, a thermometer and a camera that was tripped when the parachute came out. The considerable noise created by the rocket's firing brought much public attention, and because of this Goddard, with the help of the Smithsonian, persuaded the U.S. Army to let him use the artillery range at Camp Devens, Mass., 25 mi. from Worcester. Goddard began flight tests on December 3, 1929.

1929 October 15 Produced by the Ufa Film Co. in Germany and promoted by the Society for Space Travel (VfR), the science fiction film *Frau im Mond* (*Woman in the Moon*) had its premier at the Ufa Palast am Zoo in Berlin. The film's producers wanted Hermann J. Oberth to build a working liquid propellant rocket to feature in the film, and in collaboration with Rudolf Nebel he did so. The design was crude and the rocket never flew but the Chemisch-Technische Reichsanstalt (Reich Institute for Chemistry and Technology) provided test facilities for Oberth and Nebel, aided now by Rolf Engel, Klaus Riedel and Wernher von Braun, to redesign it into a working rocket. The *Frau im Mond* symbol from the film became a motif for German space pioneers and was painted on the first A-4/V-2 rocket successfully launched on October 3, 1942.

1929 December 3 Robert Goddard began tests with rocket motors at the U.S. Army artillery range, Camp Devens, Mass., having vacated the farmland at Auburn that he had used since the first successful flight on March 16, 1926. The tests at Camp Devens supported experiments into combustion-chamber cooling. No flights were made from this artillery range and by June 30, 1930, Goddard had conducted 16 static firing tests with curtain cooling (in which liquid flows around the inside of the combustion chamber), regenerative cooling and improved guidance systems.

1929 Russian theoretician Yuri Vasilyevich Kondratyuk published *Conquest of Interplanetary Space,* a major work on rocketry and space travel based upon a decade of research and writing. In this work Kondratyuk refines problems associated with space travel and discusses acceleration, attitude control, stability and the technical matters associated with liquid propellant rockets. He goes on to explore the problems associated with developing space stations and planetary bases and lays down a set of incremental steps for manned moon landings, including Earth-orbit rehearsal and circumlunar reconnaissance followed by manned landings. Kondratyuk died in 1942 at the age of 45.

1930 January The German army established a *Versuchsstelle* (experiment station) for rocket research at a place called Kummersdorf-West, approximately 17 mi. south of Berlin. Set up at the instigation of Col. Karl Emil Becker, chief of the Ballistic and Munitions Department of the Army Weapons Board, it was to be run by Capt. Walter R. Dornberger and munitions expert Capt. Ritter von Horstig. Becker charged Dornberger with producing solid propellant rockets with a range of up to 6 mi. The design also had to allow for the need for large-scale production and field use in war. Dornberger was also required to develop a liquid rocket motor for a ballistic missile with a range in excess of 20 mi. To that end a 60-lb. thrust combustion chamber was produced at Kummersdorf-West, but it weighed 400 lb. and was useless for missile application.

1930 March 21 A small group of rocket enthusiasts met in New York City to found the American Interplanetary Society, which formally came into existence on April 4 and published its first "Bulletin" in June. Founding members included the chemist William Lemkin, writers and newspaper people C. W. van Devander, C. P. Mason, Laurence Manning, Fletcher Pratt and Nathan Schachner, and engineer Clyde J. Fitch. David Lasser, editor of the magazine *Wonder Stories*, was the first president and writer G. Edward Pendray became vice-president. The reclusive Robert Goddard did not join but occasionally wrote to correct errors in the "Bulletin."

1930 April 11 Announcing its move from Breslau to Berlin, the Society for Space Travel (VfR) held a public meeting in the auditorium of the General Post Office adjacent to a room containing the original rocket designed by Hermann Oberth. Among those present was the ubiquitous and aged Hans Ganswindt, whose claim to fame lay not only in his 1891 proposal for a reaction rocket but in his airships, horseless carriages, fire engines and remarkable progeny of 23 children. Johannes Winkler gave the principal speech, and Max Valier and Willy Ley attended also. The VfR also exhibited at the Aviation Week displays at Potsdamer Platz and at the nearby Wertheim department store.

1930 May 4 Famous for his tailless aircraft, the German designer Gottlob Espenlaub flew for the first time under rocket power the aircraft specially built for him by Swiss engineer A. Sohldenhoff. Designated E 15, the tailless aircraft

was not a success, and only a few flights took place. Utilizing Sander solid propellant rockets, the aircraft suffered from the limitations of rockets of this type. Thrust was difficult to control and the duration of burn limited to less than a minute. Failures such as this helped accelerate the development of liquid propellant rocket motors.

1930 May 17 During a static-firing test of a motor he had developed for one of his rocket powered cars, Max Valier was killed in an explosion, becoming the first person to die while experimenting with liquid propellant rocket technology. Valier had switched from alcohol to kerosene fuel with a liquid oxygen oxidizer for tests that were to have resulted in successful calibration for his rocket-powered car. A champion of rocket propulsion for land and air transport, Valier wrote many articles and books in which he envisaged the coming age of jet aircraft for mass transport of people across the world.

1930 June Members of the German Society for Space Travel (VfR) began construction of a Minimum Rocket, or Mirak, in a shack on farmland at Bernstadt, Saxony, owned by Klaus Riedel's grandparents. Rudolf Nebel, with Riedel and Kurt Heinisch, put the Mirak I together in an arrangement similar to that of a powder rocket or firework. Formed from long, thin aluminum tubing, the "stick" part of the rocket was the gasoline tank with a cone-shaped combustion chamber at the top offset to one side. Liquid oxygen could be put in at the nose to a jacket that surrounded, and hopefully cooled, the combustion chamber; carbon dioxide was used as the pres-

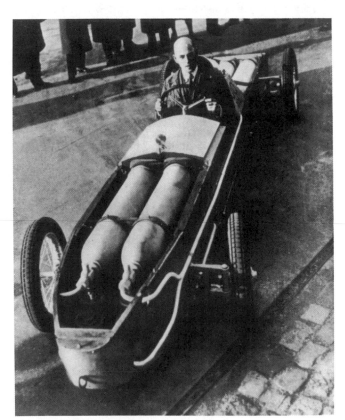

Two days before his death in May 1930, Max Valier poses in the driver's seat of his liquid propellant rocket car, which produced approximately 350 lb. of thrust.

surant. Mirak I weighed 6.6 lb. and had a thrust of up to 9.9 lb., but it exploded due to inadequate cooling.

1930 July 23 Designed and built by Society for Space Travel (VfR) members Hermann Oberth, Rudolf Nebel, Rolf Engel, Klaus Riedel and Wernher von Braun, a small liquid propellant rocket motor was successfully test-run at the Chemisch-Technische Reichsanstalt (Institute for Chemistry and Technology). It produced a thrust of approximately 15.4 lb. for 90 sec., consuming 13.2 lb. of liquid oxygen and 2.2 lb. of gasoline. Dubbed *Kegeldüse* (cone jet) because of its cone-shaped steel and copper combustion chamber, the rocket was witnessed by Dr. Franz Hermann Karl Ritter of the institute. Dr. Ritter was so impressed with its performance that he recommended that the Deutsche Notgemeinschaft (German Emergency Organization), a source of funds for sponsored projects, finance further work at the VfR.

1930 July 25 Just four months after the death of Nikolai Tikhomirov, the Leningrad Gas Dynamics Laboratory he had done so much to establish passed into the hands of the military in a decree confirming that the GDL was to concentrate its research on military weapons. On June 19, 1931, M. N. Tukhachevsky was appointed deputy chairman of the Revolutionary Military Council in charge of armaments, and on August 15, 1931, the GDL was placed under his responsibility.

1930 July Financial support from the philanthropic Daniel Guggenheim enabled Robert Goddard to move his home and work from Massachusetts to the Mescalero Ranch in Roswell, N.Mex. Here, he could work on his liquid propellant rocket motors in relative privacy and without the public attention his launches had brought around the farm in Auburn. Goddard had been visited by aviation pioneer Charles A. Lindbergh, who had written to Guggenheim encouraging him to support Goddard with funds from the Daniel and Florence Guggenheim Foundation for the promotion of aeronautics.

1930 September 27 The municipal authorities in Berlin agreed to rent out to the Society for Space Travel (VfR) an abandoned army garrison of 300 acres for the sum of 10 Reichsmarks a year. Secured by Rudolf Nebel, the new ramshackle, weed-strewn site was needed for motor tests and rocket flights paid for by benefactors and money from membership fees. The new site was dubbed Raketenflugplatz (rocket flight field) and was located in the north Berlin suburb of Reinickendorf. The VfR now had a new president, retired Maj. Hans-Wolf von Dickhuth-Harrach, who did not believe the rocket had potential as a military weapon but was intrigued by the technology.

1930 December 30 The fifth flight of a Goddard liquid propellant rocket took place when its designer launched a new device from the Mescalero Ranch at Roswell, N.Mex. With a static thrust of 289 lb., the rocket was 11 ft. tall and ascended to a height of 2,000 ft. at a speed of 500 MPH. A gas pressure tank was used to force the propellants into the combustion chamber. A key measure of the effective work performed by the propellant reaction, the rocket's exhaust velocity was 5,000 ft./sec. Several more flights were conducted until April 19, 1932, evaluating liquid nitrogen pressurization systems, gyro-controlled stabilization vanes that impinged on the exhaust efflux to deflect the thrust vector, and a variety of other technical innovations. In July 1932 the Daniel and Florence Guggenheim Foundation money used by Goddard to support

work at Roswell ran out and he returned to Clark University, Mass., to resume teaching as a living.

1930 Retired German army Gen. Artur von Baumgarten-Crusius published a book with the title *Vengeance* in which he argued that a ballistic missile with a range of 7,700 mi. would become the decisive weapon of the future. Written for the most part in 1929, when the German army was focusing its interest on the possibilities raised by developments in solid and liquid propellant rockets, the book was a frank exposé of internal army thinking at that time, and there is some justification for believing that it provides a glimpse of the army's real aspirations.

1931 February 2 A specially designed gunpowder rocket put together by the Austrian Friedrich Schmiedl began the world's first regular rocket-delivered mail service. Designated V-7, for *Versuchsrakete* (experimental rocket) No. 7, it transported 102 pieces of mail the 1.5 mi. between Schoeckel and Radegund in the Austrian Alps. By March 16, 1933, when the last flight took place between Garrach and Arzeberg, Schmiedl's rockets had delivered 1,768 items of mail. His rockets were typically 5 ft. 6 in. in length having a diameter of 9.6 in. and a weight of 67 lb. and carrying up to 300 letters in one small container a maximum distance of 2 mi. For a completely unrecorded reason, the authorities destroyed Schmiedl's equipment and ordered him to stop his flights.

1931 February 21 The first flight of a liquid propellant rocket in Europe took place at Gross-Kuehnau near Dessau. Designated HW-1 (Hückel-Winkler-Astris I), it was designed by Johannes Winkler to a description published in the December 1929 edition of *Die Rakete,* the magazine of the Society for Space Travel (VfR). Powered by a mixture of liquid methane and liquid oxygen pressurized by nitrogen, the rocket was about 2 ft. long with a launch weight of 11 lb. and reached a height of only 10 ft. when it failed. It had a combustion chamber made of seamless steel tubing, and the main structure comprised three tubelike containers for the propellant and pressurant forming a triangular arrangement. A second flight on March 14 was a complete success. Winkler built a larger rocket, but it was destroyed on the second launch attempt on October 6, 1932.

1931 April In Germany, the Society for Space Travel (VfR) experimented with Mirak II, a development of Rudolf Nebel's brainchild, the Mirak I, tested unsuccessfully in June 1930. Like Mirak I, Mirak II's hot combustion chamber could not cool quickly enough and the chamber exploded. The team quickly realized that water would be needed for cooling the combustion chamber and the Repulsor I was designed with that feature. Meanwhile, also during April, the Mirak III was designed. With realignment of the tanks and the use of nitrogen as a pressurant, it had a total weight of 8.8 lb. and a thrust of about 70 lb. for 30 sec. Mirak III was never built.

1931 May 10 In a preliminary test of the Repulsor I rocket motor at the Society for Space Travel (VfR)'s Raketenflugplatz (rocket flight field), the device accidentally roared into the air to a height of 60 ft. before falling back to the ground. This was the first time a VfR rocket took to the air. Repulsor I was the second generation of VfR rockets and had been designed by Klaus Riedel; Willy Ley gave it its name from fictional space rockets in the 1897 novel *On Two Planets.* Repulsor I had a thrust of 55 lb. from a water-cooled combustion chamber and gasoline–liquid oxygen propellants using carbon dioxide as pressurant. Repulsor I was repaired and flown again on May 14, but at a height of 200 ft. it tilted over and crashed to the ground.

1931 May 23 Designed by Klaus Riedel, the Society for Space Travel's (VfR) Repulsor II rocket made a successful flight, reaching a height of 200 ft. before crashing to Earth outside the Raketenflugplatz (rocket flight field), 2,000 ft. from where it had been launched. The rocket differed from Repulsor I in that the oxygen was fed under its own pressure. After this second successful test the VfR wrote a report to the American Rocket Society and dispensed with further work on the now redundant Mirak motors being developed by Rudolf Nebel.

1931 May The assisted takeoff of a Russian aircraft equipped with rocket units took place for the first time when a U-1 biplane was boosted into the air by two solid propellant motors. The motors were attached to the upper surface of the lower wing close to the fuselage. They had been developed at the Gas Dynamics Laboratory under V. I. Dudakov. In 1933 more powerful rockets were used to boost a Tupolev TB-1 into the air, and the roll to takeoff of this heavy aircraft was reduced from 1,100 ft. to 260 ft.

1931 June In Germany the Society for Space Travel (VfR) carried out experiments at the Raketenflugplatz with the first

German rocket pioneer Klaus Riedel inspects a solid propellant, two-stick Repulsor 2 rocket in about May 1931. Note the top-mounted combustion chamber with propellant carried in the two vertical support pipes.

of the Repulsor III–series liquid propellant rockets. The two fuel tanks were placed closer together for greater aerodynamic efficiency and better grouping of the rocket's components. On its first flight early in the month it shot to a height estimated at 1,500 ft., but a parachute trigger failed to work properly and it crashed into the ground. Several more Repulsor III rockets were made and flown and a redeveloped version, Repulsor IV, was flown for the first time in August 1931. It became the mainstay of tests through May 1932, completing 87 rocket ascents and 270 static tests.

1931 August In Moscow, the Moscow Gruppa Isutcheniya Reaktivnovo Dvisheniya (Group for the Study of Reactive Motion), or Mos-GIRD, was set up by I. P. Fortikov with Fredrikh A. Tsander as a department head on November 18. In Leningrad, a Len-GIRD was formally established by Nikolai Rynin and Jakov I. Perelman with V. V. Raxumov as its head on September 13. The GIRD groups were associated with the Osoaviakhim (All-Union Society for Aviation and Chemical Warfare). Other GIRD groups sprang up in other cities, but interest centered on those in Leningrad and Moscow. At Mos-GIRD Tsander developed a 12-lb. thrust motor running on air and gasoline and known as OR-1 (for *opytnaya raketa,* or experimental rocket).

1932 January Charged with carrying out practical experiments, members of the American Interplanetary Society completed assembly of their first device, Rocket No. 1. Inspired by the German Mirak motors, it comprised a water-cooled aluminum combustion chamber 6 in. long and 3 in. in diameter located at at the front end of two tanks, 5 ft. 6 in. long and 1.5 in. in diameter, that contained liquid oxygen and gasoline. Four aluminum guide fins were at the rear, and a cone-shaped nose section contained a parachute. Work to prepare the rocket for launch began in August 1932 with the erection of a simple proving stand on a farm near Stockton, N.J.

1932 March 3 In Moscow a meeting of the Mos-GIRD rocket research group heard reports of work on reactive devices and formulated a plan to integrate activities from the several GIRDs springing up in the Soviet Union and form a central rocket research institute. This would become the Reaction Propulsion Research Institute (Reaktirnyi Nauchno Issledovatel'kii Institute, or RNII), formed in September 1933. Attendees at the meeting heard reports of rocket motor designs for aircraft, which many believed held the key to faster air travel. Emerging talent in this area as well as other innovative applications included Fredrikh A. Tsander, M. K. Tikhonravov and Yu. A. Pobedonostsev. Sergei P. Korolev was appointed in April 1932 to head up a technical team at Mos-GIRD.

1932 July Following a visit to the Society for Space Travel (VfR) at their Raketenflugplatz (rocket flight field) north of Berlin by Capt. Walter Dornberger, Capt. Ritter von Horstig and Gen. Karl Emil Becker of the German army during the spring, Wernher von Braun and others from the VfR went to Kummersdorf-West to show off a Repulsor rocket. The army had paid the VfR 1,000 Reichsmarks for the rocket and its demonstration. Dornberger met the contingent at the the test range and directed them to an appropriate launching spot where phototheodolites, cameras and chronographs were set up. At 2:00 P.M. the rocket was launched to a height of 200 ft. before it turned in a horizontal direction and crashed. Dornberger decided to employ von Braun in the Army Weapons Department, and von Braun reported for work on October 1, 1932.

1932 September   Robert Goddard resumed teaching at Clark University and received a small grant from the Smithsonian Institution enabling him to carry out tests on liquid propellant rocket technology. Over the next twelve months Goddard continued his theoretical work on rocketry and carried out a limited number of mechanical tests in his spare time. In September 1933 Goddard received a grant from the Daniel and Florence Guggenheim Foundation for the promotion of aeronautics, and for a further twelve months he studied centrifugal pumps, rocket chambers, gyroscopes, etc., before returning to Roswell, N.Mex., in September 1934.

1932 October 1   In one of the most significant appointments in the history of rocketry, Maj. Walter Dornberger hired the 20-year-old Wernher von Braun to work for the German Army at its Kummersdorf-West experimental rocket test range. Von Braun had received his degree in engineering from the Technische Hochschule, Berlin, and the Army paid for him to continue his studies at the Friedrich-Wilhelm University, from which he received a Ph.D. on April 16, 1934. Dornberger's other recruits included Heinrich Grünow, Walter Riedel from the Heylandt Works at Brietz near Berlin and, in 1933, Arthur Rudolph, who had helped the Army produce a static rocket motor in 1931.

1932 November 12   The American Interplanetary Society's Rocket No. 1, so extensively modified that it had been officially redesignated No. 2, was successfully given its first static firing from a makeshift proving stand at a farm outside

G. Edward Pendray fills the nascent American Rocket Society's first rocket with liquid oxygen prior to tests on November 12, 1932.

Stockton, N.J. As related later by AIS Vice President G. Edward Pendray, "It is impossible adequately to describe that sight, or to convey the feeling it gave us. I suppose we were excited; but there was a certain majesty about the sound and sight which made it impossible for the moment to feel excitement as such. We forgot to remain behind the shelter of our earthworks. Moreover, we forgot to count the seconds as they passed in that downward pouring cascade of fire."

1932 December 21   The first rocket motor produced by Wernher von Braun and Walter R. Dornberger at Kummersdorf-West test range flamed into life for the first time but exploded less than a second after ignition. With a 1:3 mixture ratio of liquid oxygen and pure ethyl alcohol, the pear-shaped aluminum motor was 1 ft. 8 in. in length and calculated to produce a thrust of 650 lb. The engine was designed for a test rocket called A-1 (Assembly 1) which would have a length of 4 ft. 6 in., a diameter of 1 ft. and a weight of 330 lb. Dornberger and von Braun needed to demonstrate a working rocket to convince the German army that it should allocate additional funds for a ballistic missile. Although the A-1 never flew, the A-2 had essentially the same design, the chief difference being the location of the gyroscopic platform.

1933 March 9   Rudolf Nebel and Klaus Riedel of the Verein für Raumschiffahrt (VfR) began static tests at the Raketenflugplatz with a regeneratively cooled liquid propellant rocket motor capable of producing thrust in the 330–440-lb. range. On March 25 and April 3, 1933, two motors exploded, but a series of further tests was conducted successfully. The work was, however, soon brought to a halt. Nebel had used nefarious means to obtain funds for work on his rockets, falsifying accounts and failing to pay VfR bills, and the VfR was running into serious financial problems. Willy Ley and the president, Hans-Wolf von Dickhuth-Harrach, drew up charges against Nebel, and he was suspended. With Adolf Hitler in power, Ley left Germany for America, where he spent the rest of his life. In 1934 the VfR was wound up.

1933 March 18   Fired for the first time in a test stand, the OR-2 oxygen-gasoline rocket motor, developed by the Mos-GIRD rocket research team in Moscow, delivered a thrust of 110 lb. Designed and put together by a small team under Fredrikh Tsander, the OR-2 had been built to power a glider designated RP-1 (*raketny planer,* or rocket glider). Built by C. I. Cheranovskii, this glider never flew with the OR-2 because the case burned through under test, and the design was then adapted for the GIRD-Kh rocket. The aircraft eventually flew, powered by a 22 HP ABS Scorpion engine.

1933 March 26   Operated from a test-bed at the Mos-GIRD research facility in Moscow, an experimental ramjet engine demonstrated successful operation for a 15-min. period; a pressure of 15 atmospheres in the final compression stage was recorded. Beginning in June 1933, Mos-GIRD conducted research on pulse-jet engines. Ramjet and pulse-jet propulsion was important because significantly heavier payloads could be carried by a rocket gaining its initial lift through the atmosphere from air-breathing reaction engines. There was a general belief, first put down by Konstantin Tslolkovsky, that air-breathing engines could boost rocket stages to a speed of Mach 10 in the atmosphere.

1933 May 14 The American Interplanetary Society's (AIS) Rocket No. 2 made its first flight from Great Kills in Staten Island, New York. The motor shut down after 2 seconds at an altitude of barely 250 ft. when the liquid oxygen tank burst and the rear fins separated. The combustion chamber landed 400 ft. offshore and was recovered. For the rest of the year the AIS worked on three more rockets. Rocket No. 3 was designed by Alfred Africano, Edward Pendray and Bernard Smith. It was 5 ft. 6 in. tall with a diameter of 8 in. and weighed 20 lb., its 60-lb. thrust motor operating on liquid oxygen and gasoline. No flights were attempted.

1933 August 17 Developed by a Mos-GIRD team under the leadership of Sergei Pavlovich Korolev, the first Russian liquid propellant rocket to successfully demonstrate free flight was launched from the Nakhabino test range. The rocket, designated GIRD 9, was 7 ft. 10 in. tall with a fueled weight of 42 lb. and equipped with a motor burning a solidified charge of gasoline and liquid oxygen. The rocket flew for 18 sec. and reached a height of 1,300 ft. before it fell back to Earth; its combustion chamber had burned through.

1933 September 21 In Russia, Mos-GIRD, Len-GIRD and the Gas Dynamics Laboratory (GDL) were merged into a central organization for the study and practical development of solid and liquid propellant rocketry. The new organization was known as the Reaktirnyi Nauchno Issledovatel'kii Institute (RNII), or Reaction Propulsion Research Institute. I. T. Kleimenov, a military engineer from Leningrad, was appointed to head the RNII and Sergei Pavlovich Korolev was to be his deputy. The GDL had made outstanding progress with the development of many different liquid rocket motors since 1929, and its work would form the basis for further research. The new RNII moved into a large two-story building on the outskirts of Moscow.

1933 October 13 The first formal meeting of the British Interplanetary Society (BIS) took place in Liverpool, England. Among the founding members were the science fiction writer Arthur C. Clarke and Ralph A. Smith, who would play a leading role in drawing up plans for manned flight. On October 27, 1936, the inaugural meeting of a London branch

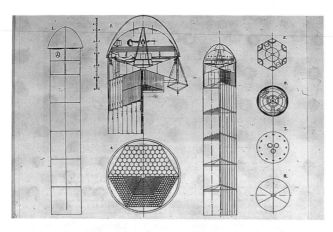

Designed by a team from the British Interplanetary Society in 1937, this spaceship was a solid propellant, multistage rocket surmounted by a pressurized compartment.

took place, and because the Liverpool group was having problems the London group took over leadership of the BIS, and a new constitution was ratified at the annual general meeting on February 7, 1937. When war broke out in 1939, the BIS was suspended, but Kenneth W. Gatland and N. H. Pantlin formed the Astronautical Development Society at Hawker Aircraft's Kingston factory. In early 1944 this and a Manchester group merged to become the Combined British Astronautical Societies in early 1944.

1933 November 25 Designed by the Mos-GIRD team of Russian rocket entrepreneurs, the GIRD-Kh design was launched from the Nakhabino range, utilizing a modified Tsander engine built originally for the RP-1 glider. This was the third flight of a liquid propellant rocket in the Soviet Union. The first two had been performed with Rocket 09, which had a length of just over 7 ft. and a weight of 66 lb. The modified OR-2 motor that powered it had a thrust of 155 lb. Like its predecessor Rocket 09, GIRD-Kh had been put together by a team which included the increasingly prominent Sergei Pavlovich Korolev.

1933 During the latter part of the year, the American Interplanetary Society (AIS) changed its name to the American Rocket Society. This name change helped remove an implied association with any ambitions of space travel, then viewed with suspicion, and it reflects how trivial the general public considered space travel to be. The AIS had decided to concentrate its efforts on designing and testing practical rocket motors. It continued to support experimental research until 1941 when it devoted its resources to the publication of technical papers and the dissemination of astronautical news. On February 1, 1963, it merged with the Institute of Aerospace Sciences to form the American Institute of Aeronautics and Astronautics.

1934 March 31 In Leningrad, the All-Union Conference on the Study of the Stratosphere began a three-day symposium at which ideas for building a sounding rocket for the upper atmosphere were circulated. Scientists wanted to gain access to the upper atmosphere for measuring cosmic rays originating in the stars, and engineers wished to expand the development of rocketry. The symposium concluded that such a rocket was needed and that it should be designed with a view to the subsequent development of a manned rocket.

1934 May 5 The first flight in the Soviet Union of a pilotless, rocket-propelled aircraft took place when the 06/I was hurled into the air and powered by the OR-2 motor which delivered a thrust of 110 lb. There was strong interest in Russia for rocket-powered aircraft and the work had been organized by Yevgeny S. Shchetinkov. The winged rocket was launched along an inclined rail, but its trajectory was erratic. A later projectile, designated 06/III, incorporated ailerons as well as elevators, the latter having been the only means of aerodynamic control on the 06/I. Over the next two years a detailed study of the dynamics of rocket-powered flight carried out by B. V. Raushenbakh led to the Model 216 winged rocket.

1934 September 9 The launch of the American Rocket Society's second projectile, Rocket No. 4, took place at Great Kills in Staten Island, New York. It was 7 ft. 6 in. tall with a diameter of 3 in. Following a 15-sec. burn of the gasoline–liquid oxygen motor, Rocket No. 4 reached a height of 382 ft.

and a speed in excess of 600 MPH, coming to Earth by parachute 1,338 ft. from the launch stand. A proposed Rocket No. 5, which would have been bigger than Rocket No. 4, was never built as the American Rocket Society decided to focus its work on the development and static testing of motors.

1934 September　Robert Goddard returned to the Mescalero Ranch at Roswell, N.Mex., where from 1930 to 1932 he had conducted live firing tests with rocket motors and flying devices. He was to remain at Roswell for seven years in one of the most remarkable displays of lone-wolf research and experimentation of the 20th century. During this period, Goddard carried out a remarkably well recorded series of test programs, culminating in pump-fed propellant systems, gas generators for turbines and flights to a height of almost 5,000 ft. The work began with A-series tests through October 1935 using rockets between 13.5 ft. and 15.25 ft. in length and equipped with pressure-fed systems and gyro-controlled vanes.

1934 December　Designed and built by a small team led by Wernher von Braun and Capt. Walter R. Dornberger at the German army experimental station in Kummersdorf-West, Berlin, two A-2 liquid propellant rockets were successfully launched from the German North Sea island of Borkum. Dubbed Max and Moritz, after two rambunctious cartoon characters, the rockets reached a maximum height of about 7,000 ft. They were powered by the 650-lb. thrust ethyl alcohol–liquid oxygen motor designed for the A-1 and operated for 45 sec. Test objectives for Max and Moritz were propulsion and guidance control.

1935 January　In developments parallel to those being conducted in Russia on winged projectiles powered by liquid propellant rocket motors, solid propellant motors from the Reaktirnyi Nauchno Issledovatel'kii Institute (RNII) were used for the first flight of the Type 48 unguided vehicle. Under the direction of M. P. Dryazgov, the solid propellant projectiles were thought to be a quick and inexpensive means of generating flight data for liquid motor projectiles (the first of which had been launched on May 5, 1934), but they quickly came to be seen as a potential weapon in their own right. A refined design called Model 217 was built and led to Model 217/II, tested on November 19, 1936.

1935 March 2　At the All-Union Conference on the Use of Jet-propelled Aircraft in the Exploration of the Stratosphere, Sergei Pavlovich Korolev presented concepts for rocket-powered aircraft in a detailed report entitled *Winged Rocket for Manned Flight*. Korolev pressed for the development of a rocket-plane laboratory specifically dedicated to the detailed design of such a vehicle. In addition to work being conducted at the Reaktirnyi Nauchno Issledovatel'kii Institute (RNII), Moscow University took up Korolev's challenge and carried out technical studies which supported the RP-318 program. V. N. Vetchinkin, Valentin Petrovich Glushko, Yu. A. Pobedonostev, M. S. Kisenko and A. V. Zagulin also read papers at the conference.

1935 March 27　During a symposium on rockets at the California Institute of Technology (Cal Tech), William Bollay, then a graduate assistant to Theodore von Karman, presented results of studies he had performed on rocket-propelled aircraft. Bollay's work had been inspired by the writings of Eugene Sänger. Publicity from this paper attracted chemist John W. Parsons and engineer Edward S. Forman to Cal Tech. On October 4, 1935, Bollay gave a lecture to the Institute of Aeronautical Sciences in Los Angeles in which he advocated development of a rocket-propelled "stratosphere plane" for meteorological research and "commercial purposes," and which affirmed that "There are thus potent reasons for the further development of the rocket plane."

1935 April 21　The first in series of static rocket motor tests conducted by the American Rocket Society took place at Crestwood, a suburb of New York City. Five motors using short and long combustion chambers were tested, but none was successful. A report of these tests was published in the June 1935 issue of the society's journal *Astronautics*. In subsequent tests carried out in August and October 1935, a great amount of information was gathered from which James H. Wyld designed a successful regeneratively cooled motor in 1938.

1935 April　The French solid propellant rocket entrepreneur Louis Damblanc published his book *Self-Propelled Rockets* and began a number of scientific tests and analyses of the combustion process involved in high-pressure motors. His tests were carried out in special rigs, and the performance of the motor was filmed in slow motion. Microphones recorded the sound so that Damblanc could correlate visual and audiophonic recordings. On June 26, 1936, he was granted a patent for his design of a multistage rocket incorporating automatic separation mechanisms. Interrupted by World War II, Damblanc's work continued after 1945 and he eventually became deputy director of the French National Space Research Center.

1935 November 22　Robert Goddard began his K-series static tests at the Mescalero Ranch in Roswell, N.Mex., refining the design of a 10-in. combustion chamber. The rocket motors resulting from these experiments demonstrated a static thrust of up to 623 lb. for 14 sec., with exhaust velocities of around 4,400 ft./sec. Ten static tests were carried out before Goddard turned once again to flight trials on May 1, 1936.

1935　The German chemist Helmuth Walter formed the Helmuth Walter Kommandogesellschaft to develop small, liquid propellant rocket motors. Prior to this, Walter had worked on the design of wakeless torpedoes for German shipyards at Kiel. His designs were based on the decomposition of hydrogen peroxide, which produces steam that quickly condenses into water. The torpedo proved impractical, but Walter's work on hydrogen peroxide with phosphate stabilizers produced a fuel known as *T-Stoff*. In 1936 Walter was given a contract to produce 90-lb. thrust motors for the aircraft industry, the first production contract awarded anywhere for rocket motors. His hydrogen peroxide systems were later adopted for the turbopumps in the V-2 rocket.

1936 February 1　The Deutsche Versuchsanstalt für Luftfahrt (German Aviation Research Institute), or DVL, signed a contract with Eugene Sänger for development work on a regeneratively cooled liquid propellant rocket motor. Construction of an experiment laboratory to support development of a 200,500-lb. thrust rocket motor began at Trauen in the Lüneburg Heath in February 1937. Sänger wanted to design rocket motors for space planes that could skip across the upper atmosphere and fly great distances, generating lift from wings. He realized that the propulsion system was the key to achieving this.

Assistants prepare a rocket for Robert H. Goddard prior to tests at Roswell, N.Mex., in November 1935. Dubbed "the father of modern rocketry," Goddard established the optimum configuration for a rocket, placing the thrust chamber at the bottom and the propellant tanks above.

1936 April 2 Members of the German army and the Luftwaffe met to agree on the joint development of a proving ground for tests with rockets and pulse-jet motors. The new test range would be located at a place called Peenemünde on the Baltic island of Usedom, suggested by von Braun's mother who recalled her husband's fishing trips there. It was to be funded jointly but run by the army. Having seen tests of liquid rocket motors developed at Kummersdorf-West by the army's Wernher von Braun, the Luftwaffe was keen not to be outdone. For its part, the army pledged increased funding for work on a ballistic missile and for a test rocket called A-3 (Assembly 3) which had been in design throughout 1935. This rocket would help develop propulsion systems, propellant feed systems, guidance units and control systems for in-flight stability.

1936 April 6 The first launch of an atmospheric sounding rocket developed by the Reaktirnyi Nauchno Issledovatel'kii Institute (RNII) took place in the Soviet Union. Developed over a period of four years in cooperation with the All-Union Aeronautic Research and Technical Society, the rocket was named *Aviavnito*. Designed by the Gas Dynamics Laboratory under the RNII, the rocket had a length of 10 ft. and a weight of 220 lb., and the engine produced a thrust of about 660 lb. The rocket was built to carry an altimeter and a barograph. Although it had been designed to reach an altitude of 35,000 ft., it never did. On a subsequent flight (August 15, 1937) it reached 1,000 ft., but during descent a parachute deployed to lower it gently to the ground tore and the rocket was damaged.

1936 May 9 In the Soviet Union, a winged, rocket-powered, unmanned projectile designated Model 216 was launched from an inclined rail for the first time. Incorporating a modified OR-2 alcohol-oxygen engine delivering a thrust of 220 lb., the rocket had a takeoff weight of 175 lb. a takeoff speed of 118 ft./sec. and a maximum speed of 590 ft./sec. (180 MPH). By November 4, 1936, four Model 216 winged rockets had been launched, two from a trolley. A later version, Model 212, flew twice in 1939 but failed to achieve predicted flight paths.

1936 May 11 Robert Goddard began the L-series liquid propellant rocket tests from his work site at the Mescalero Ranch in Roswell, N.Mex. These were conducted through August 9, 1938, and were divided into three separate but overlapping sections. Section A tests utilized the 10-in. motor developed during K-series static testing and involved seven rockets varying in length from 10 ft. 11 in. to 13 ft. 6 in. and weighing between 289 lb. and 360 lb.; these flights ended November 7, 1936. Section B tests involved eight rockets using a variety of propellants, parachute recovery devices and flight controls; they lasted from May 24, 1936, to May 19, 1937. Section C tests featured light tank construction, movable tailpieces and nitrogen-tank pressurization in flights conducted between July 28, 1937, and August 9, 1938.

1936 June 22 Robert Esnault-Pelterie was elected a member of the French Académie des Sciences in the division *"applications de la science à l'industrie"* (applications of science to industry). Esnault-Pelterie's accomplishments consisted in the way he sought solutions to engineering problems. He had proposed the use of gimbaled nozzles to facilitate flight control, designed regenerative cooling nozzles and suggested mechanisms resulting in the integrating accelerometers used in the A-4/V-2. Esnault-Pelterie was never rewarded financially by his country or his peers, and seizure of his goods and assets by the tax authorities left him penniless when he died on December 6, 1957, at age 76.

1936 October 29 Under the auspices of the Rocket Research Project at the Guggenheim Aeronautical Laboratory of the California Institute of Technology (GALCIT), the first of four rocket motor tests using a mixture of methyl alcohol as fuel and oxygen as the oxidizer took place in the Arroyo Seco north of Pasadena, Calif. Eyed with suspicion by the conservative university faculty, the GALCIT rocket team had been formed earlier in the year, original members being Dr. Frank J. Malina, Hsue-Shien Tsien, A. O. M. Smith, John W. Parsons, Edward Forman and Weld Arnold. GALCIT would conduct pioneer work with solid and liquid motors and as the Jet Propulsion Laboratory (JPL) would achieve world fame as NASA's center for planetary exploration.

1936 November 19 In Russia the first flight of the Model 217/II solid propellant winged missile took place from an inclined lattice-work ramp. With a takeoff weight of 265 lb., it had a thrust of 4,080 lb., a top speed of 850 ft./sec. (560 MPH) and a potential maximum altitude of 10,000 ft. Conceived as a research device for liquid propellant rocket aircraft, the Model 217/II firmly established the winged solid propellant projectile as a potential antiaircraft weapon and is the first clearly defined example of a rocket designed for that purpose. The political instability and uncertainty of financial support in the Soviet Union during this period inhibited further development.

1936 During the fall, Helmuth Walter conducted trials in Germany with hydrogen peroxide rocket motors fitted beneath the fuselage of a Heinkel He 72B Kadett biplane Trainer. Later trials were conducted using motors attached to a Focke Wulf Fw 56 Stösser high-wing monoplane trainer. Together, these tests helped build faith in the suitability of liquid propellant rocket motors for adding thrust to conventional aircraft during takeoff. Before the end of 1936, Dr. Adolf Baeumker, chief of the Research Department at the Air Ministry, backed Walter in the development of a throttleable hydrogen peroxide motor using a liquid catalyst.

During the year, the specification for the German Army's long-range ballistic missile was laid down by Wernher von Braun and Walter Dornberger. Designated A-4, the missile would be designed to throw a 1-metric-ton (2,205-lb. warhead 170 mi. It would need a motor burning 8 tons of alcohol and liquid oxygen, producing 50,000 lb. of thrust and reaching an exhaust velocity of 7,000 ft./sec. To permit movement via road and railway, the missile would have to be no more than 45 ft. in length, 5 ft. in diameter and 11 ft. 6 in. over the tail fins. To accommodate tests with the A-4 engine and propulsion systems for growth versions, the largest static test stand at Peenemünde would be built for motors of up to 200,000-lb. thrust.

1937 April Late in the month a Heinkel He 112 specially adapted to carry liquid propellant rocket motors flew for the first time at Kummersdorf near Berlin. Piloted by Erich Warsitz, the aircraft was supposed to carry a motor developed by Wernher von Braun and powered by liquid oxygen–methyl alcohol, but for this flight it retained its Junkers Jumo piston engine as the primary means of propulsion. The rocket motor provided a thrust of 2,200 lb. and operated for 30 sec. The aircraft crash-landed at the end of the flight due to a malfunction of the motor. It was rebuilt and flown again in June 1937 on rocket power alone.

1937 July 2 Former members of the Verein für Raumschiffahrt (VfR) Klaus Riedel and Rudolf Nebel signed a contract with the German Army Weapons Office under which they turned over all patent rights on rockets and associated devices in return for RM 75,000 (Reichsmarks) and employment. Riedel went to join von Braun at the emerging experimental test station at Peenemünde, and other VfR members were also drawn in. These included Rolf Engel, Hans Heuter, Kurt Hainisch and Helmuth Zoike, among others. German rocket research had been gathered up into the Army, and a cloak of secrecy had descended over the activities at Kummersdorf-West and at the new test site at Peenemünde. In 1935, Nazi propaganda minister Joseph Goebbels had forbidden the use of the word *rocket* in German literature.

1937 July 27 Dr. Stephen H. Smith, an Indian dentist, sent by rocket a rooster, a hen and 189 letters across the River Damodar in Bengal. In other bizarre undertakings he lobbed a small container of mice together with cheese and sugar for sustenance in what may possibly be the world's first animal flight by rocket. Smith conducted many flights with novel rocket designs of his own, including some with wings for added glide at the end of the propulsive phase. In one instance he set up a mail delivery service from the bank of a river to a small island. This activity was financed by the maharajah Sir Tashi Namgyal and sanctioned by the British Post Office.

1937 December 4 The first A-3, developed by Wernher von Braun as an experimental test rocket for guidance and pro-

pulsion systems, was launched. This first-ever rocket launch from the Army's experimental test station at Peenemünde on the Baltic took place on the island of Greifswalder Oie. A recovery parachute ejected prematurely and the motor cut off at 6.5 sec., causing the A-3 to crash 980 ft. from the launch site. The A-3 had a length of 22 ft., a diameter of 2 ft. 2.5 in., a weight of 1,650 lb. and a maximum thrust of 3,300 lb. from an engine using ethyl alcohol and liquid oxygen propellants. The rocket had a three-axis gyroscopic control system and liquid nitrogen pressurization. Subsequent flights took place on December 6, 8, and 11, 1937; all were unsuccessful. The A-3 was abandoned in favor of the A-5, an interim development design before the definitive A-4 ballistic missile.

1937 December A midshipman at the U.S. Naval Academy in Annapolis, Md., Robert C. Truax conducted his first engineering measurement of a rocket motor on test by monitoring its performance with instruments. Truax had designed a small liquid propellant motor and meticulously put it together over an eight-month period between courses at the Academy. Operating on gasoline and compressed air, the motor ran for up to 45 sec. In December 1938 further tests were carried out, with liquid oxygen being substituted for air. The tests were completed by the middle of 1939, and by that October Truax had designed a rocket for assisting flying boats into the air. This work was given formal U.S. Navy approval in 1941.

1937 Russian rocket researchers turned again to the possibility of developing a piloted aircraft powered by a liquid propellant rocket motor. Valentin Petrovich Glushko had designed several candidate motors burning alcohol, kerosene or gasoline as fuel with nitric acid as oxidant. The thrust output ranged from 175 lb. to 1,320 lb. One of these, the ORM-65, used kerosene to deliver a thrust of 385 lb., and Sergei Pavlovich Korolev suggested this motor be put into an SK-9 sailplane. Structural modifications were put under the authority of A. Y. Shcherbakov and A. V. Pallo and the project was designated RP-318. Problems with the ORM-65 resulted in two new motors being considered: the kerosene-nitric acid RDA-1–150 and the alcohol–liquid oxygen RDK-1–150.

1938 September Drop tests from a Heinkel He 111 began with small scale models of the A-5 rocket developed at the Peenemünde experimental test range in Germany. Released at 20,000 ft., the models were 5 ft. long, 8 in. in diameter and weighed 550 lb. They became supersonic at about 3,000 ft. and demonstrated the stability and effectiveness of different fin and control devices. In the absence of a supersonic wind tunnel, this was the only way of exposing the models to maximum aerodynamic pressure, a value known as max-Q, at which the missile could become unstable and lose control. These drop tests were also used to qualify parachute designs for recovering A-5 rockets from tests in which the speed did not exceed 250 MPH.

1938 October 17 Concurrent with L-series liquid propellant rocket tests at the Mescalero Ranch in Roswell, N.Mex., Robert Goddard began a series of experiments leading to the development of propellant pumps. Aware that efficient and reliable propellant delivery was an important key to successful and stable flight, Goddard concentrated on high-speed centrifugal pumps, and in more than 20 tests of 5 pumps demonstrated a variety of designs from which he selected 4 for static tests to take place between January 6 and February 28, 1939.

1938 October The Austrian engineer Eugene Sänger began work at the Deutsche Versuchsanstalt für Luftfahrt (German Aviation Research Institute) on a ¹⁄₂₀th scale model of his winged space plane. This is the first known wind-tunnel model of a reentry vehicle designed to fly back through the atmosphere and land like a conventional aircraft. The model had a long fuselage of circular cross-section supporting a low-wing lifting surface with straight trailing edge and modest sweep on the leading edge. It had conventional tail surfaces. Although a full-scale space plane was never built, wind tunnel work of this kind resulted in German aerodynamicists' adopting swept-back wings for hypersonic flight. This represented a considerable advance on theoretical studies of high-speed flight in other countries.

1938 December 28 Frank J. Malina from the GALCIT (Guggenheim Aeronautical Laboratory at the California Institute of Technology) rocket team presented a report on the use of rockets for aircraft assisted takeoff to the National Academy of Sciences Committee on Army Air Corps Research, Washington, D.C. On behalf of Dr. Theodore von Karman, Malina proposed a program of research at Cal Tech on possible motor designs and experimental demonstrations of suitable concepts. Shortly after this appeal, a $10,000 grant was awarded, and in 1939 the GALCIT Rocket Research Project was renamed the Jet Propulsion Research Project.

1938 During the fall, tests were conducted in Germany at the Peenemünde research facility with a throttleable rocket motor attached to a specially modified Heinkel He 112. A new Walter motor designated TP-1 had been developed in 1937 incorporating a liquid instead of a paste as a catalyst for the hydrogen peroxide (*T-Stoff*) propellant. Incorporating a turbine to drive hydrogen peroxide on to the liquid catalyst, variable thrust could be achieved by altering the speed of the turbine. The liquid catalyst was an aqueous solution of calcium permanganate (*Z-Stoff*). Previously, Walter motors had used a paste catalyst with a constant thrust motor. From these tests Heinkel developed the He 176 as a pure rocket research aircraft designed to be capable of exceeding existing world speed records.

1939 January 2 A top secret Section L was set up at the German aircraft factory of Messerschmitt AG in Augsburg for the development of a rocket-powered tailless aircraft designed by Dr. Alexander Lippisch of the Deutsches Forschungsinstitut für Segelflug (German Research Institute for Sailplanes). Designated DFS 194, Lippisch's delta-wing design had been selected by the German Air Ministry to have a rocket engine, and a Walter RI-203 of the type fitted to the Heinkel 176 but reduced in thrust to 660 lb. was selected. Section L was charged with producing a prototype rocket-powered interceptor designated Messerschmitt MC 163.

1939 January The January issue of the *Journal of the British Interplanetary Society* included details of a manned lunar spaceship which had been worked out over almost two years of intense effort. The design team consisted of H. Bramhill, Arthur C. Clarke, Val Cleaver, M. Hanson, Arthur Janser, S. Klementaski, H. E. Ross, R. A. Smith and James Happian Edwards. Propulsion provided by 2,490 solid propellant rocket tubes was to propel the 1,225-ton vehicle, 100 ft. tall and with a diameter of 20 ft. to a translunar trajectory. Upon

arriving at the vicinity of the moon, the three occupants were to turn the spaceship around and back it down on to the moon using four landing legs. When it was time to leave, the crew would fire off the upper section, leaving the landing section on the moon.

1939 February 11 Developed in Moscow by the Reaktirnyi Nauchno Issledovatel'kii Institute (RNII) for the RP-318 rocket-powered glider, the kerosene–liquid oxygen RDA-1-150 motor was tested for the first time. Essentially a modified version of an earlier motor designed by Valentin Petrovich Glushko (the ORM-65), it weighed 220 lb. and delivered a maximum thrust of 330 lb. RNII engineers Leonid S. Dushkin and Aleksei M. Isaev were responsible for its detailed design. An alternative engine that burned alcohol with nitric acid (the RDK-1-150) was abandoned in favor of the RDA-1-150.

1939 March 5 At an airport near the Planernaya Station outside Moscow, the first successful flight test of a ramjet-assisted rocket took place with the launch of a two-stage projectile. The first stage was a solid propellant rocket with a maximum thrust of 990 lb. that burned for 2.24 sec. with a total impulse of 573 kg./sec. The second, ramjet, stage contained solid fuel placed within the combustion chamber. A formal test program began May 19, 1939, using grains supplied by the Mendeleyev Chemical Engineering Institute. The first stage achieved a speed of 650 ft./sec. (443 MPH) and separated from the ramjet stage. In the 2.5 sec. before the ramjet ignited, speed fell to 344 ft./sec. (235 MPH) During the 5.12 sec. burn of the ramjet, speed increased to 734 ft./sec. (501 MPH).

1939 March 24 Robert Goddard began a series of eleven static tests at the Mescalero Ranch in Roswell, N.Mex., supporting development of gas generators for turbines capable of delivering liquid propellant to rocket motors. The static tests had been completed by April 28, and from July 17 to August 4, 1939, Goddard conducted a series of static firings using these turbines to produce a chamber thrust of 700 lb. for 15 sec. at an oxygen flow rate of 4 lb./sec. and a gasoline flow rate of 3 lb./sec. The gas generators worked at 180 lb./sq. in. and between May 18 and August 4, 1939, they were operated at the launching stand to support flight tests.

1939 March Experimental flights began at the German rocket research station in Peenemünde, with scale models of the A-5 development rocket. Launched without guidance, each 95-lb. model was 5 ft. in length with a diameter of 8 in. and propelled by the steam produced by the decomposition of 45 lb. of hydrogen peroxide (*T-Stoff*). With an exhaust velocity of 3,500 ft./sec., thrust was 260 lb. The models provided a quick and relatively cheap means of testing different fin and control vane configurations before flights began with the full-sized A-5. The optimum fin was broader and flatter than fins on the A-3 but set the pattern for fins on the A-4 ballistic missile. Also this month, Adolf Hitler paid his first visit to Peenemünde and was given a guided tour of the facilities by Walter Dornberger and shown cutaway models of the A-5 rocket by Wernher von Braun. He was unimpressed.

1939 June 20 Powered by a 100–1,100-lb. thrust Walter HWK RI-203 (TP-2) rocket motor, the Heinkel He 176 made its first flight from the German research center at Peenemünde piloted by Flugkapitän Erich Warsitz. The motor used turbine driven hydrogen peroxide (*T-Stoff*) and an aqueous solution of calcium permanganate (*Z-Stoff*) as a catalyst. The He 176

A two-stage rocket designed by the Soviet engineer I. A. Merkulov in 1939 seen here before takeoff. The rocket is nestled inside a structure of four guide rails.

had been specifically designed for rocket motors, but the aircraft made only a few flights and never exceeded 434 MPH. An attempt to turn the He 176 into a fighter failed because the aircraft was unstable in maneuvers and did not have the handling qualities, or the performance, necessary for that role.

1939 July 1   The GALCIT (Guggenheim Aerouautical Laboratory at the California Institute of Technology) group

received a $10,000 contract from the National Academy of Sciences to develop liquid and solid propellant rockets that could be used to augment the takeoff power of piston-engined aircraft. Thus the Army Air Corps Jet Propulsion Research Group came into force with a mandate to develop solid or liquid propellant motors for decreasing aircraft takeoff distance, increasing the rate of climb for brief periods and increasing level-flight speed when required by the pilot. This work eventually came to be known as JATO (jet-assisted takeoff). Dr. Theodore von Karman was director of the work.

1939 July 3   In Russia, the Technical Council of the People's Commissariat for the Aircraft Industry reviewed test results from experimental work on rocket-boosted ramjets conducted by the Stratospheric Committee–Jet Section. Beginning with the first successful rocket-boosted ramjet flight on March 5, 1939, further tests usefully employed rockets to accelerate experimental ramjets to operating speed. The Technical Council ordered tests involving aircraft as a next step toward the development of this form of propulsion. In August 1939 a prototype ramjet for aircraft was designed, and flight tests took place in December.

1939 July 6   Wernher von Braun submitted to the German Air Ministry a proposal for a vertically launched, rocket-powered, piloted interceptor. With a liftoff weight of 11,145 lb., the interceptor would be powered by a 22,000-lb. thrust rocket motor and reach a height of 26,250 ft. in 53 sec. using a three-axis gyrostabilizer. At this height the main motor would be shut down and power would be provided by a 1,500-lb. thrust auxiliary rocket motor giving the interceptor a top speed of 447 MPH. Gliding back down to the ground, the aircraft would land on a skid. The idea was based on successful tests with the A-3 liquid propellant rocket motor. The interceptor was never built but it inspired Erich Bachem to design the Ba 349 Natter, which flew unmanned for the first time on December 22, 1944.

1939 September 5   Maj. Walter Dornberger from the Peenemünde rocket research station in Germany accompanied Gen. Karl Emil Becker to testify before Col. Gen. Walter von Brauchitsch, commander in chief of the German Army, on developments with the ballistic missile program. Von Brauchitsch saw the potential in the A-4 and quietly approved Dornberger's plan to assign 4,000 technically qualified Army personnel to Peenemünde. These men were grouped into a new unit, Northern Experimental Command, and classed as front-line troops to render them immune from reallocation. At the end of September, groups of university professors visited Peenemünde and, sworn to secrecy, were sent away to work on specialized problems with the A-4 project. One of these was Dr. Ernst Hueter, under whom assistant professor Kurt Debus was working on voltage surges and cathode ray oscillography. When von Brauchitsch fell from favor in spring 1940, the privilege of priority status enjoyed by the Peenemünde team was withdrawn, slowing work on the A-4.

1939 September 19   In a speech in Danzig, Germany, Adolf Hitler made blatant reference to the Peenemünde rocket program when he said that if the war lasted a further four or five years Germany would have access to weapons "that will not be available to other nations." This veiled remark failed to be noticed for what it was, so effectively had the Nazis covered up the work at Kummersdorf-West and at Peenemünde. Only two years earlier, the *Journal of the British Interplanetary Society* claimed that rocket research had "ceased around the world."

1939 October Flight tests began with the A-5 missile from the island of Greifswalder Oie, part of the Peenemünde rocket research station in Germany. With a length of 24 ft. 2 in., a diameter of 2 ft. 6 in. and a weight of 1,760 lb., the missile was powered by a 3,300-lb. thrust motor burning ethyl alcohol and liquid oxygen for a maximum duration of 45 sec. Only a little larger than its precursor the A-3 and retaining only the engine of that configuration, the A-5 was in every other respect a subscale prototype for the A-4. On its first flight the A-5 reached a height of almost 30,000 ft. on propulsion, coasted several thousand feet higher and fell back. Von Braun sent the radio signal to deploy the two recovery parachutes and it was recovered from the Baltic. Some 70–80 flights took place by the end of 1942.

    At the Jet Propulsion Research Group of GALCIT (Guggenheim Aeronautical Laboratory at California Institute of Technology), Frank J. Malina completed a report on the application of liquid propellant rocket motors for auxiliary propulsion that had far-reaching implications. U.S. Navy Capt. D. S. Fahrney from the Bureau of Aeronautics took the recommendations and applied them to pilotless flying bombs. Later, he would develop these into the various guided missile projects embraced by Gorgon, the program name for a wide range of air-to-air and air-to-surface missiles developed by the Navy during the mid-1940s.

1939 November 18 Robert Goddard began a series of 24 static and live launch tests with solid propellant rockets carrying a greater propellant capacity than ever before. The rockets used were about 22 ft. in length with a diameter of 18 in., weighed up to 592 lb. including 140 lb. of liquid oxygen and 112 lb. of gasoline. These rockets used fuel pumps and combustion chambers that had been developed in earlier work. Goddard's second period of experimentation and testing at Roswell, N.Mex., ended October 10, 1941.

1939 December The first flight tests of a ramjet engine took place in Russia with a specially modified Polikarpov I-15bis piloted by Petr Y. Loginov. The biplane carried a single DM-2 ramjet under each wing but retained its radial engine for takeoff, landing and conventional flight. The ramjets worked as predicted, increasing by 12 MPH the speed of the aircraft in level flight. In all, 54 test flights were made. A bigger DM-4 engine in a special version of the I-15 designated I-153DM resulted in a 32-MPH increase in speed. A Yakovlev Yak-7 monoplane fighter was modified in 1944 to carry two DM-4C ramjets giving a speed increase of 56 MPH.

1940 February 28 Towed into the air by a Polikarpov I-5 biplane fighter, the first Soviet rocket-propelled aircraft, the RP-318, made its first flight. Pilot Vladimir P. Fedorov switched on the rocket motor which operated for 110 sec., giving the RP-318 a climb rate of 590 ft./min. The RP-318 had a length of 23 ft. 10 in., a wingspan of 55 ft. 9 in. and a maximum weight of 1,543 lb. On a future occasion aerodynamicist Aleksander Y. Bereznyak, working under Prof. Viktor F. Bolkhovitinov at the Soviet Air Force Academy, saw it perform. Aleksei Isaev told Bereznyak about a more powerful motor he was designing and Prof. Bolkhovitinov got permis-

sion from Stalin to develop a rocket fighter later designated BI, after Bereznyak and Isaev.

1940 May 28 Robert Goddard and Harry G. Guggenheim, the son of the founder of the Daniel and Florence Guggenheim Foundation that had underwritten Goddard's work in New Mexico, met with a joint Army-Navy committee in Washington, D.C. They were there to present detailed information on the potential of solid and liquid propellant rockets for military purposes. The Navy had conceived of using rockets to help aircraft get off the flight decks of aircraft carriers. The Army was generally unenthusiastic about the rocket's potential.

1940 July 1 The GALCIT (Guggenheim Aeronautical Laboratory at the California Institute of Technology) group working on development of jet-assisted takeoff (JATO) rockets received a $22,000 contract to extend and broaden its activity. Responsibility was transferred from the National Academy of Sciences to the Army Air Corps. This contract initiated the second phase of work that began July 1, 1939, and permitted the GALCIT group to move from the university campus to a site outside Pasadena on the Arroyo Seco. It was at this site that the first buildings were erected for a facility that would later be renamed the Jet Propulsion Laboratory.

1940 August Flight trials with the DFS 194 rocket-propelled research aircraft began at the Messerschmitt works at Augsburg, Germany. Powered by a derated Walter R I-203 rocket motor, it was able to achieve a top speed of 341.8 MPH. Work on the development prototype Me 163 meanwhile provided three aircraft for the new and more powerful Walter HWK R II-203b, each fitted with one manually regulated engine producing a thrust of 331–1,653 lb. The three Me 163 prototypes were completed during the winter of 1940–1941 and taken to the nearby Luftwaffe base at Lechfeld for gliding tests with Heini Dittmar at the controls.

1940 Work began in Russia on the rocket-propelled fighter designated BI, after its two principle proponents Aleksandr Bereznyak and Aleksei Isaev. Approved by Josef Stalin and the People's Commissariat for the Aircraft Industry, the BI was designed at the Soviet Air Force Academy under the supervision of Prof. Viktor F. Bolkhovitinov, and the construction of several prototypes began at a factory outside Moscow. When Germany attacked Russia in June 1941, the team was moved to an uncompleted iron foundry near Sverdlovsk, so throughout the winter of 1941–1942 all prototypes had to be assembled in buildings without roofs.

1940 December 16 The first Henschel Hs 293 air-to-surface missile—precursor to a wide range of missiles built by this aircraft company—was test-launched in Germany. The Hs 293A-1 was powered by a Walter 109–507B rocket motor with *Z-Stoff* and a thrust of 1,320 lb. Radio controlled to its target, the missile was launched by carrier-plane prior to firing the rocket motor for 10 sec. to add 120 MPH to its speed. It went into action for the first time on August 25, 1943. The Hs 293D incorporated a TV system for on-screen display in the carrier plane. Other Hs 293 variants employed delta wings and utilized the 1,342-lb. thrust Schmidding rocket motor operating on methanol *(M-Stoff)* and oxygen *(A-Stoff)*. Henschel produced a wide range of experimental rocket-boosted missiles under the designations Hs 294, 295, 296, GT 1200 and Zitterrochen.

1941 April 30 A small Argus pulse jet 265-lb. thrust was tested for the first time suspended beneath a Gotha Ga 145 D-IIWA. The German fluid dynamicist Paul Schmidt had developed the concept of a pulse-jet motor, but the Argus-Motoren-Gesellschaft was given a contract to develop a working unit. On June 19, 1942, the German Air Ministry ordered a pulse-jet powered missile which would gain infamy as the V-1 *(Vergeltungswaffe Eins),* or Retaliatory Weapon 1. Although not a missile by the definition adopted for this chronology, the undoubted success of the concept merits inclusion because of its post-war impact on the ballistic missile debate. Built by the aircraft company Fieseler as the Fi-103, the first V-1 was fired in anger on June 13, 1944, when 10 were launched against London. The last V-1 fired during the war was launched on March 29, 1945, making a total of 21,770 successful launches, 8,839 of which had been targeted against London and 8,696 against the Belgian port of Antwerp. Many, however, fell away from their targets or were brought down by guns and aircraft.

1941 May 19 The first of a series of tests with a gasoline–liquid oxygen motor began at the Fuel Oil Technical Laboratory at Fulham, London, in the first operation of a liquid propellant motor in Britain. Undertaken barely six months after the Ministry of Supply awarded a development contract to Asiatic Petroleum Co. (later renamed Shell), the 60-lb. thrust motor was a development model leading toward a jet-assisted-takeoff (JATO) rocket for the Wellington bomber. A sequence of tests was completed under the directorship of Isaac Lubbock and the ministry allowed the company to use the Flame Warfare Establishment at Langhurst, Sussex, for trials. Little use was ever made of the rockets.

1941 June 10 Engineering drawings of the proposed A-9 rocket were released by the Dornberger–von Braun team at Peenemünde, Germany. Designed as a derivative of the A-4, work had been under way for about a year on this winged version of the ballistic missile. With a terminal velocity of about 2,000 MPH, the A-4, first launched 13 June, 1942, had sufficient energy as it plunged back through the atmosphere to gain extended range by using swept wings for lift. Operating on the descending leg of its trajectory as a supersonic glider, the A-9 was projected to have a range of 340 mi., 70% greater than the A-4. Drop tests with a subscale model designated the A-7 were conducted in 1942 using a converted He 111 bomber. The A-9 was never used operationally.

1941 August 12 The first successful flight of an American aircraft powered solely by rocket motors took place at March Field near Riverside, Calif. Flown by U.S. Army Air Corps pilot Lt. Homer Boushey, the Ercoupe monoplane got off the ground in half the time and distance usually required. Developed by the Guggenheim Aeronautical Laboratory at the California Institute of Technology (GALCIT), the solid propellant rocket motors were known as GALCIT 27 and carried a 2-lb. charge of black powder producing 28 lb. of thrust for 12 sec. Three such rockets were attached to each wing tip. Because the propellant could not be stored for long periods, the GALCIT 27 rockets were not suitable for production.

1941 August 13 Test pilot Heini Dittmar successfully completed the first rocket-powered flight from the German research station at Peenemünde with the first prototype Me 163 incorporating the HWK R II-203b rocket motor. The German air force (Luftwaffe) shared the rocket research station with the Army, and the adjacent facility used by them was designated Peenemünde-West. In charge of aircraft production for the Luftwaffe, Ernst Udet was on hand for the flight and ordered a second test pilot, Rudolf Opitz, to join the program. On October 2, 1941, the Me 163 reached a speed of 623.85 MPH during a dive and Dittmar experienced compressibility near the sound barrier, cut the motor and landed safely.

1941 September Robert Goddard's team at the Mescalero Ranch in Roswell, N.Mex., was awarded contracts from the U.S. Navy Bureau of Aeronautics and from the U.S. Army Air Corps to develop rockets for jet-assisted takeoff. In July 1942 Goddard and his team were moved to the Naval Engineering Experiment Station in Annapolis, Md. There, they worked for three years on liquid propellant devices and throttleable motors, carrying out several hundred tests in the process. Robert Goddard died on August 10, 1945. Largely unrecognized in his day, he was honored years later when a major NASA research facility in Maryland was named the Goddard Space Flight Center, on May 1, 1959.

1941 December 3 The Austrian rocket engineer Eugene Sänger and his assistant Irene Bredt submitted the design for a long-range, suborbital, boost-glide rocket bomber, or *Fernbomber,* to the Luftfahrtforschungsanstalt (LFA), or Aviation Research Institute of the Third Reich. Sänger worked for the Deutsche Versuchsanstalt für Luftfahrt (VfL), or the German Aviation Research Institute, which was subordinate to the LFA. Sänger wanted to develop a space plane that could derive lift from wings and envisaged the bomber as a means of getting money from the German government. The proposal was dismissed by the LFA as too ambitious, and in August 1942 Sänger's team was disbanded. Sänger's proposal was not published until August 1944.

1941 December 16 One of the most successful entrepreneurial rocket manufacturers in the United States was established as Reaction Motors Inc. (RMI). The most notable of its four founders was the liquid rocket designer James H. Wyld, who sold his 100-lb. thrust, regeneratively cooled motor to the U.S. Navy for $5,000. Renting an old building in Pompton Lakes, N.J., the founders went to work on a 1,000-lb. thrust motor which was to be used by the Navy for jet-assisted takeoff. A second Navy contract produced a 3,400-lb. thrust motor in May 1943. In 1946 RMI, then employing 120 people, moved to Lake Denmark, N.J., and on April 30, 1958, it merged with the Thiokol Chemical Corp.

1942 April 15 A specially converted Douglas A-20, assisted by 1,000-lb. thrust liquid JATO (jet-assisted takeoff) units, took off from the U.S. Army Air Corps Gunnery Range at Muroc, Calif. It was the first time an aircraft in the United States had taken off aided by liquid propellant rocket motors. When tests ended on April 24, 44 successive takeoffs had been demonstrated. Developed by the GALCIT group, the rockets used a mixture of red fuming nitric acid (RFNA) and aniline. Earlier liquid motors from the GALCIT team used liquid oxygen and gasoline, but the liquid oxygen was changed to RFNA for ease of operation and the aniline was substituted as fuel to avoid the combustion problems associated with gasoline. Later in 1942 the Aerojet Engineering Co. received a production contract for the motors.

1942 May 15 The world's first rocket-propelled fighter, the BI-1, made its first flight in Russia with Grigori Bakhchivandzhe at the controls. With a length of 22 ft. 9 in. and a wingspan of 21 ft. 8 in., it was the third of several prototype

BIs that flew on this day, following a week of taxi tests and runway hops. Propulsion was provided by a Dushkin D-1A-1100 engine developed at the Reaktirnyi Nauchno Issledova-tel'kii Institut (RNII) and delivering a thrust of 2,425 lb. The BI-1 had provision for two 20 mm ShVAK cannons with 45 rounds per gun. Although seven prototypes had been built and testing was still under way when the war ended in May 1945, development was slow and a production batch of 50 was canceled.

1942 June 3 Austrian rocket engineer Eugene Sänger was granted a patent (Reichspatent No. 411/42) for "Gliding bodies for flight velocities above Mach 5." Sänger's proposed aircraft body was a semi-ogival shape in cross section with a wedge-shaped wing profile and was developed from work he had been conducting at Trauen in Lower Saxony, on boost-glide vehicles for hypersonic flight. Sänger wanted to design skip-glide reentry vehicles combining rocket propulsion with a lifting body capable of flight at the edge of the atmosphere.

1942 June 13 The first A-4 ballistic missile was launched from Peenemünde. It soon began rolling around its longitudinal axis and quickly came to grief as it tumbled and the motor cut off at 54 sec. The rocket fell into the Baltic Sea 96 sec. after launch and about 4,300 ft. from Test Stand VII where it had been launched. The A-4 had a length of 43 ft. 7 in., a diameter of 5 ft. 5 in. and weighed 27,500 lb. Power was provided by a 56,000-lb. thrust ethyl alcohol motor with a 65 sec. burn. A second flight on August 16, 1942, was doomed by an electrical failure 4 sec. after launch and the engine cut off at an elapsed flight time of 45 sec. when the missile had attained a height of 3,500 ft. and a velocity of 2,137 ft./sec. (1,457 MPH). The missile hit the water at 3 min. 16 sec. about 5.4 mi. from Test Stand VII.

1942 June 26 German test pilot Rudolf Opitz flew the Messerschmitt Me 163B prototype rocket fighter for the first time. He was towed into the air by a twin-engined Bf 110 and released for an unpowered glide to the ground. The production aircraft would use a more powerful Walter rocket motor

The world's first long-range ballistic missile, developed at Peenemünde, Germany, in the late 1930s, was based on a 25-ton thrust rocket motor designed by Walter Thiel. The rocket was designated A-4 but gained notoriety as the V-2.

than the one used in early Me 163s. The 109–509A-0-1 rocket motor used *T-Stoff* (80% hydrogen peroxide and 20% water) and *C-Stoff* (methyl alcohol and hydrazine hydrate) with thrust available in controlled steps between 300 lb. and 3,750 lb. The powered test program began in August 1942 and pilot training of the *Komet* began in July 1943. The Me 163B first went into action on August 16, 1944. The production aircraft had a wingspan of 30 ft. 7 in., a length of 18 ft. 8 in. and a gross weight of 9,042 lb. with 4,411 lb. of propellants. Maximum speed was 596 MPH.

1942 June Seeking a rocket fuel that could be kept for long periods and used when needed, the chief scientist at GALCIT (Guggenheim Aeronautical Laboratory at California Institute of Technology) John Parsons successfully developed a storable solid propellant comprising liquefied asphalt as fuel and an oxidizer of potassium perchlorate. Known as GALCIT 53, the propellant was the first storable solid propellant to see widespread use by the United States during World War II. It was formed into a cylinder resembling tar by cooling in a mold and was the first "cast" propellant developed anywhere.

A graduate of the Technical University in Berlin, the boyish space flight enthusiast Krafft A. Ehricke joined Walter Thiel at Germany's top-secret rocket research establishment at Peenemünde. Ehricke lacked enthusiasm for the conservative approach in design engineering and found in Thiel a kindred spirit. Frustrated by the cautious approach of Wernher von Braun and challenged by Thiel to think beyond the immediate problem, Ehricke received much valuable knowledge as he quietly worked alongside the German rocket engine builder. When Thiel was killed during the Peenemünde air raid of October 1943, Ehricke lost his mentor but never gave up his enthusiasm. When Ehricke was taken to the United States along with other Paperclip scientists in 1945, he worked first for von Braun, then for Walter Dornberger at Bell and finally for Convair. It was here, in October 1957, that he began development of the Centaur cryogenic rocket stage for which he became known as the "Father of the Hydrogen Engine."

1942 August Walter Dornberger mustered powerful arguments to justify expenditure on the A-4 ballistic missile program. Dornberger calculated that in full production it would cost RM 38,000 (Reichsmarks) to build and launch an A-4 capable of dropping a 1-ton amatol warhead on a target almost 200 mi. away. The cost of a bomber, its crew and bomb load was about RM 1.14 million, thirty times the cost of a missile. Although a bomber could drop 2 tons per flight, the Battle of Britain had shown that the average bomber had a life of six missions; thus each 1-ton bomb load dropped by a bomber cost RM 95,000 whereas 1 ton of high explosive dropped by a missile cost RM 38,000. Although the unit cost of a V-2 never fell below RM 75,000, it could have dropped to RM 35,000. These figures were important after the war when American military officials studied the case for missiles.

1942 October 3 The first successful flight of a German A-4 ballistic missile took place from Peenemünde across the Baltic Sea. The rocket motor burned for a programmed duration of 57.8 sec. and carried the missile to a height of 292,500 ft. (55.4 mi.) and a velocity of 4,875 ft./sec. (3,324 MPH). The A-4 fell into the Baltic approximately 116 mi. downrange about 4 min. 56 sec. after launch. Because it would later be arbitrarily determined that space begins at a height of 50 mi., this A-4 was the first man-made object to reach space.

1942 November 2 In response to a plea from Gen. Walter von Axthelm, inspector of antiaircraft artillery, for new means to stem the increasing numbers of enemy bombers penetrating German airspace, Wernher von Braun completed a study titled *A Guided A-A Rocket.* The priority of the A-4 prevented resources from being directed to what became the liquid propellant antiaircraft missile Wasserfall (waterfall) until August 1943, and technical problems soon appeared for what appeared at first sight to be a scaled-down A-4. But there the similarities ended: Wasserfall was in fact a completely new concept using nitric acid and a petroleum by-product called visol. These propellants are hypergolic, that is they ignite on contact, thus eliminating the need for an ignition system.

1942 Work began in Germany on the development of a multistage tactical army weapon to bridge the gap between existing long-range artillery and the extreme range (almost 200 mi.) of the A-4 ballistic missile in development at Peenemünde. The manufacturing company Rheinmetall-Borsig was asked to develop a missile that could fire 88 lb. of explosive a distance of about 100 mi. An interim development model had two stages, with six fins on the first stage and four on the second, a total length of 8 ft. 10 in. and a weight of 242 lb. delivering a 6.6-lb. test explosive. From tests with this device the definitive Rheinbote evolved, via an intermediate three-stage rocket, into a four-stage ballistic missile. Rheinmetall also started work on a surface-to-air missile, the Rheintochter.

1943 April 15 Gen. Sir Hastings Lionel Ismay, chief of staff at Britain's Defense Ministry, minuted Prime Minister Winston Churchill that intelligence information indicated that Germany had offensive ballistic missiles that could soon be used against England. This started a train of events that put Churchill's science adviser, Prof. Frederick Lindemann, in disagreement with Dr. Reginald V. Jones, head of the scientific section of M.I.6. Jones was sure the information was correct but Lindemann doubted that a rocket as big as those revealed in reconnaissance photographs could be built. Churchill backed Jones and ordered an air raid on Peenemünde.

1943 April 27 The British Anti Aircraft Command agreed to examine a plan for a rocket-powered antiaircraft missile designed to ride a radar beam to its target. Approval was eventually given to proceed with development of this device, called Brakemine, at the AA Command's Park Royal workshops. The design that emerged incorporated wings and tail fins with six solid propellant booster rockets. The first missile was fired at Walton-on-the-Naze in September 1944 but the test was unsuccessful. Over the next three years the design was refined into a credible defense weapon, but lack of funds led to its cancellation. In October 1948, English Electric was given a contract to develop its successor, later called Thunderbird.

1943 May 1 Alexander Lippisch was reassigned as director of the Luftfahrtforschungsanstalt Wien (Vienna Aviation Research Center) and was to work on a ramjet-powered supersonic aircraft. Lippisch had been instrumental in developing the Me 163 rocket interceptor. Designated LP-13a, the ramjet aircraft project was an attempt to produce a design optimized for efficient flight at subsonic speed as well as for supersonic dash capability. Propelled by powdered coal fuel combusted in a duct, the LP-13a was designed to utilize an additional liquid rocket motor for takeoff and for extra acceleration when needed, but the project was never completed.

1943 May 13 Gauleiter Fritz Sauckel, in charge of labor for the Third Reich, visited the German rocket research station at Peenemünde and discussed with Walter R. Dornberger and others the growing shortage of manpower and the needs of a production plan which envisaged delivery of several hundred V-2s a month. SS Reichsführer Heinrich Himmler wanted to use foreign labor whereas Hitler wanted only ethnic Germans working on the top-secret rocket project. After the Peenemünde air raid of August 17–18, 1943, Hitler conceded to Himmler, and plans were drawn up to use concentration camp victims as forced laborers in a production facility deep underground. Called *Mittelwerke,* it was located in the Harz mountains at Nordhausen and was to draw its labor force from the nearby Dora concentration camp. To dig the maze of underground tunnels, 60,000 forced laborers were shipped in from the Buchenwald concentration camp during September 1943. In all, 20,000 people would die at Nordhausen by the end of the war in May 1945.

1943 May 26 At Peenemünde, a German joint services commission met to discuss the future of long-range bombardment and to decide the fate of the A-4 ballistic missile project and the Fi-103 flying bomb being developed by the Luftwaffe. Minister for Armaments Albert Speer attended, as did Erhard Milch of the Luftwaffe, Admiral Karl Dönitz from the German navy and a large contingent of ministerial officials and civil servants. After demonstrations of each weapon, the A-4 giving two perfect performances but the Fi-103 failing twice, it was decided to recommend both types of missile for production. It was about this time that the A-4 became known as the V-2, or *Vergeltungswaffe Zwei* (Retaliatory Weapon 2) and the Fi-103 became known as the V-1.

1943 July 7 Walter R. Dornberger and Wernher von Braun visited Adolf Hitler at the Wolf's Lair near Rastenburg to report on progress with the A-4/V-2 ballistic missile program and to show a film of the successful launch on October 3, 1942. They displayed to Hitler a wooden model of a vast underground launch complex, designed and favored by von Braun, which the scientist wanted built for bombarding England. Dornberger opposed this, preferring the V-2 as a mobile field weapon. Hitler agreed with von Braun, saying that it could be too easily spotted and attacked from the air. Ultimately, Dornberger was proven right and the V-2 was deployed as a mobile missile launched from presurveyed locations. Hitler gave the V-2/A-4 project top priority once again and on July 26, 1943, plans were drawn up for training the first military units.

1943 August 2 Dr. Theodore von Karman, Hsue-Shien Tsien and Frank Malina of the GALCIT (Guggenheim Aeronautical Laboratory at California Institute of Technology) team in California completed a report on German rocket activities they had been asked to prepare for British Intelligence. From the limited knowledge available about these activities, the group submitted a report that was sufficiently alarming to get the attention of U.S. Army Air Force Col. W. H. Joiner, who concluded that some effort should be made to develop a ballistic missile in the United States. Col. Joiner was assigned to monitor technical developments at GALCIT and passed the report to senior Army personnel for their review.

1943 August 17–18 During the night a force of almost 600 Royal Air Force bombers dropped 1,593 tons of high explosives and 281 tons of incendiaries on the German rocket research station at Peenemünde. Most of the 800 people killed

British bombing raids partly disabled the German rocket program but could not stop it entirely. Here, damaged V-2s are seen in an assembly building at Peenemünde on the Baltic coast.

were foreign forced laborers. Among the dead was Dr. Walther Thiel, who had joined the Army rocket program in the fall of 1936 and been responsible for almost all the progress made in developing the rocket motors at Peenemünde. Dubbed Operation Hydra, the raid's objective had been to kill as many of the technical and senior staff as possible, and in this it failed. On August 21 a decision was made to move V-2 flight tests to an SS training camp centered near the Polish village of Blizna. Some development work continued at Peenemünde, but the almost total absence of firing tests led the RAF to believe its raid had been more successful than it had been.

1943 August 19 The U.S. Naval Aircraft factory at Philadelphia was authorized to develop the Gorgon aerial ramming missile which was to be powered by a turbojet engine. This program was later expanded to include pulse-jet, rocket-propelled winged missiles and development of air-to-air, air-to-surface and surface-to-surface missiles. The first rocket-powered Gorgon was launched from a Consolidated PBY-5 flying boat and achieved a speed of 550 MPH off Cape May, N.J. Using nitric sulfide acid as oxidizer and monoethylaniline as fuel, the motor produced a thrust of 350 lb., yet weighed only 10 lb. The project advanced through a wide range of winged missile projects before it was terminated in 1953.

1943 August The German company Rheinmetall-Borsig began test firing an ambitious air-to-surface missile called Rheintochter. Designated R-I, it had a tapered cylindrical body with a rounded nose supporting four servo-actuated control fins. At the rear, six symmetrically mounted swept-back fins separated six venturi ports for exhaust from the main solid propellant motor which produced a thrust of 8,800 lb. for 10 sec. Attached in tandem, a solid propellant booster produced a thrust of 165,344 lb. for 0.6 sec. By December 1944 the R-I had been abandoned in favor of the R-III which had a liquid propellant main stage and only four tail fins. The solid booster used on tests was to have been replaced by two smaller boosters attached to opposing sides of the main missile but the project was cancelled that same month.

1943 November 5 The first test launch of a V-2 rocket took place from the newly selected site near Blizna, Poland.

Ingloriously, one leg of the firing table sank into frozen ground partially thawed by the rocket's blast, sending it off at an acute angle to a destructive end 2 mi. away. Development progress was slow, and 80–90% of the missiles fired during the winter failed to fall near their test targets due to premature explosions on the way up or down. Basic research into technical problems encountered by the rockets fired on test in Poland was conducted at Peenemünde. The Polish flights ceased on August 30, 1944, after approximately 300 missiles had been fired. Evacuating Blizna in the face of the Soviet advance, 80 additional V-2s were launched on test from a site further west.

1943 During the fall a major change in the catalyst-oxidizer used for the Walter rocket engine being developed for the Me 163 tailless fighter reduced the risk of combustion instability and catastrophic explosions. Previously, *Z-Stoff* (an aqueous solution of calcium permanganate) had been used with *T-Stoff*, but the mixture clogged jets causing flow instability. A 30% solution of hydrazine hydrate in methanol known as *C-Stoff* was substituted. It would be used with *T-Stoff* in the 3,750-lb. thrust Walter 109–509A engine adopted for the Me 163B, the operational variant of the tailless fighter. *Z-Stoff* would continue to be used with *T-Stoff* for driving the propellant turbopumps.

1944 January 15 Responding to a report on German rocket developments submitted on August 2, 1943, Col. G. W. Trichel, head of the U.S. Army Rocket Development Branch at the Ordnance Department, asked the GALCIT (Guggenheim Aeronautical Laboratory at California Institute of Technology) group to design a long-range ballistic missile. GALCIT was asked to prepare financial estimates, technical plans and a development schedule. The working group would be known as ORDCIT, for Ordnance at California Institute of Technology. The Air Force was not interested in the missile idea and their antipathy to the Army's commitment to ballistic missile research continued after the war.

1944 February 5 The first successful flight of the German Wasserfall antiaircraft rocket took place from Peenemünde. The missile had the overall profile of a small V-2, with a length of 25 ft. 7 in., a diameter of 2 ft. 10 in. and a span of 9 ft. 7 in. across four equally spaced wings and 6 ft. 2 in. across the tail fins. Wasserfall was powered by a 17,600-lb. thrust nitric acid–visol engine and could attain a height of 60,000 ft. in 42 sec. at a terminal velocity of 2,165 ft./sec. (1,476 MPH). The missile came too late to be put into production, but work on the project continued until February 1, 1945.

1944 February 21 Wernher von Braun was summoned to a meeting with SS Reichsführer Heinrich Himmler, who wanted to recruit von Braun and his rocket team away from Walter Dornberger and the Army. Von Braun declined the offer and on March 15, 1944, he, Klaus Riedel and Helmut Gröttrup were arrested by the Gestapo and imprisoned in Stettin, charged with reticence over the use of the V-2 as a weapon, talk of space travel instead of war and planning to escape to England with secret rocket plans. The men were released a few days later after furious protests from Dornberger.

1944 February 28 U.S. Army Maj. Gen. G. M. Barnes at the Technical Division of the Ordnance Department received technical proposals for a long-range ballistic missile. This had been requested from the ORDCIT (Ordnance at California Institute of Technology) group in January 1944. The missile envisaged by ORDCIT would have a range of 150 mi. with a 1-ton warhead delivered with high accuracy. In June 1944 the ORDCIT group received a letter of confirmation to proceed with design in the period January 1945 through June 1946. Intermediate technical stages were to support three separate rocket projects: Private A, Private F and Corporal. At about this time the GALCIT group was renamed the Jet Propulsion Laboratory, or JPL to avoid any association in the public mind with Buck Rogers-type rocket adventures.

1944 April Tests began in Germany with the Messerschmitt Enzian ground-to-air missile, boosted into the air by four Schmidding 109–553 solid propellant rockets delivering a total thrust of 15,400 lb. for 4 sec. The sustainer engine was a Walter RI-210B operating on a mixture of Salbei (SV-Stoff), a concentrated nitric acid, and gasoline with a small quantity of furfural alcohol (Bf-Stoff) to start combustion with the nitric acid. Producing a thrust of 4,410 lb. at ignition and reducing to 2,205 lb. over the 70-sec. burn time, the Enzian missile had a length of 7 ft. 10 in. and a wingspan of 13 ft. 1 in. It was controlled in flight by radio link and joystick. Some 38 missiles were tested before the project was cancelled in January 1945.

1944 May 20 A German Army V-2 missile on test from the site near Blizna, Poland, malfunctioned and fell intact into a riverside marsh. Polish partisans rolled it into a nearby pond and put cattle in to stir up the mud, hiding the missile from view when troops arrived looking for it. On the night of July 25–26, an RAF Dakota landed in a nearby field and retrieved most of the missile which was flown out and handed over to Polish and British intelligence units. This was the most complete rocket the Allies had thus far gotten their hands on and it went a long way toward filling in the broad, sketchy picture of the German missile that was emerging on the basis of the bits and pieces from other failed tests. On June 13, 1944, a V-2 launched from Peenemünde fell into Sweden and the parts were delivered to England for analysis.

1944 June 22 The ORDCIT (Ordnance at California Institute of Technology) group received a contract from the U.S. Army Ordnance Department for the design and development of a long-range (150 mi.) rocket. Authorized expenditure of $1.6 million was followed by a definitive contract of January 16, 1945. Eventually, with contract extensions carrying the work forward to June 30, 1946, $3.6 million was spent on the ORDCIT project.

1944 June Russian Academician Vladimir Nikolaevich Chelomei was put in charge of a new design and manufacturing plant for the development of experimental cruise-missiles like the German V-1. The first such design appeared during December but it was not a success, and after the war Chelomei turned to the theoretical design of rocket motors. He went on to head a major space hardware design bureau and challenge Sergei Pavlovich Korolev about the best way to build satellite launchers and put men on the moon. Chelomei was also responsible for development of the Proton launcher and the SS-9 ICBM.

1944 July 1 Consolidation of rocket research at the Jet Propulsion Laboratory went a step further when the ORDCIT group took over responsibility for development of solid propellant rocket motors from the Army–Air Force Materiel Command. This led to contracts with the Thiokol Chemical Corp. for composite solid propellants used in Private A and F

rockets. Research on polysulfide rubber-perchlorate propellants resulted in development of the Sergeant guided missile, which entered service with the U.S. Army in 1961.

1944 July 5 Piloted by Harry Crosby, the Northrop MX-334 flying wing took to the air for the first time powered by a bipropellant liquid-fueled rocket motor burning monoethylaniline fuel and red fuming nitric acid (RFNA) oxidizer. The Aerojet XCAL-200 motor had a thrust of 200 lb. and a burn duration of 3.5 min. With a wingspan of 32 ft. and a length of 12 ft., this highly secret aircraft had been conceived in September 1942 as a rocket-powered interceptor. The test flights took place at Muroc Dry Lake, California, but the project was abandoned as the war drew to a close.

1944 July 13 Crates containing 2,500 lb. of German V-1 flying-bomb parts salvaged in Britain arrived at Wright-Patterson Field, Ohio, where the U.S. Army Air Force had ordered 13 copies for test. Within three weeks the first JB-2 had been built, and the Army drew up plans for major production. First 1,000 were ordered, then a plan for 1,000 per month emerged and by December 1944 there was a scheme to produce 5,000 per month starting in June 1945, and some were even considering the possibility of building 1,000 a day. Gen. Henry H. Arnold wanted 75,000 with a launch rate of 100 per day by September 1945 and 500 a day by January 1946. When the war ended in September 1945 only 1,385 had been built, the JB-2 having demonstrated a success rate of 78%.

1944 August 1 In Germany, the vertically launched rocket-powered Bachem BP 20 Natter interceptor was selected for development by the German Aviation Ministry. Given the official designation Ba 349, the Natter had a length of 21 ft. 3 in. and a wing span of 13 ft. 1 in. It was powered by a 3,750-lb. thrust Walter 109–509 rocket motor with four strap-on Schmidding 109–533 solid, diglycol-fuelled boosters, each producing a thrust of 2,640 lb. for 10 sec. With a takeoff thrust of 14,310 lb., the Ba 349 weighed 4,800 lb. and with a thrust-to-weight ratio of 3:1, it had a climb rate of 35,800 ft./min.

1944 August 11 The first wire-guided Ruhrstahl X-4 air-to-air missile was test-launched from a Focke Wulf Fw 190. With a length of 6 ft. 7 in. along its cigar-shaped body and a 1 ft. 11 in. span across four centrally mounted swept-back fins, set 90° apart, the X-4 was powered by a Schmidding 109–603 solid propellant rocket motor for the first ten test flights. This produced a thrust of 330 lb. for 10 sec. The BMW 109–548 for which the missile was designed operated on Tonka (R-Stoff, a self-igniting fuel composed of 50% xylidine and 50% triethylamine), and Salbei (SV-Stoff), a concentrated nitric acid that produced an initial thrust of 308 lb. which reduced to 66 lb. over the burn duration of 17 sec. The range of attack was from 4,900 ft. to 11,450 ft. Although 1,000 missile airframes were built the motors were destroyed in bombing raids, and despite full-scale production planned for January 1945 none was used in combat.

1944 August 14 Hsue-Shien Tsien at the Jet Propulsion Laboratory's ORDCIT (Ordnance at California Institute of Technology) project outlined the development program for an experimental missile that would serve as an interim step along the way towards the eventual production of a ballistic missile with a range of 1,500 mi. and capable of carrying a 1,000-lb. warhead. The rocket envisaged, designated XF36 L 20,000, was to have a gross weight of 5 tons, a diameter of 3 ft., a thrust of 20,000 lb. for 60 sec. and a range of 30 to 40 mi. The missile was to be called Corporal and special facilities to test it were constructed at the Muroc Flight Test Base, Calif. These were completed in June 1945, and a month later Dr. Theodore von Karman and Col. G. W. Trichel of the U.S. Army Ordnance Department met with representatives of the Sperry Gyroscope Co. to discuss a contract for guidance equipment.

1944 August 25 The first successful recovery of jettisonable rocket boosters took place when a German Arado Ar 234 jet reconnaissance bomber dropped its spent motors in flight. Built by Helmuth Walter one booster was mounted beneath each wing and each provided a thrust of 1,102 lb. for 30 sec. Similar to rocket motors built for the Me 163 fighters they operated through the reaction of *C-Stoff* (methyl alcohol and 30% hydrazine hydrate) and *T-Stoff* (hydrogen peroxide). Used for augmenting the installed jet engines on the Ar 234, the boosters were jettisoned shortly after takeoff and recovered by parachute. This represented the first attempt to cut the cost of booster production by recovering and reusing existing cases and motor assemblies.

1944 August Submitted on December 3, 1941, Eugene Sänger's design proposal for the *Fernbomber,* a long-range boost-glide rocket bomber, was published secretly by the Deutsche Luftfahrtforschunganstalt (LFA), or German Aviation Research Institute. The *Fernbomber* would have had a length of 91 ft., with the flat-bottom fuselage carrying a low-aspect ratio wing spanning 41 ft. 9 in. Propelled into the air by a monorail rocket-sled boost motor of 600 tons thrust, the boost-glide vehicle would have been powered by a 100-ton thrust liquid oxygen–gasoline rocket motor to near orbital speed at the edge of the atmosphere. Aerodynamic lift would maximize the range for subsonic attacks on pinpoint targets or hypersonic attack on area targets. The *Fernbomber* study formed the basis for postwar U.S. studies of orbital and suborbital vehicles.

1944 September 8 Launched at 0834 hours from a site 10 mi. south of Houffalize in the Ardennes, the first successful firing of a ballistic missile in anger was carried out by the Lehr- und Versuchsbatterie 444 (444th Training and Experimental Battery) of Army Group North. Commanded by Col. Gerhard Stegmaier, the unit fired a V-2 at Paris about 180 mi. away, hitting an area close to the Porte d'Italie and causing some damage. During the evening, the 485th Artillery Battalion fired the first missiles against London, the first falling on Chiswick, west-southwest of the City. Two days earlier the 444th had launched two V-2 rockets against Paris from a site near Fraiture, 8 mi. west of Vielsalm, but they failed due to fuel starvation.

1944 September Tests got under way at Karlshagen, Germany, with the Henschel Hs 117 Schmetterling ground-to-air missile boosted by two Schmidding 109–553 solid propellant rockets which produced a combined thrust of 7,700 lb. for 4 sec. Boosted to a speed of 680 MPH, the liquid propellant BMW 109–558 would burn a mixture of Tonka (*R-Stoff,* a self-igniting fuel composed of 50% xylidine and 50% triethylamine), and Salbei (*SV-Stoff,* a concentrated nitric acid) for 33 sec., producing a thrust of 825 lb. reduced to 132 lb. for the final 24 sec. Top speed was about 550 MPH at a ceiling of 32,800 ft. and it had a range of almost 20 mi. Production was ordered in December 1944 for commencement in March 1945, and deliveries were to top 3,000 a month by November 1945. The war ended before these plans could be realized.

Germany's A-4/V-2 missile was first used operationally in September 1944 and set the stage for future development of liquid propellant rockets for tactical and strategic purposes.

1944 November 15 The U.S. Army Ordnance Department established Project Hermes and signed a contract with the General Electric Co. five days later to initiate work to develop a series of surface-to-surface missiles. Under Project Hermes, General Electric would set up a research group to study captured German missiles and assemble a database on technical capabilities. It later dispatched a scientific group to occupied Germany for direct study of the rocket program, and designed and manufactured various test missiles. GE's activity with the Hermes projects ended on June 30, 1951. After the war, the V-2 launch program at the White Sands Proving Ground in New Mexico, which involved members of the group working under Wernher von Braun, became an integral part of Hermes.

1944 November 16 Frank J. Malina from the ORDCIT (Ordnance at California Institute of Technology) long-range rocket project at the Jet Propulsion Laboratory in Pasadena, Calif., began an 11-day visit to France with Capt. C. E. Martinson of the U.S. Army. Given the rank of colonel in the U.S. Army in case of capture, Malina was there to inspect V-1 launching ramps and, during his stay in Britain on the way over, to examine V-2 rocket motors. Technical assessment of the latest German developments with rocket and pulse-jet propulsion was a priority for the Army. On his return, Malina reported to Col. G. W. Trichel in Washington, D.C.

1944 November The German army began operational use of the Rheinmetall-Borsig Rheinbote, a tactical four-stage artillery missile, launching more than 200 against the Belgian port of Antwerp. The Rheinbote had a length of 37 ft. 5 in. and a total weight of 3,775 lb. At launch the boost stage produced a thrust of 83,790 lb. for 1 sec. Each of the first two sustained stages delivered a thrust of 12,350 lb. for 5 sec. in tandem operation. The third stage produced a thrust of 7,495 lb. for

3.5 sec. The missile threw an explosive charge of 88 lb. a distance of 136 mi. and attained a maximum height of 48.5 mi. in 4 min. 20 sec. With a terminal velocity of 3,663 MPH, the Rheinbote was Germany's fastest missile.

1944 December 1   Developed as a result of U.S. Army interest in a long-range ballistic missile, the Jet Propulsion Laboratory at the California Institute of Technology launched the first Private A rocket. Designed to qualify technical steps essential for the long-range missile, Private A had a length of 8 ft., a diameter of 10 in. and a span of 34 in. across four tail fins. With a gross weight of 600 lb., Private A burned GALCIT 61-C, an asphalt-based propellant, to produce a thrust of 1,000 lb. Boosted by four 5,500-lb. rockets that fired for 0.2 sec., Private A was launched from guide rails and achieved a maximum range of 11 mi. and an altitude of almost 3 mi. In all, 24 rockets were launched in two weeks.

1944 December 22   Following the first unmanned drop tests from a Heinkel He 111 in October, the first unmanned vertical launch of a Bachem Ba 349 rocket-powered interceptor using exclusively solid boosters took place. In all, 11 unmanned test flights took place and these were considered successful despite problems with booster performance and flight control. Modifications were made that improved flight characteristics, and the first unmanned launch with the integral Walter rocket motor installed took place on February 25, 1945.

1944 December 27   The first launch of a winged version of the German V-2 ballistic missile took place; the flight was not a complete success due to failures in the guidance system. A second, successful, flight took place on January 24, 1945. Designated A-4b, the missile was essentially the same as the projected A-9 (see 1941 June 10) and the two designations were generally interchangeable. Wernher von Braun's team had designed a much bigger rocket designated A-10 to serve as the first stage for the A-9, which would be mounted on top. This two-stage A-10/A-9 configuration would have had a launch weight of 188,090 lb., a first-stage thrust of 441,000 lb. and a range of 3,100 mi. with a 1-ton warhead; it was never built.

1945 January 16   Floyd L. Thompson and John W. Crowley Jr. of the National Advisory Committee for Aeronautics (NACA) Langley Laboratory sent a letter to NACA headquarters formally requesting permission to establish a site dedicated to high-speed missile and rocket testing. The request was backed by the highest levels in the U.S. Army and Navy. At first the intention was to use the site for high-speed projectiles which would examine the flight dynamics of supersonic objects, but later it would embrace high-altitude and high-speed rocket research. The request was approved by the NACA and by Congress on April 25, 1945. The site selected was Wallops Island, Va., and it was named the NACA Auxiliary Flight Research Station (AFRS), with Robert R. Gilruth of the Flight Research Division its first head.

1945 January 27   A liquid propellant rocket motor helped propel a special test version of the Messerschmitt 262 into the air for the first time at Lechfeld airfield. Known as the *Heimatschützer I* (Interceptor I) variant, the aircraft was officially designated Me 262C-1a. In addition to the two 1,980-lb. thrust Junkers Jumo 004C jet engines, the aircraft carried one 3,750-lb. thrust Walter HWK 509 A-2 rocket motor in the rear fuselage. Operating on *T-Stoff* (hydrogen peroxide) and *C-Stoff* (methyl alcohol and 30% hydrazine hydrate) propellants, the rocket motor was ignited at 100

MPH and propelled the aircraft into the air with phenomenal acceleration. The Me 262C-1a could reach a height of 38,400 ft. in 4 min. 30 sec. After six more flights it was seriously damaged during an air raid on March 23.

1945 January 31   German SS Gen. Hans Kammler ordered Wernher von Braun to organize the evacuation of the rocket research site at Peenemünde as the Russian army advanced on the Baltic site. The last V-2 launched from Peenemünde was fired from the Test Stand VII complex of launch pads on February 14, 1945. Between February 17 and mid-March all essential stocks, stores and personnel necessary for continued production of the V-2 were sent by road 250 mi. south to Nordhausen, and development operations and test stands were set up at nearby Bleicherode. On April 3, 1945, Walther Dornberger was ordered to select the top 500 scientists and engineers and move them from Nordhausen to Oberammergau in the Bavarian Alps. They were then dispersed among the Alpine villages, Wernher von Braun and Dornberger going to the ski resort of Oberjoch, where they surrendered to American troops on May 2.

1945 February 28   Oberleutnant Lothar Siebert was launched on the first manned test of the Bachem Ba 349 Natter interceptor. At an altitude of only 1,650 ft. the cockpit canopy became detached and the projectile nosed over, hitting the ground at great speed and killing the pilot. Six more flights were made, all successfully, before the end of the war, and some Natters were set up around Stuttgart to await the arrival of U.S. bombers. Allied tanks got there first, however, and the Natters were destroyed by the Germans to prevent them falling into enemy hands. One Natter was taken to the United States after the war and is today preserved by the National Air and Space Museum, Washington, D.C.

1945 March 16   The U.S. Army Air Force awarded the Bell Aircraft Corp. of Buffalo, New York, a contract for three XS-1 (for "experimental supersonic") rocket-powered research aircraft capable of exceeding Mach 1, the speed of sound. Each aircraft was to be powered by a single four-chamber motor providing a maximum thrust of 6,000 lb. Built by Reaction Motors Inc., the motors were designed to run on a mixture of liquid oxygen and diluted ethyl alcohol. The engine was not throttleable but each chamber could be operated independently, thus providing four increments of thrust power from 1,500 lb. The designation XS-1 was soon abbreviated to X-1, eliminating the word "supersonic."

1945 March 26   Piloted by Karl Bauer, a specially modified Me 262 jet fighter was flown for the first time with combination jet-rocket engines. Designated Me 262C-2b, and known as the *Heimatschützer II* (Interceptor II), the aircraft was powered by two 1,760-lb. BMW 003R turbojets, each motor incorporating a 2,700-lb. thrust BMW 718 rocket motor burning *S-Stoff* and *R-Stoff*. Ignited at a rolling speed of 100 MPH, the rockets produced enormous acceleration and gave the Me 262 a climb rate of 25,000 ft. in 1 min. 12 sec. Only one more flight of the Me 262C-2b was made before the war ended.

1945 March 29   The last V-2 of the war was fired against a target in Holland; the following day rocket troops were released to become general soldiers once more. Of the approximately 6,000 V-2s built a total of 3,255 had been successfully launched, 1,403 against England and 1,852 against continental targets. Of this total, all but 365 got through to their targets, representing a failure rate of 11.2%. A large number, perhaps a further 15%, detonated prematurely on reentry. A further 3,000 missiles were discovered at the end of the war, 2,100 in field storage and 250 in the *Mittelwerke* production mines.

1945 April 1 The Jet Propulsion Laboratory's ORDCIT (Ordnance at California Institute of Technology) program launched the first Private F from the Hueco Range at Fort Bliss, Tex. It had three large fins at the rear with two opposing fins spanning 5 ft. and wings spanning 2.8 ft. located near the nose of the missile. Essentially a Private A modified to generate lift, Private F was an exercise in the use of lifting surfaces for extended range. Tests revealed an instability immediately after launch, and none of the 17 rockets launched over two weeks was successful.

1945 April 11 Troops with U.S. Army Brig. Gen. Truman Boudinot's Combat Command B arrived at the *Mittelwerke* V-2 production site in the Harz mountains. They found the production facility intact with no Germans there. In the surrounding countryside the troops found 4,500 former workers and concentration camp victims released by their Nazi overlords. A plan was quickly put together by Col. Holger N. Toftoy, chief of the Ordnance Technical Intelligence team in Paris, to remove as fast as possible from the *Mittelwerke* facility as many V-2 rockets as possible for shipment out to the safety of U.S. Army units. Under an agreement signed at high level, the Russians were scheduled to arrive and take over at *Mittelwerke* on June 1, 1945.

1945 May 1 The Russian Army occupied the *Mittelwerke* V-2 production facility in the Harz mountains, Germany, and found the place had been pillaged of rockets and their components. About 3,500 personnel were eventually rounded up. Among this group, a close colleague of Wernher von Braun, Helmut Grötrupp refused a contract with the Americans and left the West, returning to Soviet-occupied Bleicherode. On May 5, the Second White Russian Army under the command of Gen. Konstantin K. Rossokovsky entered Peenemünde on the island of Usedom and found it almost completely wrecked. Maj. Anatole Vavilov took charge of the place and was ordered to destroy the structures still standing.

1945 May 5 Members of the U.S. Naval Technical Mission in Europe interrogated leading members of the German V-2 program at the small town of Kochel, Bavaria. Among those interrogating the German scientists and engineers was Dr. Hsue-Shien Tsien from the GALCIT (Guggenheim Aeronautical Laboratory at California Institute of Technology) group. He asked Wernher von Braun to prepare a summary of German rocket projects and to predict the likely path for future developments. Von Braun prepared a document entitled "Survey of Development of Liquid Rockets in Germany and Their Future Prospects," which was carried by Abraham Hyatt to the U.S. Navy Bureau of Aeronautics in Washington, D.C. Navy Lt. Robert P. Haviland used the von Braun report to stimulate interest at the Navy in an Earth satellite project.

1945 May 22 Acting on the orders of Col. Holger N. Toftoy of U.S. Army Ordnance Technical Intelligence, Maj. James P. Hamill began shipping out by rail the first of approximately 100 V-2 rockets, complete or in parts, from the *Mittelwerke* production facility. On May 27, Maj. Robert Staver organized the removal of 14 tons of documents from a mine at Dörten in the British sector, only hours before the new administrators took over. The last train load of V-2 parts departed May 31, a day before the Russians were expected to occupy the place under the terms of an agreement between the Allies. A total 341 trucks of V-2 parts had been retrieved and taken to Antwerp, Belgium. From there the crates were loaded into 16 ships, sent to New Orleans and were eventually delivered to the White Sands Proving Ground, N.Mex.

1945 May 27 Working in Allied-occupied Germany, Maj. Robert Staver received a telegram from his superior Col. Holger N. Toftoy advising him that his idea about selecting 100 top German scientists for debriefing in the United States was getting top priority in Washington and Paris. As an intermediate step, key personnel were moved by U.S. Army trucks from the zones of pending Russian occupation to the U.S. zone until a final decision about their future was made. On July 6, 1945, a U.S. War Department general staff plan for exploiting the German civilians was approved by the Joint Chiefs of Staff and by the British. A total of 350 were approved for removal to the United States, 100 of whom were to be rocket scientists. On July 24, 1945, Col. Toftoy returned to Europe to organize the selection process.

1945 May Under the title *Preliminary Design of an Experimental World-Circling Spaceship,* the RAND Corp. published a report detailing the validity of an artificial Earth-orbiting satellite. RAND envisaged a four-stage vehicle weighing 117 tons and capable of putting a 500-lb. satellite in a circular orbit of 300 mi. The propulsion system would lean heavily on the technology developed in Germany for the V-2 project and use liquid oxygen and alcohol propellants. RAND estimated a total project cost of $150 million and projected a five-year development schedule.

1945 July 3 Designed and developed by the ORDCIT (Ordnance at California Institute of Technology) team at the Jet Propulsion Laboratory, the first Baby Wac liquid propellant rocket was launched from a small tower at Camp Irwin, Calif. Baby Wac was a one-fifth scale model of the Wac Corporal atmospheric sounding rocket being developed by ORDCIT. It was put together for engineering qualification of flight characteristics. Baby Wac evaluated the use of three fins versus four and reached a height of 3,000 ft. The full-scale rocket was being developed to lob a 25-lb. package of instruments to a height in excess of 100,000 ft. and to return the scientific equipment to Earth by parachute. Baby Wac test firings were completed July 5, 1945.

1945 July 5 Soviet armed forces occupied the Nordhausen area of Germany and began to systematically reconstruct the V-2 production facilities. Under the command of Gen. Kuznetsov, Germans rebuilt elements of the plant and put back in operation the engine test stands at nearby Lehestein. In early 1946 four new facilities were added: *Zentralwerke,* where pilot production of V-2 components resumed; Nordhausen *Werke II,* where V-2 airframes and engines were assembled; Sommerda (80 mi. east of Leipzig), where component test and laboratory facilities were built; and Sonderhausen, 10 mi. south of Nordhausen, where electrical and guidance equipment was built. Helmut Grötrupp was put in charge of the *Zentralwerke.* Gen. Gaidukov, Kuznetsov's replacement, put Grötrupp in charge of all missile production in the Soviet zone.

1945 July 9 The U.S. Army formally activated the White Sands Proving Ground in New Mexico as the test site for rockets and missiles, including German V-2 rockets. It was under the authority of the Army Ordnance Department and personnel from the 69th Anti-Aircraft Battalion helped set it up. The test area, 90 mi. long and 60 mi. wide, was situated 80 mi. north of El Paso, Tex., 30 mi. east of Las Cruces, N.Mex. and 120 mi. southwest of the Mescalero Ranch, where Robert Goddard conducted his own rocket tests. (The Trinity site, where the world's first atomic bomb was tested five days later, lies in the northern part of the range.) Three hundred freight cars of V-2s and parts arrived at Las Cruces in August, and their contents were trucked to White Sands. On October 11, 1945, the First Guided Missile Support Battalion, Army Ground Forces, was activated. The German V-2 scientists began arriving at nearby Fort Bliss in October, their numbers peaking at 39 during March 1946.

1945 July 19 A U.S. Army intelligence operation was set up to gather together people previously involved with scientific projects under the Third Reich. It was originally named Project Overcast, but on March 16, 1946, it became Project Paperclip. Its mandate was to select scientists and engineers who had played an important part in developing new projects for the German war machine, including rocket scientists from Peenemünde, and offer them a six-month contract, extendable by a further six months, to go to the United States while their families remained in Landshut, Germany. On July 23, Col. Holger N. Toftoy, chief of Ordnance Technical Intelligence, received approval for his scheme to bring these people to the United States. Col. Toftoy met with the von Braun team in August 1945 and offered them contracts with the Army. By 1947, 113 German scientists had been brought to the United States.

1945 August 10 U.S. Navy Lt. Robert P. Haviland prepared a nine-page memorandum to the head of the Weapons Section of the Navy Bureau of Aeronautics recommending development of an Earth satellite referred to as Project Rex. Haviland's superior Comdr. J. A. Chambers supported Haviland and recruited Lloyd V. Berkner, head of the Electronics Materiel Branch, to persuade the bureau that it should consider the proposal. On October 3, 1945, the Bureau of Aeronautics authorized a Committee for Evaluating the Feasibility of Space Rocketry.

1945 September 6 Under the direction of Valentin Petvovich Glushko, the first V-2 engine from the newly activated *Zentralwerke* was static-fired at Lehestein in the Soviet zone of Germany. The German V-2 engineers trained Soviet military technicians under Glushko. The engine tests, however, were under the command of Dr. Joachim Umpfenbach. By this date, the Germans, under the direction of Grötrupp, had begun to assemble V-2s for test firings, and by September 1946 there were 30 V-2s ready for launching.

1945 September 8 The Combined British Astronautical Societies, which included small groups in London, Manchester, Kingston, Farnborough and elsewhere, were merged into the reformed British Interplanetary Society (BIS). The official day of incorporation was December 31, 1945, at which date membership stood at 280. Six honorary fellows were appointed: Wernher von Braun, A. V. Cleaver, O. W. Gail, Willy Ley, Guido von Pirquet and Eugene Sänger. In 1952, with membership at around 1,000, the BIS took up new headquarters in Bessborough Gardens, London, and in 1979 moved from there to its present address at South Lambeth

Road. The BIS prides itself on being the longest running astronautical society in the world, and has become a forum for technical papers and a highly respected instrument for disseminating news.

1945 September 26 The first of 10 Wac Corporal atmospheric sounding rockets developed by the ORDCIT (Ordnance at California Institute of Technology) team at the Jet Propulsion Laboratory was successfully launched. The rocket had an overall diameter of 16 ft. 2 in., a maximum diameter of 12.2 in., three tail fins and a gross weight of 665 lb. Carrying an RFNA-aniline furfuryl alcohol propellant combination, the rocket had a thrust of 1,500 lb. and a burn duration of 45 sec. Assembled by the Douglas Aircraft Co. at Santa Monica, Calif., it was boosted a few hundred feet into the air by a solid propellant Tiny Tim rocket with a length of 8 ft. and a maximum diameter of 11.75 in., a thrust of 50,000 lb. and a burn duration of 0.6 sec. Only the last six sounding rockets fired had full propellant tanks, and the fifth flight, launched October 11, 1945, reached a height of 240,000 ft. (45 mi.). The last round was fired on October 25, 1945.

1945 October 2 Under Operation Backfire, a German V-2 was launched by the British Army from a former Krupp naval gun testing range 5 mi. south of Cuxhaven, Germany. It was the first time a V-2 had been launched since the end of the war. (The last of the war was launched on March 29). The Army launched a second missile on October 14. No more V-2s were launched until the U.S. Army began its own flight program on March 15, 1946. The idea for the British shot came from Jr. Comdr. Joan C. C. Bernard, an aide to Maj. Gen. Alexander M. Cameron, Air Defense Division chief for SHAEF (Supreme Headquarters Allied Expeditionary Force). Her view was that this would be a good way of finding out how the missiles were put together and operated.

1945 October 15 Initiated as a research effort to find chemicals that would allow rocket exhaust velocities of up to 10,000 ft./sec., the first hydrogen-oxygen thrust chamber tests were conducted by Aerojet General at their Azusa test facility. Using gaseous hydrogen, the uncooled test thrust chamber completely burned out in 15 sec., but engineers measured a thrust of 45 lb. at a chamber pressure of 375 PSI with a calculated exhaust velocity of 7,280 ft./sec. For later tests, water provided cooling for chambers producing up to 500 lb. of thrust.

1945 October 29 The Committee for Evaluating the Feasibility of Space Rocketry set up by the U.S. Navy Bureau of Aeronautics on October 22 concluded that sufficient knowhow existed to proceed with a formal study of an Earth-satellite project. Advanced in its thinking, the committee suggested the use of high-energy, cryogenic hydrogen and oxygen propellants for high specific-impulse and suggested that structural complexity would be reduced if a single-stage-to-orbit (SSTO) concept was adopted. In the following month the idea gained wide support at the bureau, and confidence grew when news that Aerojet General had for the first time succeeded in firing a hydrogen-oxygen engine on October 15, 1945. On December 12, 1945, the bureau asked the Jet Propulsion Laboratory to check the committee's findings.

1945 October 31 Through its Air Technical Service Command, the U.S. Army Air Force invited proposals from industry on a 10-year program of research and development for 4 classes of missile with ranges of between 20 mi. and 5,000

mi. A leading contender for further studies, the Consolidated Vultee Aircraft Corp. (Convair) produced two competing concepts: a subsonic, pilotless cruise missile and a supersonic ballistic missile. In January 1946, Convair presented its ideas and three months later received additional funds to study how these trends might evolve over the next 5 to 10 years.

1945 December 14   The U.S. Air Force signed a contract with Bell Aircraft Corp. for a swept-wing rocket-powered research aircraft to be known as the X-2. Simple modification of the X-1 design was impossible due to a need for a completely new fuselage structure to accommodate the loads the aircraft would experience. Integrating revolutionary research results from Germany on swept-wing designs for supersonic aircraft, NACA engineer Robert Jones provided basic research work for the conceptual design of the X-2. Two X-2s were to be built, each powered by a twin-chamber Curtiss-Wright XLR25-CW-3 throttleable rocket motor. Regeneratively cooled, the motor had a thrust of 2,500–15,000 lb., the upper chamber providing up to 5,000 lb. of thrust, the lower chamber 10,000 lb. of thrust. The X-2 could carry 860.3 gal. of water-alcohol fuel and 755.8 gal. of liquid oxygen and utilized a complicated turbopump, advanced for its day.

1945 December 17   The Rocket-Sonde Research Branch of the U.S. Naval Research Laboratory (NRL) was officially constituted for the purpose of exploring the upper atmosphere with sounding rockets instrumented to carry out scientific measurements. Personnel participated in the V-2 rocket launches conducted by the U.S. Army from the White Sands Proving Ground, N.Mex. The limited number of V-2s available frustrated researchers. This led directly to the Viking rocket project, which did much to stimulate interest at the NRL in developing an artificial Earth-satellite program from which Vanguard was to emerge.

1945 December   The U.S. aircraft builder North American Aviation Inc. submitted a proposal from its technical research laboratory to the Army Air Force for a long-range cruise missile using ramjet or turbojet power and boosted into the air by a large rocket motor. From this would stem the SM-64 Navaho project and the X-10 research program. This would for the first time utilize the piggyback concept of a staged booster carrying an upper stage offset from the booster's center of gravity.

1946 January 3   Frank J. Malina of the Jet Propulsion Laboratory in California presented results of a study on satellite launchers to the U.S. Army War Equipment Board. Describing a rocket he had designed with fellow engineer M. Summerfield, Malina explained how their satellite launcher would require an initial mass of 3 million lb. in a five-stage rocket to launch a 10-lb. payload to escape velocity. Gen. Joseph W. Stilwell reflected the board's lack of interest when he wrote in his diary: "I am eminently suited to do something else and would as leif sit on a tack."

1946 January 7   The U.S. Navy launched the first of its pulse-jet flying bombs adapted from the Army Air Force JB-2 and named Loon. Two submarines were converted to carry Loon flying bombs and the first of these, *Cusk,* fired its first missile on February 18, 1947. On January 26, 1949, a Loon was launched from the converted seaplane tender USS *Norton Sound.* But the Navy terminated this program in March 1950 in favor of the more promising Regulus missile. Nevertheless, Loon did much to stimulate Navy interest in operating cruise missiles, a path that also eventually led to ballistic missile development.

1946 January 10   Following an October 31, 1945, request from the U.S. Army Air Technical Service Command for bids on a 20–5,000-mi. range supersonic missile, Convair was given funds to study unmanned weapons delivery at the extreme range. Convair began by defining two separate types of weapon: a winged cruise missile and an intercontinental ballistic missile. In December 1946 work on the cruise missile ended when Northrop was asked to proceed with the Snark, and Martin with the Matador. Convair meanwhile progressed with the ballistic missile under the designation MX-774 Hiroc and was funded to produce 10 research and development rounds. Detailed work began in June 1946 under the control of Karel J. Bossart at Vultee Field near Downey, Calif.

1946 January 15   Responding to a request from Merle A. Truve and Henry H. Porter of the Applied Physics Laboratory (APL) at Johns Hopkins University that he survey available sounding rockets in the United States, James A. Van Allen submitted a report saying that no suitable rockets existed. Impressed by a visit from Rolf Sabersky from the Aerojet Engineering Corp., Van Allen kept Aerojet informed about the report, and on February 2 the APL sent the company a letter requesting proposals to build 20 sounding rockets of a new design capable of sending a 300–1,500-lb. package of instruments to a height of at least 600,000 ft. On March 1, 1946, Van Allen suggested that the Naval Ordnance Department award a contract to Aerojet.

1946 February 27   To coordinate scientific experiments using German rockets brought to the United States, a V-2 Upper Atmosphere Research Panel was formed during an organizational meeting at Princeton University. Largely responsible for getting V-2s used for atmospheric research, Dr. Ernst H. Krause of the Naval Research Laboratory was elected chairman. Other members included Fred Whipple of Harvard University and James Van Allen, who replaced Krause as chairman in December 1947. It coordinated activities between industry, government and the universities. In March 1948 the name was changed to Upper Atmosphere Rocket Research Panel and in April 1957 to Rocket and Satellite Research Panel. It ceased to exist following the opening of the National Aeronautics and Space Administration in October 1958.

1946 March 7   Faced with an estimated development cost of up to $8 million for the engineering and design phase of its proposed satellite launcher, the U.S. Navy met with the Army Air Force to discuss possible cooperation to spread costs. The Army was generally enthusiastic and instructed Gen. Curtis E. LeMay, Deputy Chief of the Air Staff for Research and Development, to conduct negotiations with the Navy, but within a week his superior, U.S. Army Air Force Chief of Staff Gen. Carl Spaatz, had decided not to proceed. On April 9, 1946, the Navy was informed of that fact and promptly reinvigorated discussions with the Jet Propulsion Lab and Aerojet on further studies. However, during April 1946 the Aeronautical Board, which existed to integrate Navy and Air Force defense interests, asked Project RAND to prepare a satellite launcher study.

1946 March 29   The U.S. Army Air Force awarded a contract to North American Aviation Inc. for a long-range supersonic cruise missile program that would initially utilize remnants from the German rocket effort. Phase 1 would add wings to a V-2 missile, providing a range of 175–500 mi., phase 2 would substitute a supersonic ramjet engine for the rocket engine and phase 3 would put the missile on a booster powered by liquid propellant rocket engines providing a total range of 1,500 mi. The program would go through many changes and modifications, eventually becoming the SM-64 Navaho supersonic cruise missile.

1946 March   The U.S. Army Air Force issued a one-year research and development contract to Northrop for two long-range cruise missiles. One, MX-775A, was to be capable of carrying a 5,000-lb. warload 1,500 mi. while the second, MX-775B, had to carry the same load 5,000 mi. The Air Force wanted cost estimates based on a production run of 5,000 units. The shorter range missile was named Snark while the latter was named Boojam, both from the pages of Lewis Carroll. By Christmas 1946 Snark and Boojam had been cancelled, but Jack Northrop personally intervened and called Gen. Carl Spaatz, Chief of the Air Arm, to provide revised low-cost estimates. By the end of 1947 Snark was back in the budget and Northrop proceeded to prepare 10 vehicles for flight test. The first Snark flight attempt was not made until December 21, 1950.

During the month, plans were set out at the Jet Propulsion Laboratory for design and construction of an improved Wac Corporal to be designated Wac Corporal B. A lighter rocket motor and lighter propellant tanks were to be used to achieve improvements in performance. Several rockets of this type were launched at the end of 1946 and in early 1947. The success of the Wac Corporal program resulted in a contract for development of a successor, the Aerobee.

The U.S. Army Air Force began development of the W-5 atomic warhead for application in Project Mastiff, an experimental 300-mi. range air-to-surface missile. Northrop, Bell, McDonnell and Republic submitted designs for the missile but it was replaced by the Rascal, for which Bell Aircraft Corp. was eventually awarded a contract.

1946 April 16   The first V-2 rocket launched by the U.S. Army as part of a research program on German rocket technology ascended from White Sands Proving Ground, N.Mex., at 14:47 MST. Just 19 sec. after liftoff the emergency cutoff signal was sent to shut down the motor when it was observed to behave erratically. It reached a height of 18,000 ft. but flew east until it broke apart and crashed to the floor of the desert 5.3 mi. from the launch site. The second V-2 flight at 14:15 MST on May 10, 1946, was staged as a press demonstration flight. It reached a height of 70 mi. and impacted 31 mi. north. After the first few flights the missile would be used in a program of atmospheric research by the Army and the Navy and, later, in combination with a Wac Corporal developed by the GALCIT (Guggenheim Aeronautical Laboratory at the California Institute of Technology) team under the Bumper-Wac program.

1946 May 2   Douglas Aircraft completed study SM-11827 entitled "Preliminary Design of an Experimental World-Circling Spaceship," prepared in response to a request from the Aeronautical Board. It endorsed the viability of an earth-orbiting satellite program, believing it would cost some $150 million to launch a 500-lb. mass to a circular 300-mi. orbit with a 10-day lifetime. RAND emphasized the scientific value of such a project but said that "such a vehicle will undoubtedly prove to be of great military value." Released in February 1947, a second classified study refined the idea around a three-stage hydrazine–liquid oxygen rocket with a launch weight of 81,585 lb. Because it set up Project RAND, this Douglas study was subsequently referred to as a RAND report.

1946 May 17   The U.S. Navy Bureau of Ordnance awarded a contract to Aerojet Engineering Corp. for detailed design and manufacture of the Aerobee sounding rocket. The Douglas Aircraft Co. would provide aerodynamic engineering support and manufacture some of the parts. Officially known as XASR-1, the project was directed by James A. Van Allen at the Applied Physics Laboratory at Johns Hopkins University. Van Allen made up the name from Aero, for Aerojet Corp., and bee, from Bumblebee, which was the overall name of the Navy rocket program.

1946 June 28   The first V-2 that the U.S. Naval Research Laboratory had fully instrumented for research into the physics of the upper atmosphere was launched to a height of 67 mi. at the White Sands Proving Ground, N.Mex. It carried a Geiger-counter, spectrograph, pressure gauges, temperature gauges and radio transmitters. Major areas of interest lay in the study of cosmic rays, variations in atmospheric temperature and pressure with altitude, absorption measurements in the ionosphere, meteors and meteorology, sound wave propagation and the earth's magnetic field. In addition, there was interest in obtaining pictures of the earth's surface which, when published, did much to stimulate public interest in space. The use of V-2 rockets for scientific research laid the foundation for satellite proposals during the 1950s.

1946 July 1   Following studies on a potential satellite launcher that were carried out for the U.S. Navy at the Jet Propulsion Laboratory, the Bureau of Aeronautics awarded a contract to Aerojet General for studies of a test stand that could be used to test high-energy hydrogen-oxygen rocket motors. The Bureau wanted to know if it was feasible to design a 300,000-lb. thrust motor and to build and test one of 1,000-lb. thrust. Other contracts were awarded for solar engines producing electrical power and for guidance equipment to maintain control during the unpowered coast phase of an elliptical trajectory prior to circularization of the orbit. This began the engineering design phase of the Navy satellite project.

North American Aviation submitted *A Preliminary Study on the Use of Nuclear Power in Rocket Missiles*. Nuclear propulsion seemed ideal because it provided a means of heating a fuel to working temperatures without the need for an oxidizer and a conventional combustion chamber. Because the specific impulse (Isp) of a rocket is inversely proportional to the square of the molecular weight of the propellant, doing away with the relatively heavy oxidizer greatly reduced the molecular weight and significantly increased the Isp. Moreover, carrying only a working fuel without the comparatively heavy and bulky oxidizer enabled smaller rocket stages to do more work.

1946 July 9   A subcommittee of the U.S. Guided Missiles Committee of the Joint Chiefs of Staff, set up to examine the need for a long-range proving ground, recommended that the search begin for a suitable site. On October 7, 1946, the Joint Research and Development Board set up the Committee on Long Range Proving Ground and on June 20, 1947, this body recommended the El Centro area of the Gulf of California as a prime site, with the Banana River area near Cape Canaveral, Fla., as second choice. A V-2 rocket launched by the U.S. Army from White Sands on May 29, 1947, went in the wrong direction and landed in a cemetery south of Juarez, Mexico. Although not the sole reason, this incident was strong justification for a safer place from which to launch rockets. In January 1948 the president of Mexico refused to seek permission from his government for the development of the El Centro region and as a result Cape Canaveral was selected.

From this remote and sparsely populated region, missiles and rockets could be fired out across the Atlantic in relative safety.

1946 August 1 The U.S. Congress passed a bill prohibiting the exchange of atomic information between the United States and any other country which, in January 1947, became law along with the establishment of the Atomic Energy Comission. Brought by the Democrat freshman from Connecticut, Brian McMahon, it sought to prevent the proliferation of atomic weapons by stopping the distribution of information in the naive belief that other countries were incapable of producing atomic weapons. In fact, much of the work on the Manhattan project to produce the first atom bombs had been contributed by scientists from continental Europe and Britain. European scientists became U.S. citizens, but British scientists cooperated because their government had an agreement with the United States to do so. After the United States rescinded the exchange of information, Britain found itself badly behind because it had not sought to duplicate U.S. work. It was now cut off from the pool of information it had previously been given access to in exchange for the cooperation of British scientists.

1946 August 21 The Glenn L. Martin Co. received a contract to build an atmospheric research rocket for the U.S. Naval Research Laboratory. Reaction Motors of California was chosen to produce the motor, the first large pump-fed liquid propellant engine developed in the United States. It would have a thrust of 20,000 lb. and achieve flight control through the use of gimbals, whereby the entire engine was swung from side to side on commands from a guidance system, rather than with inefficient steering vanes. Originally named Neptune, it was eventually called Viking, a name chosen by Thor Bergstrahl, one of the scientists working on the project.

1946 September 26 Responding to the award of a July 1, 1946, contract on the proposed U.S. Navy single-stage-to-orbit (SSTO) satellite launcher, North American Aviation (NAA) submitted its initial report to the Bureau of Aeronautics. NAA recommended an ogive structure for the launcher, which had been called the High Altitude Test Vehicle (HATV), 86 ft. long and with a maximum diameter of 16 ft. With a maximum weight of 130,000 lb. and a thrust of 233,000–308,000 lb. from nine clustered motors, the HATV would achieve orbital speed at a height of 140 mi., considered about the minimum altitude. The HATV's theoretical mass fraction—a ratio of structural weight to launch weight—was 0.05; in other words the structure weighed no more than 5% of the launch mass, which in 1946 was a dream not realized before the Atlas Inter-Continental Ballistic Missile (ICBM) of the late 1950s.

1946 October 12–16 Approximately 20,000 German scientists and technicians were moved from the Soviet zone of Germany to Russia in 92 trains. Among them were 200 rocket engineers transported to the island of Gorodomlya in Lake Seliger, 150 mi. northwest of Moscow. At 3:00 AM local time on October 22, Helmut Grötrupp and other former senior V-2 engineers from Germany were gathered and moved to Moscow, arriving there six days later. Grötrupp was put to work at Nauchnii isledovatelskii instituit Nii-88 (Scientific Research Institute 88).

1946 November 6 The first details of a proposed man-carrying suborbital rocket were published in the *Daily Express,* an English newspaper. Devised by R. A. Smith and H. E. Ross, the rocket was based on existing A-4/V-2 missiles and substituted a manned capsule for the warhead. Dubbed Megaroc, the project envisaged strengthening a V-2 and increasing its diameter to extend the burn time to 148 sec. Smith and Ross designed a 1,300-lb. alloy cabin that would contain a rotating chair for a single human occupant. Megaroc was capable of sending an astronaut on a ballistic trajectory to a height of 225 mi., with deployment of a recovery parachute during descent. Megaroc was submitted to the British Ministry of Supply on December 23, 1946, but was rejected.

1946 November 29 The Soviet rocket engineer Mstislav V. Keldysh was appointed chief of the Scientific Research Institute No. 1. at a time when major rocket proposals were beginning to get the attention of Kremlin leaders. Under the tutelage of Keldysh, this branch of the RNII (Reaktirnyi Nauchno Issledovatel'kii Institute) would grow into the leading institute for rocket technology in the USSR. Elsewhere, the Gas Dynamics Laboratory Design Bureau (GDL-OKB) was completing development of the RD-100 liquid propellant motor for the 1RA-E geophysical research rocket and beginning the design of RD-101 engines which would be used to power the V-2-A medium-range missile, adapted in 1949 for geophysical research in the upper atmosphere. The V-2-A was developed into a research launcher capable of lifting a 4,850-lb. payload to a height of 132 mi.

1947 January 23 The first successful flight of a V-2 instrumented with telemetry was launched from the White Sands Proving Ground, N.Mex. During development flights at Peenemünde engineers could monitor at best four parameters sent down by radio signal during the flight. This marked a step toward the heavily instrumented rocket and missile flights of the 1960s for which several thousand parameters could be read simultaneously. This was also the first V-2 to carry a General Electric automatic pilot with a steering device capable of controlling the vehicle's attitude during flight. Experience with this system would result in remote-controlled missiles.

H. S. Seifert and M. M. Mills of the Jet Propulsion Laboratory at the California Institute of Technology issued Memorandum No. 3–4 *Problems in the Application of Nuclear Energy to Rocket Propulsion.* In it, they argued that chemical rockets had insufficient potential to achieve the Earth escape speed of 37,000 ft./sec. but that nuclear rockets would have a potential terminal velocity in excess of 90,000 ft./sec. They argued for the use of hydrogen as a working fluid in a nuclear heat engine where combustion would be replaced by going directly from the gaseous to the atomized condition producing very high exhaust velocities. In time, this became the principle upon which the NERVA (nuclear engine for rocket vehicle application) program was based.

1947 February 7 U.S. Air Force interest in nuclear rocket motors was focused by a report entitled *Feasibility of Nuclear Powered Rockets and Ramjets,* investigating the viability of atomic engines for ballistic missiles. The Air Force wanted to know whether nuclear rockets or ramjets were the best choice for a missile project known as MX-770, later called the X-10, which became a part of the Navaho program. In September 1948 the Air Force chose turbojets, and for a while the prospects for nuclear space propulsion dimmed.

1947 February 20 The first flight in the U.S. Air Force Project Blossom took place when a V-2 was launched from the White Sands Proving Ground, N.Mex., carrying experiment packages for the Air Materiel Command. Blossom involved 10 V-2 flights carrying a wide variety of scientific equipment for measuring biological reactions to high-speed, high-altitude flight. On the first flight the V-2 reached a height of 68 mi. and successfully tested an instrument-capsule recovery system designed for live animal tests. The ejection of the canister at a height of 68 mi. was followed by deployment of a ribbon parachute 8 ft. in diameter to check the rate of descent. A second parachute, 14 ft. in diameter, was deployed at a height of 30 mi. The canister took 50 min. to descend and was recovered 9 mi. from the blockhouse. The last Air Force flight under Project Blossom was launched on June 28, 1951.

1947 February   U.S. Army Maj. Gen. Gladeon M. Barnes, Chief of Research and Development, Office of the Chief of Ordnance, paid a visit to Fort Bliss, Tex., and discussed several ideas for long-range ballistic missiles with Army personnel at White Sands and with Wernher von Braun. In April, the Hermes contract with General Electric was expanded to included design and development of ramjet-propelled missiles, Hermes B, and large multistage missiles, Hermes C. Hermes A series vehicles were to be test rounds, the A-1 having a range of 50 mi. and the A-3 a range of up to 150 mi. Hermes A, B and C were part of the Hermes I program. Prior to flights with Hermes B, Hermes II was to be an adapted version of an existing V-2 carrying a prototype ramjet motor as a second stage. Hermes C was essentially the A-10 conceived at Peenemünde and was designed to have six 100,000-lb. thrust engines, a 100,000-lb. thrust second stage and a glider as the third stage. Alternative concepts included a V-2 as second stage and a Viking as third stage.

1947 March 14–15   During a meeting of rocket designers and aircraft manufacturers in Moscow, the USSR's deputy prime minister, Georgi Malenkov, defined future missile requirements when he said that production of the German V-2 and its basic derivatives was insufficient for Soviet needs: "Our strategic needs are predetermined by the fact that our potential enemy is to be found thousands of miles away." At a meeting of the Politburo and the USSR Council of Ministers in the Kremlin on the second day, Stalin asked that a state commission be set up to study intercontinental ballistic missiles. The original members were Col. Gen. I. A. Serov, the first deputy minister of the NKVD (the secret police); Prof. Col. Grigory A. Tokaty-Tokaev, chief scientist and deputy chairman of the Soviet air force; Prof. Mstislav V. Keldysh, Ministry of Aircraft Production; and Maj. Gen. Vasily I. Stalin, the Soviet leader's son.

1947 April 14   Leading officials from Soviet aircraft manufacturers and technical institutes met in Moscow to discuss the design concept for rocket-propelled gliders proposed by the Austrian engineer Eugene Sänger. Among those present were M. A. Voznesensky, the head of the State Planning Commission, Col. Grigory A. Tokaty-Tokaev, a rocket specialist who would later defect to the West, Air Marshal K. A. Vershinin, from the air force, and aircraft designers Aleksandr S. Yakovlev and Artem I. Mikoyan. Most technical judgments were negative, criticizing the advanced engine concept which most believed was unattainable in the short term. When briefed on the concept the following day, Stalin set up State Commission No. 2 to study the concept, and an order went out from the Kremlin to find Sänger and bring him to Russia.

1947 April 24   A French missile test area in the Sahara desert was officially inaugurated as the Centre Interarmées d'Essais d'Engins Speciaux (Inter-Service Test Center for Special Weapons). Located near the town of Colomb-Béchar, Alge-

ria, the CIESS was built up over the next 10 years with additional facilities at Hammaguir, 60 miles southwest of Colomb-Béchar. From here, flights up to 1,900 mi. were possible. Ostensibly under the control of the French Ministry of National Defense, the CIESS was also used for scientific tests with sounding rockets and, in time, would become the site from which French satellite launchers were tested.

1947 May 22   Developed by the GALCIT (Guggenheim Aeronautics Laboratory at California Institute of Technology) team the Corporal E test missile was launched for the first time from White Sands Proving Ground, N.Mex. With a weight of 12,000 lb., a length of 40 ft. and a diameter of 2 ft. 6 in., the Corporal E carried aniline-RFNA propellants and produced a thrust of 20,000 lb. On its first flight the missile impacted within about 2,500 ft. of its designated target 62 mi. from the launch site. Manufactured by the Firestone Tire and Rubber Co., the Corporal ballistic missile, the first to serve with the U.S. Army, was developed in 1951 from Corporal E. In that role, designated SSM-A-17, it had a length of 46 ft. and a range of up to 86 mi. with a velocity at motor burn-out of Mach 3. Several hundred rounds were deployed between 1953 and the early 1960s, when it was replaced by the Sergeant. The British Army operated Corporal between 1957 and 1966.

1947 May 29   Designated round #0 of the Hermes II ramjet development program, a modified V-2 was launched from White Sands Proving Ground on a relatively low altitude trajectory. Control problems caused the vehicle to deviate from the planned flight path and it impacted 47 mi. south of the launch site almost hitting the city of Juarez, Mexico. The four flight tests of Hermes II involved a modified V-2 first stage and a ramjet second stage designated Organ. The ramjet stage was 17 ft. 10 in. long with a diameter at its widest point of 4 ft. 2 in. and a square-shaped wing surface with a span of 15 ft. 3 in. A small forward elevator surface in the nose had a span of 4 ft. 11 in. Intakes for air to mix with the hydrocarbon fuel were set in the wing leading edges. For stability the four V-2 tail fins were greatly increased in surface area. The Hermes II program was influenced by the team working under Wernher von Braun out of Fort Bliss. The last launch came on November 1, 1950, and all Army ramjet work was halted in 1953.

1947 May   A top secret report from the British Joint Chiefs of Staff to the cabinet office, entitled *Future Defence Policy,* gave prophetic warning of future Soviet ambitions. It declared that "All our intelligence sources indicate that Russia is striving, with German help, to improve her military potential; and to catch up technically and scientifically. We must expect that from 1956–57 Russia will be in a position to use some atomic bombs that she may have developed, probably with German advice and technical assistance, rockets." From this came Britain's resolve to "use weapons of mass destruction on a considerable scale from the outset."

1947 June 13   The first successful firing of a liquid hydrogen–liquid oxygen rocket chamber took place at Ohio State University. The firing produced a thrust of 106 lb. with an exhaust velocity of 7,890 ft./sec. An additional 118 firings were made with the same engine, and from September 1948 an additional 38 runs were made with a chamber five times larger. Tests at Ohio State ended May 29, 1950. The two other groups in the United States studying liquid hydrogen rockets were Aerojet, which fired up its first test chamber on January 20, 1949, and the Jet Propulsion Laboratory, which ran its first test on September 21, 1948.

1947 June 26 Aerojet General successfully tested a gaseous hydrogen-oxygen combustion chamber with a thrust of 1,230 lb., the first time a thrust in excess of 1,000 lb. had been achieved with an engine using these propellants. Utilizing a convection-cooled injector recirculating the water for transpiration cooling, the test run lasted 190 sec., and the engine produced a specific impulse of 309 sec. This and other test chambers of the same size and performance were based on a 300,000-lb. thrust hydrogen-oxygen engine designed by Aerojet for the U.S. Navy High Altitude Test Vehicle (HATV) single-stage-to-orbit (SSTO) satellite launcher.

1947 June 30 The Glenn L. Martin Co. and Aerojet reported to the U.S. Navy Bureau of Aeronautics on design studies for the High Altitude Test Vehicle (HATV) single-stage-to-orbit (SSTO) satellite launcher on which it had been working since July 1, 1946. Based on an Aerojet design of a 300,000-lb. thrust hydrogen-oxygen engine, the Martin rocket concept had a length of 77 ft., a diameter of 14 ft. 5 in., a gross launch weight of 104,648 lb. with a payload of 1,450 lb. Exhibiting an astonishingly low mass-fraction of 0.028, less than 3%, the rocket would be powered by the one hydrogen-oxygen engine.

1947 July 1 The U.S. Army Air Force cancelled the MX-774 ballistic missile development contract being worked up into 10 test vehicles by Karel J. Bossart at Convair. Some funds remained, however, for test flights with three missiles and the first static firing took place at Point Loma, San Diego, on November 14, 1947. Bossart introduced novel features that would remain an integral part of later missiles. He proposed separating the reentry warhead so that structural weight could be reduced to a minimum, and gimbaled motor nozzles for flight control to replace vanes. Directional vanes in the exhaust flow, like those used in the V-2, cut thrust by 17% because they slow the exiting gases. Bossart also proposed that the propellant tanks should form the outer skin of the missile in an integral sidewall-tank assembly of pure monocoque construction such that propellant tank pressure would inflate the structure to its cylindrical form, thus saving weight. These features formed the basic design of the Atlas ICBM.

1947 July 26 President Harry S. Truman signed the National Military Act creating the Department of Defense and an Air Force independent of the Army. This disrupted Air Force and Navy proposals for a satellite launcher. The Joint Research Development Board, which had been a component of the now defunct War Department, was replaced by the Research and Development Board, and for the remainder of 1947 the launcher proposals languished. A Technical Evaluation Group met March 29, 1948, to consider the proposals but decided there was no clearly defined mission for the launcher and referred all future studies to Project RAND.

1947 July The U.S. Air Force added development of a supersonic ramjet to its contract with North American Aviation for development of a supersonic cruise missile. This gave the vehicle a range of 1,500 mi. In March 1948 the Air Force restructured the program into a three-pronged research effort: a 1,000-mi. range range test vehicle known as the X-10; a 3,000-mi. test vehicle under the XSM-64 designation; and an operational, 5,000-mi. range rocket-boosted SM-64 missile capable of delivering a thermonuclear warhead. The first X-10 was tested on October 14, 1953, powered by two Westinghouse turbojets. By the time the last test took place on January 26, 1959, 11 delta-winged X-10 missiles had made 27 flights.

1947 August 26 The German rocket engineer Helmut Grötrupp was taken by train from his chief workplace at the Nii-88 research institute in Moscow to the village of Kapustin Yar, 75 mi. from Stalingrad, where several German V-2 engineers under the authority of Marshal Nikolai N. Voronov joined Grötrupp. Using engineering equipment brought from the Nordhausen area in their zone of Germany, the Soviets were preparing to begin rocket flights from launch sites built to accommodate modified V-2 rockets. Some advanced experiments were conducted at the rocket motor test and launch area about 9 mi. to the northeast, including an attempt to measure the attenuation of radio waves as they passed through rocket exhaust.

1947 August Detailed performance specifications were drawn up for a U.S. Navy cruise missile capable of carrying a 3,000-lb. warhead a maximum distance of 500 NM at a speed of Mach 0.85. Called Regulus, the missile would have a length of 33 ft., a wing span of 21 ft. and a launch weight of 14,500 lb. Sustained in the cruise by a 4,600-lb. thrust Allison J33-A-14 or J33-A-18A turbojet, the missile was boosted into the air by two 33,000-lb. thrust solid propellant boost motors, one each side of the rear fuselage. Built by the Chance Vought company, Regulus began flight trials in 1951 and was developed for use from submarines and surface ships. It became operational in 1955 and served aboard diesel- and nuclear-powered submarines, cruisers and aircraft carriers before going out of production in 1959 and out of service in 1964. The Navy launched about 1,000 Regulus, many recoverable, and definitive versions could send a 3.8 MT warhead 575 mi. A developed version, Regulus II, made its first flight on May 29, 1956.

Prof. Col. Grigory A. Tokaty-Tokaev from the Soviet Zhukovsky Academy of Aeronautics in Moscow and Maj. Gen. Vasily I. Stalin, Marshal Stalin's son, were sent to the Soviet-occupied zone of Germany in search of Dr. Eugene Sänger. Impressed with Sänger's work on rocket motors, the Soviets wanted to pick his brains about rocket-powered aerospace vehicles having wings and sufficient lifting capacity to carry bombs half way round the world. But in Germany they found that the Austrian rocket engineer had already left for Paris to work for the French. A concerted effort into long-range ballistic missiles, begun this year at the instruction of Marshal Stalin, resulted in many study submissions for ballistic rockets.

1947 September 6 A V-2 missile was launched from the deck of the aircraft carrier USS *Midway*. The test took place in the Atlantic Ocean off Bermuda to determine the impact a missile launch would have on normal carrier operations and on how effectively the flight preparation and launch phase could be conducted on a ship under way. The missile exploded at a height of only 6 mi. but the rest of the exercise went well and aircraft were taking off as arranged immediately after the launch.

1947 September 25 The first Aerobee sounding rocket was launched when a live booster lobbed a dummy rocket across the test range at White Sands, New Mexico. The first Aerobees were 19 ft. long, weighed 1,600 lb. and utilized the same RFNA-furfuryl alcohol propellants used by the Wac Corporal. A 6-ft.-long solid propellant booster was used to accelerate Aerobee to a velocity of 1,000 ft./sec. (682 mph) and early versions of the sounding rocket could lift a 150 lb. package of scientific instruments to a maximum altitude of 81 mi. The first all-live Aerobee launch took place on November

24, 1947. Later versions of what would become the workhorse for US sounding rocket experiments could lift a weight of 100 lb. to a height of 1,365 mi. or send 1,000 lb. to a height of 250 mi.

1947 September In the Soviet Union, Prof. Grigory A. Tokaty-Tokaev completed the basic design details of Project TT-1, a long-range ballistic missile utilizing three liquid propellant stages. Inspiration for the fuel-cooled expansion chamber adopted for the engine came from the work of Dr. Eugene Sänger. Each stage was powered by a single engine operating on ethanol–liquid oxygen propellants contained in single, divided tanks. Engines in the first two stages were pump-fed and in the third stage a pressure-fed system from a gas generator was employed. Liftoff thrust was approximately 220,000 lb. and cruciform fins on first and second stages provided aerodynamic stability during ascent, exhaust vanes providing directional control outside the atmosphere. Political infighting and intrigue involving competing design teams caused chaos in the many different rocket proposals. Some scientists, like Tokaty-Tokaev, whose project was rejected, fled to the West to escape imprisonment while others like Prof. M. V. Keldysh remained aloof from the power play.

1947 October 1 The U.S. Department of Defense received proposals for atomic warheads that could be carried to their targets by unmanned missiles. In reality, however, the sheer size of these weapons made missile application impractical. The atomic bombs then available weighed 11,000 lb. and were 5 ft. in diameter and almost 11 ft. long. By early 1949, bomb technology had improved to the extent that a marriage between the nuclear weapon and surface-to-surface missiles became a possibility. On April 29, 1949, the Atomic Energy Commission's division of Military Application asked the Military Liaison Committee to study that possibility.

1947 October 14 U.S. Army Air Force test pilot Maj. Charles E. "Chuck" Yeager exceeded the speed of sound in level flight for the first time during the 50th air-launched drop test of the Bell X-1. The X-1 reached a speed of Mach 1.06 at an altitude of 43,000 ft. The power plant was a four-chamber Reaction Motors XLR11-RM-5 with a maximum thrust of 6,000 lb. run on liquid oxygen and diluted ethyl alcohol. The chambers could be fired singly or in groups. The X-1 was an important first step toward the development of the first fully operational hypersonic, rocket-powered research aircraft designated X-15. It was also the start of a 21-year period of almost unbroken high speed aerodynamic research using rocket-propelled aircraft.

1947 October 18 The first rebuilt German V-2 rocket fired from Russian soil was launched by a team under the direction of Helmut Grötrupp from a stand 9 mi. northeast of the town of Kapustin Yar, 75 mi. from Stalingrad. Much of the equipment used to launch the rocket as well as the tracking cinetheodolite stations to record the progress of the missile in flight had been brought from the Nordhausen area in Soviet-occupied Germany. About 20 rebuilt V-2s were launched before Grötrupp returned to Moscow on December 1, 1947.

1947 October Various studies and research projects on satellites and space vehicles conducted by the U.S. armed services were coordinated in a Committee on Guided Missiles set up in the Joint Research and Development Board. The JRDB succeeded the War Department's Aeronautical Board, defunct since the establishment of the Department of Defense in July.

Over the next two months a technical group under Clark B. Millikan concluded that an Earth-orbiting satellite project was feasible. On January 15, 1948, James V. Forrestal, the first secretary of defense, concurred that satellites should be developed when the time was right but would not give the effort priority at this time.

1947 November 8 The first rocket chamber to use the highly toxic but powerful oxidizer liquid fluorine fired into life at North American Aviation. Devised by experimental chemist William L. Doyle, the engine ran on hydrazine fuel. Fluorine was found to decompose Teflon seals commonly used in rocket engines and, combined with moisture and impurities in the fluid, to attack and burn metal. Doyle's *Heath Robinson* test rig in the NAA parking lot gave rise to serious concern on health and safety grounds. In 1949 he joined Ohio State University to work on hydrogen-fueled engines.

1947 December 12 The German V-2 scientist Helmut Grötrupp briefed officials of the Soviet Scientific Council on modifications to the V-2 that would greatly extend its performance and enhance its capabilities for military applications and for scientific research. The small cadre of German scientists and engineers with Grötrupp in the USSR had developed the design of the German missile to use radar guidance and achieve a maximum range of 550 mi. The Soviets then incorporated such innovative features of the Grötrupp team's designs into their own rocket, which had been developed simultaneously and without the knowledge of the Germans. In 1948 Grötrupp and his team were sent to an island in Lake Seliger near the source of the Volga to work on rocket motors and missile design.

1947 December 16 Aerojet General received permission from the U.S. Navy to build a liquefier for producing liquid hydrogen (LH2) which the company wanted to use in its first liquid hydrogen–liquid oxygen combustion chamber test. Aerojet had worked out it could build its own plant and produce the LH2 it needed for less than the quoted price from the lowest commercial bidder. By the end of 1948 the company was running a three-shift operation and had produced more than 1,000 lb. of LH2.

1948 January 15 The U.S. Air Force issued an Earth-satellite policy in which it claimed it had "logical" responsibility for taking the prime role in this arena. Since the newly formed Department of Defense had made the Air Force, Army and Navy independent entities, there had been fierce competition among the services to seize the initiative and each sought as much of the research and development budget as possible. The Air Force believed its position was justified by the strategic implications of satellites and launchers and announced that research would "be pursued as rapidly as progress in the art justifies . . . [making] recommendation of the development phases of the project at the proper time."

1948 February 20 The Russians moved Helmut Grötrupp and his family from the comforts of Moscow to the bleak subsistence of Gorodomlya island where the bulk of the Germans brought from Nordhausen had been sent. Sergei Pavlovich Korolev had been trying to encourage the German engineers to work with him on a stretched variant of the V-2, but they were not interested. Other design concepts were pursued by the Grötrupp group but they were not taken up by the Russians. On December 21, 1950, Grötrupp was relieved of his duty as head of the German engineers.

1948 May 13   The first in a series of eight launches with a Wac Corporal sounding rocket attached to the top of a V-2 rocket took place from the White Sands Proving Ground, N.Mex. Called Bumper-Wac, the combination had a height of 62 ft. and a launch weight of 14.5 tons. In this first multistage launch of a major rocket system, the altitude reached was only 79 mi. due to a premature cutoff of the Wac Corporal. The next three flights were either partial or total failures, but the objective of sending a package of instruments to record altitude was achieved on February 24, 1949, when a height of 244 mi. was recorded. The sixth flight was a failure but the next two, on July 24 and July 29, 1950, were successfully launched from Cape Canaveral, Fla.

1948 June 11   The first of five V-2 flights made with live animals occupying a bio-instrumented recovery capsule in the nose of the rocket was launched from White Sands Proving Ground, N.Mex. The V-2 reached a height of 39 mi. but the monkey died of breathing difficulties before hitting the desert floor when the recovery system failed. The 9-lb. Rhesus monkey named Albert gave its name to the series of five flights, although the last launch on August 31, 1950, was made with a mouse. None of the animals survived and all five recovery capsules failed to deploy their parachutes. A part of Project Blossom operated by the Air Force Air Materiel Command, the biomedical flights were controlled by the Aero-Medical Laboratory at Wright-Patterson Air Force Base.

1948 July 13   The first of three MX-774 test missiles built by Convair for the U.S. Air Force made its first flight from the White Sands Missile Range, N.Mex. Intended to reach a height of 100 mi., the missile fell back to Earth from 6,200 ft., impacting just 650 ft. from the launch position. On September 27, 1948, the second MX-774 ascended to a height of 10 mi. and coasted on to a maximum altitude of 29 mi., but the rocket exploded when gas pressure in the tanks built up to an excessive level. The last MX-774 was launched December 2, 1948, but it reached a height of only 30 mi. MX-774 had a length of 32 ft. 7 in. and a diameter of 2 ft. 6 in., with power provided by four uprated Thiokol alcohol–liquid oxygen motors similar to those used in the Bell X-1, each delivering a thrust of 2,000 lb.

1948 September 1   The U.S. Air Force took over the Banana River Naval Air Station near Cape Canaveral, Fla., in the first step to build up the planned Joint Long Range Proving Ground for rockets and missiles, which was activated on October 1, 1949. In February 1950 the U.S. Coast Guard opened up its small patch of land at Cape Canaveral to missile use, and the government obtained the rest of the area from private ownership. On May 16, 1950, its name was changed to Long Range Proving Ground Base, and on August 1, 1950, it became Patrick Air Force Base. By clearing underbrush and pulling up the palmettos, construction workers had the first small concrete pad completed by June 20, 1950. It was a 100-ft. strip of concrete laid over sandy soil from where a mated V-2 and Wac Corporal would make the first successful launch on July 29, 1950.

1948 September 15   The Committee on Guided Missiles of the Joint Research and Development Board, successor to the Aeronautical Board at the U.S. War Department, recommended that Project Hermes study the possibility of launching Earth satellites. This was the first reference to the U.S. Army–Wernher von Braun team being employed to evaluate the prospect of a space program.

1948 September 17   Soviet scientists attempted the first launch of the R-1 rocket, a direct copy of the German V-2. A control system failure prevented the R-1 from making a successful flight on this occasion. The rocket first flew successfully on October 10, 1948; this was followed by several launches over the next few months. The R-1 formed the basis for developing rocket propulsion techniques and provided valuable practical experience in assembling, handling and firing what were then large rockets.

1948 September 21   The Jet Propulsion Laboratory carried out its first run with a liquid hydrogen–liquid oxygen rocket chamber four months after Ohio State University had conducted the first such firing in the United States. The JPL was studying the use of high-energy hydrogen for the U.S. Navy Bureau of Aeronautics and wanted to look at the possibility of using a nuclear reactor to heat hydrogen, thus avoiding the need to carry oxygen. This was the origin of nuclear rocket propulsion concepts which would eventually lead to a development program between NASA and the U.S. Atomic Energy Commission, a predecessor of the Department of Energy. Meanwhile, interest in cryogenic propellants waned because they were difficult to handle and could not be stored. The Army and the Navy preferred using storable propellants in rockets with short-notice launch capability.

1948 October   Britain's Ministry of Supply, a government department responsible for contracts and procurement, awarded a contract to English Electric for the design and development of a ground-to-air guided missile for the defense of the United Kingdom against enemy aircraft. The company was already forming its guided weapons division and work proceeded toward what would be known as the Thunderbird. Before that, test missiles were built under the Red Shoes program. Red Shoes and Thunderbird were part of the combined program to give the British Army an air defense weapon. In a parallel program, the Royal Air Force would get the Bloodhound air defense system.

1948 November 13   Harry E. Ross presented a paper to a meeting of the British Interplanetary Society in London describing a manned space station in Earth orbit which could serve as an astronomical observatory and conduct research on zero gravity. The Ross station comprised a toroidal (doughnut-shaped) unit housing a 24-man crew as well as telescopes and research instruments. Earth-based shuttles were to resupply the station every three months. Ross also presented his idea of moon landing expeditions using Earth-orbit rendezvous to assemble a large vehicle which would then be transported to moon orbit from where it would be lowered to the surface.

1949 January 19   Designed to a specification drafted in August 1945 and developed by the Glenn L. Martin Co. beginning March 1946, the first XB-61 Matador pilotless flying bomb was, unsuccessfully, launched for the first time from Holloman Air Force Base, N.Mex. Designated TM-61A, the missile had a length of 39 ft. 8 in. and a wingspan of 27 ft. 10 in. with a weight of 12,000 lb. Powered by a 4,600-lb. thrust Pratt and Whitney J33-A-37 turbojet, Matador could carry a 3,000-lb. warload at 650 MPH to a maximum range of 620 mi. Initial boost came from a single 50,000-lb. thrust solid propellant rocket motor under the tail. A later version, the TM-76 Mace, had a length of 44 ft. and a wingspan of 22 ft. 11 in. With a weight of 18,000 lb., Mace had a 100,000 lb.-thrust motor providing launch boost. More than 1,000 missiles were produced. Matador was declared operational in 1955, followed by Mace in 1959. Both were retired in 1969.

1949 February 21 Project RAND report D-405 identified key reasons for government-run space programs. Believing satellites and space travel would have a major impact on world opinion, the report cited what it considered to be the principal reasons for a U.S. space program: "As a demonstration of U.S. technological superiority; as a device for communication; as a reconnaissance instrument; and as an instrument of political strategy." These were the first formal declarations of the primary reasons why the USSR and the United States would inaugurate space projects and use ambitious goals like the moon landing for stimulating national acclaim. Post-Sputnik reaction around the world would clearly identify the desired link between technological prowess and perceived national greatness.

1949 February 25 As part of a continuing program of aerodynamic research, the no. 3 Douglas D-558-II Skyrocket flew for the first time on rocket power. Three Skyrockets had been built from an original specification calling for jet-powered transonic research with swept-wing aircraft. The no. 3 was the first to carry a Reaction Motors XLR-8-RM-5 engine delivering a maximum thrust of 6,000 lb., but, deployed to supplement the Westinghouse J-34-40 turbojet engine, it carried only 362 gal. of diluted ethyl alcohol–liquid oxygen propellants. In October 1949 the no. 1 Skyrocket began flight tests with a rocket engine installed in addition to its turbojet. Without a jet engine fitted, the no. 2 Skyrocket carried an almost identical rocket motor but 723 gal. of propellant and made its first powered flight on January 26, 1951. The last Skyrocket flight was with the no. 2 aircraft on December 20, 1956.

1949 March 2 An instruction from the U.S. Navy to Aerojet General nearly brought an untimely halt to work on liquid hydrogen rocket motors at this company when the Bureau of Aeronautics ordered a switch to anhydrous hydrazine. The Navy wanted Aerojet's liquefier equipment for the first hydrogen bomb test in the Pacific, though this was unknown at the time. Ironically, Aerojet had produced the liquid hydrogen sold to the Jet Propulsion Laboratory which they used to fire the world's first LH$_2$-LOX (liquid hydrogen–liquid oxygen) combustion chamber on September 21, 1948. Aerojet would eventually produce the RL-10 engines for the world's first LH$_2$-LOX upper stage, the Centaur, first launched May 8, 1962.

1949 March 15 The French government's Délégation Ministérielle de l'Armement (DMA) formally agreed to begin development of an independent national nuclear force and associated delivery vehicles, including ballistic missiles. This decision initiated research at French army laboratories into all forms of solid and liquid propulsion. The Laboratoire de Recherches Balistiques et Aérodynamiques (LRBA), located at Vernon, near Paris, developed a liquid propellant missile based on the German A-4/V-2, about which the French had learned a great deal since the end of World War II. Their missile, the Véronique, first flew on August 2, 1950.

1949 April 29 The U.S. Military Liaison Committee examined the possibility of marrying the atomic bomb and long-range missiles. By this date warhead technology had advanced so much that atom bombs could weigh as little as 3,000 lb., bringing them within the capabilities of unmanned cruise-type

missiles. In January 1950, the U.S. Navy got permission from the secretary of defense to attach the W-5 to the Regulus cruise missile. On December 18, 1950, the Army got permission to put the W-7 warhead on the Corporal tactical missile. In October 1951, the Air Force planned to mount the W-13 warhead on its Snark cruise missiles. Thus began an evolving process that would result in the first nuclear-tipped ICBMs a decade later.

1949 April The Soviet Scientific Council reviewed proposals for 10 advanced rocket and missile projects developed by Soviet design teams and by the collective of German former V-2 engineers brought to Lake Seliger on the Volga. Some proposals were for missile designs of intermediate range and potentially capable of a range of up to 1,800 mi. It was during this period that the Soviets developed a wide range of small, medium and large rocket designs and engine concepts, and the gradual consolidation of this effort contrasts markedly with the absence of a formal set of missile programs in the United States. Grötrupp proposed a 159,000-lb. missile carrying a 6,000-lb. warhead and utilizing a single pump-fed rocket engine of 220,000-lb. thrust and gimbaled for flight control. This design produced technical concepts incorporated later in rockets developed by teams of indigenous Russians.

1949 May 3 The U.S. Naval Research Laboratory's first Viking atmospheric research rocket was launched from White Sands Proving Ground, N.Mex. It had a length of 45 ft. and a diameter of 2 ft. 8 in. with four swept fins at the base spanning 9 ft. 2.5 in. With 7,000 lb. of ethyl alcohol–liquid oxygen propellant the launch weight was 10,000 lb.; propulsion was provided by a 20,000-lb. thrust pump-fed engine with a turbine driven by steam from decomposition of hydrogen peroxide. Viking was designed to carry a 500-lb. payload of scientific instruments to a height of 100 mi. Viking No. 1. reached a height of only 50 mi. because the turbine leaked, causing premature cut-off. Twelve Viking rockets were launched in two phases. After the flight of the seventh, which reached a height of 136 mi., on August 7, 1951, the rocket was modified for improved performance but not until its second flight (*Viking* 9) did the improved rocket make a successful ascent.

1949 May Working with meager funds at Convair in California, Karel Bossart continued his work on a ballistic missile concept (begun under U.S. Air Force project MX-774) and formulated a radical concept that would create America's first ICBM. Seeking ways to avoid the need for his rocket to ignite a second stage at extreme altitude, an event always likely to cause failure, Bossart decided it would be safer to incorporate several engines in a row at the base of the rocket and jettison all but the central sustainer engine on the way up. Leaving the remaining rocket motor to continue burning propellant from the main tanks, at the point in the trajectory where a second stage would normally ignite, he dispensed with the excess weight of rocket motors in what would become known as the 1½ stage concept. When resurrected as MX-1593 on January 16, 1951, the Convair ICBM project would be called Atlas, in recognition of Convair's major shareholder, the Atlas Corp.

1949 June A three-man ad hoc committee set up by the U.S. Air Force examined the possibility of putting atomic or nuclear warheads on guided missiles. This was deemed feasible and led to the judgment that a wide range of missiles was justified only if they all carried atomic warheads. It endorsed the Air Force Rascal missile which, on December 6, was deemed to be capable of carrying a 3,000–5,000-lb. atomic

Beginning in 1949, Soviet engineers began flights with a modified V-2 missile carrying a recoverable nose section instrumented for scientific measurement of the upper atmosphere.

warhead. By 1955, Rascal was being designed to carry a 3,000-lb. warhead 105 mi. at Mach 2.95. In 1959 Rascal was replaced by Hound Dog.

The newly formed Gesellschaft für Weltraumforschung (GfW), or the German Association for Space Travel, passed a resolution recommending "an international meeting of all (astronautical) societies to explore the possibilities of forming an international association." In addition to Germany, there were already space societies in England, France, the United States, Canada, Denmark and Argentina. The GfW contacted in England the British Interplanetary Society and in France the Groupement Astronautique Française, or French Space Group, and received their support. The First International Congress was held in Paris in 1950 but not until the Second Congress in London, September 4, 1951, was the International Astronautical Federation established.

1949 August 26 In a major step forward in the design of rocket motors for high-energy propellants, Ohio State University ran the first test of a regeneratively cooled combustion chamber running on liquid hydrogen propellants. The motor delivered a thrust of 100 lb. at a chamber pressure of 188 lb./sq. ft. and, with an oxidizer-to-fuel mass ratio of 4.1, produced an exhaust velocity of 10,466 ft./sec. A total of 33 runs were successfully made without a failure and one achieved an exhaust velocity of 10,728 ft./sec. while another fired for 159 sec. Regenerative cooling using hydrogen helped the efficiency of the motor because it eliminated heat loss and allowed lower-density hydrogen to enter the combustion chamber and help stabilize the combustion cycle.

1949 August 29 The first Soviet atomic device was detonated at a remote location about 65 miles west-southwest of the city of Semipalatinsk in the southern Soviet republic of Kazakhstan. The device had a yield of 20 KT. It had been just over three years since the decision to make atomic weapons had been ratified by including it in the list of projects for the 1946–50 five-year plan, which the Politburo requested be compiled on August 8, 1945, two days after the atom bomb dropped on Hiroshima. In the intervening period Russian scientists got some help from British scientists working as spies and from drawings supplied by Soviet scientists working in the United States. On hearing of the first Soviet atomic test, physicist Edward Teller capitalized on national fears and rushed through a program to build a U.S. thermonuclear (hydrogen) bomb. It was first detonated on November 15, 1952.

1949 October 1 The Joint Long Range Proving Ground Base at Cape Canaveral, Fla., was activated under the organizational responsibility of the Joint Long Range Proving Ground (JLRPG). On May 16, 1950, the the JLRPG became the Long Range Proving Ground Division, and on June 30, 1950, it was again renamed to Air Force Missile Test Center. On May 15, 1964, it became the Air Force Eastern Test Range, finally changing to Eastern Test Range on February 1, 1977, when it went under the control of the Space and Missile Test Center.

1949 November 18 President Harry S. Truman appointed a special committee to help him decide whether to authorize development and testing of thermonuclear (fusion) weapons, or hydrogen bombs. The committee consisted of Secretary of State Dean Acheson, Secretary of Defense Louis Johnson and Atomic Energy Commission Chairman David Lilienthal. Theory indicated that fusion weapons had unlimited explosive yield, unlike fission (atomic) weapons of the type developed in the Manhattan Project. In a theoretical analysis of thermonuclear weapons, the hydrogen bomb was named "Super." Two members of the committee recommended a start of the Super but Lilienthal did not. Nevertheless, on January 31, 1950, President Truman announced that he was approving development of thermonuclear weapons. The fact that their explosive yield would imperil conventional bombers releasing such weapons provided further impetus for the development of strategic ballistic and cruise-type missiles.

1950 January 31   The first U.S. Navy N-7 Terrier surface-to-air missile was delivered for tests at the beginning of one of the most prolific and fruitful industrial programs in air defense missile history. From Terrier came a veritable host of missiles and shipboard air defense systems resulting in Tartar missiles for small ships and a completely new family of replacement missiles in the 1970s called Standard. The first Terrier sea launch took place in late 1951 from the converted seaplane tender USS *Norton Sound* and the first successful downing of a target drone occurred in 1953. The Terrier BW-1 was deployed with the U.S. Navy beginning June 15, 1956.

1950 April 15   Personnel from the U.S. Army Ordnance Department working on missile development at Fort Bliss on the Texas–New Mexico border began a move to Redstone Arsenal in Huntsville, Ala. Five-year contracts given to the Germans from Operation Paperclip were renewable, and although some elected to return to their native country, most stayed on to work for the new, consolidated U.S. Army Ordnance Guided Missile Center. Col. Holger N. Toftoy had been instrumental in getting approval for the new facility, which would also make use of the adjacent Huntsville Arsenal previously used for chemical production. The move was completed in October 1950. Army Maj. James Hamill was the director and Wernher von Braun was appointed technical director, with Kurt H. Debus acting as his special assistant.

1950 May 11   Under Project Reach, the U.S. Naval Research Laboratory launched Viking No. 4 from the deck of the USS *Norton Sound*, a seaplane tender modified into a guided missile research ship. Carrying 900 lb. of scientific instruments, Viking No. 4 reached a height of 105 mi. Accompanied by the destroyer USS *Osbourn,* the *Norton Sound* had been positioned on the geomagnetic equator off Christmas Island in the Pacific for the launch, which was intended to evaluate the ability of the rocket's guidance system to compensate for roll and pitch moments at liftoff and to research the possibility of launching ballistic missiles from ships at sea.

1950 May 19   The first of five flights of the Hermes A-1 rocket took place from White Sands, N.Mex. Employing a 13,000-lb. thrust motor developed at General Electric's Schenectady, N.Y., facility, the scaled down V-2 operated on alcohol and liquid oxygen. It had a maximum altitude of 15 mi. and a range of 38 mi. with a terminal velocity of 1,850 MPH. The last flight took place on April 26, 1951. An A-2 version powered by a solid propellant motor developed at the Jet Propulsion Laboratory and Thiokol was ground tested in December 1951 but was cancelled in October 1952 before it flew. With various thrust ratings of 18,000–22,000 lb., thirteen Hermes A-3 rockets had been flown by the time the program was canceled in 1954.

1950 June 15   Ohio State University chemist William L. Doyle carried out the first liquid hydrogen–liquid fluorine rocket chamber test. In January 1951 Doyle obtained an exhaust velocity of 14,108 ft./sec. Doyle discovered that liquid fluorine was denser than had been believed, and that provided an added value for rocketeers. Not only was fluorine the most powerful oxidizer, it would take up less volume per unit mass and reduce the rocket's structural weight accordingly. Fluorine was difficult to handle, however, and it was not widely used.

Developed by General Electric, the Hermes A-1 performed five flights between May 19, 1950, and April 26, 1951. This provided a foundation for further U.S. research into propulsion and guidance systems.

1950 June   Work on the supersonic cruise missile being developed for the U.S. Department of Defense by North American Aviation under a project designated MX-770 was redirected towards the development of a truly intercontinental weapon system to be known as WS-104A and called Navaho. It was to be a cruise missile with a length of 87 ft. 4 in. and a wing span of 40 ft. 3 in., propelled into the air by a booster 91 ft. 6 in. in length and powered by a three-chamber rocket motor delivering a thrust of 415,000 lb. At launch the total system would weigh 290,000 lb. The cruise missile would be powered by two Westinghouse turbojets providing a range of 5,500 NM with a 7,000-lb. warhead. An interim step using a reduced scale configuration (XSM-64) would make its first flight attempt on November 6, 1956.

The Ordnance Guided Missile Center at Redstone Arsenal, Ala., received an order from the Chief of Ordnance to study a surface-to-surface theater missile with a 150–500-mi. range carrying interchangeable warheads, one weighing 1,500 lb., the other 3,000 lb. The existing concept for the Hermes C-1 was adapted to this requirement and the new missile was designated XSSM-G-14. The missile incorporated a warhead designed to separate from the main body of the rocket after motor burnout, thereby avoiding the problem so many V-2 rockets had experienced when they began to tumble and frequently broke up during the ballistic coast phase. Separable warheads became a standard feature of all tactical and strategic ballistic missiles. The XSSM-G-14 was named Ursa, then Major and finally Redstone, after the place where it was designed.

The first flight from Cape Canaveral, Florida, took place on July 24, 1950, with the launch of a V-2/WAC-Corporal as part of Project Bumper.

1950 July 24　The first flight from the U.S. Air Force Long Range Proving Ground Base at Cape Canaveral, Fla., took place when the seventh of eight Bumper-Wac rockets was launched on a depressed trajectory 200 mi. downrange. An attempt to launch Bumper-Wac No. 6 five days earlier ended when the missile failed before launch due to salt corrosion of certain parts. On June 30, 1951, the Air Force changed the name of the organization operating the range from Long Range Proving Ground Division to Headquarters, Air Force Missile Test Center (AFMTC); the cape area was designated the Cape Canaveral Missile Test Annex. On July 21, 1951, an agreement with Britain secured U.S. rights to a 1,000-mi. range through the Bahamas, and a subsequent agreement extended this through Ascension Island more than 5,000 mi. from Cape Canaveral. On May 15, 1954, the AFMTC became the Air Force Eastern Test Range, and on February 1, 1977, it was renamed, simply, the Eastern Test Range.

1950 August 2　The first French ballistic missile, the Véronique, made its maiden flight. Its name derived from the place of manufacture (VERnon) and the French word for electronics (électrONIQUE). With a length of 24 ft. 10 in., the R-series, of which this was the first, had a weight of 3,165 lb. and burned diesel oil. On the first attempt it reached a height

of only 10 ft., but by February 1952, after seven more flights, it had demonstrated a maximum height of 5,900 ft. In 1952–53, 11 N-series flights took place. With 8,800-lb. of thrust and 1,455 lb. of furaline propellant some reached a height of 37 mi. In four launches during 1954, the Véronique NA sounding rocket carried 132 lb. to a height of 84 mi. Further development resulted in the Vesta rocket of 1964, capable of lifting 1,100 lb. to a height of 250 mi., and in the Vexin engine used in Eméraude and the Valois used in the Diamant launcher.

1950 December 21　Six months after the U.S. Air Force had invalidated it by increasing the performance requirement, the first flight attempt with a Northrop N-25 Snark cruise missile took place at Holloman Air Force Base, N.Mex. It failed, as did another on March 8, 1951, before the first successful test occurred on April 16, 1951. Bigger than previous flying bombs, this U.S. Air Force missile had a length of 51 ft. 11 in., a span across its thin, high-aspect-ratio swept wings of 42 ft. 6 in., and a weight of 28,000 lb. Power was provided by a 5,000-lb. thrust Allison J-33 turbojet. The N-25 had a speed of Mach 0.85 at its maximum range of 1,500 mi. and made 28 flights, the last 7, from August 29, 1952, taking place from Cape Canaveral. To meet the new specification, Northrop redesigned the missile around the more demanding requirement: supersonic dash at the end of a 5,500 NM flight carrying a 7,700-lb. thermonuclear warhead to within 1,500 ft. of the target. This became the N-69.

1951 January 16　The U.S. Air Force initiated project MX-1593, calling for a research-and-development effort for what would become the first U.S. ICBM (intercontinental ballistic missile), and on January 23, 1951, Convair received a contract to study optional engineering concepts. The San Diego, Calif., company was asked by the Air Force to resolve the argument over which system to adopt for very long-range, pilotless weapon delivery: a jet-powered cruise missile or a rocket-propelled ballistic missile. Karel Bossart was put in charge of the study and by September 1951 Convair decided in favor of a ballistic missile integrating design innovations developed under the earlier MX-774 study and more recent in-house work. By this time the name Atlas had been chosen and the recommended configuration would have lightweight propellant tanks as an integral part of the missile structure, gimbaled nozzles for flight control, jettisonable rocket motors and a separable warhead. The Air Force stipulated a range of 6,330 mi., and the first design had a launch weight in excess of 440,000 lb., with seven first-stage engines each producing more than 125,000 lb. of thrust.

1951 January　In response to a request from the U.S. Army for design proposals on a tactical battlefield missile, the Ordnance Guided Missile Center at Redstone Arsenal, Ala., recommended a single-stage ballistic missile for ranges of 500 to 550 mi. but a two-stage missile for ranges beyond that. The Army wanted a missile capable of throwing a 3,000-lb. warhead a distance of 500 mi. with an accuracy of 3,000 ft. The single-stage version was approved and Wernher von Braun's team proceeded with detailed design around the 78,000-lb. thrust ethyl alcohol–liquid oxygen Navaho propulsion unit built by Rocketdyne. In February 1951 the Army increased the throw-weight to 6,900 lb. to accommodate an atomic warhead. To expedite development and utilize the existing engine, the range would be lowered to a maximum 200 mi.

1951 March 13 The U.S. Atomic Energy Commission received a report suggesting that atomic or nuclear warheads could be used for defense against mass air attack. So desperate was the U.S. Air Force Air Defense Command (ADC) for an antiair weapon that it examined the possibility of tossing 10–20 KT, Mk.4 or Mk.7 airdrop bombs into the path of approaching bomber formations. On March 23, 1953, the ADC provided a specified requirement for warheads to be used on air-to-air rockets and on surface-to-air defense systems. The former would comprise a missile project called Ding Dong, and the latter would include Bomarc. Ding Dong evolved into the Genie missile.

1951 March 21 The Russians began repatriating German rocket engineers and technicians, sending the first group of 20 from Gorodomlya back to their homeland. Not for two years did the Russians take steps to move the bulk of the German research group out of the USSR, but after Stalin died on March 5, 1953, things changed, and by November 22 that year, all had been returned, including Helmut Gröttrup. The Russians had bled the Germans dry of any useful information they could contribute, and in the end the Soviets wanted to build rocket forces their way, without dividing the responsibility with their former enemy.

1951 May 1 The U.S. Army formally began development of its 200-mi. tactical battlefield missile which on April 8, 1952, it officially named Redstone, after the Guided Missile Center at Redstone Arsenal, Ala. Time was of the essence and the Army had stipulated a development period of only 20 months, 12 test missiles being required by May 1953 in three lots of four vehicles each. Ready by January 1953, Lot 1 would test propulsion, missile structure, control systems and roll control between cutoff and warhead separation. Lot 2 would test warhead separation, maneuverability during descent and aerodynamic heating. Lot 3 would qualify guidance, tracking, launch procedures and training.

1951 June 14 During a two-day meeting of the National Advisory Committee for Aeronautics (NACA) subcommittee on stability and control, aerodynamicist Max Hunter from the Douglas Aircraft Co. proposed a research program exploring the flight regime between Mach 5 and 10. Hunter reasoned that too little was known about the effect of the atmosphere on bodies at this speed and that future missile development would benefit from such a program. Others supported this view, a consensus among aerodynamicsts that resulted in several proposals for such a research effort. By the end of 1951 a strong lobby had emerged supporting development of a new hypersonic aircraft program. This led directly to the X-15.

1951 June 22 The Jet Propulsion Laboratory, Calif., fired the first Loki I sounding rocket, named after the Norse god of mischief, which was to be adapted for several multistage combinations. The basic Loki had been developed at JPL as a small solid propellant antiaircraft missile with an effective ceiling of 12 mi. Fin-stabilized and unguided, Loki had a length of 3 ft. and was never used operationally. As a sounding rocket it could lift a weight of 8 lb. to 57 mi. and was used in balloon launches during International Geophysical Year. Loki II, named *Hawk*, could lift 8 lb. to 75 mi. In another guise and named HASP (high altitude sounding projectile), a Loki missile modified by the Naval Ordnance Laboratory could lift 6 lb. to a height of 18 mi. It was given initial acceleration from 5-in. naval guns.

1951 July 15 The first English Electric Thunderbird surface-to-air missile was launched from the Royal Aircraft Establishment firing range at Aberporth, Cardiganshire, Wales. Following three years of development under the guidance of L. H. Bedford, this air-defense weapon emerged from a wide variety of configurations with a length of 20 ft. 10 in., a main body diameter of 1 ft. 9 in. and a span across four cruciform tail sections of 5 ft. 4 in. Four strap-on boosters supplied thrust during the early part of the ascent, taking the four fins with them when they were jettisoned. Sustained acceleration was provided by a solid propellant motor in the main body of the missile. Operated by the British Army from 1959 until 1976, Thunderbird could hit air targets at a range of almost 47 mi. A contemporary of Thunderbird, the Bloodhound surface-to-air missile was powered by two Mach 2 ramjets and boosted into the air by four solid propellant rockets. Deployed with the Royal Air Force from 1958 to 1992, it could hit air targets at a distance of more than 50 mi.

1951 July 30 The U.S. Air Force requested the National Advisory Committee for Aeronautics (NACA) to test 1/10 scale models of Convair's B-58 bomber design by firing them to Mach 2 using solid propellant rockets launched from its Wallops Island Flight Test Range. The first instrumented model was launched on September 11, 1952, by a combination of two Deacon motors placed side by side with a separation of 12 in. There had been criticism of delays in the development of military aircraft projects due to the lack of information about flight characteristics and the impact of high speed on various wing and body shapes. Rocket-powered research aircraft were unable to provide aerothermal data on a wide range of such shapes, and it fell to the NACA to carry out that work. Other model aircraft tested during 1952 included shapes for the F-100, F-102, F-104, F8U-1, F2Y-1 and FY-1.

1951 August 22 In the first major ballistic missile launch by an all-American crew, a German V-2 lifted off from the White Sands Proving Ground, N.Mex., at 12:00 noon MST and climbed to a height of 132.6 mi., the highest achieved by any V-2. Designated TF-1, for Training Flight No. 1, it was prepared and launched by the First Guided Missile Support Battalion on the first of five "green suit" launches aimed at demonstrating successful flights under realistic field conditions. The last V-2 was launched on September 19, 1952, almost 10 years after the first flight of a missile of this type, bringing to an end five years and five months of tests in the New Mexico desert. In all, 74 V-2s had been launched from White Sands, in addition to which there were two ship-launched flights and two Bumper flights from Cape Canaveral for a grand total of 78 German V-2s launched by the United States. The British also launched two in Operation Backfire.

1951 August Fifteen months after Douglas Aircraft received a contract to develop the U.S. Army Honest John short-range tactical missile, the flight test program began at White Sands, N.Mex. The first post-war U.S. missile to enter service, the solid propellant Honest John had a length of 24 ft. 10 in., a diameter of 2 ft. 6 in., a fin span of 4 ft. 6 in. and a weight of around 4,500 lb. The mobile missile was carried on, and launched from, a truck-mounted rail. Production of more than 20,000 Honest John missiles was undertaken by Emerson Electric, and the type remained in service with non–U.S. army units well into the 1980s.

1951 September 4 The Second International Astronautical Congress, and the first meeting of the International Astronau tical Federation (IAF), opened in London under the auspices of the British Interplanetary Society (BIS). The first IAF president was Eugene Sänger. The draft constitution had been drawn up by the 1950 BIS chairman Val Cleaver. The IAF quickly became the only truly international forum for space scientists and engineers from all political systems, largely because the constitution enshrined in its mandate a study of "peaceful, not military" space projects. This language permitted the communist countries to become IAF members and to provide a forum for discussion across political boundaries. The IAF was to hold an annual congress in a different country each year, and the number of space societies grew rapidly throughout the 1950s. The first congress in eastern Europe took place in Warsaw, Poland, in 1963.

1951 September 7 Kenneth W. Gatland presented a paper titled "Minimum Satellite Vehicle" to the Second International Astronautical Congress in London. Designed by Gatland, A. M. Kunesch and A. E. Dixon, the three-stage vehicle was the culmination of three years' work. Called Minimum Satellite Vehicle, it had a length of 51 ft. 1 in., a diameter of 6 ft. 4 in. and a launch weight of 62.4 tons carrying a 221-lb. payload. The concept was a breakthrough in that it did away with the idea that satellites had to be launched on giant rockets lifting heavy payloads. The proposal can be directly linked to the decision to launch a U.S. satellite. Alexander Satin from the U.S. Office of Naval Research visited London and met Arthur C. Clarke on November 25, 1952. Clarke advised Satin to read the book defining the design of the Gatland-Kunesch-Dixon launcher, and Satin returned to the United States to put together Project Orbiter, which led directly to the first U.S. artificial satellite.

1951 September 20 The first living animals to return safely from a ballistic rocket shot were recovered after a flight aboard a special capsule launched by an Aerobee rocket. The flight took place from Holloman Air Force Base, N.Mex., and involved a "payload" of one Rhesus monkey and 11 mice in a capsule which was safely recovered by parachute. An earlier attempt on April 18, 1951, had failed to recover the one monkey on board, which died when the nose section of the Aerobee rocket failed to deploy its parachute. A second successful recovery was made on May 21, 1952, when an Aerobee launched two monkeys, "Pat" and "Mike," and two mice. Worldwide criticism of the six preceding V-2 and Aerobee flights in which animals died brought a halt to U.S. flights with animals until a Thor-Able flight on April 23, 1958.

1951 October 12 The First Symposium on Space Flight was held at the Hayden Planetarium in New York. It was from this that the emergence of work carried out by Wernher von Braun's team on the design of futuristic space vehicles would begin to gain acceptance in the United States. Although most of the papers presented were sheer speculation, it was one of the first forums to bring together acknowledged rocket experts and astronomers to discuss the technical problems and the scientific benefits of space flight.

1951 October The International Council of Scientific Unions formally agreed to a proposal suggesting that the period July 1, 1957, to December 31, 1958, be designated as the Third International Polar Year. During the First International Polar Year, 1882–83, 11 countries had set up 12 research stations in the Arctic; for the Second International Polar Year, 1932–33, 20 nations conducted expeditions. In spring of 1951, Lloyd V. Berkner suggested cutting in half the usual 50-year gap between such events. In October 1952 the name was changed to the International Geophysical Year, and in the United States the National Committee for the IGY was established within the National Academy of Sciences. The IGY provided incentive for the Soviet Union and the United States to develop the first scientific satellites and proved a catalyst for a burgeoning space program. The IGY was extended for the twelve months of 1959 in what was called the International Geophysical Cooperation.

1952 March 22 In an article called "Man Will Conquer Space Soon," published in *Collier's* magazine, New York, Wernher von Braun presented a comprehensive description of an ambitious space program leading to manned lunar landings. The core project was to be a three- stage rocket with a height of 265 ft. and a first-stage thrust of 31.5 million lb. The first stage would be powered by 51 rocket motors of approximately 618,000 lb. of thrust each, the second stage by 34 motors of 116,000 lb. of thrust each, and a reusable space plane powered by five motors of 99,000 lb. of thrust each would be on top. The space plane would carry a payload of approximately 36 tons.

1952 April 17 Bell Aircraft Corp. made a formal presentation to the U.S. Air Force Wright Air Development Center on a proposed boost-glide vehicle conceived by Capt. Walter R. Dornberger, then working as consultant to Bell. Dubbed "Bomi" (for bomber-missile), it involved a family of two-stage vehicles stacked piggy-back and launched vertically. One configuration had a 4,000-mi. range, another a 6,000-mi. range, and one had a returnable upper stage which could fly back to the launch base. The suborbital bomber version would have delivered a single nuclear bomb while the orbital version with two nuclear bombs would have incorporated storable propellants to add the final push. The gross liftoff weight was to be 940,000 lb. with a combined thrust from the two stages of 1.1 million lb. Eugene Sänger was asked by Dornberger to join him on the study at Bell, but he refused. Dr. Krafft Ehricke, then with the Army Ballistic Missile Agency, did however work with Dornberger. Initial evaluation of Bomi was completed April 10, 1953, with generally negative reaction. Further studies were conducted, however, and Bell came up with a reconnaissance version.

1952 May 21 The National Advisory Committee for Aeronautics released the first of two proposals for a hypersonic research program to explore the flight regime between Mach 3 and Mach 10. Hubert Drake and Robert Carmen of the High Speed Flight Station envisaged a piggyback configuration in which a supersonic plane capable of Mach 3 would carry a smaller aircraft carrying rocket motors operating on water-alcohol and liquid oxygen for a hypersonic dash to Mach 10. The piggyback concept would form the cornerstone of the NASA Shuttle program that evolved during the 1960s. The second proposal released late in May 1952 was for a more conservative use of existing projects. David Stone of the Pilotless Aircraft Research Division (PARD) wanted to use solid propellant Sergeant rockets to boost a Bell X-2 to Mach 4.5. Both proposals converged into the NACA report of June 23, 1953.

1952 June 18 Harry Julian Allen, head of the High Speed Research Division at the NACA Ames Aeronautical Laboratory, Calif., first proposed the idea that a blunt, rather than a streamlined, body would be most effective in dissipating thermal energy generated through friction with the Earth's atmosphere. Assisted by Alfred Eggers, Allen showed that a blunt body would permit a shock wave to stand some distance in front of the structure and absorb most of the heat instead of passing it to the boundary layer attached to the vehicle. Allen proved that the blunt body generated no more heat than a pointed cone, the heat load being determined by the rate of deceleration and not the shape or mass. This had a profound effect on the design of any object built to survive reentry and led directly to the blunt after-body shape of ballistic manned space vehicles.

1952 July 10   Additional funding was provided Convair by the U.S. Air Force for the Atlas ICBM research and development project, MX-1593, and a component development program began. In December 1952 the USAF Science Advisory Board reviewed progress on the concept and early in 1953 gave its formal approval to continue. Convair selected a five-engine configuration in a missile 88 ft. in length and 12 ft. in diameter producing a liftoff thrust of 650,000 lb. When in 1954 the Atomic Energy Commission gave the company details of a breakthrough in lightweight thermonuclear warhead design, the missile changed again. In October 1954 Convair came up with a three-engine configuration in a 75-ft. missile, 10 ft. in diameter, and on December 16, 1954, this design was set in stone.

1952 July   The British Chief of the Air Staff, Sir John Slessor, visited Washington, D.C., with a recently compiled Global Strategy Paper with which he hoped to persuade the U.S. Air Force that global nuclear deterrence should be the cornerstone for future defense policy. At first it was viewed by the U.S. Joint Chiefs as a ploy for Britain to renege on its NATO commitments by substituting the threat of massive nuclear attack for expensive manpower in Britain and continental Europe. At this time the Pentagon viewed nuclear weapons as one of several systems available for offensive operations. The British argument was persuasive and laid the groundwork for studies leading to U.S. adoption of a concept of massive retaliation and the early use of nuclear weapons.

1952 August   A useful marriage of balloon and rocket helped pave the way for instrumentation that would eventually be developed for use on the first U.S. satellite in 1958 when University of Iowa physicist James A. Van Allen began sounding the upper atmosphere off Greenland and discovered soft radiation at a high altitude. Van Allen wanted to reach altitudes above 50 mi. and used a balloon to carry a Deacon rocket high into the atmosphere before igniting it for ascent with its package of instruments. Recovered later by parachute, the instruments told Van Allen that there was considerably more radiation than expected, leading to a suspicion that earth's magnetic field may have trapped particles of radiation. This stimulated Van Allen to instrument the first U.S. satellite, *Explorer I*, to search for those belts of radiation, and it eventually discovered them.

1952 September 10   The first development vehicle of a new U.S. area defense system made its first flight from a site near Cape Canaveral. Called XF-99 Bomarc (an acronym of Boeing and Michigan Aerial Research Center), it was so large that the U.S. Air Force referred to it as the "unmanned interceptor." With a length of 45 ft. 3 in. and a launch weight of 15,000 lb. the missile had an airplane shape with a wing span of 18 ft. 2 in. Housed in a protective concrete structure, operational Bomarcs would be elevated for vertical launch and hurled to a height of 30,000 ft. by an integral 35,000-lb. thrust

liquid propellant rocket motor. Two 11,500-lb. thrust Marquardt ramjet engines would take over to push Bomarc to a cruise altitude of 60,000–80,000 ft. and a maximum speed of Mach 3. Range was 230 mi., 440 mi. with Bomarc B, employing a 50,000-lb. thrust Thiokol solid propellant boost motor and 14,000-lb. thrust ramjets. Redesignated IM-99A, the first Bomarc to fly with all propulsion operating was launched in February 1955 and on October 2, 1957, a Bomarc A was successfully launched from Patrick Air Force Base, Fla., under the control of a New York air defense controller. A total of 366 Bomarc As and 349 Bomarc Bs were built. Canada bought the Bomarc B which it employed for air defense instead of the Avro Arrow interceptor. The last squadron was deactivated in 1972.

1952 October 3   The first British atomic explosion took place from HMS *Plym*, a ship moored 400 yd. off the island of Trimouille in the Monte Bello group 50 mi. off the coast of western Australia. With an explosive yield equivalent to 25,000 tons of TNT (25 KT), the device weighed 10,000 lb. This first test was given the code name Hurricane. In a little over one year, on November 14, 1953, the first operational British atomic bomb was delivered to RAF Wittering, England, ready for use by Vickers Valiant medium bombers. Given the then secret name of Blue Danube, it paved the way for the development of the British thermonuclear bomb. Further British atomic weapons tests were conducted after Hurricane, including two in 1953, six in 1956 and three in 1957. The last of twelve atomic tests in Australia took place on October 9, 1957. Later tests were conducted from islands in the Pacific Ocean.

1952 November 1   At 7:15 A.M. local time on the island of Eniwetok in the Pacific, the world's first large-scale thermonuclear device was detonated by American scientists and engineers as part of Operation Ivy. Dubbed "Mike," the device exploded with a yield of 10.4 megatons (MT). Once the Teller-Ulam theory of thermonuclear fusion was demonstrated, work accelerated on design and fabrication of an air-drop weapon popularly known as the hydrogen-bomb. This would lead to lighter and smaller bombs that could be carried in the nose cones of guided and ballistic missiles. With a potential yield orders of magnitude greater than a fission (atom bomb) weapon, the possibilities were constrained only by the imagination of the engineers.

1952 December 15   The second in a series of uprated Viking atmospheric research rockets was launched successfully for the first time, following failure of the first (No. 8) on June 6, 1952. Viking No. 9 carried a 765-lb. payload to 135 mi. and 5,795 ft./sec. (3,600 MPH) over the White Sands Proving Ground in N.Mex. The uprated Viking had a propellant capacity of 11,400 lb., 4,000 lb. more than earlier Vikings. To accommodate the extra propellant the diameter of the rocket was increased from 2 ft. 8 in. to 3 ft. 9 in. Without increase in structural weight this resulted in a (then) record propellant-to-weight ratio of 4:1, expressed as a mass ratio of 0.8. Viking bridged the technology gap between the V-2 with a mass ratio of 0.69 and the Thor and Atlas rockets with mass ratios of around 0.94. The greatest altitude achieved by a Viking was 158 mi. on May 24, 1954. The last two Vikings were launched from Cape Canaveral, Fla., on December 8, 1956, and May 1, 1957, in support of the Vanguard satellite project.

1952   In Russia, bench tests began on a 120,000-lb. thrust liquid propellant engine for a medium-range ballistic missile,

Designed to U.S. Navy specifications, the Viking sounding rocket was a direct precursor to the Vanguard satellite launcher and successfully performed nine flights between May 3, 1947, and May 1, 1957.

later designated SS-3 Shyster in the West. Burning kerosene and liquid oxygen, the RD-103 engine evolved from the propulsion system developed by German rocket engineers for the V-2. The SS-3 was the last Soviet rocket to benefit from V-2 engineering, and the choice of propellants reflected its heritage. Flight tests with the SS-3 began during 1953 and development continued through 1954.

In a book entitled *Across the Space Frontier,* Joseph Kaplan, Wernher von Braun, Heinz Huber, Willy Ley and Fred L. Whipple outlined the blueprint for the development of a reusable spaceplane and the assembly of a permanently manned, Earth-orbiting space station. Von Braun did much of the technical work regarding the launch vehicle and space plane. One of the founders of the German Society for Space Travel (VfR), Willy Ley described a rotating, wheel-shaped space station, and Harvard astronomer Fred Whipple described the celestial observations possible from space. The book was stimulated by articles which appeared in *Collier's* magazine and did much to generate popular awareness of space flight.

1953 January   The U.S. Air Force awarded a contract to North American Aviation under the title REAP (Rocket Engine Advancement Program) to examine future rocket engine technologies and to explore the feasibility of high-thrust liquid propellant motors. The Air Force wanted to know what developments could be expected in rocket engineering and whether there were physical limits to the size of motors. It also sought improvements with the Navaho engine, which NAA was able to make under this program. The combination of practical applications and theoretical studies resulted during 1954 in a recommendation at NAA to carry out a feasibility study on a 1-million-lb. thrust motor. Research on this began around March 1955, and by 1957 studies had established an optimum thrust of 1.5 million lb. The project was designated F-1.

1953 February 21   The first successful powered flight with the second generation X-1 series research aircraft took place when test pilot Jean "Skip" Ziegler flew the Bell X-1A after it was dropped from the belly of a modified Boeing B-50 which had carried it into the air from Edwards Air Force Base, Calif. The

X-1A was superficially similar to the three Bell X-1s, but fuel capacity was increased from 498 gal. to 570 gal. and oxygen capacity from 437 gal. to 500 gal. The four-chamber Reaction Motors XLR11-RM-5 rocket motor was the same as that used in the first-generation series. The first Bell X-1B flew its preparatory unpowered glide flight on September 24, 1954, and the last of 53 X-1A/B flights took place on January 23, 1958. One second-generation X-1, the X-1D, was lost after its first unpowered flight when an explosion destroyed it before it could be dropped from the B-50 on July 24, 1951.

1953 March 15   The Soviet R-5 missile was successfully launched for the first time. Powered by alcohol and liquid oxygen, it would be adapted for both military and scientific tasks. As the R-5M ballistic missile, dubbed SS-3 in the West, it flew for the first time on January 21, 1955. It was launched with a nuclear warhead for the first time on February 2, 1956, and entered service that year. The R-5 was used in several different variants for sounding rocket research and for carrying experiments to the edge of space. Designations for such flights included the R-5A, R-5B, R-5V and R-5R.

1953 April   U.S. Air Force Assistant Secretary for Research and Development Trevor Gardner asked for a reassessment of the time required to design and develop an operational intercontinental ballistic missile. Two months later, the Director for Research and Development at USAF headquarters, Gen. Donald M. Yates, proposed that Secretary of Defense Charles E. Wilson authorize a thorough examination of the missile programs of the Army, Navy and Air Force. This Air Force initiative sparked the first missile race, over control of U.S. long-range missile programs between the Army and the Air Force, a struggle that would only be resolved more than three years later on November 26, 1956.

1953 May 12   On what was to have been its first powered flight following a series of drop tests that began June 27, 1952, the no. 2 Bell X-2 was destroyed in an explosion that killed the pilot Jean "Skip" Ziegler and observer Frank Wolko flying in the B-50 carrier-plane. It was found that a leather gasket in contact with liquid oxygen had carried an organic substance that caused detonation by uniting with the oxidizer. The leather gasket was impregnated with a fifty-fifty mixture of tricresyl phosphate and carnuba wax and a total quantity of 0.45 lb. was present in the liquid oxygen tank bulkhead. When kept at room temperature the phosphate separated, but it solidified as a separate constituent when the temperature dropped in the liquid oxygen tank, thus transforming it into a solid plug of organic matter. The no. 1 X-2, yet to fly, was modified as a result.

1953 June 23   The National Advisory Committee for Aeronautics released a report summarizing conclusions about hypersonic flight research. In it, a study group under Clinton E. Brown of the Aerodynamics Committee urged restraint in developing advanced concepts. A proposal from David Stone of the Pilotless Aircraft Research Division for a rocket-boosted X-2 research effort was endorsed, effectively limiting near-term hypersonic research to Mach 4.5 (see May 21, 1952). Events, however, overtook the conservative NACA, and U.S. Air Force pressure for an advanced high-speed research program resulted in a joint meeting on July 9, 1954.

1953 July   R. W. Bussard of the Oak Ridge National Laboratory rekindled U.S. Air Force interest in atomic motors for space propulsion with his study entitled "Nuclear Energy for Rocket Propulsion." Between July and December 1954 a series of meetings at the Weapons Division of the Atomic Energy Commission's Los Alamos Scientific Laboratory, N.Mex., provided a forum for discussing the feasibility of developing nuclear rockets as missiles or launch vehicles, and on October 18, 1954, an ad hoc committee on nuclear missile propulsion, later known as the Condor Committee, was formed to study applications of nuclear power to rocket propulsion.

1953 August 6   The Northrop N-69 Snark U.S. Air Force cruise missile made its first, unsuccessful, flight attempt from Cape Canaveral. It was not until the sixth flight on June 3, 1954, that the N-69 flew its programmed 3 hr. 31 min. Completely redesigned from the precursor N-25, the N-69 had a length of 69 ft. 11 in., a wing span of 42 ft. 4 in. and a weight of 49,600 lb. Snark was launched on the incline by two solid propellant boosters, each providing a thrust of 105,000 lb. for 4 sec. Later models adopted two 130,000-lb. thrust boosters. Sustained power was provided by a 10,500-lb. thrust Pratt and Whitney J57-P-17 turbojet. Plagued with problems, numerous missiles crashed into what the engineers referred to as the "Snark-infested waters" of the Atlantic Ocean off Cape Canaveral. Designed to achieve a range of 5,200 NM, on only two flights did it exceed 4,400 , and it was 20 mi. off target at a range of 2,100 NM.

Geophysicist S. Fred Singer of the Applied Physics Laboratory at Johns Hopkins, presented a paper at the Fourth International Congress on Astronautics in Zurich, Switzerland, describing his design for a minimum-size satellite vehicle called MOUSE (Minimum Orbital Unmanned Satellite Experiment). Singer had taken the work of Kenneth Gatland, A.M. Kunesch, and A.E. Dixon from the 1951 conference and adapted it to provide a launcher capable of placing a 100-lb. satellite in orbit. On September 2, 1954, Singer and fellow geophysicist Lloyd V. Berkner got backing for the MOUSE concept from the International Scientific Radio Union. On September 20, 1954, the International Union of Geodesy and Geophysics passed a resolution calling for an Earth satellite.

1953 August 12   The Soviet Union detonated its first fusion weapon with a yield of 400,000 KT, a thermonuclear bomb with a small component of hydrogen. The fear of a communist hydrogen bomb, and persuasive arguments from British defense chiefs that sought to have the U.S. strategic arsenal become the cornerstone of a global deterrent force, encouraged a total reexamination of the policy of large standing armies. Secretary of State John Foster Dulles had defined the concept of "massive retaliation" during the first six months of the new Eisenhower administration, the policy of ready-use articulated by the President when he argued that "I see no reason why they [nuclear weapons] shouldn't be used just exactly as you would use a bullet." Fear of a Soviet nuclear bomb, combined with Republican criticism that the Democrats seemed content to wait for Moscow to make the "first move," stiffened Congressional resolve to support development of nuclear weapons that could be used for tactical as well as strategic warfare. This led to increased effort on small nuclear warheads and directly opened the way for a wider range of missile options. A full-scale Soviet thermonuclear bomb was dropped from the air on November 23, 1955.

1953 August 20   The first launch of the U.S. Army Redstone ballistic missile took place from Cape Canaveral, Fla., while engineers monitored 16 data points telemetered to the ground. Redstone had a length of 69 ft., a diameter of 5 ft. 10 in.,

weighed 62,000 lb. and was powered by a 75,000-lb. thrust motor operating on ethyl alcohol–liquid oxygen propellants. It was built to throw a W29 nuclear warhead 200 mi. but adopted the low-megaton-yield W39 which also equipped the Snark cruise missile. The U.S. Army launched 38 Redstone missiles on development flights, the last on November 5, 1958, of which 35 were successful. On the fourth test flight, honoring a tradition started with V-2s at Peenemünde, engineers taped to the lower section of the missile a good-luck cardboard artwork of a voluptuous blonde. As the missile accelerated after launch the cutout blonde flew away, distracting the tracking radar antenna, which ignored the ascending missile and followed her to the ground! No more cutout blondes were allowed to fly on Redstones. In all, 62 missiles were produced, missiles 13–17 and 30–62 being assembled by the Chrysler Corp., the remaining 25 at Redstone Arsenal.

1953 August   In two related but coincidental events, construction of the prototype Atlas A ICBM began at the San Diego, Calif., facility of Convair, and in Britain the Air Staff called for conceptual designs for a long-range ballistic missile (LRBM). Over the next eighteen months, the British government would move silently toward the development of an LRBM called Blue Streak and gain much assistance from Convair and from Rocketdyne, the manufacturer of the Atlas propulsion system, in developing its shorter-range equivalent. The demonstration of the first Soviet thermonuclear bomb on August 12, 1953, and the threat it was believed to make to world peace, made this decision seem inevitable.

1953 September 11   One of the most important missiles ever developed, the first in a long and continuing program of Sidewinder air-to-air missiles, was fired successfully for the first time. With a length of 9 ft. and a weight of 155 lb., the AIM-9 had a range of 2 mi. and incorporated an infrared heat seeker in the nose and a solid propellant motor in the main body. Characterized by eight small fins in two groups of four equally spaced around the body of the missile, the Sidewinder has been built in greater numbers than any other missile in the West. Following the basic version, of which 80,000 were produced, variants were developed with a range of up to 11 mi. Over time, Sidewinder became the short-range antiaircraft missile to partner the heavier Sparrow with a longer range and radar guidance. More than 150,000 Sidewinders have been produced.

1953 September   U.S. Air Force Assistant Secretary for Research and Development Trevor Gardner set up the Strategic Missiles Evaluation Committee (SMEC), limited to the task of "studying long-range intercontinental strategic missiles under development . . . and making suitable recommendations for improving this program." The SMEC was composed of ten leading members of the U.S. industrial community chaired by Dr. John von Neumann from the Institute for Advanced Studies. Within weeks, the committee heard news that a breakthrough in nuclear weapon sizing made possible the delivery of thermonuclear warheads by existing missile technologies that merely needed a political and military commitment to realize.

1953 October 30   The ground was laid for an expansion of U.S. missile programs with the completion of National Security Council Paper 162/2, *Basic National Security Policy*. It began by defining the Soviet Union as the fundamental problem for U.S. national security and by asserting that any

Developed by the U.S. Army team led by Wernher von Braun, the Redstone medium-range ballistic missile entered field service in 1958 and led directly to the Jupiter-C satelite launcher and, indirectly, to the Saturn C-1.

Soviet move of apparently peaceful intent was nothing other than a gesture. It claimed that the threat from Soviet aircraft carrying thermonuclear weapons on one-way missions was real but did not believe the USSR would deliberately launch all-out war. It stipulated a requirement for the United States to build up a strong military posture "with emphasis on the capability for inflicting massive retaliatory damage by offensive striking power." Enshrined within the policy statement was the recommendation to reach "a stage of atomic plenty and ample means of delivery." This effectively authorized the Pentagon to determine for itself what "plenty" and "ample" meant within the overall determination of national security interests and the ability to execute "massive retaliation."

1953 October A U.S. Air Force contract for analytical studies of long-range ballistic missiles was placed with the Ramo-Woolridge Corp. The organization was formed by Simon Ramo and Dean E. Woolridge, two members of the Strategic Missiles Evaluation Committee, and comprised a team of 170 people accomplished in missile technologies and management techniques. A Guided Missile Research Division was set up in 1954. It was forbidden from taking production

contracts and in 1957 was renamed Space Technology Laboratories, or STL, and given technical direction of the Thor, Atlas and Titan missile programs. In 1958 Ramo-Woolridge merged with Thompson Products to become the Thompson-Ramo-Woolridge Corp., more popularly known as TRW, of which STL was a subsidiary.

At a special meeting of a group known as the Gassiot Committee, members of the British Royal Society met in London to ponder the value of a sounding rocket program for atmospheric research. Stimulated by a lecture given in Oxford during the summer of 1952 when American scientists described how they used sounding rockets for research, the British scientists asked the Royal Aircraft Establishment to design and develop a family of solid propellant instrument carriers. From this emerged the Skylark family of sounding rockets.

1953 November 19 Personnel of the NACA Pilotless Aircraft Research Division (PARD) of the Langley Research Center conducted the first live firing of the Nike-Deacon sounding rocket. The Deacon originated as one of a family of solid rocket motors bearing ecclesiastical names developed by the National Defense Research Council toward the end of World War Two. The NACA acquired about 300 surplus solid propellant grains and used them for launching aerodynamic models from Wallops Island Test Station, Accomack County, Va., in 1947. Nike-Deacon was adopted as a sounding rocket project by the Upper Atmosphere Rocket Research Panel in 1954. The Aeronautical Engineering Department of the University of Michigan carried out conversion work funded by the Air Force Cambridge Research Laboratories to transform the PARD Nike-Deacon into a full-fledged sounding rocket. It was launched in that role for the first time on April 8, 1955. Combined with the solid propellant booster of the Nike I guided missile, the Nike-Deacon could propel a 50-lb. payload to a height of 69 mi.

1953 December   The world's first surface-to-air guided missile defense site was activated near Washington, D.C., with the U.S. Army Nike Ajax. The system was massive, fixed in concrete emplacements and aimed at providing air defense to the North American continent. The missile took the form of a canard with four delta-shaped wings and a large 59,000 lb.-thrust Hercules boost motor with three clipped triangular fins. Nike Ajax had a total length of 34 ft. 10 in. and a weight of 2,455 lb., of which about 50% was the booster. The boost motor fired for 2.5 sec. after which the 2,600-lb. liquid propellant sustainer took over. The system had a range of 25 mi. and by the time production ended in 1958 about 16,000 rounds had been delivered for air defense forces in the U.S. and many other countries. From 1958 it was progressively replaced by the more sophisticated Nike Hercules, which had a range of almost 90 mi. and a ceiling of 30 mi. The Hercules served with the U.S. Army until it was replaced by the Patriot in 1978.

1953   The largest geophysical rocket used in the Soviet Union came into use during the year. Like the 750-mi. range SS-3 Shyster ballistic missile from which it was adapted, the V-5-V utilized a single 120,000-lb. thrust RD-103 liquid propellant engine burning a mixture of kerosene and liquid oxygen with a specific impulse of 245 sec. As a research rocket, it was capable of lifting a scientific payload of 2,865 lb. to a maximum height of 320 mi. The V-5-V had a length of just over 75 ft. and a diameter of 5 ft. 5 in. It was used extensively for biomedical flights with dogs and other animals. Sometimes the dog, attired in pressure suit and helmet, would be ejected from a descending capsule to return through the lower atmosphere dangling from a parachute.

In a book entitled *Man on the Moon,* Wernher von Braun, Fred Whipple and Willy Ley described in detail the manned exploration of the moon. It did much to stimulate interest in the colonization of earth's nearest celestial neighbor. According to the authors, three moonships assembled at an Earth orbiting space station would carry 50 scientists and technicians to the lunar surface. The space station, and the reusable spaceplane that would be used in its assembly, had been described in an earlier book, *Across the Space Frontier,* published in 1952. The initial expedition was expected to remain on the moon for six weeks before returning lunar samples to Earth.

1954 February 10   Set up by the U.S. Air Force in September 1953, the Strategic Missiles Evaluation Committee (SMEC) completed its review of U.S. ballistic missile programs. It specifically examined the Snark, Navaho and Atlas missiles, believing all three to have outdated specifications. In particular, it wanted to relax the Air Force performance requirement

for an accuracy of 1,500 ft. over intercontinental range. The SMEC believed that with the enhanced yield of thermonuclear warheads accuracy could be relaxed to 2 to 3 NM. The SMEC believed an operational ICBM could be available in six to eight years but asked for further studies on the vulnerability of land basing. It linked Soviet air defenses and the vulnerability of U.S. bombers as the primary justification for ICBMs.

1954 March 1   RAND issued a two-volume top-secret report titled *An Analysis of the Potential of an Unconventional Reconnaissance Method.* Summarizing the results of Project Feedback, a compilation of studies funded by the U.S. Air Force Air Research and Development Command, it recommended "the earliest possible completion" of a military reconnaissance satellite as a "vital strategic interest to the United States." RAND suggested two concepts: one in which television pictures taken by the satellite were stored on tape and broadcast to the ground over friendly territory, and a conventional camera taking pictures which could be processed on board and retransmitted by electrical signals. Designated Project 1115 (Pied Piper) in May 1954, the Pentagon gave final approval two months later. The unclassified designation was weapon system WS-117L, the "Advanced Reconnaissance System."

Availability of thermonuclear warheads for ballistic missiles came a step closer during the month when nuclear tests in the Operation Castle program demonstrated the imminent availability of "emergency capability" bombs carried by B-36 bombers. In the first of seven tests conducted in the Pacific, the 15 MT Shrimp device demonstrated an explosive yield 2 ½ times the estimated yield of 6 MT. The need to reduce the size of thermonuclear bombs for smaller aircraft stimulated work on warheads capable of being carried by long-range missiles such as Atlas and Titan. This validated the ICBM concept by relaxing the weight-lifting (warload) requirement and moving extant technical capability closer to the mission objective.

1954 March 27   President Dwight D. Eisenhower convened a meeting of the Office of Defense Management's Science Advisory Committee and at the instigation of atom physicist J. Robert Oppenheimer held a discussion on the threat to U.S. security from Soviet ambitions. In an important move that was to fuel Eisenhower's concern about the strength of U.S. military preparedness, committee member James R. Killian Jr. was directed by the President to set up a Technology Capabilities Panel. The TCP completed its report in February 1955 and concluded that "because of our air-atomic power we have an offensive advantage and are vulnerable to surprise attack." The report encouraged the view that big missile programs were an investment in maintaining strategic capabilities and not an expansion in military strength per se.

1954 April 1   Based on work carried out following a major presentation on April 17, 1952, Bell Aircraft Co. received a U.S. Air Force contract for work on a reconnaissance version of its proposed Bomi boost-glide vehicle. From this evolved an 800,000-lb., three-stage rocket-boosted vehicle with each upper stage riding piggyback and the first stage manned for fly-back capability. Now designated Weapon System MX-2276, the boost-glide vehicle would have had a maximum altitude of 259,000 ft., a top speed of 15,000 MPH and a range of 12,200 mi. The study expired in May 1955 but Bell continued to work on the concept using its own funds. Interest from the Air Force waned as ICBM development expanded, and Bomi's nuclear delivery role was superseded by missiles.

1954 June 12 The British and American defense secretaries signed an agreement whereby the United States would provide Britain with information on its ballistic missile development program. In particular, North American Aviation was to allow Rolls Royce access to the development work on liquid propellant rocket engines conducted at Rocketdyne in California. Moreover, General Dynamics Convair Division was to work with De Havilland at Stevenage, England, on the detailed design of a British long-range ballistic missile (LRBM). This was the single most important step in bringing large ballistic missile technology within the range of British financial and technological capabilities.

1954 June 25 At the Office of Naval Research (ONR) in Washington, D.C., an important meeting took place which was to directly stimulate military interest in heading up a satellite program. Using details from the Gatland-Kunesch-Dixon proposal of 1951 for a Minimum Satellite Vehicle, Fred Durant, president of the International Astronautical Federation, Dr. Fred Whipple of the Smithsonian Astrophysical Observatory, Dave Young from Aerojet General and Fred Singer met with Comdr. Hoover of the ONR and Wernher von Braun from Redstone Arsenal in Huntsville, Ala. Accepting that his own grandiose rocket proposals of 1950–52 were too ambitious, von Braun seized the initiative and proclaimed that a 5-lb. satellite could be launched into a 200-mi. orbit using a Redstone missile with Loki rockets as an upper stage.

1954 June The U.S. Department of Defense approved development of a surface-to-air missile capable of reaching targets below the effective acquisition altitude of the Nike radar system. Under a program managed by the Army Ordnance Corps, the Raytheon Manufacturing Co. produced a compact missile with a radar in the nose called the Hawk, its name an acronym for Homing All the Way Killer. The first training unit was deployed with Hawk in June 1959. Each missile had a height of 16 ft. 6 in., a diameter of 1 ft. 2 in., a launch weight of 1,295 lb. and a range of 22 mi. In 1960, Hawk intercepted Honest John, Little John and Corporal missiles. An improved Hawk appeared in 1968. More than 50,000 rounds have been sold to 21 customers, making this the most prolific U.S. strategic air missile (SAM) produced to date.

1954 July 9 A joint NACA/Air Force/Navy committee met to discuss plans for a hypersonic, rocket-propelled research aircraft program. NACA had been dragging its feet on a Mach 10 vehicle, but the Air Force and the Navy wanted vital flight data about the performance of vehicles at this speed. Under Project 1226 the Air Force defined the specification during late 1954, resulting in the designation X-15. The requirement was formalized at a meeting of the NACA Aerodynamics Committee on October 5, 1954. The X-15 was to be capable of Mach 7 and an altitude of 250,000 ft. A bidders conference was held January 18, 1955, and proposals were to be in by May 9, 1955.

1954 July 15 Appointed on June 1, 1954, to take over U.S. Air Force ballistic missile programs, Brig. Gen. Bernard A. Schriever set up the Western Development Division of the Air Research and Development Command (ARDC) and established his office at Inglewood, Calif. Brig. Gen. Schriever's boss was Maj. Gen. James McCormack, vice commander of the ARDC. The Western Development Division was established in response to a call from Trevor Gardner, assistant secretary for Research and Development to Air Force Secretary Clark B. Millikin, for enhanced priority on ICBMs. It was endorsed by the Strategic Missiles Evaluation Committee, and this reorganization was the first practical step toward speeding up missile development as recommended by Von Neumann's committee. In June 1957 the Western Development Division would be renamed the U.S. Air Force Ballistic Missile Division (BMD).

1954 July 26 The British cabinet secretly approved development of a thermonuclear (hydrogen) bomb for operational use by the Royal Air Force. The first Soviet thermonuclear explosion on August 13, 1953, and the March 1, 1954, test of a semi-operational U.S. thermonuclear bomb under Operation Castle persuaded the Cabinet that Britain could no longer vacillate on a decision. During earlier secret cabinet meetings there had been minority opposition to the development of the hydrogen bomb on moral grounds. Prime Minister Winston S. Churchill, who had been returned to office on October 25, 1951, successfully argued that Britain could not maintain her place as a world power without the ability to manufacture her own thermonuclear weapons. This decision would lead to development of the Blue Streak long-range ballistic missile.

1954 August 3 Wernher von Braun and Comdr. George Hoover from the U.S. Office of Naval Research, discussed with Gen. H. N. Toftoy, the commanding officer of Redstone Arsenal, Ala., prospects for launching a 5-lb. satellite using an Army Redstone booster with Loki upper stage rockets. Gen. Toftoy spoke with Gen. Simon, chief of Army Ordnance, who in turn spoke with Rear Adm. Furth, chief of Naval Research. Both agreed to go ahead. The Navy would assume responsibility for the satellite, tracking and logistical support and the Army would provide the rockets and launch facilities. It was called Project Orbiter and the title of a von Braun report linked it to the 1951 Gatland-Kunesch-Dixon proposal: *The Minimum Satellite Vehicle Based Upon Components Available from Missile Development of the Army Ordnance Corps.*

1954 October 4 Responding to rising international support for an Earth-satellite program, the Special Committee for the International Geophysical Year (known by its French acronym CSAGI, for Comité Spécial de l'Année Géophysique) announced that "CSAGI recommends that thought be given to the launching of small satellite vehicles, to their scientific instrumentation, and to the new problems associated with satellite experiments, such as power supply, telemetering, and orientation of the vehicle." CSAGI was acting on recommendations from the International Scientific Radio Union and the International Union of Geodesy and Geophysics which thought that the International Geophysical Year would be an appropriate event for the peaceful exploration of near-Earth space.

1954 October 18 The U.S. Air Force Scientific Advisory Board convened an ad hoc committee on nuclear missile propulsion to study applications of reactors to missiles and launch vehicles. Industry participated by making presentations on various forms of reaction propulsion, including turbojet, ramjet and nuclear rockets. Frank Rom of the NACA Lewis facility joined H. F. Bunze from the Wright Air Development Center to brief Air Research and Development Command on the possible design configuration of a nuclear-powered ICBM, and the Air Force commissioned a report from the Atomic Energy Commission. In turn, the AEC asked Los Alamos Scientific Laboratory and Livermore Radiation Laboratory to carry out parallel six-month studies on reactor design. Their reports were completed and presented in March 1955 to the second meeting of the Air Force Scientific Advisory Board. Los Alamos and Livermore set up nuclear rocket research divisions on April 15, 1955.

1954 November 27 The U.S. Air Force published its secret System Requirement (No. 5) for a military reconnaissance satellite program, the first U.S. satellite system to be given official approval. Designated the Advanced Reconnaissance System (WS-117L), the General Operational Requirement (No. 80) was released on March 16, 1955. It defined the objective as a surveillance system for "preselected areas of the Earth" so as to "determine the status of a potential enemy's warmaking capability." Design proposals were requested from Lockheed, Martin and Bell with project management through the Air Force Western Development Division.

1954 In Britain, the Ministry of Supply put together the specification for a long-range ballistic missile (LRBM) capable of delivering nuclear warheads to targets in the Soviet Union from bases in the United Kingdom. The ministry specified a range of up to 3,000 mi. and wanted the missile to be protected during the preparation for flight by housing it in a silo. Britain was keen to play as full a role as possible in the alliance of non-communist states and saw the nuclear-tipped ballistic missile as a means by which it could, as British politicians would say later, maintain a "place at the table." Moreover, it was thought ballistic missiles would allow reductions in other, more expensive, means of nuclear weapons delivery and permit cuts in the defense budget.

The operational requirement for a rocket-powered thermonuclear stand-off bomb was formally issued by Britain's Royal Air Force. A stand-off bomb is one which has its own propulsion, allowing it to be dropped some distance from the target. The Royal Aircraft Establishment, Farnborough, and the Avro aircraft company began studies on such a weapon, and in March 1956 Avro received a development contract for the stand-off bomb, called Blue Steel. The Mk.1 missile had a length of 34 ft. 9 in., a diameter of 5 ft. 8 in., and a wingspan of 12 ft. 11 in. With canard foreplanes and ailerons on short and stubby wings, Blue Steel was to be carried by Vulcan and Victor bombers. Powered by a Stentor liquid rocket motor using kerosene–hydrogen peroxide propellants, it had a maximum speed of Mach 2 and a range of up to 110 mi. The first firing tests with two-fifths scale models took place in 1957.

In a massive program involving huge resources in men and material, the Soviet Union began deploying large numbers of surface-to-air missiles and with this single stroke that seriously threatened manned aircraft, did more to alter the strategies and defense design and procurement policies of the West than any other action. First in service was what NATO code-named the SA-1 Guild, a liquid propellant missile with a range of 20 mi. Big by any standard, SA-1 had a length of 39 ft. 5 in., a diameter of 2 ft. 3 in. and a fin span of 9 ft. 2 in. Still in the final stages of development, its stablemate the SA-3 Goa was a more compact and efficient two-stage solid propellant missile with a length of 22 ft., a maximum diameter of 2 ft. 3 in. and a maximum fin span of 4 ft. Goa was produced in large numbers for many countries and a navalized version was put to sea. Still in final design and test, the SA-2 Guideline would enter service during 1956, providing cover to a range of 31 mi. The two-stage SA-2 had a total length of 35 ft. 2 in., a maximum diameter of 2 ft. 3 in. and a maximum fin span of 7 ft. 2 in. At peak the Soviets alone deployed 4,000 SA-2 launchers, but the missile was widely exported and remains in use today.

In the Soviet Union, an intensive program to develop a Mach 3 cruise missile got under way. In what was considered a direct competition between ballistic missiles and cruise weapons, the Burya project would build for the Soviet Union a broad base of research experience in supersonic projectiles and ramjet engines. Like its R.7 ICBM competitor, Burya was designed for a range of 5,000 mi. and would be launched vertically by two powerful liquid propellant boosters. Burning kerosene–nitrogen acid propellants, each booster had four 44,100-lb. thrust rocket motors. Total liftoff thrust with all eight engines running was 352,800 lb. Launch weight was 278,000 lb. When accelerated to speed, a ramjet would take over for sustained cruise at Mach 3.2 and the boosters would be jettisoned. Burya had a length of 62 ft. and a wing area of 645 sq. ft. Tests began in 1957 and five flights were made, but the success of the less vulnerable R.7 ICBM led to the cancellation of Burya.

In the USSR, Mikhail Kuzmich Yangel, a former deputy of Sergei Pavlovich Korolev, was put in charge of his own new rocket design bureau in Dnepropetrovsk which was given responsibility for producing the first Soviet silo-launched ICBM, the SS-7. Later, this would form the basis for a satellite launcher, the SL-7. Yangel's greatest achievement was the design and development of the RS.20, which became known in the West as the SS-18, a massive ICBM with a throw-weight of 15,000 lb. Yangel died in 1971 and was succeeded by Vladimir Fedorovich Utkin.

Science fiction writer, space prophet and founding member of the British Interplanetary Society, Arthur C. Clarke described expeditions to the lunar surface in a book published in London called *The Exploration of the Moon*. Based on work carried out for the BIS by Harry Ross and illustrated with paintings by R. A. Smith, it speculated on the design and layout of moon cities populated by resettled Earth people living in giant domes that entrapped a comfortable atmosphere.

1955 January 4 The U.S. Air Force Air Research Development Command issued System Requirement 12, which defined the perceived need for a missile-launched reconnaissance vehicle capable of reaching an altitude of 100,000 ft. and a range of 3,300 mi. SR-12 was derived from work done at Bell Aircraft on the Bomi and Project MX-2276 and at Boeing on a similar high-speed, boost-glide vehicle under project MX-2145. SR-12 was later subsumed under General Operational Requirement 12 issued May 12, 1955, which outlined a piloted, high-speed reconnaissance requirement to be met by 1959.

1955 January 10 The proposed U.S. Army-Navy Project Orbiter satellite launching plan was presented to Assistant Secretary of Defense Donald A. Quarles in the first stages of getting formal approval to proceed from the Pentagon and the White House. The Naval Research Laboratory had not been consulted in the formative stages of Orbiter and quickly came up with an alternative scheme based on the Viking rocket. Late in the month the National Academy of Sciences' International Geophysical Year committee set up the Technical Panel on Rocketry and at its first meeting established the Subcommittee on the Technical Feasibility of a Long Playing Rocket, known as the LPR committee. Public dissatisfaction with talk of expensive space rockets forced the academy to downplay the concept until it was approved.

1955 January 12 Development of major Soviet launch facilities began at a location north of the small town of Tyuratam east of the Aral Sea when the first 30 workers arrived by rail. Work on the sprawling site centered on a new rail spur cutting 12 mi. north from the main line linking Moscow with the eastern towns. The old town of Tyuratam was built up into the city of Leninsk — so named on January 28, 1958 — where thousands of engineers, technicians and scientists gathered to live and work. The focus of early site preparation was the launch complex to be used for the first Soviet ICBM. Designated R.7, and later known in the West by its NATO code name SS-6 Sapwood, the rocket would be launched from a position just north of the rail spur. Over time, what the Russians refer to as Baikonur was built into a massive military-industrial area running 85 mi. from east to west and 50 mi. north to south.

1955 January 14   The U.S. Air Force formally authorized a start on America's first ICBM and awarded contract AF04(645)-4 to Convair for development of the Atlas missile, which was designated SM-65 under Weapon System 107A-1. The missile had a diameter of 10 ft., but length was to vary between 75 ft. 10 in. and 82 ft. 6 in., depending on the type of reentry warhead carried. Atlas was powered by the Rocketdyne MA-2 propulsion system that burned kerosene–liquid oxygen propellants from the two main tanks. MA-2 comprised a central 57,000-lb. thrust LR105 sustainer engine flanked by two 150,000-lb. thrust LR-89 motors in a skirt with a maximum diameter of 16 ft. The combined thrust of 357,000 lb. (387,000 lb. in the Atlas E) was maintained for just over 2 min., after which the two booster engines were jettisoned. The single sustainer continued to burn for a further 3 min., sending the warhead on its ballistic trajectory.

1955 January 24   Working to a requirement identified in late 1954 for a high Mach-number research rocket, the U.S. Air Force issued a contract to Lockheed Aircraft Co. for the X-17. Designed to provide data on reentry vehicles moving at a speed of Mach 15 and with Reynolds numbers up to 24 million, 26 X-17s were built by Lockheed, with a length of 40 ft. 5 in., a span across the first stage fins of 8 ft. 6 in. and a launch weight of 12,000 lb. Flight tests began with the first of three quarter-scale models on May 23, 1955, when a speed of Mach 3 was achieved. The first of a series of three half-scale vehicles was launched on June 23, 1955, followed by a series of seven full-scale development vehicles beginning August 26, 1955. Several technical problems delayed the start of research, which did not get under way until July 17, 1956.

1955 February   In response to recommendations from the U.S. National Security Council, the Army and the Navy began studies of intermediate range ballistic missiles. It recommended deployment of IRBMs with a range of 1,500 NM on friendly territory overseas and on surface ships at sea. The Army responded quickly with a design that would be called Jupiter A, and the Navy studied a wide range of deployment methods which would influence the design of such missiles. The National Security Council made its formal recommendation with the approval of the President on September 13, 1955, clearing the way for the Army to conduct discussions with the Navy about a joint venture based on the Jupiter liquid propellant missile.

The U.S. Army Ordnance Corps requested Bell Telephone Laboratories to carry out an 18-month study into defense against high-speed cruise and ballistic missile threats that would probably emerge between 1960 and 1970. Some Bell engineers were already working with their counterparts at the Douglas Aircraft Co. on a solid propellant air defense missile.

The system that evolved from this study was Nike II, and during June 1955 defending against the ballistic missile threat became a priority. Army Ordnance at Redstone Arsenal received an interim briefing on December 2, 1955, defining a feasible antiballistic missile (ABM) system using low-yield nuclear warheads to disable incoming missiles. The Army general study received the Nike II study in October 1956, and in February 1957 the Army selected Western Electric and Bell as primary contractors on what was now an exclusively anti-ICBM system named Nike-Zeus.

1955 March 10   The subcommittee on the Technical Feasibility of a Long Playing Rocket, known as the LPR committee, presented its recommendations on a satellite project to the U.S. National Committee (USNC) for the International Geophysical Year. It proposed three alternative concepts: a large rocket stage carrying a number of smaller rockets for orbital injection; a two- or three-stage rocket carrying a larger satellite instrumented for scientific measurements; a very powerful rocket, such as the Air Force Atlas, forming the basis for a new satellite launcher with heavy payload capability. The USNC selected the second concept and opted to request a program involving 10 launches of a vehicle capable of lifting 30–50-lb. satellites, expecting to orbit at least five at a total program cost of little more than $7 million and involving 25 scientists.

1955 March 16   The U.S. Air Force issued General Operational Requirement No. 80 calling for proposals from industry on the design and development of a photographic satellite reconnaissance system. Bids came in from RCA, Martin and Lockheed. The last proposed optional direct-TV transmission and film-recovery systems, and from these studies emerged the three primary observation systems which would be employed: imagery transmitted by radio signal, images returned to Earth in recoverable capsules, and an early warning surveillance system transmitting observations by radio.

1955 March 17   The U.S. Department of Defense authorized full-scale development of the Genie air-to-air missile, to be equipped with the W-25 atomic warhead. With a weight of only 219 lb., the warhead had a yield of 2 KT; with a length of 9 ft. 7 in. and a maximum diameter of 1 ft. 5 in., the 830-lb. Genie was powered by a 36,000-lb. thrust Thiokol SR-49 solid propellant rocket motor. Genie was unguided and accelerated to a maximum velocity of Mach 3 at a maximum range of 6 mi. The warhead had an effective radius of 1,000 ft. The first live missile was fired from a Northrop F-89J Scorpion on July 19, 1957. Several thousand had been produced before production ended in 1962.

1955 April 15   The Soviet government announced that a permanent commission for "interplanetary communications" had been established in the Astronomics Council of the Soviet Academy of Science. This followed a former announcement that a Tsiolkovsky medal would be awarded every three years for outstanding work in interplanetary communications. Couched as it was the vague language of international statement, this was considered by many to be an announcement of intent to proceed with space projects and satellites. It did much to stimulate concern at the U.S. National Security Council about the possible consequences of the Soviet Union launching a satellite before the United States.

The First Air Division of the U.S. Air Force was activated at Offutt Air Force Base, Nebr., to operate photographic reconnaissance balloons designed to fly across the Soviet

Union and be recovered over the Pacific. Techniques developed in the recovery of these balloons were adopted for the recovery of Discoverer reconnaissance satellite pods. Released from sites in West Germany, the 24-ft.-diameter balloons were nicknamed "Moby Dick" and carried the WS-119L recovery pod, comprising a camera with two lenses, film for 450–500 pictures and a recovery package. Fairchild C-119 cargo aircraft carrying trapeze-like fixtures were used to snag descending parachutes and recover the film. Between January 10 and March 1, 1956, 516 balloons were launched, but only 40 returned pictures of communist territory. A total of 13,813 images were taken, covering 8% of the Russian and Chinese landmass.

Nuclear rocket divisions were set up at the two competing nuclear research facilities, Los Alamos and Lawrence Livermore. They were the best equipped laboratories to handle preliminary investigation of nuclear propulsion and to conduct tests, without which the theoretical projections could never be examined in practical applications. The U.S. Air Force Scientific Advisory Board ad hoc committee Condor made a formal recommendation on October 18, 1955, to begin work on a nuclear rocket.

1955 April An industrial team headed by De Havilland Propellers was put together for the United Kingdom's long-range Blue Streak ballistic missile, and in August 1955 an agreement was announced whereby Rolls Royce would receive technical information on rocket engines from the Rocketdyne Division of North American Aviation. In 1956 the British studied silo-launched methods and were able to help the U.S. Air Force with technical details. This proved of value during development of the Minuteman silo-launched ICBM concept. In 1957 the Convair Division of General Dynamics agreed to provide technical help to De Havilland on the main structure of the missile. Blue Streak had a length of 61 ft. 6 in., a diameter of 10 ft., a launch weight of 199,000 lb., and a theoretical range of 2,800 mi. which was increased during development to 3,000 mi.

1955 May 6 The U.S. National Committee of the International Geophysical Year provided a detailed satellite proposal to the National Science Foundation. It proposed a 10-vehicle program at an estimated cost of $9,734,500, of which $2,234,500 was for satellite development and the balance for

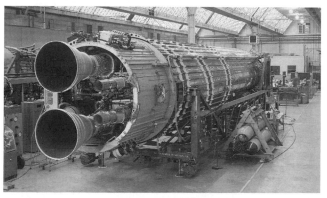

Through technology acquired from North American Aviation and Rocketdyne in the United States, Britain's Blue Streak missile was developed between 1955 and 1960 as a nuclear deterrent but canceled when it became vulnerable to enhanced Soviet long-range missiles.

the rockets. The committee wanted to retain funding for the rockets within the satellite project, thus delineating the precise lines of demarcation between the Department of Defense and the scientific community. The Army and the Navy had already proposed Project Orbiter, and the Air Force was known to be working on secret reconnaissance satellite projects. IGY scientists were concerned about military control of their own project and wanted to maintain an independent position, believing the best way to achieve that was jurisdiction over the launchers for the IGY satellites.

1955 May 20 James A. Lay Jr., executive secretary of the U.S. National Security Council, issued a secret report entitled *U.S. Scientific Satellite Program.* The report noted that "a group of Russia's top scientists is now believed to be working on a satellite program." It commented that "considerable prestige and psychological benefits will accrue to the nation which first is successful in launching a satellite." The report advised the NSC that the technology and knowledge gained would benefit "missile and anti-missile research and development programs of the Department of Defense," and it endorsed plans for a scientific satellite as well as concurrent, but highly secret, work on a military reconnaissance satellite.

1955 May 26 Expanding upon a series of biological flights to the edge of the atmosphere and near-space that began during 1951, scientists in the Soviet Union started using more powerful rockets to reach altitudes in excess of 150 mi. Launched from the Kaupstin Yar facility on the Volga near Volgograd, these included the SS-3 Shyster MRBM (medium-range ballistic missile) adapted to carry dogs and other small animals on ballistic trajectories. The animals were to experience several minutes of weightlessness as they fell back to Earth from the apogee of their ballistic trajectory. These flights helped prepare the way for orbital flights that would begin before the end of 1957.

1955 May During the month U.S. Assistant Secretary of Defense for Research and Development Donald A. Quarles submitted a secret evaluation of the various satellite proposals to the President's Special Assistant Nelson A. Rockefeller. Rockefeller added the following: "I am impressed by the costly consequences of allowing the Russian initiative to outrun ours through an achievement that will symbolize scientific and technological advancement to people everywhere. The stake of prestige that is involved makes this a race that we cannot afford to lose." The report and added comments went to the Secretary of the Treasury, the Central Intelligence Agency, the Bureau of the Budget, the Joint Chiefs of Staff and the National Security Council.

In Britain, the Royal Aircraft Establishment, Guided Weapons Department, issued proposals for a research rocket in support of the Blue Streak LRBM program. A contract for its development was awarded on July 29 to the Saunders Roe aircraft company. Called Black Knight, it was to have a single-stage, liquid propellant rocket powered by a four-chamber, 16,400-lb. thrust, Armstrong-Siddeley Gamma 201 kerosene–high-test peroxide motor. With a length of 33 ft. 6 in. and a diameter of 3 ft., Black Knight had a launch weight of 12,000 lb. and carried four fins, spanning 4 ft. 8 in., for aerodynamic stability. It could throw a 250-lb. instrumented reentry vehicle to an apogee of 500 mi. and had a speed of 12,000 ft./sec. In 1959 a combination Blue Streak/Black Knight satellite launcher was considered.

The U.S. Army reported good progress with development

of the Little John tactical battlefield missile at Redstone Arsenal. Designed as a simpler, neater and more compact equivalent of the more cumbersome Honest John, the missile had a protracted gestation, receiving no field evaluation before 1958 and no in-service use before 1962. Little John had a length of 14 ft. 6 in., a diameter of 12.5 in., a launch weight of 780 lb. and a range of up to 10 mi. with a 1–15 KT W-45 nuclear warhead. Little John was retired by 1970, 500 having been built.

1955 June 8 A secret directive from U.S. Secretary of Defense Charles E. Wilson to Assistant Secretary of Defense for Research and Development Donald A. Quarles assigned him "responsibility for coordinating implementation of the scientific satellite program within the Department of Defense." The U.S. National Committee for the International Geophysical Year wanted to keep scientific satellite work away from the military, but all it managed to secure was total control over the satellite shell and its instrumentation. The Department of Defense would develop the launcher and the National Science Foundation was to coordinate affairs between the National Academy of Sciences and the Pentagon.

1955 June 15 An agreement was signed by the United States and Britain for the exchange of information about nuclear weapons, warheads, launch crew training and military planning in the use of missiles. The way had been cleared by an amendment to the McMahon Act, signed by the President on August 30, 1954, (originally prohibiting the transfer of atomic weapons information to any other country) which President Dwight D. Eisenhower had requested on February 17, 1954. The McMahon Act had imposed restrictions on Britain's ability to develop nuclear weapons and ballistic missiles, and by this amendment and the 1954 Atomic Energy Act, effective August 30, 1954, President Eisenhower opened a door to the transfer of information about the external characteristics of these weapons. The latest agreement permitted disclosures about the detailed design and engineering of the nuclear components. This led directly to the design of warheads for Britain's Blue Streak long-range ballistic missile (LRBM).

1955 July 1 The Project Orbiter satellite proposal from the U.S. Army/Office of Naval Research team that had presented its design to the Stewart Committee offered a reworked design considerably modified from the one incorporating Loki rockets in upper stages to the Redstone booster. Now, Wernher von Braun had adopted a suggestion from the Jet Propulsion Laboratory to use Baby Sergeant rockets instead, and this configuration would be used in the Jupiter-C program. Project Orbiter was said to cost $17.7 million for eight launchers and their satellites, about $8 million more than the U.S. National Committee estimate presented May 6, 1955, and both would eventually appear woefully inadequate.

1955 July 6 The Stewart Committee, named after its chairman, Homer J. Stewart of the Jet Propulsion Laboratory and set up to determine the type of satellite launcher to use, began a series of meetings at the Pentagon in Washington, D.C. They met with Project RAND personnel to hear of various satellite proposals. On July 7 they were briefed by the U.S. Army and the U.S. Air Force on their own project proposals followed on July 8 by a presentation from the Viking design team at the Glenn L. Martin plant in Baltimore, Md. On July 9 Wernher von Braun gave a two-hour presentation on Project Orbiter followed by further briefings from other Redstone Arsenal specialists.

1955 July 27 Alan T. Waterman, director of the National Science Foundation, and Under Secretary Herbert Hoover Jr. met with President Dwight D. Eisenhower and convinced him of the need to announce publicly the U.S. satellite program. The intelligence community was concerned that the Soviets would soon announce their own satellite program. Without waiting for the Stewart Committee to formally decide on the specific type of launcher to use, the President issued the following statement on July 29 through his press secretary James Hagerty: "I am now announcing that the President has approved plans by this country for going ahead with the launching of small Earth-circling satellites as part of the U.S. participation in the International Geophysical Year." On August 2 the USSR announced that it too would launch satellites.

1955 July The U.S. Army Guided Missile Development Division at Redstone Arsenal completed preliminary design of an intermediate range ballistic missile (IRBM) capable of throwing a 2,000-lb. warhead a distance of 1,500 NM. The liquid propellant IRBM would be powered by the S-3 engine developed by the Rocketdyne Division of North American Aviation. With a thrust of 150,000 lb. and operating on RP-1–liquid oxygen propellant, the S-3 was almost identical to the Rocketdyne LR-79 adopted for the U.S. Air Force Thor IRBM. The Army stipulated a maximum miss distance of less than 1 mile at the target. Named Jupiter, this missile would be the basis of an Army-Navy joint project for the Navy sea-based IRBM.

1955 August 3 Established to choose between three contenders for launching the first U.S. satellite, the Stewart Committee came out in favor of a proposal from the Naval Research Laboratory for a launcher based in part on the Viking atmospheric sounding rocket. The NRL had offered two proposals: one, called the M-10, would use a modified and more powerful Viking as its first stage, with two solid propellant upper stages; the second, called M-15, would use a liquid propellant Aerobee-Hi second stage and a spin-stabilized solid propellant third stage. M-15 would take longer to develop but the Committee preferred this design. On September 9, 1955, the Department of Defense authorized the Naval Research Laboratory to take charge of the project, and on September 16, 1956, Milton Rosen's wife selected the name "Vanguard" for the project. Milton Rosen had been a key figure in the Viking rocket program.

1955 August 6 A year after the Allied Occupation Forces lifted a ban on the development of rockets in Japan, Prof. Hideo Itokawa from Tokyo University carried out the first in a series of tests with small powder rockets. Called Pencil, they were 9 in. long and contributed to the Baby series with a length of 12 in. and a weight of 0.44 lb. In August, Prof. Itokawa's team began testing these rockets on the Akita Range, firing them across the Sea of Japan. Soon, two-stage Babies were tested, each with a length of up to 1 ft. 6 in., carrying telemetry and camera equipment. This work resulted in the University's Kappa series with a length of 18 ft. 4 in. and capable of carrying 44 lb. of instruments to a height of 31 mi. First fired in 1956, it was Japan's contribution to the International Geophysical Year. The biggest in a long line of Kappa rockets was the two-stage, 5,500-lb thrust Kappa 10 of 1965, with a length of 32 ft. 6 in. and capable of lifting 375 lb. to a height of 155 mi.

1955 August 16 A U.S. Navy McDonnell F2H-2 Banshee successfully launched a rocket to an altitude of 180,000 ft. off Wallops Island Test Station in the first air-launched rocket flight for high-altitude atmospheric research. The rocket, a FFAR (folded-fin aerial rocket), was launched to evaluate the possibility of utilizing relatively small projectiles to carry packages of instruments to an altitude of approximately 180,000 ft. It paralleled similar efforts using balloons, but whereas the balloon-launched rockets, dubbed "rockoons," were successful, the "rockaire" concept was never formally adopted as a method of obtaining atmospheric data. The Air Force did, however, try out the technique in 1956.

1955 September 7 Following an invitation to bid dated February 4, 1955, Reaction Motors, Inc., received a contract to build the XLR99 rocket engine for the X-15 research aircraft. Using an 18,000-lb. supply of anhydrous ammonia and liquid oxygen contained in separate tanks in the fuselage, the 1,025-lb. engine would produce a maximum thrust of 57,000 lb. Thrust was throttleable between 30% and 100%, with a maximum propellant flow rate of 13,000 lb./min. exhausting the onboard supply in 85 sec. Propellant would be delivered by a high-speed turbopump powered by a decomposing, 90% hydrogen peroxide concentration, monopropellant fuel in a gas generator. The XLR99 had an exhaust nozzle diameter of 39.3 in. The first flight of an X-15 powered by the XLR99 took place on November 15, 1960.

1955 September 21 The Bell Aircraft Co. boost-glide vehicle defined under the U.S. Air Force System Requirement 12 was developed further in a contract for advanced studies incorporated in Special Reconnaissance System 118P. In this work, Bell would design a vehicle capable of Mach 15 and high-altitude photographic reconnaissance operations. On December 19, 1955, the Air Force added to this the need for a manned, hypersonic, rocket-powered bombing and reconnaissance system. It was not unusual for the Air Force to combine these two roles in a single system requirement: the North American B-70 was, for example, defined both as bomber and reconnaissance platform.

1955 September 30 North American Aviation was informed it would receive a contract to build the hypersonic X-15 rocket-powered research aircraft. Aircraft builders Bell, Douglas and Republic had submitted unsuccessful bids. The first of three X-15s was rolled out of the NAA factory at Downey, Calif., on October 15, 1958. It had a length of 50 ft., a wing span of 22 ft., and a gross weight of 33,300 lb. On flight tests it would be carried aloft under the starboard inner wing section of a converted B-52. The X-15 was to be powered by a throttleable, 57,000-lb. thrust rocket motor, but an interim power plant comprising two 4,000-lb. thrust XLR11 motors was installed in the first two aircraft.

1955 September The U.S. Army ordered the Redstone rocket modified into a test vehicle for reentry bodies built to protect warheads on ballistic missiles. Called the Composite Reentry Test Configuration, the 12 Redstones assigned to the RTV (Reentry Test Vehicle) program would be designated Jupiter-C but only three would be fired. The first stage was lightened and lengthened to contain more propellant and water-alcohol fuel was replaced with UDMH (unsymmetrical dimethyl hydrazine), increasing thrust to 83,000 lb. The second stage comprised 11 scaled-down Sergeant solid propellant rockets called Recruit arranged in a drum-shaped circle, each motor 4 ft. in length with a diameter of 6 in. and a thrust of 1,500 lb.

for a total stage thrust of 16,500 lb. The third stage comprised three uprated Baby Sergeant solids nested within the center of the 11 Recruits, each delivering a thrust of 1,800 lb. for a total thrust of 5,400 lb.

1955 October Acting on the recommendations of the Strategic Missiles Evaluation Committee, the U.S. Air Force awarded the Martin Co. a definitive contract for design and development of a two-stage liquid propellant ICBM designated SM-68 under Weapon System 107A-2. Asked by the Air Research and Development Command to select a suitable name, Martin public relations official Joe Rowland thought of the gargantuan task prescribed for the missile and chose the name Titan. Serving as a back-up to Atlas, Titan would pick up technical improvements that came too late to be incorporated in its competitor while retaining the more conservative two-stage design. The green light was given in 1957 and Titan I emerged, 90 ft. 6 in. in length with a first-stage diameter of 10 ft., a second stage diameter of 8 ft. and a launch weight of 220,000 lb. Power was provided by a 300,000-lb. thrust Aerojet LR87-1 with twin gimbaled chambers in the first stage and a single 80,000-lb thrust Aerojet LR-91 in the second stage. Both stages used RP-1-liquid oxygen propellants in stiffened tanks and the missile had a throw weight of 4,000 lb.

1955 November 2 The U.S. nuclear rocket program got under way with the formal agreement to begin Project Rover, a joint activity between the Atomic Energy Commission and the U.S. Air Force. Los Alamos National Laboratory and Lawrence Livermore would begin separate USAF research on reactor designs under the management of the Propulsion Laboratory at Wright Air Development Center. In the spring of 1956 the Air Force assigned some of its property in Nevada for tests: Los Alamos operated from Jackass Flats and Livermore from Cain Spring. Before its cancellation in 1972, the United States would invest $1.5 billion on nuclear rocket research and development.

1955 November 8 The Pentagon instructed that a joint Army-Navy ballistic missile committee be established to proceed with development of the Jupiter A IRBM for land- and sea-based deployment. The Army's Guided Missile Development Division had completed the preliminary missile design defining a length of about 90 ft. with a diameter of 7 ft. 11 in. Under the terms of the joint agreement, the Navy would design and develop shipboard systems for Jupiter. Increasingly concerned, however, with the relatively large size of the missile—a consequence of the liquid propellant design selected—the Navy proposed a resizing of the missile to a maximum length of 50 ft. and a diameter of 10 ft. Secretary of Defense Charles F. Wilson intervened to order a compromise length of 58 ft. and a diameter of 8 ft. 9 in. for the common missile.

Rocketdyne was established as a separate division of North American Aviation (NAA) with headquarters at Canoga Park, Calif., and test facilities in the Santa Susana Mountain region. Ten years after it had first stripped down V-2 engines shipped from Germany and rebuilt them with U.S.-standard parts, Rocketdyne had grown into one of America's leading rocket propulsion companies. The first contingent of 850 administration staff moved into Canoga Park on November 14, 1955, and a further 1,700 had been drawn together from separate facilities by the end of the year. On October 1, 1959, Rocketdyne acquired the former Astrodyne, Inc., which had been formed by NAA and Phillips Petroleum to manufacture complete propulsion systems.

1955 November 18 The first powered flight of the Bell X-2 took place when the no. 1 aircraft was dropped from the belly of a modified B-50 from Edwards Air Force Base, Calif. Piloted by U.S. Air Force Col. Frank K. Everest, the aircraft remained subsonic on this flight but achieved Mach 1.4 (924 MPH) on its third powered flight April 25, 1956. The highest speed successfully achieved by the X-2 was Mach 2.8706 (1,900.34 MPH) on May 22, 1956, on a flight piloted by Col. Everest. On September 27, 1956, U.S. Air Force Capt. Melburn Apt was killed and his aircraft destroyed, when the sole remaining X-2 lost attitude control at Mach 3.196 (2,094 MPH, the fastest speed achieved by a rocket-powered research aircraft) and crashed.

1955 November 28 The Navy issued its long-term objective to develop a solid propellant missile for deployment aboard surface ships and submarines concurrent with its joint agreement with the Army for a ship-based version of the liquid propellant Jupiter A. On December 2, 1955, Rear Adm. William F. Raborn Jr. was appointed head of the Special Projects Office set up to run what was now called the Fleet Ballistic Missile (FBM) program. The FBM office coordinated Navy studies on the problems associated with using liquid propellant missiles at sea: difficulties with storing and handling cryogenic propellants, the several hours needed to prepare the missiles for launch and the relatively slow initial acceleration off a heaving deck rendering the missile liable to directional error as it left the pad. On March 20 1956, the Navy approved detailed study of a solid propellant missile.

1955 December 1 Bell presented a detailed proposal to the U.S. Air Force for a three-phase development program in support of System 118P. The Bell design included a two-stage rocket-boosted glider capable of a speed of Mach 15 and a maximum altitude of 165,000 ft. The vehicle proposed for the first phase of the program would have a range of 5,000 mi., and the intermediate phase vehicle would have a range of 10,000 mi. The final phase was expected to produce a vehicle capable of going into Earth orbit. The Air Force accepted the proposal and on March 20, 1956, awarded Bell a contract for studies of a manned high-altitude reconnaissance platform known as Reconnaissance System 459L.

1955 December 2 At a National Security Council meeting attended by President Dwight D. Eisenhower, the highest national priority was accorded the four major U.S. missile programs currently under way: the Atlas and Titan ICBMs, and Thor and Jupiter IRBMs. Eisenhower was convinced the United States should have an effective missile force so as to present the threat of massive and immediate attack as a deterrent, and emphasized that the country had to have a reliable missile system, "even if I have to run it myself!" He was never convinced, however, that missiles should be the primary means of fighting a war. At this date, the United States had about 5,000 MT in nuclear weapons, but by 1956 that figure jumped to 14,000 MT, and by 1960 to 20,000 MT. The number of bombs or warheads had increased from 5,000 in 1956 to 17,000 in 1960. Thanks to the major expansion programs of the Kennedy administration, there would be 33,000 U.S. nuclear bombs and warheads by 1967.

1955 December 15 The first powered flight with the reworked no. 2 first generation Bell X-1 took place with Joseph Walker at the controls following an air-drop from a specially modified B-50 carrier-plane. Designated X-1E, it was built from the airframe of the second X-1 after its 54th flight on

America's first intercontinental ballistic missile (ICBM), the General Dynamics Atlas had integral skin and propellant tank walls, the whole kept rigid by the fluids and a pressurizing gas.

October 23, 1951, specifically to test a thin-section wing and a new low-pressure fuel turbopump for the Reaction Motors LR-8-RM-5 engine. Delivering the same thrust as the XLR-11 had previously, the new engine was the same power plant type used in the Douglas D-558-II and was tested in the hope it could replace high-pressure nitrogen systems. The 26th and last X-1E flight was flown on November 6, 1958. This also marked the last flight of a Bell X-1.

1955 December 19 The U.S. Air Force asked industry to carry out preliminary design of a hypersonic manned bomber. Boeing, Convair, Douglas, McDonnell, North American and Republic submitted proposals, and on June 12, 1956, System Requirement 126 was released defining a formal study of a rocket-bomber (dubbed Robo) capable of carrying a payload of up to 25,000 lb. Robo was required to double as a reconnaissance platform and to integrate boost-glide studies conducted by Boeing and Bell as part of the Bomi and Brass Bell work.

1955 December 28 The Douglas Aircraft Co. received a contract from the U.S. Air Force Air Research and Development Command for detailed design and development of an intermediate range ballistic missile one month after the Department of Defense gave permission for the Air Force to develop an IRBM. Officially known as SM-75, a part of Weapon System 315A, the missile was called Thor as was its Army counterpart, Jupiter, it had a specified range of 1,500 NM with a 1.4 MT W-49 thermonuclear warhead. The 1,450-lb. Mk.2 reentry cone was selected for Thor, in 1958 the only operational heat-sink reentry vehicle. Manufactured from beryllium, it was replaced by lightweight ablative reentry vehicles on Atlas and Titan.

1955 December The Propulsion Group of the Western Development Division, U.S. Air Force Air Research and Development Command, began a detailed research program on the application of storable solid propellants to large ballistic missiles. The Air Force was concerned about both the amount of time it took to prepare liquid propellant missiles for flight and the seemingly unnecessary size they required for a given throw weight. In February 1956 the WDD completed initial design details of an advanced solid propellant IRBM called Project Q. In April 1957 the WDD let the first contract for solid propellant motor design go to the Air Research and Development Command.

1955 Soviet armed forces began to deploy what NATO would later designate the SS-3 Shyster and the SS-N-4 Sark. A single-stage liquid propellant missile, the SS-3 utilized alcohol–liquid oxygen propellants in its RD-103 motor, but from 1959 switched to kerosene fuel. The SS-3 had a length of 68 ft. 10 in., a diameter of 5 ft. 6 in. and a launch weight of 57,300 lb. and provided a range of up to 750 mi. with aerodynamic control by four fins and vanes in the exhaust stream. Transported on a wheeled vehicle, it could carry conventional or nuclear warheads. The SS-N-4 Sark was the world's first submarine-launched ballistic missile (SLBM) and two were carried on seven converted Zulu-class submarines. Beginning in 1957 three missiles were carried by each of 22 specially designed Golf-class submarines, and from 1958 by 15 Hotel-class vessels as well. The Sark had a length of 49 ft., a diameter of 5 ft. 11 in. and a weight of 44,100 lb. Its range was a modest 375 mi. Ejected from a surfaced submarine by four solid propellant charges, the missile had two liquid propellant stages and a single 1 MT warhead. Fears about a Soviet breakthrough in SLBMs did much to spur decisions in the U.S. Navy over the proposed Jupiter S and Polaris programs.

During the year studies began on one of the most bizarre forms of rocket propulsion envisaged. Dr. Stanislaus Ulm of the Los Alamos Scientific Laboratory in New Mexico proposed the use of thermonuclear fusion to propel gigantic spaceships across the solar system. Thousands of nuclear "bombs" would be detonated against a huge shield, the opposite side of which would support the spaceship structure, and push the vehicle to enormous speed by acting on a sprung plate against which the shock waves would thrust. In October 1958 the Advanced Research Projects Agency awarded a $1 million contract to General Dynamics for work on what it dubbed Project Orion. The agency extended this by a further $300,000 contract in August 1959. In June 1963 General Dynamics received another contract to study the concept for interplanetary manned flight, but the nuclear test ban treaty of 1963 stopped this work.

1956 February 1 The U.S. Army Ballistic Missile Agency (ABMA) came into existence, taking over the Guided Missile Development Division at Redstone Arsenal. Maj. Gen. John B. Medaris, a highly accomplished and aggressive officer, was placed in command and Wernher von Braun was named director of the Development Operations Division. Kurt Debus, one of the Paperclip engineers from Peenemünde, was put in charge of the Missile Firing Laboratory and controlled launches from Cape Canaveral. He later became director of the NASA Kennedy Space Center, serving until he retired in 1974. The ABMA was the Army's attempt to achieve parity with the increasingly influential Air Force missile development effort, which threatened to seize control of all long-range weapons from the Army.

1956 March 14 The first Jupiter A missile was launched on a test flight from Cape Canaveral, Fla. Essentially a Redstone fitted with several modified components for the development of the Jupiter IRBM program, it was a useful adaptation of an existing vehicle for project development. In all, 19 Redstones

converted into Jupiter A configurations were launched over the next two years in support of technical development with the IRBM. In these Redstones the Army flew systems and components designed at Redstone Arsenal, Ala., in configurations they would adopt with the new missile.

1956 March 18 In the first of a series of moves to consolidate ballistic missile programs, U.S. Air Force Headquarters gave Strategic Air Command and the Air Research and Development Command responsibility for operating and training Thor missiles and crews. In July 1956, SAC announced it was entering the planning phase for all future missile programs and was primarily interested in the Thor IRBM, an ICBM, Navaho and Snark cruise missiles as well as in defense suppression missiles carried by manned penetrating bombers. During this planning process the U.S. Air Force reached wide-ranging decisions about the missile force it wanted and from mid-1956 made concerted efforts to secure primary responsibility for medium, intermediate and long-range missiles.

1956 March 20 The U.S. Air Force and Bell Aircraft completed negotiations on a contract defining a study for Reconnaissance System 459L, also known as Brass Bell. By the end of 1956 Bell came up with a rocket-boosted glide vehicle with a maximum altitude of 170,000 ft. and a range of 6,000 mi. Envisaged as a two-stage vehicle with propulsion units based on Atlas ICBM engines, Brass Bell was theoretically capable of supporting a third stage that would give it a maximum speed of 15,000 MPH and a range of 11,500 mi.

1956 March 30 The U.S. Department of Defense asked the Atomic Energy Commission to study an atomic warhead for the Hughes AIM-47A Falcon GAR-11 air-to-air rocket. The missile had a length of 7 ft. and a diameter of 11 in., with a range of 5 mi. and a terminal speed of Mach 2. On June 23, 1958, the W-51/54 warhead being developed for the Falcon missile was also applied to the Davy Crockett (nuclear) recoilless rifle. When the nuclear Falcon became operational, it had a yield of 0.01–1 KT, and almost 2,000 were produced by the end of production in 1955. Falcon was withdrawn from service by 1972.

1956 March The U.S. Air Force began Project 7969, "Manned Ballistic Missile Research System," based on preliminary activities with recoverable capsules in what eventually became the Discoverer program (code name Corona) and led to the development of a manned spacecraft. Industry was asked to submit proposals and the first arrived by the end of this year. The launch vehicle was to have been an Atlas together with an Agena upper stage. The Air Force wanted to limit acceleration forces on the pilot to a maximum 12 g and stipulated that the capsule should point forward at all times. This required a gimbaled seat for the pilot. Eventually, the Air Force received 11 industry proposals, and regular meetings on this project were held with the NACA over the next two years. Some of the studies, particularly of life support systems, were fed into NASA's Mercury program.

1956 April 11 Lockheed Aircraft Corp. was awarded a contract by the U.S. Navy to examine the feasibility of a submarine-based solid propellant missile capable of throwing a nuclear warhead 1,500 NM. The study assumed a warhead weight of 3,000 lb. and selected a two-stage design incorporating a cluster of six motors, each 3 ft. 4 in. in diameter, surrounding a similar motor which would operate as the second stage. Length was 41 ft. 4 in. with a diameter of 10 ft., conforming to a Navy specification that the missile be contained within an envelope 50 ft. in length and 10 ft. in diameter. With a launch weight of 162,000 lb., the Jupiter S missile was approximately 50% heavier than the liquid propellant Jupiter designed by the Army Ballistic Missile Agency. Preliminary design showed that an 8,500-ton submarine could carry four missiles of this type.

1956 May 18 The U.S. Air Force awarded a contract to Aerojet General Corp. for supporting work on the Project Rover nuclear rocket program. On July 15, the Rocketdyne Division of North American Aviation received a similar contract. On December 28, 1956, the Armed Forces Special Weapons Project team was charged by the Department of Defense with examining the military advantages of a nuclear-powered ICBM, and determining whether these would justify the enormous cost of development. The group reported that there was no immediate need for a nuclear-powered ICBM, but recommended that the program should continue for possible application to future requirements. On March 6, 1957, the Jackass Flats, Nev., site was chosen for the Project Rover tests, the first of which was conducted on July 1, 1959.

1956 May 28 The RAND Corp. issued a report on the technical possibility of a special carrier for lunar scientific exploration based on an Atlas missile adapted for space research. Technical studies indicated that a retro-rocket on the missile would be used to decelerate it to a relatively low speed, but that a long spike on the forward part of the vehicle would absorb the impact velocity, calculated at around 50 ft./sec. (approximately 340 MPH). Instruments lowered to the lunar surface would then, said RAND, transmit data to Earth on a wide variety of scientific measurements. This was the first in a series of reports for the U.S. Air Force which helped circulate topics for debate within the service.

1956 May 29 The first Chance Vought SSM-N-9 Regulus II cruise missile took to the air boosted by a 115,000-lb. thrust Aerojet solid propellant motor. Completely different from the Vought Regulus I, this missile had a length of 67 ft. 2 in., a wingspan of 20 ft. and a weight of 22,500 lb. Capable of carrying a 2,920-lb. W27 nuclear warhead 800 NM at Mach 2, the missile was sustained in flight by a 15,000-lb. thrust General Electric J79-3A turbojet. The Navy fired a Regulus II from the submarine USS *Grayback* on September 16, 1958, and planned to deploy the missile in 23 submarines. Erroneously believing cruise missiles to be obsolete, the U.S. Navy cancelled Regulus on December 18, 1958, as it prepared to take on a bigger role in strategic nuclear deterrence with Polaris. Moreover, nuclear-powered aircraft carriers capable of handling heavier and more flexible aircraft carrying miniaturized nuclear weapons led to an unfavorable view of pilotless flying bombs.

1956 June 12 The U.S. Air Force Air Research Development Command issued System Requirement 126 outlining requirements for the Robo rocket-bomber. It was to be a suborbital, world-circling bomber operating at a minimum altitude of 100,000 ft. Contracts were issued to Convair, Douglas and North American Aviation for the detailed design of such a vehicle. The payload was to be a maximum 25,000 lb. and the vehicle was to incorporate lessons learned from the U.S. Air Force Bomi and Brass Bell studies. On June 20, 1957, a special committee of the Air Force was convened to evaluate the Robo concept.

1956 June A highly classified report entitled *Physical Recovery of Satellite Payload: A Preliminary Investigation* was issued to the U.S. Department of Defense by the RAND Corp. Where once it had been thought impossible to recover objects safely from space without considerable damage from heat as they passed back through the atmosphere at high speed, engineers now planned a broad range of reentry tests using specially designed nose cones fired on ballistic missiles. Adapting this technology to protect film shot from satellites in space, RAND developed the concept of recoverable film capsules. This added a third intelligence recovery method to the two embraced by the U.S. Air Force's WS-117L Program which envisaged alternative wet-film scanning and TV transmission of images captured through on-board cameras.

1956 July 6 The first flight of the Nike-Cajun sounding rocket took place from Wallops Island Test Station. The Cajun upper stage for the Nike I guided missile booster adapted to scientific tasks was inspired by the National Advisory Committee for Aeronautics decision to improve the performance of the Nike-Deacon first test-fired on November 19, 1953. Thiokol Chemical Corp. provided an improved propellant grain for what was essentially a Deacon rocket case, 8 ft. 8 in. in length with a weight of 166 lb. The first live firing of the Cajun, named by Joseph G. Thibodaux of New Orleans, took place at Wallops Island on June 20, 1956. The Nike-Cajun could send a 50-lb. package of scientific instruments to a maximum altitude of 104 mi. It was used extensively during the International Geophysical Year.

1956 July 17 The full-scale flight test program of the Lockheed X-17 reentry research vehicle began with a successful flight from Cape Canaveral, Fla. Developed in response to urgent requirements for information about the effects of the atmosphere on high-velocity reentry vehicles, Lockheed produced a three-stage research rocket capable of boosting instrumented nose cones to a maximum speed of 9,500 MPH. The first stage comprised a 48,000-lb. thrust Thiokol XM-20 Sergeant capable of operating for 28 sec., during which it would boost the X-17 to a height of 90,000 ft. The second and third stages were Thiokol XM-19 Recruit rockets that burned for 1.53 sec. There was a cluster of three, each rated at 33,900 lb. thrust, in the second stage and one rated at 35,950 lb. thrust in the third stage. Research results obtained from the series of 24 launches through August 22, 1957, provided valuable data for the new generation of ICBMs. Three classified X-17s were launched beginning August 27, 1958, in support of a clandestine series of very high altitude nuclear tests.

1956 July Following a 10-year development program originally called Project Hot Shot, the Sperry Gyroscope Sparrow I air-to-air guided missile became operational with the U.S. Navy Atlantic and Pacific fleets. An outstanding weapon of great potential and long service with armed services around the free world, Sparrow quickly became partnered with Sidewinder as the basic air-to-air weapons of combat aircraft. The AIM-N-2 Sparrow I had a length of 11 ft. 8 in. and a fin span of 3 ft. 3 in. with a modest range of 5 mi. Later versions under the new designation AIM-7 introduced in 1962 had a range of 25–31 mi. One, the AIM-7F, had a range of 62 mi. In all, more than 40,000 Sparrows were built over the next 40 years and the weapon remains the primary choice for U.S. combat aircraft.

1956 September 4 U.S. Navy Adm. Arleigh A. Burke, chief of Naval Operations, was told by physicist Edward Teller from the Lawrence Livermore National Laboratory, Calif., that progress in the design of thermonuclear weapons presaged a revolution in miniaturization. When confronted with designs for the 162,000-lb. weight solid propellant Jupiter S missile, Teller asked, "Why are you designing a 1965 weapon system with 1958 technology?" He elaborated by asserting that impending developments with new solid propellants and thermonuclear warheads that weighed no more than 900 lb., meant that a submarine-launched (1,500-NM range) ballistic missile less than 20% the size of the Jupiter S proposed in April 1956 could be built and made operational by 1965 at the latest.

1956 September 20 The first modified Redstone missile in the Reentry Test Vehicle (RTV) program took place when a Jupiter-C propelled a small reentry cone 682 mi. into space and 3,335 mi. downrange from Cape Canaveral, the greatest distance achieved to date by a ballistic missile. On the second launch, May 15, 1957, a small Jupiter intermediate-range ballistic missile nose cone was flown, but a guidance malfunction in the Jupiter-C took the cone out of the target area and prevented recovery. Telemetry was received which indicated that the cone survived the temperatures and pressure of reentry. The third and final RTV flight took place on August 8, 1957, with a one-third scale model of the Jupiter nose cone. The missile reached an altitude of 600 mi., recovery taking place 1,200 mi. downrange. This was the first man-made object recovered from space. Now in the Smithsonian Institution, Washington, D.C., the cone carried a letter to Gen. John B. Medaris, chief of the Army Ballistic Missile Agency, as the first item of space mail and was publicly displayed by President Eisenhower on November 7, 1957.

1956 October 23 The Pentagon's Science Advisory Committee (not the same organization as the President's SAC), recommended that the U.S. Navy "receive top priority, equal to that of the other IRBM [Jupiter] program" for its solid propellant sea-launched missile. The SAC had been encouraged by the reworked design around low-weight thermonuclear warheads. This put the missile, now called Polaris, in the 30,000-lb. class versus the 162,000-lb. design baselined on April 11, 1956. Launched from firing tubes on a submerged submarine, the Polaris system would avoid the vulnerability of land-based missiles and bombers and form the third leg of a strategic nuclear triad. Assisted by Lockheed, the Navy put together a study group to define the parameters of the missile and fix its physical dimensions. This effort, completed in March 1957, enabled the Navy to develop both missile and submarine simultaneously.

1956 October 29 Lockheed received a contract from the U.S. Air Force to build first generation reconnaissance satellites under the program designated WS-117L, also known as the Advanced Reconnaissance System. Code-named Pied Piper, WS-117L envisaged both recoverable and nonrecoverable satellites. Cameras and surveillance equipment were to be an integral part of a payload launched by Thor with a Vanguard second stage. Development was slow because of doubts at the Pentagon, and not until *Explorer I* and Vanguard satellites were launched in 1958 did the military provide adequate funding. The satellite carrying a recoverable capsule that evolved from WS-117L was operated by the CIA under the name Discoverer. The one that would radio-transmit its pictures was originally code-named Sentry by the Air Force, then SAMOS (satellite and missile observation system).

1956 November 6 Originally scheduled for September 1954, the North American Aviation XSM-64 Navaho supersonic cruise missile made its first rocket-boosted flight attempt, but failed 26 sec. after launch when a rate-gyro malfunctioned at an altitude of 9,000 ft. The cruise missile itself had a length of 67 ft. 10 in. and a span of 27 ft. across its delta wing. The interim booster used in early tests had a length of 76 ft. 3 in., a diameter of 5 ft. 10 in. and two liquid propellant rocket motors delivering a total thrust of 240,000 lb. The fully operational configuration was to have been a bigger version of the XSM-64 combination cruise missile–booster. Four more unsuccessful flight attempts were made (on March 22, April 25, June 26 and August 12, 1957) before the first success.

The U.S. Air Force Air Research Development Command, Research and Target Systems Division, initiated a research program for a rocket-boosted glider. Known as HYWARDS (hypersonic weapon and research and development supporting system), or System 455L, the research work was conceived as a technology support program for the Air Force's Brass Bell and Robo rocket-boosted glide vehicles. Formalized through Systems Requirement 131, HYWARDS was expected to produce a research vehicle for testing various technologies, initially from an air-launched platform like the X-15, then from a rocket-boosted flight on a converted ICBM.

1956 November 26 In an attempt to eliminate competition between the services, U.S. Secretary of Defense Charles Wilson published a memorandum establishing operational limits on the missiles the Army and the Air Force would be allowed to develop. The Air Force was to be responsible for all ICBMs (intercontinental ballistic missiles) with a range in excess of 1,500 mi. and IRBMs (intermediate-range ballistic missiles) with a range over 200 mi. The Army was allowed to develop tactical battlefield missiles with a range of up to "about 200 mi." This had a serious effect on the Jupiter program. The Army was permitted to continue its development, but the Air Force was to take over and operate the Jupiter missile. The Navy was allowed to develop and operate sea-based IRBMs, a class of missile which was later designated SLBM (submarine-launched ballistic missile).

1956 November Soviet scientists and engineers met to discuss details of a space program which would include manned orbital flights and missions to the surface of the moon. With imminent availability of a launch vehicle derived from the R7 ICBM and a payload capability of several tons, prospects for manned flight became a realistic possibility. It was agreed that design of a manned spacecraft should proceed at a low level until verification of the launch vehicle had been completed and orbital flight demonstrated by a satellite program. The ability of living things to exist in space would be demonstrated by the launch of a dog in Russia's second satellite. When these things had been accomplished, a further review would determine the way forward. An ultimate goal was to conduct manned landings on the moon.

1956 December 5 During the trouble-plagued Northrop N-69 Snark program a test missile fired from Cape Canaveral, Fla., went dramatically off course and was lost, heading in the general direction of South America. Snark test flights frequently ended far short of their mission objectives due to technical failures. On this occasion, missile serial number 53-8172 N-3309 failed to respond to guidance commands that should have steered it on a test flight across the Atlantic Missile Range. It also ignored destruct command signals. Radar tracking indicated it had reached the South American jungles, prompting one Miami newspaper to borrow verse from Henry Wadsworth Longfellow: "They shot a Snark into the air, it fell to Earth they know not where." They found out where in 1982 when it was discovered by a Brazilian farmer.

1956 December 8   The U.S. secretary of defense authorized the Navy to proceed with full-scale development of the Polaris Fleet Ballistic Missile system, thus formally marking the start of the program. The Navy was also instructed to cancel its participation in the Army's Jupiter A program, and ten days later the Joint Army-Navy Ballistic Missile Committee was dissolved. Work was to concentrate on the development of a high performance, two-stage, solid propellant missile with a length of 28 ft. 6 in., a diameter of 4 ft. 6 in. and a gross weight of about 28,000 lb. Development of the W-47 warhead accelerated under the control of the Atomic Energy Commission and eventually produced a 600 KT yield weapon, 800 KT in a developed version, both of which weighed only 600 lb. System accuracy was about 1 NM at a range of 1,200 NM.

The first flight associated with Project Vanguard, the U.S. satellite program, took place from Cape Canaveral when the 13th Viking rocket was launched carrying instruments to check tracking equipment that would be carried by the satellite. Designated TV-0 (Test Vehicle-Zero), it reached a height of 16.5 mi. and a downrange distance of 97.6 mi. A second test launch, the last flight of a Viking rocket, took place on May 1, 1957, when TV-1 took a prototype of the Vanguard third stage on a ballistic trajectory 121 mi. above the Earth and 450 mi. downrange.

1956 December 13   A U.S. Air Force F-86D Sabre successfully fired a Rockaire Deacon rocket to an altitude of 28 mi. after carrying it into the air from Holloman AFB, N.Mex. attached to the aircraft's fuselage. Following a U.S. Navy test on August 16, 1955, this was only the second time an aircraft had launched a sounding rocket. The Deacon rocket had a diameter of 8 in. and a total length of 8 ft. 9 in., of which the forward 4 ft. 2 in. was the instrument section, and was capable of lifting 40 lb. of experiments to a maximum height of 40 mi. Three other flights were conducted on December 14, 17 and 19.

1956 December 14   At a high-level presentation to the Air Ministry, British scientists and senior defense officials briefed the Air Staff on progress with development of the Blue Steak long-range ballistic missile. While projected milestones at this early stage in engineering design were on schedule, the briefing exposed many potentially insoluble problems. These included decisions on a basing mode, whether or not to group the missiles in fixed sites or disperse them individually, whether to program the missiles to hit specific targets or allow them to be retargeted at random and what the cost growth was likely to be. There were already concerns about peripheral expenditure for additional support projects such as Black Knight, the rocket which would be used for warhead reentry tests.

1956 December 28   The U.S. Armed Forces Special Weapons Project completed a review of nuclear-powered ICBM concepts, finding little use for such a combination. Nevertheless, it recommended continued study of the idea for future applications. On January 12, 1957, tne Department of Defense recommended cuts in the level of work on the nuclear rocket program and the reactor called Rover, and Los Alamos began defining their first nuclear propulsion reactor, Kiwi-A. The military was gradually losing interest in nuclear space propul-

sion and on March 18, 1957, the Atomic Energy Commission assigned Rover to Los Alamos.

1956 December   U.S. radar installations picked up the first in a series of test firings associated with development of the Soviet R-7 intercontinental ballistic missile. Stage separation and altitude trials were conducted over approximately 40–50 test launches to evaluate propulsion, guidance, trajectory control and command sequences. Because Soviet warhead engineers required a throw weight of 15,000 lb., the R-7 was designed to generate a thrust of 1.1 million lb. By this date, construction of the R-7 launch site had been completed and checkout of the transport, fueling and support equipment began.

1956   During the year, in a book entitled *The Exploration of Mars,* Willy Ley and Wernher von Braun explained how the first manned missions to Mars might be conducted. Believing Mars to have a relatively dense atmosphere, the authors envisaged a winged rocket gliding down to the surface. From there, the cylindrical "fuselage" of the glider would be rotated to a vertical position, thus converting it into a conventional rocket for the return to Earth. The book did much to bring public awareness of the possibilities inherent in contemporary research on rockets and missiles and of the possibility of interplanetary travel.

1957 January 17   The NACA Langley Aeronautical Laboratory issued its report on the HYWARDS (hypersonic weapon and research and development supporting system) boost-glide vehicle studies conducted for the U.S. Air Force. Headed by John Becker, the Langley group had examined possible configurations for the research vehicle and recommended aiming for a rocket-boosted glide vehicle capable of Mach 18. The preferred configuration was a flat-bottomed shape with the fuselage on top to minimize hot areas. It was the first attempt at optimizing the configuration of a boost-glide vehicle using aerothermal design analysis. All previous designs, such as Bomi, Brass Bell and Robo, were dictated by simple aerodynamic rules and relied on substantial heat-shielding to protect aluminum structures from high temperatures.

1957 January 25   Less than six months after completion of engineering design, the U.S. Air Force Thor IRBM made its first, albeit unsuccessful, ascent from Cape Canaveral, Fla. The missile rose a few inches from the pad and collapsed in a ball of fire. The second Thor was also a failure when faulty range-safety instrumentation destroyed the missile, and two more, in May and August 1957, were failures also. The first successful Thor launch took place on September 20 when a missile flew 1,300 NM as programmed, and on another flight it flew an extended range of 2,300 NM. The first Thor with guidance was launched on December 19, 1957. Built by Douglas, the missile had a length of 65 ft., a maximum diameter of 8 ft. and a liftoff weight of 110,000 lb. Thor was powered by a single 150,000-lb. thrust Rocketdyne LR-79 propulsion system burning kerosene–liquid oxygen propellants.

1957 January   Construction work started on a rocket test range at Spadeadam Waste, a remote part of East Cumberland, Scotland. Although U.S. rocket engine test stands in the Santa Susana mountains, Calif., could have been used to test the motors being developed by Rolls Royce for the Blue Streak long-range ballistic missile, Britain wanted its own test range and the Spadeadam Rocket Establishment was selected. It would support tests of individual engines as well as of complete Blue Streak rocket stages, although no launches would be made from the United Kingdom. The Woomera Test Range in Australia would be used for launches.

1957 February 14   The first Skylark sounding rocket was test-fired from the Woomera test range in South Australia. Conceived in 1955 as Britain's contribution to the International Geophysical Year, it was designed at the Royal Aircraft Establishment, Farnborough. Skylark was essentially a Raven solid propellant rocket motor producing a thrust of between 8,900 lb. and 15,600 lb. A typical Skylark at the high end of the thrust range could lift 300 lb. of instruments to an altitude of almost 100 mi., or 450 lb. when boosted by a solid propellant Cuckoo motor attached to the bottom of the Raven. Skylark became the standard British sounding rocket and was developed over the years to provide a maximum capability of lifting 400 lb. to an altitude of 500 mi., or 165 lb. to 930 mi. More than 400 Skylarks had been launched by 1993.

1957 February 21   Los Alamos Scientific Laboratory, N.Mex., completed a report, *Metal Dumbo Rocket Reactor,* for the U.S. Atomic Energy Commission. Based on an advanced high thrust-to-weight ratio design, Los Alamos proposed development of an efficient, very high density Dumbo reactor for application as a reusable single-stage-to-orbit shuttle. Taking off and landing like a conventional airplane, the vehicle would require special shielding to prevent radioactive products from leaking to the atmosphere. Dumbo was a competitor to the less technically sophisticated NERVA (Nuclear Engine for Rocket Vehicle Application) and lost out in a funding battle.

1957 March 1   The U.S. Army Ballistic Missile Agency fired the first Jupiter missile on test from Cape Canaveral, Fla. With a length of 58 ft. and a diameter of 8 ft. 9 in., the missile weighed 110,000 lb. and was powered by a Rocketdyne S-3D engine delivering a maximum thrust of 150,000 lb. The first launch ended at 74 sec. and a height of 48,000 ft. when it broke up after overheating in the tail section. On the second flight, April 26, 1957, the missile became unstable when sloshing propellant upset the guidance at an altitude of 60,000 ft. Engineers resolved the problem by placing perforated cans in the propellant, and slosh baffles were added later to break up the wave-action in the tanks. The third test flight on May 31, 1957, was a success, and the test program was completed on February 4, 1960, after more than 25 Jupiter missiles had been fired. A new telemetry system permitted simultaneous measurement of up to 211 parameters during flight.

1957 March 4   Checkout of the first Soviet R-7 ICBM began at Tyuratam, soon to be known as the Baikonur Cosmodrome. The R-7 had a core stage 92 ft. in length with a maximum diameter of 9 ft. 8 in. surrounded by four tapered strap-on boosters, each with a length of 62 ft. 4 in. and a maximum diameter of 9 ft. 9 in. The core stage was powered by four RD-108 engines linked as a single propulsion system using a common turbopump and delivering a thrust of 211,650 lb. Four vernier chambers provided attitude control and steering. Each booster comprised four clustered RD-107 engines, fed by a single turbopump, with a thrust of 224,910 lb. and two steering nozzles. This gave the first Soviet ICBM 5 clustered elements, 20 main combustion chambers and 12 steering motors. The first R-7 was moved to the launch pad on May 5.

Built for the USSR's first satellite launcher and ICBM, the RD-107 engine comprised four combustion chambers and two small vernier motors fed by one turbopump. One RD-107 was attached to each of four boosters on the main stage.

1957 April 4   Several rocket-engine builders got interested in advanced propulsion concepts, seeing them as an area within which they could compete with each other. Perry W. Pratt (no relation to Pratt and Whitney) and C. Branson Smith made a presentation to United Aircraft on the possibility of getting involved with cryogenic hydrogen-fueled rocket engines. Pratt and Smith were engineers with Pratt and Whitney, owned by United, and were responding to the company's desire to expand its line of products to include both aircraft engines and rocket motors. At this time both Aerojet and Rocketdyne were each heavily involved in the manufacture of liquid propellant rocket motors and the two companies believed that to compete industrially they would have to come up with something special. They decided to enter the esoteric world of supercold high-energy propellants. Smith calculated that a cryogenic upper stage would add 50% throw weight to an ICBM without increasing its gross mass, thus allowing the ICBM to put a satellite in orbit. This finding encouraged the company to fund further research and on March 4, 1958, Pratt and Smith proposed development of a 15,000-lb. thrust motor.

1957 May 15   From a launch site near the town of Tyuratam in Kazakhstan, the Soviet Union began long-range tests with the R-7 intercontinental ballistic missile. Among those present were: Sergei Pavlovich Korolev, chief designer; Valentin Petrovich Glushko, engine designer; Nikolai Pilyugin, responsible for the guidance equipment; Viktor Kuznetsov, in charge of gyroscopes and attitude control; and Mikhail Ryazansky, chief designer of radio and communication systems. The first flight failed after 50 sec., but in eight flights, the last of which was flown August 3, 1957, the R-7 demonstrated a range of up to 4,000 mi. On August 26, 1957, the TASS news agency announced that "it is now possible to send missiles to any part of the world." This stimulated a dramatic increase in the pace of U.S. missile programs, and an "emergency capability" warhead was authorized for Polaris.

The first British thermonuclear (hydrogen) bomb was dropped from an RAF Vickers Valiant (serial number XD818) over Malden Island in the Pacific Ocean, 1,600 mi. south of Hawaii and 200 mi. south of Christmas Island. With a yield of approximately 1 MT, the device was a prototype of the operational bomb known as Yellow Sun. The developed bomb would weigh 7,000 lb. but was preceded in service by an interim H-bomb known as Violet Club, available for use from March 1958. Availability of a thermonuclear device gave credibility to the decision to build a long-range ballistic missile (Blue Streak) and a stand-off weapon (Blue Steel).

1957 June 7 Seventeen months after the decision had been taken to create an operational ballistic missile testing area, an old Army facility known as Camp Cooke in Calif., north of the Santa Ynez river, was named Cooke Air Force Base by the U.S. Air Force. Here missile crews would be trained on Thor, Atlas, Titan, Minuteman and MX missiles. The base would also support polar orbit satellite launches and ballistic rocket flights during development of warheads and reentry vehicles. The area south of the Santa Ynez river became the Naval Missile Facility, Point Arguello, where crews for Polaris, Poseidon and Trident were to be trained. On October 4, 1958, the name was changed to Vandenberg Air Force Base, and in 1964 the two portions of the site became known as North Vandenberg AFB and South Vandenberg AFB.

1957 June 11 The first launch of an Atlas A series missile took place from Cape Canaveral, but the flight was terminated 60 sec. after liftoff when the prototype missile went into a series of violent loops after one of the two boost motors failed. The A series carried no central sustainer engine and was limited to a theoretical "range" of 600 mi. A second flight attempt on September 25, 1957, also failed, but Atlas finally performed as planned on December 17, 1957, impacting 600 mi. downrange from the Atlantic Missile Range. Another success followed on January 10, 1958, but the last four Atlas A flights were a mixed bag rated between "success" and "partial success," the final series A flight being flown on June 3, 1958. It was followed by the first series B launch on July 19, 1958, which was a failure.

1957 June 18 The chief of U.S. Naval Operations, Adm. Arleigh A. Burke, formally approved the characteristics of the Fleet Ballistic Missile submarine that would carry the Polaris SLBM (submarine-launched ballistic missile). The pace of the program had been accelerated due to Soviet tests of its SS-6 ICBM, which defense experts said rendered U.S. air bases and launch sites vulnerable to attack. For this reason, the first five ballistic missile submarines were converted Skipjack hulls. With a 128-ft. missile launch tube section amidships, five *George Washington*–class submarines were ordered in February and July 1958. The first three had Albacore hulls while

the last two had whale-shaped hulls. Each boat was about 380 ft. long with a displacement of 6,700 tons and a crew of 110, 10 officers and 100 enlisted men. The program envisioned initial operational capability by January 1, 1963, and the first operational patrol actually began on November 15, 1960.

1957 June 20 A review of the U.S. Air Force Robo studies being conducted by Convair, Douglas and North American revealed differing views about the configuration. Douglas had selected a three-stage boost-glide vehicle, while North American selected a powered glider propelled by a combination rocket-turbojet but launched by a two-stage rocket. Funding its own studies, Boeing chose an unmanned boost-glide vehicle, and Republic offered its XF-103, also a contender, with a supersonic ramjet (scramjet) propulsion system. The Air Force ad hoc committee that had convened to evaluate the Robo (rocket-bomber) concept issued its final conclusions and recommendations on August 1. Exceeding its mandate, the committee reviewed the technical problems associated with the boost-glide concept and concluded that "with moderate funding, an experimental vehicle could be tested in 1965, a glide missile in 1968, and Robo in 1974." Thus was born the basis for Dyna-Soar, the design concept of which was formally sealed on October 10, 1957.

1957 July 11 The U.S. Department of Defense canceled its XSM-64 Navaho supersonic cruise missile program after four successive failures following the first flight attempt on November 6, 1956. More than 11 years after the program began it still didn't work and technical developments with ballistic missiles had rendered Navaho obsolete. Seven more flights were authorized, and they took place between August 12, 1957, and November 18, 1958; none was a complete success. However, the Navaho program made a marked contribution to the development of rocket motors and developments of the Navaho propulsion system evolved into the motors used in Jupiter, Thor and Atlas missiles.

1957 July The Air Force Ballistic Missile Division, successor to the Western Development Division, decided to upgrade the solid propellant missile program it had been working on since December 1955 to ICBM status, and in September it established a weapon systems office at Inglewood, Calif. On February 15, 1958, the AFBMD completed its first plan for the development of an ICBM, called Minuteman, after the Minute Men militia mobilized by the Continenal Congress in October 1774. Like its namesake, the Minuteman missile was designed to be launched within 60 sec. of an alert.

1957 August After a searching examination of its reconnaissance satellite programs, the U.S. Air Force decided to abandon one of three concepts in development under the WS-117L program. Accepting technical advice that real-time TV transmissions of images from space were almost impossible with existing capabilities, that satellite concept was dropped. Believing it would be difficult to locate and retrieve a small canister of film, the Air Force also became disheartened over the film-return concept and dropped it by the end of the year. This left only the third concept under WS-117L: wet-film images processed on board the satellite and scanned by light beam for transmission on radio signals. This would be code-named Sentry, later SAMOS (satellite and missile observation system).

1957 September 18 The first successful flight of an XSM-64A Navaho test missile took place from Cape Canaveral,

Fla., when the winged missile flew downrange for 17 min. 55 sec. The next flight, on November 13, ended when the missile had to be destroyed 1 min. 15 sec. after launch. A near perfect flight was completed on January 10, 1958, when the Navaho flew 1,240 mi., but the test missile was destroyed on the next flight, February 25, 1958, just 20 sec. after launch. At this point the Navaho test program came to an end, but two more flights were conducted for the RISE (Research In Supersonic Environment) program.

1957 October 4   The world's first artificial Earth satellite was launched by the Soviet Union at 22:28:04 local time (19:28:04 UT) into a 141 × 585 mi. orbit at an inclination of 65.1° with a period of 96.2' min. It was launched from the Baikonur launch site either by an adapted R-7 ICBM (NATO code name SS-6 Sapwood) or an SL-1 satellite launcher. This was the seventh attempt to orbit the first satellite. After the sixth, unsuccessful, attempt, Premier Nikita Khrushchev had canceled the effort. Sergei Pavlovich Korolev, convinced he could achieve success, disobeyed orders and prepared a final attempt. Called *Sputnik 1,* this satellite weighed 184.3 lb. and comprised a 22.8-in. sphere to which were attached four radio antennas for transmitting a 1-W signal on frequencies of 20.005 and 40.002 MHz. Power was provided by silver-zinc accumulators with a life of 14 days. The satellite was instrumented to measure its internal density and temperature. *Sputnik 1* remained in orbit 92 days, burning up in the atmosphere on January 4, 1958. Also placed in orbit, the sustainer stage burned up on December 1, 1957. After the success of *Sputnik 1,* Khrushchev reversed his opinion and gave major funds for space activities.

1957 October 8   When asked by President Dwight D. Eisenhower why the Vanguard project had not yet put up a satellite, Deputy Secretary of Defense Donald Quarles said the Army could have launched a satellite in 1955, but that the Pentagon wanted to keep the project under civilian control. Eisenhower was told that a reconnaissance satellite program was under way but when informed about petty rivalry between the Army and the Air Force on ballistic missiles, the President called for "a 'Manhattan Project'–type approach in order to get forward in this matter." (Manhattan was the code name for the atomic bomb project.) Eisenhower ordered the Army to go ahead with a satellite launch attempt.

Army Ballistic Missile Agency chief Gen. John B. Medaris authorized the use of an Army Redstone rocket, modified to the Jupiter C test configuration, to launch a U.S. satellite. When asked how long he would need to put a satellite in orbit, Wernher von Braun had indicated that 90 days would be sufficient. In effect, approval for the Army to go ahead with a satellite attempt turned the effort into the first space race—one between Army Jupiter C rocket engineers and civilian scientists on the Vanguard program. The Army program of satellites would be called Juno, Juno I comprising modified Redstone rockets in the Jupiter C configuration for 15–30-lb. class satellites, and Juno II equipped with Jupiter rockets for space probes (Phase I) and Earth satellites in the 100-lb. class (Phase II). Both Juno I and Juno II used the same upper-stage cluster of Sergeant solid rockets from the Jet Propulsion Laboratory.

1957 October 10   Acting on U.S. Air Force ad hoc committee on Robo recommendations issued August 1, 1957, the Air Research Development Command announced a three-step development program consolidating three projects—HYWARDS, Brass Bell and Robo. Designated System 464L, it

was to be called Dyna-Soar, an acronym for dynamic soaring. Dyna-Soar I was a suborbital glide vehicle scheduled for aerodynamic evaluation in 1966. Beginning with flights in 1969, Dyna-Soar II was a suborbital glide vehicle with bombing and reconnaissance capability. Dyna-Soar III was to be the definitive version of Dyna-Soar II, capable of going into orbit on rocket-boosted flights beginning 1974. All three variants were to carry at least one crew member.

1957 October 11   General Dynamics Convair Division consultant engineer Krafft A. Ehricke asked Rocketdyne applications engineer A. G. Negro to provide details of what his company could do to produce a small cryogenic rocket motor burning liquid hydrogen and liquid oxygen. Ehricke was fired with enthusiasm for developing a satellite launcher and wanted to improve existing launchers by adding high-energy upper stages. This work received the support of General Dynamics and they submitted a proposal to the Air Force in December 1957 entitled "A Satellite and Space Development Plan." Optimized to match the Atlas first stage, the upper stage Ehricke proposed was to be powered by four 7,000-lb. thrust engines. The U.S. Air Force supported research and Convair discovered that Pratt and Whitney had done pioneering work on pumps for cryogenic engines. The Air Force and Advanced Research Projects Agency put Convair and Pratt and Whitney together for what would emerge as the Centaur stage.

1957 October 14   In further reaction to the launch of *Sputnik 1,* the American Rocket Society proposed that a civilian-controlled body should be set up within government for managing an expanded space science program. They were not alone in seeking a consolidated and coordinated space science program for the nation. The conservative National Advisory Committee for Aeronautics (NACA) established a committee to review space technology, and under the chairmanship of Dr. James A. Van Allen, the Rocket and Satellite Research Panel of the National Academy of Sciences proposed on November 21, 1957, that a National Space Establishment be set up. These two groups made a joint proposal on January 4, 1958, and 10 days later the NACA issued a staff study calling for a national program for space combining the NACA, the Department of Defense, the National Science Foundation and the National Academy of Sciences.

1957 October 16   The first man-made objects to be fired to escape velocity and sent on an interplanetary trajectory were put into space atop an Aerobee sounding rocket from Alamagordo Air Force Base, N.Mex. A brainchild of Fritz Zwicky of the California Institute of Technology, the project involved powders of titanium (Ti) and carbon (C) compressed in a conical insert within a shaped charge. When detonated, TiC droplets "exploded" with a velocity of 9.3 mi./sec., or 33,480 MPH. Carried to a height of 53 mi. by the Aerobee and detonated, the droplets exceeded the escape velocity of the Earth by some 8,000 MPH. They thus became the first artificial "planets" of the sun.

1957 October 20–23   During a series of meetings held at Fort Bliss, Tex., Gen. John B. Medaris and Wernher von Braun discussed with the U.S. Army Scientific Advisory Panel ABMA (Army Ballistic Missile Agency) studies on a large, clustered booster with a first-stage thrust of 1.5 million lb. Eventually designated Juno V, the booster was presented as a heavyweight satellite and spacecraft launcher underpinning an ambitious national space program managed by the Army. The fact that the Army should see itself as the architect of space exploration was due in the main to Wernher von Braun's persistent efforts to apply rockets to space transportation but also to a perceived vacuum in U.S. plans for exploiting rocket propulsion for space travel. Since von Braun worked for the Army it was natural they would want to exploit his ideas. The meetings briefed the advisory panel on the substance of an ABMA plan formally released to higher authority on December 10, 1957.

1957 October 23 The first ballistic flight of the Vanguard satellite launcher took place from Cape Canaveral, Fla., with a live first stage carrying dummy upper stages. The primary objective of the TV-2 (Test Vehicle-2) mission was to test the first stage powered by the 28,000-lb. General Electric X-405 rocket motor that burned kerosene–liquid oxygen propellants for 146 sec. TV-2 reached a height of 109 mi. and impacted 330 mi. downrange. Although the upper stages were inert, the external shape of TV-2 was identical to a fully operational launcher. Vanguard had a height of 72 ft. and a diameter of 3 ft. 9 in. with a launch weight of 22,600 lb. On a satellite launch where all stages were live, the second stage would burn UDMH/IWFNA (unsymmetrical dimethyl hydrazine/inhibited white fuming nitric acid) propellants producing a thrust of 7,700 lb. for about 120 sec. and separate from the third stage and its satellite. After coasting for about 4 min. 40 sec. the solid propellant X-405 third stage would ignite to produce a thrust of 2,300 lb. for approximately 30 sec.

1957 November 3 The Soviet Union launched the world's second artificial Earth satellite carrying the dog Laika in a pressurized cabin. Called *Sputnik 2,* the total payload weighed 1,102 lb. but remained attached to the central sustainer stage of the SS-6 Sapwood (SL-1) launch vehicle, weighing about 16,500 lb. empty of propellant. Laika was wired to provide basic biological data which was telemetered to the ground, but there was no provision for recovery of the dog. Placed initially into a 132 × 1,031 mi. orbit inclined 65.3° to the equator, the combination spent rocket stage and payload reentered earth's atmosphere and burned up on April 14, 1958, some time after the dog had expired. Subtracted from the weight of the rocket stage to which it was attached, the great weight of *Sputnik 2*— compared with Vanguard and *Explorer I* satellites being prepared by the United States—aroused fears that the USSR had much bigger rockets. This was interpreted as a military threat because the SS-6 had been designed as an ICBM. In reality, the SS-6 was three times more powerful than Atlas and had a throw weight of 15,000 lb.

1957 November 7 An ad hoc civilian group under the chairmanship of H. Rowan Gaither presented President Eisenhower and the National Security Council with a grim report that did much to fuel major expansion of U.S. missile programs. Entitled *Deterrence and Survival in the Nuclear Age,* it was prepared in response to a request from the Federal Civilian Defense Commission for $40 billion for the construction of nuclear shelters. The Gaither Committee reported that the government did not appreciate the extent of the Soviet threat and that what mattered was not U.S. force but the ability to ride out a surprise attack and strike back. It warned that by the early 1960s the USSR would be capable of destroying U.S. retaliatory forces, and it urged that a second-strike force be made operational. Several members of the Gaither Committee later worked on John F. Kennedy's strategic policies which dramatically increased missile development and production in the erroneous belief that the USSR had a lead in ballistic missiles.

Launched on November 3, 1957, in Sputnik 2, *the dog Laika was the first living thing placed in orbit. Here, technicians deliver the back-up canister to an exhibition in Brussels. Note the spherical dome for Laika's head.*

Responding to a wave of post-Sputnik panic that seized many Americans with fear about Soviet supremacy in missiles and space technology, President Dwight D. Eisenhower broadcast a public speech in which he declared U.S. defenses sound. He announced that the position of Special Assistant to the President for Science and Technology had been established, that James R. Killian would fill it, and that the Science Advisory Committee had been moved from the Office of Defense Mobilization to the White House. From now on it would be known as the President's Science Advisory Committee, or PSAC. The President also announced that a "breakthrough" had been made in that an object had been returned to Earth intact, and he then displayed the Jupiter-C nose cone recovered from a ballistic shot on September 20, 1956. At a time when tests with reentry vehicles and anything to do with nuclear weapons was a closely guarded secret, it was difficult for the American people to judge whether or not the nose cone recovery was the "breakthrough" proclaimed.

At celebrations to mark the 50th anniversary of the October 1917 revolution, a large military parade in Red Square, Moscow, displayed publicly for the first time two important Soviet missiles that would be widely used throughout the coming decades. Developed from captured German V-2 rockets, the SS-1B Scud A surface-to-surface missile had a range of 50 to 90 mi. and a launch weight of 12,125 lb. It had a length of 36 ft. 11 in. and a diameter of 2 ft. 9 in. The mobile missile took about one hour to prepare and fill with UDMH/RFNA (unsymmetrical dimethyl hydrazine/red fuming nitric acid) propellants; once launched guidance was internal but a radio command secured engine cutoff. Capable of carrying conventional or nuclear warheads, Scud A served in large numbers with Soviet and client-state customers until the early 1970s. It was replaced by the 100–175-mi. range Scud B in the early 1960s, a missile weighing 13,900 lb. The Scud B was widely deployed in Warsaw Pact countries and in several Middle East states including Egypt, Iraq, Libya and Syria. Also displayed in Moscow for the first time, the solid propellant Frog (free-rocket-over-ground) series included an evolving line of tactical battlefield missiles still in wide use in many countries today. Frog-1 had a length of 32 ft. 9 in., a diameter of 2 ft. 10 in., a weight of 6,600 lb. and a range of 230 mi. Later Frog versions had a range of up to 37 mi. and all could lob conventional or nuclear warheads.

1957 November 12 A top-secret RAND report entitled *A Family of Recoverable Satellites* was issued to selected recipients. Compiled by a team led by Merton E. Davies and Amrom H. Katz, it described how a Thor rocket would launch a modified Aerobee second stage carrying a recoverable film capsule. After taking pictures of selected targets, the film canister would be brought out of orbit and caught in a special net carried by aircraft or recovered from the sea. This idea resulted in Discoverer (code name Corona), the first of which was launched on February 28, 1959, by Thor-Agena. Agena was a cylindrical structure to which a Bell Hustler rocket motor was attached at one end and the payload at the other. It would eventually be developed into one of the most versatile workhorses of U.S. civil and military space programs. The 15,000-lb. thrust Bell LR81-BA-1 Hustler had been designed for the MA-1C air-to-surface missile intended for the Convair B-58 bomber.

1957 November 15 The U.S. Air Force formally approved the development plan for the System 464L Dyna-Soar boost-glide program announced October 10. The project was to receive $3 million in fiscal year 1958, and studies were to be conducted on both manned and unmanned versions. The Air Force was not yet convinced that the Dyna-Soar needed a crew. On December 21, 1957, the U.S. Air Force issued System Development Directive 464L for the development cycle of the initial boost-glider, the first formal approval given to a manned space project.

The U.S. Information Agency's Research and Reference Service began a confidential survey, concluded on December 1, 1957, resulting in two reports. The first, *The Impact of Sputnik on the Standing of the United States versus the USSR,* was released during December; the second, *Post-Sputnik Attitudes toward NATO and Western Defense,* appeared in February 1958. Throughout Britain and mainland Europe, an average of 97% of people polled had heard of the two satellites. French popular opinion favored development of a national nuclear deterrent. At the same time, while in Britain the popular vote for siding with the United States increased from 10% to 30%, mainland Europe remained lukewarm and in France there was a decided reversal, the majority favoring a political alliance with the USSR. The surveys affirmed that "Great Britain is the only country showing an increase in solidarity with the West."

1957 November 22 The main committee of the National Advisory Committee for Aeronautics voted to set up a Special Committee in Space Technology. Under the chairmanship of H. Guyford Stever, dean of the Massachusetts Institute of Technology, the committee comprised leaders in space technology and science, including Dr. James Van Allen, William Pickering and Wernher von Braun. This was the first committee established by the NACA to examine astronautics and define the NACA's role in the definition and exploitation of space technology and exploration.

1957 November 25 Chaired by Sen. Lyndon Johnson, the Preparedness Investigating Subcommittee of the U.S. Senate Committee on Armed Services started the Inquiry into Satellite and Missile Programs. It held hearings that lasted until January 23, 1958, and about 70 Department of Defense witnesses were heard in testimony that produced 2,300 pages of transcript. Sen. Johnson quickly asserted his interests and increasingly took the lead in Senate debate on how best to mobilize national resources for a major expansion in space activities. He would orchestrate the establishment of a Senate space committee and play a leading role in determining America's direction in civilian projects.

1957 November The U.S. Army proposed the Nike-Zeus solid propellant antiballistic-missile (ABM) missile as the basis for an antisatellite system. This was reiterated in January 1960 when it was linked with the ABM proposals. In March 1961 when the ABM system was under threat of cancellation, the Army requested permission to develop an antisatellite system with Nike-Zeus, a move thought by some to be a pretext for keeping it in the budget. Permission was granted in May 1962 and Bell Laboratories was asked to begin work on converting the missile to an antisatellite role. This work was conducted under a program originally code-named MUDFLAP and later redesignated Program 505. The first test launch took place on December 17, 1962.

1957 December 5 The Department of Defense announced it would be setting up an Advanced Research Projects Agency (ARPA) to handle military space programs and antimissile missile-research. It was intended to handle all space matters until legislation established a national space agency. In fact, it became a central department of defense research body for handling new and innovative technology applications matched to military roles and requirements. The directive formally establishing ARPA was issued by the secretary of defense on February 7, 1958. The general mood in support of an expanded space program was pressed by the U.S. Air Force Scientific Advisory Board's ad hoc committee on space technology. Completed December 9, 1957, its report recommended a vigorous effort at an early moon landing, asserting that "Sputnik and the Russian ICBM capability have created a national emergency."

1957 December 6 The first attempt to place a U.S. satellite in orbit failed 1 sec. after liftoff when the Vanguard launch vehicle had risen only 4 ft. from the launch pad. Designated TV-3, the mission failed when the Vanguard fell back in a ball of flame, throwing to one side the nose cone carrying a small test satellite that would have been placed in orbit had everything worked as planned. It was not a scheduled launch attempt because the second stage had never been test-flown, but a 6.4-in. sphere weighing 4 lb. was attached to the top of the third stage just in case all three stages worked and it would be claimed that a satellite had been launched. A second attempt on February 5, 1958, also ended in failure when a control system malfunctioned, causing the Vanguard TV-3BU to break up at a height of 1,500 ft., 57 sec. after launch.

1957 December 10 The Development Operations Division of the Army Ballistic Missile Agency released "A National Integrated Missile and Space Vehicle Development Program." It covered the clustered booster concept of a 1.5-million-lb. thrust launcher powered by four 380,000-lb. thrust Rocketdyne E-1 engines and called Juno V, asserting that it was "the key to space exploration and warfare." Juno V would eventually adopt eight existing rocket motors developed from the S-3 used in the Jupiter missile. Various near-term

The Vanguard satellite launcher stands ready for flight at Cape Canaveral, Florida. Beaten to space by the Soviet Union's Sputnik and eclipsed by the U.S. Army's Jupiter-C launcher, Vanguard successfully carried three satellites into space.

launchers were proposed for a quick start on space activity, including Juno I and Juno II. Proposed Juno III and Juno IV configurations were developments of the Jupiter IRBM but never got off the drawing board. Seeking to capitalize on a perceived willingness to give the country an expanded role in space, the Army dusted off ambitious rocket projects from Wernher von Braun's team at Huntsville, Ala., and from Gen. John B. Medaris, commander of the Army Ballistic Missile Agency, which for several years had been seen as too futuristic and lacking a stated mission.

1957 December 20 The Redstone Jupiter-C rocket for the U.S. Army's first satellite attempt arrived by Douglas C-124 cargo plane from Huntsville, Ala., and was kept under guard at Patrick AFB, Fla., overnight. On January 13 the rocket was moved by truck to Cape Canaveral's Hangar D where it was prepared for launch. On January 24 under cover of darkness and without searchlights the rocket was assembled in a vertical position on Pad 26. A gantry made from sections of an old oil derrick was wheeled into position around the vehicle, and the solid propellant, spin-stabilized cluster of Jet Propulsion Lab upper stages was attached together with the *Explorer I* satellite. Canvas shrouds draped over the structure hid everything except the tail of the rocket.

1957 Very soon after the successful launch of *Sputnik 1*, Soviet Premier Nikita Khrushchev reacted to world-wide acclaim for the achievement by ordering Sergei Pavlovich Ko-

rolev to develop a recoverable capsule capable of carrying cameras and film for reconnaissance and surveillance from space. Known as Project Zenith, the design evolved into the Vostok configuration, which was used to carry the first human into space on April 12, 1961. The design owed its origin to newly won support from a hitherto uninterested Khrushchev who had merely tolerated Korolev's ambitions for launching satellites until he recognized the great technological and ideological tool his engineers had developed for the Soviet Union.

1958 January 2 A series of one-fifth scale tests of the Polaris underwater launch concept began in the Lockheed underwater missile facility at Sunnyvale, Calif. These tests simulated realistic sea conditions for over 3,000 dummy launches to provide data on hydrodynamic performance. A full-size dummy test launch program began in March 1958. In these tests, redwood logs cut to size were launched into San Francisco Bay from a test facility at the Naval Shipyard. The underwater launch concept required the vertical launch tube to be pressurized to a level above that of the surrounding water with a diaphragm over the top through which the missile could pass without admitting water. Also this month, the first 6,700-ton Polaris submarine USS *George Washington* was created from a keel which had been laid down as USS *Scorpion*. A 130-ft. midships section comprised 16 launch tubes in two parallel rows of eight, an arrangement known as "Sherwood Forest," after the legendary home of the mythical English outlaw Robin Hood.

1958 January 6 After almost 14 months of meetings and deliberation, the report *International Security: The Military Aspect* commissioned by the Rockefeller Brothers' Fund reported its findings. Under the direction of Henry Kissinger, it asserted that the United States must build a "powerful, instantly ready, retaliatory force" to "make our response fit the challenge." Coupled with the Gaither report of November 7, 1957, the Rockefeller findings were a powerful tool for lobbyists on Capitol Hill to wag accusing fingers at what many congressmen asserted was a lame-duck Republican administration presiding over a country "rapidly losing its lead in the race of military technology." Kissinger's report urged increased funding for Polaris, more IRBMs and ICBMs and bases made less vulnerable to attack. In 1961, the incoming Kennedy administration used the Gaither-Rockefeller reports to build a missile expansion program.

1958 January 15 The 864th Strategic Missile Squadron (SMS) of the U.S. Air Force Strategic Air Command was activated at the ABMA's (Army Ballistic Missile Agency) Army Ordnance Guided Missile School, Huntsville, Ala., with the Jupiter Intermediate Range Ballistic Missile (IRBM) developed by the Army. It was charged with training Air Force units that would operate the missile. Overseas deployments began with a squadron of 15 missiles operated by personnel of the Italian Air Force. The first Italian-launched Jupiter was fired from Cape Canaveral on April 22, 1961. A total of 30 missiles was eventually deployed by the 864th SMS in southeastern Italy. Later, a single squadron of 15 missiles was operated by the 865th SMS for the Turkish Air Force from a base in Turkey. Operationally, Jupiter carried a 1.44 MT W-49 warhead. The missile was withdrawn following a deal with the Russians after the Cuban missile crisis of 1962.

1958 January 21 Acting upon a decision to rescind the range limitation on U.S. Army missiles stipulating that they neither develop nor operate missiles with a range of more than 200

Launched from Cape Canaveral, Florida, a Jupiter in the Juno II satellite launcher configuration pitches over prior to slamming back into the ground. Failures came thick and fast to all major rocket programs during the late 1950s and early 1960s.

mi., the army held discussions with the Atomic Energy Commission (AEC) on a suitable warhead for the Pershing battlefield missile. Now that development was cleared to go ahead, the Army could reestablish its interest in acquiring a short-to-medium range capability. For a while the AEC chose to combine Army requirements with those of the Air Force for the Minuteman ICBM, but on December 15, 1958, the definitive concept of a selectable-yield weapon was agreed by all parties. Production of the warhead, designated W-50, began in March 1963, and approximately 280 had been manufactured when production ended in December 1965.

1958 January 22 National Security Council Action Memorandum 1846 gave satellite reconnaissance systems the highest priority. Allen Dulles at the CIA and James Killian had successfully persuaded President Dwight D. Eisenhower that intelligence-gathering satellites would be a vital part of national security and that the military should not be given exclusive control over their development and operation. Accordingly, because the Air Force was not enthusiastic about some of the concepts proposed thus far, the CIA took over the film-recovery vehicle being developed by Lockheed. It was given the top-secret code name Corona, but publicly it would be known as the Discoverer program for "technology research." The Air Force would be required to operate the system, as well as its own program, SAMOS (satellite and missile observation system).

1958 January 25 The U.S. Air Force completed examination of 111 industrial bidders for the contract to build the Dyna-

Soar boost-glide vehicle. Ten companies had been selected to receive formal requests for proposals, including Boeing, Chance Vought, Convair, Douglas, General Electric, Lockheed, Martin Marietta, North American and Western Electric. Three more companies (McDonnell, Northrop and Republic) applied to be included and were accepted, but several contractors teamed together, reducing the number of industrial contenders to nine. Proposals for design and manufacture were presented to the Air Force in March 1958. Three teams opted for a "satelloid" approach whereby Dyna-Soar would be boosted to 17,400 MPH as a temporary earth-orbiting satellite. Six teams chose the lift-body approach after the Sänger antipodal bomber utilizing higher lift-drag in a winged glider. Various boosters were studied, including Minuteman, Atlas and Titan.

1958 January 29 At a conference held at Wright-Patterson AFB, Ohio, U.S. Air Force and NACA officials presented their respective proposals for manned orbital vehicles. Air Force proposals included early design concepts from Northrop, Martin, McDonnell, Lockheed , Convair, Aeroneutronics, Republic, Avco, Bell, Goodyear and North American. Only McDonnell's proposal closely matched an NACA concept involving a ballistic, high-drag capsule featuring a heat shield, a supine position for the pilot and recovery effected by parachute. Maxime A. Faget of the Langley Aeronautical Laboratory had presented one of the two NACA proposals, the other being a Langley proposal for a flat-bottomed lifting body. The Faget/McDonnell concept marks the origin of the Mercury spacecraft design.

1958 February 1 *Explorer I,* the first U.S. artificial earth satellite, had been launched from Pad 6 at Cape Canaveral, Fla., by a Juno I rocket (serial number RS 29) at 03:48 UT on February 1 (22:48 EST, January 31). Adapted from the Jupiter-C reentry test vehicle, Juno I was identical to its military counterpart except for the addition of a fourth stage atop the third-stage cluster of three Baby Sergeant solid propellant rockets. The combined fourth stage and instrument section was a cylindrical tube 6 in. in diameter, 6 ft. 8 in. in length; the bottom 3 ft. 10 in. comprised the solid propellant fourth stage and the upper 2 ft. 10 in. formed the payload section painted with white stripes on black for thermal balance. The modified Redstone first-stage cutoff came at 2 min. 36.7 sec. followed by separation of the upper stages at 2 min. 42.3 sec. Spin-stabilized at 750 RPM, the cluster coasted upward for 4 min. 1.45 sec., until the second stage ignited at an elapsed time of 6 min. 43.75 sec. Firing for about 5 sec. each, with a few seconds between for separation, stages 2, 3 and 4 propelled the fourth stage and payload into a 225 × 1,594 mi. orbit inclined 33.24°, burnout occurring at about 7 min. 8 sec. After depletion of propellant the empty fourth stage weighed 12.67 lb. and the instrument section, which constituted the weight of the payload, weighed 18.13 lb., although the satellite was said to weigh the 30.8 lb. of the total orbited weight. Designed by the Army Ballistic Missile Agency's Josef Boehm, one of the German V-2 engineers, *Explorer I* carried a geiger counter, provided by Dr. James A. Van Allen and transmitting data on 108 MHz at 10 mW and two micrometeoroid detectors, provided by Dr. E. Manring of the U.S. Air Force Cambridge Research Center and transmitting on 108.3 MHz at 10 mW. All three experiments weighed a total 10.83 lb., and the geiger counter discovered regions of trapped radiation known thereafter as Van Allen belts. Mercury batteries provided power, the last data being transmitted May 23. *Explorer I* reentered the atmosphere and burned up on March 31, 1970. Following launch on March 5, 1958, the second Juno I (RS 26) failed to place the *Explorer II* satellite into orbit when the fourth stage did not ignite, and fell to earth 2,000 mi. downrange.

1958 February 6   The Senate created the Special Committee on Space and Astronautics, chaired by Sen. Lyndon B. Johnson. It would be Johnson's job to steer legislation on a national space agency and to mastermind many of the bureaucratic channels through which the decision-making process would pass. When the House Select Committee on Astronautics and Space Exploration was set up on March 5, with majority leader John W. McCormack serving as its first chairman, the pace was set when George Feldam, chief counsel and director said that he would "like to get the jump on the Senate Committee, so our Committee might in the public mind be the leader and not the follower." The House Committee on Science and Astronautics was established on July 21, 1958, followed by the Senate Committee on Aeronautical and Space Sciences three days later.

1958 February 8   The council of the National Academy of Sciences (NAS) passed resolutions expressing deep concern at the possible biological contamination of other worlds in the solar system by spacecraft bearing microorganisms from earth. As a result, an ad hoc committee, Contamination by Extraterrestrial Exploration (CETEX), was established; it was later renamed Committee on Extraterrestrial Exploration. A code of conduct to prevent contamination was recommended on May 12–13, 1958, and on July 6–8, 1959, the biology panel of the Space Science Board of the NAS met to consider ways of preventing contamination through spacecraft sterilization, a method that was adopted by NASA.

1958 February 20   The 705th Strategic Missile Wing, U.S. Air Force Strategic Air Command, was activated at RAF Lakenheath, England, in support of Thor Intermediate Range Ballistic Missile (IRBM) operated by the Royal Air Force. Thor training for RAF crews began in August 1958 and was provided by the 392nd Missile Training Squadron activated on September 15, 1957. Under an agreement between President Dwight D. Eisenhower and Britain's Prime Minister M. Harold Macmillan signed February 28, 1957, Britain would take over ownership of missiles located on its territory. Deployment of Thor consisted of 60 missiles at four bases located in eastern England and operated by Royal Air Force crews trained by the Air Force. Each base had 15 missiles with 3 missiles per squadron.

1958 February 26   A meeting was held at the U.S. Army Ballistic Missile Agency (ABMA) for discussions with the Jet Propulsion Laboratory (JPL) about development of Juno launch vehicles. In addition to Juno I (Jupiter-C) and Juno II (Jupiter) launchers, a Juno III was conceived to employ a cluster of solid propellant upper stages, called Red Socks, atop a Juno II Jupiter Intermediate Range Ballistic Missile (IRBM). Larger than those used on Juno I and II, the new upper stages would be developed by the JPL under a plan to upgrade the Army launch capability. Because spin-stabilized upper stages would not provide the accurate guidance and trajectory control demanded by the Army's list of planned lunar missions, Juno III was abandoned by the Advanced Research Projects Agency on March 27 in favor of the Air Force Thor-Able launcher. Thor-Able had three-axis stabilization and pointing control.

1958 February   The U.S. Advanced Research Projects Agency (ARPA) took over control of a military weather satellite, equipped with television cameras, proposed by RCA. Developed in the Army Ballistic Missile Agency (ABMA) and originally scheduled for launch by a Juno I in the spring of 1958, it was called Janus and doubled as a reconnaissance satellite. Under ARPA, the meteorological aspects of the RCA satellite were emphasized to prevent duplication with existing U.S. Air Force programs, and in March 1958 it was reassigned to a Juno II flight and its name changed to Tiros, an acronym for Television Infra-Red Observation Satellite. By early 1959 it had moved launch vehicles twice more, first to the Juno IV and, when that was cancelled, to Thor-Able. On April 13, 1959, Tiros was transferred to NASA's Goddard Space Flight Center and the first satellite launched on April 1, 1960.

1958 March 4   C. Bernard Smith, an engineer from the U.S. aircraft engine company Pratt and Whitney, sent a proposal to the Air Research and Development Command (ARDC) for a 15,000-lb. thrust rocket motor operating on cryogenic, hydrogen-oxygen, high-energy propellants. Smith knew that Lockheed was studying advanced versions of the WS-117L Agena stage and that this motor would more than double its performance. The ARDC was aware that Convair consultant engineer Krafft A. Ehricke had a design proposal for using cryogenic engines in an upper stage to the Atlas launch vehicle and decided the time had come to improve the ICBM's satellite launching capability. On November 14, 1958, Pratt and Whitney received a contract to design and develop what was later designated the RL-10, two of which would power the Centaur upper stage.

1958 March 5   President Dwight D. Eisenhower approved a recommendation from his Advisory Committee on Government Organization that "the leadership of the civil space effort be lodged in a strengthened and redesignated National Advisory Committee for Aeronautics." Key recommendations in getting the NACA as lead agency for restructuring into a spacefaring body came from the President's science adviser James Killian, Budget Bureau chief Percival Brundage and NACA chairman James Doolittle. Eisenhower wanted a clear demarcation between civil and military space programs, clearing the way for open publicity without fetters from the Pentagon and leaving the military to quietly get on with national security interests. But he also wanted to keep the scientists on his side and knew they would smart under Pentagon authority. In this way he hoped to eliminate any contest between the two factions while laying the foundations for broad debate on specific programs and priorities.

1958 March 6   The National Security Council approved a nuclear explosion in space to test a theory put forward by Dr. Constantine Christofilos of the University of California Radiation Laboratory that electrons produced from a nuclear explosion would be trapped by the earth's magnetic field. If that proved to be so, the guidance systems of enemy missiles could be disabled in flight. Called Operation Argus, three undisclosed launches with solid propellant missiles took place during late August and early September from the USS *Norton Sound,* 1,100 mi. southwest of Cape Town, South Africa. Detonations were in the 1–2 KT range at altitudes of 125–300 mi. The *Explorer IV* satellite measured the results, which proved Christofilos correct.

1958 March 10   A three-day working conference on the "Man in Space Soonest" (MISS) concept was held at the Air Force

The 21.5-pound Vanguard I *satellite sits atop the third stage of the Vanguard launcher, its four antennas extended, at Cape Canaveral, Florida.*

Ballistic Missile Division (AFBMD) in California. Gen. Bernard Schriever announced that Advanced Research Projects Agency head Roy Johnson had requested the AFBMD to report on a definitive concept for getting a U.S. airman into space as quickly as possible. Of three programs favored by the Air Force, one was the blunt body capsule concept developed at NACA's Langley Laboratory, another involved the development of some form of lifting body and a third laid down a space objective culminating in a space station or a flight to the moon.

1958 March 17   The world's fourth, and America's second, artificial earth satellite, the 3.2-lb., 6.4-in.-diameter *Vanguard I,* was successfully placed in a 406 × 2,465 mi., 34.2° orbit by the three-stage Vanguard launch vehicle. Mission TV-4 finally achieved the program objective 32 months after it had been announced by President Dwight D. Eisenhower and six weeks after the the U.S. Army/JPL Jupiter C had put the first U.S. satellite, *Explorer I,* in orbit. Before the Vanguard program ended, two more satellites *(Vanguard II and III)* were successfully orbited by this type of launcher. On six further attempts (April 28, May 27, June 26, September 26, 1958, and April 13 and June 22, 1959), Vanguard's second stage failed, resulting in the loss of the satellite. The last Vanguard launch successfully put the *Vanguard III* satellite in orbit on September 18, 1959. Of 11 Vanguard launch attempts since December 6, 1957, only three had been a success, all of which were still in orbit during January 1994.

1958 March 18   A NACA report entitled *Preliminary Studies of Manned Satellites, Wingless Configuration, Non-Lifting* was released by Maxime A. Faget, Benjamin Garland and James J. Bulgia. This eventually became the blueprint for the specification on the Mercury development program and would be issued as NASA Technical Note D-1254 in March 1962. In a related event, at a conference on high-speed aerodynamics held at the Ames Aeronautical Laboratory, Calif., March 18–20, more than 500 representatives from the armed services, NACA, government and industry heard 46 papers read by scientists and engineers from every NACA facility describe current thinking on space flight. Some dealt directly with the prospects for manned flight, drawing upon work carried out on reentry vehicles as well as ballistic capsules.

1958 March 20   At a meeting with staff from the Jet Propulsion Laboratory (JPL), Calif., U.S. Army Ordnance Missile Command chief Gen. John B. Medaris announced the formal start of a three-stage launch vehicle based on the Jupiter Intermediate Range Ballistic Missile (IRBM) and called Juno IV. Conceived at the Army Ballistic Missile Agency (ABMA) Development Operations Division, JPL completed preliminary design of the Juno IV on April 4, 1958, proposing a basic 150,000-lb. thrust Jupiter IRBM as the first stage, a 45,000-lb. thrust second stage and a 6,000-lb. thrust third stage. Both upper stages would use storable nitrogen tetroxide hydrazine propellants. JPL Director William H. Pickering began studies of a 350-lb. Mars spacecraft which eventually resulted in the Vega-Ranger spacecraft. During September 1958 a JPL-ABMA review deferred introduction of the second stage and adopted a six-vehicle plan whereby the first three would use the first and third stages only. Said to be no better than existing launchers, Juno IV was cancelled on October 9, 1958.

1958 March 26   President Eisenhower endorsed a report from the Purcell Panel that firmly set the basic guidelines for future United States military space activities. Headed by Edward

Purcell, the panel had been set up early in 1958 by the President's Science Advisory Committee to frame the outline of a U.S. space program. It addressed scientific and military issues and largely endorsed the President's view that space should not became another battleground and that military activities outside earth's atmosphere should be limited to intelligence-gathering and support systems for earth-based activities. Although policies would change with each president, the basic tenet of American planning, that no weapons of mass destruction should be placed in orbit, became a fundamental constraint upon military options.

The third Juno I launch vehicle (RS 24) successfully placed the 31-lb. *Explorer III,* America's third satellite, in a 117 × 1,740 mi., 33.5° orbit following launch at 12:45 local time from Pad 6 at Cape Canaveral, Fla. *Explorer III* carried experiments similar to those installed on *Explorer I,* with the added advantage of a tiny magnetic tape recorder capable of storing data acquired during one complete orbit and dumping it when the satellite passed over a suitable receiving station. The high eccentricity of the orbit resulted in a shorter-than-planned life for *Explorer III,* which reentered earth's atmosphere and burned up on June 28, 1958, 12 days after it stopped transmitting data.

1958 March 27   Secretary of Defense Neil McElroy approved America's first moon program when he endorsed plans for the U.S. Air Force Ballistic Missile Division (AFBMD) to launch three probes on Thor-Able launchers and for the Army Ballistic Missile Agency (ABMA) to launch two probes on Juno II launchers in Operation Mona. Under way since mid-1957, planning for Operation Mona called for the Space Technology Laboratories (STL) to design and build the AFBMD probes and the Jet Propulsion Laboratory (JPL) to do the ABMA probes, all five of which were to be called Pioneer. An earlier Army/JPL plan to adopt a Juno III "Red Socks" configuration incorporating more advanced solid propellant upper stages on a Jupiter Intermediate Range Ballistic Missile (IRBM) was dropped as too ambitious to get early results.

1958 March 31   In efforts to consolidate and streamline its rocket and missile programs, the U.S. Army Ordnance Missile Command (AOMC) was formed under the command of Gen. John B. Medaris. It would have its headquarters at Redstone Arsenal and was given command of the Ballistic Missile Agency, the White Sands Missile Range, Redstone Arsenal (later named the Army Ordnance Missile Support Agency) and the Army Rocket and Guided Missile Agency, formed a day later. On April 23, the AOMC assumed control of Army military communication satellites and, on August 15, 1958, was given its most ambitious program when the Advanced Research Projects Agency (ARPA) authorized a start on the Juno V clustered launch vehicle. On October 21, 1959, President Eisenhower announced that the AOMC would lose von Braun's Development Operations Division, Army Ballistic Missile Agency (ABMA) to NASA, and on March 14, 1960, both it and the Juno V, then named Saturn, joined the civilian space agency.

1958 April 1   The U.S. Air Force awarded a $160,000 contract to the University of Chicago for a new photographic lunar atlas to be compiled by Yerkes Observatory with Gerard P. Kuiper as principal investigator. The side of the moon visible from earth was to be divided into 44 sectors, each represented by four or more pictures taken from U.S. observatories and from the Pic du Midi Observatory in France. The original contract ran until March 31, 1959, but subsequent increases extended this to April 30, 1960. On November 1, 1958, the Air Force contracted with the University of Manchester, England, where Z. Kopal was principal investigator, for topographic lunar maps for the Air Force Aeronautical Chart and Information Center. The Air Force said it wanted the maps for "intelligence purposes." The contract expired on April 30, 1960. During 1958, the U.S. Army Map Service began its Lunar Analysis and Mapping Program. Completed in spring 1960, it formulated a plan for lunar orbit missions to complete the job, but these aspirations were only realized when NASA flew its Lunar Orbiter mission in 1966–67.

Wernher von Braun sent the Working Group on Vehicular Program, of the NACA Special Committee on Space Research, a second, revised edition of *A National Integrated Missile and Space Development Program,* which had been in preparation since April 1957 and first presented to the U.S. Army chiefs on December 10, 1957. The report outlined a U.S. national space program based on 11 existing and proposed launchers, the smallest of which was the Vanguard and the largest of which was a recoverable 3-million-lb. booster comprising two F-1 engines and a delta wing. The report also identified 14 separate engines to support the launcher program, including high-energy upper-stage motors and advanced propulsion systems for the 1970s. It marks the transition of attention from high U.S. Army authority to the civilian NACA, von Braun having quickly recognized that it was within this latter organization that the seeds of a future U.S. space agency had been sown. A further revised report was submitted to the NACA on July 18, 1958.

1958 April 2 Draft legislation establishing a National Aeronautics and Space Agency was sent to Congress. The bill was attached to a National Aeronautics and Space Act drawing clear guidelines for the new agency, and hearings began in the House of Representatives on April 15. President Eisenhower directed both the NACA and the Department of Defense to review "programs currently under way within or planned by the Department [of Defense and to recommend] . . . which of these programs should be placed under the new Agency." Discussions between NACA Director Hugh L. Dryden and Deputy Secretary of Defense Donald A. Quarles set in motion a realignment of military facilities that would be brought under the control of the new space agency. When passed by Congress, the proposed space agency would become the National Aeronautics and Space Administration.

1958 April 17 In testimony before the U.S. House Select Committee on Astronautics and Space Exploration, Maj. Gen. John B. Medaris justified development of large rocket engines by claiming "the Army must make long-range plans for the transport of small combat teams by rocket. I also believe that cargo transport by rocket is economically feasible. The major breakthrough which must be effected before any national space program can gain substantial impetus is the development of engines of much greater thrust. I am convinced that we must have a well-tested capability of a minimum of 1 million pounds' thrust, either by a combination of multiplex engines or by a single engine."

1958 April 23 The first Thor-Able nose cone test launcher was fired from Launch Complex 17A at Cape Canaveral, Fla. A derivative of the U.S. Air Force Thor Intermediate Range Ballistic Missile (IRBM) the new vehicle was designated type LV-12 and had been designed to test ablative and heat-sink reentry vehicles for the Atlas ICBM. It would double as a

satellite and space probe launcher. Called Able, the LV-12's second stage comprised the 7,500-lb. thrust Aerojet General AJ-10-41 liquid propellant motor, originally developed as the second stage for the Vanguard satellite launcher, and gave Thor-Able a height of about 76 ft. The payload for this launch, first of the Thor-Able 0 series, was a mushroom-shaped, General Electric heat-sink reentry vehicle also carrying a mouse, the first living animal carried on a U.S. rocket flight since two monkeys and two mice on May 21, 1952. Intended to fly 6,300 mi. down the Atlantic Missile Range, the vehicle exploded at 2 min. 26 sec. when the Thor liquid oxygen turbopump failed. Two successful Thor-Able 0 shots were, however, carried out on July 9 and 23, 1958, when test cones were lobbed 5,500 mi. and 6,000 mi., respectively. These two flights carried a live mouse for biological tests. Neither was recovered.

1958 April 25 The U.S. Air Force Ballistic Missile Division completed its first development plan for a long range space program. Based on studies conducted over the past year it put a case for concurrent development of four broad programs: Man in Space Soonest (MISS) would put an astronaut in orbit as soon as possible; Man in Space Sophisticated (MISS) would develop earth orbital operations for continuously manned stations; Lunar Reconnaissance (LR) envisaged the exploration of the moon's surface by television and through the use of unmanned spacecraft; and Manned Lunar Landing and Return (MLLR) would send the first human explorers. All this was to be accomplished by 1965 at a cost of $1.5 billion.

1958 April The British aircraft firm of Short Brothers and Harland, based in Northern Ireland, received a contract from the government for development of the world's first shipboard missile air defense system designed to replace the rapid-fire gun. Called Seacat, the missile was tested and evaluated between 1960 and 1962 and subsequently purchased by 16 navies. The 143-lb. Seacat had a range of 4 miles.

1958 May 6 In response to a request from NACA headquarters to prepare a space technology program for the new U.S. civilian space agency, Langley Aeronautical Laboratory requested, and received, funds to develop a four-stage solid propellant launch vehicle capable of placing a 150-lb. payload in a 500-mi. orbit. The Pilotless Aircraft Research Division (PARD) at Langley had wanted to launch five small-scale recoverable orbiters for investigating problems associated with manned flight and saw this as their chance to develop the launcher. A very large solid propellant motor, 30 ft. in length, 3 ft. 6 in. in diameter, weighing 22,650 lb. and producing a thrust of 100,000 lb., had been developed by Aerojet for the U.S. Army-Navy Jupiter Senior program, and this was proposed as the first stage, with a modified Sergeant as second stage and two X-248 stages produced either by the Allegheny Ballistics Laboratory (ABL) or the Hercules Powder Co.

1958 May 13 The first U.S. Lunar and Planetary Colloquium was held in Downey, Calif. Called to debate an increasing range of serious proposals for moon bases and unmanned planetary flights, it was sponsored by the RAND Corp., North American Aviation and the California Research Corp. The colloquium served as a platform for scientific papers on selenology and selenography and as a forum for scientific debate. The second colloquium was held at RAND's facilities in Santa Monica, Calif., on July 15, and at the third meeting, held at the Jet Propulsion Laboratory (JPL) Calif., on October 29, 1958, Dr. Harold Urey of University of

California at La Jolla presented a paper entitled "The Chemistry of the the Moon." Dr. Urey became a staunch supporter of lunar studies and would eventually help build the U.S. lunar science program.

1958 May 15 In a study of future space science missions conducted by NACA's Langley Aeronautical Laboratory, telescopes placed in orbit above earth's dense atmosphere received high priority. Astronomical observatories could not only obtain clearer images of the universe free from atmospheric aberrations, but portions of the spectrum invisible to astronomers on earth would also broaden the window of research. Langley suggested the new civilian space agency prepare a vigorous program involving stabilized telescopes in space capable of receiving ultraviolet light, x rays and gamma rays. In October, NASA set up a working group under Nancy Roman to explore the feasibility of launching large observatories, and in April 1959 the long-range space science plan included stable orbiting telescopes to observe in visible, infrared, ultraviolet and x-ray regions of the spectrum. Langley also proposed an orbiting platform for earth studies, later to mature as the Orbiting Geophysical Observatory (OGO) series.

Launched by an SL-1 from Baikonur, *Sputnik 3* was placed in a 135 × 1,158 mi. orbit inclined 65.18° The 2,926-lb. satellite took the shape of an extended cone with hemispherical end plate, 11 ft. 8 in. tall and with a base diameter of 5 ft. 8 in. Packed with scientific equipment for monitoring the earth's ionosphere, newly discovered radiation belts and cosmic rays, it transmitted to earth on 20.005 MHz via external antennae and was the first satellite to carry solar arrays and silver-zinc batteries. This was the fourth and last satellite launch with the basic SL-1; future flights with launchers of this type carried an upper stage. The *Sputnik 3* launch followed the loss on February 3, 1958, of an SL-1 and its *Sputnik 3–1* payload, which had a weight of 2,756 lb. *Sputnik 3* continued to orbit the earth until April 6, 1960, when it burned up in the atmosphere.

1958 May 16 The U.S. Army 217th Field Artillery Battalion conducted the first troop launch of the Redstone missile from Cape Canaveral. The second troop launch took place at White Sands Proving Ground, N.Mex., from where all subsequent practice firing took place. The missile went operational with the 40th Field Artillery Missile Group in July 1958, deployed in Europe as a component of the Seventh Army. The 46th Field Artillery Missile Group joined it on April 25, 1959. It remained on operational duty until replaced by Pershing in the early 1960s. As the Army's first medium-range ballistic missile it was a success, although much of its genesis leaned heavily on V-2 experience and it was cumbersome and unwieldy as a military weapon.

1958 May 20 The NACA and the U.S. Air Force signed a Memorandum of Understanding concerning the Dyna-Soar boost-glide vehicle. Under its terms, the NACA would assist the Air Force with technical matters and provide test facilities for the program. The memo recognized the Air Force as having primary responsibility for the program except for the test and evaluation phase, where the two organizations would share decisions and responsibility.

1958 May A communication satellite committee was set up in the Advanced Research Projects Agency of the Department of Defense. By the end of the year a preliminary plan had evolved for development of a geostationary satellite system

capable of linking U.S. armed forces around the world. Satellites placed at 20° W and 180° W would connect U.S. armed forces between the continental United States and the Far East. In May 1959, three separate projects had emerged: Steer (a Strategic Air Command polar system), Tackle (an advanced polar system), and Decree (the stationary global system). In February 1960 these separate elements were grouped into Project Advent.

1958 June 16 From a group of nine industrial contenders offering proposals to construct the Air Force–NACA Dyna-Soar boost-glide vehicle, the Air Force selected the Martin-Bell and Boeing-Vought teams for a two-contractor competition, at the end of which a single team would be selected to build the vehicle. Proposals from the other companies included earth orbiting reentry glider and suborbital glider, with range a function of the power of the booster. The Martin and Boeing proposals envisaged a suborbital glider but the Air Force wanted to leave the orbital capability as an option. In the competition that was to last 18 months, each company would prepare detailed technical and cost proposals.

1958 June 23 The U.S. Air Force awarded a contract to North American Aviation, Rocketdyne Division, for preliminary design of the F-1 kerosene liquid oxygen engine potentially capable of producing up to 1.5 million lb. of thrust. Other contractors were asked to bid, but Rocketdyne had seized the clear lead through the amount of work already done in-house and was thus selected on December 17, 1958, for a definitive contract to develop the F-1. By this time, NASA had, in accordance with the National Aeronautics and Space Act, absorbed the Air Force project, and it was the civilian space agency that signed a contract with Rocketdyne on January 19, 1959. At this date NASA had no mission for the engine but it was clear that Army advanced launch vehicle studies carried out by von Braun's team at Huntsville, Ala., would benefit from the engine.

1958 June 25 The first of nine attempts by the Soviet Union to send scientific instruments to the vicinity of the moon with first-generation Luna space probes failed when the SL-3 launcher malfunctioned during the ascent. Culminating in a failed launch on April 18, 1960, these attempts were unique in utilizing a second stage never used for any other purpose, a rare occurrence in Soviet rocketry. The second stage comprised a kerosene–liquid oxygen Kosberg propulsion system with a thrust of 11,025 lb. through one nozzle and four verniers. Uprated to a vacuum thrust of 1.1 million lb., the main sustainer and four strap-ons were essentially the same as the SL-1 launcher, but the total launch mass with the second stage was 615,000 lb. SL-3 had an overall length of 109 ft. 11 in. Each Luna space probe weighed about 770 lb. and was encapsulated by a protective shroud atop the second stage, which was intended to propel it direct to escape velocity without first going into orbit. By doing this the challenging in-orbit engine restart was avoided. The only successful first-generation Luna space probe flights were launched on January 2 *(Luna 1),* September 12 *(Luna 2)* and October 4, 1959 *(Luna 3).*

1958 June 27 More than 12 years after Northrop received a contract to develop the SM-62 Snark and more than 7 years after its first successful flight, the 556th Strategic Missile Squadron, Strategic Air Command, became the first U.S. Air Force unit to fire the missile. It was launched from Patrick Air Force Base, Fla., and would be deployed operationally by the 702nd Strategic Missile Wing (SMW) at Presque Isle Air Force

Base, Maine, from May 1959. In November 1959 the Strategic Air Command recommended cancellation of Snark because it absorbed resources needed for other, more promising, strategic programs, but Air Force Headquarters refused, and on March 18, 1960, the first SM-62 was put on alert. Two months after his inauguration, on March 28, 1961, President Kennedy announced that the program would be abandoned and on June 25, 1961, the 702nd SMW was deactivated. Between December 21, 1950, and December 5, 1960, a total of 118 Snark N-25 and N-69 launch attempts were made.

1958 June 30   New tasks for the WS-117L were itemized in a launch schedule that envisaged the first flight in August 1960. As well as carrying photographic equipment, SAMOS (satellite and missile observation system) was also required to carry instrumented electronic intelligence gathering (elint) equipment as precursor test runs to dedicated elint satellites. Only the first three SAMOS satellites carried elint equipment and since only the second of these was successfully placed in orbit, only one operational launch was a success. Nevertheless, this did prepare the way for piggyback, or subsatellite, launches to take place from August 29, 1963.

1958 June   After three years of design studies, full-scale development of a rocket-powered torpedo was authorized in a U.S. Department of Defense contract to General Electric. Called Subroc, the device took a further five years to bring to operational status and was to have been deployed first with the submarine USS *Thresher,* but when that vessel was lost with all hands on April 10, 1963, it was deployed with the attack submarine USS *Permit.* Subroc had a launch weight of 4,085 lb., a length of 20 ft. 6 in. and a diameter of 1 ft. 9 in. With a 35-mi. range, it carried a solid propellant rocket which powered it to supersonic speed on leaving the water. It then dived down and submerged, its 1–5 KT W-55 warhead homing in on its target. Production was completed in 1978 and Subroc was withdrawn during the 1990s.

1958 July 4   The first test in a U.S. Navy program designed to put an air-launched satellite in orbit failed when two solid propellant motors fired from a ground facility at the Naval Ordnance Test Station, China Lake, Calif., exploded 1 sec. after liftoff. Called Project Pilot, the program envisaged a four-stage solid propellant rocket dropped from beneath the port wing of a Douglas F4D-1 Skyray. The rocket had a length of 14 ft. 4.5 in. and a span across the four tail fins of 5 ft. 5 in. The first two stages comprised pairs of Hotroc rockets, each pair with a combined thrust of 28,200 lb. for 4.8 sec. fired 12 sec. apart. The third stage, an ABL X-241, had a thrust of 2,720 lb. for 36 sec. The fourth stage produced a thrust of 1,155 lb. for 5.7 sec., while the fifth stage produced a thrust of 172 lb. for 1 sec. In a second ground test on August 8, 1958, the rocket blew up at the point of ignition.

1958 July 8   A two-day meeting began in Washington, D.C., between the NACA and the U.S. Air Force concerning a joint endeavor to fund and develop the four-stage solid propellant Scout launch vehicle. The Air Force wanted to test 100-lb. reentry cones from an altitude of 2,000 mi. and combined its requirements with those of the NACA to agree on a common specification. As revised, Scout would be designed around existing rocket stages to place a 100-lb. object in a 400-mi. polar orbit or 130 lb. in an equatorial orbit. The first stage would comprise an 86,000-lb. thrust Algol I derived from the Jupiter Junior, an alternative to the Jupiter Senior from Aerojet. The second stage would comprise a 55,000-lb. thrust

Castor I, an adapted version of Sergeant. The third stage would be the uprated ABL X-248, designated X-254 Antares, producing a thrust of 14,500 lb. The fourth stage would be the 3,060-lb. X-248 named Altair.

1958 July 18   Under the chairmanship of Wernher von Braun, the Working Group on Vehicular Program of the NACA Special Committee on Space Research sent NACA Headquarters a revised projection of rocket and launch vehicle development first issued on December 10, 1957 under the title *A National Integrated Missile and Space Vehicle Development Program.* In it, von Braun's group proposed a space program based on five succeeding generations of launcher, the smallest being Vanguard and the largest comprising a winged recoverable booster with a first-stage thrust of up to 6 million lb. The space plan envisaged a manned orbital flight in July 1960, unmanned lunar soft landing in February 1961, the first Juno V launch in September 1961, first winged aerospace plane flight in August 1962, a four-man space station in November 1962, first manned circumlunar flight and return in July 1964, a twenty-man space station in September 1964, a 1,000-ton manned moonship in July 1965, a fifty-man space station in September 1967, large scientific lunar base in 1972, a permanent moon base from 1973 to 1974 and the first manned Mars landing in 1977.

1958 July 25   The first attempt to place an air-launched satellite in orbit had indecisive results. As part of the Project Pilot operation conducted from the Naval Ordnance Test Station at China Lake, Calif., a five-stage solid propellant rocket was dropped from beneath the port wing of a U.S. Navy F4D-1 Skyray with the aircraft at an altitude of 41,000 ft. and a pitch-up angle of 58°. The pilot, Comdr. William W. West, observed ignition of the Hotroc first stage over the Santa Barbara Channel but lost sight of the rocket thereafter. Tracking stations reported weak signals but the poor quality of the telemetry made confirmation impossible. In a second orbital attempt on August 12, the Hotroc blew up at ignition, and another indecisive air launch took place 10 days later. This time weak signals were picked up at a tracking station in New Zealand from what may have been the satellite, which, if successful in orbit, would have comprised the spent fifth stage of the rocket weighing 1.25 lb. with a diameter of 8 in. Four more orbital attempts were made, the last on August 28, 1958; none was successful.

1958 July 26   Launched at 10:06 local time, the 37-lb. *Explorer IV* satellite was successfully placed in a 163 × 1,373 mi., 50.1° orbit by a Juno I launcher RS 44 from Cape Canaveral, Fla. A new fourth stage solid propellant motor was carried which permitted an increase in payload capacity, which for this satellite was devoted entirely to Dr. James A. Van Allen's research into artificially produced radiation from Operation Argus. The satellite continued to provide Argus data until transmissions ceased on October 6. *Explorer IV* reentered the atmosphere on October 23, 1958. This was the third and last satellite successfully launched by a Juno I rocket. Two more, unsuccessful, attempts were made, both in 1958: RS 47 with *Explorer V* on August 24, when the first stage shunted the upper stage cluster following separation, and RS 49 with *Beacon I* (an inflatable sphere for measuring atmospheric density, originally called *Explorer VI*) on October 23, when the payload broke away from the fourth stage during launch. Juno I had successfully orbited payloads on three out of six attempts, a success rate of 50% compared to Vanguard's 27.3%.

1958 July 29 The U.S. National Aeronautics and Space Act was signed into law. The Act (Public Law 85-568) declared that (a) it was the policy of the United States that "activities in space should be devoted to peaceful purposes for the benefit of all mankind", that (b) nonmilitary space activities should be the responsibility of a civilian agency and military activities should be managed by the Department of Defense, and that (c) U.S. aeronautical and space activities should: expand human knowledge; improve the performance of space vehicles; develop appropriate launch vehicles for unmanned and manned flight; set up long-range studies for future activities; preserve its role "as a leader in aeronautical and space science"; direct information gleaned from space to appropriate government agencies; cooperate with other nations in space; and utilize the scientific resources of the United States. On August 8, 1958, President Eisenhower appointed Dr. T. Keith Glennan, president of the Case Institute of Technology, as the first administrator of the National Aeronautics and Space Administration, with Dr. Hugh L. Dryden, director of the outgoing NACA, assistant administrator.

1958 July 30 In an experimental test of a specially designed contour couch built to an idea from the NACA's Maxime A. Faget, Carter C. Collins, a volunteer, demonstrated he could withstand a 20-g load on the centrifuge at Johnsville, Penn. This demonstrated that the human body could absorb the force of deceleration on reentry and that a contour couch where the occupant lay supine with his or her back to the axis of deceleration was the best method for protecting the body. Accordingly, this method was chosen for the NACA manned spacecraft but it would not be used with later manned space vehicles because the forces of acceleration and deceleration were not as severe. The highest forces of deceleration would be experienced during a ballistic flight on a Redstone missile where astronauts would experience almost 12 g on reentry. By comparison, Apollo reentry never exceeded 7 g and Shuttle flights reach only about 3 g.

1958 July 31 The first live nuclear device launched by the United States on a ballistic missile was detonated at an altitude of 252,000 ft. (47.7 mi.) with a yield of 2 MT in a test dubbed Teak, part of the Hardtack series of nuclear weapons tests. Launched on a Redstone missile fired from the southwest end of Johnston Island in the Pacific, the W-39 warhead produced a flash seen from Hawaii, 800 mi. away, and a luminous red sphere around the 18 mi. fireball that eventually grew to a diameter of 600 mi. and continued to glow for 5 min. All high-frequency short-wave communications across Australia was blacked out for 8 hr. and for 2 hr. over Hawaii. A second test on August 11, 1958, dubbed Orange, was of similar yield but did not cause radio blackout. Detonated at 141,000 ft. (26.7 mi.), the W-39 produced a fireball and a gray radioactive cloud seen from Hawaii. Most birds on a nearby island wild-life refuge were kept grounded by a complex water-sprinkling system, but a few intrepid terns were singed by the detonations.

1958 July The first public attempt to present estimates of relative missile strength between the United States and the USSR was published by leading newspaper columnist Joseph Alsop. His alarmist predictions were based on leaked information from mischief-makers in the CIA and the Air Force, leading Alsop to project a terrifying monopoly for the USSR. He contended that in 1959 the USSR would have 100 ICBMs, while the United States would have none; that in 1960 they would have 500, versus 30 U.S. missiles; in 1961 they would

have 1,000, versus 70 in the United States; in 1962 they would have 1,500, versus 130; and in 1963 they would have 2,000, while the United States would still have 130 ICBMs.

1958 August 1 Dr. Hugh L. Dryden, director of the NACA, spoke disparagingly about Wernher von Braun's scheme for launching a manned capsule by Redstone missile during Congressional hearings on the Space Act prior to the inauguration of NASA on October 1. Von Braun believed Redstone could be used to send a manned capsule to space during a ballistic flight from Cape Canaveral. Although this was in fact accomplished less than three years later, Dryden's statement that such a flight would have "about the same technical value as the circus stunt of shooting a lady from a cannon" denied him the opportunity to serve as head of NASA. Head of NACA for 10 years, Dryden was considered by some too conservative for the space age and this outburst seemed to endorse that view.

1958 August 2 The first successful Atlas B prototype ICBM incorporating all three liquid propellant rocket motors in the Rocketdyne MA-2 package was launched 2,500 mi. down the Atlantic Missile Range from Cape Canaveral, Fla. System qualification tests continued with the B series in a total of 10 flights including the first 5,000-NM design range demonstration on August 28, 1958. The last Atlas B flight took place on February 4, 1959; six of these flights had been successful, including the first full performance flight of 6,325 NM on September 28, 1958. Reentry vehicle tests were the responsibility of the Atlas C flights, the first of six being launched on December 23, 1958. Only three were a total success, one of which was the first recovery of an Atlas nose cone at a range of 5,500 NM on July 21, 1959. The last Atlas C was launched on August 24, 1959.

1958 August 7 On the day that Presidential Science Adviser James R. Killian telephoned T. Keith Glennan and offered him the job of running NASA, Abe Silverstein, associate director of the NACA's Lewis Propulsion Laboratory (LPL), chaired the first meeting of an ad hoc committee he had formed to study government plans for launch vehicles and propulsion systems. Adelbert O. Tischler from LPL was tasked with gathering all available information on hydrogen-oxygen and hydrogen-fluorine propulsion as potential applications for high-energy upper stages. At the second meeting one week later, U.S. Air Force work on a Bell Aircraft hydrazine-fluorine engine of 12,000-lb. thrust and a theoretical Aerojet General hydrogen-oxygen engine of 100,000-lb. thrust was studied. At the third meeting, on August 28, 1958, it was decided to back hydrogen-oxygen engine development, and an Advanced Research Projects Agency (ARPA) order one day later sanctioned a high-energy upper stage for the Atlas ICBM, increasing its potential as a satellite launcher.

1958 August 15 The Advanced Research Projects Agency (ARPA) authorized the Army Ordnance Missile Command (AOMC) to proceed with full-scale development and funding of the clustered Juno V launch vehicle. The schedule envisaged a "full-scale captive dynamic firing by the end of calendar year 1959." The AOMC assigned development of Juno V to the Army Ballistic Missile Agency (ABMA) Development Operations Division, Redstone Arsenal, Huntsville, Ala. On September 10, 1959, NASA's pending administrator Dr. T. Keith Glennan visited the AOMC and received a briefing on Juno V, and thirteen days later the ARPA directive was changed to include development of Saturn into a multistage space

Atlas incorporated a single sustainer engine and two boost engines which were jettisoned during ascent, leaving the remaining engine to push the stage to terminal velocity. Seen here at launch on August 2, 1958, is an Atlas B.

launcher "capable of performing advanced missions." Von Braun had been reluctant to depart from a cluster of four E-1 engines, believing along with Maj. Gen. John B. Medaris that integrating eight S-3D engines was impossible, but Richard Canright from the ARPA stuck out for an eight-engine booster, citing early availability and cost savings. Von Braun was told firmly that if he could not integrate eight engines, a contractor would be found who could. On January 6, 1960, the ARPA-NASA Large Booster Review Committee decided that Juno V was the optimum route for the United States to

achieve large-booster capability, and NASA began planning to utilize the launcher.

1958 August 17   The first of three U.S. Air Force Thor-Able I flights carrying Pioneer moon orbit probes took place from Launch Complex 17A at Cape Canaveral, Fla., but ended in a ball of flames only 1 min. 17 sec. after liftoff when a turbopump failed. Thor-Able I was a three-stage version of the two-stage Thor-Able 0 used earlier in the year for tests with heat-sink reentry vehicles. To the Thor-Able 0 configuration was added a 2,500-lb. thrust solid propellant Thiokol X-248 third stage motor. The Pioneer 0 (Able I) probe comprised a conical structure 2 ft. 6 in. in diameter, weighing 84 lb. at launch, and was to have been inserted into lunar orbit by a 3,000-lb. thrust solid propellant Thiokol TX-8 motor protruding from the top. It was surrounded by eight small Atlantic Research Corp. solid vernier motors, fired and jettisoned immediately after third stage burnout, reducing probe weight to 75 lb. After retrograde stage burnout, Pioneer 0 would, in moon orbit, have weighed 51 lb., of which 39.6 lb. were instruments.

1958 August 18   Under pressure from the military to begin a wide range of space weapons projects, President Eisenhower approved National Security Council directive NSC 5814/1 *Preliminary United States Policy in Outer Space,* which defined the way space would be used by the military. U.S. Air Force Gen. Bernard Schriever said that America's "safety as a nation may depend upon our achieving space superiority," citing "space battles" as the deciding engagements of the future. Air Force Chief of Staff Gen. Thomas D. White said that supremacy in space would ensure "the capability to exert control of . . . Earth." NSC 5814/1 claimed that reconnaissance satellites would aid arms control efforts by helping to compile a more accurate picture of global activity. It urged the use of reconnaissance satellites "in a political and psychological context most favorable to the United States." Pied Piper was cancelled, the Air Force SAMOS (satellite and missile observation system) and MIDAS (missile defense alarm system) programs were strengthened and placed in greater secrecy and the Advanced Research Projects Agency was given control of the Discoverer (Corona) spy satellites being operated for the CIA.

1958 August 27   The first of three, three-stage solid propellant Lockheed X-17 research rockets carrying nuclear devices under the clandestine Operation Argus was fired from the missile ship USS *Norton Sound* located in the South Atlantic, 1,100 NM southwest of Cape Town, South Africa. Carrying a very low-yield device of 2 KT, the X-17 reached an altitude of 300 mi. before the nuclear charge was detonated over a position 38.° S × 11.5° W. Launched on August 8, a second X-17 detonated its 1–2 KT device at 49.5° S × 8.2° W, followed by a third at 300 mi., over 48.50° S × 9.7° W, on September 6, 1958. The exo-atmospheric effects were monitored by the *Explorer IV* satellite. A report given to President Eisenhower on November 3, 1958, pointed out that x rays from the explosion could penetrate ballistic warheads in space and disable electronic control systems, that high energy electrons could generate radio noise and that delayed radiation from fission could block radio communication.

1958 August 28   After two years of deliberation about possible high-energy upper stages for the Atlas launch vehicle, General Dynamics' Convair Division was awarded a contract by the Advanced Research Projects Agency (ARPA) to develop the Centaur hydrogen-oxygen stage carrying two Pratt and Whitney LR-115 (later redesignated RL-10) engines, each delivering a thrust of 15,000 lb. The ARPA asked the Air Force Research and Development Command to oversee the work and Pratt and Whitney received the RL-10 contract on November 14, 1958. On July 1, 1959, the Centaur was transferred to NASA's Marshall Space Flight Center. One year later, the Atlas-Centaur was proposed as the launch vehicle for the Venus and Mars probes scheduled for flight in 1962. The first successful test flight did not take place until November 27, 1963.

1958 August 29 A colloquium on space law was held at the Houses of Parliament, The Hague, the Netherlands, attended by lawyers and jurists from around the world. Chaired by Andrew G. Haley, a prominent U.S. legal expert, the colloquium debated matters arising from the first 10 months of the space age. The question of national sovereignty over the moon and the planets was discussed and concern expressed over the use of satellites for spying. Lawyers envisaged a mass of damage claims from people affected by falling objects and the United Nations was urged to examine this matter. The resulting legislation held launcher countries responsible for damage caused by their rockets or satellites.

1958 August Various proposals by several design bureaus in the Soviet Union were evaluated by senior scientific and engineering staff with a view to deciding on a development plan for manned or unmanned space flight. By November 1958 a decision had been made to proceed with the design of a manned spacecraft called Vostok at the expense of sophisticated unmanned satellites. Initial design details of the Vostok spacecraft were completed by March 1959. The value of manned space flight in demonstrating technological prowess was not lost on the Soviet hierarchy. While unmanned satellites for military surveillance and the scientific exploration of space were given approval, attention in the short term was to focus on getting a man in space and on developing unmanned space probes for the exploration of Venus and Mars. For this reason, not until March 1962 would the USSR launch another scientific satellite.

1958 September 5 At an altitude of 40,500 ft. over Cape Canaveral, a Convair B-58 Hustler supersonic bomber released a Lockheed air-launched ballistic-missile (ALBM) on test at a speed of Mach 1. Powered by a 50,000-lb. thrust solid propellant Thiokol XM-20 similar to that used in the X-17, the ALBM had a length of 30 ft. and a diameter of 2 ft. 7 in. The motor was expected to burn for 29 sec., but after ignition the missile behaved erratically and 33 sec. later it went into the Atlantic Ocean. Developed under Project 199C (High Virgo), the project began as an attempt by Convair to find extra jobs for the B-58. Only four ALBM flights took place; the last, on September 22, 1959, was in support of a satellite inspection and destruction test. Nicknamed "King Loftus IV," it was planned as an interception of the *Explorer IV* satellite, but when orbital parameters were declared suspect, the target shifted to *Discoverer V*. Launched at 37,500 ft. and Mach 2, all communication was lost at 30 sec. and the heavily instrumented missile carrying 13 cameras was never found.

1958 September 8 The Black Knight reentry vehicle test rocket made its first flight from the Woomera test range in South Australia, a site cleared for Britain's missile tests. On its first flight, Black Knight reached a height of 300 mi. and a downrange distance of 60 mi. The first two-stage Black Knight

was launched on May 24, 1960, carrying as second stage a solid propellant Cuckoo, a boost motor for the Skylark sounding rocket. Later, more powerful four-chamber, first-stage propulsion was provided by the 21,000-lb. thrust Gamma 301 motor. Eight flights were made with this engine. The two-stage Black Knight could lift a 250-lb. instrumented reentry vehicle to a height of 2,000 mi. When fired downwards from peak altitude, a 100-lb. head could be pushed back through the atmosphere at a speed of 17,000 ft./sec., simulating the reentry of a ballistic warhead. The last of 22 Black Knight flights took place on November 25, 1965, but the technology was applied to the Black Arrow satellite launcher, which adopted an uprated version of the Gamma engine.

1958 September 11 The Army Ballistic Missile Agency (ABMA) signed a contract with North American Aviation, Rocketdyne Division, for development of the S-3D rocket motor, eight of which would be clustered in the Juno V (Saturn I) launch vehicle. The S-3D was the latest of the line originally developed from research on Navaho and used as the single engine powering Jupiter and Thor missiles. What the ABMA wanted was an off-the-shelf motor that could be made more compact and simplified for use in the Saturn first stage. It would be designated H-1.

1958 September 23 U.S. Advanced Research Projects Agency (ARPA) director Roy W. Johnson, Maj. Gen. John B. Medaris and other ARPA and Army Ordnance Missile Command (AOMC) personnel established a baseline development and test program for the Juno V (Saturn I) program, expanding the flight test schedule from one to four launches. By this time the booster was designed around clustered Redstone and Jupiter missile tanks to save time and money. A full-scale captive firing was scheduled before the end of 1959, with a flight test in September 1960 followed by a second propulsion test and two more flight tests. Funding at this stage did not include development of upper stage configurations, the selection of which would become a lengthy, complex and much contested process as vested interests in government and industry vied to build up the clustered 1.5-million-lb. thrust first-stage into a viable heavyweight satellite launcher. On October 13, 1958, an AOMC document set the launch schedule with the first four flights projected for September to October 1960, January 1961, June to July 1961 and October to November 1961.

1958 September 24 Flight testing of Polaris SLBM (submarine-launched ballistic missiles) prototypes got under way with the launch of the first AX-series flight test vehicles. Several problems were encountered, including unexpectedly high base heating, and not before AX-6 was launched on April 20, 1959, did a successful test take place. The AX-series ended in October 1959 after 17 tests. In September of that year, the first of 40 A1X tests began, and over the following 10 months, 30 flights took place from the Atlantic Missile Range off Cape Canaveral, during which the success rate increased sharply. Preparations for live underwater test launches from the fleet ballistic missile submarine USS *George Washington* followed its launch on June 6, 1959. On that date the Department of Defense had authorized 9 submarines, each equipped with 16 Polaris missiles, but the total would increase to 41 submarines and 656 missiles operational by April 1967.

In accord with the U.S. National Aeronautics and Space Act, the National Aeronautics and Space Council (NASC) came into being. It was written into the legislation under the aegis of Sen. Lyndon Johnson, in efforts to keep both civilian and military emphasis strong and at the same time guarantee that nothing vital to national interest would get lost or overlooked. Civilian and military space interests being separate, NASA and the Department of Defense did not operate within the same framework, but Johnson recognized that many national interests would benefit from a unified oversight and policy guidance. Under the chairmanship of the U.S. president, the NASC would include the secretaries of state and defense; the chairman of the Atomic Energy Commission; the NASA administrator; the president of the National Academy of Sciences; the vice president of Shell Oil and president of the NACA (once established), James Doolittle; and the investment banker William Burden.

1958 September 26 Dr. Kurt Debus briefed U.S. Air Force Gen. Donald B. Yates, the commander of the Air Force Missile Test Center, Cape Canaveral, on the Juno V launcher and anticipated requirements for launch pads and a new control complex. Juno V, soon to be officially named Saturn, would have a potential blast yield equivalent to 240 tons of TNT, and safety requirements demanded a ground radius of 5,400 ft. A team led by James Deese selected a site 1,000 ft. north of launch complex 20 (LC-20) a Titan ICBM test pad. The Air Force protested and in mid-January 1959 the Advanced Research Projects Agency (ARPA) selected a new area 2,330 ft. north of LC-20 for the new launch complex 34. A reinforced concrete control center similar to that at LC-20 could be located just 1,050 ft. from the pad and support 130 launch personnel. A service tower on rails straddling the launch pedestal, to be used to vertically stack the rocket stages, moved back 600 ft. for launch.

1958 October 1 The National Aeronautics and Space Administration (NASA) formally came into being, replacing the old National Advisory Committee for Aeronautics (NACA) Jurisdiction for two U.S. Air Force lunar probes, two (ABMA) lunar probes and three earth satellites in the Juno II program passed to NASA as well as the Rocketdyne F-1 engine, for which the Air Force had a study contract with North American Aviation. NASA inherited the following staff and facilities (with percentage of funded space work in parentheses): 170 at Headquarters, Washington, D.C.; 3,200 at the Langley Aeronautical Laboratory, Va. (40%); 80 at the Pilotless Aircraft Research Station, Va. (90%); 1,450 at the Ames Aeronautical Laboratory, Calif. (29%); 2,700 at the Lewis Flight Propulsion Laboratory, Ohio (36%); and 300 at the High Speed Flight Station, Calif. (42%). In total, 7,900 people and a 1958 budget of $100 million were administered from the Dolly Madison House, Washington, D.C. Within seven years it would employ more than 30,000 people and reach a peak annual budget of $5.25 billion, paying for 376,700 jobs in science and industry. T. Keith Glennan remained its administrator until January 20, 1961, with Dr. Hugh L. Dryden as deputy administrator until his death on December 2, 1965.

1958 October 3 The NACA manned spacecraft plan was presented to the the Advanced Research Projects Agency, and on October 6 personnel from the Langley Research Center visited the Army Ballistic Missile Agency to discuss the procurement of Redstone and Jupiter missiles for ballistic test shots with the capsule. During October 7 a spacecraft briefing was given to the NASA administrator, T. Keith Glennan, whose response was "Let's get on with it." Even by this early date the external appearance of what would be named the Mercury spacecraft bore all the characteristics of the definitive design. About 40 potential manufacturers were given a

briefing on November 7, 1958, specifications for the manned spacecraft were issued seven days later and final copies were mailed to about 20 prospective bidders three days later.

1958 October 5 The Space Task Group (STG) was set up at Langley Field, Va., with the specific purpose of implementing the NASA manned spacecraft plan. Robert R. Gilruth was named project manager with Charles J. Donlan his deputy. A total 35 people from Langley were assigned to the STG, among whom was Maxime Faget who had originated the ballistic capsule idea, with a further 10 from the Lewis center, among whom was Glynn S. Lunney, later to be an Apollo flight director and future director of the Johnson Space Center. These men formed the nucleus of what would become the NASA Manned Spacecraft Center in Houston, Texas, famous throughout the world as "mission control" for all U.S. manned space flights beginning with *Gemini IV,* launched on June 3, 1965.

1958 October 9 One of 14 aerospace companies that submitted bids, the Boeing Airplane Co. was awarded a U.S. Air Force contract to assemble and test the Minuteman ICBM. Thiokol was selected to design and fabricate the first stage, then the largest solid propellant rocket motor. Tested for the first time on April 13, 1959, the first stage had a length of 24 ft. 2 in., a diameter of 5 ft. 11 in. and a filled weight of 45,000 lb. The motor produced a thrust of 200,000 lb. Built by Aerojet General, the second stage had a length of 10 ft. 5 in., a diameter of 3 ft. 8 in., a propellant weight of 11,000 lb. and was fired for the first time on April 14, 1959. Both first and second stages incorporated four exhaust nozzles. Hercules Powder Co. was responsible for the third stage, which had a length of 7 ft. 6 in., a diameter of 3 ft. 1 in. and was fired for the first time in May 1959. Overall, the LGM-30A Minuteman I had a length of 53 ft. 9 in., a diameter of 6 ft. 2 in. and a launch weight of 65,000 lb. With a slightly larger second stage, the LGM-30B version had a length of 55 ft. 11 in.

1958 October 11 The first launch under the auspices of the newly established National Aeronautics and Space Administration took place at 08:42 UT from Cape Canaveral, Fla., when the second Thor-Able I sent the moon probe *Pioneer I* into space. The first Pioneer launch had been conducted by the U.S. Air Force, but on October 1 the civilian space agency took over all military space probes and scientific satellites. All three stages functioned but the final velocity imparted to the 84-lb. satellite was 34,425 ft./sec., a mere 825 ft./sec. below the minimum required to achieve escape velocity. Accordingly, the probe reached a maximum distance of 71,700 mi. at 11:42 UT, October 12, and fell back toward earth. In a desperate effort to push *Pioneer I* into a 20,000 × 80,000 mi. orbit of the earth, engineers tried to fire the solid propellant retro-rocket designed to put *Pioneer I* into moon orbit. At 1.6° C, the batteries were too cold to send current for ignition and *Pioneer I* fell into the atmosphere at 03:46 UT the next day.

1958 October 21 At a special briefing called by Langley Research Center, potential bidders on a new solid propellant test rocket for the NASA manned spacecraft were briefed on requirements. NASA wanted a rocket that could throw a 3,942-lb. mass, the spacecraft, on ballistic test shots where the recovery sequences could be flight-tested in simulation of the capsule's return from space. The rocket, called Little Joe, was conceived by Maxime Faget and Paul Purser to have four Pollux solid propellant rocket motors fired sequentially in pairs, but this was later changed to four 55,000-lb. Castors, burning for 25 sec., and four 33,000-lb. thrust Recruits burning for 1.5 sec. In this configuration, a maximum liftoff thrust of 242,000 lb. was produced by ignition of two Castors and four Recruits, cut to 110,000-lb. thrust at 1.5 sec., followed by ignition of the remaining two Castors at 20 sec. In another configuration, four Castors and two Recruits ignited at liftoff for a combined thrust of 286,000 lb. On December 29, 1958, North American Aviation was awarded a contract to build Little Joe.

1958 October 23   A Juno I was launched from Cape Canaveral, Fla., carrying the 9-lb. satellite *Beacon Explorer 1.* Designed as the first geodetic satellite, NASA's *Beacon 1* failed to reach orbit when the upper stage separated from the first stage at 7 min. 4 sec., 1 min. 2 sec. prior to burnout, leaving it with insufficient velocity. *Beacon 1* consisted of an inflatable 12 ft. diameter sphere packed in a cylinder 4 ft. 2 in. long with a diameter of 7 in. NASA's second geodetic satellite, *Beacon Explorer 2,* launched August 14, 1959, on another Juno I, also failed when upper stages separated.

1958 November 5   In Britain the government's Cabinet Defense Committee met to consider defense expenditures in general and the Blue Streak ballistic missile program in particular and raised questions about alternatives to the British rocket. Consideration had been given to a developed version of the U.S. Thor missile and it had been suggested that its range could be extended from 1,700 to 2,300 mi. and its lifting capacity increased, thereby filling the requirement met by Blue Streak. The Pentagon had expressed willingness to carry out the necessary technical modifications to Thor to increase its performance but warned that costs, which would have to be borne by Britain, could be high. The defense committee also discussed the desirability of using Polaris missiles in place of Blue Streak and noted that Britain was already cooperating with the United States in development of this solid propellant sea-launched ballistic missile.

1958 November 7   NASA's director of Space Flight Development, Abe Silverstein, received a proposal from the Jet Propulsion Laboratory (JPL), Calif., to conduct a Space Flight Program Study, approval being granted on November 18. The JPL set up a NASA Program Study Committee for the $1.3 million contract, designated NASw-6 and comprising seven people, each of whom formed a working group with specialists from various departments at the JPL. The study was to formulate a five-year program for deep space exploration beginning with the moon, Venus and Mars. It was completed and presented to NASA on April 30, 1959. Thus began the JPL's long association with unmanned planetary exploration.

1958 November 8   The third and last Thor-Able I launch vehicle lifted off from Launch Complex 17A, Cape Canaveral, Fla., at 07:30 UT carrying the 84-lb. *Pioneer 2* moon probe with attached retro-rocket and verniers identical to *Pioneers 0* and *1.* The first two stages fired as expected but the third stage did not, and the probe reached a height of only 963 mi. before falling to destruction in the earth's atmosphere at 08:12 UT. This was the last of the three U.S. Air Force Pioneers toward which so much hope and too much ambition had been directed. The goal of putting a probe into orbit around the moon had simply been too great for the technology and knowledge of the late 1950s, and almost eight years would

pass before that feat was achieved when *Lunar Orbiter I* was launched on August 10, 1966.

1958 November 14   The U.S. Air Force made a significant change to the official direction of the Dyna-Soar boost-glide vehicle, emphasizing it as a research and development effort which might lead to a proposed weapon system, and signed a cooperative agreement with NASA. Air Force Vice Chief of Staff Gen. Curtis E. LeMay was advised by Air Force Asst. Sec. for Research and Development Richard C. Horner "that if a strong weapon system program were offered to Department of Defense officials, Dyna-Soar would probably be terminated." At the same time the three-step development program announced October 10, 1957, was simplified to a two-step one: The first step would be to produce a research vehicle, the second to develop a weapon system if needed.

1958 November 19   The U.S. Advanced Research Projects Agency (ARPA) announced that a large communications satellite would be placed in geostationary orbit 22,300 mi. above the equator late in 1960 by a Juno V (Saturn) launch vehicle. At this date the launcher comprised a clustered first stage of eight H-1 engines, a second stage comprising an adapted Atlas or Titan ICBM and a cryogenic Centaur third stage. On January 20, 1959, ARPA awarded separate work orders to the Air Force and the Army for an active repeater communications satellite. In hearings before the House Committee on Science and Astronautics on February 5, 1959, von Braun described the Saturn-launched system comprising three geostationary satellites at 120° intervals around the earth and providing continuous communications between any two points on the globe.

1958 November 20   Administrator T. Keith Glennan, with Deputy Secretary of Defense Donald Quarles, organized the transfer of the Vanguard satellite program from the Naval Research Laboratory to NASA, the first of a series of acquisitions that would strengthen the civilian space agency. NASA also obtained the Jet Propulsion Laboratory, owned by the Army but staffed and run by the California Institute of Technology, which it formally took over on December 31, 1958. Further acquisitions included the Army Ballistic Missile Agency's Development Operations Division on July 1, 1960, and its Missile Firing Laboratory, which became the Launch Operations Directorate (LOD) in charge of NASA launches at Cape Canaveral, Fla., and Vandenberg AFB, Calif. On November 29, 1963, the LOD at Cape Canaveral became the John F. Kennedy Space Center.

1958 November 24   The NASA Space Task Group placed the first order for a launch vehicle in support of manned orbital space flight when one Atlas was reserved by the Air Force Ballistic Missile Division, Calif. NASA asked the U.S. Air Force to test launch an Atlas C to qualify the aerodynamics of the manned capsule design in a mission which would be known as Big Joe. It was a logical progression from the solid propellant booster for ballistic tests Little Joe, a name derived by Maxime Faget from blueprints which showed four holes up (for the solids) reminiscent of the crap game throw of a double deuce on the dice.

1958 November 26   Suggested by Abe Silverstein, NASA's director of Space Flight Operations, during the fall, the name Mercury was formally approved by Administrator T. Keith Glennan and Deputy Administrator Hugh L. Dryden, for America's first manned space flight program. It was first used publicly on December 17, 1958. Silverstein chose the name because Mercury was the winged messenger of the ancient Roman gods. Silverstein had served as associate director of NACA's Lewis Propulsion Laboratory and was to be a key administrator at NASA during the next three years. The Office of Space Flight Operations was a visible element in NASA and his vigorous leadership, meticulous attention to detail and daily oversight helped minimize delay.

1958 December 6 The first of two Pioneer moon probes launched by Juno II launch vehicles ascended from Cape Canaveral, Fla., at 00:44:52 local time. This marked the first flight of Juno II, a modified U.S. Army Jupiter missile with propellant tanks lengthened by 3 ft. and the solid propellant upper stage cluster used on Juno I. With such a powerful first stage—150,000-lb. thrust versus 83,000 lb. with Juno I—the launcher would have been optimized with more powerful upper stages. Juno II had a length of 76 ft. and a diameter of 8 ft. 9 in. The 13-lb. *Pioneer III* probe comprised a cone 1 ft. 8 in. in length with a base diameter of 9 in. and a 3-in. spike antenna on top. Wrapped around the base were two 5-ft. wires released from the spinning fourth stage after separation to unwind under reverse centripetal force and slow the rate of rotation from 415 RPM to about 11 RPM. Due to premature cutoff of the Juno II first stage, *Pioneer III* failed to reach escape velocity. After ascending to an apogee of 63,580 mi., it fell to Earth 38 hr. 6 min. after launch.

1958 December 9 Three aeromedical specialists, Maj. Stanley C. White, U.S. Air Force, Lt. Robert B. Voas, U.S. Navy, and Capt. William Augerson, U.S. Army, were named as a special team selected to nominate 150 candidate astronauts by January 21, 1959. The plan envisaged 36 short-listed for further tests which would reduce the candidate list to 12, 6 men eventually being selected for a nine-month training prior to flight. At the end of the month this first attempt at a biomedical astronaut screening program was cancelled. On January 5, 1959, astronaut qualifications were laid down: age less than 40 years; height less than 5 ft. 11 in.; excellent physical condition; bachelors degree; graduate test pilot; minimum 1,500 hr. flying time; qualified jet pilot. In all, 508 service records were searched, out of which 110 pilots qualified. A special committee reduced this to 69 briefed and interviewed in Washington, of which 53 volunteered and 32 were picked for further tests. The 7 Mercury astronauts were selected on April 2, 1959.

1958 December 11 Of the 20 or so aerospace companies mailed with final specifications for the Mercury spacecraft, 11 companies responded with proposals. These included AVCO, Chance-Vought, Convair, Douglas, Grumman, Lockheed, Martin, McDonnell, North American Aviation, Northrop and Republic. After an exhaustive technical examination and selection process, negotiations were started with McDonnell on engineering and legal questions prior to a definitive contract to build America's first manned spacecraft.

1958 December 13 Launched on a ballistic fight aboard a U.S. Army Jupiter missile, a primate named Gordo was sent from Cape Canaveral down the Atlantic Missile Range in a test to measure physiological reactions. Telemetry data provided confirmation that Gordo was weightless for about 8 min. 20 sec., experiencing 10 g during the boost phase and 40 g during deceleration as the nose cone decelerated in the atmosphere from a speed of 10,000 MPH. It was not recovered.

1958 December 15 At an interagency meeting on launch vehicles, NASA proposed development of the Atlas-Vega, incorporating a standard 360,000-lb. thrust Atlas first stage, a second stage powered by the 12,500-lb. thrust General Electric GE 405H-2 RP-1/LOX and an optional 6,000-lb. thrust solid propellant Jet Propulsion Laboratory third stage. Overall, the launch vehicle would weigh 297,000 lb. and have a payload capacity of 4,800 lb. to a 345-mi. orbit. Intended as an interim vehicle pending availability of the Atlas-Centaur, Atlas-Vega was envisaged as a general purpose launcher for a variety of early payloads. The Jet Propulsion Laboratory received a contract to develop the third stage on January 30, 1959, and Convair got the primary contract for Vega development, and General Electric was contracted for the second stage on March 18, 1959. The first flight was set for August 1960, but on December 11, 1959, the vehicle was dropped in favor of Atlas-Agena B sponsored by the Department of Defense.

1958 December 16 The first missile fired from Vandenberg AFB, Calif., also the first Thor Intermediate Range Ballistic Missile (IRBM) fired by an operational crew, was launched by the First Missile Division, Strategic Air Command, from what later became known as SLC-2E on a qualification flight out over the Pacific. The second missile launched from Vandenberg, a Thor-Agena A, placed the first Discoverer reconnaissance satellite in orbit on February 28, 1959, and it became the first satellite placed in polar orbit. Vandenberg was used on an almost fifty–fifty basis by the U.S. Air Force for development testing of ballistic missiles and by NASA and the Department of Defense for launching satellites.

1958 December 17 The Rocketdyne H-1 engine completed its first full-power firing test. The engine was a developed version of the Rocketdyne S-3D but with the turbopump moved from the top of the engine to the side. The first H-1 for the Saturn launcher arrived at the Army Ballistic Missile Agency (ABMA) on May 3, 1959, and was first fired May 21 before a long-duration (2 min. 31.03 sec.) run on June 2, 1959. Early versions would produce a thrust of 165,000 lb., the same as the S-3D. The fully operational H-1 would produce a thrust of 188,000 lb., further development increasing this to 205,000 lb. The H-1 would eventually be adapted for the first stage of the Delta.

1958 December 18 In an audacious publicity exercise aimed at sending clear signals to the Soviet Union that the United States now had an ICBM, an Atlas B prototype was launched by the U.S. Air Force into orbit from Cape Canaveral, Fla., in an exercise known as Project Score. After 7 Atlas failures in 16 missile test attempts, success was not a foregone conclusion, but Atlas 10B fired itself into a 115 × 922 mi., 32.3° orbit. Covered by a protective nose cone on top of the missile was 150 lb. of communication equipment which broadcast a prerecorded Christmas message from President Eisenhower every day for thirteen days. Weighing a total 8,750 lb., the inert missile reentered the atmosphere and burned up on January 21, 1959. Compared to *Sputnik 3*'s weight of 2,926 lb., *Score*'s payload was hailed by public relations officials as the world's heaviest. But this was not the case because *Score*'s declared weight included the mass of the launcher, whereas that of the Soviet satellite was limited to the payload proper. When the SS-6 *Sputnik 3* launcher is included, which like Atlas remained attached to the orbited payload, the Soviet weight in orbit becomes 16,150 lb.

1958 December 19–20   A total of 19 U.S. scientists met at the Massachusetts Institute of Technology to discuss problems associated with the detection of life on other planets. They called themselves the East Coast Panel on Extraterrestrial Life, or EASTEX, and were sponsored by the Armed Forces Committee on Bioastronautics. Chaired by Melvin Calvin, professor of chemistry at the University of California, Berkeley, the first meeting concerned itself with possible life forms that might be encountered by unmanned spacecraft searching for evidence of living things. Among those present was Wolf Vladimir Vishniac, born in Berlin of Latvian parents who fled the Russian revolution. In March 1959 Vishniac was given the first NASA grant for work in biological sciences, and he designed an instrument known as the Wolf Trap for detecting microbial life. The instrument was never used but stimulated successors which helped form the experiments proposed for NASA's Voyager and Viking Mars projects.

1958–1959   Developed as a replacement for the SS-3 Shyster, the Soviet SS-4 Sandal was tested during 1958 and brought to operational status during 1959. This medium range ballistic missile (MRBM) was the first Soviet weapon to adopt storable liquid propellants, giving it a quicker reaction time than missiles which first have to be fueled for flight. Its GDL RD-214 motor used kerosene and red fuming nitric acid (RFNA) to produce a thrust of 163,100 lb. The SS-4 had a length of 73 ft. 6 in., a diameter of 5 ft. 5 in., a launch weight of 61,000 lb. and a range of up to 1,100 mi. with a 1-MT warhead. A maximum of about 500 SS-4s were deployed between 1963 and 1976. After this the type was gradually withdrawn, the last being deactivated in the early 1990s.

1959 January 2   The first space probe successfully launched to Earth-escape velocity was sent toward the moon by a Soviet SL-3 launcher which put *Luna 1* into an orbit of the sun, imparting to it a speed of approximately 25,200 MPH. Weighing 796 lb., this space probe was the fifth Luna-type spacecraft launched by an SL-3, all earlier attempts failing to meet their objectives. Only probes successfully launched were given a number, each with a specific mission objective. *Luna 1* sped past the moon at a distance of 3,727 mi. at 14:59 UT on January 4, slowed by the gravitational pull of the Earth to a speed of 5,100 MPH. From there the probe dropped into an elliptical orbit about the sun of 91.5 million × 122.5 million mi. Radio contact with *Luna 1* was lost at 07:00 UT on January 5 when the probe was at a distance of 371,000 mi. from Earth.

By the end of 1958 approval had been given for Wernher von Braun to begin development of the Juno V launcher, later called Saturn. America's first heavyweight launcher comprised a cluster of tanks and engines strapped together.

A staff report from the U.S. House of Representatives Select Committee on Astronautics and Space Exploration gathered contained projections from leading experts on the next 10 years in space. Wernher von Braun from the Army Ballistic Missile Agency believed that a manned circumnavigation of the moon would occur within 8–10 years, followed by a manned lunar landing a few years later. NASA administrators believed that circumlunar flight would take place within the decade of the 1960s and that instrumented landing vehicles would carry scientific equipment to the surface. Extensive investigation of the lunar surface by unmanned spacecraft was predicted for the 1960s by Roy K. Knutson, chairman of the Corporate Space Committee, North American Aviation.

1959 January 6   An undecided matter concerning protection of the Mercury manned spacecraft from the searing heat of reentry was the topic discussed at a special meeting in NASA headquarters. Two types of shield were studied: an ablative shield which would char or burn away, carrying heat with it as the material eroded, and a beryllium heat-sink designed to absorb and dissipate heat. It was decided to proceed with a capsule that could receive either type, and 12 ablation and 6 heat-sink shields were ordered accordingly. The ablation material selected was a fiberglass bonded with modified phenolic resin which tests proved had good structural qualities after exposure to heat. On reentry from orbit, the heat shield would experience temperatures of about 3,000°F and the front of the shock wave, just inches ahead of the shield, would reach a temperature of 9,500°F.

1959 January 7   In France a Committee on Space Research (Comité National des Recherches Spatiales) was established by the government with responsibility to the prime minister. France had a strong interest in space research, having developed a broad base of solid propellant sounding rockets since the Second World War. The committee was enlarged on February 10, 1962, and transformed into a Space Council. In April 1962 this was put under the Ministry of State in Charge of Scientific Research, Atomic and Space Affairs. Charged only with creating policy, the committee lacked the muscle to set up planning and research activities. On December 19, 1961, a new organization, the Centre National d'Etudes Spatiales (CNES) was established to manage national space projects.

1959 January 8   NASA's Milton Rosen and Richard Canright of the Advanced Research Projects Agency (ARPA) completed a concise, two-page document summarizing testimony from government and industry teams on formulating a national launch vehicle program. The two authors were co-chairmen of a Joint ARPA-NASA Committee on Large Clustered Booster Capabilities and heard evidence from Aerojet General, Rocketdyne, Convair Astronautics, Douglas Aircraft, Martin, the Army Ballistic Missile Agency (ABMA) and the U.S. Air Force. Acknowledging the eight-engine clustered Juno V as feasible and ABMA as the "best qualified from an engineering standpoint" to develop the rocket, the document criticized program organization, claiming its costs and development time had been underestimated. An alternative, suggested the authors, would cluster three Atlas ICBMs and three Centaurs as the second stage. Milton Rosen teamed with Eldon Hall to prepare a recommended national launcher program submitted on January 27, 1959.

1959 January 15   The NASA Beltsville Space Center formally came into existence at Greenbelt, Md., in accordance with an edict from Congress that a "space projects" facility be set up. On May 1, 1959, NASA announced that the new facility would be known as the Robert H. Goddard Space Flight Center (GSFC), in honor of the first person to fire a liquid propellant rocket. It was officially dedicated on March 26, 1961, the 35th anniversary of that event, with Mrs. Goddard in attendance. The GSFC would be responsible for scientific and applications satellites and for several cooperative satellite ventures with other countries as well as managing the Delta launch vehicle program. Although located at Langley Research Center, the Space Task Group managing manned flight was initially the responsibility of the GSFC.

1959 January 20   President Eisenhower expressed impatience with continuing demands for more nuclear-bomb-making capacity when he was asked by John McCone, chairman of the Atomic Energy Commission, for expanded reactor capacity to meet military needs. Of the Pentagon, Eisenhower said, "They are trying to get themselves into an incredible position of having enough to destroy every conceivable target all over the world, plus a threefold reserve. The patterns of target destruction are fantastic . . . they talk about megaton explosions as though they are almost nothing." On February 18, Eisenhower asked Gordon Gray, his national security assistant, to carry out a study of the effect on the atmosphere and the environment of an all-out nuclear war. It was never done.

1959 January 23   The first in a series of six U.S. Air Force Project Bravo ablative nose cone test flights began from Launch Complex 17A, Cape Canaveral, Fla., with the first flight of the Thor-Able II, which ended prematurely when the second stage failed to fire. In a program directed by the Ballistic Missile Division, General Electric (G.E.) and Avco ablative reentry vehicles for the Atlas ICBM were to be tested by the same launcher configuration as that first employed for Project Able beginning April 23, 1958. The second launch on February 28, 1959, was also a failure when the RVX-1 reentry vehicle was not recovered, and the same fate befell the third flight on March 20. The fourth flight, on April 7, 1959, was a success when the Thor-Able II lobbed the Avco RVX-1-5 ablative nose cone 5,000 mi. down the Atlantic Missile Range in a 25-min. flight which reached 15,000 MPH. The fifth flight carried a G.E. nose cone 6,000 mi. on May 21, 1959, and was also recovered, but the last flight on June 11, 1959, ended in failure when the G.E. reentry vehicle was lost.

1959 January 27   In attempting to bring order and structure to a widening variety of existing and prospective launch vehicles, NASA formulated a National Space Vehicle Program plan with the central theme that one launcher should be developed for each series of missions. This, it said, would allow specific launchers to build a history of reliability through the use of common hardware, only the payload needing to vary with the objectives of the mission. NASA identified four launchers that it said would form the core of its inventory: Atlas-Vega (4,800 lb. to Earth orbit), Atlas-Centaur (8,500 lb. to Earth orbit), a five-stage Juno VA and VB, (20,000 lb. to Earth orbit) and a five-stage, 6-million-lb. thrust launcher called Nova (175,000 lb. to Earth orbit). As defined at this date, Nova comprised four 1.5-million-lb. thrust F-1 engines in the first stage, one F-1 in the second stage, cryogenic, high-energy 320,000-lb. and 80,000-lb. thrust third and fourth stages and a 20,000-lb. thrust solid propellant fifth stage.

1959 February 2   The U.S. Army proposed that the Juno V launch vehicle should be renamed Saturn, because Saturn was the next planet out to Jupiter. Since Jupiter had been the most recent space launch vehicle developed at the Army Ballistic Missile Agency's Development Operations Division, Saturn seemed a logical choice. On the next day, Roy Johnson, director of the Advanced Research Projects Agency (ARPA), approved the change. The name fixed, Johnson next sought to resolve the dilemma of a burgeoning variety of upper-stage configurations for the big launcher and met with Maj. Gen. John B. Medaris, chief of the Army Ordnance Missile Command, to press the urgency of an agreement between ARPA and NASA on what these should comprise.

1959 February 6   The first Titan I (missile A-3) was launched successfully for the first time from Cape Canaveral, Fla., with a dummy upper stage containing a 45,000-lb. water ballast. This was the first time a ballistic missile had performed flawlessly on its first flight. The second launch, also with inert second stage, was successfully accomplished with missile A-5 on February 25, as was the third flight with missile A-4 on April 3. The separation of first and second (inert stages) was first demonstrated in flight with missile A-6 on May 4. The fifth flight on August 14, 1959, ended in disaster when the hold-down supports released the missile prematurely. The sixth flight on December 12, 1959, failed when a one of the command relays inadvertently destroyed the missile in flight.

After a concerted scrutiny of bids, NASA awarded a formal contract for the Mercury spacecraft to McDonnell Aircraft Corp. calling for design and development of 12 models. An additional eight spacecraft were ordered later as well as two procedural trainers, an environment trainer and seven checkout trainers. McDonnell delivered the first production Mercury spacecraft to the Space Task Group (STG) on January 25, 1960, less than one year after the formal contract had been signed. Merely a structural shell devoid of internal systems, it was fitted out by the STG and became the spacecraft for the Mercury Atlas-1 mission flown on July 29, 1960.

1959 February 13   The French began rocket launch operations from their Sahara Test Center in Algeria. At first, launches were conducted from Colomb-Béchar, then from Hammaguir as well. The range extended 1,860 mi. to the southwest but when orbit flights began they were launched eastwards. Beginning in 1964, the Centre d'Essais Guyanais at Kourou in French Guiana was built up for orbital launches starting in 1970. Operations in Algeria terminated on July 1, 1967.

1959 February 17   Designated *Vanguard II,* the first satellite designed to conduct a cloud survey of the Earth was successfully placed in a 346 × 2,063 mi., 32.88° orbit with a period of 125.9'. It was the second satellite launched by a Vanguard rocket and comprised a sphere 20 in. in diameter and weighing 23.7 lb. Because the satellite was placed in orbit together with the third stage of the Vanguard rocket, the total orbited mass came to 71.5 lb. Two photocell units were attached to the satellite, which revolved at 50 RPM and transmitted data to the ground from power supplied by mercury batteries. The transmitters lasted 27 days but the orbital lifetime of the satellite is 150 years. Cloud data could not be received due to instability in the satellite's spin rate.

1959 February 19   Dr. Homer E. Newell, assistant director of NASA's Office of Space Flight Development, defined 12 science objectives for lunar flights using the Atlas-Vega launch vehicle. To gather detailed information about the moon itself, a lunar satellite swinging to within 50–100 mi. of the surface would measure solar radiation and magnetic fields and view the surface in visible and infrared. Soft-landing spacecraft would provide seismological data and measure bearing strength, structure and composition of the surface. The lunar satellite eventually became the Lunar Orbiter and the soft lander became the Surveyor spacecraft. On April 20 Dr. Robert Jastrow urged early adoption of the Lunar Orbiter concept, and on April 27, in its earliest formal remark concerning the post-Mercury manned flight program, NASA declared that answers to the more profound questions regarding the moon "will have to wait until man himself has reached the moon and can conduct detailed surveys on the spot."

1959 February 28   A Thor-Agena A launched the world's first military reconnaissance satellite from Vandenberg AFB, Calif. *Discoverer I* comprised the Lockheed 2205 Agena A stage with a 10,000-lb. thrust Bell Hustler 8048 rocket motor, but without the reentry capsule that on future missions would be used to recover film shot from space. Agena A/Discoverer had a length of 19 ft. 2 in., a diameter of 5 ft. and an on-orbit weight of 1,700 lb. *Discoverer I* was placed in an incorrect, 101 × 601 mi., orbit due to a malfunction with the attitude control system. It burned up in the atmosphere five days later. Over the three years to February 27, 1962, 38 satellites named Discoverer were launched. Of this total, 12 failed to make it into orbit, 4 did not carry any capsules, 5 carried capsules that failed to return, 5 carried capsules that returned but were never recovered and 12 returned capsules that were recovered.

1959 February   Technical specialists from the United States visited Australia to select a site for a tracking antenna 85 ft. in diameter and similar to one already in operation at Goldstone, Calif. The team was a joint effort from the Advanced Research Projects Agency and NASA and examined seven areas close to the Woomera test range built up by Britain in anticipation of its Blue Streak long-range ballistic missile program. An agreement between the United States and Australia was signed on February 26, 1960, and the 85-ft. dish became operational on November 1, 1960, as part of NASA's Deep Space Instrumentation Facility. In related developments, the Johannesburg, South Africa, tracking station was dedicated on September 8, 1961. With this in place, at least one 85-ft. antenna would be in direct line of sight with interplanetary spacecraft at all times.

1959 March 2   At a meeting with President Eisenhower and members of the National Aeronautics and Space Council on national launch vehicle plans, Advanced Research Projects Agency (ARPA) representatives described their conception of upper-stage configurations for the Saturn launcher. Saturn A would comprise the clustered first-stage booster with eight H-1 engines. Saturn B would add an RP-1–liquid oxygen second stage, and Saturn C would replace that with a cryogenic liquid hydrogen–liquid oxygen second stage. On March 13, 1959, the Army Ordnance Missile Command (AOMC) presented its "Saturn Systems Study" to ARPA and screened 1,375 proposed configurations, giving detailed study to the preferred 14. Atlas and Titan were selected as optimum upper stages, and on April 30 ARPA received an AOMC study of the two-engine Titan first stage as second stage for the Saturn. ARPA leaned toward the cryogenic Centaur as Saturn's third stage and, on May 19, 1959, informed the AOMC that this would be the configuration for military satellite launches.

1959 March 3   At precisely 05:10:56 UT, a Juno II was launched from Cape Canaveral, Fla., carrying the 13-lb. *Pioneer IV* moon probe. Identical to *Pioneer III* (launched on December 6, 1958), it was designed to pass within 20,000 mi. of the moon, measuring the space environment before and after its closest encounter and triggering a photoelectric sensor carried along to qualify part of a photographic system designed for a later mission. Due to low velocity at direct insertion and a minor attitude error in pitch and azimuth, *Pioneer IV*'s closest approach to the moon was 37,300 mi., too far to trigger the photoelectric sensor, but the probe was tracked for 82 hr. 4 min. to a record distance of 407,000 mi. as it slipped into an orbit of the sun, the first U.S. probe to do so. Thus the first U.S. lunar program—the five Pioneer flights—ended with no significant success. This was also the second and final Juno II-launch in the Juno II/Phase II program for launching Pioneer moon probes for the U.S. Army. Earth-orbital, Juno II/Phase I flights for NASA began on July 16, 1959.

1959 March 6   A thrust of more than 1 million lb. was achieved for the first time in the United States in a test stand at the U.S. Air Force–Rocketdyne Propulsion Laboratory in Santa Susana, Calif., when a thrust chamber designed for the F-1 was successfully static-tested. It was also the first time a single chamber had produced this level of thrust. While the thrust chamber was not configured as a working motor, in August 1959 successful tests with an injector pattern for the F-1 were carried out at the same facility, resulting in a thrust exceeding 1 million lb. The injector would deliver fuel and oxidizer to the combustion chamber in precise quantities and at a fixed mixture ratio. On February 10, 1961, a thrust level of 1.55 million lb. was attained at the NASA High Thrust

For Advanced Saturn launch vehicles, a type later designated Saturn V for Apollo moon missions, NASA developed the F-1 liquid propellant engine which produced 1.5 million pounds of thrust.

Area at Edwards AFB, Calif., and two months later the engine demonstrated 1.64 million lb. of thrust. A full-duration test lasting 2 min. 30 sec. was conducted on May 26, 1962.

1959 March 11   In tests conducted at the Wallops Island test range by NASA Langley's Pilotless Aircraft Research Division, the first full-scale simulation of a Mercury spacecraft pad-abort was carried out by firing the escape rocket. This was a 52,000-lb. solid propellant rocket in a cylindrical canister 5 ft. 10 in. long and 1 ft. 3 in. in diameter, attached to the top of the spacecraft by a lattice tower. Designed to wrench the capsule free of its launch vehicle if the latter showed signs of exploding, the escape motor would burn for less than 2 sec., carrying the spacecraft to a maximum height of 2,500 ft. for aborts off the pad. It would also constitute the primary means of escape for the astronaut in his capsule during ascent to orbit. The escape tower would normally be jettisoned about 20 sec. after Atlas booster-engine separation.

1959 March 12   Formed officially in London during November 1958, the first full meeting of the International Council of Scientific Unions' Committee on Space Research (COSPAR) began a two-day session at The Hague in the Netherlands. COSPAR evolved from a United Nations resolution signed by 19 countries that specifically called for the creation of a special committee to discuss international matters relating to space research. At the first meeting, the U.S. representative Richard Porter conveyed an assurance that NASA would willingly launch scientific experiments from other COSPAR members. With the USSR a member, COSPAR became the channel through which U.S. participation in international space endeavors would forge links across the political divide.

1959 March 20   A U.S. Army task force began to develop a plan for an outpost on the moon. Under the direction of Army Ordnance Missile Command chief Maj. Gen. John B. Medaris, the study, called Project Horizon, drew heavily on the advanced thinking of Wernher von Braun at Huntsville, Ala. When the report was finished and delivered to Army Chief of Staff Gen. Maxwell D. Taylor on June 8, 1959, it comprised 808 pages in 4 volumes and read like a compendium of von Braun's ideas published by Collier's magazine in March 1952. Wildly disproportionate to national resources, it envisaged a $6 billion program, involving 149 Saturn rockets, to build a lunar outpost. Between 1964 and 1967, 229 Saturn launchers would be needed to equip and replenish the base for 12 men (no women were mentioned). Project Horizon was the last concerted effort by the U.S. armed forces to get a foothold in space exploration.

Professor James A. Van Allen, whose name had been given to the radiation belts around Earth discovered by his instruments on *Explorer I,* provided additional information about these zones. Based on data from the *Pioneer IV* satellite, the upper radiation belts extended from 8,000 mi. to 55,000 mi., a greater distance than previously known. The inner belt extended from 1,500 to 3,000 mi. It was feared that the intensity of the radiation would prohibit astronauts spending long periods within the lower zone.

1959 March 26   The USSR Academy of Science announced that it would continue the full exploration of the moon and the planets in the seven-year program which would follow the existing program extending from 1958 to 1965. This was communicated by NASA's Dr. Robert Jastrow to Homer E. Newell, and on this date an additional funding supplement for

Atlas-Vega third stage development was made available to the Jet Propulsion Laboratory. JPL director Dr. William H. Pickering had emphasized the need for extra money to prevent slippage in Vega development that would compromise a planned mission to Mars in 1960 and close the gap between Vega and Atlas-Centaur. If that happened, said Dr. Pickering, Vega might be seen as unnecessary and cancelled. As it turned out, his fears were justified.

1959 April 1 NASA's Director of Aeronautical and Space Research John W. Crowley Jr. announced that a research steering committee on manned space flight would be set up under the chairmanship of Harry J. Goett of Ames Research Center, Calif., with a mandate to look beyond Mercury and Dyna-Soar and recommend long-term goals. Nominated members included representatives from every NASA facility. Known hereafter as the Goett Committee, it met for the first time on May 25–26, 1959, and drew up a tentative list of priorities, including Mercury orbital flights, manned space stations, lunar reconnaissance, manned moon landings, Mars-Venus reconnaissance and Mars-Venus landings. In what may have been the origin of a basic flaw in future Apollo planning, the committee recommended that each NASA center study a moon landing plan, but as an end mission and not as part of a sustained series of objectives. The lack of a broader purpose for Apollo after the first landings would doom the project to cancellation.

1959 April 2 Representatives from 20 companies attended a briefing by NASA on a worldwide tracking network it wanted for the Mercury manned space program. At this time, 14 stations were envisaged through which engineers could receive telemetered data from the spacecraft and voice and biomedical data from the astronaut. Moreover, each site would be linked with a teletype service and would provide the Cape Canaveral control center with trajectory information and landing area predictions. On July 20, 1959, negotiations for construction were begun with the Western Electric Co. and their subcontractors Bendix, IBM, Bell Laboratories and Burns & Roe, a final contract being signed 10 days later. All Mercury tracking stations were declared operational on March 31, 1961.

1959 April 4 In an attempt to put together a fast-paced lunar exploration program utilizing unmanned spacecraft, NASA issued the first Atlas-Vega launch schedule that anticipated eight flights beginning in August 1960 and proceeding at two-month intervals through October 1961. Over the next three months the schedule for the first three flights was modified to provide for a vehicle test launch in August 1960, a lunar impact mission in October and a Venus mission in January 1961. On August 4, 1959, the Vega schedule had slipped to a test flight on January 10, 1961, a lunar fly-by on March 7, a weather satellite on June 5 and a lunar mission on October 22, 1961. A communications satellite was scheduled for launch on January 1, 1962, followed by a lunar orbital mission on March 5. The remaining two Vegas would be 1962 stand-by flights for the last two assigned missions. Further delays were announced on August 20 and on September 30, 1959, the final precancellation Vega changes scheduled the eight flights between January 1961 and July 1962: *Vega*s *1, 2* and *3* were to be test flights, *Vega*s *4, 6* and *8* lunar satellite launches and *Vega*s *5* and *7* weather and communication satellite flights, respectively. Vega was terminated December 11, 1959.

1959 April 5 The Bullpup air-to-surface missile became operational with the U.S. Navy. Developed over five years by the Martin Co., Bullpup answered the need for a precision stand-off bomb capable of being launched by carrier-based aircraft. The missile had a length of 10 ft. 6 in., a diameter of 1 ft. and a span of 3 ft. 1 in. across four tail fins. Bullpup was powered by an Aerojet solid propellant motor providing a terminal speed of Mach 1.8 and a range of 7–10 mi. A variant for the Air Force was produced in large numbers and export and license-built versions were sold to NATO countries. Bullpup could carry a variety of nuclear (W-45) and nonnuclear warheads.

1959 April 6 Admiral William F. Raborn said the U.S. Navy would eventually like to build 45 fleet-ballistic-missile submarines for its Polaris missiles. With each vessel carrying 16 missiles, this would bring to 720 the number of SLBMs (submarine-launched ballistic missiles) operational with the Navy. This figure of 45 vessels remained remarkably consistent over the next two years, although opinion as to the deployment rate varied. Some admirals wanted all 45 boats in service by 1964, others wanted to ease the schedule back by three years. The Navy eventually got 41 boats by 1967.

1959 April 9 The seven Mercury astronauts were formally announced by NASA administrator T. Keith Glennan at a press conference in Washington, D.C. The chosen men, along with their branch of service and the responsibilities assigned each four months later, were: Lt. Comdr. Alan B. Shepard Jr., Navy (tracking and recovery); Capt. Virgil I. Grissom, Air Force (attitude-control systems); Lt. Col. John H. Glenn Jr., Marines (crew layout); Lt. Malcolm Scott Carpenter, Navy (navigation systems); Lt. Comdr. Walter M. Schirra Jr., Navy (life-support systems); Capt. Donald K. Slayton, Air Force (Atlas launch vehicle); Capt. Leroy Gordon Cooper Jr., Air Force (Redstone launch vehicle). They reported for duty on April 27, 1959.

1959 April 14 The first Atlas D was launched from Cape Canaveral, Fla. It was a failure, as were the second and third on May 18 and June 6, respectively. The first successful Atlas D flight was accomplished on July 28, 1959. Atlas D was the first operational version developed from precursor prototypes employed for testing various elements of the missile's performance and capabilities.

1959 April 15 Responding to a request from the Department of Defense–NASA Saturn ad hoc committee, the Army Ordnance Missile Command (AOMC) informed the Advanced Research Projects Agency (ARPA) that the 430,000-lb. thrust first stage of the Titan ICBM should be the upper stage for the Saturn launch vehicle. On May 19, 1959, ARPA chief Roy W. Johnson concurred with the recommendation and authorized the AOMC to begin discussions with Martin and Aerojet over procurement of an adapted Titan first stage. Also on May 19, the 30,000-lb. thrust Centaur cryogenic stage was designated the third stage for Saturn. This was the final configuration change before Herbert F. York, director of the Pentagon's Defense Research and Engineering, cancelled work on the Titan stage for Saturn in June, saying that the Air Force's Titan-C could do all the work envisaged for Saturn.

1959 April 16 The first RAF launch of a Thor Intermediate Range Ballistic Missile (IRBM) took place from Vandenberg AFB, Calif. Under an agreement between the United States and the United Kingdom, Thor would be based in England,

NASA's first seven astronauts were selected in April 1959. Left to right, front row: Walter M. Schirra, Donald "Deke" Slayton, John H. Glenn Jr. and Scott Carpenter; back row: Alan B. Shepard Jr., Virgil I. Grissom and L. Gordon Cooper Jr.

and the first missile was delivered on September 19, 1958. Deployment of all 60 missiles was completed on April 22, 1960, when the first RAF squadron went operational. All squadrons were operational by 1961, but they were withdrawn as part of a reciprocal agreement with the Soviet Union after the 1962 Cuban missile crisis. The last launch complex in Britain was deactivated on August 15, 1963. Reliability of the missile had increased measurably. By the end of the program a total 71 Thor IRBMs had been launched on test, of which 49 (69%) were successes. A further 43 flights in support of ballistic tests not associated with the IRBM program had a success rate of 86%. As space launch vehicles, almost 550 flights (to 1994) with Thor and Delta derivatives displayed a reliability of more than 96%, Delta alone achieving more than 98%.

1959 April 20 NASA placed an order with Douglas for 12 3-stage Thor Delta launch vehicles, each comprising a 150,000-lb. thrust DM-19 first stage, a modified Vanguard second stage carrying the 7,500-lb. thrust Aerojet AJ-10-142 and the third stage, a solid propellant 3,000-lb. thrust ABL X-248 Altair. Very similar (in external appearance identical) to the Thor Able, this new launcher was destined for development into a major launch system. With a payload capability of 200 lb. to low Earth orbit, the first Thor Delta was launched on May 13, 1960. Numerous variants, adaptations and growth versions would keep Delta going as a satellite launcher until the next century. The most recent version produces a thrust of almost 1 million lb. and is thus capable of putting a payload of 11,110 lb. into low Earth orbit.

1959 April 23 In what would eventually be seen as a fatal switch in the categorization of the Dyna-Soar boost-glide project, Dr. Herbert F. York, director of the Pentagon's Defense Research and Engineering, established its key objective as the "nonorbital exploration of hypersonic flight." The Air Research Development Command had objected to developing Dyna-Soar primarily as a potential weapon system, thus denying it an Earth-orbit objective. Eventually, Dyna-Soar would become exclusively orbital in its projected role, but its lack of a clear mission status would ultimately bring about its demise.

1959 April 24 The U.S. Air Force amended the Agena development contract with Lockheed to incorporate design and fabrication of a new and improved Agena B. On December 11, 1959, NASA cancelled the planned Atlas-Vega launch vehicle in favor of the Atlas-Agena B, but the first combination to fly with the new stage was the Thor-Agena B. This adopted the uprated DM-21-improved Thor stage of 170,000-lb. thrust, carrying the Lockheed Agena B with a restartable Bell Hustler 8096 (XLR-81-Ba-11) motor rated at 15,000-lb. thrust. The stage itself had a length of 23 ft. 7 in. – 4 ft. 3 in. longer than Agena – and carried more propellant for increased burn time. Thor-Agena B could place a 3,000-lb. payload in a 115-mi. orbit. When used with Atlas, the combination adopted the Bell Hustler XLR-81-Ba-9 and could place a 5,800-lb. satellite in a 345-mi. orbit, send a 750-lb. probe to the moon or a 450-lb. package to Mars or Venus.

DeMarquis D. Wyatt, assistant to Abe Silverstein, director of NASA's Space Flight Development, explained to the House Committee on Science and Astronautics why his agency was asking for $3 million to support orbital rendezvous studies. Wyatt explained that NASA believed a manned vehicle to succeed Mercury should be capable of demonstrating an ability to find and link up with a vehicle already in orbit. Rendezvous, he said, would become the fundamental objective of future manned operations in Earth orbit, leading to docking with a space station for the transfer of crew and supplies. In citing the possible need for a two-man Mercury spacecraft to carry out tests with orbital rendezvous and docking, Wyatt was describing what three years later would result in the two-man Gemini, the precursor to Apollo.

1959 April 30 Capturing for its own plans all eight Atlas-Vega missions planned by NASA at this date, the Jet Propulsion Laboratory completed and delivered to NASA the study "Exploration of the Moon, Planets and Interplanetary Space."

It proposed 12 flights between August 1960 and March 1964, beginning with the first of seven Atlas-Vega missions sending a spacecraft on a lunar fly-by, followed by a Mars fly-by launch in October 1960, a Venus fly-by in January 1961, a Lunar hard-lander in June 1961, a lunar satellite in September 1961, Venus orbiter and Venus entry-probe launches in August 1962 and a Mars entry-probe launch in November 1962. Also in November 1962, the first of four Saturn I launches would lift a Mars orbiter into space, followed by a lunar orbit and return mission in February 1963, a lunar soft-lander in June 1963, and a Venus soft-lander in March 1964. This report formed the basis for NASA's lunar and planetary research program, and when Atlas-Vega was cancelled in December 1959 the missions were distributed among Atlas-Agena flights.

In Britain, the Air Council agreed to base Blue Streak long-range ballistic missiles in underground silos widely dispersed throughout the countryside. The RAF, which would operate the weapon, expected to have about 100 missiles based in underground concrete structures for protection from surprise attack. Engineers wanted at least 300–500 ft. of solid rock within which to locate the silos and that was lacking throughout eastern England. It had been thought desirable to put them on the eastern side to derive maximum value from the missile's range in reaching targets in the Soviet Union.

1959 April In budget hearings for fiscal year 1960 funding requests, NASA described a research and development project for an advanced meteorological satellite. Work on this project – designated Nimbus, from the Latin for *cloud* – began at NASA during August 1959, and on March 8, 1960, the U.S. Weather Bureau (later the National Oceanographic and Atmospheric Administration, NOAA) solicited proposals for an engineering study of an infrared spectrometer which could be carried into space by a weather satellite. The Weather Bureau's Panel on Observations over Space Data Regions endorsed the need for a research satellite beyond the Tiros family, and in fall of 1960 NASA issued a request for proposals from industry to build a Nimbus satellite.

In the first formal proposal for a life sciences program, NASA's Office of Space Science described a goal of launching a recoverable capsule containing living beings to be exposed to the space environment before returning to Earth for analysis. In November 1960 NASA produced a working document defining the framework for such an effort, suggesting a flight project involving flights with frog eggs, germinating seeds, bacteria and algae. The Space Science Board of the National Academy of Sciences suggested possible experiments examining the effects of day-night cycles and radiation on living things in space. In July 1962 NASA announced that six centers were examining the plan for 3–6 orbital flights dedicated to biological research. Life science work at NASA was in disarray, proper management practice among life scientists was nonexistent and Congress repeatedly refused to fund research.

1959 May 9 In what amounted to the first concept for the future Ranger lunar impact probe, NASA's chief of Space Vehicle Propulsion, Milton W. Rosen, sent Space Flight Development chief Abe Silverstein a memo proposing a simplified lunar exploration program to start with camera-carrying spacecraft taking high-resolution pictures prior to the impact of a lunar probe. This concept matured into a TV-equipped impacting spacecraft that sent back to Earth images up to the point it impacted the lunar surface and was destroyed. Although only one of several scientific tasks eventually selected for Ranger, this was in fact the only one that worked.

1959 May 12 A long-awaited statement on British government policy on space research was issued by Prime Minister Harold Macmillan in the House of Commons. It said that design studies were in hand which would determine the feasibility of converting military rockets into tools for space research, that scientific instruments were to be developed for use in space and that possible cooperation with the United States would be examined. Professor H. S. W. Massey from University College, London, would visit the United States for talks on the use of American rockets for launching British instruments into space. The government firmly denied that it was planning to launch animals or humans into space.

1959 May 20 Two months after the Soviet Union resumed testing ICBMs and in response to a request from Gen. Thomas S. Power, head of Strategic Air Command, the U.S. secretary of defense approved acceleration of the Minuteman ICBM program as part of the revised defense plans of the Eisenhower administration. Gen. Power became an advocate of a stronger ballistic missile force and quickly applied Air Force intelligence to his argument, frequently against the conclusions drawn by Army, Navy and CIA intelligence sources, which all said that the Soviets were not as far along in ICBM deployment as Gen. Power thought.

1959 May 27 The first living things carried on a U.S. rocket and recovered safely were lobbed 300 mi. downrange from Cape Canaveral, Fla., by a Jupiter missile launched at 11:35 UT. This second test involving primates—a 7-lb. Rhesus monkey named Able, and a 1-lb. squirrel monkey, Baker—also carried yeast, corn, mustard seed, fruit fly larvae and human blood. Another test involving the eggs and sperm of sea urchins examined fertilization in weightlessness. The mammals' heart rate, pulse rate, muscle reaction and respiration was measured. This time the mammals were recovered after the 15-min. flight and 9 minutes of weightlessness, although Able died on the operating table several days later. Baker lived for almost another 30 years, but died on November 27, 1984, of kidney failure.

1959 May In the Soviet Union, preliminary design of the Vostok manned spacecraft was completed. The spacecraft would weigh approximately 10,400 lb. and consist of two elements: a spherical capsule, called a descent module, 7 ft. 6 in. in diameter and weighing about 5,400 lb. and equipped with an ejection seat for the single, strapped-in occupant, and an instrument module with a length of 7 ft. 7 in., a diameter of 7 ft. 6 in. and a weight of about 5,000 lb. With descent and instrument module integrated and held together by four equally spaced metal straps, overall length was 14 ft. 5 in. Vostok would have a pressurized atmosphere of oxygen and nitrogen, and up to 16 spherical gas containers were attached to a ring forming an upper ledge to the instrument section. An umbilical carried consumables, including electrical power from batteries, from the instrument section to the descent section. Shrouded by the conical base of the instrument section, the TDU-1 hypergolic retro-rocket designed by A. M. Isayev developed a thrust of 3,500 lb.

1959 June 4 Robert Gilruth, director of NASA's Space Task Group (STG), suggested a study should be made of a post-Mercury manned flight program using vehicles capable of aerodynamic maneuvers during descent for land landings at selected areas. On August 12, the STG's New Projects Panel met for the first time and Alan B. Kehlet from the Flight Systems Division was asked to start a program aimed at conceptualizing a spacecraft capable of carrying three men, maneuvering in the atmosphere and making a land landing as a backup to water landing. The panel received from McDonnell Aircraft Corp. recommendations on how to develop the basic Mercury spacecraft to meet these needs, including in-orbit maneuvering and controlled-descent capability.

1959 June 9 Dr. Herbert F. York, director of the Pentagon's Defense Research and Engineering, approved Advanced Research Projects Agency (ARPA) funds for development of the cryogenic Centaur, but cut all funds for Saturn. Dr. York, formerly director of the Lawrence Livermore Laboratory and chief scientist at ARPA, had long wanted to get the ABMA's (Army Ballistic Missile Agency) Development Operations Divisions out of the Army and into NASA, and had become frustrated with the directed way the Army was apparently committing valuable resources to the Saturn heavyweight-satellite and spacecraft launcher. He cabled Roy Johnson, ARPA's director, claiming that there was "no military justification" for Saturn and that a proposed Titan-C from the Martin Co. would accomplish all the Army wanted from Saturn. Titan-C had been proposed as a launcher for Dyna-Soar and comprised four first-stage engines delivering a thrust of 600,000 lb. and two engines of the same type in the second stage. Milton Rosen from NASA and Richard Canright from ARPA drafted a memorandum arguing for continued funding until a special review committee had examined all related issues. Dr. York agreed and from September 16 to 18, 1959, a booster evaluation committee sat in judgment.

1959 June 11 The Advanced Research Projects Agency (ARPA) issued a $600,000 study to Radio Corporation of America (RCA) for studies of satellite interception techniques. Called SAINT, the project originated in the Air Research Development Command during 1956, but ARPA took the work on after the post-Sputnik reorganization of space responsibilities. The RCA study was under the control of Dr. Robert C. Seamans, later secretary of the Air Force. In August 1959 the Air Force Ballistic Missile Division proposed a coorbiting vehicle that could approach and rendezvous with a potentially hostile satellite for inspection purposes and on April 5, 1960, applied for permission to set up a demonstration program. SAINT was formally approved on August 25, 1960.

1959 June 12 Reacting both to growing concern at the slow progress made on the proposed Atlas-Vega launch vehicle and to a request for help with an unmanned lunar exploration program based on its Agena upper stage, Lockheed submitted to NASA a four-flight plan to send four separate probes to the moon during the second half of 1960: two lunar satellites and two soft-landers each weighing about 750 lb. Lockheed pointed out that by using Atlas-Agena with its inertial ascent guidance no midcourse maneuver would be necessary en route to the moon, thus saving spacecraft weight and complexity. It would, said Lockheed, be possible for Atlas-Agena to send a spacecraft to within 300 mi. of the lunar aim point without postinsertion trajectory corrections. Coupled to the Jet Propulsion Laboratory (JPL) study completed April 30, 1959, Lockheed's program was instrumental in shaping NASA's moon exploration program. Following cancellation of Atlas-Vega in December, JPL completed evaluation of the Lockheed proposal on January 13, 1960, and endorsed adoption of Agena for all JPL spacecraft missions.

1959 June 22 The first U.S. Air Force ballistic missile, the Thor IRBM (Intermediate Range Ballistic Missile), was deployed in England, the first of 60 operational missiles that would all be deployed by April 22, 1960. The Thor units had been activated for some months. Thor was fitted with a 1.44 MT W-49 warhead fitted to an Mk.2 reentry vehicle. With the appearance of a flattened cone, the heat-sink vehicle had a weight of 1,200 lb., a diameter of 5 ft. 3.6 in. and a height of 5 ft. 3.5 in. The Mk.2 was also the first operational heat-sink reentry vehicle. The missile and its warhead had a range of 1,700 mi. Britain's Royal Air Force deactivated the Thor force in August 1962 and the U.S. Air Force Strategic Air Command withdrew the warheads, which it had continued to own and maintain.

1959 June 25–26 At the second meeting of the NASA Goett Committee, Bruce T. Lundin from the Lewis Research Center described propulsion requirements for manned moon missions based on a 10,000-lb. Earth-return vehicle (having, as it turned out, almost exactly the weight of the Apollo command module). From NASA headquarters, John Disher pointed out that the Army Ballistic Missile Agency (ABMA) was developing advanced moon plans for its huge Saturn launchers and that this organization should be embraced by NASA. The U.S. Air Force had already surrendered the 1.5-million-lb. thrust F-1 to the civilian space agency. In discussing moon mission plans, Disher said studies should continue on direct-ascent as well as Earth-orbit rendezvous modes. In the former, a giant Saturn or Nova launch vehicle would send a heavy moon lander direct from the surface of the Earth to the moon; in the latter, a series of separate flights with smaller rockets would assemble a manned vehicle in Earth orbit first. From Langley, Laurence K. Loftin Jr. described a projected manned space station and urged that stations and Earth-orbiting laboratories be accepted as an integral part of plans for a lunar landing.

1959 June 30 A review by the Advanced Research Projects Agency of U.S. military space projects identified key areas for expansion. Space-based military communication systems were to be developed in a program called Notus, consisting of four elements: Courier, Steer, Tackle and Decree. Courier comprised a 510-lb. satellite, and the first, *Courier 1B,* was successfully launched into a 601 × 755 mi. orbit inclined 28.3° by a Thor-Able-Star launch vehicle on October 4, 1960. It was the first active-repeater communication satellite and operated for 17 days. Steer was the secret code for a strategic polar communications satellite system which was to have comprised four polar-orbiting satellites for continuous communication between Strategic Air Command aircraft and ground stations in the United States. Tackle was the code name for an advanced polar communication satellite system to evaluate advanced technologies and microwave communication. Decree was to have been a global communications satellite for instantaneous message relay comprising seven geostationary satellites for broadband point-to-point communications with military forces worldwide. In 1960, Steer, Tackle and Decree folded into the single research and development project called Advent. It was never developed.

1959 June During the month, the Dyna-Soar source selection board concluded its examination of the proposals submitted by Martin and Boeing. Secretary of the Air Force J. H. Douglas questioned the cost of the program and called for a reassessment. On November 2, 1959, the Weapons Board of the U.S. Air Force headquarters approved restoration of a three-step plan comprising suborbital, orbital and weapon systems development first mooted on October 10, 1957. Step I would be a 6,570–9,410-lb. manned glider launched by Titan I on suborbital test flights, Step II a bigger Titan for orbital tests and Step III a full orbital military capability.

1959 July 1 The first operational reactor test in the Atomic Energy Commission (AEC) research program to produce a nuclear rocket propulsion system, the Kiwi-A reactor was run for the first time. Named after the flightless bird from New Zealand, since Kiwi could never leave the ground, the graphite core reactor was fired for 5 min. at a temperature of 3,000°C and generated 70 MW of thermal energy. Two more Kiwi-A tests were conducted, on July 8, 1960, and October 19, 1960, to check reactor design. During that year, the AEC and NASA formed a working agreement to proceed with development of a space nuclear propulsion system.

1959 July 13 Dr. Homer E. Newell of the Office of Space Sciences in NASA's Office of Space Flight Development set up the Lunar Science Group (LSG). Initial members included George M. Low, Robert Jastrow and Milton Rosen. This group replaced the earlier ad hoc Working Group on Lunar Exploration chaired by G. Schilling and would establish the course of early NASA lunar exploration objectives, shaping goals and mission plans around the systematic reconnaissance and scientific probing of the moon. On December 3 and 4 the LSG submitted recommendations to NASA administrator T. Keith Glennan endorsing a program based around the Atlas-Agena B launcher.

1959 July 16 The first Juno II/Phase II launch took place from Cape Canaveral, Fla., at 17:37:03 UT, but the vehicle was destroyed by the range safety officer only 5.5 sec. later after a malfunction in the electrical system caused the guidance equipment to fail, tipping the vehicle over. The 91-lb. *Explorer S-31* satellite was to have studied earth's radiation. Another Juno II launched on August 14, 1959, failed to put into orbit the 10-lb. *Beacon 2* satellite comprising a 12-ft. inflatable sphere stowed inside a cylinder 7 in. in diameter. The third Juno II/Phase II launch on October 13, 1959, successfully placed *Explorer VII* in a 345 × 676 mi. orbit where it recorded cosmic- and x-ray radiation and heat balance information. Of the 5 further attempts culminating in a failure on May 24, 1961, only *Explorer VIII* (November 3, 1960) and *Explorer XI* (April 27, 1961) were successfully launched. Of 10 Juno II/Phase I and II flights, only 3 were a complete success.

1959 July 22 NASA selected the B. F. Goodrich Co. to design and manufacture the Mercury astronaut pressure suit. During orbit and reentry, the Mercury spacecraft would be pressurized with oxygen to 5.5 lb./sq. in. above ambient pressure. During ascent, the cabin would bleed down to an absolute pressure of 5.5 lb./sq. in. In this way there would always be a 5 lb./sq. in. difference between the spacecraft and the external environment in the vacuum of space. Should anything go wrong with the environmental control system or the wall of the spacecraft rupture or be punctured by a micrometeoroid, the suit would be required to keep the astronaut alive at this pressure during the return to Earth. The seven Mercury astronauts were fitted with prototype suits for the first time at the company's Akron, Ohio, factory on November 5, 1959. Between this date and January 1960, 10 development suits were delivered for evaluation. There were to be 8 operational suits, 1 for the "eighth astronaut:" the astronauts' physician Dr. William K. "Bill" Douglas, whose lot it was to duplicate every test the astronauts themselves carried out.

1959 July 23   The redirection of NASA's lunar and planetary program to achieve political objectives and gain maximum international credit for the United States was set up during a lengthy meeting between top space officials, the Office of the President and the National Security Council. The NSC had called for the use of unmanned deep-space missions on schedules contrived to demonstrate U.S. capabilities in science and technology, but when NASA administrator Dr. Glennan advised that launch windows to the moon came at monthly intervals whereas launch windows to Mars and Venus occurred once every two years, lunar missions were ordered as priority flights. Following this meeting, a revised flight schedule was established authorizing a total of nine lunar missions and one interplanetary mission in the period from October 1958 through December 1960.

1959 July 27   At the Pentagon, Dr. Herbert York, director of Defense Research and Engineering, directed the Air Force and Advanced Research Projects Agency (ARPA) to seek a common upper stage that could be used for the Saturn launch vehicle and the launcher that would be used to carry the Dyna-Soar boost glider. Two days later, Dr. York suspended work at ARPA on the Saturn second stage pending the outcome. On July 31, 1959, that order was amended, limiting it to work on Titan adaptation as the Saturn second stage. This effectively removed ICBMs as candidate upper stages for Saturn—to the relief of NASA and the Army Ballistic Missile Agency. Work could now begin on designing new upper stages for the clustered booster instead of compromising on performance by using converted missiles. The first of these plans emerged with the report of the Saturn Vehicle Team on December 15, 1959.

1959 July 29   Following a visit to the United States by a British delegation headed by Prof. Massey of University College, London, the government of the United Kingdom announced it had reached an agreement whereby British satellites would be launched on NASA Scout rockets beginning in 1961. This announcement marks the formal beginning of Britain's national space program and was expected to result in several launches over a three or four year period with scientific results released to the international community.

1959 July 30   NASA awarded a contract to Western Union for tracking and ground control systems for use in the Mercury manned spacecraft program. With experience on the Real-Time-Computing-Complex (RTCC) for the Vanguard Earth-satellite program, IBM received a contract in 1959 for Mercury ground computers and software. The spacecraft would not itself carry digital computers, but IBM 7090 mission control computers and software formed the core for future man-machine interface on an unprecedented scale and forged a uniquely successful link between NASA and IBM that would underpin future success in flight operations.

1959 August 7   One of the most successful of NASA's early science satellites, the 143-lb. *Explorer VI* was successfully placed in a 152 × 26,345 mi. orbit inclined 47°. The first Explorer launched by NASA and the only one with a Thor-Able III, it had a 153,000-lb. thrust Thor first stage, a 7,500-lb. thrust Aerojet second stage and an Altair third stage based on the Vanguard upper stage. A 5-lb. thrust Atlantic Research Corp. 1KS 420 injection stage was attached to the

satellite for tweaking up the apogee if necessary, which was not. Built by Space Technology Laboratories, *Explorer VI* was an irregular spheroid with a diameter of 2 ft. 2 in. and a height of 2 ft. 5 in. and was the first satellite to carry solar paddles (each 20 × 22 in.), four of which were attached to struts at 90° intervals around the satellite's waist, presenting a total span of 7 ft. 2 in. Each of the eight paddle surfaces carried 1,000 solar cells and fed NiCd batteries. *Explorer VI* sent to Earth the first complete, televised cloud-cover picture, obtained as the satellite passed over Mexico City. Data was transmitted until October 9, 1959, and the satellite decayed on July 1, 1961.

1959 August 12   NASA's Space Task Group, established to manage the Mercury manned spacecraft program, held the first meeting of its New Projects Panel, a think tank for developing strategy on long-term goals and objectives. If there was a single point at which the Apollo program began, it was during this meeting chaired by H. Kurt Strass. With the ultimate objective of a manned moon landing, intermediate steps were outlined involving a new, three-man spacecraft capable of two-week missions in Earth orbit for Earth environment studies. Alan B. Kehlet was charged with setting up a program to design the spacecraft, which had to be capable of returning to Earth from the vicinity of the moon. At a second meeting six days later the panel decided that the three-man spacecraft should be designed first and that 1970 was a valid target date for the first manned moon landing.

1959 August 17   A document describing NASA's 10-year program for space science endorsed development of an Orbiting Solar Observatory (OSO), for which negotiations had begun with Ball Brothers Research Corp. By September 30, 1959, Ball had completed initial design studies for a series of spacecraft weighing 300 lb., each equipped with instruments

Wind shaft tests with a Mercury spacecraft configuration gave NASA engineers the opportunity to test drogue parachute configurations for safe retrieval of the manned vehicle.

built to observe the sun free from the obscuring effects of earth's atmosphere. Because each spacecraft would carry an interchangeable suite of instruments carried as passengers on a common spaceframe, OSOs became known as "streetcar" satellites. This pioneered the idea that a common platform, or "bus," built on a small production line would be more economic than separate, purpose-built designs. The first OSO contract was signed with Ball in October 1959 and OSO-1 was launched on March 7, 1962.

1959 August 19   President Eisenhower's new science adviser, George Kistiakowsky, was tipped off about chaos within the wide range of highly classified space projects run by the U.S. Air Force. The SAMOS (satellite and missile observation system) program was singled out as being particularly troublesome and, intending to simplify the objective, Kistiakowsky in the spring of 1959 overruled an Air Force decision not to adopt the recovery method for image retrieval. It was a short-lived decision, however. By fall of 1960, the Air Force got its way when concern in Congress over the loss of Gary Powers and the U-2 spy plane during an overflight of the USSR injected a massive increase in funds for research and development. This enabled scientists to overcome technical hurdles with the radio transmission concept and retain it as the method used for SAMOS data.

1959 August 21   The first live hardware test of the Mercury program involving a rocket flight failed to achieve its objective when transient electrical signals prematurely fired the Mercury launch escape systems rocket 30 min. before the planned liftoff time. The boilerplate Mercury spacecraft was attached to the first Little Joe rocket, LJ-1, for a test of the launch escape system under severe dynamic conditions. Situated close to the shore at Wallops Island, Va., LJ-1 was to have fired its boilerplate payload on a ballistic trajectory when, instead, the escape motor tore the capsule free and sent it skyward. At an apogee of 2,000 ft., the clamp that held the escape tower to the capsule severed and the tower jettison rocket fired.

1959 August 24   An internal memorandum circulated among staff at the Bell Telephone Laboratories told of an ad hoc group that had been set up to study the possibility of developing an active communications satellite for the American Telephone and Telegraph Co. (AT&T). On October 21, 1960, AT&T formally requested approval from the Federal Communications Commission for an active experiment on the Bell satellite. Approval was granted on January 19, 1961, with allocation of frequencies permitting the satellite to receive signals on 6390 MHz and to retransmit on 4170 MHz. Agreement was reached with NASA for the launch of two 170-lb. Telstar satellites to elliptical orbits using a Delta launch vehicle. This was the first reimbursable, commercial launch for NASA.

1959 August 26   Firing trials with an interim design of the Nike-Zeus surface-to-air missile began at White Sands, N.Mex. under the guidance of the primary contractor, Western Electric, and direction of the U.S. Army. Nike-Zeus was the first U.S. attempt at producing a missile designed to destroy incoming nuclear warheads fired by ballistic missiles. The ABM system would operate in conjunction with BMEWS, ballistic missile early warning system, radar installations which had been set up in Alaska and in the United Kingdom and equipped with sophisticated acquisition radars built by Bell Telephone Labs. The radar installations that arose from the Nike-Zeus program were the largest of their kind to date

and advanced radar technology by leaps. Nike-Zeus was the first to use a discrimination radar which by mechanical and electronic means could separate warheads from decoys. The first Nike-Zeus launch occurred on December 16, 1959, but the definitive missile was fired for the first time on April 28, 1960.

1959 August 27–29   At the Commonwealth Spaceflight Symposium hosted in London by the British Interplanetary Society, Geoffrey K. C. Pardoe, chief engineering coordinator of the United Kingdom's Blue Streak long-range ballistic missile, presented a paper explaining how the missile could be developed into a launcher capable of placing a 1,000-lb. satellite in a 300-mi. orbit. A configuration was suggested in which Blue Streak would be the first stage to a multistage launcher developed in cooperation with other European countries. The second stage, suggested Pardoe, could be a twin-motor arrangement within the protective fairing of a payload shroud on top of the missile.

1959 August 31   An innovative method of braking a satellite for descent through the earth's atmosphere was presented at the 10th Congress of the International Astronautical Federation, which opened in London on this day. The so-called drag-brake method envisaged an inflatable parasol deployed by a satellite in low orbit, so that traces of atmosphere would slow it sufficiently for reentry. Dispensing with weighty retro-rockets, this method would gain credibility and eventually become known as aerobraking, although no practical tests had been carried out by the mid-1990s.

1959 August   Engineers and scientists at the Mathematical Institute of the Soviet Academy of Sciences completed trajectories and mission plans for planetary space flights to Venus and Mars. They deduced that the optimum flight plan was based around an interim Earth parking orbit, where the spacecraft and a rocket boost stage were first placed in Earth orbit, from whence the rocket stage would fire at the appropriate time to accelerate the payload to escape velocity. The earliest launch windows were calculated to be September 1960 for a flight to Mars and January 1961 for a flight to Venus. Plans were made for hasty preparation of a spacecraft, launched during the window of opportunity beginning January 1961, to impact Venus.

1959 September 1   NASA received a report from McDonnell Aircraft Corp. detailing how the one-man Mercury spacecraft could be adapted to carry a special heat shield capable of withstanding entry into earth's atmosphere from lunar distance. In returning from Earth orbit, a capsule would reenter at a speed of about 17,000 MPH, but from the moon the Earth would pull a spacecraft to an entry speed of 25,000 MPH. In a typical plan proposed by this company, a Centaur upper stage would boost a combined third stage and spacecraft to an elliptical Earth orbit with an apogee of 1,200 mi., whereupon the third stage would fire to drive the modified Mercury into Earth's atmosphere at 25,000 MPH. The idea was not taken up by NASA, but it surfaced again during the Gemini program when that spacecraft was proposed for a similar job.

1959 September 9   At precisely 08:19 UT, the first major test in NASA's Mercury manned space flight program began when an early production Atlas D was launched from Cape Canaveral carrying a boilerplate spacecraft 95 mi. into space and 1,496 mi. down the Atlantic Missile Range in a 13 min. flight. Programmed to reach a speed of 16,800 MPH and throw the

Mercury spacecraft 2,000 mi. downrange, the two Atlas booster motors failed to separate at 2 min. into the flight and, encumbered by the additional weight, the vehicle achieved a speed of only 14,857 MPH. The spacecraft experienced 12 g during reentry, and when it was recovered the next day, inspection of the ablation material revealed that less than 30% of the coating had eroded, despite higher-than-planned temperatures.

In what can be considered the beginning of the Jet Propulsion Laboratory's (JPL) lunar reconnaissance project, later assigned the name Ranger, the JPL released the functional design details of its Vega V-1 spacecraft concept. Designed for suitability as a lunar satellite as well as in fly-by missions, it was based on the assumption that Atlas-Vega would be assigned to launch it. At this date, Vega was increasingly seen as an unnecessary duplication of effort and the shift toward Atlas-Agena was well under way. In December, when Vega was cancelled, the JPL modelled the V-1, by then redesignated S-1, on the new upper stage.

The first U.S. ICBM was declared operational when the 576th Strategic Missile Squadron, 1st Missile Division, U.S. Air Force, fired its first missile, an Atlas D, from Vandenberg AFB, Calif. The missile traveled a distance of 4,950 mi. across the Pacific. This was only the seventh Atlas D launch, of which four had been failures. The first Atlas D to go on alert was activated at Vandenberg on October 31, 1959, and from that date the United States officially had an ICBM it could use in war. Deployed on fully exposed, above-ground launch pads known as soft sites, or early semihard coffins, a total of 30 Atlas Ds were declared operational with four squadrons on March 30, 1961, located (chronological order) at Warren AFB (Wyoming), Fairchild AFB (Washington), Forbes AFB (Kansas) and Offutt AFB (Nebraska).

1959 September 12   The first man-made object to impact the moon was launched at 12:36 UT from the Baikonur launch site in the Soviet Union by an SL-3. Called *Luna 2* (sometimes *Lunik 2*), the 860-lb. space probe was tracked by Britain's Bank radio telescope when signals became weak near the moon. The probe was similar to *Luna 1* and departed the Earth in a direct ascent, that is, it was not first placed in an orbit of the Earth. Transmitting on frequencies of 19.993, 19.997, 20.003, 49.986 and 183.6 MHz, *Luna 2* struck the moon approximately 268 mi. from the visible center at 22:04:24 UT on September 13. There was no attempt to decelerate the probe and it fell freely under the moon's gravity and was destroyed when it struck the surface at 7,380 MPH after having first been slowed by the earth's gravity and then accelerated by the pull of the moon.

1959 September 15   Loaded with just enough propellant to get it airborne, the first full-scale U.S. Air Force Minuteman test missile was fired from a silo at Edwards AFB, Calif. With inert upper stages and dummy warhead, several tests like this demonstrated the ability of Minuteman missiles to be hot-launched from silos. With a depth of 80 ft. and a diameter of 12 ft., Minuteman silos were considerably smaller than Atlas and Titan silos, eliminating the need for blast deflectors and exhaust channels. During this time development of a modestly upgraded variant of Minuteman I, designated LGM-30B, was begun. It would have a first-stage increased in length by 2 ft. 2 in., extending the operational range from 4,600 to 5,500 mi.

1959 September 16–18   Dr. Herbert York from the Pentagon's Defense Research and Engineering and NASA's Deputy Administrator Dr. Hugh L. Dryden cochaired a Booster

Launched September 12, 1959, the USSR's Luna 2 *became the first object fired from Earth to hit the moon. Note the communications antenna and the small box-shaped array of solar cells.*

Evaluation Committee to resolve differences concerning the applicability of proposed Saturn and Titan-C launch vehicles. The Saturn proponents carried the day but only on condition that the ABMA (Army Ballistic Missile Agency) be shifted from the jurisdiction of the AOMC (Army Ordnance Missile Command) to NASA. Von Braun stipulated full development of Saturn as his price for moving to NASA. When the committee heard a broad range of NASA missions that Saturn could perform and its competitor could not, Titan-C was canceled. Three NASA propulsion experts, Eldon Hall, Francis Schwenk and Alfred Nelson, set to work on a detailed study of Saturn and candidate upper stages, presenting a report of their conclusions on October 23, 1959.

1959 September 17   The world's first navigation satellite took off from Launch Complex 17A, Cape Canaveral, Fla., when a Thor-Able IV lifted off with the 265-lb. *Transit 1A*. Developed by the U.S. Navy for the Polaris Fleet Ballistic Missile submarines, Transit used the radio-Doppler navigation method and grew out of *Sputnik 1* observations made by scientists at the Applied Physics Laboratory, Johns Hopkins University, Md. The first satellite was lost when the third stage of the Thor-Able IV failed to fire. Three solar-powered Transits were launched in 1960, three in 1962 and two in 1963.

1959 September 18   *Vanguard III* was successfully placed into a 317 × 2,326 mi. orbit, inclined 33.35° to the equator with a period of 129.55'. The three-stage Vanguard rocket was designated TV-4BU and incorporated an Allegheny Ballistics

Laboratory X-248 solid propellant third stage in place of the usual X-405 stage. The spherical satellite weighed 52.25 lb. including scientific instruments for measurement of the Earth's magnetic field and the lower elements of the Van Allen radiation belts. Equipment for the satellite had been provided by the NASA Goddard Space Flight Center and the U.S. Naval Research Laboratory and included a 26-in. magnetometer boom attached to a sphere 20 in. in diameter. This was the last Vanguard satellite and the last Vanguard rocket launch.

Secretary of Defense Neil McElroy assigned booster development for the Department of Defense to the U.S. Air Force, leaving the Army without a mandate for Saturn. A memorandum giving technical direction of the Saturn to NASA was signed by Administrator Glennan on October 21, 1959. Nine days later, on November 2, 1959, it was approved by acting Secretary of Defense Thomas Gates and by President Eisenhower. McKinsey and Co., a firm of private management consultants who had helped set up NASA the previous year, aided in the monumental task of transferring the Development Operations Division of the ABMA (Army Ballistic Missile Agency) to the civilian space agency. NASA assumed full technical and managerial control of the Saturn program on March 16, 1960.

1959 September 21 Research began at the Aviation Medical Acceleration Laboratory, Johnsville, Penn., to measure the ability of a pilot under sustained acceleration to control a space vehicle. Undertaken in support of the NASA Mercury program, a variety of sidearm controllers were used. Instead of using a conventional center stick for flight control, a small hand unit at the extremity of the arm rest enabled the pilot to control his spacecraft under high g-force by rotating or tilting his wrist. A three-axis system controlling roll, pitch and yaw was found most suitable, and this form of attitude control was adopted for Mercury spacecraft.

1959 September 23 The Department of Defense announced that responsibility for all military space transportation was to be vested in the U.S. Air Force. The Advanced Research Projects Agency relinquished interest in space systems with the exception of antimissile defenses. The decision eliminated Army and Navy work on satellite launchers. NASA would get the Saturn launch vehicle program developed at the Army Ballistic Missile Agency and continue it as a civilian launcher. The Air Force was then funding three study requirements on space options: SR-182 on a strategic interplanetary strike force; SR-183 on a lunar observatory keeping Earth under surveillance; and SR-192 on a strategic military base on the moon for control of the solar system.

1959 September 24 The first flight-ready Atlas-Able launch vehicle was destroyed in a massive explosion on Launch Complex 12 shortly after a static firing test when a propellant line ruptured. The vehicle, designated Atlas-Able IV A, had a conventional 360,000-lb. thrust Atlas first stage, a 7,500-lb. thrust Aerojet AJ10-101 second stage and an ABL X-248 Altair third stage. In this configuration, NASA did to the Atlas what the Air Force had done with Thor by taking the upper elements of the (relatively) tried-and-tested Vanguard to upgrade an existing single-stage missile and transform it into a launch vehicle. Atlas-Able IVA was being prepared for the launch of the first in a series of three Pioneer moon probes managed by NASA to complement and extend the previous five Pioneers from the Air Force and the Army. Not one was a success.

1959 September 25 Hughes Aircraft made an informal approach to NASA, offering to develop a geosynchronous communications satellite. Not knowing that the Department of Defense was developing what was expected to be the United States' first geosynchronous satellite of that type under the name Advent, Hughes made a formal presentation on February 4 and 17, 1960. Frustrated over NASA's polite but seemingly indifferent attitude, Hughes considered building the satellite with its own money and buying a Scout vehicle to launch it. On March 1, 1960, the company authorized its engineers to begin work on the satellite, named Syncom, without outside funds. Another presentation on August 16, 1960, this time to NASA Administrator T. Keith Glennan, brought the suggestion that Hughes should perhaps turn its efforts to a low-altitude satellite. Between October and December 1960 Hughes tried in vain to garner support from the Air Force, several U.S. companies and even the British. Recognizing that Advent would take years to get ready, on June 23, 1961, the Department of Defense gave its blessing to NASA's developing a geosynchronous communication satellite of its own and Hughes received a contract on August 10, 1961.

1959 October 2 Science and engineering personnel at the Jet Propulsion Laboratory met to consider future unmanned lunar and planetary missions and to discuss one topic: What kind of a space spectacular might the United States perform that the Soviet Union was incapable of doing? This was the first of two occasions when the same question was asked. This first time, at least one attendee argued that a photographic

Launched on October 4, 1959, the second anniversary of Sputnik 1, Luna 3 *became the first spacecraft to photograph the far side of the moon and transmit those pictures to Earth.*

mission to the far side of the moon would undoubtedly require a level of technical capability denied to scientists in the USSR. Two days later *Luna 3* was launched and took pictures of the far side. The question was repeated, with an uncanny similarity of phrase, after Yuri Gagarin became the first man to orbit the Earth on April 12, 1961, and the answer would result in the Apollo program and its goal of a moon landing.

1959 October 4   *Luna 3* was successfully launched toward the moon by an SL-3 from Baikonur in the USSR. Weighing 614 lb., the probe carried camera equipment for viewing the moon's far side. Placed into a highly elliptical orbit of the Earth (25,040 × 296,000 mi.), *Luna 3's* trajectory brought it to within 4,350 mi. of the lunar surface at 14:16 UT on October 6. As it continued to drift toward apogee, *Luna 3* neared the most distant part on its elliptical path when it passed directly between sun and moon. Utilizing sensors at one end to lock on to the sun, the probe took 35-mm photographs from the other end through 200-mm and 400-mm focal length lenses aimed at the fully illuminated disk of the moon. Taken in a 40-min. period, 17 pictures were then transmitted to Earth, providing the first views of the far side of the moon. *Luna 3* reentered Earth's atmosphere and burned up on April 29, 1960.

A Little Joe solid propellant rocket carrying a boilerplate Mercury spacecraft was launched from Wallops Island in a flight to test the system and spacecraft–launch vehicle integrity. Designated Little Joe-6, the test flight lasted 5 min. 10 sec. The spacecraft reached a height of 37 mi., a downrange distance of 79 mi. and a speed of 3,075 mi. The aerodynamic characteristics of the spacecraft were considered sound. A second test flight, LJ-1A on November 4, 1959, carried a Mercury spacecraft on an abort test where the escape rocket fired to simulate an emergency. The test was not a complete success because pressure in the escape motor built up too slowly and did not realistically duplicate an abort. The flight lasted 8 min. 11 sec. and achieved a height of 9 mi., a downrange distance of 11.5 mi., a speed of 2,022 MPH and a maximum load of 16.9 g.

1959 October 6   The search for an effective reentry cone for U.S. intercontinental ballistic missiles moved a step nearer with the successful flight and recovery of a General Electric Mk.3 warhead atop an Atlas D. The launch took place from Complex 11 at Cape Canaveral, Fla., and was the ninth of an Atlas D since the first of this type on April 14. The Mk.3 reentry cone was fabricated from ablative materials, which are lighter than heat-sink materials. Ablative cones do not leave telltale trails of ionized gases, easy to track on radar, and do not slow down the warhead as much as heat-sink types such as the 1,375-lb. Mk.2 employed by the Jupiter missile.

The first full-scale Blue Steel stand-off bomb was tested at the Aberporth test range in Wales. Dropped from a Vickers Valiant bomber, the missile was powered by a Double-Spectre rocket motor and full range tests were conducted later from the Woomera range in Australia. The Avro Vulcan bombers of 617 Squadron, Royal Air Force, went operational with Blue Steel in September 1962, followed by Vulcans of 27 Squadron and 83 Squadron and the Handley Page Victors of 139 Squadron and 100 Squadron. Released from high altitude, the Blue Steel had a range of 200 mi., or 110 mi. from low level. The weapon was withdrawn from service between 1973 and 1975.

1959 October 13   The first successful satellite-intercept demonstration took place under the Bold Orion project when a highly classified Martin Co. two-stage missile designated

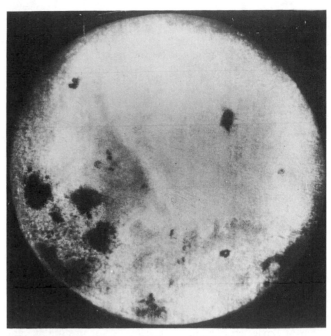

The moon's far side, as photographed by Luna 2. *Immediately apparent is the absence of vast lava sheets, characterized on the moon's near side by dark basaltic surfaces.*

ALBM 199B was fired by a Boeing B-47 bomber in flight using the *Explorer VI* satellite as a target. The missile passed within 4 mi. of the satellite, close enough to disable it had the interceptor detonated a low-yield nuclear charge. Under Project Bold Orion, designated U.S. Air Force 7795, the Air Research and Development Center had given the Martin Co. a contract in March 1958 to develop a 12-shot program using one- and two-stage rockets. The first airdrop from a B-47 had been conducted on May 26, 1958. Bold Orion was Martin's failed contender for the GAM-87 Skybolt requirement.

A Juno II launcher placed the 91.5-lb. NASA scientific satellite *Explorer VII* into a 342 × 680 mi., 50.3° orbit. With conical upper and lower sections attached to a narrow ring and equal height and diameter of 2 ft. 6 in., the satellite was built by the Army Ballistic Missile Agency under the guidance of Dr. Ernst Stuhlinger, one of the German scientists brought to the United States after the war. *Explorer VII* carried instruments for measuring the sun's radiation, the intensity of x rays and ultraviolet rays, heavy cosmic rays and the composition of the earth's ionosphere as well as for recording micrometeorite impacts. It continued to provide data until August 24, 1961, and remains in orbit.

1959 October 23   NASA propulsion analyst Eldon Hall sent Abe Silverstein, head of Space Flight Development, a summary of potential Saturn launch vehicle configurations he had prepared with the help of colleagues Francis Schwenk and Alfred Nelson. The group recommended a Saturn B-1 consisting of a 400,000-lb. thrust Titan and 30,000-lb. thrust Centaur upper stages. Saturn B-2 would replace the Centaur with an enlarged stage and have uprated first-stage engines. A Saturn C-1 configuration would comprise an upgraded clustered first stage, a 600,000-lb. thrust second stage and the enlarged Centaur from the B-2. This initial study by NASA of what it might do with the recently acquired Saturn program contrasted with configurations engineered at the ABMA

(Army Ballistic Missile Agency), which was soon to be absorbed into the civilian space agency. The ABMA proposals were presented on November 2, 1959. A Saturn Vehicle Team chaired by Abe Silverstein was put together at NASA on November 27 to converge these various configurations into a definitive series of Saturn launchers and their report was completed on December 15, 1959.

1959 October 28   The first of a series of test launches using high-altitude sounding rockets to check deployment of inflatable balloons in space took place when NASA launched Project Shotput. Fired to a height of 250 km. by a Sergeant-Delta, the experiment canister held a folded balloon which was designed to inflate to a diameter of 100 ft. On the first launch the sphere ruptured, preventing deployment, as it did on the second launch January 16, 1960. The first successful deployment occurred on the third attempt, February 27, 1960, and radio transmissions were reflected via the sphere from Holmdel, N. J. to Round Hill, Mass. A second attempt on April 1, 1960, was also successful and inflation occurred at a height of 236 km., clearing the way for an attempt to orbit a balloon called *Echo I* on May 13, 1960.

1959 October   The preliminary list of candidates for the first team of Soviet cosmonauts was screened for unsuitable candidates at the Bordenko military hospital in Moscow. The head of cosmonaut training was Col. Gen. Nikolai P. Kamanin assisted by his deputy, Col. E. A. Karpov, who specialized in aviation medicine. All candidates were from the Soviet air force since pilots were considered to have the right qualities for space flight. The screening process included personal interviews, medical examinations and physical tests to judge their suitability for resisting physical stress. The final selection of 20 pilots was completed by February 25 and they all reported for duty on March 14, 1960.

Rationalizing several different proposals put forward by the U.S. Air Force Ballistic Missile Division (BMD) and the Strategic Air Command (SAC) for the ideal number of Minuteman ICBMs to deploy, an official Air Force compromise was reached on 1,200 missiles. The BMD had wanted to deploy a single "missile field" of 1,000–1,500 solid propellant Minuteman missiles which could be launched all at once in a salvo of such ferocity that it would, according to the theorists, have obliterated all enemy opposition. For its part, the SAC wanted many more missiles—up to 10,000—although the official Air Force figure was 2,000–3,000, closer to that of the BMD.

1959 November 2   With the help of NASA's Abe Hyatt and Adelbert Tischler, Eldon Hall completed a summary of Saturn configurations proposed by the ABMA (Army Ballistic Missile Agency). The configuration of Saturn B initially reviewed, with Titan and Centaur upper stages 10 ft. in diameter, had now been given a Titan of specially increased diameter (13 ft. 6 in.) to cut bending loads from aerodynamic forces afflicting the original long and slender configuration. The four-stage B-1 had a four-engine 400,000-lb. thrust Titan second stage of 18 ft. 6 in. diameter, a cryogenic third stage of the same diameter with four RL-10 engines and a total thrust of 80,000 lb. and an uprated cryogenic Centaur fourth stage of 40,000-lb. thrust. The B-2 was identical to the B-1 but with six RL-0 engines for a third stage thrust of 120,000 lb. The Saturn B-3 had an optional first-stage cluster of eight H-1s, or four H-1s and one 1.5-million-lb. thrust F-1. The second stage was the same as the B-1, but the cryogenic third stage had two new engines for a combined thrust of 300,000 lb., and the fourth

stage was identical to the third stage of the B-1. A proposed B-4 had eight H-1s in the first stage, new cryogenic second and third stages of 600,000-lb. and 300,000-lb. thrust, respectively, and the same stage four as the B-3. A Saturn C had the same configuration as the B-3 but with a new cryogenic second stage of 900,000-lb. thrust.

1959 November 9   The U.S. Air Force declared the Boeing-Vought team the winner of the competition for detailed design and assembly of the Dyna-Soar aerospace vehicle. Martin Marietta would build the Titan launch vehicle used to boost it into space. Eight days later the Air Force officially designated Dyna-Soar as System 620A and on April 27, 1960, ordered ten production vehicles (serial numbers 61 2374 through 61 2383) of which eight would be used for space flights: two to be delivered in 1965, four in 1966 and two in 1967. The other two were for static and drop tests. Air-launched test drops were scheduled to commence in July 1963.

1959 November 10   A very large five-stage sounding rocket named Strongarm made its first flight from Wallops Island. It had been developed by the U.S. Army Ballistics Research Laboratory with the cooperation of the University of Michigan. Strongarm used an Honest John as the first stage, two Nike missiles as second and third stages, a modified Recruit rocket for the fourth stage and a reduced-scale Sergeant for stage five. Strongarm could lift a 150-lb. load of experiments to a maximum altitude of 1,000 mi.

1959 November 24   Assistant Secretary of the Air Force Dr. J. V. Charyk rejected funding requests for the fiscal years 1959 and 1960 for the Dyna-Soar boost-glide program, citing concern from the Air Force Scientific Advisory Board about the "inadequacy of technical knowledge in the areas of aerodynamics and structures." Dr. Charyk outlined a Phase Alpha research program before formal development and funds were restored. Phase Alpha began December 11, 1959. By January a more conservative program envisaged formal go-ahead in mid-1960, drop-tests of Dyna-Soar glider models in 1963 and unmanned and manned flights in 1964.

The U.S. Department of Defense asked the Atomic Energy Commission to team with the Navy on the development of a clustered atomic warhead (CLAW), later known as a multiple reentry vehicle (MRV), for the Polaris A3. The CLAW concept envisaged nuclear bomblets from a single warhead dispersed against a target like buckshot. The Department of Defense was concerned about the development of Soviet antiballistic-missile missiles and the proliferation of targets. Designated W-58, the warhead was to replace the W-47 in earlier versions of the missile and in 1962 prototypes designated XW-58 were tested in Operation Dominic. Production of the 200-KT W-58 began on March 19, 1964, and the first Polaris A3 missiles with W-58 warheads arrived on station aboard the USS *Daniel Webster* in October of that year.

1959 November 26   The 372-lb. *Pioneer P-3*, which should have been sent to the moon from Launch Complex 14, was destroyed when the payload shroud from Atlas-Able IVB separated 45 sec. into flight and the ascending vehicle broke up 25 sec. later. The vehicle had been launched at 07:26 UT. Pioneer carried two midcourse correction rockets which could be fired en route to the moon on command from either the Jodrell Bank observatory, England, or an observatory in Hawaii. The probe was then to have injected itself into a circular orbit of the moon, 3,000–4,000 mi. above its surface, and to have transmitted crude outlines of the surface via a 2-lb. scanning device.

1959 November An outgrowth of WS-117L, a plan to develop a missile defense alarm system (MIDAS) was transferred from the intelligence community to the U.S. Air Force. Initial plans scheduled ten flights, with two in 1960 and four each in 1961 and 1962. The fully operational system was expected to comprise eight satellites, with four equally spaced in each of two orbits. Comprising infrared sensor packages attached to an Agena, the Hustler engine would be used to maintain a proper separation between satellites.

The Canadian government received a proposal from Bristol Aerojet of Winnipeg for a family of powerful sounding rockets covering a variety of different payload configurations. Contracts for development were awarded the following month for the Black Brant series of rockets. Named after a species of geese in the genus *Branta,* the rockets evolved from a series of Black Brant I and II test rockets and research vehicles. Black Brant III would be capable of lifting a 40-lb. package to 110 mi., the two-stage Black Brant IV would carry 40 lb. to 600 mi. and a single-stage Black Brant V would lob 150 lb. to a height of more than 200 mi.

1959 December 1 NASA held a briefing for 150 potential industry participants in its Orbiting Astronomical Observatory (OAO) program to provide information on planning and scheduling. In February 1960 technical management of the OAO program was placed with the Goddard Space Flight Center, and between April and September of that year NASA examined eleven industry proposals to build the satellite. A contract was signed with Grumman Aircraft Corp. on October 10, 1960, for a 3,000-lb. satellite which would be launched by an Atlas-Agena D. By June 1961 Grumman had completed negotiations with a team of subcontractors and the Goddard Space Flight Center was managing the instruments program involving Princeton, the University of Wisconsin, the Smithsonian Astrophysical Observatory and the University of Michigan Observatory. In January 1962 Kollsman Instrument Co. was contracted to provide the primary mirror.

1959 December 3 NASA's Jet Propulsion Laboratory released its first ten year plan that out a vigorous program of lunar and planetary exploration "which will result in the United States' regaining and maintaining a position of world leadership in this area of technology." In a proposed fifteen flights, the JPL identified four lunar missions in 1961, two of which were survivable capsule impact flights, and two planetary missions using a common spacecraft bus in 1962. Capsule and probe flights to Mars and Venus would be flown in 1964, followed by a total of five soft landings on Venus and Mars in the period 1965–1969. An interplanetary probe was proposed for 1965, followed by a comet mission in 1966, a probe to the planet Mercury in 1967 and another to Jupiter in 1968, followed by a solar probe in 1969. The key element which prevailed throughout the next 20 years was commonality of design, utilizing a similar bus for different science missions.

1959 December 4 The first of two Mercury spacecraft abort tests involving primates launched on Little Joe boosters took place from NASA's Wallops Island launch site. Designated LJ-2, the flight demonstrated an initiated abort at 59 sec. elapsed time, an altitude of 18.2 mi. and a speed of Mach 5.5 (4,466 MPH). The spacecraft was boosted by the escape motor to a height of 53 mi. and a downrange distance of 194 mi., and a rhesus-monkey named Sam experienced a 14.8 g reentry

without ill effect, landing in the Atlantic Ocean 11 min. 6 sec. after liftoff. The second primate test, *LJ-1B,* was launched on January 21, 1960, when another rhesus monkey, Miss Sam, attained a maximum altitude of 9 mi., a speed of 2,022 MPH and a downrange distance of 11 mi., experiencing 4.5 g in simulation of a low-altitude abort.

1959 December 8 In the first major reorganization of NASA Headquarters, Abe Silverstein's Office of Space Flight Development was renamed Office of Space Flight Programs, retaining its three separate divisions: Advanced Technology, Space Science and Space Flight Operations. Out went the Propulsion division, replaced by a new Office of Launch Vehicle Programs to be headed from January 1 by U.S. Air Force Gen. Don R. Ostrander from the ARPA (Advanced Research Projects Agency). It incorporated Wernher von Braun's ABMA (Army Ballistic Missile Agency) team and had three divisions: Propulsion, Vehicles and Launch Operations. Presaged by the acquisitions of the JPL and the impending transfer of the ABMA to NASA, this was a major realignment of administrative control, but proved only temporary. In May 1960 the Office of Technical Information and Educational Programs was established and the use of more sophisticated management practices became universal.

1959 December 8–9 During the third meeting of the NASA Goett Committee, Langley Research Center's H. Kurt Strass, charged by the New Projects Panel with looking into the requirements of a three-man spacecraft, reported on the Space Task Group's thinking about the steps leading to manned moon flights. Lunar circumnavigation and an unmanned supply base on the surface were considered preliminary steps essential to a manned landing. STG thought that a 10,000-lb. cylindrical laboratory module attached at one end to a 6,600-lb. conical reentry module would be optimum for a lunar orbit mission. Seymour C. Himmel from the Lewis Research Center, Ohio, reported on a six-stage Saturn launcher capable of making the Earth-moon flight by direct ascent, but Herman H. Koelle from the ABMA urged the wisdom of using Earth-orbit rendezvous techniques.

1959 December 15 Abe Silverstein, chief of NASA's Office of Space Flight Development, submitted the report of the Saturn Vehicle Team which assembled the recommended configurations into classes A, B and C. All but a proposed C-3 adopted the eight clustered H-1 engines as first stage, now designated S-I. The A-1 configuration adopted the slender Titan and Centaur as upper stages, A-2 had clustered IRBMs (intermediate-range ballistic missiles) to replace Titan as second stage, and B-1 had four H-1 engines in a second stage, four RL-10 engines in stage three and a Centaur as stage four. All C-series had cryogenic upper stages. C-1 stage two, called S-IV, had four RL-10 engines and stage three was an adapted Centaur called S-V. These became stages three and four respectively for the C-2, with a new stage two, called S-III, powered by two 150,000-lb. thrust engines. The C-3 had a first stage cluster delivering 2-million-lb. thrust and a cryogenic second stage, S-II, with four 200,000-lb. thrust engines. S-III and S-IV comprised stages three and four. Payload capability was a maximum 54,000 lb. for the C-3. This arrangement of proposed launchers was the first use of the building-block principle whereby a few upper stages could build a variety of configurations tailored to mission needs.

1959 December 16 NASA issued its first Long Range Plan defining space goals and objectives for the next decade. As a benchmark on the civilian space program defined under the Eisenhower administration it provides a useful foundation for measuring the dramatic changes brought by President Kennedy beginning in early 1961. For 1960 the Long Range Plan envisaged the first suborbital manned Mercury mission, the first weather satellite and the first flights of the Scout, Thor-Delta and Atlas-Agena B launchers. In 1961 the first manned Mercury orbit and Atlas-Centaur flights were to take place. In 1962 flights to Venus and Mars would take place, in 1963 the first flight of the multistage Saturn launcher as well as the start of unmanned lunar soft-landings and the first orbiting astronomy observatories. For the second half of the decade, NASA envisaged manned circumlunar flight, unmanned reconnaissance flights to Mars and Venus and an Earth-orbiting station. Beyond 1970 a manned moon landing was anticipated. All this was to be accomplished with funding levels increasing from a projected $524 million for the fiscal year 1960 to $1.45 billion in 1964 and $1.6 billion by 1968. In reality, NASA's authorization for fiscal year 1960 was $490 million while authorization for the fiscal year 1964 would reach a peak of $5.35 billion.

1959 December 21 NASA's chief of the Office of Space Flight Development, Abe Silverstein, sent a letter to the JPL (Jet Propulsion Laboratory) director, Dr. William H. Pickering, authorizing five Atlas-Agena B lunar flights in 1961 and 1962 as well as two Atlas-Centaur flights to Mars and Venus. This effectively started the JPL lunar and planetary exploration program and projects that would become known as Mariner (Mars and Venus), Ranger (lunar impact), Lunar Orbiter and Surveyor (lunar soft lander). Lunar reconnaissance was put on high priority and Silverstein emphasized that high resolution pictures obtained "in the period immediately preceding impact" were an early requirement.

1959 December 31 NASA Administrator T. Keith Glennan formally approved the recommendation of Abe Silverstein's Saturn Vehicle Team, clearing development of the Saturn C-1 in a three-stage configuration. The first stage (S-I) would have eight H-1 engines producing a total thrust of more than 1 million lb., the second stage (S-IV) would have four 20,000-lb. thrust cryogenic RL-10 derivative engines and the third stage (S-IV) would carry two 15,000-lb. thrust RL-10s in a modified Centaur. On January 18, 1960, Saturn received a DX rating putting it on highest national priority. On January 26, 1960, an S-IV bidders conference was held where 37 contractors were invited to submit proposals by February 29. No firm commitment had been made for upper stages to equip proposed Saturn C-2 and C-3 configurations, but on February 3, 1960, potential bidders were briefed on the S-II, conceived as a cryogenic stage with a diameter of 18 ft. 4 in. and four 200,000-lb. thrust engines. This array of new stages would underpin NASA's selection of heavy launchers for Apollo.

1959 Tests with the the SS-5 Skean, the first Soviet Intermediate Range Ballistic Missile (IRBM) got under way. The SS-5 was in some respects a scaled-up SS-4 and had a length of 82 ft., a diameter of 8 ft. and a launch weight of 130,000 lb. Propulsion was provided by a twin-chamber RD-216 engine producing a thrust of 388,000 lb. with storable UMDH-RFNA propellants. Control was maintained with jet vanes in the rocket exhaust of each nozzle with guidance internal. The SS-5 could throw a 3,500-lb., 1-MT warhead 2,600 mi. with a 50% probability of getting within 1.4 mi. of the target. The missile was deployed in 1961 and maintained a peak deployment of

100 rounds until 1973, whereafter 90 were maintained through 1977, with final retirement in 1983.

1960 January 11–15 The first International Space Science Symposium was held at the Centre Universitaire Méditerranéen, during which scientists from around the world exchanged information on research into the upper atmosphere conducted by sounding rockets and satellites. Organized by COSPAR (Committee on Space Research) the symposium provided the first truly international forum for discussion of earth's newly discovered radiation belts, the magnetic field of the planet and lunar photography.

1960 January 18 At the start of a new decade, NASA presented its budget for year 1961 to Congress, requesting an appropriation of $802 million. With various transfers and amendments, this was brought to $966.7 million for the year beginning July 1, 1960, an increase of 85% over fiscal year 1960. This was the last year of the Eisenhower administration, whose conservative planning was the hallmark of modest expansion. Fiscal year 1961 was also the first year that civilian space expenditure exceeded military space funds, which this year stood at $814 million. Not until fiscal year 1982 would military expenditure on space once more exceed the NASA budget. In relative terms, annual defense space spending would remain relatively flat until 1980, only reaching $2 billion in fiscal year 1969. NASA's budget peaked at $5.25 billion in fiscal year 1965 and fell dramatically thereafter. During the decade of the 1960s, NASA spent $36 billion and the Pentagon spent $14.7 billion on space, 29% of all space spending. When NASA funds are added to other space spending government departments, the United States spent $51.6 billion on space projects, less than 4% of the $1,301.5 billion in total federal expenditures during the decade. In the 1970s that dropped to 1.7%, when of total federal expenditures of $3,323 billion, the U.S. government spent $55.443 billion on space, of which NASA got $34 billion and the Pentagon $20.5 billion, or 38% of total U.S. expenditures for space. In the 1980s total space expenditure of $215.3 billion accounted for 2.4% of the $8,816 billion spent by the government, of which NASA got $76.8 billion, 36%, and the Pentagon received $113.3 billion, 52% of total U.S. government space money.

1960 January 21 NASA established the first launch schedule for its lunar Atlas-Agena B missions and anticipated five flights between February 1961 and February 1962. Flights 1 and 2 were expected to be test flights to escape velocity and flights 3 to 5 were to be lunar impact missions with the S-series JPL (Jet Propulsion Laboratory) spacecraft shortly to be renamed Ranger. During this month the JPL decided to let industry contract for lunar spacecraft after Ranger and to build all planetary spacecraft in-house. Ranger A-1 was slated for launch on the first Atlas-Agena B in a test configuration weighing 630 lb., while the three impact spacecraft would each weigh about 800 lb., including a 300-lb. survivable capsule. On February 13, 1960, all five flights were converted to impact missions, and the launch schedule slipped by three months.

1960 January 26 An unnumbered National Security Council document finally killed efforts by elements in the armed forces to develop space weapons when it proposed that the United States should endorse the freedom of any nation to launch "orbital space vehicles or objects not equipped to inflict injury or damage." It affirmed the administration's belief that "space is freely available for exploration and use by all." It defined acceptance of "peaceful" space activities and to outlaw hostile space activities by other nations submitted on March 16, 1960, to the ten-nation Committee on Disarmament in Geneva, a proposal that the orbiting of "weapons of mass destruction shall be prohibited." Little progress was made on a formal agreement although no "weapons of mass destruction" were ever put in orbit.

1960 January 28 NASA's Associate Administrator Richard E. Horner briefed the House of Representatives Committee on Science and Astronautics on the space agency's 10-year plan. In 1960 NASA expected to launch the first weather satellite, carry out a manned suborbital Mercury flight and inaugurate Scout, Thor-Delta and Atlas-Agena B launchers. In 1961 it expected to put a man in orbit, launch a lunar-impact probe and fly the Atlas-Centaur for the first time. In 1962 probes would be launched on fly-by missions to Venus and Mars. 1963 and 1964 were scheduled for the first flight of a two-stage Saturn rocket, the first soft landing on the moon, the first unmanned circumlunar flight and return and the first reconnaissance of Venus and Mars. NASA hoped to begin manned circumlunar flight in the period 1965–70, with a manned landing envisaged for sometime after 1970.

1960 January 29 In closed testimony before the Senate Aeronautical and Space Sciences Committee, Allen Dulles, director of the CIA, presented a revised National Intelligence Estimate that said the Russians had a three-to-one lead over the United States in ICBMs. He claimed the Soviets already had 10 ICBMs operational to the United States' 3, that by June the Russians would have 35 on hand and that by mid-1963 they would have 350–640 ICBMs. He mentioned two sites presently being used for missile tests: Baikonur and Kapustin Yar. In fact, the Soviets would only ever have a maximum 4 SS-6 ICBMs operational at one site, Plesetsk, and not before the SS-7 appeared in 1962 did the USSR possess even a modest ICBM force.

1960 February 2 The first Titan I launch, with both first and second stages live and rocket motors operating, took place successfully from Cape Canaveral, Fla., on the first attempt. On February 24, a Titan I conducted a full-range flight demonstrating reentry vehicle separation as it would be required to perform if launched in anger. Carrying an Avco Mk.4 reentry vehicle simulating a 4-MT thermonuclear warhead, the missile flew 4,900 mi. down the Atlantic Missile Range. Titan I would eventually be raised to the top of its silo for launch, and the first live firing to demonstrate this technique took place on September 23, 1961.

The annual Senate Preparedness Hearings (known as the Joint Hearings) to examine relative U.S.-Soviet missile programs began in Washington, D.C., and continued on February 3, 4, 8 and 9. Fears of Soviet supremacy in ballistic missiles, fueled by some senior members of the armed forces and by the CIA, were countered by officials from the Eisenhower administration. Secretary of Defense Thomas Gates had a broader picture and testified that there was "no deterrent gap." The debate spilled over into the public domain, and during this election year, party-political positions polarized around opposing interpretations.

1960 February 5 NASA's Jet Propulsion Laboratory (JPL) briefed fourteen potential contractors for the Ranger lunar impact spacecraft, requesting bids for a six-week study of the associated technical problems and their solutions. Industry had 10 days to respond. On February 9 the JPL formally

proposed adoption of the name Ranger, which until that date had been an informal appellation, and 8 days later a new flight schedule was published naming 5 flights between April 25, 1961, and April 1962. On February 15 NASA received proposals for a six-week study of the hard-landing Ranger capsule and 11 days later a JPL evaluation committee selected North American Aviation and Ford Motor Company's Aeroneutronic Division for this work. These two companies and Hughes Aircraft Co. were awarded design study contracts for the capsule on March 1, 1960. On March 25 the contract for a lunar impact TV camera was awarded to RCA's Astro-Electronics Division.

1960 February 25 Built by the Martin Co., the first U.S. Army Pershing tactical ballistic missile was launched down the Atlantic Missile Range from Cape Canaveral, Fla. With a length of 34 ft. 9 in. and a diameter of 3 ft. 3 in., *Pershing I* had a launch weight of 10,141 lb. and a selectable range of 100–460 mi. carrying a W-50 warhead. The W-50 featured three yields of 60 KT, 200 KT and 400 KT in three separate 697-lb. warhead sections between 11 ft. and 12 ft. 3 in. in length. The 26,290-lb. thrust Thiokol TX-174 first stage had a burn time of 38.3 sec., while the 19,220-lb. thrust Thiokol TX-175 second stage burned for 39 sec. The Pershing system was transported in four vehicles and highly mobile. Raised to the vertical for launch, it was fired remotely. Pershing entered service in July 1962 and by 1964 was deployed to Germany where it was operated by U.S. and Luftwaffe forces.

1960 March 1 NASA established an Office of Life Sciences Programs charged with carrying out research in biotechnology and in basic medical and behavioral sciences. It would conduct research into the effects of space and planetary environments on living organisms and into the question of extraterrestrial life. It would also monitor the contamination of unmanned

By July 1962 the U.S. Army's Pershing tactical battlefield missile had become fully operational; operational deployment to Europe took place two years later.

planetary spacecraft and advise on sterilization methods for inhibiting the transfer of Earth-based organisms to alien or inert environments. The office was also tasked with developing a life sciences research program, and over time it would carry out much research on the data from planetary exploration insofar as it related to the question of life in other worlds.

1960 March 3–5 At a NASA staff conference in Monterey, Calif., the JPL outlined the refined Ranger lunar exploration schedule. Flights RA-1 and -2, launched in April and July 1961, would test spacecraft performance on highly elliptical trajectories. Flights RA-3 to RA-5, in October 1961 and January and April 1962, would send individual Ranger spacecraft on direct impact flights to the moon. During the approach phase, high-resolution TV pictures would be sent back to Earth up to the point of impact. Some distance from the moon, rough-landing capsules would be released carrying instruments designed to survive the hard impact. On May 7, 1960, the Ranger schedule was republished showing a three-month slip.

1960 March 7 At the U.S. Air Force Wright Air Development Center, NASA's Mercury astronauts began an indoctrination program into free-floating weightlessness using a modified Convair C-131B transport aircraft. The rear of the aircraft was cleared and padded so that, when flown on a parabolic path, it simulated weightlessness for a period of about 15 sec. This was considerably less than the 40 sec. which astronauts had been flying in North American F-100 fighter aircraft, but it provided opportunity for free movement whereas the F-100 prevented movement and was useful only for hand activities such as feeding and drinking. About 90 parabolic C-131B flights had been flown by March 10. Training in a converted Boeing C-135 began September 21, 1960, affording 30 sec. of simulated weightlessness. The differing times are a function of structural limits on the different aircraft when held in parabolic profiles.

1960 March 11 A Thor-Able IV successfully launched the 95-lb. solar satellite *Pioneer V* onto an escape trajectory by imparting a velocity of 36,480 ft./sec. With the same configuration as *Explorer VI* launched on August 7, 1959, the uprated Thor launcher utilized the Rocketdyne MB-3 Block II propulsion system with a thrust of 165,000 lb., an improved Aerojet AJ10-101A second stage, and an uprated X-248-A4 third stage. *Pioneer V* carried the first digital telemetry system used on a U.S. satellite, with power provided by 4,800 solar cells on four paddles. Injected to a heliocentric orbit inclined 3.35° to the ecliptic, *Pioneer V* continued to transmit to a distance of 22.5 million mi., reached on June 26, 1960. The perihelion of 74.967 million mi. was reached on August 10, 1960, and aphelion of 92.358 million mi. on January 13, 1961.

1960 March 14 Twenty Soviet air force pilots named as members of the first Russian cosmonaut team reported for duty at the M. V. Frunze Central Airport, Moscow. The team consisted of: Igor N. Anikeyev; Pavel I. Belyayev; Valentin V. Bondorenko; Valery F. Bykovsky; Valentin I. Filatyev; Yuri A. Gagarin; Viktor V. Gorbatko; Anatoli Y. Kartashov; Yevgeny V. Khrunov; Vladimir M. Komarov; Aleksei A. Leonov; Grigori G. Nelyubov; Andrian G. Nikolayev; Pavel R. Popovich; Mars A. Rafikov; Georgi S. Shonin; Gherman S. Titov; Valentin S. Varlamov; Boris V. Volynov; and Dmitri A. Zaikin. Only 12 would make space flights. Kartaskov left the group in 1960 because of a hemorrhage. Anikeyev, Filatyev and Nelyubov were dismissed in 1961 for drunken

brawling. Bodorenko was killed in 1961 (see entry for March 23, 1961). Rafikov retired in 1962 for unspecified reasons. The remainder either made flights or were on back-up crew duty. For a while the cosmonaut candidates stayed at a house in Leninsk Prospect but they all moved to Zvezdny Gorodok in April 1960.

1960 March 15 Mstislav V. Keldysh, the vice president of the Soviet Academy of Sciences, approved mission plans for spacecraft to Venus and Mars. An attempted flight to Mars was scheduled to begin with a launch attempt on September 27 of a spacecraft designated M-1. It was to have been placed on course for a 3,100–18,600-mi. flyby of Mars when it would transmit to Earth a number of pictures through a 750-mm telephoto lens resolving detail as small as 1.75–3.5 mi. across. Three merchant ships, named *Il'chevsk, Krasnoda* and *Dolinsk* and modified as floating tracking stations, were moved to the South Atlantic to cover parts of the mission. Technical problems delayed the launch attempt until October 10.

The U.S. Army Ballistic Missile Division's Development Operations Division at Huntsville, Ala., was transferred lock, stock and barrel to NASA along with all its projects and programs. The Huntsville facility was named the George C. Marshall Space Flight Center (MSFC) by President Eisenhower, and Wernher von Braun was named director when the formal transfer took place on July 1, 1960. The MSFC had just over 5,200 employees, more than any other NASA field center, and raised total NASA employment by 50%, which stood at 15,682 by the end of the year. The MSFC would manage all the Saturn launch vehicle programs and serve as lead center for the Skylab program as well as developing and testing Shuttle propulsion systems and advanced rocket motors.

1960 March 18 The first in a series of tests with solid propellant rockets launched from a floating, vertical position took place at the Naval Missile Center, Point Mugu, Calif. Believing that the cost of land-based ICBMs could be reduced by eliminating costly ground facilities and silos, the Navy research program aimed to prove the feasibility of basing missiles in vertical positions afloat. *Hydra 1* comprised a 4-ft. wooden vehicle powered by a solid motor. In a second test, *Hydra 1A,* a 105-ft. telegraph pole was propelled from the sea by the solid rocket motor of a Genie missile.

1960 March 28 The first firing of clustered engines in a Saturn launcher test-booster took place at an ABMA (Army Ballistic Missile Agency) test stand, Redstone, Ala. Two H-1 engines in the eight-engine cluster mounted in the SA-T assembly were successfully fired for 8 sec. The second test on April 6 involved four engines in a 7-sec. firing, and all eight engines were fired for 8 sec. on April 29, 1960. Further tests increased the firing duration to 24 sec. on May 17, when a thrust of 1.3 million lb. was achieved, 35 sec. on May 26, 75 sec. on June 3, 110 sec. on June 8 and 121 sec. on June 15, 1960.

NASA's Mercury astronauts began open-water egress training to familiarize them with getting out of a spacecraft after it had splashed down. The training took place in the Gulf of Mexico off the coast of Pensacola, Fla., and was carried out with the cooperation of the U.S. Navy School of Aviation Medicine. Training took place in heavy seas with 10-ft. swell and astronauts were able to demonstrate that they could get out of their spacecraft and into a life raft within about 4 min. To do this the astronaut had to unstrap himself and maneuver his body under and behind the instrument panel, working his body up and out the top of the truncated cone that formed the exterior shape of the spacecraft.

1960 March At the Ideal Home Exhibition in London sponsored by the *Daily Mail* newspaper, an exhibit called Space Vehicle attempted to give the public an impression of what the interior of an orbiting space station would look like. Based on designs prepared by the Douglas Aircraft Co., the exhibit was 62 ft. high and 17 ft. across, allowing visitors to walk into it from a second floor entrance and inspect interior detail. An estimated 150,000–200,000 people visited the exhibit. Dramatic descriptions of what a spaceflight would be like were printed in the exhibition catalogue.

1960 April 1 The world's first civilian weather satellite, *Tiros I,* was launched from Cape Canaveral, Fla., by Thor-Able V into a 432 × 464 mi. orbit inclined 48.4°. Injected into orbit by the last of the Thor-Able family, the 270-lb. pillbox-shaped *Tiros I* was 1 ft. 7 in. high and had a diameter of 3 ft. 6 in., around which 9,200 solar cells continuously charged NiCd batteries, providing an average 19 W of electrical power. Two TV cameras were carried and a magnetic tape recorder could store up to 32 pictures. *Tiros I* had transmitted a total 22,952 pictures by the time it ceased operating on June 17, 1960. This first-generation weather satellite matured from a research and development program into a semioperational service through a further nine Tiros launches, each satellite carrying two cameras and, on approximately half the missions, an infrared radiometer for heat measurement and Earth radiation budget

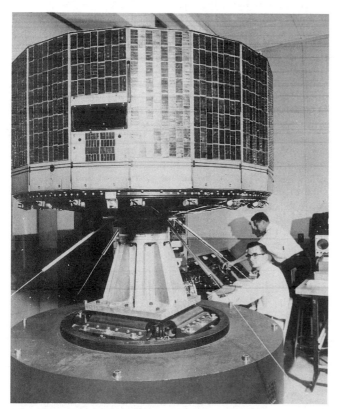

The Tiros weather satellite began a revolution in meteorology that did more to save lives, through early storm warnings, than any other application of the space age.

instruments. The last first-generation satellite, *Tiros X,* was launched on July 2, 1965, and the last to be switched off was *Tiros IX* on June 12, 1968. The longest lived had been *Tiros VII,* launched on June 19, 1963, which operated for 1,809 days and sent 125,331 pictures to Earth. In total, the Tiros program transmitted 649,077 images of the world's weather.

NASA's Space Task Group began a series of presentations to field centers describing guidelines for an advanced manned spacecraft program to follow Project Mercury. STG's Director Robert R. Gilruth defined the mission model: manned circumlunar reconnaissance but compatibility with extended duration Earth-orbit operations; 14-day flight time with a three-man crew; pinpoint landing; weight limits restricted to 15,000 lb. for the complete lunar spacecraft and 25,000 lb. for the Earth-orbit mission permitting use of the Saturn C-1 or C-2 launchers. What STG proposed was a common equipment module that could serve to house instruments for environmental studies of the Earth or reconnaissance and mapping of the lunar surface. A three-man crew had been selected on psychological grounds for work-sharing with one person acting as mission "commander."

1960 April 13 Britain's Minister of Defense Duncan Sandys informed the House of Commons that the Conservative government had decided to cancel the Blue Streak long-range ballistic missile, saying that a deterrent based on fixed sites would be quickly outdated. In announcing that Britain had already spent £100 million on Blue Streak, the minister claimed that the cost at completion would be £500–600 million. Proposals to convert Blue Streak into the first stage of a satellite launcher had already been made. Minister of Aviation Peter Thorneycroft asked Commonwealth countries to support Britain financially in this application, but with the exception of Australia, all declined to participate.

The second *Transit 1B* navigation satellite, the first to reach space, was carried to a 232 × 465 mi. orbit with an inclination of 51.3°. This was also the first launch of the Thor-Able Star, which had a modified Aerojet AJ10-104 second stage using UDMH-IRFNA propellant and producing a thrust of 7,890 lb. The complete vehicle had a length of 79 ft. 3 in. and a launch weight of 120,000 lb. A malfunction in the launcher left the satellite in a lower orbit than planned. The 265-lb. *Transit 1B* consisted of a sphere 3 ft. in diameter with two stable oscillators and dual frequency transmitters operating on 216/162 MHz and 324/54 MHz. It transmitted until July 12 and decayed out of orbit on October 5, 1967.

1960 April 14 At a four-day conference that concluded at the Langley Research Center, a joint U.S. Air Force/NASA conference on the Dyna-Soar boost-glide program heard officials review the recently concluded Phase Alpha study and Boeing describe the reentry vehicle. It was projected to weigh 9,283 lb., carry 1,000 lb. of scientific equipment and a pilot, carry a skid landing gear and be reusable on up to four flights. Boeing estimated a length of 34 ft. 8 in., a span of 19 ft. 8 in. on a deep lifting body launched by a modified Titan booster. With an L/D (lift/drag) of 4.5, it had a projected 2,000-NM crossrange. A variety of lifting-body shapes was also being considered for a program increasingly referred to as a research tool for manned hypersonic weapons systems. In another, significantly related, event this day, Air Force Chief of Staff Gen. Thomas D. White sent letters to Gen. Landon and Gen. Wilson stressing the need for quiet cooperation with NASA because he was convinced that the civilian space agency would "eventually be combined with the military."

1960 April 15 The final study reports on the NASA Ranger rough-landing capsule were submitted by North American Aviation, Ford-Aeroneutronic and Hughes Aircraft. On April 27, 1960, Ford-Aeroneutronic was awarded a contract to develop five capsules, and on May 23, 1960, the first scientific experiment was approved: a seismometer developed by the California Institute of Technology. On July 1, 1960, Ford-Aeroneutronic conducted the first helicopter drop test to determine the energy absorption properties of natural balsa wood, which had been recommended as a possible impact-absorbing material within which the scientific instruments would be housed.

1960 April 26 NASA formally announced negotiations with Douglas Aircraft on design and development of the cryogenic S-IV second stage for the Saturn C-1 launch vehicle. The cryogenic S-IV was to carry four 17,500-lb. thrust LR-119 engines operating on liquid hydrogen–liquid oxygen propellants. Pratt and Whitney would provide the LR-119, an upgraded version of the 15,000-lb. thrust LR-115, which in its RL-10 form was the propulsion for Centaur. A contract was signed with Douglas on May 20, 1960, and Pratt and Whitney signed the LR-119 contract on August 8, 1960. The S-IV and its successor the S-IVB would be used as the terminal stage of every Saturn launcher except for the last Saturn V flight in 1973 which carried the Skylab space station into orbit.

1960 April 28 The first launch of a full-scale Nike-Zeus ABM (antiballistic missile) took place from White Sands Missile Range, N.Mex. The first stage consisted of a 450,000-lb. thrust Thiokol solid propellant motor, the most powerful built to date. The second stage was also built by Thiokol, while the spherical third-stage motor was the first space bus designed to guide the warhead to the incoming nuclear weapon it was built to destroy. This technology led to maneuvering ICBM reentry vehicles which could deploy separate warheads on individual trajectories to their specific targets. Nike-Zeus had a length of 63 ft. 3 in., a diameter of 5 ft. and a span across the wedge-shaped first-stage fins of 8 ft. 2 in. The second stage carried large canard control surfaces. Launch weight was 40,000 lb. and the missile had a range of 250 mi. Live interception of an Atlas warhead took place over the Pacific on July 19, 1962.

1960 April One of the most important contributions to NASA's advanced manned spacecraft concept was presented in a report from Dr. Charles Stark Draper, director of the laboratory that carried his name. Draper showed that it would be possible for the spacecraft to carry its own on-board navigation systems, giving it a level of independence from Earth and providing stronger confidence in back-up capabilities. On-board navigation would be important, for instance, if all communications between the spacecraft and ground tracking stations were lost. The complexities involved in celestial navigation by spherical trigonometry utilizing reference points derived from onboard instruments would, said Dr. Draper, pose no serious limitations to solving the complex equations that would at least provide crude data for course-correction maneuvers enabling a safe return to Earth. Draper's work formed the basis for guidance and navigation in the Apollo program.

1960 May 1 U.S. pilot Francis Gary Powers was brought down over the Soviet Union during an intelligence gathering overflight of the USSR and created an incident that had wide and far-reaching repercussions on U.S. military space projects. Already concerned about delays and changes in the SAMOS (satellite and missile observation system) program, President Eisenhower asked Secretary of Defense Thomas Gates to investigate and reevaluate the program. The inevitable halt to overflights and a need to increase the capabilities and frequency of launch with reconnaissance satellites stimulated a revision in intelligence gathering. The review was part of an effort driven by science adviser George Kistiakowsky which involved the former presidential science adviser James Killian and Dr. Edwin Land of the Polaroid Corp.

1960 May 6 The first physiological data collected and recorded on board an aircraft in flight was gathered during the 14th flight of the rocket-propelled X-15 research aircraft. U.S. Air Force Capt. Robert M. White was wired to provide basic body data. Physiological data would be a vital element in measuring human response to high-stress work environments and become an essential element in monitoring the reaction of astronauts to spaceflight, space walking and general work tasks inside and outside spacecraft.

1960 May 9 The first production Mercury spacecraft from McDonnell was used in a beach-abort test. Launched at NASA's Wallops Island facility, the test involved the launch escape system's lifting the spacecraft to a height of 0.5 mi. and carrying it 1 mi. downrange at a maximum speed of 976 MPH. The flight lasted 1 min. 16 sec. and was scheduled to explore anomalies in the performance of the solid rocket motor observed in test data from the Little Joe series of abort simulations. It also served to further qualify the recovery system. Great care was taken to explore every aspect of the abort process. Launch vehicles were converted ballistic missiles with a miserable performance record, and there was little expectation that a manned space program could be performed without at least a few aborts. In reality, the "man-rating" of launch vehicles improved reliability and no abort was ever carried out on any Mercury, Gemini or Apollo mission.

1960 May 13 The first NASA Thor Delta launch vehicle was launched from Cape Canaveral, Fla., carrying a deflated plastic satellite called *Echo A-10*. Although the first two fired as planned, attitude control jets on the second stage failed and the upper stages and payload fell back down through the atmosphere. The *Echo* satellite comprised a 100-ft. sphere which was to have been deployed and inflated in orbit for experiments in passive communication using the satellite as a "moon" to bounce radio signals back to Earth. Despite this inauspicious start, the Delta launch vehicle was developed further and in 1962 the first stage was upgraded to a thrust of 160,000 lb. while the second stage was given retro capability, denoting a Delta A model. With a 170,000-lb. thrust DM-21 first stage, the 3-ft. longer tankage of the second stage and a longer burning third stage, Delta B appeared later in the same year with a capability of lifting 830 lb. to low orbit or 150 lb. to geostationary transfer orbit. Delta C appeared in 1963 with a larger (X-258) third stage which increased low-orbit payload to 900 lb. Between 1960 and 1969 NASA launched 36 Delta derivatives. Successive versions used strap-on solid propellant boosters beginning with the Thrust-Augmented Delta, or Delta D, first launched on August 19, 1964.

1960 May 15 The first launch of a prototype Vostok manned spacecraft took place from the Baikonur cosmodrome in the USSR when an SL-3 launcher placed *Korabl-Sputnik 1 (KS-1)* into a 195 × 229 mi. orbit. SL-3's second stage was the first to carry the upgraded R0-7 engine with a thrust of 12,300 lb.

KS-1 was not intended for recovery and the descent module was not covered with ablative material that would be used later for protecting capsules from the heat of reentry. It did carry a retro-rocket, but the orientation system failed and when the motor fired at 23:52 UT on May 18, *KS-1* was pointing in the wrong direction. The spacecraft went into a higher orbit instead of back through the atmosphere. Moreover, the motor burned for only 26 sec., too brief a period to de-orbit the spacecraft even had it been aligned correctly. The separated instrument module burned up in the atmosphere on May 5, 1962, followed by the descent module on October 15, 1965.

1960 May 16 In Japan, the National Space Activities Commission (NSAC) was formed in the prime minister's office to serve in an advisory role, making recommendations on policies, plans, programs and budgets. From the outset it firmly established the ethic of a peaceful exploration of space and has consistently outlawed attempts from some Japanese ministries to divert space technology to military purposes. In 1965 it assigned to the Institute of Space and Aeronautical Science (formed in the University of Tokyo in 1954) management of the solid propellant Lambda and Mu launch vehicles. In 1966 the NSAC developed a detailed plan for scientific work in space, incorporating research, technology and the development of launch vehicles. On August 16, 1968, its name was shortened to Space Activities Commission and it master-planned a long-term space program for the country.

1960 May 16–17 A conference was held at the NASA Langley Research Center on orbital rendezvous and the problems associated with developing techniques for its routine implementation. The first day was spent hearing reports from all the NASA field centers and the second day was dedicated to round-table discussions in which a unanimous consensus backed the need for early development of rendezvous and docking. The following month, the Space Task Group released a set of ground rules for manned projects, including manned circumlunar reconnaissance, and backed the idea of a modified Mercury spacecraft possessing a greater lift-drag ratio to facilitate some degree of postreentry control over the landing point.

1960 May 18 NASA formally approved a broad range of lunar and planetary missions and released names for each: lunar impact, Ranger; lunar soft-lander, Surveyor; mobile lunar soft-lander, Prospector; lunar satellite, Pioneer; Venus and Mars probes (1962), Mariner; Venus and Mars orbiters (1965), Voyager. Names with nautical connotations were chosen to signify dramatic journeys to distant places. During the month NASA approved the unmanned lunar surface exploration program comprising two parts: an orbiter for photographic coverage (eventually Lunar Orbiter) and a surface lander (Surveyor). On May 19 NASA adopted the generic name Mariner for its planetary missions.

1960 May 24 An Atlas-Agena A launched the first successful MIDAS (missile defense alarm system) satellite from Cape Canaveral, Fla., placing it in a 302 × 317 mi. orbit inclined 33° to the equator. Designated *MIDAS-2 (MIDAS-1* failed to reach orbit after launch on February 26, 1960), the satellite had been developed as a missile early-warning system powered by batteries. It failed on the second day, when the downlink system malfunctioned, and decayed back through the atmosphere on February 7, 1974. The first two satellites weighed 4,465 lb. and 5,070 lb., respectively, and were precursor test

vehicles to the operational version, beginning with *MIDAS-3* launched July 12, 1961, into polar orbit.

1960 May 31 NASA selected North American Aviation, Rocketdyne Division, to develop the 200,000-lb. thrust, high-energy cryogenic J-2 upper stage engine for Saturn launch vehicles. Burning liquid hydrogen and liquid oxygen, the J-2 would add performance denied to hydrocarbon fuels and significantly improve the lifting capacity of the big launch vehicles. A definitive contract was signed with Rocketdyne on September 10, 1960. The J-2 was more than just a rocket motor. Because it had to start and stop in space more than once, it was developed as an integrated propulsion system forming a functional part of the rocket stage from which it obtained propellant by incorporating pressurization and control systems for independent operation. Significantly more powerful than the 15,000-lb. thrust cryogenic Aerojet RL-10, the J-2 provided much useful information in the design of high-energy engines.

1960 May The U.S. Air Force Flight Dynamics Laboratory awarded a number of contracts for studies to determine the means by which it could better understand the upper atmosphere and near-Earth space where boost-glide vehicles would operate. Precise aerothermal data was essential for the design of reentry bodies coming back into the atmosphere from space. The several different studies converged into what became the ASSET (aerothermodynamic/elastic structural systems environmental tests) program. ASSET would involve test vehicles and small reentry bodies launched on ballistic rockets.

1960 June 7 A Bomarc air defense missile in ready storage (capable of being launched within 2 min.) at McGuire AFB, N.J., accidentally exploded into flames in a conflagration that destroyed the 350-lb., 10-KT nuclear warhead. The fire was caused by a high-pressure helium tank's exploding and rupturing one of the missile's fuel tanks. The high explosive that was designed to trigger the warhead did not go off and safety devices worked as designed to prevent leakage. Only a small area round the base of the missile and out along a 100-ft. drainoff channel for water used by firefighters was contaminated. This was one of only two missile accidents involving nuclear weapons.

1960 June 20 To determine the feasibility of deploying solid propellant Minuteman ICBMs on mobile railroad car launchers using existing track across the United States, a specially equipped train departed Hill AFB, Utah, to travel western and central rail routes. In four trial runs that ended August 27, 1960, the Air Force checked the ability of the nation's rail network to support missile trains and evaluated communications difficulties that could arise as well as technical problems such as vibration. In a Minuteman deployment plan that involved a total of 540 missiles operational by 1964, 150 ICBMS were to have been carried on 50% of the 250,000 mi. of railroad across the 48 contiguous United States. The balance would be placed in underground silos. The Air Force Ballistic Missile Division was pressing the White House for a plan involving a total of 1,200 Minuteman ICBMs, while some in the Air Force wanted 2,000–3,000, and Gen. Thomas S. Power personally asked President Kennedy for 10,000 when he assumed office in January 1961. In reality, the entire Minuteman force would never exceed 1,000. Secretary of Defense Robert S. McNamara canceled the Mobile Minuteman program on December 7, 1961, in favor of deploying the entire force in specially hardened silos.

Manned tests with the Mercury spacecraft environmental control system (ECS) began. Developed by AiResearch, the ECS was required to pressurize the spacecraft to 5.5 lb./sq. in. with 100% oxygen and to maintain a temperature of 80°F. A total oxygen quantity of 7.9 lb. would be carried in two storage spheres. Activated charcoal would remove odors and lithium hydroxide canisters would cleanse the atmosphere of exhaled carbon dioxide. Qualification tests involved subjects with full pressure suits simulating first a full mission immediately followed by a 12-hr. post-splashdown wait for recovery forces to arrive—a specified requirement of the suit. The purpose of these tests was to measure human tolerance.

1960 June 22 The first successful launch of two satellites on one vehicle took place from Cape Canaveral, Fla., when a Thor-Able Star carried the 223-lb. *Transit 2A* navigation satellite and the 42-lb. *Solrad 1* solar radiation satellite, both for the U.S. Navy. *Solrad 1* was also known as Greb, for Galactic Radiation Energy Background, which would provide data until April 1961. Modifications to the Aerojet second stage qualifying it as Able Star provided an orbital restart capability which was used to place *Solrad 1* in a 372 × 612 mi. orbit and *Transit 2A* in a 375 × 632 mi. orbit. Weight savings in the Transit navigation satellite, made possible by increasing the number of solar cells and reducing the number of batteries, allowed double payloads to be scheduled for future Thor-Able Star flights. The last Thor-Able Star launch took place on August 13, 1965, placing the Navy *Transit O-5* navigation satellite in orbit.

1960 July 1 The first flight of a solid propellant Scout rocket took place at NASA's Wallops Island launch site. The four-stage vehicle had a length of 71 ft. 4 in. and a weight of 36,842 lb., carrying a 193-lb. payload. The launch of *Scout ST-1* should have carried a radiation and acceleration package prepared by the NASA Langley Research Center to a height of 2,325 mi. and a downrange distance of 5,060 mi. The Algol and Castor stages fired as planned but the Antares third-stage burn was so rough that it caused problems that prevented the fourth stage from firing. The payload did, however, make measurements to a height of 1,007 mi. The next launch, of *ST-2* on October 4, 1960, went as planned and the second Scout reached a height of 4,028 mi. and a downrange distance of 6,675 mi. The first attempt to orbit a satellite, on December 4, 1960, failed when the second stage refused to fire and NASA's *Explorer S-46* fell into the Atlantic. The second attempt, on February 16, 1961, was a success.

1960 July 5 The House of Representatives Committee on Science and Astronautics declared that "A high-priority program should be undertaken to place a manned expedition on the moon in this decade. A firm plan with this goal in mind should be drawn up and submitted to the Congress by NASA." It added with some vigor: "The space program is not being pushed with sufficient energy." Just over 10 months later, President Kennedy would receive rapturous applause in setting exactly that objective before a joint session of Congress.

1960 July 9 NASA's Jet Propulsion Laboratory awarded separate, partially funded study contracts to North American Aviation, Hughes Aircraft Co., McDonnell Aviation and Space Technology Laboratories for the Surveyor lunar exploration program scheduled to follow Ranger. The JPL selected

Hughes for a definitive contract to build seven Surveyor spacecraft on January 17, 1961, and the contract was signed on March 1 of that year. At this date the program envisaged the use of soft-landers and orbiters built from a common space bus, but on December 17, 1962, JPL engineers declared they were "unable to give adequate support to the Surveyor Orbiter and recommended deletion of this project." Demands placed upon the JPL by Apollo engineers eager for lunar surface data to complete final design details of the manned lander concentrated attention on the soft-lander. The lunar satellite concept was redirected to the Lunar Orbiter project, proposals for which were sought by NASA on August 30, 1963.

1960 July 12 Beginning this date, the seven NASA Mercury astronauts started a 5-day desert survival course at Air Training Command Survival School, Stead AFB, Nev. Although designed to land on water, an in-orbit emergency could bring the spacecraft to Earth at any location, including arid regions. The course consisted of 1½ days of academic tuition, 1 day of field demonstrations and 3 days of isolated remote training. A set of basic survival tools carried in the spacecraft were used to simulate as realistically as possible the conditions the astronauts might have to survive in until recovered.

1960 July 15 NASA Administrator T. Keith Glennan approved the Mariner planetary exploration program put together between the JPL and the Office of Space Flight Development at NASA Headquarters. In the plan, Mariner A would have a weight of 1,074–1,512 lb. and comprise the hexagonal frame design of the Ranger spacecraft. It was planned for an Atlas-Centaur launch on a Venus fly-by in 1962 following one test flight to check spacecraft design and systems performance. In a revised plan dated February 1961 it was to have been used for additional Venus flights in 1964 and 1965. On August 30, 1961, Mariner A was canceled because the Atlas-Centaur was behind schedule, and the Venus mission was transferred to a Mariner R-series spacecraft that was lighter but also based on Ranger. Mariner B was proposed as an 882–1,325 lb. spacecraft scheduled for Atlas-Centaur. NASA planned to fly a test mission in 1963 and missions to Venus or Mars in 1964. Redefined several times, it was at first seen as carrying an instrumented lander but in February 1962 the Venus mission was dropped. On April 9, 1962, the Mars landing was dropped and the Venus mission reinstated. On March 14, 1963, the mission was changed to a checkout flight carrying a lander preparatory to a biological study of Mars in a program later named first Voyager, then Viking.

1960 July 20 The first functional Polaris SLBM (submarine-launched ballistic missile) was launched underwater from the fleet ballistic missile submarine USS *George Washington* off Cape Canaveral. The flight was a success, as was a second launch this day 2 hr. 53 min. later. With a range of 1,200 NM, a length of 28 ft. 6 in. and a diameter of 4 ft. 6 in., the Polaris A1 weighed only 28,500 lb. and had a first-stage sea-level thrust of 70,000 lb. and a second-stage vacuum thrust of 33,000 lb., with both stages exhibiting an Isp of 231 sec. The 600-KT W-47 nuclear warhead weighed only 600 lb. and had an accuracy of about 6,000 ft. The USS *George Washington* departed Charleston, S.C., on its first operational mission with 16 missiles on November 15, 1960, five years ahead of the original program goal laid down in late 1955. The USS *George Washington* returned from its first patrol on January 21, 1961.

1960 July 23 Presidential candidate John F. Kennedy was given the first of two national security briefings by the Central Intelligence Agency before his election. The second meeting took place on September 19, and vice presidential candidate Lyndon Johnson was given a briefing on July 27, 1960. At a third briefing, a Navy intelligence officer informed Kennedy that projections of a massive missile gap favoring the USSR were not corroborated by the best of U.S. intelligence. Kennedy ignored this information and fueled alarmist warnings about a runaway "missile-gap," making this a major element in his political campaign. After discovering in 1961 the myth of the missile-gap fear, he would claim he had not been shown top-secret 1960 estimates.

The second Soviet Korabl-Sputnik, *KS-2(1),* failed to reach orbit when the second stage of the SL-3 launch vehicle malfunctioned. In seven precursor tests with unmanned Vostoks only four were successful. This was the eleventh SL-3 flight, of which seven had failed to place their payloads in orbit. Of the remaining Vostok test flights only one other failed to reach orbit, when *Korabl-Sputnik 4 (1)* carrying dogs Shutka and Kometa fell back to Earth in an SL-3 malfunction on December 22.

1960 July 28 The first in the second series of U.S. Navy test flights with the ultimate objective of placing an air-launched satellite in orbit was successfully conducted when a McDonnell F4D-1 Skyray released a four-stage Caleb rocket off the coast of California. The missile was designed to use the three uppermost stages of the five-stage rocket used for the Project Pilot tests conducted between July 4, 1958, and August 28, 1958. Only one Hotroc rocket was employed as the first stage. The Caleb rocket had a length of 18 ft. 3 in. and a launch weight of 3,000 lb. It was theoretically capable of placing a 20-lb. satellite in a 1,000-mi. orbit. For this first test only the first stage was live, but none of the five Caleb rockets carried more than two live stages. Following a second flight on October 24, 1960, when the second stage failed to fire, the Navy switched to a McDonnell F4H Phantom as the drop plane and changed the project name to Hi-Hoe. Of the three Hi-Hoe flights only the last, on July 26, 1962, was a success, when a two-stage Caleb reached a height of 726 mi.

1960 July 28–29 NASA held its first Industry Program Plans Conference at headquarters in Washington, D.C., at which it presented an overall picture of the long-range plan as well as current needs and requirements. Deputy Administrator Hugh L. Dryden described the advanced manned spacecraft and said that approval had been granted for it to be called Apollo; the name had been suggested by Abe Silverstein in January to perpetuate the use of mythical gods in the names of manned space vehicles. George M. Low, NASA's chief of Manned Space Flight, said that Apollo would carry out circumlunar and Earth-orbit missions by 1970 and that lunar landings and a permanent manned space station would follow.

1960 July 29 The Mercury Atlas-1 flight took place from Launch Complex 14, Cape Canaveral, Fla., with a D-series Atlas converted to carry spacecraft no. 4. Planned as a test of spacecraft integrity through ascent and a simulated abort, although no actual escape system was carried, MA-1 should have carried the spacecraft to a height of 112 mi., a speed of 13,050 MPH and a downrange distance of 1,490 mi. in a 16-min. test of the heat shield and recovery system. Instead, just 59 sec. after liftoff at 14:13 UT, the adapter connecting the spacecraft to the launch vehicle collapsed and the Atlas aborted. The mission reached a height of 8.1 mi., a speed of 1,701 MPH and a downrange distance of 6 mi. The spacecraft was not recovered because the parachute deployment system was not designed to experience this combination of failures and did nothing to prevent it from slamming into the Atlantic Ocean.

1960 August 8 The U.S. Air Force Air Material Command Ballistic Missile Center requested industry participation in a program to study the feasibility of using reentry bodies to return film capsules from the SAMOS (satellite and missile observation system) satellites. While SAMOS was primarily required to transmit surveillance pictures by radio signal, a concurrent study had been conducted to determine the feasibility of recovery. The Air Force request called for design proposals on maneuvering reentry bodies and capsules.

1960 August 10 The first satellite from which an object was returned to Earth intact was launched by a Thor-Agena A from Vandenberg AFB, Calif. *Discoverer XIII* was placed in a 160 × 424.5 mi. orbit inclined 82.8° and its capsule reentered the atmosphere on the 17th orbit, just over one day later, and was recovered from the sea about 330 mi. northwest of Honolulu. The satellite was equipped with a KH-1 reconnaissance camera system built by Itek, but failed to operate as planned. Thus far, the Discoverer satellites had not achieved a great deal of success. Only eight of the 13 satellites launched had attained orbit.

1960 August 12 The world's first passive communication satellite was launched by NASA on a Delta from Cape Canaveral, Fla., into a 947 × 1,046 mi. orbit inclined 47.2° to the equator. Called *Echo I,* it was a 100-ft. aluminum-coated sphere inflated after reaching orbit by expansion of residual air and weighing 166 lb. Echo carried two 11-oz. radio beacons transmitting on 107.9 MHz at 5 mW, electrical power provided by 70 solar cells charging five NiCd batteries. A taped message from President Eisenhower was bounced off the satellite next day between Cedar Rapids, Iowa, and Richardson, near Dallas, Tex. *Echo I* was joined on January 25, 1964, by *Echo II,* a 135-ft. sphere weighing 457 lb. and launched from Vandenberg AFB, Calif., by Thor-Agena B into a 639 × 818 mi. orbit inclined 81.5°. *Echo I* burned up in the earth's atmosphere on May 24, 1968, followed by *Echo II* on June 7, 1969.

1960 August 13 U.S. Secretary of the Interior Fred A. Seaton announced that the U.S. Geological Survey (USGS) had completed the first photogeological map of the near side of the moon. It consisted of three maps, each 3 ft. in diameter. The first map named features on the lunar surface, the second provided a photogeologic map putting a stratigraphic sequence on the craters, the third mapped the prominent lunar rays and provided correlation with surface features. Undertaken as part of a program to provide lunar landing site selection data, it was the start of the USGS's long occupation with other worlds in space. During the Apollo moon landings, the USGS's Astrogeologic Center at Flagstaff, Ariz., completed thousands of moon maps and charts.

1960 August 16 U.S. Secretary of Defense Thomas S. Gates announced the formation of a Joint Strategic Target Planning Staff (JSTPS) composed of representatives from all the armed services holding nuclear weapons. With the widespread proliferation of nuclear weapons among the armed forces, the Department of Defense was putting together a Single Integrated Operational Plan, or SIOP. Instead of relying on each service to compile its own hit list of nuclear targets, SIOP would knit together tactical and strategic delivery systems, enabling the Joint Chiefs to plan a series of preplanned nuclear attacks in a coordinated and predetermined way. President Eisenhower wanted this because he felt each service was trying to obtain enough nuclear strike power to win a future war by itself. SIOP was to be the blueprint for complete integration but, paradoxically, would invigorate the campaign for more firepower.

1960 August 18 The first successful U.S. photoreconnaissance satellite was launched from Vandenberg AFB, Calif., when a Thor-Agena A carried *Discoverer XIV* to an orbit of 116 × 500 mi. and an inclination of 79.65°. Carrying an Itek KH-1 camera and 3,000 ft. of 70 mm film, the satellite was able to observe 35° to the left and right of its orbital ground track. On the 17th orbit (August 19, 1960) the film recovery capsule returned to Earth with pictures covering almost 20% of the land mass of the USSR, more images than had been taken on all 24 U-2 spyplane flights over Soviet territory. Resolution was about 50 ft., but within three years this had improved to 9 ft. and by 1967 to about 5 ft. Six Fairchild C-119s and one Lockheed C-130 flew the 200 × 60 mi. rectangle where the capsule was expected to drift down by parachute. The C-119 that snagged the 84-lb. capsule was patrolling an outer region up to 400 mi. away. *Discoverer XV,* the last in the series using an Agena A, was launched on September 13 but its capsule was lost at sea.

1960 August 19 The first successful flight of a prototype Vostok manned spacecraft took place when a Soviet SL-3 launcher put the 10,150-lb. *Korabl-Sputnik 2* into a 190 × 211 mi. orbit inclined 64.95°. Strapped to the ejection seat were two dogs named Strelka and Belka in the first of a series of flights designed to qualify on-board systems including life support equipment. Also carried were a variety of biological specimens including white mice, rats, flies, fungi, plant seeds, algae and a spiderwort plant. After about a day the retro-rocket fired as planned and brought the *KS-2* descent module safely back through the atmosphere. At a height of approximately 23,000 ft. the main recovery parachute opened and the ejection seat fired, bringing the two dogs to Earth only 6 mi. from the designated recovery point.

1960 August 25 Deputy Director of Defense Research and Engineering John Rubel gave the U.S. Air Force formal approval for a four-flight demonstration of antisatellite inspection capabilities in the SAINT (Satellite Inspector) program. The Air Force briefed contractors during September, and RCA was selected as primary contractor for the inspection system launched by Atlas-Agena B. The plan was for the Agena B to put the SAINT vehicle into an orbital path slightly ahead of a target of inflatable spheres. SAINT would use on-board propulsion to rendezvous with and inspect targets from a maximum distance of 50 ft. TV cameras and radiation sensors would be part of the diagnostic equipment. The four test flights were to have begun in March 1963 and been followed by 15 development flights toward operational deployment by 1967. When religious groups protested about the name, the Air Force changed it to Hawkeye, but when that was found to imply snooping it became first Program 621A, then Program 720.

A panel set up to examine U.S. reconnaissance satellite programs in the wake of the Gary Powers U-2 spyplane incident reported its findings to the President at a special meeting of the National Security Council. Members of the panel included Presidential Science Adviser George Kistiakowsky, James Killian and Dr. Edwin Land of the Polaroid Corp. It recommended the creation of an intelligence organization separate from the armed forces and the CIA. Under the leadership of Dr. Joseph Charyk, the undersecretary of the Air Force, the National Reconnaissance Office was set up six days later. It was authorized to take control of all Air Force and CIA reconnaissance satellite projects.

1960 August 29 NASA and the U.S. Atomic Energy Commission signed a memorandum setting up a Space Nuclear Propulsion Office under joint control and headed by Harold B. Finger. Just fourteen days earlier, Martin, Douglas, Convair and Lockheed had started work on RIFT (reactor-in-flight test), for the Marshall Space Flight Center. RIFT was a program to develop a flight rated engine and would lead to NERVA (nuclear engine for rocket vehicle application). Both RIFT and NERVA would evolve from Kiwi-A tests and a new larger reactor called Kiwi-B. NASA issued a request on February 2, 1961, for proposals to build NERVA, and in a special message to Congress on May 25, 1961, President Kennedy increased nuclear rocket funds as part of his Atoms for Peace program.

1960 August 30 Goddard Space Flight Center held a bidders conference on NASA's Orbiting Geophysical Observatory (OGO) program. It was to be another "streetcar" satellite defined for the Orbiting Solar Observatory program (see entry for April 17, 1959), comprising a rectangular box with solar cell panels and provision for 20 Earth-science experiments. This was the third, and last, of NASA's first-generation observatory-class satellites including the Orbiting Astronomical and the Orbiting Solar Observatories. On December 21, 1960, NASA issued a contract to Space Technology Laboratories for analysis and design of three OGO satellites each weighing about 1,100 lb. and launched by an Atlas-Agena B. In April 1961 the OGO design was completed and on August 3, 1962, TRW received a contract to build the satellite, six being launched between September 5, 1964, and October 12, 1979.

1960 August 31 In an organizational move that removed control of space-based reconnaissance and surveillance systems from the U.S. Air Force, the Navy and the CIA, the National Reconnaissance Office (NRO) was formally established. Set up within the Office of the Secretary of the Air Force under cover of the Office of Missile and Satellite Systems, it was in fact responsible for all U.S. military space observation systems. At one stroke this tightened the secret work under way on projects that previously had rubbed shoulders with aircraft, missile and rocket development projects, ensuring a protected enclave for specialists with highly classified activities. Air Force Brig. Gen. Robert E. Greer was put in charge of the development division at the AFBMD (Air Force Ballistic Missile Division), Los Angeles. Not for 13 years would the existence of the NRO be acknowledged.

1960 September 13 NASA's Space Task Group held a briefing at Langley Research Center for prospective bidders on the advanced manned spacecraft named Apollo. Three six-month feasibility studies were to be carried out and a formal request for proposals was issued. Bids from 14 companies were submitted from an initial list of 64 that expressed interest. They included: Boeing, Chance-Vought, Convair, Cornell Aeronautical Laboratory, Douglas, General Electric,

Goodyear Aircraft, Grumman, Guardite Division of American Marietta, Lockheed, Martin, North American Aviation and Republic Aviation. The contenders, reduced to 12 when Cornell and Guardite withdrew, were invited to provide primary contractor proposals. Technical assessment began on October 10, 1960.

In a move that initiated a long-standing relationship between NASA and the U.S. Department of Defense, the Aeronautics and Astronautics Coordinating Board was signed into existence by NASA administrator T. Keith Glennan and James H. Douglas, the deputy secretary of defense. Six panels were established covering manned spaceflight, spacecraft, launch vehicles, ground equipment, research and technology, and aviation. The NASA deputy administrator as well as the director of Defense Research and Engineering were to be cochairmen of the separate panels composed of equal numbers from each agency. NASA Administrator James E. Webb went a stage further on October 1 by asking W. Fred Boone, U.S. Navy Admiral (Ret.) to head a new Office of Defense Affairs established at the space agency's headquarters on December 1, 1962.

1960 September 14 NASA's Ames Research Center set up a team to investigate the possibility of developing a solar probe. Scientists wanted to study the sun's radiation, the solar wind and its interaction with the earth's outer atmosphere. The best way to do that was to design and build probes which could be launched on heliospheric orbits between the Earth and Venus, between the Earth and Mars or at the same distance from the sun as the Earth but some way from it. An interplanetary spacecraft in the Pioneer class to expand upon the information obtained by *Pioneer V* was proposed by Ames in April 1962.

1960 September 19 Bell Aerosystems announced that it had completed a series of firing tests with a turbopump engine fed on a combination of fluorine and hydrogen and that in the past four years it had conducted 600 engine firings during development. The tests included hydrogen, hydrazine and ammonia fuels used with the fluorine oxidizer in motors with thrusts up to 35,000 lb. The advantages of fluorine were that it was hypergolic (it would ignite on contact with the fuel) and that it had a higher specific impulse and was therefore more efficient.

1960 September 21 The first four-stage Blue Scout solid propellant launcher was fired from Cape Canaveral, Fla. Blue Scout was the U.S. Air Force version of the NASA Scout sounding rocket and small satellite launcher. The first Blue Scout was programmed to carry a 33-lb. package of experiments to a height of 16,600 mi. with impact 7,000 mi. downrange. Radio contact was lost shortly after the launcher left its pad, but the four stages apparently performed as planned. The second Blue Scout launch, on November 8, also proved a failure when the second stage burned out prematurely. The Air Force Ballistic Missile Division planned 12 hyperenvironment test system (HETS) launches.

1960 September 25 The Atlas-Able VA mission to send a 387-lb. Pioneer space probe to the vicinity of the moon failed when the second stage prematurely stopped running and the vehicle fell back into the Atlantic Ocean. Weighing 387 lb., *Pioneer P-30* was to have fired its retro-rocket 62 hr. after launch some 5,000 mi. from the moon and dropped into a 2,000 × 3,000 mi. lunar orbit. The probe carried a simple scanning device which was to have returned crude pictures of the lunar surface.

1960 September 27–28 NASA held a conference at the Marshall Space Flight Center with representatives from U.S. industry to brief them on opportunities to participate in the new Saturn launcher program and to provide specifications for the new S-II stage designed as the second stage for the Saturn C-3. The S-II was expected to have four 200,000-lb. thrust cryogenic engines. This was to be the first high-thrust, cryogenic high-energy upper stage developed for the Saturn program and would call for new and innovative technologies to provide for the large volumes of liquid hydrogen and oxygen carried in its tanks. The conference also heard that after initial fabrication for research and development flights NASA wanted to shift most Saturn manufacturing out into industry.

1960 October 4 Following the failure of the *Courier 1A* Thor-Able Star launcher on August 18, 1960, the U.S. Air Force successfully launched *Courier 1B* from Cape Canaveral, Fla., to a 604 × 751 mi. orbit with an inclination of 106.9°. The 507-lb. satellite was the first to carry active-repeater communication equipment and, with 19,200 solar cells, provide an unprecedented 62 W of electrical energy. Operating in the 1.7–2.3 GHz range, Courier had been built by Space Technology Laboratories, a division of TRW, and initial tests began on its second orbit with a message from President Eisenhower. The satellite failed after 17 days but remains in orbit.

The International Sporting Committee of the Fédération Aéronautique Internationale (FAI), originally set up to control aviation records for speed, distance, altitude and weight-lifting capacity, announced that the United States and the USSR had reached agreement on a framework for spaceflight records. Henceforth, the FAI would recognize categories for highest, furthest and greatest weight for each mission that reached a height of 62.1 mi. and exceeded the previous record by 10%. Soon other categories would be added, including number of astronauts or cosmonauts carried, speed and duration.

1960 October 10 The first Soviet spacecraft utilizing a new, three-stage variant of the SS-6 Sapwood ICBM and built to fly past the planet Mars was destroyed in a launch failure. Designated SL-6, it had the same first-stage/strap-on booster configuration of the SL-3 with a 66,150-lb. thrust second stage built by the Kosberg bureau and a 15,435-lb. thrust third stage from the Isayev bureau. Overall, the SS-6 had a launch weight of 674,700 lb. and a total height, including payload shroud, of 137 ft. 9 in. The first two stages were intended to place the third stage and payload into Earth orbit from where the third stage would propel the space probe, designated M-1, to escape velocity and on course for Mars. A second attempt with another spacecraft on October 14 also failed. During a visit to the United Nations, Soviet Premier Khrushchev had taken to New York a small model of the 1,400-lb. Mars probe that should have been put on course for that planet. He returned to Moscow with the model still in its case.

1960 October 11 The first test flight of an Atlas E took place from Cape Canaveral, Fla., but was only a partial success. Upgraded with the Rocketdyne MA-3 propulsion system providing a total launch thrust of 387,000 lb., the E series could throw a 6,175-lb. warhead a maximum distance of 5,000 mi. Atlas E and F would carry the 3.75-MT W-38 nuclear warhead in an Avco Mk.4 ablative nose cone. The Mk.4 had a length of 11 ft. 3 in., a diameter of 2 ft. 2 in. flaring to 7 ft. 4 in. at the skirt, and weighed 225 lb. empty. Instead of standing exposed on soft sites above ground like Atlas D, the 27 Atlas Es deployed were placed horizontally in semihard concrete coffins from where they would be raised 90° in 2 min. for launch 13 min. later. Three Strategic Missiles Squadrons operated nine missiles each at: Warren AFB, Wyoming; Forbes AFB, Kansas; and Fairchild AFB, Washington. The Atlas E was fully deployed on November 20, 1961.

1960 October 12   NASA announced it would launch private U.S. communication satellites for a reimbursable fee. On October 21 American Telephone and Telegraph (AT&T) submitted to the Federal Communications Commission (FCC) a request for permission to launch and operate such satellites. On January 19, 1961, the FCC granted AT&T a license to operate an experimental trans-Atlantic satellite link, to be known as Telstar. In April 1961 General Electric published similar plans for a satellite system.

1960 October 16–19   NASA held a management conference at Williamsburg, Va., and heard a paper read by the director of NASA's Office of Business Administration, Al Siepert, *A NASA Structure for Project Management*. This formed the basis for what became known as NASA General Management Instruction No. 4-1-1 and came to be considered one of the most important shifts in the way the U.S. space agency organized major projects. Instead of attempting to coordinate all field centers in a common effort, specific centers would in future be assigned management responsibility for particular projects. Separate offices at NASA Headquarters would link other centers to carry out subsidiary work in their respective areas of expertise. In modified form, it is a practice still in operation more than 35 years later.

1960 October 17   NASA's chief of Manned Space Flight, George M. Low, sent Abe Silverstein, associate administrator of the Office of Space Flight Development, a memorandum explaining how he was setting up a working group on the manned lunar landing program. Low explained that clear understanding of the mission requirements for a manned moon landing would enable Apollo contractors to define the particulars of Apollo circumlunar and Earth-orbit requirements. In addition to Low, the group comprised Eldon Hall, Oran Nicks and John Disher. It was to produce ground rules for the mission, project spacecraft weights and launch vehicle requirements and prepare an integrated development plan including hardware needs and funding requirements set against a projected schedule.

1960 October 19   NASA announced that preliminary design studies on large solid propellant boosters would be conducted by Aerojet General, Grand Central Rocket and Thiokol. The studies would last six months and examine the feasibility of heavyweight solid propellant launchers with a first-stage thrust of between 1 million and 20 million lb. To date, the ability to control the firing duration and the thrust of liquid propellant motors, coupled with their higher efficiency and performance levels, had focused attention away from solids. However, major advances in the development of solid propellant and solid motor-case technology achieved through work on Polaris and Minuteman missiles made the brute force available from big solids of increasing interest to NASA.

1960 October 21   Convair Astronautics received a NASA contract to study an adapted version of the Centaur upper stage, the S-V, which at this date was expected to be the third stage of the Saturn C-1. In early 1961 it would also find application as the terminal stage of a proposed four-stage Saturn C-2. The cryogenic S-V (Centaur) provided the design base for the S-IV, already under contracted development with Douglas Aircraft, and had two 15,000-lb. thrust Pratt and Whitney LR-115 engines. The S-IV was essentially an enlarged cryogenic stage with provision for four 20,000-lb. thrust uprated LR-119 engines. Convair submitted their design proposals for the S-V in January 1961, but late the same month Wernher von Braun began to cast doubts on the value of this stage, and on June 1, 1961, it was deleted from the first 10 Saturn C-1 research and development flights. The need for the S-V never returned and it was not developed.

1960 October 24   The worst accident in the annals of Soviet rocketry occurred on launch pad 41 at the Baikonur complex in Kazakhstan. Designed by Mikhail Kuzmich Yangel from the Dnepropetrovsk rocket assembly plant, the first SS-7 Saddler ICBM was being prepared for flight when a first-stage control valve malfunctioned. The SS-7 had been designed to use storable dimethylhydrazine and nitric acid propellants and emerged as a two-stage missile with a length of 104 ft. 4 in., a diameter of 9 ft. 3 in. and a launch weight of 225,000 lb. Pressure from Khrushchev to get the SS-7 tested and in service as the first Soviet ICBM on constant readiness for launch drove technicians beyond the point of caution. During efforts to change the valve while the missile was being fueled, an explosion engulfed the pad area taking the lives of 54 technicians and engineers as well as the chief of the Soviet Strategic Rocket Forces, Field Marshal Mitrojan Nedelin. With a range of 6,800 mi., SS-7s were deployed from 1962, building eventually to a force of 170. The missile remained in service until the early 1980s.

1960 October 25   From the 12 prospective primary contractors for Apollo, NASA selected Convair, General Electric and Martin for six-month feasibility studies. Two days later, representatives from these companies visited the Space Task Group to clarify aspects of the work and received contracts on November 15, 1960. At a review of progress on January 12, 1961, General Electric presented their elliptical cone and half-cone designs for a reentry capsule, Martin described their M-1 and M-2 lifting body designs as well as Mercury-type ballistic capsules and a variety of modified reentry cone shapes, and Convair presented five alternative configurations based on a cross section of ballistic capsules and lifting bodies. All three companies showed schematics whereby the reentry capsule was attached to the forward end of an orbital section. Final results were submitted May 15–17, 1961.

1960 October 26   The first orbital flight of an Agena B rocket stage was launched by a Thor booster carrying the U.S. Air Force *Discoverer XVI* reconnaissance system integrated with the stage itself. The mission failed when the Agena B failed to separate from the Thor. The first successful Thor-Agena B flight was the *Discoverer XVII* mission of November 12, 1960. Agena B displayed an increase in reliability over its predecessor, Agena A, reflecting technical improvements during this period. Of 19 Agena A flights between February 1960 and 1961, (15 on Thor and 4 on Atlas) only 12 (63%) had been successful. Agena B would be used sporadically on Thor and Atlas a total 71 times through 1966, of which 59 (83%) were successful. In its first 15 years as an upper stage for Thor and Atlas beginning 1962, the improved Agena D would display a 92% success rate in 98 launches.

1960 November 1 Speaking in the House of Commons, Prime Minister Harold Macmillan announced that the government had agreed to make available a permanent sheltered base for U.S. Navy Polaris-equipped ballistic missile submarines at Holy Lock on the Firth of Clyde, 30 miles west of Glasgow in Scotland. The first missile depot ship was scheduled to arrive in February 1961 and plans eventually envisaged a squadron of nine submarines to be based there. Of these, six would normally be on their eight-week ocean patrol and three would be in port or on local exercise.

1960 November 3 The first of NASA's ionospheric research satellites, *Explorer VIII,* was placed in a 253 × 1,056 mi. orbit with a 49.9° inclination. Designed to obtain knowledge about ionization in the upper atmosphere by looking down at the phenomenon (hence their nickname "topside sounders"), Explorer satellites in this series were instrumented to obtain information about the nature, density, dynamic behavior and charged particle distribution. The ionosphere is a region with a high concentration of free electrons extending from about 35 to 600 mi. The orbit of *Explorer VIII* was designed to examine the region of highest concentration at around 220 mi. Other NASA ionospheric satellites included *Explorer S-45, Beacon Explorer A, Explorer 20* and *Explorer 22,* the last of which was launched on October 10, 1964. Nine international ionospheric satellites were also launched by NASA.

1960 November 8 The first Little Joe abort test with a production Mercury spacecraft took place from NASA's Wallops Island test range. The flight of *LJ-5* was normal for 15.4 sec., at which point the escape motor on spacecraft no. 3 ignited prematurely; but the spacecraft did not detach from the launch vehicle until impact. In a 2 min. 22 sec. flight it only reached a height of 10.1 mi., a speed of 1,785 MPH and a downrange distance of 14 mi., picking up 6 g on descent. A repeat attempt was made with spacecraft no. 14 on the *LJ-5A* attempt, March 18, 1961, but the escape motor fired prematurely 19 sec. after launch and the spacecraft did not separate as expected. Instead, separation was effected using the retrorockets. In a 2 min. 48 sec. flight, the spacecraft reached a height of 7.7 mi., a speed of 1,783 MPH, a deceleration of 8 g and a downrange distance of 18 mi. The recovery sequence worked and the spacecraft was used again for a final effort to get the planned test results with *LJ-5B* on April 28, 1961. This time one Little Joe motor did not ignite until 4 sec. into flight, pitching the vehicle into a more severe environment, but the spacecraft performed as required. It reached a height of 2.8 mi., a speed of 1,780 MPH, a deceleration of 10 g and a downrange distance of 9 mi. in a 5 min. 25 sec. flight.

1960 November 12 A severe solar storm produced changes in the Earth's atmosphere, increasing its density to the point where satellites were affected and their orbits changed. Atmospheric drag on the 100-ft. Echo balloon satellite increased twofold and noticeably altered the orbit, even though the satellite was more than 600 mi. above Earth. The effect of solar radiation was to accelerate the satellite as it was "pushed" along in the stream of high-energy particles. This positive effect was used later in theories about "solar sailing," included in an Arthur C. Clarke novel.

Weighing 2,400 lb., the first Discoverer reconnaissance satellite successfully launched by Thor-Agena B was placed in a 116 × 616 mi., 81.9° orbit. Launched from Vandenberg AFB, Calif., *Discoverer XVII* released its 300-lb. capsule for reentry on the 31st orbit and was snatched in midair by a C-119, only the second time air recovery had been accom-

A Mercury spacecraft sits atop its Redstone launcher ready for the MR-1 abort test. Note the tower for the launch escape motor.

plished. Agena B was powered by a 15,000-lb. thrust Bell 8096 engine burning IRFNA-UDMH propellants and was capable of being restarted in space. The Agena B/Discoverer configuration had a total length of 25 ft. and a diameter of 5 ft. Between October 16, 1960 (when the first attempt had failed) and November 24, 1962, the Air Force launched 37 Discoverer-class satellites on Thor Agena B. Of this total, 7 failed to reach orbit. During the life of the program, Agena B/Discoverer weight increased to 2,550 lb.

1960 November 14 The U.S. Air Force Ballistic Missiles Center awarded the Martin Co. a contract for reentry body designs in support of the SAMOS (satellite and missile observation system) program. Conceived to support a plan whereby film canisters from SAMOS surveillance satellites could be recovered on Earth, the study switched in March 1961 to development of a maneuverable lifting body for aerothermal research purposes. This quickly evolved into the START (spacecraft technology and advanced reentry tests) program which involved three phases: ASSET (aerothermodynamic/ elastic structural systems environmental tests); PRIME (precision recovery including maneuvering reentry); and PILOT (piloted low-speed tests). The Martin work led eventually to the SV-5P lifting body program, redesignated X-24.

1960 November 15 North American Aviation test pilot A. Scott Crossfield flew the X-15 with its intended power plant for the first time. Fitted with the 57,000-lb. thrust XLR99 it had been designed to carry, the no. 2 aircraft reached a speed of Mach 2.97 (1,960 MPH) and an altitude of 81,200 ft. The first in-flight restart of the XLR99 took place on November 22, 1960. The last of thirty X-15 flights utilizing the interim XLR11 propulsion system carried U.S. Air Force Capt. Robert M. White in the no. 1 aircraft to a speed of Mach 3.5 (2,275 MPH) and a height of 78,150 ft. on February 7, 1961.

1960 November 21   In an attempt to launch the first Mercury-Redstone on a preliminary suborbital test of the spacecraft and launch vehicle combination before a primate flight, premature cutoff of the Redstone motor took place when vehicle MR-1 was about 1 in. off the pad. It settled back to its pad supports, but since the spacecraft had received a cutoff signal it triggered its launch escape system and recovery parachutes. The launch tower and rocket tore free, leaving spacecraft no. 2 on the launch vehicle and draped in parachutes just 3 sec. later when that component of the recovery equipment deployed itself. A second attempt, dubbed MR-1A, on December 19, 1960, was a success. The spacecraft reached a height of 131 mi., a speed of 4,909 MPH and a downrange distance of 235 mi. With a deceleration of 12.4 g, the spacecraft was safely recovered from the Atlantic Ocean.

A U.S. interagency Aeronautics and Astronautics Coordinating Board issued a statement via its Unmanned Spacecraft Panel declaring that NASA was to develop a research and development satellite for experiments in satellite communications. Because the Department of Defense had the work on a geosynchronous communication satellite called Advent, NASA was to confine its work to a research program on medium- and low-orbit satellites. Late in November 1960, Space Technology Laboratories received a design study contract for an active communications system that could lead to a commercial satellite system. The preliminary specification was drawn up by the Goddard Space Flight Center on January 13, 1961, and twelve days later NASA briefed industry on what it wanted built for what was now termed Project Relay.

A unique method of decelerating an object returning from space was tested when a rocket fired from Santa Rosa Island off the west coast of Florida carried Ballute, a combined drag balloon and parachute, to a height of 32 mi. Weighing 500 lb., the capsule descended to 75,000 ft. where it deployed a balloon 9 ft. in diameter that trailed behind the main body. Reducing descent velocity below the speed of sound, the balloon was severed at 35,000 ft. and a parachute released for soft landing.

1960 November 22   The first of five new and enlarged Fleet Ballistic Missile (FBM) submarines was launched from Groton, Conn., and named USS *Ethan Allen*. With a length of 410 ft. and a submerged displacement of 7,900 tons, the five *Ethan Allen*–class vessels each carried 16 Polaris missiles and a crew complement of 139 (12 officers and 127 enlisted men) compared with 140 (13 officers and 127 enlisted men) on each of the five *George Washington*–class submarines. The last boat in the Ethan Allen–class class, the USS *Thomas Jefferson*, was launched on February 24, 1962, and commissioned on January 4, 1963. It first put to sea on operational deterrent duty on October 28, 1963, carrying 16 Polaris A2 missiles. These were the first FBM submarines designed from the outset for SLBMs (submarine-launched ballistic missiles)

1960 November 28   At a meeting in Meyrin, Switzerland, scientists, engineers and government officials from every European country outside the Eastern Bloc met to discuss a unified program for space research and cooperation. The 11 nations represented agreed to establish a Preparatory Commission to Study the Possibilities of European Collaboration

in the Field of Space Research. The Meyrin Agreement came into force on February 27, 1961, and at meetings in Paris and Rome the framework of an international agreement to form the European Space Research Organization (ESRO) was worked out. On June 14, 1962, Belgium, France, Germany, Italy, the Netherlands, the United Kingdom, Spain, Swedenand Switzerland signed the agreement on June 14, 1962, with Denmark signing in December 1962. Ratification by member states took two years to complete and ESRO went into force on March 20, 1964.

1960 November 30   A Thor-Able Star launch-vehicle from Cape Canaveral, Fla., failed during the ascent phase, resulting in the loss of the 200-lb. *Transit 3A* navigation and 40-lb. *Solrad 2* solar energy detection satellites when the range safety office destroyed the vehicle by radio command. *Solrad 2* was also known as *Greb 2*. Pieces of the launch vehicle were found on the island of Cuba but no damage was reported.

1960 December 1   A meeting was held in Strasbourg, France, between government science and technology representatives from the United Kingdom, France, Germany, Belgium, Denmark, Italy, the Netherlands, Norway, Spain, Sweden and Switzerland with a view to building a European launcher. There was general support the United Kingdom's Blue Streak missile as the first stage and for developing a new French second stage and a German third stage. Instigated by the British, who sought a use for their canceled ballistic missile, the proposal appealed to the French, who sought independence from U.S. launchers. Discussions continued with a final set of meetings held between January 30 and February 2, 1961. Final agreement to form a European Launcher Development Organization (ELDO) came with the signing of the convention on March 29, 1962.

The fourth flight attempt of a prototype Vostok manned spacecraft was launched into a 112 × 155 mi. orbit carrying dogs Pchelka and Mushka. Designated *KS-3,* the 10,060-lb. spacecraft carried a solar-orientation system replacing an infrared system which would take its reference from the sun and not the limb of the Earth. The orbit for *KS-3* was much lower than previously adopted, demonstrating the type of orbit manned Vostoks would fly. If the retro-rocket failed, a low orbit would decay more quickly and bring the spacecraft back through the atmosphere within the lifetime of the on-board consumables, a prudent safety measure that would be used by Mercury orbit planners in the United States. *KS-3's* reentry a day after launch went wrong and its descent module burned up in the atmosphere. During the launch of *KS-4(1)* on December 22, 1960, the second stage failed but the biological payload was safely recovered.

1960 December 4   The first attempted launch of a U.S. satellite from a site other than Cape Canaveral, Fla., or Vandenberg AFB, Calif., failed when a NASA Scout launch vehicle malfunctioned after liftoff from Wallops Island, Va. The four-stage Scout was carrying the 15.5-lb. *Explorer S-56* satellite, an inflatable balloon 12 ft. in diameter when deployed in orbit. Made of aluminized Mylar, the balloon contained a small radio beacon. Tracking by visual and radio-detection means would have enabled scientists to determine the density of the earth's atmosphere, an objective thwarted when it instead ended up in the Atlantic 80 mi. downrange.

1960 December 5   The Douglas Aircraft Co. received a contract from the U.S. Air Force for design and development of Weapon System 138A, a strategic air-to-surface missile designated GAM-87A Skybolt. Essentially a two-stage solid propellant missile, Skybolt was designed to be carried under inboard wing pylons of the Boeing B-52 bomber. In a decision made public in March 1960, British Prime Minister Harold Macmillan and President Eisenhower announced during a meeting at Camp David that, having canceled the British Blue Streak ballistic missile in April 1960, the United Kingdom was to procure Skybolt for use with its V-bomber force as a strategic nuclear deterrent.

1960 December 7   The first successful U.S. photoreconnaissance satellite carrying a KH-2 Itek camera system, *Discoverer XVIII* was launched by Thor-Agena B from Vandenberg Air Force Base, Calif., to a 151 × 411 mi. orbit and an inclination of 81.5°. The first attempt to launch a KH-2 camera system, on *Discoverer XVI,* October 26, 1960, failed when the Atlas remained attached to the Agena and the combination failed to reach orbit. After 48 orbits of the Earth, *Discoverer XVIII*'s capsule was returned through the atmosphere and retrieved in a midair snatch. The KH-2 carried more film and, with Agena B, was able to remain in space twice as long as the KH-1.

1960 December 9   The U.S. Air Force asked the Los Alamos Scientific Laboratory, N.Mex., to design and develop a new lightweight thermonuclear warhead for the LGM-30A Minuteman I solid propellant ICBM only months after the 1.2-MT W-56 had been selected with an Mk.11 reentry vehicle. Indications that the initial LGM-30A version of Minuteman I would not have the required range dictated extra weight savings in warhead technology; every 1 lb. saved in throw weight provided an additional 5 mi. added to the missile's range. About 150 W-59s carried in an Avco Mk.5 warhead were produced for LGM-30A between June 1962 and July 1963, all being retired by June 1969. The Mk.11/W-56 warload weighed about 2,200 lb. and was employed on the LGM-30B version of Minuteman I and on Minuteman II, giving the latter a potential error radius of only 1,200–1,800 ft. at extreme range.

1960 December 14   NASA Associate Administrator Robert C. Seamans Jr. received a briefing from Langley Research Center engineers on orbital rendezvous as it related to national space programs. Clinton E. Brown and Ralph W. Stone Jr. described a lunar-orbit rendezvous method for supporting a manned moon landing. Instead of launching from Earth to the moon on direct-ascent or assembling parts of the spacecraft in the Earth-orbit rendezvous, they argued for launching directly from Earth to the moon with a combination of vehicles that would first enter lunar orbit and from there descend to the surface, ascending to rendezvous with an Earth-return vehicle at the conclusion of surface activity. Considerable work on lunar-orbit rendezvous had been done by Langley's John C. Houbolt, who attended the briefing.

1960 December 15   The third of three Pioneer moon orbiters launched from Cape Canaveral, Fla., was lost when the Atlas-Able VB failed to place the 388-lb. probe on an escape trajectory as planned. The second stage ignited prematurely at 1 min. 8 sec. during first-stage burn and the vehicle was destroyed 6 sec. later. *Pioneer P-31* was essentially the same as two earlier Pioneer moon orbiters which failed to leave Earth. Thus ended the first, frustrating and unsuccessful phase of sending U.S. space probes to the moon. Not until July 28, 1964, would the first completely successful U.S. moon mission begin.

1960 December 20   President-elect John F. Kennedy announced that he would delegate his chairmanship of the National Aeronautics and Space Council to Vice President-elect Lyndon B. Johnson. Eisenhower had worked hard to prevent further expansion of space activity and, as he prepared to leave office, warned that "Further testing and experimentation will be necessary to establish whether there are any valid scientific reasons for extending manned space flight beyond the Mercury Program." Johnson would be a vital part of orchestrating Congressional support for a major commitment to increase NASA's budget, broaden the range of space activities and set a central goal: a manned landing on the moon by the end of the decade.

1960 December   During the month, U.S. Air Force and contractor teams held a series of meetings on a new aerospace plane, or ASP, concept. It grew from work at several propulsion contractors, notably the Marquardt Corp., on air-breathing/hydrogen-fueled hypersonic scramjet engines. Instead of carrying oxidizer and fuel in special tanks for combustion in a rocket motor, the ASP would ingest air (oxidizer) from the atmosphere for the combustion of liquid hydrogen carried internally. Operating like an aircraft, the ASP would take off and land horizontally and gain most of its orbital speed in the atmosphere, switching its engine to internal oxidizer tanks for getting into orbit.

1961 January 3   In a move that began the process of establishing an autonomous manned spaceflight facility, NASA's Space Task Group (STG) became a separate field element reporting directly to NASA Headquarters. Previously, STG had been under the auspices of the Goddard Space Flight Center and supported administratively by the Langley Research Center. During the first half of the year plans to build an independent manned spacecraft center matured and a selection team was chosen to evaluate candidate sites. In August 1961 the team toured locations in Florida, Louisiana, Texas, Missouri, California and Massachusetts looking for an appropriate site.

1961 January 5   A manned lunar landing task group was set up in NASA under the chairmanship of George M. Low, hereafter known as the Low committee, and received its first briefings on proposals for such a mission. The Marshall Space Flight Center (MSFC) proposed a three-phase program to accomplish the first Apollo flight in 1965 by a spacecraft designated Apollo A, a manned circumlunar return flight in 1966 by an Apollo B design, manned lunar landing and return beginning July 4, 1967, semipermanent lunar science stations manned by 18–20 people from 1969 and permanent bases after 1971. To support the mission objectives, the MSFC envisaged 36–48 Saturn launches per year from 1966 and the use of nuclear-powered upper stages from 1965. Launch vehicles included Saturn C-2, C-3 and Nova configurations carrying four, six or eight (1.5-million-lb. thrust) F-1 engines.

1961 January 5–6   NASA's Space Exploration Program Council met at Headquarters in Washington, D.C., and discussed the long-term needs of a manned lunar landing program. There was unanimous agreement that a manned moon program would require sophisticated orbital rendezvous techniques that should be researched and understood before the detailed design of spacecraft for the lunar landing mission. This could only be obtained through the experience gained from a separate manned spaceflight program. McDonnell proposed a one-man space station comprising a Mercury spacecraft and a cylindrical laboratory which would provide up to 14 days of of habitability. On February 1, 1961, Space Task Group Director Robert Gilruth asked STG manager James A. Chamberlin to begin talks with McDonnell Aircraft Co. on the development of an improved Mercury for future needs.

1961 January 7   The first successful flight of the U.S. Air Force Blue Scout solid propellant research rocket took place from Cape Canaveral, Fla. Carrying a package of scientific instruments in a 392-lb. payload, Blue Scout reached a maximum altitude of 1,000 mi. and a downrange distance of 1,400 mi. Defense scientists had planned to recover a 90-lb. canister of equipment, but this was not accomplished. Instruments measured radiation levels in the lower Van Allen belts, micrometeorite levels and electrical signals below and above the ionosphere.

1961 January 9   NASA's Low committee met for the first time and heard from senior officials that a three-phase moon program was considered a logical step forward: (1) manned lunar landing and return, (2) a limited period of lunar exploration and (3) a lunar scientific base. Launch vehicles in the Nova class with first-stage thrusts of 3 million lb. and 6 million lb. were considered. A basic Lunar landing mission was judged to require a Nova-class launch vehicle capable of lifting a payload of 60,000–80,000 lb. to Earth escape velocity. The first-draft report was submitted on January 24, followed on February 7, 1961, by the committee's final report asserting that a manned lunar landing could be accomplished within the decade.

1961 January 10   Jerome B. Wiesner of the Massachusetts Institute of Technology submitted the report of a nine-member Ad Hoc Committee on Space which he chaired for President-elect John F. Kennedy. The report blamed NASA for spending too much effort on developing an in-house research establishment and giving too much emphasis to manned flight. In retrospect the report was unfairly critical of NASA, given the low level of technical experience in spaceflight and rocketry during these early days, and it suffered from too little fact-finding during preparation. Played down by some historians, it did imbue NASA and the executive branch of the government with a sense of inadequate leadership at NASA. This opened the way for President Kennedy and Vice President Johnson to take more direct control of space goals and objectives. On January 11, 1961, Wiesner was named by Kennedy as his special assistant for science and technology.

1961 January 12   A Nike-Cajun sounding rocket was launched by an Italian crew from a test range on the Mediterranean island of Sardinia. This was the first Italian/U.S. cooperation launch, one of many would take place over the following years. The Nike-Cajun reached a height of more than 100 mi., releasing a sodium vapor cloud visible for miles. The launch was part of an atmospheric research program and began for Italy a long association with space exploration.

1961 January 19   The American Telephone and Telegraph Co. (AT&T) received permission from the Federal Communications Commission to conduct experimental radio communications via repeater satellites in orbit. Repeater satellites, or active communication satellites, receive and amplify radio signals from one station for retransmission to another station. They would become the means by which global telecommunication services would expand during the coming decade. The AT&T system was to be known as Telstar.

1961 January 20   T. Keith Glennan stood down as NASA administrator pending selection of a replacement by the incoming Kennedy administration. Vice President-elect Lyndon B. Johnson wanted a political appointee capable of handling tough Congressional opponents but also of working well with scientists and engineers. Kennedy wanted to put the retired head of Army research and development, Gen. James M. Gavin, in charge of NASA, but Johnson persuaded him it would be disastrous to give a military man the job of running the civilian space agency. Jerome B. Wiesner, soon to become Kennedy's special assistant for science and technology, wanted a man with roots in science, but Johnson did not, and Wiesner had the first of his many confrontations with the Vice President. Incoming chairman of the Senate Aeronautical and Space Sciences Committee, Sen. Robert M. Kerr (D-Okla.), played an influential role in getting James E. Webb appointed on January 30. Along with Deputy Administrator Hugh Dryden, Webb was sworn in to office on February 14, 1961, and served until his resignation on October 7, 1968.

The NASA Marshall Space Flight Center issued study contracts to North American Aviation and Ryan Aeronautical for development of a paraglider recovery system for the Saturn launch vehicle. The concept was based upon a design by Francis M. Rogallo of NASA's Langley Research Center. Although many studies on Saturn booster recovery were conducted by NASA field centers and contractors, the concept was not adopted.

1961 January 24–25   NASA held an industry briefing for its Relay experimental communication satellite program. Relay was expected to test the operability of transoceanic communication via satellite, determine the radiation levels at orbital altitude and to measure the degrading effect of the space environment on solar cells and satellite equipment. NASA planned to launch four Relay satellites, two on Delta and two on Atlas-Agena B. Uplink signals would be transmitted in the 400–500 Hz band, and the satellite would retransmit at 2.2–2.3 KHz. Industry bids were due by March 6. On May 18, Radio Corp. of America (RCA) received a contract to build three Relay satellites.

1961 January 25   NASA announced that it had selected 12 women airplane pilots for special tests of their suitability for spaceflight. The tests became more an examination of the psychophysiological differences between men and women than an evaluation of individuals concerned. What these tests produced was a definition of differences in mind and body between the sexes. NASA built a large data bank on the significant differences that emerged from this study. It came to show that both men and women were uniquely equipped for different tasks, and NASA was to exploit these differences in the work it gave women to do when they were selected as astronauts.

1961 January 31   The first successful U.S. Air Force SAMOS (satellite and missile observation system) military reconnaissance satellite was launched by Atlas-Agena A from Vandenberg AFB, Calif., into a 294 × 346 mi., 97° polar orbit. Designated *SAMOS-2*, it provided images of the Earth which were transmitted to the ground by radio signal. *SAMOS-2* gave only poor results and the third and fourth satellites launched on September 9 and November 22, respectively, failed to reach orbit. SAMOS satellites weighed about 4,200 lb. Launched December 22, *SAMOS-5* remained in orbit eight months. The last six satellites, *SAMOS 6–11,* were launched during 1962, on March 7, April 26, June 17, July 10, August 5 and November 11. *SAMOS-6* remained in space for 15 months, *SAMOS-9* for seven days, *SAMOS-7* for two days, and the rest for one day.

The Mercury-Redstone 2 mission was launched from Pad 5 at Cape Canaveral, Fla., carrying a 37-lb. chimpanzee named Ham. The purpose of the flight was to qualify the spacecraft for launch aborts by firing off the escape rocket which carried the spacecraft away from the Redstone. The 4,190-lb. Mercury spacecraft no. 5 was lobbed to a height of 157 mi., a maximum speed of 5,857 MPH and a downrange distance of 422 mi. in a flight lasting 16 min. 39 sec. The primate experienced a maximum acceleration of 14.7 g due to a large overthrust with the Redstone, which threw the spacecraft to a greater distance than planned. Because of this, a booster development flight, dubbed MR-BD, was flown on March 24, 1961, with spacecraft no. 3, reaching a height of 113.5 mi., a speed of 5,123 MPH and a downrange distance of 307 mi., fully qualifying the Redstone for the first manned Mercury flight.

1961 January Continuing problems with development of the Centaur cryogenic upper stage brought about a revision in the 10 scheduled development flights, pushing back the first of these from June to November 1961. The original schedule anticipated a first Atlas-Centaur flight this month. Centaur was designated as an upper stage for Atlas and as the third (S-V) stage for Saturn C-1 launch vehicles. Under development at Pratt and Whitney, the Centaur's RL-10 engines encountered failures this month with vertical firings of two-engine sets in a typical Centaur configuration. Some 230 successful ignitions had already been made with engines in a horizontal attitude, and engineers discovered a relatively minor cause for the vertical failures which delayed development. Engine flight-rating tests were completed on November 4, 1961, and Centaur program management was transferred from the U.S. Air Force to NASA's Marshall Space Flight Center on January 1, 1962.

Soviet scientists and engineers completed plans for 1962 launch trajectories to Mars and Venus. Having failed in their attempt to send M-1-class unmanned spacecraft to Mars during October 1960, engineers working under Korolev developed a common design, designated Object MV, applicable to both Mars and Venus missions. Three flight attempts to Venus were planned, and launch attempts actually took place on August 25 (with *2MV-1*), September 1 *(2MV-1)* and September 12 *(2MV-2),* 1962. Three 1962 flights to Mars were scheduled for launch attempts on October 16 (with *2MV-4*), October 31 *(2MV-4)* and November 16, 1962 *(2MV-3).* The first two were fly-by missions but the *2MV-3* mission was to attempt a landing, on June 28, 1963, if launched on the scheduled date.

1961 February 3 Responding to a call for bids on an experimental weather satellite for research and development beyond Tiros, NASA selected General Electric to build two Nimbus satellites. Five companies had put in bids. The U.S. Weather Bureau wanted a satellite to test an infrared spectrometer and this was to be one of the instruments carried by Nimbus. In April an interagency panel recommended transforming Nimbus into an operational system, but the Weather Bureau became increasingly dissatisfied with the program's slow gestation and on October 4, 1963, announced their withdrawal.

1961 February 4 The first attempt to send a spacecraft to the vicinity of another planet took place when *Sputnik 7* was launched by the USSR from Tyuratam into a 131 × 198 mi. Earth parking orbit with an inclination of 65°. The Earth parking orbit concept calls for the terminal stage with spacecraft to be placed in Earth orbit before it is fired and boosted toward its objective. A failure in the third stage of the SL-6 launcher prevented the series-1VA spacecraft from being propelled out of Earth orbit and on course for a fly-by of Venus, as planned. This was the second failure for the new three-stage launch vehicle derivative of the SS-6. With a total mass of 14,295 lb., the combined stage and spacecraft remained in Earth orbit for 22 days until they decayed back down through the atmosphere.

1961 February 7 The final report of NASA's Manned Lunar Working Group chaired by George Low was submitted to Associate Administrator Robert C. Seamans. It was essentially a distillation of several different proposals, some highly ambitious, from the Space Task Group, the Langley Research Center and the Marshall Space Flight Center. It proposed the use of Saturn C-3 and Nova-class launch vehicles, the latter now being a generic name for any launcher using clustered (1.5-million-lb. thrust) F-1 engines. It endorsed development of an Apollo "A" vehicle for Earth orbit flight beginning in 1966, a "B" vehicle for manned circumlunar flight from 1967 and lunar landings from 1968 to 1969. Rendezvous operations would figure heavily, and it was proposed that NASA make maximum use of a classified Air Force space rendezvous program called Hawkeye.

1961 February 10 As part of the official opening ceremony of the NASA deep-space tracking station at Canberra, Australia, the voice of the deputy administrator, Dr. Hugh Dryden, was bounced off the moon. Transmitted from the deep-space tracking station at Goldstone, Calif., the signal was picked up at the new facility in Australia. Along with a similar tracking station at Johannesburg, South Africa, the Goldstone and Canberra stations would continue to support lunar and planetary missions until the station in South Africa was closed in 1974. From that date, coverage from a longitude close to the prime meridian would be taken over by a station near Madrid, Spain.

In a presentation to the U.S. Academy of Sciences' Space Science Board, John L. Sloop defined NASA's projections for the coming decade's manned and unmanned space activity using Saturn and Nova launch vehicles. Apollo Earth orbiting flights were expected to begin in 1966 utilizing a Saturn C-1 to launch command and service modules weighing 19,000 lb. Flights into and out of moon orbit with a 45,000-lb. Apollo could begin in 1967 and would use the more powerful Saturn C-2. Manned moon landings would wait until after 1970, when it was expected that the 12-million-lb. thrust Nova-class launcher would be available for direct ascent. Alternatively, said Sloop, smaller Saturn-class vehicles could utilize the Earth-orbit rendezvous mode. Unmanned spacecraft using Saturn were projected to include the Prospector moon rover and the Voyager Mars lander.

1961 February 12 Eight days after an attempt to launch a similar vehicle, the Soviet Union sent a 1,419-lb. spacecraft toward Venus in the first successful use of the Earth parking orbit escape trajectory concept. The Soviet practice of announcing and naming only those spacecraft successfully placed on Earth escape trajectories began with this vehicle. Designated 1VA, but called *Venera 1* after launch, the Venus vehicle consisted of a cylindrical structure 6 ft. 8 in. long by 3 ft. 6 in. in diameter with two solar panels attached to opposite sides and a wire-mesh parabolic communications antenna approximately 6 ft. 6 in. in diameter attached to the base. Three further antennas were located on the exterior, as was a boom carrying a magnetometer and a charged-particle detector. *Venera 1* also carried a Soviet medal encapsulated in a globe of the Earth set inside a sphere made up of pentagonal pieces. The sphere was capable of floating—just in case the surface of the planet was water—and was protected by a thermal cover to insulate it from the heat of entering the atmosphere of Venus.

1961 February 13   NASA began discussions with the McDonnell Aircraft Co. about an advanced manned spacecraft based on the Mercury design. Work began immediately on various designs and concepts. On learning in early March that extravehicular operations (spacewalking) would probably be an integral part of that requirement, Space Task Group engineer Maxime Faget raised with McDonnell the possibility of making the advanced spacecraft a two-man vehicle so that one astronaut could remain inside for safety reasons. McDonnell then drew up preliminary plans for a two-man vehicle based on the performance specification of the Atlas-Centaur, a launch vehicle that would provide sufficient lift capability.

1961 February 16   The first satellite placed in Earth orbit by a solid propellant rocket was launched by NASA from the Wallops Island Flight Test Range when a four-stage Scout put *Explorer IX* into a 394 × 1,605 mi. orbit. The satellite comprised an aluminum balloon 12 ft. in diameter, inflated by gas pressure and weighing 15 lb. It was to have provided information on the density of the Earth's atmosphere by accurate tracking to measure the degradation of the orbit. In this way the effect of the atmosphere on the satellite could be more accurately defined. Although this was made more difficult by the failure of a beacon transmitter, the satellite was usefully tracked until it decayed in the atmosphere on April 9, 1964. Through a series of improvements the Scout was considerably upgraded and by 1965 the A-1 model could place a 210-lb. payload into a polar orbit or 270 lb. into an easterly orbit. By 1979 the G-1 model offered the capability of putting 366 lb. into polar orbit or 460 lb. into easterly orbit. The 100th Scout launch occurred on June 2, 1979. Thereafter the launch rate declined. The 118th and last Scout launch took place on May 8, 1994, when the Department of Defense satellite MSTI-2 was placed in orbit from Vandenberg Air Force Base, Calif.

1961 February 17   The first attempt to launch a wide-angle area mapping camera on a military reconnaissance satellite began with the launch of *Discoverer XX* from Vandenberg AFB, Calif., on a Thor-Agena B. Placed in a 179 × 488 mi., 80.9° orbit, the Corona-class satellite carried a KH-5 camera system developed secretly under Project Argon. Designed and built by the Itek company, the camera had a 3-in. focal length lens with a 4.5 × 4.5 in. film, each frame covering a 345 × 345 mi. surface area in overlapping imagery. After reaching orbit the KH-5 camera aboard *Discoverer XX* failed and the satellite decayed back into the atmosphere on July 28, 1962, without ejecting a capsule. Two further attempts to launch a KH-5 camera system aboard Discoverer satellites were similarly unsuccessful before the first successful mission of a KH-5 package aboard a Corona satellite on May 15, 1962.

1961 February 21   In a test of maximum aerodynamic heating under conditions of a worst-case reentry, the Mercury-Atlas 2 mission was flown from Pad 14 at Cape Canaveral. Spacecraft no. 6 reached a maximum height of 114 mi. and a downrange distance of 1,432 mi. and reached a top speed of 13,277 MPH. During descent the spacecraft experienced a deceleration of 15.9 g, the highest ever achieved by a manned spacecraft, albeit without a pilot on this flight. The spacecraft was recovered about 38 min. after the 17 min. 56 sec. flight, and inspection of its exterior condition indicated it had survived well the fierce conditions of reentry.

1961 February 22   NASA's Space Task Group announced that a shortlist of three crew members had been selected from the team of seven Mercury astronauts to train for the first U.S. spaceflight, scheduled as a Mercury-Redstone mission launched on a suborbital trajectory to test spacecraft and pilot. In order of preference, they were Alan Shepard, John Glenn and Virgil Grissom. The announcement came more than a month after Robert Gilruth, director of the Space Task Group, had personally made the choice and gathered all seven astronauts together to tell them the news. Up to this point, competition among the seven men had been strong. From April 12, 1961, they would all face competition of a different kind.

A Thor-Able Star launched from Cape Canaveral, Fla., failed to place the *Transit 3B* and *Lofti 1* satellites in the correct orbit when the satellites remained attached to the second stage of the launch vehicle. Instead of being placed in a roughly circular orbit at a 500-mi. altitude, the payload and attached Able Star achieved a 100 × 623 mi., 28.4° orbit which decayed March 30. Designed by the U.S. Navy, the 57-lb. *Lofti 1* satellite was instrumented to study the transmission of very-low-frequency signals through the ionosphere. The only other satellite of this type, *Lofti 2A,* was launched along with four other satellites by Thor-Agena D on June 15, 1963. It decayed July 18, 1963.

1961 February 27   All contact with the Soviet spacecraft *Venera 1* was lost at a distance of 1 million mi. from Earth and more than 62 million mi. from Venus. Telemetry had been routinely sent to Earth by radio command from ground stations and indicated all was well with the spacecraft. Instrumentation aboard what the Soviets described as an "automatic interplanetary station" provided details of temperature and the pressure of the sealed compartment containing batteries and flight equipment for attitude control and communications with Earth. Before failing, it detected what was later determined to be the solar wind, a stream of energetic particles radiating from the sun. Inoperative, *Venera 1* passed within 62,000 mi. of Venus on May 19, 1961, the first man-made object to perform a planetary fly-by, and entered a heliocentric orbit of about 67 million × 95 million mi.

1961 March 1   Formal specifications for the Saturn launch vehicle family were announced by NASA. The C-1 would have S-I, S-IV and S-V stages. On April 7, 1961, NASA selected a configuration of six Pratt and Whitney 15,000-lb. thrust LR-115 cryogenic engines in the S-IV, rather than four 17,500-lb. thrust LR-117 engines as previously projected. On June 1, 1961, the S-V for the C-1 was deleted. The C-2 was now proposed in two configurations: a three-stage version would have S-I, S-II and S-IV stages, with an S-V attached to a four-stage version. On March 28, when President Kennedy submitted to Congress the revised NASA budget for fiscal year 1962, NASA Administrator James Webb authorized the go-ahead for Saturn C-2. On May 27 Wernher von Braun officially unveiled the Saturn C-3, which would utilize a new first stage powered by two 1.5-million-lb. thrust F-1 engines generating a total thrust of 3 million lb. at liftoff.

1961 March 6 In response to the critical Wiesner report of January 10, U.S. Secretary of Defense Robert McNamara issued a directive permitting each of the military services to develop new space technology under the director of Defense Research and Engineering. On March 28 McNamara assigned responsibility for research and development of reconnaissance satellite systems and sensors to the U.S. Air Force, although the CIA and the National Security Agency continued to be responsible for the satellites themselves and their operations.

1961 March 9 The first of two test fights with man-rated Vostok spacecraft took place from the Baikonur cosmodrome in the USSR when an SL-3 launcher carried the 10,400-lb. *KS-4* into a 114 × 155 mi. orbit. On board was one dog named Chernushka and a dummy cosmonaut as well as other biological payloads. The descent module was safely recovered after one orbit, 1 hr. 46 min. after launch. The final test prior to the first manned flight took place on March 25, 1961, when *KS-5* was placed in a 111 × 153 mi. orbit. Like its immediate predecessor it carried a dog, Zvezdochka, and a variety of small biological samples. It too was returned to Earth at the end of the first orbit, completing a 1 hr. 45 min. flight with safe recovery of the payload. These two precursor missions cleared the way for the first manned orbital flight.

A Redstone missile launched from Cape Canaveral, Fla., carried the first device for automatic arming and fusing of a nuclear weapon in flight. The purpose of this was to prevent warheads from a rogue missile from falling on friendly territory and detonating. Only when the missile had achieved flight criteria demonstrating a successful trajectory on course to its assigned target would the nuclear warhead be armed and fused. Redstone served as the development vehicle for this vital safety feature applied to all ballistic missiles.

1961 March 22 NASA Administrator James Webb took Deputy Administrator Hugh Dryden and Associate Administrator Robert Seamans to a meeting with President Kennedy, Vice President Johnson, Director of the Budget Bureau David Bell and deputies. They were there to discuss the budget request for fiscal year 1962 and amendments to Eisenhower's spending plan for the year beginning July 1, 1961. Kennedy was cautious about accelerating the Apollo program, preferring to wait for further studies and recommendations on the future direction of the U.S. manned spaceflight objectives. About launch vehicles he was more positive, fully endorsing acceleration of the Saturn program, and believed that heavy lift capability was the key to gaining ground on Soviet accomplishments in space.

1961 March 23 The Soviet cosmonaut candidate Valentin Vasilyevich Bondarenko died following a fire in the pure oxygen atmosphere of an isolation chamber. Scientists wanted to know how well the human body stood up to prolonged exposure to pure oxygen and Bondarenko was at the end of a 10-day period of isolation in a special chamber when the accident occurred. Accidentally igniting a rag on a hot ring, Bondarenko made desperate efforts to put out the fire but the flames were so intense in the oxygen atmosphere that his efforts were to no avail. Although alive when technicians finally got him out of the chamber, Bondarenko died shortly

thereafter in the hospital. He was the first spaceflight trainee of any nationality to be killed in an accident.

The Lunar Sciences Subcommittee of the NASA Office of Space Flight Programs recommended capabilities for the proposed Surveyor Orbiter, stating that it should obtain imagery with a resolution of 33 ft., extensive photography of the limb region with a resolution of at least 1.6 mi., general reconnaissance photographs of the moon at a resolution of 330 ft. or better and stereo pairs in high-resolution areas. Although use of a Surveyor Lander was attractive, NASA began looking at a separate spacecraft in the 2,100–2,400-lb. weight class launched by Atlas-Centaur. Not before July 20, 1962, however, did the space agency identify the framework for such a program. Even at this date little work had been done on appropriate camera systems for long-duration lunar orbit photography.

1961 March 24 Soviet cosmonauts Valery F. Bykovsky, Yuri A. Gagarin, Grigori G. Nelyubov, Andrian G. Nikolayev, Pavel R. Popovich and Gherman S. Titov left Moscow for the launch complex at Tyuratam where one of them would be launched into space. Next day they witnessed their first rocket launch, the KS-5 Vostok mission, precursor to the first manned flight, before returning to Moscow. On April 3, heeding recommendations from the National Academy of Sciences and the military, the Kremlin gave its formal approval for the flight to go ahead. Gagarin and Titov were singled out as the two candidate cosmonauts for the single seat in Vostok. On April 5 the six cosmonauts returned to Tyuratam and two days later the two prime candidates went through spacecraft familiarization training. On April 8 the State Commission met to decide who would fly and on April 10, two days before the flight, it was announced that Gagarin would fly the mission with Titov as backup.

1961 March 25 The first detailed measurement of the Earth's outer radiation belts was the goal of NASA's *Explorer X* satellite, launched on a Delta from Cape Canaveral, Fla. The 78-lb. satellite comprised a drum 1 ft. 7 in. in diameter surmounted by a pedestal atop which a 1 ft. 1 in. sphere was mounted, presenting a total height of 4 ft. 4 in. Placed in a 137 x 112,535 mi., 33.9° orbit, the satellite would, every 4.7 days, reach a distance from Earth equal to half the distance between the Earth and the moon. Apogee was reached on March 27 and signals ceased a day later after instruments had recorded much higher radiation levels than anticipated at that distance from Earth.

1961 March 28 In a major speech to Congress on the defense budget for fiscal year 1962, President Kennedy asked for money to double production of the solid propellant Minuteman ICBM and increase development efforts on the solid propellant Polaris A-3 SLBM (submarine-launched ballistic missile), while eliminating two liquid propellant Titan II ICBM squadrons (planned but not yet equipped with missiles), phasing out the Snark cruise missile and cancelling plans to base Polaris on the cruiser *Long Beach,* a scheme that had been proposed as a means of early missile deployment. Greater reliance on missiles hit manned aircraft programs, the U.S. Air Force B-70 supersonic bomber being canceled outright. Compared with Eisenhower's defense plans, the Air Force would still get 126 Atlas ICBMs, 108 Titan ICBMs versus 126, 600 Minuteman ICBMs versus 540 and 464 Polaris versus 304. ICBM strength would grow from 792 to 834, and total missile strength from 1,096 to 1,298. For NASA, President Kennedy requested an increase in budget from $1.109

billion to $1.235 billion, although this was lower than the $1.417 billion NASA had requested.

1961 March 31   Hugh Odishaw from the Space Science Board (SSB) of the U.S. National Academy of Sciences responded to a request from NASA's Homer Newell calling for a list of experiments that could be flown by Mariner B to Mars. At this date Mariner B comprised a flyby spacecraft, taking measurements of the planet as it passed within 6,800–9,900 mi. of the surface, and a landing capsule. About 176 lb. of scientific instruments would be carried aboard the flyby spacecraft if no lander capsule was carried, or 51 lb. of instruments could be sent to the surface with a lander. The SSB's Planetary Atmospheres Study Group proposed a set of instruments for the flyby, including a radiation package, cosmic ray detector, camera equipment with 0.6-mi. resolution, a magnetometer and infrared and ultraviolet radiometers. For the lander, they suggested a TV camera, mass spectrometer, gas chromatograph, radar altimeter and temperature and pressure measuring equipment for use during descent.

1961 March   The U.S. Navy described at congressional budget hearings its Early Spring antisatellite proposal utilizing a "minimum energy missile . . . launched vertically with just enough power to arrive at the altitude of the satellite at zero gravity. At that point it can hover and wait for the satellite to come and then by terminal guidance . . . kill it with some mechanism." Early Spring became an umbrella term for many antisatellite proposals, including some using Polaris with a modified Sparrow upper stage, but the Department of Defense did not want to impede the production of Polaris SLBMs (submarine-launched ballistic missiles) by attaching research programs to it. Another antisatellite proposal called Project Skipper involved the use of high velocity pellets shot into space by a Scout rocket, but this was never seriously studied.

1961 April 12   The world's first manned spaceflight began at 06:07 UT when an SL-3 launch vehicle carried the Soviet

The Vostok manned spacecraft was attached to the second stage of the SL-3, the white cylinder at left.

cosmonaut Yuri Gagarin in the 10,425-lb. *Vostok 1* spacecraft from its Tyuratam launch pad in the Soviet Union. Similar in design to the precursor unmanned test vehicles in the Korabl Sputnik series, Vostok comprised a 5,425-lb. pressurized descent module 7 ft. 6 in. in diameter. The cosmonaut sat on an inclined ejection seat with an instrument panel facing him on which was positioned a porthole with an optical orientation system known as the *Vzor*. Through this the pilot could make rudimentary horizon alignments in space. The descent module was attached to a 5,000-lb. double-cone instrument module with a maximum diameter of 8 ft. and a length of 7 ft. 5 in. The truncated upper cone provided a receptacle for the descent module, with a retro-rocket system at the lower end. This braking engine, designated TDU-1, had been designed by Aleksei M. Isayev and had hypergolic propellants of nitrous oxide and an amine fuel. It produced a thrust of 3,560 lb. for 45 sec., slowing the assembly by approximately 500 ft./sec. Around the top of the instrument module were 16 spherical oxygen and nitrogen containers providing a sea-level atmosphere in the pressurized descent module. Communication antennas were attached to both modules. Gagarin entered *Vostok 1* at 04:10 UT but there was a 20 min. delay to solve a problem with the hatch. After launch the SL-3 placed *Vostok 1* in a 105 × 196 mi. orbit inclined 65° to the equator. Gagarin experienced sunset over Kamchatka 42 min. into the flight, and TV pictures of him inside the spacecraft were sent to Earth. At no time during the flight did the pilot have manual control over *Vostok 1*. Fearing he might endanger his own mission, controllers had locked the manual controls, and only in the event that all contact with the ground was lost could he open an envelope containing codes to unlock them. Over Africa at 1 hr. 18 min. the retro-rocket was fired, the instrument module separated and 10 min. later, at an altitude of 23,000 ft., Gagarin ejected from the descent module. He landed on the banks of the Volga, 16 mi. southwest of Engels; his flight had lasted 1 hr. 48 min. Although each descent module carried a parachute system, all cosmonauts elected to eject from their Vostok spacecraft during descent.

NASA established the Saturn Program Requirements Panel under the chairmanship of William A. Fleming. Tasked with gathering together all potential mission applications for the Saturn series launch vehicle, the committee would correlate planning with Saturn development, procurement and production. The committee was permanent and served as a focal point for integrating all Saturn-related activity at NASA.

1961 April 14   Clearly upset by the Gagarin flight of April 12, President Kennedy was assertive and demanding when he met during the evening with NASA Administrator James Webb and his deputy, Hugh Dryden, to discuss a response to the Soviet achievement. With Kennedy were Special Counsel Theodore Sorenson, Budget Bureau Director David Bell and Science Advisor Jerome Wiesner. The President was considerably more bullish about Soviet space endeavors than he had been at the budget meeting of March 22 and asked "Is there any place we can catch them? What can we do? Can we put a man on the moon before them? Can we leapfrog?" Dryden argued for a massive technological effort and Webb diplomatically reiterated his faith in Kennedy's ability to choose an appropriate response. Kennedy ended the meeting by appealing for direction: "If somebody can just tell me how to catch up. There's nothing more important."

1961 April 18   Representatives from 30 U.S. firms attended a briefing at the NASA Marshall Space Fight Center on the S-II, scheduled as the second stage of the Saturn C-3 launch vehicle.

Initial design and development bids were received on May 11 from seven firms: Aerojet General, Chrysler, Convair Astronautics, Douglas Aircraft, Lockheed, Martin Co. and North American Aviation. From these bids Aerojet General, Douglas Aircraft, Convair Astronautics (now General Dynamics) and North American Aviation were asked to submit detailed proposals on June 15. On September 11, 1961, NASA announced that North American Aviation had been selected to develop the S-II. Although conceived to have four 200,000-lb. thrust cryogenic J-2 engines, the S-II would be built with five J-2s and comprise the second stage of the Saturn V.

1961 April 19   Prepared by John C. Houbolt from the NASA Langley Research Center on his own initiative, a circular entitled "Manned Lunar Landing via Rendezvous" that proposed sequential steps to a manned moon mission was released for internal consumption. In the manned orbital rendezvous and docking (MORAD) phase a Mercury spacecraft would demonstrate link-up to a satellite launched by Scout rocket during 1961–63. In the Apollo rendezvous phase (ARP), Atlas, Agena and Saturn launchers would demonstrate Earth-orbit assembly of space station and crew-transfer techniques during 1962–65. After that, a manned lunar landing involving rendezvous (MALLIR) program would use two Saturn C-2 launchers, one to lift a command vehicle and a lunar lander into Earth orbit and one to send the docked vehicles to lunar orbit where one astronaut would be left in the command vehicle while two descended to the surface. The upper half of the lander would fire itself back into lunar orbit, where the two astronauts would transfer to the command vehicle for return to Earth.

A test mock-up of a sea-launched ICBM named *Hydra 2* was successfully launched from the sea off San Clemente Island, Calif. With a mass of 20,000 lb., the 46-ft. test missile floated vertically and was ignited on command to demonstrate the feasibility of basing ICBMs in vertical tubes off the coast. Advocates said sea-based deployment would draw enemy attacks away from land and that the missiles would be safer from attack because they would be harder to pinpoint. It was also thought that space launchers of unlimited size could be accommodated from sea-based launch positions. None of these came to fruition.

1961 April 20   As promised during a meeting held April 14, 1961, President Kennedy sent a memorandum to Vice President Johnson seeking clear guidance on how to handle Soviet space achievements: "I would like for you as Chairman of the Space Council to be in charge of making an overall survey of where we stand in space. (1) Do we have a chance of beating the Soviets by putting a laboratory in space, or by a trip around the moon, or by a rocket to land on the moon, or by a rocket to go to the moon and back with a man?" Only five days earlier the Kennedy administration had suffered a humiliating embarrassment when Cuban exiles, using weapons provided by the CIA and supported by air attacks, failed to land in the Bay of Pigs and overthrow communist leader Fidel Castro. This irritated Kennedy, inciting him to mobilize a major space program to counter Soviet achievements and reassert U.S. prestige. When discussing the possibility of a manned moon landing, he turned to his science adviser, Jerome Wiesner, and said: "It's your fault. If you had a scientific spectacular on this Earth that would be more useful — say, desalting the ocean, or something just as dramatic — then we would do it."

Senior NASA officials from the Jet Propulsion Laboratory and the Space Task Group (STG) met at headquarters in Washington, D.C., to discuss the relationship between the unmanned Prospector and the manned Apollo programs. Prospector was expected to convey a 1,500-lb. roving vehicle to the lunar surface for extensive wide-area geophysical surveys, but the STG wanted it for landing logistics supplies close to a manned lunar lander. When the STG failed to support it in the automated survey role, the JPL switched the mission mode to an Apollo support role, but that was not enough to save the overweight, and costly, vehicle from cancellation.

1961 April 22   The first Jupiter IRBM (intermediate-range ballistic missile) launched by a non-U.S. firing team was sent on a ballistic test flight by a squadron of the Italian air force which had been receiving instruction from the 864th Strategic Missile Squadron, U.S. Air Force. Under a joint operations plan through NATO, the Italians were to take charge of 30 missiles, operating 15 in each of two squadrons. The Italians set up a central base at Puglia in the southeast of the country, but the missile was mobile and could be deployed anywhere. With a range of 1,500 mi., the IRBM would have the capability to fire warheads. During 1962 the Turkish Army took responsibility for one unit of 15 Jupiter missiles.

1961 April 24   To demonstrate the launch technique being planned for the second generation Titan ICBM, Titan II, a Titan I was ignited at the bottom of a test silo at Vandenberg AFB, Calif., to demonstrate the viability of this concept. The Martin Co. was developing the Titan II to operate on storable, hypergolic propellants that simplified the start-up process by allowing the fuel and oxidizer to ignite on contact. Reaction time was to be significantly reduced by launching the Titan II from the bottom of its vertical underground silo. Modified silos with exhaust ducts on either side were necessary so the hot gases could escape as the missile ascended.

1961 April 25   The Mercury-Atlas 3 flight began from Pad 14 at Cape Canaveral when the modified Atlas D launch vehicle lifted off on the first attempt to put a Mercury spacecraft in orbit. An earlier plan to have the MA-3 Atlas shut down 150 ft./sec. short of orbital velocity, followed by retrofire for a splash down between the Canary Islands and the African coast, had been abandoned in favor of a full orbital flight with the spacecraft making one revolution of the Earth. Just 40 sec. off the pad, at a height of 16,400 ft., the Atlas was destroyed by the range safety officer when it went out of control, triggering the launch escape system which pushed spacecraft no. 8 to 24,000 ft. The parachutes deployed and lowered the spacecraft to the Atlantic, 6,000 ft. from the pad, from where it was recovered and refurbished for MA-4.

1961 April 26   The U.S. Air Force modified the Dyna-Soar test program to include 20 air-launched tests starting in 1964. Two unmanned launch tests on modified Titan II rockets would be conducted during August 1964, prior to the first of 12 manned ballistic shots beginning in April 1965. A flight was to be made in April 1966 in which a Dyna-Soar glider would be launched eastward from Cape Canaveral, Fla., and land back in the United States at Edwards AFB, Calif., near the end of the first revolution. The Air Force anticipated the first operational capability for October 1967, when the vehicle would be ready to conduct reconnaissance, satellite inspection and space-based attack missions. The complete weapon system was scheduled to be fully deployed and operational by late 1971.

1961 April 27 The end of an era in nascent space exploration occurred when the last successful Juno II launcher developed from the Jupiter missile carried NASA's 95-lb. *Explorer XI* satellite to a 309 × 1,104 mi. orbit inclined 33.9°. From this date only second-generation missiles adapted as satellite launchers would carry spacecraft into orbit. A joint venture between the NASA Goddard and Marshall Space Flight Centers, *Explorer XI* comprised a three-section octagonal box, 7 ft. 5 in. long including the fourth stage motor, to which it remained attached, and was instrumented to search for gamma rays. It was the first satellite to detect them successfully. *Explorer XI* returned data until December 6 and remains in orbit, 170 × 207 mi. in late 1995.

1961 April 29 Vice President Lyndon Johnson sent President Kennedy a reply to his memorandum of April 20, 1961, asking for guidance on ways to outdo Soviet space spectaculars. Basing his remarks on numerous discussions and meetings with NASA, the Department of Defense, the Budget Bureau and other interested government agencies, Lyndon Johnson advised the President that nothing short of a manned moon landing would stand a reasonable chance of beating the Russians. Johnson was already at work preparing the Senate for a major decision for Kennedy, and on May 3, 1961, the senior leaders on the Senate space committee gave tacit approval for a national commitment to a moon-landing goal. A day later Johnson got similar approval from the House space committee. Johnson was leaving on a tour of Asia May 9 and, knowing that few White House advisers approved of expanded space goals, mustered as much support as possible for the decision.

1961 April At the request of the United States, Soviet double-agent Oleg Penkovsky provided detailed information on the Soviet missile program. He reported that serious problems with the SS-6 had prevented its deployment as an ICBM and that only one operational launch site at Plesetsk had been prepared for it. Only now were there between one and three SS-6s available for use in anger, and there would never be more than four. Tests of the SS-7 ICBM with storable propellants resumed in May and deployment began in 1962. CIA reconnaissance satellites of the Discoverer and Program 622A type were given the job of thoroughly searching for signs of Soviet ICBM activity. The missile-gap myth was being unraveled.

The U.S. Air Force Flight Dynamics Laboratory awarded McDonnell Aircraft Corp. a production contract for a series of unmanned test vehicles in the ASSET (aerothermodynamic/elastic structural systems environment tests) research program. The lifting-bodies were to be blunt, flat-bottomed delta shapes about 5 ft. 8 in. in length with a span of 4 ft. 7 in. weighing about 1,200 lb. and of semimonocoque construction. They would carry hydrogen peroxide attitude control thrusters, integrated gyro and flight control systems and full telemetry. All were boosted to an altitude of 225,000 ft. and a speed of 13,300 MPH by Thor or Delta launch vehicles; two vehicles were to be instrumented for aerothermodynamic tests and four for aerothermoelastic tests.

1961 May 2 NASA Associate Administrator Robert C. Seamans established an ad hoc task group for a Manned Lunar Landing Study which was to be chaired by William A. Fleming. This was the first formal in-house NASA study to decide upon a common set of criteria by which to plan a manned lunar landing program. With 1967 as a target landing date, the Fleming committee was tasked with setting guidelines, establishing cost estimates and determining the size of post-Saturn, Nova launch vehicles capable of placing 400,000 lb. in low Earth orbit. The group started work on May 8 and submitted its final report on June 16, 1961.

1961 May 3 To demonstrate the in-silo launch technique which the Titan II liquid propellant ICBM would use, a Titan I with inert upper stage became the first ballistic missile fired from a hole in the ground when it ascended from its silo, 146 ft. deep, at Vandenberg AFB, Calif. Devoid of a live second stage, the missile had reached a height of 20 mi. and a downrange distance of 40 mi. when the first stage cut off and the missile was intentionally destroyed by radio command. Titan I would continue to be deployed in silos but elevated to the surface for launch. Fueling and prelaunch preparations would take a minimum 20 min., whereas Atlas took more than 30 min. to launch. Superfast fueling and high-speed elevator hoists cut Titan I launch time, but only with the Titan II, constantly fueled and launched from the bottom of its protective silo, were reaction times cut to a few minutes. This was a vital margin when Soviet missiles could hit targets in the U.S. within 30 min. of launch.

1961 May 4 The Boeing Co. proposed streamlining the Dyna-Soar boost-glide program by eliminating suborbital tests and going direct to orbital tests with Dyna-Soar I. This was approved by the project office in June, when it was announced that probable development cost would be $967.6 million and that the first unmanned orbital flight test was estimated to take place in November 1963 using the Titan III booster. The U.S. Air Force designated Titan III as the booster for Dyna-Soar on November 14, 1961. This eliminated interim Titan boosters and saved money.

1961 May 5 The first U.S. manned spaceflight was launched from Pad 5, Cape Canaveral, when Alan Shepard rode the Mercury-Redstone 3 mission on a suborbital trajectory. Mercury had a length of 25 ft. 11.5 in. from the tip of the launch escape tower to the base of the retropackage. The 2,822-lb. spacecraft had a length of 10 ft. 11.5 in. including the retropackage, or 9 ft. 7 in. without the retropackage, and a diameter of 6 ft. 2.5 in. Launched at 14:34 UT, the Redstone fired for 2 min. 22 sec., followed immediately by removal of the launch escape tower and 10 sec. later by separation of the spacecraft using the three solid propellant posigrade rockets attached to the retropackage to shunt it away from the Redstone. At 2 min. 37 sec. the spacecraft turned around as it continued to ascend, and at 3 min. 10 sec. the automatic attitude system was switched off and the spacecraft brought under manual control. The automatic retrofire sequence began at 4 min. 4 sec., when the spacecraft was oriented in a 34° pitch and the three solid motors fired, at 5 min. 14 sec., for 10 sec. each at 5 sec. intervals. From a maximum altitude of 116.5 mi., spacecraft no. 7 began its descent, with retropack jettison at 6 min. 14 sec., atmospheric reentry at 7 min. 48 sec. and, 32 sec. later, peak deceleration of 11 g. Parachutes were deployed beginning at 9 min. 38 sec., and the spacecraft splashed down at 15 min. 22 sec., 303 mi. from the launch pad, having achieved a peak velocity of 5,134 MPH.

1961 May 8 President Kennedy received a memorandum from NASA Administrator James Webb and Secretary of Defense Robert McNamara entitled *Recommendations for our National Space Program: Changes, Policies, Goals*. It laid out a plan for a dramatically accelerated space program with manned lunar landing around 1967 as the primary objective.

Lt. Comdr. Alan B. Shepard Jr. (USN) walks from the transfer van to the waiting MR-3 launcher and its Mercury spacecraft on the morning of his flight, May 5, 1961.

The following day Kennedy met with senior government officials, questioned them on specifics related to the goal and unequivocally backed the Webb-McNamara text in its entirety. That day members of the Senate space committee began planned press leaks concerning the imminent announcement of a massive space endeavor.

NASA personnel were briefed by executives from the Martin Marietta Co. on the performance capabilities of the Titan II missile, adapted as a space launch vehicle. Martin believed the Titan would be appropriate for future lunar spacecraft, but NASA Associate Administrator Robert C. Seamans was skeptical. On hearing the submission, Seamans was impressed and directed STG's (Space Task Group) Robert Gilruth to factor it in to studies of an enlarged successor to the one-man Mercury missions then being defined for rendezvous and docking as well as extravehicular activity (EVA). At a senior NASA staff meeting on August 7, 1961, Gilruth recommended the Titan II as a "desirable booster for a two-man spacecraft."

1961 May 17 NASA's Space Task Group (STG) issued a work statement for studies of a paraglide system that could be used as an inflatable wing to give a manned spacecraft lift during descent through the lower atmosphere, allowing it to select a precise landing point. STG Director Robert Gilruth organized contracts with Goodyear Aircraft Corp., North American Aviation and the Ryan Aeronautical Company, anticipating that one of them would eventually receive a contract to develop the system. On November 20, 1961, a contract was awarded North American for design and development of a paraglider landing system for use on future manned spacecraft.

1961 May 18 The director of NASA's Space Task Group, Robert Gilruth, defined plans for Apollo A, the Earth-orbit version of the spacecraft conceived as a multifunctional three-person spacecraft capable of supporting orbital laboratory activities or circumlunar flights. A cylindrical laboratory, with a diameter of 12 ft. 10 in. and a length of 8 ft. 9 in., was to be carried between the upper stage of the Saturn C-1 launch vehicle and the Apollo equipment-propulsion module (later called the service module). Gilruth requested the Ames Research Center to prepare and forward a set of experiments that could be conducted in this laboratory section. Ames responded in a letter to Gilruth dated May 31, 1961, that proposed astronomical and Earth observation, micrometeoroid and solar observations.

1961 May 19 The U.S. Air Force Space Systems Division announced it was going to develop a manned lifting-body vehicle called SAINT II and capable of inspecting satellites in orbit. It was an outgrowth of the unmanned satellite inspection vehicle SAINT I, canceled in mid-1960. Launched by a modified Titan II carrying a fluorine-hydrazine upper stage called Chariot, SAINT II was expected to carry out missions denied to the winged Dyna-Soar boost-glider. An unmanned flight in 1964 was expected to precede 12 manned test launches. In October 1961 a highly critical report on SAINT II brought about its demise before the end of the year.

1961 May 22 During a routine meeting at NASA's Langley Research Center, informal plans were drawn up for a manned lunar landing with 10 Saturn C-1 launch vehicles. Typical of options for a manned lunar landing utilizing the smallest Saturn launcher, it envisaged the assembly in Earth orbit of four separate modular elements, including two 14,000-lb. lunar landers. A further five Saturn launchers would carry to Earth orbit the propellant necessary for sending the assembly on to the moon. An alternative concept utilizing just two Saturn C-2 launchers envisaged a single lunar lander weighing 11,000 lb. Just a month later Langley engineer John D. Bird drew a rough sketch showing how one Saturn C-3 could accomplish the manned landing using lunar-orbit rendezvous techniques.

1961 May 25 In a 47-min. address to a joint session of Congress, President Kennedy announced that a goal of landing astronauts on the moon was to be implemented immediately and that it was to be accomplished by the end of the 1960s. In a historic speech, Kennedy said, "I believe that this nation should commit itself, before this decade is out, to landing a man on the moon and returning him safely to Earth. No single space project in this period will be more exciting, or more impressive to mankind, or more important for the long-range exploration of space; and none will be so difficult or expensive to accomplish." Thus was NASA publicly committed to the most demanding mandate it would ever receive.

NASA Associate Administrator Robert Seamans set up a study group on manned lunar missions mode under the chairmanship of Bruce T. Lundin. Bringing together the directors of the Office of Launch Vehicle Programs and the Office of Advanced Research Programs, Lundin's study would supplement the work of the Fleming committee and examine all possible means of reaching the moon in the period 1967–70. Efforts were to focus on launch vehicles, and when the final report was submitted June 10, 1961, it recommended the use of two or three Saturn C-3 launchers using the Earth-orbit rendezvous mode.

1961 May 26　To pay for its ambitious program of manned lunar exploration and the requirements placed upon it by Congress, NASA requested a second budget amendment for fiscal year 1962 that increased authorization funding by $549 million. This brought to more than 60% the increases requested of the new Congress in two months. It was the beginning of a massive escalation in U.S. space budgets under the Kennedy administration. Defense space spending for fiscal year 1962 was just under $1.3 billion. After supplementals later in 1961, NASA's adjusted appropriation for fiscal year 1962 was $1.825 billion, an increase of virtually 100% over that of fiscal year 1961.

1961 May　Just before President Kennedy announced the moon-landing goal of Project Apollo, a Gallup opinion poll of a cross section of the American public asked if they thought the estimated total cost of $225 for every man, woman and child in the United States was worth it. Only 33% said "yes," and 58% said "no," while 9% had no opinion. After Alan Shepard made his ballistic Mercury flight, a U.S. news agency estimated his trip had cost every American $2.23. NASA was inundated with baskets of mail from U.S. citizens containing checks for that amount.

1961 June 1　NASA announced that the first ten Saturn C-1 test flights would adopt a two-stage, not a three-stage, configuration and that the last four of these might carry Apollo boilerplate spacecraft. The two-stage C-1 would be capable of placing 20,000-lb. payloads in low Earth orbit, comparable in capacity to the three-stage version thanks to improvements in the S-IV (second) stage and an additional 100,000 lb. of propellant that was carried in the extended S-I (first) stage from the seventh flight on. As envisaged at this time, the first three flights would carry the live first stage only.

1961 June 8　NASA's director of Space Flight Programs, Abe Silverstein, explained how unmanned lunar exploration would help support the manned lunar landing program. Impacting Ranger and soft-landing Surveyor spacecraft would be followed by Prospector, said Silverstein before the Senate Committee on Aeronautical and Space Sciences. Prospector was envisaged as a heavy lander capable of moving across the lunar surface to collect samples, set out packages of scientific instruments and detonate small explosive devices for seismic measurement of subsurface layers. Weighing 5,000 lb., Prospector was to be launched by Saturn C-1. Funds were made available in fiscal year 1962, but were denied the following year due to the project's high cost. Almost all tasks assigned to Prospector were carried out by Apollo astronauts.

Bell Aerosystems displayed its man-carrying small rocket lift device (SRLD) for the first time. Consisting of a twin-nozzle hydrogen peroxide propulsion system mounted to a glass-fiber corset, the SRLD was fully throttleable from 0–300-lb. thrust and weighed 114 lb. when loaded with 47 lb.

of propellant. The operator slipped his arms through padded lift rings and secured the mount with quick-release safety belts around the abdomen. In tests the SLRD had propelled its operator to a maximum distance of 815 ft., a height of 60 ft. and a top speed of almost 60 MPH.

1961 June 9　NASA Director of Space Flight Programs Abe Silverstein informed JPL Director William Pickering that four additional moon flights with the Ranger spacecraft could be added to the five Block I and Block II flight vehicles already approved. Scientific studies were to support the requirements for information about lunar surface hardness and general topography generated by engineers on the Apollo manned lunar landing program and, in addition, to make "a corresponding contribution to national prestige during the early phase of our lunar program." The JPL was asked to look at the possibility of equipping the four additional Rangers with high-resolution cameras or penetrometers for measuring the characteristics of the surface.

NASA and the U.S. Atomic Energy Commission selected Aerojet General to build the NERVA nuclear rocket motor and Westinghouse to build the reactor. The contract was effective July 10 and would result in the NERVA and RIFT programs integrating in 1962. At this time Kiwi-B was under development as a more powerful test bed for reactor design than Kiwi-A. The first Kiwi-B test series was completed December 7, with a power run at 300 MW for 30 sec. On that date NASA's Marshall Space Flight Center held a preproposal conference attended by 29 companies to hear plans for the RIFT program. RIFT would use a NERVA engine developed from the Kiwi-B design.

1961 June 15　In a historic memorandum President Kennedy charged Vice President Johnson with the task of setting up a telecommunications satellite service. Kennedy affirmed that he was "anxious that this new technology be applied to serve the rapidly expanding communications needs of this and other nations on a global basis, giving particular attention to those of this hemisphere and newly developing nations throughout the world." In March 1962 George C. McGhee, under secretary of state for political affairs, urged, "It would make possible a closer interlinking of remote nations. We are hopeful the Soviet Union . . . will join with us in implementing this objective." The Satellite Act of August 31, 1962, emerged from this action.

1961 June 16　The Fleming committee submitted its report *A Feasible Approach for an Early Manned Lunar Landing* to NASA Associate Administrator Robert Seamans. It recommended a target date of August 1967 and the use of one liquid or solid propellant Nova-class launcher capable of sending a total mass of 160,000 lb. to escape velocity. The liquid Nova would have eight (1.5-million-lb. thrust) F-1 engines. The Apollo spacecraft configuration was judged to weigh 150,000 lb., including the 12,500-lb. command module that was to return to Earth following direct ascent from the moon. It was believed that 12 Nova test flights would be necessary to prove reliability for the first manned moon landing. The Saturn C-3 would also be used in early qualification and dress-rehearsal flights around Earth. The program was projected to cost $11.68 billion over six years.

1961 June 20　NASA Associate Administrator Robert Seamans set up an ad hoc task group under Donald H. Heaton to evaluate program plans and resource needs for an early manned lunar landing. Endorsed by the Fleming committee, the Saturn C-3 was to form the basis of the analysis, which was completed during August 1961. The Heaton group also looked at the Saturn C-4 then being designed at the Marshall Space Flight Center with four F-1 engines in the first stage and four J-2 engines in the second stage and capable of placing 180,000 lb. in Earth orbit. It was found that either C-3 or C-4 launchers could perform the mission via Earth-orbit rendezvous and docking of several spacecraft and propulsion elements. It did not directly examine the direct-ascent Nova concept proposed by the Fleming committee but did recommend that the 200,000-lb. thrust J-2 be adopted for the second stage of the C-1 and the third stage of the C-3, and this was eventually approved.

1961 June 22 A tentative flight schedule for four additional Ranger moon flights was set with RA-6 through RA-9 slated for launch between December 1962 and June 1963. These would effectively close out the impacting Ranger program prior to the soft-landing Surveyor missions now expected to begin in August 1963. Designated Block III flights, *Rangers 6–9* were approved during August 1961 as spacecraft configured to support the engineering needs of Apollo moon landings by providing detailed information on surface conditions. Each Block III was expected to weigh around 800 lb. and carry at least three high-resolution TV cameras providing around 1,600 images from an altitude of 800 ft. to the point of impact.

The U.S. Advanced Research Projects Agency approved the Vela nuclear detection system of satellites which had been under study since September 1959. The Vela program envisaged a constellation of five satellites equipped with sensors provided by the Lawrence Livermore Laboratory. It originated as one of three means of detecting nuclear explosions: Vela Uniform, using seismic detectors to monitor underground explosions; Vela Sierra, with surface detection of nuclear detonations; and Vela Hotel, the use of satellites to detect explosions on the surface or in the lower atmosphere. In December 1961, TRW Systems Group was awarded a development contract for Vela.

1961 June 23 NASA announced cancellation of plans to develop the Saturn C-2 launch vehicle, with efforts now focusing on the C-3 and Nova concepts. C-2 had been considered too close in performance to the C-1 to justify full-scale development. The C-3 was designed around a S-IB stage 30 ft. in diameter with two 1.5-million-lb. thrust F-1 engines and cryogenic S-II (four J-2) and S-IV (six LR-115) upper stages. The C-3 would have been capable of placing 105,000 lb. in low Earth orbit. The Saturn C-4 design gradually lost out to a more powerful version with five F-1 engines in the first stage. At this date such a launcher was firmly in the Nova class, but it would become known as the C-5, and later the Saturn V.

NASA's Associate Administrator Robert Seamans asked Kurt H. Debus, director of the Launch Operations Directorate, and Maj. Gen. Leighton I. Davis, commander of the Air Force Missile Test Center, Cape Canaveral, to perform an analysis of future launch facility requirements for Saturn and Nova vehicles. The Debus-Davis study assumed an Earth-orbit rendezvous program for initial build-up in the Apollo program with direct-ascent using Nova vehicles for the manned landing. The first part of the Debus-Davis report was submitted to NASA on July 31, 1961, laying the foundation for facilities at Cape Canaveral capable of supporting a vigorous launch program of up to 50 Saturn launchers per year.

1961 June 29 The first nuclear power plant sent into space was carried by the 174-lb. U.S. Navy satellite Transit 4A

launched on a Thor-Able Star from Cape Canaveral, Fla. Designated SNAP-3A (systems for nuclear auxiliary power-3A), the spherical generator weighed 4.6 lb. in a 5-in. sphere fueled with plutonium 238, a radioactive isotope with a half-life of 89.6 years. Operating on thermionic conversion, in which heat produced by radioactive isotope decay is converted to electrical energy via thermocouples, SNAP-3A produced 2.7 W of power and supplemented solar cells. This type of nuclear power plant became known as a radioisotope thermoelectric generator, or RTG. Transit 4A was placed in a 534 × 623 mi. orbit, inclined 66.8°. Also carried, to a 534 × 634 mi. orbit, was the 55-lb. *Injun 1,* first of three radiation-measuring satellites built by the University of Iowa, and *Solrad 3.* They failed to separate from each other.

1961 July 3 The U.S. Air Force presented its preferred ballistic missile package to Secretary of Defense Robert McNamara. It asked for a total 3,316 ICBMs by 1967, of which 2,500 would be Minuteman missiles in fixed silos and 415 in the rail-mobile mode. Although allowed only 792 ICBMs under the Eisenhower plan and 834 under Kennedy's plan of March 28, 1961, the Air Force had ambitions to greatly expand its strategic nuclear strike force. The Navy was holding to a consistent recommendation for 45 fleet ballistic missile submarines equipped with 720 Polaris missiles.

1961 July 5 After intense development by Israeli scientists and engineers, a multistage solid propellant sounding rocket was launched to a height of 50 mi. off the Israeli coast. Designated *Comet 2,* it carried a sodium vapor experiment for measuring atmospheric phenomena.

1961 July 6 NASA awarded the Boeing Co. a six-month contract to study development of super-heavyweight rockets that were in the Saturn and Nova class and utilized solid propellants in lower stages and liquid propellants in upper stages. Because the NASA Marshall Space Flight Center was almost exclusively concerned with liquid propellant boosters, solids had been given little attention as candidates for launch vehicles. Over the next two years solid propellant stages would be increasingly incorporated in plans for heavyweight launchers.

1961 July 7 NASA and the Department of Defense set up the Large Launch Vehicle Planning Group under Nicholas E. Golovin. With a mandate to study the development of solid and liquid propellant launchers, it formally came into existence 13 days later and on August 23, 1961, announced that it wanted the help of selected NASA centers to examine four alternative ways of sending men to the moon: direct ascent (Office of Launch Vehicle Programs), Earth-orbit rendezvous (Marshall Space Flight Center), lunar-orbit rendezvous (Langley Research Center) and lunar-surface rendezvous, or LSR (Jet Propulsion Laboratory). The LSR technique envisaged landing several support modules and return spacecraft on the moon prior to the first manned flight.

1961 July 12 The U.S. Air Force launched the 3,500-lb. *MIDAS-3* early warning satellite into a 2,087 × 2,196 mi. orbit at 91.2° from Vandenberg AFB, Calif., by the first Atlas-Agena B configuration. *MIDAS-3* carried solar cells instead of the batteries that had been carried by *MIDAS-1* and *-2.* One MIDAS readout station was outside the United States, located at Kirkbride, Cumberland, United Kingdom. Development of a geosynchronous early warning system would await availability of the more powerful Agena D in 1966. Until then the program was cut back to a research effort known as Program 461, and in six more launches with Atlas-Agena B by July 19, 1963, only four reached orbit.

1961 July 15 NASA Space Task Group Director Robert Gilruth announced that Virgil Grissom would fly the second Mercury-Redstone suborbital flight, designated MR-4, with John Glenn acting as his backup. If the MR-4 flight went according to plan and if final qualification tests with the Atlas went well, the next manned mission would be an orbital flight. Because Shepard, Glenn and Grissom had been selected as the prime candidates for the first ballistic flight MR-3 the decision to fly Grissom on the MR-4 mission virtually confirmed John Glenn as the candidate for the first U.S. manned orbital flight.

1961 July 18–20 A NASA-industry Apollo technical conference was held in Washington, D.C., for what was considered the most prestigious space contract for the foreseeable future—the design and development of the manned moon vehicle. Approximately 300 potential contractors attended the briefings and heard detailed proposals from NASA, General Electric, Martin and General Dynamics (Convair) on the results of their feasibility studies. NASA said it would draw up final specifications for Apollo by the end of August 1961. When it did, the contract was for a command module carrying the three-man crew and a service module housing spacecraft systems and lunar orbit propulsion. The contract for a third, lunar landing, module would be awarded later based on the final determination of the mission mode: if direct ascent, it would become an integral part of Apollo; if not, it would be a separate landing vehicle carried to lunar orbit docked to Apollo.

1961 July 21 At 12:20 UT, the MR-4 mission carrying NASA astronaut Virgil Grissom lifted off from Pad 5 at Cape Canaveral at the start of a suborbital flight lasting 15 min. 37 sec. In a flight path that closely followed that flown by Alan Shepard on May 5, 1961, spacecraft no. 11, weighing 2,813 lb., reached a height of 118.3 mi., a speed of 5,168 MPH and a range of 302 mi. After landing in the Atlantic Ocean, Grissom was waiting inside his spacecraft as the recovery helicopter attached its line when the side hatch suddenly blew off. As water entered the spacecraft Grissom got out through the open hatch and the spacecraft began to sink. The heli-

Following a 15-minute ballistic flight to the edge of space on July 21, 1961, the MR-4 spacecraft Liberty Bell 7 *is hooked by a Marine helicopter; it sank only seconds after this photograph was taken.*

copter hooked up the recovery line, but the water-laden spacecraft weighed 5,000 lb. − 1,000 lb. greater than the helicopter's lifting capacity. With two wheels in the water, it was forced to abandon the attempt when its engine began to overheat. Grissom, his suit meanwhile becoming waterlogged, was in grave danger of drowning when a second helicopter winched him from the sea. Named *Liberty Bell 7,* the spacecraft sank in 2,800 fathoms.

1961 July 24 President Kennedy issued a statement on international telecommunications by satellite, affirming that "private ownership and operation of the U.S. portion of the system is favored." He invited "all nations to participate in a communications satellite system in the interest of world peace and closer brotherhood of peoples throughout the world." In this, Kennedy began to encourage participation in what would become the International Telecommunications Satellite Consortium, Intelsat.

1961 July 27–28 Officials from NASA and the McDonnell Aircraft Co. met to define the future course of work by that company on Mercury and post-Mercury requirements. McDonnell was directed to develop an 18-orbit version of the one-man Mercury spacecraft and, in a separate project, to develop a two-man version of the basic Mercury for advanced missions involving long duration, extravehicular activity and rendezvous and docking. The two-man concept became known as Mercury Mark II and was referred to as such when McDonnell engineers met with NASA's Space Task Group to prepare a draft project description for NASA Headquarters. When McDonnell submitted their detailed specification of the Mark II on November 15, 1961, they envisaged a spacecraft carrying a separate adapter for in-orbit equipment, fuel cells for electrical power production, ejection seats for emergency, launch escape rocket in tractor mode, bipropellant maneuvering thrusters and integral guidance and navigation equipment. The company interpreted the "long duration" requirement as seven days.

1961 July 28 NASA selected 12 U.S. aerospace companies to prepare bids for design and development of the Apollo spacecraft: Boeing; Vought; Douglas; General Dynamics/Convair; General Electric; Goodyear; Grumman; Lockheed; McDonnell; Martin; North American Aviation; and Republic Aviation. The work statement defined three phases of the program: Phase A, manned Earth orbit flights up to 14 days in duration using the Saturn C-1; Phase B, circumlunar and lunar orbit flight on Saturn C-3 or Nova-class launchers for rendezvous rehearsal and reconnaissance; Phase C, manned landings with Saturn C-3 or Nova. A source evaluation board was set up under the chairmanship of Walter Williams, who was replaced on November 2, 1961, by Maxime Faget of the Manned Spacecraft Center.

NASA and the American Telephone and Telegraph Co. signed a cooperative agreement for the development and testing of two communication satellites. Later named *Telstar I* and *II,* they were the world's first active repeater communication satellites, the first built by private industry and the first commercial payloads launched by a government agency. AT&T was to pay for the satellites and NASA would be reimbursed for the launch on Delta rockets. *Telstar I* was launched on July 10, 1962, followed by *Telstar II* on May 7, 1963.

1961 August 2 NASA headquarters announced that it was making a worldwide survey of potential sites for very large

launch vehicles of the Saturn and Nova class. In the interest of public safety, greater isolation than was afforded by Cape Canaveral, Fla., was considered essential for boosters with a first stage thrust that could exceed 20 million lb. NASA found potential problems with organizing the logistics of launches from foreign locations and preferred U.S. locations. In fact, very large boosters failed to materialize.

1961 August 5 A segmented solid propellant rocket motor developed by United Technology Corp. (UTC) in California produced a more than 200,000-lb. thrust for 1 min. 20 sec. The center segment contained 55,000 lb. of solid propellant, the largest single segment produced in the United States to this date. In a related test, also for the U.S. Air Force, a larger solid propellant motor produced a thrust of 500,000 lb. for 1 min. 20 sec. on December 9, 1961. These tests were part of the development process for Titan boosters.

1961 August 6 Soviet cosmonaut Gherman S. Titov, the second man to orbit the Earth, the first man to sleep in space and the first to remain in orbit for more than one day, was launched at 06:00 UT aboard the 10,430-lb. *Vostok 2* by an SL-3 launcher from the Tyuratam launch complex to a 103 × 144 mi. orbit and an inclination of 65°. Cosmonaut Andrian G. Nikolayev had been appointed his backup for a mission that had been extended from three orbits to one full day on the express orders of Soviet Premier Nikita Khrushchev. Although he suffered from nausea, dizziness and disorientation—a condition later described as space sickness—Titov managed to get 8 hr. of rest, beginning at 15:30 UT, as he passed up the eastern side of Florida. For much of the flight the temperature control system malfunctioned, the cabin interior reaching a low of 43°F (6°C). Titov was allowed to carry out some manual tasks aboard Vostok, firing attitude control thrusters on the spacecraft before retrofire and reentry. Titov ejected from his descent module at an altitude of about 21,300 ft., coming to Earth near the Volga river at an elapsed time of 1 day 1 hr. 18 min.

1961 August 8 The first flight on an Atlas F took place from Cape Canaveral, Fla., and was powered by the Rocketdyne MA-3 propulsion system first utilized on the Atlas E. Designed as a new quick-fire version, it was capable of traveling 5,000 mi. in 30 min. and was interchangeable with the Atlas E. Hardened underground silo basing was selected for the 72 Atlas F missiles, deployed with 6 Strategic Missiles Squadrons operating 12 missiles apiece at: Altus AFB, Okla.; Dyess AFB, Tex.; Lincoln AFB, Nebr.; Plattsburgh AFB, N. Y.; Schilling AFB, Kans.; and Walker AFB, N.Mex. Full deployment was completed on December 20, 1962. Each silo was 174 ft. deep and 52 ft. in diameter, the missile being hydraulically raised to the surface on a vertical elevator platform from which it was launched. The most powerful nuclear device carried by an operational Atlas ICBM would be the 3.75-MT W-38 warhead carried inside an Mk.4 ablative reentry vehicle.

1961 August 9 NASA awarded a contract to the Massachusetts Institute of Technology Instrumentation Laboratory for design and development of Apollo guidance and navigation systems. The job of putting together hardware and software for the moon landing missions was challenging and not within the capabilities of existing systems. The Apollo computer began as a derivative of the Polaris SLBM (submarine-launched ballistic missile) computer, which formed the basis for Apollo Block I spacecraft, but evolved into the Apollo Block II computer utilizing radically new technology. It was the decision to adopt the lunar orbit–rendezvous mode, with its greater complexity, that forced development of the more advanced computer specification.

1961 August 11 NASA awarded a contract to Hughes Aircraft for construction of the world's first experimental geosynchronous communication test satellite. Developed under the auspices of the Goddard Space Flight Center, the three satellites ordered were to be numbered in the Syncom series. They would each employ redundant transponders with 2-W traveling wave tubes, receiving on 7,360 MHz and transmitting on 1,815 MHz. The Delta launch vehicle would place the satellite in a geosynchronous transfer orbit which would be made circular at an altitude of about 22,000 mi. and an inclination of 33°. In that orbit it would appear to weave a figure-eight between 33° N and 33° S.

1961 August 16 The first of four NASA Energetic Particles Explorer (EPE) satellites, EPE-A (*Explorer XII*), launched by Delta from Cape Canaveral, Fla., was placed in a 180 × 47,800 mi. orbit inclined 33.3°. They each comprised an aluminum hexagon, 2 ft. 2 in. across, 1 ft. 7 in. high, with a truncated cone on top and a magnetometer on the bottom, giving a total height of 4 ft. 3 in. Spin-stabilized, with four solar paddles providing an average 16 W, EPE-A carried 10 instruments from the University of Iowa, University of New Hampshire, Ames Research Center and the Goddard Space Flight Center. Until December 6, EPE-A provided data revealing that the outer Van Allen belts are made up almost exclusively of protons, and decayed on September 1, 1963. Weighing 89 lb., EPE-B (*Explorer XIV*) was placed in a 174 × 61,190 mi., 32.9° orbit on October 2, 1962. It provided data until October 8, 1963, and decayed on July 1, 1966. The 100-lb. EPE-C (*Explorer XV*) was placed in a 194 × 10,760 mi., 18° orbit on October 27, 1962, providing data until February 9, 1963, before decaying on December 19, 1978. The 101-lb. EPE-D (*Explorer XXVI*) was launched December 21, 1964, to a 190 × 16,280 mi., 20.2° orbit. It is still in orbit, with parameters of 182 × 6,267 mi. at 19.8° in December 1993.

1961 August 23 In what was essentially a test launch of a partially complete spacecraft, an Atlas-Agena B launched *Ranger 1* to a 121 × 360 mi., 32.9° Earth parking orbit. A planned second burn of the Agena B engine to propel the 675-lb. spacecraft to an elliptical orbit with an apogee of 685,000 mi. failed to occur and it was left in Earth orbit. However, the two panels with their total 8,680 solar cells deployed after separation from the Agena B as did the high-gain antenna, giving the spacecraft a span of 17 ft. and a height of 13 ft. *Ranger 1* made 111 orbits of the Earth and burned up in the atmosphere on August 30, 1961. An almost identical situation occurred to *Ranger 2* after it was launched on November 18, 1961, with the same objectives. The Agena B failed to fire a second time, leaving the spacecraft stranded in a 128 × 169 mi., 33.3° orbit, and it reentered two days later.

1961 August 25 The Future Projects Design Branch at NASA's Marshall Space Flight Center (MSFC) completed a Nova preliminary planning document aimed at consolidating a very wide range of super-heavyweight booster concepts. With a manned lunar landing before it, NASA was convinced it would need a very large launcher for the 1970s. The MSFC Nova concept, also called Saturn C-8, had a payload capacity of 350,000 lb. to low Earth orbit and a launch weight of 9.5 million lb. Powered by eight F-1 engines arranged in a circle and 44 ft. in diameter, the N-I stage would have a liftoff thrust

of 12 million lb. The N-II stage, with a diameter of 33 ft., would have eight J-2 engines and a total thrust of 1.6 million lb. The N-III stage, 26 ft. 7 in. in diameter, would be powered by two J-2 engines for a combined thrust of 400,000 lb. When Saturn C-5 was selected for Apollo, Nova became subject to a wide range of post-Saturn studies and configurations throughout the decade. The most extravagant concept envisaged a 19,000-ton launch vehicle, 535 ft. tall and 70 ft. in diameter, with a solid propellant liftoff thrust of 54.9 million lb. and a 500-ton payload.

A Scout launcher carried the micrometeoroid satellite *Explorer XIII* from Wallops Island, Va., to a 175 × 606 mi. orbit with an inclination of 36.4°. It weighed 187 lb., of which 50 lb. was the spent motor case of the final stage and 12 lb. was the support section, with the exterior surface virtually covered with five types of micrometeoroid detector. Placed in a lower orbit than planned, the satellite reentered the atmosphere three days later and burned up. *Explorer XIII* was the first of three NASA S-55 micrometeoroid survey satellites, others being *Explorers XVI* and *XXIII*.

1961 August 26 Aerojet General fired the largest solid propellant rocket motor to date when a 70-ton segmented motor produced a thrust of just over 1 million lb. at a test stand near Sacramento, Calif. The motor, 30 ft. long with a diameter of 8 ft. 4 in., fired for 1 min. 27 sec. The test was part of a research and development program for large solids which, until recently, had lagged behind developments with liquid propellant rocket motors.

1961 August 28 In response to concerns about the availability and payload capacity of the Atlas-Centaur launch vehicle for the Mariner A spacecraft to Venus, the Jet Propulsion Laboratory (JPL) proposed a lighter spacecraft based on the already available Atlas-Agena B. Called Mariner R, the new spacecraft would weigh less than 460 lb., compared to 1,250 lb. for Mariners A and B, and carry 25–40 lb. of scientific equipment for observation of the planet during a close fly-by. Delays with Atlas-Centaur would preclude use of that launcher for early planetary missions and, by taking elements of the Mariner R and Ranger lunar probes, the JPL could make use of a 56-day launch window between July 18 and September 12, 1962, to launch two spacecraft to Venus. Two days later NASA approved Mariner R to Venus in 1962, canceled Mariner A and authorized the JPL to prepare Mariner B for an Atlas-Centaur launch in 1964 to either Mars or Venus.

1961 August 30 A Thor Agena B launched *Discoverer XXIX* from Vandenberg AFB, Calif., to a 94 × 337 mi., 82.1° orbit with the primary objective of spying on the Soviet northern missile test site near the town of Plesetsk. Intelligence from a variety of sources told the CIA that operational ICBMs were being placed at the site, transforming a test and development center into the first Soviet offensive launch complex. *Discoverer XXIX* was equipped with the first in a series of improved camera systems known as KH-3 and returned its capsule to Earth two days later. A number of SS-7 Saddler missiles with storable liquid propellants, were observed at surface launch sites.

1961 August 31 The Soviet Union announced that it was ending a moratorium on nuclear weapons testing that began March 31, 1958. In announcing a resumption of testing, the USSR said that it could deliver nuclear weapons to targets anywhere on Earth by means of "powerful rockets like those

Majs. Yuri Gagarin and Gherman Titov rode to begin their unrivaled spaceflights around the Earth." On September 1, 1963, the White House announced that the USSR had begun a new program of nuclear testing that morning. It was the first nuclear test by any power since the fall of 1958.

1961 September 2 The U.S. Air Force Space and Missile Systems Organization (SAMSO) awarded parallel study contracts to Boeing and Ling-Temco-Vought (LTV) for a solid propellant upper stage midway in capability between the Scout and the Thor-Able Star. What emerged was a program that would result in three separate stages: Burner I, Burner II and Burner IIA. Burner I was the Altair motor used as the fourth stage of Scout and adopted for Thor. Burner II and IIA were based on the Thiokol TE-M-364-2 developed as the solid propellant retro-motor for the Surveyor moon lander. This motor was also developed as an upper stage for Delta.

1961 September 5–8 Scientists from 10 countries, including the United States and the Soviet Union, attended the Seventh International Conference on Science and World Affairs held at Smuggler's Notch, Vermont. Among several topics discussed were proposals for greater cooperation in developing a worldwide system of weather satellites, a global communications satellite system, a broad exchange of space biology data and a joint exploration of the planets. Enthusiastic optimism among the scientists was not mirrored by events at a political level: Angered by a statement from President Kennedy that the United States would go to war should Russia threaten the freedom of Berlin, the Soviet Union began a series of atmospheric nuclear tests culminating in late October with the detonation of a 58-MT thermonuclear device, the largest ever demonstrated.

1961 September 7 NASA's Marshall Space Flight Center selected a tract of land at Michoud, near New Orleans, alongside the navigable waters of the Mississippi River as an assembly site for Saturn rocket stages to be known at first as the Michoud Assembly Facility (MAF). NASA officially redesignated the facility Michoud Operations on December 18, 1961. The first booster assembly at Michoud Operations began October 4, 1962, when the Chrysler Corp. started fabrication of the first stage (S-I) of the SA-8 Saturn launcher. Boeing activated the S-IC (Saturn V first stage) portion of Michoud Operations on October 22, 1962, when it started tooling up for stage fabrication. On July 1, 1965, the facility reverted to its former title of Michoud Assembly Facility. It currently produces external tanks for the NASA Space Shuttle program.

1961 September 11 U.S. Air Force and NASA officials conducted an inspection of the full-scale mock-up of the Dyna-Soar boost-glide vehicle at the Boeing Co. facilities in Seattle, Wash. At this date the vehicle had a length of 35 ft. 4 in., a span of 20 ft. 5 in. across the sharply swept delta wing, which had a surface area of 345 sq. ft., and a projected launch weight of 11,390 lb. During the month Secretary of Defense Robert S. McNamara questioned the wisdom of spending almost $1 billion on a research and development program when no clear military mission had yet been defined. In response, the Air Force stressed the need "to demonstrate manned, orbital flight and safe recovery" of a lifting-body type vehicle before the military potential could be defined.

NASA announced that North American Aviation's Space and Information Systems division would receive a contract to build the Saturn S-II stage. Designed as the second stage for the C-3 launch vehicle, the S-II was expected to carry four J-2 liquid hydrogen–liquid oxygen cryogenic engines delivering a stage thrust of 800,000 lb. On November 6, 1961, the Marshall Space Flight Center directed North American to design the S-II stage with a fifth J-2 engine for application as the second stage of the advanced Saturn launch vehicle, eventually known as the Saturn V. Bids for the advanced Saturn first stage, the S-IC, were invited September 17, and on December 15, 1961, Boeing was selected to build it at the Michoud Assembly Facility. The S-IC stage had five F-1 engines delivering a total thrust of 7.5 million lb.

1961 September 13 The first of two unmanned orbital qualification flights for the Mercury-Atlas launch vehicle and spacecraft lifted off from Pad 14 at Cape Canaveral, Fla. Designated MA-4, the Atlas had been significantly modified as a result of problems with three of the four Mercury Atlas flights to date. This time the flight was copybook perfect: The Atlas provided a peak 7.6 g and a velocity of 25,705 ft./sec., sending the 2,645-lb. spacecraft (no. 8A) and an astronaut simulator into a 98.9 × 142.1 mi. orbit. At 1 hr. 28 min. 59 sec., toward the end of the first orbit, the first retro-rockets fired in the vicinity of Hawaii, and spacecraft no. 8A began to descend toward the atmosphere. The drogue parachute, 6 ft. in diameter, deployed at 41,750 ft., followed by the main parachute, 63 ft. in diameter, at 10,050 ft. and splashdown at 1 hr. 49 min. 20 sec.

1961 September 15 Acting on the request of the U.S. Air Force, the Department of Defense Research and Engineering Office formally requested a study be made on adapting the Titan II ICBM into a standard space launch vehicle. The Air Force was given permission to proceed with this under Project 624A during December, and in May 1962 the preliminary design of a Titan III configuration was completed. On August 11, 1962, the Air Force received formal authorization to proceed with detailed design and awarded Martin Marietta a contract for systems integration. A contract for development and manufacture of Titan IIIA was awarded Martin Marietta on February 25, 1963, and the first vehicle was accepted from the manufacturer on June 30, 1964. From this first derivative evolved Titan IIIB, IIIC, IIID and 34D configurations for the Air Force and the IIIE/Centaur for NASA.

1961 September 19 NASA Administrator James Webb announced that the new Manned Spacecraft Center (MSC) would be constructed on a 1,020-acre site donated by Rice University and located southeast of Houston, Tex. Adjacent to Clear Lake, the site had been chosen from 20 candidate locations around the United States and judged against 10 qualifying criteria. NASA purchased 600 additional acres, and in April 1962 work began to prepare a 1,620-acre plot; personnel, now numbering 1,100, began occupying temporary buildings in Houston and at nearby Ellington AFB. To prevent disruption, Mercury staff remained at Langley until completion of the MA-7 mission of May 24, 1962, but all staff had relocated by that July. By the end of 1962 there were more than 2,400 personnel at the MSC, and in mid-1963 that number had grown to 3,400 in the buildup toward two-man Gemini and three-man Apollo missions. Movement from temporary buildings to the Clear Lake site began in September 1963 and was completed by February 1964.

1961 September 21 Eurospace, an association of European space-related industries, was formed. Its president was Jean DeLorme, who called the organization a European Industrial

Space Study Group. As a forum for discussing the possible, Eurospace helped stimulate space policy throughout Europe and led the way in talks with the United States on trans-Atlantic cooperation in space research. The 96 industrial organizations in the 11 participating countries represented in Eurospace were linked with 31 banks and insurance companies and 13 corresponding members in the United States.

A top secret U.S. National Intelligence Estimate (NIE) was issued by the Central Intelligence Agency reporting that fewer than 25 Soviet ICBMs had been built, that only six were likely to be available for operations within the next year and that a maximum three missiles could be fired against the United States as of this date. Based upon stunning evidence to refute Air Force claims about a massive Soviet missile deployment, the report drew upon evidence from reconnaissance imagery returned to Earth in recovery capsules launched on *Discoverers XIV, XVIII, XXV, XXVI* and *XXIX*. Each successful Discoverer mission with KH-1 or KH-2 cameras produced more images of the USSR than had been delivered by all 24 overflights of Soviet territory with U-2 spyplanes between July 4, 1956, and May 1, 1960.

1961 September 22   Secretary of Defense Robert McNamara completed his review of U.S. strategic retaliatory forces and forwarded to the Joint Chiefs of Staff the recommended numerical levels of missile deployment. The U.S. Navy target of 45 fleet ballistic missile submarines was cut to 41, equipped with 656 Polaris missiles, and the Air Force was ordered to plan for a maximum 900 Minuteman, 129 Atlas and 108 Titan ICBMs in fixed silos. The Air Force had wanted to deploy 150 of the 600 Minuteman ICBMs it had been allowed to buy (under the March 28, 1961, plan) in a rail-mobile mode, each train carrying five missiles, but this option was canceled on December 13 for financial reasons. The Air Force and Navy had separately lobbied for a combined force of more than 4,000 missiles. In the certain knowledge that the Soviet Union had not yet begun ICBM deployment at all, that total had been cut by more than half. Nevertheless, at this date the U.S. Air Force had only about 45 ICBMs versus the Soviets' 13, and the Kennedy administration knew this as a fact when ordering up a massive expansion of strategic firepower.

1961 September 23   The formal flight schedule for the four additional unmanned Ranger moon probes recently approved was circulated at the Jet Propulsion Laboratory, Calif. *Rangers 6–9* were scheduled for flight in December 1962 and in February, April and June 1963. Each spacecraft would carry six RCA television cameras and little else so that scientists could concentrate on giving engineers what was "necessary for initial designs of manned lunar spacecraft." In what was a realistic, if not altogether complimentary, reference to the poor level of technical success at this stage, it was also justified on the basis that nine rather than six flights would "increase the probability of obtaining at least one successful lunar mission during the entire Ranger series."

1961 September 24   A major reorganization of NASA established four new program offices: the Office of Advanced Research and Technology, the Office of Space Sciences, the Office of Manned Space Flight and the Office of Applications. Robert Gilruth was also named as the first director of the Manned Spacecraft Center being built at Houston, Tex. All the field centers would report directly to Robert Seamans, NASA associate administrator.

1961 September 26   NASA held a conference in New Orleans to brief potential bidders on industrial production of the S-I stage for the Saturn C-1 launcher at the Michoud Assembly Facility which the Marshall Space Flight Center planned to build in that area. The Chrysler Corp. was selected on November 17, 1961, for a contract to build 20 rocket stages for the 1.5-million-lb. thrust booster. Signed two months later, this contract was modified in future years to reflect the changes to Saturn launch vehicle planning. Chrysler only built 2 Saturn C-1 stages, but 12 upgraded stages of the same type for the Saturn IB.

1961 September   Along with Surveyor contractor Hughes Aircraft, the NASA Jet Propulsion Laboratory selected a shortlist of thirteen scientific experiments to be conducted on the lunar surface by the unmanned soft lander. A weight allowance of 345 lb. was budgeted on the 2,500-lb. spacecraft and proposed equipment included survey TV camera, subsurface sampler, chemical analysis equipment, geophysical probes, magnetic detectors, radiation detectors and a surface sampler. In March 1962 the Atlas-Centaur was reassessed as having a lunar payload limit of 2,100 lb. This cut the science payload down to 115 lb., leaving just the TV and drill equipment, although the launch vehicle payload capability crept back up to 2,250 lb. During October 1962 a review of instrument design increased to seven the potential range of candidate tasks. However, in the attempt to concentrate development on a flagging program, only one science instrument, the survey TV camera, was integrated into *Surveyor I* when it was launched on May 30, 1966. Later a surface sampler and chemical analyzer were added. The first six Surveyors had a science load of 62 lb., with *Surveyor VII* carrying 71 lb.

1961 October 2   A U.S. Air Force Atlas missile launched on a test flight from Cape Canaveral, Fla., carried a scientific passenger pod (SPP) developed by General Dynamics, Convair Division. Given a brief ride into space as the missile flew its ballistic trajectory, the SPP had been developed by the Air Force Office of Aerospace Research through its Aerospace Research Support Program (ARSP). The ARSP had been set up to consolidate military space experiments into a single unified program. The pod, 8 ft. 10 in. long, could accommodate a variety of payloads.

1961 October 3   NASA Marshall Space Flight Center asked Douglas Aircraft to carry out a study of an S-IVB stage for advanced Saturn launch vehicles. Essentially a modified S-IV, the S-IVB would replace the six Pratt and Whitney LR-115 engines with a single 200,000-lb. thrust Rocketdyne J-2 engine, and was to be used as the upper stage for launchers tasked with circumlunar missions. Originally developed for the S-II stage, the J-2's restart capability would be essential for Earth-escape missions. NASA announced on December 20, 1961, that Douglas had been selected for S-IVB development.

1961 October 7   The U.S. Air Force once again changed the structure of the Dyna-Soar program, this time eliminating the suborbital test flights. The fifteen air-launched flights would precede the first of two unmanned orbital flights, scheduled for November 1964, and the first manned flight, now expected to occur in May 1965 on a Titan IIIC launch vehicle. Scheduled for June 1966, the ninth flight would be an unmanned test of reentry from a high apogee to demonstrate the ability of the boost-glider to reach satellites in a wide variety of orbits. The flight test program was expected to continue through the end of 1967.

1961 October 9 NASA received submissions for the Apollo spacecraft contract from five prospective bidders: General Dynamics–Avco; General Electric–Douglas–Grumman–Space Technology Laboratories; McDonnell Aircraft–Lockheed–Hughes Aircraft–Vought; Martin; and North American Aviation. Oral reports from the five bidders were presented two days later at a meeting in the Chamberlain Hotel, Old Point Comfort, Va., and eleven panels began studying the proposals. Technical assessments were completed on October 20 with a final report to the source evaluation board on November 1, 1961.

1961 October 21 An Atlas-Agena B launched by the U.S. Air Force from Vandenberg AFB, Calif., carried the *MIDAS-4* satellite into a 2,166 × 2,336 mi., inclined 91.2° to the equator. *MIDAS-4* incorporated the West Ford experiment from the Lincoln Laboratory, Massachusetts Institute of Technology, comprising 350 million copper threads (0.7 in. long with 0.001 in. diameter) which were to have been dispersed to a 5 mi. × 24 mi. belt around the Earth. The needles were to have served as a reflector for extremely short wavelengths (8,000 MHz), but the satellite failed to disperse the needles as planned. Project West Ford had been opposed by radio astronomers who claimed it would have blocked views of the universe at those frequencies.

The first attempted satellite launch from the Soviet site at Kapustin Yar, and the first operational use of the SL-7 launcher, took place when a Cosmos-series satellite was carried into space on a flight that failed in its attempt to place it in orbit. The SL-7 was derived from the SS-4 ballistic missile. It had a four-chamber RD-214 first-stage engine of 159,000-lb. thrust on kerosene–red fuming nitric acid propellants, and a four-chamber RD-119 second-stage engine of 24,255-lb. thrust on UDMH (unsymmetrical dimethyl hydrazine)–liquid oxygen. With a nominal height of 98 ft. 5 in., a diameter of 5 ft. 5 in. and a launch weight of 95,000 lb., the SL-7 could orbit a 1,100-lb. payload. Called the Volgograd station by Russians, the launch site is located between the towns of Kapustin Yar and Akhtyubinsk and is an outgrowth of the complex used to fire V-2 rockets in 1947. The last SL-7 lifted off with *Cosmos 919* on June 18, 1977. By this date 146 had been launched, of which five had failed—a success rate of 96.6%.

1961 October 23 The first submerged firing of a Polaris A2 test missile took place from the submarine USS *Ethan Allen* at sea off Cape Canaveral. The Polaris A2 had been developed in just three years into an SLBM (submarine-launched ballistic missile) with a 1,500-NM range by improvements made to the second stage of this two-stage solid propellant missile. It had initially been thought that first-stage propellant improvements would give Polaris a range advantage of 300 NM over the A1, but engineers discovered that an eightfold increase in performance would accrue from weight-trimming measures applied to the second stage. The first stage was lengthened by 2 ft. 6 in. The first operational deployment of the missile was on June 26, 1962, when the USS *Ethan Allen* departed Charleston, S.C.

1961 October 25 NASA announced its decision to build a large rocket motor test facility at a site on the Pearl River in Mississippi. It would be known as the Mississippi Test Operations, renamed the Mississippi Test Facility (MTF) on July 1,

1965. The first rocket stage tested at MTF was the S-II-T, an all-up Saturn V second-stage systems test model, which arrived October 17, 1965, and was successfully static fired on April 23, 1966. The first S-IC Saturn V first stage arrived on the barge *Poseidon* from the Michoud Assembly Facility and was tested on March 3, 1967.

1961 October 27 The first flight of a Block I Saturn C-1 began with liftoff from LC-34, Cape Canaveral, at 15:06 UT. The 162-ft.-tall vehicle weighed 925,000 lb., but only the S-I first stage was live, the dummy second (S-IV) and third (S-V) stages being filled with a total 190,000 lb. of water ballast. A Jupiter missile nose cone was attached to the top dummy stage. Block I vehicles carried eight H-1 first-stage engines with a rated thrust of 165,000 lb. providing a liftoff thrust of 1.32 million lb. In a flight lasting 6 min. 48 sec., SA-1 reached a height of 84.6 mi. and a downrange distance of 206 mi. The vehicle was instrumented for 505 in-flight measurements, of which 485 performed as planned.

NASA's Space Task Group prepared a detailed operational plan for manned spaceflight in the period 1963–65. In the plan, the two-man Mercury Mark II would be launched by Titan II and begin with an unmanned qualification flight launched in May 1963. Succeeding flights would occur at two-month intervals, the second flight being the first manned mission, envisaged as an 18-orbit shakedown flight. The next 2 flights were to be long-duration missions of up to 14 days, followed by rendezvous and docking flights using an Atlas-Agena B. The Agena B would accommodate a docking receptacle at the front to give the Mercury Mark II crew opportunity for docking and undocking tests. The document envisaged the last of 12 flights in March 1965.

1961 October Ford's Aeroneutronic Division began work on the design of a lander capsule for NASA's Mariner B planetary spacecraft concept while science staff at the space agency sifted experiment proposals. From a list of 64 candidate proposals, 8 were selected for the fly-by bus and 10 for the lander capsule. On February 19, 1962, NASA's director of lunar and planetary sciences received word that the payload-carrying capacity of the Centaur, to be used as the second stage to an Atlas for launching Mariner B, was being reduced. This severely cut the experiment load for Mariner B, but NASA hoped to continue paying for development of science equipment so as not to lose the basic research, acting in the belief that these experiments would eventually be flown.

Studies on rigid and inflatable space stations began at the NASA Langley Research Center. Under the generic term *Apollo X,* one idea envisaged an inflatable sphere attached to the command/service module of Apollo. Providing gravity-gradient stabilization, the sphere would be manned for scientific experiments. These studies continued and in early 1962 embraced several concepts for rigid stations housing up to 30 research scientists and engineers.

1961 November 1 NASA Administrator James Webb formally implemented a major and highly significant reorganization of the civilian space agency, abolishing the four major research and development offices and forming four new Headquarters program offices. Out went the Offices of Advanced Research Programs, Space Flight Programs, Launch Vehicle Programs and Life Science Programs. In their place, with considerable shuffling of personnel and suboffices from the former groupings, came the Office of Advanced Research and Technology under Ira A. Abbott, the Office of Space Sciences under Homer E. Newell (Abe Silverstein's former deputy), the Office of Manned Space Flight under D. Brainerd Holmes from RCA and the Office of Space Applications under Morton J. Stoller. On November 1, 1963, these were merged into the Office of Space Science and Applications under Newell.

In an attempt to orbit a small test satellite to check out the communications network for the NASA Mercury manned spaceflight program, a solid propellant Scout rocket was launched from Pad 18B at Cape Canaveral, Fla. Shortly after launch the rocket developed erratic motions and at 28 sec. it began to break apart. The Cape Canaveral range safety officer pushed the self-destruct button at 43 sec. elapsed time and the Scout exploded along with its small test satellite. Examination of the debris disclosed an error by a technician who had crossed two wires in the pitch and yaw gyroscopes, causing error signals to go to the opposite control function. There was insufficient time to prepare another Scout and satellite.

1961 November 2   Signor Andreotti, Italy's minister of defense, announced that his country planned to procure Scout launchers from the United States and use them to put its own satellites in orbit. Italy already had a strong program of space research using sounding rockets, and through its space committee the Italian government wanted to expand this to include satellites. Italy would eventually become a strong partner in European space organizations and participate in the Spacelab and Columbus orbital laboratory programs.

1961 November 3   At the nine-nation Western European Conference in London, England, delegates announced the historic decision to develop an independent satellite launcher for a European satellite. The British were to provide the Blue Streak as the first stage of the launch vehicle, the French would provide the Veronique sounding rocket as the launcher's second stage and the West Germans would provide the third stage. Launches would take place from the Woomera site in Australia. Delegates expected the first satellite launch to take place in 1965.

1961 November 6   NASA's director of Manned Space Flight, D. Brainerd Holmes, received a memorandum from the Office of Launch Vehicles and Propulsion describing the organization of a working party to be chaired by Milton Rosen. Taking material submitted by the Golovin, Fleming, Lundin, Heaton and Debus-Davis committees, the Rosen working party would bring together all the disparate requirements of NASA and Department of Defense large-launch-vehicle programs and recommend a strategy for future launch vehicle evolution in the Saturn and Nova classes.

1961 November 13–22   Meteorologists from 27 countries attended the International Meteorological Satellite Workshop organized jointly between the U.S. Department of Commerce Weather Bureau and NASA. Opened by NASA Administrator James Webb, the workshop included lectures on Tiros and visits to the Goddard Space Flight Center and the national meteorological center at Suitland, Maryland. The USSR, Poland and Czechoslovakia were invited, but declined to attend. Serving as a forum for international cooperation in the science of weather, it forged links that would result in the Global Atmospheric Research Program.

1961 November 15   John C. Houbolt of the NASA Langley Research Center wrote "Somewhat as a voice in the wilderness," to Robert Seamans, NASA associate administrator, concerning the lunar-orbit rendezvous (LOR) method of get-

ting a manned spacecraft to the moon. Frustrated by what he saw as a preoccupation with "grandiose plans," Houbolt appealed for "developing the simplest scheme possible," but claimed that in the numerous committees set up to evaluate optional modes "everyone seems to want to avoid simple schemes." Houbolt had "been appalled at the thinking of individuals and committees. . . . I am bothered by stupidity being displayed by individuals who are in a position to make decisions which affect not only NASA, but the fate of the nation as well. . . . Committees are no better than the bias of the men composing them." Seamans was impressed with Houbolt's forthright and unconventional approach, and as a result his LOR idea gained a broader consideration.

The last U.S. Navy experimental navigation satellite Transit was launched as one of two satellites carried into space from Cape Canaveral, Fla., on a Thor-Able Star. Powered by solar cells and the second SNAP-3A radioisotope generator sent into space, the 190-lb *Transit 4B* was placed in a 582 × 700 mi., 33.4° orbit. The second satellite, *Traac* (transit research and attitude control) was put in a 562 × 720 mi. orbit, in which it began gravity-gradient attitude control tests. Using a 100-ft. copper-tube aerial, the satellite's long axis was aligned with the Earth's center to set up a stable alignment and test the effectiveness of this technique for future satellites. Both satellites were still in orbit at the end of 1995.

1961 November 20   NASA engineer Milton Rosen submitted his launch vehicle report to NASA Associate Administrator Robert Seamans and recommended immediate development of the Saturn C-5, with five F-1 engines in the first stage, four or five J-2 engines in the second stage and a single J-2 in the third stage. The lunar landing, advised Rosen, should utilize a direct-ascent Nova with eight F-1 engines in the first stage, four M-1 engines in the second stage and one J-2 in the third stage. Originally conceived as having four F-1 engines, the addition of a fifth F-1 to the C-5 came after Rosen pressed hard for improved first-stage performance. He convinced Wernher von Braun that this would significantly add to the vehicle's capability and explained how overconservative engineering from the Marshall Center allowed for a fifth engine to be installed at the intersection of the crossbeams to which the four existing F-1s would be attached.

1961 November 24   The NASA Apollo source selection board, chaired by Maxime Faget from the Manned Spacecraft Center, completed its report on five bids to build the manned lunar spacecraft. Assessed for technical content, technical qualifications and business management, the Martin Co. came out on top, closely followed by General Dynamics and North American Aviation, tied for second, and finally General Electric and McDonnell Aircraft. The board said: "The Martin Company is considered the outstanding source for the Apollo prime contract . . . first in technical approach, a very close second in technical qualifications, and second in business management. North American Aviation is considered the desirable alternate source."

1961 November 27   The first meeting of the United Nations Committee on the Peaceful Uses of Outer Space was held. A four-nation resolution to declare space "free for exploration" and divorced from sovereign claims was subsequently introduced by the U.S. ambassador to the UN, Adlai Stevenson, on December 4. On December 11 the General Assembly's main political committee adopted a revised version cosponsored by 24 countries, including the Soviet Union and the United States.

1961 November 28 NASA awarded North American Aviation (NAA) a design and development contract for the Apollo command and service modules. In reaching this decision NASA Administrator James Webb placed emphasis on the technical content of NAA's submission and the company's qualifications for doing the job. Experience with the rocket-powered X-15 and the XB-70 experimental supersonic bomber placed NAA in a leading position for this lucrative contract. NAA's Space and Information Systems Division would build Apollo at their plant in Downey, Calif. Optimistically, NASA expected to award a lunar landing module contract within six months, but that would be delayed until November 7, 1962, when a final decision had been made about the lunar landing mission mode.

A confidential minute written by Britain's Minister of Aviation Peter Thorneycroft informed the British Cabinet that the Kennedy administration was attempting to woo the Italians away from participation in ELDO and the Europa launch vehicle program. Fearful that a successful European satellite launcher would take away the prospect of the United States' launching satellites for non-U.S. organizations, the United States made generous offers of reimbursable flights for Italian satellites. In the Europa program, Italy was to provide the satellites for orbital test flights. If Italy pulled out of Europa, mused the Kennedy administration, the ELDO program could collapse, taking with it the threat of competition. It did not collapse.

1961 November 29 The second of two orbital qualification tests of the Mercury spacecraft was successfully accomplished, clearing the way for the first U.S. manned orbital spaceflight. Designated MA-5, the flight began at 17:08 UT when an Atlas lifted off from Pad 14. The 2,866-lb. Mercury spacecraft no. 9 carried a 37.5-lb. chimpanzee named Enos and was placed in a 99.5 × 147.4 mi. orbit inclined 32.5°. As a rehearsal for the first manned orbital flight, the mission was planned to last three revolutions of the Earth, but a problem with the environmental control system caused the chimpanzee to overheat, and when attitude control problems emerged, the flight director, Christopher C. Kraft, decided to bring spacecraft no. 9 back one orbit early. The mission had lasted 3 hr. 20 min. 59 sec., and Enos was rescued from his spacecraft by the recovery ship *Stormes*, a destroyer completed in 1945.

The NASA Space Task Group formally announced that astronaut John Glenn was the prime candidate for the first U.S. manned orbital spaceflight, an event eagerly awaited since Yuri Gagarin's flight on April 12, 1961. Glenn's backup would be Malcolm Scott Carpenter, and if the first flight went according to plan, the backup astronaut would become the prime pilot for the second mission. Thus began a tradition of assigning both prime and backup crews, with the latter assisting with flight preparations and being ready to step in for the prime pilot or crew, as required.

1961 December 11 The first Soviet photoreconnaissance satellite based on the Vostok spacecraft and reentry module was launched from the Baikonur cosmodrome. An upper-stage failure on the SL-3 launcher prevented the satellite from reaching orbit and it fell back to Earth. Named Zenith, the first generation spy satellites had been developed parallel to the Vostok manned spacecraft and were capable of fully autonomous operation. This is why Soviet pilots had less command and control capability at their disposal than their U.S. counterparts; the vehicle was designed from the outset to operate independent of human intervention. The next, partly successful attempt was made on April 26, 1962.

NASA formally defined the program objectives for the Mercury Mark II program in an attachment to its contract with McDonnell Aircraft Company. In it, the spacecraft was to accomplish five specific objectives: perform long-duration flights of up to 14 days, determine the ability of astronauts to function in the space environment, demonstrate rendezvous and docking, develop simplified countdown and launch procedures and make controlled landings on land. The modified Mercury was to incorporate an adapter for orbital equipment that could be left in space when the crew module returned to Earth, rely on ejection seats for emergency crew escape and thus dispense with a launch escape rocket, utilize modular systems outside the pressurized crew compartment for rapid servicing and accomplish program objectives by October 1965.

1961 December 12 The first amateur radio satellite, Oscar (orbital satellite carrying amateur radio), was launched on a Thor-Agena B with *Discoverer XXXVI. Oscar-1* was equipped with a 145-MHz beacon, transmitting "hi" for 18 days before the 11-lb. satellite decayed on January 31, 1962. *Oscar 2* was launched June 2, 1962, operating for 18 days. *Oscar 3* followed March 9, 1965. Weighing 31 lb., it was one of eight satellites launched by a Thor-Agena D and had a transponder enabling 98 radio "hams" to send 176 messages to each other. *Oscar 4,* launched December 21, 1965, and equipped with a 432/145-MHz transponder, brought "hams" in the United States and the Soviet Union together for the first time. Amsat was formed as an international body to support "hams," and *Oscar 5,* built in Australia and launched in 1970, stemmed from this organization. Launched between 1972 and 1978, *Oscars 6, 7* and *8* operated for several years.

1961 December 15 The Boeing Co. received a NASA contract to develop the S-IB stage for advanced Saturn launch vehicles of the C-4 and C-5 class. The stage was 30 ft. in diameter when scaled to the C-4, with four F-1 engines, or 33 ft. in diameter to accommodate five F-1s in the C-5 configuration. The C-4 was canceled because the C-5 more closely matched requirements for the lunar landing mission, and Boeing's work switched to the larger-stage variant with the considerably beefier thrust structure to which its five F-1s were attached.

1961 December 19 The French government established the *Centre National d'Etudes Spatiales* (CNES), a national space agency, to define objectives and lay a framework for space research and the development of rockets and space transportation systems. With headquarters at Brétigny outside Paris, Prof. Pierre Auger was appointed the first president of the new agency, but in 1962 he moved to become director general of the European Space Research Organization (ESRO) and was replaced at CNES by Maurice Levy. The first director general of CNES was Gen. Robert Aubinière, replaced by Michel Bignier. In 1973 CNES established its main research and testing facilities as the *Centre Spatial de Toulouse* in southwestern France.

1961 December 20 United Nations General Assembly Resolution 1721 (XVI) required all spacefaring parties to register the launch of all satellites and spacecraft and furnish data about their orbital parameters. It left open the question of where space starts and stops and permitted the use of space for reconnaissance purposes, a proviso which the Soviet Union protested when, in June 1962, it began its fruitless attempts to get them banned, claiming: "The use of artificial satellites for the collection of intelligence information in the territory of foreign states is incompatible with the objectives of mankind in its conquest of space."

An Atlas F rocket carrying a Rhesus monkey named Scatback was lost at sea during a malfunction shortly after launch from Cape Canaveral, Fla. This was the last U.S. life-science mission involving live animals carried in special canisters launched on ballistic, suborbital trajectories in a program that began June 18, 1948, when the first Project Albert V-2 launch failed. In all, the United States had launched 20 life-science animal missions, of which 13 had been failures, resulting in the deaths of 26 animals (8 monkeys and 18 mice).

1961 December 21 A newly formed NASA Manned Space Flight Management Council tasked with identifying and resolving problems with manned flight programs held its first meeting. Comprising members from the Marshall Space Flight Center, the Manned Spacecraft Center and major NASA offices, the council determined the configuration of the proposed Saturn C-5, and on January 9, 1962, NASA announced that this launch vehicle would be the one adopted for the Apollo manned lunar landing program, supporting either Earth-orbit or lunar-orbit rendezvous.

1961 December In Japan, the University of Tokyo opened a proving ground for solid propellant rockets at Noshiro, Akita Prefecture. Previously having operated under a budget provided by the Institute of Industrial Science, the University of Tokyo would expand its interest in the development of solid propellant sounding rockets and form the basis for Japan's first steps toward building a satellite launcher. Research into development of solid propellant rockets in the Mu series began during April 1963.

1961 Late in the year, and stirred to action by an ambitious challenge from the United States, the Soviet government asked the Korolev design bureau to begin development of a launch vehicle with a 100,000-lb. payload capability. It was to be known as the N-1 (not to be confused with the moon landing launcher) and achieve operational status by 1965. Sergei Pavlovich Korolev was also asked to develop the N-2, with a 155,000-lb. payload capability, by 1970. The Cholomei bureau was ordered to develop a version of the Proton launcher, designated UR-500, capable of sending a manned spacecraft around the moon and back. Within six months Korolev had merged the N-1 and the N-2 into a proposal that resulted in a new, much more powerful launch vehicle also given the designation N-1.

1962 January 3 The name Gemini became the official designation of the two-man Mercury Mark II spacecraft. It was suggested by Alex P. Nagy of NASA headquarters because the twin stars Castor and Pollux seemed to characterize the two-man crew and because the astronomical symbol for Gemini (II) permitted concurrency with the Mercury Mark II designation. At this date the first of 12 flights was scheduled for July or early August 1963, with the next 3 flights to be launched at six-week intervals and subsequent flights at two-month intervals. The first rendezvous and docking mission with an Agena was provisionally scheduled for February or May 1964, with the last of the 12 missions in April or May 1965.

1962 January 15 The design of NASA's first planetary spacecraft, Mariner R, was frozen at the Jet Propulsion Laboratory, where engineers were, on a crash basis, putting together two lightweight vehicles for flyby of the planet Venus. Launched by Atlas-Agena B in July and August 1962, they would take advantage of the next launch window for Earth's nearest planetary neighbor. Launch windows are set by the relative positions of Earth and a celestial target to minimize launch energy and maximize the payload-carrying capacity of the launch vehicle. Launch windows to Venus occur at 19-month intervals when the planet is in inferior conjunction, its closest approach to Earth. If Mariner R missed the 1962 window, it would have to wait until early 1964. Three spacecraft were prepared: *Mariner R-1* and *R-2* would be launched and *Mariner R-3* would be used on the ground to troubleshoot problems with the flight vehicles.

1962 January 16 NASA's Marshall Space Flight Center (MSFC) presented a preliminary development plan for a Nova launch vehicle to the Lunar Program Planning Committee. The baseline Nova configuration recommended by MSFC incorporated eight (1.5-million-lb. thrust) F-1 engines in the N-1 first stage, one or two (1–1.5-million-lb. thrust) cryogenic M-1 engines in the second stage and optional third stages, including one or two (200,000-lb. thrust) J-2 engines or a nuclear stage. For this, NASA awarded several study contracts during 1962. Martin, General Dynamics and Boeing conducted studies on large chemical rockets. General Dynamics and Douglas studied chemiconuclear rockets. Aerojet, Space Technology Laboratories, Douglas and RAND studied sea-launched vehicles. General Dynamics, Martin, Douglas and RAND studied advanced chemical rockets with emphasis on land recovery for subsequent reuse. After May 1963 only General Dynamics, Martin and Douglas continued Nova studies.

1962 January 20 A U.S. Air Force Strategic Air Command crew fired a Titan I ICBM from Vandenberg AFB, Calif., for the first time, and the first two squadrons became operational at Lowry AFB, Col., on April 20 and May 10. Operational Titan Is would be equipped with the Mk.4 reentry vehicle containing one 3–4-MT W-38 thermonuclear warhead. The reentry vehicle carried 10 solid propellant decoys which were released after the warhead separated from the second stage of the missile, confusing radar defenses and antiballistic missile missiles. Titan I was deployed with six squadrons, each operating nine missiles in three triple complexes. Each triple complex housed 200 personnel and required the removal of 700,000 cu. yd. of earth, with 96,000 cu. yd. of concrete and 24,000 tons of steel. The 54 operational Titan Is were deployed with two squadrons at Lowry AFB, Colo., and one each at: Beale AFB, Calif.; Larson AFB, Wash.; Ellsworth AFB, S.Dak.; and Mountain Home AFB, Idaho. With a one-missile silo provided as a spare for each squadron, total force deployment was 60 missiles; full activation was completed in 1963.

1962 January 23 Analysis of potential power sources for the two-man Gemini spacecraft was completed by NASA's Manned Spacecraft Center. Options included batteries, solar cells or fuel cells that work on the principle of reverse electrolysis, in which energy is liberated in a chemical reaction producing electricity. Batteries were too heavy for the power and life required and solar cell panels would be too large, cumbersome and work only on the sunlit side of Earth. Fuel cells by General Electric offered weight and power advantages over competitors and used oxygen and hydrogen brought together over a catalyst to produce electrical energy and water as a by-product. The water could be used to supplement the spacecraft's cooling system and so help remove some of the excess heat produced by the fuel cell itself. General Electric received a development contract on March 20, 1962.

1962 January 24 NASA awarded a contract to Aerojet General for the design and development of a large-thrust cryogenic rocket motor for application in the second stage of the proposed Nova super-heavyweight launch vehicle. Designated M-1, the motor would operate on liquid hydrogen–liquid oxygen propellant, which it would consume at the rate of 170,000 lb./sec., and produce a thrust of 1.2 million lb. The motor was to have had an overall height of 23 ft. and a nozzle diameter of 15 ft. The M-1 was being considered as a replacement for the cluster of eight 200,000-lb. thrust cryogenic J-2 engines in the original Nova N-II stage. By mid-1962 it was decided to incorporate a cluster of four M-1s for a total thrust from stage two of 4.8 million lb. The first stage design retained the eight F-1s producing a liftoff thrust of 12 million lb., and an S-IVB was selected as the third stage. In this configuration, Nova could lift a payload of 400,000 lb. into low Earth orbit.

NASA failed in the first attempt to orbit five satellites on one launch vehicle when a Thor-Able Star from Cape Canaveral, Fla., fell back into the Atlantic Ocean. With insufficient thrust from the second stage, the attempt to put a 219-lb. cluster of satellites into space was frustrated. The cluster comprised the spherical 58-lb. solar x-ray satellite, *Greb 4;* the spherical 58-lb. radio-wave-propagation monitor, *Lofti 3;* the cylindrical 8-lb. surveillance research satellite, *Surcal;* the cubical 59-lb. radiation belt monitor, *Injun 2;* and the spherical range-calibration satellite, *Secor.*

1962 January 25 NASA authorized full-scale development of the Saturn C-5 launch vehicle in a configuration carrying five 1.5-million-lb. thrust RP-1–liquid oxygen F-1 engines in the (first) S-IB stage, five 200,000-lb. thrust J-2 engines in the second (S-II) stage and a single J-2 in the (third) S-IVB stage. The primary NASA mission for this vehicle was to support manned lunar landing operations utilizing rendezvous techniques. For a while this launch vehicle was referred to as the Advanced Saturn. The first stage of the C-5 was later redesignated S-IC and the designation S-IB reassigned from that of a rocket stage to the completed, uprated Saturn C-1.

1962 January 26 NASA launched *Ranger 3,* the first of the Block II spacecraft, on an Atlas-Agena B from Cape Canaveral, Fla. The 727-lb. Ranger included a hardened capsule weighing 328 lb. and resting on top of a 5,080-lb. thrust retro-rocket. Released as *Ranger 3* approached the moon, the spherical capsule contained instruments protected by balsa-wood shock absorbers. Fired at an altitude of 52,000 ft., the retro-rocket was to have burned for 10 sec., slowing the capsule from 6,000 MPH to zero just 1,100 ft. above the surface. One-sixth that of Earth's, lunar gravity would accelerate the capsule to a speed of 150 MPH at impact, allowing seismometers to survive and send reports of moonquakes. A TV system on the main spacecraft was to have returned images from a distance of 2,500 mi. to the point of impact. *Ranger 3* missed the moon by 22,862 mi. and went into orbit around the sun at approximately the radius of the Earth's orbit.

1962 January 29 NASA announced selection of three industrial contractors to prepare final proposals for the RIFT (reactor-in-flight test) upper stage that would carry a NERVA nuclear engine. General Dynamics, Lockheed and Martin

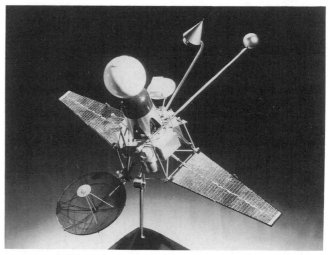

A scale model of the Ranger 3 spacecraft, built by the Jet Propulsion Laboratory, in the cruise configuration it would assume after launch on January 26, 1962.

Marietta prepared detailed proposals. On February 19 NASA and the Atomic Energy Commission assigned Jackass Flats, which had served as the test site for Kiwi reactor runs since 1959, as the development center. On May 17 NASA selected Lockheed for RIFT-stage fabrication and testing.

1962 February 7 In a special message to Congress, President Kennedy proposed the creation of a "Communications Satellite Corporation" to be financed through the sale of stock to communications companies and the general public. Comsat Corp. would have responsibility to develop, own and operate communications satellites. In time it would become the U.S. signatory for the International Telecommunications Satellite Consortium (Intelsat). The House of Representatives passed the Communications Satellite Act of 1962 by a vote of 354 to 9 on May 3.

1962 February 9 The U.S. Air Force issued Advanced Development Objective-40 (ADO-40), calling for the "demonstration of the technical feasibility of developing a nonorbital collision-course satellite interceptor system capable of destroying satellites in an early time period." Issued in response to a directive from President Kennedy to the Department of Defense for an antisatellite system "at the earliest practicable time," ADO-40 resulted in a development plan by Air Force Systems Command (AFSC) on June 30, 1962. AFSC came up with a test program based on the Thor LV-2D missile, and on February 8, 1963, the test series was designated Program 437.

1962 February 16 The U.S. Air Force awarded a contract to Martin Marietta for a study of the design criteria involved in a Titan III standardized launch vehicle comprising both liquid and solid propellant stages. The core vehicle would be the two-stage Titan II with storable propellants, to which would be attached two solid propellant boosters with a combined thrust of more than 2 million lb. Titan III was to be the launch vehicle for the X-20 and for heavy unmanned military satellites. After the X-20 was canceled, the Titan III–series launch vehicles continued in development and became the workhorses for medium-to-heavy satellites and spacecraft. Titan derivatives would remain the standardized heavyweight expendable launch vehicles for the Air Force until the end of the century at least.

NASA's Associate Administrator Dr. Robert C. Seamans described in a letter to Deputy Director for Defense Research and Engineering John H. Rubel interest in very large solid propellant motors with thrust levels of between "2.5 million and 5 million pounds and burning times of approximately 115 sec." Dr. Seamans pointed out that solid propellant motors at the higher thrust level would probably have a length of 100 ft. and a diameter of 20 ft. "with greater potential for reliability" due to the smaller number of motors required to achieve very high first-stage thrust levels of super-heavyweight launchers. The next day the largest solid propellant motor fired to date was tested when Aerojet General demonstrated 600,000 lb. of thrust from a 53-ft.-tall solid 8 ft. 4 in. in diameter that burned for 98 sec.

1962 February 18   NASA announced that a reentry test program called Project Fire would be set up and managed by the Langley Research Center in efforts to obtain detailed information about the effects of kinetic friction on selected materials in contact with the atmosphere. Velocities approximating those which would be reached by spacecraft reentering Earth's atmosphere from the moon would be achieved after launch on an Atlas D. An Antares solid propellant motor, also used as the third stage of the Scout launcher, attached to a Project Fire reentry body would boost it to about 24,800 MPH. Ground stations would track the Project Fire body and determine the effects of reentry heating on propagation of radio signals and heat insulation materials as well as provide data on heat load and heat rate.

1962 February 20   The first U.S. manned orbital spaceflight was successfully accomplished when an Atlas launch vehicle placed a Mercury spacecraft carrying astronaut John Glenn in an orbit from which it was recovered three revolutions later. Liftoff of MA-6 from Pad 14 at Cape Canaveral, Fla., occurred at 15:47:39 UT, with booster engine cutoff at 2 min. 9.6 sec., escape tower jettison at 2 min. 33.4 sec. and sustainer engine cutoff at 5 min. 1.4 sec. and a velocity of 25,730 ft./sec. The 100 × 162.2 mi. initial orbit had a period of 1 hr. 28 min. 29 sec. and an inclination of 32.5°. The 2,988-lb. spacecraft (no. 13), named *Friendship 7,* separated from the

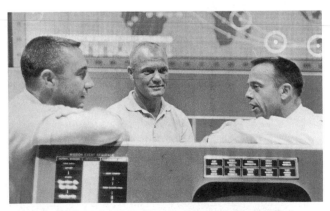

The first three American space veterans chat casually against a console at the Mercury control center, Cape Canaveral, Florida. Left to right: Capt. Virgil I. Grissom (USAF), Col. John H. Glenn Jr. (USMC) and Lt. Comdr. Alan B. Shepard Jr. (USN).

Atlas 2 sec. after shutdown. Activities went as planned, but on the third orbit the astronaut was told to leave the retropackage on during reentry. He was not told telemetry indicated (erroneously) that the landing bag had prematurely deployed and could drift loose if the straps holding the pack on were not retained. Retrofire began at 4 hr. 33 min. 8 sec., and as *Friendship 7* returned through the atmosphere 10 min. later, large chunks of the retropackage were observed by the pilot hurtling past his window. The spacecraft splashed down at 4 hr. 55 min. 23 sec.

Soviet Premier Khrushchev sent a letter to President Kennedy congratulating the American people on the first U.S. manned orbital spaceflight and offering cooperation in the exploration of space. Kennedy responded positively and pursued the offer through a letter to Khrushchev dated March 7, 1962, suggesting a worldwide weather satellite system, an exchange of tracking and data services, a joint endeavor to map the Earth's magnetic field and a cooperative project to develop international telecommunications via satellite. Khrushchev replied on March 20 with a proposal to set up an international rescue service for cosmonauts. Khrushchev also offered talks at an informal level, and Kennedy asked NASA Deputy Administrator Dryden to meet with academician Anatoli A. Blagonravov, the chairman of the Commission on Exploration and Use of Space at the Soviet Academy of Sciences. The first meeting took place in New York on March 27, 1962.

1962 February 21   NASA Administrator James Webb wrote to the president of the American Telephone and Telegraph Co. to ask Bell Systems for assistance by providing "experienced men capable of giving NASA the benefit of advanced analytical procedures . . . required for the successful execution of [the] manned spaceflight mission." Concerned that government salaries would limit the talent NASA could recruit, Webb sought outside assistance with the management expertise essential to success in the manned moon landing program. AT&T responded by setting up Bellcomm, which closely examined Apollo administration and planning.

1962 February 23   To reflect more clearly the experimental nature of the U.S. Air Force Titan–launched boost-glide vehicle, the project designation X-20 was to be substituted for the name Dyna-Soar on the orders of Secretary of Defense McNamara. Other designations had been considered, including XJN-1 and XMS-1. The Department of Defense complied on June 26, firmly categorizing it from this date as a research project within the "X" series initiated by the Bell X-1. A three-day X-20 symposium was held at Wright Field beginning November 5, 1962, and on December 19 the Air Force Systems Command ordered a review of all aspects of the program.

Twelve European nations agreed on the draft of an agreement creating the European Space Research Organization (ESRO). They were: Britain, France, West Germany, Italy, Belgium, Denmark, Norway, Sweden, the Netherlands, Switzerland, Austria and Spain. ESRO would cooperate with NASA and the International Committee on Space Research (COSPAR), of which the Soviet Union was a member. On June 14 delegates from 10 of the ESRO countries signed the convention establishing the organization; Denmark and Norway signed later. Prof. Pierre V. Auger of France was named director general. Over eight years, ESRO would spend $300 million on space experiments, including the first ESRO satellite in 1967.

1962 February 24 McDonnell Aircraft awarded a subcontract to North American Aviation for Gemini spacecraft liquid propellant orbit attitude and maneuver systems. Gemini would carry eight 23.5-lb.-thrust attitude control thrusters located in pairs at 90° intervals around the circular (10 ft. in diameter) base of the spacecraft's adapter section. Four 94.5-lb. orbit maneuvering thrusters facing radially outward were located at 90° intervals around the adapter section where it was attached to the reentry module. Two 94.5-lb. thrusters attached to the base of the adapter section at 180° intervals faced aft and two 79-lb. thrusters at the forward part of the adapter faced forward. The eight orbit maneuver thrusters allowed the spacecraft to move up, down, left, right, back or forward. All sixteen orbit attitude and maneuver system (OAMS) thrusters would be jettisoned with the adapter section at the end of the mission. For attitude control of the reentry module as it began its descent, sixteen 23.5-lb. thrusters in two redundant sets were clustered around the nose of the spacecraft.

1962 February 27 *Discoverer XXXVIII,* the last in the program, was launched by a Thor-Agena B from Vandenberg AFB, Calif., to a 129 × 212 mi., 82.2° orbit carrying a new imaging package designated KH-4. With this, a dual camera system, stereo pairs could be obtained to give vital height information for photo-interpreters analyzing the film after recovery. Vertical information was a major step forward in reconnaissance satellite data. On March 21, after 65 orbits of the Earth, the recovery capsule was separated from the satellite and retrieved in a midair snatch. Subsequent military reconnaissance satellites of this type would be designated under the Corona program. The last KH-4 satellite system was launched on December 21, 1963, by which date it had been replaced by the KH-4A.

1962 February Five female cosmonauts were selected in the Soviet Union, the second cosmonaut selection since the first group of twenty between January and February 1960. They were: Tatiana Dmitryevna Kuznetsova, Valantina Leonldovna Ponomareva, Irina Bayonovna Solovyeva and Valentina Vladimirovna Tereshkova and Zhanna Dmitryevna Yorkina. Only Tereshkova made a spaceflight.

1962 March 1 In what was designated by Mayor Robert Wagner "John Glenn Day," the city of New York hosted America's first spaceman to orbit the Earth and awarded John Glenn and STG (Space Task Group) Director Robert Gilruth the city's Medal of Honor. An estimated 4 million people lined the streets, giving the motorcade a ticker-tape welcome characteristic of the city. The next day the Mercury astronauts were guests of the United Nations and Glenn acted as spokesman during a reception hosted by Acting Secretary General U. Thant. Meanwhile, astronauts Scott Carpenter and Walter Schirra were training as prime and backup, respectively, for the three-orbit MA-7 flight. Astronaut Slayton had originally been selected as the prime for MA-7, but a minor erratic heart rate which had showed up in some tests caused him to be disqualified, and this was publicly announced on March 15.

1962 March 6 NASA selected Martin Marietta to build a radioisotope-thermoelectric generator (RTG) to provide electrical energy for the Surveyor soft-landing lunar spacecraft during its operating period on the moon. Power requirements for Surveyor indicated a need for RTGs, which would have the added advantage of providing electrical power during the two weeks of lunar night. Batteries alone would provide insufficient power and endurance, and solar cells would be unable to produce electrical energy unless sunlight fell directly on the panels. Nevertheless, RTGs were withdrawn from the Surveyor program and the spacecraft adopted a hinged solar cell panel capable of tracking the motion of the sun across the lunar sky.

1962 March 7 First of a series of Orbiting Solar Observatories, NASA's *OSO-1* was launched by Delta from Cape Canaveral, Fla., to a 343 × 370 mi. orbit inclined 32.9°. The OSO program would last 16 years and provide detailed information about the sun and its behavior. The 458-lb. *OSO-1* comprised a nine-sided spinning base with a de-spun fan-shaped top section presenting a total height of 3 ft. 1 in. The 3 ft. 8 in. spinning section carried three fiberglass balls containing pressurized nitrogen on arms 7 ft. 8 in. across. *OSO-1* measured electromagnetic radiation in ultraviolet, x-ray and gamma-ray regions until August 6, 1963. All built by Ball Brothers, the other first-generation OSOs were almost identical to *OSO-1: OSO-2* (545 lb.), launched February 3, 1965, returned data until November 1965; *OSO-C,* lost in a launch failure August 25, 1965; *OSO-3* (620 lb.), launched March 8, 1967; *OSO-4* (597 lb.), launched October 18, 1967, returned the first picture of the sun in ultraviolet; *OSO-5* (641 lb.), launched January 22, 1969; *OSO-6* (640 lb.), launched August 9, 1969. *OSO 7* and *8* were of different, later design and are entered separately under September 29, 1971, and June 21, 1975.

In testimony before a subcommittee of the House Committee on Science and Astronautics, NASA provided a detailed description of one concept for a 7.5-million-lb. thrust Saturn C-5 utilizing a NERVA nuclear engine as the third stage. The Earth-escape payload capability of the complete Saturn C-5/NERVA launch vehicle was calculated to be double that of the conventional Saturn C-5 with its three chemical stages and exceed the lift capability of the proposed 12-million-lb. thrust Saturn C-8 and Nova launchers.

1962 March 14 In the first selection of its kind, the U.S. Air Force announced the names of six candidate astronauts chosen for general manned military space missions without affiliation to a specific program. They were: Neil A. Armstrong (NASA); Capt. William J. Knight (USAF); Milton L. Thompson (NASA); Capt. Henry C. Gordon (USAF); Capt. Russell L. Rogers (USAF); and Maj. James W. Wood (USAF). On September 17, 1962, Armstrong transferred to NASA as an astronaut. On September 19, 1962, Gordon, Knight, Rogers, Thompson and Wood were reassigned to the Dyna-Soar program.

1962 March 16 The first Titan II ICBM made its first flight from Cape Canaveral, Fla. A considerably reengineered derivative of the Titan I, it had a first stage lengthened by 7 ft. and a second stage of equal diameter (10 ft.) to that of the first stage. The modified rocket motors were Aerojet LR87 derivatives of similar Titan I engines utilizing hypergolic propellants which ignited on contact: a fifty-fifty mixture of UDMH and hydrazine (Aerozene 50) as fuel and nitrogen tetroxide as oxidizer. The new engines eliminated 134 of the 245 moving parts in the original LR87s and reduced the number of power control operations from 107 to just 21, simplifying the ignition process. Moreover, the hypergolic propellants had a specific impulse of 280 sec. versus 262 sec. Sea-level thrust was 430,000 lb. in the first stage and 100,000 lb. from the uprated LR91 engine, essentially a scaled down LR87, in the second stage. Titan II had a length of 103 ft., a launch weight of 330,000 lb. with a throw weight of 7,500 lb.

A Soviet SL-7 launch vehicle carried the 695-lb. scientific satellite *Cosmos 1* from Kapustin Yar launch complex to an 84 × 127 mi., 49° orbit. So began a continuous series of satellites launched under the Cosmos designation and embracing scientific, research, applications and military duties. Just as the United States had placed critical national security satellites under a veil of secrecy, so too was the Soviet Union moving many of its forthcoming military satellites under the anonymity of the cover name Cosmos. *Cosmos 1* decayed back through the atmosphere nine days later.

1962 March 19   McDonnell Aircraft Co. awarded a Gemini spacecraft retro-rocket development contract to Thiokol Chemical Corp. As designed, four 2,500-lb. thrust solid propellant motors would be located at the intersecting corners of a crossbeam assembly in the forward part of the adapter. When the rear, equipment section of the adapter was jettisoned prior to retrofire, the forward section would be exposed. The motors would fire sequentially for 5.5 sec. each or, during an abort, in one 10,000-lb. thrust salvo. Mercury spacecraft were placed in an orbit that would decay back through the atmosphere within the time the life-support system could keep the astronaut alive, thus affording a "backup" should the solids fail. Gemini would use orbits that would not decay quickly but could use the orbit maneuver thrusters if the retro-rockets failed.

1962 March 23   Reacting to a recent spate of publicity from the U.S. Air Force for its expanding space activities, President Kennedy issued a directive through the Department of Defense prohibiting any U.S. armed service from announcing in advance the launch of military satellites. It also prohibited the use of military satellite names and the identification of payloads on launch vehicles. Conforming to the United Nations resolution on the matter, the United States would continue to announce all satellites when launched.

1962 March 27   NASA Deputy Administrator Hugh Dryden met with Soviet academician Anatoli Blagonravov in New York to discuss potential areas of space cooperation between the United States and the Soviet Union. Although enthusiastic at a personal level, the Soviet view was that current projects were too far along to introduce joint activity but they did not rule out future cooperation on new projects. There were favorable reactions in the United States and the Soviet Union, and a second meeting took place in Geneva on May 27, 1962, at which both agreed to cooperate in a worldwide weather program, a study of the Earth's magnetic field and experiments in communications using the Echo satellite. This activity was formalized in an exchange of letters on October 18 and 30, 1962.

1962 March 28   Gen. Curtis LeMay, U.S. Air Force chief of staff, predicted that "beam-directed energy weapons would be able to transmit energy across space with the speed of light and bring about the technological disarmament of nuclear weapons." Uttered only two years after the first working laser had been demonstrated, his prophetic words would underpin a sustained, but low-level effort within the Air Force defined in Project Blackeye under what were described as Directed Energy Weapon (DEW) studies. They would emerge publicly

21 years later in the Strategic Defense Initiative (SDI) announced by President Reagan on March 23, 1983, dubbed "Star Wars" by the media.

1962 March 31   NASA formally approved the final design configuration of the Gemini spacecraft, as defined and submitted by the primary contractor, McDonnell Aircraft Company. The spacecraft would consist of a pressurized crew compartment for two astronauts side-by-side in a reentry module with a maximum length of 12 ft. and a maximum diameter of 7 ft. 6 in. The forward section was 5 ft. 5 in. in length and comprised a cylinder attached to the conical rear section containing the crew compartment. The base of the reentry module was attached to the 7 ft. 6 in. adapter section with a maximum diameter of 10 ft., the same as the Titan launcher. The equipment section of the adapter would contain all the systems for orbital activity. Gemini had a length of 18 ft. 10 in.

1962 March   Soviet designer Sergei Pavlovich Korolev proposed a stretched version of the single-seat Vostok spacecraft, called Sever and capable of carrying two crew members on a circumlunar flight. Sever would have rendezvoused and docked to three upper stages, each launched by a separate SL-3. When the three stages had been docked in tandem by the *Vostok 7* cosmonaut, the Sever spacecraft would be docked to the opposite end of the rocket stages. Vostok would then undock and return to Earth, leaving the assembled orbital complex and the manned Sever to be fired toward its objective. Korolev wanted to use existing technologies for a manned circumlunar flight, but the limitations of the basic spacecraft were too great. Sever formed the basis for the Soyuz spacecraft.

1962 April 2   A meeting to review the proposed lunar-orbit rendezvous (LOR) mode for Apollo moon landing operations was held at NASA Headquarters. Involving representatives from most NASA field centers, the meeting heard that one Saturn C-5 could send to the moon an Apollo command-service module and lunar excursion vehicle (LEV) combination capable of landing two men on the moon and, after a day's stay on the surface, returning them to the orbiting Apollo, from where they would be returned to Earth. On April 24, 1962, NASA engineer Milton Rosen recommended adopting the Saturn C-5/LOR mode. On May 25, 1962, NASA's director of Manned Space Flight, D. Brainerd Holmes, asked for agency-wide submissions on costs and schedules for direct-ascent, Earth-orbit rendezvous and LOR modes. For direct ascent a Saturn C-8 with eight F-1 engines in the first stage, eight J-2 engines in the second stage and one J-2 engine in the third stage was projected. The Marshall Space Flight Center proposed a Nova with the same respective first and third stages but a second stage with two M-1 engines.

1962 April 6   In an air-launched test of a potential antisatellite weapon, a U.S. Navy McDonnell F-4 Phantom successfully dropped a Caleb rocket at high speed and an altitude of approximately 38,000 ft. The two-stage rocket was a scaled-down version of the four-stage solid propellant Caleb designed as satellite launcher. With a length of 16 ft. 6 in., a diameter of 2 ft. and a weight of 3,000 lb., it could lift a 10–20-lb. payload into low Earth orbit. Managed by the Naval Research Laboratory, the flights took place from the Pacific Missile Range, Calif. A second flight on July 25, 1962, enabled the two-stage rocket to reach a height of 1,000 mi. Both demonstrated that Caleb-sized missiles could be launched on satellite intercept missions, but no target was identified and there was no attempt to provide in-flight tracking.

1962 April 9 The Reuters news agency reported that the Tower of London, England, had been asked by Air Research Manufacturing Co. of Los Angeles, Calif., to provide details of articulated joints on the suit of armor worn by King Henry VIII. The company wanted to know if there were any innovative mechanical techniques employed by armorers of old that might prove useful in the design of space suit limb joints.

McDonnell Aircraft Co. awarded a Gemini spacecraft ejection-seat development contract to the Weber Aircraft division of Walter Kidde and Co. Unlike Mercury, the relatively low g-levels of Gemini missions precluded the need for contour couches. Crew escape during powered ascent was planned to be through the use of individual ejection seats to an altitude of 70,000 ft. (13 mi.) or by firing the retro-rockets to lift the reentry module away from the Titan launch vehicle above it. In reality, lack of confidence in crew survival during the worst possible Titan failure restricted use of the ejection seats to a maximum altitude of 15,000 ft. (2.8 mi.) and reliance on the retro-rocket escape method thereafter to an altitude of 522,000 ft. (99 mi.). Nevertheless, a special ejection rocket was built by Rocket Power under a contract awarded May 15, 1962.

1962 April 19 The first air-drop test of a two-stage GAM-87A Skybolt solid propellant air-to-surface missile fell short of its objective when the second stage failed to ignite after being released from the underwing pylon of a B-52 bomber. The next three flight tests also failed to achieve their objectives. A full-range flight was, however, demonstrated on December 19, 1962. With a length of 38 ft. and a launch weight of 11,300 lb., the missile had a design range of 1,000 mi. carrying a 600–700-lb. warhead or 600 mi. carrying a 1,600-lb. warhead. The warhead selected was the 2-MT W-59, weighing 565 lb. in a package 4 ft. in length and 2 ft. in diameter. On December 7, 1962, President Kennedy confirmed that he intended to cancel Skybolt and informed the British the following day. Britain had committed to buying Skybolt for its V-bomber force, each Vulcan carrying two missiles.

NASA awarded a contract to IBM for the world's first digital computer to be fitted to a manned spacecraft, the two-man Gemini. Measuring 1 ft. 2.5 in. × 12.75 in. × 1 ft. 6.9 in. and weighing just 59 lb., the unit included an auxiliary tape memory that increased memory storage capacity to 1.17 megabytes. To achieve the necessary reliability, IBM increased error rates from 1 bit in 100,000 to 1 bit in 1,000,000,000, considerably advancing the state of the art. The tape memory was first carried on the *Gemini VIII* spacecraft.

NASA announced plans to put John Glenn's Mercury spacecraft no. 13 on a world tour of 20 places on every populated continent. Known as the "fourth orbit of *Friendship 7*," the publicity tour was organized by the United States Information Agency, to whom the spacecraft was on loan. Accompanied by William Bland of the Mercury Project Office, it would serve to publicize not only the feat itself but also the broader objectives of the U.S. civilian space program popularized within the ambitions of the Apollo moon-landing venture.

1962 April 20 The first U.S. Air Force Strategic Air Command Titan I ICBM squadron became operational with its nine missiles at Lowry AFB, Colo., followed by a second squadron at Lowry on May 10. Like Atlas, Titan I was compromised in operational efficiency by the use of liquid oxygen and kerosene propellants that could only be loaded shortly before launch. Full Titan I deployment of all six squadrons of nine missiles each was completed in early 1963. The other Titan I squadrons were based at Beale AFB, Calif.; Larson AFB, Wash.; Ellsworth AFB, S.Dak.; and Mountain Home AFB, Idaho.

The Air Force announced the names of eight candidate astronauts, the second group selected for general military space duties on future programs. They were: Capt. Charles C. Bock Jr. (USAF); Capt. Albert H. Crews (USAF); Lt. Comdr. Lloyd N. Hoover (USN); Maj. Byron F. Knolle (USAF); Capt. Robert H. McIntosh (USAF); Capt. Robert W. Smith (USAF); Maj. Donald L. Sorlie (USAF); and Capt. William T. Twinting (USAF). Capt. Crews would be transferred to the Dyna-Soar program on September 19, 1962.

1962 April 23 Launched from Cape Canaveral, Fla., with the identical flight objectives of its predecessor, the 730-lb. moon probe *Ranger 4* was placed on a highly accurate course for lunar impact by its Atlas-Agena B launch vehicle. Shortly after injection, communication through the high-gain antenna was rendered impossible by a fault in the spacecraft timer controlling a preprogrammed sequence that should have prepared the spacecraft for ground commands. *Ranger 4* was observed to be tumbling and unable to achieve attitude control, and the spacecraft impacted the far side of the moon at an elapsed time of about 64 hr. and a speed of 5,963 MPH. This was the second of five Ranger Block II spacecraft and the fourth consecutive failure for the program.

1962 April 25 The second Block I Saturn C-1 launch vehicle (SA-2) was launched from Cape Canaveral at 09:00 UT. With a liftoff weight of 927,000 lb., SA-2 had a thrust of 1.32 million lb. from its eight H-1 engines. The first stage carried a partial, 620,000-lb. load of propellant and fired for 1 min. 55 sec. to reach a speed of about 5,500 ft./sec. Controllers sent a destruct signal to the vehicle at 2 min. 43 sec. when SA-2 was at a height of 65 mi., releasing 190,000 lb. (22,900 gal.) of water ballast in dummy second and third stages. Known as Project High Water I, this experiment helped physicists understand the constitution of the ionosphere by upsetting the water content. The vapor cloud formed into an ice cloud 4.6 mi. across and ascended to a maximum height of 90 mi.

1962 April 26 Designated *Cosmos 4,* the first partly successful Soviet reconnaissance satellite was launched by SL-3 from the Baikonur cosmodrome to a 177 × 197 mi. orbit inclined 65°. First-generation, low-resolution photographic satellites called Zenit were adapted Vostok spacecraft equipped with four cameras in place of an ejection seat or human occupant. The film was recovered from the inside of the descent module when it returned to Earth on April 29, but there had been problems with the attitude control system and the cameras. Not before *Cosmos 7,* launched on July 28, 1962, and recovered August 1, 1962, was the first totally successful Soviet reconnaissance satellite flown. The Soviets launched 41 photoreconnaissance satellites of the first-generation class: 5 in 1962; 6 in 1963; 9 in 1964; 7 in 1965; 9 in 1966; and 5 in 1967. The last of its type, *Cosmos 157,* was launched on May 12, 1967. Almost all weighed about 10,150 lb. and had a lifetime of eight days.

Postponed from April 10 because of a technical problem with the launcher, the Anglo-American scientific satellite *Ariel 1* was launched from Cape Canaveral, Fla. and placed in a 242 × 754 mi., 53.9° orbit by Delta. Named after the airy spirit in

Shakespeare's *The Tempest*, the 132-lb. satellite (designated S.51) was built at NASA's Goddard Space Flight Center and equipped with British instruments for studying the ionosphere. It comprised a fat cylinder, 1 ft. 11 in. in diameter and 1 ft. 8 in. tall, with four solar paddles and two 4-ft. instrument booms extending from the lower section. *Ariel 1* provided data until November 1964 and decayed back into the atmosphere on May 24, 1976.

1962 May 1　　The U.S. Air Force Space Systems Division awarded a contract to Lockheed for development of the standard Agena D restartable upper stage into a Gemini Agena Target Vehicle (GATV) that would be placed in orbit by Atlas launchers with which Gemini could rendezvous and dock. Modifications over and above the standard Agena, with its 16,000-lb. thrust Model 8247 Bell engine, included a secondary propulsion system, or SPS. This included two attached modules, one on either side of the primary propulsion system, each carrying a 200-lb. thrust rocket motor capable of at least 20 firings and a 16-lb. thrust motor capable of firing 90 times. At the opposite end, a conical target docking adapter (TDA) with associated visual cue lights would serve as a receptacle into which the nose of the Gemini spacecraft would be inserted for docking practice.

1962 May 6　　For the first and only time, a missile from the U.S. strategic arsenal was launched on test and its live nuclear warhead detonated in simulation of a wartime firing under the code name Operation Frigate-Bird. The event took place in the Pacific Ocean when the Fleet Ballistic Missile submarine USS *Ethan Allen* launched a Polaris A2 SLBM (submarine-launched ballistic missile) to its full range of 1,500–1,700 mi. and detonated the 600-KT warhead in a 10,000–15,000 ft. air burst near Christmas Island. The shot was part of the Dominic series of nuclear tests involving 36 nuclear explosions between April 25 and November 4, 1962, of which 29 were from bombs dropped by B-52 bombers. These were the last U.S. nuclear tests before the 1963 atmospheric test ban.

1962 May 8　　The first of 31 new and enlarged fleet ballistic missile submarines, the USS *Lafayette,* was launched from the Groton, Conn., slipway of the General Dynamics' Electric Boat Division and named by Jackie Kennedy, wife of the President. With a length of 425 ft. and a submerged displacement of 8,250 tons, the USS *Lafayette* began deterrent patrol duty on January 4, 1964. The USS *Lafayette* had a crew of 147 (14 officers and 133 enlisted men) and provision for 16 Polaris A3, Poseidon C3 or Trident C4 SLBMs (submarine-launched ballistic missiles). The last *Lafayette*-class boat, the USS *Will Rogers,* was launched on July 21, 1966, and began its first operational patrol on October 3, 1967, with 16 Polaris A3 missiles. Between February 1969 and March 1975, all 31 boats were retrofitted with the Poseidon C3. Between September 1978 and December 1982, 12 *Lafayette*-class boats were retrofitted with the Trident C4.

NASA launched the first of 10 scheduled Atlas-Centaur development flights 17 months after the original scheduled date. The Centaur had a length of 28 ft. 6 in. (42 ft. with payload shroud), a diameter of 10 ft. and two 15,000-lb. thrust cryogenic RL-10 engines. Atlas-Centaur had a total height of 105 ft., a continuous diameter of 10 ft. and a launch mass of about 300,000 lb. It could put 8,500 lb. in low Earth orbit, send 2,300 lb. to the moon, or 1,300 lb. to Mars or Venus. For this first shot the vehicle was instrumented with about 540 telemetry channels, 384 of which were on the Centaur. The flight plan called for the Atlas to throw the

Centaur and its payload shroud 1,175 mi. downrange and to a maximum altitude of 300 mi. Loaded to 40% capacity with liquid hydrogen–liquid oxygen propellant, the Centaur was required to fire for 25 sec. about 13 min. into the flight, 8 ½ min. after Atlas cutoff. All went well until the weather shield on the Centaur came off and the hydrogen tank ruptured, causing an explosion at 55 sec. A thorough program review took place and in September 1962, when the Marshall Space Flight Center proposed canceling the vehicle, Centaur development was moved to the NASA Lewis Research Center.

1962 May 10　　A Thor-Able Star was launched from Cape Canaveral, Fla., carrying the first of two geodetic research satellites, *Anna 1A*. The 7,900-lb. thrust Able Star failed to ignite, preventing *Anna 1A* from reaching orbit. A joint venture between the U.S. Army, Navy, NASA and Air Force (the acronym of which formed its name), Project Anna sought to achieve a celestial triangulation point for accurate distance determination on Earth. The satellite was an aluminum sphere 3 ft. in diameter, with a band of solar cells extending the midsection diameter to 4 ft. The 355-lb. *Anna 1A* carried two sets of xenon lights designed to produce five light flashes twenty times a day. Range rate measurements were to be made using the 162–324 MHz transmitters.

In the Lunar and Planetary Sciences division of NASA's Office of Space Sciences, concerns surfaced about the availability of the Centaur upper stage for launching planetary spacecraft to Mars in 1964. A flight test of *Mariner B* scheduled for 1963 was canceled and a *Mariner B* flight to Venus planned for 1964 was deferred until 1965. Launch windows to Mars occur at 25-month intervals. *Mariner B* could not be sent to Mars in November 1964 as planned, which meant a delay of more than two years since the next window would not come until December 1966. In concert with the Jet Propulsion Laboratory, NASA developed a plan to launch a lightweight spacecraft to Mars called *Mariner C*. Preliminary design studies around the Atlas-Agena B began, but the successful flight of an Agena D on June 28, 1962, encouraged use of that stage, with its capacity to launch a 575-lb. spacecraft to Mars versus Agena B's 495 lb.

1962 May 11　　General Dynamics Convair Division received a NASA contract for design and development of the Little Joe II, a rocket which would be used to test the Apollo spacecraft on unmanned suborbital flights. The booster would comprise a cluster of seven Algol 1-D solid propellant rockets generating a maximum liftoff thrust of 311,000 lb. Tests would take place from the Army White Sands Missile Range in the New Mexico desert north of El Paso, Texas, testing the Apollo spacecraft design for maximum dynamic pressure and for abort using the launch escape rocket.

1962 May 15　　The first successful photoreconnaissance mission using the KH-5 geodetic mapping camera began with the launch of a Corona-class satellite from Vandenberg AFB, Calif. Placed in a 189.5 × 394 mi., 82.3° orbit by a Thor-Agena B, the 3,300-lb. satellite remained in space for just over three days before releasing its film capsule for recovery. Conceived as a means of providing coordinates on the location of military targets in the USSR, the photographs allowed analysts to improve target location accuracy from several miles to around 1,000 ft. and, by the end of the 1960s, to within 450 ft. Prelaunch target accuracy was vital for missile guidance systems programmed with latitude and longitude. Unlike bombers, the missiles could not visually identify the target and adjust for position errors. The last of seven successful KH-5 mapping camera flights out of twelve attempts took place with a Corona satellite launch in August 1964.

1962 May 18 McDonnell Aircraft Corp. awarded Northrop Ventura a development contract for the Gemini spacecraft parachute landing system. At this date only the first Gemini mission was expected to land by conventional parachute, subsequent flights using the inflatable paraglider for gliding descent to a land landing. As development evolved, the paraglider system was dropped and the parachute system was the only method of safe recovery used operationally. After changes over the next year, the design eventually adopted incorporated a drogue parachute 8 ft. 4 in. in diameter and manually released below 50,000 ft., an 18 ft. 4 in. ringsail pilot parachute released at 10,000 ft. and an 84 ft. 2 in. ringsail main parachute. Seconds after final parachute deployment the spacecraft is tipped forward on a two-point suspension. This allowed the reentry module to slice into the water and dissipate the shock of impact.

1962 May 23 NASA's Ames Research Center began wind tunnel tests of a 50% scale model of the paraglider being developed for the Gemini spacecraft. The purpose of these tests, which ended July 25, 1962, was to evaluate loads and obtain aerodynamic data on deployment, inflation and gliding characteristics. North American Aviation began a series of drop tests to integrate the parachute and paraglider recovery systems. As envisaged at this date, the plan called for the forward section of the reentry module to be jettisoned at 60,000 ft., followed by deployment and inflation of the parawing for a controlled glide. After landing, the inflatable wing would be immediately jettisoned to prevent it toppling the spacecraft.

1962 May 24 The second U.S. manned orbital spaceflight was accomplished when astronaut Malcolm Scott Carpenter piloted the Mercury spacecraft named *Aurora 7*. Liftoff occurred at 12:45:16 UT and the 2,974-lb. spacecraft (no. 18) was inserted into a 100 × 166.8 mi., 32.5° orbit at an elapsed time of 5 min. 10 sec. Experiments included the observation of liquids in weightlessness and deployment and inflation of a small, tethered balloon. As with all three-orbit Mercury spacecraft, the automatic attitude control system had 35 lb. of propellant and the manual system had 25 lb., but Carpenter made extensive use of both and inadvertently used the two simultaneously. To compensate, the spacecraft was allowed to drift for 1 hr. 17 min. in addition to drift incorporated in the flight plan. Carpenter fired the retro-rockets 3 sec. late and with the spacecraft off attitude by 25° in yaw, resulting in a splashdown 250 mi. beyond the planned location at a duration of 4 hr. 56 min. 5 sec. The astronaut waited in his spacecraft for 2 hr. 59 min. until recovery forces arrived.

1962 May The Planetary Programs Office at the NASA Jet Propulsion Laboratory commissioned a study of future planetary exploration missions from the Advanced Planetary Spacecraft Study group under Philip K. Eckman. The group examined several different concepts that had been under way for the previous two years at the JPL and in NASA Headquarters, including the Voyager Venus/Mars spacecraft and Navigator, a probe for studying the sun, comets, Mercury and Jupiter. The JPL wanted to manage an advanced Voyager orbiter capable of carrying a lander on later missions to Mars. The Apollo moon landing goal convinced planetary scientists that manned flights to Mars would be the next goal, and the JPL wanted to be at the center of this activity. In November

1962 NASA held a review of Voyager and concurred with the JPL's proposal to bring in outside contracts. Several companies had been studying the concept for up to a year.

1962 June 4 At the U.S. Air Force School of Aviation Medicine, Brooks AFB, Tex., two volunteers began a 14-day simulated long-duration Gemini spacecraft mission. During that period they breathed 100% oxygen at a pressure of 5 lb./sq. in., as Gemini crew members would in space. The men suffered no ill effects and provided useful data for physiologists to plan medical experiments on the two-man flights.

1962 June 7 The director of NASA's Marshall Space Flight Center, Wernher von Braun, sent a memorandum to Dr. Joseph F. Shea, deputy director (systems), Office of Manned Space Flight, endorsing the lunar-orbit rendezvous (LOR) mode for Apollo moon landing operations. He also recommended the development of a special one-way logistics vehicle launched to the moon by Saturn C-5 to support the manned landings. Endorsement by von Braun was an important step in getting formal commitment from NASA for the LOR mode. On June 22, 1962, the Manned Space Flight Management Council met and recommended that the LOR be adopted for Apollo and that the Saturn C-5 should be developed for this role. It also endorsed the logistics vehicle and advised a six-month study be started.

1962 June 8 NASA and the Soviet academician Anatoly Blagonravov announced that the United States and the Soviet Union had reached agreement on a "world weather watch" using satellites to coordinate global monitoring of the Earth's meteorology by satellite. Information would be provided to the World Meteorological Organization (WMO), an international body used for the distribution of global weather data. The agreement grew out of correspondence between President John F. Kennedy and Premier Nikita Khrushchev that began when Khrushchev sent a congratulatory message to Kennedy on the flight of John Glenn, February 20, 1962.

1962 June 12–13 Leading representatives from NASA Headquarters, the Manned Spacecraft Center, the Marshall Space Flight Center and private industry met at Cape Canaveral to discuss the launch mode for Saturn C-5 launchers. The three contenders for design of what would be designated Launch Complex 39 envisaged an offshore launch facility accessed by barges carrying the launchers or a surface pad accessed by either a rail conveyance or a crawler way. The fixed land site with crawler way linking the pad to a vertical assembly building was selected as the cheapest, quickest to build and most flexible for increased flight rates.

1962 June 15 NASA's Office of Manned Space Flight (OMSF) sent the Office of Space Science requirements for information about the lunar surface to support the manned landing goal. The OMSF wanted data about the moon to support Apollo design decisions by the end of 1965 and information for modification of spacecraft hardware by the end of 1966. The OMSF stressed the need for engineering data over science and implied an importance within NASA which was never authorized. The attitude taken by OMSF personnel imparted a sense of dominance over other NASA departments, such as those for space science and space applications; the OMSF seemed to envelop the others' decision making and subsume it under its own aegis. This infuriated scientists, and the first confrontation took place on March 26, 1963, when Dr. Homer E. Newell, director of the Office of Space Science, sent a memo to NASA Administrator James Webb complaining about the disinterest in science at Houston's Manned Spacecraft Center.

The first Canadian Black Brant III sounding rocket was fired from Wallops Island, Va., carrying a 93-lb. payload to a height of 69 mi. and a speed of 5,618 ft./sec. With a height of 36 ft. and a launch weight of 3,100 lb., Black Brant III was a formidable tool for atmospheric research. The two-stage Black Brant IV made its first flight in June 1964 as a specially configured high-altitude rocket. The Black Brant series continued in use through the 1990s and were used by a variety of research institutes in many countries.

1962 June 18   The first in a series of dedicated U.S. electronic intelligence gathering (elint) satellites was launched by Thor-Agena B from Vandenberg AFB, Calif., to a 230 × 255 mi. orbit inclined 82.1°. The 1,500-lb. satellite was a test mission for operational satellites based on the Agena D, beginning with the second satellite launched on January 16, 1963. Launched at six-month intervals, the next nine satellites were sent up on TAT-Agena D vehicles into near-circular orbits about 320 mi. above Earth at 82° inclinations. The last of these was launched on January 17, 1968, succeeded by a second generation beginning on October 5, 1968.

1962 June 19   The U.S. Air Force Systems Command announced it was developing a "self-maneuvering unit" (SMU) for astronauts outside their spacecraft. Weighing about 160 lb., the pack would enable astronauts to propel themselves independent of a tether or restraint, using small gas thrusters to provide reactive force. The SMU would contain its own power, control and stabilization systems, enabling an astronaut to remain in close proximity to a work station without manual control of the SMU. The name was changed to modular maneuvering unit and then to manned maneuvering unit before its application in the Gemini program.

1962 June 27   NASA's director of Manned Space Flight announced that based on the results of the first two three-orbit U.S. manned spaceflights, the next manned mission would attempt to remain in space for six orbits. Piloted by Walter Schirra, with Gordon Cooper as backup, the MA-8 mission would be the ultimate utilization of the basic Mercury spacecraft design. Work on an eighteen-orbit version of the spacecraft was already under way for mission MA-9, but for six orbits, considered a prudent intermediate demonstration, the unmodified design would do. MA-8 and MA-9 would, however, carry 5.3 lb. of lithium hydroxide for cleaning carbon dioxide from the cabin atmosphere. Changes for MA-9 would include five 3,000-WH batteries and one 1,500-WH battery, a 22% increase in total over previous Mercury spacecraft; 75 lb. of propellant for the automatic and manual attitude systems; 3 oxygen bottles for a total supply of 12 lb. instead of 2 bottles for a total 8 lb.; and 10 lb. of drinking water instead of 5.5 lb.

1962 June 28   The first Agena D upper stage made its first flight atop a U.S. Air Force Thor launch vehicle and launched a Corona/KH-4 military reconnaissance satellite into a 131 × 430 mi., 76° orbit. Thor-Agena D had a payload capacity of 1,600 lb., about the same as the Thor-Agena B, but the new upper-stage model had a secondary propulsion system capable of making up to five orbital adjustments. The primary propulsion system was the 16,000-lb. thrust Bell XLR-81-BA-11 engine. With increased propellant capacity, Agena D provided greater mission flexibility over a longer period in orbit than the Agena B. Agena D would be used with the Atlas launch

vehicle and form the basis for a target vehicle in the two-man NASA Gemini program.

1962 June 29   Responding to requests for more data about the lunar surface at a faster rate, members of the Jet Propulsion Laboratory, Calif., met with NASA's Office of Space Science to present its plans for extended Ranger and Surveyor missions in support of Project Apollo. On July 20 NASA Headquarters released guidelines for adding five Ranger flights during 1964. Block IV (*Rangers 10–14*) would carry TV cameras similar to those of Block III, but the rough-landing capsules of Block II spacecraft (*Rangers 3–5*) would carry none. The Block IV plan was approved on October 3, 1962, but modified on March 1, 1963, when Block IV was reduced to three spacecraft (*Rangers 10–12*) and six Block V spacecraft (*Rangers 13–18*) were added. Block V was expected to carry rough-landing capsules with TV cameras for surface shots. Later that month Block V was doubled to 12 spacecraft (*Rangers 13–24*).

1962 July 8   As part of the Dominic series of U.S. nuclear weapons tests, a Thor missile detonated a 1.4-MT thermonuclear device at an altitude of 248 mi., 19 mi. from Johnston Island in the Pacific. The shot was known as Starfish Prime, one of the Fishbowl tests devised to measure the effect of thermonuclear explosions on radio communications. The pulse of electromagnetic radiation created an electrical surge through the distribution network on Hawaii, 800 mi. away, burning out fuses, opening circuit breakers over a wide area, disabling 30 strings of street lights and triggering several hundred burglar alarms. Within a few days electrons from the explosion trapped by the earth's magnetic field damaged solar cell panels on weather and communication satellites. NASA's Space Task Group studied data from sounding rockets which indicated that at peak the man-made radiation belt was 400 mi. wide and 4,000 mi. deep, but that by the time of the MA-8 Mercury mission in October 1962 it had dissipated.

1962 July 10   *Telstar I,* the world's first active repeater communication satellite, the first commercial satellite and the first reimbursable launch for NASA, was lifted by Delta from Cape Canaveral. The 170-lb., 2 ft. 10.5 in. sphere carried a helical antenna attached to the top of the spaceframe, received on 6,390 MHz, transmitted on 4,170 MHz at 2.25 W and could provide 600 one-way voice channels or one TV channel. The 15 W of electrical power was provided by 3,600 solar cells. The three-stage Delta had a liftoff weight of 112,000 lb. and placed *Telstar I* in a 582 × 3,513 mi. orbit inclined 44.8°. The ground station at Maine could communicate with the satellite for 4 hr. 10 min. each day and link up with France and England for 1 hr. 42 min. each day. *Telstar I* transmitted until February 21, 1963.

NASA Associate Administrator Robert Seamans officially notified the director of the Office of Manned Space Flight, D. Brainerd Holmes, that the lunar-orbit rendezvous (LOR) mode had been formally approved for Apollo. Administrator James Webb, Deputy Administrator Hugh Dryden and Seamans had made the final decision, which by now was merely a formality. The direct-ascent mode would require a Saturn C-8 that would not be available before the end of the decade, and the earth-orbit rendezvous mode called for too many launches for each mission. LOR was the best method in terms of time, cost and potential success, although direct ascent was less complex and posed fewer potential problems due to its inherent simplicity. A news conference was held next day to announce the decision to the public.

1962 July 16 Edward O. Olling from the Space Station Program Office at NASA's Manned Spacecraft Center, Houston, Tex., completed and submitted a study of a number of logistics vehicles needed to support a space station. He examined a logistics vehicle based on the Apollo spacecraft, an enlarged version of Apollo called Diana capable of carrying six astronauts, the Gemini spacecraft and a reusable vehicle capable of carrying 12 people. Olling showed economic advantage in total program costs involving a new, reusable transportation system.

1962 July 17 The fifth set of astronaut's wings awarded to a U.S. pilot were won by USAF Capt. Robert M. White when he flew the no. 3 X-15 to a height of 314,750 ft., or 59.6 mi. Arbitrarily selected as the boundary between earth's atmosphere and space, an altitude of 50 mi. had been accepted as the qualifying height for "space" flight. Other U.S. pilots who would qualify for astronaut wings on X-15 flights included: Joseph A. Walker, the first civilian (January 17, 1963; July 19, 1963; and August 22, 1963); USAF Maj. Robert A. Rushworth (June 27, 1963); USAF Capt. Joseph H. Engle (June 29, 1965; August 10, 1965; October 14, 1965); civilian John B. McKay (September 28, 1965) civilian William H. Dana (November 1, 1966; August 21, 1968); USAF Capt. William J. Knight (October 17, 1967); and USAF Maj. Michael J. Adams (November 15, 1967). In all, 8 pilots flew 13 spaceflight missions in the X-15.

1962 July 19 For the first time, an antiballistic missile system successfully demonstrated the interception of a simulated nuclear reentry vehicle launched by an ICBM. An Atlas D launched from Vandenberg AFB, Calif., lobbed an inert reentry vehicle (RV) across the Pacific Missile Range toward Kwajalein Atoll, 4,300 mi. westward, where acquisition, discrimination and target-track radars were located together with a Nike-Zeus ABM. The 16,000 MPH reentry vehicle was acquired by the ABM radars and identified in 20 sec., followed by a 30-sec. period during which discrimination analysis decided if it was a real warhead or decoy. The target-track radars tracked the RV for 10 sec. before the launch command was given to the Nike-Zeus. Interception occurred 100 sec. later. In a live action, the missile would detonate a kiloton-yield warhead to prematurely disable the incoming reentry vehicle.

1962 July 20 NASA headquarters assembled a brief for the tentatively proposed Surveyor Orbiter program expected to include five missions in moon orbit. The launch vehicle was expected to be the Atlas-Centaur carrying a spacecraft expected to weigh about 1,700 lb. and incorporating as much technology as possible from the Surveyor Lander. The Jet Propulsion Laboratory was tasked with preparing a list of design requirements by September 1 and a project plan by November 30, 1962. By September, pessimism about early availability of the cryogenic Centaur led NASA to switch to Atlas-Agena and, with JPL, conduct a feasibility study on a lightweight orbiter capable of being sent to the moon by this smaller launch vehicle.

NASA Administrator James Webb announced the need for a real-time computer-complex (RTCC) at the new Manned Spacecraft Center, Houston, Tex., unifying existing computers with the Space Task Group at Langley Research Center, Goddard Space Flight Center and Cape Canaveral's Mercury Mission Control. On October 15, IBM was given the job of building up the new RTCC at the MSC. Originally scheduled to be operational with *Gemini VI,* delays in the

two-man program enabled parallel computations to be run during the *Gemini II* flight on December 9, 1964, final tests to be conducted during the manned *Gemini III* flight of March 23, 1965, and transfer of mission control from Cape Canaveral to Houston for the *Gemini IV* flight of June 3–7, 1965.

1962 July 22 The 446-lb. *Mariner R-1* spacecraft was launched by Atlas Agena-B from Launch Complex 12 at Cape Canaveral, Fla., the first of two NASA planetary spacecraft designed to send back to Earth information about the planet Venus during a close fly-by. At the end of a countdown that began at 04:33 UT, July 21, the Atlas-Agena B lifted off the pad at 09:21:23 UT. The Atlas was programmed to fire for approximately 4 min. 59 sec., but halfway through the burn the vehicle began to deviate in yaw and a self-destruct signal was sent by the range safety officer at 4 min. 53 sec. to prevent its erratic path taking it into shipping lanes. There was no destruct package on the Agena. The radio transponder on *Mariner R-1* continued to transmit for 1 min. 4 sec. The second spacecraft was prepared and assembled on its Atlas-Agena B for launch on July 25. This was canceled due to stray voltage in the Agena batteries and the final count began on August 26 at 22:37 UT.

1962 July 25 NASA's Manned Spacecraft Center invited 11 companies to bid for a contract on the lunar excursion module (LEM). These firms were Lockheed, Boeing, Northrop, Ling-Temco-Vought, Grumman, Douglas, General Dynamics, Republic Aviation, Martin-Marietta, North American and McDonnell Aircraft. Recently renamed from lunar excursion vehicle, the LEM would be carried to moon orbit, docked to the front of an Apollo spacecraft and used by two astronauts to reach the lunar surface. For launch on the Saturn C-5, the LEM would be carried in a special adapter between the third stage of the launcher and the Apollo spacecraft. En route to the moon, the Apollo would turn 180° and dock to the LEM, where it would remain until it separated for descent from moon orbit to the surface.

1962 July A special task group at the Soviet Academy of Sciences recommended immediate development of a large launch vehicle capable of placing 165,000-lb. payloads into low Earth orbit. This project was to replace the original N-1 and N-2 launch vehicles which by this time had been canceled. During the remainder of the year Soviet space engineer Sergei Pavlovich Korolev and propulsion engineer Valentin Glushko worked up the basic design of such a launch vehicle. It was eventually developed as a (redesignated) N-1 that would form the basis for the Soviet moon-landing launcher. For the time being the N-1 program was subordinated by the firmer commitment to a manned circumlunar program which would result in two competing concepts: one, from the Chelomei bureau, authorized on August 3, 1964, and the other, from the Korolev bureau, approved as the official program on December 25, 1965.

Intelligence information reaching the United States revealed Soviet ICBM activity. In the wake of the cancellation of the S-6 as an ICBM, the SS-7 was just coming into service, with about 20 missiles operational. Plans for "hardening" ICBM launch sites by placing a new design of missile in a concrete silo like Titan II were discussed. Designated SS-8 Sasin in NATO coding, the missile had a length of 80 ft., a diameter of 9 ft., a launch weight of 170,000 lb. and a range, with a 5-MT warhead, of 6,500 mi. The SS-8 was operational from 1963, but only 19 of this "lightweight" ICBM were ever deployed at SS-7 sites. The 209 SS-7s, each capable of carrying a 25-MT warhead, and SS-8s comprised the only Soviet ICBM force until the introduction of the massive SS-9 in 1965. By that time the US had 880 ICBMs. SS-7 and SS-8 were withdrawn in 1977.

1962 August 2 Based on successful tests with large solid propellant rocket motors, Aerojet General Corp. proposed a massive satellite launcher called Sea Dragon. With a takeoff weight of 20–100 million lb., it would be assembled in drydocks and towed to a point of launch. Sea Dragon would be launched from water without the use of pads, gantries or servicing towers. Several times larger than the Nova concept put forward by NASA, Sea Dragon would be capable of lifting payloads of up to 2,000 tons. Few could anticipate the need for such a behemoth and none was ever found.

1962 August 5 In the first test of a new photoreconnaissance system designed to obtain high-resolution pictures of special interest areas, an Atlas-Agena B carried the 4,100-lb. KH-7 satellite from Vandenberg AFB, Calif., to a circular orbit of 127 mi. and an inclination of 96.3°. This mission, which was essentially a test of the new top secret Eastman Kodak camera system, was the first launch under Project Gambit, which aimed to provide the CIA with highly detailed pictures using telephoto lenses. A second Atlas-Agena B orbital test flight launched on November 11 preceded the first operational KH-7 flight with an Atlas-Agena D on July 12, 1963.

1962 August 8 The House Committee on Science and Astronautics published a report on the Project Anna geodetic satellite program. Criticizing its management by the Department of Defense when it was essentially a scientific research program, the committee recommended that Anna be moved to NASA. This gave NASA the mandate to proceed with its own geodetic program, which was to emerge in 1964 in the Explorer series.

1962 August 11 Cosmonaut Andrian G. Nikolayev was launched aboard the 10,412-lb. *Vostok 3* spacecraft from a Tyuratam launch pad at 08:30 UT. His backup was cosmonaut Bykovsky. From an initial orbit of 103 × 136 mi. inclined 64.98°, the spacecraft remained in space for 3 days 22 hr. 22 min., and during its time in orbit was joined by the *Vostok 4* spacecraft launched on August 12. The cosmonaut ate breakfast just 50 min. after launch, and "lunch" at an elapsed time of 12:00 UT. After more than 9 hr. in orbit, Nikolayev transmitted television images to Earth just one hour before eating his last meal of the "day" and shortly before starting his first rest period. The food consisted of preprepared meals. At an elapsed time of 23 hr. 32 min. cosmonaut Pavel Popovich was launched aboard *Vostok 4* and the two manned spacecraft continued to orbit the Earth for almost two days before Nikolayev returned on August 15, six minutes before Popovich at the end of the world's first dual space mission.

1962 August 12 In the second manned spaceflight within 24 hr., Soviet cosmonaut Pavel Popovich was launched aboard the 10,425-lb. *Vostok 4* spacecraft at 08:02 UT to a 105 × 138 mi. orbit inclined 64.95°. His colleague cosmonaut Komarov had been backup for the SL-3 launch until an illness brought his replacement by Volynov. When *Vostok 4* entered orbit it was only 4 mi. from *Vostok 3* carrying cosmonaut Nikolayev. Because the Vostok was unable to modify, or change, orbit in space, the two spacecraft could not be brought closer together, nor could either prevent a gradual separation as the natural perturbations acted upon each vehicle. At the time of retro-

fire, the two spacecraft had drifted 1,770 mi. apart, *Vostok 4* returning to Earth August 15 at an elapsed time of 2 days 22 hr. 57 min.

1962 August 21 Interpreting the dual flights of *Vostok 3* and *Vostok 4* as indicative of a satellite inspection capability, President Kennedy telephoned Secretary of the Air Force E. Zuckert and asked for a rundown of where the United States stood with respect to military space programs, implying that the Soviet Union was in danger of gaining superiority. Zuckert responded the next day by reminding Kennedy of his preelection promises about military strength, implying that the President should do more. Kennedy replied On August 27 by admonishing the secretary and asking for a forward strategy, to which Zuckert answered on September by requesting "a stimulant." This encouraged Kennedy to put more money into intelligence and antisatellite capabilities.

1962 August 23 A four-stage solid propellant Scout launcher carried a 198-lb. satellite from Vandenberg AFB, Calif., to a 388 × 526 mi. orbit with an inclination of 98.6°. Launched by the Air Force, the satellite carried sensors that would help in the development of military weather satellites with their unique requirements for atmospheric data. These technologies were developed under Program 417 and would combine in the Defense Meteorology Satellite Program (DMSP) satellites.

1962 August 25 The Soviet Union attempted to launch a spacecraft to Venus in their first attempt during the 1962 launch window. Designated *Sputnik 19,* the spacecraft and terminal stage failed to leave the 107 × 165 mi. Earth orbit where they had been placed by an SL-6 launch vehicle. The 2MV-1 spacecraft probably weighed about 1,960 lb. and was intended to reach the surface of the planet and effect a landing. Additional attempts at sending spacecraft to Venus during the 1962 launch window were made with the lander *Sputnik 20* (another 2MV-1) on September 1 and *Sputnik 21* (2MV-2), a flyby mission, on September 12. On each occasion the terminal stage failed to accelerate the spacecraft out of Earth orbit.

1962 August 27 At 06:53:14 UT, an Atlas-Agena B carrying NASA's *Mariner R-2* spacecraft was successfully launched from Complex 12 at Cape Canaveral, Fla. The Agena B took over from Atlas and in a burn lasting 2 min. 27 sec. placed itself and Mariner (designated *Mariner II* after launch) in a 116 × 123 mi. Earth parking orbit at an elapsed time of 8 min. 12 sec. After coasting for 16 min. 17 sec. Agena fired for 1 min. 34 sec. to send the 447-lb. spacecraft on its way to Venus at an initial speed of 25,420 MPH, 26 min. 3 sec. after liftoff. The spacecraft separated from Agena at an elapsed time of 28 min. 39 sec., and the upper stage turned around and vented excess propellant to shift its trajectory away from that of the spacecraft. *Mariner II* was a variation of the hexagonal design employed for Ranger, with six circumferential modules for equipment and systems and a liquid propellant course-correction motor. Two panels carrying 9,800 solar cells would provide 148–222 W of electrical power and attitude control was provided by 10 jets supplied from two titanium bottles containing 4.3 lb. of nitrogen gas. A circular high-gain communications antenna was deployed after separation. *Mariner II* had a height of 11 ft. 11 in., a base diameter of 5 ft. and a span of 16 ft. 6 in. across the solar panels.

1962 August 31 President Kennedy signed into law the Satellite Act, which provided for a private U.S. Communications Satellite Corp., Comsat. The general public was to own 50 percent, with the remaining shares in the hands of international telecommunications operators. Comsat was envisioned as the pivotal organization for a new global telecommunications satellite system, with national subscribers paying in proportion to their use of the system. This would form the basis for the Intelsat structure and provide funds for the development and launch of satellites linking almost every country on Earth.

1962 September 3 Dr. D. L. Mordell at Canada's McGill University announced plans to launch a ballistic sounding rocket by gun from the island of Barbados. Under Operation Harp (high altitude research project), a 470-lb. Martlet missile would be fired to approximately 200 mi. altitude by one of two 16-in. guns provided by the U.S. Navy. The program was in cooperation with the U.S. Army, and the first such firing took place on January 26, 1963, when a 475-lb. instrumented capsule was shot 15 mi. into the air and fell back into the sea 9 mi. off southeast of Barbados. This work inspired further studies on a supergun capable of firing satellites into space.

1962 September 4 Proposals to design and build the lunar excursion vehicle (LEM) for NASA's Apollo lunar landing program were received from nine companies. Of those invited to bid on July 15, 1962, only McDonnell Aircraft and North American Aviation failed to present submissions. On September 5 NASA announced it would award two three-month studies of an unmanned lunar logistics spacecraft for consolidating lunar research after the first landing. Industry presentations on the primary LEM contract were held September 13–14, and one-day visits to bidding companies were made September 17–19, 1962.

1962 September 5 At 00:24:35 UT the *Mariner II* spacecraft fired its 50.5-lb. thrust liquid propellant midcourse correction motor for 28.3 sec., imparting a velocity change of 102 ft./sec. *Mariner II* was at this time 1.492 million mi. from Earth. The motor used a monopropellant, anhydrous hydrazine, with nitrogen tetroxide as the starting fluid and aluminum oxide pellets as the catalyst. The system could execute a maximum 200 ft./sec. velocity change. The maneuver was necessary to reduce the Venus fly-by distance from 240,000 mi. to about 20,000 mi. and was calculated by precisely tracking the spacecraft. Before this flight engineers lacked definitive data on the exact orbit and location of the planet with respect to Earth and there were major uncertainties in the calculations. By tracking *Mariner II* as it moved closer to its target, flight dynamicists obtained the data they needed, first to define more precisely the relative orbits of the two planets and then to calculate the mass of Venus. In all, 16 orbit computations were made between this date and December 7, 1962, when the gravitational effect of Venus was first felt by the spacecraft.

At the annual dinner of the Society of British Aircraft Constructors in London, Mr. Julian Amery, the minister for aviation, described how he had encouraged the Air Ministry to examine the concept of an aerospace plane capable of taking off and landing like an aircraft but of reaching orbit with a useful payload. He said that studies were being carried out and that three concepts were under analysis: a Mach 7 delta strike and reconnaissance system; a second stage carried by the delta to reach Mach 14 and operate at an altitude of 150,000–180,000 ft; and a small vehicle carried by the first two, capable of reaching orbit and returning to Earth.

1962 September 7 NASA headquarters held a review of proposed Mars exploration plans for the 1964 launch window at which 28 participants reviewed existing Mariner C and Mariner B programs and heard proposals from the Jet Propulsion Laboratory for advanced Voyager spacecraft. Mariner C would be a lightweight spacecraft launched by Atlas-Agena D for a fly-by carrying a television camera and an infrared spectrometer to detect organic molecules. The prospect of searching for life on Mars had driven plans for a lander capsule, which now could not be launched during the 1964 window due to delays with Atlas-Centaur. The competition between an increasingly delayed 1,300-lb. Mariner B and the prospective 6,000-lb. Voyager, with sophisticated biological life-detection instruments in the lander, eventually led to the cancellation of Mariner B.

1962 September 9 The first of six Atlas F ICBM squadrons became operational at Schilling AFB, Kans. Other bases included Altus AFB, Okla., Dyess AFB, Tex., Lincoln AFB, Nebr., Plattsburgh AFB, N.Y., and Walker AFB, N.Mex. Each squadron had 12 missiles and all were operational by December 20. Each Atlas F was deployed to a vertical silo from where it would be raised to the surface for launch. This completed Atlas deployment of 129 missiles, including three test and training pads at Vandenberg AFB, Calif., which could be used in time of war.

1962 September 10 A fire which broke out in the pure oxygen atmosphere of a space simulation chamber at the U.S. Air Force School of Aerospace Medicine, Brooks AFB, Tex., seriously injured one of two officers inside at the time. The fire occurred on the 13th day of a 14-day experiment aimed at investigating the physiological effects of breathing pure oxygen for long periods. On November 17 another fire broke out in a pure oxygen cabin at the U.S. Naval Air Engineering Center, Philadelphia, Penn. The fire hazard of a pure oxygen atmosphere was already of concern to Apollo engineers designing the pressurized cabin of the command module to accommodate three astronauts in a pure oxygen, 5 lb./sq. in. atmosphere.

1962 September 17 The names of nine Group 2 astronauts selected by NASA were announced by Robert Gilruth, director of the Manned Spacecraft Center. They were: Neil A. Armstrong (NASA test pilot); Maj. Frank Borman (USAF); Lt. Charles "Pete" Conrad Jr. (USN); Lt. Comdr. James A. Lovell Jr. (USN); Capt. James A. McDivitt (USAF); Elliot M. See Jr. (General Electric test pilot); Capt. Thomas P. Stafford (USAF); Capt. Edward H. White II (USAF); and Lt. Comdr. John W. Young (USN). With the exception of Elliot See, killed in an aircraft crash on February 28, 1966, all would make flights in space. Armstrong and Young would get to walk on the moon, White would be the first American to walk in space, but die in the January 27, 1967, Apollo 204 fire, and Young would be the only astronaut to fly Gemini, Apollo and Shuttle.

The first in a series of tetrahedral research satellites built by TRW for the Department of Defense was launched as a piggyback payload on a KH-4 reconnaissance satellite mission from Vandenberg AFB, Calif. Known also as *ERS-2* (Environmental Research Subsatellite-2), *TRS-1* was one of the smallest satellites launched. With four triangular surfaces on a tetrahedral shape measuring only 6 in. across the base, *TRS-1* weighed 1.47 lb. and had 28 solar cells which produced 1.2 W. It carried a 3 ft. 4 in. long antenna and was to have measured radiation from the Starfish nuclear detonation. *ERS-2* failed to eject from the KH-4/Agena assembly and was unusable. *ERS-3* and *-4* were launched aboard an Atlas-Agena B/MIDAS on December 17, 1962, but that failed to reach orbit. *ERS-5* and *-6* (TRS-2/3) were successfully ejected from a MIDAS launch on May 9, 1963.

1962 September 19   The U.S. Air Force chose six astronauts from previous selections to serve as engineering test pilots for the manned Dyna-Soar program. They were: Capt. Albert H. Crews (USAF); Capt. Henry C. Gordon (USAF); Capt. William J. Knight (USAF); Capt. Russell L. Rogers (USAF); Milton O. Thompson (NASA test pilot); and Maj. James W. Wood (USAF). Capt. Crews transferred to the Manned Orbital Laboratory program in 1965, Capt. Knight went to the X-15 and lifting-body research programs in 1964, following Thompson who had preceded him on that career path in 1963.

1962 September 21   Oran W. Nicks, NASA director of Lunar and Planetary Programs in the Office of Space Sciences, asked USN Capt. Lee R. Scherer, on assignment to NASA, to form a working group to study adaptations of the Atlas-Able (Pioneer) 5 and Ranger lunar spacecraft to a lunar orbiter photographic role. Space Technology Laboratories, the Able 5 contractor, had sent Nicks a detailed proposal on such a vehicle based on an Atlas-Agena D launcher. It was to have utilized a 100-in. focal length spin-scan camera using film similar to one developed for the U.S. military reconnaissance program by Merton E. Davies at RAND in 1958. The shift from Centaur to Agena and the basic photographic system concept that arose from this study formed the basis for NASA's Lunar Orbiter.

1962 September 24   The Space Technology Center of the General Electric Co. proposed an escape system for astronauts unable to return to Earth in their spacecraft. Called MOOSE (man out of space easiest), it involved use of a prepackaged rescue assembly with plastic sack, foldaway heat shield, retro-rocket pack and containers of foaming plastic. Clambering out of his spacecraft, the astronaut would zip himself into the sack, fire the retro-rocket and inflate the heat shield with expandable foam, and return to Earth on a parachute deployed at 30,000 ft. MOOSE was never adopted.

1962 September 26   NASA completed preliminary plans for the development of the Mississippi Test Facility (MTF) where Saturn launch vehicle engines and rocket stages would be tested. The 13,500-acre MTF was located in southwestern Mississippi, 35 mi. from NASA's Michoud Operations (formerly Michoud Assembly Facility), where the Saturn stages were built. Barges would take the stages to the MTF for static firing on test and deliver them to Cape Canaveral for stacking and launch.

1962 September 27   A first-generation Soviet photoreconnaissance satellite, *Cosmos 9,* was launched by SL-3 from Baikonur to a 181 × 215 mi. orbit inclined 65°. Its purpose was to observe possible U.S. reaction to the transfer of medium-range ballistic missiles to Cuba. It returned after four days and was followed by the launch of *Cosmos 10* on October 17, which reentered on October 21. One day after it returned, President Kennedy announced the missile build-up to a shocked American public. Thus began a crisis that would bring the United States and the Soviet Union close to all-out war, brinkmanship that would have been unthinkable for the President had he not known the true nature of the missile

gap—that the Soviets had far fewer ICBMs than the United States.

1962 September 28   A meeting on possible space station work was held in Washington, D.C., with participation from NASA's Office of Manned Space Flight, Office of Advanced Research and Technology, and three field facilities: Manned Spacecraft Center, Marshall Space Flight Center and the Langley Research Center. It was noted that emphasis on the Apollo program had preoccupied NASA and that there was little funding for conceptualization. Maxime Faget emphasized the need for a low-cost station design, saying that a station would serve most usefully as a stepping stone to manned Mars missions. Others disagreed, believing that basic scientific research from Earth orbit was the primary reason for having a station in space.

1962 September 29   The first satellite designed and built by a nation outside the United States or the Soviet Union was launched by NASA from Cape Canaveral, Fla., when Canada's *Alouette I* was successfully placed in a 616 × 637 mi. orbit inclined 80.5° to the equator. NASA had launched the satellite using a Thor-Agena B. Named after a Canadian lark, *Alouette I* had been built by De Havilland Aircraft Co., the Canadian Research Council and the Canadian Defense Research Telecommunications Establishment. It weighed 320 lb. and was an oblate spheroid, with a height of 3 ft. 6 in. and a diameter of 2 ft. 10 in., equipped with two unfurlable dipole antennas for measuring the ionosphere. The backup spacecraft was launched by NASA and designated *Alouette II* on November 29, 1965.

1962 September   NASA Flight Research Center Director Paul F. Bikle approved construction of a manned M2 lifting-body glider. Several engineers at the Ames Research Center had conducted preliminary evaluations of the half-cone M2 lifting-body that had been conceived in the late 1950s. It was felt that now was the time to build a small model and fly it. Put together by volunteer engineers and the help of Gus Brieglab from a local sailplane company (Sailplane Corp. of America), the M2 gradually took shape in a hangar at the FRC. Outside hung a notice: *Wright's Bicycle Shop.* Designated M2-F1, the lifting-body was finished early in 1963.

The NASA Manned Spacecraft Center completed a preliminary Apollo flight development plan leading to manned moon landings. The launch escape system would be qualified in two simulated tests from a flat pad in tests designated PA-1 and PA-2. Little Joe II would be used for three suborbital, in-flight, escape system tests designated A-001, A-002 and A-003. System components would be tested on Saturn C-1 flights SA-6 through SA-10. Four unmanned Saturn C-1B flights would qualify the launcher for manned flights and rehearsal of lunar mission techniques using the SA-200 series of numbers. Six unmanned Saturn C-5 were deemed necessary to qualify the launcher for manned operations, and each flight was given a SA-500 series designation. In this system, the A denoted Apollo, the first digit stood for launch vehicle type and the last two digits denoted the flight sequence.

In a technical memorandum compiled from 11 papers by staff at the Langley Research Center, NASA revealed the wheel-shaped space station was both feasible and practicable. Langley had conducted more than two years of study and North American Aviation had performed a six-month evaluation of the preferred concept: a 171,000-lb. structure with a diameter of 150 ft. and provision for up to 38 people. Langley claimed the station could be used to learn what astronauts could do in space, develop closed-loop life support systems and perfect rendezvous techniques.

1962 October 3 The third U.S. manned orbital flight was successfully completed by astronaut Walter Schirra when he rode the MA-8 Mercury spacecraft named *Sigma 7* on a six-orbit mission lasting 9 hr. 13 min. 11 sec. Launched from Pad 14, Cape Canaveral, Fla., at 12:15:11 UT, the 3,029-lb. spacecraft no. 16 was initially placed in a 100 × 175.8 mi. orbit, inclined 32.5° and with a period of 1 hr. 28 min. 55 sec. The flight was the most successful Mercury mission to date and conducted in textbook style. The astronaut performed to a very high level and conserved large quantities of propellant through the judicious use of thrusters in a skillful manner. The spacecraft landed 275 mi. northeast of Midway Island, the first to come down in the Pacific Ocean, just over 5 mi. from the recovery force. This mission was the first to be relayed live to Europe by TV signals sent via the Telstar communications satellite.

1962 October 18 *Ranger 5* was launched by Atlas-Agena B on an Earth escape trajectory toward the moon from Launch Complex 12 at Cape Canaveral, Fla. The 755-lb. spacecraft was the last of three Block II spacecraft carrying an instrumented rough-landing capsule and TV cameras for images in the lunar approach phase; its mission objectives were identical to those of its two predecessors. After separation from Agena and deployment of its appendages, a short circuit occurred at 1 hr. 13 min. after liftoff, with total loss of solar cell power. The batteries were exhausted about 7 ½ hr. later, but shortly before this an unsuccessful attempt had been made to make a midcourse correction using the onboard thrusters, and the spacecraft swept past the moon at a distance of 450 mi. on its way to solar orbit. Next day, Dr. Homer Newell set up a board on inquiry drawn from members outside JPL to analyze completely the reliability of the entire program.

1962 October 22 Responding to intelligence information that the Soviet Union was shipping SS-4 medium-range missiles to Cuba and preparing to install longer-range SS-5 IRBMs (intermediate-range ballistic missiles), President Kennedy ordered an arms quarantine around the island. On October 26 Kennedy sent Soviet Premier Khrushchev a message demanding the immediate withdrawal of the SS-4s and a halt to shipments of SS-5. For a time the peace of the world hung in the balance as both the United States and the Soviet Union put their armed forces on a war footing. In long-range missile power, the Soviets had no more than 25 SS-7 ICBMs ready to fire. Soviet SLBMs (submarine-launched ballistic missiles) had a relatively short range. The United States had just over 100 Atlas and Titan I ICBMs and 144 SLBMs on fleet ballistic missile submarines, a total of around 250 missiles. On October 27 Khrushchev agreed to remove SS-4 and SS-5 missiles from Cuba and Kennedy agreed to withdraw the 15 Jupiter missiles from Turkey; armed conflict had been averted.

The U.S. Air Force announced the names of ten candidate astronauts selected to support general manned military space programs, the third such group selected. They were: Capts. Alfred L. Atwell; Neil R. Garfield; Edward G. Givens; James A. Roman; Charles A Bassett; Michael Collins; Joe M. Engle; Francis G. Neubeck; and Alfred H. Uhalt; and Maj. Tommie D. Benefield. Unlike Groups 1 and 2, they were not specifically seconded to the Dyna-Soar program. Capts. Givens, Bassett, Collins and Engle later transferred to NASA as astronauts.

1962 October 24 An attempt to send an unmanned spacecraft on a flyby of the planet Mars began when a Soviet SL-6 launch vehicle lifted off from the Tyuratam launch complex. The 23-ft.-long, 14,300-lb. second stage and spacecraft were placed in an interim Earth parking orbit of 112 × 301 mi., inclined 49° pending re-ignition for Earth escape velocity. This failed to take place, and an explosion five days later left 24 pieces temporarily in orbit. In the absence of a successful Earth departure, this satellite was designated *Sputnik 22.* Similar to *Mars 1,* successfully launched on November 1, 1962, the 2MV-4 spacecraft weighed about 1,950 lb. A similarly unsuccessful Mars attempt was made on November 4, 1962, when *Sputnik 24* (2MV-3) failed to leave Earth orbit for its planned Mars landing.

1962 October 25 Identified as a Surveyor program engineer, USN Capt. Lee R. Scherer completed for NASA a confidential report entitled *Study of Agena-based Lunar Orbiters* in which he examined two industry proposals for a lightweight lunar orbiter. Scherer supported a concept from Space Technology Laboratories for a 705-lb. spin-stabilized spacecraft to photograph the entire lunar surface from a polar orbit 1,000 mi. above the moon. Imaging resolution with a 100-in.-focal-length film scan system was calculated to be 60 ft., but if an equatorial orbit of 25 mi. was adopted, resolution would be less than 20 in. An alternative system from RCA would utilize a 440-lb. three-axis stabilized spacecraft injected to lunar orbit from a Ranger bus, but its resolution would be 100 ft. at best.

1962 October 26 A U.S. Air Force Thor-Agena D carried the 2,425-lb. Starad satellite from Vandenberg AFB, Calif., to a 120 × 3,440 mi., 71.4° orbit. Instrumented to measure the radiation caused by the nuclear Starfish detonation in space on July 9, the satellite was maneuvered to monitor the artificial field at various locations. It continued to return data to Earth until January 18, 1963, and the satellite decayed back through the atmosphere on October 5, 1967.

1962 October 29 In reply to a request from President John F. Kennedy for information on the cost of advancing the moon landing schedule by one year, NASA Administrator James Webb sent a letter outlining a revised budget to the White House. NASA had scheduled the first moon landing for late 1967, but advancing it to late 1966, said Webb, would call for an additional $1.7 billion in the next two fiscal years. To achieve a late 1966 landing, NASA would have to advance the first manned Apollo CSM (command and service module) flight from May 1965 to November 1964 and move the first manned Saturn C-1B flight from May 1966 to October 1965. The first Saturn C-5 would have to move up from April 1966 to September 1965, with the first manned Apollo/C-5 mission in June 1966 instead of June 1967.

1962 October 31 A Thor-Able Star launched from Cape Canaveral, Fla., carried the 350-lb. geodetic satellite *Anna 1B* to a 673 × 731 mi. orbit with an inclination of 50.1°. Like its predecessor, *Anna 1A,* the satellite carried a flashing beacon, a radio ranging device called SECOR (sequential collation of range) and a Doppler ranging system using a transmitter with a frequency controlled by a quartz crystal oscillator. *Anna 1B's* beacon flashed at 5.6-sec. intervals 20 times a day. Only the SECOR system failed to operate correctly.

1962 November 1 The Soviet Union had successfully launched the first spacecraft to fly past the planet Mars when an SL-6 launch vehicle lifted off from Tyuratam carrying the

1,970-lb. *Mars 1*. Placed initially in an Earth parking orbit, the second stage of the launch vehicle fired again to propel *Mars 1* (a 2MV-4 class spacecraft) to Earth escape velocity and a flight path that would carry it past Mars on June 19, 1963. The spacecraft was a complete redesign from its 1960 predecessors and became the standard for future Mars and Venus vehicles. It comprised an assembly of two barrel-shaped modules welded end-to-end for a length of about 9 ft. and a diameter at the widest point of 3 ft. 7 in. At the top, a propulsion system would maintain attitude and perform one course correction. Two solar cell panels, each about 3 ft. 4 in. wide and 4 ft. long, were attached like wings either side for a total span, including hemispherical radiators on each end, of 13 ft. 1 in. A high-gain antenna, 7 ft. in diameter, for communication with Earth was attached to the side of the main modules.

1962 November 5 United Technology Corp. (UTC) announced that it had developed a new technique for throttling liquid propellant rocket motors. By injecting gaseous helium into the fuel, thrust would increase or decrease according to the ratio of the gas to the liquid. UTC claimed that a 10,000-lb. thrust rocket motor could be "throttled" down to 100-lb. thrust using this method.

1962 November 6 NASA approved a project proposal from the Ames Research Center for interplanetary Pioneer spacecraft to explore the solar wind and radiation far from Earth. NASA wanted to know the level of solar radiation in deep space not only to expand the scientific understanding of the solar system, but to quantify the radiation threat for future astronauts far from Earth. On January 29, 1963, a request for bids to build Pioneer was issued to U.S. industry, and on February 1, 1963, NASA asked scientists to propose experiments for the Pioneer series. An initial set of experiments was selected by July 23, 1963.

1962 November 7 NASA announced that the Grumman Aircraft Engineering Corp. had been selected to build the Apollo lunar excursion module (LEM), and with it the last major contract was awarded for the manned moon landing program. Late in getting under contract due to uncertainty about the landing mode, the LEM would be the pacing item right up to completion of the manned landing goal. The early LEM design displayed a descent stage with five legs supporting a pressurized ascent stage for two astronauts, with a circular, forward-facing egress hatch, two forward and two downward-looking windows in front and four clusters of attitude control thrusters, one on either side, one in front and one at the back. NASA signed the LEM contract on March 11, 1963.

1962 November 16 The third Block I Saturn C-1 (SA-3) was launched from LC-34 at Cape Canaveral, at 17:45 UT, the first U.S. launch vehicle to exceed 1-million-lb. mass at liftoff. SA-3 carried the full design load of 750,000 lb. RP-1–liquid oxygen first-stage propellant, increasing vehicle weight to 1.1 million lb. The first stage produced 1.32 million lb. of thrust for 2 min. 21 sec., at which point the four inboard engines were shut down, leaving the outer four engines to produce a combined thrust of about 680,000 lb. for a further 8 sec. SA-3 performed Project High Water II, in which the two dummy upper stages were broken open at a height of 104 mi., 131 mi. downrange, releasing 190,000 lb. (22,900 U.S. gal.) of water, 4 min. 52 sec. after liftoff.

By the end of 1962, NASA had decided the moon landing plan for Apollo and settled the configuration of the spacecraft. While a lunar module landed on the moon, the command/service module would orbit the moon. Here, the CSM is given scale by test engineers.

1962 November 17 At 21:00 UT research assistant Whilden P. Breen entered a small isolation chamber at the University of Maryland's space research laboratory at College Park, Md. He was there to test his reactions to being alone for five months as a lone astronaut might be on a long-duration spaceflight. The chamber consisted of one 12-ft.-square room for general living activities and two 5-ft.-square alcoves. The subject's "day" comprised several different programs of tasks and duties. Closed-circuit TV was used to observe Breen, who emerged from isolation at 15:45 UT, April 17, 1963, having survived 150 days 19 hr. 45 min., apparently none the worse.

1962 November 27 NASA awarded a contract to Ling-Temco-Vought for a velocity boost package that would propel an instrumented package into Earth's atmosphere for high-speed measurements of reentry conditions. No measurements currently existed on the effects of reentry at speeds which would be experienced by an Apollo spacecraft returning from the moon. In this test, called Project Fire, a cone shaped like the command module would be launched to an altitude of 400,000 ft. and driven back down into the atmosphere at about 25,000 MPH.

1962 November 30 The final NASA report of the Ranger Board of Inquiry set up after the fifth consecutive failure with a spacecraft of this type was completed and released on December 5, 1962. Chaired by Dr. Albert J. Kelley, it criticized the JPL for poor management, too many projects and a confused set of priorities, resulting in a technically flawed design. The board believed that the JPL had not achieved "the high standards of technical design and fabrication" essential for successful mission operations and recommended postponement of the first Block IV (Ranger 6) spacecraft, then scheduled for a January 1963 launch, until a combined design review and testing process had been completed. On March 11, 1963, Northrop Corp. was put under contract to act as systems contractor for Ranger Block IV and V. The Ranger 6 spacecraft was subjected to rigorous examination and thirteen 66-hr simulated lunar missions totaling 860 hr. in the JPL space chamber.

1962 November Encouraged by the highly successful *Mariner II* mission, NASA tentatively approved plans for a *Mariner C* flight to Mars in 1964 using an Atlas-Agena D launch vehicle. Now called Mariner-Mars 1964, the mission would involve three spacecraft designated *Mariner C, D* and *E*. The project approval document was signed on March 1, 1963, and in May of that year *Mariner E* was dropped due to budget pressures, leaving two spacecraft assigned to the Mars launch window beginning November 1964. They would be known as *Mariner III* and *Mariner IV* after launch. The pressure to prepare lightweight Mariners for launch at the 1964 window was driven almost exclusively by the desire to achieve planetary success ahead of Soviet scientists. U.S. planetary scientists had by this date become enthused with the challenge of beating their counterparts in the USSR in a manner similar to that which encourages rival teams in any research and development effort. It was a mistake of legislators in Washington, D.C., that they believed the scientists to be supporting a race between political systems rather than a race for knowledge between scientists.

1962 December 3 The U.S. Air Force canceled the satellite interception and inspection program originally called Program 720 (formerly SAINT), citing loss of role and a lack of clear support for the concept. There was opposition from NORAD (North American Air Defense) because they said it could be overwhelmed with a large number of decoys that could invalidate early warning by saturating the radar and it would be better not to provoke the Soviet Union into that development. No flights took place, but the technology that had been developed by RCA for SAINT was applied to the Air Force Program 437 antisatellite system (using Thor boosters) which had emerged from Advanced Development Objective 40 on February 9, 1962.

Acting on behalf of NASA, the U.S. Army Corps of Engineers issued a contract to four New York architectural firms for design of a Vertical (later renamed Vehicle) Assembly Building (VAB) at Cape Canaveral for assembly and erection of the Saturn C-5 launch vehicle. The Corps selected Morrison-Knudson Co., Perini Corp. and Paul Hardeman, Inc., to build the VAB. With a volume of 129.482 million cu. ft., it would have a height of 525 ft. 10 in., a length of 716 ft. 6 in. and a width of 518 ft. Covering 7.5 acres, it would rest on 4,225 steel pilings driven 160 ft. into bedrock. Consisting of a main high-bay area, where up to four Saturn C-5s could be assembled simultaneously, and a low-bay area for stage checkout, the VAB would contain 10,000 tons of air conditioning to prevent clouds forming inside and rain falling on the workers.

1962 December 4 NASA asked the U.S. Army Corps of Engineers to conduct a six-month study on the feasibility of constructing a research station on the moon. The Corps was to evaluate soil movement and excavation techniques, define budget and resource requirements and prepare a schedule of construction. NASA at this time had no program for the "manned lunar laboratory" hypothesized in the study outline, but wanted the Corps of Engineers to provide preliminary conclusions so that it could integrate these in future planning.

1962 December 7–8 Following a series of failures to Kiwi-B test reactors in the nuclear propulsion for space program, President Kennedy was given a briefing at Los Alamos. He allowed funding to continue but emphasized the need for success. Changes were made and cold-flow tests replaced hot reactor tests. These were successful, the first, on May 15, 1963, confirming vibration to have been the cause of earlier problems. As a result of budget problems, Rover was downgraded from a flight to a technology research effort and RIFT was canceled in December 1963. Lockheed's contract to develop the 80 ft. × 36 ft. 6 in. nuclear upper stage was terminated in February 1964. NERVA development was to continue.

1962 December 11 The first Minuteman I solid propellant ICBM went operational with U.S. Air Force Strategic Air Command (SAC) at Malmstrom AFB, Mont. Adjustments to the deployment plan provided SAC with a total 800 Minuteman missiles, the last of which became operational on June 15, 1965. Of these, 150 were of the LGM-30A specification based at Malmstrom AFB. The 650 LGM-30B missiles were in blast-resistant underground silos at Ellsworth AFB, S.Dak., Minot AFB, N.Dak., and Whiteman AFB, Mo., each with 150 missiles, and Warren AFB, Wyo., with 200 missiles. LGM-30B differed from LGM-30A in having a slightly larger second-stage motor made of titanium instead of steel. A single missile wing comprised three or four squadrons of 50 missiles, divided into five flights of 10 missiles each. Each flight had a single launch control facility (LCF) manned by two people. The silos were dispersed on farmland, the 1.5 acres for each silo being purchased from farmers at a price of $2,500. By May 1966, LGM-30C Minuteman II was being deployed and the build-up continued toward a total force of 1,000 missiles.

1962 December 13 The first successful launch of five satellites into orbit on one vehicle took place when a Thor-Agena D carried a combined payload weighing 304 lb. The five satellites were all placed into slightly different orbits with average parameters of 143 × 1,727 mi. and an inclination of 70.3°. Only one satellite, the 114-lb. *Injun 3,* was named, the others anonymous. The first decayed out of orbit on July 1, 1963, and the last decayed on February 9, 1967.

A NASA Delta launched from Cape Canaveral, Fla., successfully placed the experimental communication satellite *Relay I* in a 818 × 4,624 mi. orbit inclined 47.5°. *Relay I* comprised a 172-lb. tapered octagonal cone with a height of 2 ft. 8 in. and a diameter of 2 ft. 5 in. Transponders received at 1,727 MHz and transmitted at 4,170 MHz, providing two-way telephone links. The satellite also carried equipment for evaluating the degradation of solar cells in relatively low orbit, measuring radiation levels in that orbital path and supporting technology research for communication satellites. Over 500 tests were conducted. The second of the two satellites, *Relay II,* weighed 188 lb. and was launched on January 21, 1964, and operated for 20 months, compared with *Relay I*'s 26 months.

1962 December 14   At 19:59:28 UT the *Mariner II* spacecraft made its closest approach to the planet Venus, passing within 21,645 mi. of its cloud-shrouded surface at a speed of about 15,100 MPH. This was the first successful planetary encounter by a spacecraft from Earth. *Mariner II* carried six instruments: microwave and infrared radiometers for determining atmospheric composition and temperature; magnetometer for determining the strength and direction of interplanetary and planetary magnetic fields; high-energy radiation detector; solar plasma detector for determining the strength, energy and direction of charged particles emitted by the sun; and a cosmic dust detector. Spacecraft systems and science experiments were controlled by a central computer and sequencer (CC and S) and data returned to Earth by the transmitter at 8⅓ bits/sec. on a 3-W signal. *Mariner II* was last heard from at 07:00 UT, January 3, 1963, 65.8 million mi. away. It remains in a 65.5 million × 98.06 million mi. orbit of the sun. *Mariner II* made the astonishing discovery that Venus has a surface temperature of almost 800°F with a dense carbon dioxide atmosphere, and the phrase "greenhouse effect" was thus coined.

1962 December 16   The 222-lb. NASA science satellite *Explorer XVI* was launched by Scout from Wallops Island, Va., to a 466 × 733 mi. orbit with an inclination of 52°. This was the third NASA micrometeoroid satellite, previous ones having been a failed attempt (S-55) on June 30, 1961, and *Explorer XIII. Explorer XVI* was made up of 160 pressurized cells wrapped around the Altair (fourth) stage of the launcher (which comprised a cylinder with a length of 6 ft. 4 in. and a diameter of 2 ft.) 60 foil gauge detectors, 46 copper wire grids wound on melamine cards and a variety of materials and detectors. The satellite continued to operate until July 22, 1963. It remains in orbit.

1962 December 17   The first Nike-Zeus modified by the U.S. Army as an antisatellite test vehicle was launched from White Sands Missile Range as part of Project 505 (MUDFLAP). The three-stage rocket reached an altitude of about 115 mi., and a height of 175 mi. was achieved during the second flight on February 15, 1963. Tests then switched to the project's base on Kwajalein Atoll in the Pacific Ocean, but two intercept attempts at simulated targets flown on March 21 and April 19, 1963, were unsuccessful. The third test from Kwajalein was successful, however, when a Nike-Zeus passed within close range of a preorbited Agena D stage equipped with radar and range sensors on May 24, 1963.

1962 December 18   The world's first attempt to test an ion propulsion system in space was made when the U.S. Air Force launched a Scout rocket from Vandenberg AFB, Calif., carrying a prototype electric engine. The test was not a complete success due to a minor malfunction of some equipment, but it represented a first step on the path to a new and innovative form of propulsion. Electric engines work by catalyzing a fuel paste in a stream of charged particles and can operate for long periods at low thrust.

Prime Minister Harold Macmillan of Britain and President Kennedy began a series of four-day meetings at Nassau in the Bahamas to discuss Britain's strategic deterrent in the aftermath of a decision by the United States on December 7 to cancel Skybolt, which the United Kingdom had agreed to purchase. It was agreed that Britain could buy the Polaris SLBM (submarine-launched ballistic missile), effectively transferring the British nuclear deterrent from the Royal Air Force to the Royal Navy. Britain would manufacture the warheads and the submarines but purchase the missiles from Lockheed. To avert accusations that it was helping with the proliferation of nuclear weapons, the United States sought to have Britain agree on a multinational force with French and German participation. This was rigorously protested by the British, who maintained their claim to an "independent" nuclear deterrent. The Macmillan government planned to build five submarines, each equipped with 16 missiles. When the Labour government came to power in October 1964 they canceled one vessel.

1962 December 19   The first in a series of operational U.S. Navy Transit navigation satellites was launched by Scout to a 184 × 450 mi., 90.6° orbit from Vandenberg AFB, Calif. Designated *Transit 5A,* the satellite was scheduled as the first of an operational Transit series, but a power failure on the first day in orbit rendered it useless. *Transit 5A* decayed on September 25, 1986, and was succeeded by *Transit 5A3* on June 16, 1963. Thereafter, Transit satellites were veiled in secrecy, only two or three being launched each year until replaced by the Nova series.

1962–1963   During 1962 development was finalized on the Soviet SS-8 Sasin ICBM, a two-stage derivative of the SS-5 Skean intermediate range ballistic missile. The liquid propellant Skean had a length of 80 ft., a maximum diameter of 9 ft. and a second stage with smaller diameter. With a launch weight of about 170,000 lb., the missile had a range of 4,300 mi., a throw weight of 3,500 lb. and a 3-MT warhead yield. SS-8 entered service in 1963 and numbers peaked at just 19 before it was withdrawn in 1977. The Sasin had an accuracy of about 6,000 ft. across full range.

1963 January 2   Oran W. Nicks from NASA's Office of Space Sciences asked the director of the Langley Research Center, Floyd L. Thompson, if he would like to consider managing a lunar orbit photographic mission. Problems with getting the Ranger impact probe to work, management of Surveyor and planetary missions and a general overload of work cast serious doubt on the ability of the Jet Propulsion Laboratory to take on this additional project. Within two months Langley had produced a report on its ability to handle the project and provided NASA with a set of recommendations about the sort of photographic equipment needed. Meanwhile, Lee Scherer and Eugene M. Shoemaker, on loan from the U.S. Geological Survey, had completed a thorough analysis of what could be salvaged from Ranger and Surveyor to cut time and costs on the orbiter.

1963 January 3   Secretary of Defense Robert S. McNamara authorized priority development of an ABM (antiballistic missile) system based on advanced electronic, radar and missile technologies. Stimulated by the Nike-Zeus ABM program, which had been highly successful as a seven-year research program but inadequate for operational deployment, the new program would first be called Nike-X, then Safeguard, and would evolve into a two-tiered defense system using new missiles: the long-range Spartan and the super-fast short-range Sprint. Bell Laboratories became the primary system contractor and integrator. Development of the 5-MT-yield W-71 Spartan warhead began at Lawrence Livermore National Laboratory in March 1968 and was first tested underground in a shaft on the island of Kamchatka in the Aleutian chain on November 6, 1971, the largest underground nuclear test conducted by the United States. The enhanced-radiation, low-kiloton yield W-66 warhead for Sprint went into development at Los Alamos National Laboratory in 1966.

1963 January 4 The Soviet Union attempted to send the first of its second-generation Luna spacecraft to a soft landing on the surface of the moon, but failed when the SL-6 launcher left it stranded in Earth orbit. Placed in a circular 94-mi. orbit, the spacecraft and terminal stage decayed into the atmosphere after one day. A second attempt, also with an SL-6, on February 2, 1963, failed even to get the spacecraft into Earth orbit. The SL-6 was to become the standard launch vehicle for the second-generation Luna spacecraft, which typically weighed up to 3,600 lb., about four times the weight of the first-generation Soviet moon spacecraft. It was to serve the Soviet Union in this capacity for the next five years.

Rice University in Houston, Tex., announced it had established the first Department of Space Science in any college or university in the United States. Dr. Alexander Dressler was appointed chairman and professor of space science. Rice had previously donated land for the construction of the Manned Spacecraft Center, and this would also become the site of the Lunar Receiving Laboratory for the quarantine and initial examination of astronauts and moon samples and for the Lunar and Planetary Science Institute.

1963 January 11 A new group of Soviet cosmonauts reported for training duty at their facility outside Moscow, constituting the third group selected for spaceflight. They were a more mature and experienced group than their predecessors and included: Yuri Petrovich Artukhin, Lev Stepanovich Demin, Anatoli Vasilyevich Filipchenko, Aleksei Aleksandrovich Gubarev, Pyotr Ivanovich Kolodin, A. Kuklin, Vasily Grigorievich Lazarev, Obratsov, Vladimir Aleksandrovich Shatalov, Boris Borisovich Yegorov and Vitaly Mikhailovich Zholobov. The most famous of these was Shatalov, who would perform the first rendezvous and docking in space during the *Soyuz 8* flight launched on October 13, 1969, and later take charge of the cosmonaut training facilities.

The U.S. Air Force announced that some Thor missiles, returned from the United Kingdom as a reciprocal agreement with the Soviet Union for its pledge not to place ballistic missiles on Cuba, would be converted to satellite launchers. At least 15 of the 64 Thor missiles from the United Kingdom would be used in this way, but all 45 Jupiter missiles returned from Turkey and Italy would be dismantled. The U.S. Air Force had satellite launch pads for the Thor, but none for the Jupiter, and there were no further requirements for the latter's being used as satellite launchers. With conversion of the Thor production line to Delta satellite launchers, it made sense to recycle Thor missiles for that purpose.

1963 January 18 Concerned that the X-20 boost-glider might not be the appropriate manned vehicle for the U.S. Air Force, Secretary of Defense McNamara ordered a study be conducted of a military version of the two-man Gemini spacecraft called Blue-Gemini. The Air Force favored the one-man X-20, but McNamara was dubious about the role such a winged boost-glider could perform, stating his belief that the Air Force would be hard-pressed to demonstrate a need for any manned space vehicle.

1963 January 26 Special assignments to the widening range of U.S. manned spaceflight programs were announced for NASA astronauts, as follows: Cooper and Shepard, pilot phases of Mercury; Grissom, Gemini; Glenn, Apollo; Scott

Carpenter, lunar excursion training; Schirra, Gemini and Apollo operations training; Slayton, astronaut activities coordinator (assigned to this position in September 1962); Armstrong, trainers and simulators; Borman, launch vehicles; Conrad, cockpit layouts; Lovell, recovery systems; McDivitt, guidance and navigation; See, electrical and mission planning; Stafford, communications; White, flight control systems; Young, environmental control and survival equipment.

1963 January 30 In a report to the House Armed Services Committee, Secretary of Defense Robert S. McNamara announced that the Air Force was in the process of starting basic research on "an advanced hypersonic manned vehicle" as successor to the X-20/Dyna-Soar program. Under the designation Aerospace Plane, the new project would use hydrogen-fueled air-breathing engines to fly through the atmosphere like a conventional aircraft, switching to internal oxygen supplies to combust hydrogen for orbital flight.

Grumman selected four contractors to build elements of the Apollo lunar excursion module: the Rocketdyne division of North American Aviation for the throttleable descent engine; Bell Aerosystems for the fixed-thrust ascent engine; Marquardt Corp. for the reaction control system and thrusters; and Hamilton Standard for the environmental control system. When development problems plagued descent and ascent engines, Grumman would change contractors for those elements. The reaction control system would comprise four quads of 100-lb. thrusters operating on Aerozene 50–nitrogen tetroxide propellants cross-fed from two redundant systems. The LEM design would eventually put the four quads on the four corners of the ascent stage, providing attitude control and stabilization during descent and ascent. Environmental control would ensure a pure oxygen atmosphere at 5.5 lb./sq. in.

1963 February 1 NASA announced that it had awarded a study contract to Republic Aviation Corp. for a Synchronous Meteorological Satellite (SMS) system. Administered by Goddard Space Flight Center, the four-month study would define the technical requirements for a weather satellite placed in geostationary orbit at a fixed longitude and capable of observing the world's weather systems on a continuous basis across almost an entire hemisphere. The SMS would complement the Tiros series of low-altitude satellites. In April, four-month study contracts were awarded to RCA, on camera requirements, and to Hughes Aircraft, on ways to modify Tiros.

The American Institute of Aeronautics and Astronautics (AIAA) officially came into existence through a merger of the American Rocket Society (ARS) and the Institute of the Aerospace Sciences (IAS). Dr. William Pickering of the Jet Propulsion Laboratory became its first president. The AIAA became the leading technical society for U.S. aerospace interests and attracted membership from space professionals in the United States and abroad.

1963 February 2 Using a tracking antenna 82 ft. in diameter at Goldstone, Calif., the Jet Propulsion Laboratory successfully bounced signals off the surface of Mars for the first time. Transmitted at 100 kW, the radio signals took 11 min. 6 sec. to make the 62-million-mile round trip from Earth. The return signals indicated that the surface was rough and uneven in places but smooth in others, rather like Earth. Radar scans were useful for getting an indication of surface texture.

1963 February 7 From this date the terminology for identifying Saturn launch vehicle types changed, simplifying no-

To rehearse procedures considered vital for Apollo moon missions, NASA developed the two-man Gemini, seen here in scale with the one-man Mercury spacecraft.

menclature by deleting the "C" prefix and shifting from Arabic to Roman numerals. Thus, the Saturn C-1B became the Saturn IB and the Saturn C-5 became the Saturn V. This terminology was maintained throughout the life of the Saturn launch vehicle programs.

1963 February 12   NASA selected the Marion Power Shovel Co. to build the 3,000-ton crawler-transporter which would be used to move assembled Saturn Vs on their mobile launch platforms (MPLs) from the Vehicle Assembly Building to the launch pad more than 3 mi. away. The crawler was to be 131 ft. long, 114 ft. wide, with a height of 20 ft., equipped with four double-tracked crawlers, each 41 ft. long and 10 ft. high. Each track link weighed 1 ton. Powered by two 2,750-HP diesel engines driving four 1,000-kW generators for 16 traction motors, the crawler had a maximum loaded speed of 1 MPH. The crawler transporter would also be used to move the mobile service structure (MSS) from its park position to the pad and back again prior to launch.

1963 February 14   A Delta launched from Cape Canaveral, Fla., placed the 86-lb. *Syncom I* communication satellite in a geosynchronous transfer ellipse from where the onboard solid propellant apogee motor was fired to circularize the orbit at

about 22,000 mi. Just 20 sec. after the ground command to fire the motor, all contact with *Syncom I* was irretrievably lost. Tracking indicated the satellite to be in a 21,371 × 22,830 mi. orbit inclined 33.3° with a period of 23 hr. 46 min. *Syncom 1* was the first object launched to geosynchronous orbit.

1963 February 25   NASA Langley Research Center hosted a major planning and review meeting of a proposed lunar orbiter photographic project that would assist the Office of Manned Space Flight in determining the engineering needs of the lunar excursion module's landing gear. Representatives from the OMSF, Office of Space Science, Bellcomm and Space Technology Laboratories were tasked with preparing studies necessary to define the job a lunar orbiter could do within the overall program of lunar exploration and to identify reliability and mission capabilities. It was confirmed that an orbiter using existing technology could photograph and identify a landed Surveyor spacecraft and that there was a 93% probability in achieving one successful mission in five and an 81% chance of two successes.

1963 February 28   An attempt to launch the first thrust-augmented Thor-Agena D took place from Vandenberg AFB,

Calif. It failed to reach orbit, as did a second TAT-Agena D on March 18. In both cases the payload comprised a Corona reconnaissance satellite, successor to the Discoverer series. The launcher was a Thor-Agena D with three 54,000-lb. thrust solid propellant Castor I strap-on boosters attached to the main stage. Each booster had a length of 19 ft. 6 in. and a diameter of 2 ft. 7 in. Launch weight was 138,000 lb. and liftoff thrust a total 335,000 lb. By adding strap-ons, total payload capability was increased from 2,600 lb., with the unaugmented Thor-Agena D, to just over 3,500 lb. The first successful TAT-Agena D flight took place on June 15, 1963.

1963 March 1 NASA Ames Research Center, Calif., asked industry for feasibility studies on building a satellite capable of carrying living things into orbit and being returned to Earth with the specimens intact. Up to six recoverable biological laboratories capable of operating in orbit from three to thirty days were envisaged. After the study period, General Electric was selected as primary contractor of the Biosatellite program. Cut to three satellites, the series was initiated when the first satellite was launched on December 14, 1966.

The NASA Manned Spacecraft Center reported conclusions from studies conducted on the feasibility of a manned orbital space station, emphasizing that existing Apollo hardware could be used to assemble such a facility. Comprising three arms, the station would house eighteen astronauts, with an initial crew of six people launched by a modified Apollo spacecraft, followed by two more six-person crews for permanent occupation. Successive revisits would be made using a reusable shuttle-type space vehicle.

1963 March 5 NASA issued a request for proposals for its Voyager advanced planetary spacecraft to 21 U.S. aerospace companies specifying a program objective for missions to the inner planets between 1967 and 1975. Voyager was expected to be a combination orbiter-lander, but the Jet Propulsion Laboratory was reluctant to include a lander, believing that a large orbiter would best serve the needs of planetary science. NASA wanted to fly a Venus mission in 1967 and a Mars mission in 1969, but the JPL thought this would be rushing things. Flight weight was set at 4,795–7,000 lb. with the Saturn IB launch vehicle, but contractors were also asked to look at a smaller 3,970-lb. spacecraft launched by Titan III and at growth versions up to 59,500 lb. for Saturn V. A total of 37 organizations attended a briefing on March 11, with delivery of proposals on March 25, 1963. Two days later the selection process began, as a result of which General Electric was given a study contract in April.

1963 March 6–7 In testimony before the House Committee on Science and Astronautics, Subcommittee on Manned Space Flight, D. Brainerd Holmes, NASA director of Manned Space Flight, described combustion instability problems with the powerful F-1 engine being built for the Saturn V. In 250 test firings over two years, instability had been observed on seven occasions. This rate was unacceptable and indicated that fuel and oxidizer were surging as they entered the injector, a device which sprayed the propellants into the combustion chamber. On November 19 NASA confirmed that this potential threat to the Saturn schedule had been resolved.

1963 March 7 At a meeting of the NASA Gemini Program Planning Board, it was agreed that an ad hoc study group be formed to integrate Department of Defense experiments into the operational flight phase of the two-man program. This group met in continuous session between March 25 and April

26 and compiled a report presented to the planning board on May 6 and recommending a series of experiments. The group recommended additional Gemini flights be added to the already scheduled 12 to accommodate extra tasks the Department of Defense desired, but this was denied. The Department of Defense set up a liaison office at the Manned Spacecraft Center, Houston, Tex.

Grumman engineers completed a technical study of options for electrical power production on the Apollo lunar excursion module (LEM). It was based on a projection that the standard LEM moon mission would require 24-hr. continuous operation from undocking with the Apollo command module to rejoining it after ascending from the surface. For this mission a power requirement of 61.3 kWh was established, all of which three fuel cells in the descent stage would provide. Batteries would provide power for the ascent stage as it carried the two-man crew back into moon orbit. Pratt and Whitney was subsequently selected to provide the fuel cells, but the decision to use this type of electrical energy production was changed on March 2, 1965, when an all-battery LEM was adopted instead.

1963 March 11 NASA and the French government announced an agreement whereby payloads provided by the French National Center for Telecommunications would be launched on sounding rockets from Wallops Island, Va. Aimed at investigating the propagation of very low frequency (VLF) radio signals at heights between 46 and 62 mi., the initial phase of the tests would, if successful, be followed by a second phase in which a French satellite would be launched by NASA on a Scout to investigate VLF characteristics.

1963 March 14 Due to concerns about the development of the descent engine for the Apollo lunar excursion module, a bidders conference was held at Grumman for a mechanically throttleable engine to be built concurrent with the helium-injection engine contracted to Rocketdyne. This second-source contracting would be a feature of critical elements in the Apollo program and provide backup systems which would help keep key flight events on schedule. The Rocketdyne engine employed an untried design concept, and on July 3 Space Technology Laboratories received a contract to develop a more conventional motor using proven concepts.

1963 March 18 The first launch of a new panoramic photoreconnaissance system for targets of special interest took place with the flight of a thrust-augmented Thor Agena D from Vandenberg AFB, Calif. The launch vehicle and the Corona satellite payload failed to reach orbit. Developed in secret under Project Lanyard, the KH-6 Itek camera carried a 5 ft. 6 in. focal length lens for panoramic shots onto 4.5-in. film. A second KH-6 flight on May 18, 1963, placed the payload in orbit but the Agena stage failed. The first and only success with a KH-6 came with the third and last launch on July 31, 1963, although the camera malfunctioned after 32 hr. Project Lanyard was tied in with Project Gambit, the CIA's high-resolution photoreconnaissance system, which began tests with the first KH-7 payload launched on August 5, 1962.

1963 March 21 Contact with the Soviet spacecraft *Mars 1* was irretrievably lost less than three months before it would fly past the planet it was named for. A routine communication session with the spacecraft, then 66 million mi. from Earth, indicated that all internal systems were functioning as expected but that the spacecraft was experiencing difficulty in maintaining attitude hold. The solar cells had to be correctly aligned with the sun to produce electrical energy. Moreover, the fixed parabolic dish antenna was highly directional, and appreciable deviation from an Earth-aligned attitude would result in loss of signal. This is what happened, and the spacecraft flew to within 120,000 mi. of Mars on June 19, 1963, in silence. It remains in an elliptical orbit of the sun with a perihelion of 86 million mi. (just inside Earth's orbit) and an aphelion of 149 million mi.

At a meeting at the NASA Manned Spacecraft Center, Houston, Tex., guidelines were laid down for extravehicular activity (EVA) in the Gemini program. The basic pressure garment was considered sufficient to sustain life outside the spacecraft, but with supplementary thermal outer layers including gloves and special boots. A tether incorporating 12 nylon-encapsulated communication wires and long enough to allow access to the rear of the spacecraft adapter section was to be developed. Special maneuvering devices were to be developed to provide stability and a method of moving about outside the spacecraft. A baseline 30 min. of EVA was established, with modifications supporting space walks from spacecraft no. 4.

1963 March 26 The U.S. Air Force awarded a supplemental contract on the X-20 boost-glider, calling for conversion of a Boeing B-52C bomber for air-drop tests of the glider as well as the preparation of Launch Complex 40 at Cape Canaveral for flights of the Titan IIIC/X-20 combination. Over the next few months the Air Force restructured the program to include six flights of an X-20A to demonstrate reconnaissance and satellite inspection capabilities. An X-20B was proposed with the capacity to carry out antisatellite missions, and a bigger X-20X satellite inspector/destroyer version, with a 14-day endurance and a maximum altitude capability of 1,000 mi., was mooted. When Deputy Defense Secretary Harold Brown favored a permanently manned orbiting space station, the X-20 was canceled on December 10, 1963.

1963 March 28 In the first test of its kind ever conducted during a live launch, NASA sent the fourth Saturn C-1 (SA-4) on its way into space from Cape Canaveral. It was programmed to test a simulated engine failure during ascent. Launched at 20:12 UT, SA-4 carried 625,000 lb. of propellant for a liftoff weight of 940,000 lb. At 1 min. 40 sec., one of four inboard H-1 engines shut down, simulating an unplanned cutoff. The remaining seven engines continued to fire until an elapsed time of 1 min. 53 sec., when the three live inboard engines shut down, followed 7 sec. later by the outer four H-1s. This demonstrated that the guidance system would continue to power the launcher on seven engines should one fail in flight. The last of four Block I Saturn C-1 flights, SA-4 reached an apogee of 81 mi. and a downrange distance of almost 130 mi.

1963 April 2 The Soviet spacecraft *Luna 4,* the third example of the Soviet Union's second-generation spacecraft for lunar exploration, was launched by an SL-6 to an Earth escape trajectory after a brief period in Earth orbit. *Luna 4* weighed 3,135 lb. and incorporated a propulsion system which would be used for course corrections and for decelerating the spacecraft to almost zero relative to the surface of the moon, ejecting a 220-lb. spherical lander for a hard touchdown on the lunar dust. The spacecraft failed to adjust its trajectory for lunar contact and passed within 5,280 mi. of the surface at 01:24 UT on April 6, deflected by the lunar mass into a barycentric orbit of the Earth with initial parameters of 55,900

× 435,000 mi. Led by Andrei Severny, a group of scientists in the Crimea had optically tracked *Luna 4* to a distance of 87,000 mi. Due to perturbations of the Earth, moon and sun, the spacecraft would eventually shift to a heliocentric orbit.

NASA announced it had signed a definitive contract with McDonnell Aircraft Co. for thirteen Gemini spacecraft, one of which was to be used exclusively for ground tests. In addition, the company was to provide two mission simulators (one for the Manned Spacecraft Center, Tex., the other for Cape Canaveral, Fla.), a docking simulator trainer, five boilerplate spacecraft and three static test articles for vibration and impact evaluation.

1963 April 3 NASA launched the first in its series of Atmosphere Explorer satellites, AE-A *(Explorer XVII)* on a Delta from Cape Canaveral, Fla. Placed in an orbit of 158 × 570 mi., with an inclination of 57.6°, the spherical satellite had a diameter of 2 ft. 11 in. and weighed 405 lb. Powered by batteries, AE-A was spin-stabilized at 90 RPM, carried eight instruments from the Goddard Space Flight Center and discovered a band of neutral helium surrounding the Earth before it ceased operating on July 10. AE-A decayed on November 24, 1966. Second in the series, AE-B *(Explorer XXXII)* was placed in an incorrect orbit of 180 × 1,687 mi., inclined at 64.5°, on May 25, 1966. It had solar cells to supplement battery power and a tape recorder, and it returned data for eight months. AE-B decayed on February 22, 1985.

1963 April 11 Representatives from the NASA Manned Spacecraft Center (MSC) visited the Langley Research Center to receive a briefing on their Manned Orbital Research Laboratory (MORL) concept. MSC was keen to support a serious study of space station requirements and considered the Apollo spacecraft a suitable logistics and ferry vehicle. Over time the station concepts grew in size, until launch and resupply costs became a major factor and the Apollo spacecraft appeared inadequate for the task. MSC cautioned against making the station too big, opting for small resupply and ferry vehicles cheaper to launch than Apollo.

1963 April 17 The U.S. Air Force awarded parallel contracts to three contractors to begin development of large solid propellant rocket motors: Aerojet General Corp. and Thiokol were to conduct parallel development of separate half-length motors each 260 in. in diameter with a thrust of 3 million lb.; Thiokol would also develop a solid 156 in. in diameter with a thrust of 3 million lb. and a 156-in.-diameter solid of 1-million-lb. thrust utilizing a movable nozzle; Lockheed would develop a 156-in. solid of 1-million-lb. thrust utilizing deflectors for steering. Based on the results of the parallel Aerojet/Thiokol tests, one of these two contractors would be asked to demonstrate a 6-million-lb. thrust motor.

1963 April 25 With delays in the Ranger lunar impact program due to repeated spacecraft failures and technical problems at the Jet Propulsion Laboratory, Ranger Project Manager Harris M. Schurmeier requested a list of priorities from NASA and got word that Headquarters now seriously questioned the wisdom of proceeding with Block V hard landers, which were, in effect, backup for the more capable Surveyor soft landers expected to reach the lunar surface during 1965. On May 22 NASA formally cut the Block V program from twelve to six spacecraft, making funding available for the Lunar Orbiter program in fiscal year 1964; it then canceled the three Block IV spacecraft on July 12, 1963. At this date the six Block V spacecraft were to fly between April 1965 and August 1966, but on December 13, 1963, these were also canceled, leaving just the four Block III spacecraft *(Rangers 6–9)*, which had yet to fly.

1963 April 29 Changes to the Gemini program flight schedule were approved by NASA headquarters. Instead of one unmanned qualification flight there would now be two: the first an orbital mission to demonstrate launch vehicle and spacecraft compatibility without separation of the Gemini from the second stage or recovery, and the second a suborbital test of the heat shield and recovery systems. These two flights were now scheduled for December 1963 and July 1964. The first manned flight would be a three-orbit shakedown flight in October 1964, with subsequent flights at three-month intervals through January 1967. Water landing by parachute was now planned for the first six flights, followed by paraglider landings on flights seven to twelve.

1963 May 1 The Future Projects Office at NASA's Marshall Space Flight Center completed a summary of Nova studies carried out by U.S. aerospace companies. From this came follow-up studies conducted during 1963 by General Dynamics, Martin, and Douglas on launch vehicle traffic needs between 1975 and 1990 for launchers with a 1-million-lb. (500-ton) single-flight orbital payload capability. Missions envisaged by contractors included a 50-person lunar base from 1975, manned Mars exploration from 1981, heavy unmanned planetary spacecraft from 1979 and space stations from 1980. In addition, the Martin Co. envisaged a Nova military strike force beginning in 1976 and building to 15 flights a year by 1981, as well as a global transportation system starting with 52 flights in 1980 building to 242 annually by 1990. None of these concepts, some involving imaginative launchers with thrust levels of up to 55 million lb., were pursued.

1963 May 2 In the first major test firing of a West German rocket since 1945, Dr. Berthold Seliger fired a three-stage solid propellant rocket to a height of 63 mi. from a firing range at Cuxhaven. The rocket had been designed to reach a height of 93 mi. and the terminal stage was returned to Earth by parachute. Without proof, Seliger claimed he had been involved in the development of the German V-1 and V-2 and intended that the three-stage rocket should be used for space research.

NASA and the Brazilian Commission for Space Activities (CNAE) announced that they had signed a memorandum of understanding for "a cooperative program in space research of mutual interest for peaceful scientific purposes." The CNAE wanted NASA's help to investigate the equatorial ionosphere by monitoring satellites and was building three ground stations to do that.

1963 May 3 The Rust Engineering Co. of Birmingham, Ala., began the design of a 4,900-ton mobile servicing structure (MSS), which would be used for the Saturn V launch vehicles as they were being prepared on the pad at Cape Canaveral, Fla. Construction began on February 21, 1965, the structure was topped out on November 9, 1965 and the MSS was completed and moved to Complex 39A on test for the first time on July 22, 1966. With a total height of 402 ft. and a base 135 ft. × 132 ft., it provided five platforms for servicing the Saturn V and would be moved by the crawler transporter from

its park position to the pad, where personnel could gain access to work areas on the launch vehicle. It would be moved back to the park position prior to launch.

1963 May 4 Provisional agreement was reached in Washington, D.C., by eight NATO countries for a force of 25 surface ships, each carrying eight Polaris missiles and operated by multinational crews. The countries were the United States, the United Kingdom, West Germany, Italy, Holland, Belgium, Turkey and Greece. The United States wanted to retain missile-firing control, and under the terms of the agreement each NATO country would have right of veto over the use of the missiles in anger. Final agreement was not reached and the idea was dropped.

1963 May 6 The continually delayed Mariner B planetary spacecraft concept was canceled and approval tentatively given for a combined NASA flyby/probe mission to Mars utilizing the 1966 launch window. In this plan, Mariner spacecraft E and F would each be equipped with a small atmospheric entry probe and launched by Atlas-Centaur. Project approval was formally granted by NASA on December 19, 1963, but initial discussions about a more ambitious mission began a month later. NASA wanted a combined orbiter/lander flight, believing it would more effectively serve as a precursor to Voyager surface missions beginning in 1971. At this time it became clear that NASA would focus on Mars after the manned Apollo program, and attention shifted dramatically to the unmanned exploration of that planet.

In classified testimony before the House of Representatives Military Appropriations Subcommittee, Dr. Harold Brown, director of Defense Research and Engineering, said that the MIDAS missile defense alarm system was being all but terminated. He said that unforeseen difficulties in detecting missile exhaust plumes by infrared energy had rendered the system virtually useless. He further testified that, of the $423 million spent so far, almost half the development money had no useful return because the science and the physics of the atmosphere had not yet been understood well enough to devise operationally useful technology.

1963 May 7 NASA launched *Telstar II* on a Delta launcher for the American Telephone and Telegraph Co. (AT&T) from Cape Canaveral, Fla., to a 600 × 6,716 mi. orbit inclined 42.7°. Similar to *Telstar I*, the active-repeater communication satellite was the second of two Telstars used to transmit color and black-and-white television programs from the United States to England and France. It continued to transmit experimental programs until May 1965.

1963 May 8 President Kennedy gave approval for development of an antisatellite system, "at the earliest practicable time," based on the Thor missile. Known as Program 437, it was a successor to SAINT, and in a state department memorandum it was defined as a defensive system against Soviet orbital bombs. On July 6, 1963, it was given highest national priority. Tests were scheduled to take place from Johnston Atoll in the Pacific.

1963 May 9 An experimental MIDAS early warning satellite was launched by Atlas-Agena B from Vandenberg AFB, Calif., and in six weeks of successful operation demonstrated the feasibility of missile detection and observation. From a 2,239 × 2,287 mi. orbit with an inclination of 87.4°, the satellite detected nine Titan II, Atlas E, Minuteman ICBM and Polaris SLBM (submarine-launched ballistic missile) launches. Designed to observe the hot exhaust of liquid propellant, the infrared sensors also detected the relatively cooler exhaust from the solid propellant Minuteman and Polaris — a particularly notable feat. MIDAS also carried two 1.5-lb. Tetrahedral Research Satellites (TRS-2 and -3), built by TRW and called *ERS-5* and *ERS-6,* and dispersed 400 million copper needles into a belt surrounding the Earth in an experiment called Project West Ford. Placed in a 2,241 × 2,683 mi. orbit at an 87.3° inclination to the equator, the ring provided a surface off of which radio waves were bounced to demonstrate passive global communications for military forces. Another MIDAS, launched on July 18, 1963, detected Soviet missile flights.

1963 May 10 Soviet spacecraft designer Sergei Pavlovich Korolov approved a plan to fly cosmonauts around the moon in a manned spacecraft, Soyuz, comprising three separate modules: Soyuz-A, Soyuz-B and Soyuz-V. The 14,200-lb. Soyuz-A would consist of a pressurized bell-shaped descent module carrying two or three people, 6 ft. 6 in. in diameter with a height of 7 ft. 10 in., attached at the base to a drum-shaped instrument module, 8 ft. 6 in. in length and 8 ft. 2 in. in diameter. The instrument module had a rocket motor at the back for rendezvous, docking and course corrections. A habitable, cylindrical orbital module, 10 ft. 2 in. long with a diameter of 7 ft. 6 in., was to have been attached to the forward end of the descent module. Soyuz-B was to have been an unmanned rocket stage, 15 ft. 5 in. in length and 8 ft. 2 in. in diameter, carrying 27,500 lb. of liquid oxygen–kerosene propellant but launched empty. Propellant would be carried into orbit by three Soyuz-V tanker modules, 13 ft. 9 in. in length and 8 ft. 2 in. in diameter, and then transferred to Soyuz-B in space. Soyuz-B would have sufficient propellant to push the manned Soyuz-A out of Earth orbit to a circumlunar trajectory. The whole idea was later dropped in favor of a manned landing plan.

1963 May 15 The last Mercury mission, the fourth U.S. manned orbital flight, lifted off of Pad 14 at Cape Canaveral, Fla., at 13:04:13 UT, beginning a 22-orbit flight that would last 34 hr. 19 min. 49 sec. Initially, the spacecraft was placed in a 100.3 × 165.9 mi. orbit inclined 32.5°. When the Mercury program gathered momentum during 1959, an 18-orbit flight was set as a goal for future missions. In planning MA-9, flight engineers selected a slightly longer mission. Pilot Gordon Cooper slept fitfully during periods of the flight and carried out several tests and experiments as well as assembling biomedical data which would serve as a baseline for the long-duration flights of the two-man Gemini program. Named *Faith 7* by Cooper, the 3,034-lb. spacecraft performed flawlessly and splashed down in the Pacific Ocean May 16, less than 4 mi. from the recovery ship. The only anomaly during reentry occurred when a faulty connector forced Cooper to take over manual attitude control.

1963 May 22 A solid propellant Scout rocket carried a 480-lb. nuclear reactor mock-up to an altitude of 800 miles before falling back to the Atlantic Ocean southwest of Bermuda. Designated RFD-1 (reentry flight demonstration 1), the purpose of the flight was to obtain test data on how a nuclear reactor could be designed to break up harmlessly in the upper atmosphere. Supplied by the Atomic Energy Commission, the nonradioactive SNAP (systems for nuclear auxiliary power) space nuclear reactor had dummy fuel rods.

1963 May 23 A joint space program between NASA and the New Zealand National Space Research Committee got under way with the launch of an Arcas sounding rocket from Birdling's Flat on Canterbury Plains. Instrumented to measure the electrical potential of the ionosphere, the rocket payload had been designed and built in New Zealand by Dr. John B. Gregory of the University of Canterbury. Two further launchings of Arcas rockets took place with NASA assistance later in the year.

1963 May 29 On the 19th test launch of the Titan II qualification program, flight vehicle N-20 was launched from Cape Canaveral carrying "Pogo," a specially modified propulsion system designed to eliminate vertical oscillations of up to 2.5 g observed in test flights since the first Titan II launch on March 16, 1962. N-20 failed after 55 sec. but not because of the Pogo suppression equipment, which included oxidizer standpipes and fuel accumulators. The oscillations were not important when the missile operated in the ICBM role for which it had been designed, but they could impair the ability of astronauts to respond to emergency situations. Pogo suppression devices were tested on three additional Titan qualification flights.

1963 June 1 The NASA Manned Spacecraft Center awarded contracts for space station studies to Lockheed and Douglas Aircraft. The baseline requirement was for proposals on a 24-person Earth-orbiting laboratory with a five-year mission life. The contractors had a free hand in proposing supply and logistical systems, although NASA stipulated a cyclical revisit for exchanging crews and consumables as well as scientific experiments. The contracts were completed and studies presented in January 1964.

1963 June 6 Officials from the NASA Manned Spacecraft Center, Houston, Tex., briefed Administrator James Webb on the benefits of continuing the Mercury program to include at least one more orbital flight. Engineers believed the spacecraft could fly longer missions than the 22-orbit flight of Gordon Cooper on May 15–16, 1963, and pointed to the advantage of maintaining an operational momentum prior to the start of the two-man Gemini program. At this date the first manned Gemini missions were scheduled for late 1964. Technicians were already working on modifications to spacecraft no. 15A for a possible MA-10 flight, but Webb decided the Mercury program should stop at this point to focus attention on the Gemini and Apollo projects. On June 12, 1963, James Webb informed the Senate space committee, "There will be no further Mercury shots."

Mr. Adlai Stevenson, U.S. ambassador to the United Nations, sent a letter to UN Secretary General U Thant complaining that the Soviet Union had failed to register the launch of six satellites between August 25, 1962 and January 4, 1963. The letter was prompted by complaints from the Soviet Union that the United States was conducting "dangerous activities" in space, drawing attention to the Project West Ford experiment. Under agreements signed by the USSR, all satellites must be registered with the UN. All flights cited by the United States were Mars, Venus or lunar probes that failed to leave Earth orbit.

1963 June 8 The first Titan II squadron became operational with the U.S. Air Force Strategic Air Command (SAC) at Davis-Monthan AFB, Ariz. The sixth and last Titan II squadron became operational on December 31. Squadrons were established in pairs at each base, others being McConnell

AFB, Kan., and Little Rock AFB, Ark. By the end of 1963, when all Titan IIs were in place, SAC had a total of 237 liquid propellant ICBMs, of which 108 (Titans) had storable propellants and could be launched rapidly. The 54 Titan IIs on alert were each equipped with the General Electric Mk.6 reentry vehicle, a cone-shaped ablative structure with a length of 10 ft. 2 in., a diameter of 4 ft. 11 in. and an empty weight of 2,076 lb. It housed the 9-MT W-53 warhead, a cylindrical device with a length of 8 ft. 6 in. and a diameter of 3 ft. weighing just under 7,000 lb. With this warhead, the most powerful deployed on a U.S. ICBM, Titan II had a range of up to 9,325 mi. By the end of the year SAC also had 13 Minuteman squadrons operational with 650 solid propellant missiles, bringing to 887 the total number of operational ICBMs. At this date, the Soviet Union had fewer than 200, building toward a total of 209 SS-7s and SS-8s by the end of the year.

1963 June 13   NAA (North American Aviation) signed a contract with the David Clark Co. for Gemini space suits after evaluating them against competing suits from the B. F. Goodrich Co. Some elements from the latter's design would be utilized in the David Clark suit, however. The G3C would be used for intravehicular operations and incorporated an outer layer of temperature-resistant nylon over a link-net material to control ballooning. The pressure layer was made of neoprene-coated nylon lined with a nylon comfort layer. About 60% of the oxygen delivered to the suit was directed through boots and gloves, flowing up arms and legs and across the torso, with 40% coming through a helmet neck ring and up across the faceplate. The extravehicular G4C suit provided added meteoroid and thermal protection.

1963 June 14   At 12:00 UT the Soviet cosmonaut Valery F. Bykovsky was launched in the 10,408-lb. *Vostok 5* to a 101 × 130 mi. orbit inclined 65° to the equator. His backup was Boris V. Volynov. In the absence of advance information on the purpose of this flight, observers in the West believed it to be a repeat of the dual flights of *Vostok 3* and *Vostok 4* launched in August 1962. When the spacecraft had been in orbit for two days without further news of a second launch, expectations faded until *Vostok 6,* carrying the first woman cosmonaut, was launched on June 16. Bykovsky remained in space until June 19, returning to Earth after a flight lasting 4 days 23 hr. 6 min., still the longest record for a solo space mission.

1963 June 15   The first launch of six satellites on one launch vehicle took place when a Thor-Agena D carried 3,589 lb. of payload to orbit from Vandenberg AFB, Calif. The primary payload comprised a 3,307-lb. classified military satellite based on the Agena D; a very low frequency 57-lb. test satellite, *Lofti 2A ;* an 86-lb. solar radiation satellite, *Solrad 6*; a 55-lb. radiation dosimeter payload, Radose; a 77-lb. classified satellite; and a 7-lb. surveillance calibration satellite, *Surcal 1C*. All six satellites decayed during July and August 1963.

1963 June 16   At 09:30 UT the Soviet cosmonaut Valentina Tereshkova was launched from the Tyuratam complex in the 10,392-lb. *Vostok 6*—the first flight carrying a woman into space. There have been unconfirmed reports that the prime crew member was an unnamed woman who was replaced by her backup, Tereshkova, making cosmonaut Irina Solovyov backup to Tereshkova. Ascending on an SL-3 launch vehicle, the spacecraft was placed in a 104 × 135 mi. orbit inclined 65° to the equator. At orbit insertion, *Vostok 6* was about 3 mi.

Yuri Gagarin, the first man in space, chats with Valentina Tereshkova, the first woman in space, launched aboard Vostok 6 *on June 16, 1963.*

away from *Vostok 5,* launched two days earlier. Over the years there have been persistent but unconfirmed reports that Tereshkova became ill during the flight and was in great pain for portions of the mission. *Vostok 6* returned to Earth on schedule June 19 after a flight lasting 2 days 22 hr. 50 min. Tereshkova married cosmonaut Andrian Nikolayev on November 3, 1963, and on June 8, 1964, they had a daughter, Yelena Andrianovna. As the first child of two people who had been into space, the girl was given frequent medical examinations to determine whether there had been any medical effects ascribable to the experiences of her parents. None was found.

1963 June 18   Philip Bono, chief advanced project engineer at the Future Systems Division of the Douglas Missile and Space Division, described a radical new space transportation concept. Called ROMBUS (reusable orbital module-booster and utility shuttle), the concept envisaged a single-stage manned space cargo carrier capable of being used at least 20 times. ROMBUS would weigh 14 million lb., and its propulsion system would develop 18 million lb. of thrust at launch from propellant in eight jettisonable tanks strapped around the sides of the vehicle. It would have a height of 95 ft. and a base diameter of 80 ft. Carrying 500-ton loads into orbit, it would descend through the atmosphere protected by a heat shield and back itself onto a landing pad with legs for stability. ROMBUS was never developed but it remains a landmark in single-stage designs.

1963 June 21   The U.S. Air Force announced contracts worth a total $1.5 million placed equally with General Dynamics Astronautics, North American Aviation and Douglas Aircraft Co. for investigation of the Aerospace Plane (ASP) concept of an air-breathing space plane powered by scramjet. Various concepts had been examined and a variety of different propulsion systems were considered. There was, however, a difference of opinion about the military need for such a vehicle and concern among some that the Air Force was funding basic research of questionable value. The ASP was to be powered by an air collection and enrichment system (ACES) engine that extracted and liquified oxygen from the atmosphere—a concept developed by the Marquardt Corp. and General Dynamics.

1963 June 24 NASA's Langley Research Center announced that Boeing and Douglas Aircraft had been selected for parallel studies on a Manned Orbital Research Laboratory (MORL). The station conceived at Langley comprised a four-person workshop with resupply and logistics flights changing crew members. The experiment's plan envisaged one astronaut remaining in space for up to a year to conduct physiological experiments and perform biomedical research on the effects of weightlessness on the human body.

1963 June 26 The NASA Manned Spacecraft Center signed a contract with the Boeing Co. for the research project *Study of Lifting, Re-Entry Horizontal Landing Type Logistics Spacecraft for Manned Orbital Space Station.* Boeing was instructed to study the M-2 lifting-body developed at the Ames Research Center and the HL-10 designed at the Langley Research Center and to utilize these as the baseline design for a spacecraft which could be launched atop Saturn IB launch vehicles.

1963 June 27 The first of two Hitchhiker satellites instrumented to measure radiation in the Earth's Van Allen belts was launched by thrust-augmented Thor-Agena D from Vandenberg AFB, Calif. The 176-lb. satellite was ejected from the aft section of the Agena D, which formed the boost stage for a Corona military reconnaissance satellite placed in an orbit of 122 × 246 mi. at an inclination of 81.6°. *Hitchhiker 1* was ejected near apogee and was propelled by its own injection motor to an orbit of 201 × 2,571 mi. at an inclination of 82.1°. It returned data for three months and remains in orbit. *Hitchhiker 2* got a ride on the back of a KH-7 launched by Atlas-Agena D on August 14, 1964, to a 163 × 2,332 mi., 95.6° orbit. It decayed March 8, 1979.

1963 June A one-person space scooter was tested at North American Aviation's Space and Information Systems Division, Calif. Designed by Jack Bell, director of Lunar and Planetary Systems, and John W. Sandford, research specialist in Advanced Systems, the scooter was capable of hovering or maneuvering in any axis. It could be used to hop across craters, put a lunarnaut down inside a crevasse, scale steep cliffs or transport crew members between orbiting vehicles. It was never adopted for operational use.

1963 July 1 NASA selected the Boeing Co. to conduct a four-month study of a base on the moon that could be built after initial manned lunar landings. This was the first in a series of studies to determine the feasibility of using Apollo hardware to establish a tentative foothold in lunar exploration. The Apollo lunar excursion module (LEM) would stay on the moon for only a limited period, but the Apollo and Saturn launch vehicles could form the basis for a series of follow-up missions using elements of the LEM to build a more permanent facility. The belief that Apollo represented the start of a permanent colonization of the moon was, for NASA, a grave error of judgment; when the first landings had been accomplished, Congress was no longer interested in expanding that capability.

1963 July 3 NASA announced that Republic Aviation Corp. would build the Advanced Orbiting Solar Observatory (AOSO) for the Goddard Space Flight Center. A 1,250-lb. satellite designed to point 250 lb. of scientific instruments accurately at the sun to measure solar radiation from a near-polar orbit 300 mi. above Earth, the AOSO was to have observed x rays, gamma rays and ultraviolet emissions. It was to have extended the work carried out by NASA's Orbiting Solar Observatories, but the program was canceled for budgetary reasons on December 16, 1965, less than four years prior to launch.

1963 July 4 NASA's Marshall Space Flight Center awarded a study contract to Space Technology Laboratories (STL) for definition and design requirements of an operational nuclear propulsion system for the 1970s. The Saturn V and Nova conventional launch vehicles were specified as candidates for lower stage elements, the nuclear upper stage operating only in the vacuum of space. Parallel study contracts awarded to General Dynamics and Douglas Aircraft Co. were to examine a completely new chemical/nuclear launch vehicle configuration capable of placing 500 tons in Earth orbit.

1963 July 12 The first Atlas-Agena D launched the first 4,400-lb. KH-7, U.S. Central Intelligence Agency close-look reconnaissance satellite into a circular 102 mi., 95.4° polar orbit from Vandenberg AFB, Calif. This was also the 100th launch of an Agena upper stage. Launched to collect imaging data on areas of special interest sometimes disclosed by pictures from the Corona/KH-4 series, operating lifetimes were short, sometimes just one day. On one occasion a KH-7 dipped to within 76 mi. of the Earth to get the theoretically best picture resolution. KH-7s released film recovery capsules snatched midair by JC-130B aircraft operating out of Honolulu; their contents were delivered to the National Photographic Interpretation Center in Washington, D.C. The last of 34 Atlas Agena D/KH-7 missions was launched on May 22, 1967, its role superseded by the KH-8 launched by Titan IIIB.

1963 July 16 NASA's Manned Spacecraft Center completed a review of alternative docking mechanisms for linking together the main Apollo spacecraft with the lunar excursion module and settled on an extensible probe and drogue device. The apex of the conical command module carrying the three-man crew would comprise a tunnel to the end of which would be attached a telescoping probe. With the probe extended, three small capture latches on the nose of the probe would engage the lip of a conical drogue in the upper docking tunnel of the LEM. Once engaged, the probe would be retracted, mating the two circular docking rings on respective vehicles via twelve main latches on the command module and creating an air-tight seal. With probe and drogue removed and respective hatches opened, crew access to the LEM would be made through the short tunnel connecting this vehicle to the command module.

1963 July 20 The first static firing of the biggest solid-propellant production rocket motor took place at Coyote, Calif. For the first time a full-size five-segment solid propellant motor built by United Technologies Corp. for the Titan IIIC satellite launcher was fired, producing more than 1 million lb. of thrust for 2 minutes. Two 75 ft. motors, each with a liftoff weight of 500,000 lb., would power the Titan off its launch pad. Payload capability would be 25,000 lb. to low Earth orbit.

1963 July 23 NASA announced that George E. Mueller, vice president for Research and Development at Space Technology

During 1963 development of large, solid propellant boosters for the Titan launch vehicle got under way. Here, a test booster is prepared for firing in a vertical position.

Laboratories, would take over as director of the Office of Manned Space Flight, succeeding D. Brainerd Holmes, who was to resign September 1, 1963. Mueller brought fresh thinking to manned flight planning and instituted several changes that, over time, were to improve the chances of landing a man on the moon by the end of the 1960s. Most notable of these was the concept of "all-up" testing that significantly cut the number of Saturn development flights.

1963 July 25 Representatives from the United States the Soviet Union and the United Kingdom signed an agreement banning nuclear weapons tests in the atmosphere, underwater and in space. This effectively opened the door on a greater level of scientific cooperation between the United States and the USSR, a move urged by British astronomer Sir Bernard Lovell, director of the Jodrell Bank radio telescope in England. Sir Bernard had visited the USSR as a guest of the Soviet Academy of Sciences, and upon his return wrote to NASA Deputy Administrator Dryden to express his belief that the Soviets had postponed plans for a manned lunar landing on the basis that the risks were too great and the returns too small.

1963 July 26 The first successful geosynchronous satellite to operate as planned was launched by NASA from Cape Canaveral, Fla., when a Delta launch vehicle carried the 86-lb. *Syncom II* communication satellite to a geosynchronous transfer ellipse. About 6 hr. after launch, its 1,000-lb. thrust solid propellant apogee motor fired, increasing speed by 4,696 ft./sec. and placing the satellite in a 22,112 × 22,801 mi. orbit

inclined 33.1°. Onboard hydrogen peroxide jets adjusted the orbit so the satellite could drift to 55° W longitude. *Syncom II* described a figure-eight pattern in the sky and was used to test the transmission of 1 two-way voice channel or 16 teletype channels.

1963 July 27 NASA and a group of Swedish scientists began a series of four sounding rocket flights from a launching site at Kronoberg, Sweden. Four Nike-Cajun rockets with rocket grenade payloads were launched between this date and August 8 to study noctilucent clouds, faintly luminous, very high and fast-moving clouds only visible at twilight for brief periods in summer at high latitude. Direct sampling of the clouds would enable scientists to measure temperatures, pressure, density and atmospheric changes.

1963 August 1 The first antisatellite system deployed by the United States was declared operational from its base at Kwajalein Atoll. The U.S. Army's Program 505 incorporated the powerful three-stage Nike-Zeus equipped with a nuclear warhead designed to detonate near enemy satellites causing damage to the electronics. A test on January 6, 1964, passed within 700 ft. of a simulated satellite at an altitude of 91 mi. and a slant-range from launch of 109 mi. Another Nike-Zeus in April 1964 launched communication components for exoatmospheric beam intercept, and three successful flights out of four attempts in June and July 1965 evaluated missile warhead and antisatellite equipment. Finally, a missile fired on January 13, 1966, qualified the launch crew to operationally prepare and fire a Nike-Zeus when needed. Secretary of Defense Robert S. McNamara canceled Program 505 on May 23, 1966, due to duplication with the U.S. Air Force Program 437 and his belief that the Air Force should have control over antisatellite activities.

1963 August 5 Newly selected astronauts and two Mercury astronauts began a 5-day survival course at Stead AFB, Nev. The course was designed to prepare them for emergency landings in remote desert regions of the Earth. The course was split into three phases including 1½ days of academic study, 1 day of field demonstrations in survival techniques and 22 days during which pairs of astronauts were left alone in the desert to apply their lessons to field situations.

NASA selected TRW to build five interplanetary Pioneer spacecraft to investigate the sun's radiation and the solar wind in deep space, seen by NASA as vital platforms for harvesting data about the solar system. The final design review on the TRW spacecraft was held in April 1964 and the first, *Pioneer 6,* was delivered to Cape Canaveral on December 5, 1965. On February 22, 1966, NASA canceled the fifth spacecraft as a budget-cutting measure. On April 28, 1966, TRW was informed that a fifth spacecraft would be built after all, from spares. It was eventually named *Pioneer-Helios,* only to be canceled again in 1973.

1963 August 16 The first flight of the M2-F1 lifting-body took place at the NASA Flight Research Center, Calif. Piloted by Milton Thompson, it was towed into the air at Rogers Dry Lake behind a Douglas C-47. Attached to the converted DC-3 by a 1,000-ft. cable, the lifting-body was taken to a height of 10,000 ft. and released. Built as an unpowered glider, a precursor to powered successors, the M2-F1 was half conical in shape without wings. It had a length of 20 ft., a height of 10 ft. and a width of 13 ft. incorporating vertical rudders, two trailing-edge flaps and two elevons. The M2-F1 had made 100 towed drop flights when tests ended in August 1964.

NASA and the Soviet Academy of Sciences announced agreement on a range of cooperative space projects between the United States and the Soviet Union. Meteorological data was to be exchanged and joint communications experiments using the *Echo II* satellite were to be conducted. In addition geomagnetic data were to be exchanged from the World Magnetic Survey of 1965, for which satellites would help expand understanding of the Earth's magnetic field.

1963 August 21 Titan II development flight N-24 was launched from Cape Canaveral carrying for the first time a specially developed malfunction detection system (MDS) in support of the NASA two-man Gemini program. Because Gemini would incorporate abort systems for safely removing the crew from a launcher run amok, special equipment to report potentially dangerous situations was needed. The MDS included an electronic package that integrated with spacecraft abort equipment. This flight was a failure unconnected with the MDS, and four more Titan development flights verified the system, qualifying the Titan II for manned flight.

1963 August 23 The U.S. Air Force awarded two contracts for development of solid propellant rocket motors: Thiokol was to produce a half-length motor of 260-in. diameter producing a thrust of 3 million lb. and a two-segment demonstration motor of 156-in. diameter also producing 3-million-lb. thrust; Aerojet General would demonstrate a half-length motor of 260-in. diameter producing a thrust of 3 million lb. Based on results from these demonstrations, the Air Force planned to select one contractor to develop a full-scale 260-in. motor.

1963 August 25 The first in a series of significantly enhanced U.S. military reconnaissance satellites was launched by a thrust-augmented Thor-Agena D from Vandenberg AFB, Calif., to a 100 × 199 mi., 75° orbit. Designated KH-4A, the Corona satellite carried an Itek camera system and had provision for two recovery capsules instead of only one with KH-4 and KH-5 systems. Only one film capsule was recovered from this mission and from the next KH-4A mission four days later. On the third mission, launched February 15, 1964, both film capsules were recovered for the first time. Effectively doubling the picture-taking capability of the Corona satellite, KH-4A replaced the standard KH-4, which made its last flight on December 21, 1963.

1963 August 28 The first Little Joe II launcher was fired for the first time from White Sands Missile Range, N.Mex., carrying a dummy Apollo command and service module to a height of 4.5 mi. and a downrange distance of 9 mi. Little Joe had a height of about 29 ft. and a diameter of 13 ft. to which was attached the 52-ft.-tall dummy Apollo and launch escape system. Four fins at the base provided aerodynamic stability and the vehicle had a launch weight of 56,500 lb. Six solid propellant Recruit rockets, each with a thrust of 34,500 lb. for 1.5 sec., boosted Little Joe II into the air, while a 103,200-lb. thrust solid propellant Algol sustainer rocket carried Little Joe II on its ballistic trajectory. Algol was built to fire for 42 sec., but on this flight a ground command at 34 sec. cut open the motor case to limit the flight.

The director of the NASA Flight Research Center, Paul F. Bikle, responded to a request from Maxime A. Faget of the Manned Spacecraft Center for flight tests to be conducted on the HL-10 lifting-body conceived at the MSC. Bikle was discouraging citing existing tests with the M2-F1 as sufficient research on this class of vehicle, but offered test facilities for the HL-10 if the MSC came up with $75,000. MSC wanted to maintain its hold on manned spaceflight and believed lifting-body research to be the key to future projects.

1963 August 29 The first in a long line of U.S. piggyback electronic intelligence-gathering (elint) satellites was launched along with a first-generation photoreconnaissance satellite on a Thor-Agena D from Vandenberg AFB, Calif. The reconnaissance satellite entered a 181 × 201 mi. orbit while the subsatellite was placed in a 193 × 268 mi. orbit, both orbits at an 81.8° inclination. The 130-lb. subsatellite decayed on September 28, followed by the 2,200-lb. reconnaissance satellite on November 7, 1963. Beginning with the next elint satellite on October 29, 1963, piggyback satellites with Corona vehicles were launched by thrust-augmented Thor-Agenas.

1963 August 30 The Lunar Orbiter program was formally approved by NASA; Lee Scherer was appointed program manager and requests for proposals to build five flight spacecraft were issued to industry. Langley Research Center would manage Lunar Orbiter, which was the first program specifically established to supply engineering information to the Apollo Program Office at NASA Headquarters. It was also the first NASA program to include cost, delivery and technical performance incentives in its contract to the successful bidder, signaling a new awareness at NASA of spaceflight engineering criteria and of industrial team performance. Thus the old research and development ethic of the National Advisory Committee for Aeronautics had been fully replaced by vigorous management of hardware design and fabrication of operational spacecraft.

1963 September 1 The U.S. Air Force awarded General Dynamics a contract to develop a standard, low-cost satellite. Called Orbital Vehicle One (OV1), the "orbital pods" were intended to be sent aloft on Atlas missiles carrying advanced ballistic reentry systems (ABRES) nose cones. Each pod was officially designated a Satellite for Aerospace Research (SATAR) for the Air Force Aerospace Research Support Program (ARSP). The pod was 12 ft. long with a diameter of 2 ft. 6 in., capable of carrying a 300-lb. payload to a 500-mi. orbit. OV1 consisted of a cylindrical orbital section and a jettisonable propulsion module powered by a 5,800-lb. thrust X-258 motor. The Atlas missile would fly a ballistic reentry trajectory, releasing the OV1 pod at apogee for orbit insertion.

1963 September 12 NASA announced that it would adopt a new sterilization technique to prevent contamination of the lunar environment from unmanned spacecraft. The failure of Ranger spacecraft to achieve their planned mission objectives was attributed in part to the high temperatures used in earlier sterilization methods. The new technique would employ special "clean-room" facilities, in which dust and microbial contaminants were kept to a minimum, followed by the use of germ-killing substances. Development of these techniques did much to improve hospital clean-rooms such as surgical theaters and open-wound treatment centers, reducing the number of post-operative cases and, in some cases, deaths.

1963 September 17 The U.S. Air Force ballistic missile early-warning system (BMEWS) network of ground-based radar stations was officially declared complete as the station at Fylingdales, England, became operational. Located on the Yorkshire moors in northern England, Fylingdales comprised three dish antennas of 84-ft. diameter, each encapsulated by a radar-transparent ball 140 ft. in diameter. The system was capable of detecting approaching missiles on ballistic trajectories and would remain operational for 30 years.

1963 September 18 The first of six ASSET (aerothermodynamic/elastic structural systems environmental tests) shots was successfully conducted from Cape Canaveral when a Thor missile fired the blunt reentry body to a height of 195,000 ft. and a speed of Mach 12 (10,900 MPH). The booster lobbed *ASSET-1* about 1,000 NM downrange, but a search failed to find it. The vehicle was instrumented for aerodynamic tests, and good data were accumulated from the 20 min. it spent maneuvering on reentry. The various materials attached to the vehicle were subject to temperatures as high as 4,000°F. The second test, on March 24, 1964, was only a partial success because the second stage of the Delta launch vehicle failed to ignite. The third flight, on July 22, 1964, successfully tested a vapor-deposited disilicaide coating, while the fourth, on October 27, 1964, was the first to measure structural and elastic effects. The fifth ASSET was launched on December 8, 1964, repeating the objectives of its predecessor, while the sixth, on February 23, 1965, propelled a similar load 2,700 mi. downrange and to a speed of 19,000 ft./sec. (12,950 MPH).

1963 September 20 Addressing the 18th General Assembly of the United Nations, President Kennedy invited the Soviet Union to join the United States in a "joint expedition to the moon," asking, Why should man's first flight to the lunar surface "be a matter of national competition?" Increasingly concerned at political tension between the two countries, Kennedy sought to use space as a means of bringing the USSR to the international community of free nations. NASA had mixed views on the issue and Congress resisted the idea, although Vice President Johnson fought hard to keep options open. Kennedy was assassinated before he could gather support for this idea. NASA Deputy Director Hugh Dryden, the other proponent of U.S.-Soviet space cooperation, died on December 2, 1965. The matter was not revived with any great enthusiasm until after the moon landing had been accomplished in July 1969.

1963 September 28 For the first time three satellites were launched on the same vehicle when a Thor-Able Star carried the *Transit 5BN1* navigation satellite and two classified payloads from Vandenberg AFB, Calif., to a 669 × 687 mi. orbit with an inclination of 89.9°. *Transit 5BN1* was the first satellite to rely exclusively on a nuclear power source for electrical energy and carried a SNAP-9A radioisotope thermoelectric generator (RTG) providing 25 W of electrical power. The satellite tumbled in orbit and stabilized inverted, unable to communicate with Earth. *Transit 5BN2,* also carrying a SNAP-9A, was successfully launched on December 5 to a 665 × 690 mi. orbit with an inclination of 90°.

1963 October 14 A Gemini Mission Planning Coordination Group at the NASA Manned Spacecraft Center, Tex., concluded that a rendezvous at first apogee after launch was to become a test objective of the two-man program. In the Apollo lunar landing mission plan, an ascending lunar excursion module ascent stage would lift off from the surface of the moon and rendezvous with the Apollo command and service modules on the first lunar orbit. In preparation for that activity, Richard R. Carley had suggested that Gemini should practice such an operation around Earth. A first apogee rendezvous was achieved on *Gemini XI,* launched September 12, 1966.

1963 October 17 United Nations Resolution 1884 (XVIII) prohibiting the orbiting of nuclear weapons or any device of mass destruction was passed. It called upon all spacefaring powers to refrain from using space as another medium for the extension of offensive warfare. The effort made by the Kennedy administration to get the resolution through the UN General Assembly was matched by successful efforts by the United States to fend off persistent attempts by the Soviet Union to get reconnaissance satellites banned.

The first two Vela nuclear detection satellites were launched by Atlas-Agena D from Cape Canaveral, Fla. Built for the U.S. Air Force by TRW, each 485-lb. satellite could detect a nuclear explosion to a distance of 100 million mi. and would be used to police test-ban agreements. The spacecraft were placed initially in a highly elliptical 230 × 57,000 mi. orbit inclined 38°. About 18 hr. after launch, a command was sent to fire a kick motor on *Vela 1,* circularizing its orbit at 57,000 mi. A similar operation was performed with *Vela 2* on October 19, but on the opposite side of the Earth. Each Vela took the form of a regular icosahedron, 4 ft. 8 in. in diameter, supporting 14,000 solar cells. *Vela*s 3 and 4 were launched on July 17, 1964, followed by *Vela*s 5 and 6 July 20, 1965. A new generation of Velas began with *Vela*s 7 and 8 launched on April 28, 1967.

1963 October 18 NASA announced the names of the 14 astronauts it had selected for training as Group 3 of the recruitment cycle: Maj. Edwin E. Aldrin Jr., Capt. William A. Anders, Capt. Charles A. Bassett II, Capt. Michael Collins, Capt. Donn F. Eisele, Capt. Theodore C. Freeman and Capt. David R. Scott from the U.S. Air Force; Lt. Comdr. Richard F. Gordon Jr., Lt. Alan L. Bean, Lt. Eugene A. Cernan and Lt. Roger B. Chaffee, from the USN; Capt. Clifton B. Williams Jr., from the USMC; R. Walter Cunningham, a RAND research scientist; and Russell L. Schweickart, MIT research scientist. Bassett was killed in an aircraft crash on February 28, 1966, Freeman was killed in an aircraft crash on June 7, 1967, Chaffee died in the Apollo 204 fire on January 27, 1967, and Williams died in an aircraft crash on October 5, 1967. The rest flew at least once in space, and Aldrin, Scott and Bean walked on the moon. Eisele died of a heart attack on December 2, 1987.

1963 October 24 Exactly three years after 54 technicians lost their lives in a pad explosion at Baikonur, 7 Soviet engineers were killed in another disaster that occurred while a launcher was being prepared for a nonmilitary mission. In nearby Leninsk, where the engineers and technicians lived, monuments were erected to the dead of each incident, with a silver rocket atop a crumbling block of concrete. Individual stone tablets were laid with the name of each person.

1963 October 26 The first submerged launch of the Polaris A3 SLBM (submarine-launched ballistic missile) took place when the submarine USS *Andrew Jackson* launched a missile of this type from underwater off Cape Canaveral. Development of the A3 with a 2,500-NM range included a redesign of 85% of the A2 and considerable changes and improvements to propellant, chamber materials and control systems. While the shape of the missile changed, the overall dimensions (length 31 ft., diameter 4 ft. 6 in.) remained the same, although weight increased to 36,000 lb. Polaris A3 carried three 200-KT W58 warheads. The A3 went operational when the USS *Daniel Boone* departed Apra Harbor, Guam, December 25, 1964. By October 1965 all Polaris A1s had been withdrawn in favor of the A2 and A3 variants, and by 1967, 13 of the 41 fleet ballistic missile submarines carried the A2 and 28 carried the A3. The last Polaris A3 patrol ended October 1, 1981, when the submarine *Robert E. Lee* returned to Pearl Harbor, Honolulu.

1963 October 30   Based on recommendations from a study conducted by Bellcomm, NASA cut Apollo flight test missions and introduced the system known as "all-up" testing. This envisaged the integration of complete spacecraft and launch vehicles from the outset, instead of launching a large number of flights to test individual elements or modules separately. It would cut the number of launches but put more reliance on each mission working correctly. Previous test philosophy had been dictated by the overcautious German rocket engineers at Huntsville. Manned Space Flight director George Mueller suggested cutting "dead-end" tests of equipment that would not fly operationally without major modification, thereby maximizing the value of each test launch. Mueller abandoned plans for four manned Saturn I flights, ending flights of that vehicle with SA-10 and shifting the first manned Apollo flights to Saturn IB, albeit at the expense of nine months' delay.

1963 October   The U.S. Department of Defense Scientific Advisory Board this month concluded in a report that the Air Force had not made an adequate case for continuing with the Aerospace Plane (ASP). It asserted that too much emphasis had been placed on the possible application of such a plane and that too little work had been done on the unique propulsion system and on the demanding aerodynamic requirements. Congress voted to delete all funds for the ASP for fiscal year 1964, but work continued on supporting technologies under a study defined under System Requirement 651.

1963 November 1   A Soviet SL-5 launch vehicle, an obscure variant of the SL-3, launched the first of two maneuverable satellites, *Polyot 1.* From the Baikonur cosmodrome, the 1,325-lb. satellite was placed first in a 211 × 368 mi., 58.8° orbit. A day later it was maneuvered by ground command to an orbit of 213 × 893 mi. with an inclination of 58.92°. Other, minor maneuvers were made and the satellite finally decayed on October 16, 1962. The second satellite in the series, *Polyot 2,* was launched from Baikonur on April 12, 1964, to a 150 × 301 mi., 59.92° orbit. From there it also changed its orbit. On May 10, 1964, it was observed by U.S. tracking stations in a 188 × 298 mi., 58.06° orbit. *Polyot 2* decayed back into the atmosphere on June 8, 1966.

1963 November 7   The first Apollo test took place at White Sands, N.Mex., when command module boilerplate no. 6 was used for the first pad abort test. Designated PA-1, the test involved the firing of the 147,000-lb. thrust launch escape system (LES) solid propellant rocket motor. Similar in concept to the escape system selected for Mercury, the Apollo LES was designed to wrench free from a malfunctioning launch vehicle the command module and its three-man crew. crew. The escape motor would fire for 3.2 sec., followed by separation of the command module for a recovery by parachute. Under normal flight conditions, the LES would be jettisoned at an altitude of 56 mi. A boost protective cover was added to protect command module windows from soot. The escape tower separated 15 sec. after liftoff and the module landed at 2 min. 45 sec.

The MIDAS satellite early warning program was brought to a halt and completely reoriented to focus on submarine-launched and intermediate-range ballistic missiles. In an elaborate program, a network of satellites in circular orbit 7,480

mi. above the Earth was to have demonstrated missile detection and a capability to pinpoint the launch site. Set in two separate phases involving 18 launches scheduled to begin in January 1966, the first and second parts of the first envisaged 11 satellites in 2,000-mi. orbits. Satellites in the third part of the first phase would be put in 6,900-mi. orbits, while the second phase would set up a constellation of operational satellites at 7,480 mi.

1963 November 11   The first Soviet Zond interplanetary spacecraft was launched by SL-6 launch vehicle from the Tyuratam complex. The purpose of the flight was a lunar fly-by and imaging mission to obtain pictures of the far side. The 1,960-lb. spacecraft remained in an Earth orbit of 113 × 134 mi. after the terminal stage failed to fire and left it stranded. The Zond series of spacecraft were designed to test technologies necessary for the future development of unmanned planetary spacecraft.

1963 November 15   The Boeing Co. completed a major study for NASA and submitted findings in the self-descriptive report *Initial Concept of Lunar Exploration Systems for Apollo (LESA)* that laid the foundation for much of NASA's planning over the next five years. The study envisaged modules weighing 25,000 lb. transported to the moon by Saturn V and supporting 2–18 persons for semipermanent habitation. LESA was the culmination of several separate studies on what would emerge as post-Apollo planning.

1963 November 16   The first second-generation Soviet photoreconnaissance satellite, *Cosmos 22,* was launched by SL-3 from Baikonur to a 119 × 237 mi. orbit inclined 64.9°. It was a developed version of the adapted Vostok spacecraft, which comprised the first-generation series and had an increased weight of 10,540 lb. due to better optics, high resolution and a larger film supply. Flights of the second-generation series were interleaved with continuing flights of the first-generation series. On June 8, 1966, the first of a new low-resolution variant of the second-generation series was launched under the designation *Cosmos 120.* The Soviets launched 102 second-generation satellites: 1 in 1963; 3 in 1964; 10 in 1965; 12 in 1966; 17 in 1967; 24 in 1968; 25 in 1969; and 10 in 1970. The last in the series, *Cosmos 355,* was launched on August 7, 1970, and reentered eight days later.

1963 November 19   The Blount Brothers Construction Co. of Alabama and the M. M. Sundt Construction Co. of Arizona, began work on a 3.5-mi. crawler way for the Saturn V Launch Complex 39 at Cape Canaveral. The initial contract covered the crawler way and the first pad, Complex 39A, which was to be 18,159 ft. from the Vehicle Assembly Building (VAB) where Saturn V launchers would be erected. The crawler way had to withstand a maximum load of 6 tons per sq. ft. and would comprise two 40-ft.-wide lanes with a 50-ft. median strip. The crawler way was built using conventional highway-type methods. A 24-ft. service road bordered the crawler way on its south side. A second contract, awarded to the George A. Fuller company of California on November 30, 1964, covered preparation of an 11,300-ft.-long spur to the crawler way that started 12,000 ft. out from the VAB and linked it to the second pad, Complex 39B.

1963 November 22   The first live transmission of television signals across the Pacific Ocean was accomplished via the *Relay I* communication satellite. From the U.S. ground station in the Mojave desert, signals were sent to Japan's new

Space Communication Laboratory north of Tokyo. Taped messages from Japanese ambassador Ryuji Takeuchi and NASA Administrator James Webb were mixed with pastoral and cultural scenes in a presentation put together by ABC-TV and NBC-TV. A taped message from President Kennedy was deleted when news came of his assassination only hours earlier.

Only a few hours before his assassination, President Kennedy made what were to be his last public statements on the missile and space programs he had done so much to install as permanent features of American politics and world history: "What we are trying to do in this country . . . I believe is quite simple: and that is to build a military structure which will defend the vital interests of the United States. And secondly, we believe that the new environment, space, the new sea, is also an area where the United States should be second to none."

1963 November 24   North American Aviation completed a study for the NASA Manned Spacecraft Center on an extended-duration Apollo spacecraft. NAA had been contracted by the MSC to define the capabilities of the Apollo spacecraft for long-stay missions in space. NAA selected an Earth-orbiting science role for the spacecraft and described systems that could be modified to support lengthy missions. It was believed that the Apollo spacecraft could remain in orbit for up to one year conducting observations and scientific research.

1963 November 25   Proposals for design and development of a life-support package for EVA (extravehicular activity) astronauts in the Gemini program were received at the NASA Manned Spacecraft Center, Tex. The equipment envisaged the supply of high pressure oxygen for astronauts working outside their spacecraft and the integration of regulators and flow control valves in an autonomous package that could be attached to the astronaut's space suit during EVA. Initial EVA operations were envisaged to last 10–15 min. and required the crew members to be attached to the spacecraft by a tether. The Garrett Corp. was awarded a contract in January 1964 to develop the package and provide flight hardware.

1963 November 27   The first of ten scientific satellites designed to study the Earth/sun radiation budget over a full 11-year sunspot cycle was launched by Delta from Cape Canaveral, Fla., when Interplanetary Monitoring Platform-A *(Explorer 18)* was placed in a 125 × 122,800 mi. orbit inclined 33.3°. The 128-lb. IMP-A carried 35 lb. of experiments which revealed the presence of high-energy radiation beyond the Van Allen belts around Earth. The satellite transmitted data until May 1965 and reentered in December 1965. Other Earth-orbit IMPs included: IMP-B *(Explorer 21)* launched on October 4, 1964, but to an apogee of only 59,400 mi. instead of 126,500 mi. and capable of gathering data only from within the magnetosphere; IMP-C *(Explorer 28)* launched on May 29, 1965, returned data until July 4, 1968; IMP-D *(Explorer 33)* launched July 1, 1966, was planned for lunar orbit but entered a 9,880 × 270,560 mi. path around Earth, returning good data (see entry for that date); IMP-F *(Explorer 34)* launched May 24, 1967, operated from a 154 × 131,187 mi. orbit; IMP-G *(Explorer 41)* was launched on June 21, 1969; IMP-I *(Explorer 43)* launched March 13, 1971 operated until it decayed October 2, 1974; IMP-H *(Explorer 47)* launched September 23, 1972 orbited Earth at a distance more than halfway to the moon (see entry). IMP-E (July 19, 1967) was placed in lunar orbit and is entered separately, as is IMP-J (October 26, 1973).

NASA launched the second Atlas-Centaur (AC-2) from Launch Complex 36 at Cape Canaveral, Fla., successfully placing the 10,200-lb. upper stage in a 338 × 1,055 mi., 30.4° orbit with a lifetime of about 100 years. This was the first successful flight of a launch vehicle using cryogenic propellants. AC-2 had a total height of 109 ft. Modified since the AC-1 failure of May 8, 1962, the two cryogenic Centaur engines burned for more than 6 min. The AC-3 flight on June 30, 1964, was to have been a repeat of AC-2 but the Centaur engines shut down 2 min. 7 sec. prematurely, 4 min. 13 sec. after ignition, when the stage began an unplanned roll that shifted propellants away from the outlets at the bottom of the tank. With insufficient velocity, the Centaur fell into the South Atlantic from a height of 345 mi. and at a maximum speed of 11,425 MPH.

1963 November 28   In honor of the late President John F. Kennedy, who had been assassinated six days earlier, the NASA Launch Operations Center at Cape Canaveral would be known as the John F. Kennedy Space Center, said President Lyndon Johnson. The next day Johnson signed an order to that effect and also changed the name of Cape Canaveral to Cape Kennedy. Although this received the approval of Florida Governor Farris Bryant, it brought complaints from local residents, and in 1974 the geographic area known as Cape Kennedy reverted to being called Cape Canaveral.

1963 December 2   NASA selected Douglas Aircraft Co. for a study and evaluation of the manned orbital laboratory concept. The company had completed a three-month study which formed the basis for a further period of analysis and evaluation lasting up to nine months. NASA wanted the studies to center on a six-person station comprising a cylindrical module launched by a single rocket.

1963 December 9   The University of Tokyo opened the Kagoshima Space Center on the southern tip of Kyushu, the southernmost of the four main islands of Japan. Kagoshima was situated opposite the island of Tanegashima where the National Space Development Agency (NASDA) would have its launch facilities. Kagoshima would achieve recognition as the launch site for Japan's other space organization, the Institute of Space and Aeronautical Sciences (ISAS). Mu and Lambda series launchers would be fired from Kagoshima.

1963 December 10   Acknowledging that $410 million had already been spent on the project, Defense Secretary Robert S. McNamara announced the cancellation of the X-20 boost-glider, asserting that it had limited objectives and no defined military mission. He said that some funds would be diverted into the Manned Orbiting Laboratory (MOL), a military space station to be launched by Titan III and routinely visited by astronauts using a developed version of the Gemini spacecraft known as *Gemini X*. The Department of Defense had already signaled its intention to increase its efforts on reentry technology and research through the START (spacecraft technology and advanced reentry tests) and ASSET (aerothermodynamic/elastic structural systems environmental tests) programs.

North American Aviation, Inc., completed a confidential study on the single-stage-to-orbit (SSTO) concept for heavy launch vehicles. Postulating a payload requirement of 1 million lb. to low Earth orbit, the study proposed a conical structure, 174 ft. tall and with a base diameter of 160 ft., with a liftoff weight of 30 million lb. NAA showed that by incorporating very large air-augmentation (scramjet) engines that dramatically reduced the oxidizer load, the liftoff weight could be as little as 2.1 million lb. for the same payload capability. Studies like these helped support the theoretical advantages inherent in SSTO vehicle designs.

1963 December 12 The first Lambda 2 solid propellant sounding rocket from the University of Tokyo, Japan, made its first flight from the Uchinoura range. Developed as a successor to the Kappa built for the International Geophysical Year, Lambda 2 had a length of 54 ft., a weight of 15,400 lb. and a first-stage thrust of just over 90,000 lb. The 22,000-lb. thrust second stage had been the first stage of Kappa 8, first flown in 1960, and the two-stage rocket could lift 400 lb. to a height of 310 mi. A three-stage Lambda 3 appeared in 1964, and a developed version in 1966 could lift 220 lb. to an altitude of more than 1,200 mi. The Lambda series contributed much to the Mu-class orbital rockets, elements of which were tested beginning in 1965, although the first Japanese satellite was placed in orbit (on February 11, 1970) by a Lambda 4S-5.

1963 December 15 An ad hoc working group on Apollo experiments recommended that Apollo's principal objectives should be to examine physical and geological properties of the area surrounding the landing site, perform geologic mapping, conduct investigations of the moon's interior, study the lunar atmosphere and perform radio astronomy from the surface. Chaired by Charles P. Sonett from NASA's Ames Research Center, the working group amplified guidelines sent from NASA's Office of Space Science to the Manned Spacecraft Center. The reception given to these recommendations bordered on hostility. MSC considered this an intrusion, with added tasks they believed would compromise, and jeopardize, their engineering priorities.

1963 December 19 NASA launched the second of its air-density balloon satellites, AD-A *(Explorer XIX),* on a Scout launcher from Vandenberg AFB, Calif., to a 367 × 1,488 mi., 78.8° orbit. Essentially similar to *Explorer IX,* it differed only in having some solar cells to power a beacon transmitter broadcasting on 136.62 MHz. The 15-lb. satellite decayed on May 10, 1981. Second in the series, AD-B *(Explorer XXIV)* was launched by a Scout on November 21, 1964, to a 326 × 1,552 mi., 81.4° orbit, together with Injun-B *(Explorer XXV),* in the first dual-launch NASA payload arrangement. AD-B decayed on October 18, 1968. The 90-lb. Injun-B obtained radiation data until December 1966 and remains in orbit.

1963 December 20 From five bids submitted for the NASA contract to design and build five Lunar Orbiter spacecraft, the Boeing Co. was selected for a definitive contract. Submissions from Hughes, TRW/STL, Martin Marietta and Lockheed were rejected largely on technical grounds, and Boeing won with its design for a 794-lb., three-axis-stabilized spacecraft utilizing an Eastman Kodak film processing system developed in 1960 for a military reconnaissance satellite program. The system worked by passing film through a focal plane optical imaging system on to a semidry chemical developer in a process called Bimat. The negative images would be passed across a photomultiplier which converted light passing through a scanner into an electrical signal whose strength would vary with the intensity of the film emulsion. The signal would then be transmitted to Earth and reassembled into a picture. Eastman Kodak was awarded a contract in October 1964.

1963 December 22 The Comsat Corp. issued an invitation for bids on the design of a global communication satellite program. At this stage, Comsat envisaged either 12–18 satellites orbiting 5,800–13,800 mi. above Earth, or 6 geosynchronous satellites at 22,000 mi. It estimated that the intermediate orbit system could begin launching in 1966, with global coverage from 1967, or a geosynchronous system could begin launches in 1967, with a global service from 1968. In March 1964 Hughes was awarded a contract for the Early Bird *(Intelsat I)* precursor satellite.

1963 December 23 The U.S. Air Force Systems Command submitted a proposal designated Program 437 X for a satellite inspector version of the Program 437 antisatellite weapon. The intelligence community wanted to know more about Soviet satellites via a Corona/Discoverer camera system in the General Electric Mk.2 reentry nose cone of a Thor missile. Five to seven photographs of a target satellite could be taken at altitudes between 70 and 420 mi., after which the film would be moved to a recovery capsule and returned through the atmosphere for airborne recovery by a C-130 from Hicham AFB, Hawaii. Program 437 X was conceptual successor to SAINT (for Satellite Interceptor).

1963 December 24 NASA and the U.S. Atomic Energy Commission announced major changes to Project Rover, the development of nuclear-powered space propulsion systems. The RIFT (reactor-in-flight test) program was to be canceled and NERVA (nuclear engine for rocket vehicle application) flight objectives were deferred until technology from the Kiwi NERVA programs were "satisfactorily established." Ground tests and experimental engine work were to be emphasized.

1963 December 26 The director of the Marshall Space Flight Center, Wernher von Braun, sent to Apollo Spacecraft Program Manager Joseph Shea a proposal for using the Saturn V in an extended exploration of the lunar surface. In a program called Integrated Lunar Exploration System (ILES), a combination of modified launchers and spacecraft could be used to fill the gap between the initial lunar landings and more permanent bases on the moon. A dual Saturn V mission with rendezvous around the moon could deliver an integrated lunar taxi/shelter to the surface, where the crew could live and base geological expeditions.

1963 December 31 NASA's Manned Space Flight director, George Mueller, completed a revision of Saturn-Apollo flight schedules. Saturn I flights would end with SA-10, planned for June 1965, the first Saturn IB flight (SA-201) would be in the first quarter of 1966 and the first Saturn V flight would be scheduled for the first quarter of 1967. Mueller also significantly reduced the number of unmanned launch vehicle demonstration flights, assigning the earliest manned Apollo mission to the third Saturn IB flight in the third quarter of 1966 and the first manned Saturn V launch to the vehicle's third flight in the third quarter of 1967. With the lunar excursion module as the pacing item, Mueller cautioned that the first flight-rated LEM might not be ready for the first Saturn V flight.

U.S. Air Force Brig. Gen. Samuel C. Phillips was appointed deputy director of NASA's Apollo Program Office, assisting the director, George Mueller. Because Mueller was also associate administrator for Manned Space Flight, increasing responsibility would fall on the shoulders of "Sam" Phillips. Over time, he would assume great responsibility for much of the progress achieved in the Apollo program. When the January 27, 1967, Apollo 204 fire brought the flight test program to a halt, it was Phillips who put the program back on track and expedited the successful return to flight. Phillips was brought back to help NASA reestablish the Shuttle program after the loss of Challenger on January 28, 1986.

1963 December The Ingalls Iron Works of Birmingham, Ala., began work on a mobile launch platform (MLP) for the Saturn V launch vehicles. The Saturn V would be assembled on a 6,000-ton MLP parked in the Vehicle Assembly Building (VAB). The MLP would comprise a flat platform 160 ft. long, 125 ft. wide and 25 ft. high. At one end of the platform would be a 398-ft.-tall umbilical tower with nine moveable service arms to connect with the rocket stages. A central cutout on the other side of the platform would permit the passage of exhaust gases from the Saturn V's five F-1 engines. Four hold-down arms, each weighing more than 20 tons, would support the Saturn V until it was released for flight. To move the MLP a crawler transporter would be positioned underneath the mobile launcher, set atop six 22-ft. pedestals in the VAB, jack it up and, with a total weight of more than 9,000 tons, roll to one of two Complex 39 launch pads.

1963 The Soviet navy began deployment of the SS-N-5 Serb, the first Soviet SLBM (submarine-launched ballistic missile) capable of being fired from underwater. With a length of 42 ft. and a diameter of 6 ft. 7 in., the two-stage solid propellant missile had a launch weight of 37,000 lb. and carried a single reentry vehicle with a 1-MT warhead. Serb had a range of 900 mi. but a comparatively poor c.e.p. accuracy of only 9,000 ft. at full range and was considered the second-generation Soviet SLBM after the SS-N-4. Serb was first deployed on Golf II–and Hotel II–type submarines, each vessel carrying three missiles. A maximum deployment of 60 SS-N-5s was achieved in 1973, and almost all had been withdrawn by the late 1980s.

1964 January 7 The first power tool developed for use in space was demonstrated by Black and Decker. Designed under a contract to the Martin Co., the tool was similar to home power drills but had 99.97% less reactive torque than the domestic model. This was accomplished by making the case rotate in the opposite direction to de-spin the motion of the chuck. In weightlessness, an astronaut operating a standard power tool would turn around the chuck as the bit remained still. Black and Decker was subsequently given the job of designing the first moon drill and it was used on *Apollo 15, 16* and *17* to drive core stems into the lunar surface.

1964 January 11 The first in a series of 13 U.S. Army geodetic satellites, *Secor 1,* was launched as one of four piggyback satellites on a Thrust Augmented Thor-Agena D from Vandenberg AFB, Calif., to a 563 × 578 mi. orbit inclined 69.9°. The primary payload was a 1,550-lb. classified U.S. Navy satellite. Other piggyback satellites included GGSE-1, an 86-lb. gravity-gradient stabilization experiment, the 100-lb. *Solrad 7A,* and an unnumbered Greb. Built by the Cubic Corp., the 40-lb. *Secor 1* carried a beacon transmitter and its name, an acronym of "sequential collation of range," describes the satellite's geodetic function. *Secors 2–6* were launched the following year, on March 11, March 9, April 3, and August 10, 1965, and on June 9, 1966. Three advanced derivatives, *EGRS 7–9,* were launched on August 19 and October 5, 1966, and June 29, 1967. *Secors 10–13* were launched on May 18 and August 16, 1968 (two satellites), and on April 14, 1969; only the latter made it to orbit. *Solrad 7A* was the first of two larger Solrad satellites built by the Naval Research Laboratory. The second flew piggyback on March 9, 1965.

1964 January 15 North American Aviation applied load to the first fuel cell electrical production unit for the Apollo command/service module. It had been delivered by Pratt and Whitney, the prime contractor for the fuel cells. The Apollo service module would carry three fuel cells, each capable of producing 1.4 kW of electrical energy at 27–31 V Each cell was about 3 ft. 8 in. tall, 1 ft. 10 in. in diameter, and weighed 245 lb. The cryogenic hydrogen-oxygen reactants would be stored in four separate spherical pressure vessels. As reactants were depleted during a normal mission, heaters and fans inside the vessels would maintain pressure by slightly warming and stirring the contents. This would turn a portion to gas, which requires greater volume, thereby retaining the overall pressure.

1964 January 21 North American Aviation provided details of its proposed Block changes to the Apollo spacecraft configurations. Block I command and service modules would carry no rendezvous or docking equipment but would provide an early means for qualification of basic and essential spacecraft systems. Block II command and service modules would be delivered later and incorporate weight-saving measures while carrying all equipment necessary for operations with the lunar excursion module and for supporting lunar landing operations.

1964 January 22 The first of 25 full-scale Gemini spacecraft paraglider landing system tests began with wing deployment, inflation, glide and radio-controlled maneuvering. Not until the sixth flight on April 30, 1964, was a complete deployment sequence successfully demonstrated, but problems were encountered on the next four flights. The next successful flight, the twelfth (July 29, 1964), was followed by ten problem flights, but the last three (October 23, November 6 and December 1, 1964) successfully demonstrated the deployment sequence. On February 20, 1964, NASA Administrator for Manned Space Flight George Mueller, proclaimed that all 12 Gemini flights would use water landings, and on April 29 it was decided to phase the paraglider out of the Gemini program, a public announcement to that effect being made August 10, 1964.

1964 January 29 NASA launched the first of six two-stage Block II Saturn I launch vehicles, the first vehicle of this class to lift a payload into orbit, at 16:25 UT. It was the first launch from Cape Canaveral's Complex 37. With a height of 164 ft. and a liftoff weight of 1,121,680 lb., SA-5 was the heaviest U.S. flight to date. Improvements to the first stage provided four stabilizing fins and a propellant capacity of 850,000 lb. for the eight H-1 engines, now delivering full-rated output for a stage thrust of 1.504 million lb. Inboard engine cutoff came at 2 min. 21 sec., followed 6 sec. later by the four outboard engines. The six 15,000-lb. thrust RL-10A3 engines (previously designated RL-115s) in the S-IV ignited at 2 min. 29 sec. and ran for about 8 min., putting it into a 163 × 478 mi. orbit inclined 31.5°. Eight recoverable cameras were ejected from their ascending stack; seven were retrieved. During flight a total 1,183 readouts were sent to tracking stations by telemetry, twice the previous record total of a Block I Saturn. Total mass in orbit was 37,700 lb., of which 19,500 lb. was the spent S-IV and an instrument unit for controlling SA-5 during powered flight. The stage decayed into the atmosphere on April 30, 1966.

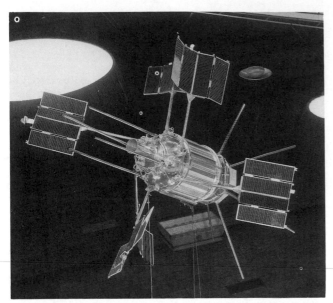

In the first Soviet dual launch, two science satellites were placed in Earth orbit by a single SL-3 on January 30, 1964. Here, a replica of Elektron 1 *displays solar panels and instrument apertures for measuring the inner Van Allen radiation belts.*

1964 January 30   A Soviet SL-3 launch vehicle carried *Elektron 1* and *Elektron 2* from Baikonur cosmodrome to Earth orbit, the first time the Soviet Union had put two satellites on one launcher. The 725-lb. *Elektron 1* was placed in a 252 × 4,412 mi., 61° orbit, while the 980-lb. *Elektron 2* went to an orbit of 285 × 42,377 mi., inclined 61°. Each satellite carried instruments to conduct simultaneous measurements of Earth's radiation belts. *Elektron 1* carried six solar cell paddles on a cylindrical body with sensors at opposing ends. *Elektron 2* carried solar cells around its cylindrical body section and on a flared "skirt" at the base. The second pair were launched by SL-3 on July 10, with *Elektron 3* placed in a 251 × 4,365 mi. orbit, and *Elektron 4* in a 285 × 41,076 mi. orbit, both at 61°. All four are still in orbit as of 1995.

At precisely 15:49 UT, NASA launched *Ranger 6,* the first of four Block III lunar impact probes, into space by Atlas-Agena B from Launch Complex 12 at Cape Canaveral, Fla. The 804-lb. spacecraft was to have transmitted up to 3,000 pictures from six TV cameras with 25-mm and 75-mm lenses operating during the final 10 min. before lunar impact. Two wide-angle and four narrow-angle cameras weighing 59 lb. were to have begun operating 900 mi. from the surface, transmitting their images live to ground stations on Earth. Modified considerably from earlier Rangers, Block III spacecraft had a total height of 10 ft. 3 in. and a solar wingspan of 15 ft. across panels carrying a total 9,792 cells and delivering 200 W. Launched on a course that would miss the moon by only 495 mi., *Ranger 6* conducted a 92.2 MPH midcourse maneuver on January 31. Just 1,290 mi. before lunar impact the TV cameras were warmed up, but no images were sent and the spacecraft struck the Sea of Tranquillity at 09:24 UT, February 2, 1964.

1964 January   NASA completed plans for extravehicular activity in manned Gemini flights. The Flight Crew Operations Directorate at the Manned Spacecraft Center planned for a crew member to open a hatch and stand up during GT-4,

the second manned flight, followed by a full EVA (extravehicular activity) to the rear of the spacecraft adapter on GT-6. On GT-7 and GT-8, tethers and hand-holds were to be evaluated, while on GT-9 an astronaut maneuvering unit was to be tested. Advanced EVA procedures and techniques were to be evaluated on GT-10, -11, and -12, the last three flights. As the program progressed, EVA activities were deleted from the long-duration *Gemini 7* mission due to the need for a full pressure suit, considered unsuitable apparel for a 14-day mission.

1964 February 3   The Boeing Co. issued the final report of a study it had conducted for NASA, *A Lifting Re-entry Horizontal-Landing-Type Logistic Spacecraft.* The concept envisaged the launch of a lifting reentry body on a Saturn IB to a manned Earth-orbiting space station. The vehicle was required to carry twelve people, up to 15,000 lb. of cargo, and be capable of remaining in space for up to six months before returning to Earth and a horizontal landing. Boeing also examined six-person versions of the Apollo spacecraft. Nine days later Lockheed, responding to a similar NASA requirement, submitted its report *Study of a Ballistics Reentry Type Logistics Spacecraft.*

1964 February 4   Representatives from the NASA Manned Spacecraft Center and the Marshall Space Flight Center met to discuss trajectory requirements for important tests of the Apollo heat shield. Fabricated by the Avco Corp., Apollo's ablative heat shield covered the complete exterior of the command module and comprised a phenolic epoxy resin impregnated into a stainless steel honeycomb matrix. The exterior temperature of up to 5,000°F, incurred as the spacecraft reentered Earth's atmosphere at speeds up to 25,000 MPH, would be reduced to 70–150°F on the interior of the spacecraft. Shield tests would be the primary objective of the first two Saturn IB flights (missions 201 and 202).

1964 February 13   NASA's Manned Spacecraft Center requested proposals from 50 companies for studies on how the lunar surface should be explored and how points on the surface could be located for selected navigation. On March 20, 1963, NASA's Office of Space Science and Applications began organizing groups of scientists to define the science objectives of Apollo. Increasingly, the Apollo program was being seen as a useful hook upon which to hang other activities, and the scientific application of manned landings to the overall space program was perceived as a more meaningful long-term investment in keeping alive a public interest in manned moon shots.

1964 February 14   The first of four demonstration tests under the U.S. Air Force Program 437 antisatellite program took place when a Thor missile launched from Johnston Island in the Pacific successfully intercepted a rocket body left over from the launch of *Transit 2A* on June 22, 1960. The second test took place on March 2 and the third on April 21, all three tests conducted by contractors. The fourth test, on May 28, was carried out by personnel from the 10th Air Defense Squadron, but failed. The system was declared operational on June 10, and a series of combat test launches began on November 16. The system could intercept targets up to an altitude of 230 mi. and a slant range of 1,725 mi. with a 1.5-MT Mk.49 warhead.

1964 February 19   The Soviet Union marked the beginning of the 1964 Venus launch window with an unsuccessful attempt to send a spacecraft to that planet. Launched by an SL-6, the Zond spacecraft failed to reach Earth orbit. A similar attempt on March 1, 1964, also resulted in an upper-stage failure before the terminal stage and Zond reached Earth orbit. On March 27, 1964, another Zond failed to leave its 104 × 123 mi. Earth parking orbit and the 14,400-lb. assembly, including terminal stage, decayed back into the atmosphere the next day. Designated *Cosmos 27* by the Soviets, *Zond 1,* the first successful Zond-series spacecraft, was launched April 2, 1964.

1964 February 20 The U.S. Air Force announced it was about to award contracts to Douglas Aircraft, Martin Marietta, and General Electric for space station study contracts. The three aerospace companies were to design a manned orbital laboratory (MOL) the Air Force was planning to launch on a Titan for basic research into the use of astronauts on military duties. The MOL would also be used to test in space equipment for use in unmanned military space programs and, as Air Force Chief of Staff Gen. Curtis E. LeMay said, "to acquire information essential to determining accurately the threat from space." He is not on record as describing what he thought that threat might be.

Edward Z. Gray, NASA director of Advanced Manned Missions, explained during testimony before the House Committee on Science and Astronautics' Subcommittee on Manned Space Flight the agency's plans for exploiting Apollo hardware after the initial lunar landing. The Apollo logistic support system (ALSS) would consist of an unmanned lunar module with the ascent stage replaced by an LEM (lunar excursion module) truck, or cargo module. The LEM truck would place on the lunar surface a 7,000-lb. module with a volume of 2,000 cu. ft. capable of supporting two people for 14 days. Another LEM truck could carry a roving vehicle capable of carrying two people for distances of up to 300 mi. in 14 days, perhaps to rendezvous with another habitability module launched to the place the astronauts would drive to. Later, the lunar exploration system for Apollo (LESA) would replace the LEM on a Saturn V with a 25,000-lb. module capable of supporting three people on the lunar surface for 90 days. Three interlinked modules would support eighteen people for 2 years.

1964 February 24 The European Launcher Development Organization (ELDO) formally came into being with agreement from 11 countries (the United Kingdom, France, Germany, Belgium, Denmark, Italy, the Netherlands, Norway, Spain, Sweden and Switzerland) to develop the ELDO A *(Europa I)* launcher. Worked out between Hawker Siddeley in Britain and SEREB, a French military and industrial organization, ELDO A would have the United Kingdom's 272,000-lb. thrust Blue Streak first stage, a 61,700-lb. thrust French Coralie second stage, and a 49,600-lb. thrust German third stage called Astris. The two upper stages would burn N_2O_4–UDMH. ELDO A would be capable of placing a 2,800-lb. payload in a 310-mi. orbit. Two other versions were proposed: two-stage ELDO B-1, with cryogenic second stage, capable of putting 4,600 lb. to a 125-mi. orbit; three-stage ELDO B-2 with a new cryogenic second stage using ELDO B-1's second stages as its third stage.

1964 February 25 Following technical evaluation of the prospects for ending Apollo missions on land or water, North American Aviation recommended water landing as the primary medium for recovery. Land landings for the 10,000-lb. command module would be difficult to achieve without sustaining damage or some personal injury to the crew. It was felt some damage might be incurred even in water landings,

and the conical design of the command module gave it the unfortunate characteristic of being stable in either apex-up or apex-down positions. A self-righting system comprising three inflatable air bags, each 22 cu. ft. in volume, was provided in the area around the forward tunnel close to the parachute housing.

1964 February 26 Studies on space station designs performed by the Lockheed Aircraft Co. were submitted to NASA. Lockheed concluded that a Saturn V could place a 24-person facility in orbit as early as 1968. Crew visits and logistics flights would take place using conventional launch vehicles.

1964 February Chief Soviet rocket designer Sergei Pavlovich Korolev began work on a second-generation Vostok called Voskhod. Aware that NASA was preparing the two-man Gemini for space-walking, docking and two-week flights, Korolev stretched Vostok to improbable limits. By removing the ejection seat and putting the crew in track suits, he could squeeze in three cosmonauts for the world's first multicrew mission. But there would be no means of emergency escape and the crew would have to remain inside their spacecraft through touchdown. A wire suspended beneath the spherical descent module would contact the ground in the final seconds prior to landing and trigger a braking rocket situated on top, firing downward to arrest the 32 ft./sec. (22 MPH) descent rate to a maximum 6 ft./sec. (4.1 MPH) and cushion the impact. Korolev's team was horrified at the absence of safety features, but they were persuaded to adapt Vostok into Voskhod when told that one of the three seats would be given to an engineer.

1964 March 2 NASA announced that Hughes Aircraft had been selected to design and build the five applications technology satellites (ATS) which would serve as orbital technology test beds for a variety of space applications. Included in the orbit evaluations would be stabilization and control techniques for meteorological and communication satellites. The first satellite, ATS-1, was launched on December 7, 1966. A sixth, much larger satellite was later contracted to Fairchild.

NASA and the U.S. Department of Defense set up a National Space Station Planning Sub-Panel of the NASA–Department of Defense Astronautics Coordinating Board. The NSSP would make recommendations to the Board on future space station developments for the period after the Apollo moon landing and U.S. Air Force Manned Orbital Laboratory programs had been completed. The Air Force planned to launch six manned orbital laboratories over a period of 18 months beginning in late 1967 or early 1968. During this month Brig. Gen. Joseph S. Bleymaier was appointed to the newly established post of deputy commander for Manned Systems of the Air Force Systems Command Space Systems Division.

1964 March 12 Edward Z. Gray, NASA Advanced Manned Missions director, requested from the Langley Research Center a project development plan for their proposed MORL (manned orbital research laboratory) space station. MORL was not an approved NASA program, but the field centers wanted to be ready with a full complement of studies and designs if and when the agency moved forward in the post-Apollo period with plans for manned stations. The Manned Spacecraft Center was also asked to submit plans for an Apollo X concept that would use the spacecraft as the base for a plug-on research module, an Apollo orbital research laboratory (ORL), and a large orbital research laboratory (LORL).

On June 18, 1965, the Douglas aircraft company received a study contract from Langley integrating the Apollo extension system as a precursor to MORL.

1964 March 16   NASA issued a request for proposals from industry for a 9-month feasibility study on an Apollo logistic support system, or ALSS, incorporating an unmanned LEM (lunar excursion module) truck for carrying a variety of cargo to the surface of the moon for the use of astronauts landed by a separate LEM. Also to be examined was the concept of a stay-time extension module, or STEM, incorporating "saddle-bags" carrying 3,000–5,000 lb. of supplies attached to the LEM. With the added propellant for landing, this would raise total Earth-payload launch weight by 15,000 lb. Saturn V was believed to have sufficient growth capability to incorporate this. The equipment would make it possible to set up a lunar base for two people for 7–10 days using the LEM with STEM cargo.

As an expansion of its ASSET (aerothermodynamic/elastic structural systems environment tests) program, the U.S. Air Force took steps to formally set up the START (spacecraft technology and advanced reentry tests) research effort. It was intended as a unifying program to encompass a broad range of research and development associated with lifting reentry bodies. On July 8, 1964, the Air Force authorized the Martin Marietta Corp. to begin work on a suitable research vehicle, which eventually emerged as the SV-5, part of the PRIME (precision recovery including maneuvering entry) program.

1964 March 19   A Delta launched from Cape Canaveral, Fla., carried the 120-lb. *Beacon Explorer A* satellite toward space but failed to place it in orbit when the 3,000-lb. thrust X-248 third stage motor burned for only 22 sec. instead of 40 sec. Octagonal in shape, 1 ft. 6 in. across, the 120-lb. satellite had four blade-like solar panels, each 5 ft. 6 in. × 10 in., and two 5 ft. whip antennas. A truncated pyramid on top, the main body held 360 fused silica reflectors used for laser tracking from a telescope at Wallops Island in geodetic experiments. In addition, coded radio signals were to provide data on the ionosphere. Two internal bar magnets were installed for passive attitude stabilization with Earth's magnetic field.

The U.S. Department of Defense announced that the first Army units equipped with the Pershing tactical missile were about to deploy to West Germany in support of the Fourth Battalion, 41st Artillery, Seventh Army. Known as the MGM-31A, the Pershing I was entering widespread use with the U.S. Army, and was also to be deployed with conventional warheads by the West German Luftwaffe.

1964 March 20   The European Space Research Organization (ESRO) formally came into being. Membership included Belgium, Denmark, France, Germany, Italy, the Netherlands, the United Kingdom, Spain, Sweden and Switzerland, with Austria and Norway having observer status. ESRO had been set up to "provide for, and promote, collaboration among European States in space research and technology, exclusively for peaceful purposes." It was charged with developing scientific satellites and space probes, buying the use of existing launchers from organizations like NASA, promoting space science among member states and disseminating results and information. Facilities included a European Space Technology Center (ESTEC) at Noordwijk, the Netherlands, a research laboratory ESLAB located near ESTEC, sounding rocket launch pads at ESRANGE near Kiruna, Sweden, a data center at Darmstadt, Germany, and a European Space Research

Institute (ESRIN) to conduct physics research. ESRO was abolished in 1975 at the formation of the European Space Agency.

1964 March 24–26   The first formal inspection of the Apollo lunar excursion module (LEM) test mock-up TM-1 was conducted at Grumman's Bethpage, N.Y., facility. The LEM now had four landing legs and a cylindrical ascent stage with offset ascent engine propellant tanks on either side. The LEM would be connected to the Apollo spacecraft at the top hatch and docking interface, but was designed so that when it returned from the lunar surface to meet the command/service modules, it would approach and dock utilizing the forward facing hatch that the crew had previously used to egress to the lunar surface. The mock-up review indicated that this was a bad idea, and on April 24, 1964, a meeting at the Manned Spacecraft Center decided to delete this option and put in an overhead window to preserve the top docking interface as the only one.

1964 March 25   NASA's Lewis Research Center awarded contracts to Aerojet General Corp. for development of the M-1 liquid hydrogen–liquid oxygen cryogenic rocket engine. One contract was for assembly of the engine and the other covered design of suitable test facilities. As now envisaged, the M-1 was to have a thrust of 1.5 million lb. On March 24, 1966, the Lewis Research Center performed initial tests on the turbopumps developed for the M-1. At 200,000 HP, the gas generator was the largest ever built for handling propellants in high-energy rocket engines. Lacking a clear application, the M-1 was eventually canceled without major development.

1964 March 27   A four-stage Scout launch vehicle carried the second satellite built for Britain, *Ariel II,* from Wallops Island, Va., to a 112 × 524 mi., 51.6° orbit. Assembled by Westinghouse, the 150-lb. *Ariel II* was a modified version of *Ariel I* and carried British science instruments from the universities of Cambridge and Manchester and the Air Ministry Meteorological Office. *Ariel II* observed ozone in the outer atmosphere, galactic noise, and measured micrometeoroid impacts. The satellite was powered by 5,400 solar cells carried on four paddles and deployed a 130-ft. dipole antenna for the astronomy experiment as well as two loop antennas. *Ariel II* returned data until November, and decayed back into the atmosphere on November 18, 1967.

1964 March 31   In a frank and open admission of faults within the Ranger program at the Jet Propulsion Laboratory, NASA Administrator James Webb sent Rep. George P. Miller, chairman of the House Committee on Science and Astronautics, a long explanatory letter describing why the *Ranger 6* spacecraft had failed to send TV images of the lunar surface two months earlier. Administrator Webb admitted that "the two video systems were more complex than required and were not completely redundant. . . . The possibilities of failure were increased as a result of practices in the design and construction of the spacecraft. . . . Tests did not simulate true spaceflight conditions." This frank and almost self-condemnatory approach was a hallmark of James Webb and helps explain why Congress maintained faith in NASA during times of crisis such as when, less than three years later, three astronauts were killed in a ground test of the Apollo spacecraft.

1964 March   Cosmonauts selected as candidates for the first Voskhod mission began training. To eclipse U.S. two-man

Gemini missions expected to begin at the end of 1964, *Voskhod 1* would follow one unmanned test flight in early October and carry three cosmonauts on a flight lasting one day. The candidates included Vladimir M. Komarov, Konstantin P. Feoktistiv, and V. G. Lazarev in the first team, and Boris V. Volynov, G. Katys, Boris B. Yegorov and A. Sorokin in the second team; physicians Yegorov, Lazarev and Sorokin had been specially recruited for Voskhod flights. Design engineer Gay Severin convinced Sergei Pavlovich Korolev that *Voskhod 2* could carry two suited cosmonauts and an inflatable airlock to accomplish the world's first space walk ahead of the Gemini astronauts. Four candidate cosmonauts were selected for that April 1964 mission: Pavel I. Belyaev, V. V. Gorbatko, Y. V. Khrunov and A. A. Leonov. Gorbatko was eventually replaced by Dmitri Alekseyevich Zaikin.

Lockheed conducted a study for the NASA Manned Spacecraft Center on a space station launched by Saturn V into low Earth orbit. The concept emerged as an MSC recommendation on one way the United States could maintain preeminence in space after the Apollo lunar landing program. The station was to rotate and have a lifetime of up to five years, housing 24 people. There was to be a zero-g laboratory for scientific research and crew duty cycles of between three months and one year. A modified Apollo spacecraft carrying 6 people was proposed as a logistics and crew supply vehicle. The core Gemini and Apollo programs were believed to provide sufficient technology research to build the station with little development.

1964 April 1   The Institute of Space and Aeronautical Science (ISAS) was formed at the University of Tokyo, Japan. Development of the Lambda series of solid propellant rockets continued at the university, and flight activities were transferred to the Kagoshima Space Center. On July 11 a Lambda 3 rocket was launched to an altitude of 528 mi. On June 20, 1965, the university announced a program of scientific satellites which would be managed by the ISAS. Beginning with fiscal year 1980, commencing April 1, 1980, the word "Aeronautical" in ISAS was changed to "Astronautical" and that terminology is retained in this chronology.

1964 April 2   A Soviet SL-6 launch vehicle carried the *Zond 1* spacecraft to an Earth parking orbit and from there to escape velocity and a flight path that would carry it past Venus. *Zond 1* weighed 1,960 lb. and incorporated several technical advances over its predecessors in the Venera series. It was designed after the launch, November 1, 1962, of the *Mars 1* spacecraft, which had set new standards in planetary spacecraft. Communications with *Zond 1* failed during the latter half of May 1964, but the spacecraft passed Venus at a distance of 600,000 mi. on July 19, 1964. Another Zond failed to reach orbit following launch on an SL-6 on June 4, 1964.

1964 April 8   The first flight in the Gemini program began with the launch of GT-1 (Gemini Titan-1) at 16:00:01 UT from Complex 19 at Cape Canaveral, Fla., to demonstrate the structural integrity of the Titan II launch vehicle and the two-man spacecraft. GT-1 had a height of 109 ft. and a liftoff weight of 341,000 lb. The spacecraft weighed 7,026 lb. and equipment pallets were placed in the two ejection seat positions to send information on structural measurements by telemetry. Acceleration peaked at 5.5 g just before first stage shut down 2 min. 14 sec. into the flight, and at 7.3 g at second stage shut down 5 min. 40 sec. after liftoff. The second stage and spacecraft remained attached in their 100 × 199 mi. orbit inclined 32.6° to the equator. Tracking stations continued to

A modified Titan II carried the first, unmanned Gemini spacecraft to space on April 8, 1964. A total of ten manned Geminis would be placed in Earth orbit during 1965–66.

obtain engineering information until 4 hr. 50 min., when the mission was considered over. The 13,785-lb. assembly reentered the atmosphere 4 days later.

1964 April 13   NASA's Manned Spacecraft Center Director Robert Gilruth announced that astronauts Virgil Grissom and John Young would be the prime crew members for the first manned Gemini flight, GT-3. Their backups would be astronauts Walter Schirra and Thomas Stafford. Because NASA planned to launch Gemini spacecraft at three-month intervals and because each mission required extensive preparation and training, backup crews would rotate to become prime crews three missions on. Thus, the backup crew for GT-3 would be named as the prime crew for GT-6, and so on. Following analysis of data from the unmanned GT-1 flight, NASA, on April 15, 1964, scheduled GT-2 for August 24 and GT-3 for November 16, 1964.

1964 April 14   The first of two 200-lb. Project Fire spacecraft was launched from Cape Canaveral by a redundant Atlas D ICBM converted as a launcher. Boosted to a height of more than 500 mi. above Earth, an Antares II solid propellant motor fired for 30 sec. to push the spacecraft back toward the atmosphere at a speed of 37,800 ft./sec. (26,000 MPH), simulating a capsule returning from the vicinity of the moon. This was the highest reentry speed achieved by a man-made object to date, and the Project Fire capsule returned to impact in the Atlantic 32 min. after launch and 5,200 mi. from the Cape. In association with this launch, NASA sent a Nike-Apache sounding rocket to a height of 97 mi. from Ascension Island to measure precisely the atmosphere at the time of reentry. The second 32-min. test, *Fire II,* took place on May 22, 1965, with an impact 5,130 mi. from Cape Canaveral.

Eleven engineers and technicians were injured, three critically, when the third stage of a Delta launch vehicle inexplicably ignited in a checkout building at Cape Canaveral, Fla. The X-248 solid propellant motor was being attached to an Orbiting Solar Observatory satellite prior to spin testing for balance when it suddenly ignited and rose from the stand, hitting the ceiling, knocking off the satellite and careering across the building to the opposite side. Sidney J. Dagle, a Ball Brothers technician, died April 17 from his injuries, and John W. Fassett died a day later from burns.

1964 April 15 Brig. Gen. Joseph S. Bleymaier, USAF, reported to NASA senior management on several problems with detailed design of the manned orbital laboratory (MOL) The crew would be launched in a modified Gemini spacecraft situated on top of the cylindrical laboratory itself. Difficulties were being encountered in designing an intravehicular access from one to the other. Several methods of transfer were being considered. Astronauts in the MOL would conduct about 36 experiments from the canceled X-20 program. Bids for the contract to build the MOL were to be sought by December 1, 1964, with hardware development under way by mid-1965.

1964 April 16 Westinghouse successfully cold-tested a nuclear reactor designed for the NERVA (nuclear engine for rocket vehicle application) engine. Designated NRX-AL (NERVA Reactor Experiment Al), it confirmed that earlier failures with the Kiwi-B series had been successfully diagnosed and corrective measures incorporated in the more refined design. On May 13 the Kiwi-B4D reactor was hot-fired for the first time, producing 1,000 MW for 1 minute. That test was extended to 10 min. 30 sec. at the same power level on August 28. The NRX-A2 reactor was tested at high power for 6 min. on September 24. On May 28, 1965, the NRX-A3 reactor ran at medium to low power levels for 45 min.

1964 April 21 The U.S. Air Force launched a Thor-Able Star from Vandenberg AFB, Calif., carrying *Transit 5BN3,* the third U.S. Navy satellite powered by a SNAP-9A nuclear generator containing 2.2 lb. of plutonium 238. The launch vehicle failed to reach orbit and fell back into the Earth's atmosphere together with the satellite and its toxic fuel. The satellite broke up at the edge of the atmosphere and the plutonium was dispersed across a wide region of the southern hemisphere. The Air Force judged there was no danger to life, but the adverse publicity brought a halt to nuclear-powered Transit satellites. The series reverted to a solar power source with *Transit 5C* June 4, 1964.

1964 April 22 Michael I. Yarymovych, director of NASA's Manned Earth Orbital Mission Studies, outlined the four different kinds of space station being considered with the Department of Defense: Extended Apollo, Apollo Orbital Research Laboratory (AORL), Medium Orbital Research Laboratory (MORL) and Large Orbital Research Laboratory (LORL). All but the LORL, sent into orbit by a Saturn V, were expected to be launched by Saturn IB. Yarymovych said it was "becoming increasingly clear that the Extended Apollo is an essential element of an expanding (manned) Earth orbital program" and that the Department of Defense's Gemini B/Manned Obital Laboratory might form the core of the Orbital Research Laboratory program.

1964 May 1 At the May Day parade in Moscow, a new surface-to-air missile trundled through Red Square was given the NATO code designation SA-4 Ganef. With a length of 29 ft. 6 in., a diameter of 2 ft. 7.5 in. and a launch weight of about 5,000 lb., the missile was boosted by four solid propellant rockets and propelled in flight by a ramjet engine. Stability and control was effected by four large tail fins, with a span of 8 ft. 6 in., and four forward fins set at 45° to the others. Ganef had a range of 47 mi.

In accordance with a statement on the future of Strategic Air Command issued by U.S. Secretary of Defense Robert S. McNamara on December 1, 1961, the deactivation of America's first ICBM, the Atlas D, began with the removal of the first missile from alert status at Vandenberg AFB, Calif. The last Atlas D was removed October 1, 1964, and on November 19, 1964, McNamara announced a decision to phase out Atlas E and F along with Titan I. Some of these missiles had been operational for less than two years, their redundancy indicative of rapid advances made in missile technology. The last Atlas F was deactivated on April 12, 1965. Over time, numerous Atlas ICBMs were converted into ballistic test and satellite launchers.

1964 May 7 The Boeing Co. received a contract for five Lunar Orbiter spacecraft with delivery dates between May 7 and December 18, 1966. In comparison with the Ranger program conducted in-house at the Jet Propulsion Laboratory, it was a model of efficiency and economy. Lunar Orbiter cost NASA less money than the Ranger program, providing a 100% success record with five spacecraft compared to a 33% success rate out of nine Ranger flights. The first Lunar Orbiter was delivered only six weeks after the original contract date and all five successful flights were accomplished within a 12-month period beginning August 10, 1966.

1964 May 11 The U.S. Air Force Systems Command proposed development of a nonnuclear antisatellite system known as Program 437 Y. An earlier designation, Program 893, had been abandoned when fears emerged that funds would not be forthcoming for a completely new system, so it was veiled under the old Program 437. In March 1965 it was given the designation Program 922 and contracts were let to Hughes, LTV, and Northrop. In June 1967, LTV was chosen to build the system but it was canceled in 1969.

1964 May 13 Designated A-001, the second Little Joe II flight took place from White Sands, N.Mex., in a test of the launch escape system at transonic speed and high aerodynamic pressure. Launched at 12:59:59 UT, the Little Joe II was propelled into the air by six Recruit motors, which fired for only 1.5 sec., after which an Algol sustainer continued to fire for a further 40.5 sec. It carried Apollo boilerplate no. 12 to a height of 15,400 ft., where the escape system fired, carrying BP-12 away from the self-destructed booster at 44 sec. elapsed time to an apogee of 29,772 ft. The only event to mar an otherwise perfect test was when one of the three parachute risers broke, collapsing one of the three main parachutes. BP-12 touched down at a descent rate of 26 ft./sec., 22,400 ft. from the launch point, 5 min. 50 sec. after launch. The Apollo recovery system included two conical-ribbon drogue parachutes 16 ft. 6 in. in diameter, three ring-slot pilot parachutes, 7 ft. 2 in. in diameter, and three diameter ringsail main parachutes of 83 ft. 6 in.

A Kiwi B-4D reactor was operated at Los Alamos Scientific Laboratory for about one minute at the Jackass Flats, Nev., test facility. The reactor used liquid hydrogen as propellant and coolant. This configuration was representative of a flight system that might be used as the upper stage of a launch vehicle for use in the vacuum of space. The test had been planned to last eight minutes, and although it was cut short the test was considered a success.

1964 May 20 NASA awarded contracts to Boeing and the Bendix Corp. for separate nine-month feasibility studies on an Apollo logistics support system (ALSS) using the Saturn V/Apollo configuration for extended lunar surface activity. Payload would be delivered to the lunar surface by LEM (lunar excursion module) truck. In some NASA studies of the concept, the Saturn IB would carry a Centaur third stage for launching unmanned logistic supply vehicles or cargo during the build-up phase of a lunar surface base.

1964 May 28 SA-6, the first of five Block II Saturn I launch vehicles configured to carry a simulated Apollo on top the S-IV stage, was launched from LC-37 at Cape Canaveral as Apollo mission A-101 carrying Boilerplate no. 13. With a height of 190 ft. (succeeding Block II launchers had a height of 188 ft.), weight and performance was generally the same as the first Block II launched January 29, 1964. But instead of the open-loop guidance of earlier Saturns, SA-6 had an ST-124 active guidance system which determined continually during flight the most efficient steering commands for the gimballed motors. SA-6 placed a total mass of 37,300 lb. into a 111 × 127 mi. orbit which decayed into the atmosphere on June 1, 1964. Remaining Block II Saturn C-1s were: mission A-102 on SA-7 (BP-15), September 18, 1964; A-103 on SA-9 (BP-16), February 16, 1965; A-104 on SA-8 (BP-26), May 25, 1965; and A-105 on SA-10 (BP-9), July 30, 1965, the last three with Pegasus micrometeoroid detector arrays.

Lockheed Propulsion Co. fired the world's first 156-in.-diameter solid propellant rocket motor. With a height of 75 ft. and a weight of 450,000 lb., the motor burned for 1 min. 49 sec. at Potrero, Calif., delivering a maximum 900,000 lb. of thrust. This was the first of 10 tests in the U.S. Air Force Space and Missile Systems Office Program 623-A series to evaluate large-thrust solids for application to ballistic missiles and satellite launchers. A 2 min. 15 sec. firing took place on September 30 when a Lockheed 156-in. solid with 650,000 lb. of propellant produced a thrust of 1.2 million lb. A new nozzle was also tested which would bring economies to the production of solid propellant boosters for Titan rockets.

1964 May The U.S. Air Force began awarding small contracts in support of the proposed Manned Orbital Laboratory program. The first of six four-ton laboratories was scheduled to be launched in 1968 on a Titan III-C for a flight lasting one month. Each MOL would carry approximately 6,000 lb. of scientific experiments. Brig. Gen. Bleymaier said he envisaged NASA flying candidate MOL crew members on Gemini and Apollo missions, giving them spaceflight experience before they were orbited in the laboratory. On June 9, 1964, NASA's George Low emphatically denied that NASA would fly any astronauts other than those selected for its own manned programs, implying that the Air Force would have to provide its own flights.

1964 June 2 NASA's Flight Research Center awarded two $1.2 million contracts for alternative lifting body design concepts from the Norair Division of the Northrop Corp. The two design concepts had been developed at NASA Ames and Langley and would be tested in the wind tunnel at the Flight Research Center. They were both classed as heavyweights and would be designated M2-F2 and HL-10. Each would be carried aloft under the wing pylon of a converted B-52 mother plane and dropped for glide tests leading to powered flight utilizing a single 8,480-lb. thrust Thiokol XLR11-RM-13 rocket motor installed in the fuselage. In addition, four 400-lb. thrust throttleable hydrogen peroxide thrusters would be fitted to the M2-F2 to assist flare-out on the landing approach. The HL-10 would utilize two 500-lb. thrust hydrogen peroxide thrusters.

1964 June 4 The first Blue Streak launch took place from the Woomera Rocket Range in Australia (June 5 local time). Designated F1, the first of three Europa I test flights originated at launch pad 6A at Range E, 300 mi. NNW of Adelaide. The vehicle had a total length of 69 ft. 3 in., a diameter of 10 ft., and a liftoff weight of 205,4543 lb. The two Rolls Royce RZ.2 liquid propellant engines produced a combined thrust of 260,000 lb. At launch, the stage contained 133,007 lb. of liquid oxygen (forward) and 57,840 lb. of kerosene (aft). F1 flew a ballistic trajectory more than 900 mi. northwest of Woomera, reaching a speed of almost 9,000 ft./sec. Programmed to fire for 2 min. 33 sec., the engines shut down 6 sec. early when they sensed oscillations caused by propellant sloshing in the tanks. The second test, F2, was successfully conducted on October 19 (October 20 Australian time), followed by F3 on March 22 (March 21 Australian time), 1965.

The *Transit 5C* U.S. Navy navigation satellite powered by solar cells was carried by Scout to a 531 × 594 mi. orbit with an inclination of 90.3°. This was the first navigation satellite launched after the launch failure of *Transit 5BN3* on April 21, which had spewed its radioactive contents across the southern hemisphere. Most Transit satellites weighed 130 lb. and comprised a sphere 3 ft. in diameter. Beginning with a launch on October 6, Transit satellites were numbered sequentially in a 0 designation series, that satellite being numbered 0-1. Four Transit satellites were launched each year in 1965, 1966 and 1967, with one in 1968 and one in 1970. From 1967 technologies were developed for its successor, and the last two Transits were launched in 1985.

1964 June 8 Commenting on NASA's claim that the Soviet Union was in a race to land the first men on the moon, *Newsweek* magazine claimed that "No evidence has appeared that the Soviet Union is building a larger rocket to go to the moon. Many observers believe that the United States is racing itself." In fact, the Soviet government had just made the decision to attempt a manned moon landing before the Americans by using the N-1 heavy-lift launcher. Disagreement over the best way to accomplish that feat had delayed firm plans for two years, since Sergei Pavlovich Korolev first proposed an early Sever configuration sent round the moon by a rocket stage fueled in space by three SL-4 launchers. In electing to go for a lunar-orbit rendezvous plan like NASA, the Russians envisaged precursor steps involving unmanned

(L-1) and manned (L-2) circumlunar flights, and lunar landing and return missions (L-3).

NASA began an industry design competition for an automated biological laboratory (ABL) which would bring together experiments for the search for life on Mars as part of the proposed Voyager program. The ABL would be set down on the surface of Mars where a range of life-detection instruments would sample the atmosphere and the soil for signs of organic or inorganic activity. The competition would result in proposals from which a one-year effort to review scientific requirements would begin.

1964 June 14   The first French Emeraude sounding rocket was launched from Hammaguir, Algeria, but suffered a guidance malfunction which destroyed the vehicle. Emeraude was the single-stage L-13 vehicle designed to serve as the first stage of the Diamant satellite launcher. Burning WFNA (white fuming nitric acid) and turpentine in its Vexin engine, the Emeraude was a scaled-up version of the Vesta sounding rocket. Emeraude had a liftoff weight of 39,000 lb., a thrust of 61,750 lb., and could lift an 880-lb. payload to 125 mi. The second launch two days later, and the third on October 20, also failed. The fourth attempt, on February 27, 1965, was a success, as was the fifth and last on May 13, 1965. Flights with the Saphir second stage attached to Emeraude began on July 5, 1965.

1964 June 17   The first flight of the French Rubis test rocket took place from Hammaguir in Algeria. Incorporating the P-2.3 Agate test vehicle as a first stage and a P-0.6 second stage, Rubis reached an altitude of 1,118 mi. It had been developed as part of the "precious stones" series of launchers that also included Agate, Topaze, Emeraude and Saphir. Rubis had a length of 31 ft. 6 in., a liftoff weight of 7,500 lb. and a launch thrust of 42,000 lb. It was capable of lobbing a 77-lb. payload to a height of 1,500 mi. and had been developed as a sounding rocket and also as the third stage of the Diamant satellite launcher. The P-0.6 stage would double as the third stage of Diamant. Test launches in support of Diamant came on May 31, June 3 and June 5, 1965.

1964 June 23   The U.S. Army Map Service announced that it had completed its topographic survey of the moon. The map that resulted from this work was the first to show variations in elevation over the visible face of the lunar surface. The map covered an area of approximately 8 million sq. mi. with more than 5,000 named surface features. The mapping survey had been conducted with the cooperation of NASA. Five days later the U.S. Air Force Aeronautical Chart and Information Center completed what lunar geologists termed "the best lunar maps ever made" when they released a set at a scale of 1:506,880. Future, more detailed maps made from images returned by spacecraft would build upon this early work.

1964 June 28   In the first overseas deployment of any British missile, the United Kingdom Ministry of Defense announced that RAF V Bloodhound Mk.2 surface-to-air missiles would be sent to the Far East. The missile would be used to defend RAF bases in Malaysia. The Mk.2 had a much higher performance than the Mk.1 and was a more mobile system than its predecessor. The Mk.1 was about to be retired from RAF service because it was considered redundant. Designed as a defense against manned aircraft, these were no longer considered the main threat to the United Kingdom, unlike other areas where the Mk.2 was to be deployed.

1964 June   Based on work conducted at the Royal Aircraft Establishment (RAE), Farnborough, England, NASA's Manned Spacecraft Center adopted the water-cooled undergarment as the foundation garment for advanced space suits which might be worn by Apollo astronauts. The idea involved plastic tubing sewn into long underwear for the purpose of carrying water round the body to remove heat. Dr. John Billingham of the RAE had developed the idea during 1962 and joined the Manned Spacecraft Center in 1963. The water-cooled undergarment would remain a feature of space suit design from Apollo on.

1964 July 1   A National Space Development Center (NSDC) was established in Japan's Science and Technology Agency (STA) to execute national space planning strategies for the National Space Activities Commission working directly under the prime minister's office. The NSDC was also directed to bring more dynamic activity to space research and technology. The existing Institute of Space and Aeronautical Science at Tokyo University had limited scope for advancing the capabilities of Japan's expanding interest in aerospace. In May 1966 the NSDC decided to develop a national launch site for satellites at a specially built range on the fishing island of Tanegashima, one of the southernmost islands. Agreements with fishermen, however, would limit rocket flights to two periods each year — for fear the noise and vibration would frighten away their livelihood.

1964 July 2   Using giant rocket-powered vehicles to move large numbers of troops rapidly around the globe was one use put forward by Philip Bono and George C. Goldbaum of Douglas Aircraft for their Ithacus concept. This vehicle would be launched from giant nuclear-powered aircraft carriers of the USS *Enterprise*-class carrying up to 1,200 troops to any place on Earth in under 45 minutes. Ithacus grew from an earlier concept called Icarus and utilized several aerospike, or plug-nozzle, engines arranged around the bottom circumference of the massive launcher, which would be shaped like a tapered cylinder.

1964 July 6   The first U.S. electronic intelligence-gathering (elint) satellite carried piggyback on a second-generation high-resolution reconnaissance satellite was carried on an Atlas-Agena D from Vandenberg AFB, Calif. The KH-7 satellite entered a 75 × 215 mi., 92.8° orbit while the 130-lb. elint subsatellite went into an orbit of 184.5 × 234 mi. inclined 92.9°. A total of eight subsatellites were flown on second-generation KH-7 reconnaissance satellites, the last on September 16, 1966.

1964 July 10   NASA Administrator Charles W. Mathews reported at a senior staff meeting that the second manned Gemini flight, GT-4, would last four days and use battery power, but that the GT-5 mission would be open-ended for up to seven days and use cryogenic fuel cells. Power requirements dictated that batteries alone could not support a two-man mission lasting more than five days. GT-6 was to be a two-day flight demonstrating rendezvous with an Agena target vehicle, and GT-7 was to be a long-duration flight of up to fourteen days. On July 27, 1964, astronauts James McDivitt and Edward White were announced as the primary crew for GT-4, with Frank Borman and James Lovell as their backups. Two days later Gemini Program Manager Kenneth S. Kleinknecht confirmed that an astronaut on GT-4 would open a hatch and "step into space."

1964 July 15 The U.S. Department of Defense began a program to develop its own global Defense Satellite Communication System (DSCS) following the breakdown of negotiations with the Comsat Corp. about the use of an international satellite system for military purposes. The Pentagon was concerned about the degree of control non-U.S. members of such a system would have on military use of these satellites. A global defense communications system was made more necessary by the expansion of strategic and tactical nuclear forces and by the extensive use of foreign bases by the U.S. military.

1964 July 20 The first successful operation in space of an electric motor took place during the ballistic flight of a Scout rocket launched by NASA from Wallops Island, Va. Called *SERT I*, for Space Electric Rocket Test, the four-stage rocket carried the 375-lb. payload to a height of 2,500 mi., from where it fell back into the Atlantic Ocean about 2,000 mi. downrange, 48 min. after launch. *SERT I* carried two ion motors, one of which, built by the Lewis Research Center, was run for a total 30 min., generating an ion beam which produced a thrust of 0.0055 lb. Built by Hughes Aircraft, the other engine did not produce thrust, but the flight was considered so successful that a second test was canceled.

1964 July 21 NASA announced that it was inviting scientists from around the world to participate in manned and unmanned flights by submitting proposals for experiments that could be carried by lunar spacecraft. The submissions would be handled by the Office of Space Science and Applications and envisaged the participation of individuals and foreign governments. This effort would grow during and after Apollo and bring to NASA a strong flavor of international participation in lunar and planetary exploration. For a long time it was the primary means of contact with space scientists in the Soviet Union, and the exchange of information from planetary programs became a cornerstone of international cooperation.

1964 July 24 At an international conference in Washington, D.C., 13 participating nations including the United States agreed to establish a global satellite communications system on commercial grounds. Five other nations agreed to join before August 19. For a further six months after that date, any nation belonging to the International Telecommunications Union would be free to join. The U.S. signatory was the Communications Satellite Corp. (Comsat), which would own 61% of the shares and operate the system on behalf of all the members. The name of this organization was to be the International Telecommunications Satellite Consortium (Intelsat).

The McDonnell Aircraft Corp. was awarded a contract from the U.S. Air Force to conduct a detailed study of Gemini B, a manned vehicle based upon the concept of a second-generation Gemini spacecraft. The Air Force wanted Gemini as a crew vehicle for the proposed Manned Orbital Laboratory but was already looking toward development of a unique ferry vehicle that could move crew and supplies to and from the orbiting facility. The Air Force planned to use START technologies to begin the design of a manned maneuvering reentry vehicle. This would lead, said the Air Force, to a larger, winged shuttle vehicle capable of supporting a permanently manned military space station.

1964 July 28 At 16:50:07 UT, NASA launched the first successful U.S. moon probe when *Ranger 7* was sent on its way to lunar impact by an Atlas-Agena B from Cape Cana-

veral, Fla. The 806-lb. spacecraft was virtually identical to its predecessor and was propelled out to an initial Earth orbit of 119 mi. by a second burn of the Agena B. A 50-sec., 67-MPH midcourse maneuver was completed at 10:27:59 UT the following day, and *Ranger 7* impacted in the area of the moon's Sea of Clouds just south of the crater Copernicus at 13:25:49 UT on July 31, 1964. In the final 17 min. 12 sec. to the time of impact, *Ranger 7* transmitted 4,316 images of the moon; the last was transmitted from a height of 1,740 ft. with a resolution of about 1 ft. 6 in. just 0.2 seconds to destruction at a speed of 5,853 MPH. Its impact coordinates were 10.7° S latitude by 20.7° W longitude, just 8 mi. from the aim point.

The Mariner-Mars 1966 mission involving two fly-by/probe flights to Mars were tentatively canceled by NASA pending the completion of final documentation on a more ambitious Advanced Mariner 1969 mission. Project approval for Advanced Mariner was granted by NASA on August 2, 1964, and termination of the Mariner-Mars 1966 mission was officially acknowledged on September 4, 1964. Advanced Mariner was conceived as a precursor to the ambitious Voyager biological survey of Mars, at this date scheduled for a flight in 1971. Advanced Mariner 1969 survived only a few months, being canceled on November 20, 1964. The Mariner E spacecraft designation, one of two originally assigned to the Mariner-Mars 1966 mission, was applied later to the Mariner-Venus 1967 mission, which was then called Mariner V.

1964 July 30 NASA Headquarters asked the Gemini Program Office to study a range of missions between the 12 currently scheduled. In the *Advanced Gemini Missions Conceptual Study* a further 16 flights were projected, including a mixture of NASA and U.S. Air Force missions. Among those proposed were a preliminary space station flight, a satellite intercept mission and a lifeboat rescue mission demonstrating the ability of the two-man spacecraft to return the crew of a disabled space vehicle to Earth in an emergency. More advanced concepts suggested a circumlunar flight and a lunar orbit mission. Neither of these missions were approved.

1964 August 3 In an historic order, number 655-268 of the Central Committee of the Communist Party, Soviet engineers were commanded to conduct a circumlunar flight and perform a landing on the lunar surface. An increasingly influential rival to chief Soviet spacecraft designer Sergei Pavlovich Korolev, Vladimir Chelomei approved the final proposals from his design engineering team for a circumlunar mission. Based on Chelomei's Proton launcher (designated UR-500), the mission envisaged a 4,400-lb. spacecraft called LK-1 carrying one cosmonaut on a flight around the moon. In a configuration similar to NASA's Gemini spacecraft, the conical LK-1 module would have been attached to a cylindrical equipment module, which in turn would have been attached to a propulsion module, the total mass being about 37,000 lb. Electrical power was to be provided by solar cells. Despite being adapted to carry two cosmonauts, the UR-500/LK-1 was canceled late in 1965 when Korolev's UR-500K/L-1 program was adopted. It formed the basis for TKS and Almaz manned ferry ship and space station designs.

1964 August 8 President Johnson announced that the U.S. Air Force was to launch a series of 24 military communication satellites, 8 at a time on Titan IIIC launch vehicles, to geostationary orbit. The first launching was scheduled for 1966. This was emphasized as an "interim" system and would be known first as the Initial Defense Communication Satellite Program (IDCSP) and later as the Initial Defense Satellite

Communication System (IDSCS). It would be followed by more advanced systems known simply as DSCS. On February 2, 1966, the United Kingdom signed an agreement with the U.S. Department of Defense to provide ground stations for cooperative tests.

1964 August 17   The Titan 2 launch vehicle was on Complex 19 at Cape Canaveral, Fla., in preparation for the GT-2 flight when the pad area was struck by lightning during a severe electrical storm. Inspection to examine the launcher for damage was completed September 22. Meanwhile, on August 27, 1964, the second stage was removed and stored as Hurricane Cleo swept through the Cape Canaveral area. Re-erected on September 1, preparations for GT-2 resumed, only to be halted September 8 when both stages had to be removed from the pad as Hurricane Dora struck the region. When Hurricane Ethel also threatened to pay a visit, the stages remained stored until finally reerected September 14, 1964. The schedule then envisaged a GT-2 launch on November 17, 1964, followed by GT-3 on January 30 and GT-4 on April 12, 1965.

The NASA Manned Spacecraft Center's Spacecraft Integration Branch laid down objectives for the proposed Apollo X spacecraft conceptualized as an Earth-orbiting laboratory launched by Saturn IB to a 230-mi. orbit. Apollo X Configuration A was for two astronauts and no laboratory module; Configuration B incorporated a single laboratory module for 45-day missions; Configuration C would accommodate a double-laboratory module for three astronauts on a 45-day flight; and Configuration D would support three astronauts for 120 days and provide an independent laboratory module. Laboratory modules would be accommodated in the spacecraft adapter atop the Saturn IB and retrieved in a transposition and docking maneuver similar to that used by Apollo to extract the lunar module.

1964 August 18   The first launch of a new Soviet satellite launcher, the converted SS-5 Skean IRBM (intermediate-range ballistic missile) known as the SL-8, took place from Baikonur. This was also the first triple-launch of a Soviet payload into orbit. Designated *Cosmos 38–40,* the three satellites, each weighing 110 lb., were placed in separate orbits with a perigee of 128 mi. and apogees of 478 mi., 496 mi. and 460 mi., respectively, at an inclination of 56.2°. With orbital periods of 95.2 min., *Cosmos 38–40* were precursor tactical military communication satellites; they decayed, respectively, on November 8, November 17 and November 18. The SL-8 first stage was powered by the twin-chamber RD-216, developed by the GDL-OKB, producing a liftoff thrust of 330,750 lb. and a burn time of 2 min. 50 sec. The second stage has a 35,300-lb. thrust engine burning for 6 min. 15 sec., and both stage burn UDMH4-N2O4 propellants. Typically, the SL-8 has a length of 105 ft., a weight of 220,000 lb., and can lift 3,750 lb. to low orbit. Later named Cosmos, the SL-8 had by mid-1995 been launched on almost 420 flights, of which 96% had been successful.

1964 August 19   The first thrust-augmented Delta (TAD) launch vehicle successfully placed the world's first geostationary satellite in orbit when *Syncom III* was sent on its way from Cape Canaveral, Fla. The TAD had three strap-on solid propellant motors providing a total first-stage thrust of 333,550 lb., with the normal 7,500-lb. thrust second stage and 5,700-lb. thrust X-258 third stage. It was more powerful than the launcher used with *Syncom I* and *II* because it was required to steer the satellite's course into the plane of the Earth's equator and this required more energy. *Syncom III*'s

solid propellant apogee motor placed the satellite in a 21,246 x 22,539 mi. orbit with an inclination of less than 1° to the equator. This eliminated the figure eight double ellipse prescribed by the two previous Syncoms. *Syncom III* drifted to 180° longitude, where it was used for communication tests between the Philippines and California and to transmit coverage of the Tokyo Olympic Games.

1964 August 20   The International Telecommunications Satellite Organization, Intelsat, came into force for an interim period of five years, after which it was established as a permanent structure. Beginning with 16 member countries in 1965, the system grew to include 60 countries by 1970, 107 by 1975 and 154 by 1980. Capacity to handle communications traffic by the orbital system expanded from 75 simultaneous telephone circuits in 1965, to 2,100 by 1970, 6,650 by 1975 and 20,300 by 1980. Annual charges to telecommunications entities for a single circuit decreased from $32,000 per annum in 1965 to $5,040 in 1980.

1964 August 22   Two Soviet precursor tactical military communication satellites were placed in elliptical orbit by SL-7 launched from Kapustin Yar. Designated *Cosmos 42* and *Cosmos 43,* the 220-lb. satellites operated from orbits of 143 × 692 mi. and 141 × 683.5 mi., respectively, at 49° and with orbital periods of 97.8 min. This was the first double Cosmos launch and both satellites decayed in 1965: *Cosmos 42* on December 19 and *Cosmos 43* on December 27. The technology development inherent in this mission may have been related to covert store-dump communications missions beginning with *Cosmos 158,* launched by SL-7 on May 15, 1967. All other tactical military communication missions continued to use the SL-8, first exploited for this role on August 18, 1964.

1964 August 25   A four-stage Scout launch vehicle carried the 97-lb. ionospheric measuring satellite, *Explorer XX,* from Vandenberg AFB, Calif., to a 540 × 634 mi., 79.9° orbit. Built by Airborne Instruments Laboratory in the United States, Ionosphere Explorer A took the form of a cylinder with truncated cone on the bottom and a cone with sphere on top. It had a height of 3 ft. 10.5 in., a diameter of 2 ft. 2 in. and carried three deployable dipole sounding antennas, two 62 ft. in length and one 122 ft. long. *Explorer XX* continued to transmit data on 136 MHz until July 1966 and the satellite remains in orbit.

1964 August 28   A Thor-Agena B launched from Vandenberg AFB, Calif., put the 830-lb. *Nimbus 1* experimental weather and environment satellite in orbit, but in an elliptical path of 263 × 579 mi. inclined 80° instead of the planned circular orbit. Designed as a test platform for new sensors, detectors and monitoring equipment, eight Nimbus satellites were launched by NASA. *Nimbus 1* comprised a lower ring section 4 ft. 9 in. in diameter with a hexagonal upper section containing control equipment. It carried 10,500 solar cells on two panels and had a total height of 10 ft. 10 in. Built by General Electric, the satellite carried three 1-in. vidicon cameras and took 27,000 pictures before it failed on September 23 when the solar panels were unable to track the sun. Essentially the same, *Nimbus 2* was launched on May 15, 1966, to a 684 x 734 mi. sun-synchronous orbit inclined 100°; it operated until January 18, 1969, by which time it had transmitted 210,000 images.

An SL-3 launch vehicle carried the first in a series of Soviet experimental weather satellites, *Cosmos 44,* from Baikonur cosmodrome to a 384 × 534 mi. orbit inclined 65°. With a weight of 3,400 lb., it carried sensors that were being developed for an operational system. Other experimental weather satellites included *Cosmos 44* (September 13, 1964), *Cosmos 58* (February 26, 1965), *Cosmos 65* (April 17, 1965), *Cosmos 92* (October 16, 1965), *Cosmos 100* (December 17, 1965), *Cosmos 118* (May 11, 1966), *Cosmos 122* (June 25, 1966), *Cosmos 144* (February 28, 1967), *Cosmos 156* (April 27, 1967), *Cosmos 184* (October 24, 1967), *Cosmos 206* (March 14, 1968) and *Cosmos 226* (June 12, 1968), all of which weighed 4,730 lb.

1964 August 31   The names of four flight directors for the Gemini and Apollo programs were announced by NASA. Manned Spacecraft Center Director Robert R. Gilruth appointed Christopher C. Kraft, John D. Hodge, Eugene F. Kranz and Glynn S. Lunney to these positions. The role of the flight director would be to control and manage the mission from launch to recovery and report directly to a mission director. Both Kraft and Lunney would eventually be appointed as directors of the MSC. In 1983 John Hodge would head a task force on space station plans in the run-up to the commitment to build Space Station Freedom, announced by President Reagan on January 25, 1984.

1964 September 1   The U.S. Air Force launched its first Titan IIIA standardized space launch vehicle from Cape Canaveral's Complex 20 at 15:00:06 UT. The flight of this three-stage vehicle proceeded normally through stage three (Transtage) ignition, but a premature shutdown caused the stage and dummy payload to reenter and impact the Atlantic Ocean near Ascension Island. The Titan IIIA comprised the first and second stages of a Titan II plus a 16,000-lb. thrust Aerojet General Transtage with twin thrust chambers burning a blend of hydrazine and UDMH (unsymmetrical dimethyl hydrazine) for fuel and nitrogen tetroxide for oxidizer. Transtage had a length of 15 ft. and a diameter of 10 ft. ensuring a common diameter for the entire length of the vehicle. Transtage carried 24,000 lb. of propellant for a maximum burn duration of 6 min. 40 sec. with multiple restart capability accommodating payloads up to 50 ft. in length and 20 ft. in diameter.

NASA's new Electronic Research Center was formally opened at 129-acre site in Cambridge, Mass., when Winston E. Kock took the oath of office as director. The center was formed from a grouping of personnel from Headquarters and from the Northeastern Office at Cambridge. It set out several research efforts in systems, components, guidance and control, instrumentation and data processing, and electromagnetic research. Design and construction of the new facility was completed by 1969. Facing a budget squeeze, NASA closed the ERC in 1970 and the buildings were transferred to the Department of Transportation.

1964 September 5   The first of NASA's six Orbiting Geophysical Observatories, *OGO-1,* was launched by Atlas-Agena B from Cape Canaveral, Fla., to a 175 × 98,287 mi., 31° orbit. Carrying 20 experiments, more than any previous U.S. satellite, in a package of instruments weighing 190 lb., *OGO-1* comprised a rectangular box, 5 ft. 7 in. long and 2 ft. 7 in. square. Two large solar panels supporting 32,000 cells were attached to opposite sides of the satellite, and two 22-ft. and four 6-ft. experiment booms deployed from opposite ends. Fully deployed, the 1,046-lb. *OGO-1* had a length of 59 ft. and a span of 20 ft. across the solar panels. The data handling system could store up to 86 megabits of information and transmit to Earth at 128 kilobits/sec. Two booms failed to deploy properly and the satellite was operated in a spin

stabilized mode at 5 RPM, which allowed 14 experiments to work as planned. *OGO-1* was turned off on November 25, 1969.

1964 September 9   The United Kingdom's Minister of Aviation, Julian Amery, announced his government's decision to develop the Black Knight experimental rocket into a national satellite launcher. It would have three stages, the first being powered by a cluster of eight BS.606 engines burning hydrogen peroxide and kerosene, the second having a single two-chamber BS.625A engine using the same propellants, and the third stage employing a solid propellant rocket motor to place the experimental satellite in orbit. As a satellite launcher, Black Knight would be capable of placing 440 lb. in polar orbit.

1964 September 24   The first LGM-30F Minuteman II solid propellant ICBM was successfully launched down the Atlantic Missile Range from Cape Canaveral, Fla. With a length of 59 ft. 10 in. and a launch weight of 70,000 lb., it had a 34% increase in propellant capacity compared to the Minuteman I and improved guidance. The first-stage propellant comprised a combination of ammonium perchlorate and polybutadiene acrylic acid producing a thrust of 200,000 lb., the 61,000-lb. thrust second stage had a polyurethane and ammonium perchlorate propellant, while the 35,000-lb. thrust third stage used a nitrocellulose, nitroglycerine and ammonium perchlorate mixture. Minuteman II was deployed with the Mk.11 warhead containing a 1.2-MT W-56 warhead which, with new guidance equipment, had a target probability error no greater than 1,200–1,800 ft. at the missile's full range of 7,000 mi.

Dr. Nancy Roman, head of NASA's astronomy programs, outlined current plans for an orbiting astronomical telescope with a mirror 100-in. in diameter. The Manned Orbiting Telescope (MOT) would operate automatically, but periodic visits by trained astronauts would be made to replace science instruments and collect photographic film. An interim link between the unmanned Orbiting Astronomical Observatory and the MOT would be a 50–55 in. telescope "that could be used manned or unmanned." Dr. Roman said the MOT was being considered for launch "in 15 to 20 years." This program would mature into the Hubble Space Telescope, launched by the Discovery Space Shuttle on April 24, 1990.

1964 October 6   The first, and only, unmanned test model of the Soviet Voskhod spacecraft was launched by SL-4 from the Baikonur cosmodrome into a 108 × 238 mi. orbit inclined 64.8° to the equator. The SL-4 utilized the same four boosters and first stage employed for the Vostok flights, but the first stage carried an additional 6,625 lb. of propellant that increased burn time by 10 sec., and the upper stage was bigger and more powerful. With a length of 29 ft. 6 in. and a diameter of 8 ft. 9 in., it had a single 66,150-lb. thrust RD-461 liquid oxygen–kerosene engine with a burn time of 4 min. The more powerful upper stage raised overall launcher height to 147 ft. 4 in., liftoff weight to 677,500 lb. and payload capacity to 12,500 lb. To maintain secrecy over the new spacecraft, it was designated *Cosmos 46* but remained in orbit only one day. Adapted from the one-person Vostok, the multiperson Voskhod had a weight of 11,690 lb. and incorporated a backup retro-rocket situated on top of the spherical descent module, as well as a braking rocket for a softer touchdown, changes which added almost 1,000 lb. to the weight of the descent module.

1964 October 9 A nine-day test and qualification of the expanded NASA worldwide tracking network began in preparation for manned Gemini missions. The old Mercury tracking network had been reworked over a period of two years into the Manned Space Flight Network (MSFN) to cater for the more demanding communications traffic generated by the two-man Gemini and three-man Apollo missions. With more complex tasks producing larger volumes of telemetry and real-time engineering data, the Mercury network had been insufficient for the job.

1964 October 10 NASA's program of geodetic and gravimetric measurement began with the launch of Beacon Explorer B *(Explorer 22)* on a four-stage Scout from Vandenberg AFB, Calif., to a 549 × 669 mi. orbit inclined 80°. The Scout was similar to that employed for *Explorer XX*, except for the 5,800-lb. thrust X-258 (Altair) stage. Also the last of five satellites in the first phase of NASA's ionospheric monitoring program, the 116-lb. Beacon Explorer B was almost identical to its predecessor, Beacon Explorer A. It was the first satellite to provide geodetic ranging data through reflected laser beams transmitted from the ground; it remained in orbit as of 1995. The 132-lb. Beacon Explorer C *(Explorer 27),* last of the original NASA geodetic series, was launched on April 29, 1965. Its battery failed in 1968 but it remains in orbit.

1964 October 12 Carrying cosmonauts Vladimir Mikhailovich Komarov, Konstantin Petrovich Feoktistov and Boris Borisovich Yegorov, the 11,730-lb. *Voskhod 1* spacecraft was launched from the Baikonur cosmodrome at 07:30 UT on the world's first multicrew spaceflight. They were also the first people to be launched into orbit without space suits. The backup crew were cosmonauts Boris Valentinovich Volynov, Georgi Petrovich Katys and Vasily Grigorievich Lazarev. The SL-4 launch vehicle was the same configuration as that which had been used to orbit the *Cosmos 47* precursor test mission six days earlier; it placed *Voskhod 1* in a 111 × 254 mi. orbit. Designed for one cosmonaut, the interior of the spherical descent vehicle was cramped, providing little opportunity for experiments or movement. When the descent vehicle returned to Earth after a flight time of 1 day and 17 min., it was the first time Soviet cosmonauts had remained inside their spacecraft from launch to landing; Vostok cosmonauts had ejected during the descent. Also, they returned to a new leadership in the Kremlin. In their brief absence, Soviet Premier Nikita Khrushchev had been removed and replaced by General Secretary Leonid Ilyich Brezhnev and Deputy Prime Minister Aleksey Nikolayevich Kosygin.

1964 October 17 The David Clarke Co. delivered to NASA's Crew Systems Division the first prototype EVA (extravehicular activity) space suit. It was assigned to astronaut James McDivitt for tests in the Gemini mission simulator. On October 19, 1964, astronaut Russell Schweickart began an eight-day evaluation of the G3C suit in tests of the biomedical harness and the operability of body sensors under the suit. During this period, when Schweickart wore the suit continuously for eight days, he went through a simulated four-day mission and several zero-gravity simulations on parabolic aircraft flights. The first prototype G4C EVA suit was delivered to Crew Systems Division on December 28, 1964.

1964 October 30 NASA test pilot Joseph A. Walker carried out the first flight of the lunar landing research vehicle (LLRV). Built by Bell Aerosystems, the LLRV comprised a 3,710-lb. open framework lunar landing training vehicle with a vertically mounted turbofan engine providing 83% of the vehicle's weight to simulate lunar weight (the moon's gravity is about 17% that of Earth's gravity). Two 500-lb. thrust Bell rocket motors supported the other ⅙th Earth weight, allowing simulated Apollo lunar excursion module descent paths to be flown. Two LLRVs were built and the first was delivered to the Manned Spacecraft Center on December 13, 1966, for astronaut training.

1964 October 31 Theodore C. Freeman, the first U.S. astronaut to lose his life in the line of duty, was killed after his T-38 struck an 8-lb. snow goose with a 2-ft. wingspan and lost power during a landing approach at Ellington AFB, Tex. Regaining control of the aircraft, Freeman made a second approach, but realizing he would not make the landing strip, he put the aircraft on course for a crash landing in fields, avoiding buildings. By this time the aircraft was down around 100 ft., and although Freeman ejected, he was too low for his parachute to deploy properly. Freeman had been selected as an astronaut with the Group 3 team on October 18, 1963.

1964 November 5 At 19:22:05 UT an Atlas-Agena D launched *Mariner III* from Complex 13 at Cape Canaveral, Fla., into an Earth parking orbit and from there to an Earth escape trajectory which should have put the spacecraft on course for Mars fly-by. At the end of the second Agena D burn, tracking data indicated a velocity deficit of 650 MPH, and just over 5 min. after the spacecraft separated, telemetry disclosed that the four solar panels had not deployed. The low injection velocity was what would be experienced if the 250-lb. payload shroud had failed to jettison during ascent. Instruments were turned off to conserve battery power and plans made to fire the midcourse correction motor to shake off the shroud. Power ran out at an elapsed time of 8 hr. 43 min. before this could be made to happen. Examination of payload shroud design showed that air would build up between the honeycomb shroud structure and its thin fiberglass skin. The subsequent buckling would create a deformed structure that would jam when called upon to jettison. NASA designed and built a new, 297-lb. magnesium-thorium shroud for *Mariner IV* and delivered it to Cape Canaveral on November 22. Still in its shroud, *Mariner III* was stuck in a 92 million × 122 million mi. orbit of the sun.

As defined through an internal NASA memorandum, the designation of what had previously been known as the Apollo space suit assembly was henceforth to be called the extravehicular mobility unit (EMU). This was to distinguish the moonwalk suit more properly from a less sophisticated design used as a pressure garment inside the spacecraft, eventually called the intravehicular pressure garment assembly. The EMU would comprise a liquid-cooled garment (LCG), an extravehicular pressure garment assembly, a portable life support system (PLSS), an oxygen purge system (OPS), a lunar EVA visor and EVA gloves. The LCG would be worn as an undergarment, with tubing to circulate water for body cooling. The PLSS backpack would provide oxygen and cooling and remove exhaled gases on the lunar surface. The OPS would sit on top of the PLSS and provide a 30-min. oxygen supply for emergency. Overall the EMU would weigh just over 30 lb. on the moon, equivalent to 183 lb. on Earth. International Latex Corp. received a contract on November 5, 1965, to fabricate the Apollo space suit.

1964 November 6 NASA launched a 295-lb. science satellite, *Explorer XXIII,* on a four-stage Scout from Wallops Island, Va., to a 289 × 607 mi., 51.9° orbit. The launch was scheduled to encounter winds in excess of 70 MPH between 25,000 ft. and 40,000 ft., so as to measure loads on the Scout for future engineering reference. This was the third and last in a series of S-55 micrometeoroid satellites, others including *Explorers XIII* and *XVI*. Put together by the Langley Research Center and the Goddard Space Flight Center, the experiment package for *Explorer XXIII* was contained in a cylinder 2 ft. in diameter and 7 ft. 8 in. long which, in the first year of operation, recorded 122 micrometeoroid hits. Data was transmitted through July 1966 and the satellite decayed on June 29, 1983.

1964 November 10   Denis Healey, Britain's defense minister, announced that he was setting up a review committee to examine the potential role of satellites and space systems in the nation's defense programs. Headed by Prof. Hermann Bondi, from Kings College, London, the committee would look at the possible application of communication satellites. This review would begin a process leading to the Skyney military communication satellites.

1964 November 19   In a weekly management report from NASA's Apollo Spacecraft Program Office, it was noted that serious thought was being given to using radioisotope thermo-electric generators (RTGs) to power groups of scientific instruments left on the surface of the moon by Apollo astronauts. Grumman was asked to study this, and the SNAP-27 RTG was eventually selected to power the Apollo lunar surface experiments package (ALSEP) comprising a variety of instruments left at landing sites when the crew returned to Earth. Plutonium 238 dioxide fuel would be carried to the moon in a special heatproof and shock-resistant graphite cask attached to the back of the LEM descent stage. An astronaut using a special handling tool would extract the fuel core on the lunar surface and insert it into the generator, where it would provide about 60 W of electrical energy and power the instruments.

1964 November 20   The NASA Manned Spacecraft Center awarded a contract to McDonnell in which it was required to examine four different logistics vehicles for an orbital research laboratory, or space station. The four alternate concepts McDonnell was to evaluate included an Apollo-type ballistic design, a lifting-body such as the M-2 or HL-10 shape, a variable-geometry (swing-wing) type and a high-lift, fixed-wing shuttle configuration. The launch vehicle was to be either a Titan IIIC or a Saturn IB capable of putting the logistics vehicle in a 200–300 mi. orbit. This study was the first serious attempt by NASA to determine the optimum design concept for a reusable logistics vehicle serving manned space stations. The findings were submitted August 27, 1965, in a detailed report in which McDonnell evaluated optional configurations within each of the four categories and did not make trade-offs between concepts.

1964 November 21   In NASA's first dual satellite flight, *Explorers XXIV* and *XXV* were launched together on a four-stage Scout from Vandenberg AFB, Calif., with the purpose of providing complex air-density data from the upper atmosphere. Weighing 19 lb., *Explorer XXIV* entered a 344 × 1,551 mi., 81.4° orbit, where it deployed as a balloon 12 in. in diameter and covered with 4,000 white dots for passive thermal control. The 90-lb. *Explorer XXV* was placed in a 345 x 1,547 mi., 81.4° orbit, from where it would monitor particle density and radiation flux. *Explorer XXV* comprised a spherical aluminum shell with 40 sides and a 12 in. diameter

incorporating a tube on top from which the deflated *Explorer XXIV* was ejected. The orbital geometry was set so that the perigee point would arrive at the north polar region around the winter solstice and migrate around the Earth once every six months. *Explorer XXIV* decayed on October 18, 1968, but *Explorer XXV* remains in orbit, inoperable.

1964 November 28   At 14:22:01 UT an Atlas-Agena D successfully launched the *Mariner IV* spacecraft from Launch Complex 12 at Cape Canaveral and placed it in a 107 × 114 mi. Earth parking orbit 8 min. 37 sec. later. At an elapsed time of 40 min. 52 sec. the Agena D burned a second time, for 1 min. 35 sec., pushing *Mariner IV* to a speed of 25,598 MPH on course for Mars. The spacecraft separated from the Agena D 2 min. 41 sec. later with the combination in Earth's shadow. At 52 min. 59 sec. *Mariner IV*'s central computer and sequencer (CC & S) began deploying solar panels and a scan platform of science instruments. The 575-lb. *Mariner IV* was built around an eight-sided magnesium structure containing 7 electronics compartments forming the main spaceframe 1 ft. 6 in. high and 4 ft. 6 in. across the flats. Four 19-lb. arrays, each 5 ft. 11.5 in. long by 2 ft. 11.5 in. wide carrying a total 28,224 solar cells, were attached to the top of opposing octagon flats and produced 640 W near Earth and 310 W at the distance of Mars. A single solar sail at the extremity of each array would be used to help maintain attitude. Data would be transmitted at 33.33 bits (binary digits)/sec., or 9.33 bits/sec. at the distance of Mars, via a fixed ellipsoidal high-gain antenna, 3 ft. 10 in. by 1 ft. 9 in., on top of the spacecraft. A low-gain tube antenna 3.9 in. in diameter extended 7 ft. 4 in. above the spaceframe. *Mariner IV* had a span of 22 ft. 7.5 in. and a height of 9 ft. 6 in. to the top of the low-gain antenna.

1964 November 30   A Soviet SL-6 launch vehicle successfully placed the 1,960-lb. *Zond 2* spacecraft on course for Mars following a brief period in Earth parking orbit. The 3MV-4A spacecraft did not carry TV cameras although its successor, *Zond 3*, did. *Zond 2* carried electric thrusters for attitude control, the first time plasma jets had been used. During January 1965, *Zond 2* was tracked by Britain's Jodrell Bank radio telescope and during a visit to that facility on February 17, 1965 the president of the Soviet Academy of Science, Mstislav Keldysh, confirmed that Soviet scientists might use its telescope in future for communicating with deep space missions. *Zond 2* suffered from a failure in one bank of solar cells and contact was lost in April 1965, four months before it flew past Mars.

1964 December 5   NASA's Mars bound *Mariner IV* spacecraft performed a 20 sec. (55.7 ft./sec.) course correction burn at 16:09:11 UT, 1.264 million mi. from Earth, to shift the closest fly-by distance from 153,125 mi. to just over 6,000 mi. from the surface by the time it arrived July 15, 1965. The course correction propulsion system was similar to that carried by *Mariner II* and had provision for total velocity corrections of up to 267 ft./sec. in one or two separate burns. Attitude control of *Mariner IV* in pitch, roll and yaw was by pairs of twelve cold gas jets located at the extremities of the four solar panels and fed from two 2.5-lb. nitrogen spheres. The jets were linked by logic circuitry to three gyroscopes, the sun sensor and the Canopus star sensor. This was the first time a star sensor had been used.

1964 December 7   The chairman of the Senate Committee on Aeronautical and Space Sciences, Sen. Clinton P. Anderson, wrote to President Johnson suggesting that the U.S. Air Force

Manned Orbiting Laboratory (MOL) should merge with the proposed Apollo X concept. Sen. Anderson wanted the MOL canceled and funds transferred to the NASA project. A week later Anderson said he had received an explanation from the Air Force and from NASA that satisfied him as to the value of maintaining independent programs. It was agreed by both agencies that the objective of MOL and Apollo X should be to achieve a truly universal space laboratory as an end product.

1964 December 8   Apollo mission A-002 carried boilerplate-23 command and service modules on a launch escape system test at White Sands, N.Mex. Its Little Joe II launcher incorporated two Algol and four Recruit solid propellant motors. The escape system carried BP-26 to an apogee of 50,360 ft. and contained the first test of a canard system in the nose of the escape motor which helped deflect the simulated spacecraft away from Little Joe II. Also flown for the first time was a soft boost protective cover (BPC) designed to protect the exterior surface and windows on the command module from exhaust gases of the escape motor. This was torn free shortly after the escape system fired. The Apollo parachute system began to deploy at an altitude of 23,500 ft., and BP-23 touched down at 24 ft./sec., 32,800 ft. downrange.

1964 December 9   Britain's Prime Minister Harold Wilson set up a committee to review the United Kingdom's future in space. The committee would examine the merits of developing an independent satellite launcher such as the Black Knight as well as its commitment to the European Launcher Development Organization (ELDO). The Labor Party had a majority in Parliament and was concerned about the amount of money these programs might attract and the true value to the economy of the technology involved.

1964 December 10   The first fully successful U.S. Air Force Titan IIIA standardized launch vehicle was launched from Cape Canaveral, Fla., and a dummy payload placed in a 97 × 105 mi., 32° orbit, decaying in less than three days. Total orbited mass was 9,000 lb., of which 3,750 lb. was payload. This payload capability was close to the TAT-Agena D and the LTTAT-Agena D, and only two more flights of the Titan IIIA were made. On February 11, 1965, the second launcher carried LES-1 (Lincoln Experimental Satellite-1) into a circular 115-mi. orbit, and through a sequence of three Transtage maneuvers raised it to a 1,729 × 1,742-mi. orbit. After separating from the Transtage, a solid propellant boost motor on LES-1 failed to fire, preventing the 68-lb. communications satellite from raising its apogee to a planned altitude of 10,900 mi. The last Titan IIIA flight on May 6, 1965, was a repeat of the previous mission and sent LES-2 into a 1,739 × 9,200 mi. orbit. In all, Transtage was fired four times, the last after ejecting LES-2 and a second radio calibration satellite weighing 75 lb.

1964 December 11   The Atlas-Centaur AC-4 launch vehicle was launched from Complex 36 at Cape Canaveral, Fla., carrying a 2,070-lb. mass model of the Surveyor spacecraft. The launch vehicle had a total height of 112 ft., and the Centaur stage burned for almost 6 min. to put itself into a 102 × 111 mi., 30.6° orbit. A planned restart of the cryogenic RL-10 engines took place, but a premature shutdown prevented the stage from achieving a planned apogee of 3,100 mi. and it reentered the atmosphere the next day. By the time AC-5 was launched on March 2, 1965 (unsuccessfully, due to a premature Atlas shutdown), the Centaur development program had been cut from 10 to 8 flights. The AC-6 flight on

August 11, 1965, simulated a Surveyor launch by sending the 2,100-lb. dynamic model of Surveyor to a 102 × 510,870 mi. orbit on a direct ascent trajectory, one into which the spacecraft would be launched to Earth escape velocity without first going into an Earth parking orbit.

Landing rockets were tested by NASA's Manned Spacecraft Center using a boilerplate Gemini spacecraft dropped from a Fairchild C-119 cargo aircraft. Released at a height of 7,000 ft. above Trinity Bay, Tex., the boilerplate carried two 12,000-lb. thrust solid propellant rockets that ignited 10 ft. above the water for 1.5 sec. Parachutes lowered the spacecraft at 27 ft./sec. and the rockets slowed it to a descent rate of 9–10 ft./sec. The deceleration at impact was reduced by rocket-braking from 7–8 g to about 2.5 g. The concept of rocket-braking was thought potentially useful for land-landing manned spacecraft. Although the Russians adopted it for their Soyuz spacecraft, it was never used by any U.S. manned vehicle.

1964 December 12   The first demonstration of thrust vector control by movable nozzle on a large solid rocket motor took place at the Thiokol Chemical Corp.'s Wasatch, Utah, facility. In a 2 min. 10 sec. firing of a 156-in. solid, the nozzle was moved to demonstrate how the line of thrust could be vectored to maintain flight control during an ascent to space. In the last 30 sec. of the test a plastic extension on the nozzle skirt blew off and reduced thrust by 8%. At peak the solid had produced a thrust of more than 1.3 million lb. This was the first test firing by Thiokol in the U.S. Air Force Program 623-A research program to evaluate large-diameter solids in which Lockheed was also testing 156-in. motors.

1964 December 15   A satellite designed and built in Italy was launched by a NASA-trained Italian crew when a four-stage Scout rocket was fired from Wallops Island, Va. The *San Marco I* (SM-I) satellite weighed 254 lb. and had a diameter of 2 ft. 2 in. Placed in a 124 × 490 mi. orbit inclined 37.8° to the equator, SM-1 carried air density and ionospheric measurement instruments and transmitted to Earth on 136.53 MHz and 20.005 MHz. The launch was a test of launch crews and operations prior to equatorial satellite launches from a specially built platform in the Indian Ocean off the coast of Kenya.

1964 December 19   NASA Associate Administrator Robert C. Seamans Jr. signed the Voyager project approval document which finally put the advanced planetary program on a formal footing. In study for three years, Voyager as now envisaged would include two spacecraft to Mars in 1971 and two in 1973. NASA had just canceled an Advanced Mariner 1969 mission to make way for Congressional funding for Voyager. At this date the first astronauts were confidently expected to land on the moon during 1968. President Johnson's special assistant for science, Donald F. Hornig, wanted an unmanned Mars mission for 1969 to maintain momentum in space exploration, so NASA agreed to fly a Voyager test spacecraft on a fly-by mission that year. On January 15, 1965, NASA requested industry proposals for a Voyager Phase IA study, and in April 1965 awarded contracts to Boeing, General Electric and TRW.

1964   During the year, Soviet rocket engineers selected the configuration for the N-1 manned moon launcher. The Valentin P. Glushko design bureau wanted to apply engines of high thrust using cryogenic or storable propellants while Korolev preferred a larger number of conventional engines with lower thrust. Under Glushko, the GDL-OKB designed and developed the 1.4-million-lb. thrust RD-270 engine operating on UDMH-N2O4 propellants. Glushko wanted to cluster eight RD-270s in the first stage of the N-1; but Korolev preferred to cluster twenty-six, then thirty, NK-33 engines which operated on liquid oxygen and kerosene. In the bitter conflict that ensued, Glushko blocked use of established rocket engine design bureaus. To build the N-1 motors, Korolev had to recruit the Kuznetsov aircraft engine bureau, which had never built rocket engines before. Glushko, meanwhile, worked with the Chelomei bureau to come up with a competitive launcher designated UR-700; this was proposed on November 16, 1966, as a simpler means of getting cosmonauts on the moon. The eventual failure of the N-1 launcher in flight trials (1969–1972) brought about the dismissal in 1974 of Vasili Mishin as head of the Soviet space program, a position he had held since the death of Korolev on January 14, 1966.

During the year, the British Aircraft Corp. completed feasibility and preliminary design details on a reusable logistics vehicle capable of carrying satellites and cargo from the surface of the Earth to low Earth orbit. Called MUSTARD, for multiunit space transport and recovery device, the concept envisaged three almost identical lifting body vehicles clustered in a back-to-back arrangement so that the outer two sandwiched the third. Designed by Tom Smith, MUSTARD would have had a launch weight of about 992,250 lb. carrying a payload of almost 7,000 lb. Burning cryogenic liquid hydrogen–liquid oxygen propellants, the outer two elements were boosters carrying propellant to feed their own engines as well as the rocket motor in the third element. Before separating, the boosters would pump propellant to the center element, leaving it to continue on into space. Carrying a crew of three to six people, the orbiter could be used up to 50 times. The British government showed no interest and the project was abandoned.

1965 January 8   Philip Bono, a space engineer with Douglas Missile and Space Systems Division, applied for a patent on the design of a recoverable, single-stage space launcher called ROMBUS (reusable orbital module-booster and utility shuttle). It would have the capacity to lift 1 million lb. of cargo to a 175-mi. circular orbit and be returned to Earth and used again up to 20 times. ROMBUS would have integral propulsion for getting into orbit, de-orbiting, landing retrothrust and landing with legs. Eight strap-on liquid hydrogen fuel tanks would be jettisoned during ascent. The concept had been developed under a contract with NASA's Marshall Space Flight Center.

1965 January 14   NASA's Office of Manned Space Flight received recommendations from a special task force on procedural changes that, if implemented, could reduce the gap between manned Gemini missions from three months to two months. Most changes would have to occur at the McDonnell plant in St Louis, Mo. Testing conducted in the factory, rather than after delivery to Cape Canaveral, would reduce the time that integration and launch crews would need to prepare vehicles for flight. The task force said that accelerated schedules could be implemented by GT-6. Many operational tasks that would be required of Apollo were to be rehearsed by Gemini, and the Apollo spacecraft was on track for flights to begin in late 1966.

1965 January 18   In a special message to Congress, President Lyndon B. Johnson announced development of a new class of submarine-launched ballistic missile to supplement existing variants of Polaris. To be called Poseidon C3, the missile would be larger than the Polaris A3 and weigh 77% more and it would have twice the throw weight of the A3. The missile was designed to carry the multimegaton W-67 warhead, development of which got under way this year with a plan to put it on both Poseidon and Minuteman III missiles, but the weapon was canceled in December 1967. Instead, Poseidon had a set of variable warhead options. It would usually carry ten 40–50-KT W-68 warheads in single multiple reentry vehicles, the accuracy of which was such that 50% would fall within 1,800 ft. of the target. Up to 14 W-68s could be accommodated, and the range of the Poseidon C3 varied, according to warhead, up to 2,800 NM.

1965 January 19   The first Thor-Altair launcher carried a 331-lb. experimental military weather satellite from Vandenberg AFB, Calif., to a 292 × 511 mi. orbit inclined 98.8°. The Altair had been developed into the Burner I stage, for which systems evaluation contracts had been awarded to LTV and Boeing on September 2, 1961. There were to be five more Thor-Altair flight attempts in support of the evolving Defense Meteorology Satellite program (DMSP), one of which, on January 6, 1966, was a failure. Others launched successfully included flights on March 18, May 20 and September 10, 1965, and on March 30, 1966.

The second and final unmanned Gemini mission began when a Titan II lifted off from Complex 19 at Cape Canaveral, Fla., carrying the 6,884-lb. GT-2 spacecraft to a ballistic trajectory for tests of heat-shield and spacecraft-recovery systems. Launch occurred at 14:03:59 UT, with staging at 2 min. 34 sec. and second stage shut-down at 5 min. 36 sec. With a velocity of 25,605 ft./sec. and an altitude of 103.5 mi., the spacecraft separated from the launch vehicle at 5 min. 52 sec. and reached a peak altitude of 106 mi. before jettisoning the equipment section of the adapter. The retro-rockets fired at 6 min. 54 sec. and the retrograde section of the adapter separated 45 sec. later. Spacecraft reentry began at 9 min. 20 sec. and maximum deceleration reached 9.9 g before parachute deployment and splashdown in the Atlantic Ocean, 2,143 mi. downrange, at 18 min. 16 sec.

1965 January 21   The first in a series of U.S. Air Force Orbital Vehicle One piggyback packages, *OV1-1*, was launched on Atlas missile 172D from Vandenberg AFB, Calif. *OV1-1* was contained in a canister attached to the side of Atlas. It failed to separate from the launcher five minutes into flight as planned and was lost along with its seven experiments. *OV1-1* was to have been propelled to a 350 × 1,500 mi. orbit by a 5,800-lb. thrust solid propellant X-258 motor as the Atlas reached the 365-mi. apogee of its ballistic trajectory. The OV1-series orbit module was a 2 ft. 6 in. sphere with 5,000 solar cells producing 22 W of power, and two whip antennas. Launched on May 28 and October 5, 1965, respectively, *OV1-3* and *OV1-2* also failed to reach orbit, the first OV1 success being achieved on March 30, 1966, although by that time *OV2-1* had already flown.

1965 January 25   NASA and the U.S. Department of Defense announced an agreement on the U.S. Air Force Manned Orbital Laboratory proposed for full-scale development. NASA would provide to the Department of Defense information on Apollo and Gemini spacecraft configurations to enable the Air Force to conclude studies about which type of vehicle to adopt for crew transport. Comparative analysis of the different requirements of the two parties would allow a future decision to be made about which type of space station to build in the future, said NASA.

1965 January 28   NASA conducted the first in-flight test of the high-altitude ejection system designed for the Gemini spacecraft when a simulated escape was conducted from an F-106 flying at 15,000 ft. and a speed of Mach 0.72. Both the seat and the dummy occupant were recovered, although the aneroid responsible for initiating deployment failed. It also failed during the second and final test, conducted on February 12, 1965, at 40,000 ft. and a speed of Mach 1.7, when the parachute also failed to deploy. Beginning February 17, 1965, a series of jumps using the personnel parachute system took place, the last on March 13, 1965. In ten dummy and five live jumps, only two tests failed when the manual override was used.

1965 February 5   The Service Propulsion System (SPS) engine designed for the Apollo service module was tested for the first time in a 10-sec. firing at White Sands, N.Mex. The SPS would provide propulsion primarily for placing the docked Apollo LEM (lunar excursion module) in moon orbit and then, at the conclusion of surface activities, fire again to return the crew to Earth. It would also be used for translunar and trans-Earth trajectory corrections and for orbital changes around the moon. The service module would contain 41,000 lb. of SPS propellant in four cylindrical tanks, two for the 50–50 mixture of hydrazine and UDMH fuel and two for the nitrogen tetroxide oxidizer. The 20,500-lb. thrust Aerojet General engine had many redundant parts and was gimbaled for flight control.

1965 February 8   NASA announced selection of astronauts Gordon Cooper and Charles "Pete" Conrad for the flight of *Gemini V*, with astronauts Neil Armstrong and Elliot See as backups. The third manned flight of the Gemini spacecraft was planned as a seven-day mission and would, for the first time in space, use fuel cells for electrical energy. This was the first crew assignment for Armstrong, who would make three flights in space, serve as commander of the first manned moon landing and become the first person to walk on the lunar surface.

1965 February 11   A U.S. Air Force Titan IIIA put the first Lincoln Experimental Satellite *(LES 1)* into a 108 × 128 mi. orbit inclined 32° attached to a Transtage. Developed by the Massachusetts Institute of Physics' Lincoln Laboratory, the satellite carried special equipment for evaluating operational tactical military communications. From the initial orbit, the Transtage shifted to a 116 × 1,766 mi. orbit and then to a 1,721 x 1,737 mi. path. The 69-lb. *LES 1* separated from the Transtage but failed to fire its solid propellant apogee motor to put it in a planned orbit of 1,725 × 11,500 mi. After launch on a Titan IIIA on May 6, *LES 2* was placed in its planned 1,753 × 9,385 mi. orbit. On a double flight launched by a Titan IIIC on December 21, *LES*s *3* and *4* were left in a 121 × 20,900 mi. orbit instead of the planned geostationary orbit. The 428-lb. *LES 5* was placed in geosynchronous orbit after launch on a Titan IIIC July 1, 1967. *LES 6* followed on another Titan IIIC on September 26, 1968, with the launch of *LES*s *8* and *9,* each weighing 1,000 lb., on March 15, 1976.

1965 February 16   The first of three Pegasus micrometeoroid payloads was placed in a 307 × 455 mi., 31.7° orbit by a Saturn C-1 SA-9 launched from Cape Canaveral. Pegasus comprised two 14 × 48 ft. winglike detector panels folded for launch into seven hinged frames on top of the S-IV stage. Four separate panels carried 25,000 solar cells for electrical power; the *Pegasus I* payload weighed 3,126 lb. in boilerplate no. 16

of the Apollo service module. Orbital weight was 23,000 lb. for the complete assembly, including the S-IV stage to which the payload remained attached until it decayed into the atmosphere on September 17, 1978. *Pegasus II* was launched by an SA-8 on May 25, 1965, decaying on November 3, 1979, and *Pegasus III* by an SA-10, the last Saturn I, on July 30, 1965. It reentered on August 4, 1969. All three Pegasus payloads played an important role in helping to define the density of micrometeoroid particles surrounding Earth and in characterizing the types and sizes of particle which might threaten manned vehicles.

1965 February 17   At 17:05:01 UT, an Atlas-Agena B successfully launched the 809-lb. *Ranger 8* from Cape Canaveral, Fla., into an Earth parking orbit of 115 mi. from which the Agena B successfully fired again to boost the spacecraft toward the moon. Correcting a lunar trajectory miss distance of 1,136 mi., a 59-sec. (72.5 MPH) midcourse maneuver began at 10:27:09 UT the next day, bringing the impact point to within 15 mi. of the assigned target in Mare Tranquillitatis. The TV system was turned on at 08:47:30 UT, February 20, and the first picture received from an altitude of 1,560 mi. at 09:34:29 UT, just 23 min. 8 sec. before impact. In all, *Ranger 8* returned 7,137 pictures, the last from a height of only 1,540 ft. *Ranger 8* impacted at 09:57:37 UT, at a position 2.59° N by 24.77° E, 888 mi. from *Ranger 7* and 141 mi. from *Ranger 6.*

1965 February 18   In testimony before the House Committee on Science and Astronautics, NASA's associate administrator for Manned Space Flight, George Mueller, outlined the agency's plans for exploiting Apollo hardware. Embracing all potential applications of the spacecraft, including Earth-orbiting laboratory and extended lunar surface exploration, Mueller explained that this program was now called Apollo extension system, or AES. It would, said Mueller, "follow the basic Apollo manned lunar landing program and would represent an intermediate step between this important national goal and future manned spaceflight programs."

1965 February 22   A Soviet SL-4 launch vehicle lifted off from the Baikonur cosmodrome, placing *Cosmos 57,* the unmanned precursor to *Voskhod 2,* in a 109 × 318 mi. orbit. Maintaining the tradition of conducting at least one test flight with the configuration of a new spacecraft, *Cosmos 57* was to qualify the systems and design innovations applied to the basic Vostok to transform it into a two-person vehicle from which a space walk could be performed. Essentially the same as *Voskhod 1,* it weighed about 12,530 lb. and incorporated a test airlock and pressure suit. Shortly after reaching orbit the spacecraft was destroyed when a self-destruct system, incorporated to prevent its falling into foreign hands, was unintentionally detonated by ground command. Only the airlock and suit had been evaluated and no further tests were conducted before the manned *Voskhod 2* flight.

1965 February 24   In testimony before the House Committee on Science and Astronautics NASA described the new "phased project planning" which it hoped would smooth the process of moving new projects from concept to operation. Under the new scheme, four distinct phases would be adopted: Phase A, conceptual and feasibility studies; Phase B, preliminary definition; Phase C, final definition and preliminary design; Phase D, detailed design, development and operational evaluation. This phasing process would be applied to future projects.

1965 February 25 NASA's Manned Spacecraft Center and the David Clark Co. reached an agreement whereby the latter would supply modified Gemini space suits for Block I, Earth-orbit Apollo flights. These would suffice for early Apollo missions until the Hamilton Standard extravehicular mobility unit became available for Block II missions and lunar surface activity. The Apollo 204 crew (Grissom, White and Chaffee) were wearing adapted Gemini suits when they were killed on January 27, 1967. When manned Apollo flights began in October 1968, crews wore the Hamilton Standard suit.

1965 February 27 The Thiokol Chemical Corp. static fired its full-scale solid propellant rocket motor with a 156-in. diameter, the largest yet fired, to produce a thrust of 3 million lb. for 1 min. 4 sec. Weighing 900,000 lb. and containing 800,000 lb. of propellant, the motor was placed in a vertical position down a 12-story pit. Based on this test, Thiokol's President Harold B. Ritchey predicted that motors of 7-million-lb. thrust could be produced within 30 months. The U.S. Air Force test was part of a development program transferred to NASA during 1965 and managed by the Lewis Research Center. NASA envisaged a motor 260-in. in diameter, and a half-length motor of that size was fired for the first time on September 25, producing 3.5-million-lb. thrust.

1965 March 1 In Project Long Life, missileers from the 44th Strategic Missile Wing, Ellsworth AFB, S.Dak., fired a Minuteman I on test to demonstrate the ability of a Launch Control Facility to launch, when necessary, if given the coded command to fire their missiles in anger. Because the missile silo was on farmland, it was given a flight duration of only 7 sec., which carried it about 2 mi. from the launch point, thus preventing catastrophic damage if the guidance malfunctioned. On February 24, 1966, crews of the 341st Strategic Missile Wing at Vandenberg AFB, Calif., simultaneously launched two Minuteman I (LGM-30A) ICBMs to demonstrate that, when called upon, salvo launches could be accomplished by a two-person Launch Control Facility. This is the only time that two U.S. ICBMs have been launched simultaneously.

1965 March 2 Managers at the NASA Manned Spacecraft Center decided to delete fuel cells from the Apollo lunar excursion module and adopt an "all-battery" LEM configuration. Within two weeks Grumman had developed the power supply and distribution concept that would be retained through production. The descent stage would carry four 400 A/hr. silver zinc batteries and the ascent stage would carry two 296 A/hr. silver-zinc batteries, respectively, providing a nominal 28-V and 30-V direct current to the power distribution system. The seven batteries weighed a total 790 lb.

Less than five months after Soviet Premier Nikita Khrushchev had been deposed by Leonid Brezhnev and Aleksei Kosygin, and following an intensive review of space and missile programs, the Ministry of General Machine Building (MOM) was formed out of an influential department in the State Committee for Defense Technology. Sergei A. Afanasayev was appointed its first minister, whose place in the hierarchy of Soviet administration allowed him to report direct to Leonid V. Smirnov, head of the powerful Military-Industrial Commission (the VPK). Smirnov himself reported directly to the Soviet prime minister. The MOM was given control of almost all Soviet space and missile programs and jurisdiction over all the separate design bureaus.

1965 March 9 The largest number of satellites carried into space by a single launch vehicle was flown on a Thor-Agena D out of Vandenberg AFB, Calif. Weighing a total 300 lb., the eight satellites were: *Greb 6* (88 lb.) in a 564 × 584 mi. orbit; *GGSE 2* and *3* (9 lb. each, gravity gradient stabilization experiment); *Solrad 7B* (103 lb., solar radiation); and *Secor 3* (40 lb., geodetic monitor) in orbits of 562 × 583 mi.; *Oscar 3* (31 lb., amateur radio) in a 564 × 585 mi. orbit; *Surcal* (11 lb., surveillance calibration) in a 565 × 585 mi. orbit; and *Dodecahedron* (9 lb., surveillance calibration for *Surcal*) in a 563 × 586 mi. orbit. All orbits had an inclination of 70.1°. *Oscar 3* transmitted for 16 days, *Surcal* reentered on March 27, 1981, the rest remain in orbit.

1965 March 12 The Soviet Union launched *Cosmos 60*, designated such after it failed to leave Earth orbit and head for the moon as planned. The latest in the Luna series of hard-lander capsules, *Cosmos 60* was the first attempt in almost two years to put instruments on the lunar surface. It was placed in a 121 × 154 mi. orbit by an SL-6 launcher and decayed back down through the atmosphere on February 26, 1968. The difficulty experienced by the United States in getting lunar spacecraft to function as planned was mirrored by the lack of success that followed Soviet efforts. Having begun to master propulsion techniques and the operation of launch vehicles, engineers now faced challenges in successfully operating increasingly complex spacecraft by remote control or programmed flight computers.

1965 March 18 At 07:00 UT an SL-4 carried cosmonauts Pavel Belyaev and Aleksei Leonov in the *Voskhod 2* spacecraft to a 107 × 308 mi. orbit. The 12,500-lb. *Voskhod 2* had an inflatable airlock comprising thick layers of material folded like a concertina against the side of the spherical descent module for launch. There was little room inside the descent module. Leonov was between Belyaev, on his right, and the 550-lb. airlock assembly. In orbit it was inflated by pressurization to form a cylinder projecting 8 ft. from the side of the descent module, with an inner diameter of 3 ft. 3 in., and an outer diameter of 3 ft. 11 in. After donning an emergency life support pack, which he did only with the help of Belyaev, Leonov opened the hatch and squeezed into the airlock, shutting it behind him before connecting up a 16 ft. 5 in. umbilical. Sealed in the airlock, Leonov released a valve to vent the atmosphere and opened the 2 ft. 4 in. diameter outer hatch at the opposite end. At 08:30 UT he moved outside, watched by two TV cameras mounted to the exterior, and became the first person to "walk" in space. Leonov had difficulty trying to get back in due to pressure in his suit causing it to balloon, only squeezing back in by opening a suit valve to reduce pressure to 3 lb./sq. in. With the outer hatch shut, the airlock was repressurized using gases carried in spherical tanks on top of the descent module, and Leonov rejoined Belyaev after his 20-min. space walk. The airlock was jettisoned prior to reentry. When commanded, the primary retro-rocket at the base of the instrument module failed to fire, and reentry had to be delayed by one orbit. The solid propellant backup retro-rocket had to be used, but it only had sufficient impulse to decelerate the descent module after the instrument module had separated. Then the 26,500-lb. thrust motor fired for 1.7 sec., slowing the 6,285-lb. spacecraft by 200 ft./sec. (133 MPH). Far off course, the 1 day 2 hr. 2 min. flight ended in snow-laden forests.

1965 March 20 At the McDonnell Aircraft Co. in St. Louis, Mo., five tests of the *Gemini IV* spacecraft began in the altitude chamber. The first run was unmanned, and on the second run a simulated mission took place with crew but without depressurization. The third run repeated the second, while the fourth run simulated a mission in vacuum and the fifth run repeated the fourth test for the backup crew. During these altitude simulations the primary and backup pilots practiced stand-up EVA (extravehicular activity) procedures through the right hatch in which the astronaut would stand up on his seat with head and torso outside the spacecraft. On May 15, 1965, qualification of the G4C EVA suit was completed. It incorporated additional protection against micrometeoroid penetration and two overvisors, one for physical impact protection and one with a gold coating for attenuating solar light.

1965 March 21 At 05:37:02 UT an Atlas-Agena B launched *Ranger 9,* the last NASA lunar impact probe in the series, from Cape Canaveral, Fla., and in two Agena D burns sent it to a lunar impact trajectory just 400 mi. from the planned impact point, the crater Alphonsus. The midcourse maneuver, usually programmed for 16 hr. after liftoff, was delayed 18 hr. to get better opportunities for impact accuracy. At 12:30:09 UT on March 23 the motor began a 31-sec. (40.6-MPH) burn shifting the impact point to within 2.76 mi. of the target. The first picture was received from an altitude of 1,405 mi. and the last of 5,814 images was taken at a height of 2,006 ft. before *Ranger 9* impacted at 14:08:20 UT. It rests at a position 12.91° S by 2.38° W, 590 mi. from *Ranger 8* and 338 mi. from *Ranger 7.* In all, *Ranger*s *7, 8* and *9* had sent to Earth 17,259 pictures.

1965 March 23 NASA launched the first U.S. multiperson space mission and conducted the first orbital maneuvering with a manned space vehicle during the 4 hr. 52 min. 31 sec. flight of *Gemini III.* Launched from Cape Canaveral's Complex 19 at 14:24 UT, GT-3 took just 5 min. 34 sec. to put astronauts Grissom and Young in a 100 × 139 mi. orbit inclined 32.5° to the equator. The 7,111-lb. spacecraft had been named *Molly Brown,* after the "unsinkable" Denver heroine of the *Titanic* disaster, when NASA refused Grissom permission to call it the Titanic in mockery of the sinkable qualities of his Mercury spacecraft, *Liberty Bell 7.* At an elapsed time of 1 hr. 33 min., a 1 min. 14 sec. (50 ft./sec.) orbit maneuver put the spacecraft on a 98 × 105 mi. path followed on the second orbit by a 10 ft./sec. plane-change, adjusting the inclination by 1.2 arc seconds. At 4 hr. 21 min. the final orbit change lowered perigee to just 52 mi., sufficient to bring the spacecraft into the atmosphere should the retrorockets fail. The crew were recovered by helicopter after splashdown 58 mi. short of the Atlantic aim point.

1965 March 29 At an informal meeting in the office of NASA Manned Spacecraft Center Director Robert Gilruth, the possibility of conducting a full EVA (extravehicular activity) outside the *Gemini IV* spacecraft raised for the first time. Gilruth's superior, George Mueller, was briefed on April 3 and began a chain of events that would result in approval for the *Gemini IV* mission to include a full EVA for the pilot, astronaut Edward White. He would be supported by an umbilical and tether attached to the spacecraft and use a hand-held maneuvering unit containing 0.7 lb. of gaseous oxygen ejected on demand at 120 lb./sq. in. for propulsion. All *Gemini IV* EVA hardware qualified by May 19. Acceleration of the EVA plan was predicated by the successful space walk of Soviet cosmonaut Alexei Leonov from the *Voskhod 2* spacecraft on March 18, 1965.

1965 March The U.S. Atomic Energy Commission began a major retrofit program, replacing existing W-47 warheads on Polaris A-1 and A-2 SLBMs (submarine-launched ballistic missiles) with the improved W-47-Y2 Mod 3. The existing warhead had a yield of 600 KT while its updated replacement had a yield of 800 KT. All 288 missile tubes on the 18 U.S. Navy Fleet Ballistic Missile submarines with Polaris missiles were retrofitted by late 1967.

1965 April 1 The first stage of the first Saturn IB launch vehicle was fired for the first time at Huntsville, Ala. The eight uprated H-1 engines produced a total stage thrust of 1.6 million lb. On April 23, 1965, the Marshall Space Flight Center announced that it had passed all qualification tests and that it was ready for flight. The Saturn IB utilized a modified Saturn I first stage with an S-IVB second stage powered by the 200,000-lb. thrust cryogenic J-2 rocket motor. The S-IVB would be used as the third stage of the Saturn V, reducing the number of new stages fired for the first time when that launcher made its initial flight.

1965 April 3 An Atlas-Agena D launched from Vandenberg AFB, Calif., carried the first SNAP-10A nuclear reactor into space. Known as *Snapshot,* the 970-lb. conical structure had a length of 12 ft. and a base diameter of 5 ft., with a 250-lb. reactor containing uranium 235 fuel and 37 moderator rods. Cooled by a liquid metal alloy of sodium and potassium, SNAP-10A's reactor was on top with a lithium hydride shield that protected the structure from neutron radiation. Placed initially in a 805 × 826 mi., 90.2° orbit, SNAP-10A produced 500 W of electrical energy for 43 days. A secondary payload, the 88-lb. *Secor 4,* was placed in an orbit of 797 × 816 mi., but failed to operate as planned. Both remain in orbit, inoperative.

1965 April 5 NASA's Manned Spacecraft Center announced astronauts Walter Schirra and Thomas Stafford as the primary crew for the planned rendezvous and docking mission of *Gemini VI.* This was scheduled as the first flight to attempt a docking with the modified Gemini-Agena Target Vehicle (GATV), adapted from the Agena D upper stage and launched by an Atlas approximately 1 hr. 40 min. before the Gemini spacecraft. Astronauts Virgil Grissom and John Young were to be the backup crew for this mission.

1965 April 6 The world's first commercial communication satellite, *Intelsat I (Early Bird),* was launched by a thrust-augmented Delta from Cape Canaveral to a geosynchronous transfer ellipse. From there an apogee kick motor circularized the orbit at geostationary altitude and the satellite was positioned over the Atlantic Ocean at 27.5° W. Capable of handling 240 two-way telephone circuits or one TV channel with two 6-W transmitters, *Intelsat I* was a spin-stabilized cylindrical drum 2 ft. 4 in. in diameter and 1 ft. 11 in. high, covered with 6,000 solar cells and weighing 85 lb. in orbit. It began operation June 28 and was switched off in January 1969, but was reactivated between June 29, 1969, and August 13, 1969, to help handle busy television traffic during the first moon landing. It remained in orbit through 1995.

1965 April 9 NASA's associate administrator for Manned Space Flight, George Mueller, formally announced the transfer of control over manned flights from the Kennedy Space Center to the Manned Spacecraft Center at Houston, Tex. Beginning with the *Gemini IV* mission launched June 3, all future manned spaceflights would be controlled from

Houston; the Kennedy Space Center would control countdown and launch and relinquish authority once the launch vehicle cleared the launch tower.

1965 April 12   Conforming to a directive from U.S. Secretary of Defense Robert McNamara, the U.S. Air Force Strategic Air Command deactivated the last Atlas and Titan I ICBM. With this phase-out program, SAC entered the second phase of its missile deployment plan, in which it would rely on 54 Titan II and 1,000 Minuteman ICBMs for the strategic retaliatory force. Atlas and Titan I suffered from the use of liquid oxygen, a cryogenic oxidizer, which meant these missiles could only be fired after a fueling process that left them vulnerable to attack.

1965 April 16   NASA's Marshall Space Flight Center conducted the first firing of the S-IC stage, the first stage of the Saturn V launch vehicle. The stage operated for 6.5 sec. as planned. On a normal launch, the stage would burn for about 2 min. 40 sec. With a length of 138 ft., a diameter of 33 ft. and a span across the base fins of 63 ft., the S-IC generated 7.5 million lb. of thrust from the five F-1 engines at its base. The S-IC would remain the largest single rocket stage until tests with the Soviet N-1 moon booster began on February 21, 1969. Since the N-1 was unsuccessful as a launch vehicle, the S-IC remains the biggest and most powerful operational rocket stage ever built. Empty, the S-IC weighed 290,000 lb., but fully fueled for launch with RP-1–liquid oxygen propellants it weighed 5 million lb.

1965 April 23   The Soviet Union launched its first communication satellite, *Molniya 1,* on an SL-6 from Baikonur cosmodrome, to an elliptical orbit of 334 × 24,421 mi. inclined 65° to the equator. With apogee over northern latitudes, military and civil communication between Soviet ground stations was conducted via four 40-W receiver/transmitters and two antennas 3 ft. in diameter. The 3,860-lb. satellite carried six wing-like solar panels providing 500–700 W of electrical power. The 441-lb. thrust restartable orbit-adjustment engine had a total firing time of 1 min. 5 sec. TV images were relayed at 625 lines/frame on 3.4–4.1 GHz, with voice and data at 0.8–1.0 GHz. Initially, Molniya orbit planes were spaced at 120° and then, with more satellites, at 90°. Set to last more than 30 years, the *Molniya 1* system supported the launch of 87 satellites, the last on December 22, 1993.

1965 April 24   At its static test facility at Santa Susana, Calif., North American Aviation fired the first S-II stage, the second stage for the Saturn V launch vehicle. With a length of 81 ft. 6 in., a diameter of 33 ft. and five J-2 engines producing a stage thrust of 1 million lb., the S-II had an empty weight of 90,000 lb. and a fueled weight of just over 1 million lb. With cryogenic liquid hydrogen–liquid oxygen propellants, the S-II would take over from the first (S-IC) stage and burn for about 6 min., carrying the upper stage and payload from a height of more than 40 mi. to an altitude of more than 100 mi., and from a velocity of almost 9,000 ft./sec. to 23,000 ft./sec.

1965 April   Cosmonauts for multicrew Voskhod missions were selected in the Soviet Union. *Voskhod 3* was planned as a two-week flight scheduled for late 1965 or early 1966, and for that mission cosmonauts Vladimir Aleksandrovich Shatalov, Georgi Timoteyevich Beregovoi, Georgi Stepanovich Shonin and Boris Valentinovich Volynov were selected as candidates. *Voskhod 4* was scheduled to carry the first journalist to space, and in July Yaroslav Golovanov and Yuri

Letunov were selected from those ranks for training. *Voskhod 5* was to have been a two-week biological flight, and for that three candidate scientists were selected, also in 1965: Prof. Yevgeni A. Illyin, Yuri A. Senkevich and Aleksandr S. Kisilev. The flight was canceled before pilot cosmonauts had been chosen to fly with them. These three candidates comprised the first scientists chosen by either the United States or the Soviet Union to train for spaceflight. It is believed that Voskhod 6 would have provided an opportunity for a female cosmonaut to conduct a space walk. None of these additional Voskhod missions were flown.

1965 May 3   NASA contractors Boeing, General Electric and TRW began work on Phase IA of the Voyager planetary program, scheduled for flights to Mars in 1971 and 1973 on Saturn IB launchers. As *Mariner IV* results came in from the first fly-by of Mars, atmospheric density of 4–7 millibars was discovered to be much less dense than the 10–30 millibars expected or the 100 millibars some scientists had thought. This overturned work to date on the concept of aerobraking (use of the atmosphere to reduce speed) for the lander, placing greater demands on rocket propulsion to slow the lander for a controlled touchdown. With a maximum Saturn IB load capacity of 7,000 lb., weight for braking motors would have to be taken from the science payload.

1965 May 4   From the NASA Manned Spacecraft Center, Maxime Faget submitted to headquarters a detailed outline for the Apollo extension system (AES) program. The ultimate objective was to hone Apollo hardware for extended stays of up to six weeks in Earth orbit and up to two weeks on the surface of the moon. Improved or alternate hardware development was expected to increase Earth-orbit duration to three months. This was deemed necessary as an intermediate step between short-duration flights with ballistic spacecraft and permanent manned space stations. The AES and launch vehicle development activity was integrated into a Saturn/ Apollo Applications Office at NASA Headquarters during August 1965.

1965 May 9   The Soviet Union successfully launched the 3,255-lb. *Luna 5* spacecraft toward the moon using an SL-6 launch vehicle sent into space from the Tyuratam launch complex. The spacecraft was designed to fire a retro-rocket close to the lunar surface, decelerating the vehicle from an approach speed of around 7,500 ft./sec. to zero just above the moon. At that point the 220-lb. hard-landing capsule would be ejected for a free-fall to the surface. After a successful translunar flight, the retro-rocket failed to fire and the spacecraft struck the moon and was destroyed, the last signal being received just after 19:09 UT on May 12, about 5 min. prior to the predicted lunar contact time.

At the May Day parade in Moscow, the Soviets publicly revealed two new ICBMs, the solid propellant SS-13 Savage and the liquid propellant SS-10. With open trusswork separating the three stages, Savage had a length of 65 ft., a maximum diameter of 5 ft. 7 in. and a launch weight of 78,000 lb. Produced by a relatively obscure Soviet missile design team, the Aleksandr D. Nadiradze bureau, its upper stages were adapted as the two-stage SS-14 Scapegoat. With a range of almost 6,000 mi. and an accuracy of 6,000 ft., the SS-13 carried one 750-KT nuclear warhead. In service from 1969, the SS-13 reached peak deployment of 60 missiles in 1972, all around the Yoshkar Ola region between Gorki and Kazan, until their removal under the terms of START (Strategic Arms Reduction Talks). The two-stage SS-14 was present but not

Seen here in the foreground during its public unveiling on May 9, 1965, the Soviet SS-10 Scrag is pulled into Red Square, Moscow.

revealed. Hidden in a cylindrical container attached to a rebuilt IS-3 tank chassis, the missile had a height of 34 ft. 9 in., a diameter of 4 ft. 7 in. and a range of 2,500 mi. The SS-10 Scrag missile utilized nonstorable (kerosene and liquid oxygen) propellants in a three-stage configuration with a length of 124 ft. 8 in., a diameter of 9 ft. and a launch weight of 375,000 lb. It was never deployed.

1965 May 19   The third Little Joe II Apollo abort simulation from White Sands, N.Mex., failed in its primary objective of a high altitude test to 120,000 ft. when the booster disintegrated 25 sec. after launch. Liftoff of Apollo boilerplate 22 on mission A-003 occurred at 13:01 UT when four solid propellant Algol motors ignited to generate a thrust of 412,800 lb. At just 2.5 sec. the vehicle went out of control and eventually broke up before the two Algol motors comprising a second stage could fire. The launch escape system fired at a height of 12,400 ft., and under violently adverse conditions, with the command module apex forward, the parachute system successfully righted the simulated command module and brought it to Earth.

1965 May 23   The Life Sciences Committee of the U.S. National Academy of Sciences' Space Science Board informed NASA that it recommended astronauts returning from the moon be quarantined for a period of three weeks. They felt this was necessary to prevent the possible contamination of Earth with microbial life forms from the lunar surface. In the report *Potential Hazards of Back Contamination from the*

Planets, the board was mindful of the risks taken with future missions to Mars, a more likely source of life. NASA would concur with the recommendations, and quarantine became standard for Apollo crews returning from the lunar surface. The scheme was discontinued on April 26, 1971, by which date it had been determined that no bioorganisms would survive on the moon.

1965 June 3   NASA launched the 7,879-lb. *Gemini IV* spacecraft carrying astronauts James McDivitt and Edward H. White II and conducted the first U.S. space walk. Launch occurred at 15:15:59.6 UT from Cape Canaveral's Complex 19, and the GT-4 Titan II put the spacecraft in a 101 × 175 mi. orbit inclined 32.53°. Beginning at 8 min. 47 sec. elapsed time, a total of 74 orbital maneuvers within 1 hr. 20 min. were performed in efforts to stay close (station-keep) to the spent second stage. This was aborted at the end of the first orbit due to excessive use of propellant. Preparations for EVA (extravehicular activity) began at 1 hr. 40 min. but the space walk was suspended for one revolution. Pilot White opened the spacecraft hatch at 4 hr. 18 min. and stood up on the seat to assemble the hand-held maneuvering unit (HHMU). He spent 22 min. moving around outside tethered by a 25 ft. umbilical, but experienced difficulty closing the hatch, finally repressurizing the spacecraft at 5 hr. 6 min. *Gemini IV* landed in the Atlantic at 97 hr. 56 min. 12 sec.

1965 June 8   The Soviet *Luna 6* spacecraft was launched toward the moon by an SL-6 launch vehicle from the Tyuratam complex in Kazakhstan. With a weight of 3,135 lb., the spacecraft was intended to put a hard-lander capsule on the surface of the moon and was fired toward the moon by the terminal stage after the assembly had first been placed in an interim Earth parking orbit. The day after launch, Soviet engineers commanded the liquid propulsion to fire for a course correction. The motor fired but failed to shut down as commanded and moved the spacecraft to a flight path that missed the moon by 100,000 mi. three days after launch.

1965 June 15   The NASA M2-F2 lifting body was rolled out from the Hawthorne, Calif., plant of the Northrop Corp. The wingless research vehicle had a convex underbody, flat top and two vertical fins sweeping up from the afterbody. It had a length of 22 ft. 2 in., a width of 9 ft. 7 in. and a gross weight of up to 10,000 lb. The first of 14 glide flights was conducted at Edwards AFB, Calif., on July 12, 1966, and on November 21, 1966, it was grounded for installation of its Thiokol XLR11 rocket motor. The first glide flight with the engine in was performed on May 2, 1967, but on the second glide flight on May 10, 1967, pilot Bruce Peterson lost control and made a heavy crash-landing attempting to avoid a helicopter. He was badly injured and suffered severe facial injuries. The film of his M2-F2 rolling end over end across the desert floor was used, some would say distastefully, as the lead into the film *Six Million Dollar Man.* The M2-F2 was rebuilt as the M2-F3.

1965 June 18   Titan IIIC, the new U.S. Air Force heavyweight launch vehicle and the first rocket to demonstrate a liftoff thrust greater than 2 million lb., ascended into space from Cape Canaveral, Fla., carrying a 21,098-lb. simulated payload to a 104 × 120 mi., 32.2° orbit. Titan IIIC was the basic uprated launch vehicle from which would be developed a broad family of heavyweight launchers for the U.S. military space program. It had a 73 ft. first stage and a 23 ft. 4 in. second stage, each with a diameter of 10 ft., essentially the same as the core Titan IIIA. Two solid propellant boosters, each with a length of 85 ft. and a diameter of 10 ft., attached to opposite sides of the first stage provided a liftoff thrust of 2.3 million lb. Only when the jettisonable boosters carried the core stages to a downrange distance of 30 mi., 2 min. after launch, did the first stage fire, producing in the vacuum of space a thrust of 526,000 lb. for 2 min. 17 sec. The second stage produced a thrust of 102,000 lb. for 3 min. 27 sec. A Transtage provided extensive orbital maneuvering capability for the payload. Development of Titan IIIC extended the length of the stages and increased burn times. Payload capability was 29,000 lb. to low Earth orbit or 3,400 lb. to geosynchronous orbit. Titan IIIC was used for a wide range of military satellites placed in geostationary orbit.

1965 June 25   The U.S. Atomic Energy Commission tested for the first time a new generation of graphite core reactor, called *Phoebus 1A,* in a full power run for 10 min. 30 sec. Phoebus was developed as a test-bed for nuclear rocket reactor technology and not as a flight engine. It was capable of producing 1,000–1,500 megawatts of power, but its successor *Phoebus 2* was almost four times as powerful. It was cold-tested for the first time on July 12, 1967.

1965 June 29   NASA announced the names of six scientist-astronauts, selected by NASA and the National Academy of Sciences, constituting the fourth group of U.S. astronauts Owen K. Garriott (Stanford University); Edward G. Gibson (Applied Physics Labs, Philco Corp.); Duane E. Graveline (flight surgeon); Lt. Comdr. Joseph P. Kerwin (flight surgeon, USN); Frank Curtis Michel (asst. prof. of space sciences, Rice University, Tex.) and Harrison Schmitt (astrogeologist, U.S. Geological Survey). Garriott would fly on the second manned Skylab mission (SL-3) in 1973 and on the Shuttle mission STS-9 launched on November 28, 1983. Gibson flew on the third Skylab mission (SL-4) in 1983–84. Kerwin accompanied Garriott on SL-2. Schmitt became the first scientist to work on the moon when he accompanied Eugene Cernan to the lunar surface on the last Apollo flight, *Apollo 17,* in December 1972.

The second of two pad abort tests with an Apollo boiler-plate command and service module took place from Complex 36 at Cape Canaveral, Fla. The simulated spacecraft was the refurbished BP-23 previously used for the A-002 mission on December 8, 1964, with a Little Joe II. The launch escape motor ignited at 13:00 UT and carried the command module to an altitude of 9,258 ft. The test qualified launch escape and pitch control motors, boost protective cover and canard surfaces. The BP-23A command module landed 7,600 ft. from the launch site.

1965 June 30   A Lunar Landing Research Facility was put into operation at the NASA Langley Research Center. Comprising a steel girder framework 250 ft. high and 400 ft. long, the LLRF provided a means by which 83% of the weight of a lunar landing test vehicle could be suspended by pulleys weight, thus simulating the weight it would have on the moon. In this way, pilots could use simulated Apollo lunar excursion module terminal descent trajectories to rehearse control techniques and develop procedures for actually landing on the lunar surface.

1965 July 1   NASA announced that it had selected astronauts Frank Borman and James Lovell as the primary crew for the 14-day *Gemini VII* long-duration flight then scheduled for the end of the year. Astronauts Edward White and Michael Collins would be their backup crew. The *Gemini VII* mission was essentially a biomedical data gathering flight to support the planned Apollo lunar landing flights which would last approximately 10–12 days. To date, no flights of greater than 5 days' duration had been conducted by either the United States or the Soviet Union, but results from the 4-day *Gemini IV* mission indicated serious bone-demineralization.

1965 July 2   An attempt to test unmanned the *Voskhod 3* spacecraft configuration ended in failure when the SL-4 launch vehicle malfunctioned and the 12,600-lb. spacecraft was destroyed. What would have been designated *Cosmos 71* was to have been a qualification flight for the two-week *Vostok 3* mission, planned to eclipse the scheduled fourteen-day flight of *Gemini VII.* If anything, the failure emphasized the vulnerability of a crew unable to escape during ascent to orbit. The Vostok design from which Voskhod evolved was unable to support more than one ejection seat, so none was carried. It may have been for this reason that further manned Voskhod missions were canceled. Already in advanced stages of development, the Soyuz spacecraft had an escape system.

1965 July 5   The French Saphir sounding rocket in its Diamant configuration was tested for the first time in a successful flight from the Sahara Test Center in Algeria. Utilizing the Emeraude rocket as the first stage and the Topaze P-2.3 rocket as the second stage, Saphir represented the first two stages of the Diamant satellite launcher. It had a liftoff weight of 40,200 lb. and could lob a maximum payload of 930 lb. to an altitude of 600 mi.

1965 July 8   The first of two Sparrow missiles launched by F4B Phantom aircraft as part of an atmospheric research program was released in midair but failed when the second stage refused to fire. Designated Sparoair III, the device was a converted two-stage antiaircraft missile capable of carrying a 30-lb. load of instruments to a maximum altitude of 200 mi. The second and last attempt, conducted on May 26, 1966, also failed. In another series of air-launched atmospheric tests called ALARR (air-launched air-recoverable rocket), modified Genie rockets fitted with air sampling gear were fired from an F4B Phantom and recovered by a C-130 transport aircraft as they fell back though the atmosphere. On one notable occasion, November 16, 1966, an ALARR Genie collected fragments from the Leonid meteor shower as they hurtled through the atmosphere. Meteors are remnants of fragmented comet nuclei that burn up upon entering the atmosphere, thereby creating shooting stars.

1965 July 15   NASA's *Mariner IV* made the first successful flyby of the planet Mars and returned 21 complete pictures of the surface. Images were taken between 00:17:21 and 00:43:45 UT. The spacecraft made its closest encounter at 01:00:58 UT when it passed 6,118 mi. above the planet. The spacecraft disappeared behind Mars for a period of 53 min. 53 sec. beginning at 02:19:11 UT before picture playback to Earth began across a distance of 135 million mi. beginning at 12:49:54 UT. Due to the distance, it took 12 min. 4 sec. for the radio signal to reach Earth. A Cassegrain telescope with a 12-in. focal length and a 1.62 in. mirror focused light on a single TV camera recording pictures on tape. Each picture was made up from 200 lines of 1,250 digits each, or 250,000 binary digits in all, which was sent to Earth at 8.33 bits/sec. via a 3-W transmitter. The camera was on a scan platform that could move to track planets. *Mariner IV* carried instruments for measuring solar wind, galactic cosmic rays, radiation, magnetic fields and dust.

1965 July 16 The Soviet Union launched what was, at this date, the world's most powerful satellite launcher from its Baikonur cosmodrome. Later designated SL-9, the two-stage launcher had been developed by the Chelomei bureau. The first stage comprised six clustered tanks containing UDMH propellant surrounding a core tank carrying N_2O_4, a configuration similar to that used for the U.S. launcher Saturn I. At the base of the SL-9 were six RD-253 engines from the GDL-OKB, each delivering a thrust of 368,200 lb. for 2 min. 10 sec. With a height of 66 ft. 3 in. and a diameter of 24 ft. 3 in., the first stage had a liftoff thrust of 2.209 million lb. The second stage had a thrust of 540,000 lb. from four engines burning UDMH-N_2O_4 propellants for 3 min. 28 sec. This stage had a height of 45 ft. and a diameter of 13 ft. 7 in. The satellite launched by the first SL-9 was a 26,900-lb. test vehicle incorporating heavy cosmic-ray experiments named Proton, a name that became synonymous with both the launcher and the satellite. *Proton 1* was placed in a 118 × 390 mi., 63.5° orbit, from which it decayed on October 11.

The first Soviet five-satellite payload was lifted from Baikonur on an SL-8 launcher. Designated *Cosmos 71–75*, the satellites were part of a precursor development program for tactical military communications. Each weighed 110 lb. and was placed in a 56.1° orbits with a period of 95.5 min. Perigee values varied between 324 mi. and 335.5 mi., with apogee values between 336 mi. and 400 mi. The first to decay, *Cosmos 71* reentered on August 11, 1970, with the last, *Cosmos 74*, falling on December 13, 1979.

1965 July 18 A Soviet SL-6 launched the 2,115-lb. *Zond 3* (vehicle number 3MV) spacecraft on to a heliocentric orbit. Late in getting off the pad, it had missed the October-November 1964 Mars opportunity and was relegated to a test of several new technical systems being developed for later missions. Unlike its predecessor *Zond 2,* the spacecraft carried a TV camera, and it took 25 pictures of the moon's far side after passing 5,700 mi. from the surface. The orbit of *Zond 3* extended at aphelion to a distance equal to the mean radius of Mars's orbit (141 million mi.) but Mars was not at that point in its orbit when the spacecraft passed along that part of its own, heliocentric path. *Zond 3* was the last planetary spacecraft designed by the Korolev bureau, as the OKB (design bureau) was now heavily involved with the N-1/L-3 manned lunar program. The Giorgi Nikolaevich Babakin bureau took over planetary spacecraft design, dropped the Object MV concept and on March 27, 1969, performed the first launch attempt with its new Mars design. The Zond name was applied to a later program that had nothing to do with *Zond*s *1–3.*

1965 July 19–31 At a NASA summer conference at Falmouth, Mass., scientists met in the first major effort to draw up a 10-year program of manned lunar research beginning with the first landing. Defined as a part of the Apollo Extension System (AES), one manned Lunar Orbiter mission per year for five or six years was proposed for comprehensive mapping and geophysical surveys. Five or six AES manned lunar surface expeditions would employ a local scientific survey module (LSSM) capable of carrying two astronauts and 600 lb. of equipment on traverses of 5–9 mi. and a lunar flying vehicle able to transport a 300-lb. payload a distance of 9 mi. Post-AES recommendations included three-person stay-times of at least two months, with traverses of up to 500 mi. from base beginning in 1975.

1965 July 21 R. Wayne Young of the NASA Manned Spacecraft Center prepared a memorandum defining the requirements of the Apollo lunar surface experiments package (ALSEP). The ALSEP would comprise a group of scientific equipment carried to the moon aboard the Apollo lunar excursion module, deployed by the astronauts and left on the lunar surface in operating condition when the crew returned to Earth. Powered by a SNAP-27 RTG (radioisotope thermoelectric generator), the instruments would continue to transmit information to Earth long after the crew left the moon. Some scientists thought the RTG would contaminate instrument readings, and Apollo engineers were concerned about the amount of heat the plutonium 238 fuel core would radiate between launch and the lunar deployment.

Bell Aerosystems described a rocket-powered lunar flying vehicle (LFV) the company had designed to allow astronauts to explore the surface of the moon far from their landing site. Nicknamed "Hopper," the LFV would be able to travel 50 mi. above the lunar surface without stopping. Technology from this design was being worked into a manned flying system also under study at Bell. The MFS would be capable of transporting an astronaut and 300 lb. of equipment for up to 15 mi. Neither system was developed.

NASA announced that it had dropped plans to develop a lunar surface roving vehicle carried to the moon by the unmanned Surveyor spacecraft. Although it received little attention, due to a need to keep work on Apollo's primary goal of landing two men on the moon as the primary focus, in-house studies continued at NASA on ways to exploit Apollo hardware for extended scientific surveys of the moon. There were many ideas, some logical but complex and others ambitious but far too expensive. One idea was to use the descent stage of the manned lunar excursion module as a soft-landing platform for delivering up to 5 tons of equipment the astronauts could use after they landed in a separate LEM.

1965 July 22 Simulated countdowns rehearsing the synchronized launches of Gemini Titan-VI and the Atlas-Agena D carrying the Gemini Agena Target Vehicle were conducted at Cape Canaveral, Fla. A full dress rehearsal involving astronauts Thomas Stafford and Eugene Cernan, demonstrated that the two vehicles could be simultaneously prepared on their respective launch pads. The countdown stopped 18 sec. before ignition of each vehicle as they were sequentially brought to that point in preflight preparations, 95 min. apart.

1965 July 26 NASA confirmed that it was still examining the possibility of using the Gemini spacecraft for a circumlunar flight using a modified spacecraft carrying added thermal protection to insulate it from the heat of the high-speed reentry it would experience when it entered the atmosphere at 25,000 MPH. The mission would use a Titan II to place a manned Gemini spacecraft in Earth orbit and a Titan III-C to dock with it and propel it to a circumlunar trajectory. In a letter to Congressional Representative Olin E. Teague on September 10, 1965, NASA Administrator James Webb affirmed, "our main reliance on operating at lunar distances . . . is the Saturn V/Apollo system."

1965 July 30 Lockheed delivered to NASA's Manned Spacecraft Center the final report on a modular space station program evolving from utilization of basic Apollo hardware through the orbital assembly of a permanently manned facility. Six missions were used as the baseline for this study: a 45-day flight with three crew members; a 12-month mission with three to six people in a two-compartment laboratory; a 90-day mission with three to six people in polar orbit; a 90-day mission for three to six people in geosynchronous orbit; a 1–5 year mission for six to nine crew members; and a 5–10 year mission utilizing a twenty-four- to thirty-person Y-shaped modular configuration.

1965 July   Grumman completed its NASA-funded study of the Apollo Extension System (AES) for Earth-orbit space station and extended lunar surface missions. Grumman incorporated U.S. Air Force requirements in the Manned Orbital Laboratory program and proposed the use of a modified lunar excursion module (LEM) for Earth-orbit science missions. The descent stage, not necessary for Earth-orbit missions, could be used to house scientific instruments, and the pressurized ascent stage would afford crew space for scientists working away from an Apollo laboratory. Grumman added an unsolicited review of moon applications and suggested an LEM shelter for extended stays on the surface, an extended LEM providing a personnel carrier and an LEM truck without ascent stage for landing large payloads.

1965 August 3   At 21:21:53 UT the last readout of 21 pictures of Mars from the *Mariner IV* spacecraft was received on Earth. Two complete picture tape loads had been transmitted from Mariner to ensure no data was lost. As the spacecraft moved from north to south, the swath of pictures moved from 37° N by 173° E to 44° S by 279° E. When examined, the pictures revealed a surface with a high incidence of craters, and tracking data from signal occultation when the spacecraft passed behind Mars shortly after encounter indicated a 4–7 millibar atmospheric pressure (0.4–0.7% that of Earth), which superficially indicated a planet less likely to have life than had been previously thought. The spacecraft was now 150 million mi. from Earth and 5 million mi. beyond Mars in a permanent orbit of the sun from 103.05 million mi. at perihelion to 146.23 million mi. at aphelion. Telemetry was last received at 22:05:07 UT on October 1, 1965, when *Mariner IV* was 192.2 million mi. away and more than 19 million mi. beyond Mars. Periodic tracking and communications continued over the next 17 months.

Three companies were named by NASA to receive $500,000 contracts to design prototype Apollo lunar surface experiments packages (ALSEP). Bendix Systems Division, TRW Systems Group and Space-General Corp. would each perform six-month studies at the end of which NASA would award a development contract to one of them. On February 14, 1966, the Manned Spacecraft Center was advised that the Office of Space Science and Applications had selected seven experiments: passive lunar seismic experiment; lunar tri-axis magnetometer; medium-energy solar wind detector; suprathermal ion detector; lunar heat-flow experiment; low-energy solar wind experiment; and an active lunar seismic experiment. On March 16, 1966, Bendix was selected for ALSEP development.

1965 August 7   Indonesia began a series of ten flights using Japanese Kappa 8L two-stage meteorological sounding rockets. The flight took place from a site near Bandung, West Java, and the 136-lb. rocket reached a height of 210 mi. Indonesia was interested in studies of the upper atmosphere and the Earth-space margin for determining better weather prediction methods. Rain and water run-off is important for the rice crop and vital to the economy of the nation.

1965 August 8   The Indonesian government announced that it had successfully launched a solid propellant sounding rocket to space for the first time. Designated K-81, the rocket reached an unrecorded altitude above the nominal air-space boundary of 50 mi. The K-81 was later said to be capable of carrying a useful scientific payload of instruments to a maximum altitude of 210 mi.

1965 August 9   Douglas Aircraft Co. static-test fired the first S-IVB stage for the first time. The captive firing came as the company prepared the S-IVB for its first application as the second stage of the Saturn IB launch vehicle. Developed by Douglas as one of the most versatile upper stages ever built, the S-IVB had a length of 58 ft., a diameter of 21 ft. 8 in., an empty weight of 29,700 lb. and a fueled weight of 259,000 lb. Providing 200,000 lb. of thrust from one restartable J-2 engine, the S-IVB remained the upper stage for both Saturn IB and Saturn V launch vehicles. The S-IVB was also used as the basic structure for the Skylab space station and among its many secondary tasks was to be steered to lunar impact on several moon missions to provide vibrations for seismometers left on the surface by astronauts.

An explosion in a Titan II ICBM silo at Complex 4 near Searcy, Ark., killed 53 out of 55 workmen. Located about 50 mi. from Little Rock, Ark., the silo was 170 ft. deep. Immediately after the explosion a fire broke out which asphyxiated most of those who died. Most workers were unable to use a ladder to escape up the side of the silo when two workmen became jammed trying to get out through a limited-egress route. As a result, changes facilitating speedier egress were made to silo design.

1965 August 10   A new version of the four-stage solid propellant Scout launcher carried the 53-lb. *Secor 5* U.S. Army geodetic satellite from Wallops Island, Va., to a 702 × 1,503 mi., 69.2° orbit. An acronym for sequential collation of range, Secor-series satellites were designed and developed by the Army Map Service of the Army Corps of Engineers to pinpoint the location of land bodies separated by large expanses of ocean and to determine the correct shape of the Earth. Distances up to 1,000 mi. apart could be charted to an accuracy of less than 100 ft. by using signals from three accurately surveyed ground stations to determine mathematically the exact position of a fourth station on an uncharted island. The Scout had a new Castor 2A second stage producing 58,000 lb. of thrust and a new FW-4 fourth stage producing 5,600 lb. of thrust. *Secor 5* remains in orbit as of 1995.

1965 August 12   NASA decided to fly a new, lightweight space suit on the 14-day *Gemini VII* mission. Called G5C, it weighed 9 lb. and comprised a lightweight structure similar to that of the G4C EVA (extravehicular activity) suit but without restraint layer, and a soft helmet with integral helmet and visor. Concern about the physical health of the crew wearing a full pressure suit on this long-duration flight led to NASA's authorizing use of the G5C. Under consideration was a plan allowing the crew to remove their suits on orbit during periods of benign activity, donning them only for orbital maneuvers and reentry. This was approved on November 29, and when *Gemini VII* was launched on December 4, 1965, mission rules permitted one astronaut to remove his suit at any given time. Eight days into the flight, rules were changed allowing both crew members to leave their suits off at the same time.

1965 August 13   The first seven-satellite payload carried on a single launcher lifted off from Vandenberg AFB, Calif., on the last Thor-Able Star. The satellites in this all-Navy flight included five Surcal and two classified payloads. It was an academic, rather than a technical, feat since an eight-satellite cluster had already been orbited on March 7. One Surcal failed to separate from the second stage but a second, of the 9-lb. Dodecahedron-type, successfully deployed twelve 25-ft. antennas. Each orbit was different, but they averaged 670 × 733 mi. at 90° inclination. All remain in orbit.

1965 August 20 Design work on a scheme to convert the spent second stage of a Saturn IB launch vehicle into an orbital workshop got under way at NASA's Marshall Space Flight Center. Utilizing the S-IVB as a near-term space station, the 11,000 cu. ft. interior of the depleted liquid hydrogen tank seemed an attractive use for this otherwise redundant hulk. The concept envisaged astronauts converting the S-IVB for occupancy. An airlock would occupy the position taken up by a lunar excursion module when that stage was used with Saturn V for moon missions. The airlock would link an Apollo CSM (command/service module) with the interior of the hydrogen tank, affording access to the interior of the hydrogen tank for habitability and experiments.

1965 August 21 At 13:59:59.5 UT, NASA launched the 7,947-lb. *Gemini V* spacecraft carrying astronauts Gordon Cooper and Charles Conrad. After separating from the launch vehicle at 5 min. 57 sec., *Gemini V* was in a 100.5 × 217 mi. orbit destined for an eight-day mission powered by electrical energy from two fuel cell sections. At 2 hr. 7 min. 15 sec. the crew ejected a small package called a radar evaluation pod (REP) from the back of the adapter module and tracked it as it moved away at a speed of 5 ft./sec. Shortly thereafter the crew reported falling oxygen pressure in the fuel cell reactant dewar, or pressure vessel. Only by powering down all unnecessary equipment could the crew keep the spacecraft in orbit for the planned duration. The spacecraft splashed down in the western Atlantic to complete a successful mission after 7 days 22 hr. 55 min. 14 sec. The REP decayed out of orbit on August 27, two days before *Gemini V*'s splashdown.

1965 August 24 The first joint flight of a sounding rocket by the United States and Brazil took place when a Nike Apache was fired from Wallops Island, Va., to a height of 101 mi. The rocket carried a 60-lb. payload of instruments to measure the lower regions of the ionosphere and obtain data to be compared with data obtained from further launchings from a site near Natal, Brazil. The hardware was provided by NASA, but the launch was performed by a team of Brazilian engineers. The first flight from Natal took place on December 15 and was conducted by the Brazilian Space Activities Commission.

The U.S. Department of Defense and NASA set up the joint Aeronautics and Astronautics Coordinating Board (AACB) to bring together technical studies of reusable launch and logistics vehicles. The AACB set up the Subpanel on Reusable Launch Vehicle Technology comprising joint Department of Defense–NASA chairmen, eight members from the Department of Defense and ten members from the civilian space agency. Their report, submitted on September 14, 1966, categorized three levels of development: Class I X-20-type vehicles, said the report, would be available by 1974 to perform simple orbital operations; Class II lifting bodies launched by reusable cryogenic boosters would be available by 1978 for more complex operations; advanced Class III reusable, single-stage logistics vehicles incorporating supersonic ramjet propulsion were projected to be operational by 1982.

1965 August 25 President Lyndon B. Johnson announced he had approved full-scale development of the U.S. Air Force Manned Orbital Laboratory (MOL). The Douglas Aircraft Co. was to build the laboratory, 25 ft. long with a 10-ft. diameter, and General Electric would plan, develop and integrate the scientific experiments. Unmanned flight tests of preliminary hardware would begin in 1967, with the fully equipped laboratory being launched by Titan III-C in 1968. The first of five crew visits to the MOL would take place later that year. About 15 Air Force astronauts would be trained for the MOL, all selected from the Aerospace Flight School at Edwards AFB, Calif.

1965 September 3 An SL-8 launched from Baikonur carried the first preoperational tactical military communication satellites to Earth orbit. Designated *Cosmos 80–84,* each of the five 110-lb. satellites were placed at an altitude of 932 × 932 mi. with an inclination of 56.1°. One of the satellites had a radioisotope thermoelectric generator (RTG) for electrical power. A second quintuplet launch took place on September 18, with the orbits of the satellites *(Cosmos 86–90)* averaging 857 × 1,050 mi. All ten satellites remain in space.

1965 September 10 NASA's Saturn/Apollo Applications Office officially redesignated the Apollo Extension System (AES) to the Apollo Applications Program (AAP). This was effectively the date NASA declared it would extend the use of Apollo hardware in Earth-orbit and lunar-surface missions before starting development of post-Apollo programs. While some managers wanted NASA to move quickly with a follow-up to moon missions, developing new vehicles for ambitious objectives, others preferred a more cautious approach to see what future goals might be defined through the extended use of existing hardware. The latter view would prevail and lose for NASA the initiative of setting new targets and maintaining momentum after the initial moon landings.

1965 September 20 NASA announced crew selections for the *Gemini VIII* mission, planned as a rendezvous and docking flight incorporating extensive EVA (extravehicular activity). Astronauts Neil Armstrong and David Scott would be the primary crew, with Charles Conrad and Richard Gordon the backups. The space walk, which could last up to 1 hr. 35 min., was planned around a backpack carrying an independent supply of oxygen that would permit Scott to operate without an umbilical connected to the spacecraft. It was called an extravehicular support package (ESP) and was to have fed oxygen to a chestpack called an environmental life support system (ELSS), standard equipment on the last four Gemini flights. On *Gemini VIII,* the astronaut was to be tethered by a 75-ft. line allowing free and independent maneuvering using a hand-held propulsion unit. Because *Gemini VIII* returned early, the ESP was never used and the one chance for a Gemini astronaut to operate independent of a life-support umbilical was lost.

1965 September 25 Aerojet General Corp. test fired a half-length 260-in. diameter solid propellant rocket motor for 2 min. at its facility in Dade County, Fla. The largest solid propellant rocket in the United States, the SL-1 developed a thrust of 3.6 million lb. and was followed on February 23, 1966, by the first firing of the second motor, SL-2, which developed a thrust of 3.5 million lb. for 2 min. 6 sec. In both instances the motor was nose down in a 150 ft. deep pit. Both firings were part of the national solid rocket motor development program started in 1963 by the U.S. Air Force and transferred to the Lewis Research Center in 1965.

1965 October 4 The Soviet Union launched the 3,320-lb. spacecraft *Luna 7* in an attempted to hard-land a 220-lb. capsule equipped with a TV camera on the lunar surface in the Oceanus Procellarum. Flight events went as planned and the spacecraft switched on a radar altimeter which, at a lunar altitude of 47 mi., would trigger a timer for delayed ignition of the retro-propulsion system. In a malfunction, the timer started the engine immediately without counting down a programmed interval. Although it fired almost exactly as planned, the engine shut down at too great an altitude for the capsule to survive the resulting impact and the spacecraft was destroyed October 7 at 22:08:24 UT.

1965 October 5 The first successful launch in a long series of U.S. Air Force Orbiting Vehicle satellites was carried by Atlas D-Altair 2 from Vandenberg AFB, Calif., to a 250 × 2,151 mi. orbit with an inclination of 144.3°. Designated *OV1-2*, the 190-lb. satellite carried six radiation research experiments integral to the Air Force Weapons Laboratory's (AFWL) Manned Orbiting Laboratory program. Mounted, for the first time, to the nose of the launcher with a simulated engineering payload, it gave Atlas a unique spade-blade appearance. *OV-1-2* remains in orbit. The last of 20 successful OV1-series satellites was launched on January 31, 1972.

1965 October 11 William B. Taylor, director of the Apollo Applications Office at NASA Headquarters, presented a paper at a meeting convened by the American Institute of Aeronautics and Astronautics explaining NASA plans for the use of basic Apollo hardware after initial moon landings. Under the Apollo Applications Program (AAP) the lunar orbit capability could be extended, said Taylor, by adding consumables (fluids and gasses essential for continued spacecraft operation) to the Apollo CSM (command/service module) for 28-day missions during which probes could be sent to the surface from lunar orbit. Lunar surface activity could be expanded from 36 hr. to 14 days through the use of separate lunar excursion modules (LEMs) converted for shelter and taxi use. Launched first, the shelter would be unmanned and carry 6,000 lb. of exploration equipment for the crew to use when they arrived later in the LEM taxi.

1965 October 14 The second of NASA's six Orbiting Geophysical Observatory satellites, *OGO-2* was launched by TAT-Agena D from Vandenberg AFB, Calif., to a 258 × 943 mi., 87.3° orbit. Identical to *OGO-1*, it weighed 1,147 lb., including 210 lb. of experiments and instruments to measure the near-Earth space environment and magnetic fields. Two experiments failed shortly after launch. Difficulties with the Earth-lock horizon scanners depleted attitude control gas by October 23. When the satellite began to tumble, five experiments were rendered useless and six were degraded but useable. Battery failure by April 1966 left only eight experiments operational, and *OGO-2* was put on standby in November 1967 after returning 72,000 hr. of data. All operations ceased on November 1, 1971, and it decayed on September 17, 1981.

1965 October 15 The first U.S. Air Force Force OV2-series satellite package, *OV2-1*, was launched from Cape Canaveral on the second Titan IIIC. Carrying a 130-lb. payload incorporating 14 experiments, the 375-lb. package was intended to separate into a 435 × 4,350 mi. orbit after the third of 10 planned Transtage burns. However, on the second burn one of the two Transtage main engines failed to fire, throwing the whole combination into a destructive tumble. OV2-series

satellites grew out of the Advanced Research Environment Test Satellite (ARENTS) program and all were cube-shaped, 2 ft. × 1 ft. 10 in., with four solar paddles carrying 20,160 cells and spanning 7 ft. 6 in. to produce 63 W of power. Out of five OV2 flights only one was successful. *OV2-1* decayed on July 27, 1972.

1965 October 25 NASA attempted the first rendezvous and docking mission in the manned Gemini program when it simultaneously prepared an Atlas-Agena target vehicle on Cape Canaveral's Complex 14 and the GT-6 vehicle on Complex 19 in support of *Gemini VI*. The Atlas-Agena was launched at 15:00:04 UT, but 6 min. 16 sec. later the Agena propulsion system failed and the target vehicle broke up before it reached orbit. Astronauts Walter Schirra and Thomas Stafford were in their spacecraft participating in countdown checks toward a planned launch about 1 hr. 41 min. later. As the astronauts left their spacecraft on the launch pad — their target now resting on the bottom of the Atlantic — Walter Burke from the McDonnell Aircraft Co. suggested using the *Gemini VII* spacecraft on its long-duration mission as a target for the *Gemini VI* spacecraft.

1965 October 28 The White House formally announced that NASA would try to launch the *Gemini VI* spacecraft while *Gemini VII* was in space on its 14-day mission so that it could demonstrate rendezvous techniques worked out in theory but not yet performed in orbit. Some managers had wanted to place a docking cone in *Gemini VII*'s adapter module, enabling *Gemini VI* to insert its nose section as though it were an Agena target vehicle. Concern about vulnerable systems equipment in the back of the adapter mitigated against that, however. Experience with five previous Gemini launches gave confidence that pad personnel could erect and launch a second Gemini within 8 days of the first.

1965 November 2 The second Soviet SL-9 heavy-lift satellite launcher carried the 26,900-lb. *Proton 2* from Baikonur cosmodrome to a 119 × 396 mi., 63.5° orbit from where it continued to operate until it decayed on February 6, 1966. The payload weight included instruments for measuring heavy cosmic rays and four large solar paddles deployed for electrical energy production. A third Proton launch attempt on March 24, 1967, ended in failure with the SL-9. Designated *Proton 3*, the third successful satellite in this series was launched by SL-9 on July 6, 1966, to a 115 × 363 mi., 63.5° orbit. It decayed on September 16, 1966.

1965 November 6 The first of three Geodetic Earth Observation Satellites, GEOS 1 *(Explorer 29)*, was launched from Cape Canaveral, Fla., in heavy rain by the first thrust-augmented improved Delta to a 692 × 1,414 mi. orbit inclined 59.4°. Also designated Delta E, the new launcher was similar to the TAD (or Delta D) introduced in 1964 but with an improved, large-diameter AJ-10–118E second stage, a U.S. Air Force–developed 5,600-lb. thrust FW-4 third stage in place of the X-258, and a bulbous fairing from the Agena. It could put a 450-lb. satellite in geosynchronous transfer, or 1,620 lb. in low Earth orbit. The 385-lb. octagonal satellite was topped by an eight-sided truncated pyramid 4 ft. across the flats and 2 ft. 8 in. high. A 2-ft. hemisphere attached to the underside carried a spiral antenna and a 60-ft. boom could be extended from the top for gravity-gradient stabilization. Five experiments were carried for establishing a three-dimensional geodetic reference system.

1965 November 7 At the military parade in Red Square, Moscow, the Soviets displayed the SS-15 Scrooge land-mobile missile for the first time. Carried inside a cylindrical container mounted to a converted IS-3 chassis, the missile remained an enigma, most deployments being to the Chinese border. The container had a length of about 65 ft. and a diameter of almost 7 ft. It was equipped with jacking devices for elevation to a vertical position. The SS-15 was the largest mobile weapon system displayed to date.

1965 November 8 NASA announced that astronauts Elliot See and Charles Bassett had been selected as the primary crew members for the upcoming *Gemini IX* mission. Their backup would comprise astronauts Thomas Stafford and Eugene Cernan. Scheduled for the third quarter of 1966, the mission would be a two- to three-day rendezvous and docking flight, with EVA as another objective. Expected to last up to 2 hr. 25 min., the space walk would make use of the astronaut maneuvering unit (AMU) stowed in the back of the adapter module. This would enable Cernan to maneuver using thrusters operated through hand controllers attached to arms on the large backpack. Connected to the spacecraft by a 125-ft. tether, Cernan was to evaluate the AMU for the U.S. Air Force.

1965 November 12 The Soviet Union launched *Venera 2* to Venus. Placed initially in an Earth parking orbit by SL-6 launcher, the 2,123-lb. spacecraft was the first of four launched by the Soviet Union during the 1965 Venus launch window. The 2,116-lb. *Venera 3* was launched on November 16 and successfully boosted out of Earth orbit on course for Venus. Designated *Cosmos 96* when it failed to leave Earth orbit, a third spacecraft was sent into space by an SL-6 launcher on November 23; a fourth Venera-series spacecraft failed to reach orbit on November 26. A 71 ft./sec. course correction maneuver was performed by *Venera 3* on December 26, 1965, transferring the 37,625-mi. fly-by trajectory to an impact course with Venus. Communication was lost halfway to the planet. No course corrections were made to *Venera 2*, which passed the planet at a distance of 14,900 mi. on February 27, 1966. Following these flights and the death of Korolev in 1966, all planetary spacecraft design was transferred to the Georgi Nikolaevich Babakin design bureau.

The U.S. Department of Defense selected the first eight astronauts for the U.S. Force Manned Orbiting Laboratory program. They were: Maj. Michael Adams (USAF); Maj. Albert H. Crews (USAF); Lt. John L. Finley (USN); Capt. Richard E. Lawyer (USAF); Capt. Lachlan Macleay (USAF); Capt. F. Gregory Neubeck (USAF); Capt. James M. Taylor (USAF); and Lt. Richard H. Truly (USN). At this date the MOL was scheduled for launch in 1969 or 1970, a delay caused by the rising costs of the war in Vietnam. On July 20, 1966, Maj. Adams was transferred to the X-15 rocket research aircraft project to replace Capt. Joe Engle, who had transferred from assignment on the X-15 to become an astronaut with NASA.

1965 November 17 Christopher Kraft, NASA Gemini program flight director and future director of the Johnson Space Center (as the Manned Spacecraft Center was renamed), Houston, escaped death aboard an airliner en route from New Orleans to Miami when a teenage boy pointed a gun at his head and pulled the trigger. The gun misfired and the would-be assassin was overpowered by a fellow passenger and Paul Haney, NASA public affairs chief. The gunman, 16-year-old

Thomas Robinson, had wanted to go to Cuba to aid anti-Castro rebels.

1965 November 19 The product of a joint NASA/U.S. Naval Research Laboratory project to study solar × rays during a quiet period of the sun's 11-year cycle, a four-stage Scout carried the 126-lb. *Explorer 30* science satellite from Wallops Island, Va., to a 440 × 548 mi., 59.7° orbit. Essentially two hemispheres 2 ft. in diameter separated by a narrow band, the 125-lb. satellite carried ion chambers and geiger counters to monitor solar x-ray and ultraviolet emissions. Two low-thrust ammonia vapor jets were used to maintain attitude, and solar cells provided 6 W of power. Data was transmitted to the United States and foreign stations in 13 countries. *Explorer 30* remains in orbit.

1965 November 22 Apollo Program Director Samuel C. Phillips began a detailed survey of work under way at North American Aviation on the Apollo command/service modules and the S-II, the second stage of the Saturn V launch vehicle. On December 19, 1965, he informed John Leland Atwood, NAA's president, that he was "definitely not satisfied with the progress and outlook of either program." Phillips was concerned that slack management practices he had detected at North American threatened the pace of the program and might lead to concerns about safety. Atwood reviewed the comments and implemented changes which did much to accommodate Phillips' concerns.

1965 November 26 France successfully launched her first satellite on a French launch vehicle from Hammaguir, Algeria, thus becoming the third nation after the Soviet Union and the United States to place a satellite in orbit on its own. The three-stage Diamant launcher had a 69,700-lb. thrust liquid propellant Emeraude first stage, a 33,700-lb. thrust Topaze solid propellant second stage and a 11,690-lb. thrust Rubis third stage. With a total height of 62 ft., Diamant put the A-1 *Asterix* satellite in a 328 × 1,099 mi. orbit inclined 34.65° to the equator. The 92-lb. satellite carried a radio transmitter but no scientific equipment. Designed to operate for two weeks, the A-1 ceased broadcasting after two days but remains in orbit. The primary purpose of the flight was to qualify the Diamant launcher.

McDonnell Aircraft Co. proposed building a cheap alternative to the Gemini Agena Target Vehicle for use as a standby in case modifications being made to the Agena were not completed in time for the *Gemini VIII* mission, the next in line for a rendezvous and docking flight. The so-called augmented target docking adapter (ATDA) concept incorporated only the docking end of the Agena together with stabilization thrusters for a total weight of 1,354 lb. Within the orbital payload capability of an Atlas launch vehicle, the ATDA could be used as a passive docking target, but unlike the Agena it could not be maneuvered in space. NASA approved the concept on December 9, 1965, and McDonnell began building the ATDA five days later.

1965 November 29 Two Canadian satellites were launched by NASA on a Thor-Agena B from Vandenberg AFB, Calif. The 319-lb. *Alouette 2* was placed in a 314 × 1,856 mi., 79.8° orbit and the 218-lb. *Explorer 31* was put into a 314 × 1,851 mi. orbit, also at 79.8°. *Alouette 2* was the *Alouette 1* backup model and took the form of an oblate spheroid with a diameter of 3 ft. 6 in. and a diameter of 2 ft. 10 in. With two sets of beryllium dipole antennas, one pair 70 ft. long and one pair 240 ft. long, there were in addition five whip antennas and two electrostatic probes. *Explorer 31* took the form of an octagon, 2 ft. 6 in. across and 2 ft. 1 in. high, with a 1 ft. 9 in. spectrometer extending from the top. The cooperative venture was also known as ISIS-X, for International Satellites for Ionospheric Studies. Both remain in orbit.

1965 November 31 Marshall Space Flight Center briefed NASA Headquarters on the S-IVB orbital workshop (OWS) concept for the Apollo Applications Program. The next day Associate Administrator for Manned Space Flight George Mueller discussed the concept and asked Wernher von Braun to draw up a program development plan and present it at the next meeting of the Manned Space Flight Management Council (MSFC). Mueller wanted experiments included, funding requirements defined and work allocation schedules prepared showing how the work would be distributed among NASA field centers. The MSFC presented these plans on February 11, 1966, and included observation instruments incorporated in a converted lunar excursion module called an Apollo telescope mount, or ATM.

1965 December 3 The Soviet Union launched the 3,422-lb. *Luna 8* spacecraft toward the moon after an SL-6 launcher had first placed it, and its terminal stage, in an interim Earth parking orbit. Fifth in a series of Soviet hard-lander flights attempted over nine months, *Luna 8* was successfully placed on a trajectory to the moon, but the retro-propulsion system fired too late and the spacecraft was destroyed when it struck the surface at too great a speed.

1965 December 4 The 8,076-lb. *Gemini VII* spacecraft with astronauts Frank Borman and James Lovell on board was launched from Cape Canaveral's Complex 19 at 19:30:04 UT and placed in a 100 × 204 mi. orbit at separation 6 min. 9 sec. later. The purpose of the flight was primarily to gather medical data from 14 days of weightlessness, with the added objective of providing a target vehicle for the *Gemini VI* spacecraft. Lovell removed his lightweight G5C suit at an elapsed time of 1 day 9 hr. and donned it again at 6 days 4 hr., when Borman removed his suit. Lovell again removed his suit at 7 days and, with the exception of the *Gemini VI* rendezvous phase and reentry, the crew flew the rest of the mission without suits. The crew performed five orbital maneuvers during the mission to increase the orbital life and place the spacecraft in the planned orbit for rendezvous with *Gemini VI*. That occurred at a *Gemini VII* elapsed time of 10 days 23 hr. 58 min. and lasted 3½ orbits. *Gemini VII* splashed down in the western Atlantic at an elapsed time of 13 days 18 hr. 35 min. 31 sec.

1965 December 5 Hamilton Standard tested a portable life support system (PLSS) designed for Apollo astronauts on the lunar surface. Weighing 65 lb., the PLSS operated for more than 3 hr. in a vacuum chamber while a test subject walked a treadmill to simulate the metabolic output of an astronaut working on the moon. All functions of the PLSS were tested, including the provision of cooling water for the astronaut's liquid-cooled undergarment, oxygen supplied at a pressure of 3.7 lb./sq. in., controlled temperature and humidity and the removal of carbon dioxide exhaled by the astronaut.

1965 December 6 NASA launched a second French satellite, *FR-1*, into a near-polar orbit of 462 × 480 mi. inclined 75.9° by Scout launcher from Vandenberg AFB, Calif. NASA provided the four-stage launcher as part of a cooperative agreement with the French national space agency Centre

Wearing the unique G-5C space suit, astronauts Frank Borman and James A. Lovell Jr. walk to the gantry prior to launch on December 4, 1965, at the start of their two-week mission.

Nationale d'Etudes Spatiales (CNES). The 135-lb. satellite was to study propagation of very low frequency radio waves and measure electron densities. It had an octagonal center-section with truncated octagonal prisms mounted on top and bottom, 2 ft. 3 in. across the corners and 4 ft. 4 in. high. FR-1 remains in orbit.

1965 December 7 The first test launch of a Thor missile equipped with satellite inspection equipment under the U.S. Air Force's Program 437 X was successfully conducted from Johnston Island in the Pacific. Also known as Advanced Program 437, or 437 (AP) for Alternate Payload, to differentiate it from the antisatellite program, three more test flights were conducted on January 18 and March 12, 1966 and July 2, 1967. U.S. Secretary of Defense Robert F. McNamara prohibited operational tests using Soviet satellites as targets, but the planning envisaged Thors being available on a 30-min. alert in case that decision was reversed during a national crisis.

1965 December 12 The first attempt to launch astronauts Walter Schirra and Thomas Stafford in their *Gemini VI* spacecraft ended dramatically only 1.6 sec. after the two Titan first-stage engines ignited on Cape Canaveral's Complex 19. Ignition occurred at 14:54 UT, but a dust cap inadvertently left over an inlet port to the gas generator starved it of oxygen and started the automatic shut-down sequence. Both astronauts had the option of activating their ejection seats or sitting tight. They chose to take the latter course. Designated *Gemini VI-A,* the mission was to have carried Schirra and Stafford to a rendezvous with *Gemini VII,* now 8 days into its 14-day flight. Had they used the ejection seats they would not have had the option of a second try because the spacecraft would have been unusable.

1965 December 14 Lockheed fired a 156-in. solid propellant motor under the program begun under the U.S. Air Force and recently transferred to NASA. With a height of 75 ft., 700,000 lb. of propellant, one motor segment containing high-burn rate propellant, a submerged nozzle buried up inside the motor case and liquid injection thrust vector control, the motor produced an average thrust of 3 million lb. thrust for 55 sec. Another test took place on January 15, 1966, when a single 34 ft. tall segment of a 156-in. motor filled with 350,000 lb. of propellant produced a thrust of more than 1 million lb. for 1 min. 5 sec. The last of five Lockheed 156-in. solid tests in this program took place on April 7, 1966, when a single-segment motor with 156,000 lb. of propellant produced 300,000-lb. thrust for 2 min.

1965 December 15 The second attempt to launch the *Gemini VI-A* mission on the first space rendezvous flight succeeded when the 7,817-lb. spacecraft lifted away from Cape Canaveral's Complex 19 at 13:37:26.5 UT. At insertion to a 100 × 161 mi. orbit, *Gemini VI* trailed *Gemini VII* by 1,238 mi. Pilots Walter Schirra and Thomas Stafford began the sequence of rendezvous maneuvers at 1 hr. 34 min. 3 sec., raising apogee to 169 mi. to put *Gemini VI* a constant 17 mi. below *Gemini VII.* At 2 hr. 18 min. 1 sec. a phasing maneuver raised perigee to 139 mi. and at 2 hr. 42 min. 8 sec. a plane change maneuver adjusted the orbital inclination from 28.97° to 28.89°. A small corrective height adjustment maneuver took place at 3 hr. 3 min. 20 sec. and a co-elliptic sequence maneuver at 3 hr. 47 min. 37 sec. produced a 167 × 170 mi. orbit, with *Gemini VI* now 198 mi. behind *Gemini VII.* A terminal phase maneuver began at 5 hr. 18 min. 56 sec., producing a 180 × 186 mi. orbit and rendezvous with *Gemini VII* 37 min. later. The two spacecraft remained together for 5 hr. 18 min., each performing a pirouette for pictures and to practice close-in maneuvers, on one occasion closing to within 1 ft. *Gemini VI* splashed down in the western Atlantic at 25 hr. 51 min. 24 sec. within 8 mi. of the aim point.

NASA announced the cancelation of its Advanced Orbiting Solar Observatory (AOSO), a successor to the highly successful Orbiting Solar Observatory satellites. Managed by the Goddard Space Flight Center, AOSO was to have been built by the Republic Aviation company and launched in 1969 into a polar-retrograde orbit using a TAT-Agena D. It was to have weighed about 1,250 lb. and carried 250 lb. of scientific instruments. AOSO was canceled to save money for other programs and in the knowledge that other, more advanced solar research projects were being developed for the Apollo Applications Program.

1965 December 16 NASA interplanetary science probe *Pioneer 6* was launched by a thrust-augmented improved Delta from Cape Canaveral, Fla., to an Earth-escape trajectory and a heliocentric path just inside the Earth's orbit of the sun. This was also the first Earth-escape mission for a Delta launch vehicle. The 140-lb. *Pioneer 6* consisted of a 3 ft. 1 in. cylinder, 2 ft. 11 in. tall, with three 5 ft. 4 in. booms deployed from the midsection, and a 4 ft. 4 in. mast carrying high-gain and low-gain antennas. Six experiments totalling 35 lb. — the highest payload-structure ratio of any interplanetary space-

craft — were divided into three groups collecting magnetic field data, measuring the solar wind and detecting cosmic rays. *Pioneer 6* resides in a heliocentric orbit between 76 million and 92 million from the sun inclined 0.1695° to the ecliptic. In 1995 *Pioneer 6* was still returning data.

Responding to a request for definition of goals for the Apollo Applications Program (AAP), Marshall Space Flight Center Director Wernher von Braun defined Earth observation for better human understanding of natural resources as the primary objective. He believed that spin-offs from technology research were not a proper justification for space activity and that NASA should bring solid results and benefits to U.S. taxpayers. Von Braun commented on the growing problem of expanding world population, on the need for better management and distribution of Earth's limited resources, and judged that Americans should take up the challenge of pioneering in this field. In this way, he said, the various proposals under the AAP would become elements combining to support this broad objective.

1965 December 22 The Voyager Mars mission based on the use of two Saturn IB-Centaur launch vehicles to carry combination flyby/landers was deferred due to the cancelation of the Saturn IB-Centaur. Moreover, there were no funds for a test flyby in 1969 or two missions in 1971. Planners would quickly focus on the possibility of carrying out dual flights in 1973 and 1975 on Saturn V. It brought a dilemma to the mission planning teams because, whereas Saturn IB-Centaur had been just capable of carrying the heavy spacecraft, Saturn V was too powerful and costly, providing a Mars mission payload capability of more than 70,000 lb! NASA explained the position to Voyager contractors from January 3 to 5, 1966, a week before managers began to plan a revised program for the next two years. The White House had wanted a Mars mission in 1969, and project documentation for the 1969 Mariner-Mars fly-by mission utilizing Mariners F and G was signed on March 28, 1966. In flight, these spacecraft would be known as *Mariners VI* and *VII,* respectively.

1965 December 25 Responding to serious concerns about the pace of the U.S. manned flight program, Soviet space officials met to consider Korolev's proposal for a manned circumlunar mission using a stripped-down version of the Soyuz spacecraft. The plan was a more considered rework of the Sever proposal of March 1962, envisaging only the descent and instrument modules of Soyuz launched by Chelomei's Proton (UR-500) launcher with an added upper stage. Designated UR-500K, this launcher would send the stripped down Soyuz (L-1), weighing 11,400 lb., on unmanned circumlunar test flights. The UR-500K/L-1 program was later given block designations in the Cosmos and Zond series. UR-500K/L-2 was to be the manned phase utilizing what were to have been designated Soyuz-L (lunar) spacecraft.

NASA formally approved the Mariner E mission for a fly-by of Venus in 1967 utilizing the spare spacecraft left over from the 1964 Mariner-Mars program. Grounded due to funding constraints, it now would serve as a stop-gap until the more powerful Atlas-Centaur was available to launch bigger spacecraft to Venus. There had been reluctance to develop a more advanced Venus spacecraft until it was clear Atlas-Centaur could be relied upon to be ready for a specified launch window, and for the next two years it would be committed to the Surveyor program. The flight was made possible by deferral of the Voyager-Mars mission from 1971 to 1973, opening a funding gap which the 1967 Mariner-Venus could fill.

1965 December 28   The first in a new class of Soviet electronic intelligence (elint) gathering satellites was launched from Baikonur when an SL-8 carried the 1,930-lb. *Cosmos 103* to a 294.5 × 307.5 mi. orbit inclined 56°. This first test satellite demonstrated essential systems being developed for the operational design, but it was March 24, 1967, before an SL-8 carried the second test satellite, *Cosmos 151*, to a similar orbit. The first operational Soviet elint satellite, *Cosmos 189*, was launched on October 30, 1967.

1965   Presidential approval for the U.S. military Manned Orbiting Laboratory (MOL) on August 25 brought a decision in the Soviet Union to develop a Soviet equivalent. Known as Orbital Piloted Station (OPS), its design and assembly was assigned to the Chelomei bureau. Also known as Almaz, the cylindrical OPS had a length of 47 ft. 9 in. and a maximum diameter of 13 ft. 7 in. A forward section of narrower diameter supported a manned reentry capsule. The OPS was to have been launched by the Chelomei UR-500K Proton launcher, as was a transport/logistics ship (TKS) with a length of 57 ft. 6 in. and a maximum diameter of 13 ft. 7 in. The TKS was to be manned at launch with a three-man crew in a Merkur capsule on front with forward-facing windows for docking at the OPS. Access back into the TKS cargo section was via a hatch in the heat shield of the reentry capsule. By 1969 the OPS was ready for flight, but priority was given to the N-1/L-1 lunar landing plan and space station Almaz waited.

Following almost 10 years of development, one of the most enduring Soviet ICBMs was made ready for deployment in large numbers around the borders of the USSR. The SS-11 Sego was a two-stage, storable liquid propellant missile with a length of 62 ft. 4 in., a diameter of 7 ft. 10 in. and a weight of 105,000 lb. Three versions were developed: Mod 1 with a single 950-KT warhead and an accuracy of 4,600 ft. at a range of 6,000 mi.; Mod 2 with a 1-MT warhead, penetration aids, an accuracy of 3,600 ft. and a range of 8,000 mi.; Mod 3 with three 200-KT multiple reentry vehicles, an accuracy of 3,600 ft. and a range of 6,500 mi. Peak deployment of 1,030 silo-based missiles was reached in 1974, but numbers declined slowly thereafter, reducing to 520 by 1984 and withdrawal under the 1990s START (Strategic Arms Limitation Talks) agreement.

The U.S. Department of Defense began a study of future ICBM survivability in light of expanding Soviet missile strike power. The United States feared improvements in accuracy, and the fielding of MIRV warheads could threaten Minuteman and Titan silos in a preemptive strike. Under Strategic Systems X, it sought to avoid being drawn into a "launch under attack," or LUA, strategy where the entire U.S. missile inventory would have to be fired in the 15 minutes or so of warning missileers would get before Soviet warheads began to fall, disabling the retaliatory force. These studies went on for six years as the Soviet ICBM force expanded.

1966 January 7   In a telephone conversation with Oran Nicks, NASA's director of Lunar and Planetary Programs, Deputy Director of the Manned Spacecraft Center George Low outlined plans for a Lunar Sample Receiving Laboratory. It was to be a curatorial facility where moon samples returned from the lunar surface would be kept and analyzed, where selected samples would be distributed to investigating scientists and where laboratory services would be provided for guest investigators. It was also to incorporate a quarantine section for moon materials and, for three weeks after their return from the lunar surface, Apollo astronauts. Located

adjacent to the Manned Spacecraft Center, the Lunar Receiving Laboratory would eventually mature into the Lunar and Planetary Institute.

1966 January 14   Sergei Pavlovich Korolev, probably the most important figure in the annals of Soviet space achievements, died in Moscow at the age of 59. Unknown to the Russian people until the announcement of his death, Korolev had always been referred to simply as "the chief designer." Born December 30, 1906, in the town of Zhitomir, Korolev had been the driving force in mobilizing support among the Soviet hierarchy and in motivating support and loyalty from his staff. With the death of Korolev, the Soviet space program lost momentum and became fragmented.

1966 January 16   The Space Science Board of the U.S. National Academy of Sciences issued a report, *Space Research: Directions for the Future*, which detailed recommended research objectives in lunar and planetary exploration for the next 15 years. Expressing dissatisfaction with the priorities placed at NASA on manned spaceflight, the SSB placed the unmanned exploration of Mars as a clear priority, followed by a similar survey of the moon and Venus. The report represented the views of the majority of scientists, who saw little science coming from the Apollo program and its proposed successor projects.

1966 January 18   The NASA HL-10 lifting body was rolled out of Northrop's Hawthorne, Calif., factory and delivered to the Flight Research Center at Edwards AFB after a period of wind tunnel testing at NASA Ames. With a launch weight of 6,000 lb., a length of 22 ft. 2 in. and a width of 15 ft. 1 in., it took the form of a thick, delta-shaped body with flat underside swept up at the nose and a rounded upper body area supporting a tall dorsal fin at the rear. Two outward-canted rear fins swept up from the flat underbody. The first flight of the HL-10, an unpowered glide from 45,000 ft. following release from the B-52 mother-plane, caused major problems for test pilot Bruce Peterson because there was much less lateral control and flow separation than predicted. After landing, which was accomplished safely, the HL-10 was modified by changing the shape of the fins.

1966 January 20   The first item of Apollo command/service module Block I flight hardware was tested when the last Little Joe II carried CSM-002 to an altitude of 15 mi. where the launch escape system took over and demonstrated safe activation from a simulated tumbling booster with power on. The mission, designated A-004, also confirmed the structural integrity of the launch escape tower under conditions of extreme stress. For this flight, the adaptable Little Joe II was equipped with two 103,200-lb. thrust solid propellant Algol motors in the first "stage" and two Algols in the second "stage" which, co-located inside the cylindrical structure with the other two Algols, were ignited at 37 sec. elapsed time, 3 sec. before the first two cut off. At liftoff, five 33,395-lb. thrust Recruit motors provided an additional 167,000-lb. of thrust for 1.5 sec. The command module landed 21.5 mi. downrange of the launch site. This was the last Little Joe II/Apollo flight, all features of the launch escape system having been cleared for manned operations with flight hardware.

1966 January 25   In its weekly report to the White House, the U.S. Department of Defense announced that an alert network to monitor Soviet reconnaissance satellites was now operational. Called SATRAN (satellite reconnaissance advance notice), the network was jointly operated by the Defense Intelligence Agency and the Air Force Systems Command. The increasing pace of Soviet photoreconnaissance satellite activity (7 in 1963, 12 in 1964, 17 in 1965) brought fears that Soviet intelligence could routinely spy on U.S. military equipment. SATRAN provided warnings to bases within optical range to cover up sensitive equipment during the satellite's pass.

NASA named astronauts John Young and Michael Collins as the primary crew members for *Gemini X*, with James Lovell and Edwin Aldrin as their backups. Young would be the first astronaut to make a second Gemini flight. *Gemini X* was planned as a docking flight with rendezvous on the fourth apogee using a trajectory and maneuver sequence similar to that employed for *Geminis VI-A* and *VIII*. Maneuver sequences for *Geminis IX-A* and *XII* would achieve rendezvous at third apogee. *Gemini XII* demonstrated a first-apogee rendezvous simulating a lunar ascent envisaged for Apollo operations at the moon.

1966 January 28 NASA Headquarters asked the Apollo Program Office at the Manned Spacecraft Center to evaluate the impact of a joint rendezvous and docking mission on the third manned Apollo CSM (command/service module) flight. Two flights would be involved, both involving Saturn IB launches. Designated mission AS-207, a Block II CSM would be launched first to a circular, 300 mi. orbit, followed by the unmanned LEM (lunar excursion module) on flight AS-208 into a 126-mi. orbit. Rendezvous would occur within 4.5 hr., followed by docking and extraction of the LEM from the spacecraft/LEM adapter on top of the S-IVB stage. It had the advantage of testing in Earth orbit the LEM, the Block II CSM and Apollo rendezvous and docking before the Saturn V could fly an all-up mission. About this time it became a convention to reverse the letter designations for the Saturn launches: SA-200 series Saturn IB flights became AS-flights and SA-500 series Saturn V missions were numbered in the AS-series.

1966 January 31 At 11:41:37 UT the Soviet Union launched *Luna 9* on course for the moon. Sent on its way by an SL-6 launch vehicle, the 3,490-lb. spacecraft performed a trajectory correction maneuver at 19:29 UT, February 1, 145,000 mi. from Earth and 118,000 mi. from the moon. At a lunar altitude of 47 mi., with just 45 sec. to surface contact, a radar altimeter on *Luna 9* started a timer which ignited a retrorocket after separating redundant elements of the spacecraft. Just before touchdown a spherical, 220-lb. capsule about 2 ft. in diameter was ejected to a hard landing cushioned by crushable, shock-absorbing material protecting a 3.3-lb. TV camera, transmitter and antennas. The landing took place at 18:45:30 UT, February 3, and 4 min. 10 sec. later four petal covers forming the top half of the spherical capsule hinged open at 90° intervals to stabilize the small capsule on the surface. The open petals had a spread of 6 ft. 3 in. The first TV pictures were broadcast in a transmission beginning at 01:45 UT on February 4, and were picked up by the British radio telescope at Jodrell Bank, England. Jodrell Bank released the first pictures that day and they were shown on British television the same evening and in newspapers the following day, ahead of the Russians. On the surface at 7.13° N by 64.5° W, *Luna 9*'s capsule continued to transmit data and TV panoramas until 22:55 UT on February 6, when the batteries ran out.

1966 January The Douglas Aircraft Co. submitted a report to NASA summarizing its work on the proposed Manned

Orbiting Research Laboratory. Douglas' MORL studies began with a Phase I review of system options in the period June–September 1963, continued with a Phase IIa study of optimized designs in the December 1963–November 1964 period and concluded with a Phase IIb development study in December 1964–February 1966. Launched by a Saturn IB into a 230-mi. orbit inclined 28.7° to the equator, MORL was to have been an orbiting laboratory with a diameter of 21 ft. 7 in. and the flexibility to accommodate an expanding range of scientific tasks. It would have received crews delivered by Apollo logistics vehicles launched by Saturn IB.

1966 February 1 The retro-rocket system designed to slow NASA's Surveyor spacecraft for a controlled descent to the surface of the moon was successfully tested by Hughes Aircraft and the Jet Propulsion Laboratory. The descent propulsion system comprised a 9,500-lb. thrust solid propellant retro-rocket weighing 1,400 lb. which would burn for about 40 sec. and decelerate Surveyor from a speed of just over 5,800 MPH, 47 mi. from the surface, to about 270 MPH, 5.4 mi. from the surface, with just over 2 min. to landing. The 30 throttleable, 104-lb. thrust, liquid-propellant vernier motors, one adjacent to each of the three Surveyor landing legs, would together perform midcourse maneuvers, maintain attitude during retro-rocket burn and jettison and reduce the speed of the spacecraft from 270 MPH to 3.5 MPH just 14 ft. above the lunar surface. These maneuvers would be in response to signals from a radar altimeter and doppler velocity sensor. The 181 lb. of propellant would consist of monomethylhydrazine-monohydrate fuel and MON-10 (90% nitrogen tetroxide/10% nitrogen oxide) oxidizer. A Surveyor spacecraft successfully demonstrated a soft landing under vernier propulsion on May 11, 1966.

1966 February 3 The first weather satellite in the Tiros Operational Satellite (TOS) series was launched by Delta E from Cape Canaveral, Fla., to a 432 × 521 mi. orbit inclined 98°. Launched for the Environmental Science Services Administration (ESSA), TOS was an operational version of the Tiros weather satellite program. Identical to Tiros satellites, the 305-lb. *ESSA I* was to provide daily cloud cover pictures of the entire sunlit portion of the globe and transmit these images to ESSA ground stations in Alaska and Virginia. Its companion, the 290-lb. *ESSA II*, was launched February 28 by a thrust-augmented Delta to a retrograde orbit of 843 × 885 mi. inclined 101°. *ESSA II* was to provide automatic picture transmission (APT) readout for a given area at the same time each day to any APT receiver worldwide. Other APT satellites were *ESSA IV* (launched January 26, 1967), *ESSA VI* (November 10, 1967) and *ESSA VIII* (December 15, 1968). Ground station readout satellites included *ESSA III* (October 2, 1966), *ESSA V* (April 20, 1967), *ESSA VII* (August 16, 1968) and *ESSA IX* (February 26, 1969).

The first breadboard tests of the NERVA (nuclear engine for rocket vehicle application) began at the U.S. Atomic Energy Commission's plant at Jackass Flats, Nev. (Breadboard tests are made with all the equipment of an operational system but arranged in a test configuration.) In the first such test, a part of the Rover series, the reactor developed 440 MW versus its design output of 1,100 MW, and an exhaust temperature of 2,000°F versus the design temperature of 3,500°F. The breadboard tests ended on March 25, demonstrating a power output of 1,130 MW and a temperature of 3,715°F. In 10 tests the breadboard engine had been run for 1 hr. 50 min., 29 min. at full power.

1966 February 17   A three-stage French Diamant rocket launched from Hammaguir, Algeria, placed the instrumented science satellite *Diapason I* (D-IA) into a 313 × 1,711 mi. orbit inclined 34° to the equator. This was the second French satellite and the first instrumented for scientific investigations from space. Signals transmitted from orbit were received at the Brétigny Tracking Center near Paris. The main body of *Diapason I* had a diameter of 1 ft. 7 in., a height of 1 ft. 8 in. and four rectangular solar cell panels, each 1 ft. 4 in. × 8 in. The 42-lb. D-1A measured the Earth's magnetic field and tested the ability of French tracking stations to control a satellite in orbit.

1966 February 22   A Soviet SL-4 launch vehicle from the Baikonur cosmodrome sent the 12,600-lb. *Cosmos 110* into a 116 × 562 mi. orbit inclined 51.9° to the equator. The spacecraft was the last of the Voskhod series and carried two dogs, Veterok and Ugolek; this was the first in a series of biological spaceflights involving animals and insects under the direction of Prof. Yevgeni A. Illyin. Designated a cosmonaut for a brief period when Sergei Korolev was planning to send scientists on a two-week biological research flight, Prof. Illyin had continued his work in the unmanned Cosmos program when manned Voskhod missions were canceled during the second half of 1965. The two dogs made 330 orbits of the Earth and landed in good condition after 21 days 18 hr. in space.

1966 February 26   At 16:12:01 UT the first 1.6 million-lb. thrust Saturn IB launch vehicle was launched from Complex 34 at Cape Canaveral carrying Apollo CSM-009. Designated AS-201, the mission was a test of launcher/spacecraft compatibility and a suborbital demonstration of the Apollo spacecraft. At launch the assembly stood 224 ft. tall and weighed 1.318 million lb. The four inboard H-1 engines shut down at about 2 min. 21 sec. elapsed time, followed by the four outer H-1 engines 5 sec. later. The first (S-IB) stage separated, and 3 seconds after first-stage cutoff the S-IVB fired for about 7 min. 33 sec. After the 33,805-lb. Apollo CSM (command/service module) separated from the 3,691-lb. adapter, the service propulsion system (SPS) engine fired for 3 min. 4 sec., increasing speed by approximately 5,000 ft./sec. Erratic performance on the second, 10-sec. SPS burn reduced command module reentry velocity by 730 ft./sec. to 26,481 ft./sec. Maximum aerodynamic heating occurred at 27 min. 12 sec., with maximum deceleration of 14.3 g at 27 min. 20 sec. The command module splashed down at 37 min. 20 sec., close to Ascension Island and was retrieved by the USS *Boxer* 48 min. later. The most heavily instrumented vehicle yet launched by NASA, 545 measurements were telemetered from the first stage, 470 from the second stage, 300 from the instrument unit controlling the launcher and 607 from the Apollo spacecraft.

1966 February 28   *Gemini IX* crew Elliot See and Charles Bassett were killed in a Northrop T-38 they were flying into St. Louis municipal airport during rain and fog. Cleared for an instrument landing, the T-38 turned toward the McDonnell Aircraft Co. buildings at the edge of the airfield, 1,000 ft. from the runway, and hit the roof of the building where the two spacecraft *Gemini IX* and *X* were being assembled. The T-38 bounced into an adjacent courtyard and exploded as part of the roof showered down upon spacecraft and workers. Back-up crew Thomas Stafford and Eugene Cernan were following in a second T-38 and landed safely, immediately being named as the primary crew for *Gemini IX*.

1966 February   The British government distributed a memorandum to member states of the European Launcher Development Organization (ELDO) expressing concern at escalating costs of the Europa I launcher and the planned Europa II. Of primary concern was the ELDO-PAS (Europa II) vehicle, standing for ELDO-Perigee/Apogee Stage, designed as a shorter route to achieving geosynchronous orbit capability than the ELDO B–series proposed earlier. ELDO-PAS would have comprised a Europa I with fourth (perigee) and fifth (apogee) stages capable of placing 400 lb. in geostationary orbit.

Engineers at Lockheed examined a block of the company's LI-500 ceramic thermal protection material, which had been put together for possible use as internal insulation, and determined that it could be used as protection against excess heat on the exterior of reusable lifting-body spacecraft. Effectively turning the heat shield inside out, this discovery revolutionized theory about the applicability of different materials as heat-sink insulators. The mechanical properties of the ceramic insulation would form the basis for heat shield technologies employed later on the NASA Shuttle.

1966 March 1   The Soviet spacecraft *Venera 3* became the first man-made object to reach the surface of another planet when it crashed into Venus at 06:56 UT, two days after *Venera 2* had flown past at a distance of 14,900 mi. Communication with these spacecraft ceased shortly before they encountered Venus and the precise impact time was calculated from the known trajectory. *Venera 3* carried the Soviet emblem and a miniature globe of the Earth and was intended to deploy a parachute to lower a capsule to the planet's surface. At this date no one knew that the surface of Venus was devoid of water, that atmospheric pressure was 90 times that of Earth's atmosphere at the surface, that temperatures were so great that lead would remain a liquid and that powerful winds could whip up tidal currents in the dense cloud layers.

The Soviet Union launched another in their Luna series of lunar exploration spacecraft when an SL-6 placed the terminal stage and its cargo in an initial Earth parking orbit of 113 × 120 mi. The terminal stage failed to fire, leaving the spacecraft stranded in that orbit until it decayed back down through the atmosphere two days later. Designated *Cosmos 111* and having a mass in Earth orbit of 14,420 lb. (including the terminal stage), it was not acknowledged as a Luna series vehicle by the USSR.

NASA selected Fairchild, General Electric and Lockheed as finalists to bid for six-month feasibility studies of a second-generation Applications Technology Satellite. NASA wanted to follow the first-generation ATS with up to five advanced satellites capable of carrying technology development equipment supporting research into space-based data relay, deep-space tracking, navigation, data collection, aircraft-to-aircraft communication, satellite-to-satellite relay, ship-to-ship communications and broadcast satellites. Fairchild was eventually awarded a contract to build *ATS-F* (*ATS-6* when launched).

1966 March 16   NASA began the *Gemini VIII* rendezvous and docking mission with the launch of the Atlas-Agena target vehicle from Cape Canaveral's Complex 14 at 15:00:03 UT into a 183 × 186 mi. orbit. The 8,351-lb. *Gemini VIII* spacecraft carrying astronauts Neil Armstrong and David Scott was launched from Cape Canaveral's Complex 19 at 16:41:02 UT to a 99 × 169 mi. orbit, trailing the Agena by 1,220 mi. In a sequence of rendezvous maneuvers similar to those performed on the *Gemini VI-A* mission, *Gemini VIII*

Adapted from the Agena D, the Gemini Agena Target Vehicle (GATV) was used by Gemini VIII *to rendezvous and dock with another object in space for the first time, March 16, 1966.*

completed its chase at 5 hr. 54 min. when it braked to a stop relative to the Agena 150 ft. away. Docking was achieved at 6 hr. 33 min. 22 sec., the first time two separately launched vehicles had been connected together in flight. About 27 min. later the docked configuration began unexpected roll and yaw motions simultaneously. *Gemini VIII* undocked from the Agena, then thought to be the cause of the motions, at 7 hr. 15 min. 12 sec. The motions increased to a tumble rate of almost 60 RPM before the spacecraft was brought under control using the reentry thrusters, and the spacecraft stabilized at 7 hr. 25 min. 30 sec. *Gemini XIII* was returned to Earth prematurely, with splashdown in the western Pacific Ocean at 10 hr. 41 min. 26 sec. Postflight analysis diagnosed the cause of the tumble as a thruster stuck in the "on" position.

1966 March 17 A Soviet SL-3 was launched from a site known as Plesetsk, the first time a satellite had been launched from this location. The first-generation, low resolution photoreconnaissance satellite *Cosmos 112* was placed in a 128 × 339 mi. orbit inclined 72.1°. This high-inclination orbit gave the satellite a broader band of the planet to survey, its path taking it to a higher latitude than satellites launched to a 65° orbit from Baikonur. The fact that *Cosmos 112* had been launched from Plesetsk was revealed by British schoolmaster Geoffrey Perry, who monitored shortwave transmissions from Soviet satellites at a school in Kettering, England.

1966 March 19 The last of 10 in-orbit maneuvers performed by the *Gemini VIII* Agena target vehicle placed the modified Agena-D stage in a 250 × 254 mi. orbit at an inclination of 29.04°. In a 51-sec. burn that began at 12:19:50 UT, the secondary propulsion system provided a 145 ft./sec. velocity change and placed the stage in a parking orbit from which it would be used again by the *Gemini X* spacecraft as a passive target for a second in-orbit rendezvous. In all, the primary

propulsion system was fired in orbit eight times, reaching a maximum altitude of 466 mi., and the secondary propulsion system was fired twice. Before electrical power was exhausted on March 25, 1966, all attitude control gas was vented overboard to prevent a thruster mishap causing the Agena to tumble.

NASA announced flight crew assignments for the Gemini and Apollo programs. Astronauts Charles Conrad and Richard Gordon would serve as the primary crew for *Gemini XI*, with Neil Armstrong and William Anders in the backup role. James Lovell and Edwin Aldrin were moved from *Gemini X* backup to *Gemini IX* backup. Alan Bean and Clifton Williams were named as the new backup crew for *Gemini X*. Astronauts Virgil Grissom (command pilot), Edward White (senior pilot), and Roger Chaffee (pilot) were selected as the primary crew for the first manned Apollo flight, with James McDivitt, David Scott and Russell Schweickart as backup. The first flight was expected to remain in a 100–265 mi. orbit for up to 14 days.

1966 March 23 NASA completed the first schedule of Apollo Applications Program flights. It envisaged 26 flights of the Saturn IB and 19 flights of the Saturn V in support of three orbital workshops fitted out in space, called "wet" workshops, and three fitted out on the ground, called "dry" workshops, as well as four Apollo telescope mounts (ATMs). The ATMs were to be converted lunar excursion modules docked to the Earth-orbiting workshops. The first launch in the AAP was slated for April 1968.

1966 March 28 The fully successful launch and air-snatch recovery of an instrumented rocket probe took place when a McDonnell F-4C aircraft fired a modified MB-1 Genie which was recovered in midair by a Lockheed C-130B transport aircraft. Known as the air-launched air-recoverable rocket

(ALARR), the MB-1 Genie carried instruments for research into the upper atmosphere above western California. Released at 44,000 ft., it was boosted to a maximum altitude of 150,000 ft., at which point the 180 lb. instrument payload separated and descended by parachute until snatched at 5,000 ft.

1966 March 30 An Atlas D launched from Vandenberg AFB, Calif., carried two U.S. Air Force OV1 payloads from the Aerospace Research Support Program to orbit. Both satellites were cylinders with hemispheric ends, 2 ft. 3 in. in diameter and 4 ft. 7 in. long; each was attached to a 5,800-lb. thrust X-258 motor. *OV1-4* weighed 193 lb. and was placed in an orbit of 550 × 630 mi., in which zero-gravity experiments were conducted on chlorella algae and multicell duckweed specimens exposed to alternate 12-hr. periods of light and dark. The 252-lb. *OV1-5* carried experimental optical sensors and a gravity gradient experiment. Both OV-1s were placed in retrograde orbits of 144° and are still in orbit, albeit inoperable.

As a result of the problems experienced aboard *Gemini VIII* when it temporarily spun out of control March 16, NASA Associate Administrator Robert Seamans informed Director of Manned Space Flight George Mueller that he favored putting TV cameras on manned Apollo missions. This was strongly supported by Julian Scheer, NASA director of Public Affairs, who believed it would stimulate greater public interest. Representing the views of the majority of astronauts, Donald Slayton formally opposed the idea on the basis that it was an invasion of crew privacy and would trivialize in-flight operations. Thus began a battle that the astronauts lost under the greater weight of public demand for increasing coverage of manned space spectaculars.

1966 March 31 The Soviet Union successfully launched *Luna 10,* which three days later became the first spacecraft placed in orbit about the moon. With a weight of 3,521 lb., the spacecraft was similar in many respects to the Luna vehicles carrying hard-landing capsules, but with extra propellant and control systems for lunar orbit. At 18:44 UT on April 3 the retro-motor fired to slow *Luna 10* from an approach speed of 6,890 ft./sec. to 4,100 ft./sec., dropping it into an initial orbit of 217 × 632 mi. inclined 71.9° to the moon's equator. Immediately thereafter, the 540-lb. payload section separated from the propulsion pack and *Luna 10* began a survey of the moon's magnetic field, radiation environment and gravitational irregularities. The Soviets programmed semiconductors to oscillate at specific frequencies and broadcast to Earth an electronic rendition of "The Internationale," the anthem of international socialism. Powered by batteries, *Luna 10* continued to transmit data to Earth during 219 communication sessions over 460 lunar orbits before it expired on May 30, 1966.

1966 April 4 NASA announced the names of nineteen pilots selected as NASA's fifth group of astronauts: Vance D. Brand; John S. Bull; Col. Gerald P. Carr (USMC); Brig. Gen. Charles M. Duke Jr. (USAF); Maj. Gen. Joe H. Engle (Air National Guard) Capt. Ronald E. Evans (USN); Maj. Edward G. Givens Jr. (USAF); Fred W. Haise Jr.; Col. James B. Irwin (USAF); Don L. Lind; Col. Jack R. Lousma (USMC); Capt. Thomas K. Mattingly II (USN); Capt. Bruce McCandless II (USN); Capt. Edgar D. Mitchell (USN); Col. William R. Pogue (USAF); Col. Stuart A. Roosa (USAF); John L. Swigert Jr. (civilian) ; Capt. Paul J. Weitz (USN); Col. Alfred M. Worden (USAF). Duke, Irwin and Mitchell walked on the moon. Only Bull and Givens would not fly space missions.

Evans died on April 6, 1990, of a heart attack. Givens died June 6, 1967, in a car crash. Irwin died August 8, 1991, of a heart attack. Swigert died December 27, 1982, of cancer.

1966 April 7 In an attempt to demonstrate the capability of an Atlas-Centaur launch vehicle to propel a Surveyor spacecraft to the moon in a dual-burn sequence, the AC-8 flight from Cape Canaveral's Complex 36 was only a partial success. It demonstrated successful Earth orbit insertion for the 1,730-lb. Surveyor mass model but failed to fire the two RL-10 engines a second time, which was to have simulated a lunar flight path by raising apogee to 236,000 mi. Only one engine fired and that shut down after just 17 sec., leaving the stage and dynamic model in a 109 × 207 mi. orbit. Achieving an escape trajectory from Earth parking orbit was demonstrated when AC-9 propelled a 1,740-lb. Surveyor mass model to an orbit of 103 × 252,400 mi. in a second Centaur burn lasting 1 min. 47 sec., 35 min. after launch on October 26, 1966. Had the model been aimed at the moon it would have been captured by lunar gravity. Propelling a lunar or planetary spacecraft to its objective from an Earth parking orbit greatly increases the time available to reach the appropriate point in space. *Surveyors III, V, VI* and *VII* used this technique. *Surveyors I, II* and *IV* used direct-ascent.

1966 April 8 An Atlas-Agena D launched by NASA from Cape Canaveral put the first of four Orbiting Astronomical Observatories *(OAO-I)* into a 493 × 505 mi. orbit inclined 35° to the equator. With a weight of 3,900 lb., *OAO-I* was an octagonal cylinder 10 ft. long and 7 ft. wide with two solar cell arrays of three panels, each spanning 21 ft., supporting a total 74,618 cells. *OAO-I* carried four experiments: seven ultraviolet telescopes, a high-energy gamma ray detector, a gas counter for measuring x-radiation and a low-energy gamma ray detector. *OAO-I* failed shortly after its main battery overheated on the second day.

1966 April 15 Manned Spacecraft Center Director Robert Gilruth pressed NASA Headquarters for permission to proceed with development of an Apollo experiment pallet. The pallet was envisaged as a complement of scientific instruments that could be fitted to a vacant segment of the Apollo service module. Operating from orbit about the moon as a minilaboratory, the pallet would permit extended lunar science activity while two astronauts were exploring the surface. The main body of the service module had a length of 14 ft. 10 in. and a diameter of 12 ft. 10 in. and was divided vertically into six separate pie-shaped segments, one of which would contain the pallet of instruments. Manned Space Flight Director George Mueller gave permission to to proceed but rescinded this under budget pressure on August 22, 1966. The concept was resurrected later and employed for *Apollos 15-17*.

1966 April 22 A four-stage Scout launch vehicle carried the first U.S. Air Force OV3-series satellite, *OV3-1*, from Vandenberg AFB, Calif., to a 218 × 3,567 mi. orbit with an inclination of 82.5°. This third series of Orbiting Vehicle satellites were managed by the Air Force Cambridge Research Laboratories (AFCRL). Comprising an octagonal prism, 2 ft. 5 in. × 2 ft. 5 in., with 2,560 solar cells providing 30 W, the satellite could carry a 37–60-lb. payload. *OV3-1* carried six experiments for measuring particles and fields in the Van Allen radiation belts. It remains in orbit. The last of six OV3-series satellites, five of which had been successful, was launched by a Scout on December 5, 1967.

1966 April 25 The first Minuteman II squadron, the 447th Strategic Missile Squadron, became operational at Grand Forks AFB, N.Dak. In further deployments authorized by the Kennedy and Johnson administrations, the Minuteman force expanded to a total 1,000 missiles deployed in silos hardened to withstand overpressures of 300 lb./sq. in. Grand Forks AFB was added to the five bases already deployed with Minuteman I missiles. An additional force of 200 ICBMs were located at this site, in addition to the usual three squadrons at each Minuteman base, and there was a fourth squadron at Malmstrom AFB, Mont. Strategic Air Command would built its Minuteman force to a complement of 500 Mk.II and 500 Mk.III by 1975, the Minuteman II and III completely replacing Minuteman I.

1966 April 29 One of the world's largest and most powerful space tracking stations was officially dedicated at Goldstone, Calif., when the antenna, 210 ft. in diameter, was formally inaugurated into the Deep Space Network (DSN). Operated by the Jet Propulsion Laboratory, the antenna would play a primary role in tracking and communicating with NASA's interplanetary spacecraft.

1966 April 30 From the Tyuratam launch complex in Kazhakstan, the Soviet Union launched an SL-6 carrying a Luna-series spacecraft. The launch vehicle failed to reach orbit and the payload was destroyed. Aware that NASA was about to launch *Lunar Orbiter I*, equipped with cameras, the Soviets were working hard to get a camera-equipped spacecraft to lunar orbit first. Problems with the SL-6 delayed the flight of a camera-carrying Luna spacecraft for almost four months, by which time NASA had received the first pictures back from *Lunar Orbiter I*. No further SL-6 flight was attempted until the successful launch of *Luna 11* on August 24, 1966.

1966 May 12 NASA Headquarters issued a memorandum to all the field centers reporting that the Project Designation Committee had changed the name of the Apollo lunar excursion module to lunar module, or LM, and that henceforth the Saturn IB would be known as the uprated Saturn I. On December 2, 1967, James Webb rescinded this last change and the launcher reverted to being known as the Saturn IB. (To avoid confusion, it has been retained in that form throughout this book.)

1966 May 13 After a lapse of almost fifteen months Thiokol Chemical Corp. ground-test fired another 156-in. solid propellant motor at its Wasatch Facility, part of the NASA research program transferred from the U.S. Air Force during 1965. The firing tested a filament-wound case that promised lighter structural weight and greater stage efficiency for operational solid propellant rocket motors, delivering a thrust of 300,000 lb. for 2 min. A second firing on May 26, 1967, produced a thrust of 1 million lb. and evaluated a flexible seal for movable thrust vector control nozzles. The last of five tests in the program took place on June 25, 1968, when a Thiokol 156-in. motor fired for 1 min. 58 sec.

1966 May 17 At 15:15:03 UT the *Gemini IX* Atlas-Agena target vehicle lifted away from Cape Canaveral's Complex 14, but 2 min. 10 sec. later the no. 2 Atlas booster engine gimbaled to the full pitch-down position causing the ascending vehicle to lose control and crash into the Atlantic Ocean 123 mi. from Cape Canaveral at an elapsed time of 7 min. 36 sec. On Complex 19, meanwhile, *Gemini IX* astronauts Thomas Staf-

ford and Eugene Cernan climbed out of their spacecraft as all hope of getting a launch dissipated with the demise of the target vehicle. Rescheduling the launch attempt for June 1, 1966, and designating it *Gemini IX-A,* NASA decided May 18 to use the augmented target docking adapter instead of a full-size Agena at the next attempt.

1966 May 19 As a result of a recent fire in an environmental control unit being built by AiResearch Co. for the Apollo command module, the NASA Manned Spacecraft Center (MSC) sent primary contractors North American Aviation a telex requesting removal of combustible items from the spacecraft interior. The MSC said it appeared that some materials in the Block I spacecraft would have to be modified to conform to standards limiting fire propagation but that these modifications would have to be followed by complete eradication of all fire propagation sources in Block II spacecraft. The rapid spread of fire in the pure oxygen atmosphere of a pressurized cabin was of great concern to astronauts, engineers and managers alike.

1966 May 23 The first of four test flights in the second phase of the Europa I launch vehicle program took place when a live Blue Streak stage carried dummy upper stages on a ballistic flight from launch pad 6A at Woomera, Australia (May 24 local time). Designated F4, the vehicle's two first-stage Rolls Royce RZ.2 engines had a combined upgraded thrust of 300,000 lb. The Europa I vehicle had a height of 104 ft., of which 60 ft. 4 in. comprised the first stage, 18 ft. the second stage and 12 ft. 6 in. the third stage. For this flight the upper stages were dummies. Blue Streak fired for 2 min. 15 sec., at which point the range safety officer cut the flight short by 8 sec. when he saw indications that the stack was off course. It was not, and telemetry later showed that it had been on course at a height of 30.3 mi. Nevertheless, the test was considered a success. The second flight with inert upper stages (flight vehicle number F5) was flown on November 14 (November 15 local time), when the first stage fired for 2 min. 6 sec. with impact 510 mi. downrange at 6 min. 50 sec.

1966 May 25 Exactly five years after President Kennedy set the moon landing goal, the first full-scale Apollo Saturn V launch vehicle and spacecraft combination was rolled out the Vehicle Assembly Building (VAB) at Kennedy Space Center and down the crawler way to Launch Complex 39A. Designated AS-500-F, the inert facilities mock-up vehicle weighed 500,535 lb. and comprised a shell conforming to the exact dimensions of a Saturn V and Apollo spacecraft combination. The facilities mock-up was used to check interface with support pedestals in the VAB and at the pad and with the mobile launch platform and its swing-arm assemblies.

1966 May 30 The 2,193-lb. *Surveyor I* lunar soft-lander was launched by Atlas-Centaur from Launch Complex 36 at Cape Canaveral, Fla., at 14:41:01 UT to an Earth-escape trajectory. Centaur main engine cutoff occurred at an elapsed time of 11 min. 27 sec., followed by extension of the landing legs 28 min. later and deployment of the antennas 10 min. after that. Spacecraft separation came at 14:53:38 UT, and the spacecraft's sensors locked onto the sun and the star Canopus for attitude control during coast. Electrical power was provided by 3,960 solar cells arranged across a 9 sq. ft. panel attached to a hinged arm producing a mean raw power output of 89 W. A 20.8-sec. (66.5 ft./sec.) midcourse correction burn with the three vernier motors began at 06:45:03 UT on May 31. *Surveyor I* carried a 28-lb. slow-scan survey TV camera

Surveyor, NASA's lunar soft-lander program, achieved a significant "first" when the first spacecraft in the series successfully touched down on the moon three days after its May 30, 1966, launch. The test model seen here carries a surface sampler arm and chemical analyzer used on later flights.

capable of transmitting in 200- or 600-line formats. It pointed vertically up and received the image via a mirror inside a protective hood. Also carried by *Surveyors I* and *II*, but never used, was an approach TV camera similar to the survey camera but pointing down and operating only in 600-line mode.

1966 May 31 Howard W. Tindall, the NASA manager responsible for monitoring development of Apollo computer software with the contractor, the Massachusetts Institute of Technology (MIT), reported that it would "soon become the most pacing item for the Apollo flights." The challenging requirements were so great that even on comparatively simple unmanned Saturn test flights computer programs would exceed memory capacity. To solve this, MIT resorted to complex program intermeshing for its two major lunar flight programs: COLOSSUS and LUMINARY. Memory capacity would be the limiting element in what the computer could do, restricting some operations which the spacecraft was capable of executing.

The first major study into moon-based astronomy was presented to NASA by North American Aviation in its report *Research Program on Radio Astronomy and Plasma for AAP Lunar Surface Missions*. It established guidelines for future plans and assembled a list of research priorities, highlighting the obvious advantages: skies totally clear of atmospheric aberration, a more stable surface on which to base permanent

observatories, far-side viewing devoid of reflected artificial light and greatly reduced stellar motion due to the moon's rate of rotation, 27 times less than that of the Earth.

1966 June 1 The *Gemini IX-A* augmented target docking adapter (ATDA) was launched by an Atlas launch vehicle from Cape Canaveral's Complex 14 at 15:00:02 UT and placed in a 182 × 186 mi. orbit. In this unique configuration the Atlas/ATDA had a launch weight of 264,320 lb. and a total height of 95 ft. 5 in., with the 16 ft. 1 in. ATDA and shroud placed atop a 12 ft. tall cylindrical adapter. The ATDA failed to jettison its 652-lb. shroud during ascent and the assembly entered orbit with a total weight of 2,006 lb. It was there to await the launch of the *Gemini IX* spacecraft and astronauts Thomas Stafford and Eugene Cernan, but a computer problem at T − 1 min. 40 sec. canceled the attempt for that day. Including his experiences with *Gemini VI, VI-A* and *IX,* this was the fourth time Stafford had climbed out of a Gemini spacecraft after its mission was aborted.

1966 June 2 The first soft-landing on the moon of a U.S. unmanned lunar spacecraft took place when *Surveyor I* touched down in the southwest part of the Oceanus Procellarum at 43.23° W by 2.46° S, just 9 mi. from the aim point. The solid propellant retro-motor ignited at 06:14:50 UT, 246,635 ft. from the surface, at a velocity of 8,565 ft./sec., with burnout 39 sec. later at a velocity of 428 ft./sec. The liquid propellant vernier motors continued to fire until 06:17:35 when *Surveyor I* was 12 ft. from the surface at a descent rate of 3 ft./sec., touchdown occurring 2 sec. later. With a landed weight of 649 lb., *Surveyor I* had a height of 10 ft. and described a circle 14 ft. in diameter with its three extended landing legs. The first 200-line picture was received 57 min. after touchdown, and 10,341 images in both this and the 600-line formats were taken before the spacecraft was shut down for the lunar night at 20:31 UT on June 16, 1966.

1966 June 3 Astronauts Thomas Stafford and Eugene Cernan successfully began the operations phase of the *Gemini*

A stuck shroud prevented Gemini 9 *astonauts Thomas P. Stafford and Eugene A. Cernan from docking with their target vehicle in space following a flawless launch on June 3, 1966.*

IX-A mission when they were launched in their spacecraft from Cape Canaveral's Complex 19 at 13:39:33 UT. They were placed in a 99 × 166 mi. orbit with an inclination of 28.91° and trailed the augmented target docking adapter (ATDA) by 656 mi. In three orbital maneuvers, *Gemini IX-A* caught the ATDA at third apogee and an elapsed time of 4 hr. 15 min. when the spacecraft was 100 ft. from the ATDA. The shroud covering the docking collar was clearly visible and a request by the crew to use the nose of Gemini to pry it loose was denied. A rendezvous began at 5 hr. 1 min. and took the spacecraft 12 mi. above and behind the ATDA. Cernan performed manual calculations for re-rendezvous at 6 hr. 29 min. Maneuvers for a rendezvous from above began at 7 hr. 15 min. and put the spacecraft on a separate path so that over time it drifted 96 mi. away before maneuvers for yet another rendezvous were completed at 21 hr. 28 min. The planned space walk began with a hatch opening at 49 hr. 23 min., and Cernan made his way to the back of the adapter to don the astronaut maneuvering unit. Difficulty putting it on caused his visor to fog over and he was ordered back; the hatch closed at 51 hr. 28 min. The *Gemini IX-A* mission ended at an elapsed time of 72 hr. 20 min. 50 sec., 2,300 ft. from the planned splashdown location.

1966 June 7 Britain's Foreign Office confirmed that the government had decided to pull out of the European Launcher Development Organization (ELDO). Britain was no longer willing to contribute to developing an independent European launcher by providing the Blue Streak first stage for the Europa launch vehicle. Britain agreed to continue supporting ELDO flight tests through the preliminary stages, however.

The third of NASA's Orbiting Geophysical Observatories, *OGO-3,* was launched by Atlas-Agena B from Cape Canaveral, Fla., to a 170 × 75,768 mi. orbit with an inclination of 30.9°. The 1,136-lb. *OGO-3* carried a science payload of 195 lb. and 21 instruments to measure and monitor the near-space environment, Earth's magnetic fields, the ionosphere and solar and cosmic radiation. *OGO-3* maintained three-axis stabilization for 46 days, but a failure in the attitude control system during July pushed the satellite into a permanent spin. By July 1969 data was limited to one-half of each orbit, but 15 experiments continued to function until December 1969, when operations were suspended. *OGO-3* had returned 375,000 hr. of data when the satellite was switched off on February 29, 1972.

1966 June 8 The NERVA (nuclear engine for rocket vehicle application) reactor NRX-A5 was successfully ground tested by NASA at the Atomic Energy Commission's facility at Jackass Flats, Nev. The reactor was operated at the design power of 1,100 megawatts for 15 min. 30 sec. The same reactor was restarted June 23 and operated for a further 14 min. 30 sec. at design power, equivalent to a thrust of 55,000 lb. The reactor consumed 218,000 lb. of liquid hydrogen. The tests were conducted by Aerojet General and the Westinghouse Corp. On December 15, 1967, the NRX-A6 reactor ran at design power for 1 hr., longer than required for most space missions.

1966 June 9 The reoriented U.S. early warning satellite program developed from the MIDAS concept began flight tests when an Atlas-Agena D launched the 4,400-lb. satellite from Vandenberg AFB, Calif. Instead of transferring to a circular orbit of 1,900 mi., the Agena left it stranded in a useless orbit of 108 × 2,247 mi. inclined 90.1°. Also carried on this flight were the 37.5-lb. *Secor 6* geodetic satellite, and

the 33-lb. *ERS-16* with a metal-to-metal bonding experiment. The second MIDAS follow-up was successfully launched on August 19 to a 2,287 × 2,299 mi. orbit, followed by the third launch on October 5 to a 2,288 × 2,300 mi. orbit, both at 90° inclination. All three flights carried small secondary satellites. The ambitious MIDAS follow-up was halted once more, replaced by the Program 949 geosynchronous satellite concept.

1966 June 10 A four-stage Scout launcher carried the 173-lb. *OV3-4* satellite to a 399 × 2,939 mi., 41° orbit. The satellite took the form of an octagonal cylinder, 2 ft. 5 in. across and 2 ft. 5 in. high, with solar cells covering the exterior and five 1 ft. 6 in. experiment booms deployed from the sides. The satellite was instrumented to perform spectral and depth dose measurements in the inner Van Allen radiation belt. It remains in orbit.

1966 June 15 Dr. Kurt Debus, director of the NASA Kennedy Space Center at Cape Canaveral, Fla., announced that hourly buses would carry the public on fee-paying tours of the facilities during daylight. On Sundays the public were to have the option of a free drive-through in private vehicles. In subsequent years, a Visitor Information Center opened nearby, followed by one at the Manned Spacecraft Center, Tex., as well. Public tours are a major attraction for visitors to Cape Canaveral today and form an important part of a broad program aimed at keeping the public informed about the nation's space program.

1966 June 16 The U.S. Air Force launched the first seven Initial Defense Communication Satellite Program (IDCSP) satellites from Cape Canaveral, Fla., on a Titan IIIC. Each 100-lb. satellite took the form of a polyhedron with 24 sides covered in solar cells to power the transponders, which received at 8 GHz and retransmitted at 7.2 GHz. Each satellite was placed randomly in a near-synchronous orbit on the equator, with perigees varying between 15,811 mi. and 15,829 mi. and apogees varying between 15,911 mi. and 16,126 mi. In addition, a 104-lb. gravity-gradient test satellite (GGTS) with two 52-ft. long booms for gravity-gradient evaluation was also released to orbit. An attempt to launch eight IDCSP satellites on August 26 failed when the Titan IIIC blew up. *IDCSP-8* through *-15* were successfully launched on January 18, 1967, followed by *IDCSP-16* through *-18* on July 1, 1967, and satellites *19* through *26* on June 13, 1968.

Mrs. James A. McDivitt gave birth to a daughter, the first child conceived by a U.S. astronaut's wife after her husband had been in space. This event was significant because of fears concerning genetic damage from radiation above the Earth's atmosphere. None was found and the child was normal and healthy.

1966 June 17 NASA announced the flight crew for *Gemini XII,* the last U.S. two-man space mission. Astronauts James Lovell and Edwin Aldrin were to fly the mission with Gordon Cooper and Eugene Cernan as their backups. Aldrin had made major contributions to the theory and practice of orbital rendezvous. When problems emerged on EVAs (extravehicular activities) during Gemini missions IX-A, X and XI, Aldrin worked hard to develop procedures and equipment that would overcome these difficulties. It was largely due to his efforts on the ground preflight that the EVAs conducted on the last Gemini flight were so successful.

The Department of Defense announced the names of five new astronauts for the U.S. Air Force Manned Orbiting

Laboratory program: Capt. Karol J. Bobko (USAF); Lt. Robert L. Crippen (USN); Capt. Charles G. Fullerton (USAF); Capt. Hank W. Hartsfield (USAF); and Capt. Robert F. Overmeyer (USMC). On cancellation of the Manned Orbiting Laboratory in 1969, all five transferred to the civilian space agency as part of NASA's seventh group of astronaut inductees. Each one subsequently had an illustrious career in the Shuttle program.

1966 June 24   NASA launched the *Pageos I* geodetic satellite on a TAT-Agena D from Cape Canaveral to a 2,509 × 2,610 mi. orbit inclined 85°. The 125-lb. satellite consisted of a 100-ft. inflatable sphere based on the *Echo I* design. Made from 84 panels of vapor-deposited aluminum with a 0.0005-in. mylar coating, the sphere was packed at launch in a 28-in. spherical magnesium canister. Inflated by 10 lb. of benzoic acid and 20 lb. of anthraquinone acting under solar heat, the sphere was simultaneously photographed by a network of 41 ground-based triangulation stations. Any two points on Earth could be fixed to an accuracy of 32 ft. in 3,000 mi. The sphere eventually broke up between 1975 and 1978 but remains in orbit.

1966 July 1   A thrust-augmented Delta carried NASA's *Explorer 33* from Cape Canaveral, Fla., to an escape trajectory on course for lunar orbit insertion. In NASA's first attempt to orbit the moon since the Pioneer probes of 1960, *Explorer 33* was expected to enter a lunar orbit of 800 × 4,000 mi. to observe the magnetosphere toward the sun and away from the sun. Similar to IMP-A, IMP-B and IMP-C, *Explorer 33* was also known as the AIMP, or Anchored Interplanetary Monitoring Platform. It differed from its predecessors in having a small retro-motor instead of the forward magnetometer boom. Unfortunately, the Delta performed too well, giving *Explorer 33* 70 ft./sec. too much velocity for the motor to place it in moon orbit. Instead, engineers commanded the motor to fire to place the satellite in an Earth orbit of 9,880 × 270,560 mi. at 29°. It is still in orbit and sending data.

1966 July 5   Out of numerical order in its Apollo mission designation, the second Saturn IB flight, AS-203, was launched from Complex 37 at 14:53:17 UT and achieved a 115 x 117 mi. orbit 7 min. 14 sec. later. The first stage burned for 2 min. 22 sec. and the second stage for 4 min. 50 sec. At launch the vehicle weighed 1.187 million lb. and stood 173 ft. tall. It did not carry an Apollo-type payload but a 28-ft. high nose cone which provided an aerodynamic fairing at the front end. The flight was an engineering test of S-IVB systems directly applicable to Saturn V operations. A total 1,491 measurements were taken of the vehicle during ascent, 581 of which were from the S-IVB stage tested during the first four orbits in simulation of a restart in space. At the beginning of the fifth orbit, pressure in the S-IVB's liquid hydrogen tank caused it to burst, effectively ending the tests.

1966 July 6   Contact with the U.S. soft-landing moon spacecraft *Surveyor I* in the Oceanus Procellarum was restored after the spacecraft's first lunar night, which began when it was shut down on June 16. On July 8 an unsuccessful attempt was made to fire the three vernier motors. Scientists wanted to view the lunar surface scoured by exhaust plumes to observe the effect on the soil (called regolith) and to conduct an engineering evaluation to determine whether it would be feasible to fire up the motors on a later mission and physically move the spacecraft by causing it to hop. This would permit stereo images of rocks and features in the vicinity of the lander

as well as pictures of the depressions caused by the weight of Surveyor at its original landing spot. An additional 899 pictures were taken before the spacecraft was once again shut down, bringing to 11,240 the total number transmitted. Engineering tests were conducted on subsequent lunar days, and the last data were received on Earth at 07:30 UT on January 7, 1967.

1966 July 8   At a meeting of the European Launcher Development Organization (ELDO), ministers from member states decided to establish a European Space Conference (ESC) to assemble recommendations about a more fully coordinated European space program. Both ELDO and ESRO (European Space Research Organization) were proceeding along their respective paths without clear coordination or unified direction. This factor alone held Europe back from realizing its full industrial and technical potential. From this meeting came a decision to set up a European Space Conference, or ESC.

1966 July 8–10   At a meeting in Paris, France, members of the European Launcher Development Organization (ELDO) decided to expand the rocket development program from the three-stage Europa I to a five-stage ELDO Asp launcher capable of launching 440-lb. payloads into orbit by 1971. The United Kingdom contribution was to be cut from 39% to 27% as a compromise for that country's maintaining its involvement with the ELDO program.

1966 July 14   A double payload of OV1 satellites was launched for the U.S. Air Force Aerospace Research Support Program by Atlas D from Cape Canaveral. Mounted back-to-back on the nose of the launcher, each OV1 was attached to a 5,800-lb. thrust X-258 solid propellant rocket motor to propel its payload to orbit after separation, when the Atlas would continue on its ballistic trajectory. With a weight of 260 lb. *OV1-7* carried instruments to monitor night airglow (reflected light from the sunlit side), solar x rays, cosmic rays, charged particles and electric fields, but its motor failed to fire and it fell back into the atmosphere. *OV1-8* was a 23-lb. inflatable plastic balloon with a mesh grid carried folded in a container. It was placed in a 144° retrograde orbit of 612 × 635 mi. and inflated to a diameter of 30 ft. Over time the plastic sphere broke up under sunlight, leaving just the wire grid, which provided a surface from which radio waves could be reflected as a test of passive communication systems.

1966 July 18   The *Gemini X* rendezvous, docking and EVA (extravehicular activity) mission began at 19:39:46 UT, when the Atlas-Agena target vehicle ascended from Cape Canaveral's Complex 14 to an orbit of 245 mi. Astronauts John Young and Michael Collins were launched on time at 21:20:27 UT to a 100 × 168 mi. orbit, 1,120 mi. behind the Agena. Rendezvous was accomplished at the fourth apogee and docking took place at 5 hr. 53 min. With a mass of about 15,000 lb., the docked Gemini-Agena was pushed to a record manned flight altitude when, at 7 hr. 38 min. 34 sec., the Agena primary propulsion system fired for 13 sec., increasing speed by 422 ft./sec. and adjusting the orbit to 182 × 474 mi. At 20 hr. 20 min. the Agena engine was fired for 10 sec., cutting velocity by 346 ft./sec. and lowering apogee to 237 mi. A 2-sec. burn at 22 hr. 37 min. was adjusted by the secondary propulsion system to place the combination in an orbit of 236 x 240 mi. On the dark side of the Earth, at 23 hr. 24 min., Collins opened the hatch and stood up to perform experiments, closing the hatch 48 min. later. An 18-sec. Agena burn took place at 41 hr. 4 min., followed by a 4-sec. burn just under 32 min. later, thus setting up the spacecraft for rendezvous with the *Gemini VIII*–Agena, which had been launched in March 16. Separation from the *Gemini X*–Agena came at 44 hr. 40 min., with *Gemini VIII*–Agena rendezvous at 47 hr. 50 min. At 48 hr. 41 min. Collins opened the hatch for a 39-min. EVA which took him across to the Agena docking collar to retrieve a micrometeoroid package placed on the stage at launch. The hatch was opened again at 50 hr. 32 min. to discard EVA equipment, and *Gemini X* splashed down at 70 hr. 46 min. 39 sec. The *Gemini X*–Agena performed three orbital maneuvers, on one of which it reached an altitude of 863 mi.

1966 July 19 NASA's Deputy Director Robert Seamans ordered program chiefs to conduct a detailed study of the need for a permanent space station to match future requirements. In his briefing memorandum, Seamans denied that such a facility was a logical progression from Earth-orbit missions in the Apollo Applications Program, questioning whether a permanent space station was "the best approach to achieving mission objectives." Such a study would, said Seamans, "help us to decide if such a course is desirable and when."

1966 July 20 NASA Manned Spacecraft Center Director Robert Gilruth wrote to Wernher von Braun, Director of the Marshall Space Flight Center, describing a two-year effort just completed on development of a lunar mapping and survey system (LMSS), now a part of the Apollo Applications Program. Comprising a cylindrical container with a docking drogue like that carried by the Apollo lunar module (LM) the LMSS was designed to replace the LM inside the spacecraft/launch vehicle adapter of a Saturn V. The LMSS had been conceived for lunar surface reconnaissance and site certification as a backup to Surveyor and Lunar Orbiter. It would be carried to moon orbit instead of the LM during Apollo CSM (command/service module) tests that preceded the landing. The LMSS was canceled on July 25, 1967, when Surveyor and Lunar Orbiter had achieved success.

France detonated an air-dropped atomic bomb at its Pacific test site in the Mururoa Atoll almost directly under the flight path of the orbiting *Gemini X* spacecraft and about one hour before it passed over. Mission Control at the NASA Manned Spacecraft Center advised astronauts Young and Collins aboard *Gemini X*, "Keep your heads down," and not to look at the Earth while flying over the atoll.

1966 July 29 The first of a third generation of U.S. close-look military reconnaissance satellites was launched for the CIA by the first Titan IIIB-Agena D from Vandenberg AFB, Calif. The Titan IIIB had a 78-ft. first stage with upgraded YLR-87 motors delivering a total thrust of 463,000 lb. and a 30-ft., 101,000-lb. thrust second stage. Designated KH-8, the 6,600-lb. satellite was the first to carry visible and infrared cameras, multispectral scanners and a thematic mapper. Images were returned to Earth in four ejectable pods retrieved in midair for analysis on the ground. Using the orbit maneuver capability of Agena D, KH-8s frequently dipped low, sometimes to an altitude of 69 mi., for the best resolution. Operational lifetimes grew from 7 days to around 50 days by the late 1970s and to more than 100 days by the early 1980s. Launch rates fell from 6–8 per year in the late 1960s to 2–4 per year by the mid-1980s. The last of 53 KH-8s was launched on April 17, 1984, only three having failed to reach orbit. The last of 67 Titan IIIBs was launched on February 11, 1987.

1966 August 2 Dr. George Mueller, NASA's associate administrator for Manned Space Flight, asked Deputy Administrator Robert Seamans for permission to proceed with procurement of the airlock module for the orbital workshop component of the Apollo Applications Program. Seamans got approval from Administrator James Webb on the basis that it would provide "a unique opportunity to investigate a major new manned spaceflight capability at a reasonable cost." Webb gave his approval the next day, and on August 19 NASA announced that McDonnell was to be awarded a contract to build the airlock module (AM) under the direction of the Manned Spacecraft Center. It was scheduled to fly on the second Apollo Applications Program flight in 1968, attached to a multiple docking adapter (MDA), and to receive the first crew on AAP-3.

1966 August 4 The U.S. Air Force launched the *OV3–3* research satellite by solid propellant Scout from Vandenberg AFB, Calif., to a 220 × 2,780 mi, 82° orbit. Similar in size to the two previous satellites in this class, the 165-lb. *OV3–3* carried seven instruments for monitoring particle radiation in space. It was equipped with 2,560 solar cells and standard OV3 satellite support systems. *OV3–3* remains in space, currently in an orbit of 219 × 2,237 mi.

1966 August 9 The first long-tank thrust-augmented Thor with Agena D upper stage was launched from Vandenberg AFB, Calif., carrying a KH-4A military reconnaissance satellite into a 120 × 178 mi., 100° polar orbit. Known as LTTAT-Agena D, or Thorad-Agena D, the first (Thor) stage was extended to 71 ft. in length with a constant diameter of 8 ft. up to the attachment ring for the upper stage. All previous Thors were tapered from a point halfway up the stage. Three 54,000-lb. thrust Thiokol strap-on boosters augmented the first-stage thrust of 172,000 lb. Thorad-Agena D would be used with KH-4A, allowing more film to be carried for the recovery capsule due to the extra fuel and extended burn time of the launcher.

1966 August 10 *Lunar Orbiter I* was launched at 19:26:01 UT by an Atlas-Agena D from Launch Complex 13 at Cape Canaveral, Fla. The 853-lb. spacecraft comprised a central bus with a 100-lb. thrust Marquardt axial rocket motor burning UDMH–Aerozene-50 for lunar orbit insertion and four panels at 90° intervals each carrying 2,714 solar cells for electrical power. The camera system comprised a 24-in. focal length Pacific Optical Panoramic high-resolution lens and a 3-in. focal length Schneider Xenotar medium-resolution lens, each with a fixed aperture of f/5.6 and shutter speeds of 0.04, 0.02 and 0.01 sec. *Lunar Orbiter I* had a height of 6 ft. 10 in. and a span of 17 ft. 6 in. A midcourse correction maneuver of 32 sec. (124 ft./sec.) was performed about 28 hr. into the mission but an optional second maneuver was not needed. *Lunar Orbiter I* slipped into a 117 × 1,160 mi, 12.15° moon orbit at 15:34 UT on August 14 after a 9 min. 39 sec. (2,592 ft./sec.) burn. At 09:50 UT on August 21 a 22.4-sec. burn (131.9 ft./sec.) burn trimmed the orbit to 34.8 × 1,151 mi., and at 16:01 UT on August 25 a 3-sec. (17.7 ft./sec.) burn changed the orbit to 25.2 × 1,129 mi. A total 413 photographs were taken with transmission to Earth completed 20:02 UT on September 14 before *Lunar Orbiter I* impacted the lunar surface at 13:29 UT on October 29, 1966.

1966 August 17 The second of NASA's four Pioneer interplanetary science satellites, *Pioneer 7,* was launched to a heliocentric orbit by a thrust-augmented Delta from Cape

Canaveral, Fla. The 140-lb. satellite was identical to *Pioneer 6* and carried a similar suite of experiments. Whereas *Pioneer 6* was in a solar orbit inside that of the Earth, taking only 311 days to go once round the sun, *Pioneer 7* was placed in a solar path outside that of the Earth and would take 403 days. Its orbit varied from 94 million mi. and 104.5 million mi. inclined 0.097° to the ecliptic.

1966 August 19   First in a new series of high-altitude Secor geodetic satellites for the U.S. Air Force, *EGRS-7* was launched from Vandenberg AFB, Calif., piggybacked on an Atlas-Agena D flight carrying a classified satellite. Secor was placed in a 2,287 × 2,299 mi., 90.1° orbit, where it was used for U.S. Army surveying. Built by the Cubic Corp., as were all Secor satellites, *EGRS-7* weighed 45 lb. and from the high altitude orbit enabled engineers to accurately pinpoint land masses over distances of 2,200 mi. Comprising a rectangular aluminum sphere, 1 ft. 2 in. long and 10 in. high, the satellite had electronic ranging equipment. It remains in orbit. Also launched was the 11-lb. *ERS-15* satellite with five experiments in zero-g cold-bonding, a condition where, in the absence of an atmosphere, flat surfaces tend to stick together due to the lack of lubricating molecules.

1966 August 22   NASA Manned Spacecraft Center Director Robert Gilruth asked the Jet Propulsion Laboratory to fire Surveyor spacecraft vernier engines on the lunar surface to make the spacecraft hop and provide information for determining the amount of surface erosion caused by the exhaust. The MSC wanted this information to extrapolate the effects and determine the amount of erosion which could be expected from the Apollo lunar module as it landed. On June 15, 1967, Gilruth asked Manned Space Flight Director George Mueller to press the JPL for this test. It was finally conducted during the *Surveyor V* mission, launched on September 8, 1967.

1966 August 24   The 3,612-lb. Soviet spacecraft *Luna 11* was launched by SL-6 from the Tyuratam complex into an interim Earth parking orbit and from there to a 370 × 740 mi. lunar orbit inclined 27° to the moon's equator three days later. Unlike its predecessor *Luna 10*, it carried cameras, although they failed to work. The spacecraft also carried scientific equipment for studying gamma and x-ray emissions from the surface for a crude analysis of the geochemical constituents of the lunar crust. In addition, *Luna 11* studied the moon's magnetic and radiation environment, confirming that the magnetic field is only 0.001% the strength of the Earth's magnetic field. After 137 radio sessions and 277 orbits, *Luna 11* went silent when its batteries ran down on October 1, 1966.

1966 August 25   Comprising the third Saturn IB and Apollo CSM-011 weighing a total 1.326 million lb., the AS-202 mission began at 17:55:32 UT from Complex 34 at Cape Canaveral. The 44,385-lb. spacecraft was pushed to a maximum altitude of 710 mi. with a 3 min. 35 sec. burn of the service propulsion system (SPS) engine just 11 sec. after separating from the S-IVB stage. Approaching Australia, the SPS engine fired again for about 1 min. 28 sec., followed 10 sec. later by two 3-sec. burns at 10-sec. intervals to test rapid restart. Pushed to an entry velocity of 28,000 ft./sec., the command module separated from the service module and flew a typical lunar return trajectory. Dipping on a shallow approach path to a height of 218,000 ft., the CM (command module) skipped back to a height of 264,000 ft. to begin the final phase at 23,500 ft./sec. The command module splashed down at 1 hr. 33 min. 2 sec. near Wake Island in the southwest

Pacific Ocean. During ascent a total 2,158 measurements were telemetered to the ground, 863 of which related directly to the Apollo CSM (command/service module)

After considerable problems getting the spacecraft for the first manned Apollo mission ready for delivery to the Kennedy Space Center, NASA allowed North American Aviation to ship it with many unsolved problems and an environmental control system that had uncorrected anomalies. Designated CSM-012, the spacecraft was held up due to subsystem problems and could not be tested in the vacuum chamber at Cape Canaveral until mid-December. Moreover, the crew had problems getting enough time on the simulators, and preparations for the first manned Apollo mission were delayed beyond the end of the year.

1966 August 30   Concerned at the receding opportunity for getting funds to develop further uses for Apollo hardware after the initial manned lunar landings, NASA Manned Space Flight Director George Mueller sent a confidential memorandum to Kurt Debus, director of the Kennedy Space Center, asking for a study on the "potential erosion" of Apollo capabilities and the consequences should the manned space program "go out of business."

1966 September 1   NAA's Lewis Research Center awarded Aerojet General Corp. a contract for fabrication and testing of a solid propellant rocket motor 260-in. in diameter. The new motor was to be a developed version of the two solid propellant rocket motors of the same diameter previously tested by Aerojet. With a new propellant capable of a higher burning rate, it was expected to produce a thrust of 5.25 million lb.

1966 September 2   The Soviet training center selected the team that would train for moon landings. Led by Alexei Leonov, it would include cosmonauts Yuri Gagarin, Viktor Gorbatko, Yevgeny Khrunov, Andrian Nikolayev and Vladimir Shatalov. Other cosmonauts were selected for a variety of planned flights involving new generations of manned spacecraft. A team commanded by Gagarin would fly Earth-orbit Soyuz flights and include cosmonauts Bykovsky, Gorbatko, Khrunov, Kolodin, Komarov, Nikolayev and Voronov. A team commanded by Popovich would fly military satellite missions and include Artykhin, Belousov, Gubarev, Gulyayev and Kolesnikov. Lunar fly-by missions with the Chelomei launcher would be led by cosmonaut Komarov and include Bykovsky, Dobrovolski, Kolodin, Volynov, Voronov and Zholobov. Military space plane flights would be commanded by cosmonaut Titov and include Filipchenko and Kuklin. Military Almaz space station flights would be led by Belyayev and include Beregovoi, Dyomin, Lazarev, Matinchenko, Shonin, Vorobyov and Zaikin.

1966 September 7   Formation of a new European consortium to bid for satellite design and development was announced at Britain's Farnborough Air Show. The organization was to be called MESH, an acronym taking the principle letters of the four members: Matra in France, ERNO in West Germany, Saab in Sweden, and Hawker Siddeley Dynamics in the United Kingdom. The four companies were obligated to combine their experience and pool their knowledge to compete for satellite business in what was now perceived as a growth area for commercial business.

1966 September 12   The *Gemini XI* mission got under way with the launch at 13:05:02 UT of the Agena target vehicle from Complex 14 at Cape Canaveral and of the 8,374-lb. spacecraft carrying Charles Conrad and Richard Gordon at 14:42:27 UT. At insertion, *Gemini XI* was 267 mi. below and behind the Agena, which was in a 179 × 191 mi. orbit. The rendezvous was planned to occur at first apogee, and at 49 min. 58 sec. the terminal phase maneuver took place, raising the orbit of the manned vehicle to match that of the target. At 1 hr. 25 min. the two vehicles were 50 ft. apart and stationary with respect to each other; they docked 9 min. later. A 33-min. EVA (extravehicular activity) began at 24 hr. 2 min. when Gordon attached a tether between the two docked vehicles; 1 hr. later the hatch was opened again to jettison excess equipment. At 40 hr. 30 min. 15 sec. the Agena fired its main engine for 1 min. 34 sec. increasing speed by 918 ft./sec. to produce a 180 × 853 mi. orbit, the highest ever flown by a manned spacecraft in Earth orbit. The engine was fired for 1 min. 31 sec. at 43 hr. 52 min. 55 sec. to lower apogee to 189 mi. A 2 hr. 8 min. stand-up EVA began at 46 hr. 7 min., during which Conrad and Gordon momentarily drifted into sleep awaiting sunset! The vehicles undocked at 49 hr. 55 min., and 18 min. later the crew began an evaluation of tether dynamics using the line linking the two spacecraft. The tether was jettisoned at 53 hr., and 25 min. later *Gemini XI* moved away from the Agena to begin a coincident-orbit rendezvous completed at 66 hr. 40 min. *Gemini XI* splashed down at 71 hr. 17 min. 8 sec.

1966 September 14   The Jet Propulsion Laboratory made a presentation to NASA Headquarters on the Voyager advanced Mars spacecraft. As now envisaged, two spacecraft would be launched on a single Saturn V in 1973, 1975, 1977 and 1979. Landing devices would be carried on all flights with the 6,000-lb. orbiters, though not until 1977 and 1979 would an advanced biological laboratory be put down on the surface. The basic mission plan envisaged separation of the lander and its aeroshell from the orbiter after Mars orbit insertion. A retro-rocket would decelerate the aeroshell-lander by 900 ft./sec., causing it to fall toward the atmosphere, where it would use limited aerobraking. From a velocity of 1,100 ft./sec. at 20,000 ft., retro-rockets would fire to slow it to 350 ft./sec. At this point the aeroshell would separate. Rockets on the lander would slow it for touchdown on the surface at 10–25 ft./sec.

1966 September 15   HMS *Resolution*, the first of four 8,400-ton (dived) submarines built in Scotland to carry the British nuclear deterrent in a total of 64 Polaris missiles, was launched at Barrow-in-Furness. With a complement of 143, including 13 officers and 130 ratings, the submarine performed sea trials between June 22 and August 17, 1967. It was commissioned on October 2, 1967, and departed on its first patrol during June 1968. The second submarine, HMS *Renown*, was launched on February 25, 1967, followed by HMS *Repulse* on November 4, 1967, and HMS *Revenge* on March 15, 1968. The Royal Navy's Polaris A3 missiles were initially fitted with three 200-KT W-58 multiple reentry vehicles.

1966 September 16   A new generation of U.S. Defense Meteorology Satellite Program satellites began operations with the first launch of a Thor-Burner II from Vandenberg AFB, Calif., to a 422.5 × 542 mi. orbit inclined 98.8°. Designated DMSP Block IVA, four of these military weather satellites would be launched by the Air Force, the last on October 1, 1967. Built by RCA, each spin-stabilized satellite comprised a cylinder with 16 flat sides supporting solar cells producing an average power of 20 W. Block IVA had a height of 2 ft. 6 in.,

a diameter of 2 ft. 8 in. and an orbital weight of 180 lb. including two vidicon cameras. Originally the retro-motor from the Surveyor moon lander, the Burner II stage had a height of 5 ft. 8 in., a diameter of 5 ft. 5 in., a weight of 1,780 lb. and a thrust of 10,000 lb. Other DMSP Block IVAs were launched on February 8, August 23 and October 11, 1967; May 23 and October 23, 1968; July 23, 1969; February 11 and September 3, 1970; and February 17 and June 8, 1971.

1966 September 17   Developed from the formidable SS-9 Scarp ICBM, the first Soviet SL-11 satellite launcher lifted off from Baikonur. Developed by the Yangel bureau, the SL-11 had a first stage powered by six motors with a combined thrust of 618,000 lb. and a second stage with a 199,000-lb. thrust RD-219 engine. Both stages burned UDMH/N_2O_4 propellants. The SL-11 flew a trajectory inclined 49.6°, but the payload broke up into 53 pieces. The carrier-stage remained in a 101 × 650 mi. orbit and decayed November 11. A second flight by SL-11 on November 2 was launched to a 87 × 531 mi. orbit, with the payload section breaking into 41 pieces before the carrier reentered on November 29. Also dubbed Tsyklon, the launcher would fly more than 120 times by mid 1995, exhibiting a 97% success rate.

1966 September 20   The 2,204-lb. *Surveyor II* spacecraft was launched on a direct ascent trajectory to the moon by Atlas-Centaur AC-7 at 12:32:00 UT from Cape Canaveral, Fla. The Centaur main engine cutoff came at an elapsed time of 11 min. 23 sec., and after deploying landing legs and antenna the spacecraft separated from the Centaur 1 min. 20 sec. later. All went according to plan until the 9.8-sec. (31.5 ft./sec.) mid-course burn began at 05:00:02 UT on September 21, when one of the three vernier motors failed to fire and the spacecraft began to tumble at a rate of 1.22 rev./sec. An attitude control system feeding 4.5 lb. of nitrogen gas to three pairs of jets was used by all Surveyor spacecraft. This was employed to reduce the tumble rate to 0.97 rev./sec., but with no electrical power available from the solar cells the life of the spacecraft was limited. Beginning at 07:28:25 UT on September 21, in efforts to get the third vernier working, two 2-sec. burns were made, followed by five 0.2-sec. burns at 5-min. intervals and another 2-sec. burn. The cycle of five short burns and one long burn was repeated four times followed by a 2.5-sec. burn, five 0.2-sec. burns and one 21.5-sec. burn, the 39th and last firing occurring at 08:05:12 UT on September 22. The firings had increased the tumble rate to 2.43 rev./sec. With electrical power being sapped the solid propellant retro-motor was fired at 09:34:28 UT on September 22 and all telemetry stopped 32 sec. later.

Secretary of the Interior Stewart L. Udall announced that the United States would develop a satellite called EROS (earth resources observation satellite) equipped with cameras for transmitting digital images of the Earth for remote-sensing purposes. Secretary Udall said, "Facts on distribution of needed minerals, our water supplies and the extent of water pollution, agricultural crops and forests and human habitations, can be used for regional and continental planning." Fearing the use of Earth images for spying, the Department of Defense had resisted the unrestricted distribution of remote sensing data, and a compromise was worked out restricting the resolution to greater than 150 ft.

1966 September 23   A series of associated working groups at NASA Headquarters provided recommendations on future manned moon activity. Dividing it into four separate phases, the groups judged that the first phase of lunar exploration involving Ranger, Surveyor, Lunar Orbiter and initial manned landings was well under way. Phase 2, they said, could be realized through the Apollo Applications Program and call for two 14-day lunar module surface missions, one or two Surveyors and a Lunar Orbiter flight each year from 1970. Phase 3 would consist of one 90-day lunar surface mission involving three astronauts, and Phase 4 would establish semipermanent bases.

1966 September 26   Japan's first attempt to launch a satellite failed when the solid propellant Lambda 4S-1 rocket strayed off course and was destroyed. The first stage had a thrust of 80,000 lb., the second stage a thrust of 30,000 lb., the third stage a thrust of 15,400 lb. and the fourth stage a thrust of 2,420 lb. The 55 ft. long solid propellant launcher weighed 17,000 lb. and carried a 57.2-lb. satellite instrumented to measure the ionosphere and send back data. The launch attempt was made from the Uchinoura Range on Kyushu Island for the University of Tokyo's Kagoshima Space Center. A second attempt, with Lambda 4S-2 on December 20, failed when the fourth stage refused to start, and a third attempt on April 13, 1967, was equally unsuccessful when Lambda 4S-3 failed to reach orbit.

1966 September 29   NASA announced the names of crew members for the second manned Apollo flight. Astronauts Walter Schirra, Donn Eisele and Walter Cunningham would comprise the primary crew on this 14-day Earth orbit mission scheduled for 1967. The backup crew would comprise astronauts Frank Borman, Thomas Stafford and Michael Collins. The mission would be launched by Saturn IB and serve to extend the in-orbit qualification of command and service modules started with the first manned Apollo mission.

1966 October 11   Talks in Moscow between Soviet and French space officials were concluded with an agreement to cooperate in the scientific research of space by satellites. In a specific proposal the French wanted to develop a satellite to be launched by the USSR from a Soviet launch site. Prior to that, French experiments would be designed to piggyback on Soviet Molniya communication satellites.

1966 October 20   The fourth Molniya communication satellite, *Molniya 1D*, was launched by SL-6 from Baikonur cosmodrome and was equipped for the first time with a television camera. From an orbit of 314 × 24,660 mi., at an inclination of 64.9° the satellite conducted observations of clouds using wide-angle and narrow-angle lenses and optional filters. The camera could be slewed in azimuth and follow the Earth on a step-by-step basis. On November 9 viewers in the far eastern territories of the Soviet Union watched programs relayed from Moscow by *Molniya 1D*.

1966 October 22   The 3,572-lb. Soviet spacecraft *Luna 12* was launched by SL-6 and placed in a 87 × 746 mi. orbit inclined 10° to the lunar equator three days later. The first pictures of the surface returned by a Soviet spacecraft from lunar orbit were transmitted by facsimile on October 29, 1966. Unlike *Luna 10*, *Luna 12* retained its propulsion package and side modules to facilitate orbit maneuvers and picture transmission. A large radiator covered the upper portion of the spacecraft. Each picture was composed of 1,100 lines providing a resolution of 50–65 ft., but only three pictures were made public. After 602 orbits and 302 communication sessions, *Luna 12* went dead on January 19, 1967.

1966 October 26   NASA launched the *Intelsat II F-1* (*Lani* [Hawaiian for *heavenly*] *Bird*) communication satellite by thrust-augmented Delta E from Cape Canaveral, Fla., to a geosynchronous transfer ellipse of 186 × 23,393 mi. inclined 26.4°. At ninth apogee on October 30 the kick motor fired but blew off its cone shortly after ignition, leaving the satellite in a 2,072 × 23,330 mi. orbit inclined 17°. The 357-lb. satellite weighed 192 lb. after apogee-motor jettison and comprised a hatbox cylinder with a height of 4 ft. 8 in. and a diameter of 4 ft. 8 in. The 85-W power supply was provided by 12,756 solar cells around the exterior. Capable of handling 240 voice circuits with a 15-W transmitting power, *Intelsat II F-1* was to have been placed in geostationary orbit at 175° W. On November 8 it fired hydrogen peroxide thrusters to change the orbit to 2,609 × 23,306 mi. Between December 3, 1966, and February 2, 1967, the Comsat Corp. conducted a limited commercial service eight to nine hours each day.

1966 October 28   The U.S. Air Force *OV3-2* science satellite was placed by Scout launcher in a 198 × 993 mi., 82° orbit to measure electron density and charged particle density changes during a solar eclipse visible in the southern hemisphere. *OV3-2* weighed 177 lb. and returned a considerable volume of useful information prior to its demise in the Earth's atmosphere on September 29, 1971. Only one OV3 out of six failed to reach orbit: the 207-lb. *OV3-5* launched on January 31, 1967, when the Scout malfunctioned.

1966 October   Bringing together ad hoc studies that had been going on for more than two years, the U.S. astronautics journal *Spacecraft and Rockets* published an article by Jerry M. Deerwester describing how the gravitational field of Jupiter could be used to reduce transit times to Saturn, Uranus, Neptune and Pluto. A range of technical papers appeared in journals throughout the year indicating a gathering interest in a gravity-assist mission. Launch vehicles capable of accelerating space probes to the outer planets in reasonable journey times were not available. A direct flight to Pluto, for instance, would take 41 years. By using the gravitational slingshot effect of the outer planets that time could be cut to 8–9 years.

1966 November 2   NASA's Associate Administrator for Manned Space Flight George Mueller announced to the NASA field centers that he now had sufficient information on the Apollo Applications Program to put together a flight schedule for noninterference with moon mission plans. Utilizing Saturn IB flights in 1968, AS-209 would be launched with airlock module (AM) and multiple docking adapter (MDA) and receive a three-man crew in an Apollo CSM (command/service module) launched on AS-210. AS-211 would carry the Apollo telescope mount (ATM) into orbit, and this would be docked to the assembled configuration. The S-IVB from AS-209 would be fitted out for habitation by the crew, who would remain for 28 days. After a brief interval the AS-212 crew would occupy the orbital workshop for 56 days.

1966 November 3   A U.S. Air Force Titan IIIC/Transtage launch vehicle successfully tested elements of the Manned Orbiting Laboratory and the first OV4-series satellites. Launched from Cape Canaveral, the simulated MOL (designated *OV4-3*), comprised a modified Titan II oxidizer tank with a length of 38 ft. 4 in., a diameter of 10 ft. and a weight of 21,350 lb. A Gemini spacecraft adapter was carried on front, supporting the *Gemini II* spacecraft previously flown on a ballistic reentry test and now designated *Gemini B*. Inside the adapter were three subsatellites: *OV1-6* (445 lb.), classified; *OV4-1T* (240 lb.), transmitter for the "whispering gallery" experiment; and *OV4-1R* (300 lb.), the receiver for *OV4-1T*. "Whispering gallery" was the name given to an experiment for using the ionosphere's F-layer to send high-frequency and very-high-frequency signals between satellites where one was over the horizon to the other. Following Transtage shutdown at 104 mi., *Gemini B* separated for a ballistic test of a special access door cut in the base heat shield. *Gemini B* splashed down in the Atlantic 5,500 mi. downrange, the first spacecraft relaunched and recovered. The second Transtage burn pushed the remaining stack to an elliptical path from which a third burn placed it, and *OV4-3*, in a 185 × 189 mi., 33° orbit. The subsatellites were ejected from the adapter, and a small motor on *OV4-1T* fired to separate it from *OV4-1R* and place it in a 183 × 199.5 mi. orbit, versus 181 × 185 mi. for *OV41-R*; *OV1-6* was in a 180 × 183 mi. path. Transtage put *OV4-3* into a slow tumble for even solar heating; nine on-board experiments were thus conducted; it decayed on January 9, 1967. *OV1-6* decayed on December 31, 1966, *OV4-1R* on January 5, 1967, and *OV4-1T* on January 11, 1967.

1966 November 6   At 23:21:00 UT *Lunar Orbiter II* was launched by an Atlas-Agena D from Cape Canaveral, Fla., to Earth escape velocity. An 18.1 sec. (69.2 ft./sec.) midcourse maneuver was performed at 19 hr. 54 min. on November 8, and a 10 min. 12 sec. (2,722 ft./sec.) burn which began at 20:26:37 UT on November 10 put *Lunar Orbiter II* into an initial orbit about the moon of 122 × 1,163 mi. with a selenographic inclination of 11.97°. To set up the geometry of the orbit for optimum photographic requirements an orbit-transfer burn of 17.4 sec. (92.2 ft./sec.) put the spacecraft into a 30.9 × 1,151 mi. orbit inclined 11.89°. A total 422 photographs were taken of 13 primary and 17 secondary sites, but only 411 were received due to a traveling-wave-tube failure in the transmitter on December 7. *Lunar Orbiter II* impacted the moon at 07:17 UT on October 11, 1967.

1966 November 9   For the first time, a helicopter was used to snatch in midair a package of instruments descending to Earth by parachute following a launch by sounding rocket. The event took place near Wallops Island, Va., just after an Argentinian *Orion 2* sounding rocket had fired a 35-lb. payload to a height of 60 mi. The single-stage solid propellant rocket had a length of 12 ft., a diameter of 8 in. and a launch weight of 286 lb.

1966 November 11   The *Gemini XII*–Agena target vehicle was launched from Cape Canaveral's Complex 14 at 19:07:09 UT into a 183 × 188 mi. orbit inclined 28.86°. Astronauts James Lovell and Edwin Aldrin were launched in pursuit from Complex 19 at 20:46:33 UT. After nine maneuvers the two spacecraft were together by 3 hr. 46 min., rendezvous having been achieved at third apogee, and docked at 4 hr. 14 min. At 7 hr. 5 min. 6 sec. a 44 ft./sec. (51 sec.) burn of the Agena secondary propulsion system set up the orbit for photography of a solar eclipse, and a second 17 ft./sec. (18 sec.) phasing maneuver at 15 hr. 16 min. 18 sec. kept the spacecraft on track for that event. A 2 hr. 29 min. stand-up EVA (extravehicular activity) began at 19 hr. 29 min. elapsed time when Aldrin installed a hand rail between the Gemini and Agena and took pictures. At 42 hr. 48 min. a full EVA began in which Aldrin moved to the Agena docking cone and attached a 100-ft. tether for dynamic tests before evaluating hand holds and body restraints in the adapter. The EVA ended when the hatch was closed at 44 hr. 55 min. The two spacecraft were

undocked at 47 hr. 23 min., and tether evaluations continued until it was jettisoned at 51 hr. 51 min. A 55-min. stand-up EVA began when the hatch was opened at 66 hr. 6 min. *Gemini XII* splashed down at 94 hr. 34 min. 31 sec., ending the flight phase of the Gemini program.

1966 November 16   A Soviet moon landing plan devised by the Chelomei bureau was considered at a review meeting of the N-1/L-3 program devised by Korolev. A massive four-stage booster designated UR-700, with a liftoff thrust of 12 million lb., would propel a two-person spacecraft designated LK-700 to moon orbit, from where it would execute a lunar landing. The upper component of the lunar lander would have comprised the Earth-return spacecraft. The UR-700 would have used eight 1.5-million-lb. thrust RD-270 engines in the first stage and two in the second stage, with six 132,520-lb. RD-0210 engines from the Kosberg bureau in the third stage and four in the fourth stage. Adapted from the LK-1 circumlunar spacecraft, the LK-700 concept got the support of Valentin Glushko as a quicker way of beating American astronauts to the lunar surface. It was not approved, and the Korolev bureau continued with the N-1/L-3 program.

1966 November 20   The first sounding rocket launches from the European Space Research Organization's new range at Kiruna in Sweden, took place when a French Centaure was fired to a height of 78 mi. Three others were also launched this day. Kiruna would develop into a routine launch site for scientific probing of the outer atmosphere, equivalent to NASA's facility on Wallops Island, Va., but without the orbital launch option. Kiruna would continue as the primary northern sounding rocket complex with the European Space Agency, once that was formed.

1966 November 22   In a relatively obscure reply to a written question in the House of Commons, Britain's Defense Minister Denis Healey announced that the United Kingdom would buy two IDSCS communication satellites from the United States. The satellites, said Mr. Healey, were to provide secure communications between the United Kingdom and Australia and were not to have any reconnaissance function. The United States would be paid to launch the satellites. Implicit in this was rejection of a British national launch vehicle, the development of which was considered uneconomic for the limited requirements envisaged. On December 13 this position was reversed when the government announced it was to proceed with the Black Arrow satellite launcher derived from the Black Knight test vehicle.

1966 November 26   Under the designation *Cosmos 133*, the first Soyuz spacecraft was launched by SL-4 from the Baikonur cosmodrome to a 108 × 137 mi. orbit inclined 51.8° to the equator. Soyuz spacecraft comprised three elements linked together: a drum-shaped instrument module 7 ft. 6 in. in length, 7 ft. 6 in. in diameter and weighing 5,680 lb., including 1,665 lb. of propellant and a KTU-35 propulsion system for orbital maneuvering; a bee hive–shaped descent module with a length of 7 ft. 2 in., a base diameter of 7 ft. 6 in. and a weight of 5,900 lb., as well as a pressurized oxygen-nitrogen atmosphere and three couches for cosmonauts in track suits; and a spheroid-shaped orbital module with a length of 8 ft. 8 in., a diameter of 7 ft. 4 in. and a weight of 2,200 lb. The Soyuz instrument module had a KTDU-35 propulsion system consisting of a 920-lb. thrust primary motor and a 905-lb. thrust secondary motor. The assembled spacecraft had a length of 24 ft. 4 in., weighed about 13,780 lb. and could, on top of the orbital module, carry a cylindrical docking module with a length of 3 ft. 11 in., a diameter of 5 ft. 6 in. and weighing 720 lb. The *Cosmos 133* mission lasted 1 day 23 hr. 21 min., but on returning through the atmosphere the descent module heat shield was very badly burned and the spacecraft was very nearly destroyed.

1966 November 27   Built by Nord Aviation, the French Coralie rocket made its first flight test from Hammaguir, Algeria. Developed as the second stage to the Europa launcher, Coralie was modified for off-the-pad flight tests and called Cora. In this configuration it had different nozzles. The stage burned UDMH-N_2O_4 propellants, and its L10 engine had a thrust of 61,740 lb. The flight was technically a failure because an electrical short circuit swung a motor nozzle out of alignment, causing the stage to tumble and burst. The second flight from Hammaguir was also a failure. After this the Algerians insisted that the French leave their country, and tests were moved to Landes on the French Atlantic coast.

1966 November 29   Donald Slayton, chief of Flight Crew Operations at NASA's Manned Spacecraft Center, circulated a memorandum changing Apollo crew designations. From this date the Block II spacecraft crew assignments became commander (CDR), command module pilot (CMP), and lunar module pilot (LMP), replacing the categories of command pilot, senior pilot and pilot used for Block I missions. Ordinarily, the commander would occupy the left position and the lunar module pilot the right position in the command module and the lunar module. The command module pilot would occupy the center couch in the command module for launch, but this was interchangeable with the commander's position on his approval. On June 14, 1967, the Block II designations became standard for all crew positions on all Apollo missions.

1966 December 7   NASA launched the first Applications Technology Satellite *(ATS-1)* by Atlas-Agena D from Cape Canaveral, Fla., to a 10.5° geosynchronous orbit over 150.2° W. The 1,550-lb. satellite took the form of a cylinder with a diameter of 4 ft. 9 in. and a height of 4 ft. 5 in., and weighed 670 lb. on station. *ATS-1* carried two communications experiments, two meteorological experiments, three control and stabilization devices on test and seven environmental monitoring experiments. Weight in orbit after apogee motor burnout was 775 lb. Power from 23,870 solar cells delivered 185 W, and four 2-W transmitters sent data to Earth. Designed for a life of three years, *ATS-1* finally failed on March 29, 1985, having been used extensively to link Public Health Service physicians with trained health aides in remote Alaskan bush country.

Flight assignments for lunar fly-by missions, first given to Soviet cosmonauts on September 2, were changed and would now include five civilian candidates selected by the OKB-1 design bureau headed by Vasili Mishin. The civilians were Georgi Grechko, Valery N. Kubasov, Oleg G. Makarov, Aleksandr A. Volkov and Aleksei S. Yelisayev. Military members of the fly-by team now included Georgi T. Beregovoi, Valery F. Bykovsky, Yuri Gagarin, Yevgeny V. Khrunov, Vladimir M. Komarov, Alexei A. Leonov, Andrian G. Nikolayev, Vladimir A. Shatalov and Boris V. Volynov. By mid-1967, after several more crew changes, three teams had been selected for what was known as the 7K-L1 mission: Leonov and Makarov were the primary team for the lunar fly-by, with Bykovsky and Rukavishnikov as backups and Popovich and Sevastyanov in reserve. The lunar fly-by launch had been scheduled for July 26, 1967, but the *Soyuz 1* disaster

on April 24, 1967, and subsequent launcher failures prevented this.

1966 December 11   Two OV1-series research satellites were orbited by an Atlas D sent on a ballistic trajectory from Vandenberg AFB, Calif. Both standard OV1 configurations with booster rocket motor, the 290-lb. *OV1-9* carried three radiation experiments on tissue similar to that found in humans, while the 287-lb. *OV1-10* carried eight space radiation experiments and a gravity-gradient stabilization system incorporating eight booms. *OV1-9* went into a 297 × 3,004 mi., 93° orbit where it would traverse the lower Van Allen belts. *OV1-10* was placed in a 403 × 479 mi., 93° orbit. Both remain in space.

1966 December 13   The first meeting of the European Space Conference, in Paris, France, confirmed that a new Committee of Alternatives would begin work immediately on unification of European space organizations. It was from the work of the ESC that the European Space Agency would be set up in 1974 through the merger of ESRO and ELDO activities. ESC meetings were held annually to decide on the future course of European space policy.

1966 December 14   The first of three recoverable biological laboratories was launched by NASA when a thrust-augmented Delta E carried *Biosatellite I* to a 183 × 192 mi. orbit inclined 33.5°. The 940-lb. satellite comprised a blunt cone reentry vehicle, 3 ft. 4 in. in diameter, carrying 13 experiments involving bacteria, common bread mold, flowering plants, a flour beetle, a parasitic wasp, a fruit fly, wheat seedlings, frog eggs and an amoeba in a 6 cu. ft. payload section. It was attached to an adapter section for a total length of 6 ft. 9 in. and a maximum diameter of 4 ft. 9 in. At the end of the three-day mission, the reentry capsule separated but the retro-rocket failed to fire, stranding it in orbit. Although an extensive search was made for the capsule when its orbit decayed and it reentered on February 15, 1967, it was never found.

1966 December 19   The General Assembly of the United Nations endorsed the "Treaty on Principles Governing Activities of States in the Exploration and Use of Outer Space, Including the Moon and Other Celestial Bodies," generally known as the Outer Space Treaty. Under its Articles, signatories were permitted to conduct passive military activities in space but not to orbit weapons of mass destruction, only to carry out activities that would not affect another country, and were prohibited from setting up any form of military outpost on the moon or the planets.

1966 December 21   The Soviet spacecraft *Luna 13* was launched by an SL-6 from Tyuratam, Kazakhstan, to an initial Earth parking orbit and from there to a lunar intercept trajectory. In a procedure almost identical to that followed by *Luna 9*, the 3,506-lb. spacecraft put its hard-lander capsule on the lunar surface at 18:01 UT on December 25 in the Oceanus Procellarum at 18.9° N by 62.05° W. Data transmissions began 4 min. 30 sec. later after petal covers had opened to stabilize the 230-lb. capsule and two extendible arms deployed a distance of almost 5 ft. from opposite sides of the spherical capsule. One arm carried a ground penetrometer to measure the strength of the lunar soil; the other carried a radiation densitometer. Thus did *Luna 13* become the first spacecraft to directly sample the surface of another world. A TV panorama was transmitted to Earth before the spacecraft batteries ran out at the end of the month.

A U.S. Air Force Atlas successfully launched the first of three Martin SV-5D lifting-bodies to a ballistic trajectory from Vandenberg AFB over the Pacific Ocean. Although it performed as expected during the descent, it was not recovered. The SV-5D was one of four flat-bottomed research vehicles ordered by the Air Force to study reentry and maneuverability of wingless shapes. With a length of 6 ft. 8 in. and a span of 4 ft. across upswept tail fins, the SV-5D had a weight of 894 lb. It was part of the U.S. Air Force PRIME (precision recovery including maneuvering entry) program where SV-5Ds would be subjected to 16,000 MPH reentry from a maximum altitude of 500,000 ft. The second SV-5D was launched on March 5, 1967, and successfully demonstrated hypersonic fight control. The third was launched on April 18, 1967, and was successfully recovered off Kwajalein Island in the Pacific. The fourth was never flown.

1966 December 22   NASA announced a new manned Apollo flight schedule and new crew assignments. Because of technical problems preparing Apollo CSM-012 for the first manned mission, designated AS-204, that flight carrying astronauts Virgil I. Grissom, Edward H. White II and Roger B. Chaffee would be scheduled for no earlier than February 1967. The second manned mission, originally planned as a repeat of the first and carrying astronauts Walter M. Schirra, Donald F. Eisele and Walter M. Cunningham, would be deleted in favor of moving the dual mission of a lunar module and a command/service module into that slot. Originally

designated AS-207/208, it would now be designated AS-205/208. For that dual mission, LM-2 and CSM-101, the first Block II spacecraft, would be launched on separate Saturn IBs in August 1967. The crew for that mission would be astronauts James McDivitt, David Scott and Russell Schweickart with Thomas Stafford, John Young and Eugene Cernan as backups. The third manned Apollo mission was to have been flown by astronauts Frank Borman, Michael Collins and William Anders with Charles Conrad, Richard Gordon and Clifton Williams as backups. Scheduled for the end of 1967, the mission would involve the first all-up LM/CSM (lunar module/command service module) flight in an Earth orbit lunar landing mission rehearsal with an apogee of 4,000 mi.

1966 December 23   NASA Administrator James Webb approved the establishment of a new Science and Applications Directorate at the Manned Spacecraft Center. From this date, the MSC would through this directorate focus its lunar science planning for Apollo, applications programs calling for Earth-science studies from orbiting space stations and future mission planning involving space science. This would also provide a conduit for work with Bendix on development of the Apollo Lunar Surface Experiments Package of instruments left on the lunar surface by astronauts.

1966 December 24   A Soviet state commission discussed the first formal flight plan for Sergei P. Korolev's UR-500K/L-1 manned circumlunar mission proposal and decided that be-

A simulated representation of Russia's Luna 13 *spacecraft on the moon. Key: 1) petal antennas; 2) pintle antennas; 3) extendible instrument arm; 4) bearing strength sensor; 5) radiation monitor; 6) TV camera.*

cause Chelomei's three-stage UR-500K (Proton) rocket had not yet flown, it would be prudent to run the mission on two launchers. The first would place the L-1 Soyuz derivative in orbit with the first launcher followed by the crew in a standard Earth-orbit Soyuz. The two-person crew would spacewalk to the L-1 assembly and its attached Block-D (third) stage, leaving their Soyuz to return to Earth. Block-D would then boost the L-1 to circumlunar flight. The mission schedule envisaged four unmanned UR-500K/L-1 test flights in the first half of the coming year, with the first manned flight in June 1967. The first demonstration of a Block-D stage firing out of Earth orbit came under the veiled designation of *Cosmos 146* on March 10, 1967.

1966 December 30   NASA and the European Space Research Organization (ESRO) signed an agreement under which launch services would be provided on a reimbursable basis. ESRO wanted to launch several scientific satellites and had no independent launch vehicle to send them into space, at least not until the Europa family of ELDO launchers became operational. ESRO's HEOS-A (Highly Eccentric Orbit Satellite-A) interplanetary physics satellite, scheduled for launch in late 1968, would be the first of the reimbursable launches. NASA would not agree to launch European satellites that might compete with U.S. commercial interests in communication programs, and this would eventually prompt Europe into a fully independent launch vehicle program.

1967 January 11   A thrust-augmented Delta launched *Intelsat II F-2* from Cape Canaveral to a geosynchronous transfer orbit from which it was eventually maneuvered to 178° W by February 4. Identical to the failed *Intelsat II F-1*, *F-2* was the first of a new generation of operational Intelsat carriers managed by the Comsat Corp. and served the commercial Pacific traffic between Australia, Japan, Hawaii and the Washington state as well as military traffic between Hawaii and the Philippines, Thailand and Japan. *Intelsat II F* was launched March 23, 1967, and positioned over the Atlantic Ocean for service from April 7. The third and last Intelsat II series, *F-4*, was launched on September 28, 1967, and subsequently positioned over the Pacific at 174° E for service commencing November 4.

The fiercest interplanetary magnetic storm recorded since the start of space science began with a flare erupting on the sun at 02:00 UT. The distortions and disruptions in the Earth's magnetosphere were observed by five NASA satellites at various locations. With a velocity of 435 mi./sec., the shock wave reached *Explorer 33*, 260,000 mi. from Earth, 58 hr. later. *Vela 5* recorded the onset from a distance of 75,000 mi., while *ATS-1* in geostationary orbit at 22,000 mi. measured the flattening of the Earth's magnetosphere to at least 5,000 mi. below that altitude. Other satellites involved in measuring the phenomenon were *Vela 6* and *OGO 3*.

1967 January 23   A software discrepancy in the *Apollo 204* computer program was reported, indicating that serious differences would result in some critical calculations. In the most severe case, ground-based solutions would indicate a retrofire time differing by 2 min. 18 sec. from solutions produced in the on-board computer. The preparation and validation of Apollo software had become a critical pacing item. The ultimate success of Apollo would rest more on the extra work put into the preparation and checking of software programs after the *Apollo 204* fire than anything else.

1967 January 25   The first in a series of nine Soviet depressed-trajectory warhead tests was launched to a 89 × 130 mi., 50° orbit by the third SL-11 from Baikonur cosmodrome. Designated *Cosmos 139*, the flight was a test of what the Americans called a "fractional orbit bombardment system," or FOBS. It followed two launch failures of this system on September 17 and November 2, 1966. The terminal-stage simulated warhead weighed about 11,000 lb. FOBS was a technique whereby the normal ballistic flight of an ICBM was extended by placing the warhead in orbit and commanding it down before one complete revolution of the Earth. The purpose of this was to approach the United States from the south, where radar early warning was at its weakest. Other FOBS tests took place on May 17, July 17, July 31, August 8, September 19, September 22, October 18 and October 28, 1967.

1967 January 27   Apollo AS-204's primary crew members Virgil Grissom, Edward White and Roger Chaffee were killed when a fire broke out in their spacecraft (CSM-012) atop the fourth Saturn IB launch vehicle on Complex 34 at Cape Canaveral. The three astronauts entered the command module at about 18:00 UT for a "plugs-out" test, in which the spacecraft would simulate a countdown on internal power, as a rehearsal for a planned launch on February 21. Inner pressure and outer thermal hatches on the command module were sealed at 19:45 UT. In an emergency it would take 90 sec. to undo the hatches and get out. At 23:30:55 UT an electrical short circuit down the lower left interior near the environmental control system started a fire that was reported by Grissom 10 sec. later from his position on the left couch. A further report of fire came 12 sec. after that, and 2 sec. later, at 23:31:19 UT, increased pressure burst the command module, causing fire and smoke to rush over the couches and billow out from the lower right section of the spacecraft. All voice and data ceased 3 sec. later, and 7 sec. after that (24 sec. after the first fire call) the atmosphere inside was lethal.

NASA signed the Voyager project approval document formally setting up the advanced Mars orbiter/lander spacecraft for dual launches by Saturn V. Seeking ways of distributing work to as many NASA centers as possible in following the Apollo program, the Marshall Space Flight Center was given management of the spacecraft and the Saturn V launch vehicle. The Jet Propulsion Laboratory and the Langley Research Center would work on development of the lander. On January 31, 1967, NASA requested proposals from industry for a Phase B development contract, and on March 2, 1967, proposals were submitted by Grumman, Hughes, Martin Marietta and McDonnell Douglas. On May 17, 1967, Martin Marietta and McDonnell Douglas received 90-day design study contracts to start June 1.

1967 January 31   A flash fire in a pure oxygen environment in the altitude chamber at Brooks AFB, Tex., killed Airman 2c William F. Bartley Jr. and Airman 3c Richard G. Harmon. The event was immediately transmitted to the Apollo 204 Review Board set up to investigate the fire that had taken the lives of three astronauts four days earlier. The next day NASA's Manned Spacecraft Center ordered all manned tests with pure oxygen environments suspended "until further notice."

1967 January   The European Space Research Organization (ESRO) awarded a contract for the development of two satellites to the MESH consortium. Designated *TD-1* and *TD-2,* they were to weigh approximately 880 lb. and be sent to orbit from the United States by Delta in 1970 and 1971. The two satellites would study the spectrum of stars and perform observations of the effects of solar activity on the Earth's upper atmosphere. Over the next several months the size of each satellite increased and weight grew to almost 1,000 lb. The estimated cost of the project grew threefold, and on April 25, 1968, ESRO announced that it had to cancel the project due to opposition from the Italians, who were helping to pay for what was at this time the most ambitious ESRO project on the books.

1967 February 1–2 A Gemini Summary Conference was held at NASA's Manned Spacecraft Center; 22 papers reviewed the remarkable accomplishments of the two-man space program. In 10 highly successful manned flights, U.S. experience as measured by man-hours in space had increased from the 53 hr. 55 min. 27 sec. of the Mercury program to a total 1,993 hr. 37 min. 19 sec., demonstrating flights of up to two weeks, space walking, biomedical evaluation, rendezvous and docking, and the flexibility to launch at two-month intervals and to stack and prepare spacecraft for flight within a matter of days. The Gemini program had propelled the United States from playing catch-up with the Soviet Union to a position of clear and unequivocal leadership in manned flight operations during a period when the USSR failed to launch one manned spacecraft.

1967 February 3 NASA Director of Manned Space Flight George Mueller provided details of Apollo flights rescheduled in light of the January 27 fire in CSM-012. All manned flights had been halted pending the report of the Apollo 204 Review Board. Unmanned testing of launch vehicles and Apollo hardware would continue, however. The next flight would be the unmanned lunar module mission AS-206 on a Saturn IB, followed by the first Saturn V flight, AS-501. Both launches were scheduled for the second quarter of 1967. The second flight of a Saturn V, AS-502, was to occur in the second half of the year. Both Saturn V missions would perform S-IVB (third stage) restart and Apollo service propulsion system engine burns to throw the spacecraft into a highly elliptical path from which high-speed reentry would be demonstrated. The Apollo contractors were briefed to assume that when it came, the first manned Apollo mission would utilize a Block II CSM (command/service module). On March 20, 1967, NASA announced that the first unmanned LM flight would use the AS-204 launcher.

1967 February 5 At 01:17:01 UT *Lunar Orbiter III* was launched by Atlas-Agena D from Cape Canaveral, Fla. At 15:00 UT the same day a 4.3-sec. (16.7 ft./sec.) midcourse maneuver was conducted, and lunar orbit insertion began at 21:54:19 UT on February 8, when the propulsion system was ignited for a 9 min. 2.5 sec. (2,311 ft./sec.) burn, dropping the spacecraft into a 131 × 1,120 mi., 20.94° orbit. *Lunar Orbiter III* was transferred to its final orbit of 34.1 × 1,148 mi. inclined 20.91° with a 33.7-sec. (166 ft./sec.) burn beginning at 18:13:26.6 UT on February 12. During 54 lunar orbits a total of 211 dual frame (wide-angle and telephoto) exposures were made of 12 primary and 31 secondary sites. Of the 422 frames shot, 275 were recovered, but 147 were lost due to a failure of the film advance motor. The final readout was on March 2, 1967, and from these images eight candidate Apollo landing sites were selected. *Lunar Orbiter III* impacted the moon at 10:27 UT on October 9, 1967.

1967 February 6–9 A conference was held at the U.S. Air Force School of Aerospace Medicine, Brooks AFB, Tex.,

during which Dr. Hubertus Strughold recommended that astronauts on very long distance flights undergo "prophylactic surgery" to prevent appendicitis or gall bladder attacks. Dr. Alfred C. Koestler described experiments he had been conducting on primates in which 18 chimpanzees had survived rapid decompression to a simulated altitude of 150,000 ft. with full recovery within 4 hr. During World War II, Strughold had conducted horrific experiments and tests from mobile laboratories on human beings condemned to death in extermination camps run by the Nazis. He was later brought from Germany to build a school of aerospace medicine in the United States; his past was covered up as much as possible.

1967 February 7 The second test flight of a Soyuz spacecraft began when an SL-4 launch vehicle carried the 13,900-lb. *Cosmos 140* from the Baikonur cosmodrome to a 106 × 146 mi. orbit inclined 51.7°. The flight lasted 1 day 11 hr. 32 min. Not all went as planned during the mission. A failure in the temperature control system would have rendered the descent module almost uninhabitable had the spacecraft been carrying cosmonauts. Two Soyuz configurations were being simultaneously prepared for unmanned flight tests: the Earth-orbit qualification vehicle called Soyuz, equivalent to NASA's Apollo Block I, and hardware for the definitive lunar variant which would be disguised under the program name Zond, the equivalent of Apollo Block II. Soyuz would be launched by SL-4, Zond by SL-12.

1967 February 8 A French Diamant launcher carried the 51-lb. satellite *Diadème 1* from Hammaguir, Algeria, to a 363 x 833 mi. orbit inclined 40°. Designated D-1C, this was the third of four French satellites launched on the basic Diamant, subsequently designated Diamant A, and was instrumented to conduct diagnostic measurement of the launch vehicle, qualify tracking and data acquisition stations on the ground, and to carry out limited research on the Earth's ionosphere. The satellite remains in orbit.

1967 February 15 The last of four French Diamant A launch vehicles carried the 51-lb. satellite *Diadème 2* to a 368 × 1,170 mi. orbit inclined 39.4°. This was also the last time the Hammaguir launch site was used for satellite launch attempts. The mission, designated D-1D, had been funded by the French military and comprised a satellite identical to that launched on February 8. Laser beams and Doppler measurement was used to conduct geodetic experiments with *Diadème 2* until April 5, 1967, when it was silenced by a power failure. The satellite remains in orbit.

1967 February 18 A French Vesta sounding rocket carried a monkey to an altitude of 100 mi. from the Sahara Test Center, North Africa. Vesta was a successor to the Véronique rocket, the first sounding rocket developed by the French after World War II but based on design details from the German A-4/V-2. Burning nitric acid and turpentine, Vesta had been developed for the French Army, and the national space agency, *Centre National d'Etudes Spatiales* (CNES), had bought 10 for scientific research. Following a series of flights with mice and cats, the French had moved to conduct life science research with primates. A second monkey was launched in a Vesta on March 7, 1967.

1967 February Soviet engineers at the Baikonur cosmodrome began to assemble the first N-1 launch vehicle. For carrying cosmonauts to the moon, it had five stages, a launch weight of just over 6 million lb. and a total height of 370 ft.

The first stage had the exterior appearance of a tapered cylinder, with a height of 98 ft., a maximum diameter of 55 ft. 10 in. and a minimum diameter of 33 ft. The first stage incorporated a liquid oxygen tank with a diameter of 42 ft. 4 in., a kerosene tank with a diameter of 32 ft. 10 in. and a total propellant load of 3.8 million lb. Thirty NK-33 engines of 339,400 lb. thrust were mounted in two concentric rings, twenty-four around the outer circumference of the first stage and six in the center, generating a liftoff thrust of 10,182,000 lb. for about 1 min. 50 sec. With 30% greater thrust than America's Saturn V, it was the most powerful launch vehicle ever developed. The second stage was also a tapered cylinder, with a length of 66 ft., a lower diameter of 32 ft. 2 in. and an upper diameter of 22 ft. 4 in. It carried eight NK-43 engines similar to the NK-33 and in the vacuum of space provided a combined thrust of 3,157,560 lb. for 2 min. 10 sec. of flight. The cylindrically tapered third stage had a length of 39 ft. 4 in., a bottom diameter of 22 ft. 4 in. and a top diameter of 15 ft. 9 in. It carried four NK-39 engines with a total stage thrust of 361,456 lb. for a firing time of about 5 min. The first three stages would put the 210,000-lb. payload comprising two upper stages, lunar module, attitude control module and adapted Soyuz spacecraft in Earth orbit. From there, the fourth stage, powered by a single 90,200-lb. thrust engine, would fire to propel the upper elements to the moon.

The President's Science Advisory Panel (PSAC) completed its report *The Space Program in the Post-Apollo Period*, and laid the foundation for future NASA goals. The PSAC had not attempted to define specific technology paths for future development but took instead a conceptual approach. It suggested that the long-term objective for civilian space programs should be "the exploration of the planets by man." Toward that end it recommended a permanently manned Earth-orbiting space station serviced and resupplied on a routine basis by economical, and reusable, logistics vehicles. This was the first policy statement to unite future mission goals around a space station and shuttle vehicles.

1967 March 7   Testifying about the prospects for landing astronauts on the moon before the House Committee on Science and Astronautics, NASA Administrator James Webb said, "if we get this done by the end of 1969, we will be very, very fortunate." He went on to say, "[the] possibility of doing all the work necessary is less this year than it was last. And I testified at this table last year that it was less at that time than it had been the previous year."

1967 March 10   The first test flight of the Korolev 7K-L1 circumlunar Soyuz variant was launched from Baikonur to a 114 × 180 mi. Earth orbit by the first flight version of the SL-12 Proton. Designated *Cosmos 146*, the spacecraft weighed 11,065 lb. and comprised a 6,600-lb. Soyuz descent module beefed up with much thicker thermal ablative protection and a 4,465-lb. instrument module. Later spacecraft given a designation in the Zond series would have an average weight of 11,900 lb. Two solar array wings with a span of 29 ft. 6 in. were attached to the instrument module for producing electrical energy. No orbital module was carried, but a support cone similar to that which would carry a solid propellant launch escape rocket on lunar flights was attached to the top of the descent module. The four-stage SL-12 incorporated the two stages of the SL-9 to which was added two upper stages, the third stage designed by the Chelomei bureau and the fourth stage by the Korolev bureau. The third stage had a length of 21 ft., a diameter of 13 ft. 7 in. and a 141,000-lb. thrust propulsion unit from the Kosberg bureau burning

UDMH-N_2O_4 propellants for 4 min. 14 sec. The fourth stage had a length of 18 ft., a diameter of 12 ft. and a restartable Korolev bureau propulsion unit delivering 19,200 lb. of thrust for up to 10 min. from kerosene–liquid oxygen propellants. With a liftoff weight of 1.54 million lb., the SL-12 could send a 12,000-lb. payload to the moon. The *Cosmos 146* mission was the first in a series of flights planned to test the lunar Soyuz out as far as the moon. *Cosmos 146*, however, was a test of the system and not a failed lunar mission. Before the spacecraft returned to Earth at the end of its eight-day flight, the fourth stage of the SL-12 boosted it to a high altitude for a fast reentry.

1967 March 23   The United Nations World Meteorological Organization announced in Geneva, Switzerland, that agreement had been reached on a World Weather Watch to begin in 1968. Several weather satellites were to provide images of the Earth's weather direct to computer centers in Washington, D.C., Moscow, USSR, and Melbourne, Australia. This was the first step in setting up international links with meteorological satellites to establish a global monitoring program aimed at predicting storms and providing accurate 24-hr. weather forecasts.

1967 March 24   NASA instructed field centers that the Apollo Applications Program's (AAP) orbital workshop would have separate solar cell arrays laid out on "wings" situated on either side of the converted S-IVB stage. Increasing demand on electrical requirements dictated a supplementary supply. The Apollo telescope mount (ATM) was also to have its own source of electrical production, four separate solar cell arrays arranged in a cruciform layout. Previously, it had been thought the Apollo CSM (command/service module) fuel cells could provide sufficient electrical power for systems and scientific experiments. When AAP electrical needs were defined in April 1969, the two workshop arrays were required to provide a total of 12 kW and the four ATM arrays a total of 11 kW.

1967 March 27   NASA's Manned Spacecraft Center awarded six-month study contracts to General Dynamics, Boeing and McDonnell on space station facilities for assembly in orbit during the 1970s. General Dynamics was to study a support module capable of providing electrical power and life support functions. Boeing was to study the space station module and a 38-in. optical telescope that would be launched with it on a Saturn V. McDonnell was required to study a logistics/ferry vehicle for resupply flights.

1967 March   Scientists and NASA representatives in Washington, D.C., discussed the possibility of sending spacecraft beyond the planet Mars, across the asteroid belt and on to Jupiter, the solar system's largest planet. Five times as far from the sun as Earth and more than three times the distance from Earth to Mars, Jupiter alone contains 71% of the mass in all nine planets. The enormous gravity field of Jupiter could usefully serve to throw a spacecraft on to the outer regions of the solar system. In a sling-shot boost provided by its enormous mass, Jupiter could act like a giant rocket adding enormous velocity to significantly reduce the transit time required for flights to the outer planets Saturn, Uranus, Neptune and Pluto. With the rocket technology then available, a flight to Jupiter would take about two years.

1967 April 3   Following the tradition of its annual bombing competitions, the U.S. Air Force Strategic Air Command opened its first annual SAC missile competition. Participants included two combat crews and one target alignment team from each of six Minuteman and three Titan wings. The 351st Strategic Missile Wing made a clean sweep of the competition by winning all the Minuteman awards and the Blanchard Trophy, named for Gen. William H. Blanchard, who died on May 31, 1966, while serving as Air Force Vice Chief of Staff. The 381st Strategic Missile Wing won the Best Titan Wing award.

1967 April 5   The NASA Apollo 204 Review Board submitted its final report to Administrator James Webb, finding that the fire that killed three astronauts on February 27 was probably caused by "an electrical arc" among "environmental control system power wiring." Moreover, "the command module contained many classes of combustible material contiguous to possible ignition sources." Electrical arcs were made more likely because "some areas of wiring exhibited what would be referred to as 'rats nests' because of the dense, disordered array [and] there were instances where wiring appeared to have been threaded through bundles, which added to the disorder."

1967 April 6   NASA launched the *ATS-2* (Applications Technology Satellite-2) by Atlas-Agena D from Cape Canaveral, Fla., to an elliptical orbit of 111 × 6,912 mi. inclined 28° to the equator. Because the Agena D failed to restart and circularize the orbit at 6,912 mi., *ATS-2* was unable to perform its planned gravity-gradient stabilization experiment employing four 123-ft. beryllium-copper booms deployed in an X-configuration. The 815-lb. cylindrical satellite had a diameter of 4 ft. 8 in. and a height of 6 ft. Built for the Goddard Space Flight Center, *ATS-2* also carried eight environmental monitoring experiments. It decayed into the atmosphere on September 2, 1969.

1967 April 8   The second unmanned Soyuz circumlunar spacecraft was launched by SL-12 from the Baikonur cosmodrome to a 115 × 137 mi. orbit. The 11,100-lb. spacecraft was launched under the guise of *Cosmos 154* to hide its true identity and was intended as a simulated test of a high-speed reentry from a highly elliptical orbit. However, the fourth stage of the SL-12 separated too early and failed to fire a second time to propel the descent module to high apogee, and it returned to Earth on April 10. Another launch attempt failed September 23, 1967, when only five of six first-stage engines fired. Yet another launch failed on November 22, 1967, when only three of the four second-stage engines ignited and the 11,300-lb. spacecraft was destroyed. Both latter flights are widely believed to have had the same objectives as *Zond 4*, launched March 2, 1968.

1967 April 17   The *Surveyor III* spacecraft was launched at 07:05:01 UT from Cape Canaveral, Fla., by Atlas-Centaur AC-12 and successfully placed in a brief 99 × 106 mi. Earth parking orbit exactly 10 min. later. At an elapsed time of 31 min. 57 sec. the two Centaur engines were reignited for 1 min. 51 sec., moving the 2,281-lb. *Surveyor III* to a lunar intercept trajectory. A 4.2-sec. (13.8 ft./sec.) midcourse burn, the smallest maneuver of any Surveyor mission, at 05:00:02 UT on April 18, put the spacecraft on course for a landing in the southeast part of the moon's Oceanus Procellarum. In addition to a TV camera, *Surveyor III* had a 25-lb. soil mechanics surface sampler (SMSS), a small scoop at the end of a lazy-tongs, or scissors, arm attached to Surveyor just below the camera. Capable of extending from 23 in. to 35 in. from its

mounting, the scoop arm could sweep a 112° radius of the lunar surface (about 10 ft.) between two landing legs and create a trench to a maximum depth of 1 ft. 6 in.

Officials at the NASA Manned Spacecraft Center defined a list of events that had to be accomplished prior to the first manned lunar landing attempt. Among these was the demonstration of a space-walk capability between the Apollo command module and the lunar module in space, demonstration of rendezvous techniques on the first manned Earth-orbit Apollo mission and unmanned burns of the lunar module's ascent and descent propulsion systems to qualify it for manned tests. Three Earth-orbital flights of the command/service modules and the lunar module, following one flight of the Apollo spacecraft on its own, were considered an essential prerequisite for the first tests in moon orbit.

A Minuteman II was launched from Vandenberg AFB, Calif., on signals sent from a Boeing KC-135 aircraft. Part of the Airborne Launch Control System (ALCS), it demonstrated that in-silo launch of ICBMs was possible if necessary in time of war. The ALCS became operational on May 31, 1967, as an alternative means of launching ICBMs and it was the first of several improvements and additional duties attached to the ICBM force. On October 10, 1967, the Emergency Rocket Communications System became operational. This enabled Minuteman II missiles to be used to launch replacement communication satellites should existing systems be destroyed.

1967 April 19   In testimony before the Senate Committee on Aeronautical and Space Sciences, NASA's head of manned spaceflight, George Mueller, outlined plans for the Apollo Applications Program. The flight schedule assumed that Saturn IB vehicles AS-204 through AS-207, and Saturn Vs through AS-509, would be required to support the Apollo lunar landing program. AAP flights for 1968 included: *AAP-1*, the lunar mapping and survey system module, would be launched together with a manned Apollo CSM (command/service module) for 4 days of Earth-orbit evaluation before docking it to *AAP-2*; *AAP-2*, the airlock module and multiple docking adapter, would be assembled onto the orbiting workshop, converted from the S-IVB stage by the *AAP-1* crew for 28 days' habitation; *AAP-3*, a manned CSM, would rendezvous with the Apollo telescope mount launched a day later on *AAP-4* and carry it to the workshop cluster for a 56-day stay. For 1969 there were to be four Saturn IB flights (AS-212 through AS-215) each lasting 56 days and launched at 90-day intervals, to revisit the workshop cluster. One of these flights, AS-214, would also lift the APP-A package of Earth/meteorology science instruments to the orbiting cluster. In 1970–1971 AS-216 and 217 would set up orbital workshop no. 2; AS-218 would bring up APP-B, the second Earth science package; AS-219 would deliver an ATM (Apollo telescope mount) and biological laboratory to the no. 2 workshop; and AS-220–AS-221 would be revisit missions. For 1971–1972 Mueller envisaged the launch of a fully outfitted 1-year mission module (workshop) on Saturn V AS-517) and five visits using extended capability Apollo spacecraft designated CSM-EC; there were to be four CSM-EC revisits in 1972–1973. Meanwhile, projected Saturn V AAP launches were to start in 1969, with AS-510 carrying a mapping and survey system module and manned CSM to moon orbit for 8 days of imaging. In 1970 AS-511 and AS-512 would each support a 3-day lunar-surface visit with a full-capability LM (lunar module). In 1970–1971 AS-513 was scheduled to launch a manned CSM, airlock and Earth-orbiting workshop no. 3 to a geosynchronous orbit and 45–60 days of manned habitation.

AS-514 would send a CSM revisit crew plus Apollo telescope mount for solar observations and a further period of extended research. In 1971 AS-515 would deliver an unmanned lunar shelter, drill and lunar science survey module (LSSM) to the moon's surface and conduct a 7-day survey from orbit with a manned CSM. AS-516 would deliver a two-man crew to the equipment delivered by AS-515 for a 14-day period of exploration. This extensive program was the most ambitious application of Apollo hardware envisaged. As such it represents an important stage in the evolution of manned mission plans for the period after the initial manned landing.

1967 April 20   The *Surveyor III* spacecraft landed on the moon at 23.34° W by 2.94° S, about 388 mi. east of *Surveyor I*. The solid propellant retro-rocket ignited at 00:01:18 UT at a slant range of 271,334 ft. and a speed of 8,617 ft./sec. and in a 41-sec. burn reduced the speed to 462 ft./sec. at a slant range of 36,158 ft. The three liquid vernier motors took over to lower the spacecraft to a point about 12 ft. above the moon, but instead of cutting off they continued to burn through touchdown because the radar lost lock on the surface and failed to cut thrust as programmed. *Surveyor III* contacted the surface at 00:14:16.9 with a descent rate of about 4 ft./sec. It bounced back off the moon and recontacted the surface 24.2 sec. later, 50–70 ft. down the gently sloping side of a crater 650 ft. in diameter. It bounced up a second time and landed 12.1 sec. later, 35–45 ft. further down the slope just 0.5 sec. after a command from Earth shut down the verniers. The 659-lb. spacecraft returned 6,326 TV images and, in the 18 hr. 22 min. of activity with the soil mechanics surface sampler (SMSS) that began on April 21, conducted seven bearing, four trench and thirteen impact tests. The last data was returned to Earth at 00:04 UT on May 4, 1967, at the end of the first lunar day and the spacecraft was never revived again. It was visited by the *Apollo 12* astronauts on November 20, 1969, when the TV camera was removed and returned to Earth.

1967 April 23   At 00:35 UT an SL-4 launch vehicle carried cosmonaut Vladimir M. Komarov from the Baikonur cosmodrome into a 120 × 131 mi. orbit inclined 51.6° aboard the 14,200-lb. *Soyuz 1* spacecraft. The mission had been made

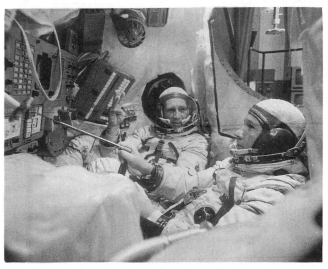

Compared to Vostok, Soyuz offered considerably greater volume, and early versions could carry three cosmonauts.

ready in a hurry. The design team and the new chief of the Korolev bureau, Vasili Mishin, had not wanted the spacecraft to carry a cosmonaut yet. Four previous orbital flights of the Soyuz/Zond design had each revealed separate technical problems, none of which had yet been completely resolved. Yuri Gagarin was backup. The mission plan involved *Soyuz 2* being launched with cosmonauts Valery Bykovsky, Aleksei Yeliseyev and Yevgeny Khrunov, 23 hr. 35 min. after *Soyuz 1*. Carrying docking modules on the front end of each orbital module, *Soyuz 2* would dock with *Soyuz 1*. Via depressurized orbital modules on respective spacecraft, Yeliseyev and Khrunov were to transfer from *Soyuz 2* to *Soyuz 1,* returning to Earth with Komarov in *Soyuz 1* during the night of April 24–25 and leaving Bykovsky to return alone in *Soyuz 2* on the night of April 25–26. Almost as soon as Komarov reached orbit, things began to go wrong. One of the solar panels failed to deploy, *Soyuz 2* was called off and attempts to stabilize the spacecraft for reentry failed on two successive orbits. Finally, on orbit 18 Komarov fired the KTDU-35 retro-motor manually, but during the descent the primary main parachute failed to deploy and the backup got tangled. It did not deploy and the descent module struck the ground 1 day 2 hr. 48 min. after launch at 280 MPH, killing Komarov. Two days later his remains were interred in the Kremlin wall. NASA astronauts Cooper and Borman had wanted to attend out of respect to a fellow spaceman but were denied access to the Soviet Union on the premise that it was "an internal matter."

1967 April 24   A new Apollo flight designation system was approved by NASA whereby the mission previously designated AS-204 would become *Apollo 1*, the first flight of the Saturn V would be *Apollo 4*, the first flight of the lunar module would be *Apollo 5*, the second flight of the Saturn V would be *Apollo 6*, and so on. There was no formal allocation of *Apollo 2* and *Apollo 3*, but it became logical to retroactively allocate those designations to the suborbital Apollo CSM flights on Saturn IB launchers (AS-201 and AS-202), although that was never formalized. Before the Apollo fire of January 27 it had been agreed that the AS-204 mission, then planned as the first manned flight, should be *Apollo 1*, and the widows of the three astronauts expressed a wish that it should remain so.

1967 April 26   A 285-lb. Italian satellite was launched from a floating platform 3 mi. off the coast of Kenya in a cooperative endeavor with the United States, Italy and Kenya. Called *San Marco 2*, the satellite consisted of a 2 ft. 2 in. sphere instrumented to obtain measurements of the air density of the ionosphere over the equator. The advantage of the platform launch derived from its position only 2.56° south of the equator, thereby permitting the Scout launcher to gain maximum advantage from the Earth's rotation. The launch pedestal consisted of two platforms fixed to the seabed on stilts. *San Marco 2* was placed in a 135 × 498 mi., 2.9° orbit that decayed on October 14.

1967 April 28   A U.S. Air Force Titan IIIC launched the first pair of a second-generation Vela nuclear detection satellite to a 67,804 × 69,991 mi. orbit inclined 32°. Designated *Vela*s *7* and *8*, the two 571-lb. satellites were placed at 180° intervals around the earth. They were modified versions of the first generation Vela and could detect detonations and yield in the atmosphere. The new Velas had a 138-lb. instrument load that included for the first time an optical flash detector called a Bhangmeter, as well as a wide variety of detectors, spectrometers and geiger counters. As a secondary payload, *ERS-18* (20 lb.), and the first two Air Force OV5-series satellites, *OV5-3* (19 lb.) and *OV5-1* (14 lb.), were deployed to a 5,357 × 69,318 mi. orbit with an inclination of 32.8°. Last of the OV satellites, the OV5s were built by TRW as an extension of their Environmental Research Satellite series first launched in 1962. They were mostly octahedrons with 816 solar cells producing 5.5 W of power. Seven OV5s were successfully launched, the last on May 23, 1969. *Velas 9* and *10* were launched on May 23, 1969, followed by *Velas 11* and *12* on April 8, 1970.

1967 May 4 At 22:25:00 UT an Atlas-Agena D launched NASA's *Lunar Orbiter IV* from Cape Canaveral, Fla., and put it on course for the moon. Because three earlier spacecraft in this series had accomplished requirements set by the Apollo program, this mission was released to conduct a photographic survey from polar orbit. At 18:21 UT on May 5, a 53-sec. (199 ft./sec.) midcourse maneuver was performed, and the spacecraft began its orbit insertion burn at 15:08:47 on May 8. The 8 min. 21.9 sec. (2,164 ft./sec.) burn placed *Orbiter IV* in a 1,682 × 3,797 mi. orbit inclined 85.48° to the lunar equator. The spacecraft obtained 199 dual-frame exposures (398 images), but 17 were not developed, some were fogged and a few were not transmitted due to film-winding problems. Nevertheless, 99% of the lunar nearside was photographed with resolution 10 times better than pictures from Earth-based telescopes, and all images had been read back to Earth by 13:28 UT on June 1, 1967. *Orbiter IV* impacted the moon on October 6, 1967.

1967 May 5 Plans for the NASA Apollo Applications Program were revised as a result of the *Apollo 1* fire. What had hitherto been designated the *AAP-1* flight involving the launch of a lunar mapping and survey system for docking to the workshop was now to be a unique mission. A tentative launch date of September 15, 1968, was now scheduled for this single mission, at the end of which the mapping module was to be jettisoned before the command module returned to Earth. It was to be followed by the first workshop flight, *AAP-1,* in early 1969, which was to be visited by the *AAP-2* crew for a 28-day mission. Later, the *AAP-3* and *AAP-4* missions would carry the Apollo telescope mount and a fresh crew to the workshop for a 56-day stay. Two additional flights in late 1969 were to visit the workshop using refurbished Apollo spacecraft from earlier flights. During 1970 there were to be two dual launches, each carrying a crew on one flight and an experiment module on the other, as well as two single-crew revisit flights using extended-capability spacecraft. In this way there would be almost continuous manning of the one workshop. In all, there were to be 25 Saturn IB and 14 Saturn V flights in the AAP schedule.

The first satellite built in Britain, *UK-3,* or *Ariel 3,* was launched by Scout from Vandenberg AFB, Calif., to a 1,681 × 3,750 mi., 85.5° orbit. The British Aircraft Corp. built the satellite, which consisted of a cylinder 2 ft. 6 in. in diameter and 10 ft. 6 in. across when four solar cell panels were deployed from the base. The 198-lb. satellite had a total height of 8 ft. and was controlled by a central electronics unit built by GEC Electronics in England. *UK-3* carried five experiments for measuring molecular oxygen, electron density and temperature in space, terrestrial radio noise, low frequency radiation and galactic radio noise. *UK-3* decayed on December 14, 1970.

1967 May 9 NASA announced the names of the crew that would fly the first manned Apollo mission in Earth orbit, designated *Apollo 7.* Astronauts Walter Schirra, Don Eisele and Walter Cunningham were the primary crew, with Thomas Stafford, John Young and Eugene Cernan as backups. Their Block II spacecraft, *CSM-101,* incorporated many modifications resulting from changes brought about by the Apollo fire of January 27, 1967, even down to noncombustible paper for flight documents. In tests to qualify the spacecraft for flight, fires were lit in a pressurized pure oxygen atmosphere to demonstrate that they would self-extinguish. Nevertheless, special fire fighting equipment was also provided. SAM was delivered to the Kennedy Space Center on May 30, 1968.

The first in a series of U.S. electronic intelligence gathering (elint) satellites carried piggyback on third-generation KH-8 reconnaissance satellites was carried into space by a long-tank thrust-augmented Thor-Agena D from Vandenberg AFB, Calif. The 4,410-lb. reconnaissance satellite was placed in a 124 × 483 mi. orbit inclined 85.1°, while the 130-lb. piggyback subsatellite was placed by the Agena D into a 345 × 503 mi. orbit inclined 84.9°. The main satellite decayed out of orbit on July 13, but the subsatellite remains in space, albeit inoperable. A total of 16 elints were launched piggyback on third-generation area-surveillance reconnaissance satellites with Thorad-Agena D launchers, the last on September 10, 1971.

1967 May 15 An SL-8 launched from Plesetsk carried the first in a series of covert Soviet store-dump communication satellites to a 459 × 510 mi., 74° orbit. Designated *Cosmos 158,* the 1,650-lb. satellite had a length of about 6 ft. 6 in. and a height of 4 ft. 6 in. and was launched as a test vehicle pending final development. The first operational store-dump satellite, *Cosmos 372,* was launched on October 16, 1970. Used primarily by the GRU (Soviet military intelligence), but at times by the KGB, satellites of this type would collect radio communication and other electronic signals as they passed over foreign embassies or military bases, store the information on board and transmit (or "dump") it to Soviet ground stations at a later time. On several occasions satellites of this type were used to transmit coded instructions to GRU and KGB field agents in foreign countries. Between one and three such satellites have been routinely launched each year to the present.

1967 May 16 A Soviet three-stage SL-6 launch vehicle placed the 9,900-lb. *Cosmos 159* spacecraft in Earth orbit for a full test of the extended-length Soyuz instrument module that, on a lunar mission, would be used to leave moon orbit and return the two cosmonauts to Earth. In a full-duration firing to simulate a trans-Earth injection burn from lunar orbit, *Cosmos 159* was pushed into a 217.5 × 37,680 mi. orbit. If conducted from lunar orbit, this would be sufficient to propel the lunar Soyuz spacecraft out of the moon's gravity and back into the Earth's sphere of influence. Two additional Soyuz propulsion tests launched on November 16, 1970, and February 6, 1970, failed to reach orbit. *Cosmos 159* reentered the atmosphere through natural orbital decay on November 11, 1977.

1967 May 30 The *ESRO 2A* satellite was launched from Vandenberg AFB, Calif., at 02:06 UT on a four-stage Scout but failed to reach orbit due to a malfunction in the third stage. ESRO 2 was the first cooperative program between NASA and the European Space Research Organization. Built by Hawker Siddeley Dynamics in England, the 162-lb. *ESRO 2A* carried seven experiments to sample radiation in the vicinity of Earth.

1967 May 31 The first of three U.S. Navy satellites built to help develop new technologies for navigation equipment, *Timation 1* was launched by Thor-Agena D from Vandenberg AFB, Calif., as part of the first cluster of nine satellites carried into space on a single launcher. Placed in a 560 × 567 mi. orbit inclined 70°, both this 9-lb. satellite and *Timation 2*, launched September 30, 1969, tested quartz clocks but found difficulties with maintaining long-term stability in a radiation environment. Other satellites included three Surcal satellites: an 84-lb. navigation satellite; a 5.3-lb., 1 ft. 8 in. calibration sphere; and a 3.3-lb., calibration sphere 1 ft. 4 in. in diameter. Two 9-lb. gravity gradient stabilization experiment *(GGSE-4 and GGSE-5)* satellites were also carried. Three satellites were classified. All nine remain in orbit.

1967 June 12 At 02:40 UT the Soviet Union launched a 2,550-lb. spacecraft called *Venera 4* carrying a hard-landing capsule designed to survive a descent through the atmosphere to the surface of Venus. The main body of the spacecraft carried scientific equipment for measuring the interplanetary environment, including a magnetometer, cosmic ray counters and hydrogen and charged particle traps. A course correction maneuver was performed on July 29, 1967, successfully aligning the trajectory with the planet. A second Venera spacecraft was launched on June 17, 1967, but this failed to leave Earth orbit and it was known simply as *Cosmos 167*.

1967 June 14 At 06:01:00 UT an Atlas-Agena D launched the 540-lb. *Mariner V* spacecraft to a 115-mi. Earth parking orbit less than 9 min. after liftoff from Complex 12 at Cape Canaveral, Fla. *Mariner V* was sent on its way to Venus after a brief 13 min. coast period before a second burn of the Agena which ended 23 min. 19 sec. after liftoff. A speed of 25,500 MPH put the spacecraft on course for a planned 46,600-mi. flypast at Venus. It was deliberately biased away from the planet to avoid any chance that either the spacecraft or launch vehicle would impact the surface. *Mariner V* was almost identical to *Mariner IV* but with different science instruments and solar panels reduced from 70 sq. ft. to 43.5 sq ft. in area and to 18 ft. in span. *Mariner V* would move toward rather than away from the sun, producing 370 W at Earth and 550 W at Venus from its 17,640 solar cells. *Mariner 5* also carried an aluminized Teflon sun shield extended 10 in. beyond the periphery of the base of the octagonal spaceframe. A 17.7-sec. (50.5 ft./sec.) course correction burn was completed at 23:08:28 UT on June 19, 1967, moving the flight path to within 3,000 of the surface of Venus.

1967 June 16 The first tactical use of U.S. military reconnaissance satellites began with the launch of a high-resolution KH-4A by a long-tank thrust-augmented Thor-Agena D (Thorad-Agena D) from Vandenberg AFB, Calif., to a 112 × 228 mi., 80° orbit. The KH-4A had been programmed to take pictures over an area extending from Egypt to Syria, covering the Six-Day War with Israel over a five-day period. When the two film capsules returned the stereo images to earth, photo-analysts were able to identify total numbers of aircraft destroyed in Egypt, Jordan and Syria, locate and identify 30 of the 35 known surface-to-air missile sites in Egypt and discover that the Suez Canal had been blocked by a passenger ship sunk 3 mi. from Port Said.

1967 June 17 The most powerful solid propellant rocket motor was fired by Aerojet General at Homestead, Fla. Third in the series of tests with its 260-in. motors, SL-3 consumed 1.6 million lb. of propellant and produced a thrust of 5.7 million lb. for 1 min. 20 sec. The only flaw in the test was the explosion of a portion of the exhaust nozzle at the end of the firing. The purpose of this latest test in NASA's Large Solid Rocket Technology program was to evaluate the performance of inert slivers to control the burnout phase and to demonstrate the performance of propellant with high burn rates.

1967 June 26 NASA awarded a contract to Martin Marietta to study the feasibility of using a Titan-Centaur combination for launching heavy payloads for the civilian space agency. NASA lacked the funds to develop a new heavyweight launcher needed to lift planetary and interplanetary spacecraft in the next several years. Reduction in NASA budget appropriations and the cost of launching Saturn vehicles weighed against the use of these ultra-heavyweight launchers, and NASA lacked intermediate rockets. On January 26, 1968, the U.S. Air Force briefed NASA officials on the Titan family, and the following month NASA decided to use a developed Titan IIIC for some planetary flights. On June 11, 1968, the Lewis Research Center awarded Martin Marietta a Titan-Centaur integration contract, and on October 30, 1969, Lewis authorized the company to develop an adapted Centaur for what was now designated Titan IIIE/Centaur.

1967 June 29 The first of three U.S. Air Force Thor-Burner II flights not dedicated to the Defense Meteorology Satellite program lifted off from Vandenberg AFB, Calif., carrying the *Secor 9* (99 lb.) and *Aurora 1* (104 lb.) satellites, the latter to investigate the Aurora Borealis. The second launch was on August 16, 1968, but a malfunction prevented the 10 satellites it carried from reaching orbit. Weighing a total 1,126 lb., these included the 395-lb. *Radcat* radar calibration satellite, the 75-lb. *Lincoln* calibration sphere for radar experiments, the 81-lb. *Orbis-Cal 1* for ionospheric measurements, the grid sphere drag experiment consisting of four 22-lb. satellites, *Secor 11* and *Secor 12* geodetic satellites, the 258-lb. *Lidos* gravity satellite and UV radiometer for shortwave tests. The last of 14 Atlas-Burner IIs carried the *STP72-1* satellite to orbit on October 2, 1972.

1967 June 30 The French Diamant B launcher program was officially approved. Designed to increase orbital capacity to 350 lb., the L-12 Emeraude first stage of the Diamant A was to be replaced by the L-17 Améthyste stage. With a thrust of 77,175 lb. from its Valois engine, the stage was a hybrid development incorporating the structural layout of Emeraude and the propulsion system of the Coralie, designed as the second stage of the Europa I launcher. It had a propellant capacity of 39,900 lb. compared to 28,150 lb. for Diamant A. The P-2.2 Topaze second stage remained essentially the same as that used by Diamant A, but the third stage was the P.68, with a vacuum thrust of 11,000 lb. Diamant B had a length of 92 ft. 8 in., a liftoff weight of 54,300 lb. and an orbital capacity of 247 lb.

The U.S. Air Force announced the names of four astronauts for the Manned Orbiting Laboratory program: Maj. James A. Abrahamson; Lt. Col. Robert T. Herres; Maj. Robert H. Lawrence Jr.; and Maj. Donald H. Peterson. The first African American selected for astronaut training, Maj. Lawrence was killed on December 8, 1967, during a routine training flight when his F-104 Starfighter crashed on landing at Edwards AFB, Calif. When the Manned Orbiting Laboratory program was canceled in 1969, Maj. Peterson transferred to NASA as part of that agency's Group 7 selection. Maj. Abrahamson subsequently went to NASA to head up the Shuttle program and later became head of the Strategic

Defense Initiative, the so-called Star Wars program. Lt. Col. Herres left in 1969 and later became the commander of Unified Space Command.

Bell Aerosystems displayed for the first time a two-man propulsion device capable of moving astronauts across the lunar surface. Essentially a rocket-powered vertical takeoff platform, the so-called Pogo (after the hopping motion of a pogo-stick) had a thrust of 600 lb. for 21 sec. With a weight of 147 lb., the device could lift one man and scientific equipment across rilles and ridges on the moon. Although limited in duration, the "flight" capabilities could be extended with additional propellant. The two-man Pogo was never adopted.

1967 July 11 Built for the U.S. Air Force PILOT (piloted low-speed tests) program, the SV-5P lifting body was rolled out from the Martin Marietta plant at Baltimore, Md., and was immediately redesignated X-24A. It was a full-scale piloted version of the diminutive SV-5D, three of which had been launched on ballistic trajectories by Atlas rockets in 1966–1967 for the PRIME (precision recovery including maneuvering reentry) program. The X-24A was designed for flights up to 100,000 ft. and Mach 2; the attribute "low-speed" was relative to the 16,000 MPH reentry of its predecessor, tests of which terminated when speed dropped below Mach 2. The wingless X-24A had a length of 24 ft. 6 in., a span of 13 ft. 8 in. across the upswept fins and a gross weight of 11,000 lb. Power was provided by an 8,480-lb. thrust Thiokol XLR11-RM-13 operating on water-alcohol–liquid oxygen fed by a hydrogen-peroxide gas generator driving a turbine. The X-24A was delivered to Edwards AFB, Calif., on August 24, 1967, and glide tests from a B-52 mother plane began on April 17, 1969.

1967 July 14 The *Surveyor IV* spacecraft was launched at 11:53:29 UT by Atlas-Centaur AC-11 from Cape Canaveral, Fla., to a direct ascent lunar intercept trajectory. The 2,295-lb. spacecraft carried a survey TV camera and soil mechanics/surface sampler (SMSS) similar to the science payload carried by *Surveyor III*. A 10.5-sec. (33.2 ft./sec.) midcourse burn began at 02:30:02 UT on July 16, and the solid propellant retro-rocket was ignited at 02:01:58.6 UT on July 17. Just 41.1 sec. later (1.4 sec. before the motor was expected to burn out), at an altitude of 49,400 ft. and a speed of 1,090 ft./sec., all signals from the spacecraft stopped without any indication as to the reason. It was estimated that *Surveyor IV*'s final resting place on the moon is in the Sinus Medii. If it continued to carry out preprogrammed landing activities and touched down as planned, it would have landed at 0.43° N by 1.61° W; if it continued on an unbraked flight path, the debris would be located at 0.47° N by 1.44° W. Efforts to find a spacecraft of that size in Apollo photographs taken over the former location have been unsuccessful.

1967 July 19 *IMP-E (Explorer 35)*, the only successful lunar orbit Explorer-class satellite in the series of ten Interplanetary Monitoring Platforms, was launched by thrust-augmented Delta from Cape Canaveral, Fla., to a translunar trajectory. The 230-lb. spacecraft was identical to *Explorer 33*, which had failed in the previous attempt to reach lunar orbit, and had four solar cell panels producing 43 W of power. On approaching the moon on July 21, an 80-lb. retro-motor was fired, producing 916 lb. of thrust for 23 sec., and *Explorer 35* entered a 497 × 7,692 mi. retrograde orbit inclined 147°. *IMP-E* carried eight experiments for measuring at lunar distance the solar wind and interplanetary fields and every

29.5 days the Earth's magnetospheric tail. It discovered a void in the solar wind on the antisolar side of the moon.

1967 July 27 For the first time, three U.S. Air Force Orbiting Vehicle satellites were launched aboard a single Atlas booster from Vandenberg AFB, Calif. *OV1-11* and *OV1-12* were carried in a dual pod on the nose of the launcher, while *OV1-86* was attached to a pod on the side. The 297-lb. *OV1-11* carried atmospheric physics experiments but failed to get into orbit due to a malfunction in its 5,977-lb. thrust FW-4S solid propellant boost motor, a type also carried by *OV1-12*. The 310-lb. *OV1-12* contained eleven experiments collectively known as the WL701 flare-activated radio-biological observatory (FARO) for biological and spectral measurements of the space radiation environment, and was placed in a 342 × 344 mi., 101.6° orbit. It decayed on July 22, 1980. The 231-lb. *OV1-86* (denoting the satellite from *OV1-8* and the 5,800-lb. thrust X-258 propulsion module from *OV1-6*) carried cosmic-ray and physics experiments and was placed in a 303 × 390 mi., 101.7° orbit. It decayed on February 22, 1972.

1967 July 28 NASA's fourth orbiting geophysical observatory, *OGO-4*, was launched by TAT-Agena D from Vandenberg AFB, Calif., to a 255 × 561 mi., 86° orbit. With a weight of 1,239 lb., the satellite carried 20 experiments for geophysical measurements of the near-Earth environment. *OGO-4* maintained three-axis stabilization for eighteen months, before the tape recorder failed on January 19, 1969. Placed in a spin-stabilized mode, the satellite was placed on standby during October 1969 and was reactivated three times during 1970 and 1971. Operations were terminated on September 27, 1971, and *OGO-4* reentered the atmosphere on August 16, 1972.

1967 July 31 The death knell sounded for NASA's Voyager Mars exploration program when the Congressional Conference Committee chopped almost in half the $70.2 million requested for work on that program in fiscal year 1968, authorizing $42 million but appropriating no funds at all. The Senate had eliminated funds completely, but the House wanted to approve continued work under a $50 million budget. Congress was concerned, however, that Voyager would stimulate large expenditure leading to calls for a manned Mars missions. In early August the Manned Spacecraft Center circulated a request for industry to participate in such a program, infuriating congressional leaders. The budget compromise was inadequate to hold together the Voyager development program, and on August 16 NASA advised the House Appropriations Committee that it would cancel the program. NASA made the formal announcement on August 30, 1967. Apart from the Mariner-Mars 1969 mission NASA had no other funded planetary program.

A two-week NASA summer study on lunar science and exploration began at Santa Cruz, Calif. Established to obtain a consensus on manned moon planning after the first landing, it advised immediate development of a local scientific survey module (LSSM), capable of manned or unmanned surface exploration, and a lunar flying unit for traversing difficult terrain. The scientists wanted ten Block II Surveyor spacecraft to carry scientific instruments to the surface between 1970 and 1975 to provide information from dispersed sites.

1967 July At the second European Space Conference held in Rome, Italy, delegates discussed nine separate ELDO launcher configurations. Most favored for development after the existing Europa I and Europa II vehicles was Europa III. This launcher would take the Blue Streak as first stage, add a new cryogenic second stage and an improved cryogenic Astris third stage to do the work of the PAS system on Europa II. A Europa IV proposal would add two Blue Streak, or solid propellant, strap-ons to the first stage, already proposed for a Europa II growth version, and apply multiple-burn capability to the third stage. Europa IV would have capacity to put a 4,410-lb. satellite in geostationary orbit.

1967 August 1   At 22:33:00 UT an Atlas-Agena D launched *Lunar Orbiter V* to escape velocity. Last in the series of photographic moon orbiters, the spacecraft executed a 26-sec. (97.6 ft./sec.) midcourse maneuver at 06:00 UT on August 3. This was also the last time NASA used an Atlas-Agena launch vehicle for lunar or planetary missions. Like its predecessor, *Lunar Orbiter V* would operate from polar orbit on primarily scientific photography, with 44 of 212 dual-frame exposures in support of Apollo site selection. The 8 min. 18 sec. (2,110 ft./sec.) insertion burn began at 16:48:54 UT on August 5, resulting in a 121 × 3,746 mi. orbit inclined 85.01° and a period of 8.5 hr. Pictures taken on orbit two secured farside imaging, and on orbit four an 11-sec. (52.4 ft./sec.) burn beginning at 08:43:49 on August 7 resulted in an intermediate elliptical orbit of 62.4 × 3,770 mi. for images at seven sites. A second, 2 min. 33 sec. (767 ft./sec.) transfer maneuver beginning at 05:08:33 UT on August 9 changed that orbit again to 61.4 × 932 mi. for the final photographic tasks. The last picture readout took place at 12:00 UT on August 28, and the spacecraft impacted the moon at 07:58 UT on January 31, 1968.

1967 August 4   The European Launcher Development Organization's *Europa I* was launched from complex 6A at the Woomera Test Range, Australia, on mission F6/1. The British Blue Streak first stage performed as planned, but the French Coralie second stage refused to ignite and the inert third stage and shroud crashed into the Simpson Desert 600 mi. from the launch site. The test had been planned to terminate with impact in the Pacific, 3,000 mi. from Woomera. The problem was traced to a stage separation mechanism. A similar fate befell the second flight with live first and second stages, F6/2, on December 6. These two flights constituted the last Phase 2 launches in the test program.

NASA announced the selection of eleven civilian scientist-astronauts constituting Group 6: Joseph P. Allen; Philip K. Chapman; Anthony W. England; Karl G. Henize; Donald L. Holmquest; William B. Lenoir; John A. Llewellyn; Story F. Musgrave; Brian T. O'Leary; Robert A. R. Parker; and William E. Thornton. Chapman, Holmquest, Llewellyn and O'Leary did not remain astronauts long enough to make a spaceflight, and O'Leary wrote a disparaging book about the space program, discrediting NASA and many key personnel at the Johnson Space Center, Houston.

A French Coralie rocket launched from the Woomera Test Range in Australia failed to carry a satellite into orbit as planned. Developed as the second stage of the Europa launch vehicle, the French wanted Coralie to place in orbit a French satellite using the stage on its own from a specially prepared launch pad. Other attempts to orbit a French satellite were made, but none was successful.

1967 August 18   NASA announced that it had awarded a contract to North American Aviation's Rocketdyne Division for development of an advanced aerospike engine. Engines of this type utilize a doughnut-shaped combustion chamber which discharges exhaust gases against the surface of a short center cone, instead of inside long bell-shaped nozzles. The concept specified by NASA envisaged an engine 8 ft. in diameter and 4 ft. 6 in. high, about 50% shorter than conventional bell-nozzle engines.

1967 September 5   Sixteen rhesus monkeys completed a two-month simulated spaceflight at the Los Alamos Scientific Laboratory, during which they were exposed to between 500 and 1,000 roentgens of gamma radiation over a 10-day period. The three monkeys given the higher dose died of radiation poisoning, but the rest survived. Scientists said this indicated that humans would be resistant to the levels of radiation expected from the maximum solar output during the period of maximum sunspot activity which occurred every 11 years. Because of the political goal of launching Apollo moon missions by the end of the decade, astronauts would be exposed to solar radiation at the next sunspot maximum in 1969.

1967 September 6   Chaired by Eugene S. Love, the Planetary Missions Technology Steering Committee at NASA's Langley Research Center met in the wake of Voyager cancellation to discuss the future of planetary exploration and asked the Jet Propulsion Laboratory, Ames Research Center and Langley to come up with proposals. With the Mars program in disarray, only the Mariner-Mars 1969 project remained intact. The JPL presented its findings October 3, 1967, calling for a 1971 Mars orbiter to obtain data on the atmosphere and photograph the surface over a long period, followed by two spacecraft to Venus in 1972 and a Titan III–launched Mars orbiter/probe mission in 1973. Two days later Langley proposed a Titan-launched Mars orbiter in 1971, Venus orbiter/probe in 1972 and larger version of the same in 1973. These studies formed the basis for the Mariner-Mars 1971 orbiter and the 1975 Viking orbiter/lander missions.

1967 September 7   NASA's second biological satellite, *Biosatellite 2*, was launched by a thrust-augmented Delta from Cape Canaveral, Fla., to a 188 × 203 mi., 33.5° orbit. The 940-lb. satellite carried 13 experiments, including seven radiation experiments isolated in a tungsten-nickel-copper sphere placed concentrically around a strontium 85 radiation source to receive one of nine selectable radiation doses from 200 to 500 roentgens. General biological experiments were carried in the back of the pod, an aluminum bowl, 2 ft. 7 in. in diameter, with 6 cu. ft. of space, mounted inside a heat shield. After separation two cold-gas jets spun the pod up to 57 RPM, whereupon a 10,000-lb. thrust solid propellant retro-rocket fired. The capsule was snatched midair by C-130 out of Hickam AFB, Honolulu, 45 hr. after launch.

1967 September 8   The 2,217-lb. *Surveyor V* spacecraft was launched at 07:57:01 UT from Cape Canaveral, Fla., by Atlas-Centaur AC-13 into a 98 × 105 mi. Earth parking orbit at a mission time of 9 min. 47 sec. At 16 min. 30 sec. the Centaur fired for 1 min. 54 sec., injecting *Surveyor V* to a translunar trajectory. After a 14.2-sec. (46.6 ft./sec.) midcourse burn began at 01:45:02 the next day, the valve controlling helium gas used to pressurize the propellant tanks failed to reseat properly, causing leakage and a 10 psi/min. pressure loss. Between 02:12:02 UT and 23:31:00 UT on September 9, five motor firings took place for a total 84.55 sec., bringing to 80.4 lb. the total amount of propellant used from a tanked supply of 181.3 lb. The purpose of these burns was to minimize the work needed from the verniers during descent by reducing spacecraft weight so that the solid retro-rocket would decelerate *Surveyor V* to about 100 ft./sec. instead of the usual 400–500 ft./sec., reducing by 75% the decrease in velocity provided by the verniers after burnout.

1967 September 11 *Surveyor V* landed on the moon in the southeast part of Mare Tranquillitatis at 1.41° N by 23.18° E on the 17° sloping inside wall of a 30-ft. crater at the end of a modified descent profile aimed at reducing the burn time for the liquid vernier motors. The main retro-rocket was ignited at 00:44:51 UT at an altitude of 147,604 ft. and a velocity of 8,487 ft./sec. It burned for 38.5 sec. and brought the spacecraft to an altitude of 4,139 ft. and a speed of just 79 ft./sec. *Surveyor V* touched down at 00:46:43 UT, 1.6 sec. after vernier motor cut-off 14 ft. above the surface. A 0.55-sec. vernier engine burn occurred at 05:38 UT on September 13 for soil erosion pictures. The 670-lb. *Surveyor V* carried a 27-lb. survey TV camera and a 27-lb. alpha-scattering instrument. Comprising a small box lowered to the surface by unreeling a nylon line attached to it, the instrument bombarded the top few microns of the lunar regolith with alpha particles and measured the energy of these particles reflected from the nuclei of atoms. In this way, scientists could determine the basic chemistry of the soil. *Surveyor V* returned 18,006 pictures during its first lunar day before it was turned off at 06:35 UT on September 29.

1967 September 15 The first in a new series of U.S. military reconnaissance gathering systems, the latest and last in the Corona family of intelligence gathering systems, was launched by thrust-augmented Thor-Agena D (Thorad-Agena D) from Vandenberg Air Force Base, Calif., to a 93 × 242 mi. orbit with an inclination of 80°. Designated KH-4B, the system carried additional film and extended useful orbital life to 19 days, at which time the recovery capsule returned to Earth with the film. The satellite decayed October 4, 1967. The last KH-4B, the last KH-4–class spy satellite and the final launch in the Corona family, was launched on May 25, 1972.

1967 September 20 NASA's Manned Spacecraft Center proposed a sequence of letter designations for Apollo flight activities: A–Saturn V/unmanned Apollo CSM (command/service module) development; B–Saturn IB/unmanned lunar module development; C–Saturn IB/manned Apollo evaluation; D–Saturn V/manned Apollo/lunar module development; E–Apollo/lunar module operations in high Earth orbit; F–lunar orbit mission; G–basic lunar landing mission; H–full-capability lunar surface mission; I–lunar mapping and survey mission; J–extended lunar landing mission using upgraded hardware. The categorization of missions according to a sequence of evolving capabilities had been put together by Owen E. Maynard, and the system was adopted by NASA Headquarters.

1967 September 29 Apollo Spacecraft Program Manager George Low identified the weight changes made necessary by modifications to the Apollo command/service modules (CSMs) and the lunar module to make the vehicles safer and fire-proof. The Apollo command module had grown in weight by 955 lb. and was now baselined at 12,524 lb., whereas the lunar module had grown by 508 lb. with a new empty weight of 9,415 lb. Modifications to the command module included a new integrated crew access hatch which could be opened in several seconds, replacing the cumbersome double hatch of the Block I CSM.

1967 October 3 Responding to a request for proposals on NASA planetary programs, the Jet Propulsion Laboratory included in its list of recommended projects a fly-by mission to Mercury using the gravity field of Venus to get it there in 1973. NASA Headquarters incorporated a mission of this type in its summary of recommendations from various field centers but pegged the mission for 1970 because the alignment of the planets that year would allow any given launch vehicle to send a heavier spacecraft than it could in 1973. Other launch opportunities for a flight to both planets occurred in 1975, 1982, 1994 and 1998. The advantage of using the gravity field of Venus to deflect a spacecraft to Mercury lay in the fact that an Atlas-Centaur could launch a 750-lb. spacecraft on a dual planet flight but could send only a 50-lb. spacecraft on a direct flight to Mercury. The Space Science Board of the U.S. National Academy of Sciences endorsed the JPL plan in June 1968.

NASA budget cuts and gloomy projections about the annual appropriation from Congress forced cuts in the Apollo Applications Program. As now envisaged, there were to be only four AAP lunar science missions, two Saturn IB orbital workshops and one Saturn V workshop, plus three Apollo telescope mounts. In all there were to be seventeen Saturn IB and seven Saturn V flights in support of the Saturn V–launched workshop. The first AAP workshop was scheduled for launch in March 1970. During negotiations over the budget for fiscal year 1969 further reductions were necessary. On January 8, 1968, it was decided to plan for a maximum six Saturn IB (AS-207 through AS-212) and three Saturn V flights (AS-513 through AS-515). One workshop would be launched on a Saturn IB and one on a Saturn V. Two lunar AAP missions were planned. The launch of the first workshop was postponed to April 1970. When NASA put its plans before Congress in early 1968, it envisaged requesting two Saturn IB and two Saturn V flights per year after AS-212 and AS-515, respectively.

The North American X-15A-2 rocket-powered research airplane achieved a world absolute speed record of 4,520 MPH (Mach 6.7), the fastest achieved by a winged, rocket-powered flying machine until the NASA Shuttle returned from orbit in 1981. Coated with an ablative heat shield material known as MA-25S, the X-15 experienced high heat loads causing the insulation to pit and char. Structural damage to the airframe was caused by a combination of high temperatures and high-speed airflow. The X-15 program provided a means of testing a wide range of different thermal protection materials designed to ablate (burn away at a controlled rate). High-speed flights in which X-15s carried test samples of thermal protection materials provided valuable data for NASA research programs on ablators.

1967 October 10 The United Nations' Treaty on Principles Governing the Activities of States in the Exploration and Use of Outer Space, Including the Moon and Other Celestial Bodies came into force. Known simply as the Outer Space Treaty, it was the first international agreement on space. It was concerned only with general principles, not specifics. On December 3, 1968, a second treaty concerning the return of astronauts and objects launched into space came into force. An agreement on international liability for damage caused to foreign property was introduced on October 9, 1973. A fourth treaty, concerning mandatory registration of objects launched into space, came into force on September 15, 1976.

1967 October 15 The *Surveyor V* unmanned spacecraft was revived at the beginning of its second lunar day when it responded to a call and reported in at 08:07 UT. alpha-scattering instrument analysis of surface chemistry resumed at 23:14 UT, and TV images were sent back in 200- and 600-line scan formats. On October 18 the spacecraft was temporarily shut down due to high temperatures as the elevation of the sun increased toward lunar noon and *Surveyor V* experienced a total eclipse of the sun by the Earth. The second lunar night began on October 23, and the spacecraft was shut down at 12:20 UT on November 1, having returned 1,048 TV images and obtained an additional 22 hr. of alpha-scattering data, bringing the total from that instrument to 124 hr.

1967 October 17 NASA accepted for core-manufacture a computer program called SUNDISK that would be used to control guidance and navigation operations on the first manned Apollo command/service module flight. Core-rope manufacture was a process whereby the computer language and program was fixed and magnetic cores carrying the relevant magnetic strength assembled for installation. It was a critical step because, once manufactured, the program could not be changed. There had been severe difficulties with the SUNDISK Earth-orbit program and if the Apollo program had not been delayed by the disastrous fire of January 27, 1967, manned flights would have been severely hampered by inadequacies in the software.

1967 October 18 The first space probe to measure conditions in the Venusian atmosphere descended toward the surface, reporting temperatures, pressures and atmospheric composition. During its approach, *Venera 4* released an egg-shaped capsule about 3 ft. in diameter and weighing 844 lb.; designed to survive 300 g deceleration. A special ablative material protected the capsule from the heat of entry, and a temperature resistant parachute was deployed to slow the rate of descent. Instruments included a barometer, a radio altimeter, an atmospheric density gauge and 11 gas analyzers. The capsule entered the predominantly carbon dioxide atmosphere at 04:34 UT and reported temperatures rising to a maximum of 536°F and a pressure of 15 Earth atmospheres. The last signals were received at 04:38 UT, and it was initially assumed the capsule had reached the surface and failed. However, analysis showed the capsule had failed at an altitude of about 15 mi., probably due to heat, and that it landed at 19° N by 38° W.

1967 October 19 The *Mariner V* spacecraft made its closest encounter with Venus and collected a comprehensive set of data. The encounter sequence nominally began at 02:49:00 UT, when the tape recorder was switched on. There were no imaging instruments on board because Venus has an opaque atmosphere. Instead, the spacecraft carried an ultraviolet photometer for measuring the upper atmosphere, a magneto-meter to determine whether the planet has a magnetic field, a trapped-radiation detector to measure any radiation belts similar to Earth's Van Allen belts, a plasma probe to measure solar wind and two investigations utilizing the spacecraft's radio signal to determine atmospheric density. Data was transmitted to Earth at 8.33 bits/sec. by the 10-W transmitter. The spacecraft reached the closest distance of 2,540 mi. at 17:34:56 UT and confirmed a dense atmosphere with a surface pressure of about 100 bars and a temperature of almost 800°F. The last data return was at 20:00:00 UT on December 4, 1967, although the spacecraft was revived between 16:35 UT on October 14 and 07:46 UT on November 8, 1968 in further tests of the communication system, and *Mariner V* was left in a 53.9 million × 68.5 million mi. orbit of sun.

1967 October 27 At 09:30 UT an SL-4 launch vehicle carried the unmanned Soyuz, designated *Cosmos 186*, from Baikonur to a 130 × 146 mi. orbit inclined 51.7° to the equator. It was to be a test of the mission sequence originally planned for the unsuccessful *Soyuz 1-2* mission when cosmonaut Vladimir Komarov had been killed. *Cosmos 186* weighed 14,100 lb. and carried a docking unit on front of the orbital module. For three days *Cosmos 186* performed maneuvers and placed itself in a 124 × 149 mi. orbit for rendezvous and docking. At 08:12 UT on October 30 another SL-4 launched an unmanned Soyuz, *Cosmos 188*, to a 124 × 171.5 mi. orbit inclined 51.7°. After maneuvering on automatic commands, *Cosmos 186* docked to *Cosmos 188* at 09:20 UT, the first time two Soviet vehicles had docked together in space. Later, TV pictures were released showing the two spacecraft linking up. The two vehicles remained docked for 3 hr. 30 min. and then separated, each maneuvering to different orbits for independent missions. *Cosmos 186* returned to Earth at 08:20 UT on October 31, and *Cosmos 188* landed at 09:10 UT on November 2.

In a test of how well spacecraft systems stood up to a lengthy sojourn in deep space, NASA's *Mariner IV* spacecraft performed a 1 min. 10 sec. firing of its course correction motor 1,062 days after launch. The burn applied a velocity change of 204 ft./sec. and modified *Mariner IV*'s orbit about the sun to 102.9 × 145.12 million mi. During August, September and October 1967 tracking stations on Earth obtained data from *Mariner IV* and *Mariner V* simultaneously to correlate observations of the solar wind from separate points in space 44 million mi. apart. *Mariner IV*'s cold gas attitude control system was depleted on December 7, 1967, but scientific data continued to be returned when requested. The last data from *Mariner IV* was received on December 20, 1967, and the mission was officially terminated, 3 years 22 days after launch.

1967 October 30 The first flight of an operational Soviet electronic intelligence (elint) gathering satellite took place when an SL-8 launcher carried the 1,930-lb. *Cosmos 189* from Plesetsk to a 325.5 × 351 mi. orbit inclined 74°. The satellite had a length of 6 ft. 7 in. and a diameter of just over 3 ft., with paddle-shaped solar array panels. Launched periodically on replacement missions, the first-generation elint satellites were used for 12 years: two in 1968; two in 1969; three in 1970; five in 1971, one of which was a booster failure; three in 1972; four in 1973; four in 1974; three in 1975; five in 1976; three in 1977; two in 1978; and one in 1979. The first-generation Soviet elint program ended with the launch of *Cosmos 1114* on July 11, 1979. The weight of this type of satellite had increased to 2,380 lb.

1967 November 3 U.S. Secretary of Defense Robert McNamara announced that beginning with the launch of an SL-11 on September 17, the Soviet Union had carried out 11 separate tests of what he referred to as the Fractional Orbital Bombardment System, or FOBS. Utilizing the power of the SL-11, the vehicle flew a trajectory which would allow it to threaten the United States from southern latitudes. The United States refrained from claiming the flights violated the UN resolution banning space weapons because its payload flew only a "fraction" of one orbit. In response, however, the Department of Defense added a counterfunction to the Nike X antiballistic missile system.

1967 November 4 In the last days of preparations for the first Saturn V launch, NASA announced a revised Apollo

During 1967 the Soviets consolidated their Molniya communication satellite system, each satellite being launched to an elliptical, 12-hour orbit.

flight schedule. In 1968 NASA planned to launch the following: AS-204 *(Apollo 5)*, an unmanned test of the lunar module (LM) in Earth orbit; the second unmanned Saturn V, AS-501 *(Apollo 6)*; the third unmanned Saturn V, AS-503; the second unmanned lunar module flight, AS-206; the first manned Apollo CSM (command/service module) flight, AS-205, on a 10-day Earth-orbit mission; and the first manned Saturn V/Apollo CSM/LM mission in Earth orbit, AS-504. In 1969 NASA planned five Saturn V/Apollo missions (AS-505 through AS-509). The first four were to be lunar distance development flights, with the first manned landing on AS-509 at the earliest. Production of Saturn V ran to fifteen launch vehicles, and after AS-509 NASA had six with which to achieve the goal set by President Kennedy on May 25, 1961.

1967 November 5 NASA launched *ATS-3 (Applications Technology Satellite-3)* by Atlas-Agena D from Cape Canaveral, Fla., to a geosynchronous orbit inclined 9.2° to the equator. The satellite weighed 805 lb. after apogee motor burnout and comprised a cylinder with a length of 6 ft. and a diameter of 5 ft., with sides covered by solar cells except for a camera opening. *ATS-3* carried nine major experiments, including several environmental and meteorological sensors as well as spacecraft technology experiments. It returned the first color photographs of the whole Earth on November 8, 1967. It was moved from its original location over 45° W to 95° W, enabling it to view the continental United States during storm season, after which it was moved back in the spring.

1967 November 7 The 2,220-lb. *Surveyor VI* was launched by Atlas-Centaur AC-14 from Cape Canaveral, Fla., at

07:39:01 UT. Placed first in an Earth parking orbit of 99 × 106 mi. 9 min. 41 sec. after liftoff, the two Centaur engines were reignited 12 min. 52 sec. later for a 1 min. 56 sec. burn that put the spacecraft on a translunar course. A 10.2-sec. (33.2 ft./sec.) midcourse burn began at 02:20:02 UT the next day. The landing site for *Surveyor VI* was the same as that selected for the failed *Surveyor IV*, a level mare area, a region of dried basaltic lava, in the Sinus Medii. It was to be the last Surveyor mission directly in support of the requirements of NASA's manned Apollo moon landing program, as the unusually high success with Surveyor landings allowed the final mission (*Surveyor VII*) to be sent to an area of purely scientific interest.

At the military parade in Red Square in Moscow, the Soviet Union publicly unveiled its heavyweight silo-based SS-9 ICBM, which was just entering operational service. With a length of 118 ft. 6 in. and a diameter of 10 ft. 2 in., the SS-9 had a mass of 420,000 lb. and storable liquid propellant combusted in six first-stage rocket motors and four verniers. But it was the throw-weight capacity of 11,000 lb. — and its 25-MT warhead — that shocked Western defense analysts. Compared to the capacity of other Soviet ICBMs (at 4,000 lb., the SS-7 had been the heaviest) and U.S. ICBMs (8,300 lb. for the Titan II, 1,625 lb. for Minuteman II and 2,400 lb. for Minuteman III), it was massive. Moreover, earlier Soviet ICBMs had an accuracy across full range of 4,600 ft. at best, but the SS-9 could get 50% of its warheads within 2,450 ft. of the target across a 7,000-mi. range. A maximum 308 SS-9s were operational by 1972. Replacement on a one-for-one basis with the still more massive SS-18 began in 1975 and was completed in 1980.

1967 November 8 Maxime Faget, director of Engineering and Development at NASA's Manned Spacecraft Center, defined options for reducing the threat of rapid propagation of fire in a spacecraft pressurized with pure oxygen. To reduce the risk of fire on the launch pad when the spacecraft was pressurized above 15 lb./sq. in., it was suggested that the interior of the Apollo command module should have a 60–40 mixture of oxygen and nitrogen before launch. As the vehicle ascended through the atmosphere, the cabin pressure would bleed down to about 5 lb./sq. in.; thereafter 100% oxygen would fill the cabin, thus satisfying the physiological requirements of the crew and reducing the threat of excessive flame propagation from pure oxygen at very high pressure.

1967 November 9 At 12:00:01 UT the 363-ft. Saturn V (AS-501) lifted off from Complex 39A at the NASA Kennedy Space Center at the start of the *Apollo 4* mission carrying the 51,591-lb. Apollo CSM-017 spacecraft and a 29,500-lb. simulated lunar module designated LTA-10R. With the 8,710-lb. launch escape system and the 3,880-lb. spacecraft/lunar module adapter, the Saturn V had a total mass of 6.12 million lb. and a payload weight of 93,681 lb. above the three stages. The Saturn V was instrumented to measure the performance of 2,894 engineering functions in flight. The first stage produced a liftoff thrust of approximately 7.5 million lb. and consumed RP-1–liquid oxygen propellants at the rate of 15 tons/sec. As the air got thinner, thrust increased to almost 9 million lb. at 2 min. 15 sec., when the center F-1 engine shut down. At 2 min. 31 sec., a speed of 8,950 ft./sec. and a height of 39 mi., the first stage (S-IC) shut down and separated 1 sec. later. Eight forward-facing retro-rockets in the fairings that covered four of the five F-1 engines fired for 0.5 sec., producing a total thrust of 606,400 lb. to shunt the giant stage back away from the ascending stack. It impacted the Atlantic

Ocean 390 mi. downrange, 9 min. into the flight. The second-stage (S-II) engines ignited at 2 min. 32 sec., and 29 sec. later the 10,650-lb. interstage adapter was jettisoned, followed 6 sec. after that by the launch escape system. The S-II fired for 6 min. 7 sec., taking the vehicle to a height of 117 mi. and a speed of 22,400 ft./sec. Four forward-facing retro-rockets produced a thrust of 139,240 lb. for 1.5 sec. to shunt the stage back away from the third stage (S-IVB) and payload when it separated at 8 min. 40 sec. The S-II impacted the Atlantic Ocean, 2,600 mi. downrange, 20 min. after launch. The S-IVB ignited at 8 min. 41 sec. and burned for 2 min. 25 sec., injecting the 81,091-lb. payload into a 113 × 118 mi. orbit. At 3 hr. 11 min. 26 sec. the S-IVB fired for 5 min., raising apogee to 10,693 mi. After separating from the S-IVB at 3 hr. 26 min. 28 sec., the Apollo SPS engine fired for 15 sec., raising altitude to 11,242 mi. At 8 hr. 10 min. a second SPS burn increased velocity for a simulated lunar return reentry at 36,000 ft./sec. and a peak deceleration force of 7.3 g. The command module splashed down in the Pacific at 8 hr. 37 min. 9 sec.

1967 November 10 *Surveyor VI* landed on the moon. The retro-rocket was ignited at 00:58:03 UT with the spacecraft at an altitude of 230,854 ft. It burned for 37.4 sec. and the verniers continued to decelerate *Surveyor VI* for a 12.5 ft./sec. touchdown at 01:01:04 UT in the Sinus Medii at 0.47° N by 1.38° W. At 10:32:01 UT the three vernier engines were reignited, causing the 661-lb. spacecraft to hop across the surface. Two cutoff commands were sent at 2 sec. and 2.5 sec. after ignition, but Surveyor responded only to the latter, causing the spacecraft to rise 10–15 ft. in the air and return to the surface 8 ft. away after a flight time of 6.1 sec. and consumption of 1.4 lb. of propellant. *Surveyor VI* carried a TV camera and alpha-scattering device for 30 hr. of active surface chemical analysis. A total 29,952 pictures were transmitted before *Surveyor VI* was shut down at 06:41 UT on November 26, 1967.

1967 November 20 NASA announced astronaut crews for the first two manned Saturn V missions, at this time scheduled as AS-504, in the last quarter of 1968, and AS-505, in the first quarter of 1969. AS-504 would be flown by James McDivitt, David Scott and Russell Schweickart, with backup crew Charles Conrad, Richard Gordon and Alan Bean. A support team of astronauts Edgar Mitchell, Fred Haise and Alfred Worden was appointed to assist with flight preparations. AS-505 would be flown by Frank Borman, Michael Collins and William Anders, with Neil Armstrong, James Lovell and Edwin Aldrin as their backups. The support crew comprised Thomas Mattingly, Gerald Carr and John Bull.

1967 November 23 The first in a series of precursor Soviet navigation test satellites was launched by SL-8 from Plesetsk to an orbit of 464 × 473 mi. at an inclination of 74°. Operating under the designation *Cosmos 192*, the 1,650-lb. satellite had a diameter of about 7 ft. and a length of about 5 ft. Five more satellites of this type were launched by SL-8 before the first-generation system was flown, the last being *Cosmos 371* on October 12, 1970. Beginning with *Cosmos 385*, launched December 12, 1970, operational satellites were placed in slightly higher orbits.

1967 November 25 A full-scale mock-up of the Soviet N-1 manned lunar landing launcher was erected on one of two launch pads built for the giant rocket at the Baikonur cosmodrome. Starting from this date, tests were conducted for three

weeks to qualify the ground facilities, conduct interface checks and provide training for servicing crews. The two N-1 launch pads were located 1,640 ft. apart, and each comprised a circular concrete section 100 ft. in diameter with three covered flame trenches. Each pad had a 410 ft. tall servicing tower and two 590 ft. tall lightning towers. Two rail tracks linked the pads with the assembly building and carried the launchers in a horizontal position on a special transporter-erector, which could raise the launchers to the vertical at the pad. The N-1 was secured to the pad by a circular ring which mated with the outer circumference of the first stage and held it down with 36 explosive bolts.

1967 November 29 Australia became the fourth country to put a satellite in orbit when the 107-lb. *WRESAT-1* (weapons research establishment satellite) was placed in an elliptical orbit of 106 × 776 mi. inclined 83.2° to the equator. The United States had provided a modified Redstone rocket having three solid propellant upper stages and called SPARTA (special antimissile research test). The satellite was a joint product of the Adelaide University Physics Department and the government Department of Weapons Research at Salisbury. It consisted of a conical structure 2 ft. 6 in. in diameter with a battery pack for a 10-day life expectancy. It was instrumented to send back information on solar radiation. *WRESAT-1* decayed on January 10, 1968.

1967 December 5 The last of six OV3-series satellites built for the U.S. Air Force was launched by Scout from Vandenberg AFB, Calif., to a 252 × 271 mi., 90.6° orbit. The 209-lb. *OV3-6* consisted of a standard OV3 satellite package and carried two basic experiments to measure the composition, temperature and electron density of the extremely tenuous upper ionosphere. It transmitted for five days and decayed out of orbit on March 9, 1969.

1967 December 7 NASA established the Aerospace Safety Advisory Panel with a mandate to monitor safety and operational plans. It was to submit regular reports to the administrator and make recommendations about technical decisions and their impact on safety and reliability. It was also expected to examine plans for future projects and advise on safety aspects. The panel was constituted of a chairman, vice-chairman and seven other members.

1967 December 13 A thrust-augmented Delta put in Earth orbit a 44-lb. satellite designed to test the Apollo tracking and communication network. Designated *TTS-1,* the TRW satellite was one of the U.S. Air Force Environmental Research Satellite (ERS) series. It was a double-pyramidal structure, 1 ft. across each side; each triangular flat carried 111 solar cells powering three VHF and one S-band antenna. The second stage and payload was placed for 22 minutes in a parking orbit of 191 × 373 mi., inclined 32.9° where the *TTS-1* was ejected from a canister on the Delta second stage to a 187 × 303 mi. orbit. That stage then fired to send the *Pioneer 8* science satellite to heliocentric orbit between 93 and 102 million mi. from the sun. *TTS-1* decayed on April 28, 1968, but in 1995 *Pioneer 8* continued to return data.

1967 December 14 In an unparalleled reactivation on the moon, *Surveyor*s *V* and *VI* responded to signals sent from Earth. *Surveyor V* responded at 06:45 UT, reactivating on its fourth lunar day since touchdown on September 11, 1967. *Surveyor VI* was reactivated by a signal on its second lunar day at 16:41 UT, but data was erratic and the last contact with that spacecraft occurred at 19:14 UT, nothing more being heard from it. In a sequence of communication sessions with *Surveyor V*, which continued through 04:30 UT on December 17, an additional 64 TV images in 200-line format were transmitted to Earth, bringing to 19,118 the total picture count for that mission.

1967 December 18   The European Launcher Development Organization (ELDO) reported that it had decided to back development of a Europa III launch vehicle following development of Europa II. The new launcher would have the Europa I first stage (Blue Streak) and a liquid hydrogen and liquid oxygen second stage with two 14,000-lb. thrust engines. Europa III would be capable of placing a 6,000-lb. payload in low Earth orbit, or 1,200 lb. into geostationary orbit. Europa III could also acquire two 420,000-lb. solid propellant strap-on stages, raising the geostationary payload to 4,000 lb. ELDO said that it was studying high specific-impulse systems for transferring payloads from low to high Earth orbit and that a small nuclear engine might be considered for this task.

1967 December 19   NASA established a new Lunar Exploration Office within the Apollo Program Office at Headquarters. Headed by Lee Scherer, it would coordinate efforts on modifying Apollo hardware for extended surface activities. From a proposed 39 lunar explorations sites, 9 would be selected for single-launch missions; the proposed dual-launch missions involving lunar module taxi and shelter vehicles for extended stays were considered a long way off. Instead, full utilization of Apollo hardware was expected to provide maximum science payloads of 750 lb., lunar stay-times of up to three days with four EVAs (extravehicular activities) and a roving vehicle capable of transporting astronauts and tools to a variety of geological sites in the vicinity of the lunar module. Landings within 3,300 ft. of the designated spot were to be achieved as a priority.

1967 December 27   An SL-11 launcher carried the first Soviet ocean surveillance satellite, *Cosmos 198*, from Baikonur to a 165 × 175 mi. orbit inclined 65°. Weighing 10,000 lb., it carried radar capable of resolving objects as small as 75 ft. long, and it was powered by a thermionic nuclear reactor fitted to one of three sections. The other two sections comprised the propulsion unit and the radar elements. With its own tiny boost motor, the reactor section would be pushed to a higher, parking orbit at the end of the mission so that it would not contaminate the atmosphere with radioactive waste products. Operational capabilities were limited by poor performance and spasmodic launches. The four satellites of this type to be successfully placed in orbit through December 25, 1971, operated for two, six, eight and ten days, respectively.

1967 December   French scientists began flight tests, with the first stage of a two-stage submarine-launched ballistic missile being developed for deployment aboard French submarines in the 1970s. Under the MSBS *(Mer-Sol Balistique Stratégique)* program, five submarines would be armed with 16 M-1 missiles; a sixth was added later. The M-1 had a length of 34 ft. 1 in., a diameter of 4 ft. 11 in. and a launch weight of 39,700 lb. The first stage was designated P10 and had a thrust of 100,000 lb. for 50 sec. Designated P4, the Rita 1 second stage had a thrust of 39,700 lb. for 55 sec. M-1 had a range of 1,490 mi. carrying a single 500-KT warhead.

1967   Nine cosmonauts began specialized training for the manned moon landings that the Soviets hoped to accomplish by the end of the decade. They comprised five military commanders (Valery F. Bykovsky, Georgi T. Dobovolsky, Pavel R. Popovich, Pyotr I. Klimuk and Aleksei A. Leonov), a military engineer (Yuri P. Artyukhin) and three civilian engineers (Georgi M. Grechko, Oleg G. Makarov and Nikolai N. Rukavishnikov). Training for the candidates was directed by cosmonaut Vitaly I. Sevastyanov, until he was replaced by Pavel I. Belyaev at the end of 1967. Because lunar trajectories made it less likely that cosmonauts would return to Earth on land, water recovery techniques were tested in the Black Sea. The team began to disband and go to other assignments during mid-1969 and had been completely absorbed into the training program for Earth-orbit application of Soyuz/Salyut space stations by the end of that year.

At the end of the year the Soviet navy took delivery of the first of 34 Yankee-class ballistic missile submarines, the first Soviet submarines capable of carrying 16 SLBMs (submarine-launched ballistic missiles); the United States had achieved that capability seven years earlier. Two rows of eight vertical launch tubes were incorporated in the hull aft of the sail. Based on the design of the five U.S. *Ethan Allen*–class (Polaris) submarines launched between 1960 and 1962, the Yankee submarines had a 9,600-ton submerged displacement and, beginning in 1968, were equipped with 16 third-generation SS-N-6 Sawfly SLBMs. Produced by a design team led by Viktor P. Makayev, the two-stage liquid-fueled SS-N-6 had a length of 31 ft. 7 in., a diameter of 5 ft. 5 in. and a weight of 42,000 lb. The Mod 1 version had a range of 1,500 mi. and a 1-MT warhead. In all versions, peak deployment of 548 SS-N-6s was reached in 1976, with a gradual decline thereafter to 192 by the early 1990s, when the missile was removed under terms of the START (Strategic Arms Reduction Treaty).

1968 January 6   NASA headquarters received a briefing from Lockheed Aircraft Corp. on the STAR (space transport and recovery) Clipper conceived as a private company venture. The STAR Clipper spacecraft itself comprised a delta-planform lifting body with upswept vertical stabilizers inclined outwards from each side of the rear and capacity for up to 22,000 lb. of cargo. The initial version had a length of 82 ft. and a span of 52 ft., with power provided by three cryogenic LO-LH$_2$ (liquid oxygen–liquid hydrogen) engines situated at the rear. The bulk of the propellant for ascent to orbit was contained in two cylindrical, jettisonable tanks connected at the nose in the form of an inverted "V" nesting the delta-shaped space plane and producing a total length of 134 ft. Future versions were bigger and capable of placing 50,000 lb. in Earth orbit. This and alternative configurations designed by McDonnell Douglas were among the last reusable space vehicles proposed by industry prior to NASA's integral launch and reentry vehicle contracts awarded October 30, 1968.

1968 January 7   The 2,289-lb. *Surveyor VII* spacecraft was launched at 06:30:01 UT by Atlas-Centaur AC-15 to a 99 × 104 mi. Earth parking orbit achieved 9 min. 53 sec. after liftoff from Complex 36 at Cape Canaveral, Fla. on a lunar landing mission. The Centaur-spacecraft combination coasted for 22 min. 26 sec. before the main engines were ignited a second time, for 1 min. 56 sec., putting *Surveyor VII* on course for the moon. Guidance was so accurate that the Centaur stage put Surveyor on course that, if uncorrected, would have been only 48 mi. from the target coordinates. The spacecraft would head for its landing site to the south, but the Centaur stage would be made to vent excess propellant, causing a change in its flight path so that it missed the moon by 12,200 mi. — a maneuver typical of all Surveyor missions. An 11.4-sec. (36.4 ft./sec.) midcourse correction burn was conducted at 23:20:09 UT, setting up the spacecraft trajectory for a landing in a hilly highland area of the moon north of the crater Tycho on January 10, 1968.

1968 January 10 *Surveyor VII*, the last of NASA's unmanned lunar landers, successfully touched down on the moon's surface at 40.92° S by 11.45° W, far to the south of the Apollo landing areas to which the first six spacecraft in this series had been targeted. Retro-rocket ignition came at 01:02:14.5 UT, and the burn lasted 42.9 sec., leaving the three vernier motors to lower the spacecraft to within 13 ft. of the surface and at a speed of 5 ft./sec., at which point they shut down. Now weighing 674 lb., *Surveyor VII* touched down at 01:05:36.3 UT, having accelerated to 12.5 ft./sec. in 1.3 sec. of free fall. *Surveyor VII* carried a TV camera, an alpha-scattering instrument and a soil-mechanics surface sampler (SMSS). In 36 hr. 21 min. of SMSS operations, *Surveyor VII* conducted sixteen bearing, seven trench and two impact tests. It also obtained 65 hr. 12 min. of data from the alpha scatterer and transmitted 20,993 pictures during the first lunar day, which ended as *Surveyor VII* was shut down at 14:12 UT on January 26.

1968 January 11 NASA successfully launched the second of three Geodetic Earth Orbiting Satellites, *Geos 2 (Explorer 36)*, on a thrust-augmented Delta from Vandenberg AFB, Calif., to a 671 × 978 mi. orbit inclined 105.8°. The 460-lb. satellite took the form of an octagon surmounted by an eight-sided truncated pyramid, to which was attached a gravity gradient stabilization boom extendable to a maximum 60 ft., and a 2 ft. hemispheric base, to which was attached a spiral antenna. The satellite carried six geodetic instruments for measuring the size and shape of the Earth as well as strengths and variations in the Earth's gravitational field.

1968 January 22 The AS-204 *(Apollo 5)* Saturn IB was launched from Cape Canaveral's Complex 37B at 22:48:08 UT, carrying the first Apollo lunar module, LM-1, into a 138 x 101 mi. orbit 10 min. 3 sec. later. The 31,528-lb. LM-1 carried no landing gear and was flown to qualify on-board systems and to demonstrate ascent and descent stage propulsion. The planned 38-sec. burn of the 10,500-lb. thrust descent propulsion system (DPS) began at 3 hr. 59 min. 42 sec. but ended after just 4 sec. when incorrect sensor measurements shut down the engine. Switching to an alternate mission plan, the DPS was fired twice, first at 6 hr. 10 min. 42 sec. with a 35-sec. firing, of which 26 sec. was at 10% and the rest at full throttle, and again at 6 hr. 11 min. 47 sec. on full throttle for 24 sec. This was immediately followed by separation of the ascent and descent stages and a 60-sec. firing of the 3,500-lb. thrust ascent propulsion system at 6 hr. 12 min. 15 sec., giving an orbit of 107 × 597 mi. A second firing at 7 hr. 44 min. 13 sec. lasted 5 min. 47 sec. until propellant depletion. The ascent stage broke up in the atmosphere at 7 hr. 52 min. 10 sec.

1968 January 29 In his fiscal year 1969 budget address before Congress, President Johnson endorsed a revitalized Mars exploration program when he announced a funding request for a new orbiter/lander spacecraft to be launched in 1973. As a cut-down version of the canceled Voyager orbiter/lander, which would have weighed almost 10,000 lb., NASA now proposed a 1973 orbiter/lander weighing 3,400 lb. for launch on a Titan III derivative. The two 1973 orbiters would carry hard landers, each weighing 800 lb. and capable of

putting 22 lb. of scientific instruments on the surface. Later orbiter missions would carry soft landers. On February 9, 1968, NASA started up a study of Titan III launchers and examined the prospect of using the cryogenic Centaur, which would have the capability of sending 7,800 lb. to Mars. Langley Research Center was assigned project management of what in November 1968 became the Viking program.

The French and German governments issued a joint request for bids on a Franco-German communication satellite, Symphonie. France wanted to have two such satellites launched on a Europa II vehicle in time for the next Olympic Games, due to be held in Munich in 1972. The first Symphonie was expected to be placed in geostationary orbit over 15° W and have the capacity to relay 200 telephone calls, 18 radio links or several TV channels. On October 28 a consortium of European companies was chosen to build the satellites. Called the CIFAS group (Consortium Industriel Franco-Allemand pour le Satellite Symphonie, formed on April 25, 1968), they included: MBB (Messerschmitt-Bölkow-Blohm), Compagnie Francais Thomson-Houston-Hotchkiss-Brandt, Compagnie Générale de Télégraphie sans Fils, Junkers, Société Anonyme de Télécommunication, Sud Aviation and Nord Aviation.

1968 February 5 NASA design teams defined two distinct options available to engineers for the Saturn V–launched orbital workshop (OWS) for the Apollo Applications Program types. OWS-B would be a generic evolution from the Saturn IB–launched workshop (OWS-A) and incorporate the solar Apollo telescope mount (ATM) as an integral part of the hardware for launch. OWS-C would be more advanced and have habitability provision for up to nine people and an overall life of one to two years. Each Saturn V workshop was to be fitted out on the ground and launched into Earth orbit by a two-stage Saturn V, unlike the Saturn IB–launched workshop, which would first propel itself into orbit, where astronauts launched on another Saturn IB would make it habitable.

1968 February 12 Responding to a signal sent from Earth at 19:00 UT, the *Surveyor VII* spacecraft was reactivated on the surface of the moon about 132 hr. after dawn on its second lunar day. Various engineering tests were performed, and the spacecraft was in good working order to conduct limited scientific activity. This included the return of 45 TV images in 200-line format, bringing to 21,038 the total number of pictures transmitted by *Surveyor VII*. The spacecraft also transmitted an additional 34 hr. 30 min. of alpha particle scatterometer data, bringing that total to 99 hr. 52 min. Contact with *Surveyor VII* was lost at 00:24 UT on February 21, and further attempts were abandoned at 06:48 UT. Thus ended NASA's unmanned lunar lander program with a total of 87,674 TV pictures. Surveyor paved the way for complex, manned lunar operations and gave valuable experience to the Jet Propulsion Lab in managing sophisticated planetary spacecraft.

1968 February 27 Addressing the Economic Club of Detroit, Mich., NASA Administrator of Manned Space Flight George Mueller gave the first public confirmation that fully reusable vehicles were to be "the backbone" of future U.S. space activities. Mueller perceived that through the use of reusable logistics vehicles, launch costs would fall from the then current average of $2,500 per pound of payload using expendable rockets, to somewhere between $5 and $50 per pound. He went on to predict that "low-cost space transportation, . . . in time, could be competitive on a cost-per-pound-per-mile basis with other forms of transport."

1968 February 28 Testifying before the Senate Committee on Aeronautical and Space Sciences, NASA's Associate Administrator for Manned Space Flight George Mueller explained the current schedule for Apollo Applications Program flights. There had been further reductions, and NASA now expected to fly just one Saturn IB workshop for in-flight conversion and one Saturn V workshop fitted out on the ground. There were to be no AAP lunar flights. Launched in 1970, the first Saturn IB orbital workshop would be manned for 28 days by a crew launched on *AAP-1*. After a brief hiatus, *AAP-3A* would carry the second crew for a 56-day mission. In 1971 *AAP-3* would put the third crew aboard the workshop, also for 56 days, and *AAP-4* would lift the Apollo telescope mount for docking at the cluster. Also during 1971 three revisit flights *(AAP-5, AAP-6* and *AAP-7)* were planned for a total of six periods of habitation and continuous operation for up to 150 days. The Saturn V workshop was to be occupied 1972–1973.

1968 March 2 At 18:30 UT a four-stage SL-12 launch vehicle carried the 11,300-lb. circumlunar Soyuz 7K-L1 variant, designated *Zond 4*, from Baikonur cosmodrome to a 119 × 127 mi. orbit. Once there, the fourth stage fired to place it in a highly elliptical orbit for deep space communication checks and high-speed reentry. However, nothing more was heard of the vehicle; it did not achieve the planned objectives for this L-1 mission because a velocity sensor failed and attitude was incorrect for reentry, preventing the descent and instrument modules from separating. *Zond 4* was destroyed. Another lunar-type Soyuz launch under the Zond designation attempted on April 22, 1968, failed to reach orbit when the escape system fired without reason. Unlike the March 2 launch, the April 22 attempt was prepared as a circumlunar mission similar to that conducted by *Zond 5*, launched on September 14, 1968. Another try on July 28, 1968, was aborted when a propellant tank was found to be cracked.

1968 March 4 NASA's fifth Orbiting Geophysical Observatory, *OGO-5*, was launched by Atlas-Agena D from Cape Canaveral, Fla., to a 144 × 92,109 mi., 31.1° orbit. The 1,347-lb. satellite carried 25 experiments for particles and fields studies, radio astronomy and the measurement of solar emissions. *OGO-5* operated as designed for 41 months until the attitude control system failed in August 1971; it was put on standby two months later. It made the first observations of a hydrogen cloud surrounding the Earth and conducted the first detailed measurement of electric fields at the edge of the magnetosphere. Operations were terminated on July 14, 1972. *OGO-5* remains in space.

1968 March 5 A solid propellant Scout launched the 198 b science satellite *Solar Explorer-B (Explorer 37)* from Wallops Island, Va., to a 324 × 545 mi., 59.4° orbit. Intended for a circular 520 mi. orbit, the off-nominal insertion was caused by a malfunction in the Scout first stage. *Explorer 37* was a 12-sided cylinder, 2 ft. 6 in. in diameter and about 2 ft. 3 in. high. Solar cells on twenty-four 10 × 7 in. panels provided power for the instruments designed to measure solar x-ray and ultraviolet emissions. *Explorer 37* was the second of two solar-explorer missions, the first of which had been launched on November 19, 1965. *Explorer 37* decayed into the atmosphere on November 16, 1990.

1968 March 15 The first flight of the modified HL-10 lifting body took place following an air-drop from a NASA B-52 mother plane over the Flight Research Center, Calif. The fastest speed of Mach 1.86 (1,227 MPH) was achieved on February 18, 1970, followed nine days later by the highest altitude of 90,303 ft. when the basic HL-10 made its 35th flight. Its single XLR11 rocket engine was replaced with three 500-lb. thrust hydrogen-peroxide engines. These enabled the aircraft to reduce the approach path from 18° to 6° and then to flare at 200 ft. for an off-power touchdown. This demonstrated that a Shuttle landing from orbital flight would be able to make an unpowered descent and land without the use of auxiliary engines. Two flights in this configuration on June 11 and July 17, 1970, completed the HL-10 flight research program.

1968 March 21 The first of a new generation of Soviet military photoreconnaissance satellites was launched by SL-3 from Baikonur cosmodrome under the designation *Cosmos 208*. This third generation incorporated a modified service module providing increased consumables for 12–13 days. Several satellites of this type carried piggyback payloads which were separated before the descent module returned to Earth with film. Second- and third-generation satellites were used concurrently until the third quarter of 1970, when the third generation took over. In that year, the Soviets launched 18 third-generation satellites, with 28 launched in 1971, 29 in 1972, 35 in 1973, 28 in 1974 and 33 in 1975, when the fourth generation was introduced. The third-generation satellites continued to be used with reconnaissance satellites of the fourth and fifth generation.

1968 March 29 NASA headquarters asked the Manned Spacecraft Center to review options for lunar photographic tasks on Apollo missions. With the abandonment of the lunar mapping and survey system module there was no provision for using the Apollo spacecraft in lunar orbit for lunar photography and mapping during the period the lunar module was on the surface. From this requirement evolved the panoramic and mapping cameras mounted in the scientific instrument module bay on J-series *Apollos 15–17*.

1968 April 3 Queen Beatrix and Prince Klaus of the Netherlands officially opened the European Space Technology Centre (ESTEC) at Noordwijk, Netherlands. The center would be the focus for satellite and payload design and development for ESRO (European Space Research Organization) and, later, the European Space Agency. ESTEC would also provide technical support for the engineering development of satellites and it would be the center of European research on systems and subsystems. In these respects, ESTEC's function was analogous to that of NASA's the Goddard Space Flight Center.

1968 April 4 At 12:00:01 UT the 6.108-million-lb. Saturn V AS-502 (*Apollo 6*) was launched from Complex 39A at the NASA Kennedy Space Center, carrying the 55,420-lb. Apollo CM-020, SM-014 and the 26,001-lb. lunar module simulator LTA-2R. Severe oscillations built up toward the end of first stage burn, and a portion of the spacecraft–lunar module adapter broke free at 2 min. 13 sec. The center F-1 engine shutdown occurred at 2 min. 25 sec., followed 3.5 sec. later by shutdown of the five outboard F-1 engines. Following S-IC/S-II separation, the five second-stage engines ignited at 2 min. 30 sec., but 4 min. 23 sec. later one of the outer J-2 engines shut down, followed by a second outer engine 1.3 sec. after that.

Allowed to burn 58 sec. longer than planned to compensate for the other premature shutdowns, the remaining three engines shut down at 9 min. 36 sec., low on velocity, after which the S-IVB separated and ignited its engine 1 sec. later. That stage burned for 7 min. 6 sec., 39 sec. longer than planned, and put the 81,421-lb. payload in a 110 × 225 mi. orbit instead of the planned 100-mi. circular orbit. At 3 hr. 13 min. 35 sec. the S-IVB's J-2 ignited a second time but shut down 16 sec. later, well short of its planned 5 min. 26 sec. burn. The spacecraft separated at 3 hr. 14 min. 28 sec., and 1 min. 38 sec. later the SPS engine fired for 7 min. 22 sec. to push Apollo to a maximum altitude of 13,800 mi., a job the S-IVB should have done. That left sufficient propellant for a second burn of only 23 sec. so it was canceled and the command module entered Earth's atmosphere at 32,830 ft./sec. for a Pacific splashdown at 9 hr. 57 min. 20 sec. The early shutdown of the J-2s on the S-II and S-IVB was attributed to fractured fuel lines caused by excessive vibration. The second J-2 in the second stage shut down because of incorrect wiring.

1968 April 6   Carried for the first time on an Atlas F launcher retired from duty as an ICBM, two OV1-series satellites were launched from Vandenberg AFB, Calif., in a 7-ft. fairing. The 235-lb. *OV1-11* was instrumented to measure radiation and evaluate the effects of the space environment on bearings and friction surfaces between different materials; it was placed in a 341 × 5,792 mi. orbit. The 222-lb. *OV1-14* carried equipment to measure radiation in the Van Allen belts and was placed in an orbit of 343 × 6,193 mi. Both orbits had an inclination of 100°, and each satellite was powered to its location by separate 5,977-lb. thrust FW-4S motors. They remain in space.

1968 April 7   A Soviet SL-6 launch vehicle launched the 3,750-lb. *Luna 14* spacecraft to an interim Earth parking orbit from which the terminal stage boosted it toward the moon. Three days later it slipped into a 99 × 540 mi. orbit inclined 42° to the lunar equator. The 540-lb. payload separated immediately thereafter, and although no pictures were transmitted, the spacecraft examined the lunar environment, solar-charged particle interaction with the Earth-moon system and lunar mass distribution. *Luna 14*, like its orbital predecessor *Luna 12*, carried out tests with electric motors of the type being developed for the Lunokhod moon rover. *Luna 14* was the last of the second-generation lunar spacecraft first launched on January 4, 1963.

1968 April 14   At 10:00 UT an SL-4 launch vehicle carried a 14,100-lb. Soyuz spacecraft, *Cosmos 212*, to a 130 × 148.5 mi. orbit inclined 51.6°. The next day at 09:34 UT another Soyuz, *Cosmos 213*, was placed in a 127 × 181 mi. orbit also inclined 51.6°. At 10:21 UT *Cosmos 212* maneuvered to a docking with *Cosmos 213*, the second time two unmanned Soyuz spacecraft had docked automatically in space. They separated after 3 hr. 50 min., *Cosmos 212* landing at 08:10 UT on April 19, followed by *Cosmos 213* at 10:11 UT on April 20. This second automated rendezvous and docking with Soyuz spacecraft cleared the way for both the first docking with manned vehicles and crew exchange, which would take place in January 1969 with the *Soyuz 4/5* mission.

1968 April 16   Following the review of a 10-year plan for European satellite launchers, Britain withdrew from both the Europa program and from ELDO (European Launcher Development Organization). The United Kingdom agreed to maintain treaty obligations to support ELDO activities up to

On April 16, 1968, the British government announced its withdrawal from the European Launcher Development Organization (ELDO). Here, the Europa I is seen prior to flight; Britain's Blue Streak is the first stage.

1971 and said it would concentrate instead on the national launcher, Black Arrow, but make Blue Streak (the first stage of Europa I) available on a commercial basis through the end of 1976. This action stirred the French into proposing a new arrangement whereby a completely redesigned launch vehicle using extant technology would provide improved operational efficiency in a way Blue Streak, with mid-1950s technology, could not.

1968 April 27   Despite serious problems that almost caused the failure of the *Apollo 6* mission on April 4, 1968, NASA Administrator James Webb approved plans to prepare the next Saturn V for a manned flight in the fourth quarter of 1968. Engineers studying telemetry from AS-502 understood the problem was caused by vertical oscillations. Called Pogo, this had been troublesome during early test launches of the Air Force Titan II launch vehicle for the Gemini two-man spacecraft. The problem could be solved by injecting helium into the Saturn V engine systems to break up the resonant frequencies that caused the oscillations. AS-503 was scheduled to be flown by astronauts McDivitt, Scott and Schweickart.

1968 May 6   A lunar landing research vehicle (LLRV) in which astronaut Neil Armstrong was practicing lunar landing techniques crashed and was destroyed during a flight session at Ellington AFB, Tex. The LLRV had a conventional jet engine supporting five-sixths of the vehicle's weight, thus simulating the effect of the moon's gravity (one sixth that of Earth). Lift rockets were used to propel the LLRV while 16 attitude thrusters provided directional control. It was thruster failure that caused the vehicle to pitch over. Armstrong ejected at a height of 200 ft., receiving superficial injuries. A second LLRV was lost in an accident at Ellington on December 8 when NASA test pilot J. Algranti ejected.

1968 May 7   The first flight-rated N-1 launch vehicle was erected on one of two Baikonur cosmodrome launch pads specially built for the giant Russian rocket. Soviet mission planners wanted four or five successful test flights of the unmanned launch vehicle and payload before scheduling the first manned lunar landing. Although a landing on the moon before the Americans had been declared an official objective by the Soviet government, funds had been less than many deemed necessary and short cuts had been made. Vasili Mishin, head of the Soviet space program since the death of Sergei Korolev in 1966, had been denied resources to build test stands for the N-1 rocket engines. The first time they were fired would be when they were ignited for launch. During checkout of the first N-1, cracks were discovered in the first stage and the vehicle was returned to the assembly building. It was moved back to the launch pad in November 1968 and then replaced on the pad by a mock-up. It was finally taken back to its pad in mid-January 1969.

1968 May 10   NASA completed a statement of work defined after consultation with the Manned Spacecraft Center and the Marshall Space Flight Center for a six-month study of an integral launch and reentry vehicle (ILRV). The guidelines required contractors to conduct an economic analysis of alternative concepts as well as a technical examination of different engineering considerations. For the first time NASA was bringing together technical and cost factors in studies of a major new concept: reusable launch systems. This was the formal start on what would ultimately become the Space Shuttle program. On October 30, 1968, NASA issued a request for proposals from industry.

1968 May 17   The *ESRO 2B* science satellite was successfully launched from Vandenberg AFB, Calif., by a NASA four-stage Scout to a 207.5 × 674 mi., 97.2° orbit. Launched under a reimbursable agreement with the European Space Agency, *ESRO 2B* was physically identical to its predecessor, *ESRO 2A*, and carried seven instruments similar to those carried by NASA's Orbiting Solar Observatory series. Built in England by Hawker Siddeley Dynamics, the satellite was powered by 3,456 solar cells and spin-stabilized at 30–40 RPM. Six of the seven experiments operated as designed and the satellite decayed out of orbit on May 8, 1971.

1968 May 23   A U.S. Air Force Thor-Burner II carried the first of three 180-lb. *DSMP* Block IVB satellites from Vandenberg AFB, Calif., to a 509 × 560 mi. orbit inclined 98.9°. An improvement on the original Block IVA, in addition to the two vidicon cameras each Block IVB satellite carried infrared sensors for night cloud mapping. The second satellite of this type was launched on October 23, followed by the last on July 23, 1969. Each satellite operated for about two years.

1968 June 26   NASA and the Atomic Energy Commission successfully tested the Phoebus 2A nuclear rocket reactor at the Jackass Flats, Nev., test facility. During the 32-min. test the motor developed 4,200 mW and for 12 min. ran in excess of 4,000 mW. The power generated was equal to a thrust of 200,000 lb., almost three times the 75,000-lb. thrust NERVA motor which was to exhibit a specific impulse of 825 sec. The reference baseline for the NERVA upper stage assembly specified a length of 140–150 ft., a diameter of 33 ft. (for compatibility with the first two stages of a Saturn V) and a 280,000–300,000-lb. fuel capacity. Over the next year NASA drew up plans for a first NERVA flight in late 1977.

1968 June   The Space Science Board of the U.S. National Academy of Sciences conducted a study of options for the scientific investigation of the solar system and two months later issued its report *Planetary Exploration 1968–1975*. It recommended "a substantial fraction of the total NASA budget be devoted to unmanned planetary exploration" and, especially, "a program of Pioneer/IMP-class spinning spacecraft to orbit Venus and Mars at every opportunity." It also recommended a "multiple drop-sonde mission to Venus in 1975, and a major lander on Mars, perhaps in 1975." In the event, the Viking mission to Mars was launched that year, and the *Pioneer Venus-1* (orbiter) and *-2* (multiprobe) missions were launched in 1978.

1968 July 1   NASA's Office of Manned Space Flight completed a summary of advanced vehicle planning studies in "support of future manned and unmanned flight programs in the 1970 decade." Studies of an "advanced fully reusable transportation system" displayed a near-term preference for ballistic spacecraft rather than lifting bodies on the basis of cost, said the summary. Not everyone at NASA thought this was correct, and a gathering consensus rapidly evolved for projecting the low-cost economics of reusable spacecraft capable of maneuvering during reentry. NASA linked the advanced technology of such a vehicle with the virtues of reduced launch costs and propagated the view that future vehicles should be funded as the best use of government money assigned to space research and technology applications. The notion was flawed because it ignored the complexity of operating a reusable manned shuttle vehicle that would eventually prove more expensive to operate than expendable rockets.

1968 July 4   In the first of two missions to measure the frequency and intensity of radio emissions from space, NASA launched the 417-lb. *Explorer 38 (Radio Astronomy Explorer A)* on a thrust-augmented improved Delta from Vandenberg AFB, Calif. The satellite entered an elliptical transfer orbit of 398 × 3,656 mi., and on July 7 the apogee motor fired, placing it in a 3,636 × 3,641 mi. orbit at 120.8°. The cylindrical satellite had a 120-ft. dipole antenna, four 750 ft. long antennas deployed to form a giant "X" and a 630 ft. long vibration damper to help it maintain gravity-gradient stabilization at a spin of 2.8 RPM. It remains in orbit.

1968 July 11   An Atlas F launched from Vandenberg AFB, Calif., carried two OV1-series satellites to elliptical Earth orbits. The 1,040-lb. *OV1-15* was placed in a 95.5 × 1,130 mi., 89.8° orbit, in which it conducted air density/solar radiation correlation measurements. Because of the low orbit, it had a relatively short orbital life, decaying back into the atmosphere on November 6. The 1,325-lb. *OV1-16* conducted ionospheric drag measurements from its 102 × 344 mi., 89.7° orbit, which decayed on August 19.

1968 July 12 The Convair Division of General Dynamics completed a report on the application of variable geometry wings to reusable lifting bodies. Compiled for NASA's Langley Research Center, the report embraced three different configurations, all with swing-wings, vehicle lengths ranging from 26 ft. to 50 ft. and wingspans ranging from 25 ft. 5 in. to 48 ft. when fully extended. All configurations were required to be compatible with the Saturn IB launch vehicle and to carry a cargo module mounted atop the rocket to which the space vehicle would be attached. Aerothermal data from the PRIME (precision recovery including maneuvering reentry) program were incorporated into the design of thermal protection systems.

1968 July 23 NASA astronaut Michael Collins underwent a surgical operation on his spine for the removal of an arthritic bone spur. Assigned as the command-module pilot for the second manned Saturn V mission (AS-504), the third manned Apollo flight, Collins gave up his place to James Lovell, who had been the backup crew member for that position. Astronaut Edwin Aldrin, originally the backup lunar-module pilot, moved to become the new backup command-module pilot, and Fred Haise was brought in as his replacement. Haise had been a member of the support team for the Borman/Collins/Anders crew.

1968 August 1 Reduced NASA budgets and a cutback in the Apollo Applications Program forced a halt to production of the Saturn V launch vehicle after AS-515. It transpired that NASA was unable to find to fund missions for the last two Saturn Vs and they remained moth-balled for possible future use. However, on December 16, 1976, NASA Administrator James Fletcher ordered AS-514 and AS-515 to be released to museums. With a payload capability of 260,000 lb. to low Earth orbit and 100,000 lb. to a lunar trajectory, Saturn V was the most powerful launch vehicle to leave the Earth's atmosphere. It was capable of placing in orbit a payload more than four times larger than that of the future Shuttle, and it was six times more powerful than the next biggest expendable rocket to reach operational status.

1968 August 6 The first Program 949 early warning satellite was carried by Atlas-Agena D from Cape Canaveral, Fla., to a near-geosynchronous, 19,686 × 24,469 mi. orbit inclined 9.9°. From a position over the Pacific, infrared sensors observed test launches from Vandenberg AFB, Calif. Program 949 was an interim precursor to the operational early-warning satellite system known as Program 747 and was specifically aimed at providing early warning of ballistic missile attack from Soviet submarines. In all there were six Program 949/Atlas-Agena D launches, the last on June 18, 1975. Five were placed in elliptical orbits and one satellite stranded in the wrong orbit.

1968 August 7 George Low, NASA's head of the Apollo Spacecraft Program Office, asked Christopher Kraft, the Manned Spacecraft Center's director of Flight Operations, to examine the feasibility of flying the first manned Saturn V mission (AS-503, designated *Apollo 8)* to the moon. Low feared that problems with the lunar module originally scheduled for AS-503 (LM-3) would delay it until at least February 1969. He wanted to use that Saturn V to send the Apollo spacecraft (CSM-103) to moon orbit during December 1968, using available hardware to achieve maximum results. The next day Low flew to the Kennedy Space Center to examine flight preparations for AS-503, which looked like it would not

be ready before January 1969. On August 9 Low went to Houston and met with Kraft, MSC Director Robert Gilruth and flight crew chief Donald Slayton. Getting their approval, he took them with him for a meeting at the Marshall Space Flight Center with its director, Wernher von Braun, Apollo Program Manager Samuel Phillips and Kennedy Space Center Director Kurt Drubs. That evening, Low discussed the spacecraft requirement with North American Aviation. On August 15 the idea was presented to NASA Administrator James Webb, who agreed two days later to support the plan subject to a "perfect" flight with the first manned mission, *Apollo 7*, scheduled for October 1968. Until then there was to be no formal announcement.

1968 August 8 Two satellites, the 21-lb. *Explorer 39* and the 155-lb. *Explorer 40*, were launched by Scout from Vandenberg AFB, Calif., and placed in elliptical orbits to conduct a coordinated measurement of atmospheric density, particle flux and very-low-frequency emissions from the sun. *Explorer 39* was a polka-dot Mylar sphere 12 ft. in diameter and inflated by nitrogen gas in its 80.66° orbit of 416 × 1,577 mi. Also known as Injun-E, *Explorer 40* was a hexagonal cylinder measuring 2 ft. 5 in. high and 2 ft. 6 in. across, with five hinged experiment booms deployed from its 423 × 1,574 mi., 80.67° orbit. Before deploying in orbit, *Explorer 39* was carried inside a tube on top of *Explorer 40*. *Explorer 39* decayed on June 22, 1981. *Explorer 40* remains in orbit. An earlier pair of air-density measurement satellites had been launched on November 21, 1964.

1968 August 9 Apollo mission planners agreed to put the software computer program COLOSSUS, the first version of the only program that could navigate an Apollo spacecraft around the moon, into CSM-103 for the *Apollo 8* mission. This would be the first time a lunar flight program had been tested in space, although the Earth-orbit software program SUNDISK would be tested on the Apollo 7 flight. The Lunar Module software program LUMINARY would be tested by *Apollo 10* in May 1969. Each was key to the eventual success of the Apollo program.

1968 August 10 NASA's Administrator for Manned Space Flight George Mueller presented a speech to the British Interplanetary Society in London, England, and made the first reference to the words Space Shuttle as the name of the space agency's future reusable logistics vehicle. Mueller explained that "Manufacturing in space, fuel, and supply storage for deep space operations, life support for crews on board space stations, require not tons, but thousands of tons of material, to be shuttled in and out of space. . . . Therefore there is a real requirement for an efficient Earth-to-orbit transportation system. . . . The objective . . . is to find a design that will yield an order of magnitude reduction in operating costs." Charts used by Mueller indicated a cargo capability of 25,000–50,000 lb. at a launch cost approaching $5 per pound.

    NASA launched the *ATS-4* (Applications Technology Satellite-4) by Atlas-Centaur from Cape Canaveral, Fla. The first Centaur burn placed *ATS-4* in a 115 × 475 mi. orbit inclined 29°. After a 1 hr. 1 min. coast it failed to fire a second time, stranding the satellite in low orbit still attached to Centaur. The cylindrical satellite had a length of 6 ft. and a diameter of 4 ft. 8 in., with thin beryllium copper strips like a carpenter's tape extending from motor-driven drums as gravity-gradient booms. Operating like scissors, they could "open" or "close" between 11° and 30° to compensate for gravitational torque. These could not be tested due to the low orbit, but 23 hr. of operation with a pair of ion engines was carried out before the satellite decayed back into the atmosphere on October 17.

1968 August 13 Soviet Union and seven other eastern bloc countries put forward a draft proposal to the United Nations for an international commercial communications satellite system known as Intersputnik. Supported by Bulgaria, Cuba, Czechoslovakia, Hungary, Mongolia, Poland and Romania, the USSR wanted to establish a system to compete with the American-dominated Intelsat. Based on a pro-rata ownership share determined by the amount of use taken out of the system, the USSR could have received only a 2% share in Intelsat and found this unsatisfactory, choosing to head its own system where it would have a monopoly.

1968 August 16 The first LGM-130G Minuteman III solid propellant ICBM flew from Cape Canaveral. More than a simple improvement on the basic Minuteman I and II, it had a completely new 33,000-lb. thrust Aerojet/Thiokol third stage and sophisticated warload incorporating post-boost bus and multiple-independently targeted reentry vehicle (MIRV) warheads. Minuteman III has a length of 59 ft. 10 in. and a diameter of 6 ft. 2 in. with a launch weight of 76,000 lb. and a range of more than 7,000 mi. The Mk.12 reentry vehicle had two or three Lawrence Livermore W-62 warheads, depending on the desired range, each having a yield of 170 KT and weighing 700–800 lb. W-62 production started in 1970 and was completed in June 1976, when 1,725 had been built. The Bell Aerospace post-boost bus contains a 315-lb. thrust bipropellant motor for fore and aft shunting, six 22.5-lb. thrust engines for pitch and yaw control and four 17-lb. thrust motors for roll control. Target accuracy is within less than 1,200 ft. at full range.

The first flight-test model of the Poseidon C3 SLBM (submarine-launched ballistic missile) was launched from a land pad at Cape Canaveral, Fla. Twenty land-test missiles were launched by June 29, 1970, and the first submerged launch took place on August 3, 1970, from a firing tube on the USS *James Madison* (SSB(N)-627). The missile had a length of 34 ft. and a diameter of 6 ft. 2 in., an increase of 3 ft. and 1 ft. 8 in. over the respective dimensions of the Polaris A3. This was the first time the diameter of a U.S. SLBM had increased, but the missile could be accommodated in existing launch tubes modified for the purpose. Between February 1969 and March 1975 the 31 Lafayette class Fleet Ballistic Missile submarines were modified and retrofitted with the C3. The 10 *George Washington/Ethan Allen*–class boats retained the Polaris A3.

NASA launched the first long-tank thrust-augmented Delta (LTTD) , or Delta N, from Vandenberg AFB, Calif., placing the *ESSA VII* weather satellite in a sun-synchronous, 552 × 567 mi. orbit inclined 101.7°. Although launched on this flight in a two-stage configuration, including three solid propellant Castor II strap-ons, Delta N could accommodate either an FW-4 or TE 364–3 third stage and was then known as the Delta L. With the 10,000-lb. thrust TE-364-3, a modified retromotor from the Surveyor moon lander first used as the Burner II, Delta L could lift 785 lb. to geosynchronous transfer orbit or 2,200 lb. to low Earth orbit. The first Delta L launched *Intelsat III F-5* on July 26, 1969.

In Japan, the Space Activities Commission (SAC) replaced the National Space Activities Commission. Set up under the authority of the prime minister's office, the NSAC had little interaction with other government and industry departments. The SAC that replaced it would play a pivotal role in mobilizing Japanese industry toward the new technologies. The shift was part of the government's commitment to apply efficient manufacturing and marketing techniques to high-tech equipment and systems, making a direct challenge to the dominant position in the computer and software industry held by the United States throughout most of the 1960s.

An attempt by the U.S. Air Force to place 13 satellites in orbit on one launcher failed when the Atlas-Burner II malfunctioned and deposited its load in the Pacific Ocean. Launched from Vandenberg AFB, Calif., the satellites comprised a 20-lb. *OV8* carrying materials experiments; a 395-lb. *Radcat* radar calibration target; a 75-lb. Lincoln Calibration Sphere for radar experiments; four 22-lb. drag coefficient spheres; the 81-lb. *Orbis-Cal 1* satellite for measuring the effects of the ionosphere on communications; two 48-lb. *Secor* geodetic satellites; one 258-lb. *Lidos* satellite to measure Earth's sphericity and gravity field; a 112-lb. *RM 18* satellite for measurements of the Earth's infrared background; and a shortwave radiation experiment called UV Radiometer. The vehicle was officially credited with carrying 10 satellites because the four drag coefficient spheres were regarded as one payload.

1968 August 19 NASA Apollo Program Director Samuel Phillips authorized changes to flight schedules. The original Mission-D objective for AS-503 was canceled. Instead, the third, and first manned, Saturn V *(Apollo 8)* would carry CSM-103 and a lunar test article on a mission designated C-prime and tentatively scheduled for launch sometime after December 6, 1968. In reality this gave authority to prepare AS-503 for the moon orbit mission first mooted 12 days earlier. AS-504 *(Apollo 9)*, carrying CSM-104 and LM-3, would fly the Mission-D plan—a full shakedown of Apollo and the lunar module in Earth orbit. Astronauts Frank Borman, James Lovell and William Anders would be moved from Mission-E, Apollo spacecraft and lunar module test flights in high Earth orbit, and assigned to the C-prime (*Apollo 8*) mission. The backup crew comprised Neil Armstrong, Edwin Aldrin and Fred Haise. The crew originally assigned to the Mission-D profile with AS-503—James McDivitt, David Scott and Russell Schweickart—would move to AS-504 *(Apollo 9)*.

1968 August 24 After the United States, the Soviet Union, the United Kingdom and China, France became the fifth nation to explode a hydrogen bomb when a 2-MT balloon-borne device was detonated 1,800 ft. above Fangatufa Atoll in the Pacific. France had just completed tests with the submarine missile, and the thermonuclear device would result in an operational bomb within the next several years.

1968 August 28 The project approval document was signed for the NASA Mariner-Mars 1971 orbiter mission. This would involve two spacecraft which were expected to conduct a detailed survey of the planet over a period of at least 90 Mars days; Mars rotates once on its polar axis in 24 hr. 37 min. 22.6 sec. The two spacecraft would be virtually identical and weigh about 2,150 lb. at launch, 1,200 lb. after consuming propellant to enter Mars orbit. Mission A called for the launch of *Mariner H (Mariner 8* after launch) on May 7, 1971, for arrival at Mars on November 14, 1971. Mission B called for the launch of *Mariner I (Mariner 9)* on May 17, with arrival at the planet on November 24, 1971. Mission A was primarily a mapping operation, with *Mariner 8* in a 750 × 10,000 mi. orbit inclined 80° to Mars' equator. Mission B was to observe variable features in a 530 × 20,500 mi. orbit inclined 50°, the spacecraft passing over the same area at five-day intervals. NASA gave the Jet Propulsion Laboratory formal approval to start the project on November 14, 1968.

A final test of the Soyuz spacecraft began at 09:59 UT when an SL-4 launcher carried *Cosmos 238* from Baikonur to a 124 × 136 mi. orbit. The mission lasted three days and the descent module returned to Earth at 09:03 UT on September 1. It is not known if Cosmos 238 was the intended target vehicle for a docking with a manned Soyuz which was delayed, but the mission did closely resemble the next Soyuz launch, which involved a rendezvous with the manned *Soyuz 3* launched on October 26, 1968.

1968 August   Under a new NASA contracting system, separate phases in the evolution of new concepts and the development of flight hardware would be organized around a set of criteria covering separate evolutionary steps. Under the scheme, called Phased Project Planning (PPP), phase A feasibility studies would identify key design elements to meet stipulated performance requirements; phase B definition studies would fix a configuration and produce preliminary design details; phase C would generate a detailed systems and subsystems design; and phase D would build and test the vehicle. The first major program to adopt the phasing sequence was NASA's Space Shuttle, initially identified as the integral launch and reentry vehicle (ILRV), for which phase A study contracts were let on January 31, 1969. Later in that program, phases C and D were combined.

The Space Science Board of the U.S. National Academy of Sciences published the report *Planetary Exploration 1968–1975*, which called for a shift in emphasis from manned space goals to the unmanned exploration of the solar system. Accepting that priority would be given to completing the first manned moon landing, the board encouraged U.S. space planners to shift away from expensive manned flight goals to what it considered more cost-effective unmanned programs. It also wanted stronger cooperation with the USSR to avoid duplication and pool resources.

1968 September 14   At 21:42 UT an SL-12 launcher carried the 11,300-lb. lunar Soyuz variant, designated *Zond 5*, from Baikonur cosmodrome to a 120 × 136 mi. orbit inclined 51.5° on what many in the West thought was a test flight for a manned circumlunar mission. At 22:47 UT the fourth stage of the launch vehicle ignited and put the payload on course for a lunar fly-by. The fourth stage separated and the Zond propulsion system, a derivative of the Soyuz KTDU-35 designated KTDU-53, performed a course correction at 03:11 UT when the spacecraft was 202,000 mi. from Earth. The descent module carried living organisms, including turtles, to test for effects from radiation at lunar distances. *Zond 5* looped behind the moon at a minimum distance of 1,210 mi. and conducted another course correction burn at a distance of 88,860 mi. from Earth. Separated from its instrument module, the descent module entered the atmosphere at 15:54 UT on September 21 and splashed down in the Indian Ocean 14 min. later with a weight of 4,511 lb., becoming the first object returned to Earth from the distance of the moon. During reentry, it had been subject to 10–15 g on deceleration and high heat loads through atmospheric friction due to a failure in the gyroscopic system which prevented the spacecraft from making a controlled reentry. The biological experiments remained intact.

1968 September 18   The first third-generation Intelsat communication satellite, *Intelsat III F-1*, was launched by long-tank Delta from Cape Canaveral, Fla. A malfunction in the launch vehicle's pitch rate system caused the Delta to fail, and the satellite was destroyed when the launcher was blown up by the range safety officer. The 642-lb. third-generation Intelsat satellites were spin-stabilized and cylindrical in shape, with a height of 3 ft. 5 in. and a diameter of 4 ft. 8 in., carrying a de-spun antenna extending 2 ft. 10 in. above the top. A total 10,720 solar cells provided 130 W of electrical power. Operating in the 6/4 MHz bands, the two transponders could handle a total 1,200 voice circuits or up to four TV programs at one time.

1968 September 26   The only successful U.S. Air Force OV2-series satellite, *OV2-5*, was carried into orbit along with the fourth and fifth OV5-series, *OV5-2* and *OV5-4*, and *LES 6*, on a Titan IIIC from Cape Canaveral, Fla. First to be released, during the synchronous transfer trajectory, was the 22-lb. *OV5-2*, put in an orbit of 114 × 22,238 mi. with an inclination of 26.4° and a period of 10 hr. 30 min. The Transtage subsequently released the 29-lb. *OV5-4* to a geosynchronous, 22,231 × 22,237 mi. orbit with an inclination of 3° and a period of 23 hr. 56 min. *LES 6* was similarly placed in a 22,120 × 22,237 mi., 3° orbit with a period of 23 hr. 51 min. Transtage then reduced its speed and released the 450-lb. *OV2-5* to a 21,821 × 22,256 mi. orbit with an inclination of 2.9° and an orbital period of 23 hr. 39 min. *OV5-2* was a standard environmental research satellite designed to take electron and proton measurements along its elliptical orbit. *OV5-4* was an experiment in transferring liquids in zero gravity. *OV2-5* had 11 separate experiments to collect data on the geosynchronous space environment. *LES 6* was designed to perform tests between ships, aircraft, portable and mobile military communication equipment.

1968 September 30   In a memorandum circulated for internal distribution only, NASA headquarters indicated the need for a logistics vehicle to supply an Earth orbiting space station comprising "a new spacecraft of larger crew size" than Apollo, to "reduce flight hardware cost. The importance of this issue is indicated by the fact that logistics can require up to 70% of the total funds for the (space station) program." This memorandum indicated the direction NASA would go in its post-Apollo programs: permanently manned Earth orbiting space stations serviced by reusable logistics vehicles taking off like a rocket and landing like an airplane.

1968 October 3   As a reimbursable activity for the European Space Research Organization, NASA launched the 185-lb. satellite *ESRO 1 (Aurorae)* from Vandenberg AFB, Calif., to a 160 × 956 mi., 93.7° orbit. Consisting of a cylindrical body with shallow truncated cones at each end, *ESRO 1* had a height of 5 ft. and a diameter of 2 ft. 6 in. Three experiment booms extended from the top and two from the bottom of the satellite. Instrumented to measure energies of particles impinging on the Earth's polar ionosphere during magnetic storms, *ESRO 1* decayed back into the atmosphere on June 26, 1970.

The first of three lunar landing training vehicles (LLTV) was flown at Ellington AFB, Tex. Essentially a developed version of the lunar landing research vehicle, the Bell Aerosystems LLTV had a single turbofan engine in the center to provide thrust equal to 83% (0.83) of the vehicle's Earth weight. Hydrogen peroxide thrusters and two 500-lb. thrust rocket motors provided pilots with sensory simulation of an Apollo lunar module.

1968 October 5 A new generation of dedicated U.S. electronic intelligence gathering (elint) satellites was launched by a LTTAT-Agena D from Vandenberg AFB, Calif., to a 300 × 318 mi. orbit inclined 74.9°. Replacing 1,500-lb. satellites carried into space at six month intervals by the TAT-Agena D, the new satellites weighed 2,000 lb. and were placed at a slightly lower inclination. Equipped with processors to filter the electronic messages and look for new signals or coded telemetry, they were able to analyze electronic data and categorize it on the basis of language, modulation and descriptive patterns. Only three more were launched: July 31, 1969; August 26, 1970; and July 16, 1971.

1968 October 7 Four days before the launch of the first manned Apollo flight, engineers at the Kennedy Space Center hoisted Apollo CSM-103 atop Saturn V launch vehicle AS-503 in readiness for the first moon orbit mission. The following day the launch escape system was installed, and on October 9 the crawler-transporter moved the *Apollo 8* launch vehicle from the Vehicle Assembly Building to Launch Complex 39A. The final decision on the moon orbit option would be based on results from the *Apollo 7* flight.

James Webb stood down as NASA administrator and was replaced by Acting Administrator Thomas O. Paine. Appointed by President John F. Kennedy early in 1961, Webb had steered NASA through its most dramatic years of expansion and seen it grow from a fledgling space agency into an organization commanding contracts that employed 400,000 people. To a Congressman critical of NASA's budget requests, Webb once railed, "Who would you rather see come through that door, me or the Russians!" Administering the exploration of the solar system, Webb brought Apollo from an idea and a goal to the verge of achievement, and deep loyalty among NASA employees would give cause for them to mourn his departure years after he had gone. Webb died March 27, 1992, at Georgetown University Hospital, Washington D.C., of a heart attack. Appointed by the incoming Republican administration, Paine was sworn in as full administrator on March 5, 1969, and served until September 15, 1970.

1968 October 11 At 15:02:45 UT the first manned Apollo mission carrying astronauts Walter Schirra, Donn Eisele and Walter Cunningham was launched from Complex 34 at Cape Canaveral. At liftoff the vehicle weighed 1.278 million lb. and included the 32,586-lb. Apollo loaded with 9,737 lb. of propellant for the SPS engine. Saturn IB AS-205 carried Apollo CSM-101 into a 142 × 177 mi. orbit inclined 31.6° to the equator 10 min. 17 sec. after launch. Apollo separated from the S-IVB at 2 hr. 55 min. 2 sec. and performed a simulated docking and lunar module extraction maneuver from the spacecraft/LM (lunar module) adapter on top of the S-IVB. For this flight the four petal doors which would normally cover the LM were hinged to fold back. On later flights they were jettisoned. At 3 hr. 20 min. 10 sec. the SPS engine fired for the first of eight burns during the mission, setting up conditions for a separation and rendezvous with the S-IVB a day later. The shortest burn was 0.5 sec. (13 ft./sec.) in duration, the longest was 67 sec. (1,691 ft./sec.). The crew transmitted the first live, in-flight TV pictures to broadcast networks around the world, but head colds and a feeling of public intrusion brought tetchy reactions from the astronauts. The 12 sec. de-orbit burn began at 10 days 19 hr. 39 min. 16 sec., and the command module separated from the service module 4 min. 17.5 sec. later, with reentry beginning 9 min. 53 sec. after that. The command module splashed down in the Atlantic Ocean at 10 days 20 hr. 9 min. 3 sec. With this flight,

Walter Schirra became the only astronaut to fly Mercury, Gemini and Apollo spacecraft.

1968 October 19 The first successful Soviet antisatellite demonstration mission began when an SL-11 launched the 8,820-lb. *Cosmos 248* target satellite from Baikonur cosmodrome to a 295 × 337 mi. orbit inclined 62.3°. A day later a 7,320-lb. interceptor designated *Cosmos 249* was launched by SL-11 to a 312 × 1,018 mi. orbit inclined 62.4°. Within hours of launch *Cosmos 249* conducted a high-speed fly-by of *Cosmos 248* and was later detonated. Launched on November 1, the 7,320-lb. *Cosmos 252* interceptor was put in a 332 × 1,019 mi. orbit for a fly-by of *Cosmos 248* and subsequent detonation into 56 pieces. *Cosmos 248* decayed February 26, 1980. This was the last such test until October 20, 1970.

1968 October 24 The last of 199 flights with the X-15 rocket-propelled research aircraft took place when NASA test pilot Bill Dana took the no. 1 aircraft to a speed of Mach 5.04 (3,682 MPH) and an altitude of 250,000 ft. Of the total 198 powered flights, 30 had been performed with the interim XLR11 engine and 168 with the XLR99 it had been designed for. Since the first powered flight of the Bell X-1 on December 9, 1946, there had been 540 flight attempts with U.S. research aircraft powered by rocket motors and carried aloft by a mother plane. These included the Bell X-1 series, Bell X-2, Douglas D-558-II and the North American X-15.

1968 October 25 At 09:00 UT an SL-4 launch vehicle lifted away from the Baikonur cosmodrome with the 14,225-lb. unmanned *Soyuz 2* spacecraft. It was placed in a 115 × 139 mi. orbit to await the launch of the 14,500-lb. *Soyuz 3*, which duly took place at 08:34 UT on October 26, carrying cosmonaut Georgi T. Beregovoi to a 127 × 140 mi. orbit. Cosmonauts Vladimir Shatalov and Boris Volynov were backups. By the time *Soyuz 3* had been launched, *Soyuz 2* had maneuvered into a rendezvous orbit and the two were brought to within about 650 ft. of each other. Another rendezvous was conducted on October 27, and *Soyuz 2* returned to Earth at 07:51 UT on October 28. Over the next two days, Beregovoi conducted Earth surveys, took pictures of land features and broadcast TV pictures of the Soyuz interior to viewers on Earth. *Soyuz 3* landed at 07:25 UT on October 30.

1968 October 30 Through the Manned Spacecraft Center and the Marshall Space Flight Center, NASA issued a formal request to industry for proposals on a study of an integral launch and reentry vehicle (ILRV). The ILRV was to be capable of lifting a cargo load of 5,000–50,000 lb. to altitudes varying between 115 mi. and 345 mi., spending up to 180 days in orbit, accommodating up to 12 people and having the ability "to land at a preselected site located in the continental United States." The ILRV was required to have a cross-range capability of 450 mi., meaning that on the way back down through the atmosphere it could fly 450 mi. left or right of the groundtrack. NASA expected to begin ILRV flight operations in 1974.

1968 October The European Space Research Organization reinstated the TD-1 and TD-2 satellite program, which had been canceled when Italy refused to pay for what it considered would bring too little work for the required investment. In a rescoped plan the program was simplified, and the first project, now known as TD1A, would proceed toward a launch planned for 1970 and would involve a series of instruments designed to study ultraviolet and infrared emissions from stars. It was proposed that experiments originally planned for TD-2 could be flown on a U.S. observatory-class satellite.

1968 November 8　A thrust-augmented improved Delta carried the 140-lb. *Pioneer 9* and the 88-lb. *TTS-2* Apollo tracking test satellite from Cape Canaveral, Fla., to Earth parking orbit. *TTS-2* was ejected to a 235 × 584 mi., 32.8° Earth orbit, where it served for tests of the NASA Manned Space Flight Network, and *Pioneer 9* was carried to escape velocity and an orbit of the sun. Virtually identical to *Pioneers 6, 7* and *8, Pioneer 9* was placed in a heliocentric path inside the orbit of the Earth, its elliptical orbit carrying it between 70 million and 93 million mi. of the sun. From this position upstream of the solar wind, it was an effective early warning monitor of solar flares and high-energy emissions that could endanger the crews of upcoming Apollo moon missions as they left the protection of Earth's magnetosphere for the first time.

　　NASA's Marshall Space Flight Center issued a 10-month contract to Boeing for studies defining a two-stage version of the Saturn V. First (S-IC) and third (S-IVB) stages would lift a payload of 158,000 lb. to low Earth orbit. In this configuration, known as the Intermediate-20, the two-stage Saturn V could lift the Saturn I workshop, Apollo Telescope Mount, and Apollo CSM (command/service module) on a single launch. Another variant, the Intermediate-21, consisted of the first (S-IC) and second (S-II) stages only.

1968 November 10　At 19:11 UT an SL-12 launch vehicle lifted away from its Baikonur launch pad to place the 11,300-lb. circumlunar Soyuz spacecraft *Zond 6* in a 115.5 × 144 mi. orbit. At 20:18 UT the fourth stage of the SL-12 fired again to propel the spacecraft toward the moon. A course correction was performed with the instrument module propulsion system at 05:41 UT on November 12, and a 1,500-mi. fly-by across the far side of the moon occurred two days later. A second course correction burn took place at 06:40 on November 16 when *Zond 6* was 146,650 mi. out from Earth, and a final tweak burn, a small adjusting burn, was performed at 05:36 UT on November 17. The *Zond 6* descent module encountered the atmosphere at 13:58 UT on November 17, with a speed of 25,050 MPH. Instead of plunging to a direct descent, it was slowed by an initial pass through the outer atmosphere before skipping back out and then reentering at a speed of 17,000 MPH. This reduced the levels of deceleration to 4–7 g, more appropriate for a manned mission. *Zond 6* landed on Soviet territory at 14:10 UT, but it was crushed on landing because the parachutes deployed too early and were ripped to shreds.

1968 November 11　NASA's Acting Administrator Thomas O. Paine formally agreed to schedule the next manned Apollo flight, *Apollo 8*, as a lunar orbit mission launched by Saturn V AS-503. Based on results from the recently completed *Apollo 7* mission, the command and service modules were considered qualified for this next, ambitious step toward a manned lunar landing. Test flights with the lunar module would wait until LM-3 was ready for an all-up Saturn V mission in early 1969 with Apollo command and service modules as well. Soviet space activity gave some U.S. officials an indication that they might be planning to send cosmonauts around the moon on a Zond mission; getting *Apollo 8* to moon orbit by Christmas this year would be a spectacular hedge against potential problems with the manned landing objective in 1969.

1968 November 16　A new Soviet launch vehicle based on the SL-9 made its first flight from Baikonur cosmodrome carrying the 37,500-lb. *Proton 4* satellite to a 154 × 1,050 mi., 51.55° orbit. Designated SL-13, the launcher was an assembly of the first three stages of the SL-12 and provided a maximum payload capability of 43,700 lb. During the early 1980s, the RD-253 first-stage engines were uprated to a thrust of 392,500 lb., increasing liftoff thrust to 2.35 million lb. and payload capacity to 46,000 lb. *Proton 4* carried 27,500 lb. of science equipment for cosmic-ray measurements and decayed on July 24, 1969.

1968 November 22　With obvious concern about the upcoming *Apollo 8* fight to moon orbit, NASA's manned spaceflight chief George Mueller reviewed for Administrator Thomas Paine the test history of the service propulsion system (SPS) engine. Built by Aerojet General, the SPS would be the only engine to get the spacecraft in or out of moon orbit, and if it failed to fire, the crew could be stranded in space. In more than 3,200 engine starts during development, the SPS had failed to start only 4 times, each of those attributable to faulty test equipment or incorrect procedures. The SPS had been built to start 50 times and burn for as little as 0.5 sec. or as long as 8 min. Tests showed it could fire for 30 min. without failing. The engine had successfully fired on test for a total of 25 hr. and in space 17 times on 5 flights.

1968 November 23　Soviet space officials announced that *Zond*s *4, 5* and *6* had been precursor flights of the *7K-L1* circumlunar spacecraft built to carry cosmonauts around the moon. Six cosmonauts arrived at Baikonur in early December ready for a manned circumlunar flight. Each had written a personal letter to the Politburo urging an immediate circumlunar attempt. On December 6 two U.S. destroyers departed Naples for the Black Sea where they could monitor communications from such a mission. When *Apollo 8* circumnavigated the moon in December 1968, the Soviet attempt was canceled.

1968 November 29　The first in a series of three phase-three flight tests of the Europa I launch vehicle took place (08:42 local time on November 30) from complex 6A at the Woomera Rocket Range, Australia. With three live stages, mission F7 had the objective of orbiting a 551-lb. test payload, designated STV-1, for the first time. With a liftoff weight of 239,000 lb., the Blue Streak first stage produced 300,000 lb. thrust for 2 min. 30 sec. The Coralie second stage burned UDMH–N$_2$O$_4$ propellants in its 61,740 lb. four-engine cluster for 1 min. 42 sec. The Astris third stage fired its 5,250-lb. thrust for only 7 sec. and failed to push the payload into orbit. A second attempt (F8) on July 9, 1969, failed when the third stage failed to ignite.

1968 November　Soviet space engineers prepared to send a circumlunar variant of the Soyuz spacecraft, designated Soyuz L, on the manned L-2 mission for which Sergei Korolev's design team had been working since December 25, 1965. Technical difficulties with the spacecraft and failures on previous unmanned (L-1) missions prevented approval being given to go ahead with the flight. Cosmonauts Aleksei A. Leonov and Oleg G. Makarov were among a group of six cosmonauts who would probably have made the flight. Others, perhaps following on later missions, were: V. F. Bykovsky, N. N. Rukavishnikov, P. R. Popovich and G. M. Grechko. During the next launch attempt, on January 20,

1969, both second and third stages of the unmanned Proton launch vehicle failed and the spacecraft was destroyed.

1968 December 4 NASA Administrator Thomas O. Paine approved the Viking Mars orbiter/lander program anticipating dual launches in July 1973. From a wide range of candidate missions and hardware options, NASA had selected a soft lander carried to Mars orbit by a Mariner Mars 1971–class spacecraft which would continue to map the planet following the lander's separation and descent to the surface. NASA's Jet Propulsion Laboratory would build the orbiter, Langley Research Center would manage the lander, and Lewis Research Center would be in charge of the Titan IIID-Centaur launch vehicle. The project approval document was signed on February 8, 1969.

1968 December 5 A thrust-augmented improved Delta launched the 238 lb. *HEOS 1* satellite from Cape Canaveral, Fla., to a 1,502 × 138,845 mi., 28.2° orbit. HEOS, an acronym for Highly Eccentric Orbit Satellite, was built by Junkers in Germany for the European Space Research Organization. It was a sixteen-sided cylinder 8 ft. 4 in. tall and 4 ft. 3 in. across, its sides covered with solar cells. A sensor boom extended from the forward end on a tripod, and four antennas were deployed perpendicular to the spin axis to give a tripod height of 5 ft. 3 in. Equipped to measure magnetic fields, cosmic radiation and the solar wind outside Earth's magnetosphere, *HEOS 1* decayed on October 28, 1975.

1968 December 7 NASA launched the second Orbiting Astronomical Observatory *(OAO-II)* on an Atlas-Centaur from Cape Canaveral, Fla., to a 475 × 483 mi. orbit inclined 35° to the equator. The 4,446-lb. satellite comprised an octagonal aluminum cylinder 10 ft. long and 7 ft. across with two three-panel solar arrays spanning 21 ft. Two sets of experiments were installed for observing in the ultraviolet, infrared, gamma-, and x-ray regions of the spectrum: seven telescopes to make spectrophotometric measurements of ultraviolet radiation, and four large-aperture TV cameras with broad-band photometers. *OAO-II* remains in orbit.

1968 December 9 Jerry M. Deerwester from the NASA Ames Research Center compiled a series of mission profiles for flights to the outer solar system using the "sling effect" of Jupiter's gravitational field. Ames wanted to extend its Pioneer series of solar satellites by sending a spacecraft to Jupiter, expanding the region of space it surveyed in observing the heliosphere (including the solar wind) and providing an early opportunity to conduct a fly-by of the planet Jupiter. Such a mission, it was said in justification, would also serve as a useful precursor to the so-called Grand Tour of the outer solar system, during which it would be possible to fix more precisely the heliocentric coordinates and mass of Jupiter. This, in turn, would help in the design of trajectories for more complex missions.

1968 December 12 The first U.S. electronic intelligence gathering subsatellite launched to evaluate Soviet antiballistic missile radars was carried piggyback on a photoreconnaissance satellite from Vandenberg AFB, Calif. The long-tank thrust-augmented Thor-Agena D launcher first put the 4,410-lb. third-generation, area-surveillance reconnaissance satellite into a 105 × 154 mi. orbit inclined 81.0°. The Agena D then maneuvered to high altitude and placed the 130-lb. ABM elint subsatellite in a 864 × 912 mi. orbit inclined 80.3°. The primary satellite decayed on December 28, but the elint

remains in space. A second pair launched February 5, 1969, were placed in similar orbits, the two elints phased to work cooperatively. Other double elint missions began with launches in October and November 1973, June 1980 and June 1983.

1968 December 19 The *Intelsat III F-2* communications satellite was launched by a long-tank Delta from Cape Canaveral, Fla., to a transfer orbit circularized at geostationary orbit two days later over the equator east of Brazil. The first successful satellite in the third-generation Intelsat III series, it was followed by *F-3* on February 6, 1969, to a position first over the Pacific and then over the Indian Ocean. *F-4*, launched on May 22, 1969, was placed over the Pacific Ocean, completing the first global Intelsat network in time to cover transmission of *Apollo 11*'s historic moon landing. Launched on July 26, 1969, *F-5* entered an incorrect orbit at 104 × 2,085 mi. from which recovery to geostationary orbit was impossible. *F-6* was launched on January 15, 1970, *F-7* on April 23, 1970 and *F-8* on July 23, 1970. The latter failed to reach geostationary orbit due to an apogee motor failure and remained stranded in a path of 12,055 × 22,389 mi.

1968 December 20 Designated *Cosmos 261*, the first in a series of Soviet joint space missions with eastern bloc countries took place when an SL-7 launched the 765-lb. satellite from Plesetsk. Placed in an orbit of 129 × 399 mi. with an inclination of 71°, *Cosmos 261* was a joint endeavor by research institutes in the Soviet Union, Bulgaria, Hungary, East Germany, Poland, Romania and Czechoslovakia to study air density and polar aurorae. The satellite decayed out of orbit on February 12, 1969.

The United Nations adopted a resolution approving the establishment of a working group on the feasibility of technology for direct-broadcast satellite (DBS) systems. DBS offered the opportunity to beam TV programs direct to dish antennas small enough to serve hotels, motels and private homes. Moreover, from geostationary orbit, DBS satellites could transmit programs across national boundaries. Many countries feared cultural pollution of national and ethnic identities by wealthy spacefaring nations.

1968 December 21 At 12:51:00 UT the *Apollo 8* mission began with the launch of Saturn V AS-503 from Complex 39A at the Kennedy Space Center, carrying the 63,531-lb. Apollo CSM-103 with astronauts Frank Borman, Jim Lovell and Bill Anders and a simulated lunar module, LTA-B, weighing 19,900 lb. The Apollo service module contained a full load (40,836 lb.) of SPS engine propellants, and the total stack weighed 6.134 million lb. at launch. At 11 min. 35 sec. the 84,431-lb. payload was inserted into a 114 × 118 mi. Earth parking orbit, still attached to the S-IVB. A 5 min. 8.4 sec. translunar injection (TLI) burn starting at 2 hr. 50 min. 37.1 sec. increased velocity to 36,228 ft./sec. , the fastest speed at which humans had ever moved. At 11 hr. 00 min. the SPS engine fired for 2.4 sec. in the first of two midcourse correction maneuvers. At 31 hr. 10 min. the crew transmitted the first TV pictures showing the Earth as a distant sphere in space, 140,000 mi. away, from a point almost two-thirds of the way to the moon. At 55 hr. 7 min. the crew transmitted a second telecast, and 31 min. later they became the first humans to escape Earth's gravity when the moon's pull became dominant. Earth's gravity had slowed the spacecraft to 4,000 ft./sec., but now the moon's gravity began to speed it up. A second course correction burn of 13 sec. began at 60 hr. 59 min. 56 sec. The SPS engine was fired for 4 min. 7 sec. at

Launched on December 21, 1968, the Apollo 8 *flight to moon orbit and back secured the first close-up images of the lunar surface from a hand-held camera. Here, the large crater Goclenius is seen transected by vaults.*

69 hr. 8 min. 20 sec., shortly after the spacecraft disappeared around the western limb of the moon. This lunar-orbit insertion (LOI) burn had slowed the spacecraft from 8,391 ft./sec. to 5,458 ft./sec. and put it in a 68.9 × 193.9 mi. moon orbit. The first telecast from lunar orbit began at 71 hr. 38 min. and viewers on Earth saw the surface of the moon on live TV for the first time. In the United States it was 07:29 EST on Christmas Eve. After two orbits, at 73 hr. 35 min. 7 sec., the SPS engine was fired for 9 sec. to circularize the orbit at 68.7 x 69.8 mi. A second telecast began at 85 hr. 40 min. (21:31 EST on Christmas Eve) and at its conclusion the crew read the first 10 verses of Genesis. On Earth an estimated 1 billion people in 64 countries watched or heard it live, with 30 additional countries receiving a recorded broadcast later that day. A 3 min. 24 sec. trans-Earth injection (TEI) burn with the SPS engine at 89 hr. 19 min. 17 sec. (01:10 EST Christmas Day, December 25) took the spacecraft out of moon orbit and put it on course for Earth. A 14-sec. midcourse correction with the reaction thrusters began at 103 hr. 59 min. 54 sec. Two more telecasts took place, at 104 hr. 24 min. and 128 hr. elapsed time. The command and service modules separated at 146 hr. 28 min. 48 sec., and reentry began 17 min. 25 sec. later. Splashdown occurred at 147 hr. 00 min. 42 sec., December 27, 1968.

1968 December 30   Aware that they were running out of time in their attempts to place the first man on the surface of the moon, the Soviet Military-Industrial Commission (VPK) approved the launch of unmanned sample retrieval missions in an effort to snatch glory by getting the first moon samples to Earth. Designed by Georgi Babakin and designated Ye-8, the series of unmanned lunar vehicles to be used for this purpose had evolved from an earlier plan to place roving vehicles on the moon for Soviet cosmonauts to use should they need to move from their original landing site to the location of a rescue vehicle. The Ye-8-5 model would have a sample re-

trieval probe and ascent rocket for return to Earth. Following an initial failure on February 19, 1969, the first launch in this series would be *Luna 15,* launched on July 13, 1969.

1968 December   A technical report from NASA's Jet Propulsion Laboratory identified seven key missions that could be undertaken using the gravity field of Jupiter to speed spacecraft to the outermost planets in the solar system. The launch opportunities discussed were in the 1976–1979 period and included a Jupiter-Saturn-Uranus-Neptune flight beginning in 1976, 1977 or 1978, a Jupiter-Uranus-Neptune mission in 1978 or 1979, or a Jupiter-Pluto mission in 1979. Via Jupiter, these missions would reach their objectives within 10 years, whereas a direct flight to Neptune would last 17 years and a direct flight to Pluto would last 41 years.

1969 January 5   A Soviet spacecraft weighing 2,494 lb., *Venera 5* was launched toward Venus by an SL-6 launch vehicle from Tyuratam. It was intended to put a capsule down through the atmosphere of the planet and because its predecessor had failed at an altitude of about 15 mi., the 1969 design was much improved to withstand the extreme pressures and fierce temperatures of the lower atmosphere. A second spacecraft of the same type, *Venera 6*, was successfully launched by another SL-6 on January 10. Each spacecraft was a probe carrier and neither was designed to go into orbit around Venus.

1969 January 9   NASA named the flight crew for the AS-506, *Apollo 11*, mission. The primary crew would comprise Neil Armstrong, Michael Collins and Edwin Aldrin. The backup crew would be James Lovell, William Anders and Fred Haise. If all went according to plan, and *Apollos 9* and *10* achieved their primary objectives, the first manned lunar landing could be attempted on the AS-506 scheduled for mid-1969.

1969 January 14   At 07:29 UT an SL-4 launched the 14,610-lb. *Soyuz 4* spacecraft carrying cosmonaut Vladimir A. Shatalov, from Baikonur cosmodrome to a 107.5 × 140 mi. orbit inclined 51.7°. During the fifth revolution this was changed to a 129 × 147 mi. orbit in readiness for a rendezvous and docking with *Soyuz 5*. At 07:05 UT on January 15 the 14,520-lb. *Soyuz 5* carrying cosmonauts Boris V. Volynov, Aleksei S. Yeliseyev and Yevgeny V. Khrunov was launched to a 124 × 143 mi. orbit. On its fifth revolution *Soyuz 5* used its propulsion system to maneuver to a 131 × 157 mi. orbit. Another maneuver was carried out on the 32nd revolution of *Soyuz 4*, putting it in a 125 × 157 mi. orbit. At 04:37 UT on January 16 an automatic rendezvous sequence began, and 43 min. later *Soyuz 4* docked with *Soyuz 5* in a combined orbit of 130 × 155 mi. After expenditure of maneuvering propellant, the mass of the docked combination was 28,497 lb. Utilizing depressurized orbital modules on respective spacecraft, Yeliseyev and Khrunov in turn transferred by space walk from *Soyuz 5* to *Soyuz 4*. This was the first time two manned vehicles had docked and the first time crews had transferred from one spacecraft to another in orbit. Among items hand carried to *Soyuz 4* were the first mail delivered to space, letters from Earth for Shatalov. *Soyuz 4* landed with Shatalov, Yeliseyev and Khrunov at 06:53 UT on January 17, followed by *Soyuz 5* carrying Volynov at 08:00 UT on January 18.

1969 January 15   NASA Apollo Program Director Samuel Phillips sent Manned Spacecraft Center Director Robert Gilruth a memorandum expressing concern at the lack of planning for Apollo lunar surface missions after the initial landing

Launched January 14 and 15, 1969, respectively, Soyuz 4 *and* 5 *achieved the first rendezvous and docking between two Soviet spacecraft, a vital activity for operations planning for the lunar landing program. Seen here in a Moscow exhibition, full-scale replicas depict the docked configuration.*

had been accomplished. Pointing to the anticipated start of an extended lunar exploration program in 1973 that involved longer stay times and translunar transportation, Phillips asked the manned flight centers to develop program objectives and establish guidelines for the bridging period between 1969 and 1973 during which Apollo hardware would have to suffice. There never was an extended lunar exploration program, but the last three moon landings *(Apollo 15–17)* exploited previously developed hardware for expanded activity.

1969 January 22 On the day NASA launched the fifth Orbiting Solar Observatory *(OSO-5),* Administrator Thomas O. Paine announced that advanced versions, *OSO*s I, J and K (designated *OSO-8, OSO-9* and *OSO-10* in orbit) were to go ahead with a launch in early 1973. A request for proposals to build the new OSOs was issued by the Goddard Space Flight Center on April 1, 1970, and in addition to the first-generation OSO contractor (Ball Brothers), Hughes Aircraft and TRW responded with bids. In December 1970 Hughes won the order to build the three satellites. By late 1971 budget pressures dimmed prospects, and in March 1972 work on *OSO*s J and K was deferred. On June 9, 1972, NASA postponed the launch of *OSO I,* and on September 26, 1972, the other two were canceled.

Officials from the Martin Marietta company briefed officials at the NASA Manned Spacecraft Center on future space transportation system concepts and announced that it planned to submit an unsolicited phase A feasibility study to NASA along with reports from the four ILRV (integral launch and reentry vehicle) contractors. Martin briefed the Marshall Space Flight Center a day later. Their preferred concept included a 1½-stage (drop-tank) lifting body designed to have significant in-flight maneuvering capability. The unsolicited Martin Marietta studies would result in the Spacemaster concept.

1969 January 27 A meeting was held at NASA Headquarters to discuss major post-Apollo moon landing plans for Earth-orbiting space stations. Fresh budget reductions had cut plans for the Apollo Applications Program to just one orbital workshop launched by a Saturn IB in late 1971, followed by one manned visit lasting 28 days and two revisits of 56 days each. A fifth Saturn IB launch prior to the last manned revisit was to take up the Apollo telescope mount. AAP activity was to be completed by the end of 1972, about the same time as Apollo lunar exploration. Regarding the matter of a permanently manned space station NASA managers had differing views, but the general agreement was that it was better to move quickly to a permanently manned space station launched by Saturn V in 1975 and serviced by a reusable manned space shuttle.

1969 January 30 A thrust-augmented improved Delta launched the 532-lb. Canadian satellite *ISIS 1* from Vandenberg AFB, Calif., to a 359 × 2,190 mi., 88.4° orbit. Part of a joint Canadian-NASA International Satellites for Ionospheric Studies (ISIS) program, this was the third ionospheric satellite after *Alouette*s 1 and 2. It had a spherical configuration 4 ft. 2 in. in diameter, 3 ft. 6 in. high and powered by 11,136 solar cells. Two extendable antennas, 240 ft. and 62 ft. long, respectively, would sound the upper ionosphere, and an onboard data storage capability, something new for the series, improved the volume of information relayed to the ground. *ISIS 1* remains on orbit but is not operating.

1969 January 31 NASA awarded four contracts for ten-month studies of its integral launch and reentry vehicle (ILRV). These were the first such contracts to be awarded for what would emerge as the Space Shuttle. North American Rockwell, General Dynamics, Lockheed and McDonnell Douglas each received $500,000 to conduct phase A (feasibility) studies of a reusable logistics vehicle capable of carrying an average 25,000 lb. of cargo to a 310-mi. circular orbit inclined 55° to the equator. In a concerted effort to use reliable technology and to discourage contractors from suggesting exotic forms, propulsion was to consist of conventional high-energy liquid oxygen–liquid hydrogen engines.

1969 January NASA's Goddard Space Flight Center published a report entitled *A Venus Multiple-Entry-Probe Direct-Impact Mission*, a mission envisaged for a funded start in fiscal year 1973. Prepared by a consortium from Harvard University, Kitt Peak National Observatory and the University of Wisconsin as well as Goddard, it described a variety of mission options for gathering information about the atmosphere and surface conditions of Venus. Among the options examined was an instrumented balloon with suspended in atmospheric buoyancy drifting round the planet with the winds, a fly-by mission with probes released for impact, a direct impact with separate probes entering at different parts of the atmosphere and an orbiter from which probes could be released. The Goddard study concluded that the best science would be obtained from four large and three small probes ejected from a spacecraft bus. Goddard awarded AVCO a contract to study frictional heating problems for probes entering the dense atmosphere. By the end of the year NASA had merged the concept with a common spacecraft bus that could be used to carry either probes or orbiter instruments.

Soviet scientists began initial flight tests with multiple reentry vehicle (MRV) warheads fired 5,000 mi. across the Pacific using SS-9 ICBMs. Essentially a cluster of three warheads fired at the same target area, the concussive effect of MRVs had a destructive advantage useful against some targets. MRV warheads would, for instance, magnify the effect of nuclear weapons dropped at random on large cities or relatively "soft" military installations. The first series of tests ended abruptly in November 1970 and were not resumed until the Mod 4 version of the SS-9, specifically tailored to MRV warheads, appeared in January 1973.

1969 February 3 NASA completed a revised Apollo flight schedule for 1969 involving five Saturn V missions. Based on outstanding performance from hardware and personnel supporting the *Apollo 7* Earth orbit and *Apollo 8* lunar orbit missions, NASA moved aggressively to get men on the moon during 1969. AS-504, scheduled for February 28, would carry CSM-104 and LM-3 on the *Apollo 9* mission for the first manned test in Earth orbit of all the elements necessary for a lunar landing. AS-505/*Apollo 10* would carry CSM-106 and LM-4 for a manned lunar mission, the objectives of which would be based on the performance of *Apollo 9*. *Apollos 11–13* would carry consecutive numerical hardware designations and any one could perform a manned lunar landing. Cursory consideration had been given to an unmanned LM landing before risking astronauts' lives, but the spacecraft had not been designed for automatic flight and to convert it would have delayed the program and cost money. *Apollo 9* was delayed when the primary crew were struck by a mild respiratory illness.

1969 February 8 The conclusions of a two-year study by the National Academy of Science recommended increased spending on space applications. The study had been conducted by 200 people from industry and universities under the chairmanship of W. Deming Lewis, president of Lehigh University. Citing the enormous benefits from weather and communication satellites, the study said that NASA was spending too little on applications programs and should conduct research on a broader range of projects.

1969 February 9 The U.S. Department of Defense's *Tacsat I* (Tactical Communications Satellite) was launched by Titan IIIC from Cape Canaveral, Fla., to a geostationary orbit over the Pacific Ocean. Formerly known as *Tacomsat*, the 1,600-lb. satellite was the most powerful communication satellite launched to date by the United States. Capable of linking small ground units using antennas only 1 ft. in diameter and of linking ground stations with aircraft, the drum-shaped *Tacsat I* had a diameter of 9 ft. 3 in. and a height of 25 ft. Tactical communication tests in orbit were conducted on the standard military UHF band and on the SHP portion of the X-band set aside for military satellite communications.

1969 February 11 NASA invited 38 scientists to help plan experiments for the Viking lander scheduled for a 1973 flight to Mars, and on February 25 it announced the primary science teams. In terms of the amount of surface science obtained, Viking would be as significant as the manned landings on the moon. Selection of a contractor to build the lander got under way at the end of the month with a request for industrial proposals. Boeing, McDonnell Douglas and Martin Marietta responded. On April 17 the Jet Propulsion Laboratory established a Viking Orbiter project office and on May 29, NASA selected Martin Marietta to build the lander.

1969 February 13 President Richard Nixon authorized the creation of a special panel, later known as the Space Task Group (STG), to give him "the direction which the U.S. space program should take in the post-Apollo period." Members consisted of Vice President Spiro T. Agnew, chairman; Thomas O. Paine, NASA administrator; Robert C. Seamans, secretary of the Air Force; and Lee A. Dubridge, science adviser to the president. Observers to the group were to consist of U. Alexis Johnson, under secretary of state for political affairs; Glenn T. Seaborg, chairman of the Atomic Energy Commission; and Robert F. Mayo, director of the Bureau of the Budget. The STG submitted its report in September 1969.

1969 February 19 A Soviet SL-12 launch vehicle carrying an unmanned lunar roving vehicle exploded just 40 sec. after liftoff from a Baikonur launch pad and fell to earth 9 mi. away. Designated Ye-8, the lunar roving vehicle had been designed as a support vehicle for manned moon landing operations. Fearing that a cosmonaut might be stranded on the moon if his LK landing vehicle was damaged, Soviet engineers planned to put a back-up lander (LKR) on the surface one month before the manned landing attempt. Ye-8 lunar rovers would then be sent to the moon and using radio beacons for directional commands would move to the LKR where they would send TV pictures to earth. If the LKR was confirmed to be in good condition, the manned N-1/L-3 mission would get the go-ahead. The Ye-8 rovers would be used to carry the cosmonaut from his LK site to the LKR rescue lander if his own spacecraft was damaged in landing.

1969 February 21 At 09:18:07 UT the first N-1, the world's biggest and most powerful launch vehicle, lifted off from the Baikonur cosmodrome with a thrust of more than 10 million lb. The flight ended in failure when fire broke out in the first-stage engines, causing them to be shut down at 1 min. 10 sec. and a height of 19 mi. The launch escape system fired and the payload modules came down 20 mi. from the pad, the N-1 launcher crashing to Earth 30 mi. away. The mission objective had been set as a test of the first three stages in placing the assembly in Earth orbit, and of the fourth stage in sending the L-1S payload to a circumlunar trajectory. The Zond version of the Soyuz spacecraft was carried, but no lunar module. The L-1S was to have photographed the Ye-8 lunar rover which should have been launched on February 19, 1969 from moon orbit. In a lunar landing mission the fifth stage would perform translunar course corrections and fire its single 19,200-lb. thrust engine to brake the L-3 payload into lunar orbit. Performing a space walk, one of the two cosmonauts in the Soyuz would enter the lunar module through a side hatch. With a length of 18 ft. and a diameter of 12 ft., the fifth stage carried the lunar module on top and would start it on its way down to the lunar surface by firing its engine again. The stage would be jettisoned just 4,000 ft. above the moon, leaving the 26 ft. high lunar module to fire its own engine and brake to a landing. After 4 hr. on the surface and a 2-hr. moon walk for the cosmonaut, the lunar module would fire the same engine it had used for terminal descent and, leaving the legs behind, ascend to low lunar orbit where Soyuz waited. The lunar module carried an attitude control module on top and would jettison this to expose a docking probe used to link up with the orbital module on Soyuz for crew transfer. The instrument module on Soyuz, about 7 ft. 6 in. longer than the module on later Soyuz spacecraft used for Earth orbit missions, would fire its single engine to return to Earth.

1969 February 24   At 01:29:02 UT Atlas-Centaur AC-19 carrying the 910-lb. *Mariner VI* spacecraft was launched from Complex 36B at Cape Canaveral, Fla., on a direct-ascent escape trajectory to Mars. Atlas sustainer cutoff came at an elapsed time of 4 min. 35 sec., followed 11 sec. later by a 7 min. 22 sec. burn with the cryogenic Centaur stage. Spacecraft separation came at 01:42:47 UT, followed 3 min. 58 sec. later by solar panel deployment. *Mariner VI* comprised an octagonal magnesium spaceframe with a width of 4 ft. 6.5 in. divided into eight equipment bays. It supported a high-gain dish antenna 3 ft. 4 in. in diameter and a tubular low-gain antenna projecting 7 ft. 4 in. above the spaceframe. Four solar panels, each 7 ft. × 2 ft. 11.5 in., carrying a total 17,472 solar cells were attached to opposing flats on the octagon, providing a total surface area of 83.2 sq. ft. and a power output of 800 W at Earth and 449 W at Mars. The spacecraft had a span of 19 ft. and total height of 11 ft.

1969 February 28   A 5.3-sec. (10 ft./sec.) course correction burn was performed by *Mariner VI* at 00:54 UT, 782,000 mi. from Earth and at a speed of 7,670 MPH. This moved the Mars fly-by point from 5,500 mi. to about 2,000 mi. The propulsion system aboard the Mariner-Mars 1969 spacecraft was similar in concept to that used by *Mariner IV* and carried a small monopropellant rocket motor with a 2.45 in. diameter thrust chamber in a motor with a length of 7.9 in. Anhydrous hydrazine forced into the combustion chamber by a nitrogen gas pressurant squeezing a flexible bladder in the propellant tank was decomposed by aluminum oxide pellets acting as a catalyst. The motor produced a thrust of 51.3 lb. and provided a maximum delta-velocity of 192 ft./sec. in up to two burns, each of which could be as brief as 0.01 sec. for a total firing time of 1 min. 42 sec. Propellant capacity was 21.5 lb. The rocket motor nozzle protruded through one side of the octagonal spaceframe below and between two solar panels. Mariner-Mars 1969 attitude control was maintained by cold nitrogen gas operating through 12 jets from a 5-lb. tanked supply.

1969 February   NASA formally approved a plan from the Ames Research Center for a Pioneer mission to Jupiter involving two spacecraft launched in 1972 and 1973. (Launch windows to Jupiter occur once every 13 months.) For more than a year Pioneer contractors TRW had been studying the use of Pioneer technology for a deep-space mission which would continue the work of earlier Pioneers by measuring the solar atmosphere in regions never visited before. TRW received a contract to build the two spacecraft as a continuation of their existing work on current Pioneers. The two new spacecraft would be designated *Pioneers F* and *G*, or *Pioneers 10* and *11* after launch.

Gathering into focus a wide range of studies conducted by planetary scientists keen to explore the outer solar system, a technical report from the Jet Propulsion Laboratory, Calif., identified launch opportunities to Uranus and Neptune using the gravity of Jupiter to shorten the transit time. The report pointed out that an opportunity would exist between 1978 and 1980 that would not occur again for another 179 years. Before 1978 the planets were aligned in such a way that proceeding to Uranus would have required too great a bending angle in the flight path. After 1980 Jupiter would be too far along in its orbit to provide sufficient gravity for this mission. During the next six months NASA drew up plans for a so-called Grand Tour of the solar system.

1969 March 3   The *Apollo 9* mission began at 16:00:00 UT when a Saturn V AS-504 lifted off from Complex 39A at Cape Canaveral, with a total mass of 6.397 million lb., including the 59,086-lb. CSM-104, astronauts James McDivitt, David Scott and Russell Schweickart and the 32,132-lb. LM-3. In an initial orbit of 118 × 119 mi., the command/service modules (CSMs) separated from the adapter at 2 hr. 45 min., turned through 180° and docked with the lunar module (LM) 17 min. 8 sec. later. After separating from the adapter/S-IVB combination at 4 hr. 8 min., the third stage ignited at 4 hr. 45 min. 47 sec. and burned for 1 min. 10 sec., putting it in a 129 × 1,918 mi. orbit. A 4 min. 2 sec. burn at 6 hr. 7 min. 19 sec. put the S-IVB in solar orbit. A series of four SPS engine burns set up the CSM/LM orbit for a complex separation and rendezvous exercise on the fifth day. Meanwhile, McDivitt and Schweickart transferred to the lunar module at about 43 hr. elapsed time to check it out, sending a telecast at 46 hr. 28 min. The LM-3 descent propulsion system engine was fired for 6 min. 10 sec. (1,739 ft./sec.) at 49 hr. 41 min. 33 sec., using various throttle settings, putting the docked vehicles in a 126 × 310 mi. orbit. A 43.3-sec. (576 ft./sec.) SPS burn at 54 hr. 26 min. 11 sec. put the docked spacecraft in a 142 × 149 mi. orbit ready for the big rendezvous exercise. On the fourth day McDivitt and Schweickart moved across to the LM, code named *Spider*, leaving Scott in the command module, dubbed *Gumdrop*. At 73 hr. 7 min. Schweickart began a 1 hr. 7 min. EVA (extravehicular activity) from the same hatch a future astronaut would use to descend to the moon. Despite feeling nauseous the previous day he proved it was possible in an emergency to move from one spacecraft to another via exterior handholds. Scott conducted a 1 hr. 1 min. stand-up EVA from the open side hatch of *Gumdrop*. On the fifth day, at 92 hr. 39 min. 30 sec., the two spacecraft undocked and McDivitt and Schweickart in *Spider* separated from Scott in *Gumdrop* to a maximum distance of 115 mi. before rendezvous through a complex sequence of seven orbital maneuvers that involved jettisoning the LM descent stage and firing the ascent propulsion system. The two spacecraft redocked at 98 hr. 59 min., and after McDivitt and Schweickart had rejoined Scott in the Apollo command module, the LM ascent stage was cut loose. At 101 hr. 53 min. 20 sec. the ascent engine ignited and burned for 6 min. 2 sec. until propellant depletion, putting it in a 143 × 4,312 mi. orbit. For the next five days the crew conducted scientific photography of the Earth and fired the SPS twice prior to the deorbit burn, with splashdown in the Atlantic Ocean at 241 hr. 00 min. 54 sec., March 13, 1969.

1969 March 18   For the first time four U.S. Air Force OV1-series Aerospace Research Support program satellites were carried to separate orbits by an Atlas F launched from Vandenberg AFB, Calif. *OV1–17* weighed 312 lb. and carried 12 solar radiation experiments and a gravity-gradient stabilization system with three 50-ft. horizontal booms forming a "Y" and two 62-ft. vertical booms. It was placed in a 247 × 288 mi., 99.1° orbit, from which it decayed on March 5, 1970. The 487-lb. *OV1–17* propulsion module separated from the satellite and fired itself into a 109 × 192 mi., 99° path; once here it was designated *OV1–17A (Orbis-Cal 2)* for radio wave propagation tests. The 275-lb. *OV1–18* was placed in a 289.5 × 362 mi., 98.8° orbit where it observed the ionosphere for radio-wave propagation effects before decaying on August 27, 1972. The 273-lb. *OV1–19* studied trapped radiation in the Van Allen belts from a 292 × 3,528 mi., 104.7° orbit. It remains in space, but is inoperable.

1969 March 20   In the first test of NERVA XE (nuclear engine for rocket vehicle application) in the flight configuration, Aerojet General and Westinghouse successfully ran the engine at 10% of full power. At full power (1,100 mW) the engine would deliver equivalent thrust of 55,000 lb. The test was limited to three 25-sec. warm-up runs prior to a full-power run. This took place on June 11, when NERVA was operated for 13 min., 3 min. 30 sec. of which was at full power.

1969 March 24   Based on postflight results from the *Apollo 9* mission, NASA announced that the *Apollo 10* mission was scheduled for a May 18 launch carrying astronauts Thomas Stafford, John Young and Eugene Cernan. The backup crew comprised Gordon Cooper, Donn Eisele and Edgar Mitchell. The objective would be to perform a full rehearsal of a lunar landing mission up to but not including descent to the moon's surface. The lunar module would approach to within 50,000 ft. of the surface, the point where powered descent would begin on a landing flight. If all went well on *Apollo 10*, the next mission would be given clearance to land on the surface.

1969 March 26   An SL-3 launch vehicle carried the first operational Soviet weather satellite, *Meteor 1*, from Plesetsk to a 400 × 443 mi. orbit inclined 81.2°. It was an outgrowth of the experimental program that realized the launch of 13 weather satellites under the Cosmos designation beginning August 28, 1964. Meteor satellites had a cylindrical body with solar array wings on opposite sides and an attitude control system using control moment gyrowheels. Weights varied between 4,400 lb. and 8,400 lb. Meteor satellites were equipped with two cameras for daytime cloud cover pictures at a resolution of 3,000 ft. and infrared sensors for nighttime images. The Soviets launched a total of 32 *Meteor 1*–series satellites, the last on July 10, 1981.

1969 March 27   NASA launched the second of two Mariner-Mars 1969 spacecraft at 22:22:01 UT from Complex 36A at Cape Canaveral, Fla. Atlas staging occurred at 4 min. 17 sec., followed 9 sec. later by a 7 min. 26 sec. Centaur burn to put *Mariner VII* on a direct-ascent to Mars. The spacecraft was an identical twin to *Mariner VI,* launched February 24, 1969, and

5.38 million mi. from Earth at *Mariner VII*'s launch. *Mariner VII* would fly a faster trajectory, arriving at close encounter only five days after *Mariner VI*. A 7.6-sec. (14 ft./sec.) course correction burn took place on April 8, 1969, when the spacecraft was more than 2.5 million mi. from Earth. This advanced the close encounter fly-by by 15 min., moving the fly-by distance from 16,000 mi. to around 2,000 mi. as the spacecraft passed over the south pole.

The Soviet Union launched the first of two spacecraft (M-69) designed to go into an orbit of Mars, but a failure with the third stage of the SL-12 launch vehicle prevented its getting into its interim Earth parking orbit. A second failure in the first stage of another SL-12 on April 2 similarly prevented its companion from reaching space. Neither spacecraft carried entry probes and both were intended to conduct photographic surveys of the planet from elliptical orbits. There had been considerable political pressure for Soviet space scientists to claim for the USSR the historic honor of placing the first spacecraft in orbit around Mars. A strenuous effort was made to prepare orbiter/landers for the 1971 launch window, for it was known the Americans planned to launch their Mariner-Mars 1971 orbiter spacecraft to be designated *Mariner*s 8 and 9 after launch.

At a meeting to decide future plans, the European Space Research Organization (ESRO) agreed on three new projects. These were the second *HEOS* for Earth science and physics, *ESRO 1B*, to follow the successful flight of *ESRO 1A*, and *ESRO 4*, a replacement for the canceled *TD-2* Earth science satellite. Other more advanced satellite proposals were being reviewed. These included an atmospheric physics project placed in polar orbit, a spin-stabilized polar ionospheric research satellite and a spin-stabilized geostationary satellite for measurement of the Earth's magnetosphere. In less defined states of conception were two cosmic-ray satellites called *COS-A* and *COS-B*, a satellite to map the sources of celestial ultraviolet radiation, and a proposed fly-by mission to the planet Mercury.

1969 April 3   Dr. Thomas O. Paine was sworn in as the new NASA administrator by Vice President Spiro Agnew in a ceremony in the Vice President's office. Paine was particularly interested in the use of space for solving Earthly problems through better monitoring of the Earth's environment, its diminishing resources and the use of weather and communications satellites for bringing information to ordinary people.

1969 April 4   NASA and the Department of Defense formed a joint working group to "assess the practicality of a common, economical space transportation system to meet the needs of both Department of Defense and NASA, and to provide information on such a transportation system to the President's Space Task Group." The reference document required the Department of Defense to "study its own needs, present and future, for a new space transportation system." It was the first formal invitation by NASA for Department of Defense participation in determining the performance requirement of a reusable Earth-orbit transportation system that eventually became the Space Shuttle.

1969 April 10   NASA announced the names of flight crew members for the AS-507 *(Apollo 12)* mission, which at this date was scheduled for launch about four months after *Apollo 11*, expected to fly in July 1969. The primary crew would comprise Charles Conrad, Richard Gordon and Alan Bean. The backup crew would comprise David Scott, Alfred Worden and James Irwin. If *Apollo 11* failed to land on the moon in

July, *Apollo 12* would be given the task of attempting that feat.

1969 April 14 A Thorad-Agena D launched the 1,269-lb. research weather satellite *Nimbus 3* from Vandenberg AFB, Calif., into a 665 × 702 mi. orbit inclined 99.6° to the equator, along with *Secor 13*. The spacecraft bus was essentially the same as *Nimbus 1* and *2*, and the payload was identical to its predecessor, *Nimbus B*, which failed to get into orbit on May 18, 1968. A total of nine experiments were carried, including an infrared spectrometer, which took the first measurement of the atmosphere's vertical temperature, water vapor and ozone level. It carried the heaviest load of experimental research equipment ever launched on a weather satellite, but the satellite weight was exceeded by the 1,366-lb. *Nimbus 4*, launched on April 8, 1970; the 1,580-lb. *Nimbus 5*, launched on December 11, 1972; the 1,823-lb. *Nimbus 6*, launched on June 12, 1975; and the 2,000-lb. *Nimbus 7*, launched on October 24, 1978, and last in the series.

1969 April 24 The first meeting of a newly formed NASA Space Shuttle Task Group evaluated various uses for reusable aerospace vehicles. It was agreed that the primary justification centered on the use of the Shuttle as a logistics vehicle for the space station, envisaged as a large, permanently manned facility placed in orbit by a Saturn-class launcher. It was deemed that the potential application of the Shuttle as a minilaboratory should be played down in public discussion and Congressional testimony for fear it would negate the reason for building a larger space station.

1969 April 28 Maxime Faget, director of Engineering at NASA's Manned Spacecraft Center, reported to Robert Gilruth, MSC director, on preferred options for a shuttle vehicle. Gilruth had wanted an internal MSC study on the integral launch and reentry vehicle (ILRV) to parallel the work being conducted by four contractors awarded phase A study money on January 31, 1969. These studies appeared to indicate preference for a reusable two-stage system comprising a manned fly-back booster and a manned orbiter capable of staging off the top of the booster and flying into orbit under the power of its own propulsion system. The booster, meanwhile, would return to land like a conventional airplane, hence the term "fly-back." These studies were fed across to the four ILRV contractors for integration with their own studies.

NASA issued requests for proposals (RFPs) on the planned 12-person Earth orbiting space station scheduled for launch in 1975 on a Saturn V. NASA wanted a 10-year life for the station, with consumables and crew sent up on a reusable logistics/shuttle vehicle. The RFPs also stipulated conceptual design of a 50-person space base consisting of specialized modules for scientific and technical research assembled in the late 1970s or early 1980s. The proposals were due in by June 9, and NASA wanted to incorporate the data into the final station design if Congress approved the program. On May 2, 1969, NASA issued a request for experiment modules to be used with the 12-person station, due May 22.

1969 April 29 Testifying before the U.S. Senate Committee on Aeronautical and Space Sciences, NASA Administrator for Manned Space Flight George Mueller explained that the Apollo flight schedule at that time envisaged a maximum five flights in 1969, with the first lunar landing no earlier than *Apollo 11*. If that was accomplished, he said, NASA would defer the fifth flight to 1970. There was hardware for flights out to *Apollo 20*, sufficient to sustain moon exploration

through mid-1972. As envisaged at that time, *Apollo 11* would land in an eastern mare region where the surface was flat and relatively free of craters in July. The second mission would land in a western mare region in November, the third in a highland (primordial) area in March 1970, the fourth in cratered highlands near Censorius in July 1970 and the fifth in the Littrow area in the eastern region of Mare Serenitatis in November 1970. The sixth landing would be at the crater Tycho in April 1971, followed by a landing among the volcanic domes of the Marius Hills in September 1971. The eighth landing would be in Schroter's Valley in February 1972, the ninth near Hyginus Rille in July 1972 and the last near the crater Copernicus in December 1972.

1969 May 5 NASA held a briefing for contractors working on the ILRV (integral launch and reentry vehicle) phase A Shuttle studies to discuss major changes to the specification. Following consultation with the U.S. Air Force the previous week, NASA accommodated military missions, which called for a much greater payload capability than the space agency needed for its own use. The Air Force wanted a payload capability of 50,000-lb., twice that originally stipulated by NASA, with a 10,000 cu. ft. payload bay at least 22 ft. in diameter. The Air Force wanted to use the Shuttle for placing in low Earth orbit an advanced generation of military reconnaissance satellites and for retrieving satellites previously orbited. The Air Force also stipulated that the vehicle have payload doors mounted on top of the fuselage.

1969 May 7 NASA established the Space Station Task Group to coordinate internal activities at the space agency on a manned Earth orbiting facility. Both the Manned Spacecraft Center and the Marshall Space Flight Center were working space station studies with different sets of contractors. Astronaut Frank Borman was named as the field director for the space station program.

1969 May 12 The NASA Office of Manned Space Flight completed a report on the future goals of the space agency's manned flight program in the document *Integrated Manned Space Flight Program 1970–1980*. It was the most coherent blueprint for NASA plans that would emerge prior to a sequence of major budget cuts that, by the mid-1970s, would have forced abandonment of these plans. In the document, NASA advocated maximum integration of post-Apollo programs with the long-term objective of manned flights to the planets, asserting that "a long-duration manned space station and low-cost space transportation" were pivotal to future capabilities. The report recommended, by the end of the 1970s, "permanent residences in both low Earth altitudes and Earth-synchronous altitudes and in lunar orbit and on the lunar surface," utilizing a "shuttle vehicle which must carry nearly all material and men into space." This was the first official hint that expendable launch vehicles might be replaced by the Shuttle. "From this vehicle, men and materials are transferred to a nuclear shuttle and are transported to a synchronous space station or to lunar orbit. Transportation between lunar orbit and surface is achieved by a reusable chemical propulsion vehicle."

1969 May 16 The Soviet spacecraft *Venera 5* arrived at Venus carrying a probe designed to survive a maximum deceleration through the atmosphere of 500 g. Scientists and engineers had designed the capsule to survive to the surface of the planet, and in an effort to prevent it suffering the fate of its predecessor, they reduced the size of the parachute so it would descend faster and soak up less heat through the insulation. *Venera 5* was equipped with instruments similar to those sent to the planet by *Venera 4* a year before. Data from the *Venera 5* probe was returned direct to Earth for a period of 53 min. before transmissions ceased while it was still some distance from the surface. *Venera 6* arrived at Venus a day later and its capsule transmitted for 51 min. when it too failed during descent.

1969 May 18   At 16:49:00 UT the *Apollo 10* mission began with the launch of AS-505 from Complex 39B at Kennedy Space Center, the only time a Saturn V was launched from pad B. With a liftoff weight of 6.412 million lb., *Apollo 10* contained payload comprising Apollo CSM-106 (63,567 lb.) carrying astronauts Thomas Stafford, John Young and Eugene Cernan and LM-4 (30,735 lb.). The lunar module (LM) carried only 2,567 lb. of propellant for the ascent engine, versus almost 5,200 lb. for a standard lunar landing flight. Instead of firing together until about 8 min. 50 sec., the center J-2 engine in the second stage was shut down at 7 min. 41 sec., leaving the outer four J-2s to burn until 9 min. 13 sec. The phased S-II shutdown would be standard for subsequent flights. From an initial Earth parking orbit of 115 × 118 mi., the S-IVB stage reignited at 2 hr. 33 min. 25 sec. for the 5 min. 45 sec. translunar injection (TLI) burn. After extracting LM-4, the docked vehicles separated from the S-IVB/adapter at 3 hr. 56 min., leaving the S-IVB to go on for a 2,016-mi. lunar fly-by at 78 hr. 54 min. A 6.7-sec. (48.9 ft./sec.) SPS course correction burn began at 26 hr. 32 min. 56 sec., and five color TV transmissions were made during translunar coast. Ignition for the 5 min. 56 sec. (2,750 ft./sec.) lunar orbit insertion burn (LOI-1) occurred at 75 hr. 55 min. 53 sec., producing a 68.6 × 196 mi. orbit. The circularization burn (LOI-2) began at 80 hr. 25 min. 7 sec. and lasted 13.9 sec. (138 ft./sec.), resulting in a 67.8 × 70.8 mi. orbit. Cernan conducted a brief checkout of the LM at about 81 hr. 55 min. At 95 hr. 2 min. Stafford and Cernan entered LM-4, dubbed *Snoopy*, for a full rehearsal of lunar landing operations. Undocking came at 98 hr. 22 min. and *Apollo*, nicknamed *Charlie Brown*, edged away. At 99 hr. 46 min. 2 sec. the descent orbit insertion (DOI) burn took place when *Snoopy*'s descent engine was fired for 27 sec. (71 ft./sec.), dropping pericynthion to 9.7 mi. (51,200 ft. above mean lunar radius) uprange of landing site no. 2. *Snoopy* had speeded up and was ahead of and below *Charlie Brown*. As *Snoopy* began to climb back up from pericynthion, the descent engine was fired for 40 sec. (177 ft./sec.) at 100 hr. 58 min. 25 sec. This raised apocynthion to 219 mi. and carried the LM up, over and behind *Apollo*, putting it in the respective position it would have when entering lunar orbit after ascending from a landing site. At 102 hr. 45 min. the descent stage was jettisoned, but incorrect switch positions sent the ascent stage gyrating until quickly brought under manual control by Stafford. At 102 hr. 55 min. 1 sec. the ascent engine fired for 15.5 sec. (221 ft./sec.), putting *Snoopy* in a 12.9 × 52 mi. orbit 285 mi. behind *Charlie Brown*. From that position a series of rendezvous maneuvers brought the two together for docking at 106 hr. 22 min. 2 sec. After Stafford and Cernan rejoined Young in the command module, the LM-4 ascent stage was jettisoned at 108 hr. 24 min. 36 sec., and 27 min. 30 sec. later the ascent engine fired for 4 min. 6 sec. (3,838 ft./sec.), propelling it to solar orbit. After extensive photographic activity the 2 min. 44 sec. (3,625 ft./sec.) trans-Earth injection (TEI) burn began at 137 hr. 36 min. 29 sec. One 6.54-sec. (1.6 ft./sec.) correction burn was made with the RCS (reaction control system)

thrusters at 188 hr. 49 min. 57 sec., and the command module splashed down in the Pacific Ocean at 192 hr. 3 min. 23 sec.

1969 May 21   The NASA Manned Spacecraft Center established a Space Station Task Group headed by Rene A. Berglund to oversee the imminent activity associated with phase B definition studies for a permanently manned Earth-orbiting facility. With its established mandate to conduct U.S. manned spaceflight operations, the MSC had a strong interest in post-Apollo activities. It was already concerned that the Marshall Space Flight Center, with its responsibility for the Apollo Applications Program workshop, might seize control of the space station when that came along.

In the wake of further budget reductions made by the incoming Nixon administration, NASA's George Mueller met with members of the Manned Space Flight Management Council to survey options for the Apollo Applications Program. Under consideration was moving from a "wet" orbital workshop launched by Saturn IB to a "dry" workshop launched by Saturn V. Lifting all AAP hardware up in one mission would eliminate one of two Saturn IB assembly flights and permit on-orbit operations to move immediately to science and research activities. The Marshall Space Flight Center agreed that this was the best way to go, and on July 8–9, 1969, all the appropriate field centers reported to the Management Council that they also agreed.

A briefing was held at NASA's Manned Spacecraft Center on the results of internal engineering studies of a shuttle vehicle being designed at the Houston, Tex., facility. Under the guidance of Maxime Faget, the MSC envisaged a two-stage booster and orbiter configuration with a total length of 188 ft. 4 in. and a launch weight of 1,351,500 lb. The vertical configuration envisaged the orbiter attached piggyback to the forward upper component of the booster fuselage. The straight-wing booster would have a length of 175 ft., a wingspan of 134 ft. and a weight of 1,035,800 lb. and would be powered by eight 209,000-lb. thrust cryogenic engines. The orbiter would have a length of 113 ft., a wingspan of 75 ft. and a weight of 315,700 lb. and would be powered by two 250,000-lb. thrust engines. The booster would carry the orbiter, with its maximum 10,000-lb. cargo, to a speed of 7,860 ft./sec. and an altitude of 43 mi. Two 12,000-lb. thrust turbofan engines would power the orbiter in the atmosphere after reentry.

NASA announced it had issued 12 requests for proposals on the Earth Resources Technology Satellite (ERTS) remote sensing system. The responses were due June 18. The first satellite, *ERTS-A*, was expected to be launched in late 1971 or early 1972. *ERTS-A* would carry sensors for viewing in the visible and infrared portions of the spectrum. Weighing about 1,000 lb., the satellite would be placed in a 500-mi., sun-synchronous orbit and view the entire Earth in 100 mi. wide swaths in less than three weeks. On October 17 NASA selected TRW and General Electric for studies on how existing satellite designs could be adapted for ERTS.

1969 May 23   Five satellites were orbited by a Titan IIIC launched from Cape Canaveral, Fla. Along with the two *Vela 9* and *10* nuclear test detection satellites, the launcher carried the last three OV5 satellites for the U.S. Air Force. The two Velas were positioned on opposite sides of the Earth in slightly different orbits averaging 37,190 × 37,640 mi. at 32.8° by 7,600-lb. thrust solid propellant apogee motors. The 25-lb. *OV5-5 (Environmental Research Satellite-29)* had a very-low-frequency plasma detector and an orbit of 10,607 × 69,377 mi. at 33°. The 23-lb. *OV5-6* tetrahedron-shaped satellite carried eight antennas for solar flare particle detection from a 10,516 × 69,371 mi., 32.8° orbit. The 29-lb. *OV5–9* was a modified tetrahedron for solar flare and radiation monitoring from its 10,592 × 69,298 mi., 32.7° orbit. All remain in space.

1969 May 27   NASA headquarters approved development of a manned lunar roving vehicle (LRV) for use in later J-series Apollo lunar landing missions. The first LRV was expected to be ready by mid-1971, and a meeting was held June 6 at Headquarters to establish requirements for the vehicle. On October 31, 1969, the Marshall Space Flight Center awarded a contract to the Boeing Co. for the design and fabrication of four LRVs. Each vehicle was to weigh no more than 480 lb. and yet carry two fully suited astronauts and their tools. With independent electric motors for each of four wire-mesh wheels, the LRV was designed to provide precise navigation data displaying heading, distance traversed and range and bearing back to the lunar module. Stowed in a quadrant of the descent stage, it would be unstowed by an astronaut on the lunar surface pulling lanyards connected to critical deployment segments of the folded vehicle.

1969 May   Final development tests of a French solid propellant two-stage land based missile took place when a prototype S-2 was fired for the first time. Designated SSBS (Sol-Sol Balistique Stratégique), the missile program had evolved over 10 years and involved research rockets in the Agate, Topaze and Emeraude series. The S-2 had a length of 48 ft. 7 in., a diameter of 4 ft. 11 in. and a launch weight of 70,550 lb. The P16 first stage produced a thrust of 121,500 lb. for 76 sec., and the P10 second stage produced 99,200 lb. for 50 sec. Deployed from 1971, the S-2 had a range of 1,700 mi. and carried a 150-KT atomic (fission) warhead. A total of 18 missiles were deployed in silos under the command of two squadrons 3 mi. apart in the Haute Provence region of France's Massif Central.

1969 June 3   NASA's Manned Spacecraft Center recommended that if *Apollo 11* landed in an eastern mare region, *Apollo 12* should land at either the *Surveyor I* or the *Surveyor III* site. The reason for this choice was to provide a visible focus for demonstrating pinpoint landings and to provide scientists with an opportunity to examine surface material identified as interesting by the unmanned landers. Moreover, the astronauts could retrieve small items from the Surveyor spacecraft for analysis on Earth. On June 12, 1969, the *Surveyor III* site was recommended in further communication from the MSC, and NASA adopted this as the aim point for the second manned lunar landing.

1969 June 5   The last of six Orbiting Geophysical Observatories, *OGO-6* was launched by LTTAT-Agena D from Vandenberg AFB, Calif., to a 247 × 677 mi., 81.9° orbit. Heaviest of all the satellites in its class, *OGO-6* weighed 1,393 lb., including 371 lb. of scientific instruments supporting 25 separate experiments for studying measurements of the Earth's upper atmosphere, the auroral regions around the globe, geomagnetic fields and solar energy. Two experiments failed early in the mission, but *OGO-6* went on to continue monitoring the Earth-sun environment during this active period of the sun's 11-year cycle. *OGO-6* decayed on October 12, 1979.

1969 June 10   David Packard, deputy secretary of defense, announced cancellation of the Manned Orbiting Laboratory on grounds of cost savings and "advances in automated techniques for unmanned satellite systems." There was some validity to the latter claim, and objectives initially laid down for the MOL were regrouped into a new generation of unmanned military reconnaissance satellites. Developed by Lockheed for the National Reconnaissance Office as Program 612, the satellite ultimately evolved into Program 467 and was designated KH-9, the first of a generation of advanced reconnaissance satellite tied in to the future Space Shuttle program.

Gerhard Stoltenberg, the West German minister for science and technology, and NASA Administrator Thomas Paine signed a memorandum of understanding in Bonn covering the launch by NASA of three satellites for studies of the upper atmosphere and the sun. The first, an "aeronomy" satellite, was to weigh about 175 lb. and be placed in a 180 × 480 mi. orbit by a Scout launcher in 1972. The second and third were called HELIOS and would be launched by NASA as part of a joint U.S.-German venture to send two spacecraft on flights close to the sun. Seven of ten experiments on each would be provided by Germany.

1969 June 12   The NASA Space Shuttle Task Group (SSTG) completed its report on the Shuttle program, confirming selection of a baseline specification incorporating U.S. Air Force requirements. Stipulated payload capability was now 50,000 lb., to and from orbit, in a 10,000 cu. ft. cargo bay, 15 ft. in diameter and 60 ft. in length. The previous on-orbit stay time of 180 days had been cut to seven days, and the flight deck was to be designed for a crew of two. The SSTG projected initial operations beginning with 16 flights in 1975, increasing to an average of 60–65 flights annually from 1978. Of these, 34–48 each year were in support of lunar operations, placing hardware in Earth orbit for transportation to the moon. The SSTG left open the matter of configuration, selecting three classes of vehicle for comparison for estimated development cost: Class I, expendable boosters/reusable spacecraft ($2.5 billion); Class II, 1½-stage (drop-tank) spacecraft ($3.9 billion); and Class III, fully reusable vehicle ($4.3 billion).

1969 June 29   A NASA Delta N launch vehicle carried the 1,536-lb. *Biosatellite 3* life sciences laboratory from Cape Canaveral, Fla., to a 221 × 241 mi., 33.5° orbit. The reentry capsule contained a 14-lb. monkey and a pressurized nitrogen-oxygen atmosphere kept at a comfortable 75°F. The object was to study the effects of the space environment on a primate for up to 30 days with extensive monitoring of physical reactions. When the monkey showed signs of stress and anxiety due to the unnatural environment, a decision was made to return the capsule to Earth on July 6. It splashed down 25 mi. from Kauai, Hawaii, and was recovered, the last of three NASA Biosatellites. The mission was not considered a success.

1969 July 3   A second N-1 lunar launch vehicle was fired from the first of two pads prepared for it at precisely 20:18:32 UT. Just after the giant rocket cleared its launch support tower, about 8 sec. after liftoff and at a height of 600 ft., a stray piece of metal found its way into one of the turbopumps feeding the thirty first-stage engines with propellant and caused it to explode. The resulting fire severed electrical lines, which cut power and shut down all remaining first-stage engines. Before the N-1 could fall back to the pad, the launch escape system fired, removing the payload and dropping it down 3,500 ft. away. As it collapsed back on its pad at an elapsed time of 18 sec., the rocket erupted in a massive

A back-up to the Biosatellite 3 *satellite launched June 29, 1969, is tested at NASA's Ames Research Center.*

explosion which completely destroyed the no. 1 N-1 pad. The flight objective for this second N-1 was to have been a repeat of the first: a circumlunar flight to demonstrate the performance of the four main stages. Not for another two years would an N-1 fly.

1969 July 10 Robert Gilruth, director of the Manned Spacecraft Center, sent George Mueller, associate administrator for Manned Space Flight, a letter outlining the need for a "data-relay satellite system as an integral part of our long-range plans for manned spaceflight." NASA planned a Data Relay Satellite System comprising large satellites placed in stationary orbit capable of relaying data and voice communications from Shuttle vehicles to the ground. This would eliminate the need for tracking stations on foreign territory and significantly increase the amount of time during which manned space vehicles would be able to communicate with ground centers in the United States.

1969 July 13 In a desperate attempt to bring to Earth a sample of lunar soil before the *Apollo 11* astronauts, the Soviet Union launched the first of their third generation Luna series spacecraft. With a weight of 5,995 lb., the *Luna 15* lunar sample retrieval mission was launched by an SL-12 from the Tyuratam complex in Kazakhstan at 04:55 UT. After an unusually long transit period, the spacecraft disappeared around the western limb of the moon at 10:48 UT on July 17 to fire its retropropulsion system 12 min. later and achieve a lunar orbit of 34 × 126 mi. with an inclination of 126° and a period of 2 hr. 30 sec. NASA astronaut Frank Borman had successfully appealed to the Russians to publish details of the *Luna 15* trajectory to confirm that it would not intersect the *Apollo 11* flight path in lunar orbit. On July 19 the orbit was changed to 59 × 137 mi. with a period of 2 hr. 3 min. 30 sec., and on July 20 this was changed again to 9.9 × 68 mi. with a period of 1 hr. 54 min. and an inclination of 127°. At 17:47

UT on July 21, a braking maneuver began, but 4 min. later *Luna 15* crashed into the moon at 440 ft./sec. and was destroyed.

1969 July 14 McDonnell Douglas received a contract to study the NASA Manned Spacecraft Center in-house Shuttle design developed by Maxime Faget. Comprising a straight-wing orbiter carried to altitude by a straight-wing, fly-back, booster, the concept was to be evaluated at three different sizing levels: Class A, with a 12,500-lb. payload capability in a 750 cu. ft. cargo bay; Class B, with a 25,000-lb. payload in a 15 ft. × 30 ft., 5,000 cu. ft. cargo bay; and Class C, with a 50,000-lb. payload and a 15 ft. × 60 ft. cargo bay. Both booster and orbiter were to be capable of being ferried a distance of up to 2,877 mi.

NASA completed an evaluation of the operating requirements of an aerial refueling concept for Shuttle vehicles returning from space. Under the proposal, Shuttle orbiters powered by turbofan engines for use in the atmosphere would hook up to U.S. Air Force KC-135 aerial refueling tankers to augment on-board fuel supplies and extend their fly-back range. The concept would give the orbiter the option of delivering a payload from space direct to the landing site of the receiving agency instead of having it trucked from a Shuttle launch and landing base. This innovative capability was not incorporated in the definitive Shuttle design.

1969 July 15 Reporting on tests to discover the effects of weightlessness on a zero-g shower, a letter from NASA's Manned Spacecraft Center to the Langley Research Center read as follows: "MSC has some excellent films of Jack Slight in the KC-135 at zero gravity. The motion pictures of Jack showering are quite revealing—not of Jack, of the action of water at zero-gravity. The interesting point is that the water strikes Jack, bounces off in droplets, but then recollects as jelly-like globs on various parts of his body. He can brush the water away, but it will soon reattach elsewhere." NASA was developing a shower for the Apollo Applications Program orbiting workshop.

1969 July 16 At 13:32:00 UT Saturn V AS-506 lifted off from Launch Complex 39A at the start of the first manned lunar landing mission, *Apollo 11*. The flight brought 3,100 press representatives, 10,000 invited guests and 1 million spectators to the Kennedy Space Center. Carrying astronauts Neil Armstrong, Michael Collins and Edwin Aldrin, the Saturn V weighed 6.398 million lb. and lifted CSM-107 (63,508 lb.) and LM-5 (33,297 lb.) into a 118 × 119 mi. parking orbit. The TLI (translunar injection) burn toward the moon began at 2 hr. 44 min. 16 sec., boosting the assembly by 10,441 ft./sec. in a 5 min. 47.5 sec. burn. Apollo separated at 3 hr. 17 min. 4 sec., turned and docked with the lunar module (LM) at 3 hr. 24 min. 3 sec., and pulled it free at 4 hr. 16 min. 59 sec. The S-IVB stage would pass the moon at a distance of 2,693 mi., 78 hr. 50 min. 34 sec. after launch. At 26 hr. 44 min. 59 sec., the SPS engine fired a 2.9-sec. (20.9 ft./sec.) course correction burn. The crew sent a color TV transmission beginning at 55 hr. 8 min. showing LM checkout and views of the Earth. At 61 hr. 39 min. 55 sec. the spacecraft passed into the lunar sphere of influence, 214,550 mi. from Earth and 38,925 mi. from the moon. The 6 min. 2 sec. (2,917 ft./sec.) LOI-1 burn began at 75 hr. 49 min. 50 sec. on the far side of the moon, putting the docked space vehicles in a 70.5 × 194.2 mi. orbit. The 17-sec. (159 ft./sec.) LOI-2 circularization burn started at 80 hr. 11 min. 37 sec., resulting in a 61.9 × 75.6 mi. orbit. At 95 hr. 20 min. Armstrong and Aldrin entered the

In May 1969 NASA named three astronauts as the crew of the Apollo 11 *lunar landing mission. They are, from left to right: Neil A. Armstrong, commander; Michael Collins, command module pilot; and Edwin E. Aldrin Jr., lunar module pilot.*

LM, named *Eagle*, and undocked from the command/service modules, called *Columbia*, 3 hr. 52 min. later. The 30-sec. (76.4 ft./sec.) descent orbit insertion (DOI) burn began at 101 hr. 36 min. 14 sec. and lowered pericynthion to 9.8 mi. Powered descent initiation (PDI) began at 102 hr. 33 min. 05 sec., with the engine throttled at 10% for 26 sec. and then full thrust for just over 6 min. Standing but restrained by support harnesses, the crew monitored the displays as they descended, first in a heads-down/feet-forward position and then, at 3 min. 52 sec., heads-up when *Eagle* rolled 174°, allowing the downward looking radar at the rear of the LM to scan the surface. At 6 min. 25 sec. the engine throttled down to progressively slow the descent rate. *Eagle* was now 23,400 ft. above the surface, moving at 1,400 ft./sec., slowly pitching forward to a vertical position. At 8 min. 26 sec. into powered descent *Eagle* was 7,500 ft. above the moon with a forward velocity of 500 ft./sec. and a descent rate of 130 ft./sec., 4.7 mi. uprange of the landing site. As *Eagle* neared the surface, a computer alarm indicated it was getting an overload of data and might wipe itself any second. Pressing on, Armstrong visually selected a relatively clear site and put the LM down at 102 hr. 45 min. 40 sec. (20:17:40 UT/16:17:40 EDT on July 20), 11 min. 25 sec. after PDI; the descent engine had been shut down by sensors on probes, one attached to the bottom of each of the four landing legs (subsequent LMs would have only three probes, deleting the one attached to the leg supporting the descent ladder). The landed lunar module had an Earth weight of 15,897 lb. and was located in the Mare Tranquillitatis, 12 mi. south-southwest of the crater Sabine D, 15 mi. south-southeast of *Surveyor V* and 42 mi. south-southwest of the *Ranger VIII* impact point. Requesting that they conduct their EVA (extravehicular activity) before a scheduled sleep period, not after it as the flight plan dictated, Armstrong and Aldrin depressurized the LM and opened the hatch at 109 hr. 7 min. 33 sec. Armstrong descended the ladder and put his left foot to the lunar dust at 109 hr. 24 min. 19 sec. (02:56:19 UT on July 21; 22:56:19 EDT on July 20) with the words "That's one small step for man, one giant leap

for mankind." A TV camera on the side of the LM transmitted images of the event. Before starting out on planned activities, Armstrong read a plaque which contained the message "Here men from the planet Earth first set foot upon the moon, July 1969 AD. We came in peace for all mankind." Aldrin joined Armstrong on the surface 15 min. later, and the TV camera was moved to a tripod 80 ft. away. Contingency samples were obtained (in case the crew had to scurry back into the LM before they could retrieve selected samples), a U.S. flag was set up on a pole, pictures were taken and the crew talked to President Richard M. Nixon. They set up a solar wind collector and the Early Apollo Scientific Experiments Package (EASEP), comprising a passive seismometer and a laser reflector. Aldrin got back in, followed by Armstrong 10 min. later. The 2 hr. 31 min. 40 sec. EVA ended at 111 hr. 39 min. 13 sec. when the hatch was closed. The crew then rested. The 21 hr. 36 min. 21 sec. lunar surface stay ended at 124 hr. 22 min. 01 sec. when the ascent engine fired and the 10,821-lb. ascent stage separated from the descent stage, which was left on the moon. The ascent engine burned for 7 min. 20 sec. (6,070 ft./sec.) putting Armstrong and Aldrin in a 10.4 × 52 mi. orbit. In a sequence of five rendezvous maneuvers, the ascent stage rejoined *Columbia* in a 65 × 71 mi. orbit, and the two vehicles docked at 128 hr. 3 min. The crew transferred themselves and the 48.8-lb. of lunar samples to the command module, and the ascent stage, jettisoned at 130 hr. 10 min, was abandoned in lunar orbit. Trans-Earth injection occurred at 135 hr. 23 min. 42 sec. with a 2 min. 39 sec. (3,279 ft./sec.) SPS engine burn. Only one course correction was performed, an 11-sec. (4.7 ft./sec.) thruster firing at 150 hr. 29 min. 55

Astronauts Neil A. Armstrong, Michael Collins and Edwin E. Aldrin Jr., ride the Saturn V into space at the start of the first lunar landing mission, Apollo 11, *on July 29, 1969.*

Astronaut Edwin E. Aldrin paused by the passive seismic experiment and looks back at the lunar module during the lunar surface EVA of Apollo 11 *on July 21, 1969.*

sec. The command module separated from the service module at 194 hr. 49 min. 19 sec., and reentry began 13 min. 47 sec. later. Splashdown occurred in the Pacific Ocean on July 24, 1969, at 195 hr. 18 min. 35 sec. Met by President Nixon, the crew were quarantined at the Lunar Receiving Laboratory, Houston, Tex., until August 10, 1969.

During his visit to the Kennedy Space Center for the *Apollo 11* launch, Vice President Spiro T. Agnew said that "It is my individual feeling that we should articulate a simple, ambitious, optimistic goal of a manned flight to Mars by the end of this century." After the launch, to applause from KSC employees, Agnew reaffirmed that "I bit the bullet for you today as far as Mars is concerned. But on the other hand, in case any of us become too enthusiastic, I want to let you know that I may be a voice in the wilderness." Agnew was stating a personal opinion not endorsed by President Nixon, who did nothing to mobilize support for such a goal.

NASA announced the formation of the Universities Space Research Association (USRA), a national consortium of 48 (later 50) U.S. universities. It was set up to foster cooperation between educational and research organizations and to operate laboratories where space science experiments could be developed.

1969 July 18   NASA Administrator Thomas O. Paine approved the adoption of a "dry" orbital workshop, fitted out on the ground and launched by a two-stage Saturn V. This meant abandoning the idea of using a spent Saturn IB second stage made habitable in space. The S-IVB would still form the structural core of the workshop, but it would support the airlock module and multiple docking adapter on top. The Apollo telescope mount would be attached to a truss structure that could swing out 90° and move it into position for observations from Earth orbit. The mission plan envisaged the workshop launched during 1972, followed a day later by a Saturn IB's taking the first crew to the orbiting cluster for a 28-day stay. Two 56-day revisit flights were scheduled for the same year. On August 8 NASA completed a contract with McDonnell Douglas for two S-IVB stages outfitted as workshops, one to serve as backup to the other, with July 1972 scheduled as the launch date.

1969 July 23   NASA announced that McDonnell Douglas and North American Rockwell had been selected to conduct parallel phase B definition studies on a 12-man space station. McDonnell's work was to be directed by the Marshall Space Flight Center, and Rockwell's work was under the jurisdiction of the Manned Spacecraft Center. The $2.9 million contracts would last 11 months, after which NASA expected to issue phase C and D contracts for detailed design, assembly and launch. A two-stage version of the Saturn V was the candidate launcher: the INT-21, with S-IC and S-II stages of Saturn V, would be capable of placing 175,000 lb. in Earth orbit; the INT-20, using S-IC and S-IVB stages, could orbit 110,000 lb.

The space station was a key element in NASA's post-Apollo plans and was to have been integrated with the reusable Space Shuttle, for which five contractors were currently performing phase A studies.

1969 July 28 Anticipating the award of phase B definition contracts on the Space Shuttle, NASA circulated a draft specification for the work and invited comment from key personnel. It stated that the primary aims of the program were to cut launch costs below an order of magnitude below the cost of launching payloads by expendable rockets, to provide a flexible payload capability for a wide variety of space applications, to provide a commercial airline–type launch environment for passengers, and to provide for high launch-rates. The early stage of phase B was to examine candidate configurations based on a two- or three-stage, fully reusable Shuttle. At this stage NASA anticipated issuing three phase B contracts early in 1970, following completion of the ILRV (integral launch and reentry vehicle) studies.

1969 July 31 *Mariner VI* made its closest approach to Mars during a fly-by of the planet. The closest pass occurred at 05:19:07 UT at a distance of 2,131 mi. Scientific instruments were carried on a scan platform capable of moving to track the planet. Weighing 165 lb., the assembly included two TV cameras. A 50-mm, high-resolution, wide-angle camera similar to that on *Mariner IV*, but covering an area 12–15 times greater, took 50 far-encounter pictures from a distance of 770,600 mi. down to 111,950 mi. The best resolution was about 15 mi. During the near encounter phase, *Mariner VI* took a total 24 pictures from the high resolution camera and a 508-mm narrow-angle camera viewing the surface through a Schmidt-Cassegrain telescope with a surface resolution of 950 ft., as against 2 mi. for *Mariner IV*. Each picture was composed of 704 lines, each of which had 945 pixels (picture elements). Pictures were played back at a maximum rate of 16.2 kilobits/ sec.

In a supplement to its June 12, 1969, report, the NASA Space Shuttle Task Group defined additional elements which it said would be essential to getting maximum benefits from the new vehicle. These included what it described as on-orbit space tugs for moving cargo between near-Earth orbits and for transferring satellites and payloads from the low Earth orbit of the Shuttle to stationary orbit 25,600 mi. above Earth. The tugs would comprise reusable rocket stages lifted into orbit by the Shuttle and left at the station or some other suitable base in space from which they would deliver or retrieve cargo at altitudes the Shuttle could not reach. Shuttles, tugs and nuclear shuttles were incorporated in what was soon referred to as the space transportation system.

Japan's Space and Technology Agency (STA) signed an agreement with the U.S. government extending the level of cooperation in space activities between the two countries. Formed on May 19, 1956, the STA existed to encourage industrial exploitation of new technologies and to enable a broad level of cooperation between government research departments and laboratories operated by commercial manufacturers. The agreement would build on a working arrangement with McDonnell Douglas whereby Japanese companies would have access to the design of the Delta launch vehicle, carry out improvements and make that technology available to McDonnell Douglas. It would emerge in Japan as the N-series.

1969 July NASA's Marshall Space Flight Center began definition studies of system requirements for (nuclear engine for rocket vehicle application) a nuclear stage powered by a NERVA (nuclear engine for rocket vehicle application) engine. In October the study was broadened to include the concept of a nuclear shuttle designed to operate between Earth orbit and lunar orbit ferrying heavy loads. If used as the third stage of a Saturn V, translunar payload would increase from 100,000 lb. to 160,000 lb., providing sufficient cargo for stays of up to 28 days on the lunar surface. The reusable nuclear Shuttle would carry 300,000 lb. of hydrogen fuel in a stage 150 ft. long and 33 ft. in diameter.

1969 August 1 Two technicians working at NASA's Lunar Receiving Laboratory, where the rocks and soil samples from the *Apollo 11* mission were being examined, became contaminated with lunar materials and entered quarantine. Ronald J. Buffum and George E. Williams joined the three astronauts, two doctors and twelve technicians, cooks and assistants. Added on August 5 were Heather A. Owens, Chauncey C. Park, Roy G. Coons and Riley Wilson, who also became contaminated when a line carrying moon samples from a vacuum chamber broke, spilling lunar materials into the room.

1969 August 5 *Mariner VII* made its closest approach to Mars at 05:00 UT, when the spacecraft passed within 2,130 mi. of the south polar region. A total of 93 wide-angle pictures were taken during the far encounter phase and 33 wide- and narrow-angle pictures during near encounter for a total 126 on the single pass. As with *Mariner VI*, which passed by on July 31, *Mariner VII* carried infrared and ultraviolet spectrometers to determine the composition of the atmosphere and an infrared radiometer for temperature measurements of the surface and the atmosphere. Joint flight results included the discovery that Mars was not as lunar-like, devoid of anything but craters and lava, as *Mariner IV* had indicated, that surface pressures were a mean average of 6.5 millibars, that temperatures varied between −190°F at the south pole to 62°F at noon on the equator and that the atmosphere was almost all carbon dioxide.

1969 August 6 NASA announced the names of the *Apollo 13* and *Apollo 14* crews. Primary crew members for *Apollo 13* would be James Lovell (CDR), Fred Haise (CMP) and Thomas Mattingly (LMP). Their backup crew would comprise John Young, John Swigert and Charles Duke. Prime *Apollo 14* crew members would be Alan Shepard, Stuart Roosa and Edgar Mitchell. The backup crew would comprise Eugene Cernan, Ronald Evans and Joe Engle. There was irony in this choice. When the *Apollo 14* backup crew was eventually cycled in as the primary crew for *Apollo 17*, subsequent lunar landing missions had been canceled. Engle was made to stand down so that scientist-astronaut Harrison Schmitt, a geologist, could fly the last lunar landing and honor a NASA pledge that at least one scientist would get to the moon.

1969 August 7 At 23:55 UT a four-stage SL-12 carried the 11,884-lb. circumlunar Soyuz variant designated *Zond 7* to a 118 × 133 mi. orbit from which the fourth stage fired it to translunar trajectory. A course correction was performed two days later at an Earth distance of 161,560 mi., and the spacecraft looped around the moon on August 11 at a distance of 1,240 mi. Like its predecessors, *Zond 5* and *Zond 6*, it carried cameras and took pictures, but this time in color instead of black and white. Like *Zond 6*, *Zond 7* performed a double reentry before landing back on Earth August 14 in a remote region of Kazakhstan after what proved to be the only successful L-1 mission.

NASA announced it had awarded an eight-month contract to General Dynamics for studies of experiment modules suitable for use with the manned space station, currently the subject of phase B definition contracts to McDonnell Douglas and North American Rockwell. As envisaged at this time, the modules would house specific sets of dedicated science, engineering and technology development equipment to complement station activity.

1969 August 12   NASA launched *ATS-5 (Applications Technology Satellite-5)* by Atlas-Centaur from Cape Canaveral, Fla., to a geosynchronous orbit inclined 7° to the equator. The 954-lb. satellite was at first unable to jettison its spent apogee motor case because it was in an unplanned spin, adding a weight of 856 lb. On September 5, with *ATS-5* over 108° W, the satellite was under control and the motor case was jettisoned. Because it was wrongly spinning counter-clockwise, the gravity-gradient test could not be performed. This would have utilized four 124 ft. long beryllium-copper booms extended from electrically driven drums. The cylindrical satellite had a length of 6 ft., or 12 ft. to the tip of a magnetometer boom, and a diameter of 4 ft. 10 in., with 22,000 solar cells covering the exterior. Experiments included an L-band air traffic control communication system, a millimeter waveband telemetry system (for the first time) and general communications equipment for C-band tests, TV, telegraph and digital data.

1969 August 14   Following the cancellation of the U.S. Air Force Manned Orbiting Laboratory, NASA's George Mueller selected seven astronauts from that program to serve as astronauts with the civilian space agency, constituting Group 7: Karol J. Bobko (USAF); Robert L. Crippen (USN); Charles G. Fullerton (USAF); Henry W. Hartsfield (USAF); Robert F. Overmyer (USMC); Donald H. Peterson (USAF); and Richard H. Truly (USN). All were to fly on Shuttle missions. Crippen was on the first Shuttle mission to orbit the Earth and subsequently became director of the Shuttle Program and then director of the Kennedy Space Center. Truly became NASA administrator. An eighth MOL astronaut, Lt. Col. Albert H. Crews (USAF), was assigned to the Flight Crew Operations Directorate at NASA's Manned Spacecraft Center.

1969 August 21   McDonnell Douglas completed a study for NASA of a logistics vehicle derived from the Gemini spacecraft that could be used to ferry crew members and cargo to a manned space station. Called Big G, the proposed vehicle consisted of a crew module with geometric proportions similar to those of the basic Gemini spacecraft but with a base diameter enlarged to 13 ft. 9 in. and capable of carrying 9 to 12 people. Attached to a cylindrical cargo/propulsion module of the same diameter, recovery would be effected with a gliding parachute, or parawing, the personnel sitting upright. The Big G would be launched by a Titan III or a combination of the first (S-IC) and third (S-IVB) stages of a Saturn V, and launch escape would be provided by an Apollo-type tower. An alternate configuration had capacity for up to 25 people.

1969 August 27   A long-tank thrust-augmented Delta was destroyed by the range safety office at Cape Canaveral, Fla., just before first stage burnout when a sudden loss of hydraulic pressure caused the vehicle to begin tumbling. The Delta had been carrying the last interplanetary Pioneer science satellite similar to *Pioneers 6, 7, 8* and *9*. Designated *Pioneer E*, it would have become *Pioneer 10* in orbit. Also on the Delta was

the last of three TTS satellites for testing Apollo tracking and communications facilities.

1969 September 5   NASA defined the objectives and requirements of a second orbital workshop launched by the first two stages of a Saturn V. Tentatively scheduled for a launch in mid-1973, *OWS No. 2* was to have been continuously manned for 1–2 years in a 242 to 310-mi. orbit. Some time was to be devoted to artificial gravity experiments by spinning the station around its center of mass, providing an added experiment option for scientists denied such a capability with *OWS No. 1*.

1969 September 8   NASA's Marshall Space Flight Center announced it was awarding an 11-month phase B contract to McDonnell Douglas for preliminary design and planning of a 12-person space station. The station was expected to be operational in 1975 and form the basis for a 50-person space base by the early 1980s. The Manned Spacecraft Center had issued a similar contract to North American Rockwell. At the end of 11 months NASA expected to decide on a single primary contractor for phase C/D, detailed design and development.

1969 September 11   George Mueller, NASA associate administrator for Manned Space Flight, outlined the general direction of manned flight planning, which envisaged the first suborbital tests of a reusable Earth-to-orbit Shuttle in 1974, initial operations with a permanently manned space station and the Shuttle vehicle in 1975 and a lunar orbit space station in 1976. Essentially a rocket-powered upper stage, the nuclear shuttle was expected to be operational in 1978 for lunar and planetary missions, with a synchronous-orbit space station available by 1979. A 50-person space base in Earth orbit, a lunar surface base and a manned Mars landing were to be accomplished in the 1980s.

1969 September 15   The Space Task Group (STG) submitted to President Nixon the report of its conclusions about the future of the U.S. space program. A maximum effort, concluded the STG, would provide a reusable Shuttle system and space station in 1975, an Earth-orbit transfer tug and lunar orbit station in 1976, a cislunar nuclear shuttle and lunar surface base in 1978, a 50-person Earth orbit space base in 1980, a manned Mars landing in 1981 and a 100-person Earth-orbit space base in 1985. To achieve this NASA budgets would reach a peak of $10 billion in 1976, twice the level of NASA's highest funding during Apollo. By eliminating the manned Mars mission, funding levels could be reduced to an annual $5.5 billion, delaying the Shuttle and station to 1977 and Earth and lunar space bases to the late 1980s. Only by eliminating all manned spaceflight could the NASA budget be kept to around $3 billion. President Nixon largely ignored the report and only in January 1972 accepted the need for a reduced-scale Shuttle.

1969 September 18   NASA and the Atomic Energy Commission (AEC) announced completion of NERVA XE (nuclear engine for rocket vehicle application) nuclear experimental rocket engine tests at Jackass Flats, Nev. Since August 1969 the XE had completed 28 successful engine starts and accumulated 3 hr. 48 min. operating time, of which 3 min. 30 sec. had been at full power, equivalent to a thrust of 55,000 lb. NASA and the AEC wanted to develop a flight engine of 75,000-lb. thrust, and the XE engine had proven the basic viability of nuclear rocket engine design and was being managed for the Space Nuclear Propulsion Office.

1969 September 22 Another attempt to launch a Japanese satellite failed when the Lambda 4S-4 solid propellant launcher strayed off course during fourth stage burn. Launched from Uchinoura, the Lambda was to have placed a 20-lb. test satellite in low Earth orbit, from which it would have detected charged particles and conducted a modest range of tests for engineering data. This was the fourth consecutive launch failure of the Lambda.

1969 September 23 A Soviet SL-12 launched *Cosmos 300*, so designated to veil its true identity as a failed third-generation Luna-series spacecraft. The terminal stage of the launch vehicle failed to propel the spacecraft out of its 114 × 117 mi. orbit, and the assembly decayed back down through the atmosphere and was destroyed four days later. At the next lunar launch opportunity on October 22, 1969, an SL-12 launched another Luna-series vehicle into a temporary Earth orbit so low that it decayed back down within a few hours, leaving only a little orbital debris to survive two days.

1969 September 24 In a quaint link between the Old World and the Space Age, Jack Warner, an innkeeper from Shawbury, England, asked the U.S. government to grant him a license for the first public-house on the moon. He offered to give it the name "The Space Inn," or alternatively "The Lunatic Tavern." Having signed an agreement not to lay claim to ownership of the moon, the government was unable to grant his request!

1969 October 1 Formed out of the existing National Space Development Center to manage Japan's expanding space interests more effectively, the National Space Development Agency (NASDA) came into being with headquarters in Tokyo. A technology center at Tsukuba was set up in 1973, and NASDA itself assumed responsibility for the development of liquid propellant launch vehicles in the N and H series. The existing Institute of Space and Aeronautical Sciences (ISAS) would continue to develop and manage the launch of solid propellant sounding rockets and orbital launchers. NASDA would expand and extend the Tanegashima launch site, and assume a role for the nation similar to that of NASA in the United States by developing and managing new technologies for civilian space programs.

A repeat of the earlier Aurorae mission, the fourth European satellite, *ESRO 1B (Boreas),* was launched by a four-stage solid propellant Scout from Vandenberg AFB, Calif., to a 181 × 242 mi., 86° orbit. Identical to *ESRO 1A,* the 176-lb. satellite conducted ionospheric and auroral studies through eight experiments. Due to a launch vehicle malfunction, *ESRO 1A* had been placed in an orbit lower than the planned one of 248.5 × 243.6 mi. and therefore decayed on November 23, although the satellite returned more than 56 hr. of useful data despite a failure in its data recording equipment.

The French firm of *Société d'Etude de la Propulsion par Réaction* (SEPR) was formed to coordinate French development work on rocket motors. The company, its name soon shortened to SEP, would conduct work on guided missiles and rocket motors for satellite launchers and form the core around which France would develop much of the technology for its defense and civilian satellite launcher programs.

1969 October 4 The U.S. Chief of Naval Operations established an office detailed to examine future concepts in submarine-launched ballistic missiles (SLBMs). Called the Undersea Long Range Missile System (ULMS), it looked to the 1990s, when existing fleet ballistic missile submarines would need replacing, and sought to fit upcoming technologies and nuclear weapons into future requirements. Polaris developments and the Poseidon SLBM had gone as far as existing submarine technology would permit. The missiles had grown to the maximum size existing submarines could carry, and there were prospects of providing the undersea deterrent force with a missile of intercontinental range carried aboard submarines that were quieter and more efficient than existing boats.

1969 October 7 The world's largest high-vacuum chamber, the Space Power Facility, was opened at the NASA Lewis Research Center. With a volume of 800,000 cu. ft., the chamber had a height of 120 ft. and a diameter of 100 ft. It was built to test large space electric power generating systems and would first be used to test a Brayton cycle power generating system operated by nuclear isotope or solar heat.

1969 October 9 NASA provided details of tentative plans for manned lunar exploration in the 1970s. On current schedules, Apollo mission hardware would continue to support manned activity through the end of 1972. According to the level of effort and the commitment of resources, Apollo follow-up expeditions could begin in 1973 and 1974 using additional equipment procured as extensions of existing hardware contracts, providing four or five missions over two or three years. Lunar orbit stations could be operational anywhere between 1976 and 1981, depending on the commitment, while lunar surface bases could be available from 1978 at the earliest. Despite these optimistic projects, there was little belief in the ability of NASA to obtain the necessary funding for lunar exploration after *Apollo 20.*

1969 October 10 At a meeting of the Manned Space Flight Management Council, major decisions were reached about modifications to equipment and hardware for the J-series of Apollo extended lunar surface visits. Projected to remain on the lunar surface for up to 72 hr., versus the 36 hr. considered optimum for the standard lunar module, it was envisaged that the astronauts would perform four extravehicular activities, or EVAs, (although this was later cut to three), and benefit from improvements to the A-7L space suit modified into an AL-7B configuration. Life support functions would be enhanced by a PLSS capable of supporting up to 7 hr. of surface activity. Mobility would be provided by the lunar roving vehicle. *Apollos 16–20* were scheduled to perform the J-series missions.

1969 October 11 The first mission to involve three manned spacecraft in orbit at the same time began at 11:10 UT with the launch of cosmonauts Georgi S. Shonin and Valery N. Kubasov aboard the 14,500-lb. *Soyuz 6* spacecraft on an SL-4 from Baikonur. Minutes later it was in a 115.5 × 138.5 mi. orbit inclined 51.7°, but maneuvers changed this to a 120.5 × 143 mi. orbit. At 10:45 UT on October 12 the 14,485-lb. *Soyuz 7* was launched carrying cosmonauts Anatoli V. Filipchenko, Viktor V. Gorbatko and Vladislav N. Volkov to a 128.6 × 140.4 mi. orbit. At 10:19 UT on October 13 the 14,650-lb. *Soyuz 8* was launched carrying Vladimir A. Shatalov and Alexei S. Yeliseyev to a 127 × 138.5 mi. orbit. It was the first time seven people were in space at the same time. On October 15, after much maneuvering, all three spacecraft approached each other to varying distances in an orbit of 124 × 140 mi. A welding experiment was conducted in the depressurized orbital module of *Soyuz 6* using an automatic device called Vulkan. After 4 hr. 24 min. in close proximity, *Soyuz 6* separated and returned to Earth at 09:52 UT on October 16, followed by *Soyuz 7* at 09:26 UT on October 17 and *Soyuz 8* at 09:10 UT on October 18.

NASA Administrator Thomas O. Paine left Washington, D.C., on a low-key visit to London, Bonn and Paris to present the plans for post-Apollo space programs and invite European cooperation. ESRO and ELDO set up a joint working group to consider participation in the manned space projects envisaged by NASA. On December 15 Dr. Paine visited Canada, and on February 24, 1970, he went to Australia, visiting Japan on March 3, 1970, for a similar purpose. While the Europeans were enthusiastic, the other countries were not.

1969 October 14   Launched by an SL-7 from Kapustin Yar, the *Intercosmos 1* satellite was placed in an orbit of 156.5 × 389 mi., with an inclination of 81.2°. The 705-lb. satellite was built to conduct a study of the sun's corona, and the effect of solar ultraviolet and x-radiation on the Earth's atmosphere. It decayed on January 2, 1970. Intercosmos satellites carried experiments from eastern bloc countries in a joint venture aimed at coordinating research programs. In total, 23 Intercosmos satellites were launched, and all but one, unnumbered, made it to orbit. The first five were similar to the first and all were launched by SL-7. The 2,360-lb. *Intercosmos 6,* launched from Baikonur on April 7, 1972, was launched by an SL-4 and its capsule recovered on Soviet territory. The next three, similar to *Intercosmos 1*, were again launched by SL-7s, including *Intercosmos 9*, called "Copernicus 500" to commemorate the 500th birthday of Copernicus. Beginning with *Intercosmos 10*, all but the last were launched by SL-8. *Intercosmos 23*, launched on August 7, 1981, by SL-3 weighed 3,310 lb.

1969 October 24   Mstislav V. Keldysh, president of the Soviet Academy of Sciences, declared that "we no longer have a timetable for manned Moon trips." In the previous ten months, NASA had put three men into orbit about the moon, conducted a full-scale manned reconnaissance, landed two men on the lunar surface and returned to Earth samples of soil and rock. In the wake of that achievement the Soviets felt they should adopt a twin-track policy of lunar surface exploration on a more protracted time scale and develop an Earth-orbit space station in parallel. Late in 1969 Soviet authorities gave formal approval for what would be called the Salyut space station.

1969 October 30   McDonnell Douglas submitted its review of the in-house NASA Manned Spacecraft Center design for a reusable Shuttle. The concept now envisaged a 148 ft. long orbiter with a 114 ft. wingspan and lift-over-drag ratio of 8.1 carried piggyback on the forward fuselage of a manned fly-back booster 207 ft. in length with a wingspan of 162 ft. A payload of 25,000 lb. could be carried in the 15 ft. × 60 ft. cargo bay of the orbiter and propulsion would consist of nine to twelve 400,000-lb. thrust cryogenic engines in the booster and two of similar type in the orbiter. The configuration would have a total height of 226 ft. and a liftoff weight of 3 million lb., of which 2,382,200 lb. comprised the booster; by November 5, 1969, the weight had been cut to 2,854,190 lb., of which the booster accounted for 2,251,190 lb.

1969 October 31   The Convair Division of General Dynamics submitted its report on the integral launch and reentry vehicle (ILRV) study it had performed for NASA. Framed by a payload requirement of getting 50,000 lb. to a 311-mi. orbit, configurations selected by General Dynamics included the two-element FR-3A and the three-element FR-4, both featuring tandem staging and parallel burn. The FR-3A had an overall length of 235 ft. and a liftoff weight of 4.33 million lb. The FR-3A booster was 210 ft. in length, 41 ft. wide and 37 ft. high, had a V tail spanning 84 ft. and carried fifteen 400,000-lb. thrust cryogenic engines. Go-around would be provided by three 52,500-lb. thrust Rolls Royce RB-211 turbofan engines. The orbiter would have a length of 179 ft. 2 in., a width of 31 ft., a height of 26 ft. and a span of 66 ft. across the V tail. Three 21,000-lb. thrust turbofan engines would provide orbiter go-around capability. Booster and orbiter each had retractable wings deployed only for fly-back, the latter having an L/D of 8.2. The three-element FR-4, which was not considered the best design by Convair, comprised a three-engined orbiter sandwiched between two boosters, each carrying nine rocket motors, for a total length of 219 ft. and a liftoff weight of 4.92 million lb.

1969 October   NASA made a significant decision about propulsion for the reusable Shuttle when it eliminated the requirement for go-around propulsion in the vehicle's operational phase. Go-around meant the ability of the Shuttle to compensate for a relatively inaccurate reentry path by using on-board engines to fly like a conventional airplane to its desired landing strip using a nominal reserve of aviation fuel. At this stage, air-breathing engines were to be retained for the development phase until accurate landings demonstrated the ability of the Shuttle to make pinpoint landings as an unpowered glider.

1969 November 1   The Chinese made an unsuccessful attempt to launch a satellite from the Jiuquan launch site, sometimes referred to as Shuang Ch'eng Tsu, at 41° N by 100° E. Developed from the CSS-3 two-stage missile, the three-stage CZ-1 (Chang Zheng, or Long March) satellite launcher had a total length of 96 ft. 7 in., a maximum diameter of 7 ft. 5 in. and a launch mass of 180,000 lb. The first and second stages were powered by $UDMH/N_2O_4$ propellants, and the third stage had a solid propellant. The four first-stage engines produced a launch thrust of about 310,000 lb. The CZ-1 could place a 660-lb. satellite in a 273-mi. orbit inclined 70°.

NASA's Kennedy Space Center, the Boeing Co. and Chrysler Corp. began a major study of possible alternative launch modes for the Saturn IB AAP flights carrying astronauts to the orbital workshop. Because all elements above the S-IB stage of the Saturn launcher were compatible with upper levels of the Saturn V umbilical tower at Launch Complex 39, NASA was considering using one of the pads at that site to launch the smaller Saturn IB. This would enable NASA to save money by closing down Launch Complex 34, the pad from which it had been planned to launch the Saturn IBs. To raise the Saturn IB above the mobile launch platform so that the upper stage and spacecraft would be compatible with the tower at LC-39, a pedestal was proposed on which the Saturn IB would be mounted for launch. On May 15, 1970, NASA announced that this plan had been adopted. The two-stage Saturn V would launch the orbital workshop from LC-39A and the three Saturn IBs would be launched from LC-39B.

1969 November 4   NASA's Ames Research Center held a meeting for scientists, engineers and prospective contractors to fully describe plans for *Pioneer F*, the first of two spacecraft scheduled for a fly-by of the planet Jupiter. Launched by Atlas-Centaur between February 26 and March 15, 1972,

Pioneer F (designated *Pioneer 11* after launch), would take almost two years to traverse the asteroid belt beyond Mars and journey to Jupiter, the largest planet in the solar system. *Pioneer F* would carry 60 lb. of instruments to measure the heavy radiation belts to within 90,000 mi. of the surface. *Pioneer G* (*Pioneer 12* after launch) would be launched in early 1973.

1969 November 5　North American Rockwell made a preliminary presentation of their completed ILRV study based on a 50,000-lb. payload capability accommodated by a piggyback configuration, with the orbiter attached to the upper forward fuselage of the booster. The design layout originated with Maxime Faget's team at the Manned Spacecraft Center and although the baseline requirement for NAR's contract was for a 25,000-lb. capability, the higher level was optimized for the work package. The orbiter had a length of 202 ft., a span of 146 ft., a liftoff weight of 871,000 lb. and two 590,000-lb. thrust cryogenic engines as well as provision for a crew of two and ten passengers. Four turbojet fly-back engines were to be provided. The booster had a length of 247 ft., a wingspan of 244 ft. and eleven 510,000-lb. thrust cryogenic engines. The stack had a total height of 274 ft., a liftoff weight of 4.494 million lb. and a staging velocity of 11,000 ft./sec. at 43 mi. NAR also submitted a sizing estimate based on a maximum launch mass of 3.5 million lb., this design providing a 35,800-lb. payload capability.

1969 November 8　A four-stage Scout carried the 156-lb. satellite *Azur 1* from Vandenberg AFB, Calif., to a 240 × 1,957 mi., 102.8° orbit. Built by the German firm of MBB (Messerschmitt-Bölkow-Blohm), *Azur 1* comprised a 2 ft. 6 in. diameter cylinder with a truncated cone at the forward end and a height of 3 ft. 8 in. Four antennas extended from the flat aft end, and a magnetometer boom atop the conical forward section produced a total height of 6 ft. 5 in. More than 5,000 solar cells provided 33 W of electrical power for the seven experiments, designed to investigate the Van Allen radiation belts, the aurora and solar particles. It remains in orbit.

1969 November 14　The second manned lunar landing mission got under way with the launch of *Apollo 12* from Complex 39A at the Kennedy Space Center at 16:22:00 UT. The 6.485-million-lb. vehicle comprised the Saturn V AS-507, CSM-108 (63,578 lb.), code name *America*, carrying astronauts Charles Conrad, Richard Gordon and Alan Bean, and LM-6 (33,587 lb.), named *Intrepid*. At 36.5 sec. and 52 sec. into flight the ascending vehicle was struck by lightning, disabling the command module displays and throwing circuit breakers. The crew had to reset both electrical power buses before reaching orbit. The SPS engine fired a 9 sec. (61.8 ft/sec) burn at 30 hr. 52 min. 44 sec. to put *Apollo 12* on a course for the moon. The flight path taken was such that the astronauts could not be returned to Earth on a gravitational slingshot maneuver should the main engine fail on the far side of the moon. The spacecraft was placed on this nonfree return trajectory only after the lunar module had been extracted from its shroud on top of the Saturn S-IVB stage so that its descent engine could be used as a backup to get the spacecraft back on a free return course if necessary. LOI-1 (lunar orbit insertion 1) began at 83 hr. 25 min. 23 sec. followed by the circularization burn 2 orbits later. Powered descent to the moon took 11 min. 58 sec., with a landing at 110 hr. 32 min. 36 sec., 620 ft. northwest of *Surveyor III*. The first EVA (extravehicular activity) lasted 3 hr. 56 min. 3 sec., when the ALSEP instruments were deployed, including dust detector,

During the second lunar surface EVA of the second manned moon mission, astronaut Charles Conrad Jr. visits the Surveyor III *spacecraft, November 20, 1969, and retrieves the TV camera for return to Earth.*

passive seismometer, magnetometer, solar wind spectrometer, ion detector and a cold cathode gauge. The second EVA lasted 3 hr. 49 min. 15 sec. when Conrad and Gordon performed a geologic field trip which included visiting *Surveyor III* and retrieving the TV camera for return to Earth. With a surface stay time of 31 hr. 31 min. 12 sec., Conrad and Gordon departed for a rendezvous with Bean in the command module, lifting away from the descent stage at 142 hr. 03 min. 48 sec. The ascent stage was sent crashing into the moon, with impact at 149 hr. 55 min. 16 sec. At 172 hr. 27 min. 17 sec., a 2 min. 10 sec. (3,042 ft./sec.) TEI (trans-Earth injection) burn put the crew and their 75 lb. of lunar samples back on course for Earth and splashdown in the Pacific on November 24, 1969, at 244 hr. 36 min. 25 sec.

1969 November 16　The first public display of moon rock took place at the American Museum of Natural History in New York. A 21.1-gram sample from *Apollo 11* it attracted 42,195 viewers on the first day, the largest attendance in the museum's history. The display ran for 10 weeks. On November 30 NASA published plans for the worldwide display of moon material. Of the fifteen rocks weighing between 0.33 and 2.5 oz. and made available for this purpose, eight were for display in the United States and seven were released through the U.S. Information Agency for overseas showing. One stone would be on display at Expo '70 in Osaka, Japan.

1969 November 22　Britain's first-generation military communication satellite, *Skynet I*, was launched by a long-tank thrust-augmented Delta L (LTTAD) from Cape Canaveral to a geosynchronous orbit over the Indian Ocean at an inclination of 2.4°. Built by Philco-Ford, the drum-shaped spin-stabilized satellite had a diameter of 4 ft. 6 in., a height of 2 ft. 8 in. and a fixed antenna on a mechanically de-spun section operating in the 375–400 MHz band. The body was covered by 7,236 solar cells providing power for a 3.5-W transmitter, and a solid propellant apogee motor protruded from the bottom, extending total height to 5 ft. 2 in. *Skynet II*, launched August 19, 1970, remained in a 168 × 22,410 mi. orbit inclined 28° when the apogee motor failed to fire.

1969 November 25   John D. Hodge, manager of the Advanced Missions Program Office at the NASA Manned Spacecraft Center, issued a memorandum stating the need for intact recovery of the Shuttle orbiter and its payload for all abort situations. Instead of safely recovering just the crew from an unplanned abort during ascent, for which provision had been made in all previous manned spacecraft, the design of the Shuttle would have to include safe recovery of the complete orbiter and its cargo. Hodge also confirmed that there would be no capsule escape system or ejection seats available in the event of a catastrophic failure during launch.

1969 November 26   J. W. Sandford, a NASA employee, sent a confidential memorandum to an unidentified addressee reporting on discussions he had earlier this month with Lt. Gen. John J. Davis, U.S. Air Force, about the potential military uses for the Shuttle. Sandford noted that the Air Force was particularly keen on manned surveillance and satellite inspection and wanted the vehicle to possess the ability to reach a wide variety of orbits and inclinations. The Air Force also sought to have the Shuttle cut launch costs by a factor of 10 and to have it replace all current expendable launch vehicles.

1969 November   McDonnell Douglas submitted their final ILRV (integral launch and reentry vehicle) study results based around a reusable Shuttle with a 25,000-lb. payload capability in a 15 ft. × 30 ft. cargo bay. Designed primarily around an enlarged HL-10, with a length of 107 ft. and a width across the flared, upswept "wing" tips of 75 ft. 8 in., the orbiter would be lifted to altitude attached back-to-back to a carrier vehicle with a length of 195 ft. and a delta wingspan of 151 ft. The mated vehicles would have a width of 107 ft. between vertical stabilizer tips. Powered by ten 400,000-lb. thrust cryogenic engines, the 2,687,875-lb. booster would carry the 730,220-lb. orbiter to altitude before the orbiter ignited its ten 448,000-lb. thrust engines and staged off the carrier. The orbiter would have had a modest L/D (lift-over-drag) ratio of 4.0. McDonnell also studied a 1½-stage (drop-tank) design but de-emphasized this when NASA asked the company to delete that option as a candidate.

1969 December 15   NASA tentatively planned to use a hand-cart called the mobile equipment transporter (MET) on *Apollo* missions *14, 15* and *16*. In fact it was used on *Apollo 14* only because two H-series missions that would have used it were canceled, replaced by J-series missions that carried the lunar roving vehicle. The MET comprised a tubular frame vehicle 7 ft. 2 in. long, 3 ft. 3 in. wide and 2 ft. 8 in. high, with two wheels. The tires were 1 ft. 4 in. in diameter and 4 in. wide and inflated preflight to 1.5 psia with nitrogen. The MET weighed 26 lb. empty and had a useful load of 140 lb. It was stowed in one of the exterior quadrants on the side of the lunar module and carried tools and sample bags.

McDonnell Douglas made its final report on the Shuttle concept developed at the NASA Manned Spacecraft Center, sized to a 25,000-lb. payload capability. The booster now had a length of 210 ft. 10 in., a wingspan of 160 ft. 4 in. and ten 400,000-lb. thrust cryogenic engines. The orbiter had a length of 148 ft. and a wingspan of 113 ft. 6 in. Liftoff weight was now 2,854,910 lb. Adopting high-aspect ratio and straight wings, the orbiter had a low cross-range capability of only 265 mi. The MSC design epitomized the NASA payload and performance requirement whereas the ILRV studies, and future Shuttle specifications, would reflect the more demanding capabilities stipulated by the Air Force.

1969 December 22   Lockheed submitted its ILRV contender in the form of two alternate configurations, each capable of sizing to 25,000 lb. or 50,000 lb. payload requirements. The smaller configuration had a baseline orbiter with a length of 173 ft., a swept wingspan of 100 ft. and three cryogenic engines. The booster for it had a length of 211 ft., a wingspan of 184 ft. 5 in. and thirteen engines. The 50,000-lb. payload capability model had a 179 ft. 2 in. swept-delta orbiter with a span of 101 ft. and three cryogenic engines. Its swept-delta booster had a length of 237 ft., a span of 200 ft. and thirteen 15,000-lb. thrust engines. Liftoff weights for the two configurations were 2,756,637 lb. and 3,057,707 lb., respectively. An alternate assembly of twin boosters sandwiching the orbiter in a trimarese configuration had liftoff weights of 3,608,733 lb. for the 25,000-lb. payload model and 4,340,230 lb. for the 50,000-lb. payload version. As with all other ILRV contenders, boosters and orbiters had air-breathing engines for go-around.

1969 December 30   NASA approved the Mariner-Venus-Mercury 1973 mission and assigned project management to the Jet Propulsion Laboratory for what would be the first dual planet flight and the first spacecraft to use the gravity assist of one planet to reach another. Following a long period of planning and spacecraft conceptual design at the JPL, in December 1970 NASA issued a request to industry for proposals to build the spacecraft. On April 28, 1971, the Source Evaluation Board recommended that NASA give the work to Boeing, which it did. The company started work on the project on June 17, 1971, the first Mariner-class spacecraft built outside the JPL. This was the last planetary spacecraft bearing that name and was designated *Mariner J* before launch and *Mariner 10* after.

1969 December 31   NASA Administrator Thomas O. Paine decided that the proposed Viking Mars orbiter/lander mission for 1973 would be postponed to 1975. He had been asked to agree to this by the Budget Bureau because it had underestimated the projected fiscal year 1971 budget position by $225 million, and this was their preferred way to absorb a cut in federal spending. During this month the basic suits of science payloads for Viking were determined, with emphasis in the lander on the biological search for life and long-term observations of changing weather at the landing sites. The orbiter would conduct long-period mapping of the surface, measure changes over time and thoroughly examine the climate of Mars as well as regional weather patterns.

1969 December   Martin Marietta submitted its unsolicited contribution to the ILRV (integral launch and reentry vehicle) Shuttle studies commissioned by NASA. Named Spacemaster, the baseline vehicle had been designed to a maximum liftoff mass of 3.5 million lb. Utilizing a double-body, fly-back booster powered by fourteen 400,000-lb. thrust cryogenic engines, the vehicle had a liftoff weight of 2,887,077 lb. The booster had a length of 197 ft. and consisted of two cylindrical fuselage sections, 28 ft. in diameter, connected by a forward wing 70 ft. across. The orbiter was nested between the two fuselage sections and had a length of 181 ft., a span across the highly swept delta wing of 107 ft., two 400,000-lb. thrust engines and a payload capacity of 36,500 lb. Martin offered an alternate design capable of carrying a payload of 50,000 lb. It had a booster 204 ft. 2 in. in length, with a span of 156 ft., and sixteen engines. The orbiter had a length of 186 ft. 10 in., a span of 110 ft. and three engines. The bigger vehicle had a liftoff weight of 4,146,000 lb.

Space Division, North American Rockwell, revised their final submission under the NASA ILRV (integral launch and reentry vehicle) contract for examining potential Shuttle configurations. The two-stage reusable vehicle now had a potential payload capability of 60,000 lb. to a 28.5° inclination orbit, 50,000 lb. to a 55° orbit and 32,700 lb. to a polar (90°) orbit; the U.S. Air Force wanted this latter performance capability for orbit placement and retrieval of an advanced generation of reconnaissance satellites. The orbiter remained the same but the booster had grown to a length of 280 ft. and a wingspan of 244 ft., and booster liftoff weight had gone down to 3,619,872 lb. Total stack height had increased to 280 ft. and total liftoff weight reduced to 4,490,849 lb. Refined aerothermal analysis showed the orbiter would be capable of a cross-range of 1,427 mi.

1970 January 4 NASA announced that it was canceling *Apollo 20* and further production of Saturn V launch vehicles after AS-515. It planned to use one Saturn V to launch the Apollo Applications Program workshop in 1972. Lunar landing missions would be suspended for 1972 to allow flight support operations to concentrate on the AAP Earth-orbiting space station. Thereafter, moon flights would be stretched out to six-month intervals. Under this plan *Apollos 13* and *14* would fly in 1970, *Apollos 15* and *16* in 1971, *Apollos 17* and *18* in 1973 and *Apollo 19* in 1974. NASA Deputy Administrator George Low said these measures were forced upon NASA by budget constraints.

1970 January 9 The Boeing Co., which had been charged with finding ways of reducing space transportation costs by at least one order of magnitude, completed an economic analysis of fully reusable Shuttle vehicles. It found that payload costs could be reduced to just 5% of existing payload costs with expendable launchers. The report emphasized standardization and hardware commonality as the key to reduced operations and turnaround (that is, making the Shuttle ready for another mission). Boeing was asked to evaluate the economic requirements of a Shuttle program launching 100 missions per year and suggested that five vehicles would be needed to support that level of activity, each Shuttle making on average one flight every 2.6 weeks.

1970 January 14 David H. Grenshields, the chief of the Thermal Technology Branch of the NASA Manned Spacecraft Center, sent a memorandum to the U.S. Air Force Flight Dynamics Laboratory (FDL) commenting upon its evaluation of the MSC Shuttle concept. The FDL had questioned the viability of the MSC straight-wing orbiter and demonstrated to its satisfaction that the design of the vehicle was aerothermally unsound. The FDL believed the straight-wing, low cross-range orbiter would not survive atmospheric reentry. It would fuel a debate, to last throughout 1970, about the relative

merits of the straight-wing versus delta-wing (high cross-range) orbiter.

1970 January 23 The first two-stage Delta N6, launched from Vandenberg AFB, Calif., placed the *Itos 1* (improved tiros operational system, or Tiros-M) meteorological satellite in a sun-synchronous, 890 × 919 mi. orbit inclined 102° to the equator. Three Castor I rockets ignited at launch, with the other three ignited at 46,000 ft. The first of a new generation of satellite capable of twice the daily weather coverage of the ESSA (Environmental Science Services Administration) series, measuring both infrared cloud cover and cloud-top and surface temperatures every 12 hr., *Itos 1* had a weight of 682 lb. Box-shaped, with dimensions of 3 ft. 4 in. × 3 ft. 4 in. × 4 ft., *Itos 1* carried 10,000 solar cells on two panels, each 3 ft. × 5 ft., providing 250 W of electrical power. The satellite became operational on June 15, after a six-month engineering checkout by NASA.

Concerned at the burgeoning complexity and potential cost and development time of the proposed NASA Shuttle program, the Manned Spacecraft Center formalized its own in-house study of an optimum vehicle configuration. Resisting pressure, principally from the Air Force, to give the Shuttle a 50,000-lb. payload capability in a high cross-range orbiter, the MSC design team under Maxime Faget baselined their proposed configuration around a low cross-range (less than 220 mi.) orbiter with a payload capability of 10,000–15,000 lb. Soon to be dubbed the "DC-3," after the Douglas DC-3 which became the workhorse of air transport, this initial configuration was retrospectively designated MSC-001.

1970 January 28 North American Rockwell and the Convair Division of General Dynamics jointly announced they would team together in a bid for the Shuttle Phase B definition contracts, soon to be awarded by NASA. North American Rockwell was to concentrate on the orbiter and assume the role of lead contractor and systems integrator. General Dynamics would support North American Rockwell and concentrate on the fly-back booster to be used to carry the orbiter to altitude. On February 4 McDonnell Douglas announced it would team with Martin Marietta, Pan American World Airways and TRW. On February 5 Grumman announced it would compete for the Phase B study. On February 6 NASA announced that it would let "alternate Phase A studies" during 1971 for industry analysis of alternative configurations.

1970 February 4 NASA successfully launched the second Space Electric Rocket Test *(SERT II)* on a Thorad Agena D from Vandenberg AFB, Calif. Placed in an orbit of 620 × 627 mi. inclined 99.1°, *SERT II* consisted of two ion engines attached to a power structure on the forward part of the Agena. Two 5 × 19 ft. solar arrays carrying 33,300 solar cells provided 1,471 W of electrical energy. In orbit *SERT II* weighed 3,100 lb. The first ion engine was turned on February 14, brought to full thrust and shut down. The second engine was turned on, but it failed on July 23 after 3,785 hr. of operation. The first engine was turned back on a day later, but it too failed, after 2,011 hr., on October 17. *SERT II* was purely a research project and was considered a success.

1970 February 11 Japan became the fourth nation to build its own satellite and launch it into space when a four-stage solid propellant Lambda 4S-5 put the 84-lb. *Ohsumi* (named after the district from which it was launched) test satellite in a 211 × 3,193 mi. orbit inclined 31.1° to the equator. *Ohsumi* consisted of a conical instrument package attached to the fourth stage of the launcher, 1 ft. 7 in. in diameter with a length of 3 ft. 4 in. Launched from the Uchinoura launch site at 13:25 local time, the tiny satellite carried 20 lb. of instruments, including accelerometers, a thermometer and a small radio transmitter and battery with four antennas. Developed and launched by the Japanese Institute of Space and Aeronautical Sciences, the satellite was named after the region from which it was launched. It remains in orbit.

A Thor-Burner II launched the first in a new generation of U.S. military weather satellites, *DMSP Block 5A*, from Vandenberg AFB, Calif., to a 475 × 605 mi. orbit inclined 98.7°. Attitude-stabilized, the 230-lb. satellite comprised a cylinder with 12 flat sides, with a height of 4 ft. and a diameter of 3 ft., supporting solar cells producing 30 W. Replacing the vidicon cameras of the Block IV series, the Block 5A had two line-scan cameras in the visual and infrared bands. The second of its type was launched on September 3, with the last Block 5A on February 17, 1971.

1970 February 13　NASA formed the Space Shuttle Program Office (SSPO) under Robert F. Thompson, former head of the Apollo Applications Program Office. At Headquarters level the SSPO would not only manage and direct activities among the contractors and subcontractors but bring together several divergent viewpoints within NASA itself. It had already been decided to aim for a two-stage fully reusable Shuttle, rather than a 1½-stage concept in which auxiliary propellant tanks were carried part of the way into orbit. Whether the orbiter would have a high or a low cross-range and whether it would have a 10,000–15,000-lb. payload, as the Manned Spacecraft Center wanted, or the 50,000-lb. capability desired by the Air Force was still open to vigorous debate.

NASA announced that distribution of 28.6 lb. of *Apollo 12* moon samples had begun at the Manned Spacecraft Center. Samples would be sent to 139 universities and laboratories in 25 states and the District of Columbia. In addition, material would be distributed to 54 scientists in 16 foreign countries: Australia, Belgium, Canada, Czechoslovakia, Finland, West Germany, Japan, Korea, Spain, Switzerland, the United Kingdom, South Africa, Italy, France, Norway and India.

1970 February 16　A scheme to electronically track 500-lb. elk on the National Elk Refuge, Jackson Hole, Wyo., and track them by satellite was outlined by NASA. A 23-lb. transmitter attached to the animal would send a signal which would be received by the orbiting *Nimbus III* weather satellite launched April 14, 1969. The collar to which the transmitter would be attached would also carry sensors for measuring each elk's skin temperature as well as atmospheric temperature, altitude above sea level and daily position at noon and midnight. The experiment proved valuable and led to many other such schemes for learning migratory habits of wild animals.

1970 February 17　NASA announced that the name of the Apollo Applications Program had been changed to Skylab, signifying a laboratory in the sky. The name had been chosen by Donald L. Steelman, USAF, when the space agency asked for ideas about a more descriptive title in 1968. Budget problems had made NASA cautious about seeming to start up a completely separate program in times of fiscal austerity, but now that it was well under way the time seemed appropriate to signal this break with the original objectives of the Apollo program.

1970 February 20　NASA issued a request for proposals from contractors bidding for Phase B definition studies on a two-stage, fully reusable Shuttle. The work statement clearly defined how NASA wanted a team approach, with one contractor building the booster and another building the orbiter. Payload weight was not stipulated, but the cargo bay was to be 15 ft. × 60 ft. and maximum liftoff weight no more than 3.5 million lb. Orbiters with both low (230 mi.) and high (1,726 mi.) cross-range were to be defined and propulsion for both stages was to draw upon a common 400,000-lb. thrust engine. On-orbit duration was to be seven days, and the reference mission was a logistics flight to a space station. The work statement anticipated an initial operational capability in the second half of 1977 and an annual flight rate of 25–75 missions per vehicle, though each was to be capable of 100 flights.

1970 February 26　NASA Marshall Space Flight Center called for bids on a definition study of a high energy astronomy observatory (HEAO) to observe and measure high-energy radiation from celestial sources. High-energy astronomy was considered a priority by space scientists and astronomers, and 20 companies were asked to bid. On April 1 NASA held a preproposal briefing for 155 scientists and industry representatives.

1970 February　Frustrated at the lack of progress with its manned lunar landing program, senior Kremlin officials ordered the Korolev bureau, now in the hands of Vasili Mishin, to expedite a manned space station ahead of NASA's Skylab. Using design materials from the OPS Almaz program, Mishin's engineers took the Chelomei bureau's concept and fitted it with controls and flight systems from the Soyuz spacecraft, even down to the aft propulsion system. By this date the Chelomei bureau had assembled 10 Almaz stations at the Krunichev factory near Moscow and had already selected three cosmonauts for Almaz. (Neither V. Makrushin, D. Yuyukov nor A. Grechanik ever flew in space. What Mishin did was wed the proven elements of Almaz and Soyuz and with Soyuz as a ferry for crew members, prepare the first Soviet space station, eventually named Zarya. On the day *Zarya 1* was launched (April 19, 1971) its name was changed to Salyut to avoid confusion with a similar name for Soviet ground stations.

1970 March 1　Wernher von Braun relinquished his leadership of the NASA Marshall Space Flight Center to take up an administrative post at NASA Headquarters; thus ended his 20-year professional association with Huntsville, Ala. He was succeeded by Eberhard F. M. Rees, another of the German scientists who came to the United States at the end of World War II. Von Braun was sworn in as deputy associate administrator for Planning on March 13, 1970. It was a thankless position. With a rapidly declining budget, NASA was fighting hard to save existing programs and had little chance of marshalling resources for his beloved grand adventures. On May 26, 1972, Von Braun announced his intention to retire from NASA and join the Fairchild Corp. He died of cancer on June 16, 1977, at the age of 65.

1970 March 3　Associate Administrator for Manned Space Flight Dale Myers sent Robert Gilruth, director of the Manned Spacecraft Center, a letter summarizing views on the Shuttle from the Management Council in the NASA Office of Manned Space Flight. He said that key to the Shuttle's cutting space program costs was the "concept of reusing payloads, capitalizing on the Shuttle's capacity to provide for retrieval, repair, and maintenance." Moreover, he told Gilruth that the council considered military "use of the Shuttle . . . essential to the current economic justification." The rapidly falling NASA budget made cost-cutting projects essential, and the Shuttle was now viewed as a means of reducing flight costs. The point so clearly defined in this letter was lost when NASA began to believe, and told Congress, that launch costs in themselves would be cut by the Shuttle's replacing expendable rockets. No economic study was ever conducted which stated such cost reductions, and when NASA failed to provide for payload retrieval and refurbishment, it lost the mechanism to lower space program costs.

Rep. Joseph E. Karth (D-Minn.) of the House Committee on Science and Astronautics, delivered a pithy broadside to NASA during a speech to the American Institute of Aeronautics and Astronautics. In commenting upon plans for a reusable Shuttle, Rep. Karth said that the agency's estimated program costs "are unrealistically low," and, "Based upon my experience with Ranger, Centaur, Surveyor, Mariner, Viking, and even Explorer, NASA's projected cost estimates are asinine. NASA must consider the members of Congress a bunch of stupid idiots. Worse yet, they may believe their own estimates—and then we really are in bad shape."

1970 March 4 The first successful suborbital flight test of Britain's Black Arrow satellite launcher took place from Woomera in Australia, when the terminal stage was fired 1,900 mi. downrange, 15 min. after liftoff. Developed by Westland Aircraft and the Ministry of Technology, Black Arrow carried eight Rolls Royce Gamma Type 8 engines burning hydrogen peroxide (HTP) and kerosene propellants for a liftoff thrust of 52,000 lb. The second stage was powered by one Rolls Royce Gamma Type 2 engine delivering a thrust of 15,750 lb., also burning HTP-kerosene propellants. The solid propellant third stage had a thrust of 5,000 lb., and the complete vehicle had a height of 42 ft. 8 in., a diameter of 6 ft. 7 in. and a launch weight of 40,000 lb. The flight was designated R1, following the failure of the first (R0) ballistic launch on June 28, 1969, when the first Black Arrow had to be destroyed one minute after liftoff.

1970 March 7 President Nixon publicly endorsed the NASA plan to use an alignment of the outer planets to launch spacecraft on fly-by missions in a grand tour of the outer solar system. Two Mariner-class spacecraft would be launched in September 1977. Called the Grand Tour inner-ring flight, the first would fly by Jupiter in January 1979, Saturn in September 1980, Uranus in February 1984 and Neptune in November 1986. The second would fly by Jupiter in February 1979, Saturn in September 1980 and Pluto in March 1986. Launched in November 1979, the third flight would visit Jupiter in 1981, Uranus in July 1985 and Neptune in November 1988. Because the Mariner spacecraft would be powered by nuclear RTGs (radioisotope thermoelectric generators), they were called TOPS (thermoelectric outer planet spacecraft).

President Nixon issued the "Statement About the Future of the United States Space Program," seeking to explain forthcoming NASA budget cuts by asserting that space goals will "not be accomplished on a crash timetable . . . space expenditures must take their proper place within a rigorous system of national priorities." It was the first public statement by a chief executive in the White House since Dwight D. Eisenhower that reminded the nation that space projects had to compete against other imperatives. Reduced budgets and

imploding aspirations had been hard for NASA to digest, and the fight to get the Shuttle approved would dominate strategy at the space agency.

1970 March 10 The first of five Diamant B launchers carried two small test satellites into orbit from the Kourou facility in French Guiana in what constituted the first flight of a foreign payload on a French satellite launcher. Both were developed in the German DIAL (Diamant Allemande) program. The first was the 143-lb. *DIAL-1/MIKA*, a cone-shaped capsule instrumented to monitor launch vehicle performance, carried on the third stage and placed in a 194 × 1,034 mi., 5.4° orbit. The 110-lb. autonomous satellite, *DIAL-1/WIKA*, was placed in a 187 × 1,013 mi., 5.4° orbit. It was equipped with instruments for measuring the geocorona and to make alpha- and beta-particle measurements. *MIKA* decayed on September 9, 1974, followed by *WIKA* on October 5, 1978. Diamant B was essentially the same as Diamant but with a more powerful first stage that delivered a thrust of 77,175 lb.

1970 March 11 In efforts to find ways of spreading the cost of expensive space projects, NASA Administrator Thomas O. Paine testified before the Senate Committee on Aeronautical and Space Sciences about his offer to Canada, Europe, Japan and Australia for "constructive participation in" the Shuttle and space station programs. As a result of a tour conducted by Paine in these countries, ESRO (European Space Research Organization) and ELDO (European Launcher Development Organization) were to set up offices in Washington, D.C., and European space scientists and engineers were to review NASA programs and, on March 13, participate in a conference to "indicate in which areas they would like to work."

1970 March 19 The first powered flight of the U.S. Air Force X-24A lifting body was conducted over Edwards AFB, Calif., by test pilot Jerauld Gentry. Carried into the air by a B-52 mother plane, the single-rocket motor was ignited for a maximum speed of 571 MPH and an altitude of 44,384 ft. This was the 10th flight of the wingless vehicle. The first supersonic flight took place on October 14, 1970, when the X-24A reached Mach 1.186 (784 MPH) and 67,900 ft. The highest altitude achieved by the X-24A was 71,407 ft. on October 27, 1970. The fastest flight at Mach 1.6 (1,036 MPH) was performed on March 29, 1971, and the last of 28 flights was flown on June 4, 1971. At that point the aircraft was returned to its builder, Martin Marietta, and reworked into the X-24B. The objective was to provide a second-generation lifting body research vehicle to follow the M2-F2, the HL-10 and the X-24A.

1970 March 20 The first NATO communications satellite was launched to geosynchronous orbit at an inclination of 6° above the North Atlantic by a thrust-augmemted Thor Delta from Cape Canaveral, Fla. Built by Philco-Ford, the cylindrical satellite had a diameter of 4 ft. 6 in., a height of 5 ft. 3 in., an in-orbit weight of 285 lb. and 7,000 solar cells for electrical power. Carrying two transponders to link NATO headquarters in Brussels, Belgium, with other NATO countries and NATO commands on land, sea and air, *NATO 1* was joined by *NATO 2* on February 3, 1971, at an inclination of 8.7°.

Robert Gilruth, director of the Manned Spacecraft Center, wrote a memorandum to Robert Lindley, assistant to Dale Myers, NASA administrator for Manned Space Flight, outlining Shuttle economics. The note claimed that development costs for the two-stage, fully reusable Shuttle with a 10,000-lb.

payload capability would be $6.0–9.5 billion and that each mission would cost $3 million to launch, implying a payload cost of $300/lb. A Shuttle capable of carrying a 25,000-lb. payload would cost $7.0–10.5 billion and $128/lb. to fly. A 50,000-lb. capability Shuttle would cost $10.5–15.5 billion to develop and $72/lb. to fly.

1970 March 26 NASA announced the names of the flight crew selected for the *Apollo 15* lunar landing mission. Primary crew members comprised David Scott, Alfred Worden and James Irwin. The backup crew comprised Richard Gordon, Vance Brand and Harrison Schmitt. Scientist-astronauts Karl Henize and Robert Parker were to operate as their support crew. The backup crew for *Apollo 15* would have become the primary crew for *Apollo 18*, which was canceled and never flew. However, geologist Harrison Schmitt was pulled from that crew and inserted into the *Apollo 17* crew flying the last lunar mission, ejecting engineering test pilot Joe Engle, who should have flown as lunar module pilot on the last manned Apollo moon mission. Thus did NASA avert accusations that in six lunar landings it failed to send a scientist to the moon.

1970 March 27 The Space Division of North American Rockwell submitted to NASA its proposal for a Shuttle Phase B definition contract, in response to the request of February 20, 1970. As required under the terms of the work statement, North American Rockwell submitted two orbiters, a low cross-range vehicle with a length of 183 ft., a 39,737-lb. payload capability, a subsonic L/D (life-over-drag) of 6.9 and a high cross-range delta planform vehicle with a length of 192 ft. 4 in., a payload capability of 14,781 lb. and a subsonic L/D of 8.2. Each orbiter had a loaded weight of 760,000 lb. and two 475,000-lb. thrust cryogenic engines. The booster had a length of 257 ft., a launch weight of 2.74 million lb., twelve 475,000-lb. thrust engines and a subsonic L/D of 6.7. All three North American Rockwell vehicles would have four turbojets using liquid hydrogen fuel for flight in the atmosphere.

Grumman Aerospace, with team members General Electric, Northrop, Eastern Airlines and Aerojet General, submitted its proposal for the NASA Shuttle Phase B definition contract. It included a proposed study of an alternate approach utilizing existing technology. Called Design 532, it would fly earlier than the conventional Shuttle and adopt a booster put together from existing hardware. Five F-1 engines used in the first (S-IC) stage of a Saturn V would form the booster, with three cryogenic J-2 engines in the low cross-range orbiter. Because of the conventional propulsion, the orbiter would have a payload of only 12,800 lb. The Grumman team failed to get a Phase B contract.

1970 March 30 McDonnell Douglas Corp. submitted its proposal for a Shuttle Phase B definition contract in accordance with the requirements issued February 20, 1970. MDC offered both low and high cross-range orbiter/booster combinations. Utilizing folding wings for launch, the low cross-range orbiter had a length of 155 ft., a 41,043-lb. payload capability and a subsonic L/D (lift-over-drag) of 6.3. The clipped delta booster had a length of 189 ft., a subsonic L/D of 8.3 and a launch weight of 2,846,710 lb., with a cruise range of 487 mi. The high cross-range orbiter had a length of 157 ft. 7 in., a 36,200-lb. payload capability and a subsonic L/D of 7.0. An innovative twin-body booster had a length of 162 ft. 6 in., a launch weight of 2,844,577 lb. and a subsonic L/D of 7.8, with a cruise range of 472 mi.

1970 March The Space Science and Technology Panel of the President's Science Advisory Committee issued the report *The Next Decade in Space*, in which it endorsed the Shuttle, cast doubt about a permanently manned space station and argued for more equal funding of unmanned and manned space systems. Believing that insufficient evidence had been accumulated to know whether automated space systems could replace humans in space, the report stated, "The most important advances in space exploration seem to rest on the development of the space transportation system."

1970 April 1 In an overview of Shuttle program activity, Leroy E. Day, manager of the Space Shuttle Task Group, officially confirmed that "it is envisioned that the Shuttle will eventually replace essentially all of the present-day launch vehicles or their derivatives except for very small vehicles of the Scout class and the (very large) Saturn V." Day also confirmed that NASA believed that the Shuttle's reusability "will make it competitive even if it carries only a fraction of its full payload capability on particular missions." Because "acceleration during descent and reentry will be less than 3 g's. . . . The benign acoustic and acceleration environment of the Shuttle [will] allow significant reductions in the cost of payloads."

An agreement was reached whereby the United States would provide integration and launch services for the United Kingdom's second-generation military communications satellites, *Skynet II*. NASA was to mate the satellite with the Delta launch vehicle and command the vehicle during all transfer orbit stages. On May 1, 1972, the U.S. Air Force agreed to provide X-band military communications between designated Earth stations and replace *Skynets I* and *II*.

1970 April 6 A day after the countdown began for their lunar landing mission, *Apollo 13* astronauts James Lovell, Thomas Mattingly and Fred Haise underwent a comprehensive medical examination at the Kennedy Space Center to determine their immunity to rubella (German measles). Backup lunar module pilot Charles Duke had come down with rubella and the primary *Apollo 13* crew members had been in contact with him. The tests showed that Lovell and Haise were immune but Mattingly was not, and although the chances of his getting it were very low, it was decided the next day to move backup command module pilot John Swigert into his place. A final decision about whether he would fly would be made only after it had been determined whether they could work as a team and "precisely execute the few intricate time-critical maneuvers which require rapid and close coordination." On April 10 it was agreed that Swigert should fly next day with Lovell and Haise.

1970 April 8 Riding piggyback on a Thorad-Agena D from Vandenberg AFB, Calif., the U.S. Army *TOPO 1* geodetic satellite was placed in a 661 × 690 mi., 99.7° orbit. Primary payload for this mission was the 1,370-lb. *Nimbus 4*, placed in a 680 × 683 mi. orbit. *TOPO 1* weighed 48 lb. and was a typical Secor satellite which the Army used for position determination and for triangulation exercises in ground tactical positioning. Both *Nimbus 4* and *TOPO 1* remain in orbit.

1970 April 11 The *Apollo 13* mission began at 19:13:00 UT when the Saturn V AS-508 lifted off from Complex 39A at the start of what should have been a lunar landing in the Fra Mauro region of the moon. It turned into near fatal disaster when a cryogenic oxygen tank in the service module blew up halfway to the moon. Crewed by Jim Lovell, Jack Swigert and

Fred Haise, *Apollo 13* had a launch mass of 6.421 million lb. and included CSM-109 (63,812 lb.) and LM-7 (33,502 lb.). During ascent the center J-2 engine on the S-II stage shut down 2 min. 12 sec. early, causing the outer four J-2s to burn 34 sec. longer. Despite this, velocity was 223 ft./sec. low at S-IVB ignition, and the third stage burned 9 sec. longer to compensate. After TLI (translunar injection), the S-IVB conducted a maneuver so that it would impact the moon at 77 hr. 56 min. 40 sec. and provide vibrations for the *Apollo 12* seismometer. At 30 hr. 40 min. 50 sec. the SPS (service propulsion system) engine was fired for 3.4 sec. (23 ft./sec.) to put the CSM/LM (control-service module/lunar module) on a nonfree return path around the moon. As with *Apollo 12*, this optimized the trajectory for the desired landing site but meant that an engine burn had to be conducted to put it back on course for Earth. At 55 hr. 54 min. 53 sec. cryogenic oxygen tank no. 2 in the service module blew up, tearing away the panel that covered bay no. 4. One of six bays in the service module, it housed the two oxygen and two hydrogen tanks and the three fuel cell electrical production units. The explosion triggered valves that shut down two fuel cells within three min. and damaged oxygen tank no. 1. It began to leak, and within 2 hr. 10 min. there was insufficient oxygen to operate the remaining fuel cell. Meanwhile, the crew had activated the lunar module to use it as a lifeboat, all hope of a lunar landing having been abandoned. At 61 hr. 29 min. 43 sec. the LM descent engine was fired for 30 sec. (38 ft./sec.), putting the spacecraft back on an Earth-return trajectory. If uncorrected, it would loop round the moon and splash down in the Indian Ocean at 152 hr. elapsed time. To speed up the return the descent engine was fired again, at 79 hr. 27 min. 39 sec. for 4 min. 23 sec. (860 ft./sec.); this aimed to put the command module on the water at about 143 hr. From this point on it was a matter of human survival, using the LM to keep three people alive for almost 86 hr. instead of two people for 48 hr., as designed. Everything was switched off except for the environmental control and communications systems. Temperatures dropped to freezing and ice formed inside the spacecraft. When exhaled gases threatened to poison the crew they jury-rigged a filter to trap carbon dioxide. At 105 hr. 18 min. 32 sec. the descent engine fired for the third and last time, a 15.4-sec. (7.8 ft./sec.) course correction burn. A final 22-sec. (2.9 ft./sec.) trim burn with the lunar module thrusters was performed at 137 hr. 39 min. 49 sec. Uniquely for this extraordinary mission, the service module was jettisoned at 138 hr. 2 min. 6 sec. and the crew saw the blasted bay no. 4 area before getting back into the command module for reentry. The lunar module was released at 141 hr. 30 min. 2 sec. and broke up as it hit the atmosphere at 36,200 ft./sec. , the RTG (radioisotope thermoelectric generator) for its undeployed ALSEP (Apollo lunar surface experiments package) surviving intact and falling to the bed of the Pacific. It could not be recovered. Command module reentry began at 142 hr. 40 min. 47 sec., followed by splashdown at 142 hr. 54 min. 41 sec.

1970 April 15   The Strategic Arms Limitations Talks (SALT) opened in Vienna, Austria, with delegations from the United States and the Soviet Union meeting for the first time to discuss the possibility of placing limits on the number of strategic warheads and delivery systems. The talks became embroiled with opposing plans and schemes as each side sought to maneuver to its advantage. On May 20, 1971, President Richard M. Nixon announced an agreement by both sides to concentrate on "working out agreement for the limitation of antiballistic missile systems."

1970 April 24   China became the fifth nation to design, build and launch its own artificial Earth satellite and launcher. The 381-lb. satellite was launched from Jiuquan, east of Lop Nor, by a CZ-1 (Long March 1) and placed in a 272 × 1,483 mi. orbit inclined 68.5°. CZ-1 had a length of 96 ft. 8 in., a diameter of 7 ft. 4 in., two stages operating on nitric acid–UDMH propellants and a solid propellant third stage. With a launch mass of 179,600 lb., CZ-1 had a liftoff thrust of 229,320 lb. The satellite comprised a multisided sphere 3 ft. 4 in. across with four 9 ft. 10 in. antennas which broadcast telemetry and transmitted the Chinese Communist Party song "The East is red" on 20.009 MHz, from which the satellite was given its name, *Tungfanghung*.

1970 April 25   The first eight payloads in a long series of fully operational Soviet tactical military communication satellites were launched by an SL-8 from Plesetsk. With orbits of about 870 × 930 mi. and inclinations of 74°, the satellites were designated *Cosmos 336–343* and comprised the first octuplet launch by the Soviet Union. Each satellite weighed 110 lb. and bore a strong resemblance to earlier, precursor types launched since *Cosmos 38–40* on August 18, 1964. Groups of eight satellites at a time were routinely launched by SL-8s from Plesetsk, usually 2–3 launches per year at least through 1993.

1970 April 27   A two-volume report entitled *DC-3 Space Shuttle Study* described a detailed technical analysis conducted at the Manned Spacecraft Center on a Shuttle constructed from "existing hardware where possible." The MSC concept envisaged a low cross-range orbiter (design model MSC-002) with a 15,000-lb. payload capability in an 8 × 30 ft. bay, a length of 122 ft. 8 in., a high aspect-ratio wing with a modest 14° leading edge sweep and a span of 90 ft. 10 in. Power was to be provided by two 297,000-lb. thrust Pratt and Whitney XLR-129-P-1 rocket motors and air-breathing engines for go-around capability. The straight-winged booster had a length of 203 ft., featuring a modest 14° leading edge sweep on the wing, which had a span of 141 ft. Propulsion comprised eleven orbiter rocket motors and four wing-mounted turbofans for cruise. Total vehicle length was 234 ft. 11 in. Development cost was estimated to be $5.912 billion, with operational capability in 1975, two years ahead of the date projected in the Phase B work statement.

1970 April 27–28   The council of the European Launcher Development Organization (ELDO) met in Paris to decide on future policy in the aftermath of Britain's decision to withdraw financial participation. Blue Streak was dropped as a candidate first stage for advanced European launchers, beginning with Europa III. Two proposals were being considered for future decision: a two-stage vehicle burning UDMH-N_2O_4 propellants in the first stage and liquid oxygen–liquid hydrogen in the second stage and capable of placing a 1,500-lb. payload in geostationary orbit; or a launcher using cryogenic hydrogen and oxygen in both stages. The ELDO council recommended the less exotic first option for analysis.

1970 April 30   NASA selected Aerojet, Rocketdyne and Pratt and Whitney for Phase B definition contracts on the Space Shuttle Main Engine (SSME), submitted in response to a request issued on February 18. The NASA Marshall Space Flight Center would manage all three contracts, each worth $6 million, which were to last eleven months. Unsuccessful bidders for the Phase B work included Bell Aerospace, Marquardt and TRW. The Shuttle Phase B contracts for the two-stage reusable vehicle directed the contractors to employ the SSME for both orbiter and booster to reduce the costs of parallel development and achieve commonality in operations, an important factor in reduced servicing and maintenance costs.

1970 May 4   A one-tenth scale model of the Manned Spacecraft Center DC-3 Shuttle concept was drop-tested at Fort Hood, Tex., under the supervision of the Operational Test Branch of the Landing and Recovery Division. Full flight tests with scale models of the 12,500-lb. payload concept were set to begin May 27, 1970. The model had a length of 13 ft. and weighed 600 lb., incorporating gyro-stabilized flight control systems under radio control. Tests were made from a Sikorski CH-54 helicopter, hovering at a height of more than 12,000 ft.

1970 May 7   Scientists participating in NASA's *Pioneer F* and *G* missions to Jupiter gathered at the Ames Research Center, Calif., for their first project coordination meeting. Each spacecraft would measure magnetic fields with a magnetometer; the heliosphere and the solar wind with a plasma analyzer; cosmic rays, radiation belts and radio signals with a charged-particle detector and cosmic-ray telescope; Jupiter's charged particles with a geiger tube telescope and a trapped radiation detector; and Jupiter's atmosphere with an ultraviolet photometer, an imaging polarimeter and an infrared radiometer. Detectors would measure the size and abundance of tiny particles trapped in the asteroid belt, a 150-million-mile-wide band of dust, rocks and boulders trapped between the orbits of Mars and Jupiter.

NASA announced that *Apollo 14* would make a landing in the Fra Mauro region of the moon, the site selected for the aborted *Apollo 13* flight. Modifications to the Apollo spacecraft would take several months, and NASA set a provisional date of December 3, 1970, as the earliest launch window for the mission. On June 30, 1970, the scheduled launch date was postponed to no earlier than January 31, 1971, which meant that *Apollo 15* would not fly before July or August 1971. In an operations change aimed at conserving propellant aboard the lunar module (LM) descent stage, the *Apollo 14* SPS (service propulsion system) engine would be used to perform the descent orbit insertion burn around the moon so that both docked vehicles would go down to 10 mi. above the surface. Apollo would then undock from the LM and fire itself back into a circular, 70-mi. orbit while the lunar module went straight into powered descent.

Robert Gilruth, director of the Manned Spacecraft Center, provided information on possible changes to work statements on the Shuttle Phase B definition contracts. He recommended that NASA set a development cost target of $5 billion through delivery of the first vehicle. It was also suggested that the Shuttle should be developed to achieve initial operational capability in 1976 rather than 1977. There was concern at NASA about the increasing gap in manned flight operations after the last Apollo mission. It was further suggested that the liftoff weight limit of 3.5 million lb. should be changed to specify a thrust-to-weight ratio of 1.1:1, leaving open the question of vehicle mass. Finally, orbiter go-around capability was to be deleted, clearing the way for the elimination of air-breathing engines, although these remained a feature of designs for some time.

1970 May 9   NASA selected North American Rockwell and McDonnell Douglas to conduct $6 million Phase B definition studies on a fully reusable, two-stage Shuttle. The contracts would cover a period of 11 months. The Manned Spacecraft Center would manage the North American Rockwell work,

and the Marshall Space Flight Center the McDonnell Douglas study. Companies that had been unsuccessful in bidding included Boeing, Chrysler and Grumman. Thus began more than two years of intensive study, appraisal and analysis before the definitive development contract was awarded.

1970 May 21   NASA awarded a preliminary design contract to TRW for a subsatellite to be carried on *Apollo 16, 17* and *18*, and released in moon orbit, and to remain there after the manned spacecraft returned to Earth. NASA wanted to carry the subsatellite in the Apollo service module and release it after the lunar module descended to the surface. The subsatellite would be instrumented to study details of the charged particles and magnetic fields. When NASA canceled *Apollo 18*, the subsatellites were reassigned to *Apollos 15* and *16*.

1970 May 22   NASA's Marshall Space Flight Center announced that Grumman Aerospace and TRW Systems Group would be awarded one-month definition and preliminary design studies for the High Energy Astronomy Observatory (HEAO) program. Scheduled for launch in 1975, the 21,600-lb. HEAO satellites would provide a platform for astronomical observations in the high-energy end of the spectrum. On November 23, 1971, NASA selected TRW Systems Group to build two satellites, expected to be launched in 1975 and 1976 on Titan IIIE launchers. Each HEAO would have a length of 39 ft. and a diameter of 9 ft.

1970 May 26   NASA's Marshall Space Flight Center described a space tug concept being developed there and at the Manned Spacecraft Center. The tug was under consideration as a propulsion unit capable of moving men and cargo between Shuttle and space station or transferring cargo or satellites from low orbit to geostationary orbit. Of modular design, the tug would comprise a rocket-stage propulsion unit, with optional upper elements comprising a cargo module, a crew compartment or a remote manipulator system. If used from a lunar base, the tug was to be capable of landing 60,000–70,000 lb. on the surface.

1970 June 1   At 19:00 UT a Soviet SL-4 launch vehicle carried the 14,530-lb. *Soyuz 9* and cosmonauts Andrian G. Nikolayev and Vitaly I. Sevastyanov from Baikonur into a 129 x 137 mi. orbit inclined 51.6°. In a flight lasting 17 days 16 hr. 58 min. 55 sec. the two cosmonauts significantly extended the manned orbital flight record, carried out several orbital maneuvers with the instrument module propulsion system, conducted biological experiments from the orbital module and qualified the spacecraft for long-duration flight. Due to the very limited reserves of attitude-control propellant, Soyuz was unable to maintain a fixed attitude in space, so *Soyuz 9* was spin-stabilized and the crew were in poor physical condition when they returned, unable to walk properly and disoriented. Nevertheless, the mission marked a transition from using Soyuz in support of a manned lunar landing objective to employing the spacecraft as a ferry and cargo vehicle for manned space stations around Earth.

1970 June 2   Rebuilt from the wreckage of the M2-F1 that crash-landed on May 10, 1967, the NASA M2-F3 made its first drop test. It differed from its former configuration only by the addition of a dorsal fin. The first powered flight on November 25, 1970, was cut short when the Thiokol XLR11 rocket motor petered out. The highest speed reached by the M2-F3 was Mach 1.36 (900 MPH) on October 5, 1972, and the maximum altitude achieved during the program, 71,500 ft., was obtained on the last of 27 M2-F3 flights which took place on December 20, 1972.

1970 June 3–4 The European Space Research Organization (ESRO) held a conference in Paris, France, on U.S. space station plans. Attended by 350 people, the conference attracted wide interest, and ESRO set up two groups to get opinions from a broad range of scientists on the possible uses they might have for a manned orbiting facility. On June 5 NASA and ESRO had discussions about joint ventures on the station and broader European participation. Eventually, these decisions would lead to agreement on a manned pressurized module.

1970 June 4 NASA announced the award of a space tug preliminary planning study contract to North American Rockwell, envisaging the tug as a vehicle capable of transferring payloads from one orbit to another. Gus R. Babb, from the Manned Space Center's Advanced Mission Design Branch, distributed a memorandum outlining significant cost advantages in using the Shuttle to launch satellites and unmanned planetary spacecraft. Babb emphasized that a restartable kick stage, like the tug, capable of moving satellites and spacecraft from low Earth orbit to their assigned trajectories would be imperative. The cryogenic Centaur stage proposed as a candidate for this job, said Babb, could be placed in orbit along with the payload.

1970 June 5 NASA agreed to launch a scientific satellite for the Astronomical Netherlands Satellite Authority on a four-stage Scout launcher. Called *ANS*, the satellite was to be launched to a near sun-synchronous orbit from which instruments provided by British and Dutch scientists would study stellar ultraviolet radiation. The satellite would be designed and built in Holland.

1970 June 12 The final test flight of the Europa I launch vehicle (F9) took place from complex 6A at the Woomera Rocket Range, Australia. With the objective of placing a 573-lb. test satellite in low Earth orbit, the launch took place at 01:10 UT (10:40 local time) and resulted in near-perfect propulsion from all three stages: 2 min. 37 sec. for Blue Streak; 1 min. 45 sec. for Coralie; 6 min. 6 sec. for Astris. Because the 661-lb. payload shroud failed to separate as scheduled at 3 min. 20 sec., the vehicle achieved a velocity of only 23,295 ft./sec. instead of the planned 25,897 ft./sec. As a consequence, the upper stage and payload failed to achieve orbit and fell back into the atmosphere.

1970 June 15 Edgar M. Cortright, chairman of the *Apollo 13* Review Board set up to investigate the explosion aboard service module 109, submitted findings and recommendations which it advised should be implemented before the next flight. The board found that all Apollo cryogenic oxygen tanks had been manufactured with inadequate switches unable to take the high electrical loads applied during ground tests. Power from test equipment partly melted the insulation and potting, leaving it vulnerable to shorting, which was what had happened on *Apollo 13* when the crew switched on fans to "stir" the liquid oxygen. If this had occurred on *Apollo 8*, which had no lunar module to serve as a lifeboat to sustain life on the long journey back to Earth, the crew would have perished. Prior to the next flight, a third oxygen tank was added and located in bay no. 1 of the service module, opposite the bay where the fuel cells were located. A contingency lunar module descent-stage battery was added to the service module as well.

The fully reusable Shuttle configuration is depicted here in this Grumman representation of an orbiter carried to the edge of space by a manned, fly-back booster. The concept is dated June 1970.

NASA announced that it was negotiating with a Grumman/Boeing team, Lockheed Aircraft and Chrysler for 11-month Phase A feasibility studies on several alternate Shuttle concepts. Worth $4 million, the Grumman contract would be managed by the Manned Spacecraft Center and cover a 1 ½-stage Shuttle concept utilizing drop tanks, a reusable orbiter with expendable booster and a booster-orbiter combination using strap-on solids and updated cryogenic J-2 engines. The $1 million Lockheed contract covered analysis of 1 ½-stage concepts utilizing both high and low cross-range orbiters. The $750,000 Chrysler contract was for a single-stage-to-orbit (SSTO) concept. The Lockheed and Chrysler contracts were to be managed by the Marshall Space Flight Center.

1970 June 19 Strategic Air Command (SAC) accepted the first flight of Minuteman III (LGM-30G) ICBMs comprising part of the 741st Strategic Missile Squadron at Minot AFB, N.Dak. The first operational test launch of a Minuteman III took place when the 91st Strategic Missile Wing launched their first round from Vandenberg AFB, Calif., on March 24, 1971. SAC would deploy 550 Minuteman III missiles, initially equipped with the Mk.12 reentry vehicle carrying two or three 170-KT W-62 warheads. During 1972 SAC began to convert Minuteman III units to the Command Data Buffer capability which enabled launch crews to electronically reprogram onboard target coordinates from the Launch Control Facility. Hitherto, retargeting of missiles could only be carried out by a technician's physically changing the target tape in the guidance system on the missile. The new capability enabled targets to be reprogrammed during the course of an active nuclear war so that warheads could be redirected to sites missed by malfunctioning missiles. The last of 550 Minuteman IIIs was declared operational on January 26, 1975.

NASA's Space Station Task Group presented Administrator Thomas Paine with the results of the Phase B definition studies conducted over the preceding nine months by North American Rockwell and McDonnell Douglas. NASA wanted the station launched in 1977 to have a minimum operating life of 10 years, supporting 12 crew members and a variety of plug-on experiment modules. It was to serve as a prototype module for a space base carrying up to 50 people. The station had to be launched by a two-stage Int-21 version of the Saturn V and made compatible with the reusable Shuttle. North American Rockwell defined a barrel-shaped station 33 ft. in diameter, 59 ft. in length, weighing 120,000 lb. Four separate decks would be provided for mixed experiments, living quarters and control stations, with ports for attached experiment modules. A forward power boom, giving the station an overall length of 92 ft., would support two 96 ft. 6 in. × 66 ft. 2 in. solar array wings for a total span of 203 ft. McDonnell Douglas proposed a similar station, 33 ft. in diameter with a length of 111 ft., consisting of a 50-ft., four-deck pressure module and electrical power from an isotope/Brayton system, in which heat produced by plutonium 238 is used to drive generators. A forward section could be extended on a telescoped boom for gravity-gradient stabilization or artificial gravity. A key feature of the 187,000-lb. station was a central tunnel running the length of the module to which astronauts could retreat and await rescue in the event of a catastrophe.

1970 June 29 The U.S. Air Force briefed a NASA/U.S. Air Force Space Transportation System committee on its requirements from the Shuttle. The Department of Defense said that a 40,000–50,000-lb. payload capability coupled to a 15 × 60 ft. payload bay capacity, would be the "most ideal size for Department of Defense and national mission projections." If provided, such a Shuttle "would replace" all existing launch vehicles used by the Department of Defense, according to the Pentagon. Of enduring irritation to NASA was the unwillingness of the Department of Defense to contribute funds toward Shuttle development even while recognizing the problem of NASA's finding the necessary "yearly funding levels."

In seeking to cut NASA's fiscal year 1971 budget, Sen. Walter F. Mondale (D, Minn.) criticized the Shuttle and space station programs as "the first essential step toward a manned Mars landing—a project which would cost anywhere between $50 and 100 billion." In asserting that "our environment on Earth desperately needs attention," Mondale called for a redirection of resources toward Earth-based problems. Such Earth-first concerns had been heightened when astronauts transmitted TV images of a "fragile Earth" and weather satellites began to bring greater awareness of environmental issues. An internal report at the NASA Manned Spacecraft Center highlighted the "significant cost savings potential in using Earth-to-Orbit Shuttle (EOS) to deliver unmanned payloads." It asserted that expendable launch costs totaled "$400 million per year for close to 60 Department of Defense and NASA flights combined" and that "this could be handled with around 20 flights of the EOS which would cost about $200 million per year including the necessary expendable kick stages." The report also said that a large "35,000 pounds" payload capacity Shuttle would achieve this.

1970 June 30 NASA pilots involved in lifting-body flight tests stated their preference for landing an unpowered Shuttle orbiter rather than one powered by jet engines. Milton O. Thompson was quoted in the *Los Angeles Herald Examiner*, "I, as a pilot, would still prefer to make an unpowered approach in the Shuttle due to the uncertainties of air-starting [a jet engine], and initial jet engine operation, and the limited fuel available [on the Shuttle]."

1970 July 6–9 The European Launcher Development Organization (ELDO) held a meeting in Bonn, Federal Republic of Germany, attended by NASA and two U.S. contractors, on the proposed Shuttle program and how Europe might become involved. About 250 representatives attended the presenta-

tion, at which North American Rockwell and McDonnell Douglas briefed the group on their respective design proposals. ELDO had already placed two space tug study contracts, one with Hawker Siddely Dynamics in the United Kingdom and the other with Messerschmitt-Bölkow-Blohm (MBB) in Germany. Although not a member of ELDO, Britain agreed to participate in these studies and share the cost.

1970 July 10 After a year of informal correspondence between NASA Administrator Thomas Paine and Soviet academician Mstislav Keldysh, President Richard M. Nixon authorized official study of a possible Earth orbit flight involving astronauts from the United States and cosmonauts from the Soviet Union. NASA engineers recommended two possible options: a rescue mission to assist a disabled vehicle in space or a rendezvous and docking flight to demonstrate orbital link-up between manned spacecraft. NASA officials left Washington for Moscow on October 23, 1970, and during a ceremony in the Soviet capital five days later, representatives of the two countries signed an agreement to work on a common docking unit that could be used with Russian and American spacecraft.

The program development office at NASA's Marshall Space Flight Center drew up an "Alternate Configuration Modular Space Station." Anticipating that the cost of launching a fully outfitted station on a single launcher would be prohibitively expensive, engineers at MSFC reworked the station requirements into a series of separate modules, each of which could be launched by a Shuttle. At both MSFC and the Manned Spacecraft Center, work was already under way in an informal sense to anticipate the demise of the big single-launch station lifted by a Saturn V derivative.

1970 July 15 NASA announced it had selected General Electric Co. to build the earth resources technology satellite (ERTS) system under a contract managed by the Goddard Space Flight Center. Two flight spacecraft and the first six ground stations for receiving digital images from the polar orbiting satellite were planned. General Electric would employ the bus configuration from the Nimbus series of weather satellites and adapt it for ERTS.

1970 July 23 The U.S. Environmental Science Services Administration (ESSA) announced a two-year effort to complete a satellite triangulation program with stations 600–800 mi. apart linking Canada with Alaska and the 48 contiguous United States. Initiated in 1964 using *Echo I* and *Echo II* balloon satellites, it had been suspended at the launch of *Pageos I* on June 23, 1966, when a global geodetic measurement program had been operated. In the renewed ESSA program *Pageos I* would be photographed from a network of U.S. and Canadian sites, enabling land surveyors and engineers to obtain a new geodetic grid accurate to within 1 ft. in 200 mi.

Donald K. Slayton, director of Flight Crew Operations at the NASA Manned Spacecraft Center, replied to a letter from Col. F. H. Fahringer, U.S. Air Force, concerning a request for information about the possibility of RAF pilots applying to be Shuttle astronauts. Slayton said that to date all NASA astronauts were required to have U.S. citizenship and that he had "adequate astronauts on hand to support all approved programs."

1970 July 24 NASA selected Philco-Ford to develop and build the Synchronous Meteorological Satellite (SMS) series which were to become part of the National Operational

Meteorological Satellite System. SMS satellites would be operated by the Environmental Science Services Administration and provide the world's first weather observation platforms from geostationary orbit, viewing continuously almost a complete hemisphere of the Earth for cloud-cover pictures by day or night.

1970 August 14 NASA sent letters to Shuttle Phase B contractors North American Rockwell and McDonnell Douglas about cost objectives for the definitive vehicle which included a mission traffic model. Cuts in NASA's budget request for fiscal year 1971 and gloomy projections about future funds brought a reappraisal of program pace, and NASA now baselined a first flight in early 1977 instead of 1975. NASA expected to fly 10 Shuttle missions in the first year, with succeeding annual rates of 15, 20, 30, 40, 50, 60, 70, 75 and 75 flights, for a total of 445 missions in the first 10 years.

1970 August 17 The Soviet Union launched the 2,602-lb. *Venera 7* spacecraft toward the planet Venus after an SL-6 launcher first placed the terminal stage and spacecraft in low Earth orbit. This spacecraft was equipped with a modified descent capsule to enable it to transmit information from the surface of Venus after descending through the atmosphere by parachute. Five days after the launch of *Venera 7*, the Soviets tried to launch a second spacecraft of the same type. It failed when the terminal stage fired briefly and shut down far short of the velocity increment needed to push it to escape velocity. Designated *Cosmos 359*, it was stranded in an elliptical orbit of 129 × 553 mi. from which it decayed into the atmosphere on November 6, 1970.

1970 August 18 McDonnell Douglas submitted to NASA a review of progress to date in its Shuttle Phase B definition study. In its original submission, McDonnell Douglas outlined two different booster designs, but the twin-body concept had been dropped. The new canard booster of Martin Marietta design had a length of 248 ft. and a span of 160 ft. The swept wings were set atop the fuselage with outward splayed vertical stabilizers and canards set low on the forward fuselage aft of the cockpit area. Primary propulsion was twelve 458,000-lb. thrust cryogenic engines. The low cross-range orbiter, now with a length of 148 ft. 2 in., had lost the upward folding wings, which were replaced by a fixed wing. The high cross-range orbiter remained essentially the same.

1970 August 26 Debate at NASA's Manned Spacecraft Center about a "cheap" Shuttle versus the more ambitious design specified in the Phase B studies prompted a memorandum from Caldwell C. Johnson, chief of the Manned Spacecraft Center's Spacecraft Design Office, which suggested cutting the payload to 10,000 lb. in a 11 ft. × 24 ft. bay. The so-called DC-3 concept would "avoid new developments— make do with things we have." Predicting that current NASA traffic rates involving 60–75 flights per year were unrealistic, the memorandum offered the view that annual Shuttle rates of 24 were the most that could be achieved. In reality NASA never managed to fly more than 9 in one year, and in the first five years of operations (1981–1986) averaged only 5 per year.

1970 September 2 The Black Arrow R2 launch vehicle failed in an attempt to orbit a British satellite from the Woomera test range in Australia. Carrying a 29-lb. satellite called *Orba*, the Black Arrow was to have placed its payload in a 217 × 683 mi., 82° orbit. The three stages fired, but the second stage shut down 13 sec. early and the vehicle had insufficient velocity by the time the third stage had completed its burn phase. On July 29, 1971, the British government announced that, while plans to put a satellite in orbit with Black Arrow R3 would go ahead, the program would be terminated after that event. The second of two satellites designed for Black Arrow would be launched by a NASA Scout.

1970 September 3 Responding to increasingly severe budget cuts, NASA Administrator Thomas Paine announced that *Apollos* 15 and 19 were to be canceled. What had been *Apollo 16* would become the new *Apollo 15* and the remaining missions would be designated *Apollos* 16 and 17, ending the manned lunar program in mid-1972 before the start of Skylab operations, presently scheduled for November of that year. This would mean that the upcoming *Apollo 14* flight would be the last H-series mission using basic hardware and that *Apollos 15–17* would all be J-series missions utilizing enhanced hardware capabilities and with the addition of a lunar roving vehicle (LRV). Cancellation of *Apollo 19* meant that there would be flights for only three instead of four LRVs. Skylab operations would continue through June 1973.

1970 September 8 NASA announced that the British Aircraft Corp. and the German aircraft company Messerschmitt-Bölkow-Blohm (MBB) were to participate in the Shuttle Phase B definition studies. The two companies were to study structures, aerodynamics, flight test instrumentation, data handling and reaction control systems. Although the companies were officially classed as subcontractors to Phase B contractors North American Rockwell and McDonnell Douglas, the British and German governments paid for the work.

1970 September 9–11 NASA held a Space Station Utilization Conference at the Ames Research Center where representatives from several key NASA centers gave presentations on their work in connection with the planned orbiting facility. From Headquarters, the Space Station Task Force gave an overview of the program and the Manned Spacecraft Center presented the operations plan. NASA wanted to involve all the centers in a coordinated effort where their respective skill areas could be applied to the long-term plan, which envisaged orbital operations beginning in 1977.

1970 September 12 Less then 14 months after the United States delivered to Earth the first soil and rock samples gathered from the lunar surface, the Soviet Union launched its first successful moon-sample return mission on an SL-12 from Tyuratam. The 12,350-lb. *Luna 16* consisted of a four-legged descent stage with a single liquid propellant retro-rocket and two adjacent vernier motors for stability and control in the lower center. The descent stage was built up from propellant tanks supporting a toroidal structure within which nested the lower section of the ascent stage. This comprised three spherical tanks in a row with four attitude control thrusters on separate outriggers at 90° intervals. Above the middle sphere was the instrument compartment on top of which was the spherical reentry capsule which would deliver a soil sample to Earth. Below the middle sphere was the ascent propulsion motor for direct ascent from the moon to Earth. On September 13 a course correction put *Luna 16* on track for lunar orbit and on September 17 the spacecraft was injected into a circular 68-mi. orbit inclined 70° changed over the next two days to a 9.3 × 66 mi. path inclined 71°.

1970 September 15 Thomas O. Paine resigned as NASA administrator, leaving George M. Low as acting administrator until a replacement could be found. Paine had been head of NASA for almost two years. The relatively rapid change of leadership that afflicted NASA between 1968, when James Webb retired, and April 27, 1971, when Paine's replacement James Fletcher was appointed, left the agency with little influence at the White House. The budget was still falling and NASA was caught between decreasing funds and ambitious, costly projects that had originated when annual funds were more than twice what they were in 1971.

1970 September 16 NASA's Manned Spacecraft Center completed a pre–Phase A study of a Shuttle-launched modular space station. Studies indicated a series of common modules could be used for a variety of purposes and were projected to be 29 ft. long by 14 ft. in diameter. In a series of 17 Shuttle flights the modules were to be launched for sequential assembly in orbit. Some flights would carry nodes to connect one module to another while others would convey solar cell arrays for electrical power. Basic structural modules would weigh 15,000–18,000 lb. each.

1970 September 20 The first Soviet lunar soft lander put itself down onto a place in the Mare Fecunditatis at 0.68° S by 56.3° E and began the process of obtaining moon soil for return to Earth. From a lunar orbit of 9.3 × 66 mi., the main descent engine fired at perilune to bring the spacecraft out of moon orbit. Guided by radar, precision braking began at an altitude of 2,000 ft. when the retro-engine was fired again. Thrust was modulated by altitude and velocity signals aligning the actual descent path to a preprogrammed profile. At a height of 200 ft. the main engine was shut down, leaving the two vernier motors to lower *Luna 16* at 8 ft./sec. to the surface at 05:18 UT. Having expended propellant first to achieve lunar orbit and then to land on the moon, *Luna 16* now had a mass of 4,145 lb. It stood 13 ft. high within a circle 13 ft. in diameter. A drill attached to the end of a rigid arm hinged at its attachment to the descent stage was lowered to the surface where it obtained a 10-in.-deep core of material. The arm rotated vertically upward to align the core with the spherical Earth-return capsule on top where it deposited the sample through the side. At 07:43 UT on September 21, the ascent stage fired its motor and brought the capsule back toward Earth.

1970 September 24 The lunar sample–return capsule (2 ft. in diameter) from the Soviet *Luna 16* spacecraft landed back on Earth carrying the first moon soil obtained by the Soviet Union. The capsule was carried back to Earth by the ascent stage of the spacecraft, which was retained for Earth tracking purposes. It communicated with Earth on a frequency of 183.6 MHz and was separated from the spherical entry capsule at a distance of 30,000 mi. and a speed of about 36,000 ft./sec. At the end of a 69 hr. 27 min. flight from the lunar surface, during which no course corrections had been made, the capsule entered the atmosphere at 05:10 UT. A beacon transmitter was picked up 4 min. later and the capsule landed by parachute 50 mi. southwest of Dzhezkazgan in Kazakhstan at 05:26 UT. It was visually observed from nearby recovery aircraft. The 3.6-oz. (0.22-lb.) lunar soil sample was delivered to the Soviet Academy of Sciences.

1970 September 25 The first orbital launch attempt by Japan's new four-stage, solid propellant Mu-4S booster failed when the terminal stage malfunctioned. With a total height of 77 ft. and a diameter of 4 ft. 7 in., the Mu-4S had a launch weight of 96,140 lb. Eight solid propellant strap-on boosters produced a total thrust of 174,635 lb. for 8 sec., at which point they were jettisoned. The first stage would continue to produce a thrust of 161,850 lb. until 1 min. 1 sec., when the 62,840-lb. thrust second stage took over and burned for 42 sec. The 27,340-lb. thrust third stage also burned for 42 sec., with the 4,190-lb. thrust terminal stage firing for 40 sec.

1970 October 1 U.S. Air Force launch crews based on Johnston Island in the Pacific in support of the Program 437 antisatellite system were transferred to Vandenberg AFB, Calif. It would take four weeks to get the system operational from an alert. Nevertheless, for more than four years the Air Force held on to the launch pads until Program 437 was terminated on April 1, 1975.

1970 October 5 Kurt Debus, director of the NASA Kennedy Space Center, received a memorandum from Dale Myers, NASA associate administrator for Manned Space Flight, "to confirm that KSC has been assigned responsibility for launch facilities design for the Shuttle program regardless of the locations of the launch site." This referred to the decision, still pending, about where the Shuttle would be launched from.

1970 October 6 Space scientists James A. Van Allen and Thomas Gold lent their support to a move by four U.S. Senators to eliminate the entire range of U.S. manned space programs within three years. Senators Clifford P. Chase (R-N.J.), Jacob K. Javitts (R-N.Y.), William Proxmire (D-Wisc.) and Walter F. Mondale (D-Minn.) were openly hostile to America's continued commitment to manned spaceflight within the civilian space budget. The two scientists opposed funds committed to manned flight on the grounds that it took money from less expensive unmanned programs which, they said, brought a much higher return for the money spent. It was a contest that has been consistently debated and never resolved.

1970 October 8 A Soviet SL-4 launched from the Baikonur cosmodrome put the 12,570-lb. *Cosmos 368* biological satellite into a 125.5 × 238 mi. orbit inclined 65°. The spacecraft carried a variety of living things for investigations into the effects of the space environment on organisms. The reentry section of the spacecraft returned the biological specimens to Earth on October 14.

1970 October 20 The last Soviet lunar variant of the Soyuz spacecraft, designated *Zond 8*, was launched from Baikonur by an SL-12 at 19:55 UT. From an Earth orbit of 125.5 × 138.5 mi., the fourth stage propelled the 12,350-lb. spacecraft to a translunar trajectory. It passed within 695 mi. of the moon's far side four days later and on October 27 returned to Earth across the north pole—the first time that had happened. The descent module made a single high-g reentry into the atmosphere and landed in the Indian Ocean at 13:55 UT, ending two years of highly successful tests with the spacecraft originally intended to carry Russian cosmonauts into moon orbit.

After a two-year hiatus, a new series of Soviet antisatellite tests began with the launch of the 7,320-lb. *Cosmos 373* by SL-11 to a 294 × 337 mi. orbit inclined 63°. The 7,320-lb. interceptor *Cosmos 374* was launched by SL-11 on October 23 to a 329 × 654 mi. orbit inclined 62.9° for a high-speed fly-by and detonation, at a safe distance, leaving 45 pieces. Preserved for future use, *Cosmos 373* was again the target for *Cosmos*

375, launched by SL-11 on October 30 to a 351 × 618 mi. orbit. It detonated after fly-by leaving 31 pieces.

1970 October 29 Drawings of a new Shuttle orbiter design from the NASA Manned Spacecraft Center showed a configuration with a 50,000-lb. payload capability in a 15 ft. × 60 ft. bay. Designated 012, this was the first of 12 orbiter configurations with a 50,000-lb. payload, significantly higher than earlier MSC concepts in the DC-3 series. The Ling-Temco-Vought company had been conducting engineering support work for MSC on a wide variety of different orbiter configurations. The 013 series designs also had 50,000-lb. payload capability and were worked on by Ling-Temco-Vought beginning November 16, 1970.

1970 November 4 At a historic two-day European Space Conference in Brussels, French, West German and Belgian delegations agreed to participate in the U.S. post-Apollo space program involving the Shuttle, a space tug, space station and space bases. The United Kingdom declined to participate on the grounds that the cost of the program was too poorly defined and that the timing was too uncertain. Moreover, said the British, the U.S. Congress had yet to approve the ambitious NASA plans. At the close of the conference, the 12 participating countries agreed to establish a committee to examine the U.S. proposals with a view to ratifying participation.

1970 November 6 The first Program 647 U.S. geostationary early warning satellite was launched by Titan IIIC from Cape Canaveral, Fla. Instead of achieving geostationary orbit, a Transtage failure left the satellite in a 16,255 × 22,300 mi. orbit at 7.8° with a period of 20 hr. The 2,100-lb. TRW satellite comprised a cylinder 9 ft. in diameter and 9 ft. long, covered in solar cells, with four solar panels at 90° intervals around the base spanning 23 ft. to provide a total 450 W. On top was a 12-ft. Schmidt telescope for the 2,000-cell infrared detector. Offset at a 7.5° angle, the telescope rotated with the main body of the satellite at 5–7 RPM, viewing a given spot on the Earth every 8–12 sec.

1970 November 9 A NASA Scout launch vehicle carried the *Orbiting Frog Otolith* and *Radiation Meteoroid* satellites from Wallops Island, Va., to low Earth orbit. The 293-lb. *Orbiting Frog Otolith* satellite comprised a pressure-tight canister, 1 ft. 6 in. × 1 ft. 6 in., containing two male bullfrogs, each weighing 11 oz. Electrodes attached to the vestibular region of the frogs' ears would record changes induced by acceleration of a centrifuge to which they were attached. The *OFO* satellite had a width of 2 ft. 6 in. and a length of 3 ft. 11 in. and was placed in a 189 × 322 mi., 37.4° orbit. The 46-lb. *Radiation Meteoroid* satellite consisted of two cylindrical segments mounted around the Scout fourth stage. The 1 ft. 3 in. high lower cylinder carried 7 sq. ft. of solar cells providing 25 W of power, and the upper cylinder housed electronics and experiments to measure micrometeoroid flux, speed and direction. It was placed in a 188 × 327 mi. orbit. Both remain in space.

1970 November 10 The first unmanned roving vehicle to operate on the surface of another world was launched as an integral part of *Luna 17* by an SL-12 launch vehicle from the Soviet launch complex at Tyuratam. The 12,400-lb. spacecraft consisted of the descent stage previously flown as part of the *Luna 15* and *Luna 16* sample-return missions but with the upper (accent) stage replaced by a surface rover called *Lunokhod 1*. Following liftoff at 14:44 UT, the SL-12 put *Luna*

17 in an interim Earth parking orbit and from there brought it to a lunar trajectory, during which two course-correction maneuvers were performed on November 12 and 14. Following lunar orbit insertion on November 15, *Luna 17* was in a circular path of 53 mi. inclined 141° to the equator. Changed a day later to an elliptical path that brought it to within 12 mi. of the surface, the spacecraft descended to the bottom of a shallow crater in the Mare Imbrium at a position 38.28° N × 35° W and came to rest at 03:47 UT on November 17, 1970.

1970 November 12 NASA announced that the Marshall Space Flight Center and the Manned Spacecraft Center were examining proposals that the U.S. space station be launched by the Shuttle as a series of separate modules and assembled in orbit. Composed of cylindrical modules 14 ft. in diameter and 58 ft. long, the station would support 6–12 people over 10 years of continuous occupation. NASA said the station could be assembled within a few months in a circular orbit 200–300 mi. above Earth at a 55° inclination.

1970 November 14 NASA became the first U.S. government agency to convert to the SI (*Système internationale d'unités*) system of measurement in all its technical and scientific documentation. The metric system was, at this date, standard in 110 countries, and the widescale use of NASA data and documentation throughout the world made this shift a logical step. With aspirations for international cooperation in post-Apollo programs, NASA wanted to integrate with the majority of engineers and scientists around the world. In fact, more than half of NASA publications already used metric at this time.

1970 November 17 At 06:28 UT the Soviet *Lunokhod 1* moon rover descended to the lunar surface from its position on top of the descent stage of *Luna 17*, thus becoming the first remotely controlled vehicle operated on the surface of another world. Shaped like an old-fashioned bath tub, *Lunokhod 1* had eight small wheels, each independently powered by its own electric motor. Power was provided by solar cells attached to the underside of the hinged "bath tub" lid which, when folded back, exposed them to solar energy. Batteries were provided to supply power to survive lunar night. The 1,667-lb. *Lunokhod 1* carried four TV cameras for close-up and panoramic views, an x-ray spectrometer for soil analysis, thumpers to test the strength of the surface and cosmic-ray detectors. The French had provided a laser-ranging retroreflector for accurate measurement of the Earth-moon distance. *Lunokhod 1* continued to operate for five days, during which it traveled a total distance of 646 ft. and sent back 14 TV images.

1970 November 18 Searching for ways to cut Shuttle development costs, NASA officials discussed the possibility, suggested by Wernher von Braun, that the booster might be funded for development before the orbiter. Some felt that booster development should begin before the orbiter in any event, citing the commonality of booster design to low or high cross-range orbiters. It was also said by an unattributable source, "If the orbiter is delayed, the booster could be used to launch conventional upper stages and payloads in the interim." Others disagreed, saying that the booster was a key element in a total, reusable space transportation system and should not be developed separately.

1970 November 19 NASA formally changed the requirements of the Shuttle Phase B definition contracts being

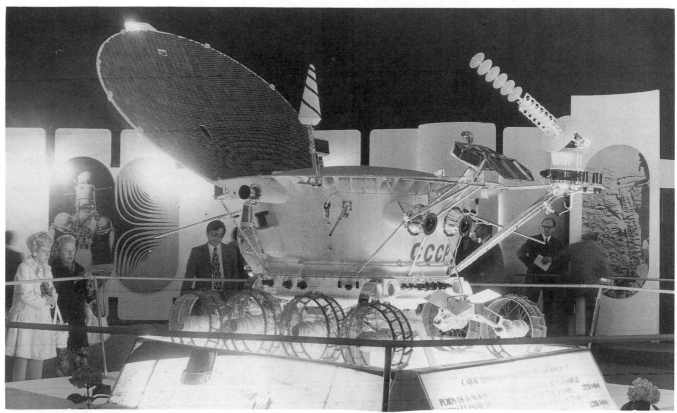

As it appeared on the moon, and as seen here at an exhibition, the Lunokhod 1 *moon rover with its solar cell array hinged back from the top of the vehicle.*

performed under teams led by North American Rockwell and McDonnell Douglas. Unspecified in the original work statement, the Shuttle payload requirement was now fixed at 25,000 lb. to a 311-mi. orbit with a 55° inclination. The payload bay dimensions were reconfirmed at 15 ft. diameter by 60 ft. length.

1970 November 20 A highly unconventional Shuttle orbiter design was circulated under the 012 series designation at the NASA Manned Spacecraft Center. Featuring a forward payload bay and aft propellant tanks for the cryogenic hydrogen and oxygen feeding the two cryogenic engines at the rear, the so-called *Blue Goose* had forward canards and a low wing. To accommodate a varying center of gravity the straight, slightly swept wing could be moved forward or back over a range of 12 ft. This configuration had many undesirable features and was never seriously evaluated.

1970 November 24 A Soviet SL-4 carried the 16,530-lb. unmanned *LK (Lunova Korobly)* lunar landing module from Baikonur cosmodrome to a 119 × 144 mi. orbit with an inclination of 51.6°. The test version of the *LK* was known as T2K, veiled under the designation *Cosmos 379*; its first flight was scheduled to simulate events associated with a lunar landing mission. The *LK* was designed with a 4,520-lb. thrust motor burning UDMH-N_2O_4 that on a manned lunar landing mission, was to lower it to the lunar surface and fire again to carry the cosmonaut back into lunar orbit. It was fired on November 25 in an 863 ft./sec. burn simulating a descent to the moon that placed the spacecraft in an Earth orbit of 122 ×

752 mi. Two days later it was fired again in a 4,980 ft./sec. burn to simulate ascent to lunar orbit, putting T2K in a 110 × 8,725 mi. orbit. It decayed on September 21, 1983.

1970 November 30 NASA launched the third Orbiting Astronomical Observatory *(OAO-B)* on an Atlas-Centaur from Cape Canaveral. The 2,400-lb. aerodynamic shroud covering the 4,680-lb. satellite failed to separate, making it impossible for *OAO-B* to achieve orbital velocity, and it fell back into the Atlantic Ocean. The octagonal *OAO-B* had a length of 10 ft. and a width of 7 ft. It was to have conducted stellar observations in the ultraviolet region of the spectrum, utilizing a 38-in. Cassegrain telescope.

1970 December 2 A Soviet SL-13 launch vehicle carried the 41,310-lb. *Cosmos 382* into space for a series of maneuvers totaling 8,488 ft./sec., leaving it in a 1,600 × 8,720 mi. orbit. *Cosmos 382* was the veiled designation for the extended propulsion section of the lunar-orbit Soyuz variant complete with upper modules. This was designed as part of the series of flight tests prior to manned lunar landing flights, which were later canceled.

1970 December 3 NASA's Marshall Space Flight Center issued a request for proposals on the preliminary design of the research and applications module (RAM). A cylindrical structure up to 58 ft. long and 14 ft. in diameter, with a weight of 20,000 lb., it would be capable of carrying a wide range of scientific equipment. The RAM would be launched in the cargo bay of the reusable Shuttle and docked to the permanently manned space station NASA hoped to have in orbit by the late 1970s. Proposals were due January 8, 1971, and three teams submitted bids: General Dynamics/North American Rockwell/TRW, Bendix; General Electric/Lockheed/Grumman; Martin Marietta/McDonnell Douglas/IBM/Honeywell.

1970 December 10 The Soviet moon rover *Lunokhod 1* was activated for its first full lunar day of operations on the western edge of the Mare Imbrium and the beginning of a concerted exploration of the area around the *Luna 17* landing site. For 13 earth days the vehicle was operated by its four "drivers" on Earth, and when it was brought to a halt for oncoming lunar night, *Lunokhod 1* had traveled a further 4,994 ft. The vehicle was operated on each successive lunar day until traversing operations ceased on September 15, 1971. Designed to operate for 3 lunar days, it had exceeded that capability threefold and moved a total distance of 34,582 ft., or 6.5 mi. A final communication session was held between September 30 and October 4, 1971, when it returned the last of some 20,000 TV images, including 206 panoramas. *Lunokhod 1* had performed 25 chemical tests and 500 mechanical tests on the lunar soil.

1970 December 11 The first operational ITOS weather satellite, *ITOS A*, was launched by a Delta N6 from Cape Canaveral into a sun-synchronous orbit of 888 × 915 mi. inclined 102° to the equator. Virtually identical to *ITOS 1* launched on January 23, 1970, the satellite was redesignated *NOAA-1* in orbit, in recognition of its operational duty for the National Oceanic and Atmospheric Administration. Built by RCA and managed by NASA's Goddard Space Flight Center, *ITOS B*, launched on October 21, 1971, failed to reach a usable orbit due to problems with the launcher. *ITOS D/NOAA-2*, orbited on October 15, 1972, carried scanning radiometers. *ITOS E* was launched on July 16, 1973, but failed to reach orbit, and *ITOS F/NOAA-3* was sent up successfully on November 6, 1973. *ITOS G/NOAA-4*, along with two other satellites, was launched on November 15, 1974. *ITOS H/NOAA-5* was successfully orbited on July 29, 1976, the last of the Tiros M series.

In a preliminary draft of the Shuttle Phase C/D work statement, the Manned Spacecraft Center stipulated a 25,000-lb. payload capability in a 15 ft. × 60 ft. cargo bay but indicated that it should be capable of "launching and landing with payloads up to 60,000 lb." The orbiter cross-range was set at 1,726 mi., which met U.S. Air Force payload and mission requirements. At this date, North American Rockwell and McDonnell Douglas were about halfway through their Shuttle Phase B definition studies.

1970 December 12 The first U.S. satellite launched by another country, and the first equipped with instruments to detect high-energy x-ray sources in space, *Small Astronomy Satellite-A (Explorer 42)* was placed in a 330 × 355 mi. orbit inclined 3° to the equator. The 315-lb. satellite was launched on a four-stage Scout by an Italian crew from the San Marco platform in the Indian Ocean off Kenya. The first of three, *SAS-A* comprised a drum 2 ft. in diameter and 2 ft. high surmounted by a 140-lb. x-ray experiment 13 ft. tall. Four 5 ft. 6 in. × 1 ft. solar arrays provided electrical power. *Uhuru*, the Swahili word for *freedom*, was adopted as the name of *Explorer 42*. *SAS-A* decayed on April 5, 1979.

The 154-lb. French satellite *Péole* was launched on a Diamant B from Kourou, French Guiana, to a 321 × 464 mi. orbit with an inclination of 15°. *Péole* was designed as precursor to the *Eole* weather satellite, which was to be launched on a U.S. Scout. Although it did not reach its planned orbit, several tests were conducted with the communications payload. The exterior surface carried laser reflectors for geodetic tests. The French national space agency, CNES (Centre National d'Etudes Spatiales), was particularly keen to develop space-based geodesy. *Péole* decayed on June 16, 1980.

The first in a series of operational Soviet navigation satellites, *Cosmos 385*, was launched by SL-8 from Plesetsk to a 605 × 608 mi., 74° orbit. Two more were launched in 1971, as *Cosmos 422* (May 22) and *Cosmos 465* (December 15), each satellite in an orbital plane displaced 120° to its neighbor. *Cosmos 475* (February 25, 1972) and *Cosmos 489* (May 6, 1972) replaced *Cosmos 385* and *Cosmos 422*, respectively. The five satellites of this first-generation system were gravity gradient–stabilized and weighed 1,500 lb.

1970 December 15 The Soviet spacecraft *Venera 7* entered the atmosphere of Venus at 05:01 UT and separated its insulated entry capsule. Unlike the egg-shaped spheroid carried by *Veneras 4, 5* and *6*, *Venera 7* was perfectly spherical in shape and had additional thermal protection. When aerodynamic braking had reduced its entry velocity of 38,000 ft./sec. to 820 ft./sec., a temperature-resistant parachute was deployed. Data on pressure and temperature was transmitted to Earth for 35 minutes, after which the signals ceased abruptly and it was assumed the capsule had failed during descent or been destroyed on impact. Analysis of the signals, however, revealed an additional 23 min. of data transmitted by the capsule at a strength of only 1% the descent level, indicating it had rolled on top of its flexible antenna. Thus did *Venera 7* become the first man-made object to survive to the surface of another planet and continue to transmit science data.

NASA's Manned Spacecraft Center requested a proposal from North American Rockwell for design and planning studies of a modular space station. With a crew of six the initial station would be assembled by Shuttle flights for operational duty in early 1978 and have the potential to grow to a twelve-person facility with additional modules launched subsequently on separate Shuttle missions. NASA wanted to know whether it could accommodate all the science and research investigations planned for the single-launch space station defined in earlier studies.

The Manned Spacecraft Center Flight Analysis Branch issued a memorandum breaking down the potential Shuttle traffic envisaged for the 445 flights predicted for the first 10 years of operation beginning in 1977. Of the total, 42.25% were for placing NASA satellites in orbit, 34.61% were for placing Department of Defense satellites in orbit, 15.73% were to be in support of space station operations and 7.41% were to be space tug flights (figures rounded to nearest second decimal place). Of this total, more than 60% of missions would be flown to an orbital inclination of 28° endorsing Kennedy Space Center as the primary launch site.

1970 December 18 The first of a much improved generation of Soviet electronic intelligence gathering satellites, *Cosmos 389*, was launched by an SL-3 from Baikonur to a 371.5 × 422 mi. orbit inclined 81.2°. The 8,380-lb. satellite was packed with sophisticated signals-monitoring equipment and data acquisition and scanning instruments. Until 1977 these satellites were launched at the rate of 1–3 each year. Beginning in 1978 the flight rate increased to an average of 4 each year. Satellite mass varied between 4,410 lb. and 8,820 lb. The last of 35 satellites in this series, *Cosmos 1441*, was launched by SL-3 on February 16, 1983.

1970 December 23   NASA decided to provide a rescue capability for the Skylab astronauts during each of their long periods of habitation aboard the orbital workshop. It called for the use of a conversion kit to temporarily equip the Apollo command module with seats for five astronauts. The *SL-1* mission would comprise the first two stages of the Saturn V, the orbital workshop, the airlock module, the multiple docking adapter, and the Apollo telescope mount. *SL-2, -3* and *-4* missions would comprise the separate visits to Skylab involving Saturn IB vehicles AS-206, -207 and -208, respectively. If called upon to retrieve a stranded Skylab crew, each vehicle scheduled for launch could be prepared for a rapid rescue carrying a two-person crew up to Skylab and returning all five astronauts to Earth. After the launch of AS-208, a spare Saturn IB (AS-209) would be prepared for roll-out to the pad in case of emergency.

NASA Associate Administrator for Manned Space Flight Dale Myers informed the Manned Spacecraft Center that as part of its Phase A alternate Shuttle contract, the Grumman/Boeing team was to study an orbiter employing externally mounted liquid hydrogen drop tanks. This was being studied because of the promise it gave of cutting orbiter weight and costs, and of relieving the technical complexity inherent with internal cryogenic tankage for engine propellant. NASA also wanted the Marshall Space Flight Center to study the use of the Saturn V S-IC first stage as an interim booster pending availability of the fully reusable booster. Earlier, NASA had rejected the notion of building the booster first in a phased development program.

1970 December 29   A European industrial consortium was formed to compete for satellite contracts from the European Space Research Organization (ESRO). Called STAR (Satellites for Telecommunications, Applications and Research), the group included British Aircraft Corp., Contraves A.G. of Switzerland, CGE Fiar and Montedel-Montecatini Edison Electronics S.p.A. of Italy, SABCA of Belgium, L. M. Ericsson AB of Sweden, Fokker VFW of the Netherlands, Dornier-System GmbH of the Federal Republic of Germany and Thompson CSF of France. AEG-Telefunken of Germany would participate on communication satellite technology.

1970   During the year a series of developments began to focus attention at NASA on a scientific investigation of the atmosphere and surface conditions at Venus. The Space Science Board of the U.S. National Academy of Sciences and the NASA Planetary Missions Board examined the scientific potential for such a mission and issued *Venus—Strategy for Exploration*, which became known as the Purple Book from the color of its cover. It proposed the use of spin-stabilized Explorer spacecraft with Delta launchers used for flights involving orbiters, atmospheric probes and landers. The report proposed two mutiprobe vehicles for the 1975 opportunity, two orbiters for the 1976–1977 opportunity and a lander mission in 1978. Because of funding constraints the plan was modified to include two multiprobe flights in 1976–1977, an orbiter in 1978 and a single multiprobe mission in 1978.

China began to deploy nuclear-tipped missiles in its northeast and northwest regions. The CSS-1 comprised a liquid propellant missile operating on UDMH/N_2O_4 propellants. With a length of 69 ft. and a diameter of 5 ft. 3 in., it had a launch weight of 57,300 lb. and could carry a single 15-KT warhead across a range of 700 mi. Only 100 CSS-1s were ever built. The CSS-2 had a length of 67 ft. 7 in., a diameter of 8 ft. 1 in. and a weight of 59,500 lb. It had a 2-MT warhead and

a range of 1,500 mi. A maximum 60 CSS-2 were deployed by 1980.

1971 January 14   The NASA Manned Spacecraft Center's Propulsion and Power Division completed evaluation of two alternate Shuttle booster concepts. One, called the "Big Dumb Booster," envisaged the use of an expendable, pressure-fed, storable propellant booster generating a first-stage thrust of 2.6 million lb. and a second-stage thrust of 1.5 million lb. The second concept was for a delta-winged orbiter equipped with overwing external hydrogen and oxygen propellant tanks for its propulsion system and strap-on solid propellant boosters. This concept approximated the configuration eventually selected: a reusable orbiter fed with propellants carried in an external tank and boosted on its way by solid propellant rockets. On April 1, 1971, NASA added external liquid hydrogen tank studies to the primary Phase B definition contracts.

1971 January 16   A delegation from NASA that included Administrator Thomas Paine arrived in Moscow for meetings with officials at the Soviet Academy of Sciences to discuss joint manned spaceflights. Until this date each side had talked about compatible docking systems for future spacecraft. On January 20 NASA proposed to their Russian hosts that each side should agree to a joint flight involving existing Apollo and Soyuz spacecraft. Launched separately, they would link up in space using a separate docking module carried into orbit by one of the two manned vehicles. Technical details were discussed during a reciprocal visit by Soviet officials to the NASA Manned Spacecraft Center that began June 20, 1971.

1971 January 19–20   NASA's Office of Manned Space Flight Management Council made important changes to the Space Shuttle Requirements Document which identified objectives to which the system should be designed. The stipulated mission requirement now was for a maximum payload of 65,000 lb. delivered to a 115-mi. due east (28° inclination) orbit with the air-breathing engines removed. This payload requirement equated to a 25,000-lb. load carried to a 311-mi. due east orbit or a 40,000-lb. payload delivered to polar orbit. A cross-range capability of 1,266 mi. was stipulated, ruling out the low cross-range concepts, and booster and orbiter engine thrust was set at 550,000 lb. at sea level. Launch and reentry accelerations were not to exceed 3 g's.

1971 January 26   The first of a new, fourth-generation series of Intelsat communication satellites, *Intelsat IV F-2*, was launched by Atlas-Centaur from Cape Canaveral, Fla., to a transfer orbit, from which it was positioned in geostationary orbit over 25.5° W. With a launch weight of 3,080 lb. (1,313 lb. in orbit), each Intelsat IV series satellite comprised a cylinder with a diameter of 7 ft. 10 in. Solar cells around the exterior provided 570 W of electrical energy to power the 12, 6/4 GHz, transponders with a total capacity for handling 4,000 telephone circuits and two TV channels. Directional antennas mounted on top of the cylinder gave the satellite a total height of 17 ft. 5 in. *Intelsat IV F-3* was launched on December 20, 1971, followed by *Intelsat IV F-4* on January 23, 1972, *Intelsat IV F-5* on June 13, 1972, *Intelsat IV F-7* on August 23, 1973, *Intelsat IV F-8* on November 21, 1974, *Intelsat IV F-6* on February 20, 1975 (destroyed in a launcher malfunction) and *Intelsat IV F-1* on May 22, 1975.

1971 January 29   George M. Low, NASA's acting administrator, publicly explained the NASA budget for fiscal year

1972 requested by the Nixon administration. Reporting a halt to the downward spiral of annual space program budget reductions, Low said that the $3.271 billion spending plan provided only $10 million for Shuttle work compared to a NASA request for $190 million. Low confirmed that engine development would go ahead but that NASA was still not sure whether it could afford to let contracts for full scale development. More significant, all funds for NERVA (nuclear engine for rocket vehicle application) were cut, although some work did go ahead on a 15,000-lb. thrust version. A year later even that work was brought to a halt.

1971 January 30 Severe criticism of NASA's proposed Shuttle program was made by former scientist-astronaut Brian O'Leary in an article in the *New York Times:* "Once again, our priorities seem to be grossly twisted. Once again, we confront a new multibillion dollar project which is peripheral to our pressing domestic problems. A number of Senators and scientists share the view that the President and Congress have acted irresponsibly." O'Leary's views galvanized informed opinion opposed to this further commitment to manned spaceflight, largely at the expense of a broader utilization of unmanned, Earth applications projects.

1971 January 31 At 21:03:02 UT the *Apollo 14* mission was launched from Complex 39A at the Kennedy Space Center. Carrying astronauts Alan Shepard, Edgar Mitchell and Stuart Roosa, the vehicle weighed 6.42 million lb. and included CSM-110 (64,464 lb.) and LM-8 (33,652 lb.). Earth parking orbit and translunar injection went according to plan, but docking with the lunar module was delayed 1 hr. 54 min. by repeated difficulties with the docking probe, finally being

Apollo 14 *astronauts Alan B. Shepard Jr. (right) and Edgar D. Mitchell rehearse procedures for the* Apollo 14 *lunar landing mission, launched January 31, 1971.*

accomplished at 4 hr. 57 min. The LOI (lunar orbit insertion) burn went off as planned at 81 hr. 56 min. 40 sec., putting the docked spacecraft into a 67.8 × 195.2 mi. orbit. Instead of circularizing that orbit at the end of the second revolution, the SPS (service propulsion system) engine fired a 21-sec. descent orbit insertion burn at 86 hr. 10 min. 52 sec., putting the vehicles in a 11 × 67.7 mi. orbit. Shepard and Roosa entered the LM (lunar module) named *Antares*, and undocked from Apollo, dubbed *Kitty Hawk*, at 103 hr. 48 min. At 105 hr. 11 min. 45 sec. Mitchell changed the CSM (command/service module) orbit to 64.4 × 73.5 mi. Powered descent began at 108 hr. 2 min. 26 sec. and touchdown occurred 11 min. 33 sec. later. The LM remained on the surface for 33 hr. 30 min. 31 sec. and two EVAs (extravehicular activities) were conducted. The first lasted 4 hr. 47 min. 50 sec. as Shepard and Roosa deployed the ALSEP (Apollo lunar surface experiments package) experiments: passive seismometer, ion gage, ion detector and charged-particle detector. It also involved an active seismic experiment where Shepard laid out a string of geophones up to 310 ft. from the ALSEP and walked along triggering a set of small explosive charges to remotely profile the surface to a depth of 500 ft. A box containing four mortars was activated after the crew returned, lobbing charges up to 5,000 ft. from the site. The second EVA lasted 4 hr. 34 min. 41 sec. and took Shepard and Roosa on a geological traverse with the handcart for tools, at the end of which Shepard produced a golf club and proceeded to tee off! *Apollo 14* returned to Earth February 9,1971, with 94.5 lb. of lunar samples, splashdown occurring at 216 hr. 1 min. 58 sec.

1971 January McDonnell Douglas completed a study of the use of a Shuttle booster to launch an expendable upper stage and payload. Being designed to lift a manned reusable orbiter from the surface of the Earth on its way into space, the booster could also double as the first stage of a cargo system capable of placing great weights in space. McDonnell Douglas found that a 12-engine fly-back booster developing a launch thrust of 4.8 million lb. could lift a payload of up to 170,000 lb. into orbit through a range of upper stages. McDonnell believed that its own S-IVB stage would be a suitable candidate for providing upper stage propulsion.

Grumman started work for NASA on the conceptual evaluation of external hydrogen tanks for a Shuttle orbiter. Initial studies revealed external overwing tanks with a diameter of 14 ft. and a length of 76 ft. Work on this concept aimed to reduce the weight, and the cost, of a reusable orbiter. In other changes, NASA now scheduled horizontal atmospheric test flights with the Shuttle orbiter for June 1976, followed by a first orbital flight in April 1978 and initial operational capability (launching satellites and performing scheduled missions) from the second half of 1977.

1971 February 9 A new and more vigorous period of Soviet antisatellite tests began with the launch of *Cosmos 394*, weighing 1,653 lb., on an SL-11 from Plesetsk to a 355 × 381 mi. orbit inclined 65.8°. This was the first time that this particular launcher and launch site had been used for Soviet ASAT flights. The 7,320-lb. *Cosmos 397* interceptor was launched by SL-11 from Baikonur on February 25 and placed in a 357 × 1,368 mi. orbit from which it performed the usual high-speed fly-by from above, detonating later and fragmenting into 82 pieces. Another target, *Cosmos 459*, was launched by SL-11 from Plesetsk on November 29 to a low, 138 × 161 mi., 65° orbit; it was and intercepted by *Cosmos 462*, launched on December 3 by SL-11 from Baikonur to a 143 × 1,118 mi. orbit, breaking into 29 pieces after the fly-by.

This was the last of the so-called Phase I Soviet antisatellite tests; Phase II began on February 16, 1976.

1971 February 16 The second Japanese satellite was launched from the Uchinoura Space Center by a four-stage Mu-4S booster. Called *Tansei*, the 138-lb. MS-T1 spherical test satellite entered a 615 × 690 mi. orbit inclined 29.7° and was designed to transmit telemetry, which it did but only for one week. The satellite was a test payload for the first Mu-4S launcher and was instrumented to observe the ionosphere, cosmic rays and solar high-frequency radio emissions. It's name, meaning "light blue," was chosen for the colors of the University of Tokyo, whose Space and Aeronautics Institute developed the project. It remains in space.

1971 February 26 A Soviet SL-4 put the 15,995-lb. T2K test version of the LK lunar landing spacecraft, *Cosmos 398*, in Earth orbit. Second of three LK test missions, it was put through a series of maneuvers totalling 5,160 ft./sec., simulating the way a lone Soviet cosmonaut was intended to land on the moon and return to the orbiting LOK (Lunova Orbitalny Korably) in moon orbit. In two burns of the main propulsion system, the T2K was put through simulation of landing and ascent burns, leaving it in an orbit of 126 × 12,368 mi. with an inclination of 51.6°.

1971 March 1 The NASA Marshall Space Flight Center issued a request for industry to bid on a Phase C/D detailed definition and fabrication of the Space Shuttle Main Engine (SSME). The request was sent to Aerojet General, Pratt and Whitney and Rocketdyne—the companies that had been working on the SSME Phase B studies. The throttleable, reusable cryogenic propulsion system was the heart of the Shuttle's capability. NASA considered it a pacing element in the program which, if delayed, could postpone the booster and orbiter programs commensurately. The traffic schedule accompanying the request for bids predicted 4 flights in the first year of operation, followed by annual totals of 6, 15, 20, 30, 40, 50, 60, 70, 75 and 75 Shuttle flights in successive years.

NASA Headquarters released the document *The NASA Program: A Forecast for the Seventies*, which laid down a baseline of activities more realistically aligned with projected budgets. Gone was the concept of an early attempt to send men to Mars, gone even the moon bases and stations. Instead, by the end of the coming decade, NASA expected to have the Shuttle, an orbital transfer stage (Centaur or Agena) for moving satellites to different orbits and a sortie module, which NASA wanted to build to use inside the Shuttle cargo bay as a minilaboratory. For the 1980s the agency projected a permanently manned space station and a space tug with which to propel satellites to synchronous orbit.

Dale Myers, NASA's associate administrator for Manned Space Flight, reported to the chairman of the Space Science Board, Charles H. Townes, his findings regarding the use of scientist-astronauts on an Apollo moon mission. Amid deep concern within the scientific fraternity that scientists were being overlooked, NASA's Associate Administrator for Space Science Homer Newell had spent several hours in Houston talking to the scientist-astronauts about their frustrations at not being assigned a crew place. Myers reported that, in discussions with Manned Spacecraft Center Director Robert Gilruth, he had agreed to assign Harrison "Jack" Schmitt, to *Apollo 17*, the last moon mission.

China launched its second satellite on a CZ-1 from Jiuquan to a 165 × 1,128 mi. orbit inclined 69.9° to the equator. Called *SKW-2 (Space Physics Experiment-2)*, it carried solar cells and several items of scientific equipment. The ascent was unintentionally witnessed by a U.S. Navy pilot off the coast of Vietnam returning to his aircraft carrier. *SKW-2* continued to transmit until it decayed back through the atmosphere on June 17, 1979. Utilizing the CZ-2 (Long March 2), the next three attempts to launch recoverable observation satellites, on September 18, 1973, July 12, 1974, and November 5, 1974, failed.

NASA announced the flight crew for the *Apollo 16* lunar landing mission. The primary crew would comprise John Young, Thomas Mattingly and Charles Duke. The backup crew would comprise Fred Haise, Stuart Roosa and Edgar Mitchell. By this date there was almost no chance of the backup crew flying a lunar mission. With only three more moon flights left, their only opportunity would come if the primary crew were disabled in some way. *Apollo 16* was at this time scheduled for March 1972.

In testimony before the U.S. House of Representatives Committee on Appropriations, NASA explained that from a peak of more than 400,000 people working on civilian space programs in 1966, there would be only 108,000 by the middle of 1971, and just under 100,000 by the end of the year. Much hope was pinned on the proposed shuttle and space station programs. No other manned flight programs had been approved, and Skylab would be completed in 2½ years. As it was, there would be a hiatus of several years before these new vehicles would be fully operational.

1971 March 5 NASA announced that it had awarded Phase B extension contracts to North American Rockwell and McDonnell Douglas for studies of a modular space station launched by the Shuttle. Effective February 1, 1970, the 10-month study was to concentrate on a station which would include a general purpose laboratory module and a crew quarters module with command and control area. Expanded versions would bring additional crew facilities and experiments in life sciences, astronomy, Earth resources and physics.

1971 March 12 Grumman presented NASA with a review of its eight-month alternate Shuttle Phase A concepts study showing clearly how external hydrogen drop-tanks tanks for the orbiter produced a "lighter, simpler, cheaper" vehicle. Gross liftoff weight was cut from 3.6 million lb. to 2.8 million lb. and projected development costs had been cut from $8.37 billion to $7.37 billion. Because external tanks would have to be replaced for each flight, operational (flight) costs had risen, but not by an amount that would offset the savings made in development.

1971 March 13 The first long-tank thrust-augmented Delta with six Castor strap-ons and the 10,300-lb. thrust TE-364-3 third stage was launched from Cape Canaveral, Fla., carrying the 635-lb. *Explorer 43 (IMP-I)* to a 146 × 122,146 mi. orbit inclined 28.75°. Designated Delta M-6, the launcher could place 1,000 lb. in geosynchronous transfer orbit or 2,850 lb. in low Earth orbit. *IMP-I* was the first in a new generation of interplanetary monitoring platforms from the NASA Goddard Space Flight Center. Comprising a 16-sided drum, it was 6 ft. tall and 4 ft. 6 in. in diameter, with three solar arrays producing 110 W. Two 12-ft. booms carried instruments, with two 5-ft. booms for stability. Twelve science instruments weighing a total 215 lb. would provide data on solar and galactic cosmic rays, electrical fields and energetic particles.

1971 March 19 A new satellite inspection technique was demonstrated for the first time by the Soviet military, beginning with the launch from Plesetsk of a 1,653-lb. target, designated *Cosmos 400,* on a converted SS-5 Skean. Placed in a 609 × 625 mi. orbit inclined 65.8°, it became the target for the 7,320-lb. *Cosmos 404,* launched April 4. From an intermediate path it moved to a 498 × 628 mi. orbit and conducted a flyby at the relatively slow rate of 390 MPH. *Cosmos 404* was maneuvered to a 105 × 496 mi. orbit and de-orbited later the same day.

1971 March 21   A U.S. Air Force Titan IIIB Ascent-Agena D launched the first of six 1,559-lb. Jumpseat satellites from Vandenberg AFB, Calif., to an elliptical orbit of 12,232 × 15,391 mi. inclined 9.9° to the equator. The Ascent-Agena D differed from the conventional Titan IIIB-Agena D in having a full-diameter shroud of 10 ft. × 58 ft. 2 in., 18 ft. 2 in. of which encapsulated the Agena D and 40 ft. the payload. Designed as communications intelligence (comint) satellites, Jumpseats were built for the U.S. National Security Agency to intercept signals transmitted to Soviet Molniya satellites. Launched on August 21, 1973, the second Jumpseat went into an orbit of 286 × 24,418 mi. inclined 63.3°, characteristics that would be typical of the remaining four Jumpseat satellites. The last was launched on February 8, 1985.

1971 March 31   More than seven years after it was announced by President Lyndon B. Johnson, the first operational patrol with the Poseidon C3 SLBM (submarine-launched ballistic missile) began from Charleston, S.C., when the USS *James Madison* put to sea with 16 rounds on board. Considerable improvements had been made in the ability of the missile's warhead to deliver nuclear charges in the threat of antiballistic missile (ABM) defenses. Soviet ABM deployments had worried the United States because of the implied vulnerability of its warheads. The first Poseidon patrol ended at Holy Loch, Scotland, on May 10, 1971, when the *Madison* pulled alongside the tender USS *Canopus.*

1971 April 1   The Canadian *ISIS 2* (International Satellite for Ionoshperic Studies) satellite was launched by a Delta E from Vandenberg AFB, Calif., to a 843 × 888 mi., 88.1° orbit. The 582-lb. satellite consisted of an eight-sided sphere, 4 ft. tall and 4 ft. 2 in. in diameter, with two cross-dipole antennas, one 240 ft. long and one 61 ft. 6 in. long. As a joint U.S.-Canadian effort to study the ionosphere, it was the third satellite in a series of three, the others being *Alouette II* and *ISIS 1. ISIS 2* carried 12 ionospheric experiments, eight from Canada and four from the United States. It remains in orbit.

1971 April 12   NASA selected a team led by General Dynamics to conduct a Phase B definition study of the research and application module (RAM) concept for use with the space station. The work was to last 15 months and provide NASA with preliminary design, operational analysis and development mock-ups of critical sections of a RAM for engineering analysis. Teamed with General Dynamics were North American Rockwell, TRW and Bendix in the United States, SAAB of Sweden, MBB (Messerschmitt-Bölkow-Blohm) and ERNO of Germany, Selenia of Italy and MATRA of France. Four RAM applications were being studied: a support module for small experiments and control equipment, a pallet for instruments exposed to the vacuum of space, a pressurized general purpose laboratory and an automated free-flying module.

1971 April 15   The French D-2A *Tournesol* satellite was placed in a 284 × 433 mi. orbit with an inclination of 46.37°. Launched from Kourou, French Guiana, on a Diamant B, the

200-lb. satellite was equipped with solar and geocoronal measuring equipment, and although it failed to reach its planned orbit, 27 months of data was returned, after which it went silent. The D-2A platform was an improvement on the D-1 series used for previous Diamant satellites, with active cold-gas nitrogen stabilization, a payload spin rate of 180 RPM and a system of sun sensors developed by Space Technology Laboratories in the United States. *Tournesol* decayed on January 28, 1980.

1971 April 19   At 01:39 UT a Soviet SL-13 lifted away from the Baikonur cosmodrome carrying the first Salyut space station into a 124 × 138 mi. orbit inclined 51.5° to the equator. Weighing 40,800 lb., *Salyut 1* took the form of three cylinders stacked one above the other, each with a greater diameter than the one below, with a total length of about 47 ft. 6 in.; the maximum diameter was 13 ft. 7 in. The station comprised a transfer compartment and two work compartments. The cylindrical transfer compartment had a length of 9 ft. 10 in. and a diameter of 6 ft. 6 in. and could double as an airlock. A Soyuz spacecraft could dock at one end and a hatch at the opposite end allowed passage into the first of two compartments, which had a length of 12 ft. 6 in. and a diameter of 9 ft. 6 in. and was equipped as a living area. It contained a table for eating, replenishable water tanks, food refrigerators, a small library of books, games, sanitary and hygiene units and sleeping bags. At the opposite end of the rest compartment was the main work compartment. It had a length of 13 ft. 6 in. and a diameter of 13 ft. 7 in. A 4 ft. long conical section provided an adapter between rest and work compartments and contained equipment for medical monitoring of the cosmonauts and for physical exercise. At the rear of the station, inaccessible to the crew, was the propulsion module. With a length of 7 ft. 1 in. and a diameter of 7 ft. 3 in., it had a KTDU-66 engine, essentially a modified Soyuz KTDU-35. The module also carried attitude control thrusters. Two sets of solar cell arrays, one on the transfer compartment and one on the propulsion module, provided electrical power from a total surface area of 410 sq. ft. Windows were provided for earth views and for visual

1971 April 20   NASA signed an agreement with the Canadian Department of Communications for development and launch of an experimental communications technology satellite, CTS. With 200-W traveling wave tubes, the satellite was to be based on technology developed for the ATS-6 satellite. With transmitting power levels up to 20 times greater than then current commercial communication satellites, CTS would pioneer the use of small ground antennas for use in remote areas.

1971 April 22   At 23:54 UT the 15,000-lb. *Soyuz 10* spacecraft carrying cosmonauts Vladimir A. Shatalov, Alexei S. Yeliseyev, and Nikolai N. Rukavishnikov, was launched from Baikonur on a rendezvous and docking mission to the *Salyut 1* space station. This was the first flight of the ferry version of Soyuz, incorporating a docking system which would permit cosmonauts to transfer from one vehicle to another without going outside. Also, the propellant load was reduced to 1,100 lb. The initial orbit of 129 × 153 mi. was too high, and at an elapsed time of 11 hr. 1 min. a series of automatic maneuvers began which brought *Soyuz 10* to within 600 ft. of *Salyut 1.* At this time the station was in an orbit of 114 × 149 mi. and at 01:47 UT on April 24 Shatalov docked his spacecraft to *Salyut 1.* Intending to remain aboard the station for up to four weeks, the crew attempted to equalize the atmospheric pressure between Soyuz and the station. They were unsuccessful and undocked after 2 hr. 31 min., redocking at 05:47 UT the same day. They remained attached for 1 hr. 30 min. and undocked at 07:17 UT, returning to Earth for the first night landing carried out by either U.S. or Soviet crew members. The mission had lasted 1 day 23 hr. 46 min. To prevent it decaying back down into Earth's atmosphere, the propulsion module on *Salyut 1* was fired on April 26, 1971, placing the station in an orbit of 157 × 171.5 mi.

1971 April 24 A Scout launch vehicle carried the 360-lb. Italian *San Marco 3* satellite from the San Marco platform in the Indian Ocean off Kenya to a 138 × 446 mi., 3.2° orbit. A sphere 2 ft. 6 in. in diameter, with solar cells for power, the structure comprised a floating inner structure attached to the outer shell by triaxial flexing members. Drag caused by atmospheric density at perigee translated into motion by the inner flexing members, providing data by which scientists could calculate the magnitude of drag at different altitudes. Also measured was the temperature and distribution of molecular nitrogen, and the neutral composition of the atmosphere. *San Marco 3* decayed on November 29, 1971.

1971 April 26 NASA Acting Administrator George Low announced that Apollo astronauts returning from the moon would no longer be quarantined. The Interagency Committee on Back Contamination would continue to operate at least through the final lunar landing mission, but it did not believe there was any reason to continue with the isolation of the astronauts and their samples. No trace of bacteria had been found and there were no grounds for believing that any life-form existed on the moon. The crew and their samples would, however, continue to be examined.

1971 April 27 James B. Fletcher was sworn in as the fourth NASA administrator, taking over from Acting Administrator George Low, who had been holding the reins since Thomas Paine resigned on September 15, 1970. Fletcher had a background in industry, having worked for the Ramo-Wooldridge Corp.'s Guided Missile Research Division before it became Space Technology Laboratories, and was the first president of Space Electronics Corp., which developed the Able Star upper stage. He came to NASA after seven years as president of the University of Utah. Fletcher would serve the agency for six years until he retired on May 1, 1977, and become the only administrator to serve twice, being called back in May 12, 1986.

1971 April The results of a feasibility study on an orbiting lunar station (OLS) were presented to NASA by North American Rockwell. Capable of supporting crews of eight on rotation for a five-year mission, the OLS was planned to operate from a 60-mi. polar lunar orbit for controlling and coordinating surface activities and as a support base for mapping, data processing and lunar satellites. North American Rockwell speculated that a reusable nuclear cislunar shuttle would be the mode of propelling the station, 12 ft. in diameter, from Earth orbit to the moon. In May North American Rockwell presented its Lunar Base Synthesis Study to NASA and for the first time identified the Shuttle, rather than a Saturn V, as the launch vehicle to initial Earth orbit.

1971 May 5 The first successful U.S. geostationary early warning satellite, the second Program 647 flight, was placed in orbit over the Indian Ocean by a Titan IIIC launched from Cape Canaveral, Fla. After the second satellite of this type was launched on March 1, 1972, to a position over Panama,

they were so successful that the system was declared operational. Although these flights were supposed to be only a Phase 1 test network, they were handed over to Air Defense Command in 1972 as watchdogs for surprise attack from Soviet submarines.

1971 May 9 At 01:11:02 UT an Atlas-Centaur AC-24 was launched from Complex 36A at Cape Canaveral, Fla., carrying the *Mariner I* spacecraft, the first of two Mariner-Mars 1971 simultaneous orbiter missions to the planet Mars. The uprated Atlas engines for this flight developed a liftoff thrust of 403,000 lb. and the total stack weighed 327,000 lb., 2,273 lb. of which was the spacecraft. Flight events went according to plan until the Centaur main engines started at an elapsed time of 4 min. 25 sec., when the stage began tumbling. A failure in the gyroscopic flight control system caused the engines to receive incorrect guidance signals, and at an elapsed time of 6 min. 5 sec. the Centaur stage shut down prematurely. The stage and its payload fell into the Atlantic Ocean at an elapsed time of 10 min., 900 mi. downrange from Cape Canaveral.

1971 May 10 A Soviet SL-12 launch vehicle, known as the 8K82, successfully placed a planetary spacecraft weighing 8,500 lb. in an Earth parking orbit of 90 × 99 mi. prior to reigniting the terminal stage for an Earth escape trajectory. Mission planners had prepared a specially lightened version of their Mars orbiter/lander combinations, two of which would be successfully launched at the 1971 planetary window and designated *Mars 2* and *Mars 3*. Political pressure to put a spacecraft in Mars orbit ahead of the Americans had been strong, but Soviet scientists calculated from the published mission plans for Mariner-Mars 1971 that the NASA spacecraft would get there first. Accordingly, they off-loaded the lander from the first spacecraft so that the launch vehicle would provide sufficient velocity to put it in Mars orbit ahead of *Mariner 8* or *9*. Unfortunately, the timer which commanded reignition of the SL-12 terminal stage had been incorrectly set and the stage failed to fire, leaving what was now designated *Cosmos 419* stranded in an orbit from which it decayed two days later.

1971 May 19 A 10,255-lb. Soviet spacecraft called *Mars 2* was placed on course for the planet Mars by an SL-12 launch vehicle from Tyuratam. The spacecraft was equipped with two 520-mm focal-length cameras with wide- and narrow-angle lenses, an infrared radiometer for surface temperature distribution, a water vapor detector and an instrument to measure surface relief by determining the levels of carbon dioxide along a sighting line. The 1,400-lb. lander was encapsulated in an aeroshell to protect it from the heat of atmospheric entry until slowed by drag. It would be separated during the approach to Mars, leaving the spacecraft to fire its retromotor and enter Mars orbit. A second spacecraft designated *Mars 3* and weighing 10,238 lb. was sent on its way to Mars by an SL-12 launched May 28.

1971 May 30 At 22:23:04 UT Atlas-Centaur AC-23 successfully launched *Mariner 9* on a direct ascent trajectory which, uncorrected, would bring the spacecraft to within 16,000 mi. of Mars. *Mariner 9* comprised a modified *Mariner 6/7* octagonal spaceframe to which were attached four 7 ft. by 2 ft. 11.5 in. panels carrying a total of 14,742 solar cells producing 800 W at Earth and 500 W at Mars. The panels had a surface area of 82.8 sq . ft. and a span of 22 ft. 7.5 in. A circular, 3 ft. 4 in. in diameter high-gain antenna was set on top and to one side of the octagonal spaceframe, and a low-gain tubular antenna with a 4-in. diameter extended 4 ft. 9 in. A horn-shaped medium-gain antenna was mounted on a solar panel outrigger. Attitude control was maintained via two sets of six cold gas jet thrusters drawing upon a 5.4-lb. supply of nitrogen gas contained in two tanks. Information could be transmitted to Earth at eight selectable speeds from 8.33 bits/sec. to 16.2 kilobits/sec.

1971 May 31 General Dynamics' Convair Division issued its Operational Site Evaluation Trade Study on potential launch sites for the Shuttle. Candidates included several U.S. Air Force bases (Clinton-Sherman, Holloman, Wendover Auxiliary and Edwards) as well as the Western Test Range (Vandenberg AFB) and the Kennedy Space Center. The study concluded that the Kennedy Space Center should be the primary launch and landing site for easterly missions and that the Western Test Range should be the launch and recovery site for polar orbit missions.

The financial analysts Mathematica, Inc. completed a report for NASA entitled *Economic Analysis of New Space Transportation Systems*. It clearly indicated that a reusable shuttle was not a less expensive system than expendables. Mathematica estimated it would cost $9.92 billion to develop a shuttle and tug but that the low cost of operating a complex reusable shuttle could never offset the cost of development and the low run-on costs of expendable launchers. Even at a prodigious flight rate of 736 missions over 13 years, Shuttle flight costs would be $18 billion, versus $15 billion for the same number of expendable launchers. Only when the costs associated with building and maintaining satellites and pay-loads were included did the shuttle/tug system show lower program costs than an equivalent space program with expendable launch vehicles.

1971 May NASA's Manned Spacecraft Center released general arrangement drawings of its 020 series orbiter design, the first from MSC to feature a centerline belly tank for liquid hydrogen and the last to show internal tanks for the liquid oxygen propellant. During the month MSC released drawings of orbiter concept 021. With all main engine propellants in a single centerline belly tank, it was the first concept to show the definitive configuration. Instead of a reusable booster, concept 021 used a single solid propellant rocket motor beneath the external propellant tank.

1971 June 4 Lockheed presented NASA with the findings of its Alternate Space Shuttle Concepts study and demonstrated that its 1½-stage LS 200–10 configuration, utilizing overwing drop-tanks, was economical and viable. Descended from its STAR Clipper concept, the LS 200–10 configuration dispensed with a booster and incorporated nine 248,000-lb. thrust engines in a blended delta wing lifting body. With two tapered cylindrical propellant tanks converging at the nose, the vehicle had a liftoff weight of 3.67 million lb.

1971 June 5 At 00:22:00 UT a 5.1-sec. (22 ft./sec.) course correction burn was performed by *Mariner 9* en route to Mars, shifting the fly-by point from a distance of 16,000 mi. to less than 800 mi. from the surface. The propulsion system comprised two spherical titanium tanks, each 2 ft. 6 in. in diameter, which dominated the top of the octagonal spaceframe. These tanks held 1,050 lb. of monomethylhydrazine and nitrogen tetroxide propellants in teflon bladders, expelled by a quantity of 31 lb. of nitrogen gas from two tanks. The 300-lb. thrust rocket motor mounted vertically upward be-

tween the propellant tanks was designed for putting *Mariner 9* in Mars orbit but also for performing up to two course corrections and two in-orbit trim burns. It was capable of a total velocity change of 5,550 ft./sec. in up to five separate maneuvers. Because *Mariner 8* had been destroyed in a failed launch attempt, *Mariner 9*'s mission would be a hybrid of objectives originally assigned to the separate spacecraft.

1971 June 6 Soviet cosmonauts Georgi T. Dobrovolsky, Vladimir N. Volkov and Viktor I. Patsayev were launched aboard the 14,970-lb. *Soyuz 11* spacecraft from Baikonur at 04:55 UT. At one stage Alexei Leonov had been selected as the mission commander. The SL-4 launch vehicle placed the spacecraft initially in a higher orbit, but this was corrected to 115 × 135 mi. and after several automatic rendezvous maneuvers the spacecraft docked with *Salyut 1* at 07:49 UT June 7. With the docked assembly in an orbit of 132 × 155 mi., Patsayev became the first cosmonaut to enter the space station, closely followed by the other crew members. Several days were spent activating the station's on-board systems and checking them out for routine operations. Several orbit adjustments using *Salyut 1*'s propulsion module raised the orbit to 161 × 175 mi. Medical experiments were conducted, scientific equipment aboard the station was used for astronomical and Earth observations and exercise routines were performed. On June 14 the crew exceeded the manned flight record of *Soyuz 9* and on June 26 they photographed the dramatic failure of the giant N-1 launcher after launch from its Baikonur pad. On June 29 the *Salyut 1* crew transferred back to *Soyuz 11* at 18:15 UT and later undocked from the station. The retrofire burn was completed at 22:35 UT and the orbital module and the instrument module then separated, leaving the pressurized descent module to return the crew to Earth. The physical shock of separating the modules inadvertently opened a pressure relief valve, and air in the descent module leaked away to space. All three crew members were dead when the descent module touched down at 23:17 UT, June 29, 1971, a mission duration of 23 days 18 hr. 22 min.

1971 June 15 The first Titan IIID heavyweight launch vehicle successfully placed in orbit the first of the fourth-generation series of U.S. area-surveillance reconnaissance satellites called KH-9 and more popularly known as Big Bird. Titan IIID was essentially an uprated two-stage Titan core launch vehicle with two solid propellant strap-on boosters similar to those used on Titan IIIC but with a combined liftoff thrust of 2.307 million lb. The flight sequence was almost identical to that of the Titan IIIC but without the third stage. The 29,300-lb. KH-9 was a dramatic improvement on its predecessor, the 4,400-lb. KH-7. Built by Lockheed, like all previous Keyhole satellites KH-9 carried a Cassegrain telescope from Perkin-Elmer with a 6-ft. primary mirror and a folded focal length of 20 ft. The focused light could be directed via mirrors to a number of optical, infrared, multi-spectral and thematic mapper sensors behind the primary mirror. Up to six ejectable capsules returned data to Earth. In addition, Kodak supplied a double-lens camera for high-resolution and wide-area low-resolution views. Images were scanned by a TV camera and stored on a recorder, transmitted via a relay satellite or broadcast direct to one of seven ground receiving locations. A total of 20 KH-9s were orbited between this date and June 25, 1984. The 21st, and last, was destroyed when its Titan 34D exploded 8.5 sec. after liftoff on April 18, 1986.

1971 June 16 NASA Administrator James Fletcher informed leading Senators that "a two-stage delta-wing reusable system in which the orbiter has external propellant tanks that can be jettisoned" was the preferred Shuttle configuration. He also pointed out that NASA was deliberating the option of developing the reusable orbiter first with an interim booster, followed by final development of a fully reusable booster. With annual budgets fixed at around $3.2 billion, a fully reusable two-stage Shuttle would cost around $10 billion with peak annual funding at $2 billion. Given that NASA had administrative and other program costs of $2 billion, the two-stage fly-back system would be unaffordable.

1971 June 20　The NASA Ames Research Center launched a four-stage solid propellant Scout rocket to test at high atmospheric entry speed a Planetary Atmosphere Experiment Test (PAET) vehicle. The PAET had been designed to show how a suitably designed entry body could retain its shape and its trajectory while scooping up constituent parts of the atmosphere for analysis. Engineers at Ames had worked up the PAET vehicle after measuring the flight characteristics of bodies entering atmospheres at up to 31,000 MPH. This worked demonstrated a capability at Ames to participate in planetary exploration programs and was important in having the center take on such activity in January 1972.

1971 June 25　North American Rockwell submitted the final report of its Shuttle Phase B definition study. The central focus of the study had been a two-stage, fully reusable fly-back system with optional high or low cross-range orbiters, but the latter had been made redundant by NASA's decision to opt for a high cross-range Shuttle. The delta-wing NR-161C orbiter had a length of 206 ft. 2 in. with a span of 107 ft. and a liftoff weight of 859,104 lb. Propulsion comprised two 550,000-lb. thrust cryogenic rocket motors paired in the tail and the vertical stabilizer provided a fin tip height of 60 ft. 10 in. The delta-wing/delta-canard Convair B9U booster had a length of 269 ft., a wingspan of 143 ft. 6 in. and a fin tip height of 80 ft. Twelve 550,000-lb. engines in the rear and twelve deployable turbofan engines in the undersurfaces provided propulsion for ascent and fly-back. Total system height in the mated piggyback configuration was 290 ft. with a liftoff weight of 5,047,327 lb.

1971 June 26　At 23:15:07 UT the third N-1 lunar landing rocket was launched from the second of two specially prepared launch pads at the Baikonur cosmodrome. Just 39 sec. into the flight the guidance system malfunctioned and the 370 ft. tall rocket began unexpected rotation around the vertical, causing the structure connecting the second and third stages to break 8 sec. later at a height of 18,000 ft. The upper section fell off and collapsed in a series of massive explosions as propellant erupted. The first two stages, meanwhile, continued to ascend until they too toppled and crashed to Earth 8 mi. away, creating a hole 40 ft. deep and 100 ft. across. The mission objective had been to carry an unmanned L-2 payload for a lunar orbit mission demonstrating all phases of the mission short of a landing. A lander mock-up was, however, carried within the payload shroud.

1971 June 29　The U.S. Senate cut by two-thirds the $30 million start-up money requested by NASA for the so-called Grand Tour mission involving three spacecraft launched to the outer planets in 1977 and 1979. On July 21, 1971, the Conference Committee restored a further $10 million, but poor congressional support made it impossible to proceed with the mission as conceived. In NASA's ambitious plan, three 1,445-lb. spacecraft, each powered by radioisotope thermo-

electric generators (RTGs) delivering up to 439 W of electrical energy, would fly by Jupiter, Saturn, Uranus, Neptune and Pluto. Carrying advanced STAR (self-test and -repair) computers able to operate largely independent of Earth, the Mariner-class spacecraft would be stabilized in three axes by gyroscopes and hydrazine thrusters. NASA planned a 1975 test flight for STAR components.

1971 June 30　McDonnell Douglas submitted the final report of its Shuttle Phase B study. The proposed high cross-range orbiter was less streamlined than its North American Rockwell equivalent and had a length of 174 ft. 9 in., a wingspan of 107 ft. 6 in. and a height of 59 ft. 8 in. from the base of the fuselage to the tip of the vertical tail fin. Two 550,000-lb. thrust engines were placed one above the other in the tail. The Martin Marietta booster carried a rear-mounted wing with upward swept tips, a length of 270 ft. 2.5 in., a span of 166 ft. and a liftoff weight of 3.764 million lb. Propulsion comprised twelve 550,000-lb. thrust cryogenic rocket motors in the tail and ten turbofan engines in thick canard foreplanes. When mated, the orbiter did not protrude beyond the nose of the booster so the stacked height was no greater than the length of the booster.

1971 June　North American Rockwell and McDonnell Douglas presented NASA with their separate findings on the supplementary contracts to study Shuttle configurations utilizing external hydrogen tanks for the orbiter. Although primarily concerned with fully reusable systems on Phase B studies, the two companies were also involved in applying this concept to their more detailed analyses. When external hydrogen tanks were incorporated, the North American Rockwell orbiter had a length of 181 ft. 8 in., a wingspan of 109 ft. 5 in. The McDonnell Douglas external tank orbiter had a length of 175 ft. and a wingspan of 115 ft. McDonnell Douglas maintained that an external tank concept would cut total Shuttle development costs by $1 billion.

1971 July 1　NASA issued Space Shuttle Phase B extension contracts to previous Phase B contractors North American Rockwell and McDonnell Douglas, and to Alternate Shuttle Concept Phase A contractors Grumman Aircraft and Lockheed Aircraft. With each valued at $2.8 million, the contracts were to run for four months, expiring on October 31, 1971. The official reason for the extensions was "to evaluate a phased approach . . . in which the orbiter vehicle would be developed first and initially tested with an interim expendable booster" such as the Titan III or Saturn IB. It also included a requirement to study orbiters carrying all the main engine propellant in external tanks, furthering reductions in size and cost.

1971 July 6　The Grumman/Boeing team working Phase A Alternate Shuttle Concept Phase A studies issued its final report to NASA. From what had begun as an evaluation of 29 alternative design concepts, the team focused on two configurations: the G-3 with integral propellant tanks and 550,000-lb. thrust engines and the H-33 with external orbiter hydrogen tanks and optimized 415,000-lb. thrust engines. The delta-wing G-3 orbiter had a length of 173 ft., a wingspan of 98 ft. 5 in. and three rocket motors. The G-3 straight-wing booster had a length of 296 ft. and a wingspan of 221 ft. 4 in. and was powered by thirteen motors. Liftoff weight was 5,240,920 lb. The H-33 orbiter had three rocket motors, a length of 157 ft. and a wingspan of 97 ft. The H-33 booster had twelve motors, a length of 245 ft. and a wingspan of 177 ft. 6 in.

1971 July 8 A joint endeavor between the U.S. Navy and NASA, the 260-lb. satellite *Explorer 44* was launched by a four-stage Scout from Wallops Island, Va., to a 269 × 393 mi., 51° orbit. The U.S. Naval Research Laboratory satellite, also designated *Solrad 10*, consisted of a 12-sided structure 2 ft. 6 in. across and 11 ft. 11 in. tall with four symmetrically placed solar cell panels hinged at the center section. *Explorer 44* was instrumented to monitor solar radiation and, on command, stellar radiation from celestial sources. It also served to warn of impending solar flares which could pose danger to Apollo astronauts. *Explorer 44* decayed on December 15, 1979.

1971 July 12 NASA awarded a full-scale development (Phase C/D) contract for the Space Shuttle Main Engine to the Rocketdyne Division of North American Rockwell. The contract was to be managed by the Marshall Space Flight Center. Rocketdyne's winning design was based on a 500,000-lb. thrust staged combustion cycle where oxidizer and fuel were in preburners from where they were fed directly to a 3,000 lb./sq. in. combustion chamber. The turbines would be driven by the preburners. With an operating pressure three times that of a conventional engine and temperatures much higher in the combustion stage, the engine would make exhausting demands on rocket engineers. The engine had to be throttleable, restartable and fly at least 55 missions before replacement.

NASA announced that the Manned Spacecraft Center had been selected to head up the Shuttle program and the Marshall Space Flight Center had been chosen to lead the Space Station and RAM (Research and Applications Module) programs. NASA still expected to develop the Space Station after the Shuttle and envisaged the RAM family of Shuttle-launched modules as common structures carrying experiments or logistical supplies. Some RAMs were expected to remain docked to the Station after being carried to orbit in the Shuttle and others were expected to remain close by as free-flying laboratories.

1971 July 26 The *Apollo 15* mission began at 13:34:00 UT with the launch of AS-510 carrying astronauts David Scott, Alfred Worden and James Irwin on the first of three extended-stay J-series missions. *Apollo 15*'s total liftoff weight of 6.408

Beginning a new, expanded, phase of lunar exploration, Apollo 15 *was launched on July 26, 1971, carrying a lunar roving vehicle built by Boeing, shown here on the moon at the Hadley-Apennine site.*

million lb.; the payload comprised CSM-112 (66,955 lb.) and LM-10 (36,255 lb.), a total mass greater than 100,000 lb. for the first time. Events occurred as planned with extra activities introduced by the modified hardware. At 74 hr. 6 min. 47 sec., about 4 hr. prior to the lunar orbit insertion burn, the crew jettisoned the door covering service module bay no. 1 housing panoramic and mapping cameras in a scientific instrument module (SIM). Dubbed *Falcon*, the lunar module undocked from the CSM at 100 hr. 14 min. and carried Scott and Irwin to a landing near Hadley Rille on the eastern edge of Mare Imbrium at 104 hr. 42 min. 29 sec. A 33 min. 7 sec. stand-up EVA (extravehicular activity) was performed by Scott from the top of the lunar module, the only mission on which this was done. The first EVA lasted 6 hr. 32 min. 42 sec. and brought the first operation of the lunar roving vehicle when the astronauts drove to the rim of Hadley Rille clocking 6.4 mi. They also deployed six ALSEP (Apollo lunar surface experiments package) instruments: passive seismometer, lunar surface magnetometer, solar wind spectrometer, ion gauge, cold cathode gauge and heat flow sensors down holes drilled in the surface during the third EVA. The second EVA lasted 7 hr. 12 min. 14 sec. and took the crew on a geological traverse to the foothills of a lunar feature known as the Apennine Mountains, during which they journeyed 7.7 mi. The third EVA lasted 4 hr. 49 min. 50 sec., during which further geological surveying was conducted and a further 3.2 mi. clocked on the LRV. After 66 hr. 54 min. 53 sec. on the surface, the crew departed and rejoined Worden in the CSM (command/service module), dubbed *Endeavour*. For the first time, the liftoff was televised live from the LRV camera. Docking took place at 173 hr. 35 min. 47 sec. The ascent stage was de-orbited to lunar impact and a small subsatellite was released to lunar orbit from the service module SIM-bay at 222 hr. 39 min. 19 sec. Trans-Earth injection (TEI) brought the CSM out of lunar orbit at 223 hr. 48 min. 45 sec. About 18 hr. later Worden performed a 38 min. 12 sec. EVA out of the command module side hatch to retrieve SIM-bay film canisters. With 169 lb. of samples, the command module splashed down on August 6, 1971, at 259 hr. 11 min. 53 sec.

1971 August 3 The Manned Spacecraft Center Space Shuttle Program Office circulated a compilation of Shuttle concepts and studies including: three North American Rockwell/General Dynamics concepts (two-stage fully reusable with two-engine orbiter, two-stage fully reusable with three-engine orbiter and a 2½-stage system with external hydrogen tanks); three McDonnell Douglas concepts (two-stage fully reusable with two-engine orbiter, a 2½-stage series burn system with external orbiter hydrogen tanks and a 2½-stage parallel burn system also with external hydrogen tanks); two Grumman/Boeing concepts (two-stage fully reusable and 2½-stage series booster system); and two Lockheed concepts (two-stage fully reusable, and a 1½-stage drop tank STAR Clipper derivative). Flight costs per mission had been calculated as $3 million for a fully reusable two-stage system and $4 million for a 1½-stage drop-tank system.

1971 August 7 An Atlas F launched from Vandenberg AFB, Calif., carried *OV1-20* and *OV1-21* to Earth orbit. Part of the Department of Defense Space Experiments Support Program, they were released at a height of 104 mi. *OV1-20* was maneuvered to a 83 × 1,216 mi., 92° orbit from which it released a 2 ft. 2 in., 800 lb., cast brass sphere called *Cannonball 2*. Designated *OAR-901 Low Altitude Density Satellite II*, it had an initial orbit of 83 × 1,224 mi. and made air density measurements until it decayed on January 31, 1972.

OV1–20 carried experiments: SAMSO-132, for measuring trapped protons, and SSD-975B, an electron density detector. *OV1–21* ejected a 135-lb. sphere 1 ft. in diameter and called *Musketball* (or, *OAR-907 Radar Tracking Density Satellite*) to an 85 × 549 mi. orbit, where it would conduct air density measurements. *OV1–21* then deployed to a 484 × 564 mi., 87.6° orbit and ejected a rigid aluminum sphere 2 ft. in diameter and called *Rigid Sphere 1,* and three inflatable 7-ft. spheres called *Balloon (Mylar), Grid Sphere 1,* and *Grid Sphere 2,* collectively designated the AVL-802 experiment. These were to measure the ballistic coefficient of grid spheres. Finally, a Lincoln Calibration Sphere-4 (LCS-4), a hollow aluminum sphere 3 ft. 8 in. in diameter, was released for calibrating total radar systems. *OV1–21* also carried three onboard science experiments.

1971 August 12   A Soviet SL-4 launched into low Earth orbit the 12,600-lb. *Cosmos 434,* the veiled designation for the third and last T2K test version of the LK manned lunar lander. It performed simulated moon mission maneuvers, including the longest T2K burn to date, totalling 5,350 ft./sec. and producing a final orbit of 112 × 7,335 mi. The flight qualified the LK for manned operations, but this was the last piece of Soviet manned lunar hardware tested in space. Shortly after this the Soviet mission to the moon was abandoned for political and economic reasons.

The West German aircraft manufacturer ERNO conducted the first in a series of flight tests of a reentry vehicle named *Bumerang.* Air-dropped from the open rear cargo doors of a Transall C.160 aircraft, the unpowered glider descended in a test of its aerodynamic qualities before a parachute was deployed to lower it gently to the ground. In work undertaken for the Ministry for Education and Science, ERNO had developed Bumerang as a potential shuttle for manned operations. Four more tests were conducted off the island of Sardinia in the Mediterranean, where two Soviet submarines observed the activity. Bumerang was never developed beyond the model test stage.

1971 August 13   NASA announced the crew assignments for the last moon mission, *Apollo 17,* then scheduled for December 1972. The primary crew members comprised Eugene Cernan, Ronald Evans and Harrison Schmitt. Joe Engle had been a part of the *Apollo 14* backup crew with Cernan and Evans, and was expected to be part of the primary crew for *Apollo 17.* Instead, he was removed from the flight list and replaced by Schmitt, who was a trained geologist. The backup crew comprised David Scott, Alfred Worden and James Irwin, none of whom was a scientist by profession.

1971 August 16   Initiating an experiment into atmospheric physics, a NASA Scout launcher carried the 186-lb. French satellite *Eole* from Wallops Island, Va., to a 421 × 564 mi., 50.2° orbit. An octagonal cylinder, 2 ft. 4 in. in diameter and 1 ft. 11 in. in height, it carried eight solar panels in windmill fashion which provided 20 W of power. Extending from the top was a 33 ft. boom for gravity-gradient stabilization. Over the next five months, 500 balloons were set free from Argentina to drift over the southern hemisphere at 39,000 ft. Signals from the balloons were received by *Eole* (named after the god of winds Aeolus) and relayed to ground stations. It remains in orbit.

1971 August 31   In an interim review of progress on Shuttle Phase B extension contracts, North American Rockwell and McDonnell Douglas reported that the preferred option for external tank configurations was to place both hydrogen and oxygen tanks in tandem in a single external tank to the side of which the orbiter would be attached. Other issues included optional boosters with solid propellant, storable propellants and liquid pressure–fed boosters. Storable liquid propellants were considered too toxic. As might have been expected, McDonnell Douglas's team partner Martin Marietta proposed their own Titan III launcher as the interim booster for the Shuttle, while Boeing proposed their S-IC Saturn V first stage as the booster.

1971 August   NASA decided to adopt the concept of Shuttle orbiter hydrogen and oxygen propellants contained in an external tank as the baseline design. The booster concept had by now moved away from the expensive, fully reusable, manned fly-back type to the utilization of existing booster technologies. To save further costs, the concept of phased orbiter development resulted in a Mk.I/Mk.II approach proposed for further analysis. The Mk.I orbiter would use ablative thermal protection, low-cost system technology and achieve a first flight with an interim booster during 1978. The Mk.II would be developed with a recoverable booster for flights beginning in 1983. The idea behind the phased approach was to lower peak-year funding.

1971 September 2   A Soviet SL-12 launch vehicle lifted off from Baikonur at 13:41 UT carrying the 12,400-lb. soft-landing spacecraft, *Luna 18.* From an initial Earth parking orbit, the terminal stage boosted *Luna 18* to the moon. After two course corrections, on September 4 and 6, the spacecraft was placed in a circular orbit of 62 mi., with an inclination of 35° and a period of 1 hr. 59 min., on September 7. After an unusually long period in lunar orbit, *Luna 18* fired its retro-motor to descend to a soft touchdown, but at 07:48 UT on September 11 the spacecraft crashed and nothing more was heard from it. The Soviets indicated that its communication equipment had become disabled when the spacecraft toppled on landing.

1971 September 9   NASA directed Shuttle Phase B extension contractors North American Rockwell, McDonnell Douglas and Grumman to terminate studies on solid propellant boosters for the Shuttle, leaving Lockheed to continue a low-level effort on a concept utilizing solids 156 in. in diameter. This left the hydrocarbon fueled F-1 engine in a Saturn V first stage derivative as a candidate liquid propellant booster. An environmental impact assessment had, however, determined that solid propellant boosters would be an acceptable alternative.

1971 September 14   U.S. Deputy Secretary of Defense David Packard approved the Navy Undersea Long-Range Missile System, which envisaged a completely new, high-technology submarine-launched ballistic missile to replace Polaris A3 and Poseidon during the 1980s. Program objectives sought to double the range of the Poseidon, deploy it in the current range of submarines, improve reliability and constrain operational costs to existing levels. Advanced development began in December 1971 and on May 16, 1972, the program became known as Trident. On December 14 President Nixon granted it Brickbat status, the highest industrial priority for defense projects. Also in 1972 development began of the W76 Mk.4 warhead, with a yield of 90–100 KT in a cluster of up to eight per missile. With full MIRV (multiple independently targeted reentry vehicle) capability, Trident would provide for the U.S.

Navy the same high-accuracy, hard-target kill potential that Minuteman 3 provided the U.S. Air Force.

1971 September 17   The U.S. Air Force Space and Missile Systems Office briefed senior officers on Project SPIKE, a new proposal for a nonnuclear, air-launched antisatellite system. The plan was to equip an F-106 fighter with a modified Standard AGM-78 antiradar missile and terminal homing warhead for a series of tests. The concept evolved from the impending shutdown of the Program 437 nuclear missile antisatellite system and the need to find a viable replacement. SPIKE was never funded.

1971 September 20   A high-altitude cloud of ionized barium formed when a cooperative effort between the United States and the Federal Republic of Germany culminated in the launch of a NASA Scout rocket from Wallops Island, Va. Carried to a height of 20,000 ft. above Central America, the rocket released 6 lb. of barium, which spread out to form a thin ionized cloud approximately 10,000 ft. long. Developed at the Max-Planck Institute in Munich, the experiment was devised to map high-altitude flux lines in the Earth's magnetic field by providing a visual track observable from the ground.

1971 September 28   At 10:00 UT a Soviet SL-12 lifted off from its Baikonur launch site and put *Luna 19* on course for the moon. The 12,300-lb. spacecraft relied heavily on technology from the previous, third-generation Luna-series spacecraft first launched on July 13, 1969, but it had been designed under the designation Ye-8LS for extensive photoreconnaissance of potential manned landing sites. The Ye-8LS utilized the bathtub structure adopted for the Lunokhod moon rovers. On October 3 *Luna 18* went into a near circular, 87-mi. orbit of the moon with a period of 2 hr. 1 min. 45 sec. and an inclination of 40.58°. Three days later this orbit was modified to a circular 79 mi. and a period of exactly 2 hr. 1 min. *Luna 19* carried 19 scientific experiments which it conducted from lunar orbit. These included studies of magnetic fields, cosmic radiation, the solar wind and meteoroids.

Japan successfully launched its third satellite from the Kagoshima Space Center when a four-stage Mu-4S placed the 138-lb. *Shinsei* (New Star) in a 540 × 1,159 mi. orbit inclined 32°. Instrumented to obtain information about cosmic waves and solar radiation, the 4-ft. tall satellite had an octahedral center structure with conical top and bottom, and 5,184 solar cells on the outer body for electrical power. This was the first fully instrumented science satellite built by Japan and it operated perfectly in orbit, where it remains.

1971 September 29   The last of NASA's seven first-generation Orbiting Solar Observatories, *OSO-7*, was launched by Delta N from Cape Canaveral, Fla., to a 201 × 355-mi., 33.1° orbit. With a weight of 1,400 lb., *OSO-7* was twice as heavy as previous OSOs and balance weights replaced mechanical arms with balls. The satellite had two sections: a sail-like, de-spun structure with solar cells and two sun-pointing experiments fixed on the solar disk, and a nine-sided spinning wheel with support systems and four experiments. Spin-stabilized at 30 RPM, the satellite had a height of 6 ft. 7 in. and a diameter of 4 ft. 7 in. The Delta also carried a 45-lb. test and training satellite, *TTS-4*, for calibrating the NASA Manned Space Flight Network of ground tracking stations. *OSO-7* decayed on July 9, 1974; *TTS-4* decayed on September 21, 1978.

1971 October 11   The Soviet *Salyut 1* space station was destroyed in the Earth's atmosphere following a planned retro-burn of its KTDU-66 propulsion system. For some time the station had been in an orbit of 165 × 186 mi., but the cumulative effects of continued orbital decay had lowered it to an orbit of 110 × 113 mi. The de-orbit maneuver brought the station back down through the atmosphere, where it was destroyed through friction and burned up over the Pacific Ocean. *Salyut 1* had been occupied by cosmonauts for more than 22 days of the *Soyuz 11* flight in June 1971.

1971 October 14   A Thor-Burner II carried the first in a series of advanced Block 5 DMSP (Defense Meteorology Satellite Program) military weather satellites from Vandenberg AFB, Calif., to a 495 × 545 mi. orbit inclined 98.9°. Designated Block 5B/C, this series comprised eight satellites that took the form of truncated cones with 12 sides, 11 of which carried solar cells producing 50 W. Each satellite measured 7 ft. in height, with a diameter of 4 ft. 4 in. at the bottom and 3 ft. 7 in. at the top. A visible/infrared line-scan radiometer provided images with a resolution of 2,000 ft. The four other Block 5Bs were launched on March 24 and November 9, 1971, August 17, 1973, and March 16, 1974, respectively. Three Block 5Cs were launched on August 9, 1974, May 24, 1975, and February 19, 1976; due to launch vehicle failure the last was placed in a 56 × 220.5 mi. orbit from which it decayed within hours.

1971 October 17   The U.S. Air Force launched a Thorad-Agena D from Vandenberg AFB, Calif., to a 480 × 499 mi., 92.7° orbit for a series of advanced technology tests. Developed under the Air Force Space Test Program, formerly the Space Experiments Support Program, the ASTEX (advanced space technology) payload comprised two primary experiments aboard the Agena D. Known as ERIS, the Earth-reflecting ionospheric sounder had 19 separate instruments for collecting data on proton, alpha and electron particles entering the upper atmosphere. The largest experiment comprised a 250-lb. rolled-up solar array. When deployed from a cylinder 10 in. in diameter, the unfurled array had a size of 5 ft. 6 in. × 32 ft. and supported 34,000 solar cells producing 1,500 W of electrical power. ASTEX remains in orbit.

1971 October 19   An agreement for cooperation in space research was signed by representatives of the Soviet government and the European Space Research Organization. Although no specific projects or programs were identified, contact between the two spacefaring entities was maintained and extended when the European Space Agency (ESA) was formed. Excepting France, which had a historic association with Russia, other ESA countries (most of them NATO members) were reluctant to develop strong ties. During the 1970s the Soviets expanded areas of cooperation in space programs, striking accords with India and Sweden.

1971 October 27   At a NASA review of options for different development paths in the Shuttle program, Charles J. Donlan outlined a variety of cost-saving configurations. These included phase development of a Mk.I/Mk.II orbiter with the early vehicle using J-2 engines, ablative thermal protection and first-generation systems. Alternate boosters included a liquid oxygen–RP-1 stage with F-1 (Saturn V) first-stage engines, and variously sized solids with diameters of 120 in., 156 in. or 260 in. Also undecided was the configuration of the booster/orbiter arrangement, whether to have booster and orbiter propulsion fire in series, as in a conventional multi-stage rocket, or to put solid boosters on the sides of the external orbiter propellant tank and fire booster and orbiter engines in parallel.

1971 October 28 Britain became the sixth nation to build and launch its own satellite when the fourth Black Arrow carried the 145-lb. *Prospero* (X-3) from Woomera test range, Australia, to a 334 × 990 mi. orbit inclined 82°. The launch took place at 04:09 UT and the first stage carried the upper elements to a height of 28 mi. The second stage burned out at 125 mi. and coasted on up to a height of 348 mi., at which point a 40-sec. burn of the third stage placed *Prospero* in orbit. Built by the British Aircraft Corp. and Marconi, the satellite had a diameter of 3 ft. 8 in., a length of 2 ft. 3 in. and carried a micrometeoroid detector from the University of Birmingham, the only science experiment on board. It remains in orbit.

1971 November 2 NASA authorized extensions of the already extended Shuttle Phase B definition contracts. Under the new six-month extensions, each worth a maximum of $7.2 million, North American Rockwell, McDonnell Douglas, Grumman Aerospace and Lockheed would "provide an initial study effort of four months, November 1, 1971, through February 28, 1972, with two additional one-month options through April 30, 1972." The study effort was to focus on the Mk.I/Mk.II concept, on the reusable fly-back S-IC stage and on evaluation of ballistic recoverable solid and liquid propellant boosters.

1971 November 3 The first two satellites in the U.S. DSCS-II (Defense Satellite Communications System-II) series were

The business end of Britain's Black Arrow satellite launcher which carried Prospero *into orbit on October 28, 1971. Britain is the only country to have developed an indigenous launcher for just one flight, only to abandon it afterwards.*

launched at 03:09 UT on a Titan IIIC from Cape Canaveral to separate geostationary orbits by Transtage. Each satellite in this new, second-generation series built by TRW comprised a spin-stabilized drum with a diameter of 9 ft. and a de-spun antenna section forming a total height of 13 ft. Weight at launch was 1,365 lb., with 520 W of electrical power provided by solar cells around the exterior of the spinning section. Two 20-W transmitters provided a total of 1,300 voice circuits in the 7–8 GHz band. Additional paired Titan IIIC launches occurred on: December 14, 1973; May 12, 1977; March 25, 1978; December 14, 1978; and November 21, 1979. The latter two had 40-W transmitters. A launch on May 20, 1975, left two satellites stranded in a low orbit from which they decayed six days later.

1971 November 3–4 At a major Shuttle program review held at the NASA Manned Spacecraft Center, optional configurations and sets of hardware were evaluated. The original, fully reusable two-stage system would have had a liftoff weight of around 5 million lb. with an orbiter more than 200 ft. in length and a staging velocity of 10,800 ft./sec. External hydrogen tanks reduced the length of the orbiter to about 123 ft., and combined external liquid oxygen–liquid hydrogen tanks further reduced that to 110 ft. Current candidate configurations could be narrowed to four, all of which carried orbiter propellant in an external tank: pump-fed F-1 engines or pressure-fed engines attached to liquid boosters in series-burn with the orbiter; or as an alternative parallel burn concept, twin pressure-fed liquid or twin solid rocket motors 156 in. in diameter. Staging velocities were about 6,000 ft./sec., except for the twin solid configuration where it was 4,400 ft./sec. Total development cost for the original concept was estimated at more than $10 billion, but only $6 billion for the parallel burn solid propellant booster concept.

1971 November 5 In the last launch of a Blue Streak, the Europa II launch vehicle failed to orbit a test satellite from Kourou, French Guiana, on the F.11 mission as planned. At an elapsed time of 1 min. 47 sec. guidance was lost and the vehicle began to diverge from its planned trajectory. At 2 min. 30 sec. the forward section of the Blue Streak first stage buckled as the launcher pitched forward and the Coralie second stage failed to separate, leaving the upper stages to fall back into the Atlantic Ocean about 250 mi. downrange. Europa II was scheduled to launch an 800-lb. Italian satellite into orbit using the Astris 2 third stage and a solid propellant fourth stage. The ground tracking equipment had been provided by Belgium, and the Netherlands contributed station equipment. Changing strategies and Britain's withdrawal from ELDO (European Launcher Development Organization) sounded the death knell for Europa.

1971 November 12 A flurry of Shuttle Phase B extension study reports began when North American Rockwell submitted its findings for the first extension period, followed by McDonnell Douglas, Lockheed and Grumman on November 15. NASA Deputy Administrator George Low issued a summary of their findings on November 22. It was clear that orbiter propellant would have to be carried in an external tank to drastically cut development costs and that the lowest cost would accrue to an Mk.I/Mk.II orbiter using any one of four solid or liquid propellant booster designs in either series or parallel burn. Development costs ranged between $4.5 billion and $6.5 billion, with launch costs varying between $6 million and $12 million per flight. The Office of Management and Budget (OMB) asked NASA to look at a space-glider concept launched by an existing rocket in what it thought might be a low-cost configuration. NASA found it would cost $3 billion to develop and $30 million to fly.

1971 November 14 At 00:17:39 UT the main propulsion system on *Mariner 9* was ignited for a burn lasting 15 min. 15 sec., reducing its speed by 5,251 ft./sec. The first spacecraft placed in orbit had a periapsis (low point) of 869 mi., an apoapsis (high point) of more than 12,000 mi., a period of 12 hr. 34 min. 01 sec. and an inclination of 64.4°. This orbit would allow the science instruments to observe selected areas of the planet under the same lighting conditions every 17 days, a compromise worked up following the loss of *Mariner 8*. *Mariner 9* carried two wide-angle and two narrow-angle cameras identical to those carried by Mariner-Mars 1969. Because *Mariner 9* was closer to the surface, the cameras could resolve objects as small as 3,280 ft. and 328 ft., respectively. *Mariner 9* also carried an ultraviolet spectrometer for atmospheric measurements, an infrared radiometer bore-sighted with the cameras for temperature studies and an infrared interferometer spectrometer to analyze atmospheric composition.

1971 November 15 A four-stage Scout launcher carried the 114-lb. NASA science satellite *Explorer 45* from the San Marco platform off the coast of Africa to a 171 × 16,712 mi., 3.6° orbit. Consisting of a 26-sided polyhedron with a diameter of 2 ft. 3 in., the satellite carried five instrument booms, one 2 ft. 6 in., two 2 ft. and two 9 ft., as well as four 2-ft. antennas. Also called *Small Scientific Satellite-A (SSS-A)*, *Explorer 45* measured aerodynamic heating and radiation damage during launch and electric fields in the magnetosphere from its eccentric orbit. The satellite decayed on January 10, 1992.

1971 November 16 At 02:37:53 UT *Mariner 9* conducted a 6.4-sec. (50 ft./sec.) orbit trim maneuver with its main propulsion system, changing the orbit to a periapsis of 861 mi. and an apoapsis of almost 11,000 mi. and the period to 11 hr. 58 min. 14 sec. This burn was essentially a time-phasing maneuver to adjust the orbit for synchronizing the spacecraft's motion about Mars with tracking stations on Earth. It transpired that the orbit was varying due to anomalies in the Mars geoid in 20-day cycles and the periapsis point was moving toward the beginning of the tracking window with the Goldstone antenna in California, the primary Earth station. For much of this early period in the observation of Mars, *Mariner 9*'s view of the surface was obscured by one of the biggest dust storms ever experienced. It had begun to envelope the planet shortly before *Mariner 9* arrived on station and would persist through January 1972. Nevertheless, the first phase of observation was spent on a planet-wide reconnaissance before full-scale mapping began.

1971 November 24 An SL-6 launched a new generation of Soviet communication satellites when it carried the first Molniya 2 from Plesetsk to an elliptical orbit of 321 × 24,579 mi. inclined 65.4°. The 4,000-lb. Molniya 2 series operated in the 4–6 GHz band with greater load capacity than the Molniya 1 series. Only 16 Molniya 2 satellites were launched, the last on December 2, 1976; they varied in weight according to the communications payload. First- and third-generation Moniyas continued in use, however.

1971 November 27 The Soviet spacecraft *Mars 2* achieved a "first" when its 1,400-lb. lander separated from the main spacecraft and reached the surface of Mars. Communication with the lander ceased during descent and it is not known whether it made a soft landing or crashed. Either way, it rests at 45° S by 58° E. Meanwhile, at 20:19 UT the main body of the spacecraft went into an elliptical 855 × 15,500 mi. orbit inclined 48.9° to Mars' equator and with a period of 18 hr. The *Mars 2* orbiter continued to send back to Earth a limited range of science data and a few pictures of the surface. Each picture was composed of 1,000 picture elements on 1,000 lines for transmission to Earth, where they were reformed on magnetic tape and on paper. The primary mission of *Mars 2* was declared to have been completed on August 22, 1972, after the spacecraft had made 362 orbits of the planet.

NASA officials arrived in Moscow for discussions about a proposed joint Apollo-Soyuz flight. During the 10-day visit, a mission outline was agreed whereby separate manned vehicles would rendezvous and dock in space using a new module attached to the forward end of Apollo. Soyuz utilized an oxygen-nitrogen atmosphere at sea-level pressure, but Apollo utilized an oxygen environment at one-third atmospheric pressure. The docking module would provide an air-lock facility for astronauts passing between vehicles. A target date of mid-1975 was set for the mission. In the ensuing months NASA worked with the White House on a formal announcement. Details about a public announcement on what had, to date, been a confidential proposal were discussed with the Russians when U.S. officials visited Moscow again in April 1972.

1971 November 30 A Joint Technical Experts Group formed of NASA and European Space Conference (ESC) representatives began a three-day meeting in Washington, D.C., to define areas of European cooperation in the U.S. post-Apollo program. The ESC was the unified voice of European spacefaring nations and operated at ministerial level. During the meeting the Europeans discussed the broad range of opportunities, including Shuttle system subcontracts, an orbit-to-orbit tug and laboratory modules such as sortie cans and research and application modules. The ESC representatives stated a clear preference for a tug contract. NASA believed that while funding constraints would prevent it having a tug ready before 1984, the Europeans could have it built by 1980.

1971 November Planetary exploration projects were discontinued at the Goddard Space Flight Center, and in January 1972 the Ames Research Center took over the emerging Venus multiprobe and orbiter project. A study team was quickly organized at Ames and the concept was absorbed by the Pioneer series of spacecraft. Pioneer-Venus, as it was now called, was still relatively undefined, and a Pioneer-Venus Science Steering Group was set up by NASA to focus the work of scientists eager to put experiments on board. The Group proposed two identical spacecraft, each carrying one large probe with parachute and three small free-fall probes, to be launched at the 1976/77 window. An orbiter was to be launched in 1978 followed by a third probe mission in 1980.

1971 December 1 The first French ballistic missile submarine, *Le Redoutable,* became operational with its inventory of 16 M-1 SLBMs submarine-launched ballistic missiles). Under the MSBS (Mer-Sol Balistique Stratégique) program, France was to build four more SNLE (Sous-marin Nucléaire Lance-Engins) submarines: *Le Terrible* (operational December 1, 1973), *Le Foudroyant* (June 6, 1974), *L'Indomptable* (December 31, 1976), and *Le Tonnant* (May 3, 1980). Each had a length of 422 ft., a beam of 34 ft. 10 in., a hull depth of 32 ft. 10 in., a submerged displacement of 8,940 tons and a complement of 135 (15 officers and 120 men). A sixth boat, *L'Inflex-ible,* became operational on April 1, 1985, ensuring that, with two in dock and one in refit, at least three submarines were constantly on patrol.

1971 December 2   The 1,400-lb. Soviet *Mars 3* spacecraft reached the surface of Mars at 45° S by 158° W in a global dust storm. Soviet scientists have made an unsubstantiated claim that the spacecraft sent telemetry from the surface for 20 sec., making it the first spacecraft to send back signals from that planet. Like its predecessor, *Mars 3* carried a small device connected by a cable to the lander. It was to have measured the bearing strength and density of soil as it was moved across the surface on skids. The *Mars 3* orbiter entered an elliptical, 930 × 124,270 mi. orbit of the planet and sent back to Earth 12 pictures of the surface. The images were almost useless because of the obscuring effect of the dust. The primary mission of the *Mars 3* spacecraft ended August 22, 1972.

1971 December 5   The fourth Diamant B launch vehicle failed to carry its payload from Kourou, French Guiana, to orbit when the P-2.2 Topaze second stage malfunctioned at 20 sec. into its burn, dropping its payload in the Atlantic Ocean. Launched from Kourou, French Guiana, the 214-lb. D-2A *Polaire* satellite was to have continued the work begun by *Tournesol* and represented the second use of the modified D-2 satellite bus. It was instrumented to conduct scientific study of the distribution of hydrogen and helium.

1971 December 11   *Ariel 4*, Britain's fourth satellite in a joint U.S./United Kingdom investigation into the interaction between charged particles, plasma and the topside atmosphere, was launched by a four-stage Scout from Vandenberg AFB, Calif., to a 296 × 368.5 mi., 83° orbit. The 220-lb. satellite, also known as *UK-4,* comprised a cylinder 3 ft. high and 2 ft. 6 in. in diameter, with two experiment dipole antennas and four telemetry antennas mounted on top and four solar cell paddles inclined outwards from the base. Project management was under the United Kingdom Science Research Council, and British Aircraft Corp. built the satellite. Experiments were provided by the Universities of Birmingham, Manchester, Sheffield and Iowa. *Ariel 4* de-orbited on December 12, 1978.

1971 December 14   Identified anonymously as 1971–110A, the world's first ocean surveillance satellite cluster was launched by the U.S. Air Force for the Navy on a long-tank thrust-augmented Thor-Agena D from Vandenberg AFB, Calif. The 1,550-lb. orbited payload comprised three separate satellites, all of which occupied virtually the same 611 × 621 mi. orbit inclined 70°. The Navy had been testing ocean surveillance sensors since early 1971 on high-resolution KH-8 photoreconnaissance satellites launched by Titan IIIB, and would continue to do so for several years. Not until April 30, 1976, would the Navy launch its first operational ocean surveillance satellites.

1971 December 15–16   The council of the European Launcher Development Organization (ELDO) agreed to fund further development of the four-stage Europa II satellite launcher and to begin development on Europa III. After the failure of Europa II F.11 on November 5, the council endorsed a plan to fly F.12 and F.13 as development rounds in mid-1972 and March 1973, respectively, and to fly the first operational mission, F.14, carrying the first Symphonie communication satellite, in October 1973. On February 17 ELDO

postponed F.12 to the "first half of 1973" to allow time for complete analysis of the F.11 failure.

1971 December 20   The council of the European Space Research Organization (ESRO) unanimously agreed to join with the United States in the development of an Aerosat (aeronautical satellite) system. Under the terms proposed and agreed to by both entities, the system of communication satellites for aircraft would be owned on a fifty-fifty basis. Over the next several months the Comsat Corp. in the United States strongly opposed such an arrangement, proposing a system more like the Intelsat telecommunications consortium whereby ownership would be based on usage. Gradually, support for Aerosat withered and it was never developed.

1971 December 27   A French payload was launched aboard the Soviet satellite *Oreol* (the French called it Auréole) on an SL-8 from Plesetsk to a 255 × 1,553 mi. orbit with an inclination of 74°. Designed to study the polar regions of the upper atmosphere, the 1,390-lb. satellite was the first French payload launched aboard a Soviet rocket. Beginning an enduring Franco-Russian space cooperation, it was followed by a piggyback satellite, *MAS* (*SRET-1* by the French), launched on the *Molniya 1–20* weather satellite carried into orbit by an SL-6 on April 4, 1972. The second *Oreol* was launched by SL-8 on December 26, 1973. *SRET-2* was carried into space with *Molniya 1–30* on an SL-6 launched June 5, 1975. All three remain in orbit, but are nonoperational.

1971 December 29   Based on results filed by Shuttle Phase B double-extension contractors, NASA Administrator James Fletcher and George Low, his deputy, met with Caspar Weinberger, director of the Office of Management and Budget, to decide upon a design configuration. Senior administrators had already been given word from the White House that if NASA could confidently recommend a Shuttle configuration with a development cost of less than $5.5 billion, President Richard M. Nixon would give the program the go-ahead. The OMB agreed to fund the 65,000-lb. payload capability orbiter designed with cryogenic propellant in an external tank. The question of booster was left open, options being either a pressure-fed liquid propellant booster or a solid propellant booster, with development costs of $5 billion and $4.3 billion, respectively. Operating costs were projected to be $7.5–9 million per flight for the liquid concept and $10–13 million for the solid configuration.

1971 December 30   *Mariner 9* conducted a second orbit trim maneuver when the main propulsion system fired for 17.3 sec., imparting a delta velocity of 137 ft./sec. This raised periapsis to 1,025 mi., lowered apoapsis and extended the mean orbit period by 1 min. 14 sec. to 11 hr. 59 min. 28 sec. This established the proper cycle for data transmission to the tracking station in Goldstone, Calif., so that systematic mapping of the planet could begin. The dust cloud that had enveloped the planet since the spacecraft arrived was starting to show signs of dissipating, and mission scientists put together a coherent set of mapping cycles, the first of which began on January 2, 1972.

1971   The U.S. Department of Defense authorized a Strategic Air Command Required Operational Capability document which outlined the need for a new ICBM capable of evading direct attack from accurate Soviet warheads. President Richard M. Nixon allowed studies to proceed and build on the Strategic Systems X analyses begun during 1965. He was not

The December 1971 concept for a reusable shuttle envisaged an orbiter riding piggyback on a recoverable ballistic booster. By this date the fully reusable system was considered too costly.

prepared to authorize work on the Advanced ICBM Technology Program (AITP) and the conceptual Missile-X, or M-X, which would counter increasingly accurate Soviet ICBMs.

1972 January 2   The first systematic mapping cycle of the planet Mars with the orbiting *Mariner 9* spacecraft began with a scheduled 32 pictures per orbit. The spacecraft was in an orbit where it made 2.056 evolutions per Mars day, or sol, and almost exactly 2 revolutions per Earth day. Because of this the tracking station in Goldstone, Calif., could pick up the spacecraft for data transmission in the same part of its orbit each day. The map cycle one was designed to cover Mars from 65° S to 20° S and continued through 39 orbits. The second map cycle covered areas between 30° S and 20° N and began January 22, followed by map cycle 3 that covered areas from 20° N to between 45° N and 60° N. N starting February 10 and ending March 8, 1972. The primary mission of *Mariner 9* was essentially over by the end of that month.

1972 January 5   President Richard M. Nixon held a press conference at the "Western White House," in San Clemente,

Calif., to officially announce the go-ahead for NASA to develop and operate the partially reusable Shuttle system. It was to comprise a delta-wing orbiter carrying cryogenic hydrogen and oxygen propellants in an expendable external tank feeding three 415,000-lb. thrust high-pressure engines. The liquid or solid propellant boosters would be attached to the sides of the external tank in a parallel-burn configuration in which all main propulsion units would fire at liftoff. As required by the Air Force, the orbiter would have a maximum payload capability of 65,000 lb. to a due-east orbit or 40,000 lb. to a polar orbit. It would have go-around jet engines in the rear for power during the terminal stages of descent.

1972 January 12   Maxime Faget, director of Development and Engineering at the NASA Manned Spacecraft Center, briefed Phase B double-extension contractor Boeing on the series 040B orbiter. In the two years it had been working on in-house orbiter configurations, beginning with the DC-3 concept of a small, low-cost Shuttle, the MSC had become the lead agency with respect to developing designs for the reusable vehicle. The series 040B orbiter was the latest in a complex and evolving line of different configurations and designs. Fourteen days later the MSC published drawings of the 040C design, which had a length of 109 ft. 7 in., a wingspan of 79 ft. 11 in., three cryogenic main engines, two hypergolic maneuvering engines and thirty-four monopropellant thrusters clustered in fin-tip and wing-tip pods.

1972 January 13   The NASA Marshall Space Flight Center issued two-month contracts to Aerojet General, Lockheed, Thiokol and United Technology Center for studies of solid propellant rocket motors for the Shuttle. Valued at $150,000, each contract was to evaluate 120- and 156-in.-diameter motors in twin and quadruple configurations in a parallel-burn mode with the winged orbiter. The studies were to converge at a March 15 deadline, when NASA expected to release a request for bids on Shuttle Phase C/D full-scale development.

1972 January 19   NASA announced the crew members for the three Skylab missions. Primary crew for the first, 28-day SL-2 mission were to be Charles Conrad, Joseph Kerwin and Paul Weitz. The backup crew would comprise Russell Schweickart, Story Musgrave and Bruce McCandless. Primary crew for the 56-day SL-3 mission were Alan Bean, Owen Garriott and Jack Lousma. Crew members for the 56-day SL-4 mission were to be Gerald Carr, Edward Gibson and William Pogue. Back-up crew for the second and third missions were Vance Brand, William Lenoir and Don Lind. The launch of the SL-1 vehicle with the unmanned Skylab cluster was scheduled for April 1973, followed by the first manned mission a day later.

1972 January 20   Once again the family of U.S. electronic intelligence gathering subsatellites shifted to new carrier-satellites and a new launch vehicle when the second KH-9 Big Bird was sent into space by Titan IIID. The photoreconnaissance satellite was placed in a 97.5 × 206 mi. orbit inclined 97°. Carried piggyback, the 130-lb. radar elint subsatellite was then positioned in a 293 × 341 mi. orbit inclined 96.6°. In all, 12 elint subsatellites were carried on 11 Titan IIID/Big Bird flights, the last on May 11, 1982. A mission launched November 10, 1973, carried two elints.

1972 January 24   In outlines of budget projections, an internal NASA document defined the spread of Shuttle development costs. Total development costs from 1971 through 1980 were projected to amount to $7.37 billion at fiscal year 1973 dollar values, with a peak year funding of $1.225 billion in 1976. In public press announcements, NASA still maintained the Shuttle would cost less than $5.5 billion.

1972 January 27 NASA's Marshall Space Flight Center announced selection of Aerojet General Corp., Lockheed Propulsion Co., Thiokol Chemical Co. and United Technology Center for studies of solid rocket motors of 120-in. and 156-in. diameter applicable as boosters for the Shuttle. The solid motors would be reused several times after recovery by parachute from a splashdown in the Atlantic Ocean following launch from Cape Canaveral, Fla., or the Pacific Ocean after launch from Vandenberg AFB, Calif.

1972 January 31 *HEOS A-2,* the second of two highly eccentric orbit satellites, was launched by a Delta L from Vandenberg AFB, Calif., to a 199 × 698 mi., 89.9° orbit. Developed by the European Space Research Organization, the satellite took the form of a 16-sided cylinder, 4 ft. 4 in. in diameter and 7 ft. 10 in. tall, carrying 67 lb. of instruments to investigate particles, micrometeorites and the northern magnetic region. *HEOS A-2* decayed on August 2, 1974.

Mathematica, Inc., issued an updated version of its *Economic Analysis of New Space Transportation Systems*, the core of which was a traffic model comprising five orbiters scheduled to run 624 flights in the first 12 years of operation. The report reiterated that the manned Shuttle in itself would make no savings over existing unmanned expendable launch vehicles: a 514-flight program between 1979 and 1990 would cost $12 billion in either case. It emphasized that only when used in conjunction with an orbit transfer tug and common payload designs would savings accrue to the Shuttle program. The report did predict that development of a new expendable launch vehicle using the latest technology would immediately begin to save costs.

1972 January McDonnell Douglas and North American Rockwell (NAR) completed their Phase B extension studies on a modular space station and reported their findings to NASA. Each company proposed an initial 6-person station with growth potential to a 12-person facility. Both companies worked to a NASA projection that authorization to build the station would not occur before 1975 and that the first launch in the station assembly sequence would not come before 1980. McDonnell Douglas proposed a 6-person station assembled from three habitable modules together with a power boom for solar arrays. Four RAMs (research and application modules) would provide experiment and work areas for the scientific investigations. NAR proposed a similar arrangement but in a different assembled configuration. Modules from both companies were generally 15 ft. × 55 ft. weighing 15,000–17,000 lb.

1972 February 11 McDonnell Douglas reported to NASA the findings of its four-month extension to the already extended Phase B Shuttle program definition study. It concluded with the description of a series-burn, pressure-fed, liquid propellant booster configuration. Grumman Aerospace presented its final briefing on February 16, focusing on a series-burn, ballistic, recoverable booster, and a parallel-burn concept with two solids of 156-in. diameter. North American Rockwell also presented its extension study findings on February 16, comparing parallel- and series-burn, and liquid versus solid propellant boosters. Lockheed reported its findings on March 15,

1972, which compared all possible candidate options for different booster configurations.

1972 February 14 A Soviet SL-12 launcher carried the 12,400-lb. *Luna 20* spacecraft from Baikonur to an Earth parking orbit and from there to a translunar trajectory. After one correction burn on February 15, *Luna 20* went into a 62-mi. lunar orbit with an inclination of 65° and a period of 1 hr. 58 min. on February 18. The next day its path was modified to a 13 × 62 mi. orbit. At 17:13 UT on February 21 the descent engine fired for 4 min. 27 sec. to bring *Luna 20* out of moon orbit. At an altitude of 2,500 ft. the engine fired again to slow the spacecraft further and bring it to an altitude of 66 ft., at which point the engine shut down. Two vernier motors continued to lower *Luna 20* to a soft landing at 17:19 UT. The landing site was at 3.5° N × 56.5° E, close to the Apollonius Mountains at the edge of the Mare Fecunditatis. A sample of surface material was obtained by a method similar to that first employed by *Luna 16* in September 1970.

1972 February 18 NASA and the U.S. Atomic Energy Commission issued notices that work on the NERVA (nuclear engine for rocket vehicle applications) program must terminate by June 30, 1972. The primary contractor, Aerojet General, and the major subcontractor, Westinghouse, had been on a suspended basis since early 1971 when funding for the 75,000-lb. thrust engine was cut. Long-lead procurement was now halted and the NERVA program, which to date had absorbed more than $1.5 billion, was abandoned.

1972 February 22 At 10:58 UT the ascent stage of the Soviet *Luna 20* moon lander blasted away from the spacecraft's descent stage, accelerating to a velocity of 8,860 ft./sec. on course for Earth. Inside the spherical reentry capsule was a 3.6-oz. (0.22-lb.) core sample from the lunar surface. On a high-velocity intercept trajectory, *Luna 20*'s ascent stage separated from the capsule in the final stages of approach to Earth. Entering at a steep angle of only 30°, versus 60° for the *Luna 16* capsule, the spherical object reached the surface at 19:12 UT on September 25, 25 mi. northwest of Dzhezkazgan, Kazakhstan. The capsule came to rest on an island in the Karakingir river during a blizzard only 75 mi. northwest of where the *Luna 16* capsule had landed.

1972 March 3 At 01:49:03 UT Atlas-Centaur AC-27 was launched from Complex 36A at Cape Canaveral, Fla., carrying the 570-lb. *Pioneer 10* spacecraft on top of a 2,510-lb. TE-M-364–4 third stage. In a direct ascent to escape velocity, Atlas carried the 132-ft. configuration to a speed of 7,900 MPH and a height of 85 mi. At 4 min. 3 sec. after liftoff, the cryogenic Centaur pushed on to a speed of almost 23,000 MPH and a height of 100 mi. at 11 min. 46 sec., and the solid propellant third stage produced a thrust of 14,800 lb. for 44 sec., accelerating the spacecraft to a new record speed of 31,400 MPH at a height of 280 mi. at 13 min. 56 sec. The spacecraft comprised a hexagonal flat box 1 ft. 2 in. deep with each of its six sides 2 ft. 4 in. long. One side supported a smaller box of the same depth whose sides were irregular hexagons. It contained most of the science instruments. A dish antenna 9 ft. in diameter dominated the top of the main spaceframe and would transmit data to Earth at a maximum 2,048 bits/sec. in the asteroid belt or 1,024 bits/sec. at Jupiter via two 8-W amplifiers. Four SNAP-19 radioisotope thermoelectric generators in pairs at the end of two 9-ft. booms would provide 155 W of electrical power at launch and 140 W at

Jupiter. Spinning at 4.8 RPM, within 11 hr. of launch, *Pioneer 10* passed the orbital distance of the moon.

1972 March 7   Bound for Jupiter, NASA's *Pioneer 10* performed a two-part course correction maneuver with its propulsion system to adjust the trajectory for a fly-by on December 3, 1973. The propulsion and attitude control system was used for maintaining a spin rate of 4.8 RPM, changing spacecraft velocity and changing the attitude of the spacecraft in space. Total velocity changes adding up to 616 ft./sec. were possible through six thrusters firing steadily or in pulses producing 0.4–1.4 lb. of thrust. Hydrazine monopropellant was carried in a 1 ft. 4.5 in. diameter pressurized tank for decomposition over a catalyst in the chamber of each thruster. For its first course correction the thrusters fired in two sequences. The first, lasting 8 min. 7 sec., began at 12:19:37 UT, and the second, lasting 4 min. 16 sec., began at 13:01:39 UT for a total velocity change of 46 ft./sec. Only four days after launch, *Pioneer 10* was almost 2.2 million mi. from Earth.

1972 March 12   The European Space Research Organization's *TD-1A* astrophysics satellite was launched by Delta N from Vandenberg AFB, Calif., to a 325.5 × 342 mi., 97.55° orbit. The 1,038-lb. satellite consisted of a crate-like structure 7 ft. high and 3 ft. at the base. Two pairs of solar cell panels extended outward from opposite sides. The 319-lb. payload included seven science experiments to measure high-energy emissions from stellar and galactic sources and from the sun. The orbit was planned to keep the satellite in continuous sunlight for six months. Data recorders began failing on April 19 but 50% data was returned for six months. It decayed on January 9, 1980.

1972 March 15   NASA announced its decision to use the parallel-burn concept for solid rocket boosters and the orbiter's cryogenic main engines on the partially reusable Shuttle. Solids had been chosen over liquids because of technical simplicity and lower cost. Previously, solid rocket motors had been considered too dangerous and uncontrollable to use in conjunction with a manned space vehicle. NASA Adminis-

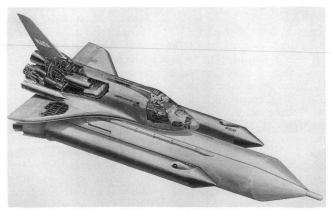

The final configuration of the Shuttle was selected during March 1972, when NASA settled on a reusable orbiter and main engine assembly which drew propellant from an external tank to which were attached two large solid-rocket boosters. This drawing depicts the winning orbiter design from Rockwell.

trator James Fletcher assured the Office of Management and Budget that the 156-in.-diameter solids could be developed faster and at a saving of $700 million in development costs, lowering total Shuttle cost to $5.15 billion.

1972 March 17   NASA issued a request for proposals on full-scale development, test and evaluation of the Shuttle. The RFP was issued to North American Rockwell, Grumman Aerospace, Martin Marietta, McDonnell Douglas, Boeing, General Dynamics, Lockheed and Chrysler. Three reference missions were described: a 65,000-lb. payload capability to a due-east, 115-mi. orbit such as might be used to launch satellites; a 25,000-lb. space station resupply flight to a 311-mi. circular orbit inclined 55°; and a 40,000-lb. payload delivery flight to a polar orbit. The payload bay was to be 15 x 60 ft. and acceleration during launch and ascent limited to a maximum 3 g. Atmospheric tests with the orbiter were to begin in early 1976, with the first manned orbital flight on March 1, 1978. Each orbiter was to be capable of 500 flights over the course of at least 10 years. NASA outlined a projected flight rate of 6 flights in the first year followed by 15, 24, 32, 40 and 60 annual flights in succeeding years, with a fixed rate of 60 per year after the sixth year.

1972 March 23   The first of a two-part course correction to the flight path of NASA's *Pioneer 10* spacecraft en route to Jupiter began at 22:02:30 UT, 10 million mi. from Earth. The maneuver thrusters fired for 34 sec. and shut down for almost 10 hr., when they were fired again at 12:02:23 on March 24, for 1 min. 4 sec. The first 5.9 ft./sec. maneuver had been performed directly away from Earth but the second 7 ft./sec. was aimed generally in the direction of Earth, 24° off the Earth-sun line. This course change moved the Jupiter fly-by point 7,860 mi. closer to the planet and aimed it for a point 87,000 mi. above the cloud tops.

1972 March 27   The Soviet Union launched the 2,602-lb. *Venera 8* spacecraft to Venus from an interim Earth parking orbit. A second spacecraft, designated *Cosmos 482,* was placed in a low Earth orbit of 122 × 134 mi. on March 31, but the terminal stage of the SL-6 fired for too brief a period and it was stranded in an elliptical orbit of 127 × 6,093 mi. These were the last Soviet Venus-bound spacecraft of the original series designed by the G. N. Babakin bureau that began with *Venera 4* launched June 12, 1967, and the last Soviet planetary missions to use the SL-6 launch vehicle. Future flights to Venus would use the more powerful SL-12. Known in Russia as the 8K82, this was first used on a planetary flight attempt for the unsuccessful *Cosmos 419* launched on May 10, 1971.

1972 March 29   A convention on the international liability for damage caused by space objects was signed in London, Washington and Moscow. It was then opened for general signature at the United Nations. Known as the Outer Space Liability Convention, it had taken several years to negotiate by the UN Outer Space Committee. Under it, states would be liable for damage caused by objects launched from their territory, and it thus covered both private and government launches.

1972 March   Engineers completed specifications on a successor to the Europa II satellite launcher, which looked increasingly as though it would be canceled. Nominally called Europa III, it had Viking motors derived from those used in the French M-20 missile in an L-150 first stage containing 330,750 lb. of propellant, and a cryogenic H-20 second stage carrying 50,715 lb. of hydrogen-oxygen propellants. With a liftoff mass of 421,500 lb., this configuration would be able to place 12,100 lb. in a low Earth orbit or 3,300 lb. in geosynchronous transfer orbit. The Germans suggested a larger first stage with Viking engines, a conventional second stage and a smaller cryogenic third stage.

1972 April 1 A significant rise in expenditure on Japanese space projects was evidenced by a 56% increase in space money for the fiscal year 1972 budget which began this date. The government would spend 23.96 billion Y, about $186 million, during the next 12 months, compared to 15.53 billion Y, or $120 million, in fiscal year 1971. These figures included all government space expenditure, of which the National Space Development Agency (NASDA) took 76% in 1972. The rapid increase continued until fiscal year 1977, when 94.91 billion Y ($736 million) was spent, of which NASA took 84%. From that date space expenditure increased at a more modest rate, reaching 117.3 billion Y ($910 million) by fiscal year 1986, remaining at about that figure in real terms through the early 1990s.

1972 April 2 The *Mariner 9* extended mission began with the spacecraft going into a series of Earth and solar occultations. Mapping started again on June 5, 1972, with the spacecraft taking pictures of the region between 40° N and the north pole. This region had been covered by the seasonal north polar hood which had receded since completion of the standard mission. This sequence ended October 16 and the last signal from *Mariner 9* was received at 22:31:00 UT on October 27, after it had operated for 349 days, completing 698 orbits of Mars. It had returned 54 million binary digits of data, 27 times the volume transmitted by *Mariners IV*, *VI* and *VII*, including 7,329 pictures mapping the entire surface of the planet. *Mariner 9* revealed Mars as a dynamic world with wide geologic variations, giant volcanoes more than 50,000 ft. high, canyons 20,000 ft. deep and a highly active atmosphere that freezes out on the surface in winter to form ice. *Mariner 9* had operated 1 year 150 days 19 hr. 18 min. It was expected to remain in orbit until about 2012.

1972 April 4 The NASA Marshall Space Flight Center established the Sortie Can Task Team to plan the definition and program plan of a scientific station carried inside the Shuttle. Evolved in parallel to the research and applications module (RAM), which began as an experiments module to be launched by the Shuttle and docked to a permanently manned space station, the sortie can would be a mini-laboratory retained in the Shuttle's cargo bay for dedicated science missions. The team was managed by Fred E. Vreuls with Hans R. Palaora as chief engineer.

1972 April 6 NASA and the National Teachers Foundation announced selection of 25 finalists from more than 15,000 applications for participation in the Skylab Student Project. An informal dinner for the 25 finalists, their teacher-sponsors and families was held on May 10. From the 25 finalists, 19 winners were selected on July 7, 1972. The purpose of the project was to stimulate interest in science and technology among students. Students in grades 9 to 13 had been eligible for submitting experiment proposals, and the winners were to work with NASA team managers at the Marshall Space Flight Center to get their experiments ready for flight on Skylab.

1972 April 14 The first in a series of Soviet solar physics satellites, *Prognoz 1*, was launched by an SL-6 launcher from Baikonur to a 139 × 277 mi., 64.9° orbit from where it was shifted to its operational orbit of 625.5 × 124,075 mi. with an inclination of 65°. The 1,859-lb. satellite comprised a squat, round-ended cylinder with four deployable solar panels, a magnetometer boom at one end and a long-wave antenna at the other. *Prognoz 1* carried 15 instruments for measuring radiation, plasma, solar winds and interactions with the magnetosphere. *Prognoz 2* was launched on June 29, 1972, to the same elliptical orbit, followed by *Prognoz 3* on February 15, 1973. After a lapse of almost three years, a heavier, 1,995-lb. generation began with *Prognoz 4*, launched on December 22, 1975, followed by *Prognoz 5* on November 25, 1976, *Prognoz 6* on September 22, 1977, *Prognoz 7* on October 30, 1978, and *Prognoz 8* on December 25, 1980. After a 30-month hiatus, the 2,340-lb. *Prognoz 9* was launched by an SL-6 on July 1, 1983, to a 236 × 447,410 mi. orbit. On April 10, 1985, the last in the series, *Prognoz 10*, was launched to a 3,170 × 121,000 mi. orbit.

NASA announced selection of the Kennedy Space Center (KSC) in Florida and the Vandenberg AFB in California as the launch sites for the reusable Shuttle. Two sites were necessary to accommodate flights to all orbital inclinations. Because the flight path into orbit was prohibited from passing over land, flights from KSC were limited to a maximum orbital inclination of 57°, whereas flights to high-inclination orbits from Vandenberg would head south over the poles. Initial flights to orbit were scheduled to begin in early 1978 from KSC, followed at the end of the decade by flights into polar orbit from Vandenberg. In reaching its decision, the Site Review Board had considered 140 locations and visited more than 40 sites.

1972 April 16 The *Apollo 16* lunar landing mission with astronauts John Young, Ken Mattingly and Charles Duke was launched from Kennedy Space Center's Complex 39A at 17:54:00 UT. The AS-511 flight vehicle had a liftoff weight of 6.430 million lb. and carried CSM-113 (67,010 lb.) and LM-11 (36,255 lb.). The total CSM-LM (command/service module–lunar module) payload mass of 103,265 lb. was the greatest weight ever sent to the moon. Following a sequence of events similar to that of the previous two J-series missions, LOI (lunar orbit insertion) burned the spacecraft into lunar orbit at 74 hr. 28 min. 28 sec. The only significant anomaly was a 6-hr. delay in landing caused by a computer problem around the moon just after the lunar module (*Intrepid*) separated from the command service module (*Orion*). Touchdown at the Descartes landing site occurred at 104 hr. 29 min. 35 sec. after a powered descent lasting 12 min. 10 sec. There were three periods of EVA (extravehicular activity). The first lasted 7 hr. 11 min. 2 sec. and involved deployment of the ALSEP (Apollo lunar surface experiments package) instruments, including heat-flow, passive and active seismometers, magnetometer, solar wind experiment, portable magnetometer and cosmic-ray detector. In walking away from the heat-flow experiment, Young caught his foot in the data cable, tearing it away from the instrument and rendering it completely inoperable. A far-ultraviolet camera was also used. The second EVA lasted 7 hr. 23 min. 11 sec. and took the crew on a geological traverse, as did the third EVA which lasted 5 hr. 40 min. 3 sec. The LRV (lunar roving vehicle) was used on all three EVAs, traversing a total distance of 16.8 mi. TEI (trans-Earth injection) was accomplished at 200 hr. 21 min. 33 sec., after the LM ascent stage had been sent crashing to the moon and the crew had ejected a subsatellite similar to that released on *Apollo 15*. A 1 hr. 23 min. 42 sec. EVA by Mattingly retrieved panoramic and mapping camera film on the way back to

Earth. Splashdown occurred On April 27, 1972, at 265 hr. 51 min. 5 sec., with 208 lb. of samples.

1972 April   Representatives of the European Space Research Organization (ESRO) met with NASA officials to discuss possible ESRO participation in the proposed Pioneer-Venus mission being developed at the Ames Research Center. A Joint Working Group was set up to define the mission's objectives and a report was issued in January 1973. *Pioneer Venus Orbiter*, emphasized the area of interest held by European scientists and focused ESRO attention toward the 1978 orbiter mission. Events were to overtake the intentions of the respective scientific communities and nothing further came of the discussions.

1972 May 1   Southern Contractors Service began dismantling the two Saturn I/IB launch sites at the Kennedy Space Center, Fla. Conceived and built to handle the first large U.S. space launchers, they had been in use for less than seven years. It was from Launch Complex 34 that the first Saturn I had lifted off on October 27, 1961, and it was from Launch Complex 37B on October 11, 1968, that the first Apollo lunar module was launched aboard a Saturn IB, bringing an end to operations at the two sites. Launch control rooms were retained for historical purposes.

1972 May 12   Eight U.S. aerospace companies submitted proposals in response to NASA's request for bids on full-scale design, development and test of the Shuttle: Grumman, Lockheed, McDonnell Douglas and North American Rockwell. Evaluation by a source evaluation board began May 15. The board was staffed by 416 people, 20 of whom were from the U.S. Air Force. Of the 396-strong NASA contingent, 269 members were from the Manned Spacecraft Center. Each bid was to be evaluated against a maximum of 1,000 points, including 400 for technical content, 250 for maintainability and 350 for program management. Each of these three areas were further divided into several specific indices. Cost proposals from the bidders were received on May 19.

General Dynamics completed the Phase B definition study of the research and applications module (RAM) which it had been conducting under contract to NASA. An added role to the four basic applications outlined in the study requirement document included the sortie mode whereby a RAM would remain attached to the cargo bay of the Shuttle for scientific research. The study envisaged the first RAM flights during 1978, when the Shuttle was expected to be available. The RAM concept served to focus study on a laboratory for the Shuttle, which would eventually emerge as Spacelab. It would also reinvigorate interest in a modular space station, which would form the basis for a complete revitalization of station studies.

1972 May 19   NASA's budget for fiscal year 1973 was approved by Congress and with it a modest start on a revised grand tour of the outer planets. Instead of launching three spacecraft to Jupiter, Saturn, Uranus, Neptune and Pluto in 1977 and 1979, two spacecraft would be launched in August 1977 to fly by Jupiter in March 1979 and Saturn in November 1980 and August 1981. The space agency had been unable to gather congressional support for the more ambitious mission plan, which NASA admitted would cost up to $1 billion. Also deleted was a planned flight test of the STAR (self test and repair) computer being designed at the Jet Propulsion Laboratory to keep control of the spacecraft far from Earth when communication time stretched to several hours.

1972 May 24   At a Moscow Summit, U.S. President Richard M. Nixon and Soviet Premier Alexei Kosygin signed the Agreement Concerning Cooperation in the Exploration and Use of Outer Space for Peaceful Purposes. The U.S. delegation was in Moscow primarily for joint signing of Strategic Arms Limitation Talks (SALT) agreement, a culmination of three years of negotiations. The agreement on cooperation in space was highlighted by formal announcement of a plan to launch U.S. and Soviet spacecraft on separate missions for a rendezvous and docking in space, called the Apollo-Soyuz Test Project (ASTP).

NASA's Manned Spacecraft Center issued a request to industry for technological development of materials capable of withstanding the severe thermal environment which would be experienced by the reusable Shuttle on its return to Earth through the atmosphere. The thermal protection systems were to be divided into two ranges of materials: those capable of withstanding temperatures of 800–1,000°F and those capable of withstanding 2,500–3,000°F. NASA had already concluded that a reusable surface insulation (RSI) was the baseline choice over ablative materials, although aerospace companies were conducting technical and cost trade-off studies between the two types of thermal protection. McDonnell Douglas, General Electric, LTV Aerospace and Lockheed were awarded contracts to study RSI.

1972 May 25   The last Corona-series military reconnaissance satellite, a KH-4B, was launched by a long-tank thrust-augmented Thor-Agena D from Vandenberg AFB, Calif., bringing to an end a historic period in the use of satellites for spying, surveillance and information gathering. Delivered to a 98×189.5 mi., $96.3°$ orbit, the satellite decayed through the atmosphere on June 4, 1972. Of the 144 Discoverer- and Corona-series satellites (covering KH-1, KH-2, KH-3, KH-4, KH-4A, KH-4B, KH-5 and KH-6 types) launched since *Discoverer I* on February 28, 1959, 102 had been successful, with an improvement in resolution from about 50 ft. down to approximately 5 ft. Since the first successful capsule recovery on August 18, 1960, 800,000 images covering an area 88 times the land mass of the USSR had been shot onto 2.1 million ft. of film returned to Earth. Responsibility for reconnaissance missions would now fall upon the KH-9 and subsequent vehicles.

1972 May 26   The United States–Soviet Union Treaty on the Limitation of Anti-Ballistic Missile Systems and the Interim Agreement with Respect to the Limitation of Strategic Offensive Arms were signed by President Richard M. Nixon and Premier Leonid Brezhnev in Moscow. Under the agreements, each country would be allowed to deploy a maximum 100 ABM missiles at only two sites. Each deployment site had to have a maximum radius of 93 mi., with one protecting a city and one protecting an ICBM field. ICBM production was to be frozen for five years or until the full treaty was negotiated, whichever was the sooner. Possession of ICBMs was to be limited to existing numbers: 1,054 for the U.S. and 1,618 for the USSR.

1972 June 5   Howard W. Tindall, director of Flight Operations at NASA's Manned Spacecraft Center, expressed discontent with the proposed method of dumping the Shuttle's external propellant tank in orbit. In the baseline plan, the tank would be separated from the orbiter and then de-orbited by firing a solid propellant rocket motor in the nose of the tank. Tindall believed the tank would not remain stable after separation and that an expensive attitude control package would have to be attached to the tank to ensure that it pointed in the right direction at retro-fire.

1972 June 14 During a visit to Washington, D.C., by a delegation from the European Space Conference, U.S. officials said that the space tug was no longer a candidate for European cooperation. Citing their belief that Europe lacked the money and the technical expertise to play a part in helping develop the intraorbit vehicle, the officials encouraged European space organizations to undertake development of the Shuttle laboratory, or sortie can, instead.

1972 June 26 At 14:53 UT an unmanned Soyuz spacecraft designated *Cosmos 496* was launched by an SL-4 from Baikonur to an orbit of 114 × 206 mi. With a weight of 14,720 lb., it was the first of a new generation of ferry vehicles designed to carry crew and a limited range of supplies to orbiting space stations. Changes had been made to some systems aboard Soyuz following the tragic loss of three cosmonauts at the end of the *Soyuz 11* flight a year before. The greatest visual change was the removal of the two solar cell arrays, which on earlier vehicles had been attached to the instrument module, and their replacement by batteries. This limited the autonomous life of Soyuz. If access to supplementary electrical power by docking with Salyut could not be achieved within two days of launch, the spacecraft would have to return to Earth. *Cosmos 496* returned after six days and may have carried extra batteries.

1972 June 30 NASA contracted with North American Rockwell for modification of Apollo CSM-111 to a configuration suitable for the Apollo-Soyuz Test Project mission and for the development and fabrication of a docking module which would be carried into space by Apollo for linking up with the Soviet Soyuz spacecraft. The docking module would comprise a cylinder, 10 ft. 4 in. long with a diameter of 4 ft. 8 in. and weighing approximately 4,300 lb. One end would carry a standard Apollo docking drogue for coupling with the Apollo command module. The opposite end would carry a different, androgynous docking unit which would be used to link up with the Soyuz spacecraft. Crew members would pass through the docking module, visiting each others' spacecraft after docking. A launch date of July 1975 was set.

1972 July 1 The European Space Research Organization handed back to Sweden the Kiruna sounding rocket launch facility called Esrange. It would retain the name, but no longer would ESRO use it as one of its facilities. ESRO had taken the decision to opt out of sounding rocket research and so had no further need for the launch base.

1972 July 6 NASA awarded a six-month contract to Arthur D. Little, Inc., for feasibility studies of a Solar Power Satellite (SPS) system using giant satellites in geostationary orbit to beam down electrical energy for use on Earth. Brainchild of the company's Peter Glaser, each station would be several tens of square miles in size and consist of flat panels converting sunlight into electrical energy transmitted in the form of microwave beams to microwave receivers, called recteners, on the ground. Managed by the Lewis Research Center, the SPS study would compare this method of electrical energy production with more conventional, Earth-based concepts.

1972 July 15 On its way to Jupiter, *Pioneer 10* entered the asteroid belt beyond the orbit of Mars, becoming the first spacecraft to chart a path through what many scientists considered a threat to the spacecraft's survival. More than 170 million mi. across, the belt is 50 million mi. thick, and any spacecraft journeying to the outer planets would have to go through this region of rocky debris left over from the formation of the solar system. *Pioneer 10* carried detectors that would record the distribution and frequency of small particles, but some scientists believed the spacecraft might be destroyed by larger fragments. *Pioneer 10*'s closest approach to any of the known asteroids was on August 2, when it passed within 5.5 million mi. of Palomar-Leyden, 0.6 mi. in diameter.

1972 July 17 NASA awarded parallel contracts to Grumman Aerospace and Lockheed Aircraft for the study of an airplane capable of simulating subsonic flight characteristics of the reusable Shuttle orbiter. Existing aircraft designs were to be considered for modification to a configuration in which they could duplicate descent and landing characteristics from a height of 35,000 ft. The selected airplane would be used for training Shuttle astronauts in descent and landing.

1972 July 22 The Soviet *Venera 8* spacecraft ejected a capsule weighing 1,100 lb. which descended to the surface of Venus, from where it transmitted data to Earth for 50 min. Entry occurred on the night side of the planet at 07:40 UT, and for 1 hr. 49 min. the capsule descended, reaching the surface at 09:29 UT. The probe had been considerably improved over its *Venera 7* precursor, and with better understanding of the Venusian atmosphere a portion of the thermal insulation could be removed, making way for more science equipment. Included was a light meter to measure the amount of sunlight reaching the surface. A gamma-ray spectrometer measured crude chemical properties of the surface and determined that the material was similar to granite on Earth. The Soviets were preparing a new and much heavier class of Venera spacecraft which would take pictures from the surface.

1972 July 23 The first U.S. Earth resources technology satellite, *ERTS-A*, was launched from Vandenberg AFB, Calif., to a sun-synchronous, 564 × 559 mi. orbit inclined 99.1°. The launch vehicle was the first Delta 900. Under the new Delta designator system, the first digit indicated the number of solid strap-ons, the second digit gave the type of second stage, and the third digit indicated whether or not a third stage was carried. *ERTS-A* would cross the equator daily at 9:30 A.M. local time, and cover the globe every 18 days. The 1,965-lb. *ERTS-A* had a height of 10 ft., a diameter of 5 ft. and solar panels spanning 13 ft. It carried 485 lb. of sensors and related equipment: a return beam vidicon system (RBVC) incorporating three overlapping cameras with green, red and near-infrared filters and a multispectral scanner (MSS) in four spectral bands, green, red and two near-infrared. The RBVC and MSS both observed the same 115 × 115 mi. area on the Earth's surface. Some 300 investigators from the United States and 31 foreign countries were involved. The satellite operated until January 16, 1978.

1972 July 26 NASA Administrator James Fletcher approved the selection of North American Rockwell to build the Shuttle. Out of a possible 1,000 scoring points, North American Rockwell had been awarded 603, edging out Grumman with 597, but well ahead of McDonnell Douglas with 489 points and Lockheed with 365. In the technical section of the scoring system, Grumman scored 240.42 to Rockwell's 195.8, but in the area of program management Rockwell scored 270.01 to Grumman's 222.95. The North American Rockwell Shuttle orbiter, based on a modified MSC-040C configuration and

Named ERT (Earth Resource Technology satellite) *at its launch on July 23, 1972, the satellite more familiarly known as Landsat ushered in a new generation of planet management.*

known as Vehicle 1, had a length of 124 ft., a wingspan of 79 ft. 6 in. and a maximum weight of 253,000 lb. carrying a 67,000-lb. payload (2,000 lb. more than the mission requirement). The external propellant tank (ET) would have a length of 183 ft. 4 in. and a diameter of 26 ft. 6 in., and the two solid rocket boosters (SRBs) would each have a length of 146 ft., a diameter of 13 ft. and a thrust of 3.521 million lb. The assembled Shuttle would have a length of 206 ft. and a liftoff weight of 5.24 million lb. The combined output of the two boosters and the three shuttle main engines would generate a liftoff thrust of 8.45 million lb. for 1 min. 49 sec., when the SRBs would separate at a velocity of 4,035 ft./sec. Abort solid rocket motors (ASRMs) attached to the rear of the orbiter would be jettisoned on the way up. North American Rockwell planned to start horizontal flight tests using air-breathing engines in August 1976, followed by unmanned vertical flights beginning in December 1977 and manned orbital flights from March 1978. NASA and North American Rockwell expected no greater interval than 14 days between flights for each orbiter.

NASA astronauts Robert Crippen, Karol Bobo and William Thornton began a 56-day period of isolation in an altitude chamber at the Manned Spacecraft Center. They were there to conduct the Skylab medical experiments altitude test (SMEAT), aimed at collecting medical data and evaluate medical equipment in support of the Skylab program. The altitude chamber (20 ft. in diameter) was pressurized to 5

lb./sq. in. with a 70/30 mixture of oxygen and nitrogen. The test was concluded September 20, and the report issued on October 18, 1972, identified several equipment problems that would have to be corrected before flight.

1972 July 29   A Soviet SL-13 launch vehicle lifted off from its launch pad at the Baikonur cosmodrome carrying the 39,700-lb. *Salyut 2* space station. A failure in the second stage, however, prevented its getting into orbit. It is believed that cosmonauts Vinogradov, Vitaly Sevastyanov and Valery Kubasov were in training for the first manned visit. After the failure the Korolev station was designated *Salyut 2–1*. There was a delay in launching a replacement Korolev station because only two of these hurriedly adapted facilities were available. Not before December 1974 would the next Korolev Salyut become available for habitation. Although called Salyut by observers in the West, the next station launched was the first Almaz from the Chelomei bureau, which had formed the basis for the Salyut design.

1972 August 9   NASA and North American Rockwell signed a letter contract for the development and construction of the reusable Shuttle in four incremental steps. The first would include the first two years of detailed design for the first two orbiters up to the preliminary design review (PDR) scheduled for mid-1974. The second would include the critical design review (CDR), tentatively scheduled for spring 1976, and would continue through the first year of flight tests. The third would cover the production of three additional orbiters and retrofitting of the two development orbiters into operational vehicles. The fourth was to have covered the operational phase beginning in the second year of manned orbital flights.

1972 August 13   A Solid propellant Scout launcher carried NASA's *Explorer 46* satellite from Wallops Island, Va., to a 306 × 504 mi., 37.7° orbit. The 370-lb. satellite consisted of a windmill-like structure with meteoroid bumper panels deployed from a hexi-cylindrical bus section. Each panel was 10 ft. 6 in. long and the satellite remained attached to the fourth Scout stage. One pair of meteoroid bumper wings failed to deploy, but a full set of data was returned from the remainder. *Explorer 46* decayed on November 2, 1979.

1972 August 16   The first in a series of second-generation Soviet navigation satellites was launched by SL-8 from Plesetsk. Designated *Cosmos 514*, the 1,500-lb. satellite was placed in an orbit of 593 × 600 mi. with an inclination of 83° instead of the 74° used by first-generation navsats. Moreover, instead of 120° orbital spacing, the satellites of the second-generation series, comprising *Cosmos 574* (June 20, 1973) and *Cosmos 586* (September 14), were spaced 60° apart. Replacement satellites in repeat orbit planes included: *Cosmos 627* replacing *Cosmos 514* on December 29, 1973; *Cosmos 628* (January 17, 1974) and *Cosmos 663* (June 27, 1974) for *Cosmos 586*; *Cosmos 689* (October 18, 1974) for *Cosmos 574*; and *Cosmos 729* (April 22, 1975) for *Cosmos 627*. On February 3, 1976, *Cosmos 800* was launched as the replacement for *Cosmos 663* but in an orbit plane 180° away to shift it from the orbit planes of the third-generation Soviet navsats, the first of which had been launched on December 26, 1974. Six more second-generation replacements were launched, the last being *Cosmos 1027* on July 27, 1978.

1972 August 19   Japan launched its fourth satellite, the 165-lb. Denpa (radio wave), on a solid propellant Mu-4S from the Kagoshima Space Center to a 144 × 3,786 mi. orbit with an inclination of 31°. The satellite was the second science satellite launched by the Institute of Space and Aeronautical Sciences of Tokyo University. It was instrumented to measure plasma waves and density levels, electron flux in the outer atmosphere/near space region and geomagnetism. *Denpa* decayed into the atmosphere on May 19, 1980.

1972 August 21 The Soviet nuclear-powered ocean surveillance satellite program entered a new and more vigorous operational phase with the launch from Baikonur of *Cosmos 516* by SL-11 to a 159 × 172 mi. orbit. Coming almost five years after the launch of the first of its type, *Cosmos 198*, on December 27, 1967, *Cosmos 516* was the first to operate for more than one month, albeit for only 31 days. Sent up on December 27, 1973, *Cosmos 626* worked for 46 days, while *Cosmos 651*, launched on May 15, 1974, operated for 71 days. Operating in pairs passing over the same region 20–30 min. apart, they monitored the movement of merchant and naval shipping. In all, 33 nuclear powered ocean surveillance satellites were launched by the USSR; the last was *Cosmos 1932*, on March 14, 1988.

NASA launched the fourth Orbiting Astronomical Observatory *(OAO 3, Copernicus)* on an Atlas-Centaur from Cape Canaveral, Fla., to a 453 × 459 mi. orbit inclined 35° to the equator. The 4,900-lb. satellite had a length of 10 ft. and a width of 7 ft., supporting two sets of three-panel solar arrays. *Copernicus* carried a 10-ft.-long ultraviolet telescope, a 32-in. mirror, an ultraviolet spectrometer and special sensors for the telescope guidance system. The *OAO 3* mission was to study the absorption of hydrogen, oxygen, carbon, silicon and other constituents in the interstellar gas.

1972 August NASA made a decision to limit the proposed Pioneer-Venus program to two missions only. A multiprobe spacecraft would be launched at the 1976/77 opportunity and an orbiter would be launched at the second opportunity in 1978. A month later scientists were invited to participate in the multiprobe mission and in April 1973 the preliminary payload was selected. An invitation for experiments on the orbiter mission was not issued before August 1973 and payloads for that mission were selected in June 1974.

1972 September 2 A Scout launcher carried an experimental U.S. navigation satellite from Vandenberg AFB, Calif., to an orbit of 445 × 536 mi. with an inclination of 90°. Called *Triad OI-1* or *TIP-1* (for transit improvement program), the 207-lb. satellite carried a new station-keeping system called DISCOS (disturbance compensation system) which engineers hoped would help counteract the disturbing effects of atmospheric drag and solar radiation on the orbit. *TIP-1* carried a new type of nuclear generator, but *TIP-2*, launched on October 12, 1975, reverted to solar cells which failed to deploy after insertion to a 281 × 507 mi. orbit. They were successfully deployed by the last in the series, *TIP-3*, launched on September 1, 1976.

1972 September 19 The first Soviet early warning test satellite, *Cosmos 520*, was carried into space from Plesetsk by an SL-6 launch vehicle. With a weight of 4,000 lb., the satellite was placed in an elliptical orbit of 388 × 24,443 mi. inclined 62.8°. In this orbit *Cosmos 520* made one revolution of the Earth in just under 12 hr. With an apogee slightly greater than the altitude of geosynchronous orbit, *Cosmos 520* spent most of that time relatively high above Earth where its infrared sensors could observe large portions of the hemisphere. Other test satellites included: *Cosmos 606* (2,750 lb.) on November

2, 1973; *Cosmos 665* (2,750 lb.) on June 29, 1974; *Cosmos 706* (880 lb.) on January 30, 1975; and *Cosmos 862* (2,750 lb.) on October 22, 1976.

The Advanced Missions Design Branch, Planning and Analysis Division, of the NASA Johnson Space Center circulated a plan changing the way the Shuttle external tank was to be separated and dumped. Instead of carrying the tank into orbit, separating it from the orbiter and then firing a retrorocket to de-orbit the inert cylinder, engineers proposed a scheme whereby the tank would be jettisoned at suborbital velocity. The tank would follow a ballistic trajectory and burn up over the Indian Ocean. The orbiter meanwhile would use its orbital maneuvering system (OMS) engines to add the necessary velocity for orbital insertion.

1972 September 23 NASA launched an Interplanetary Monitoring Platform, *IMP-H (Explorer 47),* from Cape Canaveral, Fla., to a 124,965 × 146,400 mi., 17.2° orbit. The 860-lb. satellite comprised a drum-shaped, 16-sided body with boom-deployed experiment packages and instruments to monitor particles, plasma and fields in the region between Earth and moon. *Explorer 47* had an orbital period of 12 days 6 hr. 30 min. and at that altitude could remain in orbit for more than a million years.

1972 October 2 NASA awarded study contracts for the Pioneer-Venus missions to Hughes, TRW, AVCO and General Electric. As defined at this date, a multiprobe mission to Venus in 1976/77 would comprise an 840-lb. spin-stabilized spacecraft carrying one large probe and three small probes. The probes would separate from the bus 10–20 days before encounter with the large probe containing 60 lb. of science instruments descending by parachute through the atmosphere containing 60 lb. of science instruments. Each small probe would carry 3 lb. of science equipment for a free-fall descent to the surface 1 hr. 15 min. after entry, while the bus would follow along carrying 25 lb. of instruments and transmit data until it burned up. An orbiter mission was tentatively planned for 1978.

1972 October 24 Martin Marietta delivered the X-24B lifting body research vehicle to Edwards AFB, Calif. The X-24B had been built up from the X-24A to conform to aerodynamic shapes developed by the U.S. Air Force Flight Dynamics Laboratory, and it was considered superior to first-generation lifting bodies (M2-F2, HL-10 and X-24A). The new aircraft had a greatly extended nose section, a more bulbous center body, extended outer fins at the rear and a small dorsal fin added. The X-24B had a length of 37 ft. 6 in., a span over the fins of 19 ft. 2 in. and a gross weight of 13,000 lb. Power was to be provided by the same 8,480 lb. thrust Thiokol XLR11-RM-13 used in the aircraft when it had the X-24A configuration. Drop tests with the X-24B began August 1, 1973.

1972 October 31 NASA informed the British Aircraft Corp. that it had decided against offering European companies subcontracts on the reusable Shuttle program. The policy had been that European aircraft manufacturers would be allowed to compete for subcontract work only if the Europeans had decided by this date to develop the Shuttle sortie lab. Since no such decision had been forthcoming, the opportunity had been forfeited.

1972 October Concerns about potentially high operating costs for NASA's reusable Shuttle was the focus of an engineering study conducted by Allen F. Donovan, senior vice president of Aerospace Corp. Noting that NASA had now formally dropped the concept of using air-breathing engines stowed in the rear of the cargo bay as supplementary propulsion during descent, Donovan noted that further cuts in recurring costs might be made if the solid propellant boosters were deleted in favor of two liquid propellant boosters, each using two F-1 engines from the Saturn V program. Although NASA Administrator James Fletcher had confidentially expressed interest in making "substantial changes in the configuration" to lower operating costs, a study at NASA revealed the F-1 engines would not be an improvement.

Soviet scientists began flight tests with a Mod 2 version of the SS-N-6 submarine-launched ballistic missile (SLBM). With an improved range of 1,850 mi., Mod 2 could reach any target in the United States from the 100 fathom (600 ft) line offshore, the first SLBM to have that capability. Replacing Mod 1 missiles aboard Yankee-class submarines from 1973, the Mod 2 was succeeded during 1974 by the Mod 4, equipped with two 500-KT multiple reentry vehicle (MRV) warheads. With a best accuracy of only 4,300 ft., however, the weapon was not suitable for use against hardened military targets.

1972 November 10   Canada's *Anik 1 (Telesat A)*, the world's first geostationary domestic communications satellite, was launched at 00:14 UT by the first extended long-tank Delta 1914 from Cape Canaveral, Fla., to a transfer ellipse of 117 × 22,661 mi. inclined 26.9°. The first digit, *1*, in the designation "Delta 1914" indicated that the first stage extended by 8 ft.; the *9* indicated nine strap-on Castors; the second *1* indicated a standard second stage with 10,000 lb. of propellant and the same uprated AJ-10–118F engine; and the *4* denoted introduction of the TE-364–4 third stage with 2,300 lb. of solid propellant, compared with 1,440 lb. for the TE-364–3 first used on the Delta L. The Delta 1914 could put 1,400 lb. in geosynchronous transfer or 4,050 lb. to low Earth orbit. Named after the Eskimo word for brother, *Anik 1* was moved to a geosynchronous slot at 114° W at 20:55 UT on November 13. Weighing 1,238 lb. at launch and 600 lb. on station, the drum-shaped Hughes satellite had a diameter of 6 ft. 3 in. and a height of 11 ft. 2 in., with 20,448 solar cells producing 300 W power. A de-spun section supported the main, 5-ft.-diameter 6/4 GHz antenna, which could handle 7,000 telephone calls or 12 color TV channels. *Anik 2 (Telesat B)* was launched on April 20, 1973, to 109° W, followed by *Anik 3 (Telesat C)*, the last of the first-generation series on May 7, 1975, to 104° W.

NASA issued guidelines for fixing a flight cost on the reusable Shuttle. Because the space agency expected to operate 439 flights with the Shuttle between 1978 and 1988, the recurring costs would form a high percentage of NASA budget requirements. NASA worked on the assumption that each partially-reusable solid rocket booster (SRB) would make 10 flights and that 1% of the motors would fail. The NASA traffic model current at this time projected 6 flights in 1978, 15 in 1979, 24 in 1980, 32 in 1981, 40 in 1982 and 60 thereafter from 1983 through 1987, followed by 28 in 1988.

1972 November 13   NASA conducted the Shuttle preliminary requirements review (PRR), the first major milestone in finalizing the configuration after contract award in July 1972. It noted in particular considerable growth in the projected size of the vehicle. The orbiter (Vehicle 2 configuration) now had a length of 125 ft. 9 in., a wingspan of 84 ft., an empty weight of 170,000 lb. and a maximum landing weight of 215,000 lb. The 50° swept wing had an area of 3,220 sq. ft., and maximum subsonic lift/drag had been cut from 5.7 to 5.1. The

external tank now had a length of 189 ft. 10 in., a diameter of 25 ft. 4 in. and capacity of 1.72 million lb. of liquid oxygen–liquid hydrogen. The solid rocket boosters had a length of 191 ft. 8 in. and a diameter of 13 ft. 6 in. The abort propulsion system had been abandoned, but each SRB (solid rocket booster) was now projected to deliver a thrust of 4.15 million lb. The total vehicle now had a length of 214 ft. 4 in. and a liftoff weight of 5.246 million lb.

1972 November 15   The second of three Small Astronomy Satellites, *SAS-B (Explorer 48)* was launched by an Italian crew on the four-stage Scout from the San Marco platform off Kenya in the Indian Ocean to a 276 × 393 mi., 1.9° orbit. The 410-lb. satellite comprised a cylinder 1 ft. 11 in. in diameter with a height of 1 ft. 8 in. supporting four solar paddles with a span of 13 ft. A bulbous spark chamber was mounted atop the cylindrical control section. Instrumented to measure extragalactic gamma rays, *SAS-B* decayed on May 1, 1979.

1972 November 21   The NASA Marshall Space Flight Center issued a request for proposals from industry on space tug studies. Work already under way presupposed that the early tug would make temporary use of an existing rocket stage. NASA wanted to award contracts for studies on a cryogenic tug using liquid hydrogen and liquid oxygen propellants and on a tug using storable propellants. The tug would serve as an "upper stage" to the Shuttle, transferring people, payloads or cargo to higher orbits than could be reached by the reusable space plane.

The United States and the Soviet Union began the second round of Strategic Arms Limitations Talks (SALT II) in Geneva, Switzerland. Inspired by the May 26, 1972, accord limiting deployment of antiballistic missile systems, the second round was working toward comprehensive limits on strategic weapon delivery systems.

1972 November 22   A four-stage Scout carried the European Space Agency's *ESRO 4* satellite from Vandenberg AFB, Calif., to a 152 × 729 mi., 91.1° orbit. The 253-lb. satellite comprised a cylinder 2 ft. 11 in. high and 2 ft. 5 in. across with 70 lb. of science instruments for investigation of the polar ionosphere, the near regions of the magnetosphere and auroral and solar particles. Built by Hawker Siddeley in the United Kingdom, the satellite carried experiments from Britain, Sweden, the Netherlands and Germany. It decayed on April 15, 1974.

1972 November 23   The fourth and last Soviet N-1 launch vehicle lifted off from the Baikonur cosmodrome at 09:18:07 UT. It had some modifications over previous N-1s, including more powerful first-stage engines, a better guidance and control system and improvements to the propellant delivery system. At 1 min. 30 sec. into the flight, the six inner first-stage engines shut down as planned, leaving the outer twenty-four engines to continue burning until a planned time of 1 min. 56 sec. At 1 min. 36 sec., however, several fires started among the active engines due to fractured propellant lines severed by the shock of the inner engines shutting down. A fire extinguishing system failed, and at 1 min. 45 sec. explosions began, 2 sec. before the control system shut down all remaining engines. The launch escape system activated and the N-1 was destroyed by the range safety engineers at 1 min. 48 sec. with the vehicle at a height of 25 mi. Two more N-1 launchers were prepared, the first to have been launched in August 1974, the second in December 1974. Earlier that year Vasili Mishin had been replaced as head of the space program by Valentin Glushko, a bitter opponent of the N-1 lunar launch vehicle who promptly canceled the program and cleared the way for the Energia. As late as 1993, there were still 94 pristine engines from the N-1 program in addition to 50–60 experimental test engines available for sale from a program which until the early 1990s remained a closely guarded secret.

1972 December 6   NASA announced that 90 scientists from the United States, France, Sweden, Germany and the United Kingdom had been selected to participate in the Mariner-Jupiter-Saturn 1977 (MJS-77) mission. Representing 32 institutions, the scientists were to begin development of the broad range of equipment to be carried on the Mariner-class spacecraft. Although not presented as a firm plan, NASA expected to fly the spacecraft on to rendezvous with either Uranus or Neptune or both, but only if the primary mission objectives at Jupiter and Saturn were met in full.

1972 December 7   Before a record 3,503 representatives of the world press *Apollo 17*, the last manned mission to the moon, began at 05:33:00 UT with the spectacular night launch of Saturn V AS-512 carrying astronauts Eugene Cernan, Ronald Evans and Harrison ("Jack") Schmitt. With a liftoff weight of 6.445 million lb., the payload comprised CSM-114 (66,953 lb.) and LM-12 (36,279 lb.), the heaviest lunar module of them all. The LOI (lunar orbit insertion) burn took place at 86 hr. 14 min. 23 sec. and the landing of lunar module *Challenger* occurred at 110 hr. 21 min. 57 sec. after a 12 min. 4 sec. powered descent. Evans in CSM (command/service module) *America* conducted detailed science from lunar orbit with panoramic and mapping cameras. Cernan and Schmitt deployed the ALSEP (Apollo lunar surface experiments package) on their first 7 hr. 11 min. 53 sec. EVA (extravehicular activity) and traversed a total distance of 2.2 mi. on the LRV (lunar roving vehicle). Equipments included a heat-flow experiment, seismic profiling equipment, an atmosphere composition experiment, lunar ejecta and micrometeorites experiment, gravimeter experiment, neutron probe, cosmic-ray detector and a traverse gravimeter. The second EVA lasted 7 hr. 36 min. 56 sec., the longest conducted on the lunar surface, and the crew conducted a 12.7-mi. traverse with the LRV during an extensive geological survey of the Taurus-Littrow region. On the third EVA, which lasted 7 hr. 15 min. 8 sec., Cernan and Schmitt completed their field tour with a 7.5-mi. traverse on the rover. They logged a record 22 hr. 3 min. 57 sec. exploring the landing site, collecting a record 243.5 lb. of surface material. Before climbing back in *Challenger*, Cernan unveiled a plaque which read: "Here man completed his first exploration of the moon, December 1972 AD. May the spirit of peace in which we came be reflected in the lives of all mankind." After a 74 hr. 59 min. 40 sec. surface stay, another record, the ascent stage lifted off and rendezvoused with *America*. Another two days were spent on orbital science tasks before TEI (trans-Earth injection) at 234 hr. 2 min. 9 sec. Evans conducted a 1 hr. 5 min. 44 sec. EVA to retrieve SIM-bay film cassettes, and splashdown occurred on the thirteenth day at 301 hr. 51 min. 59 sec., ending the longest Apollo mission of all.

1972 December 8   At a meeting in Bonn, Federal Republic of Germany, Germany's Aerospace Minister Klaus von Dohnanyi and Britain's Aerospace Minister Michael Heseltine agreed to cease contributions to the European launcher program, effectively abandoning the Europa III. The French proposed a new vehicle which they called the L3S, incorpo-

rating the L-150 stage from Europa III, a new second stage called L-30 and a scaled-down version of the Europa III H-20 second stage as the new cryogenic third stage, with a thrust of 13,000 lb., called H-6. Both first and second stages would have Viking 2 engines burning UDMH/N_2O_4 while the third stage would have an HM-4 engine burning hydrogen and oxygen. Liftoff thrust would be 529,200 lb. During January 1973 the second stage was revised to an L-35 size (35,000 kg, or 77,175 lb. of propellant) with a thrust of 152,000 lb.

1972 December 11   A Delta 2910 launched the fifth Nimbus weather and environmental research satellite from Vandenberg AFB, Calif., to a 675 × 683 mi., 99.7° orbit. The 1,705-lb. satellite carried six instruments for making the first vertical temperature and water vapor measurements of the atmosphere as well as providing data on the temperature of the Earth's crust and an infrared map of the Gulf Stream. *Nimbus 5* was the first in the series to begin the process of mapping and measuring ozone levels in the upper atmosphere. Because it screens Earth from solar ultraviolet radiation, the ozone is important for the health of living things. Scientists were interested in measuring changes that might indicate a thinning in the ozone screen. Further measurements were made by *Explorers 51*, *54* and *55* indicated that ozone levels were being depleted.

1972 December 15   Responding to weight and size growth in the projected design for the reusable Shuttle, NASA's Requirements Change Board approved a major downsizing in the overall size of the vehicle. The orbiter gross weight was fixed at 150,000 lb. and the return payload capability cut from 40,000 lb. to 25,000 lb. As noted in a briefing chart, the 20,000-lb. savings in orbiter gross weight was "basically a means of lowering cost-per-flight," since the total vehicle would be much lighter and smaller. This triggered a change to a Vehicle 2A configuration from the primary orbiter contractor, North American Rockwell. The company estimated the downsized orbiter would cut total Shuttle liftoff weight from 5.246 million lb. to little more than 4 million lb. and reduce cost-per-flight from $12 million to $10–10.5 million.

1972 December 20   Major decisions concerning the future of European space activity were made at the European Space Conference in Brussels attended by representatives from 12 nations. A commitment was given to merge the European Space Research Organization (ESRO) and the European Launcher Development Organization (ELDO) into a single agency by January 1, 1974. Although its inauguration was delayed, the European Space Agency (ESA) permitted each member country to participate in selected programs. Agreement was reached to develop an autonomous European satellite launcher, the French L3S, using advanced technology and a cryogenic upper stage. Full backing was also given to continue talks with the United States over participation in the post-Apollo program.

1972   Development tests with a new generation of Soviet SS-N-8 submarine-launched ballistic missile (SLBM) were successfully concluded prior to deployment during 1973. The liquid propellant missile had a length of 42 ft. 6 in., a diameter of 5 ft. 5 in. and was equipped with one 1-MT reentry vehicle delivered with an accuracy of 5,000 ft. across a range of 5,000 mi. The Mod 2 version appeared in 1974 and featured two 800-KT reentry vehicles and an increased range of 5,600 mi. The missile was fitted to Hotel-, Golf- and Delta-class submarines. Peak deployment of 292 missiles was achieved during 1979 and the missile was eventually retired during the 1990s under the terms of the START (Strategic Arms Reduction Talks) agreement.

1973 January 4 The Federal Communications Commission approved a request from Western Union Telegraph Co., to build the first domestic communications satellite system in the United States. Each of the three satellites, named Westar and built by Hughes Aircraft, would carry 12 transponders and would be capable of relaying 1,200 voice circuits, one color TV channel with audio or data at 50 megabits/sec. The FCC had six other requests under review and announced that it would do all it could to encourage commercial satellite operations by licensing all qualified applicants. *Westar 1* was launched on April 13, 1974.

1973 January 5 Budget restrictions forced NASA to redefine the size and scope of its High-Energy Astronomy Observatory (HEAO) program. Instead of two 21,600-lb. satellites launched by Titan IIIE, the HEAO program was to be suspended for one year and redirected toward three smaller satellites, each weighing less than 6,600 lb. During the year the program was rescoped and Atlas-Centaur selected as the launch vehicle. On April 10, 1974, NASA announced that TRW Systems, Inc. would be awarded a contract for development of the three HEAO satellites.

1973 January 8 At 06:55 UT a Soviet SL-12 launcher lifted off from Baikonur and sent *Luna 21* and the moon rover *Lunokhod 2* on a translunar trajectory. Following two course corrections, the spacecraft entered a 56 × 68 mi. orbit inclined 60°, with a period of 1 hr. 58 min., on January 12. In orbit change maneuvers over the next two days, flight controllers reduced perilune to 9.9 mi., from which the spacecraft descended to a landing at 22:35 UT on January 15. *Luna 21* landed with a vertical velocity of only 6.5 ft./sec. inside the crater Lemonnier, 112 mi. north of the *Apollo 17* site. The 1,852-lb. *Lunokhod 2* descended to the surface 3 hr. 39 min. later to begin a comprehensive sequence of maneuvers. Much improved over *Lunokhod 1*, it could move at almost twice the speed and had a high-mounted panoramic camera as well as four redesigned stereo cameras.

1973 January 18 European Space Research Organization (ESRO) members agreed unanimously to back development of the Sortie Lab, a pressurized laboratory designed to be carried inside the Shuttle cargo bay and operated by up to six scientist-astronauts in space. Participating countries included West Germany, Italy, Belgium and Spain. Other ESRO member countries were expected to join later. NASA wanted the Sortie Lab delivered by 1979, coincident with the planned inauguration of Shuttle flights. Development cost was estimated at $250–300 million.

1973 January 20 Responding to a requirement from NASA that the Shuttle design be reduced in size while retaining the basic payload capability, Rockwell completed preliminary design of the Vehicle 2A configuration. With a maximum empty weight of 150,000 lb., the orbiter now had a 45°/79° double-delta wing with a surface area of 2,690 sq. ft. and a span of 78 ft. 1 in. As a result, subsonic lift/drag ratio had dropped from 5.1 to 4.9. Landing weight was reduced to 179,800 lb., but minimum touchdown speed had increased by 15 knots to 165 knots. The air-breathing engines fitted for ferry and atmospheric flight tests had been relocated from above the rear fuselage to under the wings. The external tank

was resized to a length of 165 ft. 10 in. and a diameter of 27 ft., and the solid rocket boosters to a length of 145 ft. 1 in. and a diameter of 11 ft. 10 in. The complete assembly now had a height of 192 ft. 3 in. and a liftoff weight of 4.116 million lb. Anticipated budget problems in the next few years had moved the first manned orbital flight from March to the third quarter of 1978.

1973 January 25 The U.S. National Oceanic and Atmospheric Administration (NOAA) announced a Global Atmospheric Research Project (GARP) Atlantic Tropical Experiment scheduled to run from June 15, 1974 to September 30, 1974. During this period, one-third of the Earth's surface would be observed with the aid of 6 weather satellites, 38 ships, 65 instrumented buoys, 13 aircraft and a network of almost 1,000 land stations to observe phenomena from the top of the atmosphere to an ocean depth of 4,900 ft. GARP was designed to improve tropical forecast techniques for the equatorial belt and develop means of assessing pollution.

1973 January 30 NASA announced the U.S. crew for the joint Apollo-Soyuz docking flight (Apollo-Soyuz Test Project, or ASTP) naming Thomas P. Stafford, Vance D. Brand and Donald K. Slayton as the prime crew. Only Stafford had flown in space before, and Slayton, one of the original seven Mercury astronauts, had been grounded since 1962 after physicians detected a mild heart condition. Slayton's persistence in proving himself fit for spaceflight paid off when doctors reexamined his medical status. ASTP backup crew members were Alan L. Bean, Ronald E. Evans and Jack R. Lousma, with astronauts Robert Overmyer, Bob Crippen and Karol Bobko as the support team. The Soviet crew members—Alexei Leonov and Valery Kubasov as primary members, and Vladimir Dzhanibekov and Yuri Romanenko as backups—were announced on May 24, 1973.

1973 January Tests began with a Mod 4 version of the SS-9 Soviet heavyweight ICBM. Utilizing three 3.5-MT multiple reentry vehicle (MRV) warheads, the Mod 4 was the first Soviet ICBM to carry MRVs. Unlike previous MRV tests utilizing the SS-9, the new RVs were recovered by parachute; in tests conducted later this year some steps were taken to develop independent targeting, so that some warheads from the same missile could be redirected to separate targets. These MIRV (multiple independently targeted reentry vehicles) would be deployed first on the SS-17 and SS-19. The SS-9 did not carry MIRVs.

1973 February 6 The NASA Marshall Space Flight Center set up a Large Space Telescope (LST) Task Force to coordinate planning and the preliminary design phase of the mantended optical observatory scheduled for launch by the Shuttle in the 1980s. Beginning April 23, a committee of 35 astronomers representing 7 observatories and 10 universities and NASA facilities began a three-day meeting at the MSFC to review possible experiments for the LST.

1973 February 13 At the NASA Marshall Space Flight Center, a Space Tug Task Team was set up to coordinate activities on the in-orbit vehicle designed to move payloads from one orbit to another or propel spacecraft to interplanetary trajectories. On February 14 NASA selected General Dynamics, Convair Aerospace and McDonnell Douglas to study tugs with cryogenic propulsion, while Grumman and Martin Marietta were to study tugs utilizing storable propellants.

1973 February 15 NASA's *Pioneer 10* completed its traverse across the asteroid belt between the orbits of Mars and Venus without any damaging impacts from the tiny particles that occupy the region. *Pioneer 10* was now 250 million mi. beyond the Earth's orbit around the sun and 200 million mi. beyond the orbit of Mars. For the first time, a spacecraft was able to sample the environment of the solar system far from Earth and take measurements of the solar wind. *Pioneer 10* observed that the strength of the sun's magnetic field, the intensity of the solar wind and the numbers of solar high-energy particles all decline approximately as the square of the distance from the sun.

1973 February 17 The NASA Manned Spacecraft Center was renamed the Lyndon B. Johnson Space Center in honor of the late President who had died on January 22, 1973. Johnson had been the architect of NASA and played a leading role in placing the idea of a manned lunar landing before President John F. Kennedy in April 1961. Mobilizing congressional support for Apollo, Johnson paved the way for the expanding budgets and resources that gave the civilian space agency the means to carry out that mandate. Dedication ceremonies were held August 27 to coincide with the 65th anniversary of Johnson's birthday.

1973 February 28 The U.S. Air Force selected Philco-Ford to build second-generation military communications satellites for the North Atlantic Treaty Organization. They would be used to handle NATO's long-range communication requirements and supersede the first-generation (NATO I and II) satellites. Designated NATO III, the first was launched on April 2, 1976.

1973 March 6 The first U.S. geostationary electronic intelligence gathering satellite, *Rhyolite*, was launched by Atlas-Agena D to a position above the Horn of Africa. The 610-lb. *Rhyolite* (named for a gray volcanic ash containing crystals of quartz and feldspar) had been 10 years in development at TRW Systems, Inc. With a folded length of 20 ft. and a diameter of 4 ft. 6 in., *Rhyolite* deploys several solar-cell panels, a wire mesh antenna 70 ft. in diameter and several smaller dish antennas. The satellite collects telemetry signals from Soviet and Chinese ballistic missile tests, as well as radio, radar, microwave and telephone traffic. Other Rhyolites were launched on May 23, 1977 (over Borneo), with spares on December 11, 1977, and April 7, 1978.

1973 March 10 IBM received the first Shuttle software contract from NASA. IBM also received a contract from Rockwell International for Shuttle computers. The Shuttle was the first aerospace vehicle in which every function was controlled by a suite of avionics. Five AP-101 computers would provide an unprecedented level of safety through redundancy and self checks. Two mass memory units, each capable of containing 8 million 16-bit words, could accommodate three times the maximum software necessary for Shuttle operations.

1973 March 19 In responding to a query from Senator Lowell E. Weicker, NASA Administrator James Fletcher said that the agency was reluctant to proceed with the second Skylab orbital workshop because it might detract from funds for the Shuttle and other programs. Consideration had been given to a Skylab B mission, and McDonnell Douglas had prepared a second orbital workshop from a converted S-IVB as part of earlier plans to launch it for a series of expanded Earth-orbit visits. Some NASA people had pressed for Skylab B to fly between the Apollo-Soyuz mission in 1975 and the Shuttle flights beginning at the end of the decade.

1973 April 2 NASA requested McDonnell Douglas, Boeing, Chrysler and Martin Marietta to bid for the contract to build expendable external tanks for the reusable Shuttle. The contract was for a total of 445 tanks by 1988 in three increments, with delivery of 3 test and 6 flight tanks by early 1979, 54 tanks in the second increment between 1978 and 1981, and 385 in the third between 1981 and 1988. Production was in support of an estimated 6 flights in 1978, 15 in 1979, 24 in 1980, 32 in 1981, 40 in 1982, 60 between 1983 and 1987, and 28 in 1988. NASA asked the bidders to work up costs for an alternate-pace program involving 242 flights between 1978 and 1988. Flights from Vandenberg AFB were to begin in 1980, and with a maximum of 60 flights per year, NASA estimated 40 would be from the Kennedy Space Center and 20 from Vandenberg.

1973 April 3 At 09:00 UT a Soviet SL-13 launch vehicle successfully placed a 41,700-lb. Almaz military space station in a 134 × 161.5 mi. orbit following launch from the Baikonur cosmodrome. Designed by the Chelomei bureau, this first Almaz was confusingly referred to as *Salyut 2*, the name "Salyut" having also been applied to the Korolev station. Almaz had a length of 47 ft. 9 in., a maximum diameter of 13 ft. 7 in. and a massive high-resolution reconnaissance camera with a 19 ft. 8 in. focal length that shot 1 ft. 7.7 in. square pictures. During on-orbit systems test an electrical fire broke out, increasing internal pressure and causing the hull to burst. On April 18 the Soviets announced that the mission of *Salyut 2* had ended. After partial disintegration of the station, 24 separate pieces were tracked in orbit. What remained of *Salyut 2 (Almaz No. 1)* decayed on May 28 after only 55 days in space.

A team of scientists in the United States selected the landing sites on Mars for NASA's Viking spacecraft, scheduled for launch in 1975. *Lander 1* would head for a site at 19.5° N by 34° W in the region known as Chryse, located at the northeast end of a 3,000-mi. long canyon. A backup site was selected at Tritonis Lacus, located at 20.5° N by 252° W. *Lander 2* would head for a site in the Cydonia area of the Mare Acidalium region at 44.3° N by 10° W. Its backup was at the Alba region, 44.2° N by 110° W. Each site had to have minimal chance of surface obstructions that could threaten the stability of the lander, but with the best pictures showing nothing smaller than 100 ft., that was largely educated guesswork.

1973 April 6 At 02:11:00 UT Atlas-Centaur AC-30 launched the 570-lb. *Pioneer 11* onto a flight path to the planet Jupiter, where it would arrive on December 4, 1974. The spacecraft was almost identical to its predecessor, *Pioneer 10*, which was now 390 million mi. from Earth and 111 million mi. from Jupiter. A course correction maneuver was accomplished on April 11, 1973, through a 125 ft./sec. burn of the spacecraft's thrusters. *Pioneer 11* was launched with the option of making either a close pass over Jupiter's equator, making a close pass across the south pole or using the gravity field of Jupiter to deflect its trajectory back across the solar system for a rendezvous with Saturn. Much would depend on how well *Pioneer 10* performed when it encountered Jupiter at the end of 1973. *Pioneer 11* traversed the asteroid belt between August 18, 1973 and March 12, 1974.

1973 April 11 A consortium of European manufacturers was appointed to build the first in a series of European Communication Satellites. To be known as Orbital Test Satellites (OTS), they would conduct research into superhigh-frequency (SHF) communications via satellite. SHF frequencies have a greater capacity per given bandwidth than lower frequencies, but SHF signals are badly affected by adverse weather. European telecommunications engineers wanted a test satellite to provide signals to measure these effects and trade the obvious merits of high capacity against the much more difficult operational problems. Britain's Hawker Siddeley Dynamics would be prime contractor for OTS, which was to be launched on a U.S. launcher.

1973 April 13 Prime Minister Edward G. Whitlam and NASA Administrator James Fletcher opened a dish antenna 210 ft. in diameter at Tidbinbilla Deep Space Communications Complex near Canberra. As part of the Deep Space Network, it would join the station at Goldstone, Calif. (210 ft. in diameter) and a similar antenna at Madrid, Spain. When complete, the global network would permit at least one station to be in sight of any given interplanetary spacecraft at all times.

1973 April 16 A two-day meeting on the Shuttle and Sortie Lab programs at NASA's Marshall Space Flight Center was attended by 20 representatives of the European Space Research Organization and European industrial consortiums headed by MBB (Messerschmidt-Bölkow-Blohm) and ERNO. The group received briefings on these programs and then presented NASA with preliminary determinations concerning Sortie Lab, which the Europeans preferred to call Spacelab. On May 3–4, ESRO (European Space Research Organization) representatives met with U.S. space officials in Washington, D.C., and signed an agreement to develop the $300–400 million laboratory. Europe was to pay for Spacelab development and for assembly and delivery of the first module, to be followed by four more which the the United States would buy. In reality, costs soared to more than $1 billion and the United States bought only one more, for $250 million.

1973 April 23 NASA informed the General Accounting Office of new projections for Shuttle traffic rates, increasing from 581 flights to 779 the number of missions it estimated would be required for the period 1978–1991. The GAO was conducting an assessment of Shuttle economics for Congress. NASA said that about one-third of the Shuttle flights would be taken up with Sortie Lab missions. The basis for a projected $16 billion saving in overall space program costs during this period was that the Shuttle would stimulate the use of reusable payloads and modularized satellites that could be repaired or refurbished in orbit for less than the replacement cost. When the GAO issued its report to Congress on June 1, it cast doubt on the NASA figures and recommended that justification of the Shuttle be made on "other than economic grounds."

1973 April 27 The development of the Europa II launch vehicle was formally abandoned at a council meeting in Paris when the European Launcher Development Organization (ELDO) learned that France and Germany were not prepared to invest more money. The F.12 test flight had been scheduled for October 1. Cancellation also brought the dissolution of ELDO, which had spent $732 million since 1963 on a European satellite launcher which had yet to emerge. Thus far $690 million had been spent on Europa I and II. Only $40 million

had been spent on the Europa III concept, which was expected to go ahead under a different designation when the newly agreed European Space Agency was set up.

NASA decided to delete the "thrust termination" system on the Shuttle solid propellant rocket boosters (SRBs). The system comprised a series of sensors which would shut down the solid rocket motors in the event that they detected an anomaly in an SRB. They would do this by blowing off the nose cap of each SRB, thereby extinguishing booster thrust by allowing exhaust gases to escape from both ends. This would allow time for the orbiter to separate from the external tank and fly back to the launch site. The structural loads encountered by the orbiter in such a maneuver would, however, necessitate an additional 19,600 lb. of additional strengthening, and the thrust termination concept was eliminated as a weight saving measure.

1973 April Further refining the basic design, engineers at the Space Division of Rockwell International and the NASA Johnson Space Center developed the Shuttle orbiter Vehicle 3 configuration. Within a month the design of *Vehicle 3A* had appeared, an orbiter with a length of 122 ft. 10 in. and a (45°/81°) double delta-wing shape with a span of 78 ft. 1 in. Smaller and lighter than the Shuttle vehicle originally proposed when North American got the contract in July 1972, the overall vehicle now had a length of 181 ft. 4 in., including an external tank (ET), with a length of 155 ft. 5 in. and a diameter of 27 ft., and two solid rocket boosters (SRBs), each with a length of 145 ft. 1 in. The Shuttle now had the capability of landing with a 32,000-lb. payload in the cargo bay. The first orbital flight was still scheduled for the third quarter 1978. Further orbiter design refinements introduced the Vehicle 4 configuration with a length of 122 ft. 4 in. and several minor internal changes.

U.S. Navy and Air Force satellite navigation experimental research programs were combined and the Air Force was given jurisdiction over a program to develop a highly accurate operational system. The Air Force brought in their System 621B studies, which had been under way for four years. It called for three satellite constellations of four satellites in orbits at 120° intervals; one satellite in each constellation would be in geostationary orbit, three in elliptical, slightly inclined 24 hr. orbits. The Navy had been working on its Timation series of satellites. When combined, the joint effort would be called the Navigation Technology Satellite program.

1973 May 7 A five-day meeting to plan the U.S.–West German Helios project began at the Kennedy Space Center, Fla., with 200 members of a joint working group. The Helios project envisaged two West German spacecraft launched by NASA on an interplanetary orbit that would bring them to within 28 million mi. of the sun, closer than any other spacecraft. Named for the Greek god of the sun, Helios would penetrate the outer solar corona where charged particles receive their final acceleration. *Helios A* was launched on December 10, 1974, followed by *Helios B* on January 15, 1976.

1973 May 11 At 00:20 UT a Soviet SL-13 launch vehicle carried a 41,700-lb. Salyut space station from the Baikonur cosmodrome into an Earth orbit of 133 × 151 mi. inclined 51.6° to the equator. Shortly after reaching orbit the station began to malfunction, and plans to launch a crew to dock with it were abandoned. The Soviets chose to give it the designation *Cosmos 557*. Had *Salyut 2 (Almaz 1)* performed as planned and *Cosmos 557* worked successfully (thus being recognized as

Depicted in this artist's representation, NASA's Skylab space station comprised a converted S-IVB rocket stage (bottom), with attached solar array "wings," airlock module (inside the truss structure), and multiple-docking adapter (to which Apollo is docked). Off to one side is the Apollo telescope mount.

Salyut 3) the Soviets would have achieved their objective of having two manned space stations in orbit simultaneously when NASA launched Skylab on May 14, 1973. In fact, the whole effort was a fiasco, with neither station working. *Cosmos 557* decayed naturally back through the atmosphere eleven days after launch and burned up.

1973 May 14 The Skylab workshop cluster was launched from Kennedy Space Center's Complex 39A at 17:30 UT. With a liftoff weight of 6.211 million lb., the two-stage Saturn V AS-513 was programmed for a unique trajectory. S-IC center-engine cutoff occurred at 2 min. 20 sec., followed by the remaining four outer engines at 2 min. 38 sec., at an altitude of 52 mi. and a velocity of 9,129 ft./sec. S-II ignition came at 2 min. 41 sec., center-engine cutoff at 5 min. 14 sec. and outer-engine cutoff at 9 min. 48 sec. The 195,052-lb. orbital workshop cluster separated from the second stage 2 sec. later and entered a 266.9 × 269.3 mi. orbit with an inclination of 50°. About 63 sec. after liftoff one of the two Skylab solar array wings had been torn free from the side of the workshop by air pressure, carrying with it a micrometeoroid shield that doubled as thermal protection for the workshop. Soon after reaching orbit, internal temperatures began to rise, eventually stabilizing at a maximum 125°F. Moreover, debris caused by the separation of one of the two solar array wings snagged the second array, jamming it in its folded position against the workshop. Virtually uninhabitable due to high temperatures and with almost no electrical power from the remaining orbital workshop solar array, plans to send up

the first crew a day later were canceled. Teams at the Manned Spacecraft Center and the Marshall Space Flight Center immediately began developing procedures whereby the first crew could attempt to deploy both the snagged solar array and a thermal shield.

NASA reported that a total of $5,548 had been sent in by well-wishers to finance 19 student science experiments on the Skylab program. The donations ranged from $1,500 to a single contribution of $0.35 from an eight-year-old boy who had heard news that the space agency "didn't have enough money." Other contributions included the sum of one day's pay contributed after each manned flight by a Navy chief petty officer in memory of his son who had been killed in the Korean War.

1973 May 15 In pursuit of its aspiration to be a space-faring nation, India fired a two-stage rocket from its Sriharikota range. The first stage used solid propellants and produced a thrust of 5,950 lb., but the second stage was a liquid propellant rocket producing a thrust of 1,320 lb. Both stages had a diameter of 10 in. The first stage was a sounding rocket, but the domestically developed and more complex liquid propellant second stage was part of India's program to build a satellite launcher.

1973 May 21 Two French satellites were lost when the Diamant B launcher carrying them toward space was de-

Two French satellites, Castor *and* Pollux, *are prepared for launch into space by a Diamant rocket in late 1972.*

stroyed after the third stage failed to fire following launch from Kourou, French Guiana. The 79-lb. satellite designated D-5A *Pollux* was to have tested a hydrazine thruster for seven months, and the other satellite, the 168-lb. D-5B *Castor,* was to have tested orbital perturbations over a six-year period. Both satellites were to have been placed in a 186 × 802 mi. orbit. This was the last planned launch of a Diamant B.

1973 May 23–24   A review of various thermal shield designs to be carried by the first Skylab crew was held at the NASA Marshall Space Flight Center. Intended to replace the thermal protection torn away during the launch of Skylab on May 14, three from the wide range of designs were selected for the crew to carry with them and deploy in space. Two could be deployed by a space-walking astronaut laying out blanketlike covers over the hull of the workshop, and one, taking the form of a folded umbrella, could be pushed out through a small airlock from the inside of the workshop and opened to provide a shade.

1973 May 25   The first manned flight to the Skylab space station began at 13:00 UT from Complex 39B at Kennedy Space Center with the launch of Saturn IB AS-206 carrying CSM-116 and astronauts Charles Conrad, Paul Weitz and Joseph Kerwin. With a liftoff weight of 1.29 million lb., the launch vehicle for this SL-2 mission placed the 36,007-lb. spacecraft in a 92.6 × 218.8 mi. orbit. Through a sequence of orbital maneuvers, the CSM (command/service module) rendezvoused with Skylab on the fifth revolution at 7 hr. 30 min. elapsed time. Reporting that the remaining Skylab solar array boom was snagged by wire, permission was given to attempt to dislodge the debris using a 10 ft. pole. A 37-min. stand-up EVA (extravehicular activity) performed by Weitz through the open side hatch of the command module failed to free the boom. Initial attempts to dock at the multiple docking adapter failed, and not before an elapsed time of 14 hr. 50 min. was the CSM latched up. After a sleep, the crew entered the workshop at 1 day 3 hr. 45 min. and deployed an umbrellalike shade through the small airlock in the workshop wall at 1 day 11 hr. 30 min. The high temperatures to which the Skylab interior had been exposed since shortly after launch quickly began to fall, and by mission day four the atmosphere was down to 90°F. Starting at 15:23 UT on the 14th day in orbit (June 7), Conrad and Kerwin performed a 3 hr. 30 min. EVA and freed the stuck solar array boom. Another EVA was performed on the 26th day (June 19). Lasting 1 hr. 44 min., it began at 10:53 UT when Conrad and Weitz started retrieving film cassettes from the Apollo telescope mount. In preparation for return to Earth, undocking occurred at 08:55 UT on June 22, 1973, followed by the de-orbit burn at 13:11 UT and splashdown in the Pacific Ocean at an elapsed time of 28 days 49 min. 49 sec.

1973 June 3   Soviet scientists announced that the *Lunokhod 2* moon rover had completed its work on the lunar surface at the eastern edge of the Mare Serenitatis. In five consecutive lunar days, the vehicle had covered a total distance of 121,390 ft., or 23 mi., taken 80,000 TV images, of which 86 formed comprehensive panoramas, and made 740 mechanical tests of surface materials. In total, *Lunokhod 1* and *Lunokhod 2* had rolled 156,000 ft. (29.5 mi.) across the lunar surface. Soviet scientists justified unmanned exploration of the moon and the planets as both cheaper and more flexible than maned exploration.

Launched June 10, 1973, Radio Astronomy Explorer-B *is encapsulated by the Delta launcher payload shroud which will protect it in its ascent through Earth's atmosphere.*

1973 June 10   A Delta 1913 carrying nine solid propellant strap-on motors launched the 723-lb. *Explorer 49*, *Radio Astronomy Explorer-B (RAE-B)* satellite, at 14:14 UT from Cape Canaveral, Fla., to a translunar trajectory. Designed to study low-frequency radio signals from galactic and extragalactic sources, *Explorer 49* fired its lunar orbit insertion motor for 22 sec. at 07:21 UT on June 15, slipping to a 697.8 × 828.9 mi., 61.3° lunar orbit. On June 18, after insertion motor jettison, an on-board hydrazine propulsion system trimmed this to a 653.7 × 660.5 mi. orbit inclined 38.7°. The 3-ft. cylindrical satellite, 2 ft. 7 in. tall and now weighing 442 lb., extended a 120-ft. dipole antenna, a 675-ft. gravity gradient stabilization boom and four 750-ft. radio antennas to form a giant "X" with a span of 1,500 ft. Four fixed solar array panels provided 38 W of power.

1973 June 12   The first in a series of improved U.S. geostationary Defense Support Program early warning satellites was launched by Titan IIIC from Cape Canaveral, Fla. Later in the decade the second-generation DSP satellites were launched. With a weight of 3,690 lb., they each carried larger solar-cell arrays than the initial Block 647 satellites and produced 650 W of electrical energy. In the early 1980s the U.S. Air Force developed a special detector array with 6,000 infrared cells. Intended for use in the third-generation Block 14 DSPs, they were retrofitted to the last two basic Block 647 satellites launched. In all, the Air Force flew 13 DSP satellites, the last by Titan 34D on November 29, 1987.

1973 June 15   A Soviet SL-4 launch vehicle carried a 14,700-lb. unmanned Soyuz spacecraft, designated *Cosmos 573*, from the Baikonur cosmodrome into an Earth orbit of 119 × 194 mi. This mission closely followed the profile of its predecessor, *Cosmos 496* launched on June 26, 1972, but remained in space for only six days. In doing so it simulated later Soyuz space station ferry missions and conformed to the now conventional two-day "life" of the spacecraft dictated by its reliance on batteries instead of solar cells for electrical power.

1973 June 29 Soviet scientists created, for the first time, an aurora in the upper atmosphere in an experiment known as *Zarnitsa*. A sounding rocket carried an electron accelerator into the upper atmosphere between 60 mi. and 100 mi., and electron energies of 7,500–9,000 eV were achieved in a 4 kW beam. Free electrons formed the artificial aurora, mimicking the natural phenomenon caused by ionization of molecules struck by electrons from solar disturbances.

1973 June NASA issued a request for proposals from industry to design and develop the Pioneer-Venus spacecraft. Originally expected to consist of a multiprobe launch in 1976/77 and an orbiter in 1978, both launched on Delta launchers, the mission now envisaged the launch of these two spacecraft at the 1978 planetary opportunity on Atlas-Centaur launchers. NASA assessed the cost of miniaturizing components of the multiprobe as at least equal to the extra cost of shifting the flights to the bigger launcher, but with less involvement of high technology and the assurance of a proven launch vehicle; no planetary mission had ever been launched on a Delta.

1973 July 2 The prospect of a Soviet shuttle-type space plane was discussed during a Radio Moscow broadcast to West Germany. The concept envisaged a winged fly-back booster carrying an orbiter to a height of approximately 19 mi. and a speed of 4,900 MPH before separating and returning to Earth. The space plane would use its own propulsion to reach a 124-mi. orbit and return to Earth using air-breathing engines from a height of 19–25 mi. at the end of its mission. In September, cosmonaut Gherman Titov provided additional information, saying that the two-stage vehicle would takeoff from a runway.

1973 July 8 Ten Soviet cosmonauts arrived at the NASA Manned Spacecraft Center to begin flight training for the planned 1975 Apollo-Soyuz Test Project (ASTP) mission. They were given briefings on the Apollo spacecraft, visited Rockwell International, where CSM-111 was being prepared for the mission, and returned to the Soviet Union after two weeks in the United States. In November 1973 the NASA astronaut flight crew assigned to the ASTP mission spent 11 days examining the Soyuz simulators at the Zvezdny Gorodok cosmonaut training facility in the Soviet Union. Subsequently visits and meetings of the ASTP working groups preparing for the flight became more frequent.

1973 July 18 NASA issued requests to Aerojet, Thiokol and United Technology Corp., for bids on the partially reusable Shuttle solid propellant rocket motors. Bids were to be in by the end of August, based on a contract built around three incremental procurement phases: the first to deliver by September 1979 sufficient hardware for 6 development flights; the second, beginning early 1978, for the production of new and refurbished motors for 54 flights; the third, beginning July 1980, for new and refurbished units for 385 flights. The solid rocket motor cases were to be recovered from the ocean and reused.

RCA announced that in association with its subsidiaries RCA Global Communications and RCA Alaska Communications, it had signed a contract with McDonnell Douglas for the modification of the Delta launch vehicle enabling it to carry heavier communication satellites. This was the first time a commercial company had paid for the upgrading of a launch system, and under the agreement the Delta payload capacity would increase by 30%.

1973 July 21 A Soviet SL-12 launch vehicle successfully sent the first of four spacecraft dedicated to the exploration of Mars on its way following a brief Earth parking orbit. Designated *Mars 4,* it was followed into space by *Mars 5* launched on July 25, *Mars 6* on August 5, and *Mars 7* on August 9. Each spacecraft weighed about 8,030 lb., less than the optimum 10,500 lb. of their design, because the 1973 launch window was particularly unfavorable for the performance of the launch vehicle. Each spacecraft had weight for either propellant to decelerate into Mars orbit or for a lander that would make a direct entry and descend to the surface through a combination of parachute and retro-rockets. *Mars 4* and *Mars 5* were intended to orbit the planet, but *Mars 6* and *Mars 7* were landers. The plan was that the *Mars 4* and *Mars 5* orbiters would relay data to Earth from the landers *Mars 6* and *Mars 7*, although the landers were capable of direct data transmissions to Earth.

1973 July 28 Following 36 days of unmanned operations, NASA's Skylab space station received its second crew when Alan Bean, Owen Garriott and Jack Lousma were launched on the SL-3 mission from Kennedy Space Center's Complex 39B at 11:10:50 UT. With a liftoff weight of 1.29 million lb., Saturn IB AS-207 placed the 36,073-lb. CSM-117 in a 93.2 × 138 mi. orbit 10 min. 2 sec. later. Rendezvous with Skylab occurred at 8 hr. 27 min., and the crew entered the multiple docking adapter about 24 min. later. For the first few days the crew suffered from nausea caused by initial disorientation (space sickness), but low work loads and an adjusted diet eased them back into the schedule by the fifth day. The first EVA (extravehicular activity) lasted 6 hr. 29 min., beginning at 17:30 UT on August 6, when Garriott and Lousma erected a twin-pole sunshade, more permanent than the umbrella shade deployed by the SL-2 crew. The second EVA began at 16:24 UT on August 24, when Garriott and Lousma spent 4 hr. 30 min. outside replacing gyroscopes for the attitude control system in the Apollo telescope mount. The third EVA lasted 2 hr. 45 min., starting at 11:18 UT on September 22, when Bean and Garriott retrieved ATM film cassettes. The Apollo CSM undocked from Skylab at 19:34 UT on September 25, 1973, with splashdown at 22:20 UT after 59 days 11 hr. 9 min. 4 sec.

1973 August 1 At a European Space Conference in Brussels, ministers from 11 countries agreed to go ahead with development of the Spacelab pressurized laboratory for use inside the Shuttle cargo bay. Spacelab would also comprise a pallet, or platform, carried either aft of the pressure module or as a string of pallets to which unpressurized equipment or experiments not affected by exposure to the vacuum of space could be attached. At this stage, West Germany agreed to provide 52.65% of the funds for Spacelab; France accounted for 10%, Britain 6.3% and the other eight countries for the balance. Also agreed was the recommendation to proceed with the L3S launcher, which was to be called Ariane, with France paying 62.5% of the cost, Germany 18.8%, Belgium 5%, Britain 2.65%, Spain 2% and the other countries the remainder.

Development of MAROTS (maritime operational technology satellite) was also agreed, Britain paying 56%% and Germany 20%, with the balance undecided.

1973 August 9 The NASA Marshall Space Flight Center awarded contracts to Itek Optical Systems Division and the Perkin Elmer Corp., for competing designs and definition of the Large Space Telescope. The LST was to use a primary mirror 120 in. in diameter and be capable of resolving objects 10 times smaller and 100 times fainter than objects seen through ground-based telescopes. The multipurpose optical facility was to be serviced by the Shuttle and routinely reequipped with upgraded instruments and equipment.

1973 August 16 NASA selected Martin Marietta for the design, development, test and evaluation of the Shuttle external tank. The ET would carry 1.5 million lb. of liquid oxygen–liquid hydrogen propellant in a structure 158 ft. long and 27 ft. in diameter, feeding oxidizer and fuel to the Shuttle orbiter via two pipes 17 in. in diameter. Each tank would be jettisoned just before the Shuttle reached orbit, falling back into the atmosphere over the Indian Ocean and burning up. The orbiter would propel itself into orbit using orbital maneuvering engines on either side of the vertical tail. ETs would be manufactured at the NASA Michoud Assembly Facility in New Orleans.

NASA decided to delay the launch of the third crew to the Skylab space station and synchronize orbital operations with the arrival of the comet Kohoutek. Discovered on March 7, 1973, by Czech astronomer Dr. Lubos Kohoutek from his observatory in Hamburg, West Germany, the comet would reach perihelion on December 28. To be in position to use the Apollo telescope mount to conduct observations of the comet as it swept to within 8.3 million mi. of the sun at 240,000 MPH, SL-4 would be launched on November 9, 1973, for an 84-day mission. Previously, it had been scheduled for launch in mid-October.

1973 August 27 During preparations for the launch of SL-4, the third manned occupancy of the orbiting Skylab space station, cracks were found in one of the forgings at the base of the Saturn IB launch vehicle AS-208. This was repaired by September 3, but during preparations for the launch on November 6, cracks were discovered in the aft attachment fittings on all eight fins at the base of the first stage. Despite consensus that the fins would not fail during launch, all eight were replaced. During the replacement activity on November 12, still more cracks were found in reaction beams linking the first and second stages. These were deemed not to be a threat to the mission.

1973 September 12 The Federal Communications Commission approved applications for domestic satellite network licenses from five U.S. companies and groupings: Fairchild Industries and Western Union International, joint owners of American Satellite Corp.; a team request from RCA Global Communications and RCA Alaska Communications; GTE Satellite Corp. teamed with National Satellite Services, a subsidiary of Hughes Aircraft Co.; American Telephone and Telegraph Co.; and Communications Satellite Corp. In addition, Communications Satellite Corp. was given permission to build satellites for lease to AT&T.

1973 September 21 Given the interim name of Vega, the French L3S proposal for a satellite launcher was brought to its definitive specification as requests for contractor bids were

announced and the tentative allocation of costs was agreed. It was expected that France would provide 62.5% of funding, West Germany 20%, Belgium 5%, Britain 2.5%, the Netherlands and Spain 2% each, Switzerland 1.5% and Italy 1%, with Sweden and Norway given the option of providing the other 3.5%.

1973 September 24 The agreement to proceed with development of Spacelab, the laboratory module and pallet assembly for use with the Shuttle, was signed by NASA Administrator James Fletcher on behalf of the US, and Director General of the European Space Research Organization Alexander Hocke for nine European nations. The cost of the laboratory module would be divided among the participating countries: Belgium, Denmark, France, West Germany, Italy, the Netherlands, Spain and Switzerland. The estimated cost of Spacelab development at 1973 prices was $369.6 million. Germany would pay 52.55% of this, but Britain's contribution had fallen to 6.3%, supporting only the pallet assembly. From this date the American term Sortie Lab was abandoned in favor of the European name, Spacelab, and a preliminary design meeting at the Marshall Space Flight Center October 9–19 was attended by 37 delegates from ESRO.

1973 September 25 NASA invited Lockheed Aircraft Corp. to examine the prospect of using a converted C-5A cargo airplane for ferrying the Shuttle orbiter piggyback between launch and landing sites. Early flights with the Shuttle would take off from Kennedy Space Center, Fla., and land at Edwards AFB, Calif. Previously, Shuttle design engineers had planned to use air-breathing engines attached to the underside of the wings for ferrying the delta orbiter. By using a carrier plane, savings could be made in development costs for the engine and there would be no need for underwing attachment points that penetrated the reusable thermal protection material.

1973 September 27 An SL-4 launch vehicle successfully placed the 14,817-lb. *Soyuz 12* spacecraft in an orbit of 112 × 142 mi. with an inclination of 51.6°. From there it was moved to an orbit of 202.5 × 214 mi. It was the first Soviet manned space mission since the deaths of the *Soyuz 11* crew in June 1971 and included cosmonauts Vasily G. Lazarev and Oleg G. Makarov. The spacecraft carried a docking system made redundant when the two previous attempts to launch unmanned Salyut stations (*Salyut 2/Almaz 1* and *Cosmos 557*) failed earlier in the year. The crew performed some Earth observations, took photographs and then prepared to return to Earth. Without solar panels, the endurance of the spacecraft was limited, and *Soyuz 12* landed at an elapsed time of 1 day 23 hr. 16 min.

1973 October 23 Further studies into the suitability of women for spaceflight were conducted at the NASA Ames Research Center when a five-week experiment began. Eight of twelve U.S. Air Force nurses had two weeks of total bed rest, while four acted as controls. Dr. Harold Sandler was head of the study, initiated to evaluate more fully the physiology of women. As with earlier studies, tests proved that women, like men, were uniquely suited to specific tasks and had certain aptitudes not found in the opposite sex. Specifically, women proved to have enhanced perspicacity, the ability to distinguish key data points from a broad base of information. NASA wanted to exploit those differences to enhance crew performance in gender-oriented tasks. This did much to enhance the role of women in the astronaut corps.

1973 October 26   Tenth and last in a series of NASA satellites designed to investigate the Earth/sun radiation environment over the 11-year sunspot cycle, *Explorer 50* (Interplanetary Monitoring Platform-J) was successfully launched by a Delta 1604 to an elliptical orbit of 122.7 × 142,174.7 mi. inclined 28.7°. At an elapsed time of 74 hr. 34 min. an apogee kick motor was fired, placing IMP-J into an 87,727.7 × 179,487.4 mi. orbit inclined 28.7°. From this 12-hr. orbit, the 818-lb. satellite activated 12 experiments, including a magnetometer on a 10-ft. boom and four 200-ft. antennas. The first NASA satellite designed to measure Earth/solar radiation budgets had been *Explorer 18*, launched on November 26, 1963.

1973 October 30   An SL-8 launcher from Plesetsk carried the 1,210-lb. satellite *Intercomsos 10* to a 162 × 904 mi., 74° orbit. The satellite carried instruments supplied by Czechoslovakia and East Germany. This was the first of 12 consecutive Intercomsos satellites launched by SL-8, the last of which, *Intercomsos 21,* was launched on February 6, 1981. The Intercosmos program provided opportunities for eastern bloc countries to participate in space science experiments and for the Soviet Union to benefit from the equipment these communist countries provided for Soviet space research.

1973 October 31   A Soviet SL-4 launched the 12,130-lb. *Cosmos 605* biological satellite to a 133 × 249 mi. orbit inclined 62.8°. The satellite carried white rats, steepe turtles, insects, fungi and special equipment to monitor their reaction to the conditions of space. The purpose of the flight was to evaluate the response of the living cargo and to test the lift-support system on board. Special emphasis was placed on the examination of individual cells of living organisms to determine the effects of radiation on their genetic properties. *Cosmos 605* was recovered on November 22, 1973.

1973 November 3   At 05:45:59 UT NASA launched *Mariner 10* on an Atlas-Centaur AC-34 from Complex 36B at Cape Canaveral, Fla., to an Earth parking orbit. A second burn of the Centaur put it on an escape trajectory, bringing it to within 41,600 mi. of Venus on the sunward side February 5, 1974. *Mariner 10* took several pictures of the moon, beginning 21:45 UT as it passed within 68,000 mi. of the lunar surface. The 1,108-lb. *Mariner 10* design was based on the octagonal spaceframe, with the same dimensions, that had been used for *Mariner 9.* Two solar panels, each 8 ft. 10 in. long by 3 ft. 2 in. wide, with a total area of 54.9 sq. ft. and carrying 19,800 solar cells, were attached to opposing sides of the octagon and provided 500 W at Earth and 800 W at Venus and Mercury. Attitude control was via 12 cold jets and an 8-lb. supply of nitrogen gas. Communications went via a tubular low-gain antenna which protruded 9 ft. 4 in. above the octagonal structure and a high-gain dish reflector 4 ft. 6 in. in diameter. Data could be transmitted at various rates between 8.33 bits/sec. and 2.45 kilobits/sec.

1973 November 10   For the first time, two subsatellites were launched piggyback on a U.S. photoreconnaissance satellite. Carried by a Titan IIID, the 29,300-lb. *Big Bird* photoreconnaissance satellite was placed in a 101 × 165 mi. orbit inclined 96.9°. The first subsatellite, a standard 130-lb. radar electronic intelligence gathering elint, was placed in a 303 × 314 mi. orbit inclined 96.3°. The second subsatellite, a 130-lb. elint for monitoring Soviet antiballistic missile systems, was put into a 881 × 906 mi. orbit inclined 96.9°. The first subsatellite decayed on November 13, 1973, followed by *Big Bird*, the seventh of its type to be launched and the first to remain in space for more than 100 days, on March 13, 1974, and the ABM elint on December 26, 1978.

1973 November 14   At 00:41:49 UT *Mariner 10* conducted a 20-sec. (25.5 ft./sec.) velocity adjustment to move the fly-by distance at Venus from 41,600 mi. on the sunward side to about 4,800 mi. on the dark side. The *Mariner 10* propulsion system comprised a 51-lb. thrust monopropellant rocket motor with 63 lb. of hydrazine combusted over a Shell 405 catalyst. The hydrazine was expelled from its tank by a 0.5-lb. tanked supply of nitrogen pressurant. The system could make a total velocity adjustment of just over 400 ft./sec. in up to 5 burn periods. A second trajectory correction maneuver was conducted at 19:14 UT on January 21 when the motor fired for 3.7 sec., altering velocity by 4.5 ft./sec. and shifting the miss distance back toward Venus by approximately 1,300 mi.

1973 November 15   Test pilot John Manke flew the U.S. Air Force X-24B lifting body research vehicle for the first time in powered flight, achieving a speed of 597 MPH at a height of 52,764 ft. The aircraft achieved its top speed of Mach 1.76 (1,163 MPH) on October 25, 1974, and made its last rocket-propelled flight on September 23, 1975. The 36th and last flight of the X-24B program was carried out on November 26, 1975. The last six drops were unpowered glide flights to give different pilots experience in flying a lifting body. These were the last in the U.S. Air Force/NASA lifting body program, which also involved the M2-F2, the HL-10 and the X-24A. Among those glide flights were two by Capt. Francis Scobee, who became a NASA astronaut in 1978 and was killed in the Challenger disaster of January 28, 1986.

1973 November 16   The final revisit mission to the Skylab space station began at 14:01:23 UT with the launch of astronauts Gerald Carr, Edward Gibson and William Pogue on the SL-4 mission from Complex 39B at Kennedy Space Center. The AS-208 Saturn IB had a liftoff weight of 1.3 million lb., the heaviest yet, carrying the 34,291-lb. Apollo CSM-118 into a 92 mi. × 138 mi. orbit. Docking was completed at 8 hr. 1 min. on the third attempt. The first EVA (extravehicular activity) began at 17:42 UT on November 22 and lasted 6 hr. 33 min., during which Gibson and Pogue carried out repairs to external equipment, installed film in the Apollo telescope mount and set up experiments. The second EVA began at 16:00 UT on December 25 and lasted 7 hr. 1 min., during which Carr and Pogue performed experiments and changed more film. The third EVA started at 17:00 UT on December 29 for film exchange and photographs of the comet Kohoutek. The final EVA began at 15:19 UT on February 3, 1974, and accomplished a broad range of experiment activities and film retrieval before it ended 5 hr. 19 min. later. Undocking from Skylab came at 09:28 UT on February 8, 1974, with splashdown at an elapsed time of 84 days 11 hr. 16 min. 00 sec. On all three periods of habitation, Skylab astronauts had accumulated 941 hr. of solar astronomy, 569 hr. of Earth observations, 274 hr. of astrophysics experiments, 824 hr. of life science investigations, 156 hr. of the comet Kohoutek observations, 29 hr. of student experiments and had performed 41 hr. 46 min. of EVA.

1973 November 20   NASA selected Thiokol Chemical Corp. for design, development and production of the Shuttle solid rocket booster (SRB) units, two of which would power each Shuttle partway into orbit before separating and falling back on parachutes to splashdown, recovery and reuse. Lockheed, the losing bidder, contested the decision, but the judgment was made final on June 27, 1974. NASA's Marshall Space Flight Center was conducting drop tests with 1/12 scale models of the Shuttle SRB in the Tennessee River. Released at a height of 200 ft., each model was 11 ft. in length with a weight of 105 lb. and was lowered to the water by three parachutes, each 11 ft. 6 in. in diameter. The drop tests were to determine whether full-size SRBs could be towed back to port with their parachutes attached.

1973 November 26 While still four million miles from Jupiter, *Pioneer 10* encountered the bow shock wave of the planet's enormous magnetosphere. A day later the spacecraft crossed the boundary between the shock wave and the magnetic field of Jupiter, 3.54 million mi. from encounter. Like an amorphous jelly fish, Jupiter's magnetosphere would wobble under pressure from the solar wind. Storms on the sun compressed the magnetosphere closer to the planet as the intensified solar wind pushed against the bow, but when the solar wind calmed down, the strength of the magnetic field would push the bow wave farther out. *Pioneer 10* confirmed that Jupiter's magnetic field is reversed, with "north" at the south. The first ultraviolet scans of Jupiter took place on November 30.

1973 November 30 A 14,720-lb. unmanned Soyuz spacecraft was launched at 05:20 UT by SL-4 launch vehicle from the Baikonur cosmodrome. Designated *Cosmos 613*, it was a long-duration space-soak test to qualify the Soyuz design for lengthy periods in space docked at Salyut stations. Over time the extreme cold of space affects materials used in spacecraft construction, and tests to demonstrate that engineers had given Soyuz adequate protection were an essential part of preparing the equipment for flights in excess of 2–3 weeks. After six days of maneuvering, *Cosmos 613* was placed in an orbit of 155 × 248.5 mi. and then powered down. After two months in space *Cosmos 613* was powered up and returned to Earth after a mission lasting 60 days 9 min.

1973 December 4 At 02:25:05 UT *Pioneer 10* made its closest encounter with Jupiter when it passed within 81,000 mi. of the planet at a speed of 82,000 MPH, the first time a spacecraft from Earth had visited an outer planet in the solar system. Scientific data was returned to Earth at a rate of 1,024 bits/sec.; traveling at the speed of light, the signal took 46 min. to reach Earth. Of particular interest to astronomers were photopolarimeter images of the Great Red Spot, a gaseous vortex three times the size of Earth floating in the atmosphere.

1973 December 13 NASA selected Grumman to convert two Grumman Gulfstream aircraft as training vehicles for Shuttle orbiter crews. Conventional aircraft usually approach the airfield on a glide slope (angle from the horizontal) of only 2–3°. Returning to the lower atmosphere from orbital reentry, Shuttle orbiters would descend on a glide slope of 24°, necessitating special training for the pilots. The two Gulfstream aircraft were to be modified to allow them to simulate the handling characteristics of a Shuttle orbiter returning from space. The aircraft were to be delivered by mid-1976.

1973 December 16 Launched out of sequence, the first of three second-generation NASA Atmosphere Explorer satellites, AE-C *(Explorer 51),* was launched by a Delta from Vandenberg AFB, Calif., to a 96 × 2,670 mi., 68.1° orbit. Weighing 1,450 lb., AE-C comprised a cylinder 4 ft. 5 in. across and 3 ft. 9 in. tall, with solar cells on top and around the cylinder providing 100 W of power to batteries. Carrying 18 instruments from NASA, the U.S. Air Force and seven U.S. universities, AE-C carried 370 lb. of hydrazine propellant to alter perigee by up to 21 mi. in efforts to determine energy transfer, molecular processes and chemical reactions in the upper atmosphere. It decayed on December 12, 1978. AE-D *(Explorer 54),* launched by a Delta on October 6, 1975, went to an 88 × 1,922 mi., 90° orbit from which it too could dip into the outer atmosphere, but at latitudes AE-C could not reach. With a weight of 1,488 lb., AE-D was a drum-shaped polyhedron, 4 ft. 5 in. diameter and 3 ft. 9 in. high, with 12 instruments and 120 W from solar cells. It decayed on March 12, 1976. Identical, AE-E (Explorer 55) was launched on November 20, 1975, to a 278 × 279 mi, 19.6° orbit, from Cape Canaveral, Fla.

1973 December 18 Soviet cosmonauts Pyotr I. Klimuk and Valentin V. Lebedev were launched at 11:35 UT in their 14,470-lb. *Soyuz 13* spacecraft. In space at the same time as the Skylab SL-4 crew, it was the first time U.S. and Soviet cosmonauts were in orbit at the same time, although the two vehicles did not come anywhere near each other. Placed in a 138 × 158 mi. orbit inclined 51.6°, the crew performed astrophysical observations with a telescope called Orion 2 attached to the front of the orbital module. They also activated and operated a biological experiment called Oazis 2 which cultivated various organic compounds. In carrying solar cell arrays, the configuration of *Soyuz 13* was identical to the design of spacecraft the Russians would use for the joint docking flight with Apollo in June 1975. *Soyuz 13* successfully returned to Earth at an elapsed time of 7 days 21 hr. 15 min.

1973 December 19 The Soviet Union agreed to India's setting up a satellite telemetry receiving station outside Moscow to obtain information from its satellite Aryabhata, which the USSR was shortly to launch. India was building a satellite which would weigh 661 lb. and carry three science experiments. Commands would originate at India's Sriharikota facility. The cooperative program had been formalized during a recent visit to India by General Secretary Leonid Brezhnev. The chairman of India's space commission, Prof. S. Dhawan, was in charge of the project.

1973 December 25 Peter E. Glaser, vice president for Engineering Sciences at Arthur D. Little, Inc., was granted a U.S. patent for solar power satellites, a concept whereby large arrays of solar cells in orbit beam to Earth on microwave links electrical energy converted from sunlight. The patent recognized that a belt 3 mi. wide around the Earth would produce 200 times the world's electrical energy consumption.

1973 A team of Soviet designers under the leadership of ex-cosmonaut Konstantin Feoktistov began work on the third-generation "civilian" Salyut space stations, *Salyut*s 6 and 7, involving a totally new concept in operating manned space stations. Instead of carrying aboard the station at launch all the stores, equipment and maneuvering propellant that visiting crew members would need on orbit, unmanned Progress cargo tankers adapted from the Soyuz spacecraft would be used to resupply Salyut. A special docking port at the rear would facilitate Progress service missions, leaving the forward docking port for Soyuz crew visits. Equipment designed into *Salyut 6* included the multispectral MKF-6M Earth resources camera, the KATE-140 wide-angle, stereoscopic camera for topographic mapping and the BST-1M infrared-ultraviolet atmospheric telescope. With a weight of 1,430 lb., the BST-

1M was also used for astrophysical research. *Salyut 6* would also carry the Yelena gamma-ray telescope and materials processing equipment such as the Splav-01 and Kristaïl furnaces. Third-generation Salyuts were preceded by one station of second-generation design: *Salyut 4.*

The biggest Soviet heavyweight ICBM ever deployed reached the end of its development phase, ready for operational deployment from 1974. Designated RS-20, it was known in the West as the SS-18 Satan and in its Mod 1 version could throw a 25-MT warhead 9,000 mi. with an accuracy of 1,400 ft. The two-stage storable liquid propellant missile had a length of about 121 ft. 4 in., a diameter of 10 ft. and a launch weight of 480,000 lb. A total 308 missiles replaced the SS-9, peak deployment being reached in 1980. A Mod 2 version with 8–10 1-MT multiple independently targeted re-entry vehicles (MIRVs) was deployed from 1976 and a Mod 3 version, with a single 20-MT warhead and a range of 10,000 mi. was deployed from 1977. Beginning in 1979 these were replaced with the Mod 4.

1974 January 8   At 18:40 UT the first domestic communications satellite service linking east and west coasts of the United States began when conversations from New York City were routed to Alaska via the *Anik 2* satellite. The service was operated by RCA Global Communications and RCA Alaska Communications, who offered a private leased line for $1,700 per month compared to $2,298 per month from American Telephone and Telegraph using landlines. RCA hoped to have three satellites of its own operating by 1976.

1974 January 9   The Lockheed Propulsion Co. contested the award by NASA of a contract to Thiokol Chemical Corp. for

A NASA Delta 1914 stands ready April 20, 1973, to launch Anik 2, *Canada's second domestic communications satellite.*

the Shuttle solid rocket motors. Lockheed asked the General Accounting Office to examine the figures used by Thiokol to substantiate their cost estimates, which it said were unrealistic, and to look again at the technical aspects of the design, which it believed to be flawed. On June 24 the GAO said it could not fault the contract and left NASA to make a final judgment; three days later it confirmed the Thiokol deal.

1974 January 14   The NASA Marshall Space Flight Center received six study reports on the proposed space tug. The final reports were submitted by Grumman Aircraft, Martin Marietta, McDonnell Douglas, General Dynamics and Lockheed, covering a standard tug using storable propellants, a tug utilizing cryogenic propulsion and a growth version. These reports were the culmination of a Phase A feasibility analysis and presaged the onset of Phase B definition work. On May 15 the MSFC issued requirements for two additional tug contracts, one to develop schedules and planning data, and the other to recommend avionics and subsystem requirements.

1974 January 16   Plans for an Arab communication satellite were presented to the Arab League in Cairo, Egypt. The proposed network would provide radio, TV and news links between the Arab states for educational, cultural and national development programs. Fifty ground stations were proposed, to be built by member states, with costs apportioned according to use. From September 15 to 22 the Arab League voted to set up the system and considered three industry proposals: Hawker Siddeley Group from the United Kingdom proposed a satellite based on the Orbital Test Satellite (OTS); Messerschmitt-Bölkow-Blohm (MBB) put up the three-axis stabilized satellite; and a consortium comprising Hughes Aircraft, Nippon Electric of Japan and Thompson CSF of France proposed a design similar to Canada's Anik series.

1974 January 19   Launched from Cape Canaveral, Fla., at 01:38 UT, a Delta 2313 failed to place the first in a new generation of military communication satellites, *Skynet IIA,* in orbit. First in the improved 2000 series, this Delta had three (52,000-lb. thrust) Castor strap-ons and a 205,000-lb. thrust Rocketdyne H-1 first-stage engine left over from the Saturn I program. Instead of the Aerojet AJ-10 series second stage, the Delta 2000 series had a modified (9,800-lb. thrust) version of the TR-201 lunar module descent engine in a stage of equal diameter to the first stage. The third stage comprised a 14,100-lb. thrust TE-364–4. Instead of placing the satellite in a geosynchronous transfer ellipse, a problem in the Delta second stage electronics 22 min. after launch steered *Skynet IIA* to an eccentric orbit of 59.5 × 2,116 mi. inclined 37.6°. Built by Marconi Space and Defense Systems in the United Kingdom, with technical direction from Philco-Ford, *Skynet IIA* had a cylindrical main structure 6 ft. 3 in. in diameter with a height of 6 ft. 10 in. including antenna and weighed 959 lb. at launch; it was to weigh 517 lb. at apogee motor burnout, which on this flight did not take place. *Skynet IIA* decayed into the atmosphere on January 25.

1974 February 4   NASA awarded a contract to Hughes Aircraft for the conceptual design of two Pioneer-Venus flight spacecraft to be sent to Venus in 1978 on separate Atlas-Centaur launch vehicles. The contract contained an option for further work leading to final design and development, which was realized in May. In August 1974 Congress approved a formal start on the Pioneer-Venus mission, and further discussions with Hughes resulted in a final hardware contract in November 1974. Detailed design began the following month.

NASA's Ames Research Center was responsible for management of the orbiter and multiprobe spacecraft, and a start was made in March 1975 on the detailed design of scientific instruments carried aboard these vehicles.

1974 February 5   At 18:01 UT *Mariner 10* passed the planet Venus at a distance of 3,500 mi. from the dark side as viewed from the sun, 25.5 million mi. from Earth. Although Mercury was the primary mission objective, *Mariner 10* conducted a full survey of Venus. The spacecraft carried 172 lb. of science equipment, including two TV cameras which, unlike previous Mariners, were identical and could be used through 1,500-mm narrow-angle or 64-mm wide-angle lenses. Each picture was composed of 700 lines carrying 832 pixels (each coded by 8 binary digits) for 582,400 picture elements. TV images of Venus varied in resolution between 330 ft. and 81 mi. A total of 4,165 images had been taken when camera operations ended on February 13. *Mariner 10* also carried a magnetometer on a 20-ft. boom, infrared radiometer for temperature data, two ultraviolet spectrometers for atmospheric measurements and a charged particle telescope. The gravity field of Venus had slowed *Mariner 10* by 9,900 MPH and bent its trajectory in toward the orbit of Mercury.

1974 February 10   The first of the four Soviet spacecraft launched to Mars between July 21 and August 9, 1973, arrived at the planet. Programmed to fire its retro-rocket as it approached its point of closest approach, the *Mars 4* propulsion system failed to ignite and the spacecraft sped past the planet at a distance of 1,370 mi. Two days later *Mars 5* successfully placed itself in an elliptical orbit of 1,100 × 20,200 mi. inclined 35° to the equator and with a period of 25 hr. *Mars 4* and *Mars 5* were to have relayed data from the *Mars 6* and *Mars 7* landers. *Mars 5* conducted a wide range of scientific tasks and each orbiter carried two cameras, one called Zufar, with a 350-mm focal length, f/2.5 lens, and one called Vega, with a 52-mm focal length, f/2.8 lens; both were fitted with red, blue, green and orange filters. *Mars 4* and *Mars 5* returned a total of 60 pictures. The spacecraft also carried 3.5 cm and 8–32 cm radio telescopes, an infrared radiometer and three photometers for measurements of the upper and lower atmosphere.

1974 February 11   Officials at the NASA Marshall Space Flight Center described a solar-electric propulsion system (SEPS) that could be used to move satellites and cargo from low orbit to high orbit and around the solar system. Utilizing six to nine ion thrusters and two lightweight solar arrays producing 25 kilowatts of power, a SEPS-equipped space tug could move a 5,500-lb. load from low Earth orbit to geosynchronous orbit in 100 days, three times the capacity of a tug using conventional chemical propulsion. Although the electric thrusters would produce very low levels of thrust, they would operate for several months, slowly accelerating the vehicle to high velocities.

The first test launch of a Titan IIIE/Centaur launch vehicle took place from Launch Complex 41 at Cape Canaveral, Fla., but failed when the Centaur's oxygen pump did not set up proper conditions for ignition. In almost all respects the solid propellant boosters and the first and second stages were similar to the equivalent components of a Titan IIIC, but with the addition of a cryogenic Centaur third stage. The payload for this test flight comprised a series of test packages, a 210-lb. satellite called SPHINX (an acronym for space plasma high voltage interaction experiment) and a 7,744-lb. dynamic simulator of a Viking spacecraft, which vehicles of this type were

to launch on August 20 and September 9, 1976. Total vehicle height was 160 ft. Titan IIIE/Centaur could lift 38,000 lb. into low Earth orbit, 7,000 lb. to geosynchronous transfer orbit or send 8,000 lb. on Earth escape missions.

1974 February 12   Soviet military photoreconnaissance planners exercised a new technique for getting better pictures of selected targets by putting third-generation satellites into elliptical orbits with a perigee of only 114 mi. Launched this day, *Cosmos 632* entered a 114 × 207 mi. orbit and was returned to Earth after 14 days. Although perigee varied flight to flight, the more elliptical orbit introduced a more flexible phasing schedule for selected regions of the Earth.

1974 February 13   NASA and the European Space Research Organization moved a step nearer cooperation on major interplanetary and planetary missions when officials met in Paris to consider a menu of proposed projects. Included in the discussions was an out-of-ecliptic mission to study the north and south poles of the sun, flights to Jupiter and Mercury, and European experiments placed aboard NASA flights to Mars and Venus.

1974 February 16   Using a modified three-stage version of the MU-4S booster designated MU-3C, Japan launched the 143-lb. *Tansei 2*, MS-T2, test satellite from Uchinoura to a 176 × 2,009 mi. orbit inclined 31.2° to the equator. The MU-3C was the first guided booster used by Japanese rocket engineers, dispensing with the MU-4S third stage and adopting an enlarged version of the MU-4S fourth stage as the new third stage. The satellite, named *Light Blue*, for the colors of the University of Tokyo, was equipped to "test the control devices and not for scientific observation." It decayed out of orbit on January 22, 1983.

1974 February 18   The Italian satellite *San Marco 4* was launched by solid propellant Scout from the San Marco platform off the coast of Kenya to a 143.5 × 565.5 mi., 2.9° orbit. The 362-lb. satellite was instrumented to measure the constituents of the upper atmosphere and to obtain data that could be correlated with that from *Explorer 51* launched on December 16, 1973. The satellite decayed back into the atmosphere on May 4, 1976.

1974 February 27–28   The development program for the L3S Europa satellite launcher was signed in Paris, France, by contributing nations. To be known as Ariane, the launcher would have an L-140 first stage with four Viking 2 engines, an L-33 second stage with one Viking 4 engine and an H-8 cryogenic second stage with an HM-7 engine. To keep the high-technology third stage as simple as possible, it would have only single-burn capability and place its payloads in geosynchronous transfer orbit via direct injection rather than with two burns separated by a coast period. The plan was to have Ariane operational by 1980, with the French space agency (Centre National d'Etudes Spatiales, or CNES) managing the program and Aérospatiale as system architect and contractor for the first two stages. The French company SEP (Société Européenne de Propulsion) would provide all motors, Germany's MBB (Messerschmitt-Bölkow-Blohm) building the third-stage combustion chamber.

1974 February   NASA held a Preliminary Design Review on the reusable Shuttle and authorized minor changes. This led to the Vehicle 5 configuration, close to the definitive design which would emerge for flight. Vehicle 5 was noticeable for eliminating the forward fairings on the orbital maneuvering system pods so that they no longer overlapped the payload bay doors. Vehicle 5 had an empty weight of 187,900 lb. and a minimum touchdown speed of 171 knots. NASA deleted the jet engines for aerodynamic flight trials in the atmosphere and for ferry flights, preferring instead to have each orbiter carried on the back of a specially modified Lockheed C-5A or Boeing 747 aircraft converted for the purpose. Due to budget constraints the first horizontal test flights in the atmosphere were scheduled for the second quarter of 1977, a delay of six months, and the first manned orbital flight was not now expected before the second quarter of 1979.

1974 March 9 The Soviet *Mars 7* spacecraft failed to enter the atmosphere of Mars and descend to a soft touchdown, as planned, when a malfunction caused it to miss the planet by 808 mi. Instead of decelerating through friction with the atmosphere, the lander went past Mars and remained in heliocentric orbit. On March 12 *Mars 6* successfully commenced its descent toward the surface of Mars, but contact with the spacecraft was lost 2 min. 28 sec. after the parachute was deployed. It is not known if it reached the surface intact, but the calculated landing point was at 24° S by 25° W. Thus, of four Soviet spacecraft sent to Mars in the 1973 launch window only the orbiter *Mars 5* was a success.

NASA launched the United Kingdom's experimental satellite *Miranda* (UK-X4) on a four-stage Scout D from Vandenberg AFB, Calif., to a 443 × 569 mi. orbit inclined 97.8°. Designed and built for the United Kingdom by Hawker Siddeley Dynamics Ltd., *Miranda* was a rectangular structure 2 ft. 9 in. in height and 2 ft. 2 in. wide with a weight of 203 lb., powered by a solar array of 1,800 cells, 8 ft. long. The United Kingdom reimbursed NASA for the integration and launch of *Miranda*, which conducted scientific investigations and demonstrated innovative attitude control systems. It remains in orbit.

1974 March 16 *Mariner 10* conducted a course correction maneuver at 11:55:42 UT when the main propulsion system was fired for 51 sec. executing a 58 ft./sec. velocity change. This moved the projected fly-by point at Mercury from 6,200 mi. on the sunlit side to less than 500 mi. on the antisolar side of the planet. The exact magnitude of the maneuver was 1% less than precisely required, which put the fly-by distance 125 mi. closer to the surface than planned. Had *Mariner 10* passed Mercury on the sunlit side, ideal for TV images, the trajectory would not have permitted occultation of the radio signal which was important for measuring any atmosphere the planet might have had. The first of 2,363 TV images of Mercury at the first encounter were sent to Earth on March 23, 1974, when the spacecraft was 3.3 million mi. from the planet.

1974 March 19 NASA announced that it had decided to use the gravity field of Jupiter to bend the flight path of *Pioneer 11* toward Saturn. With a close pass of just 26,000 mi. above the cloud tops of this giant planet, more than 80,000 mi. in diameter, the spacecraft would collect enough energy to deflect itself in a planetary slingshot some engineers dubbed "interplanetary billiards." On April 9, 1974, *Pioneer 11* fired its thrusters for 42 min. 36 sec., expending 17 lb. of propellant and thereby adding 210-ft./sec. (143 MPH) to its velocity of 28,700 MPH. This allowed it to make a close pass around Jupiter's south polar region on December 2, 1974. At that time the gravitational pull from Jupiter would have increased *Pioneer 11*'s speed to 108,000 MPH. *Pioneer 11* would rendezvous with Saturn on September 1, 1979.

1974 March 26 The first Soviet satellite destined for geostationary orbit was launched by SL-12 from Baikonur to a 110.5 × 143 mi., 51.54° orbit. Designated *Cosmos 637*, it was a prototype engineering precursor to the operational Statsionar series of satellites, already more than three years late. Toward the end of the first orbit, the terminal stage fired to put the satellite in a highly elliptical orbit of 140 × 22,750 mi., at 49.73° inclination. This was circularized at apogee to a 21,991 × 22,233 mi., 0.25° orbit at geostationary altitude. The 4,400-lb. satellite was used for communication and engineering tests.

1974 March 29 At 20:46 UT NASA's *Mariner 10* spacecraft made its closest approach to the planet Mercury, 3,045 mi. in diameter, when it passed within 437 mi. of the surface. Because Mercury is relatively close to the sun and cannot readily be seen at night from Earth, scientists had no idea what the planet looked like. Pictures which had been transmitted from the spacecraft for several days showed an astonishingly moon-like surface with no signs of recent geological activity, on a planet devoid of atmosphere. Occultation of the radio signal to Earth began at 20:48 UT and ended 11 min. later when the spacecraft reappeared from behind the planet. The science equipment concluded its first observations of Mercury on April 2, 1974, but a technical problem consumed large quantities of nitrogen gas. After encounter, spacecraft attitude had to be adjusted using a combination of cold-gas jet firings and solar-sailing, using the solar panels and high-gain antenna to tack the spacecraft against the solar wind.

1974 April 1 Work began at the NASA Kennedy Space Center, Fla., on a special runway, close to Launch Complex 39, to be used as a landing strip for the Shuttle. Located northwest of the Vehicle Assembly Building, the grooved runway would have a length of 15,000 ft., a width of 300 ft. and a microwave scanning-beam landing system. Nevertheless, cost-saving measures would leave the runway below U.S. Air Force standards and draw criticism for its relatively primitive facilities. The concept of landing Shuttle orbiters in that part of Florida, where weather is unpredictable and fog can roll in within a few hours, would be criticized by astronauts.

1974 April 3 At 07:31 UT an unmanned Soyuz spacecraft was placed in Earth orbit by an SL-4 from Baikonur. The 14,500-lb. spacecraft was designated *Cosmos 638* and conducted extensive orbital maneuvering. The spacecraft configuration was almost identical to that of the Apollo-Soyuz Test Project, but the mission profile of *Cosmos 638* did not follow the flight plan for that international mission. It conducted orbital maneuvers quite different from those which would be employed for the ASTP mission. The spacecraft successfully returned to Earth after 9 days 21 hr. 34 min.

Indonesia announced that it would buy and operate a domestic communications satellite program in a cost-saving measure aimed at linking the nation's 13,000 islands without the expense of landlines or microwave links. Managed by the Indonesian Post and Telecommunications Directorate, the program would require two satellites, each capable of handling 144 telephone circuits or 12 color TV channels. The second satellite would serve as backup to the first. U.S. and European satellite makers were hopeful that this was only the first in a series of satellite orders from developing countries. Malaysia and the Philippines also showed interest in joining the Indonesian system.

1974 April 13 The first U.S. domestic communications satellite, Western Union's *Westar 1*, was launched from Cape Canaveral, Fla., at 23:33 UT on a Delta 2914 with nine Castor solid propellant strap-ons. Although one of the spent Castors hung up and failed to separate until the first stage was jettisoned, the TR-201 second stage steered the stack to a 143 × 22,520 mi. geosynchronous transfer orbit inclined 24.7°. The apogee motor was fired at 21:21 UT on April 16, placing the satellite in a geostationary orbit from which it was moved to its final position at 99° W. The 674-lb. spin-stabilized satellite had a diameter of 6 ft. 3 in. and a total height of 11 ft. 3 in., including a spoon-shaped, de-spun mesh antenna, receiving in 6 GHz frequencies and transmitting in the 4 GHz band. Each of 12 transponders could carry 1,200 voice circuits. *Westar 2* was launched on October 10, 1974, and positioned at 123.5° W. On August 10, 1979, Westar 3 was placed over 91° W.

1974 April 18 A joint proposal to the Federal Communications Commission (FCC) from American Telephone and Telegraph and GTE Satellite Corp. requested a licence to operate a domestic satellite service for the 50 states and Puerto Rico. Three satellites owned by the Comsat General Corp. were to be used and provide a joint system capable of handling 28,800 simultaneous telephone calls and cut from ten to seven the number of ground stations involved with the existing, separate systems.

1974 April 22 The U.S. Department of Agriculture's Soil Cartographic Division completed a photo map of the 48 contiguous U.S. states for NASA. Prepared as a mosaic of images from the *ERTS-1* satellite, it provided a compilation of regional investigations and for the first time a synoptic overview of the entire continental United States, revealing lineaments, faults and features never seen in one picture before. This map allowed the United States for the first time to set up a national water inventory, generate a national vegetation map and form the basis for planning a national water drainage network.

1974 May 1 NASA's Kennedy Space Center awarded a contract to Seelye Stevenson Value and Knecht, Inc., for design of an Orbiter Processing Facility (OPF) where winged Shuttle vehicles could be processed before stacking with external tank and solid rocket boosters in the Vehicle Assembly Building (VAB). It would also provide a place where very large payloads, such as Spacelab, could be installed prior to the vehicle being moved to the VAB. Two OPFs would be built for processing two orbiters simultaneously.

1974 May 6 A four-day conference on self-sufficient colonies in space began under the auspices of Princeton University. Dr. Gerald K. O'Neill outlined a colony for 2,000 people at the equigravisphere, a point between the Earth and the moon where their gravitational fields balanced out. Built in 15–20 years from go-ahead, the station would require 9,000 tons of building materials, and more materials from the moon could be mined by surface vehicles. O'Neill said the solar system had "virtually unlimited" resources and believed that most "dirty industries" could be moved off the surface of the Earth by the middle of the 21st century.

1974 May 9 The Federal Communications Commission (FCC) signed an agreement clearing the way for an international undertaking with the European Space Research Organization (ESRO) on development of an aerosat system providing airline communications via satellite. After 10 years of study and analysis, U.S. agencies were moving to head an international organization providing continuous telephone and data services to airliners en route to their destinations. Formed to organize the effort, the Aerosat Council was told on September 15, 1977, that the Carter administration was withdrawing funds on the grounds that it could not afford to lead this effort, thus effectively terminating the program.

At 20:05:00 UT *Mariner 10* performed a 3 min. 15 sec. (164 ft./sec.) trajectory correction maneuver to set up the orbital conditions for a second encounter with the planet Mercury. The spacecraft would swing round the sun out to a distance just beyond the orbit of Venus and then return to the same position in space where it first encountered Mercury. The planet, meanwhile, would have made two complete orbits of the sun to return again to the position where *Mariner 10* would arrive after its single "year" around the sun. A second maneuver took place at 20:05:40 UT on May 10, when the engine fired for 2 min. 19 sec., changing velocity by 90.2 ft./sec. This two-part maneuver was made necessary because the rocket motor could not be allowed to fire longer than 3 min. without a 24 hr. cool-down period. It moved the spacecraft to a flight path which would rendezvous with Mercury at a distance of 21,000 mi. on the sunlit side of the planet. An 18-sec. (10.8 ft./sec.) motor firing at 20:09 UT on July 3, 1974, shifted the fly-by point to a distance of 31,000 mi., setting up the trajectory for a possible third encounter.

1974 May 17 NASA launched the world's first weather satellite designed to operate from geostationary orbit when it sent *SMS-A* (*Synchronous Meteorological Satellite-A*) into space from Cape Canaveral, Fla. With a launch weight of 1,385 lb., the satellite had a diameter of 6 ft. 3 in. and a total height of 8 ft. 7 in. to the tip of a magnetometer boom. Built by Philco-Ford, *SMS-A* carried 15,000 solar cells around the exterior of its drum-shaped structure. Weight decreased to 630 lb. in orbit after depleting apogee motor solid propellant, and the satellite was spin-stabilized at 100 RPM after being positioned over 45° W. The weather camera comprised a

Built by Ford, the Synchronous Meteorological Satellite (SMS-1) *is preapred for shipment to Cape Canaveral, Florida, for launch on May 17, 1974.*

Hurricane Carmen is imaged by NASA's Synchronous Meteorological Satellite *from high above South America.*

spin-scan radiometer observing clouds with a resolution of 1.2 mi. in visible light and 5.8 mi. in infrared. The second preoperational satellite, *SMS-B*, was launched to geostationary orbit on February 6, 1975, and placed at 135° W.

1974 May 18 After the United States, the Soviet Union, the United Kingdom, France and China, India became the sixth nation to acquire nuclear weapons when it detonated a 10–15-KT device at an underground test facility. The chairman of India's Atomic Energy Commission, H. N. Sethna, said the test was carried out to determine the cratering effect of the Earth's surface and to measure the cracking effect of rocks beneath the soil.

1974 May 27 The final unmanned test flight of the Soyuz space station ferry vehicle began with the launch of *Cosmos 656* on an SL-4 at 07:25 UT from Baikonur. Soyuz ferry variants had batteries instead of solar panels, and the 14,720-lb. *Cosmos 656* was of that type. The spacecraft was placed initially into an orbit of 121 × 226 mi., inclined 51.6° to the Earth's equator. It was maneuvered by ground command during the first day in space but returned to Earth after a flight lasting only 2 days 25 min.

1974 May 29 At 08:57 UT a Soviet SL-12 launcher lifted off from Baikonur carrying the *Luna 22* moon orbiter, the last of two Ye-8LS orbiters originally developed as lunar landing site survey vehicles. The first *(Luna 19)* had been launched on September 28, 1971. Four days later the 8,800-lb. *Luna 22* braked into a 137-mi. circular orbit inclined 19.6° with a period of 2 hr. 10 min. On June 9, 1974, the spacecraft maneuvered into a 15.5 × 151.6 mi. orbit for high-resolution TV imaging of the surface and the remote sensing of geochemical characteristics. On June 13, after completing the photographic work, *Luna 22* was maneuvered to a 112.5 × 186 mi. orbit to continue the detailed study of the moon first carried out by *Luna 19*, launched on September 28, 1971.

Maneuvering propellant was exhausted on September 2, 1975, after 15 months of lunar study.

1974 May 30 NASA launched *ATS-6 (Applications Technology Satellite-6)* on a Titan IIIC from Cape Canaveral, Fla., to a geosynchronous orbit inclined 8° to the equator and a position over the Earth at 94° W longitude. The 3,090-lb. satellite was 26 ft. tall and supported two structural arms to which were attached semicircular solar cell arrays. A 2,000-lb. Earth-viewing module was attached by a tubular truss to a parabolic antenna 30 ft. in diameter. *ATS-6* was designed to serve an international community as a special broadcasting station in space and carried more than 20 science and technology experiments. Designed to operate for two years, *ATS-6* was finally retired in August 1979 when its thrusters were used to boost it to a higher path above geosynchronous orbit.

1974 May Responding to a Required Operational Capability document issued by the U.S. Air Force Strategic Air Command in 1971, the proposed M-X ICBM moved into the Concept Definition Phase. Several optional basing modes were studied in efforts to give the missile some means of protection from attack by accurate Soviet MIRV warheads. One was called the dash-on-warning mode, whereby a missile on its launcher would wait at the center of a network of spokes radiating to dispersed hard stands from which the missile could be fired. Another option envisaged basing missiles on the southern flank of large mountains, compromising Soviet targeting over the North Pole. Yet another envisaged a 3,000-mi. tunnel along which M-X could be moved at will.

1974 June 3 A five-stage Scout launcher incorporating a BE-3A solid propellant fifth stage carried the 59-lb. *Explorer 52* satellite from Vandenberg AFB, Calif., to a 319 × 78,853 mi., 89.8° orbit. Also known as Hawkeye, the satellite comprised an eight-sided truncated cone with a base diameter of 2 ft. 6 in., a top diameter of 10 in. and a height of 2 ft. 6 in. Two extendable booms unfurled to a length of 75 ft. for detecting electric fields, and a 5-ft. fluxgate magnetometer boom was deployed. *Explorer 52* was instrumented for investigating the interaction of the solar wind with Earth's magnetic field and for studying the topology of the magnetic field over the Earth's polar regions. It decayed on April 28, 1978.

1974 June 4 Construction of the first reusable Shuttle orbiter began at Rockwell International with a start on the structural assembly of the crew module for orbiter vehicle 101 (OV-101), later to be named *Enterprise*. Flight orbiters were designated by numbers in the 100-series. OV-101 would only be used for air-launched tests in the atmosphere. On June 14, 1976, Rockwell began assembling OV-099, assigned to engineering tests on the ground and not intended for flight. Later it would be modified as a flight-rated orbiter and carry the name *Challenger*. Structural assembly of OV-102 (later named *Columbia*), the first Shuttle to fly in space, began on June 28, 1976.

1974 June 5 The European Space Research Organization (ESRO) decided to award the Spacelab a six-year design and development contract to the team headed by VFW-Fokker/ERNO Raumfahrttechnik GmbH. One fully equipped Spacelab pressure module and associated pallet were to be delivered by April 1979, ready for installation of flight experiments for *Spacelab-1 (SL-1)* on a Shuttle flight in 1980. The pressure module was to provide a shirtsleeve environment for up to six scientists on as many as 50 missions each of between 7 and 30 days.

1974 June 6 The third French nuclear-powered ballistic missile submarine, *Le Foudroyant*, became operational with 16 M-2 missiles. Equipped with a more powerful P6 Rita II second stage having a thrust of 70,550 lb. for 50 sec., the M-2 had a range of 1,920 mi. compared with 1,490 mi. for the M-1. The two French ballistic missile submarines already launched, *Le Redoutable* and *Le Terrible*, were subsequently retrofitted with the M-2.

1974 June 11 A giant nylon parachute built by Goodyear Aerospace Corp. to provide experience in handling large parachutes was demonstrated at the NASA Kennedy Space Center, Fla. It had a base diameter of 130 ft., a base canopy circumference of 394 ft. and a weight of 2,075 lb. Three parachutes of this size were considered necessary for recovery of each solid rocket booster (SRB), two of which would be separated after the first 2 min. of powered ascent by the Shuttle.

1974 June 20 At a National Security Council meeting with President Richard M. Nixon and Secretary of State Henry Kissinger, Secretary of Defense James Schlesinger argued strongly for major cuts in both U.S. and Soviet ballistic missiles. Believing that the large number of Soviet MIRVed warheads on test would, if deployed, seriously threaten the survivability of the U.S. deterrent, Schlesinger wanted Nixon to tell Secretary General Leonid Brezhnev that a new U.S. ICBM would be developed if the Soviets continued to expand their own missile forces. Amid an atmosphere of conciliation and détente, Schlesinger was one of the few who championed development of the proposed M-X ICBM as replacement for older missiles. By the end of the year, after Nixon had resigned from office, President Gerald Ford had authorized full-scale planning for M-X development.

1974 June 21 The U.S. Air Force issued a contract to Rockwell International for design and development of three prototype satellites in the Navstar global position system (GPS) navigation satellite program. Navstar had been conceived through work in the Defense Navigation Satellite Development Program as the definitive position-fixing aid. With three equally spaced orbital rings of six satellites, each satellite 60° ahead of its neighbor on that orbit (inclined 63° to the equator), coded receivers on land, at sea or in the air, could fix position to within 5 ft. or less. With six in-orbit spares, a constellation of 24 satellites was planned.

1974 June 24 The first successful OPS/Almaz space station was launched at 22:38 UT from Baikonur. A Soviet SL-13 launch vehicle carried the 41,700-lb. *Almaz* space station, its identity disguised as *Salyut 3*, to a 136 × 168 mi. orbit inclined 51.5°. Several orbital maneuvers were conducted by remote command from the ground until the station was in an orbit of 166.5 × 169 mi. on July 2, 1974. One of the classified "military" stations developed by the Chelomei bureau, *Salyut 3* had a useful life of only 5–6 months and was intended for surveillance and military research. Unlike "civilian" Salyuts, however, it comprised a shortened main work area only 7 ft. in length. In "military" Salyuts this was used for all crew activities because the smaller forward section was full of equipment. A large recoverable capsule was attached to the front of the smaller forward section, occupying the space usually taken up with the transfer module, absent on "mili-

tary" stations. Visiting Soyuz spacecraft would dock at the rear of the station, a facility not shared by first-generation Salyuts. The propulsion module had an adapted motor with two nozzles, one on each side of an axial docking unit. Docking at the rear would be provided on "civilian" stations beginning with *Salyut 6*.

1974 July 3 At 18:51 UT cosmonauts Pavel R. Popovich and Yuri P. Artyukhin were launched aboard their 14,995-lb. *Soyuz 14* spacecraft from Baikonur by an SL-4 that placed them in a rendezvous orbit with the previously launched *Almaz/Salyut 3* space station. Two days later, in an orbit of 165 × 171.5 mi., the two vehicles were docked together and the cosmonauts entered *Salyut 3*, the first time in three years that a Soviet station had been manned. For more than two weeks the crew worked at tasks aboard their "military" Salyut and returned to Earth after a mission lasting 15 days 17 hr. 30 min., the first totally successful Soviet space station flight.

1974 July 13 To recognize the fifth anniversary of the first manned landing on the moon, President Richard M. Nixon proclaimed the period July 16 through July 24 as United States Space Week. "The knowledge to be gained from space," he said, "will lead to scientific, technological, medical and industrial advances which cannot be perceived today. In time man may take for granted in the heavens such wonders as we cannot imagine—just as superhighways across America would amaze the Puritans of 1620 or transatlantic flights would astound those who passed on the legend of Icarus. But we know that a beginning has been made that will affect the course of human life forever."

1974 July 14 The first Navigation Technology Satellite, *NTS-1*, experimental navigation satellite was launched on an Atlas F from Vandenberg AFB, Calif., to a 124.7° retrograde orbit of 8,352 × 8,552 mi. The 646-lb. satellite had been built by the Naval Research Laboratory under the name *Timation 3*, but after the merging of Navy and Air Force navigation programs, it was renamed. *NTS-1* carried receivers and transmitters to broadcast its exact position together with signals from a rubidium clock. Problems with the satellite tumbling under solar wind pressure delayed the start of experimental research.

1974 July 16 A four-stage solid-propellant Scout carried the 280-lb. German satellite *Aeros B2* from Vandenberg AFB, Calif., to a 139 × 542 mi., 97.45° orbit. Built in West Germany, *Aeros B2* carried five experiments, four German and one American, for measurements of the Earth's upper atmosphere. Comprising a cylinder 3 ft. in diameter, the satellite had a height of 2 ft. 4 in., with a conical shell at one end from which a 5 ft. 11 in. probe extended. Solar cells provided the 35 W of power needed by the bus and its payload. *Aeros B2* decayed September 25, 1975.

1974 July 18 To deliver Shuttle orbiters from manufacturing plant to launch site and to ferry them between landing and launch sites, NASA purchased a Boeing 747-123 (registration number N9668) from American Airlines. The 86th aircraft of the Boeing 747 production line, it had been delivered to American on October 29, 1970, and had logged 8,999 hr. on 2,985 flights, most between New York and Los Angeles. The aircraft would be modified with a yaw oscillation damper in the nose, strut arrangements on top of the fuselage and vertical stabilizers at the tip of each horizontal tail unit.

1974 July 26 NASA's Marshall Space Flight Center issued space tug and interim upper stage (IUS) study contracts to General Dynamics, Pratt and Whitney, Martin Marietta, McDonnell Douglas and IBM. The U.S. Air Force had been developing the concept of an orbit-to-orbit stage (OOS) which would fill the gap for an orbital maneuvering unit until the tug became available around 1983. The interim upper stage was conceived as a two-stage, solid propellant OOS capable of placing a 5,000-lb. payload in geostationary orbit.

1974 August 12 At 06:24 the 14,500-lb. unmanned Soyuz designated *Cosmos 672* was launched on a full dress rehearsal of the Apollo-Soyuz Test Project (ASTP) flight to further qualify the hardware design and to demonstrate that fully automated control and command operation could accomplish the mission as a backup to manual control. An important part of the mission was to qualify the coordination of ground control activities during the flight and to demonstrate that emergency backup procedures would work. The Soviets were acutely aware that any serious problem with their equipment on the international flight could harm future cooperation with the United States. *Cosmos 672* returned to Earth after 5 days 22 hr. 38 min., clearing the way for a manned dress rehearsal of ASTP with *Soyuz 16* on December 2, 1974.

1974 August 26 At 19:58 UT cosmonauts Gennady V. Sarafanov and Lev S. Demin were launched aboard their 14,905-lb. *Soyuz 15* spacecraft to low Earth orbit from which they could perform a rendezvous with the *Almaz/Salyut 3* space station. After rendezvous with the station, the crew made repeated, but unsuccessful, attempts to dock. Because of the limited endurance of their spacecraft, which relied on batteries rather than solar cells for electrical energy, they had little time to resolve the difficulties. Finding it impossible to correct the docking problem, the crew of *Soyuz 15* returned to Earth at the end of a mission which had lasted only 2 days 12 min.

1974 August 30 A four-stage NASA Scout launched the *ANS-1* (Astronomical Netherlands Satellite) from Vandenberg AFB, Calif., to a 160 × 729 mi. orbit inclined 98°, significantly different from the planned circular orbit due to a malfunction in the first stage. The 265-lb. spacecraft was three-axis stabilized, the first such to be launched by a Scout, and comprised an eight-sided prism 3 ft. 9 in. high, 2 ft. 5 in. deep and 2 ft. wide, supporting two solar panels and 1,980 solar cells supply 35 W of power. The 84-lb. payload comprised instruments from the Netherlands and the United States to study ultraviolet and x-ray sources. Because of the improper orbit, x-ray emissions could be studied only 25% of the time and ultraviolet emissions for only 70% of each orbit. *ANS-1* decayed on June 14, 1977.

1974 September 6 The first mail sent by satellite was transmitted between New York and Los Angeles via Western Union's *Westar 1*. The combination telegram and letter was in advance of a new service being offered by Western Union which would utilize the advantages of satellite communications to significantly cut costs. A 60-word mailgram sent by *Westar 1* was to cost $1.35 compared to $6 for a standard telegram sent by conventional means. One-business-day transmission of mailgrams would begin in late September with a New York–Los Angeles service and was to be followed later in the year with connections between the East Coast and Texas and parts of the West Coast.

1974 September 21 At 20:59 UT *Mariner 10* made its second rendezvous with the planet Mercury and passed within 31,000 mi. of the surface on the sunlit side. During an encounter sequence that included three days of observations and the transmission of 500 TV images, the spacecraft acquired pictures which extended the coverage from 50% to 75% of the illuminated hemisphere. The planet's synchronized orbit and rotation period dictated that each time *Mariner 10* returned for a close encounter, the same hemisphere of Mercury would be lit by the sun, thereby preventing images of the other (unlit) side.

1974 September 23 A recoverable module containing equipment and film from the "military" *Almaz/Salyut 3* space station was returned to Earth at the end of the primary mission phase. Although hosting only the *Soyuz 14* crew since its launch on June 24, 1974, the station had continued to send information to tracking stations on Earth. Some equipment on board was in support of new technologies for military surveillance missions, and some was monitoring the performance of the station itself. The Soviets were about to embark on their first long-duration station program, the science research station *Salyut 4*, and operational experience with earlier Almaz and Salyut stations was essential. An extended period of remote control operation of station elements continued until *Almaz/Salyut 3* was brought out of orbit and destroyed in the atmosphere on January 24, 1975.

1974 September 27 Because of budget pressure, NASA could no longer afford to build a third Viking orbiter and decided to put the second production spacecraft in storage and rework the proof-test model into a flight vehicle. This would be

Encapsulated in an aeroshell, the Viking lander sits atop the orbiter with folded solar panels.

redesignated Viking *Orbiter 1* for launch aboard a Titan III-Centaur. NASA also canceled plans for a third Viking lander. The original contract called for three orbiter/lander pairs when the mission originally scheduled for launch in 1973 included two hard landers followed by a soft lander in 1975. This had never been changed. NASA had retained the contracted hardware in the hope of eventually getting funds for a third (Viking C) mission. Martin Marietta delivered the first lander to the Kennedy Space Center, Fla., on January 4, 1975, and the Jet Propulsion Laboratory delivered the first orbiter on February 11, 1975.

1974 September   The British Interplanetary Society published early results of a study on the feasibility of an interstellar spacecraft launched from Earth to the star Barnard, six light years away. The engineers and scientists involved in the study, called Project Daedalus, proposed the use of small nuclear-fuel pellets of helium 3 detonated at the rate of 250 per second, with each bomblet producing energy equivalent to 90 tons of TNT. The thrust phase would last five years and accelerate Daedalus to 31,700 mi./sec. (114 million MPH). During the fly-by of Barnard 50 years later it would release 10–20 probes to search for life and send back information.

1974 October 10   NASA's Lewis Research Center announced that an electric rocket thruster that failed on the *SERT II* satellite launched February 3, 1970, had been restarted in space and operated for six weeks. In that period the thruster had been operated 17 times for brief periods, delivering 80% of maximum output. Engine technicians said this indicated that ion thrusters could survive long periods between use and would be applicable to missions needing several years of powered flight.

1974 October 15   A four-stage Scout launcher carried the 284-lb. satellite *Ariel 5* (UK-5) from the San Marco platform off the coast of Kenya to a 313 × 341 mi., 2.88° orbit. Built at the Appleton Laboratories, England, this fifth British satellite was cylindrical in shape, 2 ft. 4 in. long and 2 ft. 5.5 in. in diameter, instrumented with six experiments to measure the spectrum, polarization and features of celestial, non-solar x-ray sources. *Ariel 5* decayed on March 14, 1980.

1974 October 18   NASA announced that it planned to land Shuttle orbiters at the Flight Research Center, Edwards AFB, Calif., for the first few missions. Providing much longer runways and better weather conditions than were generally available from the Kennedy Space Center, Fla., Edwards would ensure greater margins of safety until confidence in landing Shuttle orbiters had been acquired.

1974 October 22   A Soviet SL-4 from the Baikonur cosmodrome launched the 12,130-lb. biological satellite *Cosmos 690* to a 134 × 227.5 mi. orbit inclined 68.8°. Carrying white rats, steppe tortoises, insects and fungi in a study of the effects of weightlessness and radiation, the satellite contained a cesium 137 gamma-radiation source to provide an average daily dose of 32 rads/hr. on command from the ground. From the 10th day in orbit, the rats were exposed to a maximum dose of 1,000 rads/24 hr. After return to Earth on November 12, the rats were found more passive than control rats on the ground, an effect attributable to radiation.

1974 October 24   A U.S. Air Force Lockheed C-5A Galaxy transport aircraft carried a Minuteman ICBM to an altitude of 8,000 ft. and dropped it out the rear cargo ramp doors to demonstrate an air-drop launch technique. The missile assumed a vertical position as it began to descend by parachute. It then fired its first-stage motor for 10 sec., reaching a height of 20,000 ft. before burnout. In looking for ways to make the future M-X ICBM invulnerable to attack from Soviet warheads, the Air Force wanted to demonstrate the feasibility of the air-launch option. This was but one basing option being examined for the M-X and was never adopted operationally.

1974 October 28   At 14:30 UT a Soviet SL-12 carried the 12,400-lb. *Luna 23* soft lander to an initial Earth parking orbit, and from there to the moon. The spacecraft went into a 58 × 64.6 mi. lunar orbit inclined 138°, with a period of 1 hr. 57 min. on November 2. From November 4 to 5 further maneuvering changed this to a 10.5 × 65 mi. orbit prior to a descent to the surface. *Luna 23* successfully soft-landed in the southern part of the Mare Crisium at 05:37 UT on November 6, 1974. The modified arm and drill assembly, designed to obtain a core sample from 8 ft. below the surface, was damaged on landing and could not obtain any material. The ascent stage was not fired back to Earth, and communications with *Luna 23* ended on November 9, 1974.

1974 October 30   The *Mariner 10* propulsion system was fired for 1 min. 13 sec., producing a 47.7 ft./sec. velocity change and successfully correcting the flight for a third encounter with Mercury. Uncorrected and as a result of that second encounter, *Mariner 10* would have passed 171,000 mi. from the planet's southern hemisphere. The corrected trajectory would bring it to within 2,700 mi. of the surface. Careful tracking confirmed that this had been a satisfactory maneuver, but precise alignment was tweaked up by yet another maneuver. On February 13, 1975, a 12.5-sec. (6.6 ft./sec.) burn was conducted, and on March 7, 1975, *Mariner 10*'s rocket motor was fired for the ninth and last time when it burned for a duration of 3.1 sec., executing a 1.6 ft./sec. velocity change to precisely set up the trajectory for a close fly-by. During the mission the spacecraft had conducted 409.8 ft./sec. in velocity adjustments.

1974 November 7   Heading for a close encounter with Jupiter, *Pioneer 11* crossed the orbit of the giant planet's outermost moon, Hades, 14 million mi. from the planet. The spacecraft entered Jupiter's magnetosphere at 06:00 UT, 4.3 million mi. from the surface. At 18:20 UT on November 27, however, the solar wind stiffened up and compressed the magnetosphere closer to the planet. The spacecraft's instruments reported it popping into interplanetary space again for 5 hr. 30 min. before the solar wind relaxed and the magnetosphere bounced back. These repeated bow shock-wave crossings confirmed the model of the Jovian magnetosphere that likens it to an unstable soft balloon buffeted by the solar wind and often squeezed in on the side facing the sun. At 01:27 UT on December 2 *Pioneer 11* crossed the orbital path of Callisto, outermost of Jupiter's four large moons, 1.125 million mi. from the planet.

1974 November 15   The first satellite built by Spain, called *Intasat*, was launched on a Delta 2310 from Vandenberg AFB, Calif., at 17:11 UT, as a secondary payload to the *NOAA 4* weather satellite; a third, amateur radio satellite was also carried. Just 24 min. after launch Intasat was released to a 912-mi. orbit inclined 101.7°. The 45-lb. satellite comprised a 12-sided, spin-stabilized polyhedron with a beacon transmitter for measurements of the ionosphere.

1974 November 21 An SL-6 carried the first third-generation Molniya Soviet communication satellite to an elliptical orbit of 1,157 × 23,920 mi. inclined 64.8°. The 3,300-lb. satellite had greater power and larger solar array panels than those flown on Molniya 1- and 2-series satellites. By this time, the Soviet Orbita ground station building program had realized more than 60 operational facilities capable of handling Molniya traffic. Molniya 3 satellites were generally flown at the rate of two each year, 45 having been launched by the end of 1993. Molniya 1 handled civil broadcasts, Molniya 3 military traffic.

1974 November 23 Britain's second operational military communication satellite, *Skynet IIB*, was launched by Delta 2313 from Cape Canaveral, Fla., to a geosynchronous orbit over the Indian Ocean at 7.7° inclination. *Skynet IIB* had a cylindrical main structure 6 ft. 3 in. in diameter, with a height of 6 ft. 10 in. including antenna, and weighed 959 lb. at launch and 517 lb. at apogee motor burnout. Spin-stabilized, the satellite had been built by Marconi Space and Defense Systems, England. One of two first-generation Skynet satellites had failed, as had *Skynet IIA* after launch on January 19, 1974.

1974 December 2 At 09:40 UT cosmonauts Anatoly V. Filipchenko and Nikolei N. Rukavishnikov were launched aboard the 14,995-lb. *Soyuz 16* spacecraft. Following an inaccurate orbital injection by the terminal stage of the SL-4 launcher, the crew placed *Soyuz 16* in a simulated ASTP (Apollo-Soyuz Test Project) prerendezvous orbit and followed closely the flight plan for that international mission. The Soviets had already flown two unmanned Soyuz spacecraft in the ASTP configuration *(Cosmos 638* and *Cosmos 672)* and spared no resources not only in qualifying the hardware for their own test requirements but in demonstrating to critics in the United States that Soviet spacecraft were safe. Opponents of the joint Apollo-Soyuz docking flight planned for July 1975 said that Soviet hardware was unsafe and that ground controllers could not be relied upon in emergencies. *Soyuz 16* returned at an elapsed time of 5 days 22 hr. 24 min.

1974 December 3 GEC-Marconi, the United Kingdom aerospace company, was awarded a contract from the European Space Research Organization for design and development of the MAROTS maritime communications satellite payload section. A consortium led by the United Kingdom company Hawker Siddeley received a parallel contract for the spacecraft bus, the main structure and supporting systems to house the payload. MAROTS (maritime orbital test satellite) was a test and development satellite for ship-to-shore communications.

1974 December 4 *Pioneer 11* made its closest encounter with Jupiter when it flew just 26,725 mi. above the cloud tops in that part of the radiation belts with the greatest intensity. *Pioneer 10* had measured radiation levels and magnetic fields considerably higher than scientists had expected, but whereas *Pioneer 10* did not go as deep into those belts, its total exposure to radiation was cumulatively higher because it passed the planet at a relatively much slower speed. *Pioneer 11* was able to add valuable scientific data to *Pioneer 10*'s more distant sampling. As viewed from Earth, *Pioneer 11* went behind Jupiter at 05:02 UT, but signals did not disappear for 40 min. because it took that long at the speed of light for the last radio waves to get back. The spacecraft reached its minimum fly-by distance over a point on Jupiter at 50° S at

05:22 UT and reappeared at 05:44 UT, 40 min. before radar stations on Earth received word.

1974 December 10 The West German solar probe *Helios 1* was launched at 19:11 UT from Cape Canaveral, Fla., on a Titan IIIE-Centaur, the first successful test of this launcher. The 815-lb. spacecraft comprised a central body with a 16-sided experiment compartment 5 ft. 8 in. in diameter and 1 ft. 10 in. tall. Conical solar arrays extended from each end, giving the structure a spool-shaped appearance with a height of 7 ft., to which a 6 ft. 8 in. antenna mast was surmounted. From an initial parking orbit a complex sequence of Centaur maneuvers placed *Helios 1* in a heliocentric orbit of 91.6 million × 30 million mi. The second of two, *Helios 2* was launched by another Titan IIIE-Centaur on January 15, 1976, to a heliocentric orbit of 89 million × 27 million mi., the closest any spacecraft had approached the sun.

1974 December 19 A Delta 2914 launched the 886-lb. experimental French–West German communication satellite *Symphonie 1* to a geostationary transfer orbit of 250 × 23,657 mi. inclined 13.2°. At 23:00 UT on December 21 a liquid propellant apogee motor, the first of its kind, fired to place the satellite in a geostationary orbit, from which it migrated to its final position at 11.5° E. *Symphonie I* had a diameter of 6 ft., a height of 1 ft. 8 in. and three solar panels extending 8 ft. 4 in. from the satellite. Developed under an agreement signed June 1967, France's national space center, Centre Nationale d'Etudes Spatiales (CNES), managed the satellite and it was built by the Consortium Industriel Franco-Allemand pour Symphonie (CIFAS). Symphonie could carry 1,200 voice circuits or two color TV channels. *Symphonie 2* was launched on August 27, 1975, and placed at 11.5° W.

1974 December 24 The first Soviet ocean electronic intelligence satellite, *Cosmos 699*, was launched from Baikonur by SL-11 to a 271 × 282 mi. orbit inclined 65°. Powered by solar cells, unlike Soviet nuclear-powered ocean surveillance satellites, it was intentionally blown up on April 17, 1975, at the end of its mission so as to prevent components that could survive reentry falling into non-Soviet hands. As flights built up, the Soviets established a pattern of launching a solo elint between pairs of nuclear-powered radar surveillance satellites. A total of five ocean elints were launched through April 18, 1979, before the launch of *Cosmos 1094* on April 18, 1979, and the start of a new operating procedure.

1974 December 26 At 04:15 UT a Soviet SL-13 launch vehicle lifted off from Baikonur with the 41,700-lb. *Salyut 4* space station and placed it in an orbit of 136 × 168 mi. and an inclination of 51.6°. *Salyut 4* was the first successful Soviet long-duration research station designed from the outset for detailed scientific work in materials processing, Earth observation, biomedical research and astrophysical sciences. It represented the sole second-generation "civilian" design, incorporating three pivotable solar-cell arrays attached to the outside surface of the living area. *Salyut 4* would remain in space for more than two years and during seven months of manned operations play host to two teams of visiting cosmonauts.

Cosmos 700, the first in a new series of Soviet navigation satellites, was launched by SL-8 from Plesetsk to a 595 × 619 mi., 83° orbit. The plane of the orbit was offset by 20° from the orbit plane of *Cosmos 627*. The second third-generation Soviet navsat, *Cosmos 726*, was launched on April 11, 1975, to an orbit plane 120° from *Cosmos 700*. An intermediate satellite, *Cosmos 755*, was launched on August 14, 1975. The next satellite, *Cosmos 778*, was sent up on November 4, 1975, to a position between *Cosmos 726* and *Cosmos 700* while *Cosmos 789* was launched on January 20, 1976, to a plane between *Cosmos 700* and *Cosmos 755*. Thus was established the first Soviet six-satellite navigation system in separate orbit planes displaced 30° from each other. Replacement satellites were launched at frequent intervals.

1974 December 31   Under the terms of a bilateral agreement signed on September 30, 1971, a Soviet ground station provided a special hot-line linking Washington, D.C., and Moscow via satellites. Communications were routed via Intelsat and Molniya satellites in an effort to provide a fail-safe means of contact between political leaders in the two countries.

1974   The first version of the Soviet SS-19 Stiletto ICBM began to enter service during the year as a replacement for the SS-11. With the Soviet designation RS-18, the missile had a length of about 88 ft. 6 in., a diameter of 8 ft. 2 in. and a weight of 172,000 lb. With a throw weight of 7,525 lb., the Mod 1 version carried six 550-KT MIRV warheads, the first Soviet ICBM MIRVs to enter service, across a range of 5,000 mi. with an accuracy of 1,150 ft. In 1977, when 100 Mod 1s had been deployed, the Mod 2 appeared with a single 6-MT warhead, a range of 5,500 mi. and an accuracy of 1,000 ft. From a peak deployment of 60 in 1978–1979, it was withdrawn by 1980, being replaced by the Mod 3, which first appeared that year. Of all versions, build-up peaked at 360 in 1983 and was maintained through the early 1990s.

1975 January 9   NASA announced a new space applications program aimed at providing detailed information about the world's oceans. Called Seasat, the satellite would accumulate scientific data on the curvature of the oceans, ocean circulation, the transport of mass, heat and surface nutrients, and provide data on the interaction of the sea and the air and its effect on weather. On November 20 the Jet Propulsion Laboratory awarded Lockheed Missiles and Space Co. a contract to develop the Seasat satellite, which would circle the Earth 14 times and view 95% of the surface every 36 hr.

1975 January 10   Cosmonauts Alexei A. Gubarev and Georgy M. Grechko were launched at 21:43 to the *Salyut 4* space station aboard the 15,050-lb. *Soyuz 17* spacecraft by an SL-4 from Baikonur cosmodrome. Rendezvous was automatic to a distance of 330 ft., after which *Soyuz 17* was manually directed to a perfect docking at the forward axial port on the station's transfer module at 01:25 UT on January 12. *Salyut 4* was to be maintained at an average altitude of 217.5 mi., a higher orbit than used by previous Salyuts and one which would need less orbit-adjustment propellant because there were fewer atmospheric particles to degrade the orbit. The cosmonauts began operations with a solar telescope called OST-1 and maintained contact with the ground via teleprinter, a new feature for *Salyut 4*. On January 17 the crew began operations with the Filin X-ray astronomical telescope and returned to Earth in the *Soyuz 17* spacecraft after a mission lasting 29 days 13 hr. 20 min.

1975 January 14   From this date, NASA's Earth Resources Technology Satellite (ERTS) was renamed Landsat. *ERTS-A* would be retrospectively named *Landsat 1* and *ERTS-B* would be known as *Landsat 2*. A more user-friendly name, it would

standardize the naming of space applications satellites; the ocean surveillance program had already been named Seasat.

1975 January 22   *Landsat 2*, the second U.S. Earth resources satellite, was launched by Delta 2910 from Vandenberg AFB, Calif., to a sun-synchronous orbit of 567 × 570 mi. inclined 99.1°. Similar to *ERTS-A (Landsat* 1), the 2,075-lb. satellite carried the same return beam vidicon system and multispectral scanners providing independent images of an area 115 × 115 mi. By January 25 these instruments were working satisfactorily. Problems developed on November 5, 1979, and the satellite was retired on January 22, 1980, only to be restarted for a limited time beginning on May 5 that year.

1975 January 26   The U.S. Minuteman Force Modernization Program was completed by Boeing when Strategic Air Command replaced the last Minuteman I with Minuteman II and Minuteman III, 500 of each being deployed at six air bases in five states. A conversion program to replace 50 Minuteman II missiles at the 341st Strategic Missile Wing, Malmstron AFB, Mont., with 50 Minuteman IIIs began January 20, 1975, and was completed July 11, with the result that there were now 450 Minuteman IIs and 550 Minuteman IIIs. The last Minuteman III research and development flight took place on March 27, 1980, ending an effort that began with the first launch of a Minuteman I missile on February 1, 1961. In a modification program aimed at increasing the missile's survivability, additional hardening to the top of the silos allowed them to withstand an overpressure of up to 1,000 lb./sq. in.

1975 February 5   Rockwell International demonstrated a prototype of their Low Cost Modular Spacecraft (LCMS) conceived as a common platform for several different satellites. Designed by the NASA Goddard Space Flight Center, LCMS comprised a spacecraft mounting assembly and an instrument ring and could fit inside the payload bay of the Shuttle. Interest in a common satellite bus had been generated when cost studies on recoverable versus expendable boosters revealed a potential advantage in designing satellites for recovery and refurbishment.

1975 February 6   The first Diamant B/P-4 launch vehicle carried the 104-lb. satellite *Starlette* to a 500 × 688 mi. orbit with an inclination of 49.8°. An acronym for Satellite de Taille Adaptée avec Reflecteurs Laser pour les Etudes de la Terre, *Starlette* had a uranium core and laser retroreflectors for research into the Earth's geoid, part of France's development of geodetic satellites. The Diamant B/P-4 differed from the Diamant B in having an increased first-stage burn time of 1 min. 58 sec., a P-4 Rita 1 second stage from the MSBS nuclear missile and a slightly larger P.68 third stage. The P-4 had a thrust of 39,700 lb. and a burn time of 2 min. 27 sec. The launcher could put 440 lb. into low Earth orbit.

1975 February 7   NASA issued a request for proposals on the procurement of telecommunications services for the Tracking and Data Relay Satellite System (TDRS) being developed by industry. Consisting at first of two specialized relay satellites in geostationary orbit and two ground stations in the United States, the system would provide 85% coverage for all Earth-orbiting satellites and spacecraft, including the Shuttle, compared with 15% using existing networks.

1975 February 14   Indonesia selected the Hughes Aircraft Co. to build a satellite communications system designed to connect 120 million inhabitants of 13,000 islands with telephone and TV. Hughes was to build two HS-376-series spin-stabilized satellites, the master control system, and nine ground stations. An additional 30 ground stations would be built by another U.S. contractor. The satellites would be called Palapa and the first was scheduled for launch in late 1976.

1975 February 24 Japan's third scientific satellite, *Taiyo*, was launched by Mu-3C from Uchinoura to a 154 × 1,944 mi. orbit inclined 31.5°. The 190-lb. satellite was instrumented by the University of Tokyo to observe solar x-rays, ultraviolet radiation and the Earth's ionosphere. It decayed through the Earth's atmosphere on June 29, 1980.

1975 March 13 Rocketdyne Division of Rockwell International completed the first Shuttle main engine, known as the integrated subsystems test bed (ISTB). It was not to be used for flight but would perform test firing to qualify component design. The ISTB was fired for the first time on low power levels from test stand A-1 at the National Space Technology Laboratories, Miss., on July 23, 1975. The engine was fired at 65% thrust for 42.5 sec. on March 12, 1976. The ISTB was used to refine many design aspects of the basic engine, and firing tests were also conducted by Rocketdyne at their facility in Santa Susana, Calif.

1975 March 15 The French–West German *Helios 1* solar probe passed, at 09:13 UT, to within 28.6 million mi. of the sun at the perihelion of its orbit—closer than any other spacecraft. Despite temperatures ten times greater than those at the distance of Earth (93 million mi.) *Helios 1*'s instruments measured the solar wind, magnetic fields, micrometeoroids and zodiacal light. The spacecraft again passed through perihelion on September 21.

1975 March 17 At 22:39:24 UT *Mariner 10* made its third and final encounter with the planet Mercury when it passed within 203 mi. of the surface. An additional 300 TV images were transmitted. The spacecraft finally ran out of attitude control gas on March 24 and at 11:54 UT lost lock on the sun and its reference star, Canopus. The command to switch off the transmitter was sent 6 min. later and Earth stations lost lock on *Mariner 10* 21 min. after that. *Mariner 10* had been the first dual-planet mission (Venus and Mercury), the first to visit Mercury, the first to perform three fly-bys of the same planet on one mission and the first to conduct four planet fly-bys on one flight.

1975 April 5 A 15,060-lb. Soyuz spacecraft carrying cosmonauts Vasily G. Lazarev and Oleg G. Makarov was launched by SL-4 from the Baikonur cosmodrome at 11:02 UT on a flight that was to bring about the first launch abort in the history of manned spaceflight. All went as planned until after the four strap-on boosters had separated, at 2 min. 2 sec., and the first stage had completed its planned burn duration of 5 min. 14 sec. At that point pyrotechnic bolts should have fired to cut free the redundant first stage. Only half the bolts fired and the first stage was still attached when the second stage ignited. Sensing the launcher was out of control, the abort system took over and fired the launch escape motor, wrenching free the Soyuz spacecraft from the top of the upper stage. The cosmonauts were accelerated to a peak of 14–15 g and an altitude of 119 mi. before returning safely in the descent module to land just 200 mi. short of the Chinese border, 21 min. 27 sec. after launch. The mission was designated *Soyuz 18-1*, the first suborbital flight into space by Soviet cosmonauts.

1975 April 6 NASA's Goddard Space Flight Center issued a contract to Beckman Instruments for a special ozone mapping instrument to be flown on the last Nimbus weather research satellite, *Nimbus 7*, planned for launch in late 1978. Gathering evidence for changes to the ozone level had been accumulated by weather satellites for several years. Beginning September 8, the Senate Aeronautical and Space Sciences committee held a series of hearings into stratospheric ozone depletion. This led to increased resources being given to scientists for research on the problem and resulted in a new generation of instruments and sensors for detecting and monitoring these changes.

1975 April 9 The NASA Johnson Space Center awarded a contract to Martin Marietta for an investigation into current technologies for manned maneuvering units (MMUs). After selecting the most promising concept, Martin Marietta was to provide a detailed design and high-fidelity mock-up and build a prototype controlled unit. The MMU was a spin-off from the unit built for use inside the orbiting Skylab space station. NASA wanted a developed version for use with the Shuttle.

1975 April 10 Marking the first launch in NASA's new Earth and Ocean Physics Applications Program (EOPAP), a Delta carried the 750-lb. *GEOS 3* satellite to a 512 × 528 mi. orbit inclined 114.96°. Deriving its name from the acronym for geodynamics experimental ocean satellite, *GEOS 3* was not an Explorer-series satellite in the National Geodetic Satellite Program like its predecessors. It consisted of a 4 ft. 5 in. wide octagon with a truncated pyramid below the base and a height of 2 ft. 8 in. Extending toward Earth from beneath the base, a 15 ft. 7 in. boom held the end mass for a gravity gradient system. *GEOS 3* carried a radar altimeter for measuring ocean features, C-band transponders for geometric and gravimetric measurements, and laser retroreflectors for accurate range data. It remains in orbit.

Microparticle detectors left on the moon by the *Apollo 17* astronauts had recorded mini–dust storms caused by static electricity charging minute particles on the surface. Scientists at the NASA Goddard Space Flight Center reported that dust particles were being shifted from high to low areas. More than 20 years earlier geologist Thomas Gold had suggested that electrostatic transportation would smooth out jagged lunar features and shift dust and fine particles into low-lying areas. Rejected by most, his theory appeared to have been proved.

1975 April 19 A Soviet SL-8 launch vehicle launched from Kapustin Yar placed the first Indian satellite, *Aryabhata*, in a 353 × 380 mi. orbit inclined 50.7°. Named for a fifth century Indian astronomer and mathematician, the 794-lb. satellite carried instruments for detecting x rays from galactic sources, neutrons and gamma radiation from the sun, and for measuring properties of the Earth's ionosphere. *Aryabhata* was built by the Indian Space Research Organization (ISRO) and comprised a multifaceted structure, 4 ft. 10 in. in diameter and 3 ft. 7 in. tall. It carried solar cells and a passive thermal control system.

1975 April 24 Gordon Woodcock, a development engineer with the Boeing Co., presented a paper at a conference on solar energy held in Los Angeles describing the solar power satellite (SPS) concept. Utilizing billions of separate solar cells laid out across a platform covering 50 square miles of space, each SPS would produce 10,000 MW, twice the energy of

America's Grand Coulee dam, and beam it to Earth on microwave links. A constellation of 45 SPSs would provide all the electrical needs of the United States. Preliminary analysis indicated the system would be cost effective by 2025. On January 19, 1976, Boeing presented the findings to the Senate subcommittee on aerospace technology and national needs.

1975 April 28   Having first detected the presence of an ozone layer, then the depletion of that protective layer by man-made pollutants, NASA announced the beginning of a major stratospheric research program to investigate this part of the Earth's atmosphere. The survey would coordinate measurements from aircraft, sounding rockets and satellites. Other federal agencies, institutions and private industry would be involved in one of the largest and most extensive environmental research programs to date.

1975 April 29   NASA announced it was inviting scientists to propose experiments for a mission to Uranus using the gravitational slingshot effect of a close fly-by of Jupiter. With a proposed launch date of November 1979, the Mariner-class spacecraft would be almost identical to the one planned for launch to Jupiter and Saturn in 1977, thereby saving additional development costs. Passing within 15,000 mi. of Jupiter in April 1981, the spacecraft would go on to visit Uranus just over four years later. NASA was unable to get approval from Congress for a separate mission, and the objectives were eventually embraced by the 1977 missions. During 1976 NASA formally adopted the name Voyager for its Jupiter-Saturn 1977 mission.

1975 May 6   NASA announces that Canada will build the remote manipulator system (RMS) for the reusable Shuttle. Attached inside the cargo bay of each orbiter vehicle, the RMS would comprise a 50 ft. long articulated arm with elbow and wrist joints capable of movement in a wide range of axes to grapple payloads with a mass of up to 65,000 lb. in or out of the Shuttle in space. A universal end effector ("hand") would attach to specially designed grapple attachments on satellites built for refurbishment or return to Earth. This essential feature of Shuttle program economics was, in fact, included in very few satellites, most notable being the Solar Maximum Satellite, *Landsat D*, and the Hubble telescope.

1975 May 7   The third of three Small Astronomy Satellites, *SAS-C (Explorer 53)* was launched by a Scout launch vehicle from the San Marco platform off the coast of Kenya. *SAS-C* was placed in a 310 × 316 mi. orbit inclined 3°. The primary objective of *SAS-C* was to measure discrete x-ray sources as a basis for more detailed studies with the High Energy Astronomy Observatory. The 430-lb. NASA satellite measured 2 ft. 2 in. in diameter and 2 ft. in height with four foldable solar arrays, and was managed by the Goddard Space Flight Center. *SAS-C* decayed on April 9, 1979.

1975 May 17   France became the first country after the United States and the Soviet Union to launch two satellites on the same launch vehicle. The second Diamant BP.4 from Kourou, French Guiana, put the 77-lb. satellite called D-5A *(Pollux)* in a 167.8 × 797.3 mi. orbit and the 170-lb. satellite named D-5B *(Castor)* in a 169.6 × 788.5 mi. orbit, both with an inclination of 30°. Both carried technology development equipment. *Pollux* carried a hydrazine microthruster delivering a thrust of 0.75 lb. *Pollux* decayed on August 5, 1975, followed by *Castor* on February 18, 1979.

1975 May 22   In Moscow for final technical reviews of spacecraft hardware, NASA's George Low and the Soviet Academy of Sciences' Vladimir Kotelnikov cochaired the Apollo-Soyuz Flight Readiness Review. The culmination of three years' hard work by engineers from the United States and the Soviet Union, the review board examined test records for respective Apollo and Soyuz manned spacecraft and the docking module that would link the two vehicles in space. Working groups reported on progress with specific elements of the hardware and on integrating flight control functions for the joint mission. The flight would be controlled from two flight control centers, one in each country.

1975 May 24   The 15,050-lb. *Soyuz 18* spacecraft carrying cosmonauts Pyotr I. Klimuk and Vatily I. Sevastyanov was launched at 14:58 UT by SL-4 from the Baikonur cosmodrome to a rendezvous mission with *Salyut 4*. The two cosmonauts had been the backup crew for the aborted *Soyuz 18–1* mission launched April 5, 1975. *Salyut 4* was in a 214 × 221 mi. orbit and at 18:44 UT on May 25 *Soyuz 18* docked at the space station. *Salyut 4* began serious deterioration during the two months of occupation, the manned utilization of the station having been delayed by the postponed second visit occasioned by the launch abort to *Soyuz 18-1*. Nevertheless, the crew spent 11 days studying Earth's natural resources, 9 days on biomedical research, 8 days operating the x-ray telescope, 6 days each on solar observations and technical experiments and 2 days on operating atmospheric observation instruments. Displaying the capability of Soviet ground stations to control two missions simultaneously, the cosmonauts remained aboard *Salyut 4* throughout the Apollo-Soyuz Test Project flight, a period when seven spacemen were in orbit at the same time. *Soyuz 18* returned its crew to Earth at an elapsed time of 62 days 23 hr. 20 min.

1975 May 31   The European Space Agency, ESA, came into operation one day after the convention was adopted and opened for signing by the member states of the former European Space Research Organization (ESRO): Belgium, Denmark, France, West Germany, Italy, the Netherlands, Spain, Sweden, Switzerland and the United Kingdom. Members would fund administration costs and core programs with participation in discretionary programs according to choice; about one-third of the budget was for obligatory programs. The approved budget for 1976 was about $600 million, a 40% increase over ESRO/ESA spending for 1975 and about $380 million more than ESRO and ELDO had for 1974.

1975 May   NASA's Langley Research Center issued feasibility analysis contracts to Boeing and Martin Marietta for a single-stage-to-orbit (SSTO) reusable launch vehicle capable of lifting 65,000 lb. into Earth orbit. The studies were to anticipate operational availability in the period 1990–2010 and be launched either from a horizontal position by sled, from a vertical position or from a carrier plane in midair.

1975 June 8   The Soviet Union launched *Venera 9*, the first in a new series of very heavy spacecraft specifically designed and built to explore the surface of Venus and launched by the USSR's heaviest operational launch vehicle. The *Venera 9* spacecraft weighed 10,844 lb. and was launched by a three-stage SL-12. This was heavier than the *Mars 2* and *Mars 3* spacecraft of 1971 and almost four times the weight of the *Venera 7* and *Venera 8* spacecraft of 1970 and 1972. A second spacecraft of this type, the 11,098-lb. *Venera 10*, was launched by another SL-12 on June 14, 1975. Each spacecraft was a combination orbiter/lander with a wide range of scientific equipment and cameras on the landers for taking images from the surface. The Soviets proudly announced that for the first time, these spacecraft carried digital computers for autonomous command. The United States had been using digital computers in space since 1965.

1975 June 21   NASA's sole second-generation Orbiting Solar Observatory, *OSO-8,* was launched by Delta 1910 from Cape Canaveral, Fla., to a 338 × 348 mi., 32.9° orbit. With a weight of 2,346 lb., *OSO-8* was more than three times the weight of the standard first-generation OSO and had a scientific payload of 823 lb. The satellite consisted of two sections: a nonspinning sail, 7 ft. 8 in. tall and 6 ft. 10 in. across, providing power from solar cells and a platform for two sun-locked instruments; and a cylindrical spinning section, 5 ft. in diameter and 2 ft. 4 in. high, carrying the six main science instruments. Solar cells and batteries provided 110 W of power and the satellite was spin-stabilized at 6 RPM. *OSO-8* decayed on July 9, 1986.

1975 June 24   A milestone in the development of the Space Shuttle main engine occurred when a test combustion chamber was fired for 1 sec., demonstrating ignition using propellants carried through the injector. The test took place at the National Space Technology Laboratories in Missouri and was the first of several short-run firings planned for the next several months designed to culminate in a run at about 20% of the engine's rated thrust of 470,000 lb.

1975 June   During NASA budget hearings for fiscal year 1976, the House of Representatives voted to significantly cut

A pre-launch view of the Orbiting Solar Observatory-8 (OSO-8), *prior to launch on June 21, 1975.*

funds for the Pioneer-Venus mission involving multiprobe and orbiter spacecraft scheduled for launch in 1978. Rallying to the call for funds to be reinstated, leading meteorologists and climatologists from the United States and other countries urged the House to think again. The mission was, they said, vital to a better understanding of the Earth's rapidly changing weather and climate patterns. By knowing more about how the atmospheres of Mars and Venus had evolved, scientists could put together a more precise picture of what was happening to the Earth's weather. In September 1975 the House-Senate Conference Committee restored the money requested by NASA.

NASA's Marshall Space Flight Center awarded a contract to Boeing for studies on heavy-lift launch vehicle (HLLV) options for each of five separate payload ranges: Class 1, 132,000–198,000 lb. to low Earth orbit; Class 2, 198,000–298,000 lb.; Class 3, 298,000–441,000 lb.; Class 4, 441,000–661,000 lb.; and Class 4, 661,000–992,000 lb. Traffic levels capable of handling between 1.1 million and 275 million lb. of cargo per year were to be considered. Furthermore, the study was to consider support of four scenario programs ranging from space station support through to assembly and operation of solar-powered satellites. The final study report was issue on June 22, 1976.

1975 July 11   An SL-3 launch vehicle carried the first in a new generation of Soviet weather satellites, *Meteor 2-1,* from Plesetsk to a 532.5 × 553.5 mi. orbit inclined 81.2°. The 6,200-lb. satellite was the first in a new generation with essentially the same design arrangement as the Meteor 1 series but with better equipment and higher resolution. The next Meteor 2–series satellite was launched on January 6, 1977, and thereafter at more frequent intervals interspersed with Meteor 1– and Meteor 3–series launches. Through the end of 1994 there were 21 flights of this second-generation weather satellite.

1975 July 13   Authorities in India reported that preparations for a unique experiment using the most powerful broadcast satellite yet launched, *ATS 6,* were nearing completion. An estimated 5 million people in 2,400 rural villages were being supplied with television sets equipped to receive educational and health programs via the NASA satellite. Under what was called the Satellite Instructional Television Experiment (SITE), electrical power was brought to villages that previously had none, village TV-caretakers were selected, and locations such as schools selected for community gatherings. More than 1,200 hr. of diverse program content was to be broadcast in the year-long SITE program beginning August 1.

1975 July 15   At 12:08 UT Soviet Apollo-Soyuz Test Project cosmonauts Alexei Leonov and Valery Kubasov were launched aboard their *Soyuz 19* spacecraft from the Baikonur cosmodrome to a 115.8 × 137.9 mi. orbit with an inclination of 51.78° at the start of the world's first international space mission. The first of two maneuvers to circularize the orbit took place at the end of the fourth revolution, at an elapsed time of 5 hr. 19 min., resulting in a 119.6 × 143 mi. orbit. At 19:30 UT the Apollo CSM-111 spacecraft was launched by Saturn IB from Complex 39B carrying astronauts Thomas Stafford, Vance Brand and Donald Slayton in to a 96.1 × 107.7 mi. orbit with an inclination of 51.75°. The *Apollo* spacecraft (CSM-111) weighed 28,054 lb. and the docking module 4,436 lb. Three orbital rendezvous maneuvers conducted by *Apollo* preceded a Soyuz circularization burn which put the Soviet spacecraft in an orbit of 142.7 mi. at an elapsed time of 24 hr. 24 min. Apollo performed five more maneuvers before achieving rendezvous with *Soyuz 19* at a Soyuz-elapsed time of about 51 hr. 15 min. Docking began at 51 hr. 49 min. and was completed 4 min. later. Stafford and Slayton moved into the docking module, closed the hatch behind them and opened the hatch into *Soyuz 19* at 54 hr. 57 min., the two commanders shaking hands in orbit for the first time two min. later. After 7 hr. 29 min. of transfer time Stafford and Slayton were back in Apollo. Beginning at 68 hr. 45 min. a series of three crew transfers began which resulted in each crew member's visiting the other spacecraft for a total duration of 11 hr. 55 min. The two spacecraft undocked at 95 hr. 42 min. elapsed time and redocked 32 min. later, with the Soviet spacecraft the active vehicle. Final undocking came at 99 hr. 6 min. and the Soyuz descent vehicle landed 7 mi. from its target point in Kazakhstan at 142 hr. 31 min. Apollo remained in orbit for a further three days, returning to Earth on July 24, 1975, at an elapsed time of 217 hr. 28 min. 23 sec. Thus ended the last flight of a U.S. manned ballistic spacecraft. In 31 manned flights, including 29 orbital missions, NASA astronauts had accumulated 22,503 hr. 48 min. 58 sec. of spaceflight experience from launch through landing. The next U.S. manned space vehicles would be reusable Shuttles, the first of which was launched on April 12, 1981. It would be 20 years before the next joint United States–Russian docking flight took place in 1995.

1975 July 17　The International Telecommunications Satellite Organization, Intelsat, requested firm bids for design and development of a fifth-generation communication satellite. The board of governors agreed a need for seven Intelsat V satellites with options for up to eight more. Although the Intelsat IV-A series had yet to be launched, the Intelsat V family was scheduled for launches beginning in 1979. They would have increased capacity of 12,000 two-way voice circuits plus one color TV channel, in the 6/4 GHz band with antenna beam separation and dual polarization, and would have pioneered the use of the 14/11 GHz band for limited use in high-traffic regions.

1975 July 21　Fourteen Arab League nations joined with six other Middle East states to participate in implementation of the Arab satellite communications network, Arabsat. The Broadcasting Union of the Arab states determined a need for 15 radio channels for direct broadcast of educational and cultural programs, with as many as 1,060 telephone and telegraph channels needed by 1976 and up to 6,200 by 1984. Arabsat would provide communications for up to 95% of the Arab world.

1975 July 24　In attempting to explain the apparent apathy to the nation's space program, veteran U.S. media anchor man Walter Cronkite said that it would be hard to maintain "the high intensity of feeling about the space program, the race to the moon, [and the] landing on the moon. We don't have an attention span in the modern world, with so much going on, that permits us to remain at that high level of excitement." With an end to manned ballistic reentry flights and almost six years before the first Shuttle flight, NASA was at a watershed in manned spaceflight, punctuated by a long hiatus in operations.

1975 July 26　A Chinese CZ-2 launcher placed the 2,440-lb. *SKW-3* satellite in a 115 × 286.5 mi. orbit inclined 69°. This third successful Chinese satellite launch was also the first successful flight of the CZ-2, which had been developed from the CSS-4 ICBM. The CZ-2 had a length of 103 ft. 8 in., a diameter of 11 ft. and a launch weight of 421,150 lb. It was powered by four YF-2 engines in the first stage delivering a thrust of 617,400 lb. and a single YF-2 in the second stage delivering 171,300-lb. thrust, with UDMH-N_2O_4 propellants. The FB-1 could place 4,410 lb. in orbit, and other flights with this vehicle included *SKW-5* on December 16, 1975; *SKW-6* on August 30, 1976; a launch failure on July 30, 1979; and a triple launch of *SKW-9A/B/C* on September 19, 1981, the last FB-1 flight.

1975 July　The U.S. Air Force announced operational plans for converting Space Launch Complex 6 (SLC-6, pronounced "slick-six") at the Vandenberg AFB, Calif., into a launch complex for the reusable Shuttle. Shuttle flights with orbital inclinations greater than 55° would take place from this facility and not the Kennedy Space Center, Fla., which would be used of flights to lower inclination orbits. SLC-6 had been built for the canceled Manned Orbiting Laboratory program. The Air Force planned to begin Shuttle flights from Vandenberg AFB in December 1982, with a second pad ready by 1986.

Boeing received a NASA contract to study a proposed large lift vehicle (LLV) made up from Shuttle elements. Without an orbiter, the three main liquid propellant engines, the external tank and the avionics package could be used to lift heavy loads into space or carry an upper stage for translunar missions. The LLV would be more efficient and much less expensive than the Saturn V because it would use recoverable components that could be reused many times.

1975 August 9　The first European Space Agency satellite, *COS-B*, was launched by a Delta 2913 from Vandenberg AFB, Calif., to a 214 × 62,069 mi. orbit with an inclination of 90.3°. Orbital dynamics changed this to 1,118 × 60,667 mi. at 91.9° by the end of the year. With a weight of 617 lb., *COS-B* was the first European scientific satellite to focus on a single experiment, a 260-lb. gamma-ray telescope for studying gamma rays from celestial sources. The satellite had been scheduled for launch on a Europa II, but with the cancellation of that vehicle the satellite was reconfigured for a U.S. launcher. *COS-B* was built by MBB (Messerschmitt-Bölkow-Blohm) in West Germany, with significant subsystems from the United Kingdom, Spain, Italy, France and Denmark. It consisted of a cylinder 4 ft. 7 in. in diameter, 5 ft. 7 in. tall, with solar cells providing 130 W of power. *COS-B* remains in orbit, but is inoperative.

1975 August 11　Attempts to launch the *Viking 1* orbiter/lander combination on Titan III-E/Centaur TC-3 were scrubbed at T − 115 min. when a faulty thrust-vector control valve, essential for directional control during powered flight, was discovered on the launch vehicle. Two days later engineers discovered that the Viking orbiter batteries had been drained from 37 volts to 9 volts by accidental movement of a rotary switch. The 7,784-lb. *Viking 1* comprised a 5,133-lb. orbiter and a 2,613-lb. lander. The orbiter was built around an octagonal ring 1 ft. 6 in. high and 7 ft. 10 in. wide with alternate 4 ft. 7 in. and 1 ft. 8 in. sides containing 16 electronics bays, 3 ft. on each long side and 1 ft. on each short side. Eight 5 ft. 2 in. × 4 ft. solar panels hinged in pairs to produce four "wings" with a span of 32 ft. and a surface area of 165.8 sq. ft. carried 34,800 solar cells producing 1.4 kW at Earth and 620 W at Mars. Communication was via a 4 ft. 11 in. circular high-gain dish antenna or an omnidirectional low-gain antenna. Data rates of up to 16 kilobits/sec. were available through the 20-W transmitter. Nitrogen gas jet thrusters on the solar panels would maintain attitude control using a 32-lb. supply. The orbiter had a total height of 10 ft. 9.5 in.

1975 August 20 At 21:22 UT a 1.413-million-lb. Titan III-E/Centaur carrying the *Viking 1* orbiter/lander was launched from Complex 41 at Cape Canaveral, Fla. With a liftoff thrust of 2.4 million lb. from the two solid propellant boosters, the vehicle climbed to an altitude of 24 mi., at which point, 1 min. 50 sec. after liftoff, the liquid propellant core stage ignited to produce a thrust of 520,000 lb. The solids were jettisoned 12 sec. later and the core stage cut off at 4 min. 16 sec., followed immediately by ignition of the 100,000-lb. thrust second stage, which burned until 7 min. 40 sec. At 7 min. 56 sec. the 30,000-lb. thrust Centaur stage fired for 2 min. 7 sec. to put the assembly into Earth parking orbit. A second Centaur burn lasting just over 5 min. pushed the 7,784-lb. *Viking 1* toward Mars.

1975 August 27 The *Viking 1* spacecraft conducted a course correction maneuver 1.741 million mi. from Earth, moving the Mars encounter distance to within 3,450 mi. of the planet. The 12-sec. firing of the Viking orbiter's 300-lb. thrust main rocket motor used 12.5 lb. of propellant from the two cylindrical tanks. Each had a length of 4 ft. 7 in. over hemispheric end-domes and a diameter of 3 ft., and contained a total 3,144 lb. of monomethylhydrazine and nitrogen tetroxide propellants. Helium gas from a tank 5 ft. 4 in. in diameter pressurized the propellant for delivery to the motor. The propulsion system dominated the top of the orbiter on the opposite side of the spacecraft to the attached lander. The system could accommodate four course correction burns, the main orbit-insertion burn and up to 20 orbit trim burns with a total velocity adjustment capability of 4,856 ft./sec.

1975 August NASA released *Outlook for Space*, a report that emphasized the need for a permanently manned space station both to derive maximum benefits from space for people on Earth and to utilize most effectively the reusable Shuttle. The station should be used, said the study, to improve Earth observation to the point where it could play a significant role in better managing Earth's resources and improving crop yield as well as offering a means of research into better vaccines and medicinal products in the unique weightless environment. Prophetically, the study indicated that NASA would be unable to achieve operational availability of such a facility before the end of the century at the earliest.

1975 September 4 Responding to a perceived need for an orbit-to-orbit tug with which it could move satellites and spacecraft from low Earth orbit to higher orbits, the U.S. Air Force decided on solid propellant for its interim upper stage (IUS). The U.S. Air Force Space and Missile Systems Organization (SAMSO) wanted the IUS to have a 3,500–5,000-lb. payload capability. Bidders conferences were held during October and December, and proposals were received from industrial competitors in March 1976. The prime candidate for a contract to build the IUS was Boeing, with Lockheed, UTC (United Technologies Corp.), Thiokol and Aerojet candidates for providing the solid propellant.

1975 September 5 The first of a new generation of Soviet photoreconnaissance satellites was launched under the guise of *Cosmos 758* on an SL-4 from Baikonur. The 8,820-lb. satellite was an adapted variant of the Soyuz spacecraft, the first major change in the basic vehicle employed for photorecon-

naissance satellites since *Cosmos 4* was launched on April 26, 1962. *Cosmos 758* was placed in a 108 × 202 mi. orbit inclined 65.2°. With high-resolution optics, it represented a major step up in capabilities but got off to an inauspicious start when it exploded on the ninth day. Only two were launched in 1976, of which one also exploded, followed by a successful flight in 1977 for the full duration of 29.5 days. One was launched in 1978 and four in 1979, but not before 1980 did the type become operational, with 5 flights that year, 10 in 1981, 8 in 1982, 11 in 1983, 5 in 1984 and 4 in 1985, by which time the 40-day version (introduced in 1982) was standard.

1975 September 9 At 18:39 UT a Titan III-E/Centaur launched *Viking 2* on course for Mars from Complex 41 at Cape Canaveral, Fla., in a launch sequence similar to that followed by the *Viking 1* launcher on August 20. Delayed first by problems with *Viking 1* and then by a spacecraft problem in *Viking 2*, the flight began on the last day of the launch window allowing the full mission objective to be achieved, and only 3 min. before heavy rainstorms would have aborted the launch. At 8 min. into the flight all telemetry from *Viking 2* was lost for 6 min., causing minor consternation among flight controllers. *Viking 2* performed a course correction maneuver when the main propulsion system fired for 20.5 sec., consuming 21.5 lb. of propellant, on September 19, 1975.

The first Japanese N launcher lifted off from its pad on the island of Tanegashima, Japan, carrying the 185-lb. Engineering Test Satellite-1 (ETS-1) *Kiku* to a 607 × 686 mi. orbit inclined 47°. The N-series launcher had been developed from the McDonnell Douglas Delta and included an MB-3 first engine modified from the U.S. Thor, three Thiokol Castor 2 solid propellant strap-on boosters, an LE-3 second stage designed by Japan's National Space Development Agency and a Thiokol TE-364–4 solid propellant third stage. N-1 had a total height of 107 ft., a launch weight of 198,500 lb. and a payload capacity of 287 lb. to geostationary transfer orbit. The satellite was instrumented to measure ascent vibrations and temperatures in space.

1975 September 12 Brooks AFB in San Antonio, Tex., was chosen by NASA as a second location for 10–20% of the lunar samples returned to Earth by Apollo astronauts. They would constitute a cross-section of all types of material returned from the moon and provide assurance that some samples would remain intact for analysis should a catastrophe befall the prime location at NASA's Johnson Space Center in Houston, Tex.

1975 September 22 NASA and the U.S. Air Force Space and Missile Systems Organization (SAMSO) jointly contracted for an advanced weather satellite. Built by RCA, it would be known at NASA as Tiros N and at the Air Force as the Block 5D series of the Defense Meteorology Satellite Program. The initial contract was for 12 satellites, nine of which would go to NASA. Attached to its Burner II upper stage, the assembly would weigh 5,950 lb., reduced to 1,060 lb. after motor burnout. Deployed, the satellite would have a length of 12 ft. 6 in., a width of 7 ft. 6 in., and 100 sq. ft. of solar arrays producing 290 W. Line-scan imaging equipment would produce visible-infrared images with a resolution of 1.7 mi., or 1,900 ft., of local areas on direct transmission.

1975 September 26 The first in a series of upgraded Intelsat communication satellites was launched when an Atlas Centaur carried *Intelsat IVA F-1* from Cape Canaveral to a geostationary transfer orbit. With a launch weight of 3,340 lb. and an on-station weight of 1,820 lb., the satellite was similar to the Intelsat IV series but with capacity to handle 6,000 telephone calls through 20 transponders. *Intelsat IVA F-1* had a total height of 22 ft. 9 in. Other satellites in the Intelsat IVA series included *F-2* on January 30, 1976, *F-4* on May 26, 1977, *F-5* on September 29, 1977 (destroyed in a launch vehicle failure), *F-3* on January 7, 1978, and *F-6* on March 31, 1978.

1975 September 27   The last of three Diamant BP.4 launchers carried the 253-lb. satellite D-2B *Aura* from Kourou, French Guiana, to a 310 × 440 mi. orbit with an inclination of 37.1°. The first French satellite built and managed entirely by industry, *Aura* was equipped with instruments for studying solar far-ultraviolet emissions and atmospheric absorption of solar ultraviolet radiation. This was the last French national satellite carried into space aboard an independent French launcher. Since November 26, 1965, France had successfully carried 12 satellites to space on 10 launches: four on Diamant A; four on Diamant B; and four on Diamant BP.4. *Aura* decayed on September 30, 1982.

1975 October 16   The first operational Geostationary Operational Environmental Satellite, *GOES-1*, was launched by Delta 2914 from Cape Canaveral, Fla., to a geosynchronous orbit inclined 6.7° at 50° N. It was the first operational version of the Synchronous Meteorological Satellite (SMS), two of which had been launched for engineering tests and preoperational tests. *GOES-1* had a launch weight of 1,375 lb., reduced to 650 lb. at apogee motor burnout. The drum-shaped satellite had a diameter of 6 ft. 3 in., a height of 8 ft. 10 in. and solar cells providing 200 W of power. A visible-infrared spin-scan radiometer provided day and night global weather pictures. *GOES-2* was launched on June 16, 1977, to a position over 75° W. *GOES-3*, launched on June 16, 1978, was placed at 135° W.

1975 October 18   A U.S. Defense Support Program early warning satellite in geostationary orbit detected a very strong source of infrared energy which blinded the sensors. This occurred five times in the following six weeks, on one occasion disabling the satellite for four hours. Since DSP satellites would be the first to observe a surprise missile attack, Department of Defense sources were deeply concerned and concluded initially that Soviet laser beams were being directed at the infrared "eyes" to hide an attack. Reconnaissance satellite images revealed the real reason to be infrared light from fires at Soviet gas pipeline breaks.

1975 October 22   Two days after they had separated for their independent missions, the *Venera 9* lander descended through the atmosphere of Venus while the orbiter slipped into an elliptical path about the planet, the first spacecraft to orbit Venus. Encapsulated in a spherical heat shield (8 ft. 2 in. in diameter) for the initial braking phase, the 3,440-lb. lander entered the atmosphere of Venus at 03:58 UT traveling at a speed of 35,100 ft./sec. Separated from its shroud at a height of 40 mi. and a speed of 820 ft./sec., the lander deployed first three drogue parachutes and then three main parachutes. These were jettisoned at 31 mi., and the lander descended in free fall for a touchdown at 05:13 UT. The 6 ft. tall lander was sitting in the Beta Regio area at approximately 33° N by 293° W. Two minutes after landing, a cover over the single camera was ejected and at 05:28 UT the one picture sent to Earth was transmitted showing a block-strewn surface. Signals were sent back from the surface for 23 min. The main spacecraft, meanwhile, decelerated to a 808 × 69,597 mi. orbit with a

period of 48 hr. 18 min. Spectrometers, radiometers and photopolarimeters were used to study the atmosphere, the cloud layers and sun-planet interactions with the solar wind.

1975 October 25   The Soviet *Venera 10* lander and orbiter arrived at Venus. The lander entered the Venusian atmosphere at 04:02 UT and touched down at 05:17 UT, having gone through a sequence identical to that followed by its companion, *Venera 9*, 1,400 mi. away. The lander rested on the surface at 16° N by 129° W and operated for 1 hr. 5 min. during which it sent to Earth a picture of the surrounding material very different from that of *Venera 9*. Like its companion, the *Venera 10* lander carried a large range of scientific equipment for measuring the atmosphere during descent. The main *Venera 10* spacecraft fired its main propulsion system to enter a 870 × 70,840 mi. orbit with a period of 49 hr. 23 min.

1975 November 12   NASA announced that it had decided to eliminate all launch vehicles except the Scout after the reusable Shuttle became operational. The U.S. Air Force decided against adopting this policy for its own launch strategy, preferring to wait and see how the Shuttle performed. Although it backed NASA in funding requests to Congress, the Air Force was not prepared to order payloads designed only for the Shuttle and kept its launcher options open.

1975 November 17   At 14:37 UT the 15,000-lb. unmanned *Soyuz 20* was launched by SL-4 from the Baikonur cosmodrome to a 210 × 224.5 mi. orbit. Two days later, at 19:20 UT, it docked to the forward port of the *Salyut 4* space station in an orbit of 212.5 × 218 mi. The primary purpose of this flight was to qualify a completely new, totally automatic rendezvous and docking system of an unmanned Soyuz with an unmanned Salyut. This was in preparation for the utilization of Soyuz spacecraft stripped of their life-support systems and used as cargo transporters, an application that would begin in 1978 with Progress spacecraft. An additional flight objective was to carry biological specimens for analysis back on Earth. A further requirement was to space-soak the Soyuz spacecraft for three months to qualify it for docking at Salyut while cosmonauts remained aboard the station for extended periods. *Soyuz 20* returned to Earth on February 16 after a mission lasting 90 days 11 hr. 47 min.

1975 November 20   The first satellite launched with a prime objective of using its instruments to monitor the Earth's ozone layer for signs of depletion from man-made causes was launched by NASA on a Delta from Cape Canaveral, Fla. Fifth and last of the Atmosphere Explorer satellites, *AE-E (Explorer 55)* was placed in a 278 × 279 mi., 19.6° orbit from which it could monitor the upper layers of the ozone region. Carrying 311 lb. of scientific instruments, it was essentially the same as *AE-D*, adapted for this important task after instruments piggybacked on weather satellites revealed concern about the state of the ozone layer, which had itself been discovered by satellites. *AE-E* decayed on June 10, 1981.

In deep space, NASA's *Pioneer 11* spacecraft made its first observations of the ringed planet Saturn beginning 16:20 UT. Although the satellite was still some 800 million mi. from Saturn, scientists wanted to observe the atmosphere from a position at a tangent to the Saturn-sun line. After passing Jupiter on December 4, 1974, *Pioneer 11* flew a course that turned its flight path more than 90° and flung it back across the solar system in an arching trajectory that would take it 100 million mi. above the ecliptic, the plane in which most of the planets orbit the sun.

A satellite was used for the first time to transmit facsimile pages of a newspaper. The Intelsat IV satellite serving the Atlantic region was used to transmit pages of the *Wall Street Journal* from Massachusetts to Florida, where a regional edition for the southeast was produced. The newspaper owners Dow Jones claimed the satellite relay was 70% cheaper than conventional means utilizing telephone landlines and microwave links. Gradually this type of service would expand to cover international editions of foreign newspapers.

1975 November 25 A Soviet SL-4 launched from Plesetsk put *Cosmos 782*, the first satellite providing artificial gravity for biological specimens, in a 141 × 252 mi. orbit inclined 62.8°. A small spinning table to which specimens were attached provided a 1-g environment. The 8,820-lb. satellite carried four experiments from the United States as well as samples from Russia, Czechoslovakia, France, Hungary, Poland and Romania, and was recovered December 15. Due to the lack of an equivalent program, this had been the first time since 1969 that the United States had been able to conduct biological experiments in space and the launch of *Cosmos 782* provided the first opportunity for U.S. experiments to fly on an appropriate satellite.

1975 November 26 The first successful launch of a Chinese recoverable satellite took place from Jiuquan when an improved FB-1, designated CZ-2C (Long March 2C), carried the 5,512-lb. satellite *SKW-4* to a 111 × 301 mi. orbit inclined 63°. Weighing about 4,000 lb., the beehive-shaped descent vehicle returned to Earth after six days, China being only the third country, after the United States and the Soviet Union, to safely recover an object from space. Carrying military photo-reconnaissance equipment, additional CZ-2 fights with satellites of this type were launched on December 7, 1976; January 26, 1978; September 9, 1982; August 19, 1983; and September 12, 1984.

The extraordinary cost-effectiveness of using communications satellites rather than ground links for high-density routes proved an embarrassment to the U.S. Comsat General Corp. Deemed by the Federal Communications Commission to be making too much money, it was ordered to reduce its international rates by 35%. Essentially a wholesaler of communications services, Comsat had been selling satellite circuits to domestic telephone and telegraph companies that connected the service to subscribers.

1975 December 5 Two inflatable spheres, called *Explorers 56* and *57*, were destroyed when the four-stage Scout carrying them to space from Vandenberg AFB, Calif., failed to reach orbit. Identified as the Dual Air Density (DAD) Explorer program, the mission was a continuation of similar air-density measurements conducted by *Explorers IX, XIX, XXIV* and *39. DAD-A (Explorer 56)* had a weight of 78 lb. and a diameter of 2 ft. 6 in., while *DAD-B (Explorer 57)* had a weight of 79 lb. and a diameter of 12 ft. Made of Mylar, they were to have been deployed to elliptical orbits from which they would have measured differences in air density with temperature, time and position. These were the last standard Explorer missions funded by NASA, although the name "Explorer" would emerge periodically in the 1980s for selected satellites.

1975 December 10 NASA issued a request for proposals from industry for a space station systems analysis study with

bids due by January 26, 1976. Two parallel contracts, one managed by the Marshall Space Flight Center and one by the Johnson Space Center, would last 18 months and study low-orbit and geosynchronous orbit stations. NASA now envisaged that the facility would be used to test solar power satellite concepts and as a place where new methods of assembling structures in space could be evaluated. The request included a requirement for contractors to examine synchronous-orbit stations 22,000 mi. above Earth.

1975 December 13 RCA's *Satcom 1*, the second U.S. domestic communication satellite after Western Union's Westar, was launched by Delta 3914 to a transfer trajectory and from there to 135° W. This was the first Delta 3000 series launch, essentially a 2000 series with nine Castor IV strap-ons, each 36 ft. 7 in. long with a thrust of 74,000 lb. Delta 3914 could lift 2,065 lb. to geosynchronous transfer as against only 1,550 lb. for a Delta 2914. *Satcom 1* weighed 1,913 lb. with apogee motor and 1,020 lb. on station. It comprised a three-axis stabilized box 5 ft. 4 in. × 4 ft. 1 in. × 4 ft. 1 in. with 24 transponders operating in the 4–6 GHz band, each of which could handle 1,000 voice circuits. *Satcom 2* was launched on March 26, 1976, to a position above 119° W. Launched on December 7, 1979, *Satcom 3* was lost when the apogee motor exploded shortly after ignition. Its replacement, *Satcom 3R*, was launched by Delta 3910 on November 20, 1981, to 132° W for cable users. Last in the first-generation series, *Satcom 4* was launched on January 16, 1982, to 83° W.

1975 December 19 *Pioneer 10* completed a course correction maneuver, increasing speed by 98 ft./sec., to tweak the trajectory aimed at Saturn so that mission planners could preserve two precise targeting options for the September 1, 1979, fly-by. Whether to fly the spacecraft between Saturn's innermost ring and the cloud tops or come in under the rings and pass upward outside them. A later maneuver would have to be made after a decision had been made as to which trajectory to adopt. Different viewpoints were considered by the mission planners and the science targeting teams who would program the spacecraft for a close look at the planet.

1975 December 22 Under the international designation *Statsionar*, the first Soviet geosynchronous communication satellite *Raduga 1* was launched by SL-12 from the Baikonur cosmodrome. *Raduga 1* was placed in geosynchronous orbit with an inclination of 3.5°. The 11,000-lb. satellite operated in the 4–6 GHz band and would handle government communication services across the Soviet Union and between Moscow and Soviet embassies in foreign countries. Varying in weight between 4,400 and 11,000 lb., Raduga satellites would be launched at an average rate of almost 2 per year, 32 having been sent to geosynchronous orbit by the end of 1994.

A newly formed domestic communication satellite company, Satellite Business Systems (SBS), filed an application with the Federal Communications Commission for an all-digital service for industrial and government users. SBS was a partnership formed from subsidiaries of Aetna Life and Casualty, Comsat General Corp. and IBM, each owning one-third. Using 14/12 GHz frequencies on two satellites, one a spare, the new system was aimed at remote customers networked to a single private line using small antennas at the customers' locations anywhere in the 48 contiguous states.

1975 The Soviet SS-17 Spanker ICBMs began operational service. Replacing the SS-11, and slightly longer, the SS-17 had a throw weight of 6,025 lb., and in the initial version had a range of 5,500 mi. with four 750-KT warheads and an accuracy of 1,450 ft. The SS-17 and SS-19 were the first Soviet ICBMs to carry MIRV reentry vehicles. Deployment of the first version, Mod 1, peaked at 130 in 1980–1981 but all were withdrawn by 1983. Total SS-17 deployment of all three versions reached 150 in 1980 and remained at that level until the late 1980s, when numbers dropped to 75 by 1990 with complete elimination of the type by 1995. With one 4-MT warhead, a range of 5,500 mi. and an accuracy of 1,400 ft., 20 missiles of the Mod 2 version were deployed between 1978 and 1982. The definitive Mod 3 appeared in 1982.

1976 January 8 The NASA Flight Research Center at Edwards AFB, Calif., was renamed the Hugh L. Dryden Flight Research Center in honor of the pioneer aeronautical researcher. Dr. Dryden served as the NACA director from 1947 and then as NASA's deputy administrator from its inception on October 1, 1958, until his death in 1965.

1976 January 17 The world's most powerful communications satellite, Communications Technology Satellite (CTS), was launched at 23:28 UT from Cape Canaveral, Fla., on a Delta 2914 to a geostationary transfer orbit. Following an apogee motor firing at 20:41 UT, CTS was placed in a geostationary orbit over 116° W by January 29. The satellite was built by the Canadian Communications Research Center and comprised a cylinder 6 ft. 2 in. high and 6 ft. across with two wing-like solar arrays delivering 1.25 kilowatts of power. CTS was equipped with a 200-W transmitter for the first tests conducted with a direct-broadcast capability, transmitting to relatively small antennas.

1976 January 18 NASA's Johnson Space Center reported that the Apollo Lunar Surface Experiments Package (ALSEP) left on the moon by the *Apollo 14* astronauts in February 1971 had abruptly ceased transmissions. On February 19 the ALSEP suddenly came on again but just as abruptly shut down once more on March 19. This was the only ALSEP station to behave so erratically and none of the four others, placed at the lunar surface by Apollos *12, 15, 16* and *17*, failed to work properly.

1976 January The Sigma Corp. conducted a study for NASA on replacement of the two Shuttle solid rocket boosters (SRBs) with a single liquid rocket booster. Known as EDIN05, the concept envisaged either three or four Rocketdyne F-1 engines from the Saturn V program attached below a liquid oxygen-kerosene propellant module and a stretched external tank. In some respects it can be said that the liquid boost module (LPM) of later derivation evolved from this concept. With NASA committed to getting the SRB-equipped Shuttle into operation, the EDIN05 was shelved.

1976 February 10 NASA's *Pioneer 10* spacecraft crossed the orbit of the planet Saturn on its way out of the solar system. Still reporting back details of the interplanetary environment, the spacecraft could not observe Saturn because the ringed planet was about 100° around in its solar orbit and had moved on from that location in space in June 1969. *Pioneer 10* was now 860 million mi. from the sun and 892.21 million mi. from Earth. On March 20, 1976, *Pioneer 10*, now 430 million mi. beyond Jupiter, detected the magnetospheric tail of Jupiter's radiation belts, thus testifying to the enormous strength of that planet's radiation belts.

1976 February 12 A second series of tests with the Soviet satellite inspection system began with the launch of a 661-lb. target satellite, *Cosmos 803*, on an SL-8 from Plesetsk to a 340 × 386 mi. orbit inclined 65.8°. Four days later the 4,410-lb. *Cosmos 804* was launched by SL-11 from Baikonur to a 349 × 384 mi. intercept-inspection orbit from which it was returned to Earth the same day. The test was part of Soviet military exercises with simulated strikes by naval and long-range aircraft on February 17 and a simulated mass launch of ICBMs two days later. *Cosmos 803* was also the target for the 4,410-lb. *Cosmos 814*, launched on April 13 by SL-11 from Baikonur to a 345 × 382 mi. orbit, performing a rapid fly-by and reentry on the first orbit.

1976 February 19 *Marisat 1*, the first privately owned communication satellite designed to provide ship-to-shore communications, was launched by Delta 2914 from Cape Canaveral, Fla., toward a geostationary orbit at 15° W. Operated by Comsat General Corp., the satellite had been built by Hughes Aircraft and was owned by a consortium comprising Comsat (86.29%), RCA Global Communications (8%) and ITT (2.3%). With its antennas, the satellite was 12 ft. 6 in. tall, had a diameter of 7 ft. 1 in. and carried 7,000 solar cells providing 330 W of power. Three UHF channels were provided as well as one L-band and one C-band channel, each 4 MHz wide. Launched on June 10, 1976, *Marisat 2* was positioned at 176.5° W. Sent to 73° E after launch on October 14, 1976, *Marisat 3* was initially used only by the U.S. Navy.

1976 February 24 NASA announced the crew for the Shuttle approach and landing test (ALT) flights which would involve the first orbiter (OV-101) being carried into the air on the back of a Boeing 747 and released for free flight down to a runway at Edwards AFB, Calif. Two crews were named: astronauts Fred Haise and Charles Fullerton, and Joe Engle and Richard Truly. Haise and Fullerton were scheduled to make the first ALT flight, with the second crew making the second flight. ALT flights would demonstrate subsonic aerodynamic handling qualities of the orbiter.

1976 February 29 A Japanese N-1 launcher carried the 306-lb. ionospheric sounding satellite (ISS) *Ume* (Plum Blossom) from the Tanegashima launch site to a 616 × 630 mi. orbit inclined 69.6°. The satellite was instrumented to observe the worldwide distribution of ionospheric radio noise. The satellite took the form of a cylinder, 3 ft. 1 in. in diameter, surmounted by solar cells, and was designed to provide research into natural phenomena that could hamper short-wave broadcasting from space. About one month after launch the satellite developed power problems and became unusable. It remains in orbit.

1976 March 5 NASA selected two aerospace companies to conduct 18-month studies on low orbit and geosynchronous orbit space stations and construction bases. The Marshall Space Flight Center awarded a contract to Grumman Aerospace Corp.; the Johnson Space Center awarded one to McDonnell Douglas. Both contracts would begin April 1, 1976, and end on October 1, 1977, and define a space station built with modules launched by Shuttle. It was to be capable of operating as a research laboratory with growth potential to a space construction base serving the anticipated requirement for a small solar power station test facility. Unlike earlier space station studies, this explored the idea of using the Shuttle and its external tank as an assembly base with first flights in 1983.

1976 March 15 A U.S. Air Force Titan IIIC carried two large Lincoln Experimental Satellites, *LES-8* and *LES-9*, from Cape Canaveral to geosynchronous orbits with inclinations of 25°. Each satellite had an on-orbit weight of about 1,000 lb. and formed part of the continuing LES program. The primary objective was to demonstrate the ability of one satellite to communicate with another, a technique known as cross-linking, as well as to handle communication between small, mobile ground terminals on land and at sea. After depositing the LES satellites in geosynchronous orbits more than 22,000 mi. above Earth, the Transtage fired again to put two solar radiation satellites in orbit: *Solrad 11A* in a 73,560 × 74,060 mi. orbit, and *Solrad 11B* in a 71,910 × 72,485 mi. orbit. Drum shaped and about 4 ft. high, the two satellites separated into two cylindrical elements, each weighing about 400 lb. They remain in orbit.

1976 March 16 The first calibration of the *Viking 2* orbiter TV cameras prior to the spacecraft entering Mars orbit began with pictures of the distant planet Jupiter. With *Viking 1* more than 115 million mi. from Earth and 13 million mi. from Mars, spacecraft science equipment was being tested for accurate measurements of how the journey so far had affected their performance. *Viking 2* was 112 million mi. from Earth and 17 million mi. from Mars. Each Viking orbiter carried two identical slow-scan 475-mm TV cameras equipped with six color filter wheels. When an image was formed on the 1.5-in. vidicon tube, it was scanned by an electron beam that neutralized the image and converted it into an electronic signal. One picture could be taken every 4.48 sec. and each image consisted of 1,056 lines made up of 1,182 pixels. *Viking 1* cameras were calibrated on March 23. The cameras were mounted on a scan platform along with the other orbiter science instruments, an infrared thermal mapper and an atmospheric water detector also calibrated on that day. Science instruments on the Viking orbiters weighed 161 lb.

1976 March 17 NASA asked scientists to submit proposals for scientific experiments on the first Spacelab mission aboard the reusable Shuttle. On February 16, 1977, 222 scientists representing the United States and 14 other countries were short-listed. NASA chose 86 of the scientists, 81 from the United States and one each from India, Japan, Canada, France and Belgium. The European Space Agency selected 136 including two from Austria and Norway.

1976 March The U.S. Defense Systems Acquisition Review Committee discussed three optional basing modes for the M-X ICBM: air-drop, multiple shelters and trench deployment. The latter two were deceptive basing modes designed to give M-X higher survivability than silos vulnerable to accurate, hard-target MIRV warheads. The Air Force proposed several M-X design options: upgrading solid propellant stages allowing the Minuteman to carry more warheads and superhardening the silos; developing a new missile 6 ft. 11 in. in diameter and compatible with Trident II; or developing a new 7 ft. 10 in. diameter ICBM capable of carrying 10 MIRVs. The committee instructed it to build the latter. In early 1977 the 3,000-mi. trench concept was changed to a series of 20-mi. trenches, 15 ft. in diameter and 5 ft. underground, one for each M-X. That proved more vulnerable than the silos because blast propagation would disable the missile wherever it was in the tunnel.

1976 April 1 The Boeing Co. announced it had received a 12-month contract to define Solar Power Satellite (SPS) concepts, whereby electrical energy produced by billions of solar cells would be transmitted to Earth on microwave links. The Boeing contract was managed by the NASA Johnson Space Center. In August a four-month contract was given to Rockwell International by the Marshall Space Flight Center for a similar study. A typical SPS would consist of a relatively flat surface constructed in space and placed in geostationary orbit. The platform would be approximately 12 mi. long and 3 mi. wide, taking the form of a flattened "W" in cross-section, accommodating 50 sq. mi. of solar cells across all surfaces facing the sun. During 1977 NASA decided to team with the Department of Energy for SPS definition studies.

1976 April 22 *NATO IIIA*, first in a new generation of communications satellites for the North Atlantic Treaty Organization, was launched by a Delta 2914 from Cape Canaveral, Fla., to a geostationary orbit over 18° W. Built by Aeronutronic-Ford, the drum-shaped satellite had a diameter of 7 ft. 2 in., a height of 7 ft. 4 in. and a weight on station of 830 lb. Operating with three 20-W transmitters in the 7–8 GHz frequency band, *NATO IIIA* was followed by *NATO IIIB* on January 28, 1977, placed over the Pacific and then moved to the Atlantic, and *NATO IIIC* launched on November 19, 1978, and placed at 50° W.

1976 April 30 The first preoperational U.S. ocean surveillance satellite system was launched by Atlas F from Vandenberg AFB, Calif. Called White Cloud, the system had been developed at the Naval Research Laboratory and involved a "parent" carrier and three separate satellites, each 8 ft. × 3 ft. × 1 ft. and covered with solar cells. The carrier (NOSS-1, or National Oceanic Satellite System-1) was placed in a 678 × 701 mi. orbit. The first main satellite *(SSU-1)* was positioned in a 679 × 701.5 mi. orbit, the second *(SSU-2)* in a 679 × 702 mi. orbit and the third *(SSU-3)* in a 673 × 707.5 mi. orbit, all three at an inclination of 63.4°. The three satellites were in parallel orbits within 30 mi. of each other and used interferometry to amplify low-strength signals from surface vessels up to 1,800 mi. away. They remain in orbit, but are inoperative.

1976 May 2 The European Space Agency completed a report on a mission to send spacecraft over the poles of the sun, proposed as a collaborative venture with NASA. Most of the planets lie in the ecliptic, that is the plane of the sun's equator, and spacecraft from Earth had always observed the sun, and measured the solar wind, from this latitude. Scientists wanted to put instruments on a spacecraft over both poles of the sun. Dubbed the out-of-ecliptic (OOE) mission, it envisaged the simultaneous launch of two spacecraft to Jupiter, one from NASA and one from ESA, using that planet's enormous gravity field to turn the flight paths through 90°. The resulting orbit change would send both spacecraft back across the solar system and over opposing poles of the sun. In April 1977 NASA solicited proposals for scientific experiments aboard the mission, soon called Solar Polar.

1976 May 4 NASA's *Lageos 1* laser geodynamic satellite was launched by a Delta 2913 from Vandenberg AFB, Calif., to a 3,632 × 3,691 mi. orbit inclined 109.8° to the equator. Resembling a golf ball in shape, the 906-lb. satellite comprised a solid sphere devoid of any active parts, electronics or power. With a diameter of 2 ft., *Lageos 1* had a solid brass core with aluminum exterior carrying an array of 426 prisms called cube-corner reflectors. A stable reflector for laser beams directed at it from Earth, the satellite was used to help

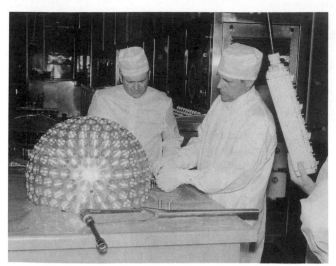

Technicians at NASA's Marshall Space Flight Center prepare the LAGEOS *geodynamic satellite for launch on May 4, 1976.*

measure the motion of tectonic plates and maintain geodetic reference positions. It remains in orbit.

1976 May 5   The French government and the European Space Agency reached agreement on the use of the launch facilities at Kourou, French Guiana, for the Ariane launch vehicle. Close to the equator, the launcher would gain the advantage of the Earth's rotational speed when placing satellites in geosynchronous transfer orbit. Owned by Centre Nationale d'Etudes Spatiales, the French national space agency, the existing Europa launch facilities would be adapted and modified for Ariane.

1976 May 13   The first of four domestic communication satellites for the Comsat General Corp., *Comstar D-1* was launched by an Atlas-Centaur from Cape Canaveral, Fla., to a geostationary orbit over the equator at 128° W. Aimed at providing spot-beam coverage of Hawaii, Alaska and Puerto Rico, the spin-stabilized satellite comprised a cylinder with a height of 9 ft. 6 in. and a diameter of 7 ft. 9 in., extended to a total height of 20 ft. 9 in. by two antennas. It weighed 3,348 lb. at launch and 1,746 lb. on station. *Comstar D-1* had 17,000 solar cells producing 760 W of power. Operating in the 4–6 GHz band, providing 14,400 voice circuits through 24 transponders, *Comstar D-1* also had two horn antennas for tests in the 19 and 28 GHz bands. *Comstar D-2* was launched on July 22 to 95° W; Comstar D-3 on June 29, 1978, to 87° longitude; *Comstar D-4* on February 21, 1981, to 127° W.

1976 May 20   NASA reached an agreement whereby the Universities Space Research Association and the Langley Research Center would develop experiments through the National Academy of Sciences for use in the Long Duration Exposure Facility. The LDEF was a multipurpose, Earth-orbiting experiment canister to be placed in space by the Shuttle and retrieved on another mission 6–12 months later. On January 9, 1978, NASA announced selection of 16 space technology experiments for the LDEF.

1976 June 2   A U.S. Air Force Titan IIIB Ascent-Agena D launched the first Hughes Satellite Data Systems (SDS) satellite from Vandenberg AFB, Calif., to a 232 × 24,607 mi. orbit inclined 63.3°. With perigee over the South Atlantic and apogee at high northern latitude, each SDS would relay coded digital image data from KH-11 photoreconnaissance satellites in real time to the United States. The second SDS was launched on August 6 in readiness for the first KH-11, launched on December 19. Hughes was contracted to build seven SDS satellites, the last of which was launched on February 11, 1987.

1976 June 10   The first of two *Viking 1* approach course correction maneuvers was performed when the orbiter's main propulsion system was fired for 2 min. 5 sec., adjusting velocity by 166 ft./sec. This burn had originally been planned for June 9 to change the speed of *Viking 1* by just 12 ft./sec., but a helium leak into the propellant tank that started June 7 had increased tank pressure to a dangerous level. It was necessary to burn off more propellant than needed to provide an increased volume in the tank, in the hope of successfully reseating the helium valve. A 2 min. 24 sec. (197 ft./sec.) burn on June 15 accomplished this. However, the flight path had also been affected, so instead of going into a 930 × 19,840 mi. orbit of Mars, Viking would have an elliptical orbit extending to a height of 31,370 mi. with a 42.6-hr. period rather than 24.6. This orbit could be adjusted to the correct values later.

1976 June 18   Designed to test a principle laid down by Albert Einstein in 1907, NASA launched Gravitational Probe-A (GP-A) on a Scout from Wallops Flight Center. The 226-lb. probe would determine the effect of gravity on time by comparing the rate of an on-board "clock" to an identical clock on Earth. The "clock" comprised a set of atomic-hydrogen masers (microwave amplification by stimulated emission of radiation) accurate to 0.005 sec. The Scout boosted GP-A to a height of 552 mi. in a flight lasting 1 hr. 56 min. during which high quality telemetry returned data from the atomic-hydrogen maser "clock," proving that, as Einstein predicted, time runs faster in the absence of gravity.

1976 June 20   In a burn lasting 38 min. and consuming 2,330 lb. of propellant, *Viking 1* orbiter put itself and its attached lander in a 941 × 31,255 mi. orbit with a period of 42.4 hr. and an inclination of 37°. The burn began at 22:38 UT and slowed Viking by more than 3,600 ft./sec. from an unbraked approach velocity of almost 13,000 ft./sec. An orbit change maneuver began at 10:44 UT on June 21 and lasted 2 min. 12 sec., lowering apoapsis to the desired 20,381 mi. and synchronizing the orbital period with the planet's rotation of 24.6 hr. In this orbit, the periapsis point would be over the Chryse Planitia touchdown site for the *Viking 1* lander at 20° N by 34° W and remain so for the same time each Mars day, or sol. A comprehensive period of site certification began using the orbiter TV cameras to conduct a reconnaissance for the final selection of the precise lander aim point. It had been hoped to put the lander down on July 4, but increasing concern over the appearance of the surface delayed the landing while the orbiter sent back detailed pictures.

1976 June 22   The 41,700-lb. *Almaz/Salyut 5* "military" space station was launched from the Baikonur cosmodrome at 18:04 UT by an SL-13 launcher, almost two years to the day since the launch of the last "military" Salyut. Like its Almaz predecessors, *Salyut 2* and *Salyut 3*, this latest surveillance and research laboratory accommodated a large reentry capsule at the forward end. From an initial orbit of 136 × 161.5 mi., *Salyut 5* was eventually maneuvered to an orbit of 161.5 ×

168 mi. where it awaited the crew launched by *Soyuz 21* on July 6, 1976.

Boeing submitted its final report on a heavy-lift launch vehicle study it had performed for the Marshall Space Flight Center. Tasked with examining four separate payload classes in four annual traffic models and four different space program scenarios, Boeing found that Shuttle systems could be applied to an HLLV. It also evaluated new booster concepts including single-stage-to-orbit (SSTO) vehicles. For class 4 payload capability, Boeing projected a SSTO-HLLV consisting of a truncated conical base with cylindrical upper section, a total height of 249 ft., a diameter of 134 ft. 6 in. and a liftoff weight of 23 million lb., including a 502,000 lb. payload. Power was to be provided by 48 engines in two concentric rings around the periphery of the base: 24 liquid oxygen–liquid hydrogen engines of 1-million-lb. thrust each and 24 liquid oxygen–kerosene engines of 500,000-lb. thrust each. The combined launch thrust was 36 million lb.

1976 June 25   NASA announced the names of 71 scientists selected to work in eight study areas proposed for a lunar polar orbiter spacecraft conceptualized at the Jet Propulsion Laboratory, Calif. Proposed for a low-cost Delta launch, the spacecraft would be placed in a 60-mi. circular polar orbit of the moon to work cooperatively with a subsatellite placed in a 3,000-mi. orbit. Designed to operate for a year, the polar orbiter would help scientists map the gravitational irregularities of the lunar interior, the subsatellite keeping track of the orbiter during its passage across the far side of the moon. The lunar polar orbiter was repeatedly denied funds and the project was never started.

1976 July 6   Cosmonauts Boris V. Volynov and Vitaly M. Zholobov were launched aboard the *Soyuz 21* spacecraft at 12:09 UT by SL-4 from Baikonur cosmodrome. Their mission was to occupy *Almaz/Salyut 5* and conduct military activities in Earth surveillance tests of new sensor equipment and selected technology tasks in the development of future spacecraft. From an initial orbit of 120 × 157 mi. the crew maneuvered their spacecraft to a rendezvous and docking at the rear of *Salyut 5*, which was in a 164 × 170 mi. orbit at an inclination of 51.6°, at 13:40 UT on the next day. A sudden, severe illness suffered by Zholobov caused the two-month flight to be cut short. *Soyuz 21* landed at 18:33 UT on August 24, after a mission lasting 49 days 6 hr. 23 min.

1976 July 8   At 23:31 UT a Delta 2914 launched the Indonesian communication satellite *Palapa A* from Cape Canaveral, Fla., to a geosynchronous transfer ellipse. At 20:30 UT on July 11 an apogee motor boosted the satellite to geostationary orbit over 83° E. Identical to the Anik and Westar satellites, the spin-stabilized Hughes HS 333 satellite had a height of 11 ft., a diameter of 6 ft. 2 in. and an in-orbit weight of 654 lb. It provided 12 color channels or up to 6,000 telephone circuits through its 12 transponders. The name of the satellite came from a 14th century ruler's vow not to eat the popular delicacy, palapa, until the whole of Indonesia was united. Identical to its predecessor, *Palapa B* was launched on March 10, 1977, and positioned at 77° E.

A Soviet satellite interceptor demonstration of a new "pop-up" technique for high altitude targets began with the launch of an 882-lb. target, *Cosmos 839*, on an SL-8 from Plesetsk to an elliptical orbit of 569 × 1,276 mi. inclined 65.9°. Launched from Baikonur by an SL-11 on July 21, the 4,410-lb. *Cosmos 843* interceptor was placed in a 82 × 215 mi. orbit. Instead of accelerating to perform a fly-by of the target

at high altitude, the interceptor was returned through the atmosphere within a few hours of launch. *Cosmos 839* remains in orbit.

1976 July 9   At 00:58 UT the *Viking 1* orbiter performed a 41-sec. (85.3 ft./sec.) maneuver, taking the spacecraft out of synchronism with the rotation of Mars so that the periapis of its orbit would slowly migrate from 34° W to 51° W by July 16. In the intervening period, the orbiter would take pictures of potential landing sites further west in the hope of finding a less rock-strewn surface. The camera look-angle enabled images to be seen as far as 56° W, and it was decided that increasingly rough ground would be encountered further west. On July 13 it was decided to halt the migration at 47° W, and a 4-sec. (9 ft./sec.) burn the following day resynchronized the spacecraft's periapsis at that longitude. The final touchdown site for the *Viking I* lander was fixed at 22.4° N by 47.5° W.

1976 July 13   The Intercosmos countries (Bulgaria, Czechoslovakia, Cuba, East Germany, Hungary, Mongolia, Poland, Romania and the Soviet Union) signed a new agreement setting out terms and conditions for a more generous exchange of technical information from space projects. Under its terms, member countries could make a contribution in equipment or services to the activities of the Soviet space program and participate in the preparation and coordination of new ventures. The agreement covered areas such as space biology, meteorology, communications, the practical study of the upper atmosphere and the physics of space.

1976 July 16   The Soviet Union agreed to launch a group of cosmonauts from Intercosmos countries on missions to the *Salyut 6*, the first third-generation scientific space station. Designed to operate for up to five years, *Salyut 6* was scheduled for launch in 1977. Long-duration occupation of *Salyut 6* by individual crews would be interspersed with visits by short-stay crews, one of which would be from one of the Intercosmos countries. The first candidates were selected in December 1976 and included Vladimir Remek and Oldrich Pelczak from Czechoslovakia, Miroslaw Hermaszewski and

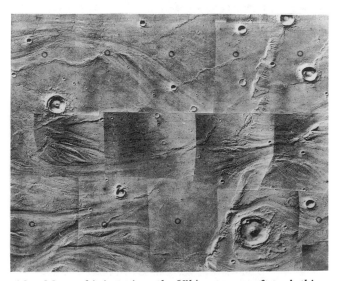

After Mars orbit insertion, the Viking spacecraft took this high-resolution image of the planned lander touchdown region in Chryse Planitia, an area 155 miles by 120 miles.

Taken August 3, 1976, on the surface of Mars, this striking picture from the Viking 1 *lander reveals a boulder- and rock-strewn dune field transected by the spacecraft's meteorology boom.*

Zenon Jankowski from Poland and Sigmund Jähn and Eberhard Köllner from East Germany. Cosmonaut candidates would be trained at Zvezdny Gorodok in return for contributions in equipment and services from their respective countries.

NASA selected Hamilton Standard Division of United Technologies Corp. to develop and produce space suits for Shuttle astronauts. Because of the expectation that female astronauts would fly on the Shuttle, the suit, or extravehicular mobility unit, would be suitable for crew members of either sex. It was to be manufactured in three adjustable sizes, incorporating a hard torso with flexible arms, and lower body and leg elements with a circular waist body lock connecting the two segments. Helmet and gloves were to be separate and a liquid-cooled undergarment was to be worn next to the body. A life support system was to be integral to the upper torso of the suit.

1976 July 20   The *Viking 1* lander successfully touched down in the Chryse Planitia region of the planet Mars. The lander had a width of about 10 ft. and a height on the surface of 7 ft. supported by three legs, each with a pad 1 ft. in diameter. Power was provided by two 35-W SNAP-19 radioisotope thermoelectric generators. The lander had two facsimile cameras 3 ft. 3 in. apart, a biology instrument housing three experiments, an x-ray fluorescence spectrometer and a meteorology boom. Encapsulated in a flat cone-shaped aeroshell heat shield 11 ft. 6 in. in diameter, the lander had separated from the orbiter at 08:51:15 UT and at 09:58:16 fired eight deorbit engines for 22 min. 16 sec., reducing speed by 360 MPH. At 11:54:08 UT the lander was oriented for entry at a pitch angle of about 20°, and at 12:03:08 UT a radar altimeter started terminal events when it registered an altitude of 151 mi. The aeroshell acted on the Mars atmosphere to reduce its speed from about 10,000 MPH to 600 MPH. At 12:10:06 UT and an altitude of 19,376 ft., a parachute 53-ft. in diameter deployed 98 ft. above the lander; 7 sec. later the aeroshell was jettisoned, followed 12 sec. after that by deployment of the three landing legs. Within 1 min. the parachute had taken the lander to a height of just 4,600 ft., slowing it to 130 MPH. Three 62–638-lb. thrust hydrazine monopropellant rocket motors throttled by a terminal descent/landing radar to a velocity of 8 ft./sec. *The Viking 1* lander touched down on Mars at 12:12:07 UT at 22.38° N × 47.49° W. It weighed 1,350 lb. on the surface. The first TV pictures appeared 5 min. after touchdown. (Note: The above are Earth receive times, just over 19 min. later than times at Mars due to the 212-million-mi. distance covered by the radio signal traveling at the speed of light.)

1976 July 28   The *Viking 1* lander began the first search for life on Mars when a surface sampler arm retrieved the first soil samples and deposited them in the 33-lb. biology instrument. With a scoop on the end, the retractable arm could extend from 3 ft. to a maximum 10 ft. from its housing just in front of one of the two lander TV cameras and sweep a radius of 120° between two footpads. Data from the lander were relayed direct to Earth at rates of up to 1,000 bits/sec. via a 2 ft. 6 in. parabolic high-gain antenna or via the orbiter at rates of up to 16 kilobits/sec. on a 30-W UHF signal. The biology instrument conducted three tests simultaneously: the labeled release test looked for signs of metabolism, the pyrolytic release test checked for micro-organisms that function by photosynthesis and the gas exchange experiment measured gas

The Viking 1 *lander's soil sample collector arm pushed a rock (left) on October 8, 1976, and left a visible track revealing dark material.*

changes in the closed environment of the test chamber. A separate gas chromatography experiment performed organic analysis on retrieved soil.

1976 July 30 NASA announced that studies were being conducted at the Marshall Space Flight Center on the feasibility of a "tethered" satellite trolled through the upper atmosphere at the end of a long cable attached to a Shuttle orbiter in space. The purpose would be to investigate with instruments the region of the outer atmosphere least known, extending between 50 mi. and 75 mi. above the Earth. On its own, a satellite would not long survive in the upper atmosphere. Trolled through that region by a more massive object in higher orbit, the instrument package could be used on a repetitive basis. A request for proposals to build such a satellite was issued by NASA on May 26, 1977, with Martin Marietta being selected but, in a cooperative program, Italy built the satellite and Martin Marietta the support hardware. The tethered satellite system (TSS) was launched aboard Shuttle mission STS-46 on July 31, 1992.

1976 August 1 Having completed India's Satellite Instructional Television Experiment (SITE) from its geostationary position over the Indian Ocean, NASA's powerful communication satellite *ATS 6* began a four-month migration westward. During this time it would conduct a set of demonstrations for the U.S. Agency for International Aid (AID) in a program known as AIDSAT. In demonstrations scheduled to last through August 26, films on the use of satellites for developing countries would be transmitted to Thailand, Pakistan, Bangladesh, United Arab Emirates, Oman, Jordan, Kenya, Yemen, Libya, Sudan and Morocco.

1976 August 6 The U.S. Air Force announced that the Boeing Co. had been selected to develop and manufacture the solid propellant interim upper stage, or IUS, a precursor to the space tug for transporting payloads from low orbit to geostationary orbit. Prospects of getting funding for the space tug diminished, and at the end of 1977 the Air Force renamed it the inertial upper stage, retaining the acronym IUS but also indicating that it was designed for a definitive, and not transitory, purpose. As designed, the IUS had autonomous control with a master sequencer.

1976 August 7 Preceded by a 26-sec. (29.5 ft./sec.) approach course maneuver performed at 01:19 UT on July 28, the *Viking 2* orbiter/lander combination was placed in Mars orbit. A 39 min. 30 sec. burn with the orbiter's main propulsion system began at 11:49 UT and cut the speed of the vehicles by 3,610 ft./sec., putting them in an elliptical orbit with a periapsis of 944 mi., a period of 27.6 hr. and an inclination of 55°. On August 9 this orbit was trimmed by a 6-sec. (13 ft./sec.) burn, cutting the orbital period to 27.3 hr.; the orbit was adjusted again to 27.41 hr. at 08:50 UT on August 14 by a 3-sec. (6 ft./sec.) burn which moved periapsis to 933 mi. In this precise sequence of maneuvers, *Viking 2* had gone directly for a migrating orbit, carrying the periapsis point 40° westward on each revolution for orbital reconnaissance photography of alternate *Viking 2* lander touchdown sites.

1976 August 9 An SL-12 launcher lifted off from Baikonur carrying the 12,400-lb. *Luna 24* spacecraft to a translunar trajectory. Following course correction maneuvers en route, *Luna 24* slipped into a circular 71.5-mi. orbit, with an inclination of 120° and a period of 1 hr. 59 min., on August 12. After maneuvering into the elliptical orbit typical of Soviet

soft-lander operations, the spacecraft touched down at 06:36 UT on August 18. Situated at 12.75° N × 62.2° E in the southeast quadrant of Mare Crisium, *Luna 24* carried a modified drill designed to obtain several separate samples from a maximum depth of 8 ft. 6 in. With a 6-oz. (0.37-lb.) collection of material, *Luna 24*'s ascent stage blasted away from the descent stage at 05:25 UT on August 19 and delivered the reentry capsule to Earth at 17:55 UT on August 22. It landed 124 mi. southeast of Surgat, Kazakhstan. This was the last flight of the third-generation Soviet lunar spacecraft. In the three successful sample return flights, 13.1 oz. (0.8 lb.) of lunar material had been delivered to Earth.

1976 August 25 At 17:47 UT *Viking 2* fired its rocket motor for 1 min. 18 sec., adjusting orbital velocity by 147 ft./sec., to halt and reverse the westward migration of the spacecraft's orbit about Mars. The orbit now had a periapsis of 893 mi. and a period of 23.98 hr., less than that of the planet. Thus the planet never completed an entire revolution for each full orbit of the spacecraft so that its orbit would appear to move east. Using orbital photography, scientists had selected a *Viking 2* lander site at 47.9° N × 225.8° W and wanted *Viking 2* moved eastward, at the rate of 9.5° longitude each day, to put it over that location for the descent. At 20:46 UT on August 27 another burn lasting 19 sec. (39 ft./sec.) halted the eastward migration and adjusted the periapsis altitude to 925 mi., fixing it over the selected landing site with a period of 24.62 hr. synchronized with the rotation of Mars. Vertical pictures of the selected site were taken on August 29 and weather pictures taken two and three days later.

1976 August 30 NASA invited scientists to propose experiments for the second Spacelab mission carried inside the Shuttle cargo bay. Unlike the first scheduled Spacelab mission, the second mission would use a train of pallets and not the pressure module. Large experiments exposed to the vacuum of space would be attached to the pallets and operated by scientist-astronauts from the flight deck of the Shuttle orbiter.

1976 September 1 NASA selected Science Applications Inc. of Los Angeles to carry out a study of opportunities for space industrialization and commercial production of materials for processing products aboard a permanently manned space station. The agency wanted to obtain a clear line of development from concept evaluation to in-orbit demonstration and manufacturing facilities. Experiments on U.S. and Soviet manned missions had already shown that under weightless conditions better vaccines could be produced, and that semiconductor crystals could be grown with higher levels of purity than on Earth.

1976 September 2 The U.S. National Oceanic Administration announced successful use of *Landsat 1* to locate fish off the coast of Louisiana in a cooperative program begun by NOAA in 1975. Water turbidity detected by Landsat would indicate the distribution of fish and surface vessels working in coordination with spotter airplanes would directed to the largest shoals. NOAA wanted this information for better resource inventories.

1976 September 3 At precisely 22:58:20 UT (09:45:05 local time) the *Viking 2* lander safely touched down on Mars at a site called Utopia, located at 47.97° N × 225.71° W, about 6 mi. from the center of the elliptical target site. A minor systems failure aboard the orbiter forced a long, but partial, communications blackout with Earth just as the lander was preparing to descend. Instead of relaying blow-by-blow the sequence of automatic events similar to those previously conducted by the *Viking 1* lander during descent, the first indication that it had landed came from scratchy communications with the orbiter reporting that the lander was sending it 16 kilobit/sec. post-landing data. Normal service was resumed by early next morning. The *Viking 2* lander deployed its seismometer, which the *Viking 1* lander had been unable to do, and biological tests began on September 12 when the surface sampler retrieved the first soil from the *Viking 2* lander site.

The International Maritime Satellite Organization (Inmarsat) was chartered in London after a four-year period of study to create a ship-to-shore satellite telephone and telegraph system. The formation of Inmarsat came on the third day of an international conference in London, where delegates from 47 countries and observers from Yugoslavia, as well as delegates from 23 international agencies, agreed to open the charter for membership. Inmarsat became a permanent organization on July 16, 1979.

1976 September 11   The U.S. Air Force launched the first of its Block 5D-1 Defense Meteorology Satellite Program (DMSP) satellites on a Thor-Burner 2 from Vandenberg AFB, Calif. The satellite was placed in a 509.5 × 525.5 mi. orbit inclined 98.6° but a power system failure prevented communications with it. It was later designated *AMS-1* as the first Advanced Meteorology Satellite, but this soon reverted to Block 5D-1 DMSP designation. The five Block 5D-1 satellites shared a common bus and payload structure with the Tiros-N series for the NOAA. The military satellites carried a scanning infrared radiometer, atmospheric density scanner, electron spectrometer gamma radiation sensor, ionospheric sounder and ion scanner. The last Block 5D-1 was launched July 14, 1980, but failed to reach orbit, leaving the one launched June 6, 1979, the last operable Block 5D-1 satellite. Although inoperable, it is still in orbit.

A 16-sec. (70 ft./sec.) engine burn performed by the *Viking 1* orbiter reduced the periapsis altitude to 927 mi. and the orbital period to 21.9 hr., causing the spacecraft to begin an eastward migration around the planet of 40° long. each day. This was because the orbital period was almost 3 hr. less than Mars took to make one revolution, so the spacecraft arrived at the low point in its orbit quicker than the planet could rotate to the same point in space. The eastward migration was equal to just over 1,200 mi. per day, but on September 20 the *Viking 1* orbiter performed a brief engine burn to slow the migration rate. Four days later, after having migrated 1.5 times round the planet in 13 days, the *Viking 1* orbiter was resynchronized with the rotating planet in another engine burn that fixed periapsis at 940 mi. and the period at 24.6 hr. The *Viking 1* orbiter was now in position to take over the *Viking 2* lander communications relay to Earth, leaving the *Viking 2* orbiter begin a major photo-mapping operation of its own. Meanwhile, the *Viking 1* lander had been put on a program of reduced science activity.

1976 September 15   Cosmonauts Valery F. Bykovsky and Vladimir V. Aksenov were launched aboard the 14,355-lb. *Soyuz 22* spacecraft at 09:48 UT. Their SL-4 launcher placed them in an orbit of 114 × 184 mi. at the start of an Earth resources mission to test equipment which was to be used in the *Salyut 6* space station. Most notable was the MKF-6 multispectral camera which occupied a part of the orbital module at the forward location where docking devices were usually carried. No dockings were planned on this flight. The

unusual orbit inclination of 64.8° had never been flown before in the manned Soyuz program and was selected to give optimum coverage for the remote sensing equipment. The crew returned to Earth at an elapsed time of 7 day 21 hr. 52 min.

Lockheed Missiles and Space Co. Inc. opened a facility for the manufacture of a silica insulation material from which Shuttle thermal protection tiles would be made. About 70% of the Shuttle orbiter's exterior surface would be covered with about 34,000 of the brittle tiles, each separated from its neighbor by a small gap which would accommodate the expansion and contraction of the alloy airframe. The heatsink tiles were so efficient that they could be held in the hand when the inside was at red hot temperatures. NASA wanted each tile to survive 100 Shuttle flights.

A culmination of six years effort by the United Nations Committee on the Peaceful Uses of Outer Space, the UN Convention on Registration of Space Objects came into force. Based on the voluntary system of registration in force since 1962, the new treaty established a mandatory system. Although the language of the treaty was not specific on issues regarding the declaration of satellite function, it did require all launcher countries to itemize debris as well as functional vehicles in orbit at that time.

1976 September 16   The NASA Marshall Space Flight Center reviewed the status of the solid-spinning-upper-stage being developed in two variants as a solid propellant perigee kick motor for satellites carried in the Shuttle. The SSUS-D would be used with Delta-class satellites, while the SSUS-A would be applicable to Atlas-Centaur–class payloads. Attached to the satellite and mounted on a spin-table in a vertical position in the Shuttle cargo bay, the spinning SSUS and its payload would be released by springs to drift away from the Shuttle. At a safe distance the SSUS would fire, raising the satellite's apogee to geosynchronous altitude. McDonnell Douglas was awarded a contract for SSUS-D on December 6 and one for the SSUS-A on February 18, 1977.

1976 September 17   The first Shuttle orbiter, OV-101, was rolled out of its fabrication facility at the Rockwell Space Division plant, Palmdale, Calif., in a public ceremony to which several hundred dignitaries had been invited. Overturning an earlier decision to call it *Constitution*, NASA succumbed to pressure from "Trekkies" and gave the name *Enterprise* to OV-101, in part after the fictional spaceship from the Star Trek TV series. It would be used for a series of approach and landing test (ALT) flights scheduled to begin in May 1977. Five flights were to be flown with *Enterprise* attached to the top of the Boeing 747 Shuttle carrier aircraft (SCA), followed by eight flights scheduled to begin in July 1977 during which *Enterprise* would separate in flight and glide to a landing.

1976 September 21   The Indian government reported outstanding success with the Satellite Instructional Television Experiment (SITE), in which 2,400 Indian villages received 1,200 hr. of educational television via the NASA *ATS 6* satellite. Prof. Yash Pal, director of the space applications center operated by the Indian Space Research Organization, said it had created "a cadre of dedicated people and the methodology necessary to sustain an ongoing program." The use of Earth resource satellites to improve crop yield and of communication satellites to expand educational programs transformed the ability of India's diverse communities to benefit from knowledge, uniting diverse cultures across the subcontinent.

1976 September 30   At 21:30 UT the *Viking 2* orbiter fired its main propulsion motor for 5 min., imparting a 1,125 ft./sec. velocity change to move the inclination of the spacecraft's orbit from 55° to 75°. This major maneuver was made to give the orbiter's TV cameras access to the north polar hood of Mars before ice began to cover it during the onset of winter in the northern hemisphere. By raising the orbital period to 26.78 hr., it also set the orbiter migrating westward at the rate of 31.7° long. each day. The final data transmission in the primary mission of the orbiter was sent on November 8, 1976.

1976 October 14   At 17:40 UT cosmonauts Vyacheslav D. Zudov and Valery I. Rozhdestvensky were launched aboard their 14,905-lb. *Soyuz 23* spacecraft by SL-4 from Baikonur. The mission objective was to occupy the *Almaz/Salyut 5* surveillance space station. A hurried evacuation of *Salyut 5* by the *Soyuz 21* crew necessitated systems checks and tests on the environmental control system, but ground controllers believed the station to be habitable. However, after successful rendezvous, the automatic approach equipment malfunctioned and *Soyuz 23* was unable to dock. The crew returned to Earth for a landing at 17:46 UT on October 16, and *Soyuz 23* became the first Soviet manned spacecraft to splash down in water when the descent module broke through the ice covering Lake Tengiz. Not for several hours could the crew be retrieved as their spacecraft floated below a thin covering of ice.

1976 October 26   An SL-12 launch vehicle carried the 4,400-lb. *Ekran* multiprogram television satellite to a geosynchronous orbit with an inclination of 7.6°. Designed to relay TV programs to remote areas, the *Ekran* system served regions in Siberia and the far north, covering 40% of the Soviet Union, where 20 million people had been denied Soviet TV broadcasts because station-to-station transmitters were too expensive. Between this date and October 30, 1992, 24 Ekran satellites were launched; 20 of these reached geosynchronous orbit. The system also included 5 satellites in the 8,820-lb. Statsionar series.

1976 November 5   The *Viking 1* orbiter sent its last data transmission to Earth at the end of its primary mission, followed by the *Viking 2* orbiter three days later. The primary mission of the Viking program, set out in the years of development preceding launch, had been completed November 15, 1976, prior to solar conjunction, when Earth and Mars are on opposite sides of the sun, on November 25. During this period the spacecraft's radio signals were tracked as solar gravity affected the carrier wave by delaying the arrival time of the signal at Earth in a manner predicted by Einstein's theory of general relativity. During this period the landers were placed on standby and no communications were conducted while planners prepared added tasks for Viking. A Viking extended mission officially began on December 13, 1976, when the period of solar conjunction had ended.

1976 November 9   The European Space Agency accepted the first solid propellant apogee motor produced in Europe. Designed by the SNIA-Viscosa of Italy in cooperation with SEP (Société Européene de Propulsion) of France, it would be used in the Geostationary Scientific Satellite and measured 3 ft. 7 in. long with a diameter of 2 ft. 3.5 in., with a weight of 672 lb. The motor would place the satellite in geostationary orbit after launch on a NASA Delta.

1976 November 29   At 16:00 UT a Soviet SL-4 launch vehicle carried the first of a new generation of modified Soyuz spacecraft into a 122 × 180 mi. orbit inclined 50.6° to the equator. Designated *Cosmos 869*, the 15,100-lb. spacecraft was the first full-scale, unmanned Earth-orbit test of the Soyuz-T series. Limited to carrying two crew members in full space suits since the deaths of the *Soyuz 11* crew on their return to Earth on June 29, 1971, the basic Soyuz had been substantially modified and given a capacity for carrying up to three suited crew members in the T-series configuration. A new 695-lb. thrust propulsion system was adopted, together with 26 additional attitude thrusters (14 × 31 lb. thrust and 12 × 5.5 lb. thrust) using UDMH-N_2O_4 propellant from a common tank supply. Chemical batteries could provide four days of autonomous flight, and new flight control electronics assisted automatic navigation and rendezvous. The *Cosmos 869* descent module returned to Earth on December 17, 1976, at an elapsed time of 17 days 18 hr. 31 min. Two more Soyuz-T test flights (*Cosmos 1001* on April 4, 1978, and *Cosmos 1074* on January 31, 1979) extended the orbital duration to 11 days and 60 days, respectively.

1976 November   Soviet scientists began sea-launch trials with a new liquid propellant submarine-launched ballistic missile, the first Soviet SLBM to carry MIRV warheads. Code-named SS-N-18 Stingray by the NATO coordinating committee, the missile had a length of 46 ft. 3 in. and a diameter of 5 ft. 11 in. The Mod 1 version, which became operational in 1977 with Delta III–class submarines, carried three 500-KT MIRV warheads with a range of 4,050 mi. The Delta III had a submerged displacement of 13,250 tons and carried sixteen missiles. The Mod 2 version had a range of 5,000 mi. carrying one 450-KT warhead and became operational in 1978. Mod 3 entered service in 1978 and had seven 500-KT MIRVs and a range of 4,050 mi. All versions were accurate to within 3,000 ft. Peak deployment of 224 SS-N-18 on 14 submarines was reached in 1982 and maintained through the START (Strategic Arms Reduction Talks) agreement of the early 1990s.

1976 December 1   Launch Complex 14 at Cape Canaveral, the site of John Glenn's historic flight into orbit on February 20, 1961, was blown up with plastic explosives operated by a U.S. Army demolition team. Rusting in the salty atmosphere of Cape Canaveral, the outdated tower atop the concrete pad was considered likely to collapse. NASA was unable to fund its preservation as a national monument due to a lack of Congressional support, but a plaque commemorating the event was set up at the approach to the pad ramp.

1976 December 9   A Soviet SL-8 launched from Plesetsk carried the 1,985-lb. *Cosmos 880* satellite interceptor target to a 347 × 383 mi. orbit inclined 65.8°. On December 27 the 4,410-lb. interceptor *Cosmos 886* was launched from Baikonur by an SS-11 to a 330 × 787 mi. orbit from which it conducted a high-speed fly-by on the descending leg and detonated, leaving 60 pieces in orbit as planned. Unlike previous interceptors, which carried radar homing systems, *Cosmos 886* employed an optical infrared homing device. *Cosmos 880* exploded into 50 pieces on November 27, 1978, and decayed back through the atmosphere on October 8, 1979.

1976 December 13   The extended mission phase of the Viking Mars exploration program got under way when the *Viking 1* orbiter received orders from Earth to resume standard operations. The *Viking 2* orbiter received a similar command a day later, the *Viking 1* lander was woken up on December 15 and

Viking 2 lander on December 17. Several days later the landers transmitted full tapeloads of weather data gathered over the preceding month. On December 20, 1976, the *Viking 2* orbiter fired its main rocket motor to lower its periapsis to 483 mi., reducing its period to 26.48 min.—just a little less than one full rotation of Mars. At the same time, this maneuver increased the spacecraft's orbital inclination from 75° to 80°, close to the optimum for orbiter science instruments.

1976 December 15 At 01:29 UT a Soviet SL-13 launch vehicle lifted away from the Baikonur cosmodrome carrying two Lapot space planes. Under the designation *Cosmos 881* and *882*, each vehicle weighed 13,230 lb. and comprised a lifting body with vertical fin and swept wings hinged to fold up to a near vertical position for atmospheric reentry and down for aerodynamic flight. *Cosmos 881* entered an orbit of 124 × 150 mi., and *Cosmos 882* was placed in an orbit of 118 × 133 mi., both at an inclination of 51.6°. Each Lapot returned to Earth before completing one revolution and was recovered on Soviet territory. Another attempt at a dual Lapot flight on August 4, 1977, ended in a launch failure. Two more Lapot test missions were completed: *Cosmos 997/998* on March 30, 1978, and *Cosmos 1100/1101* on May 22, 1979. Thereafter, scaled-down Lapot spaceplanes were incorporated in BOR-4 tests beginning with *Cosmos 1374* on June 3, 1982.

An SL-8 launcher carried the first in a new series of civil Soviet navigation satellites from Plesetsk to a 593 × 624.5 mi., 83° orbit. The fully operational system was effective with the launch of the fourth such satellite, *Cosmos 1092*, launched on April 11, 1979. Each satellite was displaced 45° from its neighbor to complete a four-satellite system with global cover. The civil and military navigation satellites continued to operate as separate systems, and replacements were provided at the rate of 5–8 per year.

1976 December 16 At a White House briefing, presidential science adviser H. Guyford Stever disclosed that, at the request of the Soviet Union, U.S. satellites had performed a satellite inventory of the Soviet grain crop to confirm a projected bumper harvest. As predicted by the imagery, the Soviets harvested 222.5 million metric tons of grain that season.

1976 December 16–17 The Council of the European Space Agency agreed to establish a coordinated European program of Earth resource–satellite data management called Earthnet. Primarily associated with the U.S. remote sensing, environmental and weather satellites, Earthnet provided the foundation upon which Europe would develop its own Earth resource program, eventually leading to the ERS program. Under Earthnet, ground receiving stations were built in Sweden and Italy to collect Landsat data, and a data processing station was built at Frascati, Italy.

1976 December 19 The first in a new series of U.S. military reconnaissance satellites was launched into a 153 × 331 mi., 96.9° polar orbit by Titan IIID launcher from Vandenberg AFB, Calif. It was the first of seven digital imaging satellites built by TRW Systems under Program 1010 and designated KH-11. With a weight of 28,600 lb., each KH-11 carried a telescope, photomultiplier tube, thematic mapper, multispectral scanners and an array of charge-couple-devices (CCDs), the heart of the new satellite. Developed at Bell Telephone Laboratories in 1970, CCDs provided the breakthrough to achieve very high resolution from televised images, approaching in quality the film returned by ejectable capsules of KH-9 satellites. Digital data was transmitted either direct to ground stations or via SDS satellites. KH-11s had a life of 2–3 years. Six were launched successfully, the last on December 4, 1984, but the seventh was destroyed on August 28, 1985, when its Titan 34D launcher shut down prematurely and plunged to Earth.

1976 December The first test flight of a new French land-based missile was successfully completed. Designated S-3, it utilized the same first stage as the S-2, operationally deployed since 1971, but adopted the 70,550-lb. thrust P06/Rita II developed for the M-20 submarine launched missile. With a range of 2,175 mi., the S-3 carried a 1.2-MT warhead. The M-3 replaced the M-2 between 1980 and 1982, being retrofitted to 18 modified silos built for the first-generation missiles of this type. Reaction time from first alert to launch is about 3 min. 30 sec.

1976 Development of three new Soviet missiles was completed: the solid propellant submarine-launched SS-N-17 Snipe ballistic missile, the solid propellant SS-21 Scarab (a replacement for the aging FROG series) and the solid propellant SS-22 (a replacement for the SS-12). The Snipe had a length of 36 ft. 3 in., a diameter of 5 ft. 5 in. and a range of 2,300 mi. Carrying a single 500-KT warhead, the Snipe had a target accuracy of only 4,500 ft. Only one Yankee-class submarine with 12 missiles was deployed from 1977, qualifying it as a Yankee II, and that vessel remained in service through the early 1990s. The SS-21 had a length of 20 ft., a diameter of 2 ft. 9 in., and a weight of 6,600 lb. carrying a 1,000-lb. warhead with a range of 75 mi. By 1990, 300 Scarabs had been deployed. The SS-22 used the SS-12 transporter and had a 500-KT warhead with a range of 550 mi. and an accuracy of 1,400 ft. The Soviets began to deploy both the SS-21 and the SS-22 in 1977.

1977 January 3 In what amounted to the largest contract issued by NASA for an unmanned space project, Western Union Communications, TRW Systems and Harris Electronics were contracted to develop the Tracking and Data Relay Satellite System. In a deal costing $79.6 million, the contract would cover the production of six satellites and components for a seventh. The specialized TDRSS relay satellites would replace almost all the ground stations around the world then operated by NASA for communicating with satellites below 3,100 mi.

1977 January 6 After almost two years of spying for the Soviet secret service, an American by the name of Andrew Lee handed the last package of materials on U.S. electronic intelligence satellites to the KGB at the Soviet Embassy in Mexico. Photographed documents obtained by TRW Systems employee Christopher Boyce told the Soviets details of the geostationary Rhyolite satellites. Eventually, Boyce and Lee were arrested and went to prison, and the Soviets developed new encryption codes for telemetry from a new generation of ICBMs.

1977 January 7 NASA announced that the first three customers had signed up for the low-cost "getaway special" payloads to fly on the Shuttle. The small, self-contained payloads could weigh no more than 200 lb. each and measure a maximum 5 cu. ft. Shaped like trash bins, each canister would contain the payload built and paid for by the customer. NASA would carry it inside the Shuttle and return it to the customer at the end of the flight. The first four were reserved by R. Gilbert Moore of Utah, a private citizen who donated half the canister to Utah State University and kept the other half for himself, Dr. L. R. Megill, chairman of the space science committee at Utah State, and Reiner Klett, a private U.S. citizen who reserved two payload slots.

The NASA Marshall Space Flight Center requested proposals from industry for a study contract on Solar Power Satellites (or Satellite Power Systems) capable of converting sunlight to electricity and beaming it to Earth on microwave links. Originally proposed by Peter Elaser of Arthur D. Little, Inc., NASA had already established the technical and economic viability of the SPS concept, but such massive structures placed in geostationary orbit would demand vast resources on a scale likely to dwarf the Apollo moon program. However, powersats had the potential to eliminate the use of fossil fuels for producing electricity.

1977 January 10–14 At the first meeting of the Inmarsat Preparatory Committee in London, delegates from 17 countries met to work out the structure of the International Maritime Satellite Organization due to be formalized in 1979. A mirror image of the Intelsat organization, Inmarsat was to be a gathering of maritime countries for mutual support and funding of a ship-to-shore satellite relay system. Kuwait became the first country to ratify the Inmarsat convention and its associated operating agreement on February 25.

1977 January 17 Presenting his last budget to Congress and the American people, President Gerald Ford announced spending plans for fiscal year 1978 that included full-scale development of the M-X ICBM. It backed the buried-trench concept for deployment. On February 22 the incoming Carter administration reversed this and cut funding, as a way of signalling to the Soviet Union their intention to negotiating strategic arms cuts. Secretary of State Cyrus Vance proposed wide-ranging limitations on new missile development. This the Soviets rejected, believing it to be a U.S. ploy to freeze a technical lead in their favor.

1977 January 18 The first flight test of the Trident C4 SLBM was successfully conducted from Cape Canaveral. Designed as a replacement for the aging Polaris A3 and Poseidon C3, the Trident I had the same physical dimensions and launch weight of the C3 but a considerably improved range and warhead capability. Whereas Polaris and Poseidon each had two stages, Trident had three, with innovative packaging of the narrow-diameter third-stage motor up inside the equipment section. The first Trident I missile to go on operational duty put to sea aboard the retrofitted USS *Benjamin Franklin* in October 1979. Trident I has a nominal range of 4,500 mi., but a stretched development of this missile, known as the D5 Trident II, would extend the range to more than 6,500 NM.

1977 January 22 The *Viking 1* orbiter fired its motor to change its orbital period about Mars to 23.05 hr., thus setting it up for a close pass of Phobos, one of Mars' two small moons. Further maneuvers on February 4 and 12 synchronized its orbit with Phobos for pictures to be taken. Close encounters took place beginning February 18, with the orbiter moving to within 60 mi. of the potato-shaped moon one day later. On March 2, 1977, the *Viking 2* orbiter made an orbit change maneuver, giving it a period of 24.73 hr. to more closely synchronize it with a daily pass across the *Viking 2* lander site so that it could relay data from the lander to Earth. On March 11 The *Viking 1* orbiter departed its Phobos-encounter orbit and dropped its periapsis to 186 mi., reducing

orbital period to 21.92 hr. for an eastward migration. Pictures taken at low altitude would resolve objects as small as 65 ft., but the view was compromised by clouds. A small trim-burn on March 24 adjusted the orbital period to 23.5 hr.

1977 January 31 The NASA Marshall Space Flight Center invited industry to bid for the contract to build the primary elements for the space telescope, the support systems module and the optical telescope assembly. The 100-in. telescope was expected to weigh about 20,000 lb. and be carried to a 310-mi. orbit with an inclination of 28.8° in the Shuttle payload bay. On July 25, NASA announced that Lockheed Missiles and Space Co. Inc. would build the support systems module and main structure, and that Perkin-Elmer would be responsible for the optical assembly. In reality, the space telescope was an "astronomically civilianized" version of the KH-11 military reconnaissance satellite.

NASA's Shuttle *Enterprise* was moved by road the 36 mi. from Palmdale, Calif., where the orbiter was built, to the Dryden Flight Research Center at Edwards AFB Calif., where approach and landing test (ALT) flights would be conducted. The *Enterprise* was mated to pylons on top of the Boeing 747 Shuttle carrier aircraft (SCA) on February 8, and taxi tests were conducted on February 15, prior to the first "captive-inactive" flights in which the orbiter would be carried atop the SCA.

1977 February 2 The Soviet *Salyut 4* science research space station was de-orbited and burned up in the atmosphere over the Pacific Ocean. Unmanned since the return of the *Soyuz 18* crew on July 26, 1976, the station had been routinely monitored by flight controllers on the ground. It had supported six man-months of occupancy and provided a docking facility to test the ability of the unmanned *Soyuz 20* spacecraft to remain in space for three months, be reactivated and return to Earth. *Salyut 4* was the sole Soviet second-generation "civilian" space station, providing invaluable data for the design and operation of the long-duration stations *Salyut 6* and *7* and *Mir*.

1977 February 7 At 16:12 UT an SL-4 carried the 15,000-lb. *Soyuz 24* spacecraft containing cosmonauts Viktor V. Gorbatko and Yuri N. Glazkov into orbit from where a rendezvous and docking with the *Almaz/Salyut 5* "military" space station was completed at 17:38 UT on February 8. Wearing breathing masks due to an acrid odor reported by the departing *Soyuz 21* crew, Gorbatko and Glazkov confirmed that the environmental control system was working as expected and that normal operations could resume. This was the last visit to *Salyut 5*, and the last time an Almaz station would be manned. The crew returned to Earth on February 25, completing a mission that had lasted 17 days 17 hr. 26 min. The next day the large recoverable capsule carried by all "military" Salyuts was jettisoned and reached the ground at 09:28 UT bringing with it film and data.

1977 February 15 Boeing submitted its final report, revised 13 days later, on a study of airline and shipping operations as they might relate to the operation of a large fleet of space freighters (heavy lift launch vehicles) each with a payload capacity of 600,000 lb. and a projected use rate of 750 flights per year for 15 years. The Class 4 vehicle selected would consist of two recoverable ballistic stages with a liftoff weight of 14.3 million lb., the first stage powered by nine liquid oxygen–RP-1 engines with a combined liftoff thrust of 19.2 million lb. and the second stage by seven cryogenic SSMEs (Space Shuttle main engines) for a combined vacuum thrust of

3 million lb. Launched from offshore pads, the vehicle would perform 11,250 flights over 15 years at a flight cost of $6.5 million per flight, a defrayed freighting charge of just over $10/lb. of cargo.

1977 February 18 The NASA Shuttle *Enterprise* was taken into the air for the first time on the back of the Boeing 747 shuttle carrier aircraft, with pilots Fitzhugh L. Fulton Jr., and Thomas C. McMurtry and flight test engineers Victor W. Horton and Thomas E. Guidry Jr., aboard the SCA. First in a series of five "captive-inactive" flights in which the orbiter remained inert, the flight lasted 2 hr. 5 min. and was followed by a 3 hr. 13 min. flight on February 22, a 2 hr. 28 min. flight on February 25, a 2 hr. 11 min. flight on February 28 and a 1 hr. 39 min. flight on March 2. A maximum altitude of 30,000 ft. was reached.

1977 February 19 The 295-lb. Japanese MS-T3 test satellite *Tansei-III* (*Light Blue*) was launched by a solid propellant Mu-3H launcher from Uchinoura to a 495 × 2,374 mi. orbit inclined 65.8°. This was the first flight for the Mu-3H, which was a stretched upgrade of the Mu-3C. It retained the eight 21,390-lb. thrust strap-on boosters but incorporated a more powerful 214, 545-lb. thrust first-stage engine and extended stage length for longer burn time. Second and third stages remained the same as the Mu-3C. The launcher stood 78 ft. tall, with a diameter of 4 ft. 7 in. and a launch weight of 107,400 lb. The satellite was instrumented by the University of Tokyo to monitor and record data about launch vehicle performance and to obtain engineering data from the satellite itself. This launch began a busy year for Japan, seeing four satellites successfully placed in orbit. *Tansei-III* remains in orbit.

1977 February 23 Japan launched its first geostationary satellite on an N-1 from Tanegashima to a position at 130° E. The 287-lb. engineering test satellite *Kiku-2* (Chrysanthemum) took the form of a cylinder with a diameter of 4 ft. 7 in. It was equipped with instruments to help design tracking and control systems for future operational satellites in geostationary orbit and to check the functioning of attitude control technologies.

1977 February 25 Hughes Aircraft was selected to design and fabricate a thematic mapper (TM) for the fourth Landsat Earth resources satellite. Designed to provide higher-quality remote sensing information than the return beam vidicon cameras flown on previous Landsats, the TM required more costly ground equipment and a more expensive processing system, opposed by many developing countries at a time when they had just completed ground stations and equipment for the RBV system. NASA, however, saw itself as a research and development agency and imposed the TM on Landsat users because it advanced the state of the art.

1977 February Soviet cosmonauts Anatoly Berezovoy and Mikhail I. Lisun were selected for a flight to the *Almaz/Salyut 5* station in *Soyuz 25* during March. When the *Soyuz 24* crew returned to Earth on February 25, it was determined that extra time was needed to prepare the *Soyuz 25* spacecraft and that it could not be ready before June. That was too late because propellant aboard the *Almaz/Salyut 5* station would be expended by maintaining it in an Earth-pointing attitude long before the spacecraft could be launched. It was decided to abandon *Almaz/Salyut 5* and commit resources to a second-generation Almaz, weighing 77,000 lb. and placed in orbit by a new heavy-lift launcher beginning 1979/80. Early in 1978

cosmonauts S. Kondratyev, B. Morozob, L. Tararin, S. Chelomei, A. Chekh and S. Chuchin were selected to train for flights on both Soyuz and TKS ferry ships, but there were delays and none of them went into space. Unmanned test flights with TKS took place beginning July 17, 1977, and the advanced Almaz was canceled in 1980.

1977 March 4 The Jet Propulsion Laboratory, Calif., announced that it had awarded six contracts to build a spacecraft powered by solar wind. Under the Solar Sail Development Project, a spacecraft would be put together using thin metallic sails to capture solar photons and propel it around the solar system. Known as "solar sailing," this concept had been around for some time and NASA now wanted to build a prototype that might be used to rendezvous with the next appearance of Halley's comet in 1986. Calculations suggested that an 8,611 sq. ft. solar sail would carry a spacecraft weighing 1,900 lb. on such a mission. Although studied in depth, the solar sail was never developed and no tests were actually carried out.

1977 March 7 The NASA Marshall Space Flight Center reported on the use of a Shuttle external tank (ET) as the core element of an orbiting space station. Developed by James E. Kingsbury at MSFC, the idea was to use the spent ET by carrying it into orbit and fitting it out for habitation after residual propellants had been vented. Believing it would not get funds for the space station it had planned to service with the Shuttle, NASA contrived to use the ET or a Skylab-Apollo telescope mount to extend the life of the Shuttle, to launch unmanned platforms that could visit later, or to modify Spacelab as an independent, free-flying laboratory.

1977 March 16 A significant milestone in the development of the Space Shuttle main engine (SSME) occurred when engineers at NASA's National Space Technology Laboratories, Miss., successfully throttled engine number 0002 to its rated power level of 470,000 lb. thrust. The ability to throttle the SSME in flight is a key feature of this reusable engine, and one which is not common in rocket motors. The 24-sec. firing took place on test stand A-2, bringing to more than 3,500 sec. the accumulated firing time on SSME engines since May 1975. On March 21, 1977, another engine ran successfully for 9 min., more than enough to power a Shuttle orbiter into space.

1977 March 25 The surface sampler arm on the *Viking 2* lander stopped operating at the surface of Mars as temperatures of −190°F began to affect the spacecraft. Never designed to operate in such cold, the lander was exposed to the approaching winter at its northern Utopia site. Winter would last from mid-April through mid-October. On April 2 the sampler arm started working again and on April 14 controllers sent the lander instructions on a winter-survival configuration in which it would periodically give out taped weather data and pictures. On April 18, 1977, the *Viking 2* orbiter fired its motor to slow the orbital period to 22.73 hr. so that it would pass over the *Viking 2* lander site once every 12 days and relay the taped data to Earth. On October 19, 1977, the *Viking 2* orbiter passed within 14 mi. of the tiny moon Deimos.

1977 April 11 The first operational Soviet early warning satellite, *Cosmos 903*, was launched from Plesetsk by SL-6 to a 391 × 2,496 mi. orbit inclined 62.8°. The 2,750-lb. satellite was positioned so that perigee was over the southern hemisphere, which extended the coverage over the important northern hemisphere from where the Soviets feared a missile attack. The system became fully operational when *Cosmos 917* (June 16) and *Cosmos 931* (July 20) had been launched to similar orbits, the three satellites spaced at 80° intervals. Two replacement satellites were launched in 1978 and two in 1979.

1977 April 14 NASA awarded McDonnell Douglas and General Electric design study contracts for a probe capable of surviving enormous pressures deep in the atmosphere of Jupiter. Conceptually developed at NASA in 1976, the Jupiter-Orbiter-Probe (JOP) was tentatively assigned a flight date in January 1982 as the first planetary Shuttle launch. In the mission plan, the probe would be carried toward Jupiter by an orbiter but separate from it 55 days prior to encounter. The orbiter would conduct detailed observation of Jupiter and its satellites after the probe had entered the atmosphere and transmitted data for about 30 min. during descent. On August 29, 1977, NASA announced selection of 114 scientists chosen to participate in the JOP mission. On October 5, 1977, West Germany signed an agreement with NASA to participate in the joint development of the probe.

1977 April 20 A Delta 2914 from Cape Canaveral, Fla., launched *Geos 1* for the European Space Agency, the first purely scientific satellite designed to operate from geostationary orbit. Built by the STAR consortium comprising 14 European aerospace companies headed by the British Aircraft Corp., *Geos 1* had a weight of 1,260 lb. and comprised a cylindrical structure with appendages for experiments to investigate the Earth's magnetosphere. A failure in the launch vehicle placed *Geos 1* in a 150 × 7,276 mi. orbit inclined 26° instead of the geostationary transfer ellipse with an apogee at 21,700 mi. Using on-board propellant, flight controllers raised the orbit to 1,324 × 23,922 mi., but the scientific return was minimal. *Geos 2* was launched by NASA on a Delta 2914 on July 14, 1978, and successfully placed in geostationary orbit at 6° E.

1977 April 25 Women successfully destroyed another bastion of male domination when Trudy Tiedemann, a former public information specialist at the NASA Dryden Flight Research Center, was selected as a commentator on the broadcast public affairs circuit for the upcoming Shuttle orbiter flight tests from Edwards AFB, Calif. Hitherto, only male public affairs officers had been heard giving transmitted commentary during NASA missions.

1977 April The Centre National d'Etudes Spatiales (CNES), the French national space agency, asked the Ballistic Systems Division of Aérospatiale to perform a feasibility study of a manned space vehicle launched by an advanced version of Ariane. Named Hermes, the vehicle was to take the form of a small space plane with a maximum projected weight of 14,330 lb. A second study to investigate the possibility of developing a small lifting body was completed in December. In 1978 further studies showed the advanced Ariane would be capable of lifting 22,000 lb. to low Earth orbit, and Hermes was designed to accommodate that weight, having a length of 40 ft. and a wingspan of 24 ft.

1977 May 6 Dr. Alan Lovelace, acting NASA administrator, and Anatoly P. Aleksandrov, president of the Soviet Academy, signed an agreement on a joint manned flight scheduled for 1981 during which a NASA Shuttle would dock to a Soviet Salyut space station for several days of experiments and joint activities. NASA felt it had little hope to getting Congress to fund a space station before the Shuttle proved itself. This was

the only way to carry out research aboard an orbiting facility for at least another 10 years. The joint flight planned for 1981 never took place. When Ronald Reagan became president-elect in November 1980, all joint U.S.-USSR ventures were put on hold.

1977 May 13 NASA requested proposals from McDonnell Douglas and Boeing for procurement of six spinning solid upper stage (SSUS) boost motors for moving satellites from low Earth orbit to geosynchronous orbit. The request was instigated by a need to find a perigee stage for the Intelsat V–type satellites that NASA wanted to launch on the Shuttle. NASA wanted to fly the first SSUS-A (Atlas payload equivalent) on a Shuttle demonstration flight in February 1979, optimistically believing the reusable launch vehicle would be ready for orbital flight tests by then.

1977 May 17 A German firm called Orbital Transport und Raketen-Aktiengesellschaft (OTRAG) fired its first rocket designed to test low-cost methods for launching satellites. The rocket was a single-stage device of simple design feeding a cluster of four 6,600-lb. motors from four propellant tanks, two containing WFNA (white fuming nitric acid) and two kerosene. The OTRAG rocket reached a height of 65,600 ft. as planned. Later flights were expected to cluster up to 40 such combinations for inserting small satellites into orbit and 300 four-engine clusters were devised with a lifting potential equal to Atlas Centaur. Tests took place from Zaire in an agreement with President Gen. Joseph Mobutu. Lutz T. Kayser, president of OTRAG, attracted investment from a wide range of interests, including countries that saw in his plans the means to acquire missiles capable of launching small nuclear bombs. At one time, Kurt H. Debus, former director of NASA's Kennedy Space Center, was chairman of the OTRAG board.

The NASA Marshall Space Flight Center announced it had investigated the unique properties of an ion propulsion system for sending a spacecraft to fly by Halley's comet in 1986. Tested for several years at the Lewis Research Center, a solar electric propulsion (SEP) system would aim solar energy concentrated by reflectors at solar cells which would power eight ion engines. A mission to the comet would benefit from a high fly-by velocity, and the long operating time of an ion engine would allow NASA to achieve this. The concept was in competition with a solar sailing method proposed by the Jet Propulsion Laboratory.

1977 May 18 Agreement between the United States and the Soviet Union on an international satellite rescue service was formalized. Known as COSPAS/SARSAT (an acronym assembled from the Russian words for "Space System for the Search of Vessels in Distress" and the U.S. "Search and Rescue Satellite-Aided Tracking"). The permanent establishment of the joint system was signed on November 29, 1979. Canada joined on August 24, 1979. France, Norway, Japan, Sweden and the United Kingdom joined later. The system worked by flying transponders on existing U.S. weather satellites and Soviet Cosmos satellites. Tuned to 121.5 and 243 MHz, early experiments were succeeded by a dedicated frequency of 406 MHz. Ships, boats, small aircraft and vehicles could carry distress transmitters for signals to be relayed via the COSPAS/SARSAT system.

A solid propellant Castor strap-on motor weighing 23,000 lb. fell off a Delta launch vehicle being readied on its pad at Cape Canaveral, Fla., for a planned mid-June launch of the European Space Agancy's high-frequency communication test satellite, CTS. In falling 12 ft., the 36 ft. motor with a 4-ft.

diameter plunged through a platform and struck the Delta's first-stage liquid oxygen tank. A structural failure in the securing linkage was responsible for the accident.

1977 May 19 The first successful Soviet high altitude, "pop-up" antisatellite test began with the launch from Plesetsk of the 1,100-lb. *Cosmos 909* target on an SL-8 to a 610 × 1,312 mi. orbit inclined 65.9°. Four days later the 4,430-lb. interceptor *Cosmos 910* was launched by an SL-11 from Baikonur to an intermediate orbit of 92 × 314 mi. orbit from which it was to have shot up to a high speed fly-by. It failed and plunged back through the atmosphere just hours after launch. On June 17 *Cosmos 918* was launched to a 77 × 122 mi. orbit from where it successfully conducted its "pop-up" maneuver and immediately reentered.

1977 May 27 The Marshall Space Flight Center issued requests for proposals to build a tethered satellite system for launch on the Shuttle, designed to investigate the region of space between 50 mi. and 75 mi. Too high for research aircraft and too low for free-flying satellites, a tethered satellite trolled through the lower reaches of space at the end of a 64-meter cable attached to a Shuttle in Earth orbit would provide valuable scientific data. On September 30 Martin Marietta and Ball Brothers were selected to conduct parallel studies.

1977 May 30 The gathering of data essential for the search for life on Mars ended when the biology instrument on the *Viking 1* lander was turned off two days after the instrument on the *Viking 2* lander stopped operating. In more than 10 months of operations, scientists on Earth had received information which left them confused about reactions observed in the three test chambers on each spacecraft. In several cycles of soil sampling, some indications seemed to mimic life, but no organisms were found and the consensus among astrobiologists was that some form of exotic chemistry, perhaps a process that links living and nonliving things, had been seen at work on Mars. Tests would continue in laboratories around Earth, but no absolute conclusion would be reached.

1977 June 16 Robert M. Frosch was sworn in as NASA administrator six weeks after James Fletcher retired from that post. Frosch came to NASA from his position as associate director for Applied Oceanography at the Woods Hole Oceanographic Institute, which he had held since 1975. For two years prior to that he had served as assistant secretary general of the United Nations. Frosch served NASA for three years and seven months, retiring on January 20, 1981. He was the only administrator within whose tenure no U.S. manned spaceflight took place.

1977 June 18 NASA astronauts Fred Haise and Gordon Fullerton were aboard Shuttle *Enterprise* when it was carried into the air for the first time by a Boeing 747 Shuttle carrier aircraft from Edwards AFB, Calif., on a flight lasting 55 min. 46 sec., achieving a height of 14,970 ft. and a maximum speed of 208 MPH. Astronauts Joe Engle and Richard Truly were aboard *Enterprise* for the second of three "captive-active" flights on June 28, which lasted 62 min. and reached 22,030 ft. and a speed of 310 MPH. Haise and Fullerton flew the third flight on July 26, reaching 30,292 ft. and a speed of 311 MPH in a flight lasting 59 min. 53 sec. These successfully cleared *Enterprise* for free flights beginning August 12, 1977.

1977 June 23 The second of two Navigation Technology Satellites, *NTS-2*, was launched by Atlas F to a 12,145 ×

12,544 mi. orbit inclined 63.2°. Built to refine the accuracies in atomic clocks to be used in the Navstar global positioning system, the 950-lb. *NTS-2* had an octagonal main body, 2 ft. 7 in. high and 5 ft. 5 in. across, with two solar cell wings providing 480 W and a 65-ft. gravity-gradient stabilization boom. A test array carried 14 different solar cells for performance analysis and the primary payload comprised a cesium beam atomic clock with an accuracy of one second in three million years. It remains in orbit with a projected life in excess of one million years.

1977 June 30 The European Telecommunications Satellite Organization (Eutelsat) was created by agreement with 17 telecommunications entities, all members of the European Conference of Postal and Telecommunication Administration. The purpose of Eutelsat was to set up a regional European telecommunications satellite system. The agreement established an Interim Eutelsat structure pending final operating agreements. The organization would provide a set of requirements, the first of which was met through experimental operations with the European Communication Satellite (ECS).

1977 July 14 Japan's first Geostationay Meteorological Satellite (GMS-1) *Himawari* (*Sunflower*) was launched by Delta 2914 from Cape Canaveral, Fla., and eventually positioned at 140° E. Designed and developed by Hughes Aircraft for the Nippon Electric Co., under contract to Japan's National Space Development Agency, the cylindrical, spin-stabilized satellite had a height of 8 ft. 10 in., a diameter of 7 ft. 1 in. and

The viewing mirror for Japan's GMS-1 weather satellite is examined by a technician prior to launch on July 14, 1977.

a weight of 1,476 lb. at launch and 670 lb. on station. The primary instrument comprised a visible/infrared spin-scan radiometer producing high-resolution images every 30 min. day and night. The satellite continued to operate until June 30, 1989.

1977 July 17 At 08:57 UT, a new type of Soviet spacecraft, the Chelomei bureau's TKS manned ferry ship, designated *Cosmos 929*, was launched from the Baikonur cosmodrome by an SL-13 Proton launcher. *Cosmos 929* weighed 41,900 lb. and incorporated an 8,000-lb. conical reentry module, 9 ft. 2 in. across, capable of carrying two cosmonauts or cargo. Looking like the conical section of a NASA Gemini spacecraft, the reentry module had been adapted from the Chelomei LK-1 circumlunar descent module. Unmanned, *Cosmos 929* had a length of almost 43 ft., including a cylindrical section adapted from the smaller work compartment of Almaz, and two rotatable solar cells arrays. External propellant tanks were attached to the cylinder and a docking unit was carried at the aft end opposite the reentry module. The habitable area was approximately half that of the Salyut space station. *Cosmos 929* entered an orbit of 133 × 162 mi. inclined 51.5° to the equator and performed several maneuvers before the reentry capsule was returned to Earth on August 16. *Cosmos 929* continued to maneuver, reaching a maximum apogee of 278 mi. before it was de-orbited on February 2, 1978, over the Pacific Ocean. Derivatives of *Cosmos 929* were launched as *Cosmos 1267*, *1443*, and *1686*.

1977 July 18 The first firing of the largest solid propellant rocket motor developed for flight was conducted by Thiokol's Wasatch Division near Brigham City, Utah, when Shuttle solid rocket motor (SRM) demonstration model, DM-1, was successfully fired for 2 min. The SRM produced a thrust of 2.75 million lb., as against the 1.15 million lb. of thrust for the SRMs used with the Titan launch vehicle. Later firings included DM-2 on January 18, 1978, DM-3 on October 19, 1978, and DM-4 on February 19, 1979. Qualification firings included QM-1 on June 13, 1979, QM-2 on September 27, 1979, and QM-3 on February 13, 1980. When assembled into a solid rocket booster, each unit had a length of 149 ft. 2 in., a diameter of 12 ft. 2 in., a launch weight of 1.257 million lb. and a vacuum thrust of 3.31 million lb. with an Isp of 264.8 sec.

1977 August 8 The Soviet *Almaz/Salyut 5* space station, unmanned since the return of the *Soyuz 24* crew on February 25, 1977, was de-orbited into the Pacific Ocean. Its final orbit prior to retrofire with the on-board propulsion system was 113 x 115 mi. Last of the "military" stations, *Salyut 5* had been plagued with failures and problems. Of three stations of this type launched, only two had been operable, and of five crews launched in Soyuz spacecraft to occupy an Almaz/Salyut, only three had been successful. One mission, *Soyuz 21*, had to be terminated early due to illness. As the United States had previously concluded, a military justification for manned spaceflight was hard to find.

1977 August 12 More than 60,000 people at Edwards AFB, Calif., watched the first free-flight descent of the Shuttle *Enterprise* as it lifted away from the top of the Boeing 747 Shuttle carrier aircraft 48 min. 28 sec. after takeoff. Piloted by astronauts Fred Haise and Gordon Fullerton, *Enterprise* separated from the Boeing 747 at a height of 24,100 ft. and a speed of 310 MPH. It touched down 5 min. 21 sec. later at a speed of 213 MPH. The second free flight in the approach and

landing test (ALT) series took place when astronauts Joe Engle and Richard Truly lifted away from the Boeing 747 at 26,000 ft. and 310 MPH on September 13 for a 5 min. 28 sec. descent and landing at 225 MPH. The third flight took place on September 23 from a height of 24,700 ft. and a speed of 290 MPH, carrying astronauts Haise and Fullerton to a landing 5 min. 34 sec. later at 221 MPH. These three flights were made with a tail cone attached to reduce buffeting from the blunt afterbody on *Enterprise*. The last two flights, beginning October 12, were made with the orbiter tail cone removed to simulate the steeper 22° glideslope orbiters would fly when returning from space.

HEAO-1, the first of three High-Energy Astronomy Observatory satellites, was launched by Atlas-Centaur from Cape Canaveral, Fla., to a 266 × 278 mi., 22.76° orbit. The 5,665-lb. satellite comprised an octagonal equipment module, standard for all HEAOs, with a 10 ft. 6 in. high hexagonal experiment module on top. HEAO-1 had a total height of about 19 ft. and a diameter of 7 ft. 8 in. A rectangular solar array deployed from the top and two other arrays were attached to one side of the experiment module, the three providing up to 460 W of electrical power. HEAO-1 carried four large instruments for conducting the most advanced x-ray survey of the sky to date. It continued to return useful data until the attitude control gas ran out in January 1979, increasing from 350 to nearly 1,500 the number of known celestial x-ray sources. HEAO-1 decayed out of orbit on March 15, 1979.

1977 August 20 Titan IIIE/Centaur TC-7 launched NASA's *Voyager 2* spacecraft at 14:29:45 UT from Complex 41. The 1,820-lb. spacecraft was attached to a 2,690-lb. propulsion module which, together with the spacecraft, separated from the Centaur. This 15,300-lb. thrust solid propellant motor burned for 43 sec. to put Voyager on course for a fly-by of Jupiter in July 1979 and Saturn in August 1981. The three-axis-stabilized spacecraft was built around a 10-sided aluminum electronics frame 5 ft. 9 in. across and 1 ft. 6 in. high. A parabolic reflector 12 ft. in diameter was attached to the top of the spaceframe capable of transmitting data at adjustable rates between 40 and 2,560 bits/sec. via 9.4-W or 21.3-W amplifiers. Three nuclear radioisotope thermoelectric generators attached to a deployable boom provided 423 W of electrical energy at Earth and 384 W at Saturn fly-by. A computer command system with two 4,096 data-word memories controlled on-board events. A 7 ft. 6 in. boom attached to one side of the spaceframe below the parabolic antenna carried the 236-lb. movable scan platform and most of the science instruments. On the opposite side, a 43-ft. deployable boom held two magnetometers.

1977 August 25 Italy's first experimental domestic communications satellite, *SIRIO* (Satellite Italiano Riceroa Industriale Orientata), was launched at 23:50 UT on a Delta 2313 from Cape Canaveral, Fla., to a 143 × 23,408 mi. geostationary transfer orbit inclined 23°. At 14:57 UT on August 27, the apogee boost motor fired to place *SIRIO* on station over 15° W by September 8. With a height of 6 ft. 6 in., a diameter of 4 ft. 7 in. and a launch weight of 485 lb., the satellite conducted super high-frequency tests in the 12 and 18 GHz bands. One of the primary experiments studied propagation characteristics of SHF radio waves transmitted through adverse weather conditions.

1977 September 5 At 12:56:01 UT, Titan IIIE/Centaur TC-6 launched the 1,820-lb. spacecraft *Voyager 1* on an Earth escape trajectory for fly-bys of Jupiter in March 1979 and

Launched on August 20, 1977, the Titan IIIE-Centaur thunders away from Cape Canaveral, Florida, carrying the Voyager 2 *spacecraft, the first of two deep-space vehicles launched toward the planet Jupiter.*

Saturn in November 1980. Although launched before it, on August 20, 1977, *Voyager 2* would be overtaken by *Voyager 1*, which was on a slightly different trajectory. Like its predecessor, *Voyager 1* carried a copper-plated 12-in. recorded disc containing "Sounds of Earth" for whoever (or whatever) might pick it up in millenia to come, a cosmic message-in-a-bottle tossed to the ocean of space. *Voyager 1* passed the moon's orbit 10 hr. after launch, 57,000 mi. from its surface.

1977 September 9   Martin Marietta rolled out the first Shuttle external tank (ET). Each ET would have a length of 154 ft., a diameter of 7 ft. 6 in. and carry 1.3 million lb. of liquid

oxygen in a forward tank and 227,000 lb. of liquid hydrogen in the aft tank. The combined total of 528,000 gal. would be consumed by the three Shuttle main engines at a rate of 65,000 gal./min. Delivered to NASA on June 26, 1979, the empty weight of the tank for the first Shuttle mission came out at 77,100 lb.; 600 lb. was saved with the third and subsequent tanks by leaving off white paint applied to the first two tanks, giving the ET a brownish color. Further weight reductions were made to the ET structure, shaving a further 6,400 lb. for the sixth, eighth and all subsequent ETs. The first lightweight ET was delivered to NASA on September 10, 1982.

1977 September 11   *Voyager 1* performed the first of two small course correction maneuvers to tweak up the trajectory for encounter with Jupiter in December 1978. The first maneuver increased the spacecraft's velocity by 8 ft./sec. and the second, two days later, by 33.2 ft./sec. Postmaneuver tracking indicated both burns were about 20% under velocity, a deficit that was corrected with a tweak burn on October 29. A hydrazine monopropellant attitude and maneuver system common to each Voyager could provide up to 660 ft./sec. in course corrections. A tank 2 ft. 4 in. in diameter contained 231 lb. of hydrazine operating through sixteen 0.2-lb. thrust jets. Four were dedicated to course corrections and the remaining twelve were divided into two redundant banks of six thrusters each for attitude stabilization. *Voyager 2's* first correction maneuver was conducted on October 11, 1977, and the second on May 3, 1978.

1977 September 12   NASA selected the Solar Electric Propulsion System (SEPS) ion drive system for the proposed Halley's comet rendezvous mission, rejecting the Jet Propulsion Laboratory's solar sail concept. The Lewis Research Center and the Marshall Space Flight Center had conducted considerable research into ion drive engines and performed several tests in space. With a cluster of mercury-ion engines, the SEPS would, NASA hoped, constitute the primary deep-space propulsion system from the early 1980s. Launched from the Shuttle in 1981, SEPS would power the Halley comet mission for a fly-by in 1986.

1977 September 13   An attempt to launch the 1,907-lb. Orbital Test Satellite (OTS) for the European Space Agency ended 54 sec. after liftoff from Cape Canaveral, Fla., when the Delta 3914 exploded. OTSs had a launch weight of 1,910 lb. at launch and 980 lb. on geostationary orbit after apogee motor burnout. Each satellite comprised a six-sided box with two solar array "wings" spanning 30 ft. 4 in. OTS had a height of 7 ft. 10 in., a length of 7 ft. and a width of 5 ft. 6 in. *OTS-A* was to have been positioned at 10° E. Built by British Aerospace, the satellite was a precursor to operational regional communications satellites Europe planned to put up for use by the European Telecommunications Union, ETU. A replacement, *OTS B*, was launched by NASA on May 11, 1978.

1977 September 18   A single picture containing both the Earth and the moon seemingly suspended in space was taken for the first time by *Voyager 1* on its way to Jupiter. At the time, *Voyager 1* was 7.25 million mi. from Earth and the picture was assembled from three images taken through separate filters. Each Voyager carried a 200-mm wide-angle and a 1,500-mm narrow-angle TV camera, each equipped with a wheel of eight color filters. Images would be transmitted to Earth at a rate of 115.2 kilobits/sec., each picture containing 800 lines of 800 pixels each, each pixel coded in 9 binary digits.

In this way a full picture load of 640,000 pixels could be transmitted to Earth in 48 sec. In comparison, it had taken more than 8 hr. to transmit the 40,000 pixels from each *Mariner IV* image of Mars in 1965.

1977 September 29 At 06:50 UT a Soviet SL-13 launch vehicle lifted off from Baikonur cosmodrome carrying the 43,700-lb. *Salyut 6* space station to an orbit of 136 × 171 mi. inclined 51.6° to the equator. From there it was maneuvered to a higher orbit to await the first crew. First of the third-generation "civilian" Salyuts, it was externally similar to *Salyut 4* but incorporated many changes. The forward transfer module had a length of 11 ft. 6 in. and a diameter of 6 ft. 6 in., with a docking drogue on the front and an extravehicular activity (EVA) hatch in the side wall. The smaller (living) compartment had a length of 8 ft. 10 in. and a diameter of 9 ft. 6 in. connected to the larger work area by a conical, 3 ft. 11 in. adapter. The large work area had a length of 8 ft. 10 in. and a diameter of 13 ft. 7 in. Instead of the inaccessible propulsion module at the rear, *Salyut 6* had a service section, with a length of 7 ft. 3 in. and a diameter of 13 ft. 7 in. It had two 660-lb. thrust rocket motors identical to the propulsion unit carried by Soyuz-T. A tunnel through the center of the service section provided access to a rear hatch and docking drogue. The rear docking port would be used by Progress cargo tankers.

1977 September 30 The operational phase of the Apollo moon landing program came to an end when NASA deliberately switched off the five ALSEP (Apollo lunar surface experiments package) science stations at the *Apollo 12, 14, 15, 16* and *17* sites to save money. The first four stations had been designed for a life of one year and the last one for two years. The *Apollo 12* station had completed almost eight years and the *Apollo 17* station almost five years, during which vast quantities of data had been acquired for analysis. More than 153,000 commands had been transmitted to the stations and more than one trillion binary digits of data received. About 10,000 moonquakes and 2,000 meteorite impacts on the moon had been recorded.

1977 September From land pads at Cape Canaveral, Fla., the U.S. Navy conducted the first of many test firings with a new British warhead for the Royal Navy's Polaris submarine missiles. Called Chevaline, the warhead was designed to counter Soviet ballistic missile defenses around Moscow by incorporating penetration aids and decoys among the live warheads. Developed at Britain's Atomic Weapons Research Establishment, Aldermaston, Chevaline comprised six 40-KT nuclear devices attached to a penetration aid carrier (PAC), essentially a maneuvering bus attached to the missile's second stage.

1977 October 9 At 02:40 UT the 15,125-lb. *Soyuz 25* spacecraft was launched from the Baikonur cosmodrome carrying cosmonauts Vladimir V. Kovalenok and Valery V. Ryumin. It was to have provided the first occupation of the *Salyut 6* space station on a mission planned to last three months. Two more cosmonauts were to have visited Kovalenok and Ryumin in November and left their *Soyuz 26* spacecraft at *Salyut 6* while they returned home in *Soyuz 25*. One of the visiting crew members was to have been a Czech cosmonaut from the Intercosmos program. Kovalenok and Ryumin were then to have moved *Soyuz 26* from the rear of *Salyut 6* to the forward port, leaving the aft port available for the first Progress cargo tanker to dock. Neither *Soyuz 25* crew member had flown

before and despite several attempts at docking, problems prevented the Soyuz spacecraft hooking up to the big space station. The mission had to be abandoned and Kovalenok and Ryumin returned to Earth at an elapsed time of 2 days 45 min.

1977 October 12 Astronauts Joe Engle and Richard Truly piloted the Shuttle *Enterprise* to a landing at Edwards AFB, Calif., in the first of two free flights in the approach and landing test (ALT) series made with a tail cone removed from the rear of the delta-winged vehicle. This was done to achieve a steep 22° glideslope and duplicate the profile that Shuttles returning from space would fly at the end of their mission. Carried to a height of 22,400 ft. and a speed of 230 MPH by the Boeing 747 Shuttle carrier aircraft, the astronauts piloted the *Enterprise* to a landing in just 2 min. 34 sec. at a speed of 278 MPH. The last free flight in the ALT series, also with the tail cone removed, was flown by astronauts Fred Haise and Gordon Fullerton on October 26, from a height of 19,000 ft. and a speed of 283 MPH to a landing 2 min. 1 sec. later at 283 MPH on a concrete runway.

1977 October 21 A new Soviet low-altitude, "pop-up", anti-satellite test began when an SL-8 from Plesetsk launched *Cosmos 959* to a 89.5 × 518 mi. orbit inclined 65.8°. The 4,430-lb. interceptor, *Cosmos 961*, was launched by SL-11 from Baikonur on October 26 to a 77.5 × 187.5 mi. orbit. Within hours it shot up to the target, simulating an intercept, and fell back through the atmosphere. *Cosmos 959* decayed on November 30, 1977. In further tests beginning December 13 with the launch from Plesetsk of an SL-8, the 1,543-lb. target vehicle *Cosmos 967* was placed in a 598 × 624 mi. orbit inclined 65.8°. On December 21, the 4,430-lb. *Cosmos 970* was launched by SL-11 from Baikonur to a 569 × 700 mi. orbit from which it maneuvered to a slow fly-by before destruction. *Cosmos 967* was also the target for *Cosmos 1009*, launched from Baikonur by SL-11 on May 19, 1978 to a 599 × 860 mi. orbit from which it conducted a close inspection prior to reentering the same day. No longer operating, *Cosmos 967* remains in orbit.

1977 October 22 A Delta 2910 carried into space two satellites designed and built at the Goddard Space Flight Center to make simultaneous measurements of near-Earth space from different distances. The first of three International Sun-Earth Explorer satellites, *ISEE-A* and *ISEE-B*, were each placed in almost the same orbit of 209 × 85,695 mi. inclined 28.95°. *ISEE-A* was a 16-sided polyhedron 5 ft. 8 in. across, 5 ft. 3 in. tall, weighing 750 lb. *ISEE-B* was a cylinder with a diameter of 4 ft. 2 in., a height of 3 ft. 9 in. and a weight of 366-lb. ISEE would measure Earth's magnetic environment, the solar wind and interactions of the two. The use of two satellites, separated by a variable distance, enabled scientists to make precise measurements of the magnetosphere's bow shock wave and tail. *ISEE-B* was built by a European consortium led by the Dornier company in Germany.

1977 October 23 The main rocket motor aboard the *Viking 2* orbiter was fired to change its orbit around Mars and lower periapsis to 186 mi., allowing the spacecraft to measure perturbations in its orbit resulting from what are termed gravitational anomalies. By carefully tracking the gravitational influence of the asymmetric distribution of mass inside Mars, scientists could calculate the way materials were distributed inside the planet. Both orbiting spacecraft were now in orbits that brought them as low as 186 mi. and there would be no more orbital changes. The Viking extended mission would end December 31, 1977, followed by a continuation mission that began January 1, 1978.

1977 October 28 The Marshall Space Flight Center announced plans for a teleoperator retrieval system (TRS) combining a low-thrust propulsion unit with approach and docking aids. Carried into space aboard the Shuttle, the TRS would be "flown" by remote control to a satellite that it would move to a different orbit as desired. Propulsion would comprise 24 cold-gas attitude control thrusters, but engineers were looking at a hydrazine system, disliked because its toxicity might prove dangerous in the Shuttle. On July 5, 1978, MSFC awarded Martin Marietta a TRS development contract. The first TRS mission was expected to boost Skylab to a higher orbit from which, on a later mission, it could be revisited by Shuttle astronauts.

1977 November 23 Europe's first geostationary meteorological satellite, *Meteosat 1*, was launched by Delta 2914 from Cape Canaveral, Fla., to a geostationary transfer orbit of 91 × 22,990 mi. inclined 27.5°. From there it was placed in geostationary orbit on the Greenwich meridian. The drum-shaped satellite had a diameter of 6 ft. 10 in., a height of 10 ft. 6 in. and a weight of 1,540 lb. at launch and 700 lb. on station after apogee motor burnout. It carried a high-resolution radiometer with a definition of 1.5 mi. in daylight and 3 mi. in infrared. Meteosats were built by a consortium of European

Europe's first indigenous weather satellite, Meteosat 1, *is attached to the terminal stage of its Delta launcher prior to its flight into space on November 23, 1977.*

manufacturers led by Aérospatiale. *Meteosat 1* was one of five geostationary satellites placed in orbit by nations participating in the Global Weather Experiment.

1977 November Intelligence information about Soviet developments with MIRV warheads reached upper levels of the Carter administration, forcing a reconsideration of the pace of development of U.S. missiles. Tests with new versions of Soviet ICBMs revealed unexpected levels of technical capability the CIA had previously believed would be applied to a future generation of missile deployed around 1985. They were now seen as applicable to SS-18 and SS-19 ICBMs already in silos. The National Security Council debated Carter's desire to placate the Soviets and trim U.S. defense spending while deciding an appropriate response to these trends. Some wanted to accelerate development of the M-X missile while others negotiating the SALT II treaty said the Soviet developments were of no importance. It did, however, play an important part in reversing Carter's policy of compromise.

1977 December 8 The second set of three U.S. Navy White Cloud ocean surveillance satellites was launched by an Atlas F from Vandenberg AFB, Calif., and placed in three closely spaced parallel orbits at 63.4°. Positioned 113° radially around the Earth from the first three White Cloud satellites launched on April 30, 1976, they would cover Soviet shipping in the Atlantic while their predecessors would cover the Pacific. The bus (NOSS-2) and the three satellites *(SSU-4, -5 and -6)* were placed in orbits at 655 × 726 mi., 655 × 726 mi., 655.5 × 725.5 mi. and 655.5 × 725.5 mi., respectively, all orbits with an inclination of 63.44°. Thus was established the first preoperational White Cloud network transmitting intelligence on 1.432 GHz with a 1-Mz bandwidth. They remain in orbit, but are inoperable.

1977 December 9 NASA's Jet Propulsion Laboratory reported on work it was conducting with the Dryden Flight Research Center to develop a reconnaissance airplane for use in the atmosphere of Mars. Developed from a design called Mini Sniffer, the airplane would use a high-aspect ratio wing and a large propeller to cruise through the rarefied atmosphere taking measurements and pictures. Such a craft would probably fly to Mars encapsulated in a protective shroud attached to an orbiter and be released into the atmosphere where it would operate for several weeks.

1977 December 10 Cosmonauts Yuri V. Romanenko and Georgy M. Grechko were launched at 01:19 UT in their 15,000-lb. *Soyuz 26* spacecraft from Baikonur cosmodrome to a rendezvous and docking at the rear of *Salyut 6* at 03:02 UT on December 12. The previous crew, aboard *Soyuz 25*, had been unable to dock at the forward port and Grechko, an experienced pilot and engineer, had been seconded to the *Soyuz 26* crew to carry out an inspection of the forward docking equipment. After entering *Salyut 6* for several days of activation and check-out, Grechko began a space walk at 21:36 UT on December 19. In 1 hr. 28 min. Grechko completed his inspection of the forward docking port, pronounced it safe to use and returned to *Salyut 6*. Romanenko had to stand by in the open sidehatch of the transfer module but briefly experienced a space walk when he drifted out and joined Grechko in an unplanned exercise. The two cosmonauts then settled down to a three-month stay aboard *Salyut 6*.

1977 December 13 The European Space Agency reported that the Société Euopéenne de Propulsion (SEP) had success-

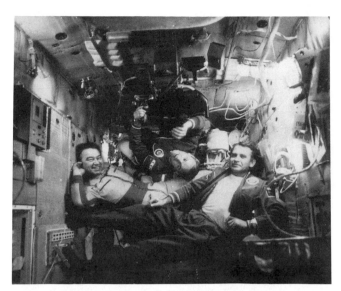

Seen aboard Salyut 6, *cosmonauts Georgi Grechko, Vladimir Dzhanibekov and Oleg Makarov give scale to the interior of the Soviet orbital space station.*

fully carried out the first test of the Ariane first stage. The four Viking II first-stage engines were fired for 1 min. 51 sec. to verify stage behavior, vibration spectra and thermal conditions at the propulsion bay. ESA was aiming for two more development tests prior to May 1978, followed by three-stage qualification tests.

1977 December 14   NASA announced its decision to fly the *Pioneer 11* spacecraft to within 18,000 mi. of Saturn's rings at encounter on September 1, 1979, and to fly down under the rings to a distance of 15,000 mi. from the planet itself. *Voyager 1* would encounter Saturn in November 1980 followed by *Voyager 2* in August 1981. If all went well at the first Voyager encounter, mission planners wanted to target *Voyager 2* for a fly-by of Saturn that would throw the spacecraft on to the planet Uranus in January 1986. *Pioneer 11* would serve as a pathfinder for that trajectory, generating the data engineers needed to know precisely the position, size and mass of the planet for a highly exact positioning needed for the *Voyager 2* option. The course correction to put *Pioneer 11* precisely on target was completed in July 1978.

1977 December 15   A NASA Delta 2914 launched Japan's domestic communications satellite (CS) *Sakura* (*Cherry Blossom*) to a geostationary transfer orbit of 103 × 22,330 mi. inclined 28.8° to the equator. From there it was placed in geostationary orbit over 135° E. The hatbox-shaped satellite had a diameter of 7 ft. 2 in., a height of 11 ft. 6 in., a launch weight of 1,490 lb. and an on-station weight of 770 lb. Designed and built by Ford Aerospace and Mitsubishi Electric, CS carried 20,000 solar cells providing 500 W of power for eight transponders in both C-band (4–6 GHz) and K-band (20–30 GHz), the first communications satellite to carry both bands. CS continued to operate until November 25, 1985.

1977 December 27   The first Soviet SL-14 launcher lifted off from Plesetsk carrying the 11,000-lb. satellite *Cosmos 972*. It placed the satellite in an orbit of 445 × 728 mi. with an inclination of 103.92°. The SL-14 Tsyklon is a three-stage

liquid propellant vehicle burning UDMH-N_2O_4. The first stage produces a thrust of 620,000 lb., the second stage a thrust of 228,000 lb. and the third stage a thrust of 17,550 lb. The SL-14 has a length of 128 ft. 10 in. and a payload capability of 12,150 lb. to low Earth orbit. A development of the SS-9 ICBM, it is the second of two satellite launchers based on that missile. The other is the two-stage SL-11.

1977 December   Boeing completed its final report on a seven-month study of Shuttle Derivative Vehicle (SDV) designs for Class 1 and Class 2 payload capabilities. Class 1 (132,000–198,000-lb. to low Earth orbit) objectives could be met by removing the orbiter and placing a cluster of three SSMEs in a recoverable module in the same relative position as they would occupy were an orbiter attached to the external tank. A 148,000-lb. payload could be accommodated in a special pod with an internal volume at least equal to that provided by a Shuttle orbiter and attached to the ET in the same relative location. A Class 2 SDV would have a similar configuration with tripropellant liquid boosters producing a total thrust of around 6 million lb. instead of SRBs and place 185,000 lb. in low Earth orbit.

1977   During the year the Soviet Union began deploying one of the most controversial missiles fielded to data. Code-named SS-20 Sabre and known in the Soviet Union as *Pioneer*, the missile was mobile and carried three MIRV warheads of 150 KT each. Designed to replace the outdated SS-4 and SS-5, the SS-20 was almost impossible to pretarget due to its versatility. Moreover, each missile threatened three targets, and replacement on a missile-by-missile basis tripled the number of targets the force could attack. At 3,100 mi., the range of the SS-20 was twice that of the aging missiles it replaced. Deployment was rapid, reaching 120 in 1979, 240 in 1981, 378 in 1983, 423 in 1985 and 441 by early 1987. At peak, on its own, this long-range intermediate nuclear force (INF) missile posed a nuclear threat to 1,323 separate targets, compared to a maximum 590 from the missiles it replaced.

A new solid propellant missile began equipping the French nuclear ballistic missile submarine force. Called M-20, it was identical to the M-2 in shape, size and performance but incorporated a new warhead carrying a 1.2-MT, MR-60 reentry vehicle with penetration aids and hardening against ballistic missile defenses. All French ballistic missile submarines were fitted with the M-20, and a new, lighter 1.2-MT warhead (the MR-61) was retrofitted to the M-20 by 1980.

1978 January 10   At 12:26 UT cosmonauts Vladimr Alexandrovich Dzhanibekov and Oleg G. Makarov were launched in their 15,000-lb. *Soyuz 27* spacecraft from Baikonur cosmodrome to a rendezvous and docking with the *Salyut 6* space station. Their mission was to dock with the troublesome forward port on *Salyut 6* and qualify it for routine operations. A docking was successfully achieved at 14:06 UT on January 12, during which *Salyut 6* occupants Yuri Romanenko and Georgi Grechko had retreated to their *Soyuz 26* spacecraft, docked to the aft port, for protection in the event of an accident. Dzhanibekov and Makarov joined Romanenko and Grechko in *Salyut 6*, bringing mail and newspapers and participating in some experiments. The two visiting cosmonauts were to return to Earth in *Soyuz 26*. Because seats were shaped for individual cosmonauts, they exchanged contour couches between respective spacecraft before undocking *Soyuz 26* at 08:05 UT on January 16. This left the aft docking port vacant for the *Progress 1* cargo tanker, launched on

January 20. Dzhanibekov and Makarov returned to Earth at an elapsed time of 5 days 22 hr. 59 min.

1978 January 20 The Carter administration directed the Department of Defense to formulate recommendations about an antisatellite system. Initially, the Air Force selected a missile-launched system using the Minuteman III with an air-launched concept in backup. The Air Force reversed itself and chose the air-launched concept as the primary system. By the end of 1978 a Boeing short range attack missile was selected as the first stage and a Vought Altair III as second stage with a miniature homing vehicle (MHV) head. The two-stage missile would be 17 ft. long and 1 ft. 6 in. in diameter, weighing 2,600 lb. and launched by a modified F-15.

At 08:35 UT a Soviet SL-4 launched *Progress 1*, the first unmanned cargo-tanker version of the Soyuz spacecraft, to an orbit from where it rendezvoused and docked with the aft docking port on *Salyut 6* at 10:12 UT two days later. The 15,480-lb. *Progress 1* retained the instrument module and the orbital module of Soyuz, but the descent module was removed and replaced with a structural section containing 2,205 lb. of propellant in supplementary tanks. A connecting interface at the aft end of Salyut linked propellant lines between Progress and the space station. The pressurized orbital module could carry up to 2,866 lb. of cargo and access from Salyut was gained via the docking hatch. *Salyut 6* cosmonauts Yuri Romanenko and Georgi Grechko unloaded oxygen bottles, water, food, sanitary supplies, film and equipment before refueling the Salyut propulsion module. *Progress 1* was loaded with waste from the station and undocked at 05:53 UT on February 6. The spacecraft was de-orbited over the Pacific Ocean on February 8, after a sequence of orbital maneuvers to test the propulsion system.

1978 January 23 NASA formally designated its Jupiter-Orbiter-Probe mission Project Galileo, after the 17th century Italian astronomer who first observed moons around Jupiter. As envisaged, the orbiter would incorporate a dual-spin design, part of the spacecraft being three-axis-stabilized and part spin-stabilized. Powered by nuclear radioisotope thermoelectric generators and carrying a high-gain antenna 16 ft. in diameter the orbiter camera would carry a 1,500-mm-focal-length lens obtaining a resolution of 98–165 ft. at the larger moons. NASA planned to launch Galileo by Shuttle in January 1982 using an inertial upper stage to boost it toward Mars where gravity-assist would throw it on to Jupiter, which it would reach in July 1985. The probe would be released 100 days prior to encounter and the orbiter would relay data to Earth as it descended through the atmosphere. The orbiter would conduct a tour of the moons on a variety of elliptical orbits shaped to mission needs.

1978 January 24 At 11:53 UT the reactor section of Soviet ocean surveillance satellite *Cosmos 954* reentered the Earth's atmosphere and broke up across the Canadian Northwest Territories showering a 500 mi. wide area with radioactive debris. Launched on September 18, 1977, the satellite was on its 2,060th orbit when it reentered, its boost motor having failed to push it to a safe orbit. In a search effort known as Morning Light, the Canadian authorities mounted a major clean-up campaign. The Soviets settled a $3 million compensation claim on December 7, 1980.

1978 January 26 NASA launched the *International Ultraviolet Explorer (IUE)* on a Delta 2914 from Cape Canaveral, Fla., to an eccentric, synchronous orbit of 104 × 28,790 mi.

inclined 28.7°. A joint NASA/ESA venture with the United Kingdom's Science Research Council, the IUE was an octagonal structure with a 17.7-in. cassegrain telescope protruding from the top, instrumented to map the ultraviolet universe in a region of the spectrum inaccessible to Earth-based telescopes. It had a height of 13 ft. 9 in. and a span of 14 ft. across extended solar-cell panels. With apogee motor, the IUE weighed 1,480 lb. including 272 lb. of scientific equipment. The motor was used to place the IUE in its final orbit of 15,985 × 28,515 mi. Transmitting to an ESA station near Villafranca, Spain, and NASA's Goddard Space Flight Center, Maryland, the IUE was the first truly international astronomy observatory. It is still operating, having made more than 9,000 observations and provided more than 250 megabytes of data.

1978 January NASA announced the names of 35 new pilot (PLT) and mission specialist (MS) Shuttle astronauts in the Group 8 selection, the greatest number ever recruited at one time: Col. Guion S. Bluford Jr. (USAF); Capt. Daniel C. Brandenstein (USN); Col. James F. Buchli (USMC); Capt. Michael L. Coats (USN); Col. Richard O. Covey (USAF); Capt. John O. Creighton (USN); Col. John M. Fabian (USAF); Anna L. Fisher; Col. Dale A. Gardner (USAF); Capt. Robert L. Gibson (USN); Col. Frederick D. Gregory (USAF); S. David Griggs; Terry J. Hart; Capt. Frederick H. Hauck (USN); Steven A. Hawley; Jeffrey A. Hoffman; Shannon W. Lucid; Capt. Jon A. McBride (USN); Ronald E. McNair; Col. Richard M. Mullane (USAF); Col. Steven R. Nagel (USAF); George D. Nelson; Lt. Col. Ellison S. Onizuka (USAF); Judith A. Resnik; Sally K. Ride; Maj. Francis R. Scobee (USAF); Margaret Rhea Seddon; Col. Brewster H. Shaw Jr; Col. Loren J. Shriver (USAF); Brig. Gen. Robert L. Stewart (USA); Kathryn D. Sullivan; Norman E. Thagard; James D. A. van Hoften; David M. Walker (USN); Capt. Donald E. Williams. Griggs died June 17, 1989, in the crash of a vintage airplane. McNair, Onizuka, Resnik and Scobee were killed on January 28, 1986, when Shuttle mission 51-L exploded.

1978 February 4 A Japanese Mu-3H solid propellant launcher carried the 278-lb. science satellite EXOS-A, named *Kyokko*, from Kagoshima to a 396 × 2,471 mi. orbit inclined 65.3°. The satellite was designed to observe the density of plasma and the temperature and composition of electron energy and to obtain ultraviolet images of aurora. This was the fifth science satellite launched by the Institute of Space and Aeronautical Sciences for Tokyo University. It remains in orbit.

1978 February 9 NASA launched the first of eight U.S. Navy Fleet Satellite Communication satellites, *FltSatCom 1*, on an Atlas-Centaur from Cape Canaveral, Fla., to a transfer ellipse from which the satellite's on-board propulsion put it in geostationary orbit at 100° W. Built by TRW, the satellite had a launch weight of 4,154 lb. and an on-station weight of 2,175 lb., consisting of two stacked hexagonal modules 7 ft. 10 in. across and 3 ft. 11 in. tall, with two solar arrays and a 15 ft. 9 in. deployable UHF antenna. There were 23 channels: 10 for the Navy, 12 for the Air Force and 1 for the Department of Defense. *FltSatCom 2* was launched on May 4, 1979, *FltSatCom 3* on January 18, 1980, *FltSatCom 4* on October 31, 1980, *FltSatCom 5* on August 6, 1981, *FltSatCom 7* on December 5, 1986, *FltSatCom 6* on March 26, 1987 (destroyed in launch failure), and *FltSatCom 8* on September 25, 1989.

The French Economic and Social Affairs Committee, chaired by Prime Minister Valery Giscard d'Estaing, agreed to begin development of an advanced remote sensing satellite called SPOT. The decision was ratified by the French government in March 1978, and a marketing organization, SPOT Image, was formed in July 1982. The box-shaped satellite, 6 ft. 6 in. × 6 ft. 6 in. × 11 ft. 6 in., was equipped with a solar cell array providing 1 kW of power. Two identical high-resolution visible (HRV) scanners would provide panchromatic images with a resolution of 33 ft. and multispectral images in green, red and near infrared with a resolution of 65 ft.

1978 February 10   Fairchild Space and Electronics Co. was selected by NASA to build a multimission modular spacecraft (MMS) for the *Landsat D* Earth resources satellite. The MMS had been designed by Fairchild as a basic spacecraft bus capable of supporting a variety of mission requirements to which different payload sections would be added for a wide range of different satellites. It was hoped that costs would come down through the use of the common bus.

1978 February 16   A Japanese N-1 launcher carried the second of two ionospheric sounding satellites (ISS-b), *Ume-2* (*Plum Blossom*), from Tanegashima to a 606 × 761 mi. orbit inclined 69.4°. ISS-b took the form of a cylinder, 3 ft. 1 in. × 2 ft. 8 in. tall with a weight of 306 lb. Built and launched by NASDA, the satellite replaced the failed *Ume-1* satellite launched February 29, 1976. It continued to operate until February 23, 1983, but remains in orbit.

1978 February 17   NASA published plans for preventing the Skylab space station from reentering the Earth's atmosphere. North American Air Defense Command and several observatories predicted this would happen as early as late summer 1979. The workshop's thruster control system was to be reactivated in spring 1978, causing Skylab to tumble and so delay the effects of atmospheric drag. NASA wanted to launch the teleoperator retrieval system around October 1979 to dock with Skylab and boost it to a higher orbit so astronauts on a later Shuttle mission could revisit the workshop.

1978 February 22   An Atlas F launch vehicle carried the first of six precursor global position system (GPS) navigation satellites, *Navstar 1*, to a 12,487 × 12,619 mi. orbit inclined 63.27°. The final boost was provided by a solid propellant motor, reducing satellite weight from 1,673 lb. to 1,005 lb. Development Navstars were also known as navigation development satellites (NDSs). Built by Rockwell International, each satellite carried three extremely accurate rubidium atomic clocks accurate to one second in 36,000 years, all of which failed on the first two satellites. Plans for 24 satellites were changed in 1980 to 18 satellites. This degraded system performance to a position accuracy of 50 ft. rather than 30 ft. Further plans for a second-generation Navstar cut first-generation flights to 11, the last launched on October 9, 1985.

1978 March 2   At 15:28 UT cosmonauts Alexei Gubarev and Vladimir Remek were launched aboard the 15,000-lb. *Soyuz 28* spacecraft to a rendezvous and docking at the *Salyut 6* space station. The Czechoslovakian Remek was the first Intercosmos cosmonaut to fly under the new Soviet plan of carrying international visitors to Salyut space stations. Czechoslovakia thus became the third country to provide a pilot for spaceflight. Docking to the rear *Salyut 6* port was accomplished at 17:10 UT on March 3, and the two visitors joined Yuri V. Romanenko and Georgi Grechko on their 83rd day in space. The next day the two long-stay cosmonauts regained from the U.S. the long-duration spaceflight record set by the Skylab SL-4 crew four years earlier. Never again would the Soviet Union lose that record. After conducting several experiments, Gubarev and Remek returned to Earth in their *Soyuz 28* spacecraft at an elapsed time of 7 days 22 hr. 16 min.

1978 March 3   The Jet Propulsion Laboratory announced that it had developed a prototype roving ball for planetary exploration. Suitable for use across the surface of Mars, it had been developed from an idea put forward by Jacques Blamont of the French national space agency, Centre National d'Etudes Spatiales (CNES). Fabricated from reinforced Mylar and Kevlar, materials produced by DuPont, the explorer ball would be powered by an internal drive system and carry up to 60 lb. of instruments a maximum distance of 100 mi. It was never developed.

1978 March 5   NASA launched its third Earth resource satellite, *Landsat 3*, on a Delta 2910 from Cape Canaveral, Fla., to a sun-synchronous, 557 × 567 mi. orbit inclined 99.1°. Similar to the two previous Landsats, the 2,116-lb. *Landsat 3* carried a modified return beam vidicon camera (RBVC) upgraded to provide panchromatic images with a resolution of 130 ft. The modified multispectral scanner (MSS) had a fifth band for monitoring natural resources, providing 262-ft. resolution images of reflected solar radiation and 790-ft. resolution images of emitted infrared radiation. Also launched with *Landsat 3* was the 60-lb. amateur radio satellite, *Oscar 8*, with an uplink frequency of 145.9 MHz and downlink frequencies of 29.4 MHz and 435.1 MHz. The Delta terminal stage carried a 75-lb. plasma interaction experiment (PIX) package which remained attached. *Landsat 3* continued providing images until March 1983, and it was turned off six months later. The Delta terminal stage and both satellites remain in orbit, albeit inoperative.

1978 March 6   A ground tracking station in Bermuda made contact with the orbiting Skylab space station and began reactivating on-board systems, reporting the workshop in good condition almost five years after launch. The Apollo telescope mount solar arrays gave power when turned to the sun. By June 19 the station had been oriented in its 242-mi. orbit to minimize the effects of atmospheric drag, delaying reentry from late summer 1979 to early 1980 to give NASA time to launch a teleoperator retrieval system on the second Shuttle flight. On December 19, due to Shuttle scheduling delays and deterioration of Skylab, NASA abandoned plans to boost the workshop and conduct a revisit.

1978 March 7   The Goddard Memorial Symposium on the International Uses of the Space Shuttle and Spacelab drew more than 400 participants to Washington, D.C., where NASA officials declared that Canada, India and the European Space Agency had reserved space aboard the reusable vehicle and that Iran and West Germany had expressed an interest in doing so. The first flight of the Shuttle was currently scheduled for June 1979. The first Spacelab flight, presently scheduled for launch in December 1980, would carry investigations from 17 research groups and be followed in April 1981 by *Spacelab 2*, which accommodated experiments from 47 scientists in the U.S. and 12 from the United Kingdom.

1978 March 13   The Boeing 747 Shuttle carrier aircraft flew the orbiter *Enterprise* from Edwards AFB, Calif., to the Marshall Space Flight Center. *Enterprise* was to be used in a mated vertical ground vibration test (MVGVT) simulating the vibration and dynamic loads associated with launch and flight through the atmosphere. Tests with the orbiter disclosed structural and design changes essential for flight-rated orbiters. Up to now, the *Enterprise* had been scheduled to be returned to Palmdale, Calif., where Rockwell would have built it up for orbital flight. It was decided instead to modify the structural test article, STA-099, and use the *Enterprise* as a fit-check vehicle at the Kennedy Space Center, Fla., and Vandenberg AFB, Calif.

1978 March 16   Soviet cosmonauts Yuri V. Romanenko and Georgy M. Grechko returned to Earth at the end of a visit to the *Salyut 6* space station on a mission lasting 96 days 10 hr. 00 min., a world spaceflight endurance record. Before returning to Earth the two crew members set up the Salyut space station for three months of unmanned operation until *Soyuz 29*, launched on June 15, brought up another crew. Romanenko and Grechko returned to Earth in the *Soyuz 27* spacecraft delivered to *Salyut 6* following launch on January 10, 1978. Their mission ended with touchdown at 11:19 UT. *Salyut 6* was in an orbit of 207 × 216 mi.

1978 March 17   NASA announced the names of four pairs of astronauts scheduled for early flights in the Shuttle program: John Young and Robert Crippen, Joe Engle and Richard Truly, Fred Haise and Jack Lousma, and Vance Brand and Charles Fullerton. Young and Crippen were scheduled to fly the orbiter *Columbia* (OV-102) on the first Shuttle flight, presently scheduled for March 1979, the first of six orbital flight test (OFT) missions to qualify the Shuttle for operational duty. Only two astronauts would fly the Shuttle during OFT missions, and the orbiter would carry ejection seats for emergency escape during ascent or descent.

1978 March 23   NASA published the findings of a blue-ribbon group of 16 U.S. scientists on detecting artificial radio signals from space in its report *The Search for Extraterrestrial Intelligence*. In it, the scientists asserted that it was both timely and feasible to conduct a search for extraterrestrial intelligence (SETI) and that the U.S. should lead the international effort to do so. To make a start, the Jet Propulsion Laboratory had requested $2 million in the fiscal year 1979 budget for an all-sky radio frequency search aimed at finding artificial transmissions from intelligent life and this sum was approved.

1978 March   Cosmonaut candidates from four Intercosmos countries were selected for training at Zvezdny Gorodok: Georgy Ivanov and Aleksandr Aleksandrov from Bulgaria, Bertalan Farkas and Magyari from Hungary, Arnaldo Tamayo-Mendez and J. Lopez-Falcon from Cuba, Jugderdemidyn Gurragcha and M. Ganzorig from Mongolia, and Dumitru Prunariu and L. Dediu from Romania. One candidate from each pair would represent his country in space during a short visit to the *Salyut 6* space station. Each cosmonaut would contribute a scientific experiment from his country and the findings would be available to Soviet space scientists.

1978 April 7   NASA launched Japan's Broadcasting Satellite Experimental (BSE) satellite *Yuri* (*Lily*) on a Delta 2914 from Cape Canaveral, Fla., to a transfer ellipse from which it was placed in geostationary orbit over 110° E. Built by General Electric, BSE comprised a box-shaped, three-axis satellite

weighing 1,494 lb. at launch and 775 lb. on station, carrying a powerful transmitter beaming to ground antennas only 3 ft. 4 in. to 5 ft. 2 in. across. In a development program managed by NASDA, the Japanese national space agency, the Japanese government wanted to provide TV to remote islands and to cities where tall buildings obscured signals from ground transmitters. *Yuri* continued to operate until January 23, 1982.

1978 April 26   *Applications Explorer Mission-A*, the heat capacity mapping mission (HCCM) satellite, was launched by a four-stage Scout D from Vandenberg AFB, Calif., to a sun-synchronous orbit of 347 × 401 mi. inclined 97.6°. The *AEM-A* was the first in a series of low-cost NASA satellites and the first attempt to measure variations in the Earth's temperature. With a weight of 296 lb., *AEM-A* had a height of 5 ft. 4 in. including antenna and a scanning radiometer obtaining visible and near-infrared images with a resolution of 1,800 ft. as the satellite covered a 435 mi. wide swath. *AEM-A* decayed out of orbit on December 22, 1981.

1978 May 11   A NASA Delta 2914 launched Europe's second Orbital Test Satellite *(OTS 2)* from Cape Canaveral, Fla., to a geostationary transfer orbit of 115 × 22,396 mi. inclined 27.4°. About 37 hr. after launch its apogee motor fired, placing the satellite over 10° E. The three-axis-stabilized satellite had been built by a European consortium headed by British Aerospace and carried transponders for 6,000 telephone circuits or up to 4,500 telephone circuits and two TV channels operating in the 11 and 14 GHz bands. OTS served as precursor to the European communication satellite (ECS) series.

1978 May 20   At 13:13:00 UT an Atlas-Centaur AC-50 carrying the 1,219-lb. *Pioneer-Venus 1* orbiter spacecraft was launched from Complex 36A at Cape Canaveral, Fla. After a brief Earth parking orbit, the Centaur reignited to propel the spacecraft to escape velocity and on course for Venus. It separated from the Centaur 24 min. 41 sec. after liftoff and performed course correction maneuvers on June 1 and November 2 totaling 16 ft./sec. The drum-shaped orbiter had a diameter of 8 ft. 4 in. and a height of 4 ft. and carried 8,850 solar cells around the 77.8 sq. ft. cylindrical exterior. The solar cells could only produce power when they faced the sun; as the spin-stabilized spacecraft rotated at 5 RPM, less than 50% of the surface area would be illuminated. The cells produced 226 W at Earth and 312 W at Venus. The top of the spacecraft had a de-spun platform rotating in the opposite direction so that it appeared to be stationary. It supported a 9 ft. 10 in. tall mast carrying a 3 ft. 7 in. parabolic high-gain antenna, the top of which comprised a low-gain antenna. Data could be sent to Earth at up to 2.048 kilobits/sec. Two axial and four radial thrusters provided attitude control from a 70-lb. supply of hydrazine stored in two tanks 12.8-in. in diameter. A 4,000-lb. solid propellant retro-rocket would consume its 398-lb. of propellant to put the spacecraft in orbit around Venus.

1978 May 31   The concept of using the weightlessness of space to produce superior medicines and materials came a step nearer as NASA announced that 17 scientists had signed contracts for materials-processing experiments aboard the Shuttle and Spacelab missions. The tests would include research into the production of better semiconductor crystals, improved electrolyte materials for batteries, pharmaceutical products with a higher degree of purity than possible in Earth manufacturing and better alloys and hybrid materials.

1978 May A study conducted for the identified 45 Soviet satellites considered priority targets for a U.S. antisatellite system. Top priority threats comprised the electronic ocean surveillance satellites used for detecting and tracking naval vessels and airborne warning aircraft. Next priority was accorded the Salyut space station used for detecting missile launches and providing ICBM target data by carrying out highly accurate position measurements. Last were the defense meteorological and early warning satellites. The Joint Chiefs of Staff were advised that the top two priority levels were to be destroyed within 48 hr. of war breaking out.

1978 June 8 The U.S. House of Representatives took up a controversial bill calling for a funded start on the development of solar power satellites, massive structures assembled in orbit for converting sunlight into electricity for use on Earth. Lobbying groups included the Sunsat Energy Corp., a non-profit organization, and the Boeing Co. With each SPS weighing around 100,000 tons, the use of a heavy-lift launch vehicle would be a key to the concept's viability. A competing technology known as the Brayton-cycle heat engine involved the use of giant dishes in space to collect solar rays and focus them to a central furnace where heated gas would drive a turbine and generate electricity. This was more efficient and less costly than using solar arrays, but more difficult to work.

1978 June 8–16 The first of three separate talks between the United States and the Soviet Union about banning antisatellite weapons was held in Helsinki. The U.S. side had not been briefed by the Carter administration before they left Washington, and received direction solely via a single sheet of paper faxed to the U.S. Embassy in Finland. For their part, the Soviets denied they had an antisatellite system and then asked for time to confer prior to the start of the second round, held in Bern, Switzerland, January 16–12, 1979. Little progress was made at this or the final round in Vienna, April 23–June 17, 1979. Soviet antisatellite tests resumed with the launch of *Cosmos 1,171* on April 3, 1980, shortly after the Soviet invasion of Afghanistan.

1978 June 10 A U.S. Defense Support Program (Program 647) early warning satellite was launched by Titan IIIC from Cape Canaveral, Fla., and placed in a near-geosynchronous orbit of 18,917 × 26,123 mi. inclined 12°. In this orbit, the satellite would execute a sweeping figure 8 over a given point on the equator in much the same manner as the experimental Program 949 satellites launched by Atlas-Agena D. Six Program 647 satellites had previously been launched by Titan IIICs into geostationary orbits, and of the five that remained to be launched (the last on March 6, 1982) only one other was placed in this unusual orbit.

1978 June 15 At 20:17 UT Soviet cosmonauts Vladimir V. Kovalenok and Alexandr S. Ivanchenkov were launched by SL-4 from the Baikonur cosmodrome to the *Salyut 6* space station in their 15,000-lb. *Soyuz 29* spacecraft. The space station was in an orbit of 211 × 220.5 mi. Docking took place at the forward transfer port at 21:58 UT on June 16, and shortly thereafter the crew moved into their new home in space, where they would remain for more than four months. The crew quickly moved ahead with scientific experiments and prepared for the first visit of an international crew aboard *Soyuz 30*, launched June 27, 1978.

1978 June 16 NASA launched the 1,385-lb. *GOES 3* geostationary weather satellite on a Delta 2914 launch vehicle, marking the 500th flight of the Thor stage. Employed as the first stage of Delta, the Thor had been developed for the U.S. Air Force as an intermediate-range ballistic missile and then adapted as the first stage of a multistage satellite launcher. This was the 142nd Delta launch by NASA, the remaining 358 having been flown by the Air Force.

1978 June 23 In the first known instance of a satellite exploding in geosynchronous orbit, a Soviet Ekran television satellite was destroyed when a nickel-hydrogen battery burst. Forewarned to anticipate the event, Soviet ground cameras observed and filmed the explosion, which clearly revealed a shower of debris at the moment of detonation. The Russians believe two other satellites launched by them had also been destroyed in low Earth orbit for a similar reason.

1978 June 27 The 5,075-lb. satellite *Seasat-A* was launched by Atlas-Agena D from Vandenberg AFB, Calif., to a 482 × 496 mi. orbit inclined 108°. Designed to provide global monitoring of oceanographic phenomena and features, *Seasat-A* was built on the Agena D stage with an attached sensor module providing a total length of about 69 ft. and a diameter of almost 5 ft. Sensors would provide wave height data, surface temperature, wind speeds and direction, and a synthetic aperture radar would produce images of 15.5 mi. resolution over a swath 60 mi. wide. *Seasat-A* operated for 106 days before contact was lost on October 9, 1978, when a short circuit drained the battery. It remains in orbit.

At 15:27 UT the international crew of Soviet cosmonaut Pyotr I. Klimuk and the Polish cosmonaut Miroslav Hermaszewski was launched aboard the 15,000-lb. *Soyuz 30* spacecraft by an SL-4 from Baikonur cosmodrome. They docked with *Salyut 6*, in an orbit of 207.5 × 213 mi., at 17:08 UT the next day and joined Vladimir Kovalenok and Alexandr Ivanchenkov in the space station. For six days the four cosmonauts conducted experiments, and Hermaszewski operated the Splav-01 furnace to test an experiment he had brought with him from Poland. *Soyuz 30* undocked from the aft port on *Salyut 6* at 10:15 UT on July 5, and returned Klimuk and Hermaszewski to the Soviet Union at an elapsed time of 7 days 22 hr. 03 min.

1978 July 7 The first of three unmanned Progress cargo tankers was launched at 11:26 UT to the *Salyut 6* space station by an SL-4 from the Baikonur cosmodrome. In an orbit of 205.5 × 210 mi., *Progress 2* docked with the aft port on *Salyut 6* at 12:59 UT on July 9. Cosmonauts Vladimir Kovalenok and Alexandr Ivanchenkov unpacked stores from *Progress 2* and on July 19 completed the transfer of propellant to the equipment module at the rear of the station. *Progress 2* brought the Kristall materials processing furnace, an improvement on the Splav-01 fitted to *Salyut 6* before it was launched. At 04:00 UT on July 29 the two cosmonauts began a 2 hr. 5 min. space walk to retrieve experiments left on the exterior of *Salyut 6*. After they returned to the transfer module, gas from *Progress 2* was used to repressurize that section of the station, conserving on-board supplies. *Progress 2* was undocked at 04:57 UT on August 2, 1978, and burned up in the atmosphere two days later.

1978 July 17 Projected demand for world telecommunications satellite traffic was announced by Intelsat at its fourth global meeting, at which 172 experts representing 93 telecommunications entities met to discuss future requirements.

Traffic was expected to double in five years, expanding from 14,105 telephone circuits at the end of 1978 to a projected 29,282 by the end of 1982. Atlantic Ocean traffic was expected to rise by 114%, Indian Ocean traffic by 98% and Pacific Ocean traffic by 91%.

1978 July 19 NASA signed its first reimbursable launch agreement with a foreign government for a satellite to be placed in orbit by the Shuttle. Administrator Robert Frosch signed the agreement with Prof. S. Dhawan, secretary of the department of space and chairman of the Indian Space Commission, for the launch of *INSAT-1*. The geostationary satellite would provide India with domestic public telecommunications, direct TV broadcasting and meteorological services.

1978 July 25 Following several leaks that had drained attitude control propellant in its path around Mars, the *Viking 2* orbiter ceased operating at 06:01 UT after 706 orbits of the planet in almost three years of continuous operation. Both orbiters had been designed to operate for a minimum 150 days in Mars orbit. Because of a technical problem with its transmitters, the *Viking 2* lander could not communicate direct with Earth so data was sent via the *Viking 1* orbiter. The *Viking 1* lander could still talk to Earth direct. The *Viking 2* lander was turned off on April 12, 1980. The *Viking 1* orbiter died on August 7, 1980, after bringing to 51,539 the total number of images returned from both orbiters. The *Viking 1* lander returned the last picture to Earth on November 13, 1982. Later that month an error in a computer instruction from Earth caused the lander's high-gain antenna to slew away from Earth. The lander was effectively useless after 6 years 4 months of Mars surface operations.

1978 July 29 A Soviet SS-18 ICBM launched from Baikonur on test was monitored by the U.S. *Rhyolite* geostationary electronic intelligence satellite positioned above the Horn of Africa. Analysts discovered that telemetry from the missile was embedded in coded transmissions to block intelligence gathering. This move had come about as a result of the documents handed to the KGB by U.S. citizens Christopher Boyce and Andrew Lee between 1975 and 1977 informing them about *Rhyolite*. Boyce and Lee also handed over details of a satellite project called Argus which was to have replaced foreign electronic listening posts in places like Iran and Turkey.

1978 August 4 Lockheed Missiles and Space Co. received a contract from NASA's Marshall Space Flight Center to develop and build a flight experiment for the Shuttle using a large solar array wing. Devised as a means of significantly increasing the amount of electrical power available, the 13 ft. 6 in. wide deployable array would extend to a maximum height of 105 ft. above the payload bay and produce 12.5 kW of electrical power. It could be used to boost power for Shuttle and Spacelab experiments and be adapted as a power array for solar-electric propulsion stages.

1978 August 7 At 22:31 UT the *Progress 3* cargo tanker was launched from Baikonur cosmodrome by an SL-4 launcher and docked to the aft port on the *Salyut 6* space station at 49 hr. 29 min. Shortly thereafter the *Salyut 6* occupants, Vladimir Kovalenok and Alexandr Ivanchenkov began moving stores into the space station. For the first time, no replenishment propellant was carried, adequate supplies having been delivered by *Progress 2*, but the orbital module was packed with provisions and recreational items, including a guitar for

Ivanchenkov! When the crew had finished unpacking all the stores from *Progress 3*, it was sealed up and undocked at 19:29 UT on August 21. After a de-orbit burn at 17:30 UT on August 23, *Progress 3* burned up in the atmosphere over the Pacific Ocean.

1978 August 8 At 07:33:00 UT an Atlas-Centaur AC-51 launched the 1,930-lb. *Pioneer-Venus 2* multiprobe spacecraft from Complex 36A at Cape Canaveral, Fla. The spacecraft was placed first in a brief Earth parking orbit and then sent on course for Venus, separating from the Centaur 31 min. 55 sec. after liftoff. The multiprobe spacecraft bus was identical to the orbiter launched May 20, 1978, but without the large antenna mast and solid propellant retro-rocket. Instead, one 695-lb. probe and three 200-lb. small probes, each shaped like a flat cone, were carried on top of the drum-shaped spacecraft. The large probe 5 ft. in diameter carried 62 lb. of science equipment to measure the composition of the clouds, atmospheric structure, pressures and temperatures, and it would deploy a parachute to slow descent, sending data to Earth via a 40-W transmitter. The small probes, 2 ft. 6 in. in diameter, carried 7.7 lb. of instruments and 10-W transmitters. All probes would send data at 64 bits/sec. to an atmospheric altitude of 19 mi. and 16 bits/sec. thereafter.

1978 August 12 The third of three International Sun-Earth Explorer satellites, *ISEE-C* was launched by Delta 2914 from Cape Canaveral, Fla., to a location known as a libration point (L1). A libration point is one in which gravitational forces balance to "hold" an object at a specific point in space, like a mass suspended between wires of different length at equal tension. *ISEE-C* was placed at L1, a point about 932,000 mi. from Earth in a direct line between the Earth and the sun but at an inclination to the ecliptic. Viewed from Earth, *ISEE-C* appeared to rotate around the sun but was in fact describing a halo around that fixed point on the Earth-sun line. *ISEE-C* would detect solar phenomena one hour before they arrived at *ISEE-A* and *-B*, which were in highly elliptical Earth orbit. Designed and built at the Goddard Space Flight Center, *ISEE-C* was a 16-sided polyhedron 5 ft. 8 in. across and 5 ft. 3 in. tall, weighing 1,034 lb. The experiments had been developed by 117 investigators from 35 universities in 10 countries.

1978 August 15 The *Pioneer-Venus 2* multiprobe spacecraft en route to Venus performed the first of three course corrections to precisely align its trajectory for the release of four atmospheric entry probes. Other maneuvers were conducted on October 20 and November 9 for a total velocity adjustment of 20 ft./sec. On December 14 the probes were checked out by radio command from Earth, and two days later the spacecraft bus was oriented and spun up to 48 RPM. At 02:37:13 UT on December 16 the large probe separated, followed by the three small probes within milliseconds of each other at 13:06:29 UT on December 20. They still had 8 million mi. to travel. All three would enter the atmosphere of Venus at about 26,000 MPH beginning 18:45 UT on December 9, 1978. A few hours prior to that, at 11:37:12 UT, the multiprobe bus adjusted its trajectory for a precise entry into the atmosphere at the desired location about 42 min. after the last probe.

1978 August 19 A workshop including scientists from European countries and the United States decided that a study should be performed on a possible comet rendezvous mission undertaken as a joint venture between NASA and ESA (European Space Agency). The NASA concept envisaged launching a 7,400-lb. spacecraft towards Halley's comet in December 1985 before it passed through perihelion. The spacecraft was to be propelled by the solar-electric propulsion system (SEPS) and carry along with it a European probe dropped off for a very close skim past the comet nucleus. The main spacecraft would go on to perform a later rendezvous with the comet Temple 2. When NASA abandoned the SEPS in January 1980, it discontinued the comet rendezvous plan, leaving the European Space Agency to proceed without U.S. involvement on a project called Giotto.

1978 August 23   The first major symposium on crop monitoring from space began at NASA's Johnson Space Center. The three-day conference on the large-area crop inventory experiment (LACIE) reported on progress with this major international effort to estimate wheat production around the globe. In a three-year effort that began in the fall of 1974, Landsat satellites had conducted continuous scans of agricultural areas, and this information had been combined with data from 8,000 weather stations around the world. Scientists had now developed techniques for recognizing wheat in satellite images and estimating the growth state and projected crop yield.

1978 August 26   At 14:51 UT an SL-4 carrying the *Soyuz 31* spacecraft and an international crew lifted off from the Baikonur cosmodrome on a rendezvous and docking mission to *Salyut 6*. Crew members Valery Bykovsky from the Soviet Union and Sigmund Jähn from East Germany docked *Soyuz 31* to the rear port on *Salyut 6* at 16:38 UT on August 27. They joined Vladimir Kovalenok and Aleksandr Ivanchenkov in the space station, where joint experiments were carried out. During his stay on *Salyut 6* Jähn helped operate the MKF-6M multispectral camera manufactured in East Germany by the Carl Zeiss company. The crews exchanged seats between the two Soyuz spacecraft on September 3 and Bykovsky and Jähn got into *Soyuz 29* for the return to Earth, leaving their *Soyuz 31* spacecraft at the aft docking port. At 08:23 UT the visiting crew undocked from the forward port on *Salyut 6* and landed at an elapsed time of 7 days 20 hr. 49 min.

1978 September 7   *Salyut 6* cosmonauts Vladimir Kovalenok and Aleksandr Ivanchenkov performed for the first time a maneuver that was to become standard for space station operations when they moved the *Soyuz 31* spacecraft from the rear docking port to the forward docking port. Left at the station by cosmonauts Valery Bykovsky and Sigmund Jähn, *Soyuz 31* occupied the rear docking port reserved for Progress tankers and had to be moved. Limited propellant reserves in the Soyuz spacecraft prevented the crew from maneuvering the manned vehicle from the back of the station to the front for redocking. Instead, they undocked from Salyut at 10:53 UT, backed away, and waited while ground controllers commanded the space station to rotate through 180°. With the forward docking port now facing them, the cosmonauts moved *Soyuz 31* back in and redocked with *Salyut 6* at 12:03 UT.

1978 September 9   The Soviet Union launched its 8,700-lb. *Venera 11* spacecraft to the planet Venus and followed it five days later with its companion, *Venera 12*, of approximately the same weight. The 1978 Veneras were lighter than their predecessors *(Venera 9 and 10)* of 1975 because the Venus launch window for this year required greater energy from the SL-12 launcher. Each spacecraft was a combination fly-by bus and lander and no component of either would attempt to enter

an orbit of Venus. There was an advantage in this because the fly-by carrier would have more time to relay communications from the landers than they would if they were going into orbit. Extensive interplanetary measurements were made during the flight to Venus.

1978 September 11   NASA selected five scientific instruments for an Earth orbiting gamma-ray observatory (GRO) tentatively scheduled for launch in 1984. Designed to study the most energetic forms of radiation in the known universe, the GRO was being managed by the Goddard Space Flight Center and all but one of the experiments were contributed by U.S. scientists; the Max Planck Institute of West Germany provided the fifth. The GRO was to be launched by Shuttle and would operate for two years from an altitude of 250 mi. NASA was unable to incorporate the GRO into its program of funded projects until fiscal year 1981.

The U.S. Navy Systems Command announced that it had awarded a contract to Hughes Communications Services for the design and development of four satellites which were to be leased for military communications. Winning over competing bids from Comsat General and TRW Systems, Hughes had offered its Syncom 4 design purpose-built for the cargo bay of the Shuttle. To be named Leasecraft, the satellites would provide communication for Navy ships as well as Army, Marine and Air Force ground-mobile forces.

1978 September 16   A satellite built and launched by Japan's Institute of Space and Aeronautical Sciences was carried by Mu-3H from Kagoshima to a 126 × 18,281 mi. orbit inclined 31.2°. EXOS-B, named *Jikiken*, was designed to measure electron density, plasma waves and other near-Earth space phenomena in a continuing program aimed at understanding the processes that could affect the operation of applications satellites. The satellite weighed 220 lb. and remains in orbit.

1978 October 3   The unmanned cargo tanker *Progress 4* was launched by SL-4 from the Baikonur cosmodrome at 23:09 UT. It docked with the rear docking port on *Salyut 6* at 01:00 UT on October 6, and cosmonauts Vladimir Kovalenok and Alexandr Ivanchenkov on board *Salyut 6* since June 1978, transferred supplies into the space station before replenishing on-board propellant tanks. *Progress 4* performed two maneuvers of the docked assembly on October 19–20, raising the orbit from 201 × 211 mi. to 223 × 225 mi. *Progress 4* undocked from *Salyut 6* at 13:07 UT on October 24 and the de-orbit burn was performed at 16:28 UT on October 26, over the Pacific Ocean.

1978 October 13   The first of the new series, *Tiros-N*, a third-generation, civilian U.S. weather satellite, was launched from Vandenberg AFB, Calif., by an Atlas F to a 528 × 537.5 mi. polar orbit inclined 98.9°. Built by RCA, *Tiros-N* had a launch weight of 1,594 lb. and in the launch configuration comprised a rectangular box 12 ft. 2 in. long and 6 ft. 2 in. across the corners. In orbit, the satellite weighs just over 1,000 lb. and deploys a flat-panel solar array. Designed to supersede the ITOS series, *Tiros-N* carried a high-resolution radiometer for obtaining data on day and night sea-surface temperatures, ice, snow and cloud conditions. It was the first satellite capable of collecting data from platforms on land, at sea or in the air on an operational basis. *Tiros-N* was subjected to rigorous engineering tests prior to the launch of the first operational satellite in this series, *NOAA-6*, on June 27, 1979.

The Nimbus 6 *weather satellite was used to track the lone journey of a Japanese explorer beginning March 4, 1978, when he left by dogsled from northern Canada on a six-month trek to the North Pole.*

1978 October 24 A Delta 2910 carried the last Nimbus weather research satellite from Vandenberg AFB, Calif., to a 587 × 595 mi., 99.3° orbit. The 2,175-lb. *Nimbus 7* carried the first dedicated ozone measuring instrument, the solar backscatter ultraviolet/total ozone mapping spectrometer (SBUV/TOMS). Continuing the work begun by *Nimbus 5,* launched almost six years before, *Nimbus 7* measured solar radiation, UV radiation and UV radiation reflected back from the surface of the Earth, as well as mapping total ozone levels. Seven other instruments were carried to measure Earth's radiation budget, humidity levels, aerosol concentrations, pollutants, thermal emissions from the land and sea and chlorophyll concentrations in coastal waters.

1978 November 2 The record-breaking *Salyut 6* cosmonauts Vladimir Kovalenok and Aleksandr Ivanchenkov returned to Earth at the end of a mission that had lasted 139 days 14 hr. 48 min., more than 43 days longer than the previous record. The landing took place at 11:05 UT and immediate medical examinations revealed that the crew had suffered no greater physical deterioration than an earlier crew that had been in space only 96 days. Kovalenok and Ivanchenkov had performed a strenuous physical conditioning program in orbit and this paid off when the time came to adapt to life back on Earth. Within two days they were taking walks and physicians cleared mission planners to prepare for even longer flights aboard *Salyut 6.*

1978 November 13 *HEAO-2,* the second of three NASA high-energy astronomy observatories, was launched from Cape Canaveral, Fla., by an Atlas-Centaur to a 323 × 336

mi., 23.51° orbit. The 6,500-lb. satellite comprised a standard octagonal equipment module and an experiment module, presenting a total length of 22 ft. and a width of 7 ft. 8 in. *HEAO-2* carried one x-ray telescope, the largest ever built, with a mirror with a 14.8-in. diameter and a focal length of 10 ft. 10 in. When spectacular results from *HEAO-2* were derived from information of unprecedented quality, scientists renamed the satellite *Einstein.* It continued to operate until the attitude control gas ran out on April 25, 1981, and it decayed into the atmosphere on March 25, 1982.

1978 November 17 NASA announced it was asking scientists to propose experiments for a planetary mission to provide the first global view of the surface of Venus. Launched by a Shuttle in December 1984, the Venus-orbiting imaging radar (VOIR) spacecraft would go into polar orbit around Venus in May 1985. Carrying a large synthetic aperture radar of the type previously flown on Seasat, the VOIR was to perform mapping activity that would provide global coverage of surface features to a resolution of about 3,000 ft. Radar is the only means by which surface features on Venus–shrouded by a dense, opaque atmosphere–can be seen from space. Although not officially approved, the VOIR project aroused intense interest among scientists because Venus is nearly the same size as Earth but has a completely different atmosphere.

1978 November 28 U.S. and Chinese officials began talks about China's buying a U.S. communication satellite in what would constitute a transfer of the most advanced technology yet sent to a communist country. A delegation from the Space Technology Research Institute of the People's Republic of China had met with representatives from NASA and the U.S. Departments of State and Commerce to discuss the deal. China said it wanted the satellite for broadcasting educational programs, and the United States said it would prohibit its use for military communications.

1978 December 4 The *Pioneer-Venus 1* orbiter successfully fired its solid propellant rocket motor, placing it in orbit around the planet nearest Earth at the start of a mission that would last almost 14 years. As viewed from Earth, the spacecraft went behind Venus at 15:51 UT and at 15:58:05 UT fired its retro-rocket for 30 sec., reducing speed by 3,400 ft./sec. When the spacecraft reappeared at 16:14 UT, ground trackers confirmed it to be in a highly elliptical orbit of 235 mi. by almost 40,000 mi. with a period of 23 hr. 11 min. 26 sec. and an inclination of 105°. Beginning December 6, a series of 7 maneuvers over 16 orbits adjusted this to 93–125 × 41,600 mi., bringing the orbital period to 24.13 hr. and thereby synchronizing it with tracking stations on Earth. Soviet scientists commanded *Veneras 11* and *12* to observe the planet when Pioneer-Venus entered orbit.

1978 December 5 The U.S. Defense Systems Acquisition Review Council met to discuss optional basing modes for the M-X ICBM, then being funded at modest levels by the Carter administration. The Air Force wanted a multiple protective shelter (MPS) system. Each missile would move around among a group of shelters, forcing the Soviets to target all shelters in the group to ensure they destroyed the missile. This was rejected by the Carter administration because it would make verification, mandatory in arms limitation talks, virtually impossible since an all but unlimited number of missiles could be hidden by shelters. The Air Force was ordered to study the next preferred mode in which missiles were dropped from aircraft.

1978 December 6 Tracking stations on Earth received the first images of cloud-shrouded Venus from the *Pioneer-Venus 1* orbiter. The instrument that took the pictures was a photopolarimeter consisting of 3.7-cm telescope with 16 filters and a prism that directed the image to silicon photodiodes enhanced to detect ultraviolet light. What they "saw" was transmitted to Earth as a picture. The orbiter also carried a 21-lb. surface radar mapper to penetrate the opaque atmosphere and produce radar "pictures" with a resolution of 12 mi. along the orbit track and 10 mi. across it. Surface elevation resolution of 300 ft. was calculated by scientists on Earth. The orbiter also carried an infrared radiometer and an ultraviolet spectrometer to measure upper atmosphere composition and temperature as well as several instruments to observe the environment of the planet and its interaction with the solar wind. It would continue to do so until it entered the atmosphere of Venus in 1992.

1978 December 8 The Intelsat board announced it would use both the NASA Shuttle and the European Ariane expendable launch vehicle for the fifth, sixth and seventh satellites in the Intelsat V series. NASA was trying to sell the Shuttle on the basis that it would be "the" U.S. satellite launch system, replacing all expendable launchers except the solid propellant Scout for small satellites. Europe was about to introduce Ariane as the major competitor to Shuttle.

1978 December 9 At 18:45:32 UT the large *Pioneer-Venus 2* probe encountered the atmosphere of Venus at an altitude of 120 mi. and a speed of 26,000 MPH and in just 38 sec. decelerated to 450 MPH, at which point it deployed its parachute and jettisoned the flat cone-shaped aeroshell protecting the 2 ft. 5 in. diameter spherical pressure vessel. At entry +17 min. 58 sec., at a height of 28 mi., the parachute was jettisoned and the probe fell through the heavy and dense atmosphere to reach the surface at 19:39:53 UT. The three smaller probes, called *North*, *Day* and *Night* for their entry points, encountered the atmosphere at 18:49:40 UT, 18:52:18 UT and 18:56:13 UT. They reached the surface after descent times of exactly 53 min., 55 min. 41 sec. and 55 min. 52 sec., respectively. *North* and the larger probe stopped functioning at the instant of impact, but *Night* operated for 2 sec. and *Day* for 1 hr. 7 min. 37 sec. The *Pioneer-Venus 2* bus encountered the atmosphere at 20:21:52 UT and burned up 1 min. 3 sec. later. The Pioneer-Venus orbiter, meanwhile, had reached its periapsis altitude of 111.6 mi. at 14:48:15 UT and was far out toward an apoapsis of more than 41,000 mi., thereby getting a good view of the planet as the probes entered.

1978 December 14 The World Meteorological Organization, along with the International Council of Ministers, announced that beginning in January 1979, 147 nations would take part in the Global Weather Experiment (GWE). Lasting 12 months, the GWE would seek to define the boundaries of weather forecasting and design a global observation system to reach them. It would be part of the Global Atmospheric Research Project (GARP), a joint effort of the two organizations and directly integrated with the World Weather Watch, which was providing 40,000 observations daily from 10 satellites, 50 research ships, 110 aircraft, 300 high-altitude balloons and 300 instrumented buoys.

1978 December 15 Intelsat announced cuts in its monthly charges from $1,140 to $960 beginning January 1979. Costs had already come down from $5,334 when the service began in 1965, allowing national posts and telecommunications agencies to lower their charges to subscribers and so encourage wider use of telephone and telephonic services. Projections showed that costs could be cut further, to less than $700 by 1983.

1978 December 16 A Delta 3914 carried the 1,956-lb. Canadian satellite *Anik B1 (Telesat-D)* from Cape Canaveral, Fla., to a geosynchronous transfer ellipse from which an apogee motor placed it in a geostationary position at 109° W. Built by RCA Astro-Electronics in New Jersey, the satellite was a replacement for the less powerful *Anik A1*. It comprised a box structure with a width of 7 ft. 1 in. and a length of 3 ft. 8 in. The two paddle-shaped, solar array panels deployed on booms had a span of 10 ft. 8 in. *Anik B1* could carry 12 commercial channels in the 14/12 GHz band.

1978 December 19 An SL-12 launch vehicle carried the 8,800-lb. Soviet communications satellite *Gorizont 1* to an elliptical orbit of 14,485 × 30,419 mi. with an inclination of 10.6° and a period of 24 hr. 14 min. Launched on July 5, 1979, *Gorizont 2 (Statsionar 4)* was placed in a geostationary orbit typical of the series. Satellites of this type relayed telephone, television and telegraph messages from central stations to remote antennas on the ground. Whereas Raduga handled government traffic, Gorizont carried "civil" and TV traffic with a special 40-W transmitter dedicated to the Moskva system comprising more than 1,000 dish antennas 8 ft. in diameter. By November 27, 1992, a total 27 Gorizont satellites had been placed in orbit. Each satellite was raised to a higher orbit at the end of its operational in order to vacate the longitudinal slot for a successor.

1978 December 21 The Soviet spacecraft *Venera 12* arrived at Venus and put the Lander down through the atmosphere to land on the surface at 03:30 UT. For 1 hr. 50 min. the lander continued to send data to Earth via the main spacecraft bus, which passed the planet at a distance of 22,000 mi. Relayed transmissions ceased because the bus went over the lander's horizon. The lander failed to send any pictures to Earth because of a failure in the camera system, but it did report environmental conditions. A temperature of 860°F was recorded and frequent and violent thunderstorms were detected. For the first time, NASA scientists worked with their Soviet counterparts to make simultaneous observations of the planet. The Pioneer-Venus orbiter performed mapping observations of the surface at the time the lander descended.

1978 December 25 The Soviet spacecraft *Venera 11* landed on the surface of Venus at 03:24 UT, 500 mi. from *Venera 12*. Like its companion, *Venera 11* failed to send any pictures from the surface due to technical problems but continued to relay data for 1 hr. 35 min. The lander reported temperatures only a little lower than those recorded by *Venera 12* and the same atmospheric density of 88 Earth atmospheres. NASA integrated its Pioneer-Venus orbiter science program with the descent of the *Venera 11* lander to provide simultaneous observation from above and within the atmosphere. This information, and that obtained with *Venera 12* on December 21, was exchanged between U.S. and Soviet planetary scientists.

1978 December The last Orbiting Solar Observatory, *OSO-8*, went silent more than three years after its launch on June 21, 1975, and two years beyond its design life. Since the first OSO had been launched on March 7, 1962, the program had contributed a wealth of important information about the sun, preparing the way for more advanced observatories like the Solar Maximum Mission launched February 14, 1980.

1978 China began to deploy its first ICBM when the CSS-3 was fielded in special silos. With a 87 ft. 10 in., a diameter of 8 ft. and a throw weight of 4,400 lb., the CSS-3 could send a 3-MT reentry vehicle a distance of almost 4,400 mi. About 30 CSS-3s were built, but only six silos were constructed. Engineers developed the missile into the CZ-1 satellite launcher.

1979 January 4 NASAs *Voyager 1* spacecraft began routine observations of the planet Jupiter, still 38 million mi. away, as it moved toward a close encounter on March 5. Voyager's science instruments had been selected to gather data about Jupiter's atmosphere and its dynamic environment, about 5 of the 13 (then) known moons and about the strength and intensity of the magnetic field. Apart from two TV cameras, each Voyager carried an infrared interferometer spectrometer and radiometer (IRIS) , and an ultraviolet spectrometer (UVS) for determining atmospheric composition of Jupiter and, possibly, its moons, as well as a photopolarimeter for studying the structure of the planet's atmosphere and a suspected sodium cloud around one of its moons, Io. Magnetometers, cosmic ray and charged particle detectors and a plasma science instrument mapped the interplanetary and Jovian environment. In addition, two 33-ft. whip antennas at 90° angles to each other "listened" for radio waves from Jupiter.

1979 January 22 For the first time since 1751, the planet Pluto moved inside the orbit of Neptune and could no longer be referred to as the outermost planet. In its orbit of the sun, which takes 248 years, the planet is outside the orbit of

Voyager 1 *took this dramatic image of Jupiter on February 1, 1979. Although shot from a distance of 20 million miles, it reveals subtle details in the atmospheric bands of this giant planet.*

Neptune for all but 20 years. It would not regain that position until March 1999.

1979 January 25 NASA announced names for the four reusable orbiters built or being assembled for the Shuttle fleet. *Orbiter 102*, scheduled to make the first manned Shuttle flight into space, was to be called *Columbia* after the ship which explored the mouth of the Columbia River in 1792 en route to becoming the first U.S.-flag vessel to circumnavigate the globe and the Apollo command module from the first manned moon landing. *Orbiter 099* would be named *Challenger* after the 19th century oceanic exploration vessel and the *Apollo 17* lunar module. *Orbiter 103* would be named *Discovery* after one of two ships in the 1770s used by Capt. Cook to discover the Hawaiian Islands. *Orbiter 104* would be named *Atlantis* after the first U.S.-operated vessel to be designed for ocean research.

1979 January 30 A joint Department of Defense/NASA program to investigate the electrical charging effects on spacecraft in high orbit, the *SCATHA* (spacecraft charging at high altitudes) satellite was launched by Delta 2914 from Cape Canaveral, Fla., to a transfer orbit from which an apogee motor placed it in a 17,527 × 26,457 mi. orbit inclined 4.9°. Weight at launch was 1,444 lb., reduced to 787 lb. at apogee motor burnout. *SCATHA* carried twelve experiments to characterize electrical and magnetic fields and the charged-particle environment of the satellite. It comprised a cylindrical structure 6 ft. in diameter and 6 ft. high from which a number of booms were deployed. Solar cells provided 300 W of electrical power.

1979 February 5 NASA signed a contract with Rockwell International formally authorizing the manufacture of two Shuttle orbiters in addition to OV-101 and OV-102. OV-101 would never fly in space. Rockwell was to deliver OV-103 and OV-104, modify OV-102 for operational flight after the first six orbital flight test (OFT) missions and convert OV-099 into a flight-rated orbiter, completing the four-vehicle fleet for space duty. NASA had wanted five space-rated orbiters in addition to the structural test article (STA-099) but had been denied funds to build more than three flight-rated vehicles and convert one to flight condition.

1979 February 6 An N-I launch vehicle from the Tanegashima Space Center placed the 287-lb. experimental communications satellite (ECS) *Ayame* (*Sweet Flower*) into a geosynchronous transfer orbit from which an apogee motor placed it in geostationary orbit. The satellite collided with the second stage just 12 sec. after it was released and all contact was lost. ECS fired its apogee motor by timer and achieved a geostationary orbit of 21,110 × 22,010 mi. The satellite was built for NASDA, Japan's national space agency, and was intended to carry out research in millimeter wave transmissions.

1979 February 8 The first of two Soviet antisatellite target satellites flown during 1979 for radar calibration tests, the 2,380-lb. *Cosmos 1075* was launched by an SL-8 from Plesetsk to a 244 × 267 mi. orbit inclined 65.8°. It decayed back through the atmosphere on October 18 and was followed by the 2,200-lb. *Cosmos 1146* on December 5, also launched by an SL-8 from Plesetsk to a 253 × 266 mi. orbit inclined 65.9°. *Cosmos 1146* decayed November 25, 1981.

1979 February 9   In a move to sound out relevant authorities on the prospects for privatizing the national remote sensing program, NASA conformed to a directive from President James E. Carter by announcing it would undertake a study on how far private industry could participate in Landsat. Since it began providing digital images, demand for Landsat data had been growing and the Carter administration was keen to see if the entire program could be privatized.

1979 February 12   An SL-14 launch vehicle carried the first in a series of Soviet Okean ocean research and applications satellites from Plesetsk to a 387 × 407 mi. orbit inclined 82.5°. The 9,920-lb. *Cosmos 1076* was used to conduct tests with a nonscanning high-frequency radiometer (designated MSU-M) for studying the water surface, obtaining images of cold and warm currents, compiling water temperature charts and mapping rough sea zones. Other satellites of this type were: the 13,935-lb. *Cosmos 1151* (January 23, 1980), *Cosmos 1500* (September 28, 1983), the 4,400-lb. *Cosmos 1602* (September 28, 1984), the 3,530-lb. *Cosmos 1766* (July 28, 1986), *Cosmos 1812* (January 14, 1987), *Cosmos 1825* (March 3, 1987), *Okean 1* (July 5, 1988), and the 4,200-lb. *Okean 2* (February 28, 1990) and *Okean 3* (June 4, 1991) .

1979 February 16   Boeing completed a study for NASA's Johnson Space Center on solar power satellite systems definition. The reference SPS design was a flat array 13 mi. × 3.3 mi. in size producing 10,000 MW of useful power from geostationary orbit. With a projected mass of 100,000 tons, each SPS would need a heavy-lift launch vehicle for carrying the structure into space economically. With each HLLV carrying only 500,000 lb., however, there would have to be 400 HLLV flights to build each SPS and 4,000 such flights to provide the United States with electricity from solar energy from space. Nevertheless, the revenue from each $9 billion SPS would make it cost-effective against coal and nuclear power. Policymakers were reminded that in 1979 the United States used 35% of the world's total power production and that if it could be cost-effective to build the 40 SPS units necessary for the United States, the other 80 or so needed to power the world would be even more cost-effective and give the United States a powerful monopoly in energy.

1979 February 18   The second applications explorer mission *(AEM-2)* was launched by a Scout from Wallops Islands, Va., to a 341 × 411 mi. orbit inclined 55°. Known as the stratospheric aerosal and gas experiment (SAGE), the 325-lb. *AEM-2* was a hexagonal structure 2 ft. 4 in. across and 2 ft. 1 in. high with two 24 sq. ft. solar arrays. Built by Boeing to measure solar radiation falling on the Earth's outer atmosphere, SAGE would help define concentrations of aerosols and ozone. By measuring the effect of these filters on sunlight in the atmosphere, the instruments could determine the quantity of dust and liquid droplets. *AEM-2* decayed on April 11, 1989.

1979 February 21   A Japanese Mu-3C solid propellant launch vehicle carried the scientific satellite Corsa-B, named *Hakucho* (Swan), from Kagoshima to a 336 × 355 mi. orbit inclined 29.9°. Weighing 212 lb., the satellite was equipped with instruments to observe x-ray stars, x-ray nebulae and other x-ray phenomena in the universe. Corsa-B decayed back into the atmosphere on April 15, 1985.

1979 February 24   The 54th mission in the U.S. Air Force Space Test Program, which began in 1967, took place when an

Atlas F-Burner II carried satellite P78-1 from Vandenberg AFB, Calif., to a 350 × 374 mi., 97.65° orbit. Named *Solwind*, the satellite had been built by Ball Brothers and borrowed much from the NASA orbiting solar observatory series. It had a cylindrical spinning section 6 ft. in diameter with a de-spun vertical portion supporting a solar array, the whole 9 ft. tall. P78-1 weighed 2,935 lb. and carried seven experiments providing information on solar wind, sun spots, electron population levels in the Earth's polar regions and the distribution of aerosols and ozone in the upper atmosphere. P78-1 served as a U.S. antisatellite target on September 13, 1985.

1979 February 25   At 11:54 UT cosmonauts Vladimir A. Lyakhov and Valery V. Ryumin were launched aboard the *Soyuz 32* spacecraft on a six-month mission to *Salyut 6* in an orbit of 184 × 192 mi. Dormant for almost three months since the return of Vladimir Kovalenok and Alexandr Ivanchenkov, *Salyut 6* was in good condition when the crew docked to the forward port at 13:30 UT on February 24 and went inside. Nevertheless, several systems components had to be replaced and an inventory was made for logistical resupply by Progress cargo tankers. Despite a problem with the propulsion system, thrusters were used to raise the orbit slightly on March 1, and on March 10 the station was placed in a gravity-gradient stabilization mode with the aft end pointing Earthwards for propellant conservation. The station had been completely prepared for operations by March 10.

1979 March 4   The first picture ever taken of a volcano actively erupting on the surface of another world was captured by the TV cameras on the *Voyager 1* spacecraft when it viewed Jupiter's moon Io from a distance of 310,000 mi. Intensive surveys of Jupiter's moons with the scan platform science instruments was a key feature of the mission. Ranging in diameter from 1,900 mi. to 3,200 mi., the four large moons Io, Europa, Ganymede and Callisto carried features dating back to the origin of the solar system. The innermost of the large moons, Io orbits Jupiter at a distance of 262,000 mi. and was observed by Voyager in a state of almost continuous eruption due to the enormous gravitational tides from its parent planet.

1979 March 5   *Voyager 1* made its closest approach to the planet Jupiter when it passed within 173,000 mi. of the cloud tops at 12:42 UT traveling at a speed of 84,000 MPH. Jupiter was 422.33 million mi. away from Earth and radio signals took 37 min. 43 sec. at the speed of light to reach ground stations. During the period around closest encounter the science instruments shifted targets from the planet itself to the innermost of Jupiter's four large moons, Io. The closest approach to Io occurred at 15:47 UT when *Voyager 1* came to within 12,752 mi. of its sulfurous surface. The closest approach to Callisto, 78,355 mi., came at 17:46 UT on March 6. By March 15 *Voyager 1* had returned more than 15,000 images of the giant planet and its satellites coded in 86.94 billion binary digits. A course correction maneuver took place on April 9 to fine-tune the trajectory to Saturn.

1979 March 7   The French Council of Ministers agreed to fund development of a reusable space glider carried into orbit by an advanced version of the Ariane launcher. Development of the space plane, called Hermes, and the launcher, Ariane 5, was expected to take about eight years. The French national space agency, Centre National d'Études Spatiales (CNES), had designed Hermes to carry a payload of 14,000 lb., attributed as either five passengers, or two passengers and

3,100 lb. of cargo, or all cargo. According to the French, a materials processing research laboratory called Minos could be carried on some Hermes missions. At this date Hermes was projected to have a length of 41 ft. with a 20 ft. long cargo or pressure module. The preliminary design featured a double delta wing with a dorsal fin and rudder.

1979 March 8 Shuttle orbiter OV-102, *Columbia,* was rolled out of its Rockwell International assembly facility at Palmdale, Calif., and delivered to the Kennedy Space Center, Fla., by Boeing 747 Shuttle carrier aircraft on March 25 in preparation for its first flight into space, now scheduled for November 9, 1979. It was placed in the orbiter processing facility (OPF) at KSC from which it was moved to the Vehicle Assembly Building on November 24, 1980. There, it was mated to the external tank and the two solid rocket boosters which would power it toward space.

1979 March 12 At 05:47 UT the unmanned *Progress 5* cargo tanker was launched by an SL-4 from the Baikonur cosmodrome to the *Salyut 6* space station, where it docked at 07:20 UT on March 14. In addition to propellant, food, water, linen dryer, extra chemical batteries and new carbon dioxide sensors, the freighter delivered a black and white television on which the crew could watch Moscow TV. On March 16 the docked assembly of Progress/Salyut/Soyuz was slowly spun around its transverse axis to permit the separation of gases and propellant in a contaminated propulsion system at the rear of the space station. Progress was then used to pump in supplementary fuel and raise the orbit to 207 × 217 mi. The freighter undocked at 16:10 UT on April 3, and was de-orbited two days later.

1979 March 15 The European Space Agency approved a plan to conduct a survey on remote sensing which was to be carried out by Denmark, France, Italy, the Netherlands, Sweden and the United Kingdom. The survey would determine the type of optical and microwave instruments for monitoring land and ocean surfaces which the ESA intended to put on its remote sensing satellites planned for the 1980s.

1979 March 19 The tenth lunar and planetary science conference began at the Johnson Space Center, Tex. Attended by more than 700 scientists from the United States, Western Europe, Africa, Australia, and the Soviet Union, the five-day conference heard the usual papers on lunar and planetary science as well as a presentation on the Pioneer-Venus mission to Venus, the *Voyager 1* encounter at Jupiter and a three-dimensional film of Mars.

1979 March 30 NASA announced it had signed an agreement with the European Space Agency for the International Solar Polar Mission (ISPM). Under the agreement, NASA and ESA would each build a spacecraft, both of which would be launched on a single Shuttle flight in February 1983. They would reach Jupiter in May 1984 and cross opposite poles of the sun in September 1986. In July 1979 NASA selected TRW Systems to build the U.S. spacecraft for the ISPM and in April 1980 allocated the two spacecraft to separate Shuttle flights in 1985. In February 1981 the incoming Reagan administration forced cancellation of the U.S. spacecraft, leaving the one spacecraft built by ESA and upsetting European plans to integrate the scientific survey of the solar environment.

1979 March 31 The U.S. Department of Defense rejected the air-launched deployment mode for the M-X ICBM, con-

cluding that too many operational uncertainties existed to rely on this method for surviving a preemptive attack from Soviet ICBMs. On May 7 President James E. Carter was briefed on the multiple protective shelter (MPS) concept which had been restored as the best basing option. Nevertheless, other concepts were reviewed, including a proposal from CIA Director Admiral Stansfield Turner that M-X be deployed in small submarines. Gen. David Jones, chairman of the Joint Chiefs of Staff, was concerned that the Soviets would find a way of detecting in which MPS shelter the M-X was located.

1979 April 4 Intelsat reported that two major new satellite communication networks were no threat to its economic well-being and that they would integrate well with the existing system of international satellites. The two new networks were the European Communications Satellite (ECS) system and the proposed Palapa B system, a joint venture of the Association of Southeast Asian Nations, including Indonesia, Malaysia, the Philippines, Singapore and Thailand. Intelsat members were obliged not to start up regional systems in competition, and there had been concerns that these two programs might contravene Intelsat regulations, but the three organizations co-existed and formed a precedent for other regional systems.

1979 April 7 The first Trident fleet ballistic missile submarine took to the water when the USS *Ohio* was launched by Electric Boat Division of General Dynamics, Groton, Conn. The *Ohio*-class had a length of 560 ft., a submerged displacement of 18,700 tons and two parallel rows of launch tubes for 24 Trident missiles aft of the sail. The *Ohio* had a complement of 164 (16 officers and 148 enlisted men). In terms of sheer targeting power, one *Ohio*-class boat could launch up to 336 (90–100-KT) warheads on separate targets. Whereas fleet ballistic missile submarines had been named after distinguished American citizens, the new generation would be named after states of the Union, an exception being when the *Rhode Island* was changed to the *Henry M. Jackson*. The 18th and last *Ohio*-class boat was funded in fiscal year 1991. As the 10 *Lafayette*-class boats remaining in service were withdrawn, the U.S. SLBM (submarine-launched ballistic missile) force was reduced to the 18 *Ohio*-class boats with their total complement of 432 Trident missiles.

Agreement was reached between the United States and the Soviet Union banning the encryption of signals transmitted from ballistic missile tests. Since an SS-18 launch on July 29, 1978, the Soviets had shrouded telemetry information in encrypted codes to hide performance and engineering details from U.S. Rhyolite geostationary electronic intelligence gathering satellites. The United States had complained that this went against the agreement on verification of missile performance enshrined within the SALT 2 treaty.

1979 April 9 NASA signed a memorandum of understanding with Indonesia for the launch of the Palapa B satellite system on separate Shuttle missions. The two new satellites would replace the Palapa A series with *Palapa B1* scheduled for launch in June 1982 and *Palapa B2* for the first half of 1983. Each satellite would be attached to a perigee kick stage which would fire after the combination had been released from the Shuttle orbiter payload bay.

1979 April 10 At 17:34 UT cosmonauts Nikolei N. Rukavishnikov and Georgy Ivanov, a Bulgarian, were launched to the *Salyut 6* space station in *Soyuz 33* by SL-4 from the Baikonur cosmodrome. As scheduled, it was to be the first of two Intercosmos visits for Vladimir Lyakhov and Valery

Ryumin. In an orbit of 210 × 220.5 mi., they observed Soyuz as it approached for docking 25 hr. 30 min. after launch, but a malfunction in the Soyuz primary engine completely prevented that. Due to the relationship of the spacecraft orbit with the Earth, Rukavishnikov and Ivanov had to wait another day before trying the backup system to get them back on the ground. At 15:47 UT on April 12 the backup engine was fired, but instead of burning for 3 min. 8 sec., it fired for an additional 25 sec. before Rukavishnikov shut it off manually. Slowed by a greater amount than planned, Soyuz entered a steeper descent trajectory which exerted 10 g on the crew instead of the usual 4 g. *Soyuz 33* touched down at an elapsed time of 1 day 23 hr. 01 min. Lyakhov and Ryumin were deeply depressed at not receiving guests to relieve their boredom, temporarily cut communications with the ground and were disgruntled for days. Their reactions, and those of other long-stay cosmonauts, were incorporated in Soviet studies on the psychological effects of spaceflight. Not before June 7, 1988, would another attempt be made to put a Bulgarian cosmonaut aboard a Soviet space station.

NASA's Boeing 747 Shuttle carrier aircraft took the orbiter *Enterprise* from the National Space Technology Laboratories, Huntsville, Miss., where it had been used for vibration tests, to the Kennedy Space Center, Fla., where it would be used to check out the facilities at Launch Complex 39. During the month it was mated to the no. 2 external tank and a pair of inert solid rocket boosters. On May 1 it was moved by a modified mobile launch platform (no. 1) to the pad at LC-39A. For the next three months it was used to fit-check pipes, umbilicals and service arms and to serve as a training aid for ground crews. The *Enterprise* was returned to the VAB (vehicle assembly building) on July 23 and de-stacked.

1979 April 18   After an absence of 15 months following the reentry of *Cosmos 954*'s radioactive core on January 24, 1978, a new pattern of Soviet surveillance and elint ocean reconnaissance satellites began with the launch of *Cosmos 1094* from Baikonur on an SL-11. Placed in an orbit of 271.5 × 284 mi. inclined 65°, the solar-powered elint was followed seven days later by another of its kind, *Cosmos 1096*. Separated by 60°, one satellite would precede the other by 15 min. over a given spot. The launches were timed for a major Soviet naval exercise in the Persian Gulf astride the Western shipping lanes.

1979 April   Two candidate cosmonauts from Vietnam were selected to begin training at Zvezdny Gorodok, the Soviet training center outside Moscow: Pham Tuan and Bui Thanh Liem. Participating under the Intercosmos program, one of the two candidates would represent Vietnam on the international flight to *Salyut 6* and carry along a scientific experiment contributed by his country. As with earlier selections involving other countries, the two candidates were serving officers in the armed forces.

1979 May 8–11   At a conference on advanced technology for future space systems, Rockwell International described an air-breathing, rocket-powered horizontal takeoff and landing vehicle called Star-Raker. The delta-winged vehicle looked like a supersonic airliner, with a length of 310 ft. and a gross takeoff weight of 5 million lb. Ten 140,000-lb. thrust, high-bypass supersonic turbofan engines in an underwing group would provide power for takeoff and cruise to the selected geographic location. Liquid hydrogen would be used in a 3.2-million-lb. thrust rocket engine to supplement the air-breathers to a height of 107,200 ft. and a velocity of 7,200 ft./sec., at which point the rocket would take over and power

Star-Raker into orbit. The vehicle would have a payload capability of up to 89,000 lb. in a bay 141 ft. long. Star-Raker was never pursued.

1979 May 13   At 04:17 UT an SL-4 launched *Progress 6* from the Baikonur cosmodrome carrying propellant and supplies to *Salyut 6* cosmonauts Vladimir Lyakhov and Valery Ryumin. Docking was achieved at the aft port at 06:19 UT on May 15, and the station crew unloaded the freighter. When *Progress 6* docked, the station was in an orbit of 201 × 210.5 mi., but its propulsion system was used to raise this to an orbit of 204 × 219 mi. *Progress 6* undocked from the aft port at 08:00 UT on June 8 and was de-orbited a day later.

1979 May 19   NASA officials arrived in China for a two-week visit during which they discussed the country's interest in purchasing a domestic communication satellite. China was also interested in receiving Landsat Earth resource data and set about establishing links with the U.S. Department of Commerce for a ground receiving station. Agrarian China was an apt choice for area surveys from space, and the value of such data did much to convince the Chinese that they needed their own remote sensing satellite.

1979 June 2   NASA launched the British high-energy astrophysics satellite, UK-6 *(Ariel 6)*, on a four-stage Scout D from Wallops Island, Va., to a 370 × 409 mi. orbit inclined 55°. Built by Marconi Space and Defence Systems, the 340-lb. cylindrical satellite had a height of 4 ft. 3 in. and a diameter of 2 ft. 3 in., with four solar array booms providing 80 W of power. Three science and two technology experiments provided data on heavy components of cosmic radiation and x-rays and on the development of new types of solar cell. UK-6 decayed through the atmosphere on September 23, 1990.

1979 June 4–5   A Presidential Review Committee gave formal approval for the full-scale development of the M-X ICBM. After several minor specification changes, the three-stage solid propellant missile was to have a length of 76 ft. 9 in., a diameter of 9 ft. 2 in., a launch weight of 195,000 lb. and a throw weight of 7,900 lb. Initially the missile was planned to have ten 335-KT W-78 warheads in Mk 12A reentry vehicles (RVs) designed for the Minuteman III, but the end result of the development, begun in February 1982 by Lawrence Livermore National Laboratory, was quite different. A restartable liquid propellant fourth stage, the "bus," could make numerous course changes, putting each RV on its own unique path to an individual target with an accuracy of 400 ft. The method by which the missile was to be deployed was deferred for further study.

1979 June 6   At 18:13 UT *Soyuz 34* was launched unmanned by an SL-4 from the Baikonur cosmodrome on a rendezvous and docking mission to *Salyut 6*. Because the scheduled *Soyuz 33* flight, launched April 10, had been unable to dock due to a propulsion system failure, the *Salyut 6* cosmonauts Vladimir Lyakhov and Valery Ryumin needed a fresh vehicle on hand to get them back to Earth at the end of their stay, which would last six months. (The *Soyuz 32* spacecraft that had carried them to *Salyut 6* on February 25 had an in-orbit life of only three months, so a replacement spacecraft was essential.) The unmanned *Soyuz 34* docked at the aft port on *Salyut 6* at 20:02 UT on June 8, just 12 hr. after *Progress 6* had vacated that location. Unmanned, *Soyuz 32* was undocked from the forward *Salyut 6* port at 09:51 UT on June 13 and returned to Earth at 16:18 UT that day. At 16:18 UT on June 14, *Salyut*

6 cosmonauts Lyakhov and Ryumin undocked *Soyuz 34* and transposed it to the forward docking port.

1979 June 7   A Soviet SL-8 launched the Indian satellite *Bhaskara* from Kapustin Yar to a 322 × 336 mi. orbit inclined 50.7°. Developed by the Indian Space Research Organization (ISRO) and named after an ancient Indian astronomer, the 1,000-lb. satellite carried television cameras and sensors for studying hydrology, meteorology and forestry. Engineers were unable to switch on the primary sensors and received data only from secondary experiments. On November 20, 1981, it was joined by *Bhaskara 2*, placed in a 319 × 346 mi. orbit inclined 50.6° by another SL-8, and decayed into the atmosphere on February 17, 1989. The second satellite decayed on November 30, 1991.

1979 June 8   The first Soviet satellite dedicated to Earth resource surveys was launched by an SL-4 from Plesetsk to a 132 × 158 mi., 81.35° orbit. Designated *Cosmos 1105*, the 12,125-lb. satellite was yet another application of the manned Vostok spacecraft. Cameras and other sensors in the visible and nonvisible portions of the spectrum were carried in the reentry module, and information was obtained via wet-film images returned to Earth at the end of the mission. *Cosmos 1105* was recovered on June 21. Three more Earth resource flights were flown this year: *Cosmos 1106* on June 12, *Cosmos 1122* on August 17 and *Cosmos 1127* on September 5. Five Earth resource Cosmos satellites were launched in 1980, followed by eight in 1981, two in 1983, eight in 1984, seven in 1985, six in 1986 and three in 1988. From May 25, 1989, the Soviet Earth resources program was identified under the title Resurs.

1979 June 12   The French space agency, Centre National d'Etudes Spatiales (CNES), signed an agreement with principal contractors on the Ariane program setting up Transpace, an organization for marketing the launcher. The Transpace agreement was signed by 29 companies from Belgium, Denmark, France, Germany, Italy, Netherlands, Spain, Sweden, Switzerland and the United Kingdom. Under an arrangement with the European Space Agency, which had overall responsibility for the program, and CNES, which managed Ariane for ESA, Transpace would evolve into Arianespace and become a highly effective instrument for attracting customers to the launcher. By the mid-1980s Arianespace would claim 50% of all commercial satellite traffic in the noncommunist world.

The first clustered firing of three Space Shuttle main engines (SSMEs) at full power took place at the National Space Technology Laboratories, Miss. Previous firings had reached only 90% of thrust, but this one achieved power levels necessary for launch. Planned to last 8 min. 40 sec., the test was aborted at 55 sec. when signals from an accelerometer on a fuel preburner erroneously indicated a malfunction. Preliminary flight certification on the SSME was completed on June 27 when engine 2004 successfully demonstrated a simulated 13 min. 43 sec. abort burn which it could perform should two of the three SSMEs fail at a critical point during ascent. However, a second clustered test on July 2 was aborted early and NASA moved the first orbital flight test back to March 1980, a decision attributable to problems with *Columbia*.

1979 June 27   The first operational third-generation U.S. weather satellite was launched by Atlas F from Vandenberg AFB, Calif., to a 503 × 514.5 mi. polar orbit inclined 98.5°. Designated *NOAA-6*, the satellite was almost identical to the Tiros-N precursor launched on October 13, 1978, and joined that satellite to provide continuous Earth coverage with an advanced high-resolution radiometer. Other satellites of the Tiros-N generation included: *NOAA-B*, launched May 29, 1980, but stranded in a 155 × 639 mi. orbit from which it decayed May 3, 1981; *NOAA-7* on June 23, 1981; *NOAA-8* on March 28, 1983 (the battery for which exploded on December 30, 1985); *NOAA-9* on December 12, 1984; *NOAA-10* on September 17, 1986; *NOAA-11* on September 24, 1988; and *NOAA-12* on May 14, 1991.

1979 June 28   At 09:25 UT the unmanned *Progress 7* cargo tanker was launched by SL-4 from the Baikonur cosmodrome to the *Salyut 6* space station in an orbit of 219 × 224 mi. It was automatically docked to the aft docking port at 11:18 UT on June 30. In a two-day operation that began July 3, *Progress 7* raised the *Salyut 6* orbit to 248 × 255 mi. The *Salyut 6* propulsion system had become suspect and there were to be no more Progress flights during 1979, so it was necessary to boost the orbit as high as possible to extend the life of the station. *Progress 7* had carried to *Salyut 6* a KRT-10 radio telescope weighing 440 lb. Folded like an umbrella, it would, when unfurled, present the form of a semirigid dish antenna with a diameter of 33 ft. Placed in the aft access tunnel with the inner hatch closed and the outer hatch open, the umbrella-like device deployed automatically as *Progress 7* undocked and slipped away from the aft docking port at 03:50 UT on July 18.

1979 June 30   The venerable NASA experimental communications satellite *ATS-6* was turned off after five years of useful work, bringing educational programs to developing countries around the world. On August 1 its remaining thruster (three of four had failed) was used to begin a series of maneuvers that would raise it several hundred miles above geostationary altitude, where it would remain indefinitely.

1979 July 9   *Voyager 2* made its closest approach to Jupiter and four of its known thirteen moons. The close encounter sequence began when the spacecraft passed within 38,500 mi. of Ganymede at 08:06 UT; 217 images were obtained. At 18:43 UT *Voyager 2* came within 128,000 mi. of Europa, and at 20:11 UT it passed tiny Amalthea at 453,000 mi. Its closest encounter with Jupiter occurred at 22:29 UT and a distance of 404,000 mi. with the spacecraft traveling at a speed of 45,400 MPH. The closest approach to Io came at 00:09 UT on July 10, at a distance of 702,000 mi. As a result of the *Voyager 1* and *2* encounters with the Jovian system and the 33,000 pictures returned to Earth, three additional moons were discovered, bringing to sixteen the total number of known satellites. The presence of a thin ring around the planet had also been observed.

1979 July 11   The Skylab space station reentered Earth's atmosphere and was destroyed over Australia at 16:37 UT (00:37, July 12, local time) with several pieces falling across an area 20 mi. wide and several hundred miles long. Due to uncertainties about the reentry point, Australian officials had only an eight-minute warning of the descending 77-ton hulk. The largest piece recovered had a length of 6 ft. 6 in. and a width of 3 ft. 3 in. It was put on display in the Kalgoorlie town hall.

Pioneer 10 crossed the orbital path of the planet Uranus, at this time the most distant planet, 1.8 billion mi. from the sun and 1.7 billion mi. from Earth. Uranus, however, was about 172° through its 84-year orbit and almost on the exact opposite side of the sun, 3.5 billion mi. from the spacecraft. Scientific instruments were still working and sending data to Earth on the interplanetary environment, and spacecraft systems were operating as well as they had during the encounter with Jupiter on December 3, 1973.

1979 July 16 The International Maritime Satellite Organization (Inmarsat) officially came into being, mandated by its 22 member nations to provide a global maritime communications network. The European Space Agency had built the first two Marecs satellites for trials with the system, and at a meeting in Paris on July 26 the Maritime European Communications Satellite Program agreed to build a third satellite, Marecs-C. The first Marecs satellite was launched on December 20, 1981.

1979 July 20 The last of three flight engines assigned to propel the first Shuttle orbiter *(Columbia)* into space passed flight acceptance testing, clearing the way for launch. By this date, the other two flight engines had been delivered to the Kennedy Space Center, Fla., and one was being installed in the orbiter. Engine tests were continually plagued with failures, however, and by the end of the year the current expectation of making the first Shuttle flight at the end of March 1980 began to evaporate. During 1980 there were two major test failures in engines being used to qualify the design.

1979 July 26 France, Germany and Italy decided to proceed with preparatory analysis of improvements to the Ariane launch vehicle. It had been proposed by the European Space Agency that the carrying capacity of Ariane could be improved by upgrading the first- and second-stage engines and by increasing propellant capacity in the third stage and adopting a more powerful cryogenic engine. The upgraded launcher would be known as Ariane 2. Taking the basic Ariane 2 and adding two solid propellant strap-on boosters to the first stage would produce an Ariane 3 version capable of placing a single load of 5,700 lb. on a geosynchronous transfer ellipse, or of lifting two 2,635-lb. satellites to the same orbit. The basic Ariane 1 could lift 4,025 lb.

1979 August 2 *Pioneer 11* began its scientific observations of the ringed planet Saturn, the first spacecraft to visit this gaseous world and the first to perform two outer planet encounters on the same mission. *Pioneer 11* was still 932 million mi. from Saturn but hurtling toward its target at 19,000 MPH. On August 16 periodic tracking by Earth stations was increased to a 24-hr. operation. On August 27 the spacecraft crossed the orbit of Phoebe, Saturn's outermost moon, 8 million mi. from the planet with speed now 20,500 MPH. At 13:00 UT on August 31 *Pioneer 11* crossed Saturn's magnetospheric bow shock wave just 898,500 mi. out from the planet, closer than the 3.5 million mi. expected. Although 100 times stronger than Earth's magnetosphere, it was only 5% the strength of Jupiter's enormous magnetosphere.

The Comsat General Corp. announced it was considering a satellite subscription television service. For a monthly fee, programs would be transmitted from powerful satellites to small antennas on the roofs of subscribers' houses. Comsat had conducted trials with the Canadian communications technology satellite (CTS) satellite and noted that tests were already being conducted by Japan with its BSE satellite in what would become known as "direct-broadcasting" transmitted straight into individual antennas.

1979 August 9 *Salyut 6* cosmonauts Vladimir Lyakhov and Valery Ryumin completed scientific observations with the

KRT-10 radio telescope (33 ft. in diameter) carried to orbit aboard the *Progress 7* cargo tanker on June 28. In attempting to jettison the antenna and its support structure from the rear docking port, the ribs of the antenna became tangled with communications antenna on *Salyut 6*. To free the jammed antenna and clear the docking port for later missions, Lyakhov and Ryumin began a space walk at 14:16 UT on August 15. In activity lasting 1 hr. 23 min., they made their way from the forward transfer module back along the exterior hull of the station, cut loose the fouled debris and returned.

1979 August 10 The first flight of India's first satellite launcher, SLV-3, ended in disaster when it tumbled into the sea 5 min. after launch from a site at Sriharikota, an island just north of Madras. The launch site was operated by the Indian Space Research organization, which also had management responsibility for the four-stage solid propellant launcher. SLV-3 had a height of 75 ft. 6 in., a maximum diameter of 3 ft. 3 in. and a launch weight of 38,230 lb. The four solid stages had thrust levels of 122,000 lb., 51,800 lb., 17,850 lb. and 4,850 lb., respectively. The first SLV-3 had been carrying the 88-lb. *Rohini RS1A* test satellite. The first successful flight of an SLV-3 took place on July 18, 1980.

1979 August 19 *Salyut 6* cosmonauts Vladimir Lyakhov and Valery Ryumin returned to Earth after six months aboard the space station. Climbing aboard the *Soyuz 34* spacecraft, which had been launched unmanned on June 6, the crew undocked at 09:07 UT and touched down 3 hr. 23 min. later. They had remained in space for a record 175 days 36 min., thus completing the primary mission of *Salyut 6*. Lifted from their spacecraft on recliners, Lyakhov and Ryumin were judged fit and well despite the psychological pressures that had made life almost intolerable at times. Denied any contact with other human beings for the full duration of their flight due to the

On the launch pad a Soviet SL-4/Soyuz combination is gripped by four restraint arms supporting the main stage just above the strap-on booster attach line, two service towers and an electrical/hydraulic/fueling structure.

failure of *Soyuz 33* to dock and begin a cycle of two Intercosmos visits, the two had at times experienced severe tension with each other. *Salyut 6* was now put in dormant mode before an extended series of manned occupations in 1980.

1979 August After five years in development at Los Alamos Scientific Laboratory, production of the W-78 warhead with a yield of 335–350 KT got under way. The U.S. Air Force planned to retrofit 300 Minuteman ICBMs with Mk.12A reentry vehicles incorporating two or three W-78 warheads, replacing the original warload of the Mk.11/W-62 combination. This conversion program began in December 1979 and was completed in 1983. From then on the land-based ICBM force comprised 450 Minuteman IIs, each with a single 1.2-MT W-56 warhead in an Mk.11 reentry vehicle; 250 Minuteman IIIs, each with two or three 170-KT W-62 warheads in an Mk.12 reentry vehicle; and 300 Minuteman IIIs with the Mk.12A. The Minuteman III/Mk.12A/W-78 combination had a target accuracy of 600–900 ft. at full range, which, with a yield twice that of the W-62 and an accuracy up to twice that of the Mk.12 warhead, gave the upgraded system truly hard-target kill capability.

1979 September 1 At 14:36 UT *Pioneer 11* crossed the ring plane of Saturn, 70,000 mi. above the cloud tops, traveling at 70,000 MPH. The spacecraft reached its closest fly-by altitude of 13,000 mi. and a speed of almost 71,000 MPH at 16:31 UT, disappearing behind the planet 1 min. later and reappearing at 17:50 UT. In approaching Saturn, *Pioneer 11* dived down around the south polar region and recrossed the ring plane at 18:35 UT. Earth was 963 million mi. away and radio signals now took 1 hr. 26 min. to cross that space. Whereas it had originally been thought that data from the spacecraft could be sent at a maximum rate of 256 bits/sec., major improvements to the Earth stations enabled data at 512 bits/sec. and sometimes 1,024 bits/sec. to be sent during the encounter. At 18:04 UT on September 2 *Pioneer 11* made its closest pass of 219,950 mi. to Saturn's moon Titan. As the spacecraft receded, the data rate was reduced progressively to 32 bits/sec. by September 6. Its flight path turned by almost 90° due to Saturn's gravity, *Pioneer 11* was heading out of the solar system in the direction almost opposite to that of its companion *Pioneer 10*. As requested by NASA, three Soviet early-warning satellites in Earth orbit were switched off to prevent interference with the signals from *Pioneer 10*.

1979 September 7 President James E. Carter announced that the United States would deploy 200 M-X ICBMs in a sheltered road-mobile system. In this concept, each missile would be carried on a transporter-erector-launcher (TEL) between 23 horizontal shelters, necessitating 4,600 shelters at 200 sites. Under the terms of the SALT agreement, the Soviets would be allowed a maximum 8,000 ICBM warheads, and since two warheads would be necessary to guarantee total destruction of each shelter, the USSR would be required to expend its entire ICBM force just to destroy one leg of the strategic triad, leaving the submarine-launched ballistic missile force and manned bombers to "savage the aggressor at will" (in the expression of the Joint Chiefs of Staff). To satisfy arms treaty verification criteria, plugs in the top of each shelter could be opened at an agreed time so that at any given moment Soviet spy satellites would confirm that only one M-X was deployed in a string of 23 shelters.

1979 September 10 NASA's Johnson Space Center announced that tests had begun on the prototype of a manned maneuvering unit (MMU) built by Martin Marietta for use with the Shuttle. It was an improved version of the maneuvering unit used inside Skylab during 1973. Astronaut Bruce McCandless was assigned to the project and tested the donning procedure, in which a suited astronaut with integral life-support backpack backs into an MMU attached to the inside wall of the cargo bay. McCandless likened the procedure to "backing into a phone booth with a large knapsack on your back." The MMU was first flown on Shuttle mission 41-B, launched on February 3, 1984.

1979 September 18 Responding to a request from NASA Administrator Robert M. Frosch, a special committee of individual consultants concluded a four-month study of the Shuttle management system with a report on deficiencies in both long-range planning and the ability to anticipate potential technical problems. Although the Shuttle was originally scheduled for launch in early 1978, NASA did not now expect the Shuttle to make its first flight into space before the spring of 1980. One member of the report team was Maj. Gen. James A. Abrahamson, former MOL astronaut and U.S. Air Force F-16 program director. He would be appointed to head the Shuttle program on October 1, 1981.

1979 September 20 The third of NASA's three high-energy astronomy observatories, HEAO-3, was launched by an Atlas-Centaur from Cape Canaveral, Fla., to a 301 × 311 mi., 43.61° orbit. The 6,000-lb. satellite consisted of a standardized equipment section attached to a box-like experiment module, with a total height of 18 ft. and a width of 7 ft. 8 in. HEAO-3 carried three instruments for gathering information on the origin and propagation of gamma and cosmic rays. The all-sky survey that resulted provided useful information on the spectrum and intensity of diffuse and discrete x-ray and gamma-ray radiation. Attitude control gas ran out on May 30, 1981, and the satellite decayed on December 7, 1981.

1979 September 25 A Soviet SL-4 launch vehicle carried the 13,200-lb. international biological research satellite *Cosmos 1129* from Plesetsk to a 135.5 × 234 mi., 62.8° orbit. The satellite contained 38 white rats and 60 fertile Japanese quail eggs and accommodated experiments from several countries, including the United States. The entire payload was devoted to biological experiments. The white rats were separated by gender and allowed to mix after several days' adapting to weightlessness, when it was hoped they would become the first animals to breed in space. They did not. Embryo development in the quail eggs was expected to begin on the eighth day. The recovery portion of the satellite was returned to Earth on October 14.

1979 October 5 NASA's Marshall Space Flight Center issued a request for proposals on a 25 kW solar power array to provide supplementary power for Shuttle Spacelab missions. The solar cell array was conceived as a means of increasing the electrical power available to Spacelab materials processing experiments, which claim large power needs by virtue of their technology. On May 7, 1980, NASA announced that McDonnell Douglas and TRW Systems had been awarded parallel contracts for design and definition of the array.

1979 October 15 NASA reported that scientists and engineers from Australia, Canada, India, Japan and the United States had met at Wallops Flight Center, Va., to compare data on ozone activity in the stratosphere. Instruments used around the world varied greatly in the calibration and precision of measurements. The meeting helped define a common standard by which to regulate observations and evaluate the data. Beginning October 21, 20 sounding rockets were launched in a two-week period to define ozone variability by region. The launch of Orion, Nike Orion, Super Arcas and Super Loki rockets coincided with satellite overflights to compare data.

1979 October 30   The third Applications Explorer Mission (AEM-3) was launched by a four-stage Scout from Vandenberg AFB, Calif., to a 220.5 × 349 mi. orbit inclined 96.8°. Dubbed *MAGSAT* (magnetic field satellite), the 400-lb. AEM-3 comprised a base module which used the same octagonal structure as the small astronomy satellite and an instrument module containing the science payload. *MAGSAT* had a diameter of 2 ft. 6 in. and a height of 5 ft. 3 in. Four solar array wings providing 160 W of electrical power extended from the structure at 90° intervals to a span of 11 ft. 2 in., and a magnetometer on a 20-ft. scissors boom protruded from the side. The satellite decayed June 11, 1980.

1979 November 2   Ending six years of reduced effort in communications satellite technology research, NASA announced a five-year development plan to help private industry develop a sophisticated experimental satellite. Lewis Research Center was to be the lead facility for this satellite, which NASA expected to launch in 1985 or 1986. This would be known as the advanced communications technology satellite, or ACTS. On June 5, 1980, parallel one-year contracts to develop the concept were awarded by NASA's LRC to TRW Systems and Hughes Aircraft.

1979 November 4   A further effort to get a full-thrust, full-duration, clustered firing with three space shuttle main engines (SSMEs) ended after only 9 sec. when sensors detected excess pressure in a critical part of the engine. Although the SSME did not set the pace for getting the Shuttle into space on its first flight, engine problems held up preparations for *Columbia*'s first flight. In early December NASA decided it could not attempt that first flight before June 30, 1980.

1979 November 8   NASA and the European Space Agency requested proposals from scientists for experiments to put aboard an international mission to comets Halley and Tempel 2. Using a solar-powered ion electric propulsion system, the spacecraft would fly past Halley in late 1985 and go on to rendezvous with Tempel 2, accompanying the latter in its orbit of the sun for more than a year. The object of the mission would be to define the chemical and physical nature of the cometary nucleus and to collect information about its tail.

1979 November 20   On the advice of his science advisor, Frank Press, President James E. Carter assigned management of all civilian remote sensing to the National Oceanic and Atmospheric Administration. The NOAA had managed three generations of weather satellite and was well placed to control the Earth resource program. The Department of Commerce would continue to encourage private sector involvement and encourage the sale of civilian data to foreign customers.

1979 November 29   NASA outlined to the Senate Committee on Appropriations changes in plans for the Galileo orbiter-probe mission. Instead of launching a single spacecraft by Shuttle in January 1982, it now wanted to separate orbiter and probe for dual launches. Early design work led to fears that a Shuttle inertial upper stage would be incapable of lifting the orbiter-probe combination; and although it could fly on a Titan-Centaur, none would be available by 1982. In the new plan, the 5,864-lb. orbiter would be launched in February 1984 using an inertial upper stage to boost it for a gravity-assist fly-by of Mars three months later. Arriving at Jupiter in mid-1986, Galileo would enter a 178,000 × 9.3 million mi. orbit of the planet. Launched by Shuttle IUS in March 1984, the probe and its carrier would enter the Jovian atmosphere in mid-1987 with data relayed via the orbiter.

The European Space Agency selected British Aerospace to build the large satellite, or L-Sat (later named Olympus), to conduct experimental broadcast television in the 20–30 GHz band. A special payload module for research into superhigh-frequency broadcasting would be built by the United Kingdom's Marconi Space and Defence Systems, with assistance from Telespazio in Italy and Philips in the Netherlands. Signals in this frequency band would carry much greater volumes of data than signals in lower frequency bands, but they were vulnerable to attenuation from rain and poor weather.

NASA's Johnson Space Center distributed a document on a space base called the Space Operations Center (SOC). The document asserted that "A compelling rationale can be drawn that it is now time to identify the next logical step in the evolution of U.S. manned spaceflight." and described an orbital base where satellites could be serviced, large space structures assembled and maneuvering vehicles berthed for transporting cargo from low to high orbit. Utilizing the Shuttle to construct a modular space base, the document described how the SOC could be expanded to a permanent manned facility for up to eight people. It emphasized the base as a place to do useful work rather than research, a support facility for many other space programs and a departure point for planetary flights.

1979 December 5   The United Nations General Assembly opened for signature the draft Moon Treaty, which had been under debate since the adoption of Resolution 2779 in late 1971. In the first treaty concerning the territory of another world, the UN made a historic first step in preserving the moon and the planets "exclusively for peaceful purposes." The United States and the Soviet Union had already announced they had no intention of laying claim to the moon or any other body, and the UN resolution forbade the deployment of "any type of weapons" on the lunar surface, in orbit around the moon or to a trajectory toward the moon.

1979 December 6   NASA selected 33 science investigations on climate and upper atmosphere research for satellites scheduled to fly in the 1980s. Of the 33, 23 experiments would be from the United States and 10 from six foreign countries, including Canada, Japan, France, the United Kingdom, West Germany and Italy. The investigations would be carried out on three platforms: a planned Earth Radiation Budget Satellite (ERBS), a seven-day Spacelab 2 flight on the Shuttle and in the form of several packages attached to NOAA (National Oceanic and Atmospheric Association) meteorological satellites.

1979 December 13   NASA's Jet Propulsion Laboratory announced that Martin Marietta and Hughes Aircraft had been selected to conduct studies of a Venus orbital imaging radar (VOIR) spacecraft to map the surface of Earth's nearest planetary neighbor. Carrying a synthetic aperture radar instrument, the spacecraft would first be placed in a 185 ×

11,800 mi. elliptical polar orbit and then lowered to a circular, 185-mi. orbit for the 120-day mapping sequence. Venus held intense interest for scientists because its dense, hot, carbon dioxide atmosphere evolved through a mechanism dubbed the "greenhouse effect." Some scientists believed the Earth might suffer the same effect in future years and they wanted to study it in detail.

1979 December 16   At 12:30 UT the unmanned *Soyuz-T 1* was launched from the Baikonur cosmodrome to the *Salyut 6* space station. During the approach phase on the second day, *Soyuz-T 1* overshot the station and had to back up for a second try. Docking was achieved at 14:05 UT on December 19. The *Soyuz-T 1* propulsion system was employed to raise the *Salyut 6* orbit, but the primary purpose of the mission was to put the new T-series Soyuz design through a full operational checkout. It remained in space for more than three months, undocking from *Salyut 6* at 21:04 UT on March 23, 1980. Postseparation maneuvers kept the vehicle busy for a further two days before it was finally de-orbited. In a procedural change, Soyuz-T series spacecraft jettisoned their orbital modules before retrofire, followed by separation of the descent module for reentry. On standard Soyuz missions, the orbital module remained attached until after retrofire.

1979 December 17   Three Space Shuttle main engines (SSMEs) completed a full-duration, full-power firing for the first time. The test took place at the National Space Technology Laboratories, Miss., and ran for 9 min. 10 sec. at varied throttle settings of 90%, 80% and 70%. This would be normal during a Shuttle launch, reducing thrust as the mass of the Shuttle declined to restrict the accelerations levels below 3 g's. The burn was terminated with just two engines running, the third having been shut down 5 sec. earlier as planned. The first Shuttle flight was now tentatively scheduled for June 30, 1980. By March 1980 that date had slipped to September/October 1980, and by June 1980 it was postponed to March 1981.

1979 December 24   The first of four test and development launches with the European Ariane 1 launch vehicle took place from Kourou, French Guiana. Ariane 1 comprised three stages in a total height of 155 ft. 6 in. and a launch weight of 463,643 lb. The first (L140) stage had a length of 60 ft. 4 in., a diameter of 12 ft. 6 in. and contained 325,385 lb. of UDMH-N_2O_4 propellants feeding four Viking V engines delivering a total vacuum thrust of 626,400 lb. The second (L33) stage had a length of 38 ft. and contained 75,175 lb. of UDMH-N_2O_4 feeding one 162,700-lb. thrust Viking V engine. The third (H8) stage had a height of 29 ft. 9 in., a diameter of 8 ft. 8 in. and contained 18,150 lb. of cryogenic hydrogen and oxygen feeding one 13,815-lb. thrust HM7 engine. The L01 mission began with first stage ignition 3.4 sec. before liftoff. The first stage shut down 2 min. 24 sec. after liftoff and the second stage ignited 4 sec. later, running for 2 min. 16 sec. Third-stage ignition came 9 sec. later and lasted 9 min. 8 sec. At 15 min. 9 min. the 3,532-lb. payload separated from the third stage in a simulated geostationary transfer orbit of 111 × 13,366 mi. The Capsule Ariane Technologique payload comprised a cylindrical module 4 ft. in diameter carrying vibration, acceleration and acoustic sensors. The third stage exploded on March 1, 1980, as expected, and the CAT decayed on November 27, 1989.

1979   During the year the Soviet Union began to reequip its SS-18 heavyweight ICBM force with the Mod 4 version.

Carrying ten 500-KT, multiple independently targeted reentry vehicle (MIRV) warheads, the Mod 4 had a range of 5,000 mi., a throw weight of 16,700 lb. (the greatest ever achieved by an ICBM) and a phenomenal accuracy of 850 ft. at full range. By 1983, earlier Mod SS-18s had been replaced by the Mod 4. The missile was deployed in SS-9 silos utilizing a cold-launch technique whereby the round is ejected from its silo by compressed gas, much like a submarine-launched missile is ejected from its launch tube. This allowed the Soviets to deploy the much more potent SS-18 in silos that did not flaunt the SALT I agreement, which limited the Soviets to missiles no bigger than existing silos could contain.

Development tests with the definitive Mod 3 version of the Soviet SS-19 (designated RS-18) ICBM were completed; it was ready for operational deployment during 1980. This version had six MIRV warheads of 550 KT each, a throw weight of 7,500 lb. and a range of 5,500 mi. With 50% of test missiles throwing their warheads to within 800 ft. of the target, this was the most accurate ICBM deployed to date. Deployment of the Mod 4 coincided with deactivation of Mod 1 and Mod 2 versions, and by 1983 peak deployment had reached 360. The missile was still in service in the early 1990s.

1980 January 25   NASA and McDonnell Douglas signed a joint venture agreement to develop a new technique for producing high quality drugs through the concept of continuous-flow electrophoresis. In a state of weightlessness, biological materials can be separated to higher levels of purity than they can in 1 g environments. Under the agreement, McDonnell Douglas would seek out a pharmaceutical company and provide a pilot production plant which NASA would launch aboard the Shuttle. This activity became known as the Electrophoresis Operations in Space (EOS) program.

1980 January 28   In its budget request for fiscal year 1981, NASA asked Congress to approve funds for a start on a gamma ray observatory designed to explore the most energetic forms of known radiation. GRO was planned as an advanced successor to HEAO-1 (High-Energy Astronomy Laboratory-1), which conducted the first low-energy gamma ray sky survey, and HEAO-3, and was to be placed in orbit by the Shuttle. Although the project was approved, Congress slashed funds and the launch date slipped from 1987 to June 1988. Design study contractors General Electric and TRW Systems were selected on February 29, 1980, with TRW eventually selected as prime contractor. Delays to the Shuttle program after *Challenger* was destroyed on January 28, 1986, delayed the launch to 1991.

1980 January   NASA announced the names of nineteen pilot (PLT) and mission specialist (MS) Shuttle astronauts constituting Group 9: James P. Bagian; Col. John E. Blaha (USAF); Col. Charles F. Bolden (USMC); Brig. Gen. Roy D. Bridges Jr; Franklin R. Chang-Díaz; Mary L. Cleave; Bonnie J. Dunbar; William F. Fisher; Col. Guy S. Gardner (USAF); Col. Ronald J. Grabe (USAF); Col. David S. Hilmers (USMC); Capt. David C. Leestma (USN); John M. Lounge; Col. Bryan D. O'Connor (USMC); Capt. Richard N. Richards (USN); Col. Jerry L. Ross (USAF); Cdr. Michael J. Smith (USN); Col. Sherwood C. Spring (USA) ; and Col. Robert C. Springer. Smith was killed January 28, 1986, when Shuttle mission 51-L exploded.

Tests began of a powerful new Soviet missile designed to equip a completely new class of nuclear powered ballistic missile submarine. Called Typhoon, the submarine would not be launched until September 1980, but the missile itself was a major step forward. Known in the Soviet Union as the RSM-52, the solid propellant Sturgeon had a range of 5,175 mi. with eight 100-KT MIRV warheads and an accuracy of 1,800 ft. Tests were not successful at first, but more flights in 1981 and 1982 were observed from the West to produce better results. The RSM-52 was given the NATO code designation SS-N-20 Sturgeon.

1980 February 13 NASA announced that it was making Delta 3910 launchers available as backup to the Shuttle during its transition to full operational status. NASA also announced that it had reached agreement with Telesat Canada for subsidized development of a new, more powerful Delta 3920 capable of putting a 2,750-lb. payload in geosynchronous transfer orbit. The Delta 3910/Payload Assist Module, scheduled for introduction later in the year, could lift 2,400 lb. NASA wanted the 3920 version to put *Landsat 4* in orbit, but Telesat wanted it as a backup to their Shuttle payload slot and was prepared to fund 20% of the cost. The Delta 3920 would have a total height of 116 ft., weigh 426,000 lb. at liftoff and produce a liftoff thrust of 525,000 lb. from the first-stage engine and five Castor IV solid propellant strap-ons. The remaining four Castor IVs would be ignited when the others burned out. The extended second stage of the 3920 would increase burn time by 30%. A PAM-D would be used as the third stage.

1980 February 14 NASA's Solar Maximum Mission (SMM) satellite was launched by Delta 3910 from Cape Canaveral, Fla., to a 303 × 306 mi. orbit inclined 28.5°. Solar Max comprised a multimission modular spacecraft (MMS) with a 5-ft. triangular framework (containing essential systems) to which a payload module was attached. The 5,200-lb. satellite, with a length of 13 ft. and a diameter of 7 ft., had two solar cell paddles delivering 1,500–3,000 W of electrical energy. Launched near the peak of the 11-year solar cycle, the satellite carried gamma-ray, x-ray and ultraviolet spectrometers, polarimeters and instruments for imaging the sun's corona. SMM was repaired by the crew of Shuttle mission 41-C on April 9, 1984, and it decayed out of orbit on December 2, 1989.

1980 February 17 A solid propellant Mu-3S launch vehicle carried the 408-lb. test satellite MS-T4, named *Tansei-IV* (*Light Blue*), from Kagoshima to a near circular orbit of 311 × 357 mi. inclined 38.7°. This new three-stage launcher had a height of 78 ft., a total weight of 109,150 lb. and a payload capacity of 570 lb. Mu-3S had the same eight solid propellant strap-ons used with the Mu-3C and Mu-3H, with an improved first-stage thrust of 229,320 lb., a second-stage thrust of 62,620 lb. and a third-stage thrust of 13,010 lb. MS-T4 was both a qualification mission for the Mu-3S and a demonstration of measuring techniques of launch vehicle performance parameters. It decayed on May 12, 1983.

1980 February 22 A Japanese N-1 launch vehicle put the second Experimental Communications Satellite, ECS-2 (*Ayame 2* [*Sweet Flower*]) into a transfer orbit from which an apogee motor placed it in geostationary position above northern New Guinea. The 573-lb. satellite was a replacement for ECS-1, lost when it bumped the second stage after launch on February 6, 1979. Contact with *Ayame 2* was lost shortly after apogee motor ignition. An abnormal burn was deemed responsible for the loss. With weight reduced to 290 lb. at apogee motor burnout, ECS-2 was placed in geostationary orbit but never used.

1980 March 3 The fully operational White Cloud ocean surveillance satellite system for the U.S. Navy was established with the launch of an Atlas F from Vandenberg AFB, Calif. From this one launch, three satellites were placed closely together in a 662 × 715 mi. orbit inclined 63.4°. Built by Martin Marietta, this third triple cluster was positioned 120° east of the first cluster, launched April 30, 1976. A replacement cluster for the first three satellites failed when the Atlas F malfunctioned on December 9, 1980. The next triple launch did not take place until February 9, 1983, followed on June 9, 1983, by another replacement set and on February 5, 1984, by a third.

1980 March 4 NASA's Marshall Space Flight Center announced contracts to Martin Marietta and Aerojet General studying the feasibility of adapting Titan launch vehicle propulsion units for a liquid boost module on the Shuttle. Overweight, the basic Shuttle was able to lift only 24,000 lb. to polar orbit from Vandenberg AFB, Calif., instead of 40,000 lb. as required by the Air Force. By fitting a cluster of four shortened Titan propellant tanks, each 10 ft. in diameter, and a pair of Titan first-stage engines to the bottom of the external tank, payload capability would be increased to 41,000 lb. Extending 24 ft. 2 in. below the dome of the external tank, the two Titan engines would ignite 5 sec. after the solid rocket boosters and burn for 3 min. 20 sec. Although feasible, it was never adopted.

1980 March 6 The European Space Agency announced a new scientific project called Hipparcos. Designed to last 2 ½ years, the satellite would be the first orbiting astrometric tool to measure positions, proper motion, parallaxes, position and displacement velocity of more than 100,000 stars. The ESA wanted the satellite launched by Ariane in mid-1986.

1980 March 12 It was decided that completion of the manned Spacelab laboratory by countries of the European Space Agency would be accomplished under a new funding structure. Cost overruns would permit expenditure of up to 120–140% of previously agreed amounts, divided between West Germany, 64.4%, France, 12.07% and the United Kingdom, 7.6%, with seven other countries contributing the balance of 15.93%.

1980 March 21 NASA's Johnson Space Center issued a contract to Boeing for a systems analysis study of the Space Operations Center (SOC). In-house studies at the JSC indicated the first modules could be launched within eight years of go-ahead and that the initial facility would comprise two habitation and work modules, two interlinking nodes (or service modules) and a logistics module to be periodically replaced by Shuttle. The SOC was also expected to support space construction by manufacturing lightweight beams and truss assemblies from flat rolls of steel alloy. All this, said JSC, could be obtained for $2.7 billion, excluding Shuttle costs.

1980 March 26 The first private company set up to run production and marketing of a major satellite launcher, the European Ariane, came into existence under the name Arianespace. Shareholders included the 36 primary European companies involved in Ariane production, 11 European banks and the French national space agency, Centre National d'Etudes Spatiales (CNES). Capital amounted to FF 120 million subscribed through 11 countries including France (59.25%), Germany (19.6%), Belgium (4.4%), Italy (3.6%), Switzerland

(2.7%), Spain (2.5%), the United Kingdom (2.4%), Sweden (2.4%), the Netherlands (2.2%), Denmark (0.7%) and Ireland (0.25%).

1980 March 27   At 18:53 UT the *Progress 8* cargo tanker was launched by SL-4 from the Baikonur cosmodrome to a rendezvous with the *Salyut 6* space station. After a docking at the aft port at 20:01 UT on March 29, Progress 8 used its propulsion system to raise the station's orbit to 209 × 216 mi. The unmanned vehicle had been launched in readiness for the first manned occupation of *Salyut 6* in 1980. After stores and supplies had been transferred to the station by *Soyuz 35* cosmonauts Valery Popov and Leonid Ryumin, launched April 9, *Progress 8* was undocked at 08:04 UT on April 25 and de-orbited the following day.

1980 April 3   Following the breakdown of antisatellite talks with the United States and the Soviet invasion of Afghanistan, tests of the Soviet antisatellite system resumed after a halt of almost two years with the launch of *Cosmos 1171*. Launched on an SL-8 from Plesetsk, the 466-lb. satellite was placed in a 602 × 623 mi. orbit inclined 65.8°. On April 18 the 7,320-lb. interceptor *Cosmos 1174* was launched by SL-11 from Baikonur to a 225 × 637 mi. orbit inclined 66.1°. The test was abandoned when the interceptor failed to close to within 5 mi. of the target, considered to be the "kill" radius. Both satellites remain in orbit but inoperable.

A Soviet SL-4 carries a Soyuz spacecraft to the Salyut 7 *space station. Note the launch escape tower and square-shaped paddles to stabilize the spacecraft in the event of an abort.*

1980 April 9   At 13:38 UT cosmonauts Leonid I. Popov and Valery V. Ryumin were launched aboard their 15,000-lb. *Soyuz 35* spacecraft from Baikonur cosmodrome to the *Salyut 6* space station. In an orbit of 209 × 216 mi., *Soyuz 35* docked to the forward port at 15:16 UT on April 10. It had been almost eight months since Ryumin returned with Vladimir Lyakhov after a six-month stay and he was now back on another mission also planned to last six months. *Salyut 6* had accomplished its primary goal of supporting three periods of occupation, but the Lyakhov/Ryumin flight of 1979 was to be repeated by the Popov/Ryumin crew. During this mission three Intercosmos crews would visit the station.

1980 April 12   A new cycle of replacement Soviet early-warning satellites began with the launch of *Cosmos 1172* on an SL-6 from Plesetsk to an elliptical orbit of 396 × 24,955 mi. inclined 62.8°. This also marked the beginning of an increase in the number of operational early-warning satellites in orbit simultaneously. With 10 launches by the end of 1981 and 3 in 1982, the system grew from three satellites situated 80° apart to nine satellites 40° apart and was completed with the launch of *Cosmos 1367* on May 20, 1982. Replacement launches included 2 in 1982; 3 in 1983; 7 in 1984; 7 in 1985; 6 in 1986; 3 in 1987; 2 in 1988; 2 in 1989 (when the satellite weight increased to 4,200 lb.); 6 in 1990 (1 stranded in the wrong orbit); none in 1991; and 4 in 1992.

1980 April 14   Engineers at the NASA National Space Technology Laboratories, Miss., successfully ran a Shuttle main engine for 10 min. 10 sec., during which a thrust level of 109% of rated power was maintained for 6 min. This engine (no. 2004) had already accumulated 11,000 sec. of test time. Shuttle engines were being overrated to 109% for contingency abort conditions under which the orbiter and external tank would have to return to the launch site shortly after launch. Achieving 109% (512,300-lb. thrust) power levels compared with a rated thrust of 470,000 lb. would also allow the Shuttle to fly different trajectories into orbit.

1980 April 25   NASA selected 26 science investigations for a planned upper atmosphere research satellite (UARS) mission. Consisting of two satellites launched to 56° and 70° orbits by the Shuttle, 2 years 6 months apart, the UARS missions were designed to obtain data on the energy input, chemical balance, chemical composition and dynamics of the stratosphere and the mesosphere. Growing awareness of the potential erosion of stratospheric ozone, vital for screening ultraviolet radiation hazardous to Earth's living things, stimulated a call from scientists for more direct measurements of these changes. UARS was reduced to one mission, delayed by the loss of *Challenger* on January 28, 1986, and was eventually launched on STS-48 on September 12, 1991.

1980 April 27   At 06:24 UT the unmanned *Progress 9* cargo tanker was launched by SL-4 from the Baikonur cosmodrome. At 08:09 UT on April 29 it docked to the aft port on *Salyut 6*, in orbit at 207 × 224 mi., only four days after *Progress 8* had vacated the same port. *Salyut 6* cosmonauts Leonid Popov and Valery Ryumin transferred stores and supplies before filling the station's water tanks on May 6 and refueling the propellant tanks. Its mission accomplished, *Progress 9* undocked at 18:51 UT on May 20, and de-orbited two days later, clearing the aft port for the first Intercosmos mission of 1980.

1980 April 29   An SL-11 launched the 9,800-lb. Soviet nuclear-powered ocean surveillance satellite *Cosmos 1176* to a 161.5 ×

164.5 mi. orbit inclined 65°. The first of a new design to reduce the risk of radioactive contamination caused by unscheduled reentry, *Cosmos 1176* would separate the core from the reactor, making it unlikely to survive in the atmosphere, which it did after boosting the core to a 542 × 599 mi. orbit after 134 days. The redesigned ocean surveillance satellite also had a much longer operational life. From March 1981 they were launched at intervals between ocean elint satellites until March 14, 1988, when the last nuclear-powered radar satellite *(Cosmos 1932)* was launched.

The French and German governments agreed to cooperate in the development of direct broadcast satellites. Each country would provide one satellite for preoperational trials. The two satellites were to be almost identical, the French satellite known as TDF and the German satellite known as TV-SAT. The satellites would be used for experimental tests and transmissions on direct broadcasting, using powerful amplifiers to send programs direct to comparatively small-dish antennas on the ground, obviating the need for large-dish antennas.

1980 May 23 The Ariane L02 flight, the second of four test flights with the new European launcher, ended in disaster when pressure fluctuations in one of the four first-stage Viking engines just 5.75 sec. after liftoff caused failures that resulted in a self-destruct signal, blowing up the vehicle at 1 min. 48 sec. With a liftoff weight of 464,100 lb., the vehicle was to have orbited a CAT payload similar to that carried by L01, a 203-lb. amateur radio satellite *(Oscar 9)*, and a 2,434-lb. plasma experiment called Firewheel. The failure was attributed to combustion instability, a very common problem in liquid propellant motors. By mid-October 37 static firings had taken place to confirm the need for modifications to the design of the injector plate bringing propellants into the combustion chamber.

1980 May 26 At 18:21 UT the international crew of Soviet cosmonaut Valery N. Kubasov and the Hungarian cosmonaut Bertalan Farkas was launched in the 15,000-lb. *Soyuz 36* spacecraft on an SL-4 from Baikonur cosmodrome. At 19:56 on May 27 *Soyuz 36* docked to the rear port on *Salyut 6* and the crew joined Leonid Popov and Valery Ryumin in the space station. After exchanging gifts and performing experiments for five days, contour couches from respective spacecraft were changed over and Kubasov and Farkas got inside the *Soyuz 35* spacecraft docked at the forward port. Leaving their own spacecraft, *Soyuz 36*, for the station crew, Kubasov and Farkas undocked from *Salyut 6* at 11:47 UT on June 3, and landed back in the Soviet Union 3 hr. 20 min. later at the end of a mission that lasted 7 day 20 hr. 46 min. The *Salyut 6* crew transferred *Soyuz 36* from the aft to the forward docking port beginning at 16:39 UT on June 4.

1980 May China began flight tests with a new generation of three-stage liquid propellant ICBMs fired into the Pacific Ocean between the Solomon Islands, Fiji and the Gilbert Islands. Designated CSS-4, the missile had a length of 142 ft., a diameter of 10 ft. 1 in. and an estimated launch weight of around 445,400 lb. Tests indicated a throw weight of 4,400 lb., a range of 9,300 mi. and a warload comprising one 5-MT reentry vehicle. Over the following year China deployed two CSS-4, bringing to a maximum eight the total ICBM complement deployed in silos. One missile has been tested with a MIRVed warhead. Chinese engineers adapted the CSS-4 into the CZ-2 and CZ-3 satellite launchers.

1980 June 5 At 14:19 UT cosmonauts Yuri V. Malyshev and Vladimir V. Aksenov were launched in the 15,100-lb. *Soyuz-T 2* by an SL-4 launch vehicle from Baikonur. This was the first manned flight of a T-series Soyuz and a final checkout of the definitive configuration for this class of Soyuz. At 15:58 UT on June 6, *Soyuz-T 2* docked at the aft port on *Salyut 6*. Solar cells had been reintroduced for station ferry missions, and although the T-series had fully automatic docking capability, a manual docking was performed on this flight. Malyshev and Aksyonov joined Popov and Ryumin, now in their eleventh week aboard the orbiting space station, for a brief visit before undocking at 09:20 UT on June 9 and returning to Earth at the end of a relatively short 3 day 22 hr. 19 min. flight.

1980 June 18 An SL-3 launch vehicle carried the 7,665-lb. *Meteor-Priroda* weather and Earth resources satellite from Baikonur cosmodrome to a 356.5 × 405.7 mi. orbit inclined 97.9°. Essentially a converted Meteor 2–series satellite, it took pictures of the Earth's surface in 10 multispectral bands with a resolution of 100–2,600 ft. in strips 100–1,250 mi. wide. Single images of up to 733 sq. mi. in area provided geologists with detailed information on faults and folds in the Earth's surface.

1980 June 29 The unmanned *Progress 10* cargo tanker was launched by SL-4 from the Baikonur cosmodrome at 04:41 UT. Bringing supplies and propellant for the space station, *Progress 10* docked to the aft port on *Salyut 6* at 05:53 UT on July 1, and cosmonauts Leonid Popov and Valery Ryumin busied themselves unpacking racks in the orbital module. New equipment for the scientific experiments was transferred, including components to extend the life of the BST-1M astrophysical telescope. Its job done, *Progress 10* undocked at 22:21 UT on July 17 and was de-orbited over the Pacific Ocean two days later.

1980 July 3 Member countries of the European Space Agency decided to proceed with development of the Ariane 2 and Ariane 3 launch vehicles with the cost divided between seven countries: France (62.41%), Germany (17.28%), Italy (16.29%), Sweden (2.09%), Belgium (0.84%), Spain (0.84%) and the Netherlands (0.25%). As with Ariane 1, development of the upgraded launchers would be under the direction the French national space agency, (CNES), Centre National d'Etudes Spatiales and the Ariane Programme Board composed of representatives from the participating countries.

1980 July 8–9 At a meeting of the European Space Agency in Paris, France, the Science Programme Committee approved a plan to send a spacecraft called Giotto on a rendezvous with Halley's comet when it next approached the sun in March 1986. The mission took its name from the 14th-century Florentine painter Giotto di Bondone, whose *Adoration of the Magi* depicted the appearance of Halley's comet in 1301. Giotto's painting is considered the first serious representation of the comet. British Aerospace became the prime contractor for development and fabrication of the spin-stabilized spacecraft, which on March 14, 1986, would make the closest fly-by of Halley's comet as part of an international program of observations.

1980 July 10 British Prime Minister Margaret Thatcher sent a letter to U.S. President Jimmy Carter expressing a desire to purchase the Trident I (C-4) submarine-launched ballistic missile as replacement for its existing Polaris missile during the early 1990s. Carter gave the approval of the U.S. government in a reply dated July 14. Later, when the United States decided to phase out Trident I, the United Kingdom government elected to buy the Trident II (D-5). It was to be equipped with eight warheads per missile rather than the 14 it was capable of carrying. Nevertheless, with the much greater range and more sophisticated penetration aids and decoys, the British Trident deterrent would be a much more potent force than the Polaris system it replaced.

1980 July 17   The Federal Communications Commission approved a joint venture between Fairchild Industries, and Continental Telephone Corp., to be known as American Satellite Corp. On December 4, 1980, the FCC authorized ASC to acquire a 20% share in Westar satellites for $31.9 million. The FCC also authorized the two joint owners of ASC to join with Western Union to form Space Communications Co., which was to launch and operate the TDRSS/Advanced Westar system; TDRSS stands for Tracking and Data Relay Satellite System.

1980 July 18   India became the seventh nation to build and launch its own satellite when the 88-lb. *Rohini RS1B* satellite was placed in a 190 × 571 mi. orbit inclined 44.7°. Launched from Sriharikota Island near Madras, the satellite was put in orbit by the SLV-3 launcher. It had solar cells producing a power level of 3 W. *Rohini RS1B* reentered the atmosphere on May 20, 1981.

1980 July 23   At 18:33 UT the second Intercosmos crew to visit *Salyut 6* cosmonauts Leonid Popov and Valery Ryumin was launched aboard the *Soyuz 37* spacecraft from Baikonur cosmodrome. Carrying Soviet pilot Viktor V. Gorbatko and Vietnamese cosmonaut Pham Tuan, *Soyuz 37* docked to the aft *Salyut 6* port at 20:02 UT on July 24. Without an infrastructure to support advanced scientific research, Vietnam had no science program to contribute to the *Salyut 6* mission, which had been customary for Intercosmos flights, but Tuan took photographs of his country from the orbiting space station. After the customary exchange of contour couches customized for each cosmonaut, Gorbatko and Tuan prepared to return in the *Soyuz 36* spacecraft docked at the forward port. They undocked at 11:55 UT on July 31 and returned to Earth at the end of a mission that had lasted 7 days 20 hr. 42 min. Popov and Ryumin moved *Soyuz 37* to the forward docking port at 16:43 UT next day.

1980 September 9   The first of a new and improved generation of Geostationary Operational Environmental Satellites built by Hughes Aircraft, *GOES-4* was launched by Delta 3914 from Cape Canaveral, Fla., to a geosynchronous orbit at 98° W inclined 3.1° to the equator. The drum-shaped satellite had a diameter of 7 ft. 1 in., a height of 11 ft. 11 in., a launch weight of 1,840 lb., or 975 lb. at apogee motor burnout. In addition to a visible/infrared spin-scan radiometer for day and night cloud cover pictures, the satellite also carried a package of instruments for monitoring the Earth-sun environment and a data collection system relaying information from remote data stations. Other satellites of this type included: *GOES-5* launched on May 22, 1981, to 85° W; *GOES-6* launched on April 28, 1983, to 135° W; *GOES-G* would have been designated *GOES-7* in orbit but its Delta was destroyed after losing power seconds after liftoff on May 3, 1986; and *GOES-7* launched on February 26, 1987, to 83° W.

1980 September 18   During routine maintenance of a Titan II ICBM silo at Little Rock AFB, Arkansas, a U.S. Air Force

technician dropped a heavy wrench socket off a work platform inside the silo where the missile was fueled and onto its launch mount. The socket struck the side of the missile and ruptured a fuel tank which caused a leak of vapors. About 8 ½ hr. later, during the early hours of September 19, the Titan II exploded, killing 1 of a team called in and injuring 21 other U.S. Air Force personnel. The 9-MT Mk.6/W-53 warhead was thrown clear by the explosion and recovered. Safety locks on the warhead prevented radioactive leakage and there was no contamination. Gen. Bennie L. Davis, commander of the U.S. Air Force Training Command, was appointed to head a Titan II executive committee that investigated the accident and concluded in December 1980 that both the missile and its handling procedures were sound.

At 19:11 UT the third international crew to visit *Salyut 6* cosmonauts Leonid Popov and Valery Ryumin was launched in the *Soyuz 38* spacecraft on an SL-4 from Baikonur cosmodrome. The Soviet pilot Yuri V. Romanenko played host to Cuban cosmonaut Arnaldo Tamayo-Mendez, the first black person to fly in space, as the *Soyuz 38* docked with the aft port on *Salyut 6* at 20:49 UT a day later. The four cosmonauts conducted joint biomedical experiments developed by Cuban physicians, although few items of equipment were added to the already burgeoning inventory of on-board hardware. The *Soyuz 38* crew returned in their own spacecraft on September 26 at an elapsed time of 7 days 20 hr. 43 min.

1980 September 28   At 15:10 UT *Progress 11* was launched by SL-4 from the Baikonur cosmodrome and docked to the aft port on *Salyut 6* at 17:03 UT two days later. Cosmonauts Leonid Popov and Valery Ryumin were preparing to leave *Salyut 6* at the end of their six-month stay and unpacked some equipment from *Progress 11* but left most of it for the next crew. With *Salyut 6* now three years into its operational life, attention to systems and equipment vital for the safety and proper running of the station was necessary. *Progress 11* carried items the next crew would need to carry out this work, as well as materials processing samples for experiments with the furnaces. *Progress 11* was eventually undocked at 10:23 UT on December 9, 1980, and de-orbited.

1980 September   The first in a new class of Soviet ballistic missile submarine was launched at Severodvinsk. Known as the Typhoon class, these giant vessels represented the biggest submarines ever built. With a submerged displacement of 25,000 tons, each vessel had a length of 557 ft. 6 in., a beam of 82 ft. and a hull depth of 42 ft. 8 in. Each Typhoon had a double hull with missile tubes for 20 SS-N-20 SLBMs in two rows of 10 forward of the sail. As each missile carried eight MIRV warheads, each Typhoon could attack 160 separate targets. The first Typhoon/SS-N-20 complement became operational in 1983, followed by a second boat in 1984. A total of six submarines were operational by 1990.

1980 October 7   NASA's Marshall Space Flight Center announced the award of contracts to Martin Marietta and Aerojet Liquid Rocket Co. for definition of a liquid boost module (LBM) for attaching to the bottom of the Shuttle external tank. NASA wanted to improve the payload carrying capacity of the Shuttle by 12,000 lb. and sought to achieve this by adapting first-stage engines from the Titan launch vehicle.

1980 October 11   Cosmonauts Leonid Popov and Valery Ryumin returned to Earth at the end of their six-month mission aboard the *Salyut 6* space station. The crew undocked the *Soyuz 37* spacecraft from the forward port at 06:30 UT, leaving the *Progress 11* cargo tanker at the aft port for the next crew. *Soyuz 37* landed in the USSR at 09:50 UT, ending a flight that began for Popov and Ryumin 184 days 20 hr. 12 min. earlier on April 9, 1980. It was another new endurance record and gave Ryumin the staggering total space time of 361 days 21 hr. 33 min. over three missions. His place on *Salyut 6* for a second long-duration flight had been made possible when the primary crew member, Valentin Lebedev, suffered a knee injury. Western observers also noted that Ryumin's mission cycle of two six-month spaceflights separated by almost eight months was about the cycle cosmonauts would experience in a flight to the surface of Mars and back. Orbital mechanics dictate that a crew would take at least six months to fly out and six months to get back, with at least six months spent on the surface of Mars. There is no evidence whatsoever to suggest this was a planned simulation.

1980 November 1 NASA announced that it had been informed by the Office of Management and Budget that President James E. Carter had approved a formal go-ahead for the so-called Venus orbital imaging radar spacecraft. Now planned for a Shuttle launch in 1986, VOIR was projected to weigh around 11,000 lb. It was to carry a synthetic aperture radar, and in November 1981 Hughes Aircraft was awarded a contract to develop this instrument. Hughes and Martin Marietta were studying the VOIR mission concept under a separate contract from NASA, but the heavy spacecraft carried a high development cost. When NASA attempted to get the project approved, VOIR ran into opposition from Congress.

1980 November 12 *Voyager 1* began its close observations of Saturn and 7 of its 12 (then) known moons. At 07:05 UT Voyager passed within 2,400 mi. of Titan, Saturn's largest moon. With a diameter of 3,200 mi., Titan orbits Saturn at a distance of just over 1.2 million mi. and is second only to Jupiter's Ganymede as the largest natural satellite in the solar system. Flying north to south, Voyager crossed the multiring plane of Saturn at 07:23 UT, the outer ring of which lies 298,000 mi. from the planet. *Voyager 1* passed within 258,000 mi. of Tethys, a moon of 650-mi. diameter 183,000 mi. from Saturn, at 11:40 UT. The spacecraft passed within 77,120 mi. of Saturn over a position at 20° S at 01:10 UT on November 13. Just 52 min. later passed within 67,300 mi. of Mimas, a moon of 240-mi. diameter 115,000 mi. from Saturn, as it headed north and upward, its flight path bent by Saturn's gravity. At 03:08 UT *Voyager 1* passed behind Saturn as viewed from Earth and 7 min. later came within 126,000 mi. of Enceladus, 310 mi. in diameter and 148,000 mi. from its planet. *Voyager 1* reappeared at 04:35 UT and 18 min. later passed 100,000 mi. from the moon Dione, 700 mi. in diameter. The spacecraft flew above the ring plane and came within 45,000 mi. of Rhea, a moon 950 mi. in diameter and 327,000 mi. from Saturn at 07:46 UT. Voyager passed within 547,000 mi. of Hyperion, 180 mi. in diameter, at 18:09 UT, and departed the Saturnian system when its scan platform instruments were turned off on December 19, 1980.

1980 November 15 First in a new series of Hughes HS 376 communication satellites, *SBS-1* was launched from Cape Canaveral, Fla., to a transfer orbit and thence to 106° W for Satellite Business Services. The launch vehicle was the first Delta 3910, incorporating a PAM-D (payload assist module-delta) in place of the TE-364–3. The payload capacity had been increased to 2,450 lb., as compared with the Delta 3914's 2,065 lb. The 2,410-lb. HS 376 series would be sold to many

Seen here with its lower, cylindrical solar cell array extended as it would be in orbit, this Hughes HS-376 satellite was built for Satellite Business Systems and launched on November 15, 1980.

customers for regional and domestic services. Spin stabilized, it comprised a cylinder 7 ft. in diameter and 9 ft. 3 in. tall. After deployment, a cylindrical solar array stowed like a sleeve around the main body of the satellite extended to a total height of 21 ft. 8 in., doubling the solar cell area and providing more than 1,000 W of electrical power. It weighed 1,300 lb. on station. The satellite carried 10 × 20 W K-band transponders making SBS the first U.S. domestic satellite company to utilize this less crowded, 12–14 GHz part of the spectrum. *SBS-2* was launched by Delta on September 24, 1981, with *SBS-3* and *SBS-4* launched by Shuttle.

1980 November 18 NASA selected Hughes Aircraft to build the Galileo probe carrier spacecraft, at this date planned for a March 1984 launch to Jupiter and arrival in July 1987. The probe would separate from the carrier 100 days prior to encounter and enter the Jovian atmosphere at a speed of 100,000 MPH. Slowed by the dense atmosphere, the probe would release a parachute to further retard its rate of descent. The Galileo orbiter, which was to be launched in February 1984 and by the time of the probe's arrival would have been circling Jupiter for a year, would relay to Earth information transmitted to it from the probe. The probe parachute system began tests on November 26, 1980, from the Naval Weapons Center Test Range, China Lake, Calif. NASA's Ames Research Center managed the probe and its carrier, with General Electric building the heat shield, while the Jet Propulsion Laboratory was responsible for the orbiter.

1980 November 25 Ball Aerospace Systems was selected to design and build an Earth radiation budget satellite, ERBS, which NASA planned to launch from a Shuttle in 1984. *Nimbus 6* and *Nimbus 7* each carried instruments for measuring the radiation budget of the Earth, but ERBS would provide budget measurements on a global basis 24 hr. a day. The satellite would be carried into space aboard a Shuttle orbiter and released to orbit using the remote manipulator arm. ERBS formed an important part of NASA's climate research program and was launched by mission 41-G on October 13, 1984.

The Department of Commerce and the Department of Defense signed a memorandum of understanding on behalf of the National Oceanic and Atmospheric Administration and the Navy, respectively, concerning development of a National Oceanic Satellite System. Conceived as a successor to the Seasat program, NOSS envisaged a joint civil-military ocean research effort with two satellites equipped to measure wind, sea surface temperature, wave state, ice, surface currents and chlorophyll concentrates. In 1981 the Reagan administration terminated NOSS to save money.

1980 November 27 At 14:18 UT cosmonauts Leonid D. Kizim, Oleg G. Makarov and Gennady M. Strekalov were launched in their 15,100-lb. *Soyuz-T 3* spacecraft by SL-4 from the Baikonur cosmodrome. Equipped with new lightweight pressure suits, they were the first three-person Soviet space crew since the tragic deaths of Georgi Dobrovolsky, Vladislav Volkov and Viktor Patsayev in June 1971. *Soyuz-T 3* docked with the forward port on *Salyut 6* at 15:54 UT on November 28. On entering the station, they activated dormant systems and prepared to complete unpacking *Progress 11*, which had been docked to the aft port since September 29. Several systems were repaired, equipment added or replaced, and considerable time was spent operating the materials processing furnace. The mission ended December 10, a day after the release of *Progress 11*, with undocking at 06:10 UT and a landing back in the Soviet Union at an elapsed time of 12 days 19 hr. 08 min.

1980 December 4 The Federal Communication Commission authorized a request from AT&T that it be allowed to put together a system of Telstar satellites, each of which would carry 24 C-band transponders. Two were to be placed in orbit and one left on the ground as a spare. AT&T wanted to free itself from the $47 million per year it was paying to Comsat General for use of the Comstar satellites. The $230 million it would cost AT&T to procure and operate the satellites would redeem the investment in 4.9 years. Each Telstar satellite would have a life of about 8 years.

A wholly owned subsidiary of General Telephone and Electronics Corp., GTE Satellite Corp., was authorized to proceed with a plan to build and launch three Ku-band satellites. The company wanted to provide data, voice, teleconferencing and image traffic via 16 transponders with 54 MHz bandwidth on each satellite, which would have a design life of 10 years. GTE wanted to compete against Satellite Business Systems. Each SBS satellite had 10 transponders, 37 MHz bandwidth and a life of seven years.

The Federal Communications Commission authorized the construction and launch of three satellites for Southern Pacific Co. Called Spacenet, the system would include C-band and Ku-band services on satellites built by RCA. Services were to be aimed primarily at the video programming industry, with the majority of users in the cable industry. The two satellites put up initially would be located at 70° W and 119° W.

The Federal Communications Commission authorized Hughes Communications, Inc., to operate three C-band sat-

ellites built by Hughes Aircraft. Called Galaxy, the HS-376 satellites would each carry 24 transponders with 6 held in reserve for rental on a preemptible basis. This meant that 18 primary users could switch to one or more of the six preemptibles if one of their transponders failed. Customers would "buy" up to 18 transponders, an innovation in communication satellite services. First to buy on *Galaxy 1* was Time, Inc., paying $8.5 million for each of six transponders. Other customers, such as Times Mirror, paid $14 million for a single transponder.

1980 December 6 The first Intelsat V high-capacity communications satellite was launched by Atlas-Centaur AC-54 from Cape Canaveral, Fla., to a 103.6 × 22,339 mi. transfer orbit. The apogee kick motor was fired two days later, placing *Intelsat V F2* at geostationary altitude over 15° E, moved in 1981 to its operational position at 338.5° E. The satellite weighed 4,251 lb. at liftoff and 1,900 lb. on station and comprised a box structure 5 ft. 5 in. × 6 ft. 7 in. × 5 ft. 9 in. Two wing-like solar arrays with a span of 51 ft. provided 1,241 W of electrical power. Intelsat Vs could handle 12,000 voice circuits and 2 TV channels, operating at both 14/11 GHz and 6/4 GHz bands. Intelsat Vs launched by Atlas-Centaur included *F1* on May 23, 1981, *F3* on December 15, 1981, *F4* on March 5, 1982, *F5* on September 28, 1982, *F6* on May 19, 1983 and *F9* on June 9, 1984, which entered an unusable orbit of 109 × 755 mi. inclined at 29.1° and decayed October 24. Intelsat Vs launched by Ariane included *F7* on October 19, 1983, and *F8* on March 5, 1984.

1980 December 18 The Intelsat board of governors decided to purchase additional satellites of the Intelsat VA type. Improved from the standard Intelsat V, the modified satellites would handle 15,000 simultaneous telephone calls, have increased reliability and greater operating flexibility. They would follow six successful Intelsat V series satellites, the last of which was launched on May 19, 1983. The upgraded capacity was needed to meet traffic demand until the high-capacity Intelsat VI series was launched.

1980 December 29 A mobile launcher platform carried the Shuttle orbiter *Columbia* and its associated external tank and solid rocket boosters from the Vehicle Assembly Building to Launch Complex 39A at the Kennedy Space Center, Fla. Modified from its Apollo configuration, the MLP had two new cutouts for SRB (solid rocket booster) exhaust but did not have the launch umbilical tower and a weight of 8.23 million lb. Aiming for a launch on March 17, NASA engineers discovered during external tank propellant loading tests on January 22 and 24 that small cracks had appeared in the insulation. This delayed the first flight until early April while repairs were made. A series of 11-day tests began on February 10 to qualify systems and procedures for the first Shuttle mission.

1980 Development tests were concluded on a new Soviet surface-to-surface missile capable of carrying high explosive or nuclear warhead. Code named SS-23 Spider, this battlefield support weapon was a replacement for the aging Scud B. Carrying a 100-KT nuclear warhead, the missile had a range of 300 mi. and an accuracy of 1,400 ft. It was gradually introduced during the decade to serve on Soviet borders.

1981 January 8 NASA announced that it would fly *Voyager 2* past Saturn in August 1981 so that it would be deflected on a slingshot course for Uranus, arriving at that planet in

January 1986. In effect, *Voyager 2* was already on such a fly-by trajectory to Saturn and the formal decision to fly past a third planet was a commitment to hold the spacecraft on its present course. *Voyager 1* had flown a trajectory selected to optimize scientific objectives at Jupiter, Saturn and the moon Titan, and because that mission had been a total success, *Voyager 2* could give up some of the potential science at Saturn to use its gravity field to visit at least one, perhaps two, more planets. If the spacecraft successfully navigated the Uranian system, it could fly on for an encounter with Neptune, which on encounter in 1989 was the solar system's outermost planet.

1981 January 16   A group of 14 universities in the Association of Universities for Research in Astronomy (AURA) was selected by NASA to set up and operate a science institute for the space telescope (ST). The facility would be set up on the Homewood campus of John Hopkins University in Baltimore and would receive data from the ST, when launched, via the tracking and data relay satellite system (TDRSS). Scientists would request the Goddard Space Flight Center to perform ST orbit attitude changes to align the telescope with selected targets.

1981 January 21   The 1981 season of Soviet antisatellite tests began with the launch of *Cosmos 1241* on an SL-8 from Plesetsk to an orbit of 606 × 628 mi. inclined 65.8°. On February 2 the interceptor, *Cosmos 1243*, was launched on an SL-11 from Baikonur to a 225 × 631 mi. orbit from which it performed its close fly-by and reentered the atmosphere on the same day. The second interceptor, *Cosmos 1258*, was launched by SL-11 from Baikonur to a 187 × 636 mi. orbit on March 14, again for a close fly-by within hours of launch and reentry on the same day. *Cosmos 1241* remains in orbit.

1981 January 24   In preparation for a new flurry of manned launches to the *Salyut 6* space station, *Progress 12* was launched at 14:18 UT by SL-4 from the Baikonur cosmodrome. It docked to the aft port two days later at 15:56 UT with Salyut 6 in an orbit of 164.5 × 199.5 mi. *Progress 12* was used to raise the orbit to 210 × 217.5 mi. in readiness for the next crew. *Progress 12* remained docked to *Salyut 6* six days after the *Soyuz-T 4* cosmonauts arrived on March 13. It was separated from the aft port at 18:14 UT on March 19, and de-orbited the following day. *Progress 12* was the last cargo tanker to visit the *Salyut 6* space station.

1981 January 29   The first mass migration of weather satellites in space began when the U.S. National Oceanic and Atmospheric Administration moved *GOES 2* from 105° W to 107° W to serve as a weatherfax relay, and *SMS 1* to a higher orbit of 22,600 mi., out of the way of other geostationary satellites. In other activity, *GOES 4* was to be moved from 98° W to 135° W as backup to the ailing *GOES 3*, and *SMS 2* would be moved from its station at 75° W to a nearby standby location after the launch of *GOES 5* on May 22.

1981 January   With urging from Congress, NASA canceled the planetary version of the solid propellant inertial upper stage and switched to the cryogenic Centaur G-prime, a standard Centaur which it said could be adapted for use with the Shuttle. This forced a change in the planned dual launches of the Galileo orbiter and probe missions to Jupiter in February and March 1984. Centaur was more powerful than the IUS, so instead of conducting two separate launches, the probe would be carried by the orbiter on a direct flight without

gravity-assist to the vicinity of Jupiter, where it would separate for atmospheric entry. Because the Centaur G-prime could not be developed in time for the 1984 opportunity, the launch of the recombined Galileo would be postponed to April 1985, with arrival at Jupiter in 1987.

1981 February 11   The first Japanese N-II launch vehicle successfully placed the fourth engineering test satellite, ETS-4 *Kiku-3* (Chrysanthemum) in a 143 × 18,682 mi. orbit, inclined 28.3° from Tanegashima. Essentially a Delta 2914 built under license, the N-II differed from its predecessor in having a stretched first stage and a new second stage, with three solid propellant strap-on boosters. Overall height was 116 ft. and the N-II had a constant diameter of 7 ft. 10 in. throughout its length, and a liftoff weight of 298,000 lb. The 1,411-lb. satellite was designed to monitor launch vehicle performance and the behavior of satellite structures in space. It was a cylinder 6 ft. 10 in. × 9 ft. 2 in. tall and continued to operate until December 24, 1984.

1981 February 20   The three liquid propellant main engines attached to *Columbia* were fired for 20 sec. while the Shuttle was on the pad at Launch Complex 39A. This flight readiness firing (FRF) came at the end of the "wet" countdown demonstration test prior to the Shuttle's first spaceflight. The tests were a final rehearsal for preflight Shuttle activity and qualified launch pad systems. The Shuttle was secured to the pad by eight tie-down fixtures, four at the base of each solid rocket booster. Alongside, a fixed service structure (FSS) comprised a lattice framework 40 ft. square and 247 ft. high surmounted by a hammerhead crane 265 ft. above the pad and a lightning tower extending to 347 ft. The FSS provided a pivot point for the rotating service structure (RSS), 102 ft. long, 50 ft. wide and 130 ft. high set between 59 ft. and 19 ft. above the pad.

1981 February 21   A Japanese Mu-3S solid propellant launcher carried the 414-lb. science satellite Astro-A, named *Hinotori*, (Firebird) from Uchinoura to a 349 × 386 mi. orbit inclined 31.3°. The third satellite to be equipped with scientific instruments for measuring x radiation, *Hinotori* carried sensors for observing and recording images of solar flares and solar particles. The satellite decayed back down into the Earth's atmosphere on July 11, 1991.

1981 March 5   The council of the European Space Agency began a series of meetings in Paris, France, to consider its position on the International Solar Polar Mission following an announcement from the United States that NASA would be forced to withdraw from the collaborative endeavor as originally conceived. Over the next several months various alternative, cheaper plans were considered, but the ESA decided to take up the U.S. offer to fly the European mission on a Shuttle/inertial upper stage as compensation for having invested in early development. Under the new arrangement, one spacecraft would be launched in May 1986, arriving at Jupiter in 1987, crossing either the south or the north pole of the sun between October and December 1989. In July 1982 NASA changed the upper stage to a Centaur when plans to launch planetary spacecraft on an IUS had to be canceled. On September 10, 1984, ISPM became Ulysses, with reference not only to Homer's hero but the Italian poet Dante's description of Ulysses' urge to explore "an uninhabited world behind the sun."

1981 March 12   At 19:00 UT cosmonauts Vladimir V. Kovalenok and Viktor P. Savinykh were launched in the *Soyuz-T 4* spacecraft from Baikonur to the *Salyut 6* space station which had *Progress 12* docked at the aft port. Savinykh was the 100th person to fly in space, a total which included 43 from the United States, 50 from the Soviet Union and seven from other communist countries. *Soyuz-T 4* docked to the forward port at 20:33 UT the next day, and the crew entered *Salyut 6* for a planned stay of 2 ½ months during which two Intercosmos visits were scheduled. With the complex in an orbit of 210 × 217.5 mi., Kovalenok and Savinykh opened up the *Progress 12* orbital module and unpacked stores and supplies for their stay. Using Progress propulsion, the complex was pushed into a slightly higher orbit ready for the first international visitors. The aft port was vacated by *Progress 12* at 18:14 UT on March 19, which then deorbited, making way for the arrival of *Soyuz 39* on March 23.

1981 March 19 John Bjornstad, a Rockwell International employee working at NASA's Kennedy Space Center, died just hours after losing consciousness in the aft section of orbiter *Columbia* as it was being prepared for its first flight into space. Along with five other people, Bjornstad was exposed to a 100% nitrogen gas purge which asphyxiated the 50-year-old worker shortly after entering the area where the three main engines are housed. All six men were given medical treatment, the remaining five being released without permanent injury.

1981 March 22 At 14:59 UT an SL-4 launcher carried the *Soyuz 39* spacecraft into space containing Soviet pilot Vladimir A. Dzhanibekov and Mongolian cosmonaut Jugderdemidiyn Gurragcha, the eighth Intercosmos crew. *Soyuz 39* docked to the aft port on *Salyut 6* at 16:28 UT on March 23, and the crew joined Kovalenok and Savinykh in the space station. Medical tests were conducted as standard practice, but Gurragcha became ill during his first weightless day and was unable to participate fully in activity aboard *Salyut 6*. *Soyuz 39* undocked at 08:20 UT on March 30 and the Intercosmos mission ended on Earth at an elapsed time of 7 days 20 hr. 43 min.

1981 April 12 Carrying astronauts John Young (Commander, or CDR) and Robert Crippen (Pilot, or PLT), on the first NASA Shuttle mission (STS-1, for Space Transportation System-1) began at 12:00:04 UT with the liftoff of *Columbia* from Launch Complex 39A at the Kennedy Space Center, Fla. Ignition of the three main engines came at T − 6 sec. with liftoff at ignition of the two solid rocket boosters (SRBs). Shock waves from the pad tore off 16 thermal protection tiles and damaged 148 more. The liftoff weight of 4,457,111 lb. included the 219,260-lb. orbiter *Columbia* of which 10,823 lb. comprised the payload consisting of development flight instrumentation in a special cargo bay pallet. The two SRBs were separated at 2 min. 10.4 sec., at a height of 31 mi., with *Columbia* and its external tank now weighing 1,476, 278 lb. The SRBs descended on parachutes, splashed down 161 mi. from Cape Canaveral at 7 min. 10 sec. and were subsequently retrieved. Meanwhile, *Columbia*'s main engines were progressively throttled to 65% of full power as the vehicle got lighter, maintaining a 3-g acceleration until cutoff at 8 min. 34.4 sec. and a height of 72.5 mi. The ET separated from *Columbia* at 8 min. 58.1 sec., followed by a ballistic trajectory and broke up over the Indian Ocean at a height of 53 mi. *Columbia* was placed in a 152.7 × 153.9 mi. orbit inclined 40.3° to the equator by two orbital maneuvering system (OMS) burns, OMS-1 at 10 min. 34.1 sec. for 86.3 sec. and OMS-2 at 44 min. 02.1 sec. for 75 sec. Shortly after the OMS-2 burn, the cargo bay doors were opened and the crew settled into two days of

The culmination of more than 10 years' work, NASA's reusable Shuttle Columbia *heads for space from the refurbished Launch Complex 36A, Kennedy Space Center, Florida, on April 12, 1981.*

systems testing. A 28.8-sec. OMS-3 burn was conducted at 6 hr. 20 min. 46.5 sec., followed by a 33.1-sec. OMS-4 burn at 7 hr. 5 min. 32.5 sec. after which *Columbia* was in a 170.2 × 170.3 mi. orbit. The 297 ft./sec. de-orbit maneuver with both OMS engines took place at 53 hr. 21 min. 31.1 sec. on the 36th orbit, and entry into the Earth's atmosphere began 17 min. 30 sec. later at a height of 75.7 mi. and a range to the runway of 5,031.3 mi. The landing, on runway 23 at Edwards AFB, Calif., came at 54 hr. 20 min. 53 sec., at a speed of 207 MPH. With a weight of 195,472 lb., *Columbia* rolled to a stop in 8,993 ft. It was delivered back to KSC by a Boeing 747 Shuttle carrier plane, arriving April 28.

1981 April 21 The Federal Communications Commission issued a memorandum setting out rules for experimental direct-broadcast satellite (DBS) systems. Several U.S. companies had made it known that they were interested in building powerful satellites capable of beaming TV programs directly to small domestic dish antennas. By July 16, 1981, 13 applications had been filed from potential DBS operators. On June 23, 1982, the FCC specified that DBS systems should operate in the 17.3–17.8/12.2–12.7 GHz regions.

1981 April 25 At 02:01 UT an SL-13 launch vehicle carried the 44,100-lb. *Cosmos 1267* unmanned TKS spacecraft from

The world's first reusable manned spacecraft returned to Earth at Edward's Air Force Base, California, when Columbia *touched down at the end of a two-day mission, carrying astronauts John W. Young and Robert L. Crippen.*

Baikonur cosmodrome to a 120 × 161.5 mi. orbit with an inclination of 51.5°. *Cosmos 1267* was similar to *Cosmos 929* (launched on July 17, 1977), and, like its predecessor, carried a reentry module developed from the Chelomei LK-1 spacecraft. A lot of orbital maneuvering took place over the following four weeks until the 12,000-lb. reentry module was returned to Earth on May 24 with *Cosmos 1267* in an orbit of 164 × 178 mi. In a further three weeks of maneuvering, *Cosmos 1267* adjusted its orbit to that of the *Salyut 6* space station and conducted an automatic rendezvous with it on June 19, although no docking took place.

1981 May 14 At 17:17 UT the *Soyuz 40* spacecraft was launched by SL-4 from the Baikonur cosmodrome carrying Soviet pilot Leonid I. Popov and the Romanian cosmonaut Dumitru Prunariu, on the ninth and final Intercosmos mission to *Salyut 6*. Docking at the aft port took place at 18:50 UT the next day and the international crew joined Vladimir Kovalenok and Viktor Savinykh in the space station. Medical experiments and technological demonstrations were performed before the visiting cosmonauts got back inside their spacecraft and returned to Earth. *Soyuz 40* undocked from *Salyut 6* on May 22, and in a procedure which by now had become routine, delivered its crew to the Soviet Union at the end of a flight that had lasted 7 days 20 hr. 41 min. It was the last time the standard Soyuz ferry would be used for space station flights; missions from now on would utilize the T-series spacecraft.

1981 May 15 A Scout launcher carried the 368-lb. *Nova 1* navigation satellite from Vandenberg AFB, Calif., to an elliptical Earth orbit. Nova was the operational version of the TIP (transit improvement program) satellite, providing more accurate position fixes than Transit with fewer updates. The octagonal satellite had a body width of 1 ft. 8 in. and a length

of 3 ft. 10 in., with a truncated attitude control system on top and four solar-cell panels. A 26-ft. scissors boom provided gravity gradient stabilization. *Nova 2* was launched on October 12, 1984, to a 715 × 744 mi. orbit with an inclination of 90.1°. The Nova satellites were replaced with the much more capable Navstar series. They remain in orbit.

NASA announced that it had set September 30, 1981, as the date for the return to space of Shuttle *Columbia* on STS-2, the second of four planned orbital flight test (OFT) missions and scheduled to last 5 days 4 hr. 55 min. On August 31, the day *Columbia* was rolled to Launch Complex 39A, the flight was rescheduled to October 9. A spillage of oxidizer propellant on September 22 damaged 360 thermal protection tiles, delaying the flight until November 4. On that date, astronauts Joseph Engle and Richard Truly entered *Columbia* but the mission was scrubbed when a hold was called at T − 31 sec. due to a malfunctioning electrical fuel cell tank.

1981 May 25 Arabsat, the Arab satellite communications organization, selected an international team led by Aérospatiale of France to build three satellites for the 22 member nations. Teamed with Ford Aerospace of the United States, Aérospatiale produced a three-axis-stabilized satellite comprising a box structure 7 ft. 5 in. × 5 ft. 4 in. × 4 ft. 11 in., with a launch weight of 2,635 lb., reducing to 1,305 lb. on orbit. Two solar array wings with a span of 67 ft. 11 in. would provide a minimum of 1,300 W to power 25 × 4–6 GHz transponders carrying 8,000 telephone circuits plus 7 TV programs, and 1 × 4/2.5 GHz transponder for one semidirect TV broadcast channel to ground antennas 10 ft. in diameter.

1981 May 26 *Salyut 6* cosmonauts Vladimir Kovalenok and Viktor Savinykh returned to Earth in the *Soyuz-T 4* spacecraft at the end of their stay aboard the space station. Their mission had lasted 74 days 18 hr. 38 min., during which time they had received two Intercosmos crews and brought an end to the manned occupation of *Salyut 6*. Kovalenok and Savinykh had few physical effects from their spaceflight and soon adapted to Earth gravity once more. Manned operations with *Salyut 6* had raised the Soviet spaceflight endurance record from just under 63 days to almost 185 days, and 29 cosmonauts had conducted 900 medical and 200 science and technology experiments, shot 19,500 frames of film and accumulated about 36,000 man-hours on board the orbiting station. By comparison, all previous Salyut missions combined had accumulated only 10,000 man-hours of occupation by 15 cosmonauts.

1981 May 31 The second satellite launched by the Indian Space Research Organization was carried from the Sriharikota range near Madras to a 115.5 × 260 mi., 46.3° orbit. Launched by the third SLV-3, the satellite, called *Rohini 2*, weighed 84 lb. and carried instrumentation to record the performance of the launcher in addition to a Landmark camera for basic Earth observation. It decayed on June 8, 1981, earlier than predicted.

1981 June 19 At 06:52 UT the 30,000-lb. *Cosmos 1267* orbital module was docked to the forward port on the *Salyut 6* space station in an orbit of 211 × 225.5 mi. The launch weight of the *Cosmos 1267* module had been reduced by extensive orbital maneuvering and by the release, on May 24, of a 12,000-lb. reentry capsule. More maneuvers were performed after *Cosmos 1267* docked to *Salyut 6*, and extensive tests were conducted over the next four months until the docked combination was in an orbit of 246 × 255 mi. Between October 1981 and July 1982 the unmanned combination decayed gradually back down toward the atmosphere until it was in an orbit of 197.5 × 201 mi. On July 28 the perigee was lowered to 138.5 mi. and a day later another maneuver brought *Cosmos 1267* and *Salyut 6* back down through the atmosphere where they burned up.

The third Ariane 1 launch vehicle took off from Kourou, French Guiana, carrying a 3,605-lb. payload consisting of the 1,541-lb. *Meteosat 2* ESA weather satellite, a 1,477-lb. experimental Indian communication satellite called *Apple* and a 587-lb. CAT capsule similar to that carried by L01. Designated L03, this was the first flight since the failure of L02 on May 23, 1980. The vehicle had a liftoff weight of 463,200 lb. and placed *Meteosat 2* in a geostationary orbit at 0° long. and *Apple* in geostationary orbit at 102° E. The *CAT-3* capsule was left in a 125.5 × 22,270 mi., 10.5° orbit and remains in space.

1981 June 21 NASA's Solar Maximum Mission satellite detected solar neutrons for the first time when there was a major solar flare producing particularly intense activity on the surface of the sun. Neutrons separate into protons and electrons within 15 minutes, and because the Earth is eight light minutes from the sun, only neutrons accelerated to at least half the speed of light will be detected by a satellite in Earth orbit. The observations were made by a gamma-ray and neutron spectrometer built by the University of New Hampshire with help from West German scientists.

1981 July 10 Almost six months after Dr. Robert Frosch retired, James M. Beggs was sworn in as the new NASA administrator. He took over from Dr. Alan M. Lovelace who had served as acting administrator since January 20. James Beggs had been executive vice president for Aerospace at General Dynamics, which he joined in 1974. Before that, he

had served four years as under secretary of transportation. Beggs had served with NASA from 1968 to 1969 as associate administrator at the Office of Research and Technology. He was to remain administrator until December 4, 1985, when he was unfairly accused of fraud while at General Dynamics.

1981 July Martin Marietta presented the results of a study into shuttle-derived vehicles, a range of configurations using Shuttle elements to increase the vehicle's lift potential. Martin provided four classes of SDV: Class 1 with the orbiter replaced by a module of three SSMEs (Space Shuttle main engines) and a cargo pod; Class 2 with a conventional orbiter and two liquid rocket boosters; Class 3 with SSME module, cargo pod and two LRBs; Class 4 with the orbiter's SSMEs removed to a module under the external tank.

1981 August 3 A Delta 3914 launch vehicle carried two Dynamics Explorer satellites from Vandenberg AFB, Calif., to separate polar orbits. Each satellite comprised a 16-sided polygon 4 ft. 5 in. wide and 3 ft. 9 in. tall. The two satellites were instrumented to study the interaction of solar energy with the outer regions of the Earth's atmosphere, to acquire data on auroras produced by energetic particles spiralling round the Earth's magnetic field lines and on the effects of solar radiation on radio transmissions and weather patterns. The 888-lb. *Dynamics-Explorer 1* was placed in a 347 × 14,475 mi. orbit inclined 89.9°. It carried 231 lb. of instruments and operated for more than nine years before retirement on February 28, 1991. The 915-lb. *Dynamics-Explorer 2* went to a 185 × 619 mi. orbit inclined 90° and performed well until it finally decayed on February 19, 1983.

1981 August 5 A privately funded rocket belonging to Space Services Inc. of Houston, Tex., blew up during a test firing of its 55,125-lb. thrust first-stage engine, destroying its launch pad at Matagorda Island, 50 mi. north of Corpus Christi, Tex. Designed and built in only six months, the rocket, called Percheron, was a low-cost attempt to produce a launcher for small satellites. The rocket had been put together with the help of GCH, a Sunnyvale, Calif., company and operated on liquid oxygen–kerosene propellants. Percheron was abandoned but its principal fund-raiser, ex-astronaut Donald "Deke" Slayton, progressed to a series of solid propellant rockets called Conestoga.

1981 August 10 The first operational flight of an N-II launch vehicle carried the first of two Japanese Geostationary Meteorological Satellites (GMS-2) from Tanegashima to an elliptical orbit from which it was placed in geostationary orbit at 140° E. Called *Himawari-2*, (Sunflower) the spin-stabilized satellite had a height of 10 ft. 2 in., a diameter of 7 ft. 1 in. and a weight of 652 lb. on station. GMS-2 carried the same instruments as GMS-1, launched on July 14, 1977, but Hughes Aircraft had built the satellite using lightweight materials to limit the weight to the capacity of the launch vehicle. It continued to operate until November 20, 1987. Of similar design, the 668-lb. GMS-3 was launched by N-II on August 3, 1984, also to 140° E.

1981 August 14 Product of a cooperative program between Canada's National Research Council and NASA, the Black Brant X sounding rocket made its first flight from Wallops Island, Va. The four-stage rocket was a development of the basic Black Brant and could carry heavier loads on a wide variety of trajectories. For this first flight, the three-stage vehicle reached a height of 390 mi. and a downrange distance of 440 mi. NASA's Goddard Space Flight Center had been responsible for development and had received help from Bristol Aerospace in Canada and Saab in Sweden.

1981 August 22   *Voyager 2* began a sequence of close encounters with the planet Saturn and its moons that would last for two weeks. Moving from above the ring plane southward, the spacecraft obtained images of Iapetus with a closest pass of 565,000 mi. at 14:56 UT. From a distance of 292,000 mi. at 14:53 UT on August 24, *Voyager 2* obtained images of Hyperion with a resolution of 5.5 mi., 10 times better than images of Hyperion from *Voyager 1*. The spacecraft came closest to Titan at 11:04 UT on August 25, but at 10 times the distance flown by *Voyager 1*. Between this event and Voyager's closest approach to Saturn, the spacecraft imaged five more moons, three of which had been discovered in images from *Voyager 1*. Flyby of the ringed planet came at 04:50 UT on August 26, when *Voyager 2* came to within 63,000 mi. of Saturn. Ten more moon imaging tasks were conducted by September 4. In flying through the Saturnian system, *Voyagers 1* and *2* had taken 18,500 images and discovered 11 new moons, bringing to 23 that planet's known complement of satellites. Altogether, the two spacecraft had returned a total 70,000 images to Earth.

1981 August 24   Representing a new class of Soviet electronic intelligence gathering (elint) satellite, *Cosmos 1300* was launched by SL-14 from Plesetsk to a 395 × 413 mi. orbit inclined 82.5°. For 18 months such elint missions were interspersed with flights of the Soviet elint satellites launched by SL-3 to an orbital inclination of 81.2°, the first of which had been flown on December 18, 1970. The new class gradually replaced the older generation (retired in 1983) with 2 launched in 1981, 2 in 1982, 4 in 1983, 4 in 1984, 6 in 1985, 6 in 1986, 3 in 1987, 4 in 1988, none in 1989, 1 in 1990, 1 in 1991, and 2 in 1992. The satellites operated in groups, each weighing 3,500–4,400 lb.

1981 August 28   Boeing released its final report on the Space Operations Center, which it had been studying under a contract from the NASA Johnson Space Center. With assembly flights projected to start in 1989, it would take two years and seven flights to make the facility operational, at an estimated cost of $4.7 billion. The initial eight-person SOC comprised two 49 × 15 ft. habitation modules side-by-side, linked at one end by a docking tunnel and at the other end by a service tunnel. The tunnels were about 82 ft. long and provided docking space for a logistics module, 26 × 15 ft. in size. The nest of modules would occupy an area 60 × 80 ft. Outside this nest, two 49 × 25 ft. hangars would be available for servicing satellites, each attached to opposite ends of the service tunnel. At 90° to the plane of the modules, two 101 × 50 ft. solar array panels would be carried on opposing booms 96 ft. long, giving the "wings" a total span of almost 400 ft. The SOC was the focus for NASA space studies over the next 18 months.

1981 August   The European Space Agency and Arianespace agreed to develop a second launch pad at Kourou, French Guiana, for the Ariane launch vehicle. Primarily configured for the Ariane 4 launch vehicle, at this time expected to be flown in 1985, the new site would be known as ELA-2 and incorporate several important improvements. With the existing ELA-1 launch pad, Arianespace could mount six launches per year. ELA-2 would increase this to twelve per year. Unlike ELA-1, ELA-2 would allow launcher assembly

and checkout prior to arrival at the pad, facilitated in the vertical position along a double rail track.

1981 September 21   A Soviet SL-14 launcher carried the French *Oreol 3* satellite into space from Plesetsk. The 2,250-lb. satellite contained instruments designed to study the magnetosphere and the ionosphere and was the latest product of Franco-Soviet space cooperation. Built in the Soviet Union, *Oreol 3* carried 220 lb. of French equipment and 155 lb. of Soviet equipment. It was placed in an orbit of 236 × 1,195 mi. at an inclination of 86.2°. It remains in space but inoperative.

1981 September 30   A Soviet SL-14 launcher carried the 1,985-lb. *Cosmos 1312* geodetic satellite from Plesetsk to a 926 × 932 mi., 82.6° orbit. First in a new series of geodetic satellites, it comprised a cylinder 6 ft. 6 in. in diameter with a length of 7 ft. A long boom extended from the forward end with the long axis pointing Earthward. Twelve rectangular solar-cell arrays hinged to the base of the satellite were deployed in a splayed configuration. Other satellites in this series launched to an 82.6° orbit included *Cosmos 1410* (September 24, 1982), *Cosmos 1589* (August 8, 1984), *Cosmos 1803* (December 2, 1986), *Cosmos 1950* (May 30, 1988) and *Cosmos 2037* (August 28, 1989).

1981 October 1   NASA announced that Maj. Gen. James Abrahamson, a former U.S. Air Force manned-orbiting-laboratory astronaut and, later, manager of the F-16 fighter aircraft development program, had been appointed to head the Office of Space Transportation. Leaving a staff position at Andrews AFB, Gen. Abrahamson brought qualities of organizational and administrative abilities Administrator James Beggs wanted for getting the Shuttle program into an operational phase. On April 16, 1982, Beggs announced that Gen. Abrahamson would be head of Shuttle operations, taking over from Stanley Weiss who had been named NASA chief engineer to succeed Walter Williams. On March 27, 1984, Gen. Abrahamson was named to head the Strategic Defense Initiative program.

1981 October 2   President Ronald Reagan announced sweeping changes to the U.S. nuclear deterrent force, authorizing deployment of the Trident II C-5 submarine-launched ballistic missile, ordering the deactivation of Titan II and approving development of the M-X ICBM with 100 missiles deployed from the late 1980s. Abandoning the Carter administration's multiple protective shelter (MPS) concept, the Department of Defense was to examine three M-X basing options: a survivable long endurance aircraft, each carrying one M-X; active defense of M-X with antiballistic missiles; or survivable locations deep underground. In an effort coded Rivet Cap, the first Titan II squadron was taken off alert on September 30, 1982, the last in July 1987. From that date, Strategic Air Command had an all-solid ICBM force.

1981 October 6   NASA launched the 963-lb. *Solar Mesosphere Explorer* satellite on a Delta 2310 from Vandenberg AFB, Calif., to a 314 × 317 mi. orbit inclined 97.6°. A second payload comprised the 115-lb. *UOSAT/Oscar 9* satellite, placed in a 298 × 301 mi. orbit. *UOSAT* was built by the University of Surrey in England, ninth in a series of amateur radio satellites launched piggyback on primary payloads. SME was cylindrical in shape, 5 ft. 6 in. long × 4 ft. 1 in. in diameter and had been designed to study ozone formation and density at altitudes between 12 and 50 mi. Five instruments would measure atmospheric constituents, water-vapor abundance and levels of solar radiation. UOSAT decayed on October 13, 1989, followed by the SME on March 5, 1991.

1981 October 28 NASA published plans to shut down the dish antennas (85 ft. in diameter) at Canberra (Australia), Madrid (Spain) and Goldstone (California), which had been part of the original Deep Space Network built to support early planetary missions. The stations would continue to operate the dish antennas of 210-ft. and 112-ft. diameter at these facilities. The loss of the smaller antennas would reduce by 30% the amount of coverage NASA could give to existing planetary missions but save $7.2 million per year.

1981 October 30 The Soviet Union launched the 11,000-lb. *Venera 13* spacecraft to Venus on an SL-12 launcher. Placed initially in a low Earth parking orbit, it was the first of two spacecraft the Soviet Union had assigned to the 1981 launch window. Working actively with information supplied by planetary scientists assimilating the data from the NASA Pioneer-Venus orbiter, Soviet engineers and scientists had developed a plan to sample surface materials on Venus. The NASA orbiter data helped in the selection of a suitable site for the Soviet lander and its companion, *Venera 14*, which was launched by an SL-12 on November 4, 1981. *Venera 13* performed a course correction maneuver on November 10, 1981, followed by a maneuver with *Venera 14* four days later which would ensure they reached the planet in March 1982.

1981 November 1–6 The first major international conference on the physics and atmosphere of the planet Venus was hosted by NASA's Ames Research Center at a hotel in Palo Alto, Calif., where 118 presentations were made by scientists from the United States, the Soviet Union and several other countries. The most visited planet in the solar system, Venus had been the target for 17 U.S. and Soviet spacecraft since *Mariner II* launched on August 27, 1962. The conference heard compelling evidence that the enormous pressures and high temperatures in the atmosphere of Venus were the result of a "greenhouse effect" and that meteorologists and climatologists had learned about threats to Earth's environment by unraveling the complexities of the Venusian atmosphere.

1981 November 9 RCA auctioned seven transponders on the *Satcom 4* communications satellite for a total $90.1 million, an average of $12.87 million per transponder compared with $3 million it cost for RCA to launch the satellite. The price hike was indicative of the burgeoning market driven by profits from telecommunication satellite operations. The Federal Communication Commission rejected the auction result as excessive, but lack of protest from the user community forced the FCC to allow the sum involved. NASA Administrator James Beggs implemented structural changes in the organization of various offices. The separate offices of Space Science and Space and Terrestrial Applications were reunited as the Office of Space Science and Applications. This reflected increasing emphasis on the two separate but distinctly related fields at NASA Headquarters.

1981 November 12 At 15:09:59 UT, after delays totaling 2 hr. 49 min. 59 sec., the STS-2 mission began with the launch of Shuttle orbiter *Columbia* from LC-39A at the Kennedy Space Center, Fla., carrying astronauts Joe Engle (CDR) and Richard Truly (PLT). About 11 min. of the launch delay was to synchronize the precise liftoff time with the orbital track of a KH-11 military reconnaissance satellite, which would be used to view the Shuttle from a distance for possible tile

damage. Liftoff weight of 4,470,308 lb. included the 230,939-lb. orbiter carrying an 18,778-lb. cargo made up of development flight instrumentation and a 5,604-lb. package of scientific instruments called *OSTA-1* (for the NASA Office of Space and Terrestrial Applications) for Earth observation. With two orbital maneuvering system (OMS) burns, *Columbia* was placed in an initial orbit of 138.1 × 143.8 mi. with a 38° inclination, but on the fifth orbit three more OMS burns changed this to 160 × 165.7 mi. Carried for the first time, the remote manipulator system (RMS) arm, built by Spar Aerospace in Canada, was tested. At 4 hr. 35 min. elapsed time, one of three fuel cells failed and flight controllers cut the mission from a planned five days to just over two days. Weighing 204,262 lb., *Columbia* landed at an elapsed time of 54 hr. 13 min. 11 sec. on runway 23 at Edwards AFB, Calif., with a roll-out of 7,711 ft. *Columbia* arrived back at KSC on November 25.

1981 November 18 In a speech to the National Press Club in Washington, D.C., President Ronald Reagan announced new arms control initiatives aimed at cutting missiles and their offensive capabilities. Empowered by what the Reagan administration called Strategic Arms Reduction Talks (START), negotiators would bargain with defense analysts in the United States and the Soviet Union. Over 11 years of talks, it would be the most significant step so far in abolishing threatening technologies for nuclear war. Adopting a twin-track policy of rigorously updating and improving the U.S. nuclear deterrent while talking down the overall threat to peace, START would carry U.S./Soviet discussions through three administrations in Washington and the collapse of communism and the Soviet Union.

The U.S. communications and electronics company RCA announced it had received a contract from NASA's Ames Research Center to carry out a preliminary design of a Mars polar orbiter capable of mapping seasonal variations and atmospheric changes. Ames and the Jet Propulsion Laboratory were separately studying the application of low-cost Earth satellite designs to a series of missions including a Mars water mission, a Mars climatology orbiter, a Mars aeronomy orbiter, a Mars geoscience orbiter and a Mars probe network. All of these varied studies eventually converged into the Mars geoscience/climatology orbiter (later named Mars Observer) proposed by the Solar System Exploration Committee in its May 1983 Core Program report.

1981 December 4 The European Space Agency handed over to NASA the first flight unit of the Spacelab pressure module during a special ceremony at Bremen, West Germany. Built by a consortium of European companies headed by ERNO in West Germany, Spacelab flight units were shipped to the United States for the first manned flight aboard the Shuttle *Columbia* on mission STS-9. Components of the second flight unit were delivered in 1982.

1981 December 7 NASA selected nine experimental teams and ten theoretical teams of scientists to take part in the Upper Atmosphere Research Satellite (UARS) program. Managed by the Office of Space and Terrestrial Applications, UARS would provide detailed information on the dynamics of the stratosphere and the mesosphere, provide data on energy input to the atmosphere from space and monitor the chemical composition of the atmosphere. The scientists would build instruments for UARS to assess the impact of human activity on the environment, and the satellite would be placed in orbit by the Shuttle. On March 6, 1985, NASA awarded a contract to General Electric for design and construction of the UARS.

1981 December 17   A Soviet SL-8 launcher carried six amateur radio satellites into orbit from Plesetsk. All six occupied slightly different orbits averaging 1,015 × 1,030 mi. at an inclination of 83°. The Soviet Union had been anxious to participate in this international endeavor, which stimulated interest among amateur radio enthusiasts. All six satellites remain in orbit.

1981 December 20   The European Space Agency's *Marecs A* maritime communication satellite was launched by the fourth Ariane (L04) from Kourou, French Guiana, to a geostationary transfer orbit of 137 × 22,258 mi. inclined 10.5°. The satellite was positioned later in geostationary orbit at 26° W. With a weight at launch of 2,218 lb., *Marecs A* took the form of a hexagonal prism, 8 ft. 2 in. high by 6 ft. 6 in. across supporting two solar panels. From May 1, 1982, *Marecs A* was operated by Inmarsat on Atlantic Ocean traffic. Ariane 1 weighed 463,398 lb. at liftoff and included a 478-lb. CAT data capsule which was left in an orbit of 135 × 22,243 mi. inclined 10.5°.

1981   Development trials of the definitive Mod 3 version of the Soviet SS-17 ICBM were completed prior to full operational deployment during 1982. This final version of the missile had a throw weight of 6,000 lb., a range of 6,000 mi. and an accuracy of 1,200 ft. with four 50-KT MIRV warheads. Replacing Mod 1 and Mod 2 versions, which were completely withdrawn by 1982, some 150 Mod 3 SS-17 were in service from 1983. With the Soviet designation RS-16, the SS-17 Spanker was gradually being withdrawn when the START agreement was negotiated in the early 1990s.

1982 January 5   The Department of Defense announced plans to deploy the first 40 of the 100 M-X ICBM missiles it wanted, in canisters in redundant Minuteman silos. For launch, the canister would be raised to the surface and the M-X ejected by pressurized gas, in the same manner as submarine-launched ballistic missiles and Soviet ICBMs using the cold-launch technique. The first 10 M-Xs were scheduled to become operational in late 1986. Titan II silos had been rejected because they were not spaced so as to be compatible with the preferred layout for antiballistic-missile defenses, one option still being examined by the Reagan administration for long-term, survivable basing. The Air Force wanted to buy 226 M-X missiles, 126 in excess of the 100 it wanted to deploy, for testing and as spares to replace faulty missiles.

1982 January 13   The decision to develop an advanced version of the Ariane launch vehicle, known as Ariane 4, was taken by the European Space Agency and would include six variants to accommodate different payload requirements. Ariane 4 would comprise a lengthened first stage similar to Ariane 3, with propellant capacity increased from 140 tons to 220 tons and the addition of liquid or solid propellant strap-on boosters. Payload capacity to geostationary transfer orbit would range from 4,190 lb. to 9,260 lb., making Ariane 4 highly competitive with U.S. Delta and Atlas Centaur launchers. The development program was funded by France (59.3%), West Germany (18.2%), Italy (6.6%), the United Kingdom (4.9%), Belgium (4.6%), Spain (2%), Switzerland (1.8%), Sweden (1.2%), the Netherlands (1.1%), Denmark (0.2%) and Ireland (0.1%).

1982 January   The U.S. Office of Management and Budget forced NASA to cancel development of the cryogenic Centaur G-prime upper stage by withdrawing funds on the grounds that not enough missions were planned to justify development costs. This brought major changes to NASA Earth-escape missions, particularly the *Galileo* orbiter/probe flight. Originally rejected in January 1981 on the grounds that it was insufficiently powerful, the solid propellant inertial upper stage (IUS) was reinstated as the escape stage for *Galileo*. Even using a more powerful IUS-2, there was insufficient energy to send the 5,600-lb. *Galileo* to Jupiter, so a 4,660-lb. thrust Star 48 solid propellant boost motor was attached to a special injection module beneath the spacecraft, and a gravity-boost trajectory was selected. Launched in May 1985 on a wide orbit of the sun, *Galileo* would gain momentum from a flyby of Earth in July 1987 and arrive at Jupiter in late 1989.

1982 February 19   Rockwell International completed a study for NASA's Johnson Space Center on the effect of the proposed Space Operations Center (SOC) on Shuttle operations. The mission model now envisaged the first of six SOC build-up flights in 1989 with operational capability in 1990. The baseline habitation module was now sized at 46 × 15 ft., the service tunnel at 50 × 15 ft., the docking tunnel at 53 × 15 ft. and the logistics module at 26 × 15 ft. The SOC was now seen as a space base and marshaling yard for large and complex payloads, a storage facility for cryogenic tug propellants, a "garage" for payloads checkout in space and a terminal for deep-space missions.

1982 February 22   NASA's Jet Propulsion Laboratory selected Martin Marietta to study options for a potential Venus radar mapping mission. NASA's existing VOIR (Venus Orbital Imaging Radar) spacecraft concept had been removed from the NASA budget request for fiscal year 1983 submitted by the Office of Management and Budget because it was too costly. Hughes Aircraft Co. would build the synthetic aperture radar for what was now called the Venus Radar Mapper, or Under the revised plan, a smaller spacecraft would be authorized as a new start in fiscal year 1984, launched by Shuttle in 1988 and placed in orbit around Venus six months later.

1982 February 26   Launched by a Delta 3910 from Cape Canaveral, Fla., the first of a new generation of communication satellites for Western Union, *Westar IV*, was placed in a geostationary transfer ellipse. From there it was moved to a geostationary position at 99° W, replacing *Westar I*. The new generation was built on the Hughes HS 376 for compatibility with either the Delta or the Shuttle. With twice the size and four times the capacity of the first Westar generation, *Westar IV* had a launch weight of 2,360 lb. and 24 transponders serving the United States from Hawaii to the Virgin Islands. Second in the series, *Westar V* was launched by Delta to 123° W on June 8, followed by *Westar VI* on Shuttle mission 51-A, February 3, 1984.

1982 February   Britain's Royal Navy successfully completed a series of flight trials with the Chevaline warhead designed to replace U.S. W-58 warheads on its Polaris missiles. During the month, HMS *Renown,* submerged 30 mi. off the coast of Cape Canaveral, Fla., launched several missiles equipped with this warhead. Chevaline carried six 40-KT warheads on a ballistic trajectory, maneuvering to confuse tracking radars, and ejected them shotgun-fashion at the target. It was not a MIRV system, where each warhead is released on a separate trajectory flying to its unique target, but would do much to confuse missile defenses. Deployment was competed by the late 1980s.

Development engineering of a new warhead for the M-X ICBM was started at the Lawrence Livermore National Laboratory. Encapsulated by an Avco Mk.21 reentry vehicle, the W-87 warhead could be fused for five different air burst or surface burst options. The standard 300-KT yield could be increased to 475 KT by an additional sleeve of enriched uranium 235. The cone-shaped Mk.21 was 5 ft. 9 in. tall and 1 ft. 9 in. across at the base. Each M-X fourth-stage bus could carry 10–12 reentry vehicles to within 400 ft. of their targets across a full range of more than 5,000 mi. The first production W-87 was ready in April 1986, and 250 had been built when production ceased in 1987.

1982 March 1   The Soviet spacecraft *Venera 13* landed on the surface of Venus at 7.5° S × 303° W equipped with imaging instruments and a device for sampling surface materials. The lander/carrier had separated February 27 and the carrier performed a course correction which took it past the planet at a distance of 22,370 mi. The 3,257-lb. lander entered the atmosphere at a speed of 36,432 ft./sec., reduced by drag to 820 ft./sec. at a height of 40 mi. when the main parachute deployed. Slowed to 500 ft./sec. by a second parachute, the lander descended to a height of 29 mi., when that too was jettisoned. A large circular plate acted as a brake in the dense atmosphere, lowering the lander to the surface at 03:57 UT and a touchdown speed of 26 ft./sec. (18 MPH). A hollow pipe with cutting blades was lowered to retrieve a 1.2-in.-long rock sample which was placed inside the lander. The lander relayed data via the carrier for 2 hr. 7 min., returning eight surface images and data on surface chemistry.

1982 March 2   NASA announced plans to improve the payload performance of the Shuttle by switching to a new lightweight solid rocket booster. SRB weight would be reduced by replacing eight of the eleven conventional steel booster segments with new filament-wound segments, chopping weight by 33,000 lb. on each booster. This would make possible a payload increase of 6,000 lb. to polar orbit. NASA wanted to begin using filament-wound SRBs in 1985, but the program was delayed by lack of funds.

1982 March 5   The Soviet *Venera 14* touched down on the surface of Venus at 13.25° S × 310° W, 600 mi. southeast of the *Venera 13* site. Like its companion, the *Venera 14* lander transmitted data via the fly-by carrier spacecraft before it continued past the planet into a heliocentric orbit. The lander also retrieved a small portion of surface material and subjected it to x-ray fluorescence analysis, a feat of unusual complexity because of the enormous pressures and 855°F temperature at the surface. Chilled down to about 86°F, an inner chamber on the lander was kept at a lower pressure for the chemical analysis. The *Venera 14* lander operated for 57 min.

1982 March 22   After a delay of just one hour to rectify a ground equipment problem, astronauts Jack Lousma (CDR) and Gordon Fullerton (PLT) were launched on the STS-3 mission by Shuttle *Columbia* at 16:00:00 UT from LC-39A at the Kennedy Space Center, Fla., to a 149.9 × 150.3 mi. orbit at an inclination of 38°. The liftoff weight of 4,468,755 lb. included the 235,556-lb. orbiter carrying a 22,710-lb. payload

The external propellant tank for the third Shuttle flight did away with the white fire-retardant latex coating. Launched on March 22, 1982, STS-3 was the first mission to sport an orange-colored tank.

comprising a wide range of experiments for NASA's Office of Space Science as well as a package of development flight instruments. Named OSS-1, the 8,470-lb. science payload incorporated packages moved around the outside of the orbiter by the remote manipulator system arm. The third of four orbital flight test (OFT) missions, STS-3 lasted 8 days 4 min. 46 sec. and ended with a landing of the 207,072-lb. *Columbia* at Northrup Strip, White Sands, N.Mex., instead of Edwards AFB, Calif., due to wet conditions. *Columbia* came to a stop in 13,737 ft. and returned to KSC on April 6, 1982.

1982 March 25 Boeing completed the study of a shuttle-derived cargo vehicle (SDCV) built up from Shuttle elements and capable of placing a payload of 86,640 lb. into low Earth orbit. Earlier studies had proposed utilization of Shuttle elements with three SSMEs (Space Shuttle main engines) in a group in the same relative position as the orbiter. The SDCV differed in that it placed a single SSME beneath a short version of the external tank with an 18 × 60 ft. cargo module on top. The two SRBs (solid rocket boosters) were to be reduced in size to a length of 122 ft. 6 in. The SDCV would have a length of 193 ft. 6 in. and a liftoff weight of 2.8 million lb.

1982 March The Department of Defense completed plans for design studies of an advanced satellite system enabling U.S. forces to communicate with each other and with commanders in the field wherever they are in the world. Called Milstar (military strategic tactical and relay), the system envisaged several high-power satellites hardened against electromagnetic impulse and radiation caused by nuclear war. In short, they were to keep the war machine running in the environment of an all-out global nuclear war. The three team bidders were: Lockheed/General Electric; TRW Systems/Hughes Aircraft Co.; and Ford Aerospace/Boeing.

1982 April 10 A NASA Delta launch vehicle put the satellite *Insat 1A* into a geostationary transfer orbit of 103 × 22,361 mi. inclined 28.4°. Boosted to geostationary altitude by apogee motor, *Insat 1A* was located over 74° E to provide limited Earth resource data and communications services, the first time both applications had been combined in one satellite. Problems with the C-band antenna and the solar array panel gave Insat a lop-sided instability problem and brought about excessive expenditure of propellant, and the satellite failed with a total loss of communication on September 4. Built by Ford Aerospace, *Insat 1A* was a complex design incorporating several separate functions not usually found in one satellite.

1982 April 15 The European Space Agency announced that it would begin development of a free-flying platform carrying scientific experiments, to be launched by Shuttle and left in space for retrieval on a later flight. Called EURECA, for European retrievable carrier, the platform would weigh about 3,300 lb. and be capable of remaining in space for up to six months, supporting a wide range of science and technology experiments. An onboard propulsion system, and power from solar arrays, would give EURECA controllers the option of moving the platform to higher orbit; the satellite would resume low orbit for recovery.

1982 April 19 At 19:45 UT a SL-13 launch vehicle lifted away from the Baikonur cosmodrome with the *Salyut 7* space station. Placed in an orbit of 132 × 162 mi. at an inclination of 51.6°, the 41,700-lb. station was essentially the same as

Salyut 6, with improvements brought about through experience with the first of the third-generation "civilian" space stations. Extra handholds were provided on the exterior to aid space walking, bigger windows were provided, better exercise equipment was installed, a broader range of medical monitoring equipment was provided, and improved eating facilities, including refrigerators and electric heating stoves, were installed. MKF-6M and KATE-140 cameras were carried for Earth observations, but instead of the BST-1M astrophysical telescope, *Salyut 7* had a XT-4M telescope and an XS-02M x-ray spectrometer.

1982 April 26 Martin Marietta submitted its final review of technology requirements for shuttle-derived vehicles. A wide variety of different booster types was examined, including fly-back boosters reminiscent of early shuttle concepts. Hybrid SRBs (solid rocket boosters), in which the forward half of the booster, containing oxygen, added to the combusting solids as they fired, were considered. Compared to a standard Shuttle orbiter cargo bay of 15 × 60 ft., an unmanned cargo module placed parallel to where the orbiter would usually be attached could accommodate loads up to 25 × 90 ft.

1982 April An entrepreneurial company called Orbital Sciences Corp. was set up to develop a solid propellant boost rocket capable of carrying Shuttle-launched satellites on geosynchronous transfer ellipses or on Earth-escape missions. A memorandum of agreement with NASA on the joint development of a transfer orbit stage (TOS) was signed on December 17, 1982. Martin Marietta became primary contractor in May 1983 and Honeywell was brought in to produce a laser inertial navigation system. TOS was capable of boosting a 13,400-lb. satellite from Shuttle orbit to geosynchronous transfer orbit. TOS had a length of 11 ft., a diameter of 7 ft. 6 in. and weighed 24,100 lb. It utilized a 21,460-lb. thrust United Technologies Orbus 21 solid propellant motor with a 2 min. 24 sec. burn time.

1982 May 13 At 09:58 UT cosmonauts Anatoly N. Berezovoy and Valentin V. Lebedev were launched in their 15,100-lb. *Soyuz-T 5* spacecraft by SL-4 from the Baikonur cosmodrome. The Igla onboard radio guidance system took them to an automatic rendezvous with *Salyut 7* and a docking was achieved at 11:36 UT on May 14. Berezovoy and Lebedev entered the station and activated systems to prepare it for manned operations. On May 17 the crew released a small 62-lb. satellite called *Iskra 2* designed for the use of amateur astronomers by members of the Moscow Sergo Ordzhonikidze Aviation Institute. It decayed on July 9. A second satellite, *Iskra 3*, was ejected on November 18 and it decayed on December 16. *Salyut 7* carried an automated systems checkout computer called Delta, and this was activated to take over attitude control commands and relieve the crew of unnecessary tasks.

1982 May 20 NASA Administrator James Beggs established a Space Station Task Force at Headquarters in Washington, D.C., to provide focus for the separate design and architecture studies being conducted at most NASA field centers. Having proven the flight capabilities and performance of the reusable Shuttle, NASA now wanted to move ahead with plans for a permanently manned station in low Earth orbit and had received approval from the White House to begin conceptual studies. Ex-Gemini program flight director John D. Hodge would head the new task force as it sought to define space station objectives, needs and capabilities. The task force would provide a bridge between the many and varied studies that had already been conducted on Shuttle-launched station concepts, and signified the firm commitment to proceed with a permanently manned facility as a national goal.

1982 May 23   At 05:57 UT a Soviet SL-4 launcher carried *Progress 13* from its launch pad at the Baikonur cosmodrome to an orbit from which it began a sequence of rendezvous maneuvers with the *Salyut 7* space station. A docking was achieved at the aft port at 07:57 on May 25. *Salyut 7* cosmonauts Anatoly Berezovoy and Valentin Lebedev opened up the orbital module and transferred to the station 1,975 lb. of scientific equipment to outfit *Salyut 7* for lengthy periods of research activity. Of this total, 533 lb. was in support of a planned visit by a French cosmonaut for which a comprehensive program of cooperative experiments had been scheduled. Two new materials processing furnaces, called Kristall and Magma-F, were installed, and 76.6 gal. of water was pumped from tanks into the *Salyut 7* Rodnik water system. The crew also monitored the supply of 1,455 lb. of propellant for the attitude and maneuvering system. *Progress 13* undocked at 06:31 UT on June 4, and was de-orbited two days later.

1982 May   The Australian national satellite system moved a step nearer reality when AUSSAT Proprietary selected Hughes Aircraft to build three communications satellites capable of direct broadcast TV, high-quality TV relay between cities and digital data transmission. The HS-376 series satellites would each provide 15 channels, four of which would carry 30-W Ku band amplifiers, while eleven would have 12-W channels. Three dish antennas would provide three receive and seven transmit beams for networking. Each AUSSAT HS-376 would provide 1,054 W of electrical power at the beginning of life and 860 W at the end. A Thiokol Star 30 solid propellant apogee motor would place each satellite in geostationary orbit.

1982 June 3   Test flights with scaled-down models of the Soviet Lapot space plane took place from Kapustin Yar when an SL-8 carried the first 2,200-lb. BOR-4 test bed to a 98 × 127 mi., 50.6° orbit. Veiled under the designation *Cosmos 1374*, the scale model was recovered after 1¼ revolutions of the Earth. As it reentered, it conducted a 375-mi. cross-range maneuver, landing 350 mi. south of the Cocos Islands in the Indian Ocean. After splashdown it deployed a radio beacon and flotation device shaped like a road-traffic cone and was retrieved by recovery forces. BOR-4 did not have folding wings like the original Lapot and was used for testing carbon leading edges and thermal protection systems designed for *Buran,* the Soviet shuttle. BOR-4 had a length of 11 ft. 2 in. and a width of 8 ft. 6 in. It was lowered by parachute at the end of flight. Similar missions were flown by other BOR-4 models: *Cosmos 1445* on March 15, 1983; *Cosmos 1517* on December 27, 1983; and *Cosmos 1614,* on December 19, 1984, which was recovered from the Black Sea away from the prying eyes of non-Soviet aircraft that had been used to photograph the space plane recovered from the earlier flights.

1982 June 6   In support of a major Soviet military exercise demonstrating a wide range of major strategic weapon systems, an SL-8 launched the 4,410-lb. *Cosmos 1375* from Plesetsk to a 608 × 629 mi. orbit inclined 65.8°. In the exercise, the Soviet Strategic Rocket Forces launched two ICBMs, two antiballistic missiles, one submarine-launched ballistic missile and one SS-20-theater nuclear delivery system. The 3,300-lb. interceptor *Cosmos 1379* was launched by SL-11 from Baikonur on June 18 to a 334 × 633 mi. orbit.

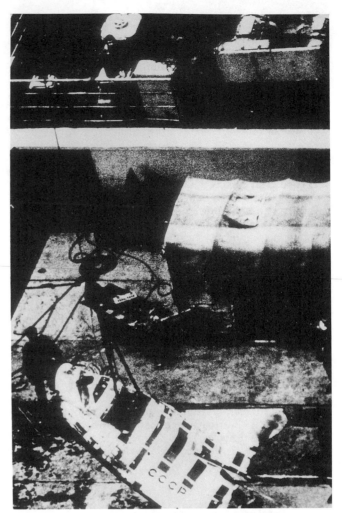

Launched on March 15, 1983, under the designation Cosmos 1445, *this Soviet space plane was retrieved from the sea by recovery ship.*

Emphasizing the importance placed by Soviet planners on antisatellite systems, the interceptor launch was immediately followed by the launch of navigation and reconnaissance satellites to simulate replacement following negation by U.S. ASAT systems. *Cosmos 1379* failed to intercept the target satellite and reentered the Earth's atmosphere the same day.

1982 June 10   Engineers at the NASA Goddard Space Flight Center commanded thrusters on the satellite *ISEE-3* (previously known as ISEE C) to fire, moving it from its halo orbit at the lagrangian point between Earth and the sun to the first in a series of five lunar fly-bys. The fly-bys took place in 1983, on March 30, April 23, September 27, October 21 and December 22, the last a scant 72 mi. above the surface. Instruments on *ISEE-3* were ideal for measuring characteristics of comet tails, and the trajectory of comet Giacobini-Zinner brought it within rendezvous distance of the satellite. Most of the energy to reach the comet was provided by the gravity-assist maneuvers at the moon, accelerating *ISEE-3* from 2,900 MPH to 5,145 MPH. *ISEE-3* was redesignated International Cometary Explorer (ICE). It would encounter the comet Giacobini-Zinner on September 11, 1985, after three course corrections.

1982 June 21 Telesat Canada signed an agreement with NASA for the launch of five Anik C- and D-series communication satellites, four on the Shuttle and one on a Delta. The flights were made available at the specially subsidized price set by NASA in competitive marketing against the European Ariane expendable launch vehicle. Telesat Canada would pay about $75 million for the five flights. Each Anik D was built around the Hughes HS 376 series spin-stabilized satellite bus.

1982 June 24 At 16:29 UT Soviet cosmonauts Vladimir A. Dzhanibekov, Aleksandr S. Ivanchenkov and French cosmonaut Jean-Loup Chrétien were launched aboard the *Soyuz-T 6* spacecraft to the *Salyut 7* space station. Thus was France the twelfth country to send a man for space travel. Docking had to be accomplished manually by Dzhanibekov and took place at 17:46 UT on June 25. About 2 hr. 15 min. later the crew transferred to *Salyut 7* and joined Berezovoy and Lebedev, making it the first time that five people had been in the same spacecraft simultaneously. (The five crew members of the 1975 Apollo-Soyuz Test Project were never in the same vehicle at once.) A broad and comprehensive range of biomedical, scientific and technological experiments were conducted over the next six days. The *Soyuz-T 6* crew withdrew to their spacecraft on July 2 and undocked from the aft port at 11:04 UT, returning to Earth at an elapsed time of 7 days 21 hr. 51 min.

1982 June 27 Astronauts Thomas Mattingly (CDR) and Henry Hartsfield (CDR) were launched aboard the Shuttle *Columbia* from LC-39A at NASA's Kennedy Space Center, Fla., at 15:00:00 UT. Both solid rocket boosters were lost when their parachutes failed to open during descent. The 4,481,935-lb. liftoff weight included the 241,772-lb. orbiter and its 24,492-lb. payload. Last of four orbital flight test (OFT) missions, STS-4 carried a classified Department of Defense payload in addition to development flight instrumentation and the first get away special (GAS), an instrumented canister flown by NASA at a special low price to encourage universities, research institutes and individuals to sponsor experiments in space. This first GAS canister contained nine experiments from the Utah State University. The remote manipulator arm was used to move a gas analyzer around the outside of the orbiter. STS-4 ended with the first landing on the concrete runway 22 at Edwards AFB, Calif., at 7 days 1 hr. 9 min. 31 sec., followed by a 9,878 ft. rollout. *Columbia* returned to KSC on July 15.

1982 June 29 The first satellite equipped with emergency distress signal transceiver equipment in the Cospas/Sarsat network, *Cosmos 1383* was launched by SL-8 from Plesetsk in Russia. A military navigation type, the 1,786-lb. satellite was followed by the first U.S. satellite in the network, *NOAA-8*, on March 28, 1983. This satellite was launched by Atlas F from Vandenberg AFB, Calif., but it failed prematurely in June 1984. *Cosmos 1574* was launched with transceivers on June 21, 1984. By October 1984, 137 incidents had been reported and 312 people had been rescued after their distress signals were relayed to emergency rescue centers.

1982 June 30 Brazil's state telecommunications company, Embratel, issued a purchase order to Canada's Spar Aerospace for two Hughes HS 376 series satellites. Under an arrangement with Hughes, Spar would assemble the HS 376s and deliver them to Guiana for launch on an Ariane 3. With a launch weight of 2,514 lb. and an on-orbit mass of 1,480 lb., the satellite would have 24 channels in the 6 GHz band and 30

amplifiers, 6 of which were for redundancy. Brasilsat was designed to last 8–10 years in orbit.

1982 July 4 The White House issued a National Space Policy document signed by President Ronald Reagan. Placing the attainment of U.S. supremacy in space as a priority second only to national security, Reagan's directive put international cooperation last. Between the two extremes, the document wanted the United States to expand its economic and scientific return from space, to bring about greater commercialization and to expand private-sector involvement. The National Space Policy had been put together through a series of analyses organized by Presidential Science Adviser George Keyworth.

1982 July 5 The second Shuttle orbiter rated for spaceflight arrived at the Kennedy Space Center, Fla., atop the Boeing 747 carrier plane and was taken to the orbiter processing facility (OPF) where it was prepared for the STS-6 mission. OV-099, named *Challenger*, was about 2,488 lb. lighter than *Columbia* due to weight-saving measures and lighter materials. It was the first orbiter configured for operational flight from the outset, requiring no ejection seats or development instrumentation. Build-up with the external tank and solid rocket boosters began October 1 on Mobile Launch Platform no. 2, all previous Shuttle flights having used MLP-1. *Challenger* was moved to the Vehicle Assembly Building on November 23 and to LC-39A on November 30.

1982 July 10 At 19:58 UT the *Progress 14* cargo tanker was launched by SL-4 from the Baikonur cosmodrome and docked to the aft port on *Salyut 7* at 11:41 UT two days later. Cosmonauts Anatoly Berezovoy and Valentin Lebedev began to move the stores across almost immediately. *Progress 14* provided an additional 1,545 lb. of propellant, 46 gal. of water and 2,425 lb. of supplies, including film for the cameras, scientific equipment, extra clothing and food. *Progress 14* was undocked from the aft port at 22:11 UT on August 10 and was de-orbited two days later.

1982 July 16 The first NASA Delta 3920 launched the 4,282-lb. *Landsat D* Earth resources satellite to a 434.5 × 435.6 ml. orbit inclined 98.2°; on station it was officially designated *Landsat 4*. The two-stage Delta 3920 differed from the 3910 in replacing the TRW-201 second-stage engine with the 9,400-lb. thrust Aerojet AJ-10–118K and enlarging the propellant tanks, within the same stage skirt, to hold 13,200 lb. of propellant. Unlike previous launch modes, in which five of nine Castor motors ignited on the pad, on this flight six were fired with the first stage, with the remaining three ignited at an altitude of 8 mi., 60 sec. after launch. Built around a multimission modular spacecraft (MMS) bus by General Electric, *Landsat 4* carried two sensors: a thematic mapper (TM) built by Hughes and a multispectral scanner system (MSS) similar to that flown on the first three Landsats. The TM consisted of a seven-band scanner collecting radiation from the Earth in a swath 115 mi. wide. The MMS bus had provision for servicing and upgrading by astronauts brought to the satellite by Shuttle, but this was not possible after NASA elected not to develop a Shuttle launch facility at Vandenberg AFB, Calif.

1982 July 22 The first launch of a production Pershing II battlefield support missile failed to achieve its objective after launch from Cape Canaveral, Fla. About 4 sec. into flight, the first stage disintegrated and the missile fell into the Atlantic Ocean. Developed as a modular and more accurate version of the Pershing IA, this second-generation missile had improved first- and second-stage motors and a terrain-matching radar guidance system. This allowed the descending warhead to make terminal corrections based on images of the approaching terrain contained in a memory section of the guidance unit. Pershing II carried a W-85 nuclear warhead with two selectable yields (5–50 KT or 10–20 KT) accurate to within 100 ft. over a full range of 1,100 mi.

1982 July 30 Cosmonauts Anatoly Berezovoy and Valentin Lebedev performed the first space walk of the *Salyut 7* mission 11 weeks after their launch. The transfer module side hatch was opened at 02:39 UT and Lebedev moved outside to retrieve canisters of amino acids and biopolymers attached to the side of the space station before it was launched on April 19, 1982, for analysis of the effects of cosmic radiation on biological substances. Assisted by Berezovoy, Lebedev tested various mechanical joints, including pipes and connectors, to evaluate their effectiveness in weightless space. The space walk ended at an elapsed time of 2 hr. 33 min.

1982 August 9 NASA identified eight U.S. aerospace companies selected for eight-month studies to identify missions and roles for a manned space station: Boeing Aerospace, General Dynamics, Grumman Aerospace, Lockheed, Martin Marietta, McDonnell Douglas, Rockwell and TRW Systems. On September 15 the Marshall Space Flight Center selected four of these to perform additional studies: Boeing to evaluate building large structures; Martin Marietta and TRW Systems to study satellite servicing and maintainability in space; and General Dynamics to study operations of an orbital transfer vehicle, or OTV. Called *Space Station Needs, Attributes and Architectural Options Study*, the analyses were to assume availability of an operational space station in 1990.

1982 August 19 At 17:12 UT cosmonauts Leonid I. Popov, Aleksandr A. Serebrov and Svetlana Y. Savitskaya were launched aboard their *Soyuz-T 7* spacecraft by SL-4 from Baikonur cosmodrome. Savitskaya was only the second woman to fly in space, more than 19 years after cosmonaut Valentina Tereshkova. *Soyuz-T 7* docked to the aft port on *Salyut 7* at 18:32 on August 20. Extensive medical experiments were carried out, with emphasis on measuring the reaction of Savitskaya's physiology to weightlessness, as well as an electrophoretic separation test on biological substances. After exchanging contour couches on August 27, the *Soyuz-T 7* crew entered *Soyuz-T 5* at the forward port and landed back on Earth at 7 days 21 hr. 52 min. At 16:47 UT on August 29, Berezovoy and Lebedev moved *Soyuz-T 7* from the aft port to the forward docking port.

1982 August 26 The first NASA Delta 3920/PAM-D launched the first in a new generation of communication satellites for Telesat Canada. The launcher differed from the basic Delta 3920 in having a PAM-D third stage, increasing the Delta 3914's geosynchronous transfer orbit capability from 2,065 lb. to 2,800 lb. Called *Anik D-1 (Telesat G)*, the satellite was an HS 376 design weighing 2,370 lb. at launch and 1,454 lb. after arriving on station at 104.5° W. With 24 transponders, each handling 480 telephone circuits, *Anik D-1* would provide Canada with TV, voice communications and data links, serving initially as backup to the three Anik As and the one Anik B launched between 1972 and 1978. *Anik D-1* had a height of 21 ft. 7 in. fully deployed and a diameter of 7 ft. 1 in.

1982 September 1 Responding to President Ronald Reagan's space policy statement of July 4, 1982, U.S. Air Force Space Command was formally inaugurated as the operating agency to consolidate all U.S. Air Force space-related operations. With headquarters in Colorado Springs, Colo., on October 1 Space Command established the Space Technology Center at Kirtland AFB, N.Mex. Space Command initially had about 6,000 civilian and military personnel worldwide. It took over the management of the North American Aerospace Defense Command and Aerospace Defense Command missile warning and space defense facilities. With the motto "Guardians of the High Frontier," Space Command initially comprised two Space Wings: the 1st located at Peterson AFB, Colo., which would manage early warning ground and space-based sensors; and the 2nd, which was to manage the Consolidated Space Operations Center at Falcon Air Force Station, Colo., to operate satellites and related systems.

1982 September 3 A Japanese N-II launch vehicle carried the 849-lb. engineering test satellite ETS-III, *Kiku-4* (Chrysanthemum), from Tanegashima to a 601 × 763 mi. orbit inclined 44.6°. The satellite took the form of a box, 2 ft. 9 in. × 6 ft. 5 in. × 2 ft. 9 in., with paddle-type solar arrays. It was equipped with instruments for measuring the efficiency of systems designed for larger satellites in the future. In particular, ETS-III measured the performance of three-axis control systems, solar array deployment and active thermal control. It remains in orbit.

1982 September 9 The fifth Ariane 1 launch (L5), the first operational flight, ended in disaster when the third-stage turbopump failed 30 sec. before scheduled cutoff, causing the vehicle to fall back into the Atlantic Ocean. With a liftoff weight of 463,398 lb., L5 was launched from Kourou, French Guiana, with the 2,218-lb. maritime communications satellite *Marecs B* and the 926-lb. Italian *Sirio-2* weather data distribution satellite on board. It performed well until the cryogenic turbopump failed 4 min. 35 sec. after third-stage ignition. Not until June 16, 1983, after modifications to the turbopump, did next Ariane launch vehicle fly.

Space Services, Inc., launched a solid propellant rocket called Conestoga I at 15:15 UT to an altitude of 196 mi. from a cattle ranch on Matagorda Island, northeast of Corpus Christi, on the south-central coast of Texas. For two years SSI had been working to develop a family of private, low-cost satellite launchers and this first successful flight carried a mock payload weighing 1,097 lb., containing 40 gal. of water ejected at peak altitude which spread out leaving a visible marker of frozen droplets.

1982 September 18 At 04:59 UT the Soviet cargo tanker *Progress 15* was launched by SL-4 from the Baikonur cosmodrome to the *Salyut 7* space station, in orbit at 187.5 × 203 mi. Docking took place at the aft port at 06:12 UT on September 20. Carrying 1,545 lb. of propellant, 40 gal. of water, spare parts for Salyut, scientific equipment, extra film for the cameras and a range of personal items for *Salyut 7* cosmonauts Anatoly Berezovoy and Valentin Lebedev, *Progress 15* was undocked from the aft port at 11:46 UT on October 14 and was de-orbited into a safe area of the Pacific Ocean two days later.

1982 September 23 The Federal Communications Commission gave Comsat General Corp. permission to begin work on a direct-broadcast satellite (DBS) system with its Satellite Television Corp. On November 5 the FCC approved seven more DBS applications: CBS, Direct Broadcast Satellite Co., Graphic Scanning Corp., RCA Americom, United States Satellite Broadcasting (Hubbard), Video Satellite Systems and Western Union. U.S. DBS systems proved notoriously difficult to get off the ground, and none of the above reached operating status as described in their original proposals.

1982 September   The U.S. Congress overturned a decision by the Office of Management and Budget in January 1982 to cancel development of the Centaur G-prime cryogenic upper stage for the Shuttle and ordered NASA to reinstate the stage for the *Galileo* mission to Jupiter. Because of the nine-month delay in its development, Centaur could not now be ready to support the scheduled launch of Galileo in May 1985, forcing a delay in the launch to May 1986. Moreover, because Centaur was more powerful than the combination IUS-2 and injection module, *Galileo* could fly direct to Jupiter without gravity assist, reaching the planet in August 1988.

1982 October 12   A new system of advanced Soviet navigation satellites was inaugurated with the launch of the first three Glonass satellites by SL-12 from Baikonur. Designated Cosmos *1413–1415*, they were placed initially in an interim orbit inclined 51.6° to the equator. Prior to separating from the terminal stage, the orbit inclination was changed to 64.8°. All three occupied orbits averaging 11,850 × 11,890 mi. and each satellite weighed 3,090 lb. Subsequent Glonass triple-sets were launched by SL-12s, with two launches per year in 1983, 1984 and 1985. Only one set was launched in 1986. One of two launch attempts in 1987 was a failure and, after two successful launches in 1988, the two in 1989 each comprised two Glonass and one Etalon geodetic satellite as the third component. Two triplet-sets were flown in 1990, one in 1991, two in 1992 and one in 1993.

1982 October 22   NASA Administrator James Beggs announced changes to the qualifying criteria for selecting payload specialists aboard Shuttle flights. Instead of allocating crew space for a payload specialist to Shuttle users taking half or more of one mission, flights from 1984 on would become available on a reimbursable basis to a wide range of customers. These would include foreign and domestic companies, international cooperative partners, members of the scientific and industrial community and the Department of Defense.

1982 October 28   NASA launched the first of a new generation of advanced Satcom communication satellites for RCA. Carried into space by a Delta launched from Cape Canaveral, Fla., *Satcom 5* (RCA-E) was placed in a transfer orbit from which it was sent to geostationary orbit at 143° W to provide a service to Alaska. Built by RCA, the three-axis-stabilized satellite took the form of a box 4 ft. 7 in. × 5 ft. 3 in. × 5 ft. 11 in. and carried 28 solid-state C-band amplifiers with 4 redundant transponders and 24 assigned to television, voice and data transmissions. This was the first all-solid-state domestic communications satellite. Later named *Alascom*, the 1,300-lb. satellite was slightly overweight and had to be launched without the added weight of telemetry on the Delta third stage. *Satcom 6*, a replacement for *Satcom 1* dubbed 1R, was launched to 139° W on April 11, 1983, and *Satcom 7*, a replacement dubbed 2R, was launched to 83° W on September 8, 1983.

1982 October 30   The first in a new series of Titan 34-D heavy launch vehicles successfully placed two satellites in geostationary orbit from Complex 40 at Cape Canaveral, Fla.: the last

Built for RCA American Communications and sold to Alaska, this RCA Satcom 5 provided long-distance telecommunications between the state and the rest of the United States after launch on October 28, 1982.

1,365-lb. DSCS-II and the first 2,476-lb. DSCS-III military communication satellite. The standard Titan IIIC utilized solid propellant boosters assembled from five stacked segments. The Titan 34D incorporated five half-segments which increased length from 85 ft. to 90 ft. 5 in. and raised the total liftoff thrust to 2.5 million lb. The first stage was increased in length from 73 ft. to 78 ft. 11 in. and burn time from 2 min. 27 sec. to 2 min. 45 sec. The second stage produced a thrust of 101,000 lb. for 3 min. 30 sec. Titan 34D carried an inertial upper stage (IUS) third stage with a length of 21 ft. 2 in. and a diameter of 9 ft. 5 in. The solid propellant IUS had two stages, the lower stage with a thrust of 42,000 lb. and the upper stage with a thrust of 17,000 lb. The 34D could place 4,200 lb. in geostationary orbit or 31,650 lb. in low Earth orbit without IUS. The last of 15 Titan 34Ds was launched on September 4, 1989. Built by General Electric, the DSCS-III was a three-axis-stabilized box, 6 ft. 9 in. × 6 ft. 4 in. × 6 ft. 5 in., supporting fold-out solar panels with a span of 38 ft. producing 1,240 W and operating on 7.9–8.4/7.25–7.75 GHz.

1982 October 31   At 11:20 UT *Progress 16* was launched by an SL-4 from the Baikonur cosmodrome. At 13:22 UT on November 2 it docked to the aft port on *Salyut 7* where cosmonauts Anatoly Berezovoy and Valentin Lebedev could retrieve the supplies it had brought. At this date *Salyut 7* was in an orbit of 218 × 227 mi. Among the items transferred to *Salyut 7* was a small subsatellite, *Iskra 3*, for amateur radio enthusiasts. Similar to *Iskra 2* launched earlier in the *Salyut 7* mission, it was placed in the *Salyut 7* airlock hatch and released by the crew to its independent orbit on November 18. *Progress 16* was still docked to *Salyut 7* when Berezovoy and Lebedev returned to Earth on December 10. It was undocked automatically from the aft port at 15:32 UT on December 13 and de-orbited a day later.

1982 November 11   At 12:19 UT astronauts Vance Brand (CDR), Robert Overmyer (PLT), Joseph Allen (Mission Specialist, or MS) and William Lenoir (MS) were launched aboard Shuttle *Columbia* from LC-39A at the Kennedy Space Center. STS-5 was the first operational Shuttle flight with ejection seats disarmed and development flight instrumentation removed. The 32,080-lb. payload comprised two commercial communication satellites. The first, for Satellite Business Systems (SBS-3), was released from the cargo bay attached to a PAM-D solid propellant perigee-kick stage at 20:18:00 UT on launch day. The PAM-D fired 45 min. after separation to put the satellite in an elliptical path from which, on November 13, its onboard apogee motor placed it in geostationary orbit. The second, for Telesat Canada *(Anik C-3)*, was released from the cargo bay at 20:24:11 UT on November 12, also with its own PAM-D, and circularized in geostationary orbit on November 16 at 117.5° W. An HS-376, it offered 14/12 GHz Ku-band transmissions for point-to-point commercial services. A planned EVA (extravehicular activity) had to be canceled when suit problems were discovered during preparation. STS-5 ended at 5 days 2 hr. 14 min. 26 sec. on runway 22 at Edwards AFB, Calif. *Columbia* was returned to KSC on November 22.

1982 November 22   President Ronald Reagan announced the proposed basing mode for the M-X missile, now known as Peacekeeper. Rejecting air-launched and deep underground basing silos 3,000–4,000 ft. below the surface, the President chose what was called closely spaced basing, or CSB. In this concept, 100 ICBMs would be located at one site, 1.5 mi. × 14 mi., near Warren AFB, Wyo. Each missile would be contained in a hardened silo no more than 1,800 ft. from its neighbor. Where previous silo dispersal had sought to keep missiles apart so that no single warhead could damage two silos, planners now relied on fratricide to protect the pack. This "dense pack" basing meant that any two Soviet warheads targeted against more than one closely-spaced silo would prematurely detonate in the atmosphere due to the electromagnetic pulse of the preceding nuclear explosion. By packing all Peacekeepers into a single location, the Soviets would be unable to knock out all 100 missiles in a preemptive strike.

1982 December 7   By 245 to 176, the House of Representatives rejected President Ronald Reagan's closely spaced basing deployment concept and cut all funds for the Peacekeeper ICBM. In January 1983 Reagan set up the Commission on Strategic Forces under Lt. Gen. Brent Scowcroft, the former assistant to the President for national security affairs, which initially was to have reported its findings to the President by February 11, 1983, and make recommendations on Peacekeeper basing modes. Broader issues of land-based missile deployment were embraced by the commission, and its report was not presented to the President and the National Security Council until April 1983.

1982 December 10   Cosmonauts Anatoly Berezovoy and Valentin Lebedev returned to Earth aboard the *Soyuz-T 7* spacecraft at the end of a seven-month stay aboard the *Salyut 7* space station, recalled prematurely due to serious psychological problems. A poor match between similar personalities

resulted in problems soon after they reached the station in May and for months they said hardly a word to each other. The situation became serious and no amount of consultation with psychologists on the ground would resolve the matter. The crew was recalled to a landing at 15:04 UT. Despite their early return, their mission established a new record: 211 days 09 hr. 05 min.

1982 December 18   Shuttle orbiter *Challenger* performed a 20-sec. test of its three main engines in a flight readiness firing (FRF), standard procedure for a new orbiter. However, engineers found excessive hydrogen in the engine compartment, and the planned STS-6 launch date of January 20, 1983, was postponed first to February 1 and then to the end of the month. Additional time was needed to prepare the vehicle for a second FRF on January 25, but concentrations of hydrogen were detected once more. The leak was eventually traced to a 0.75-in.-long crack on the no. 1 engine coolant manifold. The engine was replaced with another from the National Space Technology Laboratory, Miss., but it too displayed a fault after it arrived at the Kennedy Space Center on February 4. That engine was replaced by yet another from NSTL which arrived at KSC on March 3, 1983.

1982 December 21   The first in a new series of U.S. military weather satellites was launched by Atlas E from Vandenberg AFB, Calif., to a 504 × 511.5 mi. orbit inclined 98.7° to the equator. Designated Block 5D-2 in the Defense Meteorology Satellite Program (DMSP) series, it was based on the Block 5D-1 design, with length increased by 1 ft. 6 in., on-orbit weight increased to 1,500 lb. (1,700 lb. for later satellites in the series) and a 125 sq. ft. solar array (versus 10 sq. ft. for the 5D-1 series). Designed for a four-year life, the 5D-2 carried the 5D-1 series instruments and a microwave imager for cloud moisture measurement. The weight of DMSP instruments on each satellite had grown from 25 lb. for the Block IV of 1966 to more than 600 lb. for later 5D-2 satellites. To mid-1995, seven 5D-2s had been launched, the most recent on March 24, 1995.

1982 December 28   *Cosmos 1426*, the first of the fifth-generation Soviet photoreconnaissance satellites, was launched by an SL-4 from Baikonur to a 125.5 × 221 mi. orbit inclined 50.5°. With a weight of 14,775 lb., the satellite type was based on the Soyuz but with much longer life. *Cosmos 1426* remained in space for 67 days and transmitted digital images before returning film to Earth. Launched on May 14, 1984, the second satellite of this type, *Cosmos 1552*, remained aloft for 173 days, while *Cosmos 1643*, launched on March 25, 1985, stayed up for 207 days.

1982 December   Boeing presented a classified proposal to the U.S. Department of Defense that envisaged a single-stage-to-orbit vehicle launched along a sled. Powered by one Space Shuttle main engine (SSME), the vehicle would, in theory, have been capable of placing a 4,000-lb. payload in polar orbit. This subscale demonstrator was to have been followed by the definitive design, weighing 1.2 million lb. and capable of lifting a 20,000-lb. payload. In 1983, when Lockheed and McDonnell Douglas got involved, the project was given the code-name Science Dawn, changed to Science Realm in 1985.

1983 January 5   The Department of Defense alerted the news media to a problem that had developed in space with a Soviet ocean surveillance satellite, *Cosmos 1402*, powered by a nuclear reactor containing 100 lb. of uranium. The propulsion unit that should have propelled the radioactive part of the spacecraft to a higher orbit where it would remain for several hundred years had failed and the spacecraft was expected to reenter the atmosphere in the next few weeks. On January 7 the Tass news agency reported that *Cosmos 1402* had separated into several elements. On January 23 the main body of the spacecraft reentered, followed at 11:56 UT on February 7 by the radioactive fuel core which burned up in the atmosphere and dispersed its uranium.

1983 January 23   At 22:21 UT the radioactive nuclear reactor from the Soviet ocean surveillance satellite *Cosmos 1402*, reentered the atmosphere and broke up over the Indian Ocean about 1,100 mi. southeast of the island of Diego Garcia. Launched on August 30, 1982, the satellite was of a new design different from *Cosmos 954*, a first-generation ocean surveillance satellite that had broken up over Canada on January 23, 1978. Like *Cosmos 954*, *Cosmos 1403* could not be boosted to a higher orbit. No harmful radiation was detected. The satellite's radioactive core containing the fissile material broke up over the South Atlantic on February 7.

1983 January 26   A NASA Delta 3910 launched the 2,360-lb. infrared astronomy satellite *(IRAS)* to a 532.3 × 549 mi. orbit inclined 100.1°. The satellite comprised a cylinder 11 ft. 10 in. long and 6 ft. 6 in. across, with solar panels for electrical power and a cryogenically cooled 22-in. Cassegrain telescope to measure infrared radiation in four bands between 8 and 119 microns. To work effectively, a helium tank cooled the telescope to 2.5°F above absolute zero. *IRAS* was built in the Netherlands, the telescope was built in the United States and the United Kingdom provided tracking and data collection services. After surveying more than 95% of the sky, pinpointing the locations of more than 200,000 infrared objects for the first time, IRAS ran out of helium on November 22. It remains in orbit.

1983 January   During a period when two new and important ICBMs were being tested by the Soviet Union, the SS-X-24 and the SS-X-25, Soviet engineers began encrypting telemetry from test flights originating at the Baikonur launch facility. Also this month, the Soviets began to jam the U.S. Cobra Dane, Cobra Judy and Cobra Ball long-range and early-warning radars, and the Rhyolite electronic intelligence gathering satellites (elint) in geostationary orbit. An urgent need for an advanced intelligence gathering geostationary satellite called Aquacade was held up when the IUS (inertial upper stage) released from Shuttle mission STS-6 malfunctioned, delaying the elint for almost a year.

1983 February 1   After several years of deliberation, and on orders from the White House, NASA transferred the operational management of the Landsat system to the National Oceanic and Atmospheric Administration. There was a general belief that the Earth resources satellite program could be managed on a commercial basis and NOAA was tasked with preparing it for that. *Landsat*s *1* and *2* were no longer operating, *Landsat 3* was to be shut down later in the year, and *Landsat 4* would continue to be controlled from the Goddard Space Flight Center on a reimbursable basis. NOAA would also take over *Landsat 5*, expected to be launched in 1985 on a reimbursable basis.

1983 February 4   A Japanese N-II launch vehicle carried the communications satellite CS-2A (*Sakura,* or Cherry Blossom)

to a geostationary transfer ellipse from which the apogee motor placed it in a stationary position at 130° E. The 1,480-lb. satellite comprised a drum-shaped cylinder and de-spun antenna section 10 ft. 10 in. tall with a diameter of 7 ft. 3 in. and a weight of 770 lb. on station. *Sakura* was built by Mitsubishi with extensive technical help from Ford Aerospace. The C-band and K-band channels could handle telephone, TV and data links and were under the control of the Telecommunications Satellite Corp. of Japan. CS-2A operated until December 7, 1991. The second *Sakura* satellite, CS-2B, was launched by N-II on August 5 and placed at 135°E where it operated until January 26, 1990.

1983 February 7   The first in a series of four experimental ballistic missile intercept tests organized by the U.S. Army took place over the Pacific Ocean. Called Homing Overlay, it began with the launch of a Minuteman test missile from Vandenberg AFB, Calif., toward the Kwajalein Atoll test area in the Marshall Islands. An intercept vehicle comprising the first two stages of another Minuteman missile was launched from Meck island in the Kwajalein group to intercept the reentry vehicle (RV) launched from Vandenberg. The test was unsuccessful because the cooling system for sensors located in the head of the interceptor malfunctioned, preventing it from locking on to the RV. A second test on May 28 failed when an electronics malfunction caused the interceptor to miss its target. The third test on December 16 was partially successful, but impact was not achieved due to a problem in the onboard software. The fourth test, on June 10, 1984, was successful.

1983 February 9   The first of five Atlas H launch vehicles carried the first in a new series of upgraded White Cloud ocean surveillance satellites from Vandenberg AFB, Calif., to an elliptical orbit of 658 × 718 mi. with an inclination of 63.4°. To accommodate the heavier series of satellites the Navy chose to adopt a proposal from General Dynamics that the more powerful MA-5 engines be fitted to the SLV-3D, a model of Atlas previously developed for use with the Centaur upper stage. Developing 7,000 lb. more thrust than the SLV3-D Atlas-Centaur configuration, Atlas H could lift the new White Cloud series without an upper stage. The satellites themselves incorporated solid propellant Thiokol Star 37E kick motors for final orbit placement. The other Atlas H–White Cloud launches took place on June 9, 1983, February 5, 1984, February 9, 1986, and May 15, 1987.

1983 February 20   Japan's Institute of Space and Astronautical Sciences (ISAS) launched a Mu-3S solid propellant rocket from Kagoshima carrying the 397-lb. science satellite Astro-B *Tenma* to a 297 × 304 mi. orbit inclined 31.5°. The satellite was one in a continuing series designed to observe celestial x-ray sources. It decayed back into the atmosphere on December 17, 1988.

1983 February 26   During preflight checks on Shuttle orbiter *Challenger* prior to its first space mission, NASA engineers observed a leak in the no. 2 main engine and discovered a small crack. During checks of the no. 3 engine a similar crack was also found and both engines were removed for repairs on February 28 and March 1, respectively. The no. 2 engine was returned to the launch pad on March 8 followed by the no. 3 engine on March 10. A replacement for the no. 1 engine, discovered to have a crack in it after the flight readiness firing on December 18, 1982, was installed in the aft engine compartment on March 11.

1983 February The U.S. Air Force Space Division selected Lockheed to design and build the powerful Milstar military communications satellite. The launch of the first Milstar was scheduled for late 1987, but delays inevitably put this back for a decade. Meanwhile, in July 1984 the Air Force canceled a contract with Hughes Aircraft for the communications payload and transferred the work to TRW Systems. When weight problems jettisoned plans to launch Milstar from the Shuttle using an inertial upper stage, the Air Force switched vehicles to a Centaur stage, and when Centaur was banned from the Shuttle because of safety concerns the Air Force picked a Titan 4/Centaur launcher.

1983 March 2 At 09:37 UT a Soviet SL-13 launch vehicle lifted off from Baikonur cosmodrome carrying a 44,100-lb. TKS ferry ship designated *Cosmos 1443*. Designed by the Chelomei bureau, it was a variant of the *Cosmos 929/1267* family and carried the 12,000 lb. reentry module (by now standard) capable of returning to Earth with 1,100 lb. of cargo. In its original form, conceived in the early 1960s in competition with the Korolev bureau's Soyuz, TKS was to have carried supplies to the Almaz "military" space station. *Cosmos 1443* carried 7,700 lb. of cargo to *Salyut 7*, at this date unmanned. Access to the reentry module was gained via pressurized freight compartment, with the docking port at the rear and the reentry module forward. The aft section of *Cosmos 1443* docked with the forward port on *Salyut 7* at 09:20 UT on March 10, with the 85,000-lb. complex in an orbit of 202 × 203 mi.

1983 March 9 NASA announced that it had joined with the Department of Energy in a program called SP-100 to develop advanced nuclear reactors for large spacecraft in orbit. Research on thermoelectric and alternative power conversion systems would be conducted, as well as tests on high-temperature metals and materials, radiators for dissipating waste heat and nuclear fuels. A project office at the Jet Propulsion Laboratory would manage the research.

NASA released a request for bids on development of an Advanced Communications Technology Satellite (ACTS) designed to test radical and innovative technologies for the needs of communications satellite users in the 1990s. NASA wanted ACTS launched from the Shuttle to geostationary orbit in 1988. On August 10, 1984, RCA received a contract to design and build the ACTS, by that time scheduled for a 1989 launch. Delays to the program arose from funding cuts and reluctance on the part of Congress to use public money for what some politicians believed was research that private industry should pay for. ACTS was finally launched on September 12, 1993.

1983 March 16 Following reinstatement of the Centaur G-prime as the Earth-escape stage for the Shuttle-launched Galileo and International Solar Polar Mission (ISPM), a policy was agreed whereby both spacecraft would be launched to Jupiter within the same planetary window in 1986. Two Shuttle orbiters would be prepared for flight readiness on both Complex 39 launch pads at Cape Canaveral. The launch window would extend from May 15 to June 8, 1986, with ISPM considered the prime candidate for the first launch. It would not be possible to use a single Shuttle orbiter because the launch window was too small to turn the vehicle around for a second flight in the given period.

1983 March 23 President Ronald Reagan announced that he had authorized the Department of Defense to develop defensive technologies against ballistic missiles in a research program called the Strategic Defense Initiative (SDI) . In this effort, U.S. research laboratories and universities would determine whether it was possible to use advanced lasers and particle beams to destroy enemy missiles and their warheads in flight. If deployed, the concept envisaged directed-energy weapons placed in space, and for this reason the press coined the phrase "Star Wars" from George Lucas's film of the same name. Reagan set up a Defensive Technologies Study Team under Dr. James Fletcher, former NASA administrator, to evaluate the missile threat and draw conclusions about the possibility of mobilizing a defense. It reported in March 1984.

The Soviet Union launched an astronomical observatory called Astron from Baikonur. Carried into space by an SL-12, the 7,170-lb. satellite was equipped with an ultraviolet telescope, called Spika, built at the Crimean Astrophysical Observatory with the help of French scientists. The 1,765-lb. telescope was 13 ft. 9 in. long and comprised double-reflecting optics with a 31.5-in. primary mirror, a 10-in. secondary mirror and a focal length of 26 ft. 3 in. Astron also carried an x-ray telescope called SKR-02 with 10 spectral channels designed to study sources of radiation in the 2–25 kiloelectron-volt range. Astron operated from an orbit of 15,615 × 111,117 mi., inclined 79.8°, high above the Van Allen radiation belts.

1983 April 4 Astronauts Paul Weitz (CDR), Karol Bobko (PLT), Donald Peterson (MS) and Story Musgrave (MS) were launched aboard the Shuttle *Challenger* on its first flight from LC-39A at the Kennedy Space Center, Fla., at 18:30:00 UT on

Deployed from the Space Shuttle Challenger *on April 4, 1983, NASA's first Tracking and Data Relay Satellite (TDRS-1) relayed information from many separate satellites.*

the STS-6 mission. The 4,487,255-lb. liftoff weight included the 256,928-lb. orbiter, carrying its 46,971-lb. cargo which consisted of the NASA tracking and data relay satellite system (TDRSS) and inertial upper stage (IUS). From a circular 177 mi. orbit, TDRSS-1/IUS was deployed at 04:30:01 UT on April 5, and the first IUS stage fired 55 min. later for 2 min. 31 sec. The 1 min. 43 sec. burn of the second stage 5 hr. later, intended to place the satellite in geostationary orbit at 25,700 mi., malfunctioned and placed TDRSS-1 in an elliptical orbit with perigee 8,000 mi. lower than that. The first EVA (extravehicular activity) performed by NASA in more than nine years began at 21:05 UT on April 7, when astronauts Musgrave and Peterson opened the hatch to the cargo bay and went outside for 4 hr. 10 min. to evaluate suits and restraints. STS-6 ended at a mission time of 5 days 00 hr. 23 min. 42 sec. with a landing at Edwards AFB, Calif.

For the first time, one satellite used signals from another to navigate its position in orbit. In an announcement from the U.S. Air Force Systems Command, the Department of Defense reported that *Landsat 4* had received navigation updates from the U.S. Air Force Space Division's Navstar global positioning satellite instead of from the ground tracking stations. This technique was being developed for use with military satellites, cutting their reliance on ground transmitters that would be vulnerable in a major conflict.

1983 April 5   Eight U.S. aerospace companies reported findings from the *Space Station Needs, Attributes and Architectural Options Study* begun August 9, 1982. Boeing, General Dynamics, Grumman, Lockheed, Martin Marietta, McDonnell Douglas, Rockwell International, and TRW Systems found major cost savings in overall space activity would be gained from a manned orbital facility capable of servicing satellites and berthing space tugs and orbit transfer vehicles. They found that tugs, OTVs and provision for on-orbit servicing were essential elements in realizing the potential of a space station. They also concluded that while scientific knowledge and practical benefits would grow with microgravity research, these would not be methods for saving costs, although products might be found which could be sold by commercial concerns.

1983 April 11   The report from the U.S. Commission on Strategic Forces, headed by Lt. Gen. Brent Scowcroft, was released by the White House. Formed to reach decisions on basing modes for the Peacekeeper program, it made far-reaching recommendations about the future of the land-based deterrent. The commission advised deployment of 100 Peacekeeper missiles in existing Minuteman silos, further research into superhardening of silos against attack from highly accurate warheads, consideration of an antiballistic missile screen and engineering design of a small ICBM carrying a single warhead. The latter became known as the SICBM, or Midgetman, and was conceived by the commission as having a length of 40 ft., a diameter of 5 ft. 10 in., a launch weight of 30,000 lb. and a single warhead.

1983 April 16   NASA's Solar System Exploration Committee recommended development of two types of spacecraft for what it called a Core Program of planetary exploration. Called Planetary Observer spacecraft, low-cost applications of existing Earth satellite designs should be developed for exploration of the inner solar system, said the SSEC. More costly Mariner Mark II spacecraft, meanwhile, would be developed from the Voyager and Galileo class for a variety of missions to the outer solar system. The SSEC recommended continuation

of the Venus Radar Mapper (Magellan) mission scheduled for launch in 1988, and a Mars Observer mission for 1990 as the first of the Planetary Observer spacecraft. It was proposed that the first Mariner Mark II mission should be the comet rendezvous/asteroid fly-by (CRAF) mission tentatively scheduled for a Shuttle-Centaur launch in late 1992 followed by a Saturn Orbiter/Titan probe (Cassini) mission later in the decade.

1983 April 17   An SLV-3 launcher from the Indian Space Research Organization carried the 93-lb. satellite *Rohini 3* from the Sriharikota range near Madras to an orbit of 239 × 488 mi., at an inclination of 46.6°. The satellite was a low-technology Earth observation satellite, aimed more at engineering development than function. The satellite decayed out of orbit on April 19, 1990.

1983 April 20   Cosmonauts Vladimir G. Titov, Gennady M. Strekalov and Aleksandr A. Serebrov were launched in the 15,100-lb. *Soyuz-T 8* spacecraft by an SL-4 from Baikonur cosmodrome. In attempting to complete a rendezvous with *Salyut 7*, the crew discovered the primary tracking radar was not working. When they tried to perform a manual rendezvous, the crew overshot and ground controllers calculated that they would consume valuable propellant, reserved for retrofire, if they conducted further maneuvers. Unable to dock, they returned to Earth at the end of a mission lasting a mere 2 days 18 min. Titov and Strekalov had been specially trained to install additional solar cell "wings," carried to *Salyut 7* by *Cosmos 1443*, to the "vertical" sail attached to the smaller work module.

1983 May 2   Having been dropped in an incorrect orbit following deployment from the Shuttle on STS-6 launched April 4, 1983, the *TDRS-1* satellite, operated by ground engineers, began a series of 57 thruster firings to place it on the correct path. Using 1-lb. thrust attitude control jets on the satellite, engineers gradually raised the perigee point by the necessary 8,000 mi. to place it in geostationary orbit by June 29 and, on October 17, position it over the Atlantic at 41° W. Leased to NASA by SPACECOM, a partnership between Continental Telecom and Fairchild Industries, *TDRS-1* measures 57 ft. across its solar array panels and has two 16 ft. diameter antennas for relaying up to 300 megabits/sec. of data from up to 26 separate satellites.

1983 May 6–13   Participants from NASA, industry and the Department of Defense attended a Space Station Mission Synthesis Workshop to discuss baseline requirements for a permanent manned orbital facility. The group heard NASA's projection of the capabilities of the space station, expected to have a crew of six, to provide users with 55 kW of electrical power for experiments, give occupants more than 4,400 cu. ft. of habitation and work volume and grow to support a crew of 12–18 with 160 kW of electrical power and up to 13,500 cu. ft. of habitable volume in up to six modules. On April 11 President Ronald Reagan requested the Senior Interagency Group for Space to examine optional space station scenarios to determine whether the United States should proceed with such a facility.

1983 May 26   A NASA Delta 3914 launch vehicle carried the 1,125-lb. *Exosat* x-ray observatory satellite from Vandenberg AFB, Calif., to a highly elliptical 1,813 × 117,440 mi. orbit inclined 71.4°. Designed to study cosmic x-ray sources, the European Space Agency's *Exosat* comprised a box structure 4 ft. 7 in. × 6 ft. 10 in. supporting a 6 ft. 3 in. solar array panel and 265 lb. of science instruments. Design and development of *Exosat* had been conducted by a European consortium led by the German company Messerschmitt-Bölkow-Blohm (MBB). The satellite decayed back into the atmosphere on May 6, 1986.

1983 June 2   The Soviet Union launched the first of two 8,800-lb. spacecraft designed to conduct a radar mapping survey of Venus when an SL-12 lifted off from the Tyuratam launch complex carrying *Venera 15*. The basic spacecraft was a development of the Venera lander/carrier combinations that had been used to explore the atmosphere and surface of Venus since *Venera 4* in 1967. In the 1983 mission, however, the weight allocated to the lander portion was given over to a radar mapping device which would survey the planet from orbit about Venus. To do this the Soviet scientists had built a 20 ft. × 4 ft. 6 in. side-looking synthetic-aperture radar called Polyus-V. Solar arrays were doubled in size over previous Veneras to provide sufficient electrical power for the radar. *Venera 16* was launched by another SL-12 on June 6, 1983.

1983 June 13   NASA's *Pioneer 10* became the first spacecraft to exit the solar system, as defined by the boundary of the orbit of the most distant planet. At 12:00 UT *Pioneer 10* was 2,813,685,909 (2.8 billion) mi. from the sun, arbitrarily the point where it crossed the orbit of Neptune, traveling at a speed of 30,558 MPH. Though Pluto is normally the outermost planet, its orbit is so elliptical that it was inside Neptune's orbit and would remain so until the year 2000. At this distance it took radio signals from Pioneer 4 hr. 16 min. to reach Earth. Since launch, the spacecraft had traveled 3.59 billion mi., received over 98,900 commands and transmitted more than 126,000 million binary digits of data.

1983 June 16   The sixth Ariane 1 launch vehicle (L6) was launched from Kourou, French Guiana, carrying the first European Communication Satellite *(ECS-1)* and the *Oscar 10* amateur radio satellite. Built by a consortium headed by British Aerospace, the 2,300-lb. satellite comprised a box structure 7 ft. 2 in. wide and 7 ft. 10 in. high with two solar array wings, each 4 ft. 3 in. × 17 ft., spanning 45 ft. 3 in. producing 1,000 W of electrical power. *ECS-1* had 12 × 20-W channels and would provide telephone services and TV relay within the countries of the European Broadcasting Union from a geostationary position at 10°E. The 287-lb. *Oscar 10* was left in an orbit of 2,390 × 22,127 mi. inclined 25.9°, where it remains.

1983 June 17   The first launch of the Peacekeeper ICBM (formerly the MX) took place from Vandenberg AFB, Calif. The first stage fired after the missile had been expelled from its canister by a gas generator at the base and produced 500,000 lb. of thrust via a single movable nozzle, carrying the missile to a height of 75,000 ft. The second stage produced a thrust of 300,000 lb., taking the missile to a height of 55 mi. The 125,000-lb. thrust third stage carried the upper elements to an altitude of 132 mi. After separation, the liquid propellant fourth stage maneuvered to position the warhead dispenser for sending six reentry vehicles to separate targets. On its first flight the Peacekeeper flew 4,700 mi. downrange to the Kwajalein Atoll in the Marshall Islands. Five test flights were conducted by 1984 which confirmed the missile's functional performance, followed by a further fifteen flights to conduct flight tests and measure silo integration.

Following a failure on September 9, 1982, Ariane returns to flight on June 16, 1983, carrying Europe's ECS-1 communication satellite toward geostationary orbit.

1983 June 18   Launched precisely on schedule at 11:33:00 UT, the Shuttle *Challenger* carried astronauts Robert Crippen (CDR), Frederick Hauck (PLT), John Fabian (MS), Sally Ride (MS, the first American woman to fly in space) and Norman Thagard (MS) into orbit on the STS-7 mission. The 37,124-lb. payload comprised two communications satellites and a pallet of instruments. Released from the cargo bay at 21:02:00 UT, the Canadian *Anik C-2*, a Hughes HS-376, was boosted to an elliptical orbit by a solid propellant PAM-D and from there to geostationary orbit on June 21. Another Hughes HS-376, the Indonesian *Palapa B-1* was released from the cargo bay at 13:33:00 UT on June 19 and fired to geostationary orbit two days later. Each of the satellites' twenty-four 10-watt transponders could carry 500 telephone calls or one TV transmission. *Challenger*'s remote manipulator arm was used several times to deploy from the payload bay a package of science and technology instruments on the *SPAS-01* (Shuttle pallet satellite) built by Messerschmitt-Bölkow-Blohm (MBB) in West Germany. For several hours *Challenger* flew in formation up to 1,000 ft. away from the *SPAS-01*. Prevented by poor weather from landing at the Kennedy Space Center in Florida as planned, STS-7 touched down at Edwards AFB, Calif., two orbits later than scheduled at an elapsed time of 6 days 2 hr. 23 min. 59 sec.

1983 June 20   The standard U.S. electronic intelligence gathering (elint) subsatellite switched from Titan IIID to Titan 34D when the 40th in the series was carried into space piggyback with a KH-9 Big Bird. The 29,300-lb. KH-9 was placed in a 97.5 × 143 mi. orbit inclined 96.4° while the 130-lb. antiballistic-missile elint went into a circular 802 mi. orbit inclined 96.6°. The second, and last, in the series, was launched by Titan 34D on June 25, 1984, when a standard radar elint was put in a circular 435 mi. orbit inclined 96.5°. What should have been the last piggyback elint was lost April 18, 1986, when the launcher exploded.

1983 June 21   The U.S. government posted notice calling for privatization bids on the U.S. inventory of expendable launchers. With NASA wanting to put all its satellites and spacecraft on the reusable Shuttle and marketing the Shuttle to foreign users, there seemed no role for expendable rockets. Industry disagreed and the government opted to encourage competition with its own space agency. In the past, NASA acted as agent for Delta and Atlas-Centaur on the basis that launch costs would be reimbursed. Phase-down in production of these launchers would begin at McDonnell Douglas and General Dynamics later this year, unless launch slots could be booked for U.S. or foreign users.

1983 June 27   At 09:12 UT *Soyuz-T 9* was launched by an SL-4 from the Baikonur cosmodrome carrying cosmonauts Vladimir A. Lyakhov and Alexandr P. Alexandrov on a rendezvous mission to *Salyut 7*. Docking at the aft port took place at 10:46 UT on June 28, and the crew quickly moved inside to check out the station and unpack the large *Cosmos 1443* module launched March 2. By July 7 all stores, including an extra set of solar-cell arrays for the "vertical" solar sail, had been transferred to the station, and Lyakhov and Alexandrov settled down to a planned three-month stay aboard the space station. On August 4 the crew began to load the *Cosmos 1443* descent module with redundant items—equipment specialists wanted to examine on the ground, and failed systems from *Salyut 7*.

1983 June 28   A Delta 3920 launch vehicle from Cape Canaveral, Fla., put the *Galaxy 1* satellite in geostationary transfer orbit from which it was finally positioned by apogee kick motor at 135° W. *Galaxy 1* was the first of three wholly owned Hughes communication satellites based on the HS 376. The 1,539-lb. satellite carried 24 active and 6 spare transponders for sale or lease to the cable industry. Launched on September 22, 1983, *Galaxy 2* was placed at 74° W and would be used for large corporations and long-haul carriers. *Galaxy 3*, launched on September 21, 1984, went to 93.5° W and would serve the same customer base.

1983 July 1   Martin Marietta, builders of the U.S. Air Force Titan ICBMs and satellite launchers, moved a step nearer to commercializing their product when an agreement between them and a marketing organization called SpaceTran expired. Lack of interest from the U.S. government in keeping Delta, Titan and Atlas funded at a time when the Shuttle was being marketed as a replacement for expendable launchers led to manufacturers' setting up their own sales campaign for orders. President Ronald Reagan had approved the use of government launch sites at Cape Canaveral for privatized launchers on a fee-paying basis. SpaceTran had been operating as go-between, trying to get Titan orders for Martin Marietta.

An SL-6 launched the 2,340-lb. *Prognoz 9* science satellite from Baikonur to a highly elliptical orbit of 236 × 447,410 mi. with an inclination of 65.8°. *Prognoz 9* carried science equipment from the Soviet Union, Czechoslovakia and France, including a radio telescope with two antennas operating at a frequency of 8 mm. The satellite was designed to rotate around an axis directed toward the sun at a rate of once every 2 min. and had an orbital period of 26.7 days. Equipment for x-ray and gamma-ray studies was also carried onboard the satellite, which remains in space.

1983 July 4   The first in a series of flight tests with ⅛th-scale models of the Soviet reusable shuttle began with the suborbital flight of a BOR-5 on an SL-8. About six flights of the BOR-5 models were made prior to the first flight of the full-scale shuttle, *Buran,* on November 15, 1988. BOR-5 was a perfectly scaled replica capable of demonstrating various aerodynamic and flight control techniques as well as serving as a test-bed for *Buran* systems and thermal protection materials. The degree of effort made by Soviet engineers to qualify *Buran* systems and test shuttle-related technologies in space was unparalleled.

1983 July 18   NASA Administrator James Beggs explained to a symposium of space station planners attended by several hundred representatives from government, industry, and the military that President Ronald Reagan would within the next 12 months announce approval to launch a major space station program. He told his audience that Presidential Science Adviser George A. Keyworth had dropped his objection to the station concept and had asked NASA to plan a "grand vision."

1983 July 28   A NASA Delta launched from Cape Canaveral, Fla., put the *Telstar 3A* communications satellite into an elliptical transfer path from which it was later moved to geostationary orbit at 96° W. First of three to be operated by American Telephone and Telegraph, the Hughes 376/Telstar 3 series carried 24 × 5.5-W active and 6 spare transponders capable of relaying a total 21,600 telephone calls. The three Telstar 3–series satellites relieved AT&T of reliance on the Comsat General Corp.'s Comstar satellites. *Telstar 3C* and *Telstar 3D* were launched by Shuttle.

1983 August 3   Pressure for the White House to authorize development of a U.S. space station grew when President Ronald Reagan met heads of 11 companies and was told that a permanently manned facility was urgently needed to stimulate private investment in space. New areas of space commercialization included research into new pharmaceutical products, international cooperation with U.S. leadership and the eventual transfer of Shuttle operations from NASA to the private sector.

1983 August 11 Tests conducted by the NASA Ames Research Center duplicated critical flight aspects of the entry conditions for Galileo's Jupiter probe, scheduled for launch aboard the Shuttle in May 1986. Carried to an altitude of 98,000 ft. by a 495 ft. tall, 5.1 million cu. ft. plastic balloon, the simulated Jupiter probe was released to descend on a parachute to the desert floor of White Sands Missile Range 9 min. later. Properties at the top of the Earth's upper atmosphere are similar to those of Jupiter's upper atmosphere and accurately simulated the conditions which would be encountered by the probe when it reached the planet in August 1988.

1983 August 14 The *Cosmos 1443* logistics module which had been docked to the forward port on *Salyut 7* since March 10 was undocked at 14:04 UT. The *Salyut 7* station was in an orbit of 195 × 202.5 mi. Two days later, at 14:25 UT, the *Salyut 7* cosmonauts Vladimir Lyakhov and Aleksandr Aleksandrov undocked *Soyuz-T 9* from the station's aft port, withdrew 800 ft., waited 20 min. as *Salyut 7* flipped 180° end-over-end, and docked to the forward port. At 11:02 on August 23 the 12,000-lb. *Cosmos 1443* reentry module arrived back on Earth; the main spacecraft de-orbited to destruction on September 19.

1983 August 19 Soviet Premier Yuri Andropov informed a group of U.S. senators on a visit to Moscow that the Soviet Union was unilaterally refraining from further tests of antisatellite systems during the period the United States refrained from orbiting satellites of a similar purpose. The Soviets implicitly agreed to allow the U.S. suborbital tests and the next day Foreign Minister Andrei Gromyko presented a draft treaty to the United Nations proposing language for a ban on "space-based weapons." This followed Soviet concerns about the Reagan Strategic Defense Initiative.

1983 August 30 Just 17 min. late due to weather, Shuttle mission STS-8 began at 06:32:00 UT with the launch of astronauts Richard Truly (CDR), Daniel Brandenstein (PLT), Dale Gardner (MS), Guion Bluford (MS, the first African American to fly in space) and William Thornton from LC-39A at the Kennedy Space Center. The first night launch of the Shuttle program, STS-8 had a liftoff weight of 4,492,074 lb. comprising the 242,913-lb. orbiter including a 30,076-lb. payload comprising an Indian communication satellite and development flight instrumentation for engineering studies. The satellite, *Insat-1B*, was released from the payload bay along with its PAM-D boost stage at 07:48:54 UT on August 31. Subsequently, three apogee motor firings from a liquid propellant system placed it in geostationary orbit. The mission ended at an elapsed time of 6 days 1 hr. 8 min. 43 sec. on runway 22 at Edwards AFB.

1983 August The Swedish Space Corp. contracted with the Aérospatiale/Eurosatellite consortium for design and development of a powerful direct broadcast satellite for Scandinavia. The order was made on behalf of the Nordiskasatellitaktiebolaget (Nordic Satellite Co.), a cooperative program among Sweden, Norway and Finland. Eurosatellite was a consortium of several European satellite manufacturers: Aérospatiale of France (24%), Alcatel Espace of France (24%), Messerschmitt-Bölkow-Blohm of Germany (24%), AEG of Germany (12%), ANT of Germany (12%), and ETCA of Belgium (4%). The satellite would be known as Tele-X.

1983 September 21 NASA announced it had signed an agreement with Fairchild Industries for the development of an unmanned orbital platform called Leasecraft. Deployed from the Shuttle it would carry commercial and government payloads in remote sensing, materials research and scientific research. The platform was to measure 15 ft. × 15 ft. × 9 ft. 6 in. Leasecraft was never deployed operationally due to lack of customer interest.

1983 September 26 The first pad abort took place at 19:38 UT when Soviet cosmonauts Vladimir G. Titov and Gennady M. Strekalov were boosted away from their SL-4 launch vehicle just 90 sec. before planned ignition of the first stage engines. The cosmonauts had been waiting for launch when a fire broke out at the base of the vehicle. For an unknown reason—possibly the conflagration severed electrical cables carrying abort commands—the automatic abort system failed to operate and the crew had to fire the escape rocket manually. Experiencing a crushing 17 g's, the Soyuz orbital and descent modules were wrenched away from the instrument unit and launch vehicle and carried to a height of 3,100 ft. before the orbital module and escape rocket separated, leaving the recovery system on the descent vehicle to lower the crew to the ground, 8,200 ft. from the pad. Specially trained to install additional solar panels on the *Salyut 7* via a space walk, Titov and Strekalov were to have relieved Vladimir Lyakhov and Alexandr Alexandrov aboard the orbiting station. Their mission was retrospectively designated *Soyuz 10–1*.

1983 September 28 An SL-14 launch vehicle carried the Soviet ocean research and applications satellite, *Cosmos 1500*, from Plesetsk to a 393 × 414 mi. orbit inclined 82.5°. For the first time the satellite carried a side-looking radar covering a swath 285 mi. wide and generating data delineating water "fields" in seas and oceans, information on storm zones, tidal velocity data and ice fields. In October 1984 *Cosmos 1500* provided radar images of ice-locked ships in the Arctic Ocean, enabling oceanographers to instruct local shipping officials on best routes of access to open water. Subsequent Soviet ocean survey satellites carried radar.

Space Shuttle *Columbia* was moved to Launch Complex 39A with the 33,264-lb. Spacelab 1 installed in the cargo bay. On all previous missions, payloads had been brought to the Shuttle on the launch pad and installed via the rotating service structure. The weight and bulk of Spacelab made that procedure impossible. On October 14 a decision was made to move the Shuttle *Columbia* back into the Vehicle Assembly Building for checks on the thrust skirts at the base of each solid rocket booster. Inspection of STS-8 boosters after their recovery on August 30 revealed damage that could have caused a malfunction during launch. *Columbia* was returned to the pad on November 8.

1983 October 1 The U.S. Navy formally inaugurated its Naval Space Command, headed by astronaut Richard Truly, a Navy captain. Based at the Naval Weapons Laboratory, Dahlgren, Va., the Space Command would focus separate elements in the service that had used space-based assets for fleet and operations support. Communications, navigation and weapons targeting were all vital elements for which the Navy used satellites. There was opposition in Congress, claiming that the Air Force Space Command was all the United States needed, and even some within the Navy believed that a stronger emphasis on space assets would draw funds from traditional areas. Nevertheless, Naval Space Command proved a positive link between separate Navy users.

Intelsat began a new service for business customers via its global network of telecommunications satellites. Called the

Intelsat Business Service, it would allow small ground stations to send international messages to other ground stations via direct access through the geostationary satellites. Hitherto, subscribers had to send their traffic through landlines or domestic communication satellites. The new IBS option afforded greater flexibility, allowing companies to access the network with their own terminals.

1983 October 10 The Soviet spacecraft *Venera 15* arrived at the planet Venus and by firing a propulsion system on board placed itself in a $600 \times 40,000$ mi. orbit with a period of 24 hr. The orbit was synchronized with ground stations on Earth so that data acquired by a mapping radar on the spacecraft could be transmitted when the primary tracking station was facing the planet each day. The radar mapping device scanned a $5,600 \times 93$ mi. strip of the planet for 16 min. on each orbit. *Venera 16* arrived in orbit around Venus on October 14, two days before the radar mapping operation with *Venera 15* began. *Venera 16* started its surveys on October 20, 1983. Both spacecraft scanned the surface in the northern hemisphere across the polar region down to a lat. of 30° N at a resolution of 3,300–6,600 ft.

1983 October 20 At 09:59 UT the unmanned *Progress 18* cargo tanker was launched from the Baikonur cosmodrome to a rendezvous with *Salyut 7*. A docking at Salyut's aft port took place at 11:34 UT on October 22. With the complex in an orbit of 204.5×215.5 mi., *Salyut 7* cosmonauts Vladimir Lyakhov and Aleksandr Aleksandrov unpacked stores and supplies before beginning the transfer of 1,100 lb. of propellant. *Progress 18* remained docked to the aft *Salyut 7* port until after two space walks had been completed. The cargo tanker separated from Salyut at 03:08 UT on November 13 and was de-orbited three days later.

1983 October NASA received approval from the Office of Management and Budget to make a start on the Venus Radar Mapper (VRM), a spacecraft equipped with synthetic aperture radar designed to map the surface of Venus. The spacecraft would be assembled by Martin Marietta, but cost savings would be achieved by using some spare parts at the Jet Propulsion Laboratory left over from the Mariner, Viking and Voyager programs. The selected launch window (April 6–26, 1988) would be met by a Shuttle launching into Earth parking orbit the VRM attached to a cryogenic Centaur G upper stage. Centaur would boost the spacecraft to Venus where it would arrive on July 26, 1988, and operate at least through April 3, 1989, a full Venusian year.

NASA's Marshall Space Flight Center announced that the optical observatory being built for a Shuttle launch into low Earth orbit would be known as the Edwin P. Hubble Space Telescope. It had been so named in honor of the U.S. astronomer who died in 1953. Working with the 100-in. Mount Wilson observatory telescope, Prof. Edwin Hubble was able to show as early as 1924 that stars were gathered into galaxies and that the universe was expanding at a rate proportional to its age. The Hubble Space Telescope was considered by many astronomers to be as significant as the discoveries of Prof. Hubble.

Two major U.S. satellite manufacturers announced a new range of communication satellites. Hughes Aircraft revealed details of three spin-stabilized designs for compatibility with the Shuttle: the HS 399 with a diameter of 11 ft. 10 in., a stowed height of 11 ft. 1 in. (deployed height of 33 ft.), a weight of 2,400 lb. and 2.5 kW of power serving 16 high-power channels through 24 (50-W) traveling wave tubes; HS

394, the first satellite to combine stabilized solar array wings on the de-spun top section, providing 4 kW of power from "wings" spanning 115 ft. to 14 high-power wave tubes; and HS 399, a relatively inexpensive 12-channel satellite using 5-W wave tubes in a structure 9 ft. 2 in. in diameter and 2 ft. 11 in. tall. Ford Aerospace revealed five designs in a family called Supersat, essentially a common bus for different applications and derived from the Intelsat V design.

1983 November 1 The first of two space walks by *Salyut 7* cosmonauts Vladimir Lyakhov and Alexandr Alexandrov began at 04:47 UT, with the specific purpose of installing two additional, 49 sq. ft., solar cell arrays to opposite sides of the "vertical," 216 sq. ft. solar cell wing attached to a pivot on the top of the station. This job had been the task for cosmonauts Vladimir Titov and Gennady Strekalov, who had failed to dock following launch in *Soyuz-T 8* on April 20 and experienced an abort for their attempted launch on September 26. Working at a task for which they had not been trained, Lyakhov and Alexandrov installed only one solar array on their 2 hr. 50 min. excursion. A second space walk began at 03:47 UT on November 3, and in 1 hr. 55 min. they erected the second supplemental array.

1983 November 10 Named *Discovery*, Shuttle OV-103 arrived at the Kennedy Space Center's orbiter processing facility to begin preparations for its first flight, then scheduled for mid-1984. Third of four NASA spaceflight rated orbiters, *Discovery* weighed in empty at 147,925 lb., compared with 148,633 lb. for *Challenger* and 155,359 lb. for *Columbia*. Low temperature thermal insulation tiles over most of the upper wings and fuselage had been replaced by advanced flexible reusable surface insulation (FRSI), which saved weight. Graphite epoxy replaced some aluminum structural parts. *Discovery* was moved to the LC-39A launch pad on May 19, 1984, and a 20-sec. flight readiness firing of the three main engines was successfully conducted on June 2, 1984, in preparation for the 41-D mission, then scheduled for June 22, 1984.

1983 November 23 *Salyut 7* cosmonauts Vladimir Lyakhov and Alexandr Alexandrov ended their stay aboard the *Salyut 7* space station, now in an orbit of 199.5×209.5 mi. They returned in their *Soyuz-T 9* spacecraft, which had been in orbit since June 27, and undocked at 16:40 UT before descending to Earth at an elapsed time of 149 days 10 hr. 46 min. This was by far the longest period a Soyuz spacecraft had remained in space and returned cosmonauts to Earth. They were to have had the *Soyuz-T 9* spacecraft exchanged with the spacecraft that failed to arrive when it was aborted before launch on September 29.

1983 November 24 A Soviet SL-14 launched the first in a new series of geodetic satellites designated *Cosmos 1510* from Plesetsk. Placed in an orbit of 920×948 mi., with an inclination of 73.6°, it was operated in conjunction with a series of Soviet geodetic satellites placed in 82.6° orbits, the first of which, *Cosmos 1312*, had been launched on September 30, 1981. The weight of the *Cosmos 1510* is variously reported between 1,350 and 1,985 lb. Other Soviet geodetic satellites placed in 73.6° orbits included *Cosmos 1660* (June 14, 1985), *Cosmos 1732* (February 11, 1986), *Cosmos 1823* (February 20, 1987), *Cosmos 1950* (May 30, 1988), *Cosmos 2037* (August 28, 1989), *Cosmos 2088* (July 30, 1990) and *Cosmos 2226* (December 22, 1992).

In this artist's representation, the European Space Agency's Spacelab pressure module is attached to the Shuttle cargo bay with exposed pallet-mounted instruments at the rear.

1983 November 28 At 16:00:00 UT astronauts John Young (CDR), Brewster Shaw (PLT), Owen Garriott (MS), Robert Parker (MS), Byron Lichtenberg (Payload Specialist, or PS) and Ulf Merbold (PS), were launched aboard the Shuttle *Columbia* from LC-39A at the Kennedy Space Center, Fla., on STS-9, the first Spacelab mission. Funded by the European Space Agency, Spacelab carried 38 different sets of experiment hardware supporting 73 investigations from 14 countries. West German astronaut Ulf Merbold represented ESA on this mission. STS-9 had a liftoff weight of 4,503,089 lb. including the 247,813-lb. orbiter. It was launched onto the most northerly heading flown to date by a U.S. manned space vehicle, reaching a 155.4-mi. orbit inclined 57° to the equator. Two teams of payload specialists worked round the clock, from Spacelab activation at a mission time of 4 hr. 31 min. to deactivation at 9 days 13 hr. 5 min. The mission ended with a landing at Edwards AFB at 10 days 7 hr. 47 min. 24 sec.

1983 November NASA issued a new Shuttle flight manifest projecting 10 flights in 1984, 12 flights in 1985, 17 flights in 1986 and 24 flights in 1987. NASA introduced a new code for mission designations. Instead of progressing numerically mission by mission as had been the practice to date (STS-1, STS-2, etc.) a combination of letters and numbers would now be used to categorize each flight. Instead of being STS-10 the next mission would be coded 41-B, the first digit indicating the fiscal year of launch, the second digit indicating the launch site (*1* for Kennedy Space Center, *2* for Vandenberg AFB) and the letter suffix indicating the alphabetical sequence of launch in that fiscal year (STS-9 had been the first Shuttle launch in fiscal year 1984 so 41-B was the second).

1983 December 7–8 At a Royal Society meeting in London, England, *Technology in the 1990s: The Industrialization of Space*, Deputy Director of the NASA Space Station Task Force Robert F. Freitag summarized current plans. With a projected operational date of 1990, baseline Space Station capabilities would provide the 6–8 person crew with 75 kW of power to run the facility and power experiments, provide a 300 megabit/sec. data transmission system and cost about $8 billion to build and put together in a 28.5° orbit. A growth version would house 12–18 crew, provide 160 kW of power, berth a space-based orbital transfer vehicle and operate in conjunction with a platform in polar orbit. The growth station was predicted to cost $17–20 billion. Freitag noted that NASA hoped to involve the European Space Agency, Japan and Canada.

1983 December 9   The German manufacturer Messerschmitt-Bölkow-Blohm/ERNO and the French aerospace giant Aéro-spatiale signed an agreement to jointly design, develop and manufacture a family of communications satellites called Spacebus. Spacebus 100 was derived from Arabsat and the Spacebus 300 design evolved from two direct-broadcast satellites, *TV-Sat* and *TDF-1*, being built for German and French use, respectively. The new member, Spacebus 200, would weigh 1,500–2,400 lb. in geostationary orbit and produce 3 kW of power from solar arrays. All three types would be three-axis attitude stabilized and accommodate a range of transponders at customer choice.

1983 December 14   The Soviet Union launched monkeys into space aboard an international biology satellite named *Cosmos 1514* for the first time. The use of animals in Soviet space experiments had previously not included primates. *Cosmos 1514* was launched by an SL-4 from Plesetsk to a 133 × 161 mi., 82.3° orbit. Two macaque monkeys as well as rats and fish were on board this, the fifth, Soviet biology satellite. The 12,570-lb. satellite was one of the Vostok-type and safely returned its occupants to Earth on December 20.

1983 December 22   Dr. Edward Teller, one of the principal scientists in the development of the hydrogen bomb, sent a letter to Presidential Science Advisor George Keyworth telling him that design of an x-ray laser battle station was ready for engineering development. Some scientists close to Teller at Lawrence Livermore National Laboratory did not agree with him but were prevented by senior administrators at LLNL from sending qualifying letters. A space-based x-ray laser battle station would comprise a 150-KT nuclear bomb inside a spherical device surrounded by lasing rods. In the instant the bomb detonated, x-rays produced by the explosion would flow down the rods to separate targets before the device was consumed. The concept was highly controversial and results of underground tests were inconclusive.

1983 December 23   The U.S. government Federal Register published an announcement from NASA that it would select passengers to ride aboard the Shuttle. Reacting to a large number of written requests and telephone calls from people across the United States who wanted to fly on the Shuttle, the space agency had decided to select fit and healthy individuals who could be trained to go into space on short-duration missions. This action led to the Teacher in Space and the Journalist in Space programs.

1983 December   The West German postal and telephone agency Deutsche Bundespost commissioned a European consortium to develop a German telecommunications satellite system under the leadership of Siemens AG. Three satellites named DFS Kopernikus were to be built by a group led by Messerschmitt-Bölkow-Blohm/Erno, two for on-orbit operations and a third as ground spare. With a weight of 3,122 lb., reduced to 1,422 lb. on orbit, each satellite would comprise a central box structure with a height of 13 ft. 7 in. and a span of 50 ft. across two solar array wings providing 1,500 W. Ten transponders in the 14/11 GHz band would be available for TV, sound broadcasting, telephone channels and data, while

one 30/20 GHz transponder would be available for future applications.

The British government decided to launch its *Skynet 4* military communications satellite on the Shuttle rather than on Europe's Ariane, as had been expected. Britain was the first Ariane partner to select another launcher for its satellite and the choice brought recrimination against the United Kingdom from Ariane and the European Space Agency. In choosing the Shuttle, the United Kingdom became eligible to fly a British astronaut aboard the Shuttle that carried the satellite. A short list of candidates was drawn up from names submitted by the Royal Air Force, the Navy, the Army and the civil service.

1983   During the year, launch vehicle designers at British Aerospace together with Alan Bond, a propulsion engineer with Rolls Royce, converged on an optimized launch vehicle for future applications. Examining more than 30 different launcher configurations for low-cost launch prices and engineering efficiency, British Aerospace settled on a recoverable, winged, unmanned launch vehicle using a novel form of hybrid air-breathing/rocket propulsion system. The concept was known as HOTOL (for horizontal-takeoff-and-landing). The company wanted to play a leading role in cutting launch costs as a means of improving its satellite sales and sought a stronger role for Britain in the European Space Agency.

1984 January 10   NASA announced that it had completed negotiations with Transpace Carriers (TCI) for the commercialization of the Delta launch vehicle. NASA wanted to halt government-sponsored flights with expendable launchers to encourage payload customers to move to the Shuttle. TCI believed it could profitably market and operate Delta to satellite users, offering launches at a competitive rate to the government-subsidized Shuttle launch fees. Under the agreement, TCI would assume responsibility for the last two NASA Delta launches in 1986 and subsequent flights. The shift was delayed at first by a year, and then permanently, when commercial negotiations broke down.

1984 January 21   The first of five test launches of an air-launched antisatellite system took place when a U.S. Air Force F-15 released a two-stage missile on a test flight. Called the air-launched miniature vehicle (ALMV), it was without the miniature homing vehicle (MHV) that on later flights would guide it toward a designated test target in space. The two-stage ALMV comprised the first stage of a short-range attack missile (SRAM) and an Altair 3 as second stage. The ALMV had a length of 17 ft. 9 in. and a diameter of 1 ft. 8 in. A second test on November 13, 1984 — with an MHV attached — failed, but a spectacular success was achieved on September 13, 1985, when the Solwind satellite was destroyed 320 mi. above the Earth.

1984 January 23   A Japanese N-II launch vehicle lifted off from the Tanegashima Space Center carrying the 1,480-lb. *Broadcasting Satellite-2A (BS-2A)* to an elliptical transfer orbit. Three days later the apogee motor placed the satellite in geostationary orbit at 100° E. The box-shaped satellite had a width of 4 ft. 1 in., a height of 6 ft. 6 in. and a span of 29 ft. across the solar cell panels. The *BS-2A* was the first of two direct broadcast satellites built to provide 14/12 GHz transmissions to small antennas in remote areas where station-to-station microwave links were unavailable. Japan's National Broadcasting Corp., the Nihon Hoso Kyokai (NHK), refused to accept the satellite for commercial use when the traveling wave tube transmitters failed, downgrading it to experimental status. The *BS-2A* operated in this mode until April 12, 1989.

1984 January 25   In his annual State of the Union address, President Ronald Reagan announced his endorsement of NASA's plans for a permanent manned orbiting space station when he directed the agency to "do it within a decade." NASA Administrator James Beggs saw the space station serving as science laboratory, astronomical observatory, space manufacturing and processing center, satellite servicing facility and assembly site for large structures. Fearing that it would draw funds from military and national security space projects, Defense Secretary Caspar Weinberger and Central Intelligence Agency Director William Casey were opposed to the station. The National Academy of Sciences space science board was also opposed on the grounds that it was hard to justify such a costly facility.

1984 January 29   The first of a more powerful class of Chinese satellite launcher, the CZ-3 (Long March 3) was launched from Xichang carrying the 1,985-lb. *STW (Shiyan Tongxin Weixing,* Experimental Communication Satellite) intended for geostationary orbit. The CZ-3 was essentially a CZ-2 with a cryogenic liquid hydrogen–liquid oxygen third stage and a 9,900-lb. thrust engine producing a very credible 425 sec. Isp. CZ-3 had a total height of 141 ft. 11 in., a maximum diameter of 11 ft. and a launch weight of 445,400 lb. The first stage fired for 2 min. 9 sec., the second stage for 2 min. 8 sec. and the third stage for 7 min. 30 sec. The third stage fired again for 4 min. 51 sec., just 13 min. 30 sec. after launch, to push the payload to geostationary transfer. This it failed to do, leaving *STW* in a 180 × 286 mi. orbit inclined 31.04°. Two maneuvers were performed by the *STW*, first to a 223 × 4,023 mi. orbit, then to a 299 × 4,089 mi. orbit and an inclination of 36° where tests were performed.

1984 January 31   In the formal announcement of NASA's budget request for fiscal year 1985 the space agency asked for, and received, funds to start development of the Mars Geoscience/Climatology Observer. As part of a low-cost Planetary Observer program, the spacecraft was scheduled for launch by Shuttle in August 1990. An upper stage was necessary to propel the spacecraft from low Earth orbit to escape velocity, and in March 1986 NASA selected the transfer orbit stage (TOS) to do that job. Developed by Orbital Sciences Corp., the solid propellant TOS was the first commercially developed upper stage used by NASA in a planetary program. When NASA switched from Shuttle to Titan launch vehicle, the TOS was used as a terminal stage on the expendable launcher; the first flight was delayed until 1992.

1984 February 2   With testimony from John Hodge, director of its Space Station Task Force, NASA asked Congress to approve expenditure of $150 million in fiscal year 1985 for a permanent manned facility, now expected to be operational in 1991 at a cost of $8 billion. Over the next 10 years NASA would spend approximately $16 billion on designing and redesigning the station, at the end of which an operational capability by the year 2002 was in prospect.

1984 February 3   At 13:00:00 UT Shuttle *Challenger* was launched from LC-39A at the Kennedy Space Center, Fla., on mission 41-B with astronauts Vance Brand (CDR), Robert Gibson (PLT), Bruce McCandless (MS), Ronald McNair (MS) and Robert Stewart (MS). Liftoff weight was 4,498,443 lb., including the 250,405-lb. orbiter accommodating 33,868 lb. of

payload. Two HS-376 communications satellites built by Hughes Aircraft were on board. Western Union's *Westar VI* was released at 20:59 UT, but 45 min. later the PAM-D motor blew off its exhaust cone, stranding the satellite in a 186.4 × 754.9 mi. orbit. After discussions with controllers, Indonesia elected to proceed with deployment of its *Palapa B2*, which occurred at 15:13:16 UT on February 6. It too blew its cone and the satellite stranded in a 170.3 × 735.4 mi. orbit. At 12:10 UT on February 7, the first of two ambitious EVAs (extravehicular activities) began when McCandless and Stewart left the airlock and entered the payload bay. McCandless strapped on one of two manned maneuvering units (MMUs) and performed the first untethered space walk by moving 300 ft. away from *Challenger*. Then Stewart evaluated the MMU and the EVA ended at an elapsed time of 5 hr. 55 min. The second EVA began at 10:24 UT on February 8, and in 6 hr. 17 min. the two crew members used the two MMUs to rehearse a satellite repair planned for the next mission. The German SPAS-01 pallet was also on board *Challenger*, but a failure in the remote manipulator arm prevented it from being lifted out of the cargo bay as planned. At an elapsed time of 7 days 23 hr. 15 min. 53 sec. *Challenger* became the first orbiter to end its mission at Kennedy Space Center's specially built runway.

1984 February 8   At 12:07 UT *Soyuz-T 10* was launched by SL-4 from Baikonur carrying cosmonauts Leonid D. Kizim, Vladimir A. Solovyov and Oleg Y. Atkov. Their mission was planned as a long-duration stay emphasizing medical evaluation of the body over extended periods of weightlessness. To facilitate crew health monitoring, Atkov had been included in the crew as a fully qualified heart physician. *Soyuz-T 10* docked to the forward *Salyut 7* port at 14:43 UT on February 9, and almost immediately the crew began reactivation of the station, which had been unmanned for 2½ months since the return of Vladimir Lyakhov and Alexandr Alexandrov.

1984 February 14   A Japanese Mu-3S solid propellant launcher carried the 441-lb. scientific satellite *Exos-C (Ohzora)* from Kagoshima to a 212 × 446 mi. orbit inclined 74.6°. The satellite was equipped with optical sensing devices for measuring the upper atmosphere, the stratosphere and the mesosphere. *Exos-C* was launched for the Institute of Space and Astronautical Sciences. It decayed back down through the atmosphere on July 19, 1989.

1984 February 15   NASA announced that the Johnson Space Center was to be responsible for managing the Space Station program through five basic areas of activity: systems engineering and integration, business management, operations, customer integration and support with Headquarters in Washington, D.C. Although the decision to develop the space station program at the JSC was a natural extension of its historic role in the development of new manned space vehicles, the JSC had primarily developed space transportation or exploration systems while the Marshall Space Flight Center had historically taken the lead in Space Station work, including Skylab.

1984 February 21   At 08:46 UT the unmanned *Progress 19* cargo tanker was launched by an SL-4 from Baikonur cosmodrome. It docked to the aft port on *Salyut 7* at 08:21 on February 23 and was unpacked by cosmonauts Leonid Kizim, Vladimir Solovyov and Oleg Atkov. The usual combination of propellant, water, stores, food, clothing, equipment and scientific instruments was added to the *Salyut 7* inventory before the vehicle undocked at 08:40 UT on March 31, and de-orbited to an uninhabited area over the Pacific Ocean two days later.

1984 February 29   NASA announced that it had set up seven teams to conduct advanced technology development for the Space Station. The teams would concentrate on attitude control and stabilization systems, data management systems, auxiliary propulsion systems, environmental control and life support systems, space operations mechanisms, thermal management systems and electrical power systems. NASA wanted its centers to exploit new and innovative technologies to enhance station performance and make it as cost-effective as possible.

1984 February   Four companies were chosen by the Department of Defense to begin concept development for a hardened mobile launcher to support the Small ICBM, or Midgetman, missile recommended by a blue-ribbon panel headed by Lt. Gen. Brent Scowcroft. The companies were General Dynamics, Martin Marietta, Ball Aerospace and Boeing/Goodyear. Each SICBM mobile launcher would carry a single missile and afford protection from a nuclear explosion close by. Congress had agreed to fund studies of an SICBM on the basis that it weighed no more than 30,000 lb. and therefore posed no threat to arms reduction agreements limiting the size of new missiles.

1984 March 1   A NASA Delta 3920 launched the 4,293-lb. Earth resource satellite *Landsat D' (D-prime)* from Vandenberg AFB, Calif., to a 434.5 × 435 mi. orbit inclined 98.2° to the equator. Redesignated *Landsat 5* in orbit, the launcher also carried *Uosat 2/Oscar 9*, a 115-lb. amateur radio satellite built by the University of Surrey in England. Built by General Electric, *Landsat 5* was identical to its predecessor, *Landsat 4*. Both satellites remain in space. Due to a series of delays caused by policy changes in Landsat management, the next U.S. Earth resource satellite would not be launched until October 5, 1993, although *Landsat 6* was destroyed by a propulsion failure.

1984 March 15   Michael Heseltine, Britain's defense secretary, announced the names of four candidate astronauts from the United Kingdom who would receive training for a flight aboard the Shuttle when the *Skynet 4* military communications satellite was launched. The candidates were: Christopher Homes, a civil servant specializing in telecommunications; Lt. Col. Tony Boyle from the army; Cdr. Peter Longhurst from the Royal Navy; and Sqdn. Ldr. Nigel Wood, from the Royal Air Force. At this date, Skynet was scheduled to be carried by Shuttle *Challenger* on mission 61-D planned for launch on January 22, 1986. By that date Skynet had been remanifested to mission 61-H *(Columbia)*, scheduled to fly on June 24, 1986, but *Challenger* blew up on January 28, 1986, and all subsequent flights were canceled.

At a press conference in London, Dr. David Baker announced the world's first independent space consulting company, Space Consultants International, set up to serve international finance and insurance. With backing from insurance brokers Willis Faber and Dumas, SCI provided information and advisory services to both brokers and underwriters in the United Kingdom, Europe, India, China, Japan and the Far East. Development Manager Sarah Palmer organized conferences in London, England, Los Angeles, Calif., and Washington, D.C., to stimulate space commercialization and support the National Space Policy directive of July 4, 1982. The

company was dissolved after the *Challenger* disaster and reformed as Sigma Projects in April 1986.

1984 March 16   NASA announced selection of McDonnell Douglas and TRW Systems for parallel study contracts on space station systems definition and architecture studies. The work was to last 2 years 3 months and define the role of the station data system to characterize the environment in which both operator and user would function. The station had to serve a community of widely differing science and technology interests.

1984 March 27   Secretary of Defense Caspar Weinberger announced that Lt. Gen. James Abrahamson, then head of the NASA Shuttle program, would head the Strategic Defense Initiative antiballistic missile research program. His job would be to coordinate the many separate research programs that had been under way at the Department of Defense and the Department of Energy. Gen. Abrahamson moved to his new job on April 15 after Shuttle mission 41-C. During the month the Defensive Technology Study Team under Dr. James Fletcher reported its findings. It was generally optimistic about the SDI concept and believed it would be possible to build a low-leakage screen, in which few enemy warheads got through, by utilizing all the available laser, particle-beam and impact technologies.

1984 March 30   NASA announced plans for implementing one of the recommendations of the Solar System Exploration Committee and declared that a scientific advisory group had decided that comet Kopff should be the target for a Mariner Mark II mission in the 1990s. A team of 20 U.S. and European scientists comprising NASA's comet Rendezvous Science Working Group recommended that a spacecraft launched by Shuttle in July 1990 could rendezvous with Kopff in February 1994. The rendezvous mission would differ from a fly-by in that the spacecraft would match its orbit around the sun with that of the comet, flying in formation with it for several years. NASA incorporated into the flight an asteroid fly-by and announced that it would ask for funds in January 1986 to start development during fiscal year 1987. However, NASA deferred this and postponed the start of its Comet Rendezvous/Asteroid Fly-by (CRAF) mission for financial reasons.

1984 March   Scott Science and Technology (SST) made its first public offering to raise money for a low-cost liquid propulsion stage capable of transferring satellites from Shuttle orbit to geosynchronous altitude. The company had been formed in anticipation of a lucrative market for upper stages. The Satellite Transfer Vehicle (STV) was being designed by British Aerospace. President, and ex-astronaut, David Scott obtained an agreement with NASA for help with testing. In 1985 the design was completed around a cluster of four 900-lb. thrust bipropellant engines built by Marquardt, and three versions were projected. In 1986 SST contracted with TRW Systems for management of the STV, but a collapse of Shuttle-launched satellite business after the *Challenger* disaster made the concept redundant.

1984 April 3   At 13:09 UT cosmonauts Yuri A. Malyshev, Gennady M. Strekalov and Rakesh Sharma, the first Indian in space, were launched in their *Soyuz-T 11* spacecraft from the Baikonur cosmodrome. They docked to *Salyut 7*, in an orbit of 178.5 × 185 mi., at the aft port at 14:31 UT on April 4. When the crew transferred to the station, six cosmonauts were aboard *Salyut 7* for the first time. Because the five crew members of Shuttle mission STS-41C were also in orbit, a world record of eleven people in space at one time was established. The Indian cosmonaut, Sharma, conducted several experiments, including one which tested the value of yoga to relaxation in space. After crew contour couches had been transferred between spacecraft, the crew returned to Earth in *Soyuz-T 10*, undocking at 07:33 UT on April 11 and landing at an elapsed time of 7 days 21 hr. 41 min. On April 13 Leonid Kizim, Vladimir Solovyov and Oleg Atkov undocked *Soyuz-T 11* from the aft port and transferred it to the forward port.

1984 April 6   One of the most complex Shuttle flights scheduled by NASA, mission 41-C began at 13:58:00 UT from LC-39A, at Kennedy Space Center, Fla., when *Challenger* lifted off on its way to repair a stranded solar maximum mission (SMM) satellite. For the first time, the Shuttle was launched to a direct-ascent trajectory, where only one orbital maneuvering system burn was required to put *Challenger* in a 157.6 × 290 mi. orbit. In addition to astronauts Robert Crippen (CDR), Francis Scobee (PLT), George Nelson (MS), James Van Hoften (MS) and Terry Hart (MS), *Challenger* carried a 38,266-lb. payload including the 21,396-lb. long-duration exposure facility (LDEF) with 57 passive science, technology and engineering experiments. With a length of 30 ft. and a width of 14 ft., LDEF was lifted from the cargo bay and released at 17:19:27 UT on April 7. NASA planned to

Astronaut James Van Hoften works on the main electronics box of the Solar Maximum Mission satellite from a work platform on the end of the Shuttle's remote manipulator arm.

retrieve the LDEF during the 51-D mission in February 1985. After *Challenger* rendezvoused with the SMM satellite, Nelson and Van Hoften began a 2 hr. 57 min. EVA (extravehicular activity) at 14:18 UT on April 8, operating one of two manned maneuvering units for 42 min. in abortive efforts at connecting a forward probe to the SMM so that it could be steered back to a servicing fixture in the payload bay for repair. In contingency procedures, SMM was grasped by the remote manipulator arm at 13:52:20 UT on April 10 and placed on the servicing structure. In a 6 hr. 16 min. EVA that began at 08:58 UT on April 11, the two astronauts repaired the satellite and it was put back in orbit by the manipulator arm at 09:26:29 UT on April 12. Extended one day to April 13, the mission ended at Edwards AFB, Calif., 6 days 23 hr. 40 min. 7 sec. after liftoff.

1984 April 8   China launched its first geostationary communication satellite on the second CZ-3 from the Xichang launch site. Designated *STW-1*, the 1,985-lb. satellite was placed in geostationary orbit over the Molucca Sea at 125° E. The satellite would link remote areas across China, serving both civil and military requirements. The second geostationary communications satellite, referred to merely as *STW*, was successfully launched on February 1, 1986, to 103° E. China also had reservations pending for three slots on the Shuttle but these were never taken up.

1984 April 13   Mission controllers at NASA's Ames Research Center sent commands to the *Pioneer-Venus 1* orbiter around Venus to realign itself and make observations of the comet Encke. Tipped 37° by 200 × 0.5 sec. attitude thruster pulses, the spacecraft observed Encke with its ultraviolet spectrometer across 100 million mi. of space. The comet Encke makes a close pass around the sun, moving inside the orbit of the

NASA's first Shuttle-based rescue mission gets under way as Challenger *leaves the pad on April 6, 1984, to repair the Solar Maximum Mission satellite.*

planet Mercury once every 3.3 years and is for this reason one of the most observed comets in the solar system. *Pioneer-Venus 1* returned to observations of Venus but on December 26, 1985, began 10 weeks of observation of the comet Halley which ended March 6, 1986. After that it turned its attention once again to Venus.

1984 April 15 At 08:13 UT a Soviet SL-4 launch vehicle from the Baikonur cosmodrome lifted off carrying the Progress 20 cargo tanker. In an orbit of 172.5 × 180 mi., it was destined for the *Salyut 7* space station and docked to the aft port at 09:22 on April 17. The vehicle carried fresh supplies and life-support equipment for a planned series of space walks Leonid Kizim and Vladimir Solovyov were scheduled to perform beginning the last week in April. *Progress 20* remained attached to *Salyut 7* throughout these activities, providing replenishment air to pressurize the transfer (airlock) module on each of the four occasions the two cosmonauts went outside. The cargo tanker finally undocked at 17:46 UT on May 6, and was de-orbited next day.

1984 April 23 *Salyut 7* cosmonauts Leonid Kizim and Vladimir Solovyov began a 4 hr. 15 min. space walk at 04:31 UT to erect ladders and access equipment to check items necessary for a major repair of the propulsion module on a second space walk three days later. Having assured themselves that the equipment would work, it was folded up and left against the side of the station. Beginning at 02:40 UT on April 26, a second space walk by Kizim and Solovyov afforded opportunity to start work on the propulsion units in the unpressurized section of the aft equipment module. The crew installed a valve in a propellant conduit, and after checking it for tightness, they returned to the station. A third space walk took Kizim and Solovyov out at 01:35 UT on April 29; they installed another conduit and checked it out before ending their activity at an elapsed time of 2 hr. 45 min. A fourth space walk for these two cosmonauts began at 23:15 UT on May 3, when they took apart thermal covering over the propulsion unit and put together a second extra line conduit. It was over in 2 hr. 45 min.

1984 April 23–27 A NASA-sponsored Lunar Base Workshop was hosted by the Los Alamos branch of the Institute of Geophysics and Planetary Physics, University of California. Comprising 50 scientists, engineers, industrialists, historians and other academics, attendees concluded that the time was right for a new national effort to establish settlements on the moon as a cornerstone of future space exploration. They resolved to reconvene at a lunar bases symposium scheduled for late October to establish guidelines for setting goals and objectives.

1984 April A joint Technical Assessment Study on a U.S.-European collaborative mission to the planet Saturn and its moon Titan began. Meetings between NASA's Jet Propulsion Laboratory and the European Space Agency's ESTEC and ESOC facilities pooled ideas on the venture. Under a joint agreement to share the costs of an ambitious deep-space mission proposed by NASA as a Mariner Mark II mission, a NASA spacecraft would go into orbit about Saturn and send a European probe to land on Titan to transmit scientific information from the surface. As proposed, the spacecraft was to be launched in May 1994 for a flight around the sun before a return to Earth and gravity-assist slingshot in June 1997, fly-by of Jupiter in February 2000 and arrival at Saturn in January 2002. By 1987 a delay in getting a funded start

deferred the proposed launch date to March 1995, with Earth swing-by in April 1997, a Jupiter fly-by in May 1999, and arrival at Saturn in January 2002.

1984 May 7 At 22:47 UT the unmanned cargo tanker *Progress 21* was launched by SL-4 from the Baikonur cosmodrome. It docked to *Salyut 7*, in an orbit of 171.5 × 197 mi., at 00:10 UT on May 10, just over four days after the aft port had been vacated by *Progress 20*. That cargo tanker had supported four repressurizations of the transfer module after four planned space walks by Leonid Kizim and Vladimir Solovyov. *Progress 21* would remain docked to *Salyut 7* to support the fifth space walk, scheduled for May 18. *Progress 21* undocked at 09:41 UT on May 26, and was de-orbited the same day.

1984 May 18 Cosmonauts Leonid Kizim and Vladimir Solovyov began the fifth space walk of their stay aboard *Salyut 7* when, at 17:52, they exited the transfer (airlock) module hatch and went outside to erect two supplemental solar cell panels. They were to attach each 49.9 sq. ft. panel to opposite sides of the 216 sq. ft. main solar array wing on the right side of the station. In an operation lasting 3 hr. 5 min. the two cosmonauts completed the job, fixing first one panel before cosmonaut Oleg Atkov, inside the station, rotated the array through 180° so that the second panel could be attached from the same work station. The docked *Progress 21* cargo tanker supplied air to repressurize the transfer module.

  NASA transferred ownership of the *Viking 1* lander spacecraft on Mars to the Smithsonian Institution's National Air and Space Museum in Washington, D.C., the first time a museum had taken ownership of an object on another planet. The museum's director, Walter Boyne, accepted the spacecraft on behalf of NASA in a special ceremony where the lander was named the Thomas A. Mutch Memorial Station. Dr. Mutch had been the leader of the Viking lander imaging team but died during a climbing accident in the Himalayas in 1980. NASA agreed to carry the plaque commemorating the naming of the lander to the landing site with the first U.S. astronauts to visit the area. They also retained the option of reclaiming rights over the lander so that the astronauts could retain parts of it for scientific analysis.

1984 May 23 In the first commercial launch conducted by Arianespace, the first Ariane launch for a U.S. customer, Ariane 1 L9 carried the 2,634-lb. *Spacenet 1* communication satellite to a geosynchronous transfer ellipse from which an apogee motor placed it in geostationary orbit at 120° W. Built by RCA, *Spacenet 1* was a box structure 5 ft. 4 in. × 4 ft. 4 in. × 3 ft. 3 in. high, with a total height of 9 ft. 8 in. with dish antenna deployed. With two solar cell wings deployed, the satellite had a span of 47 ft. and would produce up to 1,220 W electrical energy. *Spacenet 1* carried 18 C-band (4–6 GHz) channels, with 14 × 8.5-W and 7 × 16-W amplifiers, and 6 Ku-band (12–14 GHz) channels. With an on-orbit weight of 1,551 lb., the satellite would provide telecommunication services to the 50 states.

1984 May 28 At 14:13 UT the cargo tanker *Progress 22* was launched from Baikonur to the *Salyut 7* space station. In an orbit of 207.5 × 222.5 mi., it docked with *Salyut 7* at 15:47 UT on May 30. Comprising additional propellant, food, water and general supplies, the cargo also included 90 lb. of extra medical equipment and 155 lb. of scientific equipment, 100 lb. of which comprised film and photographic equipment. The

Salyut 7 crew of Leonid Kizim, Vladimir Solovyov and Oleg Atkov were scheduled to conduct an extensive series of Earth observations over a period of six weeks; the inventory of equipment brought by *Progress 22* supported that program. The cargo tanker undocked from the aft port at 13:36 UT on July 15 and was de-orbited the same day.

1984 May   NASA announced the names of 17 pilot (PLT) and mission specialist (MS) Shuttle astronauts constituting Group 10: Col. James C. Adamson (USA); Ellen S. Baker; Col. Mark N. Brown (USAF); Lt. Col. Kenneth D. Cameron (USMC); Capt. Manley Lanier Carter (USN); Col. John H. Casper (USAF); Capt. Frank L. Culbertson (USN); Lt. Col. Sidney M. Gutierrez (USAF); Col. Blaine L. Hammond Jr. (USAF); Marsha S. Ivins; Lt. Col. Mark C. Lee (USAF); G. David Low; Capt. Michael J. McCulley (USN); Capt. William M. Shepherd (USN); Kathryn C. Thornton; Charles Lacy Veach; Cdr. James D. Wetherbee (USN). Carter died on April 5, 1991, in the crash of a commercial airliner while he was on NASA business.

1984 June 9   The first in a series of "stretched" Atlas-Centaur launch vehicles (AC-62) carried the *Intelsat V F-9* into space, failing to place it in the planned geosynchronous transfer orbit due to a leak in the liquid oxygen system of the Centaur. Designated Atlas G, the main stage had been lengthened by 6 ft. 9 in., accommodating additional propellant, with liftoff thrust increased from 430,500 lb. to 438,000 lb. The Centaur D-1A was equipped with pressure-fed (not pump-fed) RL-10A-3-3A engines. These changes increased geosynchronous transfer orbit capacity from 4,600 lb. to 5,100 lb. and extended the length of the launcher to 138 ft. *Intelsat V F-9* was left in a 109 × 755 mi., 29.1° orbit from which it decayed to destruction on October 24. This was the first failure of a NASA launch vehicle since 1977.

1984 June 10   The fourth and final flight in the U.S. Army homing overlay experiment (HOE) successfully intercepted a Minuteman missile fired from Vandenberg AFB, Calif. The modified two-stage Minuteman adapted as an interceptor with sensors and snare device was launched from Meck in the Kwajalein island chain of the Marshall Islands about 20 min. after the target vehicle had been launched. About 4,800 mi. along its trajectory, the reentry vehicle discharged by the Minuteman was impacted by the interceptor at a closing velocity of 20,000 MPH. The head of the interceptor comprised a set of 36 steel bands, each 7 ft. long, wrapped around a central core. When deployed prior to intercept the bands deployed like steel tape to form a set of radial spokes 15 ft. across which would snare the target.

1984 June 11   A team of scientists had been selected to provide science investigations for the Shuttle-based astronomy observatory called Astro. Denied the ability to send a spacecraft to Halley's comet in 1985, Astro would be used from the Shuttle to observe the comet from Earth orbit. The instruments included three ultraviolet telescopes and two wide-field cameras carried as a Spacelab payload in the cargo bay when the comet passed close to the sun early in March 1986.

1984 June 13   Joining the rush to develop propulsion stages capable of shifting satellites from Shuttle orbit to geosynchronous transfer orbit, RCA announced development of a Shuttle compatible orbital transfer system (SCOTS). Aimed at producing a low-cost stage, SCOTS was to have been capable of boosting a 6,000-lb. satellite. RCA wanted an upper stage available for customers selecting its series-4000 communica-

tion satellite, which could weigh anything from 4,500 lb. to 6,000 lb. The PAM-D upper stage would not have the power for satellites of this weight. Thiokol was to have provided a solid propellant motor, but the project became redundant when the Shuttle *Challenger* was destroyed in 1986.

1984 June 25   NASA conducted the first launch attempt of Shuttle *Discovery* for its maiden flight into space on the 41-D mission. The orbiter carried a 33,841-lb. payload comprising the 15,190-lb. *Syncom IV-A* communications satellite *(Leasat-1)* and its solid propellant perigee-kick motor, a 3,396-lb. unfolding solar array experiment known as OAST-1 from NASA's Office of Aeronautics and Space Technology, and the 3,560-lb. large-format camera (LFC) and truss structure capable of precise stereo photography of the Earth's surface. During the countdown to launch with astronauts Henry Hartsfield (CDR), Michael Coats (PLT), Judith Resnik (MS), Richard Mullane (MS), Steven Hawley (MS) and Charles Walker (PS), a computer malfunction appeared at T − 32 min. and again at T − 20 sec. Recycled to T − 9 min., the computer once again refused to process a launch program. The countdown was abandoned at T − 9 min. and recycled for a second attempt next day.

1984 June 26   Only one day after *Discovery*'s originally scheduled launch was aborted, NASA's Shuttle program experienced its first pad abort when *Discovery* failed to lift off on mission 41-D. Sensors detected a problem with a fuel valve in main engine no. 3 and shut down the start sequence at T − 4 sec. The engine had momentarily ignited and start commands for engines no. 1 and no. 2 had been given; the three engines are started at 0.12 sec. intervals. Engine no. 2 had only just begun the ignition sequence but engine no. 1 had not. A fire broke out under the orbiter when excess hydrogen built up, but extinguishers soon put it out. The crew (Hartsfield, Coats, Resnik, Mullane, Hawley and Walker) left the orbiter 40 min. later. The Shuttle was rolled back to the Vehicle Assembly Building on July 14 and *Discovery* was placed in the orbiter processing facility three days later. The no. 3 engine was replaced and *Discovery* was rolled out to LC-39A once more on August 9. To minimize delays to the schedule, payloads from missions 41-D and 41-F were combined into the revised 41-D mission and 41-F was deleted. As a separate mission, 41-F was to have carried a 48,363-lb. payload comprising *SBS-4*, *Telstar-3C* and *Syncom IV-2 (Leasat-2)* communication satellites as well as a science package called Spartan.

1984 June 28   NASA announced details of four field centers which would participate in Space Station development: the Johnson Space Center, already tagged as the lead center for Station development, was to select an appropriate configuration and be responsible for all systems engineering and integration as well as the structural framework to which the various elements would be attached; the Marshall Space Flight Center was to be in charge of pressurized modules, designed to a common concept for cost efficiency, as well as their outfitting, the environmental control system and orbital maneuvering vehicles; the Lewis Research Center was to be responsible for electrical power generation and distribution; and the Goddard Space Flight Center would handle definition of automated free-flying platforms.

1984 July 5   NASA's Marshall Space Flight Center announced that it was to award contracts to Martin Marietta and Boeing for studies on a space vehicle capable of moving payloads from low Earth orbit to higher orbits. Known as the orbital transfer vehicle (OTV), it was a successor to the now defunct space tug. Conceived as a relatively low-cost un-manned orbit-to-orbit propulsion stage, the OTV was to have been launched by Shuttle and left at the space station for use as necessary.

1984 July 17 At 17:41 UT cosmonauts Vladimir A. Dzhani-bekov, Svetlana Y. Savitskaya and Igor P. Volk were launched aboard their *Soyuz-T 12* spacecraft to the *Salyut 7* space station. With the orbiting complex in a path of 207.5 × 220 mi., *Soyuz-T 12* docked to the aft station port at 19:17 UT next day. Joining Leonid Kizim, Vladimir Solovyov and Oleg Atkov in *Salyut 7*, the visiting crew began a busy period of activity. On July 25 the transfer (airlock) module hatch was opened and Dzhanibekov and Savitskaya performed a 3 hr. 55 min. space walk beginning at 14:55 UT. Savitskaya thus became the first woman to make a repeat spaceflight and the first to perform a space walk. During the exercise, cutting, welding and soldering of test objects and metal plates was demonstrated, and samples were retrieved from the exterior of the space station. The *Soyuz-T 12* crew returned to Earth on July 29 at the end of a mission lasting 11 days 19 hr. 14 min.

NASA announced that microscopic latex particles for applications in medical research would soon be put on sale as the first commercial products produced in space. Beginning with STS-3, five Shuttle flights had carried a latex reactor experiment from which spheres 5–30 micrometers in size were produced. The largest latex particles produced on Earth are 5 micrometers in size. They are used for measuring the size of pores in the wall of the human intestine during cancer research, and for measuring the size of pores in the human eye in glaucoma research. The first sale was made on July 17, 1985, at a cost of $384 per unit of 30 million spheres.

1984 July 24 NASA selected LTV Aerospace, Martin Ma-rietta and TRW Systems to carry out studies on a small rocket stage carried in the Shuttle and used on orbit for retrieving satellites from high orbit so they could be repaired in space. Called the orbital maneuvering vehicle (OMV), it was smaller and had less propulsive capability than the orbital transfer vehicle. The OMV would be used for shunting payloads up and down to a low-orbit Shuttle and differ from the OTV in that it would not be based in space. Preliminary designs indicated that the OMV would be about 15 ft. in diameter and 3 ft. long.

1984 July 27 NASA Headquarters established a Space Sta-tion Office to direct the agency's development and manage-ment of the permanent manned space station. Heading the new office would be Associate Administrator Philip E. Cul-bertson, with John Hodge as his deputy. The new program office would provide policy and overall program direction to the Johnson Space Center, which would be the reporting center for all other NASA facilities working on the develop-ment of the space station.

1984 August 1 Donald P. Hearth, director of the NASA Langley Research Center, made a presentation to Congress explaining a technology research effort under way to define a successor to the partially reusable Shuttle. Called Shuttle II, the concepts ranged from two-stage, fully reusable fly-back booster/orbiter combinations to single-stage-to-orbit (SSTO) vehicles of advanced design. Langley thought a second-generation orbiter should be sized to carry a 20,000–40,000-lb. payload to low Earth orbit. A plug-in cargo module would be attached to the top of the orbiter with payloads intact.

1984 August 3 A small quantity of champagne went on a 32-sec. flight celebrating the culmination of a three-year effort to produce a solid propellant rocket capable of being launched from a vertical floating position at sea. Called Dolphin, it was fired from the Pacific Ocean off San Clemente Island, Calif. In 1981 Michael Scott, ex-Apple Computer president, and Tucker S. Thompson had persuaded a group of wealthy entrepreneurs, among them Apple cofounder Steve Wozniak, to invest in development of a low-cost competitor to Shuttle and Ariane. Called Starstruck, their company developed a simple solid propellant rocket which they expected to cluster into a configuration called Constellation. On Dolphin's one and only test flight it weighed 16,500 lb. and produced 35,000 lb. of thrust.

1984 August 4 The first launch of a more powerful Ariane launch vehicle, Ariane 3, took place at Kourou, French Guiana, carrying the 2,613-lb. *Telecom 1* communications satellite and the 2,591-lb. European Communication Satellite *ECS-2*. Ariane 3 had improved first- and second-stage engines and a third stage lengthened by 3 ft. 11 in., its propellant capacity raised to 22,930 lb., an HM7B engine producing a thrust of 14,130 lb., and burn time increased to 12 min. In addition, two 21,320-lb. solid propellant boosters, each con-taining 16,200 lb. of propellant and producing a thrust of 149,850 lb., were attached to the first stage. The boosters were ignited 4 sec. after liftoff and burned for 32 sec., generating a total vehicle thrust of more than 900,000 lb. for that period. Total vehicle height was 162 ft. 5 in. with a liftoff weight of 528,944 lb. Almost identical to *ECS-1*, *ECS-2* was placed in geostationary orbit at 7° E, while *Telecom 1A* was positioned at 8° W. *Telecom 1A* was built by Matra to provide two channels on 8/7 GHz for the French Defense Ministry, four 6/4 GHz channels for domestic telecommunications and TV services and six 14/12 GHz channels for digital data. The box-shaped satellite, 4 ft. 8 in. × 6 ft. 4 in. × 4 ft. 8 in. tall, carried 15,816 solar cells producing 1,100 W on two wings with a span of 52 ft. 6 in. The satellite had an on-orbit weight of 1,550 lb.

1984 August 6 Soviet military photoreconnaissance satellite operators tested a new in-orbit storage technique with the launch of *Cosmos 1587*, a third-generation type, from Ple-setsk on an SL-4. Placed initially in a 122 × 228 mi. orbit inclined 72.8°, the satellite was shut down and allowed to drift. After 10 days it was powered up and operated for its standard 14-day mission before returning to Earth on August 31 in a demonstration of how vital space-based assets could be stored in orbit for future activation and use. A similar demonstration began with the launch of *Cosmos 1613* on November 29, 1984, which returned to Earth after 25 days.

1984 August 8 At 08:46 UT the side hatch on the *Salyut 7* forward transfer (airlock) module was opened for the sixth space walk by Leonid Kizim and Vladimir Solovyov. Working their way to the back of the station, the two cosmonauts removed thermal blankets from the propulsion unit of the equipment section and disconnected a propellant feed line in further work to prevent the entire system from breaking down. With the blankets replaced, the cosmonauts reentered the space station at the end of an excursion that had lasted 5 hr.

1984 August 10 NASA announced that it had selected RCA's Astro-Electronics division to design and build the Advanced Technology Communications Satellite (ACTS), together with

NASA's power-tower concept for a permanently manned space station was unveiled on July 11, 1984, but would prove to be only the first in a protracted series of redesigned configurations.

team members TRW Systems, Comsat General Corp., Motorola, Hughes Aircraft and Electromagnetic Sciences. The ACTS was to be used to research superhigh-frequency bands in the 20 GHz and 30 GHz range and to make this information available to U.S. corporations, research organizations and laboratories.

1984 August 13 In a historic agreement signed between insurance underwriters Merrett Syndicates of London, England, and NASA, the U.S. space agency agreed to attempt the retrieval of the *Palapa B-2* satellite that had stranded in low Earth orbit after deployment on mission 41-B. On September 20 a similar agreement was signed for the retrieval of *Westar VI* stranded at the same time. The satellite insurers agreed to reimburse NASA for the overhead costs of $2.75 million per satellite, and Merrett Syndicates planned to sell each satellite to recoup some of the losses they incurred when settling the claim. NASA scheduled retrieval efforts as an addition to the 51-A mission.

1984 August 14 At 06:28 UT the unmanned cargo tanker *Progress 23* was launched by an SL-4 from the Baikonur cosmodrome and docked to the *Salyut 7* space station, in an orbit of 218 × 233 mi., at 08:11 UT two days later. Supplies were transferred across to *Salyut 7* by the resident crew of

Leonid Kizim, Vladimir Solovyov and Oleg Atkov. Its work accomplished, *Progress 23* undocked from the aft port at 16:13 UT on August 23 and returned to fiery destruction in the atmosphere two days later.

1984 August 16 A Delta 3924 launched from Cape Canaveral, Fla., carried three active magnetosphere particle explorers (AMPTE) into orbit for a coordinated measurement of the transfer of charged particles from the solar wind to the Earth's magnetosphere. *AMPTE 1*, the U.S. Charge Composition Explorer, took the form of an octagonal prism 3 ft. 4 in. tall and 6 ft. 8 in. across with a weight of 534 lb. placed in a 628 × 30,934 mi. orbit inclined 4°. *AMPTE 2*, the West German IRM satellite, was an irregular cylinder, with a length of 6 ft. 8 in. and a diameter of 6 ft. 8 in. and a weight of 1,334 lb., placed in a 250 × 70,726 mi. orbit inclined 27°. *AMPTE 3*, the 170-lb. British satellite UKS, was a prism 1 ft. 8 in. long and 4 ft. 11 in. in diameter placed in a 623 × 70,477 mi. orbit inclined 26.9°. *AMPTE 1* measured tracer ions released into orbit by *AMPTE 2* while *AMPTE 3* monitored disturbances in the magnetosphere produced by the ions.

1984 August 17 NASA held a Space Station Mission Requirements Workshop at which it briefed industry and potential users on what it wanted from the participating community. NASA's Johnson Space Center had generated various architectural designs for the space station, and the resulting reference configuration used in the invitation for tenders was known as the power-tower concept. This had the modules clustered together at one end of a 420 ft. tall truss assembly. Half way up the truss structure was a secondary truss at right angles forming the horizontal bar of an "H" configuration. The outside vertical bars of the "H" supported eight solar cell "wings" — four on each side. Although NASA wanted industry to use this as the baseline, it sought design originality.

1984 August 27 President Ronald Reagan announced during a ceremony at the Department of Education in Washington, D.C., that NASA would fly a U.S. educator on a forthcoming Shuttle mission. The so-called Teacher in Space program brought 11,000 applications from which the Council of Chief State School Officers selected 114 elementary and secondary level candidates. After briefings on the space program held in Washington, D.C., June 22–27, 1985, 10 finalists were selected. On July 19, 1985, Vice President George Bush announced the winners. Sharon Christa McAuliffe was the prime candidate with Barbara R. Morgan selected as her backup.

1984 August 30 After a postponement from August 29, when a computer problem aboard Shuttle *Discovery* at T − 20 min. canceled the attempt that day, the third NASA orbiter made its maiden flight from LC-39A at Kennedy Space Center, Fla. Delayed for 6 min. 50 sec. at T − 9 min., when a private plane strayed into restricted airspace, 41-D lifted off at 12:41:50 UT carrying 47,516 lb. in communication satellites, the OAST-1 extendible solar array, and experiments including an electrophoresis test for producing new vaccines. The crew was the same as that for the earlier attempt on June 25. Communication satellite *SBS-D* and its PAM-D stage were released at 20:40:18 UT, *Syncom IV-2* at 13:16:27 UT on August 31 and *Telstar-3C* and its PAM-D at 13:25:52 UT on September 1. *Syncom IV* was the first satellite designed for launch from a Shuttle and was released by the "frisbee" method, rolling out while rotating. It fired its motor to put itself in geostationary orbit at 105° W. The two Hughes 376 satellites were pushed to geostationary transfer orbit by their PAM-D motors and from there to circular orbit by on-board apogee motors. The OAST-1 array was successfully unfurled to its maximum height of 105 ft. and retracted. *Discovery* landed at Edwards AFB, Calif., at 6 days 0 hr. 56 min. 4 sec.

1984 August 31 NASA announced that it would develop a satellite to investigate celestial radiation in the extreme ultraviolet end of the spectrum. Called the extreme ultraviolet explorer (EUVE), it was designed for launch from the Shuttle in a mission anticipated for 1988. Supplied by the University of California, Berkeley, four 15.7-in. telescopes would scan the universe for emissions in the extreme ultraviolet range to complement the work carried out by the international ultraviolet explorer (IUE).

1984 August British Aerospace revealed preliminary details of its HOTOL (horizontal takeoff and landing) satellite launcher concept based on air-breathing, hydrogen-fueled propulsion. With a highly streamlined double-delta configuration, the proposed HOTOL had a length of 178 ft. and a wingspan of 56 ft. With forward canards and two outwardly canted vertical stabilizers, HOTOL would become airborne at the end of an accelerated horizontal sled run. Accelerated to hypersonic velocities by mixing ingested air with hydrogen

fuel, the vehicle would assume the conventional rocket-powered role for reaching low Earth orbit, switching to an on-board supply of oxygen at extreme altitude. Refined by Alan Bond of the United Kingdom Atomic Energy Authority, the engine design was similar to that originated by the Marquardt Corp. in the United States during the 1960s.

1984 September 11 NASA announced a new post of administrator for commercial programs, to be filled by Isaac T. Gillam IV. In accordance with the National Space Policy directive of July 4, 1982, and subsequent edicts on the need to improve the commercialization of space, NASA began to focus efforts on high-technology ventures, new commercial applications of existing technology and initiatives aimed at transferring existing space programs to the private sector. These were enshrined in a Commercial Use of Space Policy issued by NASA on November 20.

1984 September 14 NASA issued a request for proposals on the preliminary design and definition of a manned space station with the power-tower reference configuration. NASA wanted the station operational by the early 1990s and asked companies to submit their proposals by November 15, 1984. Instead of naming a single contractor to build the station, or to act as lead contractor for a team, as in previous projects, NASA wanted to divide the contracts into four separate work packages: WP1 would be managed by the Marshall Space Flight Center and include definition and preliminary design of the common pressure modules; WP2 under the Johnson Space Center would provide definition and preliminary design of the structural framework, space station systems, the habitation plan and the EVA (extravehicular activity) equipment; WP3 under Goddard Space Flight Center would define and design free-flying platforms; WP4 under Lewis Research Center would define and design the electrical power system. The specification called for a 6–8 person crew with 75 kW of electrical power operating a cluster of modules in a 345-mi., 28.5° orbit.

1984 September 21 The United States and China signed a space agreement detailing five areas of cooperation, including spacecraft environmental data, magnetic test data, thermal control technology, geodynamic data and time frequency data. It came after five years of negotiation, and proposals that a Chinese astronaut fly on the Shuttle were put forward. When *Challenger* was destroyed on January 28, 1986, the prospect of launching foreign nationals for political purposes vanished. China announced it would market its launchers for international business.

1984 September 28 First in a new series of Soviet electronic intelligence gathering satellite, the 20,000-lb. *Cosmos 1603* was launched by an SL-16 from Baikonur to a 112 × 118 mi. orbit inclined 51.6°. It maneuvered to a 528-mi. orbit inclined 66.6°, then to a 530-mi. orbit inclined 71°. *Cosmos 1603* made four daily passes across the United States, gathering elint from the same areas on a routine basis. Similar elints included: *Cosmos 1656* on May 30, 1985; *Cosmos 1697* on October 23, 1985 (which suffered booster failure); *Cosmos 1714* on December 28, 1985 (which stranded in a 275 × 530 mi. orbit); *Cosmos 1844* on May 13, 1987; *Cosmos 1943* on May 15, 1988; *Cosmos 1980* on November 23, 1988; *Cosmos 2082* on May 22, 1990; *Cosmos 2219* on November 17, 1992; and *Cosmos 2227* on December 25, 1992.

1984 October 2 *Salyut 7* cosmonauts Leonid Kizim, Vladimir Solovyov and Oleg Atkov returned to Earth at the end of almost eight months in space. The three-man crew got into the *Soyuz-T 11* spacecraft which had been in space since April 3 having displayed an endurance of six months. They undocked from the forward docking port at 08:40 UT and landed at 10:57 236 days 22 hr. 50 min. after they had been launched on February 8. When the crew had been examined on Earth and pronounced fit, albeit greatly affected after their record-breaking flight, Russian sources linked the duration of this flight to the average journey time to the planet Mars. It was further consolidation of long-term aims and goals of the Soviet space program, but it provided more tangible evidence that humans could survive weightlessness long enough to reach the nearest planet people might visit.

1984 October 5 At 11:03:00 UT Shuttle *Challenger* began the 41-G mission when it lifted off from LC-39A at the Kennedy Space Center, Fla., carrying astronauts Robert Crippen (CDR), Jon McBride (PLT), David Leestma (MS), Sally Ride (MS), Kathryn Sullivan (MS), Paul D. Scully-Power (PS) and Marc Garneau (PS). It carried a 23,465-lb. payload comprising three OSTA-3 Earth observation experiments for NASA's Office of Space and Terrestrial Applications, the 4,949-lb. *Earth Radiation Budget Satellite*, the large-format camera and several technology experiments. *ERBS* was released by the remote manipulator arm at 22:18:22 UT. At 15:38 UT on October 11, Leestma and Sullivan began a 3 hr. 27 min. EVA (extravehicular activity) evaluating a fluid transfer experiment that may be used in the future to refuel satellites. This was the first EVA conducted by a woman. The mission ended at 8 days 5 hr. 23 min. 38 sec. at the Kennedy Space Center, Fla. ERBS was part of the Earth radiation budget experiment which also involved instruments carried aboard NOAA-9 launched on December 12, and NOAA-10 September 17, 1986.

1984 October 13 President Ronald Reagan set up a National Commission on Space to be composed of fifteen members, not more than nine of whom would represent federal departments or agencies, with two members from the Senate and two from the House of Representatives in an advisory capacity. The commission was asked to examine the state of the U.S. space program and identify long-range goals and policy options for the next 20 years.

1984 October 25 Morton Thiokol successfully tested the first filament-wound solid rocket booster in development for the Shuttle. Each filament-wound booster would weigh about 28,000 lb. less than the conventional steel-case booster, thus improving Shuttle lift capacity by 4,600 lb. The tests were scheduled for completion in time for the first Shuttle launch from SLC-6 at Vandenberg AFB, Calif., presently scheduled for October 1985. The second test was completed on May 9, 1985, by which time the flight had moved to early 1986.

1984 October 29–31 "Lunar Bases and Space Activities of the 21st Century" was the running theme at a NASA-sponsored symposium held in Washington, D.C., at which 200 agency and nongovernment participants aired thoughts and philosophies on the next major space goals for the United States. A broad range of applications and reasons for setting up a moon base were discussed and practical problems and potential solutions were debated. Delegates at the symposium agreed to reconvene a year later for further discussions with the object of bringing a realistic set of program options for

political consideration of presenting moon bases as a national goal.

1984 October Astronauts in western Europe formed the Association of European Astronauts (AEA) open to anyone who had flown or been selected to train for a specific mission. The first meeting was held October 5–6, during which the purpose of the AEA was defined as an exchange of views concerning training and projects. The seven inaugural members included: ESA astronauts Claude Nicollier, Ulf Merbold and Wubbo Ockels; French astronauts Patrick Baudry and Jean Loup-Chretien; and German astronauts Rheinhard Furrer and Ernst Messerschmid.

Summing up what he described as the "worst year for space insurance," David Thompson of insurance brokers Stewart Wrightson claimed the recent loss of *Palapa B-2* and *Westar VI* had cost insurers $185 million, while the loss of *Intelsat VI-F9* would cost $103 million. Already this year, losses grossed $300 million, raising to $500 million total satellite losses to the insurance industry since space insurance began in 1965. Only $250 million had been received in premiums. From a high premium rate of 25% the insured value, premium rates had tumbled to 5–10% during the early 1980s in the belief that space transportation was becoming routine and much more reliable. During 1985 insured rates increased to 22%.

1984 November 5 Launch of a classified military satellite by the U.S. Department of Defense aboard Shuttle *Challenger* on December 8 was postponed due to a suspect bonding agent under 2,800 thermal protection tiles on the orbiter. NASA had the option of turning around the orbiter *Discovery* from its 51-A mission or flying the secret Department of Defense mission on *Challenger* when all the tiles had been replaced. When *Discovery* flew the 51-A mission on November 8, the decision was made to turn it around quickly for the Department of Defense payload and a launch date of January 23, 1985, was set.

1984 November 7 NASA announced that Senator Edwin (Jake) Garn (R-Utah), chairman of the Subcommittee on HUD-Independent Agencies of the Senate Committee on Appropriations, had accepted an invitation to fly as a payload specialist aboard the Shuttle. Elected to the Senate in 1974, Jake Garn was a former U.S. Navy pilot with 10,000 hr. in military and private civilian aircraft. Sen. Garn was assigned to fly mission 51-E, but when that was cancelled he was reassigned to mission 51-D, launched on April 12, 1985, the fourth anniversary of the first Shuttle flight.

1984 November 8 Following a 24-hr. postponement due to high winds, Shuttle orbiter *Discovery* was launched on mission 51-A at 12:15:00 UT from LC-39A at the Kennedy Space Center, Fla., on the first mission to retrieve two satellites stranded in space. Carrying astronauts Frederick Hauck (CDR), David Walker (PLT), Anna Fisher (MS), Dale Gardner (MS) and Joseph Allen (MS), the 45,336-lb. payload included the communication satellites *Syncom IV-A (Leasat-1)* and *Telesat-H (Anik D-2)* and two modified Spacelab pallets on which the retrieved satellites would be returned to Earth. At 21:04:32 UT, *Telesat-H* was released, followed by *Syncom-IVA* at 12:56:07 UT on November 9. *Discovery* then began a long sequence of rendezvous maneuvers to come up alongside *Palapa B-2*, in an orbit of 210 mi. The 6 hr. EVA (extravehicular activity) with astronauts Allen and Gardner began at 13:25 UT on November 11, and *Palapa B-2* was retrieved at 18:13 UT. The next day, *Discovery* closed the 700 mi. separating it from *Westar VI*. In a 5 hr. 43 min. EVA that began at 11:08 UT on November 13, Allen and Gardner retrieved the satellite at 14:59 UT. On both occasions, Fisher had used the remote manipulator arm to grapple the satellites into the payload bay. The mission ended at 7 days 23 hr. 44 min. 56 sec. on the Kennedy Space Center runway.

1984 November 10 The second Ariane 3 launcher (L11) carried the Inmarsat *Marecs B2* and the GTE *Spacenet 2* satellites into geostationary orbits from Kourou, French Guiana. Both satellites were almost identical to their respective predecessors, *Marecs B2* being positioned over 177.5° E to replace *Marecs B1* lost with the failure of Ariane L5. The satellite also carried an L-band (1.6 GHz) receiver as an emergency search and rescue beacon locator. *Spacenet 2* was positioned over 68° W.

1984 November 13 The U.S. Congress's Office of Technology Assessment (OTA) published its report *Civilian Space Stations and the U.S. Future in Space*, which asserted that President Ronald Reagan's commitment to a permanently manned orbiting facility could not be justified on scientific, economic or military grounds. It said the nation's goals in space were shortsighted and narrow and claimed that there was "no compelling, objective, external case" for the station.

1984 December 15 The Soviet Union launched the first of two spacecraft on a combined mission to the planet Venus and Halley's comet when an SL-12 placed the 11,100-lb. *Vega 1* on an Earth escape trajectory to the inner solar system. *Vega 2* was launched by another SL-12 on December 21. Both spacecraft were almost identical and evolved from the Venera series of Venus spacecraft. Each carried a 3,395-lb. lander and an encapsulated balloon designed to drift for two Earth days in the planet's dense atmosphere. The main spacecraft, weighing 7,700 lb. at launch, including propellant, would pass Venus and, using the gravity assist of the planet, move on to a rendezvous with Halley's comet. Both spacecraft would arrive at Venus in June 1985 and reach their closest point to the comet in March 1986.

1984 December 27 NASA Administrator James Beggs approved an extra task for the Galileo Jupiter orbiter/probe mission scheduled for launch in May 1986. On its way beyond Mars, Galileo would take a relatively close look at the asteroid named 29 Amphitrite in December 1986. About 120 mi. in diameter, the asteroid orbits the sun at a distance of about 230 million mi. and is thought by astronomers to rotate relatively slowly once every 5.39 hr. Galileo would fly within 6,000–1,200 mi. of Amphitrite, but the slightly modified trajectory would mean the spacecraft would arrive at Jupiter on December 10, 1988, instead of August 29, 1988. Moreover, the slight adjustment meant that under the new schedule, the spacecraft would be able to perform only 10 close flybys of Jupiter's moons instead of 11; so the primary mission in Jovian orbit was extended from 20 to 22 months.

1984 The French navy completed flight tests with a new generation of submarine-launched ballistic missile, the solid propellant, three-stage M-4. With a length of 36 ft. 3 in., a diameter of 6 ft. 4 in. and a launch weight of 77,175 lb., the M-4 could deliver six 150-KT (TN-70) MIRVed reentry vehicles across a range of 2,800 mi. Of all-new design, the first stage delivered a thrust of 156,500 lb., the second stage a thrust of 66,150 lb. and the third stage a thrust of 15,450 lb. Deployment began in 1985 with the submarine *L'Inflexible*,

the others being retrofitted with the exception of *Le Redoutable*.

1985 January 8 Japan's ISAS (Institute of Space and Astronautical Sciences) launched the first of two spacecraft designed to study Halley's comet during its passage through the inner solar system in March 1986. This was also the first flight of the three-stage, solid propellant Mu-3S II with strap-on boosters, a development of the Mu-3S but with augmented upper stages, a height of 92 ft. and a payload capacity of 1,700 lb. to low Earth orbit. It was also the first time a Japanese spacecraft had been propelled to escape velocity. Called MS-T5 *Sakigake*, the spacecraft comprised a cylinder 4 ft. 7 in. diameter and 2 ft. 3 in. high with solar cells providing up to 100 W of electrical power. The spin-stabilized MS-T5 carried a high-gain dish antenna on a mechanically de-spun top section and 22 lb. of hydrazine for spin and course control. With a weight of 304 lb., MS-T5 carried a plasma-wave probe for detecting turbulence in the solar wind, a solar wind instrument to measure the bulk and velocity of ions and a magnetic fields detector. Launched as a qualification model for the *Planet A* spacecraft sent on a flyby of Halley's comet, MS-T5 would observe the comet from 4.3 million mi. on March 10, 1986.

1985 January 12 NASA and the U.S. Air Force began stacking the Shuttle orbiter *Enterprise*, the external tank and two inert solid rocket boosters on Space Launch Complex-6 (SLC-6) at Vandenberg AFB, Calif. In the first full-scale Shuttle stacking to fit-check the new launch complex, engineers wanted to conduct a dry-run of facilities and work structures. Verification revealed some problems but nothing that would prohibit use of the complex. The *Enterprise* was not being considered for spaceflight.

1985 January 15 An SL-14 launch vehicle carried a set of six 90-lb. tactical military communication satellites from Plesetsk to orbits averaging 870 × 876 mi., with inclinations of 82.6° and periods of 1 hr. 53.8 min. Designated *Cosmos 1617–1622*, they were followed by a second set of six, *Cosmos 1690–1695*, on October 9 designed, as were the first, to supplement another set of tactical military satellites routinely launched by SL-8 to 74° orbits with periods of 1 hr. 55 min. Two more sets were launched in 1987, one set in 1988 and two sets in 1989. Thereafter, a new system called Locyst was launched by SL-14 beginning with *Cosmos 2090–2095* on August 8, 1990.

1985 January 17 The 1,037th and last Aerobee sounding rocket was launched from White Sands, N.Mex., carrying a 571-lb. ultraviolet light experiment to a height of 105.2 mi. The package of instruments was recovered 55 mi. downrange. The Aerobees had carried America's first scientific experiments to the edge of space, made the first map of the atmosphere, discovered the first x-ray star and carried the first mammals from the United States to space and back.

1985 January 24 At 19:50:00 UT the first dedicated Department of Defense Shuttle mission was launched from LC-39A at the Kennedy Space Center, Fla., carrying astronauts Thomas Mattingly (CDR), Loren Shriver (PLT), James Buchli (MS), Ellison Onizuka (MS) and Gary E. Payton, a payload specialist provided by the U.S. Air Force. Designated 51-C, the mission had been postponed for 24 hr. due to freezing conditions at the pad. In this secret mission, the first U.S. manned spaceflight of its kind, an inertial upper stage was used to place a 5,000-lb. signal-intelligence (sigint) satellite called *Aquacade* in geostationary orbit. Release from the orbiter *Discovery* of this much advanced successor to Rhyolite came at 11:00 UT on January 25. The mission ended with a landing at Kennedy Space Center, 3 days 1 hr. 33 min. 23 sec. after liftoff.

1985 January 25 A group of Florida undertakers called the Celestis Corp. signed a contract with Space Services Inc. for the launch of cremated human remains to a circular, 1,900 mi. orbit using a Conestoga rocket capable of placing 1,500 lb. in orbit. Utilizing the full payload capability, Celestis would pay $15 million for the launch and charge each client $5,000. Each launch could accommodate 15,000 separate cremations in capsules. SSI was planning to use the *Conestoga 2*, utilizing two Thiokol Castor IV motors for the first stage and one Thiokol Star 48 for the second stage. The flight never took place due to development problems with the Conestoga and the concept was never pursued.

1985 January 31 At a ministerial meeting, the European Space Agency authorized preparatory work on a future launch vehicle to succeed the current generation of Ariane 4 and to start development of a cryogenic engine for core stage applications. The launcher would be required to put a 15,000-lb. payload to geosynchronous transfer orbit and place 39,700-lb. space station modules, free-flying platforms or the Hermes space plane into orbit. Designated Ariane 5, the new proposed launcher would use conventional high-performance propulsion. Competing concepts included the reusable, winged HOTOL (horizontal takeoff and landing) from British Aerospace and Rolls Royce.

1985 February 4 At the end of a two-day meeting in Rome, European cabinet ministers of the European Space Agency member countries agreed to participate in the permanently manned U.S. space station. The ESA contribution would consist of an Italo-German proposal to build a pressurized experiment module called Columbus. The United Kingdom was also keen to participate and suggested it develop a free-flying Columbus platform to co-orbit with the station. Funding for Ariane 5 was also approved. Ministers agreed to Phase B funding for the French Hermes space plane, at present expected to have a length of 59 ft., a wingspan of 33 ft. and a 9,900-lb. payload capability, but Germany was not yet ready to give it full approval.

1985 February 8 An Ariane 3 launch vehicle (V12) successfully carried two communication satellites to a transfer path from which they were placed in geostationary orbit. *Arabsat 1* was positioned over 19° E but one of the two solar array panels refused to deploy. Threatened with a severe power loss that would have limited the usefulness of the satellite, ground controllers attempted an unconventional method of freeing the array by blasting it with the attitude control thrusters. It worked and the array deployed. *Brasilsat 1* was positioned over 65° W, providing 12,000 telephone circuits to urban and rural regions.

1985 February 25 Signed by President Ronald Reagan, National Security Directive 164 ordered NASA and the Department of Defense to study a second-generation space transportation system utilizing manned and unmanned systems. The study embraced a wide range of potential design configurations and had not advanced substantially when the Shuttle *Challenger* was destroyed on January 28, 1986, upsetting

plans for a Shuttle successor as the entire manned program came under review.

1985 February Soviet ground controllers lost contact with the *Salyut 7* space station shortly before the *Soyuz-T 13* spacecraft was to have launched cosmonauts for a long-duration stay lasting up to nine months. Cosmonaut commander Vladimir V. Vasyutin, flight engineer Viktor P. Savinykh and Alexandr A. Volkov had trained for the mission. Because telemetry signals from the station had been lost, it was not possible to ascertain the condition of *Salyut 7*, and initial reactions were to abandon the derelict hulk. However, keen to get a full lifetime's use out of the station, mission planners developed a daring bid to reactivate and put right the failed systems aboard *Salyut 7*. With experienced mission commander Vladimir A. Dzhanibekov and Viktor P. Savinykh assigned to a reactivation crew, preparations went ahead for a flight to the space station in June.

1985 March 1 An operating problem with the Tracking and Data Relay Satellite System (TDRSS) satellite launched by Shuttle mission STS-6 caused cancellation of mission 51-E, which was to have launched *TDRSS-B*. Engineers needed to modify *TDRSS-B* before it could be launched with a high probability of success. The *Challenger* mission had been scheduled for March 7 and was also to have carried the *Telesat-I (Anik C-1)* satellite into orbit. The assigned payload weight of 53,283 lb. would have made this the heaviest payload ever launched by Shuttle. *Anik C-1* was moved to *Discovery* for mission 51-D, launched April 12, 1985, and *TDRSS-B* was assigned to mission 51-L, which ended in disaster on January 28, 1986. With the exception of French astronaut Patrick Baudry, the 51-E crew comprising astronauts Karol Bobko, Donald Williams, Rhea Seddon, Jeffrey Hoffman, David Griggs and Sen. Jake Garn was assigned to 51-D; Baudry went to 51-G.

1985 March 3 In a "National Security Launch Strategy" directive, President Ronald Reagan ordered NASA and the U.S. Air Force to start work on a study which would lead to the joint development of a bigger and more powerful launcher than the Shuttle. Aiming for operational availability from the year 2000, the new national launch system was to accommodate payload lift needs of 75,000 lb. to polar orbit and 100,000 lb. to 500 mi. equatorial orbits. The Air Force also announced that it had agreed to book eight Shuttle missions a year from 1988.

1985 March 5 The U.S. Air Force awarded Martin Marietta a contract to begin development and manufacture of 10 advanced Titan launch vehicles. Designated Titan 34D-7, it was to be an uprated version of the Titan 34D, with an additional 1½ segments for each solid propellant strap-on booster and a 9-ft. extension in the length of the core stage. At launch the two solids would produce a total thrust of 3.2 million lb. With a Centaur G-prime as the third stage, total length would be 204 ft. A payload envelope 40 ft. long and 15 ft. in diameter permitted Titan 34D-7 to launch Shuttle-compatible payloads, giving the launcher its name of complementary expendable launch vehicle (CELV). Titan 34D-7 would be capable of placing 10,000 lb. in geostationary orbit.

1985 March 6 NASA announced that it had selected scientific investigators for the Advanced X-ray Astrophysics Facility (AXAF), an observatory that would be launched by Shuttle to complement the work of the Hubble Space Telescope, scheduled for launch in 1986, and the Gamma-Ray Observatory, scheduled for 1988. NASA wanted to begin building the AXAF by 1988 so that it could join another orbital observatory, the Space Infrared Telescope Facility (SIRTF). By the early 1990s, the space agency wanted to use these four observatories to simultaneously view the universe in infrared, visible, ultraviolet, x-ray and gamma-ray portions of the spectrum.

1985 March 11   U.S. Assistant Secretary of State Richard Burt announced a significant increase in the number of SS-20 mobile ballistic missiles moved into eastern Europe. At a meeting of NATO leaders in Brussels, he said that the total of 296 missiles installed by May 1984 had increased to 414 and that there were no signs of a halt being drawn to these offensive systems. The deployment of SS-20 and Backfire bombers had instigated the move by NATO to deploy cruise missiles and Pershing 2s in western Europe.

1985 March 13   The highest level of the policy making board at the Japanese Space Activities Commission made a series of momentous decisions about that country's future goals in space. It decided to participate in the U.S.-led international space station, already signed on to by Europe. Japan would construct an experiment module, known as the JEM, which would be permanently attached, like Europe's Columbus, to the modular facility. The commission also agreed to go ahead with a two-stage satellite launcher called the H-II. It would weigh 480,000 lb. and use cryogenic propellants in first and second stages, augmented by two solid propellant strap-on boosters.

An Atlas E carried a 395-lb. U.S. Navy oceanographic mapping satellite from Vandenberg AFB, Calif., to a 470 × 506 mi., 108.1° orbit. The satellite was designed to provide geodetic data on the southern hemisphere and the north Pacific Ocean to replace data lost by the demise of NASA's Seasat satellite. It was also tasked with measuring small variations in the height of the ocean surface. It remains in orbit.

1985 March 14   NASA announced three contractor teams for 21-month Phase B studies of the permanent manned Space Station: Boeing Aerospace and Martin Marietta Aerospace managed by Marshall Space Flight Center would handle Work Package 1; RCA/General Electric managed by Goddard Space Flight Center would perform WP3; Rockwell International's Rocketdyne Division/TRW Systems managed by Lewis Research Center would handle WP4. On April 15 NASA selected McDonnell Douglas and Rockwell International for Phase B studies of Space Station Work Package 2 dealing with the structural framework in a contract managed by the Johnson Space Center.

1985 March 20   In a move to put science firmly at the forefront of space station planning, NASA announced that as of April 15, Dr. David Black was to be the chief scientist for the Office of Space Station at Headquarters. In this post, Dr. Black would ensure that engineering design received all the necessary consideration for scientific interests and that Philip Culbertson, associate administrator for Space Station, would be well advised of the best means to give scientists affordable access to the Space Station. It was the first U.S. manned space program designed first to accommodate the needs of science rather than of engineering.

A test model of the Intelsat VA communications satellite is given scale by technicians. The first of the series was launched on March 22, 1985.

1985 March 22   The first of six improved Intelsat VA communications satellites, *VA F10*, was launched by Atlas Centaur AC-63 to an elliptical transfer orbit from where an apogee kick motor placed it in geostationary orbit. Of similar dimensions to the Intelsat V series, the upgraded satellites had a launch weight of 4,400 lb. and 2,430 lb. on station. Carrying seven antennas, each Intelsat VA could handle 13,500 two-way telephone circuits and 2 TV channels. *Intelsat VA F11* was launched by Atlas Centaur AC-64 on June 30, 1985, as was *F12* on September 29, 1985. Three Intelsat VAs were launched by Ariane: *F14* on May 30, 1986, lost when the launcher blew up; *F13* on May 17, 1988; and *F15* on January 27, 1989.

1985 March 29   Called the National Commission on Space, a 15-member team of U.S. space principals began a 12-month effort to define U.S. space policy to the year 2005. The commission was set up by President Ronald Reagan and placed under the chairmanship of ex-NASA Administrator Thomas O. Paine. Paine said that he would extend the mandate and view the next 50 years as well, incorporating bold initiatives such as international cooperation on a manned mission to Mars.

1985 March   Robert Cooper, director of the Defense Advanced Research Projects Agency (DARPA), described before the House Armed Services Committee a "transatmospheric vehicle," or TAV, that could take off and land like an ordinary airplane yet reach orbital speeds of 17,500 MPH.

The concept incorporated a cryogenic, hydrogen-fueled, air-breathing propulsion system taking in air from the ram effect of encountering air at high speed and switching to integral oxygen tanks for the final spurt into space. Development of aerospace vehicles had stopped in the 1960s, but the U.S. Air Force and the Navy were interested in the concept.

The U.S. Air Force conducted tests at the SLC-6 Shuttle facility at Vandenberg AFB, Calif., designed to suppress sound waves engineers calculated might damage the orbiter during launch. Engineers believed sound waves caused by the three orbiter liquid propellant engines and the two solids could be focused by the design of SLC-6 into a shattering curtain of sound. Testing the system with orbiter *Enterprise*, external tank and inert solid rocket boosters on the pad, 750,000 gal. of water flooded the basin area to act as an acoustic barrier. The Air Force expected to make about four Shuttle flights a year from SLC-6 beginning January 1986.

On the initiative of the Luxembourg government, the Société Européenne des Satellites (SES) was incorporated as a private company to set up a satellite system for direct broadcast TV. In October SES selected an RCA Series 4000 satellite as the bus for 16 protected transponders delivering 45 W of power in the 14 GHz band. In space the satellite, called *Astra*, would have a height of 10 ft. 5 in., a length of 6 ft. 7 in., a width of 4 ft. 11 in. and a solar array wingspan of 63 ft. 4 in. The solar cells would provide 3,576 W at the beginning of life and 2,309 W at 10 years, the satellite's nominal life. It would weigh 3,925 lb.

1985 April 1   The 8,920-ton French ballistic missile submarine, *L'Inflexible*, became operational with the new M-4 missile. The idea of building a sixth ballistic missile submarine was to enable three vessels to be operational at any one time, two of which would be on patrol. *L'Inflexible* would serve as an intermediary between the five boats of *Le Foudroyant*–class and a new, larger vessel scheduled for delivery in 1994. *L'Inflexible* was the same size as the first five vessels. The M-4 had a length of 36 ft. 3 in., a diameter of 6 ft. 4 in. and a launch weight of 77,000 lb. The three solid propellant stages had thrust levels of 156,500 lb. through four nozzles, 66,150 lb. through one flexible nozzle and 15,400 lb. through another flexible nozzle. The maneuverable bus carried six 150-KT TN-70 warheads with a range of 3,000 mi. The M-4 would be retrofitted to four other vessels, but not *Le Redoutable*.

1985 April 5   Scientists at U.S. universities were briefed on an Innovative Science and Technology program being offered by the Strategic Defense Initiative antiballistic-missile research organization. There was strong opposition to what many scientists considered was an extension of the arms race into space, and many feared that a system protecting U.S. cities and military bases from attack would give the U.S. unreasonable power to impose its will abroad. Opposition was mobilized through the Union of Concerned Scientists, a group of professional scientists against research on nuclear weapons and SDI. On May 16, 1986, they announced the names of 6,500 scientists and engineers pledged not to receive money for SDI.

1985 April 12   Following a 55-min. hold when a ship strayed into restricted waters, Shuttle mission 51-D began at 13:59:05 UT with the launch of *Discovery* from LC-39A at the Kennedy Space Center, Fla., carrying astronauts Karol Bobko (CDR), Donald Williams (PLT), Rhea Seddon (MS), David Griggs (MS), Jeffrey Hoffman (MS), passengers Charles Walker and Sen. Jake Garn. With a 35,824-lb. payload, the mission had

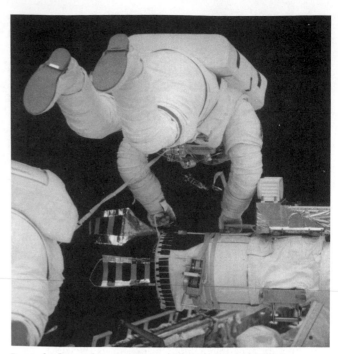

Launched to orbit on mission 51-D, April 12, 1985, Discovery *astronaut Jeffrey Hoffman rigs a "fly-swat" device to the remote manipulator as a makeshift aid to get a troubled satellilte working.*

been delayed first from March 19 to March 28 when some payloads assigned to 51-E were remanifested to this flight, and to April 12 when a work platform in the Vehicle Assembly Building dropped and damaged part of a cargo bay door. *Discovery* deployed the *Telesat-I (Anik C-1)* satellite at 23:38:38 UT, followed by *Syncom IV-3 (Leasat-3)* at 14:58:22 UT on April 13. Syncom's booster motor failed to fire as planned, and at 12:30 UT on April 16, Griggs and Hoffman performed the first contingency EVA (extravehicular activity) (3 hr. 1 min.) of the Shuttle program to attach a "flyswatter" device on the remote manipulator arm. After rendezvousing with *Syncom IV-3*, an unsuccessful attempt was made at 14:12 UT on April 17 to activate a timer switch on the satellite by striking it with the "flyswat." Leaving *Syncom IV-3* stranded in a 180 × 250 mi. orbit, *Discovery* returned to land at the Kennedy Space Center, Fla., at 6 days 23 hr. 55 min. 23 sec.

1985 April 13   The first in a new generation of heavy Soviet launch vehicles made its first flight from Baikonur in the first in-flight test of a new propulsion system that would form the first stage of the SL-16 Zenit, and the strap-on-booster for the heavyweight SL-17 Energia. Not since the appearance of the SL-9 in 1965 had the Soviets fielded so impressive a new launcher. Designated SL-16 Zenit, it was 201 ft. 6 in. tall and carried one 1.63-million-lb. thrust, four-chamber RD-170 in the first stage which burned liquid oxygen–kerosene propellants for 2 min. 24 sec. The second stage was powered by one RD-120, producing a thrust of 205,200 lb. for 5 min. Payload capability was 30,000 lb. to low Earth orbit. The first SL-16 was a ballistic test, the first orbital launch taking place on June 21, which was only marginally successful. The first operational SL-16 launch carried *Cosmos 1697*, an electronic intelligence gathering satellite, to orbit on October 22.

1985 April 15   The U.S. Air Force made a significant shift away from full use of the Shuttle for military purposes when estimates presented to the Department of Defense deleted all funds for a dedicated Shuttle mission control facility. The facility was to have been set up at the Air Force's Consolidated Space Operations Center (CSOC) , Colorado Springs, Colo. In deleting funds, the Air Force signaled its lack of interest in utilizing the Shuttle for all but satellite launches, for which NASA's mission control facilities would suffice.

1985 April 16   NASA signed a memorandum of understanding with the government of Canada covering that country's participation in the U.S. Space Station program. Under the terms of the agreement, the United States and Canada would conduct Phase B definition studies of a space construction and servicing system involving a space-erector which would be attached to the truss assembly for handling sections of the station and for moving logistics and other modules around the station exterior.

1985 April 29   Following a 2 min. 18 sec. hold to manually override a faulty computer processor, Shuttle mission 51-B began at 16:02:18 UT with the launch of *Challenger* and its 31,407-lb. Spacelab-3 payload from LC-39A at the Kennedy Space Center, Fla., to a 57° orbit. The crew was divided into two teams: astronauts Robert Overmyer (CDR), Don Lind (MS), William Thornton (MS) and Taylor G. Wang (PS) in a "gold" team; Frederick Gregory (PLT), Norman Thagard (MS), and Lodewijk Van den Berg (PS) in a "silver" team. The two teams worked round the clock on 12-hr. shifts operating 15 materials science, life science, fluid mechanics, atmospheric physics and astrophysical investigations. The mission ended at 7 days 0 hr. 8 min. 46 sec. at Edwards AFB.

1985 April   The U.S. National Security Council implemented a National Space Transportation Strategy Study to evaluate future launch concepts and articulate a strategy for future programs. In October four contractors joined the study: General Dynamics, Rockwell International, Boeing and Martin Marietta. The $24 million study was funded jointly by the U.S. Air Force, NASA and the Strategic Defense Initiative program at the Department of Defense. Each contractor had 2 years 2 months to complete their share of the work, with overall architecture evaluated in the first year followed by a year of study. The mission model for follow-up study recommendations would cover the period 1995–2010.

Following the de-regulation of telecommunication services, two of Japan's largest trading organizations, C. Itoh and Co. and Mitsui and Co., formed a joint venture with Hughes Communications in the United States called Japan Communications Satellite Co. C. Itoh owned 40%, Mitsui 30% and Hughes 30%. JCSAT procured two very powerful Hughes HS 393 satellites supporting telephone, television, facsimile and high-speed data links. This was the first time an American company shared as equity partner in a Japanese commercial satellite venture.

1985 May 8   An Ariane 3 launched the *GSTAR 1* and *Telecom 1B* communications satellites from Kourou, French Guiana, to transfer paths from where they were placed separately in geostationary orbits. *GSTAR 1* was launched for the GTE Spacenet Corp. and constituted one of two Ku-band satellites then authorized by the International Telecommunication Union. Built by RCA, the satellite comprised a box structure, 5 ft. 4 in. × 4 ft. 4 in. × 3 ft. 3 in., two solar-array wings with a span of 49 ft., a launch weight of 2,667 lb. and an on-orbit weight of 1,543 lb. Positioned at 103° W, *GSTAR 1* had sixteen 14/12 GHz channels serving all 50 U.S. states. GTE had permission to operate three Spacenet and two GSTAR satellites, with applications pending for additional satellites. Positioned at 5° W, *Telecom 1B* was almost identical to *Telecom 1A*.

1985 May 9   NASA and the government of Japan signed an agreement covering that country's participation in the U.S. Space Station. Japan was to contribute by designing and building a pressurized experiment module. One end of the Japanese experiment module (JEM) would be docked to other pressurized elements of the U.S. Space Station while the other end would connect to an exposed workdeck. In some respects this would be similar to Spacelab, providing a work area for astronauts and a pallet for instruments designed to operate in the exposed vacuum of space.

1985 May 24   NASA announced that it had reached an agreement with Hughes Aircraft to jointly develop plans for salvaging the failed *Syncom IV-3* satellite deployed from *Discovery* on the 51-D Shuttle mission launched April 12, 1985. In an agreement with the underwriters, Hughes was to pay NASA for all add-on costs to the scheduled 51-I mission associated with restarting the satellite from orbit. In the plan, NASA would rendezvous with *Syncom IV-3* and during an EVA (extravehicular activity) astronauts would secure the satellite in the payload bay and attach equipment enabling ground controllers to command it to its planned geostationary orbit.

1985 May 29   NASA and the European Space Agency signed an agreement covering cooperative participation in the U.S. Space Station. ESA was to design and build a pressurized laboratory module permanently attached to the Space Station and a man-tended free-flyer for both low- and high-inclination orbits. Called Columbus, the pressurized module would draw upon almost a decade of experience in the manned Spacelab program and provide a work area for international experiments in space. With the signing of this agreement, NASA completed agreements with the group of international partners it had invited to participate in the Space Station.

1985 May 30   Ford Aerospace announced it had been selected to negotiate for a contract to build a new generation of geostationary operational environmental satellites. Designated *GOES-I*, *-J* and *-K*, they were to be developed under the management of NASA's Goddard Space Flight Center and would provide geostationary weather watch for the decade of the 1990s. NASA had scheduled the launch of *GOES-I* (*GOES 9* after launch) in late 1989.

1985 June 5   Ground controllers at NASA's Goddard Space Flight Center announced they had performed the first of four maneuvers of the International Cometary Explorer (ICE), aligning its flight path for an intercept of the comet Giacobini-Zinner on September 11. The corrections had been calculated from telescopic measurements of the comet's trajectory as it reappeared round the far side of the sun, 149 million mi. from Earth. The last maneuver was scheduled for three days prior to intercept.

1985 June 6   At 06:40 UT cosmonauts Vladimir A. Dzhanibekov and Viktor P. Savinykh were launched aboard the *Soyuz-T 13* spacecraft by an SL-4 from Baikonur cosmo-

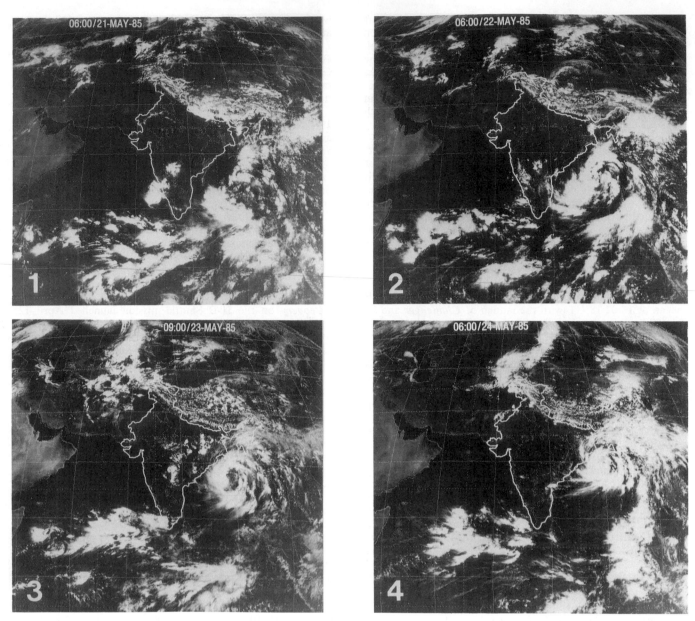

India's Insat 1B *multi-purpose satellite took this sequence of pictures showing a cyclone crossing Bangladesh between May 21 and May 24, 1985.*

drome. They rendezvoused with *Salyut 7*, in an orbit of 220.5 x 224 mi., on June 8 using manual techniques and a laser range-finder. All contact with the unmanned station had been lost in February, and the cosmonauts performed a "fly-around" inspection of the hulk, slowly tumbling due to a lack of attitude control. *Soyuz-T 13* docked to the forward port at 08:50 UT on June 8, and the crew went inside *Salyut 7*. It was dark, cold and depleted in electrical power. Within three days the crew had realigned the three exterior solar cell arrays with the sun to charge low batteries, spending rest and sleep periods back in *Soyuz-T 13*. On June 12 they had reactivated the environmental control system, odor scrubbers, heaters and the communication equipment. With contact through *Salyut 7*'s antennas restored, television was brought back on a day later, and control systems were activated on June 14. Three days after that, ground controllers checked and tested several new

components brought to the station by *Soyuz-T 13* and had *Salyut 7* back under proper management.

1985 June 10　　The Soviet spacecraft *Vega 1* passed the planet Venus on its way to a rendezvous with Halley's comet while a lander capsule carried to the planet by the main bus entered the atmosphere and descended to the surface. The capsule entered the atmosphere at an altitude of 78 mi. and deployed the first parachute at 40 mi. when the spherical aeroshell was jettisoned. Seconds later a balloon canister separated from the lander assembly and at a height of 34 mi. deployed its own parachute to slow the descent to just 26 ft./sec. The balloon was inflated at a height of 33 mi. when the parachute was released. The balloon, 10 ft. in diameter, was filled with 4.4 lb. of helium in aerostatic equilibrium with the atmosphere of Venus at an altitude of about 34 mi. Suspended 40 ft. beneath the balloon, a 14-lb. instrument gondola, 3 ft. 10 in. in length with a diameter of 5 in., carried instruments for measuring the atmosphere of Venus. Data was transmitted to Earth by a 4.5-W amplifier during the 46 hr. the balloon operated in the atmosphere. The lander sent data from the surface of the planet for 21 min., a creditable duration in the extreme temperatures and pressure of the atmosphere.

1985 June 14 The Soviet *Vega 2* spacecraft successfully deployed a lander/balloon combination to the atmosphere of Venus and continued along its heliocentric path for an encounter with Halley's comet in March 1986. Like its predecessor from *Vega 1*, the balloon was deployed on the night side of the planet and drifted in to the day side where it was warmed by the sun and destroyed. Data was sent back to Earth for a period of approximately 46 hr. The balloons had been developed in cooperation with the French national space agency Centre National d'Etudes Spatiales (CNES), although the Soviets had total control over the mission. As with previous Venera missions, data from the landers was relayed via the main spacecraft in heliocentric orbit.

1985 June 17 At 11:33:00 UT Shuttle mission 51-G began with the launch of *Discovery* from LC-39A at the Kennedy Space Center, Fla., with astronauts Daniel Brandenstein (CDR), John Creighton (PLT), Shannon Lucid (MS), Steven Nagel (MS), John Fabian (MS), French astronaut Patrick Baudry (PS) and Arab astronaut Sultan Bin Salman al-Saud (PS). The 44,507-lb. payload included three communication satellites: a Hughes HS 376 for Mexico *(Morelos-A)* deployed at 19:37:51 UT; a Ford Aerospace three-axis-stabilized satellite for the Arab League *(Arabsat-1B)* deployed at 13:56:56 UT on June 18; and an HS 376 for American Telephone and Telegraph *(Telstar-3D)* deployed at 11:20:36 UT on June 19. All three carried PAM-D perigee boost motors which fired as planned, sending their charges toward geostationary orbit. *Morelos-A* had twelve 7-W and six 10.5-W C-band and four 20-W K-band transponders. At 16:02:39 UT on June 20 the manipulator arm deployed an astrophysical satellite called *Spartan 1* and *Discovery* moved away a distance of 115 mi. for two days, closed back in and retrieved *Spartan 1* at 13:31:54 UT on June 22. The mission ended on a runway at Edwards AFB, Calif., at 7 days 1 hr. 38 min. 52 sec.

1985 June 20 The House Science and Technology Committee completed legislation for authorization of the transfer of government funds to EOSAT, the Earth Observation Satellite Co. A joint venture of RCA Corp. and Hughes Aircraft Co., EOSAT was envisaged as the organization through which the U.S. Earth resources program would be commercialized. On September 30 EOSAT announced it had been awarded a contract from the U.S. Department of Commerce for design and development of the *Landsat 6* and *Landsat 7* Earth resource satellites. EOSAT would receive $250 million over five years to build the two satellites and provide ground facilities for operating *Landsats 4*, *5*, *6* and *7* and disseminating their products.

NASA announced it would test a special can produced by the Coca Cola Co., designed to dispense the carbonated beverage in the weightlessness of space, a procedure heretofore impossible. The crew aboard Shuttle mission 51-F would test the new dispenser, which incorporated a system of bladders for pressurization and expulsion. Five days later NASA announced that a similar technology from the Pepsi-Cola Co. would also be tested aboard 51-F. Government regulations

prevented space missions appearing to endorse one product against another, hence the decision to take four cans of each.

1985 June 21 At 00:40 UT the unmanned cargo tanker *Progress 24* was launched from Baikonur cosmodrome to the *Salyut 7* space station. Docking at the aft port was achieved at 02:34 UT on June 23, and the freighter was unloaded by cosmonauts Vladimir Dzhanibekov and Viktor Savinykh. Some equipment replaced old and worn out items aboard the space station, and some of it supported a new program of Earth survey work. It also brought two additional solar cell arrays, each 49.5 sq. ft. in area, to be added to the left main solar panel during a forthcoming space walk. *Progress 24* was undocked from the aft port at 12:28 UT on July 16, and the vehicle was de-orbited the same day.

1985 June 28 The U.S. Air Force announced that a B-52 equipped with a Navstar global positioning system (GPS) flew across the North Pole navigating its way by the satellite. In a test of the system's high latitude capability, the flight also proved integration between the GPS receiver and the aircraft's on-board inertial navigation equipment.

1985 June NASA announced the names of 13 pilot and mission specialist Shuttle astronauts constituting Group 11: Jerome Apt; Capt. Michael A. Baker (USN); Col. Robert D. Cabana (USMC); Lt. Col. Brian Duffy (USAF); Lt. Charles D. Gemar (USA); Linda M. Godwin; Col. Terence T. Henricks (USAF); Richard J. Hieb; Tamata E. Jernigan; Col. Carl J. Meade (USAF); Stephen S. Oswald; Lt. Cdr. Stephen D. Thorne (USN); and Cdr. Pierre J. Thuot (USN). Thorne died May 24, 1986, in an airplane crash.

1985 July 1 The European Space Agency announced that it was to sign a contract with a consortium led by Messerschmitt-Bölkow-Blohm/ERNO in West Germany for development of the Eureca free-flying platform to be launched by Shuttle and retrieved on a later mission. The first Eureca mission was to consist of a materials and life sciences package. ESA looked upon Eureca as a flexible instrument platform that could have use during manned operations with the U.S.-led international space station.

1985 July 2 The 2,117-lb. *Giotto* spacecraft was launched at 11:23 UT to an elliptical, 124 × 22,349 mi. orbit by an Ariane 1 launch vehicle from Kourou, French Guiana. At 19:24 UT, on the third orbit, a solid propellant kick motor carrying 825 lb. of propellant fired for 55 sec. to boost the spacecraft by 4,600 ft./sec. and place it in heliocentric orbit so that it would encounter Halley's comet on March 14, 1986. With depletion of the solid propellant and the consumption of some of the 152 lb. of hydrazine attitude and course correction propellant, *Giotto* would weigh about 1,215 lb. at encounter. The spin-stabilized spacecraft comprised a cylinder, 6 ft. in diameter and 5 ft. 3 in. high, supporting a high-gain dish antenna on top, solar cells around the exterior providing a maximum 190 W of power and a tripod antenna feed for a total height of 9 ft. 8 in. A circular bumper shield would protect the under body of *Giotto* during close encounter from some of the particles vented by the comet.

1985 July 10 An international biological and life sciences flight was launched by a Soviet SL-4 from Plesetsk, Russia, when *Cosmos 1667* was placed in a 131 × 168 mi. orbit inclined 82.3° to the equator. The 12,570-lb. spacecraft carried 2 Rhesus monkeys (Verny and Gordy), 10 male rats, 10 mollusks and 1,500 Drosophila flies as well as guppy fish, corn seeds and crocuses. Adaptation to weightlessness, regeneration of amputated mollusk limbs and blood flow analysis was studied. The monkeys had been instrumented with sensors brought to the Soviet Union by U.S. scientists. Scientists from Bulgaria, Hungary, East Germany, Poland, Romania, Czechoslovakia and France were also involved in the flight, which ended July 17.

1985 July 12   The second launch-pad engine shutdown in NASA's Shuttle program occurred at T−4.734 sec. when a coolant valve on *Challenger's* no. 2 engine failed to cycle correctly and the electronic controller aborted the mission. *Challenger* was set to fly the 51-F mission, with an unmanned pallet configuration designated Spacelab-2, from LC-39A, Kennedy Space Center, Fla. Inspection of the hardware was conducted with *Challenger* on the pad, and a new launch date of July 29 was set. The first launch-pad Shuttle abort occurred to *Discovery* prior to its first flight on June 26, 1984.

NASA's Marshall Space Flight Center awarded a study contract to Martin Marietta for a unique propellant-scavenging system. The permanently manned U.S.-led international space station would need about 250,000 lb. of hydrogen and oxygen attitude propellants each year. MSFC derived a way of reclaiming the 9,000–28,000 lb. of propellants typically trapped in engine plumbing and external tanks (ETs) after each Shuttle launch. MSFC wanted to know whether a special tank with propulsion attached to the base of an ET could suck out residual propellants and detach itself from the tank before it de-orbited. An orbital maneuvering vehicle (OMV) garaged at the space station could be used to collect the scavenger module.

1985 July 16   At 13:05 UT a new variant of the unmanned Progress cargo tanker was launched by SL-4 from the Baikonur cosmodrome. Under the designation *Cosmos 1669*, the spacecraft had a weight of 15,500 lb. and would result in development of what the Soviets called the Progress M-series. When used operationally from August 23, 1989, they would have the capacity of carrying up to 5,510 lb. of cargo, an increase of more than 400 lb. over the standard Progress. This was to be the last of 13 cargo tankers sent to the *Salyut 7* space station, one more than the number of Progress tankers sent to *Salyut 6*. *Cosmos 1669* remained docked to *Salyut 7* during the August 2 space walk by Vladimir Dzhanibekov and Viktor Savinykh, but undocked at 21:50 UT on August 28, and reentered over the Pacific Ocean two days later.

1985 July 29   Following a 1 hr. 37 min. hold for correcting a software problem, Shuttle mission 51-F began when *Challenger* was launched at 21:00:00 UT from LC-39A at the Kennedy Space Center, Fla., with the 34,430-lb. Spacelab-2 payload. The crew comprised Gordon Fullerton (CDR), Roy Bridges (PLT), Story Musgrave (MS), Karl Henize (MS), Anthony England (MS), Loren Acton (PS) and John-David Bartoe (PS). The two solid rocket boosters were jettisoned as planned at 2 min. 1 sec., but at 5 min. 43.4 sec., with *Challenger* at 67 mi. and a speed of 12,640 ft./sec. (8,620 MPH), the no. 1 (center) main engine shut down due to a faulty sensor. Fullerton commanded the orbiter's computers to fly an abort-to-orbit, in which the remaining two main engines pushed on using propellant which should have been consumed by the failed engine and the two 6,000-lb. thrust orbital maneuver engines fired for 1 min. 46.4 sec. to dump 4,000 lb. of propellant and lighten the load. At 8 min. 12 sec., engine ground controller Jenny M. Howard at mission control

saw one temperature sensor in the no. 3 (right) engine fail and a second sensor start to read high temperatures which would incorrectly trigger shutdown of that engine also. She advised flight director T. Cleon Lacefield that the crew should inhibit the electrical safing circuits and press on because the engine itself was working properly though the sensors were not; this they did at 8 min. 45 sec. The two main engines shut down at 9 min. 42 sec. at a height of 70 mi. and a velocity of 25,760 ft./sec., just 114 ft./sec. lower than planned. At 33 min. elapsed time the two orbital maneuver engines were fired for 2 min. 2 sec., adding 194 ft./sec. to achieve a 122 × 162.3 mi. orbit versus the planned 122 × 214 mi. Subsequent burns put the orbiter in a circular 195.6 mi. orbit versus the planned 237 mi. orbit. Spacelab-2 comprised three pallets supporting equipment for 13 separate experiments and lasted 7 days 22 hr. 45 min. 26 sec., ending on runway 23, Edwards AFB, Calif.

1985 July 30   President Ronald Reagan announced a new Shuttle pricing policy effective October 1988 that would involve an auction process with a minimum acceptable bid of $74 million for a full payload bay. Only three Shuttle payload bay loads would be available for customers each year, but the effort was expected to drive up revenue and move Shuttle toward full cost recovery. Each Shuttle mission cost about $200 million. Concerned at the number of commercial customers signing up to fly on Europe's Ariane, the White House refused a request from the Department of Transportation for a Shuttle launch price of $129 million for a full cargo bay. Because an orbiter could accommodate three satellites each customer would pay marginally less than the $25 million charged by Arianespace.

1985 July   Testifying before congressional committees, Lt. Gen. James Abrahamson, head of the Strategic Defense Initiative (SDI) ballistic missile defense program, said that the United States would need a replacement for the Shuttle during the 1990s. Based on its ability to fly 24 Shuttle missions per year, NASA declared there would be no shortfall in launcher capacity. But the Air Force wanted a range of optional launchers, and Abrahamson announced he was spending some SDI money on studies for a Shuttle replacement.

1985 August 2   At 17:15 UT *Salyut 7* cosmonauts Vladimir Dzhanibekov and Viktor Savinykh opened the side hatch on the forward transfer (airlock) module and went outside at the start of a 5-hr. space walk. Their primary task was to attach two 49.5 sq. ft. supplementary solar cell arrays to the existing 216 sq. ft. main solar array wing on the left side of the station (as viewed with the transfer module forward). In four space walks since November 1983, cosmonauts had installed six supplementary solar array panels, increasing total solar cell area from 648 sq. ft. to 796.5 sq. ft. This did not increase the total available electrical power but compensated for degraded cell performance over time. Secondary space walk tasks included retrieving micrometeoroid and biopolymer packages left on the exterior.

1985 August 7   Japan announced the names of three astronaut candidates, one a woman, to fly in space aboard the Shuttle in a flight planned for 1988 during which Japanese scientists would carry out research in a dedicated Spacelab mission. One candidate, Takao Doi, was a NASA researcher. The second, Mamoru Mori, was assistant professor of nucleonics at Hokkaido University. The third, Chiaki Naito, was assistant at Keio University Medical School in Tokyo. NASDA (National Space Development Agency) had selected the three from 533 applicants. The Shuttle participation program was predicated on Japan's long-term commitment to the Space Station program. Shuttle flights were considered a valuable prerequisite for that endeavor.

1985 August 9 Lockheed Space and Missiles Co. announced it had joined McDonnell Douglas as a partner in Work Package 1 of the NASA Space Station program. Lockheed was to provide expertise in areas of thermal control, electrical power management and distribution among the pressurized modules, and habitability systems. It would also give input to plans for EVA (extravehicular activity) systems aboard the station.

1985 August 10 A joint announcement by NASA, the Department of Energy and the Department of Defense was made to the effect that the reactor thermoelectric power concept had been selected for the SP-100 space reactor power program. Under this program a compact and highly reliable power source for hundreds of kilowatts of electrical energy would be developed and tested. Aimed at a broad range of civil and military space power requirements, the SP-100 program was to be applicable to space power needs up to 1 MW.

1985 August 13 In San Diego, Calif., General Dynamics Space Systems Division handed over to NASA the first of seven Centaur G-prime cryogenic stages for use with the Shuttle. The wide-body Centaur had a diameter of 14 ft. 2 in., a length of 29 ft. 7 in. and would be capable of injecting a 5,302-lb. payload from low Earth orbit to escape velocity. Alternatively, the Centaur G-prime could place 10,000 lb. in geostationary orbit. Serious concerns existed about the safety of the cryogenic stage in the manned Shuttle orbiter. When *Challenger* was destroyed on January 28, 1986, the Centaur G-prime was canceled.

1985 August 18 The 307-lb. Japanese spacecraft *Suisei*, also named *Planet A*, was launched by a solid propellant Mu-3S II from Kagoshima to a heliocentric flight path that would carry it to within 94,700 mi. of Halley's comet on March 8, 1986. Identical to the MS-T5 spacecraft which acted as its precursor, *Suisei* carried an ultraviolet imaging instrument for observing the cometary hydrogen corona and a solar wind experiment for measuring the distribution of ions and electrons from the sun. Like its predecessor, *Suisei* was developed and operated by Japan's Institute of Space and Astronautical Sciences.

1985 August 20 NASA announced it had signed an agreement with Space Industries for the construction of an industrial space facility (ISF), proposed as the first privately owned habitable space structure. Designed by Space Industries' Maxime Faget, the ISF would measure 35 ft. long × 14 ft. 6 in. diameter. To be visited by astronauts on a periodic basis for changing experiments and retrieving samples, it was intended for microgravity materials processing work where even the movement of crew members bouncing off walls would disturb the delicate investigations. The ISF was never built for lack of investors.

1985 August 22 NASA selected teams to set up Centers for the Commercial Development of Space, which aimed to encourage high technology research that could make use of the Shuttle, Spacelab and the forthcoming space station. The centers were: Batelle Columbus Laboratories, Ohio; University of Alabama, Birmingham; University of Alabama, Hunts-

ville; National Space Technology Laboratories, Miss.; and Vanderbilt University, Tenn.

1985 August 27 Delayed August 24 at T − 5 min. due to thunderstorms, on August 25 when one of the orbiter's five computers failed, and this morning for 3 min. 1 sec. due to weather and an unauthorized boat in the vicinity, the Shuttle *Discovery* was launched on mission 51-I. Liftoff came at 10:58:01 UT from LC-39A at the Kennedy Space Center, Fla., the crew comprising Joseph Engle (CDR), Richard Covey (PLT), James Van Hoften (MS), John Lounge (MS) and William Fisher (MS). The 44,018-lb. payload included three communication satellites: a Hughes HS 376 for Australia, *AUSSAT-1*, deployed at 17:33:15 UT; a three-axis-stabilized satellite built by RCA for the American Satellite Co., *ASC-1*, deployed at 22:07:34 UT; and the Hughes *Syncom IV-4 (Leasat-4)* deployed in the "frisbee" mode at 10:48:55 UT on August 29. PAM-D motors on AUSSAT and ASC satellites sent them to geostationary transfer orbit and apogee motors placed them in their correct orbits on August 30 and August 31, respectively. *Syncom IV-4*'s solid perigee motor fired as planned and seven burns of the liquid apogee motor placed it in geostationary orbit, but the satellite failed due to technical problems. After rendezvousing with *Syncom IV-3*, two EVAs (extravehicular activities) by Fisher and Van Hoften totaling 11 hr. 27 min. grappled the satellite into the payload bay, attached ground command equipment and redeployed it to space. Later, controllers commanded it to its planned geostationary orbit. The mission ended at 7 days 2 hr. 17 min. 42 sec. at Edwards AFB, Calif.

1985 August 28 A U.S. Air Force Titan 34D carrying the last of seven high-resolution KH-11 digital imaging satellites was destroyed when the first-stage propulsion system shut down less than 2 min. into flight, causing the vehicle to fall back to the ground and explode in a fireball at the Vandenberg AFB, Calif. Thus began a train of events that would nearly cripple U.S. space-based intelligence gathering capability. Only one KH-11 was operating in orbit, but that had been launched on December 4, 1984, and was almost halfway through its life. The most recent KH-9 had been launched on May 11, 1982, and was no longer available. A ground-based KH-9 being used by TRW Systems to test sensors was hurriedly made flightworthy and launched on April 18, 1986, to back up the existing KH-11 already in orbit. Just 8.5 sec. off the pad, that Titan 34D exploded, destroying the last KH-9 in the process. Then, on January 28, 1986, when the Shuttle *Challenger* was destroyed in another launch accident, the manned program was put on indefinite hold, grounding the KH-12 series designed exclusively for Shuttle launches to replace KH-9 and KH-11. Not before August 8, 1989, was a reduced-capability KH-12 type launched by the Shuttle.

1985 August 30 NASA's fourth Shuttle orbiter, OV-104 *Atlantis,* was moved from the Vehicle Assembly Building at Kennedy Space Center, Fla., to LC-39A in preparation for its first mission. Last of the four spaceflight-rated Shuttles ordered by NASA, *Atlantis* had a dry weight of 171,205 lb. The three main engines were run for 22 sec. in a flight readiness firing conducted successfully on September 12. *Atlantis* was manifested to carry a classified payload for the Department of Defense as mission 51-J, scheduled for launch on October 3.

1985 September 6 The first-stage shell of a decommissioned Titan II ballistic missile was destroyed in tests of a high-powered chemical laser at White Sands Missile Range, N.Mex. Designated MIRACL (for mid-infrared chemical laser), the device was not itself a weapon but a test rig for a concept the Strategic Defense Initiative office wanted to develop. Set in a vertical position on the ground without its engines, the stage was held down by steel hawsers. When the laser hit the side of the stage, the weakened structure collapsed in what seemed an explosion, greatly impressing the media with what was little more than a faked stunt.

NASA and the U.S. Air Force awarded contracts to Boeing Aerospace, General Dynamics, Martin Marietta and Rockwell International to study space transportation architecture. The purpose of the study was to evaluate the space transportation requirements of the United States and to make recommendations about essential technologies for future space transport systems.

1985 September 11   The NASA *International Cometary Explorer (ICE)* satellite flew through the dust-filled tail of the comet Giacobini-Zinner for 45 min. beginning at 11:00 UT, during which it approached to within 5,000 mi. of the nucleus. In its seven years since launch (as *ISEE 3*), the satellite had been the first to traverse the distant regions of the Earth's geomagnetic tail, the first to conduct multiple swing-bys of the moon and the first to conduct a lunar gravity-assist maneuver. Further maneuvers would allow it to observe Halley's comet at a distance on October 31, 1985, and then from a range of only 19.5 million mi. on March 28, 1986. *ICE* would return to the vicinity of Earth in July 2012 where a lunar gravity-assist might place it in an Earth orbit from where it could be recovered.

The U.S. Air Force announced that it had selected Air Force Under Secretary Edward "Pete" Aldridge to fly aboard the Shuttle on its first launch from SLC-6 at Vandenberg AFB, Calif. Aldridge would be accompanied by Air Force Maj. John Brett Watterson along with five NASA astronauts on the March 20, 1986 mission. When *Challenger* was lost in a solid rocket motor burn-through on January 28, 1986, the mission was postponed and subsequently canceled when the SLC-6 site was abandoned and "passenger" flights halted.

1985 September 13   In the first successful test of a new weapon built by the Vought Corp., a modified U.S. Air Force F-15 fired a two-stage antisatellite missile which successfully struck and destroyed the U.S. Navy P78–1 *Solwind* satellite launched February 24, 1979. The terminal stage and impact device comprised a 35-lb., drum-shaped miniature homing vehicle containing 64 steering rockets and a refrigerated infrared sensor. The combined impact velocity of 27,800 MPH destroyed the satellite without the need for explosives. Although the Air Force had no further use of the satellite, it continued to transmit valuable scientific information about the physics of near-Earth space until the time of its destruction 320 mi. above the Earth. Two more tests with the antisatellite missile were conducted, on August 22 and September 29, 1986, against phantom targets in space. Congress halted the program after pressure from the United Nations, which sought to limit such weapons.

An Ariane 3 launch vehicle (V15) carrying the *GTE Spacenet 3* and the European Space Agency's *ECS 3* communication satellites failed to reach orbit due to a malfunction in the third-stage engine. Ignition failed to occur owing to a mixture ratio in the combustion chamber which caused the engine to stop. Without sufficient velocity to attain orbit, V15 deviated from its planned trajectory and was destroyed by the range safety officer.

1985 September 17   At 12:39 UT cosmonauts Vladimir V. Vasyutin, Georgi M. Grechko and Aleksandr Volkov were launched aboard their *Soyuz-T 14* spacecraft to a rendezvous with the *Salyut 7* space station, in an orbit of 209.5 × 219.5 mi. Docking at the station's aft port took place at 14:14 UT on the next day, and shortly thereafter the crew joined Vladimir Dzhanibekov and Viktor Savinykh in *Salyut 7*. In a crew split, Dzhanibekov would return to Earth with Grechko, leaving Vasyutin and Volkov to stay with Savinykh for the remainder of the long-duration mission. After the customary joint activities aboard *Salyut 7*, appropriate crew contour couches were transferred between respective spacecraft, and Dzhanibekov and Grechko got into *Soyuz-T 13* for the return to Earth. Undocking from the forward port at 03:58 UT on September 25, they spent a day in simulated redocking tests and landed in the USSR on September 26 at 8 days 21 hr. 13 min. Dzhanibekov had been in space for 112 days 3 hr. 12 min. *Soyuz-T 14* was then transferred to the forward port.

1985 September 23   The U.S. Department of Defense activated its first unified space command, SPACECOM, in a special ceremony at Peterson AFB, Colorado Springs, Colo. An integrated system designed to knit together all military programs in space, it would oversee and manage Air Force, Army, Navy and Marine Corps space activities, including intelligence gathering, attack alert warning, communications, navigation and unified command data transfer.

1985 September 27   At 08:41 UT a 44,100-lb. unmanned TKS ferry ship designated *Cosmos 1686* was launched by SL-13 from the Baikonur cosmodrome. It was a variant of the *Cosmos 1443* module launched on March 2, 1983, but instead of a 12,000 lb. reentry module it had a battery of astronomical telescopes weighing 2,300 lb. installed in its place. After a series of maneuvers, the aft section of *Cosmos 1686* docked to the *Salyut 7* space station, in an orbit of 208 × 219 mi., at 10:16 UT on October 2. Carrying 11,000 lb. of cargo, *Cosmos 1686* had docked to the aft port of *Salyut 7*, unlike *Cosmos 1443* which docked at the forward port. Station cosmonauts Alexandr Volkov and Viktor Savinykh unloaded some of the cargo over the next several days. *Cosmos 1686* would remain attached to *Salyut 7*, using its propulsion system to keep the station in the proper orbit; *Salyut 7*'s propulsion system could only perform attitude corrections. This was the last flight of a Chelomei TKS freighter, originally designed more than two decades earlier to carry a three-person crew and cargo to the Almaz military space stations. None of the four TKS vehicles flown carried crew members.

1985 September 28   China successfully fired its first submarine-launched ballistic missile from one of 12 vertical launch tubes on the *Xia,* submerged in the Pacific Ocean. The submarine had a length of 393 ft., a beam of 33 ft., a hull depth of 26 ft. and a submerged displacement of 8,000 tons. Designated CSS-N-3, the missile had a length of 33 ft., a diameter of 4 ft. 11 in. and a weight of about 30,400 lb. It had a range of 1,400 mi. and carried a 2-MT reentry vehicle. Only one *Xia*-class boat with 12 missiles had been deployed by the early 1990s, although three more were being built.

1985 October 3   The first flight of Shuttle orbiter OV-104 *Atlantis* on the second dedicated Department of Defense Shuttle mission was launched at 15:15:30 UT from LC-39A at the Kennedy Space Center, Fla., after a 22 min. 30 sec. hold to correct a faulty engine reading. Crew for mission 51-J comprised Karol Bobko (CDR), Ronald Grabe (PLT), Robert

Stewart (MS), David Hilmers (MS) and U.S. Air Force astronaut Maj. William A. Pailes (PS). The payload consisted of two 2,200-lb. Defense Satellite Communication System-III (DSCS-III) satellites, the second and third flown, propelled to geostationary orbit by an inertial upper stage with a reduced load of propellant. In ensuring the success of the DSCS deployment, *Atlantis* achieved a Shuttle-record altitude of 320 mi. The mission ended at Kennedy Space Center, Fla., at 3 days 1 hr. 33 min. 23 sec.

The control board responsible for approving changes to the NASA Space Station sanctioned a fundamental redesign of the structural configuration. Responding to a determination from McDonnell Douglas that the power-tower design was flawed, NASA replaced it with the dual-keel configuration. When McDonnell Douglas applied modal analysis to the power-tower concept, it discovered that oscillations in the pressure modules caused by the mass of the solar arrays on the end of the tower truss would seriously interfere with microgravity experiments. To inhibit this effect, the modules were moved to the center of the horizontal truss in an "H" configuration arrangement, and the top and bottom extremities of the vertical truss assemblies were connected with horizontal trusses. The vertical truss units were spaced 126 ft. apart and the central keel to which the pressure modules were attached supported solar cell arrays at each end for a total width of about 435 ft. The twin vertical trusses had a height of 297 ft. from top to bottom.

1985 October 7 NASA announced it had been signed up by Scott Science and Technology (SST) to provide advice and technical expertise in its development of a commercial liquid propellant upper stage for sending Shuttle payloads to higher orbit. Formed by ex-NASA astronaut David Scott, SST had been working for two years on development of a satellite transfer vehicle (STV) capable of boosting satellites weighing between 2,000 lb. and 19,000 lb. The Johnson Space Center was to monitor progress and warn of any technical problems as the STV evolved.

Burton I. Edleson, associate administrator for NASA's Office of Space Science and Applications, addressed a Global Habitability seminar in Stockholm with an appeal for more attention to environmental monitoring. Using the seminar, which was being held in association with the 36th Congress of the International Astronautical Federation, to present a major effort under way at NASA to develop an Earth-watch program, Edleson's appeal alerted scientists to a challenging requirement for advanced sensors and instruments to be carried on satellites dedicated to studying the environment and atmospheric change. The endeavor was later dubbed Mission to Planet Earth.

1985 October 11 The U.S. Department of Defense dropped plans to charge users a fee for access to the signals from Navstar global position system satellites. Preferring instead to make open and available the lower of two position-information channels, the Department of Defense said that it had come to this decision in the interests of aviation safety. The highly accurate signal would remain the sole preserve of the military user community, however.

1985 October 16 NASA canceled the mission to comet Wild 2 planned for launch on a Shuttle in 1991 for a fly-by in 1996. To be ready in time, NASA had to begin funded development of the spacecraft in 1986 but the Reagan administration refused to allocate the money. Coming only months before a convoy of European, Soviet and Japanese spacecraft flew by

Halley's comet without direct NASA participation, this latest cancellation was a severe blow for cometary scientists in the United States. NASA affirmed that it would push vigorously for permission to start a mission to comet Tempel 2 which would pass Earth in 1998.

1985 October 18 The U.S. Department of Commerce transferred the Landsat remote sensing satellite system to the Eosat Corp., a joint venture between RCA Corp. and Hughes Aircraft. The shift from government to private ownership was the continuation of a policy begun by the Carter administration. Successive delays and wrangling in Congress denied users the benefits of improved *Landsat 6* and *7* systems, which should have been scheduled for flights beginning in 1987. Under the terms of the agreement, Eosat would receive $270 million to help convert the remote sensing program into a self-financing business and to partly subsidize the next generation of satellites.

Negotiations were concluded between Aérospatiale and Dassault-Breguet, France's two leading aerospace companies, on the joint development of the Hermes space plane. Aérospatiale was given the role of primary industrial contractor while Dassault-Breguet was to be primary contractor on the aerodynamics of the reusable vehicle. The total program was estimated to cost 15 billion francs ($1.9 billion at 1985 prices), including 9 billion francs ($1.1 billion) on development and assembly, with France providing 50% and other European Space Agency countries providing the balance. There were to be two Hermes flight models launched by Ariane 5 on flights expected to begin in the mid-1990s.

1985 October 21 A CZ-2 launch vehicle carried the 5,510-lb. Chinese satellite *SKW-13* to a 107 × 245 mi. orbit inclined 63°. It was almost identical in structure to the six military photoreconnaissance satellites with recoverable capsules launched between November 26, 1975, and September 12, 1984. The return capsule from *SKW-13* returned to Earth after five days, and the main body of the satellite decayed on November 7. Other Chinese Earth resource satellites in this class were: *SKW-14* launched October 6, 1986, capsule recovered at five days, satellite reentered October 23, 1986; and *SKW-16* launched September 9, 1987, capsule returned at six days, satellite reentered October 4, 1987.

1985 October 24 NASA announced that it would fly a U.S. journalist into space aboard a forthcoming Shuttle mission. Inviting applications from candidates who felt they could use their journalistic training to communicate their experiences to ordinary Americans, NASA selected 40 nominees for final consideration by the Space Flight Participant Evaluation Committee. The destruction of *Challenger* on January 28, 1986, canceled the Space Flight Participant Program and with it plans for eventually sending an artist into space.

At the end of a two-day meeting the European Space Agency admitted to its membership Austria and Norway, bringing to 13 the number of countries that were permanent members of ESA. The two countries had been closely involved with several ESRO/ESA projects for more than 20 years and full membership status would accrue on January 1, 1987, after government review and ratification in respective parliaments.

An SL-14 launch vehicle carried the experimental precursor in a new and improved generation of Soviet weather satellite, *Meteor 3-1*, from Plesetsk to a 736 × 752 mi. orbit inclined

82.5°. The next Meteor 3 was launched on July 26, 1988, followed by the third satellite of this type on October 24, 1989, the fourth on April 24, 1991, and the fifth on August 15, 1991. All Meteor 3–series weather satellites occupied the higher orbit typified by the first in the series. Meteor 1 and Meteor 2 series satellites were placed in lower orbits. Weights varied between 3,300 lb. and 4,750 lb.

1985 October 25   NASA selected the dual-keel reference configuration for the international space station, so-called because of twin vertical booms that would provide the framework for attachment of modules and other structures. In this variation of the power-tower concept, the pressure modules would be located closer to the center of gravity. The dual-keel layout had been selected from a variety of different designs put forward by contractors. McDonnell Douglas had discovered that in the original power-tower concept the modules would not have the level of microgravity sought by materials processing scientists. After further engineering studies the dual-keel concept was publicly unveiled on May 14, 1986, as NASA's new baseline design.

1985 October 29   Society Expeditions, a company specializing in exotic vacations, announced that it was taking reservations for spaceflight trips it said would begin on October 12, 1992, the 500th anniversary of Columbus's discovery of the new world. The company claimed it had signed a $280 million deal with Pacific American Launch Systems for development of a 57 ft. long rocket called PHOENIX E. Each flight was said to cost the operators $1 million and each passenger would pay $50,000 plus $2,000 for training.

1985 October 30   At 17:00:00 UT Shuttle *Challenger* began the 61-A mission with liftoff from LC-39A at the Kennedy Space Center carrying the 31,941-lb. Spacelab pressure module. Crew for the German Spacelab D-1 mission comprised U.S. astronauts Henry Hartsfield (CDR), Steven Nagel (PLT), James Buchli (MS), Guion Bluford (MS) and Bonnie Dunbar (MS); German astronauts Reinhard Furrer (PS) and Ernst Messerschmid (PS); and Dutch astronaut Wubbo Ockels (PS) representing the European Space Agency. The D-1 mission was managed by the Federal German Aerospace Research Establishment and designed to operate 76 experiments in materials science, biology, medicine and space-time interaction. Two teams were formed: Nagel, Dunbar and Furrer in the blue team; Buchli, Bluford and Messerschmid in the red team. The two teams operated experiments round the clock on 12-hr. shifts. The mission ended at Edwards AFB, Calif., at 7 days 0 hr. 44 min. 53 sec.

1985 November 4   More than eight years after launch and a close encounter with the solar system's biggest planets, Jupiter and Saturn several years later, NASA's *Voyager 2* began observations of Uranus prior to a close fly-by on January 24, 1986. At this date the spacecraft was 1,790 million mi. from Earth with Uranus 64 million mi. ahead. Traveling at the speed of light, radio signals now took 2 hr. 25 min. to reach Earth. Because Uranus receives from the sun only 0.25% the light falling on Jupiter, the target around which Voyager's cameras had been designed, imaging exposure times had to be 16 times as long. This necessitated a special computer program to pan the cameras and prevent smear. Also, maximum data transmission rates would be 14.4 kilobits/sec. at the distance of Uranus versus 29.9 kilobits/sec. at Saturn and 115.2 kilobits/sec. at Jupiter. The far encounter phase began on

January 10, 1986, and near encounter events started 14 days later.

1985 November 5   John Young, chief of the NASA Astronaut Office at the Johnson Space Center, wrote a memorandum of weaknesses in the Shuttle design and offering a menu of changes engineers should consider for a successor. He emphasized the need for a crew escape module in the event of a catastrophe during ascent, additional strengthening in the wing to improve intact abort capabilities, and a thrust termination system for the single rocket boosters to shut them down in the event of a major anomaly. Astronauts Vance Brand and Paul Weitz also signed the memorandum.

1985 November 12   NASA's Marshall Space Flight Center issued requests for proposals to LTV, Martin Marietta and TRW Systems to compete for a contract to design and develop the orbital maneuvering vehicle (OMV). The companies were asked to respond by December 20. A simplified derivative of the old space tug concept, the OMV was to be unmanned initially and have the propulsive power to carry satellites to and from orbits as high as 1,000 mi.; the Shuttle was effectively limited to altitudes of around 300 mi.

1985 November 15   The NASA Lewis Research Center announced it had selected Sundstrand Corp., Grumman Aerospace, Boeing and the Harris Corp. for Phase B studies of the space station power system. One major uncertainty was whether to employ solar cell arrays for electrical power production or a solar dynamic (heat engine) system which would focus solar energy to a turbine. Energy needs aboard the station would be 10 times the power produced by any previous solar array system.

1985 November 21   *Salyut 7* cosmonauts Vladimir Vasyutin, Alexandr Volkov and Viktor Savinykh returned to Earth in the *Soyuz-T 14* spacecraft launched September 17. The plan had been for the three cosmonauts to remain in space until March 15, 1986, thus giving Savinykh 282 days in space, but that was abandoned when Vasyutin became seriously ill. After private, coded communication sessions with physicians on the ground it was decided to bring him home. That required the use of *Soyuz-T 14*, without which the other two crew members could not have returned in the event of an emergency. As it was, Savinykh had spent 168 days 3 hr. 51 min. in space and Vasyutin and Volkov had both spent 64 days 21 hr. 52 min. in space. This was the last long-duration stay aboard the *Salyut 7* space station, although two cosmonauts launched on March 13, 1986, would visit the station between May 6 and June 25, 1986.

U.S. Air Force Maj. Gen. Donald Kutyna announced that the Department of Defense had decided to proceed with development of a hypersonic aerospace vehicle capable of launching satellites into orbit and intended to provide low-cost access to space in support of President Ronald Reagan's Strategic Defense Initiative, the "transatmospheric vehicle." Using hypersonic ramjet engines to accelerate to a high fraction of orbital velocity, the vehicle could either deploy a separate boost stage and payload or, in a developed version, fly into space itself.

1985 November 27   At 00:29:00 UT Shuttle mission 61-B began with the launch of *Atlantis* from LC-39A carrying a 47,539-lb. payload, including three communication satellites and U.S. astronauts Brewster Shaw (CDR), Bryan O'Connor (PLT), Mary Cleave (MS), Sherwood Spring (MS), Jerry Ross

The assembly of large structures occupies Shuttle Atlantis *astronaut Jerry Boss during the second of two EVAs on mission 61-B launched November 27, 1985.*

(MS), Charles Walker (PS) and Mexican astronaut Rodolfo Neri Vela. Two HS 376 satellites were released: *Morelos-B* for Mexico at 07:46:50 UT, and *AUSSAT-2* at 01:20:33 UT on November 28. The three-axis-stabilized *Satcom K-2*, first of three built for RCA, was released at 21:57:31 UT on November 28. All three satellites deployed by *Atlantis* reached geostationary orbit. In a 5 hr. 33 min. EVA (extravehicular activity) on November 30 and a 6 hr. 41 min. EVA two days later, Spring and Ross evaluated ACCESS (assembly concept for construction of erectable space structure) and EASE (experimental assembly of structures in EVA), twice building a 45-ft. truss tower and nine times assembling and dismantling a large pyramid structure. The mission ended at Edwards AFB, Calif., 6 days 21 hr. 4 min. 49 sec. after liftoff.

Originally scheduled for March 20, 1986, the first Shuttle launch from SLC-6 at Vandenberg AFB, Calif., was postponed to mid-July 1986 in a joint NASA/U.S. Air Force decision. Designated mission 62-A, the *Discovery* carried a crew comprising astronauts Robert Crippen (CDR), Guy Gardner (PLT), Jerry Ross (MS) and Richard Mullane (MS). The Department of Defense payload was to have comprised an 11,200-lb. package of technology experiments designated AFP-675 and the 4,278-lb. *Teal Ruby* satellite AFP-888. Placed in a 460-mi. orbit, *Teal Ruby* was designed to detect aircraft flying in the atmosphere.

1985 November   McDonnell Douglas and the pharmaceutical company 3M signed a deal to work as a team to produce and market a proprietary drug to be made aboard the Shuttle. The drug, erythropoietin, would stimulate the body to produce red blood cells, a reaction of benefit to people suffering from anemia or other red blood cell disorders. An electrophoresis machine carried into orbit aboard Shuttle mission 61-B was to produce erythropoietin which would be given to human patients shortly after the return to Earth.

1985 December 2   A federal grand jury indicted three former executives of General Dynamics, including NASA Adminis-

trator James Beggs, for alleged attempts to defraud the Department of Defense in connection with a government contract. Two days later James Beggs took indefinite leave from NASA to fight his plea of innocence, and Associate Administrator Philip Culbertson became general manager in Beggs' absence. President Ronald Reagan appointed former Rand Corp. analyst William Graham, who had been with NASA only eight working days, to the post of acting administrator, but Culbertson stayed on to control daily affairs. Beggs tendered his resignation from NASA on February 25, 1986. On June 22, 1987, the Department of Justice dropped all charges against Beggs and he was exonerated.

1985 December 6   Britain and the United States signed a memorandum of agreement on cooperation in the field of ballistic missile defense under the U.S. Strategic Defense Initiative. The agreement was signed one day after Britain's Prime Minister Margaret Thatcher announced the cooperative deal in the House of Commons, the lower house of Britain's Parliament. Under the agreement, British companies would be allowed to bid for SDI work. In reality, very little work went to Britain.

1985 December 13   A solid propellant Scout launched two antisatellite target vehicles to low Earth orbit from Wallops Island, Va. In a classified flight that was run for the Department of Defense, *ASAT-1* was placed in a 195 × 482 mi. orbit, and *ASAT-2* was placed in a 195 × 480 mi. orbit. Both orbits were at an inclination of 37.1° and both satellites remain in space. *ASAT-1* decayed on August 9, 1987, followed by *ASAT-2* on May 11, 1989.

1985 December 18   In an unparalleled series of six false starts, the first attempt to launch Shuttle *Columbia* on mission 61-C was abandoned when it took too long to close out the aft compartment. Next day it was scrubbed at T − 14 sec. when a solid rocket booster subhydraulic power unit erroneously indicated it had failed. On January 6, 1986, the count was halted at T − 31 sec. when 14,000 lb. of liquid oxygen was accidentally drained from the external tank. The next day, the launch was scrubbed at T − 9 min. due to bad weather at landing sites in Spain and Senegal. The attempt to launch on January 9 was delayed when a launch-pad liquid oxygen sensor broke off and lodged in the no. 2 main engine prevalve. The January 10, 1986, attempt was abandoned due to heavy rain.

1985 December 28   An underground nuclear test in the Nevada desert researched the concept of harnessing x rays produced by a nuclear explosion into a laser weapon for use against ballistic missiles. Proposed by Dr. Edward Teller of the Lawrence Livermore National Laboratory, for whom the test was arranged, the concept envisaged an x-ray battle station in space equipped with numerous lasing rods, each pointing at a separate missile launched through the atmosphere. On command, the nuclear core would detonate, propagating powerful x rays down each lasing rod in the pico-seconds before the satellite was consumed by its own fireball.

1985   The Soviet Union began to deploy two powerful new missiles: the SS-N-23 Skiff submarine-launched ballistic missile, in Delta IV–class submarines, and the SS-25 Sickle ICBM. The liquid propellant SS-N-23 had a length of 46 ft. and could carry 10 MIRVed warheads to a distance of 5,150 mi. with an accuracy of 3,000 ft. By 1990 there were six Delta

IV vessels operational, each of which carried 16 missiles. The solid propellant SS-25, known in the USSR as the RS-12M, was the first road-mobile ICBM deployed anywhere, and by 1990 some 225 had entered service. The missile had a length of 60 ft., a diameter of 5 ft. 7 in. and a throw weight of 3,500 lb., accommodating a single 750-KT warhead with an extremely creditable accuracy of 650 ft. at a range of 6,500 mi.

1986 January 12   At 11:55:00 UT Shuttle mission 61-C began with the launch of *Columbia* from LC-39A at the Kennedy Space Center, Fla., and astronauts Robert Gibson (CDR), Charles Bolden (PLT), Franklin Chang-Díaz (MS, the first Hispanic American to fly in space), Steven Hawley (MS), George Nelson (MS), Robert Cenker (MS) and Rep. Bill Nelson (D-Fla.). The 32,763-lb. payload included the RCA *Satcom K-1* satellite, deployed at 21:26:29 UT, the Materials Sciences Laboratory-2 (MSL-2) and its three materials experiments from Marshall Space Flight Center, and Hitchhiker-G accommodating a range of technology experiments. The mission was to have landed January 17 at Kennedy Space Center but was moved to January 16 to save turnaround time. Poor weather at KSC moved it back to January 17 and then to January 18. When the weather was still unacceptable, the landing was delayed one orbit and moved to Edwards AFB, Calif., at 6 days 2 hr. 3 min. 51 sec.

1986 January 24   *Voyager 2* became the first spacecraft to fly past Uranus and its moons, gathering data which transformed scientific knowledge about this distant planet. With a diameter of 32,200 mi., Uranus and the orbital plane of its moons is tilted 95° to its orbital path round the sun so Voyager would approach the planet as though targeting the edge of a bullseye. At encounter *Voyager 2* was 1,842.6 million mi. from Earth; at the speed of light radio signals took 2 hr. 45 min. to bridge the distance. At 17:53 UT Voyager passed within 227,000 mi. of Titania, a moon 1,000 mi. in diameter which orbits Uranus at a distance of 270,000 mi. Just 1 hr. 4 min. later the 960-mi. diameter Oberon, which lies 362,000 mi. from Uranus, was passed within 293,000 mi. Closest approach to the 720-mi. diameter Ariel was at 19:05 UT, a distance of 81,000 mi. The 300-mi. diameter Miranda was passed at 20,000 mi., 43 min. later. *Voyager 2* swept past Uranus at 18:44 UT from a distance of 51,000 mi. At 11:22 UT the spacecraft went behind the planet as viewed from Earth, reemerging 1 hr. 22 min. later, 1 hr. 6 min. after passing within 202,000 mi. of Umbriel, about the size of Ariel. *Voyager 2* had photographed the rings of Uranus and added 10 new moons to the five already known.

1986 January 28   In the most public disaster to hit the U.S. manned space program, Shuttle *Challenger* was destroyed and its seven-member crew killed after liftoff from Kennedy Space Center, Fla., the first Shuttle launch from LC-39B. Carrying astronauts Francis Scobee (CDR), Michael Smith (PLT), Judith Resnik (MS), Ellison Onizuka (MS), Ronald McNair (MS), Gregory Jarvis (PS) and Christa McAuliffe (the first "teacher in space"), mission 51-L comprised a 52,685-lb. payload consisting of the *TDRS-2* tracking and data relay satellite and its IUS, the *SPARTAN-203* Halley's comet observation satellite and several technology experiments. Delayed first from January 24 to January 25 because of bad weather and then to January 27 before a 24 hr. delay for technical problems, liftoff occurred 0.25 sec. after solid rocket booster ignition which took place at 16:38:00.01 UT. Just 0.428 sec. into flight eight puffs of smoke from ignition of the booster leaked through joints in the bottom of the right-hand booster. At 35.379 sec. after solid rocket motor ignition,

flames appeared from the leaking joint and acted as a flame torch, cutting open the bottom end of the external tank. At 73 sec. the vehicle exploded as the ET became a Roman candle. Shoved aside by the force of the explosion, the two solid rocket boosters careened off in opposite directions and *Challenger* broke into major chunks. Separated from the rest of the orbiter, the intact crew compartment continued its upward trajectory from 48,000 ft. to an altitude of 65,000 ft. about 25 sec. after the breakup. It then descended, striking the ocean surface at 207 MPH with a force of 200 g's, at an elapsed time of 3 min. 58 sec., destroying the cabin and killing the crew. The two boosters were intentionally blown up at an elapsed time of 1 min. 50.25 sec. and impacted in the Atlantic Ocean.

1986 February 1   The second Chinese geostationary communication satellite, *STTW*, was carried to orbit by a CZ-3 launch vehicle from Xi Chang. The Chinese had applied to the international telecommunications authority for orbit slots at 125° E and 87.5° E. The first geostationary satellite, *STTW-1*, had been launched to 125° E but instead of occupying the second reserved slot, STW was placed at 103° E. *STTW-2*, launched on March 7, 1988, was positioned at 88° E. Launched December 22, 1988, *STTW-3* was put over 110° E. *STTW-4* was launched by CZ-3 on February 4, 1990, and placed at 98.5° E. An abbreviation of Shiyang Tongbu Tonxin Weixing, STTW satellites weighed about 2,200 lb. at launch and 975 lb. at apogee motor burnout, and comprised a cylinder 6 ft. 6 in. in diameter and 7 ft. 10 in. tall.

1986 February 10   NASA Administrator William Graham postponed the launch of *Galileo*, *Ulysses* and the *Astro-1* Shuttle payload that would have observed Halley's comet from above Earth's atmosphere. *Astro-1* should have been carried in the payload bay of Shuttle *Columbia* for the 61-E mission launched on March 6, 1986. *Ulysses* was to have been launched on May 15, 1986, by *Challenger* on the 61-F mission. *Galileo* was to have followed on *Atlantis* five days later comprising the 61-G flight. Because of its international commitment to the European Space Agency, NASA offered to launch *Ulysses* at the June 1987 launch window, although *Galileo* managers hoped to use that window too. NASA planned to modify *Discovery* to carry Centaur, thus providing two Shuttle orbiters for the cryogenic Earth escape stage. However, the 1987 window would result in *Galileo* getting to Jupiter at the same time as *Ulysses*, compromising Earth tracking capabilities. As it turned out, neither was launched in 1987.

1986 February 12   The second of two Japanese direct broadcast satellites, *BS-2B*, was launched by N-II from Tanegashima to a geosynchronous transfer orbit from where it was placed later at 110° E. The satellite was virtually identical to its predecessor, *BS-2A*, and had been launched to provide an operational service for the Nihon Hoso Kyokai (NHK) national broadcasting station. However, an attitude control problem caused the satellite to stray off its assigned longitudinal slot and by June it had drifted 20° away. Redundant systems were employed to return it to 110° E and NHK accepted *BS-2B* for service.

1986 February 19   At 21:19 UT an SL-13 launch vehicle lifted off from the Baikonur cosmodrome carrying the first of a new generation of Soviet space stations called *Mir*. With a weight of 46,100 lb., *Mir* was superficially the same as Salyut. It had a total length of 43 ft. and comprised four principle sections. The cylindrical forward transfer module characteristic of

Crew members of Shuttle mission 51-L shortly before boarding Challenger *on the fateful morning of January 28, 1986; from left to right are Ellison S. Onizuka, Christa McAuliffe, Michael J. Smith, Francis R. Scobee, Judith A. Resnik, Robert E. McNair and Gregory B. Jarvis.*

earlier "civilian" space stations was replaced by a spherical docking module reminiscent of the Soyuz orbital module. It had a diameter of 7 ft. 2 in. and incorporated a single axial docking port and four radial ports at 90° intervals, each port with a diameter of 2 ft. 7 in. The docking module incorporated lights, alignment targets and a module transfer system called Liappa. With this, large modules would dock at the axial port and deploy a short arm to latch into Liappa sockets. The large module could then be unlatched and moved 90° to reside at one of the radial ports, leaving the axial port free for further dockings. The smaller work compartment had a diameter of 9 ft. 6 in., while the larger work compartment had a diameter of 13 ft. 9 in., the combined length being 25 ft. 2 in. A service propulsion compartment, 13 ft. 9 in. in diameter and 7 ft. 5 in. long, was attached to the rear of the large work module. It incorporated two 660-lb. thrust rocket motors and a central transfer tunnel, 5 ft. 6 in. long and 6 ft. 6 in. in diameter, through which cosmonauts could pass from the aft docking port. *Mir* had 32 (31-lb. thrust) attitude control thrusters. Two large arrays spanned 97 ft. 6 in. and provided 820 sq. ft. (26% more than *Salyut 7*) of solar cells for electrical energy. Following a series of maneuvers, *Mir* settled down into an orbit of 201 × 219 mi., inclined 51.6° to the Earth's equator. *Mir* would be periodically occupied for at least the next nine years.

1986 February 22 The last of 11 Ariane 1 launch vehicles (V16) carried the French Earth resources satellite, *SPOT 1*, and Sweden's first experimental satellite, *Viking*, into orbit from Kourou, French Guiana. The 4,035-lb. *SPOT 1* satellite was placed in a 512 × 514.5 mi. orbit inclined 98.7°. The satellite began returning images within two days of launch and by the end of the second month of operation had returned 18,000 pictures. The 1,180-lb. *Viking* was placed in a 491 × 8,301 mi. orbit, with an inclination of 98.8°. The octagonal

satellite had a height of 1 ft. 8 in. and a width of 5 ft. 11 in. Eight solar panels provided 80 W of power. Four 131-ft. and two 13-ft. wire booms carried electric field sensors, and two rigid axial booms carried magnetometers for magnetospheric and auroral research.

1986 February The British government agreed to provide the sterling equivalent of $2.1 million to fund studies of the British Aerospace HOTOL (horizontal takeoff and landing) reusable satellite launcher concept. BAe was to put in an

A section of Challenger's *lower forward fuselage is retrieved from the Atlantic Ocean two days after the explosion that took the lives of six NASA astronauts and a school teacher.*

The Soviet Mir *space station is positioned ready to receive the launch shroud that will protect it during its ascent on February 19, 1986.*

additional $2.1 million of its own money. Studies indicated a development cost of $6.3 billion but flight costs of only $320/lb. of payload, about 5% that of the NASA Shuttle. HOTOL would be capable of carrying 16,700 lb. in payload to a 185 mi., 28.5° orbit. Over the next year the British government took steps to have the European Space Agency fund HOTOL definition, but France, with its Hermes space plane, proved an intractable opponent and ESA backed the latter.

1986 March 6   The first in a series of flyby observations of Halley's comet began with the Soviet *Vega 1* spacecraft which passed within 5,524 mi. of the solar side of the nucleus, a potato-shaped rock 10 mi. long by 5 mi. wide, and surrounded by a halo of dust and ice particles evaporating from the surface. *Vega 1* carried 265 lb. of scientific instruments, including one 1,200-mm, high-resolution and one 150-mm, low-resolution camera which provided 500 images in 3 hr. *Vega 1* had a span of 33 ft. across two wing-like solar cell arrays with a surface area of 107 sq. ft. Scientists expected it to receive many hits from tiny particles venting off the comet. The spacecraft suffered a 40% power loss as solar cells were destroyed by the dust cloud. The relative speed at encounter with Halley's comet was 33.5 mi./sec.

1986 March 7   The Soviet Union signed an agreement with the French government for the first long-duration research flight involving a broad range of experiments carried out by a French cosmonaut. Following the highly successful visit to the *Salyut 7* space station by Jean-Loup Chrétien, launched in *Soyuz-T 6* on June 24, 1982, the French had wanted a two-month mission for an ambitious scientific and technological experiments program. In return for the opportunity to perform a space walk, the French agreed to a mission lasting only one month. It would begin with the launch of *Soyuz-TM 7* on November 26, 1988. Chrétien was selected as the primary crew member during November 1986, with Marcel Tognini as the backup candidate.

McDonnell Douglas discontinued development of erythropoietin produced in space, a hormone believed by some researchers to be useful in fighting many forms of anemia. In the Electrophoresis Operations in Space (EOS) program, the aerospace company had been teamed with the pharmaceutical industry to fly Shuttle experiments producing a purer strain of erythropoietin than was believed possible on Earth. Subse-

quent research had shown that it might be possible to produce a comparable product on Earth. In the wake of the Shuttle disaster on January 28, and consequent uncertainties about the future of the Shuttle program, McDonnell Douglas decided to pull out.

1986 March 8   The second in a series of four spacecraft to conduct close fly-by observations of Halley's comet, the Japanese spacecraft *Suisei* passed within 94,700 mi. of the solar side of the nucleus. The spacecraft was instrumented to specifically study the comet's enormous hydrogen corona, which extends to a radius of 12 million mi., to determine the total hydrogen production rate, and to study the interaction of the solar wind with the ionosphere that surrounded the comet during its close pass around the sun. Of surprise to scientists were the two particle strikes recorded by *Suisei*, each larger than 1-mm in size, when it was expected none would be encountered at that distance. The relative fly-by velocity was 30.9 mi./sec.

1986 March 9   Third in a series of four spacecraft which passed inside the bow shock wave of Halley's comet made its close encounter with the celestial visitor when the Russian spacecraft *Vega 2* came to within 4,990 mi. of the nucleus. The spacecraft encountered fewer but larger dust particles than *Vega 1*, on March 6, losing 80% of its electrical power from damaged solar cells. Nevertheless, the spacecraft transmitted 700 pictures of the comet and neither spacecraft completely lost a scientific instrument. The relative closing velocity was 32.5 mi./sec. Data from the combined NASA/USSR tracking of *Vega 1* and *Vega 2* were passed immediately to the European Space Agency so that engineers in control of the *Giotto* spacecraft could receive updates on the precise position of the comet.

1986 March 13   At 12:33 UT cosmonauts Leonid D. Kizim and Vladimir A. Solovyov were launched aboard their *Soyuz-T 15* spacecraft by an SL-4 from the Baikonur cosmodrome. Trained originally for operations aboard *Salyut 7*, the occupancy of which had been cut short when Vladimir Vasyutin, Alexandr Volkov and Viktor Savinykh returned early on November 21, 1985, Kizim and Solovyov were to activate the *Mir* space station and then transfer to *Salyut 7* for seven weeks of activity, equipment for which had already been delivered by Progress cargo freighters. After that, they would return to *Mir* and carry on with their activities at the new station. In an orbit of 205 × 212 mi., *Soyuz-T 15*'s Igla approach system completed a rendezvous in line with the rear docking port, at which point the crew maneuvered their spacecraft round and docked to *Mir*'s forward axial port at 13:38 UT on March 15. The crew moved inside *Mir*, activated the systems and settled in.

1986 March 14   The fourth and last in a series of Halley's comet fly-by observations took place when the European Space Agency's *Giotto* spacecraft passed within 370 mi. of the nucleus, by far the closest and most perilous passage, at 00:10 UT, 87 million mi. from Earth. *Giotto* carried 11 science instruments weighing 130 lb., including a 1,000-mm multicolor camera, dust detectors, plasma analyzers, mass spectrometers, and equipment for measuring magnetic fields, ion concentrations and comet halo brightness. About 2 seconds prior to the close encounter, with *Giotto* closing on Halley's comet at 27.1 mi./sec., the spacecraft was buffeted by dust and debris which temporarily knocked it off alignment with Earth and signals stopped. Just 34 min. later the spacecraft corrected itself and strong signals were once again received until the instruments were switched off at 02:56 UT on March 15.

1986 March 19   The European Space Agency's *Giotto* spacecraft began a series of maneuvers to alter its heliocentric flight path and bring it back within 14,300 mi. of Earth on July 2, 1990, so that it could be redirected to rendezvous with the comet Grigg-Skjellerup on July 10, 1992. In a 4 hr. 30 min. burn of the hydrazine thrusters, *Giotto* changed speed by 209 ft./sec. A 4 hr. 10 min. burn the next day changed speed by a further 151 ft./sec. and a final 16-min. burn on March 21 added another 8.5 ft./sec. velocity change. Total propellant consumption for the three maneuvers was 52.5 lb.

At 10:08 UT the unmanned cargo tanker *Progress 25* was launched by an SL-4 from the Baikonur cosmodrome carrying supplies and propellant to the *Mir* space station, then in an orbit of 207 × 210 mi. Docking occurred at the aft port at 11:16 UT on March 21, and the *Mir* crew (Leonid Kizim and Vladimir Solovyov) began offloading the 4,410 lb. of supplies, including 53 gal. of water. On April 1 a test was conducted of the rigidity of the docked complex, and over the next several days *Progress 25* was used to maneuver the assembly to an orbit of 209.5 × 214 mi. The cargo tanker undocked at 19:24 UT on April 20, and was de-orbited within a few hours.

1986 March 21   NASA agreed to a proposal from Canada's Ministry of State for Science and Technology to perform definition studies of a mobile servicing center for the international Space Station. An outgrowth of work on remote manipulator systems, the MSC was conceived as a part of the station where satellites could be serviced using remote manipulators, servicing tools and dextrous devices operated from inside the pressurized modules. The MSC would also be useful in building structures on the exterior of the modules.

1986 March 28   An Ariane 3 launcher vehicle (V17) using the second, ELA-2, launch complex at Kourou, French Guiana, for the first time carried the 2,740-lb. *GTE GSTAR 2* and the 2,514-lb. *Brasilsat 2* satellites into orbit. Each satellite was virtually identical to its respective numerical predecessor. *GSTAR 2* was placed over 105° W to cover the 48 contiguous U.S. states, with Alaska and Hawaii covered by special spot-beams. *Brasilsat 2* separated from the Ariane 3 third stage two min. after *GSTAR 2* and was eventually positioned over 70° W.

1986 March 29   The U.S. National Commission on Space completed its review of the nation's space program and set ambitious goals for the future. In a 200-page report, the commission argued for a new family of space transporters, human settlements on the moon by 2017 and on Mars by 2027. New space transporters should focus, said the commission, on a low-orbit cargo vehicle by the year 2000, a passenger vehicle smaller than the Shuttle for accessing low Earth orbit and a transfer ship for moving people and cargo beyond the moon and around the solar system. Looking further ahead, it forecast for the next century, one million passengers a day in space, extraterrestrial mining and space-based "self-replicating" factories maintained by robots.

1986 March   NASA selected RCA Astro-Electronics to build Mars Observer, the first of the low-cost Planetary Observer spacecraft recommended by the Solar System Exploration Committee. Technology for Mars Observer would rely heavily on RCA's SATCOM and Tiros satellites. With a weight of 5,672 lb., including 2,961 lb. of propellant for Mars orbit insertion and maneuvering around the planet, Mars Observer would carry 343 lb. of science instruments and transmit data at up to 85.3 kilobits/sec. via a high-gain antenna 4 ft. 9 in. in diameter on a 20 ft. boom. The main body of the spacecraft comprised a box 3 ft. 3 in. high, 7 ft. wide and 5 ft. deep. A single solar-cell wing 23 ft. long and 12 ft. wide would provide 1,130 W of electrical power. The propulsion system included 24 monomethyl hydrazine–nitrogen tetroxide bipropellant and hydrazine monopropellant thrusters: 4 × 110 lb.; 4 × 4.95 lb.; 8 × 1 lb.; 8 × 0.2 lb.

1986 April 18   A U.S. Air Force Titan 34D carrying the last KH-9 *Big Bird* reconnaissance satellite was destroyed just 8.5 seconds after launch when the ascending vehicle exploded in a fireball. One of the two 1.4-million-lb. thrust solid rocket motors ruptured at a field joint causing a flame to impinge upon the core stage in a manner similar to that which caused the failure of Shuttle *Challenger*. A review board determined the cause as failure in thermal insulation in one of the booster segments. This latest mishap seriously disabled the U.S. launch vehicle fleet and temporarily prevented the launch of heavy payloads. The Air Force decided to order a further 13 Titan 34D-7 CELV launchers, bringing the order book to 23.

The White House Senior Interagency Group for Space recommended procurement of an additional Shuttle orbiter to replace *Challenger*, which had been destroyed January 28. Estimating that a replacement orbiter would cost $3.2 billion, SIG-Space also backed continued utilization of expendable launchers. It could not decide whether commercial satellite operators should be allowed to continue using the Shuttle or whether they should be forced to switch to expendables. One reason cited for the *Challenger* disaster was the need for the Shuttle to compete with expendable launchers and adhere to a schedule that left no time for fixing problems.

1986 April 21   NASA formally announced 33 possible scientific investigations for the Mars Observer mission scheduled for launch in August 1990. The spacecraft would carry seven scientific instruments for operation from a sun-synchronous, near-polar orbit of 224 mi. so that the instruments could complete a global survey of the surface and atmosphere of Mars every 56 days for 2 years. In a fight to obtain the maximum science return while keeping costs low, NASA managed to get a camera aboard the spacecraft for synoptic imaging of weather systems as well as instruments for a complete survey of the planet's environmental phenomena, which scientists considered a priority.

1986 April 23   At 19:40 UT *Progress 26* was launched by an SL-4 from Baikonur cosmodrome to a rendezvous with the *Mir* space station in an orbit of 208 × 214 mi. A docking was accomplished at the aft port at 21:26 UT on April 26, and cosmonauts Leonid Kizim and Vladimir Solovyov began to unpack the 4,500 lb. of supplies, including 440 lb. of food and 79 gal. of water. In addition, 1,100 lb. of equipment had been uploaded for transfer to the *Salyut 7* space station by the *Mir* crew who, on May 5, would transfer to the old space station for six weeks of work. *Progress 26* remained at the aft *Mir* port, automatically topping up the propellant tanks during June. At 18:25 UT on June 22, *Progress 26* was undocked and de-orbited next day.

1986 May 3   A Delta 3914 carrying the *GOES-G* weather satellite was blown up by the range safety officer at Cape Canaveral, Fla., 1 min. 31 sec. after liftoff and 20 seconds after the main engine stopped, causing the launcher to tumble and the payload shroud to break off. This was the first Delta loss since September 13, 1977; in the interim, NASA had launched 43 of this type. More significant, it was the third U.S. launch vehicle failure in five attempts this year following the Shuttle disaster and the loss of a Titan 34D. A review board attributed the loss of Delta 178 to electrical failure.

1986 May 5   In their *Soyuz-T 15* spacecraft, *Mir* cosmonauts Leonid D. Kizim and Vladimir A. Solovyov undocked from the forward space station port at 12:12 UT and began a series of rendezvous maneuvers to link up with the *Salyut 7* space station. *Mir* was in an orbit of 192 × 214 mi. and *Salyut 7* was in an orbit of 222.5 × 224 mi. To get from one station to the other, the *Soyuz-T 15* cosmonauts performed two intermediate maneuvers to set up the spacecraft for rendezvous. On the day of undocking they first moved to an orbit of 193 × 213 mi. and from there to an orbit of 191 × 212.5 mi. *Soyuz-T 15* reached *Salyut 7 on* May 6, with docking taking place at 16:58 UT. Kizim and Solovyov moved inside and reactivated the station.

1986 May 12   James C. Fletcher was sworn in as NASA administrator, replacing James Beggs, who had resigned on February 25. This was the second tenure for Fletcher, who had first served NASA between April 27, 1971, and May 1, 1977. On leaving NASA the first time, Fletcher had accepted the William K. Whiteford Professorship of Energy Resources and Technology at the University of Pittsburgh, where he remained until rejoining NASA. Deputy Administrator William Graham, who had served as acting administrator in the interim, left NASA at the end of the year. Fletcher stood down on April 8, 1989, and Dale Myers became acting administrator until a new head, astronaut Richard H. Truly, was appointed later that year. Fletcher died on December 22, 1991, at the age of 72.

1986 May 13   NASA released the conclusions of a study into an optional, man-tended approach to space station design. To save costs, an alternate development approach would be first to build a station in orbit that could be periodically visited by astronauts. Only occupied while the Shuttle was docked to the orbiting facility, the station could gradually expand to permanent habitation over 3–5 years. The man-tended design would consist of a single multipurpose laboratory module equipped with a partial environmental control system. The permanent manned facility would add a dedicated habitation module. The man-tended approach was discounted in favor of the permanent manned concept, now expected to become operational by 2002.

1986 May 15   Senior Shuttle managers told a congressional space science committee that the citizen-in-space program had been abandoned indefinitely and that it was very unlikely payload specialists would be flown from organizations and countries launching satellites from the orbiter. NASA wanted confidence in the Shuttle rebuilt before these publicity activities resumed.

1986 May 21   At 08:22 UT an SL-4 launch vehicle carried the 15,600-lb. *Soyuz-TM* spacecraft into orbit from the Baikonur cosmodrome. Launched to a rendezvous with the orbiting *Mir* space station, in orbit at 205.5 × 212.5 mi., the spacecraft was the first in a new generation of Soyuz. In appearance almost identical to the Soyuz-T series, *Soyuz-TM* was 490 lb. heavier than its predecessor. It incorporated a modified descent

module weighing 6,615 lb. and was capable of carrying either two or three cosmonauts. A standard Soyuz-T could return to Earth with 10 lb. of cargo while the TM could carry 330 lb. back to the ground. It incorporated a refined altimeter system for the retro-rocket and new parachutes for the greater weight during descent. *Soyuz-TM* docked to *Mir*'s forward axial port at 10:12 UT on May 23. It remained there until May 29, when, at 09:23 UT, it undocked and returned to Earth, touching down at an elapsed time of 8 days 22 hr. 27 min. Soyuz-TM had been qualified for manned space operations to begin.

1986 May 28   At 05:43 UT *Salyut 7* cosmonauts Leonid Kizim and Vladimir Solovyov began a 3 hr. 50 min. space walk to demonstrate the assembly of large structures in space. Using poles and equipment brought to *Salyut 7* by *Cosmos 1686*, the two cosmonauts unfolded a preformed frame assembly made by the Paton Institute in Kiev, and after successfully demonstrating erection, folded it back down again. The structure had a height of 49 ft. taking the form of a rhombus with each side 16 in. long. Sample packs measuring the impact from micrometeoroids were recovered from the exterior of the space station. A second space walk, lasting 4 hr. 40 min., began at 04:57 UT on May 31, where the rhombus was assembled again to form the base for a tower extending to a height of 39 ft. After it had been folded back up, the cosmonauts welded it closed and left instruments attached for measuring the stresses and strains on it over time.

1986 May 30   The first Ariane 2 (V18) was launched from Kourou, French Guiana, carrying the *Intelsat V F14* international telecommunications satellite. Comprising the Ariane 3 configuration but without strap-on solids, V18 weighed 482,108 lb. at liftoff. In a failure similar to that experienced by V15, the third stage failed to ignite and the vehicle was destroyed by the range safety officer. A board of enquiry recommended a complete overhaul of the HM7 igniter and more than 30 vacuum tests of a modified design were completed before Ariane flew again.

1986 May   A NASA and U.S. Air Force review of the future need for heavy-lift launchers revealed that only the Strategic Defense Initiative would need lifting capacity greater than that of the Shuttle. Gen. James Abrahamson, head of the SDI ballistic missile defense program, believed a heavy-lift launch vehicle (HLLV) capable of putting 150,000 lb. in orbit would be essential for future U.S. space needs. In four space transportation architecture studies, General Dynamics and Martin Marietta reported to NASA while Boeing and Rockwell International reported to the Air Force. In February 1987 the Air Force completed plans for what it now called an Advanced Launch System (ALS) to lift SDI payloads.

1986 June 2   NASA released results from research conducted by the Construction Technology Laboratory, Skokie, Ill., indicating that concrete made with lunar soil is much stronger than concrete made with Earth materials. This is because lunar soil contains no organic substances, considered an impurity in concrete. Considerable interest had been expressed in the potential use of lunar material and a resurgence of interest in lunar bases inspired the research, conducted by Dr. T. D. Linn, principal research engineer at CTL.

1986 June 6   Set up on February 3, the Presidential Commission on the Shuttle *Challenger* mission 51-L accident reported its findings. The 13 members included former astronaut Neil Armstrong, serving astronaut Sally Ride, physicist Richard E.

Feynman and test pilot Charles "Chuck" Yeager. The commission found that the ambient temperature at launch (36°F) prevented a rubber O-ring from sealing the gap between segments of the right hand solid rocket booster because low temperatures stiffen the rubber and make it less reactive to sudden pressure. It noted that the weakness of this joint, and its tendency to leak hot gases, had been kept secret even from astronauts flying as crew members on the Shuttle. On several flights, gases had leaked through, but senior managers had not disclosed the fact. The commission recommended this joint be redesigned to provide a greater safety margin. It also recommended that NASA develop a means of escape from the orbiter in a situation where it was unable to reach the runway during the final stages of descent.

1986 June 19   NASA formally canceled Centaur G-prime as an upper stage for spacecraft carried by the Shuttle. A broad consensus from astronauts and mission safety specialists expressed concern about carrying the cryogenic stage in the Shuttle payload bay. Cancellation made it necessary to seek alternative upper stages for missions such as Galileo, Ulysses and Venus Radar Mapper (now called Magellan) designed to use the liquid propellant Centaur. The U.S. Air Force had planned seven missions on Shuttle/Centaur. Orbital Sciences Corp. proposed their transfer orbit stage as replacement, and Boeing proposed their own solid propellant alternative.

1986 June 25   Cosmonauts Leonid D. Kizim and Vladimir A. Solovyov left the *Salyut 7* space station, having occupied it for 48 days 22 hr., and began a series of rendezvous maneuvers to get back to the *Mir* station. The *Soyuz-T 15* spacecraft undocked from the forward port on *Salyut 7* at 14:58 UT, with the complex in an orbit of 221 × 210 mi., and maneuvered to *Mir,* in an orbit of 206.5 × 227.5 mi. The spacecraft docked to the forward axial port on *Mir* at 19:46 UT on July 26, and Kizim and Solovyov resumed activities aboard the new space station.

1986 June 27   The Council of the European Space Agency approved a French resolution that the Hermes space plane be adopted as a formal ESA program. Member countries would decide individually whether they wanted to participate and how much they wanted to contribute to the financial structure of the program. This was the first time a manned spacecraft had been officially incorporated within the ESA plan, although it was only at the status of a "preparatory program." ESA expected the first Hermes launch in 1995/96.

1986 June 28   The U.S. Air Force issued a request for bids on a medium launch vehicle (MLV) capable of lifting the 2,100-lb. Navstar global positioning system satellite to orbit. The MLV requirement stipulated a lift capacity including apogee kick motor, which would raise total payload to 3,800–4,200 lb. The initial buy would be for 12 launchers, but the Air Force would eventually need up to 20. General Dynamics would put up the Atlas-Centaur, McDonnell Douglas would bid with an improved Delta and Martin Marietta would propose Titan derivatives. Air Force officials quoted an eventual need for 6,000–7,000 lb. dual launch capability, saying this may justify an MLV-2 requirement. Bids were to be in by October 20.

1986 June 30   NASA announced the appointment of Andrew J. Stofan as associate administrator for the Office of Space Station, replacing Philip Culbertson who during this month had presided over the station Systems Requirement Review, which established specifications for the dual-keel design approach. It would comprise a rectangle of truss structures, 310 × 150 ft., bisected at the shorter length by a transverse beam 610 ft. long, to the center of which the pressure modules would be attached. The U.S. habitation and laboratory modules would each have a length of 44 ft. 6 in. × 14 ft. 7 in. and be attached at right-angles to the vertical plane of the dual keel. European and Japanese modules would be attached in the same plane with common access between all modules via nodes 10 ft. 6 in. in diameter and containing six docking ports. The outer sections of the transverse beam would support solar cell arrays, four 84 × 35 ft. "wings" and one solar dynamic generator on each side. The hybrid power system would produce 75 kW of energy. The first of 14 assembly flights was scheduled for January 1993, with the last in November 1995, but delays would alter the sequence and postpone the first station launch into 1997 or 1998.

1986 July 16   *Mir* cosmonauts Leonid Kizim and Vladimir Solovyov returned to Earth at their end of their four-month stay aboard the orbiting space station. With *Mir* in an orbit of 206.5 × 210 mi., *Soyuz-T 15* undocked from the forward axial port at 09:07 UT and returned to Earth at 12:34 UT. The crew had been in space for 125 days 1 min. The events from undocking to touchdown were televised live to a Soviet audience unused to seeing such perilous endeavors as they happened. As part of the new openness, Soviet media was being invited to cover live many events in the space program that scientists, engineers, cosmonauts and politicians believed would garner support for the Soviet space program.

1986 July 17   NASA officials testifying before members of the U.S. House space science and applications subcommittee announced Shuttle flight recovery plans. Assuming the Shuttle returned to flight early in 1988, NASA expected to fly five missions in fiscal year 1988, ten in FY 1989, and 13, 14 and 13 in each of the three succeeding years. Even this proved grossly overoptimistic. The Shuttle did not return to flight before September 29, 1988, respective fiscal year (Oct. 1–Sept. 30) flight rates being 1, 4, 5, 8 and 7.

1986 July 25   A National Security Council meeting in Washington, D.C., agreed to mothball the $3.3 billion Shuttle launch complex (SLC-6) at Vandenberg AFB, Calif. Several design problems with the facility had postponed the first Shuttle flight from the West Coast launch site, and in the aftermath of the *Challenger* disaster the Air Force wanted to consolidate its future planning around expendable launch vehicles. Most significantly affected was the KH-12 series of heavyweight reconnaissance and surveillance satellites designed exclusively around the Shuttle. Each KH-12, weighing almost 40,000 lb., would have been placed in polar orbit by Shuttle and routinely revisited for refueling, recovery of data cassettes and replacement of sensors and cameras as developments allowed. KH-12 was rescaled into a KH-11 successor first launched from Cape Canaveral by Shuttle on August 8, 1989.

NASA Administrator James Fletcher visited the Johnson Space Center to discuss space station issues with managers and astronauts. He was told that, although the dual-keel concept had evolved at the JSC, there was increasing concern that the configuration was not the best and that the agency should examine alternates. Astronauts voiced concerns about the amount of EVA (extravehicular activity) work they would be required to carry out during assembly and routine servicing. NASA wanted to move from Phase B definition studies to

Phase C/D detailed design and fabrication beginning October 15, but doubts expressed about the dual-keel design would prevent that.

1986 July 28 A new series of Soviet oceanographic satellites was launched by SL-14 from Plesetsk to a 392 × 411 mi., 82.5° orbit. Given the designation *Cosmos 1766*, the 3,530-lb. satellite was the first in a precursor series to the Okean ocean satellites. The Soviets launched two more: *Cosmos 1812* on January 14, 1987, and *Cosmos* 1825, on March 3, 1987. All three remain in space. The first formally named satellite in the Okean series was launched on July 5, 1988.

1986 July 31 As a result of complaints from Congress that NASA had not consulted them in advance of its June 30 announcement about a space station configuration, Administrator James Fletcher ordered a thorough review of space station planning and configuration studies. On August 20 a Space Station Configuration Critical Evaluation Task Force was set up at NASA's Langley Research Center. It was tasked with reexamining all aspects of station architecture, in particular the amount of EVA (extravehicular activity) required during assembly and servicing work. NASA had set an upper limit of 285 man-days of EVA per year, but this would commit two astronauts to spending more than one-third of their lives working outside on assembly, repair and servicing. Many thought this was far too high and that, in any case, the dual-keel configuration would require even more EVA time.

1986 August 5 NASA announced results from the Earth radiation budget experiment which indicated the "greenhouse" effect in the Earth's atmosphere may not be as severe as some scientists had thought. Measurements of the amount of solar heat absorbed by Earth versus the amount of heat reflected back into space revealed that the Earth would be 20°F warmer without clouds. Some scientists had speculated that clouds trap heat and raise the temperature of the lower atmosphere. Launched in October 1984, the Earth Radiation Budget Satellite showed this not to be the case. While a doubling of carbon dioxide would raise Earth's temperature by 4–8°F, the increase in cloud cover caused by higher levels of evaporation might offset this effect. After further measurements, on February 6, 1989, NASA confirmed these findings.

1986 August 12 The first flight test of Japan's H-I launcher took place from Tanegashima. Using the same first stage as the N-II, a new and improved cryogenic second stage employed the Japanese LE-5 engine burning liquid oxygen and liquid hydrogen and producing 23,000 lb. of thrust. A new and bigger solid propellant third stage had also been designed but would not fly until the second H-I flight on August 27, 1987. The H-I was capable of placing 1,200 lb. in geosynchronous transfer orbit. On board the first H-I were two satellites: the 1,510-lb. *Experimental Geodetic Payload (EGP)*, placed in a 919 × 930 mi. orbit; and the first 110-lb. Japanese amateur radio satellite *(JAS-1)*, also designated *Oscar 12*, placed in a 919.5 × 930 mi. orbit. Both satellite orbits were inclined 50°.

1986 August 15 Just one week after President Ronald Reagan formally authorized procurement of a replacement orbiter for the *Challenger*, lost in a destructive explosion January 28, the White House barred the Shuttle from competing for satellite launch traffic. In the clearest and most positive move in support of the expendable launcher industry, Reagan ordered NASA to honor all existing launch commitments but to concentrate primarily on using the Shuttle for national security and civil government space services. In a stroke, and as a consequence of the *Challenger* disaster, the Shuttle was stripped of the role on which it had been sold to Congress during the 1970s.

1986 August 17 The docked assembly of *Salyut 7* and the massive *Cosmos 1686* spacecraft, which was attached to the aft docking port, began a series of maneuvers to place it in a high orbit where it was to be left unattended. From an orbit of 205.5 × 210 mi., the propulsion unit on *Cosmos 1686* moved the assembly to a circular orbit of 295 mi. by August 23. No further activities were conducted, although telemetry from the abandoned complex continued to provide data about the gradual collapse of on-board systems, allowing engineers to build a plot of its deterioration that helped with the design of future long-duration stations. The complex burned up in the atmosphere on February 7, 1991.

1986 August Martin Marietta announced plans to commercially market Titan 3 and Titan 4 heavyweight payload launchers to compete with Europe's Ariane 4 series. Titan 3 was essentially the Titan 34D without an IUS (inertial upper stage) third stage. With a weight of 1.5 million lb., a height of 150 ft. and a liftoff thrust of 2.8 million lb. from the two 5.5-segment solids, Titan 3 could lift 31,000 lb. to low Earth orbit or place 3,000–5,000 lb. in geosynchronous transfer orbit with optional upper stages. Titan 4 was essentially the Titan 34D-7, with a launch weight of 1.9 million lb. and a total thrust of 3.2 million lb. from two seven-segment solid boosters. Third stages comprised either a cryogenic Centaur stage or a solid propellant IUS, giving the launcher a maximum payload capability of 10,000 lb. to geostationary orbit.

Faced with having to select a new way of getting the *Galileo* orbiter/probe to Jupiter in the wake of Shuttle problems, Roger Diehl, head of the Jet Propulsion Laboratory's mission design team, came up with a radical solution. Launched in the October 12–November 24, 1989, window by a Shuttle-IUS combination, *Galileo* would fly past Venus at a height of 9,000 mi. on February 9, 1990, return for a 600-mi. flyby of Earth on December 8, 1990, and perform a wide loop for a second flyby of Earth on December 8, 1992, just 185 mi. above the surface. This last slingshot would throw *Galileo* to Jupiter where it would arrive on December 7, 1995, the probe to enter the atmosphere, leaving the main spacecraft to go into orbit for 22 months of observations. Before that, *Galileo* would pass within 600 mi. of the asteroid Gaspra on October 29, 1991, and within 600 mi. of asteroid Ida on August 28, 1993.

1986 September 5 A U.S. Air Force Delta 3920 was launched from Cape Canaveral, Fla., at 16:08 UT for a test in the Department of Defense Strategic Defense Initiative (SDI) program. The flight was significant in restoring faith in U.S. expendable launchers after failures earlier this year. The SDI test involved the Delta second stage and satellite entering a 137 x 138 mi., 28.5° orbit. At 45 min. mission elapsed time the satellite separated, and as it drifted away, sensors on the second stage identified it as a simulated Soviet reentry warhead. The Delta engine was fired again to achieve a 120-mi. separation with the satellite on a crisscrossing path that periodically brought the two closer together. This allowed the sensors to characterize the simulated warhead against various backgrounds. At 92 min. a combination Aries/Minuteman second stage was fired from Wallops Island, Va., as the orbiting vehicles passed over. The sensors successfully detected and locked on to the hot exhaust plumes. Later, with the Delta second stage holding a stable course, the SDI satellite maneuvered to a head-on collision with the stage, impacting it with an accuracy of 6–12 in. The SDI satellite decayed on September 28, followed by the Delta stage on November 25.

1986 September 18 NASA's Space Station Critical Evaluation Task Force completed its examination of the dual-keel configuration and recommended a change in the design of some station elements and in the build-up sequence. Connecting nodes were increased in size to the diameter of the U.S. pressure modules and outfitted with subsystems required for station management and control that previously were put on the exterior or in the pressure modules. The design also featured a considerably rescoped assembly sequence, less EVA (extravehicular activity) and early operational capability. The assembly sequence put emphasis on the transverse beam with attached pressure modules, nodes and solar power system. Only in the last 2 of 17 assembly flights would the dual-keel and horizontal beam units be built.

1986 September 30 RCA Astro-Electronics Division announced it had been awarded initial funding from the U.S. Department of Defense for development of a third variant of the Block 5D series military weather satellite. Designated 5D-3, the new series would extend operating life from four to five years, carry a greater number of sensors than its predecessor, generate higher power levels and carry improved attitude control and pointing systems. An onboard computer processor would handle the flight control system. The first 5D-3s were scheduled for launch in the late 1990s.

1986 October 4–11 At the 37th Congress of the International Astronautical Federation, Innsbruck, Austria, Hughes Aircraft and the Boeing Co. proposed a low-cost, heavy-lift launcher based on existing technology. Named after Gregory Jarvis, killed when *Challenger* blew up on January 28, the Jarvis launcher concept took two F-1 (Saturn V) engines for the first stage, one J-2 (Saturn V) engine for the second stage and a bipropellant transfer stage from the Intelsat VI program. The two-stage Jarvis would have a height of 209 ft., a liftoff thrust of 3 million lb. and place 85,000 lb. in low Earth orbit. It was proposed as a means of giving the United States a launch capability in excess of the Shuttle at less cost.

1986 October 9 In the first major move of Shuttle hardware since the *Challenger* disaster on January 28, orbiter *Atlantis,* along with its external tank and solid rocket boosters was rolled out to Launch Complex 39B at the Kennedy Space Center, Fla. While it was there engineers would conduct fit-checks of connections and umbilicals. It was returned to the Vehicle Assembly Building on November 22 after the tests had been completed.

1986 October 10 NASA notified Transpace Carriers that it was terminating negotiations about privatizing the Delta launch vehicle and opening discussions with Delta manufacturer McDonnell Douglas. Since the *Challenger* disaster and the decision to offload commercial satellites from the Shuttle and place them on expendable launch vehicles, manufacturers took a renewed interest in taking over launch operations for profit. Before the end of the year, American Satellite Co., Inmarsat and Comsat Corp., had signed contracts with McDonnell Douglas for Delta launches.

1986 October 15 French aerospace manufacturers Aérospatiale and Dassault-Breguet made formal presentations to the French national space agency, Centre National d'Etudes Spatiales (CNES), on the Hermes space plane. Of particular significance was the addition of measures aimed at crew safety. Designed to achieve orbit on the propulsion of the Ariane 5, Hermes now incorporated two 4,500-lb. thrust orbit-insertion engines and a quartet of solid propellant escape motors producing a total thrust of 352,800 lb. The escape motors would wrench Hermes free from a malfunctioning booster, providing a means by which the crew could return to Earth.

1986 October 17 NASA announced that *Nimbus 7* had produced color images enabling scientists, for the first time, to produce a complete map of the distribution of phytoplankton in the surface waters of the entire north Atlantic Ocean. Phytoplankton are microscopic plant life. *Nimbus 7*, which operated from launch on October 24, 1978, until summer 1986, produced images of coastal land regions, allowing direct correlation between abundance and distribution of microscopic life on land and water. This data enabled scientists to convert raw measurements into calculations of the amount of photosynthetic activity, and from that to define the effects of man-made changes.

1986 October 21 NASA Administrator James Fletcher appointed Thomas L. Moser to run the Space Station Program Office. Reporting to Associate Administrator Andrew Stofan, the SSPO had been recommended in a report prepared by USAF Gen. Samuel E. Phillips (Ret.). The program office would be responsible for overall technical direction and content of the space station program, including systems engineering and analysis, design of the configuration and the integration of the various elements.

1986 October 30 A meeting between officials of NASA and the European Space Agency concerning allocation of resources and responsibilities on the joint space station ended with unresolved obstacles. The ESA wanted greater autonomy and sought financial estimates of development and operations costs the Americans were unable to give. Moreover, the Europeans felt NASA was unwilling to accept them as equal partners. Space station talks with Japan began October 1. ESA Director General Reimar Lust flew to Washington on November 21 to clarify the position with the NASA Administrator James Fletcher prior to a second session with the Europeans on December 8–9 to resolve differences. NASA met with the Canadians starting December 10.

1986 October NASA selected the payload for its Comet Rendezvous/Asteroid Flyby (CRAF) mission planned for launch by Titan 4/Centaur-G in February 1993 to the comet Tempel 2. In the mission plan, a flyby of Venus in August 1993 would put the Mariner Mark II spacecraft on course for a flyby of Earth in June 1994, picking up energy at each encounter for a flyby of the asteroid 46 Hestia in January 1995. The rendezvous with Tempel 2 would occur in November 1996, close to the comet's aphelion, with the spacecraft almost as far out as the orbit of Jupiter. CRAF would accompany the comet as it moved in toward the sun, coming inside the orbit of Mars at perihelion during September 1999, with the mission ending three months later. To accomplish this mission the spacecraft would be required to make large orbital maneuvers totaling 13,100 ft./sec. NASA had hoped to ask for funds to start CRAF development in fiscal year 1989, but deferred the mission, causing a change of plan in March 1988.

1986 November 6 NASA's Space Station Critical Evaluation Task Force presented the conclusions of a review into space station assembly using expendable launch vehicles. It found that only the Titan 4 would meet all the guidelines and ground rules. Candidate transport systems included the Titan 34D, Ariane 5, Japan's H-II and the proposed Boeing-Hughes Jarvis booster. Expendable launchers would add to the complexity of station assembly and require early use of an orbital maneuvering vehicle to move station elements around. From this NASA was able to decide against the use of expendable launch vehicles.

1986 November 14 A solid propellant Scout launcher carried a museum relic into space designated *Polar Bear*. An acronym for Polar beacon experiments and auroral research satellite, *Polar Bear* was an ex-Navy transit navigation satellite that had been donated to the U.S. National Air and Space Museum in 1976 when it was deemed redundant to needs. It was removed in 1984 and, retaining its original solar panels and electronics which had never been removed, refitted with three experiments for sampling electrical particles in the Earth's polar magnetic regions. The 276-lb. satellite was placed in a 596.5 × 631 mi., 89.6° orbit where it remains.

1986 November 26 NASA announced it had selected the solid propellant inertial upper stage to replace the canceled cryogenic Centaur for its Galileo, Ulysses and Magellan planetary missions. Because the performance of IUS was less than that of the Centaur, mission profile changes were necessary. The original plan to launch Galileo and Ulysses to Jupiter in June 1987 had to be canceled and the mission trajectories redefined. Under an existing agreement, the solid propellant transfer orbit stage would be used to launch Mars Observer from a Titan.

1986 December 1 NASA announced the start of a major project using satellites to monitor environmental conditions which trigger the breeding of malaria-carrying mosquitoes. Malaria was the leading cause of disease in tropical countries, where 3 billion people lived, causing 100 million new cases and 1.5 million deaths each year. Known as the Biospheric Monitoring and Disease Prediction (DI-MOD) project, the survey would map habitats producing dense anopheline mosquitoes two months before adult mosquitoes emerge, a key factor in destroying them before maturation.

1986 December 2 The 3M Co. signed a 10-year agreement with NASA for 62 flight experiments on the Shuttle in a joint endeavor aimed at researching a wide range of commercial possibilities. 3M signed up to two flights per year in which middeck experiments would be carried aboard scheduled Shuttle flights. For its part, NASA would fly two experiments in the cargo bay per year for the first three years, followed by six in the cargo bay annually in years four through nine. Cargo bay experiments would be carried in 500-lb. getaway special canisters. 3M wanted the flights for research in organic and polymer science.

1986 December 5 The draft outline of a new U.S. policy for the military use of space was completed by the Department of Defense. Generated by a concern at an alleged increase in military space activity with the Soviet *Salyut 7* space station, the draft policy recommended a new generation of heavy-lift launch vehicles, specifically citing the proposed Hughes/ Boeing Jarvis. Recognizing that it had never managed to organize a coherent and lasting policy on manned activity, the draft acknowledged that military space activity was well advanced along an evolutionary path from research to operations. It called for a greater integration between military space activities and existing duties at sea, on land and in the air.

1986 December 11 American Rocket Co. tested the full-scale prototype of a hybrid solid/liquid rocket motor in a 20-sec. firing at the U.S. Air Force Rocket Propulsion Laboratory, Edwards AFB, Calif. The motor generated approximately 32,000 lb. of thrust. Loaded with 4,500 lb. of solid propellant, it had added capability from injected oxidizer, providing a specific impulse of more than 300 sec. The engine was part of an industrial launch vehicle which AmRoc wanted to build and fly in 1988. The project was eventually abandoned due to lack of funding.

1986 December 15 A NASA launch vehicle payload task force completed a mixed-fleet analysis of satellites and spacecraft which would need a launch vehicle between 1988 and 1995. Prior to the loss of *Challenger* on January 28, NASA had sought to shift all payloads from expendable launch vehicles (ELVs) to the manned Shuttle. The task force identified 135 payloads for the period 1988–1995, of which the Shuttle could launch only 85–100. The analysts recommended that NASA use up to 40 ELVs to accommodate the manifest and predicted the Department of Defense would need five Shuttle flights per year.

1986 December 22 The first 10 MGM-118 Peacekeeper ICBMs became operational with the 90th Strategic Missile Wing at Warren AFB, Wyo. Deployed to refurbished Minuteman III silos, Peacekeeper carried 10 Mk.21 MIRV warheads, each with a 475-KT W-78 weapon. All 50 Peacekeeper missiles had been deployed in redundant Minuteman III silos by mid-1989. Also this month, President Reagan decided to ask Congress for approval to deploy an additional 50 Peacekeepers on 25 trains in the rail garrison mode. The trains would remain in garrison on secure U.S. Air Force bases until a time of crisis, when they would be dispersed over existing railroad track.

1986 December 23 The U.S. Department of Defense authorized full-scale development of the Small ICBM (SICBM), or Midgetman. With a preliminary program plan that envisaged operational deployment in 1992, the three-stage solid propellant SICBM was expected to weigh 37,000 lb. and deliver a single 500-KT W-91 warhead in an Mk.21 reentry vehicle across a maximum range of 8,000 mi. Long-term plans envisaged a force of 500 Midgetman missiles deployed in hardened mobile launchers which Boeing had been selected to build. Each HML transporter/launcher would have a length of 100 ft., with power to 14 all-terrain steel radial tire wheels provided by two 750-HP diesel engines.

1986 The classified U.S. Department of Defense single-stage-to-orbit program, code-named Science Realm, lost favor to the winged, air-breathing/hydrogen-fueled National Aero-Space Plane (NASP) authorized by President Ronald Reagan as a Department of Defense/NASA research program. Under Science Realm, Boeing, Lockheed and McDonnell Douglas each produced their own designs of space launchers with horizontal takeoff and landing, a feature that compromised the advantages of high thrust-weight ratios of rocket engines. After NASP gained favor, rocket concepts continued in a top secret program under the NASP office, and Science Realm became Have Dawn.

1987 January 1 The European Meteorological Satellite organization, EUMETSAT, accepted responsibility for the Meteosat weather satellites. Europe was preparing to shift from experimental to operational meteorological satellites with the flight of *MOP-1 (Meteosat Operational Program 1)*. Built by Aérospatiale, each Meteosat satellite would have a diameter of 6 ft. 11 in., a height of 10 ft. 2 in., a launch weight of 1,500 lb. and an on-orbit mass of 697 lb. *MOP-1* would carry an imaging radiometer providing visible water-vapor and thermal data, as well as collecting and distributing information from data collection platforms.

1987 January 5 U.S. Secretary of Defense Caspar Weinberger requested additional funds for the fiscal year 1987 defense budget that included money for development of a heavy-lift launch vehicle (HLLV) capable of placing 100,000–150,000 lb. payloads into Earth orbit. The Department of Defense wanted the heavyweight launcher in support of operational deployment of satellites and space vehicles for the SDI ballistic missile defense program. It believed a start was needed now to have the new launcher ready by the mid-1990s when, it was envisaged, laser and particle-beam defenses would be deployed in space.

1987 January 15 The first flight test of a Trident II D5 SLBM took place from a land pad at Cape Canaveral and was completed successfully. A development of the Trident I but considerably bigger and heavier, the Trident II had a length of 44 ft. 7 in., a diameter of 6 ft. 10 in. and a weight of 133,000 lb. It had about the same range as the Trident I but with considerably increased weight-throwing capability. Each missile had capacity for up to 15 (300-KT) W87 Mk.5 warheads, originally designed for the MX/Peacekeeper ICBM. The first underwater test launch on March 21, 1989, was a failure, as was the second on August 15, traced to the first-stage nozzle assembly. The first successful underwater launch took place on December 4, 1989; the D5 became operational in March 1990.

1987 January 16 At 06:06 UT the unmanned *Progress 27* cargo tanker was launched by SL-4 from Baikonur cosmodrome. A docking at the aft axial port on the *Mir* space station was achieved at 07:27 UT on January 8. Manned operations with *Mir* had been put off for many months by delays to a family of large laboratory modules that would provide additional work areas for the *Mir* crew. Launched over several years, the laboratories would be located at opposed radial docking ports on the spherical docking module at the front of the space station. To maximize crew work load, Soviet mission planners scheduled crew visits so that *Mir* would be permanently occupied. *Progress 27* supported the start of this activity and remained docked until March 26, when it undocked at 05:07 UT and was de-orbited two days later.

1987 January 21 The U.S. Air Force awarded McDonnell Douglas a contract for production of 20 Delta launch vehicles for the Navstar global navigation satellite program. Responding to a request for a medium launch vehicle, McDonnell Douglas had been selected over competitive bidder General Dynamics for a derivative of the Atlas. Known as Delta II, the launcher would use solid propellant strap-ons made by Hercules rather than Thiokol, which had supplied solids for the Delta to date. Geosynchronous transfer orbit capability

would grow from 3,190 lb. (Delta 3920) to 3,560 lb. (Delta 7920) and then to 4,010 lb. (Enhanced Delta 2) with the introduction of stretched strap-ons. All three would feature a 4 ft. 8 in. stretch in the second stage.

1987 January 28 The Senior Interagency Group-Space reviewed the use of NASA's Space Station and the allocation of research time in the U.S. research module. It concluded that U.S. Space Station hardware should be reserved for dedicated American use and that the Department of Defense should have priority over other U.S. experimenters should a national need arise. Japanese and Canadian partners were troubled by the implication that they were not full participants and threatened to pull out.

1987 January 29 Rep. Bill Nelson, chairman of the U.S. House of Representatives' Space Science and Applications Subcommittee, directed NASA to maintain an August 1990 launch date for its Mars Observer mission despite space agency plans to delay it to the next launch window in 1992. Concerned about pressure to launch several planetary missions when the Shuttle was returned to flight status, NASA Administrator James Fletcher pressed hard for agreement to delay the Mars Observer mission. Some in Congress felt NASA should transfer the mission to a Titan 3 launch and sought funds for that to happen. Nevertheless, NASA stuck to its rescheduled September 1992 launch date for Mars Observer, although it did eventually move the spacecraft from Shuttle to Titan 3.

1987 January The Mars Study Team, formed out of NASA's Solar System Exploration Division, completed an initial report on a proposed Mars Rover–Sample Return (MRSR) mission. Envisaging two spacecraft on separate launchers, the study projected a mission scenario where the Shuttle would launch the 6,600-lb. rover mission to an aerobraking Mars orbit capture from which the rover would descend to the surface. Weighing three times as much as the rover, the sample return spacecraft would extract surface materials, lift off and rendezvous with a booster stage left in orbit and return to Earth after a round trip of nearly three years. Over the next year redefinition of the design decreased the weight of the sample return spacecraft.

The Strategic Defense Initiative, the U.S. Department of Defense antiballistic missile research effort, put out requests for bids from industry on a space-based kinetic-kill vehicle (SBKKV). The SDI office wanted a device that could be deployed to space by the late 1990s to work in conjunction with target acquisition and tracking radars and achieve destruction of enemy missiles or warheads through high-velocity impact. SDI's plan was to park 5,000–10,000 SBKKVs in garages of 6–24 in low Earth orbit so that they could be directed to attack and destroy missiles and reentry vehicles.

1987 February 1 An SL-11 launched from Plesetsk carried a new generation of Soviet ocean reconnaissance satellite to a 488 × 497 mi., 65° orbit. With a weight of 11,000 lb., the *Cosmos 1818* satellite was the first to carry an advanced nuclear reactor for space called Topaz. It provided electrical power for the powerful synthetic aperture ocean surveillance radar from an orbit more than 300 mi. higher than previous satellites. A second satellite carrying a Topaz test reactor, *Cosmos 1867*, was launched by SL-11 on July 10. The solar maximum mission satellite had been used for several years to monitor radiation emissions from Soviet nuclear-powered satellites, which had been added reason to employ a Shuttle mission on SMM repair in 1984. SMM recorded high levels of gamma radiation from *Cosmos 1818* and *1867* prompting changes to NASA's Gamma Ray Observatory to minimize disruption to its instruments.

1987 February 5 At 21:38 UT cosmonauts Yuri V. Romanenko and Aleksandr I. Laveikin were launched in their *Soyuz-TM 2* spacecraft to the *Mir* space station, in an orbit of 204 × 225.5 mi. Adopting a more fuel-efficient maneuver sequence which took a day longer than previous rendezvous trajectories, *Soyuz-TM 2* docked with the forward axial port on *Mir* at 22:28 UT on February 7, ending 7½ months of inactivity at the station. Romanenko and Laveikin began activating the *Mir* complex and unpacking cargo from *Progress 26*, docked to the aft port.

A Japanese Mu-3S II solid propellant launch vehicle carried the 926-lb. *Astro-C* satellite, called *Ginga*, from Kagoshima to a 321 × 440 mi. orbit inclined 31.1°. Measuring 4 ft. 11 in. × 3 ft. 3 in., *Astro-C* carried a low-energy x-ray telescope developed at the University of Leicester, England, and a gamma-ray detector built by the Los Alamos National Research Laboratory, N.Mex. With a design life of four years, the satellite was to study distant x-ray sources associated with black holes and pulsars. The satellite decayed on November 1, 1991.

1987 February 12 NASA met with the Office of Management and Budget to discuss the latest cost estimates on the international space station. In 1983 the space agency had told Congress the station would cost $8 billion in fiscal year 1984 prices to build. That had now increased to $15.1 billion in 1984 prices ($27.4 billion in 1987 prices), although the NASA comptroller said it could reach $16.6 billion ($30 billion). Moreover, estimates of a "bare-bones" station reduced in size and scope from the modified dual-keel baseline configuration was still high, at $12 billion.

1987 February 18 The first Earth observation satellite launched by Japan was carried from Tanegashima by an N-II to a circular, 565 mi. orbit inclined 99.9°. *Marine Observation Satellite-1 (MOS-1)* had a box-shaped structure, 4 ft. 3 in. × 4 ft. 7 in. × 7 ft. 10 in., weighing 1,645 lb. Developed by NASDA (National Space Development Agency) and Nippon Electric, *MOS-1* was a technology development project with a semioperational role, supporting a multispectral scanning radiometer for sea color data with 165-ft. resolution in four bands, visible and infrared radiometers for sea surface and temperature data, 0.6-mi. and 1.7-mi. resolution respectively, and a microwave scanning radiometer for atmospheric water vapor measurement with a resolution of 20 mi. *MOS-1* remains in orbit.

1987 February 26 Congressional opponents of the Strategic Defense Initiative antiballistic missile research program claimed at a House of Representatives foreign affairs subcommittee hearing that the Reagan administration's interpretation of the 1972 antiballistic missile treaty was "legally convoluted." The Department of Defense had wanted to conduct tests in space of kinetic-kill (impact) devices capable of knocking out missiles and warheads. The next day congresspersons were lobbied by SDI opponents in what was the start of a vigorous campaign to discredit claims that exotic weapons could provide a defense against Soviet missiles. In March, Sen. Sam Nunn, chairman of the Senate Armed Services Committee, said the tests would be illegal.

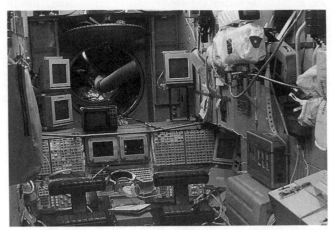

During 1987, Soviet space station activity accelerated with the Mir *facility in orbit. Seen here are control panels and crew restraint seats looking forward.*

1987 March 3 At 11:14 UT the unmanned *Progress 27* cargo tanker was launched by an SL-4 from the Baikonur cosmodrome. Docking at the aft port on the *Mir* space station took place at 11:43 UT, and cosmonauts Yuri Romanenko and Alexandr Laveikin set about unloading stores and equipment. A special crystal growth device called Korund was installed aboard *Mir*, as was a KATE-140 Earth resources camera. After a relatively brief stay at the station, *Progress 27* undocked at 05:07 UT on March 26 and was de-orbited two days later.

1987 March 6 The Space Station Office at NASA's Langley Research Center reported to Headquarters the results of a study it had performed on the use of the permanently manned facility in support of a manned mission to Mars sometime in the 21st century. NASA wanted to allow the station to expand so that it could readily support deep-space manned flights to the moon and the planets. The study also identified the heavy-lift launcher that would be required to support the orbital assembly of a Mars expedition. NASA wanted to develop such a launcher using Shuttle elements, a so-called Shuttle-derived vehicle (SDV).

1987 March 13 NASA's *Voyager 2* spacecraft conducted a course adjustment maneuver when it fired four 0.2-lb. hydrazine thrusters for 1 hr. 10 min. 27 sec. beginning at 13:06:28 UT. *Voyager 2* was heading for a close encounter with the then outermost planet, Neptune, on August 24, 1989, and this 30 ft./sec. velocity change moved up by 12 hr. the time of closest approach, putting the spacecraft on course for a 3,100-mi. flyby of the planet and a 25,000-mi. flyby of Neptune's moon Triton. Because long-time exposures would be necessary at Neptune, where the planet receives only a fraction of the light received by Jupiter and Saturn, for which the TV cameras had been designed, the spacecraft would have to be held in a very stable attitude. To test new attitude control procedures, the Jet Propulsion Laboratory conducted tests with *Voyager 1* to demonstrate that they would work on *Voyager 2*.

1987 March 14 The *Pioneer-Venus 1* orbiter began to track the newly discovered comet Wilson from its orbit around

Venus. Using its ultraviolet spectrometer to observe the rate of water evaporation from the comet's nucleus, the spacecraft continued to send data to Earth until March 21 and then conducted a second survey from March 31 until April 20, 1987. *Pioneer-Venus 1* made a further set of observations when the short-period comet once again returned to the inner solar system between June and July 1987. The spacecraft resumed its observations of Venus and on May 20, 1988, celebrated 10 years of operations since launch. At that date it had 5 lb. of attitude control gas remaining, sufficient to last five more years. The spacecraft burned up in the Venusian atmosphere on October 8, 1992, after more than 5,000 orbits.

1987 March 18 Secretary of the Air Force Edward C. Aldridge described to the House Committee on Appropriations subcommittee on defense, a recovery strategy for military satellites following the spate of launcher failures during 1986. Navstar would be offloaded from the Shuttle and moved to Delta 2 (MLV-1), and military weather satellites would also be moved from the Shuttle to refurbished Titan II ICBMs. Some of the heaviest military payloads envisaged for the Shuttle would be moved to Titan 4.

The Department of Defense announced that the Strategic Defense Initiative antiballistic missile research organization had awarded contracts to Grumman and Lockheed Space and Missile Co. for competing design and development of a boost surveillance and tracking system. The BSTS was conceived as a space-based experiment to demonstrate technologies capable of improving existing early-warning systems. It was expected to incorporate sensitive tracking and signature sensors capable of identifying ballistic missiles as they ascended through the atmosphere and discriminating them from the Earth's background.

1987 March 19 Officials from the U.S. National Security Council, the Office of Science and Technology Policy and the Office of Management and Budget agreed to plans for a space station program structure that allowed NASA to proceed with development. To counter escalating costs, NASA would break development in two. Phase One would comprise a single-keel configuration retaining the transverse beam with pressure modules and electrical power supply but without the dual-keel structure. The Phase One station would cost $10.9 billion for development and $1.3 billion for technical support, a total of $12.2 billion in 1984 prices versus $15.1 billion for the full dual-keel configuration and be assembled between 1994 and 1996. At $3–4 billion in 1984 prices, Phase 2 would add upper and lower booms and the solar dynamic power system by late 1998. President Ronald Reagan approved this plan on April 3.

1987 March 20 A Delta 3920/PAM carried the *Palapa B-2P* communication satellite from Cape Canaveral, Fla., to a transfer ellipse from which it was positioned in geostationary orbit at 113° W. This Delta was the last built before production halted in early 1984 in expectation that satellites would shift to the Shuttle. A replacement for the *Palapa B-2* stranded in low orbit following release from the Shuttle in February 1984, *B-2P* was of the same design as its predecessor, carrying telecommunications traffic on the 6/4 GHz band for Indonesia and the Association of Southeast Asian Nations. Recovered on Shuttle mission 51-A in November 1984, *Palapa B-2* was relaunched for Indonesia as *Palapa B-2R* by Delta on April 13, 1990, to 107.7° E. Fourth in the series, *Palapa B-4* was launched by Delta-2 on May 14, 1992, and positioned over 118° E.

1987 March 24 The first attempt by the Indian Space Research Organisation to place a satellite in orbit with the four-stage augmented satellite launch vehicle (ASLV) failed when the second stage malfunctioned. The ASLV had a total height of 77 ft. and a launch weight of 77,200 lb. The first stage had a thrust of 113,175 lb., but this was ignited at 49 sec. into flight after two solid propellant strap-on boosters had fired for 46 sec., producing a total thrust of 200,000 lb., and been jettisoned. After a further 50 sec., the 49,000-lb. second stage was to have burned for 36 sec. The 14,400-lb. thrust third stage was designed to fire for a further 44 sec., followed by the 5,850-lb. thrust fourth stage for 32 sec. This first ASLV carried a 330-lb. scientific satellite.

In the largest exercise of its kind, six Pershing 2 tactical battlefield missiles were fired within a three-hour period at Cape Canaveral, Fla. The missiles carried dummy warheads on programmed trajectories that lobbed the nose cones between 800 mi. and 1,200 mi. downrange into the Atlantic Ocean. The firings were conducted by U.S. Army field artillery troops based in West Germany.

1987 March 26 Atlas-Centaur AC-67 was destroyed by the range safety officer 71 sec. after liftoff from Cape Canaveral, Fla., carrying the 5,070-lb. *FltSatCom 6* military communications satellite. This was the fifth U.S. launch failure in 19 months; the failures involved every type of manned and unmanned launch vehicle. Launched into a gathering thunderstorm, several electrical strikes triggered an erroneous yaw command at 48.3 sec. which steered the Atlas hard-over into a dive, causing vehicle breakup at 50.7 sec. The review board determined that the U.S. Air Force, which provided weather data to NASA, made a poor judgment in not suggesting a halt to the launch; but it put the blame on NASA for not making a subjective decision to question the data.

1987 March 31 At 00:06 UT a 45,425-lb. unmanned observatory and logistics module destined for the space station *Mir* was launched by SL-13 from Baikonur cosmodrome. Called *Kvant*, the main experiment module weighed 24,250 lb. and comprised a cylinder 19 ft. long and 13 ft. 7 in. in diameter, with a truncated cone and docking probe at one end and apertures for a wide range of astronomical instruments at the other end. *Kvant* carried 3,300 lb. of astrophysical instruments from West Germany, the Netherlands, Britain and the European Space Agency. It also carried 5,500 lb. of cargo for the station. The logistics/propulsion module was attached to a fixture on the instruments end and comprised a stripped version of the heavy modules like *Cosmos 1443* docked to *Salyut 7*. It carried propellant and served as a space tug for maneuvering *Kvant* to a docking with *Mir*. First attempted on April 5, the operation was not a success, and a second attempt on April 9 only resulted in a partial hook-up. *Mir* cosmonauts Yuri Romanenko and Alexandr Laveikin went about their chores while ground controllers wrestled with the problem.

NASA briefed potential contractors at the Jet Propulsion Laboratory, Calif., on a proposed Mars rover/sample-return mission tentatively scheduled for launch in 1998 with samples returned to Earth in 2001. An accelerated schedule with priority funding proposed launching rover and sample-return spacecraft in 1996 on separate Titan 4/Centaur G-prime launchers. Preliminary rover designs were now based around a 1,700-lb. vehicle spending 9–18 months roaming Mars. A rover technology workshop was scheduled for April 28–30, followed by requests for industry proposals issued May 1.

1987 April 1 Addressing the U.S. Senate Subcommittee on Strategic Forces and Nuclear Deterrence, Gen. James Abrahamson, head of the Strategic Defense Initiative (SDI) ballistic missile defenses research program, described progress in directed energy weapons. Explaining how free-electron lasers could be based on the ground and directed to targets on the other side of the world by mirrors in space, he referred to a unique energy storage system capable of delivering several hundred billion watts for the fraction of a second such a laser would need to be effective against missiles.

1987 April 3   NASA established a launch schedule for the *Galileo* and *Ulysses* spacecraft, both of which would fly to Jupiter. The NASA *Galileo* would be launched by Shuttle in October or November 1989 and the European Space Agency's *Ulysses* mission would be launched by Shuttle in October 1990. The decision was based on the longer trip time of *Galileo*, which made it necessary to launch as early as possible, and on the lack of confidence in being able to launch two spacecraft on the same Shuttle within a few weeks of each other. *Galileo* would get to Jupiter in 1995, one year after *Ulysses* got its gravity slingshot back across the sun.

1987 April 7   McDonnell Douglas announced orders for nine commercial Delta II launchers. Based on its winning bid for the U.S. Air Force medium launch vehicle, the Delta II was being marketed by McDonnell Douglas to satellite customers transferred off the Shuttle. Eight organizations had ordered the Delta II: American Satellite Corp. (formerly American Satellite Co.); Hughes Communication Satellite Services; the government of India; Pacific Satellite; Inmarsat and two anonymous customers.

Secretary of Defense Caspar Weinberger wrote Secretary of State George Schultz urging that the United States retain options for military activities aboard the NASA space station without consulting its partners. Station partners in Europe and Japan were concerned about its being used for any kind of military activity. On April 13 NASA Administrator James Fletcher wrote to Frank Carlucci, White House national security adviser, warning that a wrong decision by President Ronald Reagan at this point could jeopardize the whole program, claiming that "Foreign officials . . . are accusing the United States of bad faith. Congress is questioning the appropriateness of NASA's budget being used to build a military facility."

1987 April 11   At 19:41 UT *Mir* cosmonauts Yuri Romanenko and Alexandr Laveikin left the *Mir* space station through one of the radial docking ports on a space walk to examine the aft docking interface where the *Kvant* module, launched March 31, was snagged. They found that insulation covering adjacent areas had been torn loose when *Progress 28* undocked on March 26, fouling the docking port. When the snared insulation had been cleared, ground controllers completed the docking and the two cosmonauts got back inside *Mir* after a space walk that lasted 3 hr. 40 min. The next day the service module attached to the rear of *Kvant* was undocked at 20:18 UT; but it had insufficient propellant to permit a de-orbit burn, so on April 20 it was placed in a higher orbit, of 238 × 252 mi., well away from *Mir*. Romanenko and Laveikin entered *Kvant* for the first time on April 13.

1987 April 16   During a meeting at the White House, the Department of Defense backed down in its demands that NASA's space station be used for "national security purposes" at any time and without consultation with the international partners responsible for building separate station elements. Both European nations and Canada had expressed grave

concerns at attempts by the Department of Defense to claim priority use for unspecified activities without providing any financing for the facility. The international partners were adamant that they would not participate in the program if it involved use by the Department of Defense for military research.

1987 April 21   At 15:14 UT the unmanned *Progress 29* cargo tanker was launched by SL-4 from the Baikonur cosmodrome. At 17:05 UT on April 23 it docked automatically to the rear docking port on the *Kvant* module, which itself was docked to the aft port on the *Mir* space station. *Kvant* carried a duplicate of the rendezvous and approach radar equipment attached to the rear of *Mir* so to the incoming Progress tanker it appeared to be a standard space station docking. Cosmonauts Yuri Romanenko and Alexandr Laveikin aboard *Mir* passed through the *Kvant* module and into the *Progress* orbital module to unload equipment and supplies. It contained 305 lb. of scientific equipment, 606 lb. of equipment for the operation of *Mir*, and the usual inventory of small stores and packages. *Progress 29* undocked at 03:10 UT on May 11, and de-orbited later that day.

1987 April 22   In a test to demonstrate that solid propellant Lance missiles held in storage for more than seven years could be taken out and fired at short notice, two missiles were fired simultaneously at White Sands Missile Range, N.Mex. Lance manufacturer Ling Temco Vought had been awarded a contract to validate the 10-year life guarantee on these missiles.

1987 April 23   The American Physical Society issued a report expressing deep concern over assertions from the Department of Defense that lasers and particle-beam weapons could be used to knock out enemy missiles and their warheads. It claimed that "technologies of vastly higher performance than any currently available" would be required to achieve the objectives set down by the Strategic Defense Initiative anti-ballistic-missile research organization.

1987 April 24   NASA issued a request for proposals from industry for detailed design and construction of a permanent manned space station. Proposals for this Phase C/D work were required to be in by July 21. Phase A of the station would accommodate U.S. habitation and laboratory modules, four resource nodes, the single-keel transverse truss assembly and the U.S. polar-orbiting platform. The Phase A station would provide 75 kW of power for housekeeping operations and user experiments, increased by 25 kW after discussions with congressional space committees enabling the United States to get an early start on materials processing experiments which have large power demand. Three work packages defined under the Phase B definition study contracts were to form the structure for contracts in Phase C/D. The first elements were due for launch in early 1994, with man-tended capability one year later and permanent manned capability in late 1995.

1987 April 28   The Los Alamos National Laboratory issued a request for companies that might help define applications of a nuclear-powered shuttle for carrying cargo between low and high Earth orbits beginning in the mid-1990s. The effort was to resurrect work on solid-core heat exchangers conducted in the original Rover and NERVA programs. Over the next five years several important technical breakthroughs were achieved by U.S. scientists working on this classified program. Known as the Space Nuclear Thermal Propulsion program, it defined a 75,000-lb. thrust propulsion system with a specific impulse of 1,000 sec., almost 200 Isp better than NERVA.

1987 April   The U.S. Air Force Space Division issued an announcement to industry requesting research proposals for an advanced launch system (ALS) capable of carrying unmanned payloads weighing up to 150,000 lb. Against advice from NASA, the proposal emphasized low launch cost over reliability, aiming for a factor of 10 decrease in cost-per-pound weight compared to current launch costs. Largely to accommodate Strategic Defense Initiative ballistic missile defense program payloads, the Air Force wanted to begin ALS flights in 1993. Bids were due in by May 30.

1987 May 5   The last operational Titan II ICBM was deactivated at Little Rock AFB, Ark. Taken off alert at 19:01 local time, the missile assigned to the 308th Strategic Missile Wing was deactivated at the end of a 24-hr. tour of duty by its four-man crew. Deactivation of the Titan II ended 24 years during which the missile type had been on alert for Strategic Air Command. The last Titan II silo, at Quitman, Ark., was destroyed on November 24. Equipped with a 5-MT warhead, Titan II was the most powerful strategic missile in the U.S. inventory and the decision in October 1981 to remove the 54 missiles of this type was an integral part of President Ronald Reagan's force modernization program.

1987 May 8   Claiming that the Soviet *Mir* space station was "almost totally dedicated to military purposes," U.S. Secretary of Defense Caspar Weinberger urged the House of Representatives not to prohibit the NASA space station from being used by the military. NASA, and a strong congressional lobby, wanted the station used exclusively for peaceful scientific research, but Weinberger said it would be "monumentally irresponsible" not to use the station for national security needs, claiming that Congress had been tainted by "the resonance of Soviet propaganda." On May 21 NASA Deputy Administrator Dale Myers called on the Department of Defense to "begin to think seriously about a dedicated space station of its own, possibly in polar orbit."

1987 May 15   The first successful flight of a Soviet Saturn V–class, super-heavyweight launcher took place from Baikonur cosmodrome when the first SL-17 Energia lifted off at 16:00 UT. With a total height of 192 ft. 6 in., Energia comprised a cryogenic core stage with four RD-0120 engines delivering a total thrust of 1.3 million lb. at liftoff, and four strap-on boosters each equipped with a four-chamber RD-170 engine burning oxygen and kerosene propellants for a total thrust of 6.5 million lb. The core stage fired for 12.3 sec. before the four boosters ignited, producing instant liftoff with a total thrust of 7.8 million lb. Liftoff weight was about 4.6 million lb. including a payload canister attached to the side of the launcher. The four strap-ons were jettisoned at 2 min. 20 sec., and the core stage continued to fire to an elapsed time of 7 min. 30 sec. At that point, just short of orbital velocity, the payload separated and was to have fired the two orbital insertion engines, producing a total thrust of 13,600 lb., to place it in orbit. These engines failed to fire and the payload fell back into the atmosphere. The SL-17 has a payload capability of 300,000 lb. By February 1994 only one more SL-17 had been launched, when the shuttle *Buran* was placed in orbit on November 15, 1988.

The last flight of an Atlas H took place from Vandenberg AFB, Calif., when the four satellites of a U.S. Navy White Cloud ocean surveillance system were delivered to orbit. It was the first flight of an Atlas launch vehicle since the failure of an Atlas-Centaur on March 26. The next flight of an Atlas, one of the refurbished E models, took place on June 20 when the Defense Meteorology Satellite Program *5D-2 F3* was placed in orbit. Not before September 25, 1989, would the next Atlas-Centaur fly, the last under NASA auspices.

NASA's space station operations task force completed a 10-month study of space station operations planning by recommending that control of the orbiting facility be vested in four NASA field centers: Johnson, Marshall, Kennedy and Goddard. Tasked with defining how station operations management, involving several hundred people for the 30 years the space station was to remain permanently manned, was to be coordinated, the group said that Johnson should control the systems on the station, that Marshall should control science and experiment activities, that Kennedy should be in charge of logistics and resupply, and that Goddard should control the stations' attendant unmanned platforms.

1987 May 19   At 04:02 UT the *Progress 30* cargo tanker was launched on an SL-4 from Baikonur cosmodrome. It docked to the *Mir* space station, in an orbit of 213 × 227.5 mi. at 05:53 UT on May 21. *Progress 30* supported a scheduled space walk by *Mir* cosmonauts Yuri Romanenko and Alexandr Laveikin on June 12, replenishing the atmosphere in *Mir*'s spherical docking module. *Progress 30* undocked from *Mir* at 00:20 UT on July 19 and was de-orbited the same day.

1987 May   Soviet space scientists revealed their plans for the exploration of Mars. In either 1992 or 1994, two Mars orbiters would be launched, each with either two balloons or a balloon and a small rover. The balloons were to float in the atmosphere of Mars and carry instruments round the planet. In 1994 or 1996, a large rover would be sent to Mars. Plans included a surface sample return mission before the end of the decade.

The Department of Defense sent a classified report to Congress on the potential application of Strategic Defense Initiative antiballistic missile technologies to other military missions of the next 5–10 years. It identified how SDI research was being applied to other needs of the military establishment and how the technologies might be coordinated for operational deployment of a ballistic missile defense screen.

1987 June 1   NASA Administrator James Fletcher announced the creation of an Office of Exploration, with Sally Ride serving as its acting head until August, when the veteran astronaut was due to leave NASA. Its aim was to "define missions proposed to achieve the goal of human expansion off the planet . . . [and] focus on studies of potential lunar and Mars initiatives." In order of realization, Fletcher presented four major goals: intensive study of Earth's environment in a Mission to Planet Earth program; an expanded program of unmanned exploration of the solar system; a permanently manned scientific lunar base; and the human exploration of Mars.

1987 June 9   In the first occurrence of its kind on record, lightning inadvertently launched three sounding rockets at NASA's Wallops Island, Va., firing range. The largest rocket, an Orion, crashed into the Atlantic Ocean 500 ft. from the launch pad. The other two were small calibration rockets that would have been used to test the range. Lightning struck one of the test rockets which triggered the firing mechanism and set off the others. No one was hurt in the incident.

1987 June 12   At 16:55 UT cosmonauts Yuri Romanenko and Alexandr Laveikin began a 1 hr. 53 min. space walk to install solar-cell arrays to the top of the *Mir* space station, effectively erecting a third solar-cell array to the two wing-like arrays on the station when it was launched. To save weight for the launch vehicle, the third set of arrays had not been installed but the pivotable attachment points had been. The first of two array segments was set up on this space walk. The second set was put up on a 3 hr. 15 min. space walk that took place on June 16. The new solar array assembly had a height of 34 ft. 6 in. and a surface area of 236.5 sq ft.

1987 June 22   Dr. John-David F. Bartoe was named chief scientist of NASA's Space Station program. A physicist by qualification and profession, Dr. Bartoe had been a Spacelab 2 payload specialist in 1985 and was responsible for high-resolution telescope and spectrograph experiments. Dr. Bartoe was responsible for representing the science community to the space station program and would also act as a conduit from NASA back to the scientific community.

1987 June 23   In the longest firing on record, a Space Shuttle main engine ran for a continuous 16 min. 42 sec. on test at the National Space Technology Laboratory, Miss. Of the firing, 11 min. 20 sec. was performed at 104% thrust, the remainder at 100% of rated thrust. This firing represented a firing time far in excess of the worst-case abort situation in which a single engine would be required to run on when its two neighboring engines in an orbiter had shut down prematurely.

1987 June 25   The Advisory Committee on the Space Plane, a group set up by Japan's Science and Technology Agency, submitted its findings about the desirability of that country's building a small shuttle. It advised a two-phase approach: putting a small, unmanned fly-back space plane into orbit for preliminary flight tests and then developing a manned reusable shuttle on the basis of results from the space plane. It advised that the H-II launcher being developed for test in the early 1990s should carry the small space plane into orbit. Called Hope, it was, at this date, judged to have a wingspan of almost 33 ft., a length of 39 ft. 5 in. and carry a useful payload of 6,600 lb.

1987 June   General Dynamics announced that it would use company funds to finance manufacture and assembly of 18 Atlas-Centaur launch vehicles for selling launch services. In the wake of depleted orders from NASA and the U.S. Air Force, General Dynamics wanted to exploit its 30 years of experience with the Atlas and the 21 years of Atlas-Centaur evolution to gain launch orders, leasing facilities from the Air Force and NASA. Capable of lifting 5,100 lb. to geosynchronous transfer orbit, the Atlas-Centaur D-1A was redesignated Atlas I, but a variant with bulbous payload shroud for customers with large satellites was also offered. When General Dynamics won the Air Force MLV-2 (Medium Launch Vehicle-2) contract in May 1988, it decided to build 12 Atlas I and six Atlas II versions.

1987 July 2   NASA selected Grumman Aerospace to provide Space Station program support. This vital adjunct to the government team which would manage four work-package contracts was to provide management support services as well as systems engineering and integration. Most of the work was to be conducted in the Washington, D.C., area; field offices at the four NASA centers Johnson, Marshall, Goddard and

Lewis would be assigned management of the work-package contracts.

1987 July 10   The U.S. Air Force Space Division awarded seven feasibility contracts for an advanced launch system (ALS) capable of lifting heavy payloads to space. The one-year contracts went to Boeing, General Dynamics, Hughes Aircraft, Martin Marietta, McDonnell Douglas, Rockwell International and United States Booster Inc. While providing funds for these contracts, Congress warned against using the money to push for early deployment of a Strategic Defense Initiative ballistic missile screen and urged NASA to get involved with the program. NASA had little or no need for an ALS, preferring instead that any future launch system comprise a Shuttle-derived vehicle (SDV).

1987 July 13   An access platform ruptured the last NASA Centaur stage being readied on Launch Complex 36B at Cape Canaveral, Fla., for the Atlas-Centaur AC-68 flight carrying *FltSatCom 8*, the last military communication satellite of that series. The Centaur did not have a payload or shroud on top and collapsed like a punctured balloon when the pressurized hydrogen tank was accidentally ruptured by the falling platform. NASA dismantled the Atlas-Centaur configuration by the end of August, but not before September 25, 1989, could it launch *FltSatCom 8*, using a stage from the new production batch of 18 Centaurs privately funded by General Dynamics.

1987 July 21   Air Force Secretary Edward C. Aldridge wrote to NASA seeking support for its advanced launch system heavy-lift launcher. The Air Force wanted NASA to drop its own studies of a Shuttle-derived vehicle (SDV) and focus on ALS as the heavyweight successor to Titan and Shuttle. NASA continued to argue that a future launcher based on Shuttle elements would be ready sooner and for less money than a completely new expendable launcher, citing June 1993 as an in-service target date for SDVs. As a result of different perspectives the White House began a review of space policy and hosted a meeting of the respective parties on July 28 to discuss options.

1987 July 22   At 01:59 UT cosmonauts Aleksandr S. Viktorenko, Aleksandr P. Aleksandrov and Muhammad A. Faris were launched aboard their *Soyuz-TM 3* spacecraft by SL-4 launch vehicle from the Baikonur cosmodrome. The spacecraft docked to the rear of the *Kvant* module on the *Mir* space station at 03:31 UT on July 24. Cosmonaut Faris was a Syrian participant in the broadening range of international flights surrounding Soviet manned spaceflight activity. On July 29, Viktorenko and Faris returned to Earth together with Aleksandr Laveikin, who had displayed signs of a potentially serious heart condition. They returned in the *Soyuz-TM 2* spacecraft, undocking from the forward axial port at 20:34 UT, and landed on Earth early next morning at an elapsed time of 6 days 23 hr. 52 min. Laveikin had been in space for 174 days 3 hr. 26 min.

1987 July 25   A Soviet SL-13 launcher carried *Cosmos 1870* from Baikonur to a 104 × 175 mi., 71.9° orbit. The 35,000-lb. module was in fact an Almaz space station modified by the Chelomei bureau to carry a side-looking synthetic aperture radar, with a resolution of 33–49 ft. The first of this type had been completed when further flights of the manned Almaz station were canceled in 1980. Launched in 1985, after a long and bitter dispute with the military, it was destroyed in an SL-13 launch failure. An outgrowth of manned military surveillance stations, *Cosmos 1870* was the first successful step toward an operational radar imaging satellite for remote sensing. The Soviets said the mission would concentrate on hydrology, cartography, geology and agriculture. Two days into the flight, *Cosmos 1870* was maneuvered to a new orbit of 147 × 155 mi., and over the next week it was maneuvered several times again. It decayed out of orbit on September 29.

1987 July 28 The U.S. Department of Defense completed the specification for an advanced successor to the Defense Meteorology Satellite Program Block 5D-3 series. In issuing requests for proposals, the Department of Defense wanted a polar orbiting visual/infrared imaging satellite to form a Block 6 program. Full operational status was projected for 1999, although the U.S. Navy pressed for an earlier availability. Current Block 5 series DMSP satellites transfer data to one of two ground stations which then forward it to tactical units. Block 6 was to have the capability to relay images via another satellite for direct transmission to the end user.

1987 July The U.S. Air Force Space Command's new Space Defense Operations (SDO) facility at Colorado Springs, Colo., was inaugurated. The new facility would enable urgent military space information to be collected, collated, coordinated, and categorized before transmitting it to the National Command Authority. If the United States deployed an antisatellite system, the SDO would control operations from there.

Ball Aerospace Systems received a contract from the U.S. Defense Advanced Research Projects Agency for a satellite to detect and inspect nuclear material in space. Designated P86–2, the satellite was to be called *Starscan* and carry instruments for detecting gamma rays and neutrons. The satellite was to weigh 6,675 lb. and be launched into space on a Titan II. It was to be placed in a sun-synchronous, 340-mi. orbit, by a solid propellant boost motor.

1987 August 1 Rockwell International's Space Transportation Systems Division received authority to build a fifth spaceflight orbiter for NASA's Shuttle fleet, which had been funded under a spares program authorized in 1983. Designated OV-105, the orbiter incorporated essential elements such as wings, crew module, aft fuselage and mid fuselage. In May 1989 President George Bush announced the result of a competition organized by NASA for students to select a name for the fifth orbiter and proclaimed it would be called *Endeavour*, after the ship used by the 18th-century Royal Navy officer and explorer Captain James Cook. The orbiter was unveiled on April 25, 1991.

1987 August 3 At 20:44 UT the unmanned spacecraft *Progress 31* was launched by an SL-4 from the Baikonur cosmodrome. It brought the usual inventory of propellant, water, food and sundry stores to the *Mir* station, now in an orbit of 192.5 × 224 mi. Docking was accomplished at 22:28 UT on August 5 at the back of the *Kvant* module. While *Progress 31* was docked to the complex, cosmonauts Yuri Romanenko and Alexandr Alexandrov went through a station evacuation drill, simulating a sudden emergency in *Mir*, which they were ordered to carry out without prior notice. Hurriedly donning space suits, they retreated to the relative safety of the *Soyuz-TM 3* spacecraft docked to the forward axial port. *Progress 31* undocked at 23:58 UT on September 21 and was de-orbited two days later.

Built more than 25 years previously, the first deactivated Titan 2 ICBM converted to a satellite launcher was rolled out

of Martin Marietta's facilities near Denver, Colo. Capable of placing a 4,800-lb. payload in a due-east circular orbit, Titan 2 satellite launchers would be used from Vandenberg AFB, Calif., to place, among other payloads, military weather satellites in sun-synchronous polar orbit. The first Titan 2 launched a White Cloud satellite cluster on September 5, 1988.

In the first of a series of meetings to resolve issues between NASA and the Department of Defense about future launcher needs, NASA Administrator James Fletcher met with Presidential Science Advisor William Graham to iron out differences. The Air Force wanted a new advanced launch system (ALS); NASA wanted a Shuttle-derived vehicle (SDV). The next day, Fletcher met with Vice President George Bush and Craig L. Fuller, Bush's chief of staff. On August 10 Fletcher met Bush again to discuss the space station and received support for continued studies of an SDV, now known as Shuttle-C for "cargo," within NASA.

1987 August 4 Astronaut Sally Ride submitted the results of a study she had chaired into the future of the American space program. It endorsed a vigorous approach to space exploration and recommended the moon be used as a stepping-stone to Mars, and that bases on both bodies should be a strong national goal. The lunar outpost was to be used for scientific research as well as for promoting Earth-moon transportation systems that could be used for the human occupation of Mars and the establishment of colonies.

1987 August 5 A CZ-2 launched the first Chinese materials processing experiment in the 5,512-lb. satellite *SKW-15* with recoverable capsule to a 106 × 245 mi. orbit inclined 63°. The capsule was returned to Earth on August 10, and the satellite reentered September 23. A second materials processing satellite, called *FSW-12 (Fanhui Shi Weixing-12)* launched on August 5, 1988, carried a German crystal growth experiment. This was also the last launch of the basic CZ-2 launcher.

1987 August 7 President Ronald Reagan met with senior officials from the Central Intelligence Agency to review the Soviet military space program. NASA Administrator James Fletcher advised the President on Soviet nonmilitary space developments, and Reagan discussed a strategy to restore U.S. leadership in space. Reagan was concerned that the United States needed a reinvigoration of space projects and was preparing a new space policy document for completion in early 1988.

1987 August 15 As an indication of how routine and reliable advanced weather monitoring capabilities from satellites had become, the U.S. Air Force flew its last airborne typhoon surveillance mission over the Pacific Ocean. Formed in 1947, for the last 25 years the 54th Weather Reconnaissance Squadron had been flying C-130 aircraft on typhoon patrol and was now standing down to be replaced by satellites. Parallel tests to determine the extent of satellite capabilities had revealed that in some cases the satellite provided better data than the aircraft flying in the vicinity of the typhoon.

1987 August 27 The first three-stage H-I in fully operational configuration was launched from Tanegashima, Japan, carrying the fifth engineering test satellite (ETS-5) to a geostationary orbit at 150° E. Named *Kiku-5*, the 1,212-lb. satellite was instrumented to measure the performance of three-axis-stabilized bus and solar array systems as well as providing a new data relay system for aircraft communications, navigation, and search and rescue.

1987 August 30 The first full-duration firing of a Shuttle solid rocket booster (SRB) carrying the redesigned segment joint/O-ring assembly took place at Thiokol's Wasatch facility near Brigham City, Utah. The 2 min. firing of development motor DM-8 qualified the changes brought about by findings of the national commission on the *Challenger* disaster, which increased SRB weight by more than 6,000 lb. The first of three qualification tests with full-scale SRBs, this was a crucial first step in returning the Shuttle to flight status.

1987 August The U.S. Air Force completed plans for procurement of a further 27 Titan 4 launch vehicles, bringing to 50 the total number of super-heavyweight lifters it wanted for advanced military satellites. On October 19 Martin Marietta awarded Hercules Aerospace Co. a contract to build uprated solid rocket motors for the Titan 4 launcher. Although early flights of the Titan 4 would use the seven-segment solids built by United Technologies, later launchers would use the Hercules boosters, increasing geosynchronous payload capability from 10,000 lb. to 12,500 lb.

NASA announced the names of 15 pilot (PLT) and mission specialist (MS) Shuttle astronauts constituting Group 12: Maj. Thomas D. Akers (USAF); Maj. Andrew M. Allen (USMC); Lt. Cdr. Kenneth D. Bowersox (USN); Maj. Curtis L. Brown Jr. (USAF); Lt. Col. Kevin P. Chilton (USAF); N. Jan Davis; C. Michael Foale; Gregory J. Harbaugh; Mae C. Jemison; Lt. Col. Donald R. McMonagle (USAF); Cdr. Bruce E. Melnick (USCG); William F. Readdy; Cdr. Kenneth S. Reightler Jr. (USN); Lt. Cdr. Mario Runco Jr. (USN); and Lt. Col. James S. Voss (USA).

NASA's Marshall Space Flight Center set up a special task team to evaluate Shuttle-derived vehicle (SDV) concepts as candidates for unmanned heavy-lift launchers. The concept was now exclusively sized for NASA payload requirements. The Air Force rejected SDVs as a candidate for heavy-lift launch vehicle duties because of their man-rated elements, which it believed was contrary to low-cost flight rates. The Air Force preferred to remain with the advanced launch system concept of a totally new expendable based on low-cost technologies. Thus began what would be NASA's last opportunity to develop a new unmanned launcher. Future legislation, and space policy, would give the Air Force and the launcher industry that mandate.

1987 September 1 A classified test for the U.S. Department of Defense Strategic Defense Initiative ballistic missile program was conducted by the U.S. Air Force across the Pacific Ocean. Called Aerothermal Reentry Experiment-1 (ARE-1), the test proved that a reentry vehicle only partly damaged by the glancing blow of an antimissile device would have its warhead disabled and be caused to self-destruct. A Minuteman III ICBM fired on an unrelated test from Vandenberg AFB, Calif., to the Army's Kwajalein Atoll test range demonstrated the vulnerability of incoming warheads. A device attached to the reentry vehicle imposed damage simulating the glancing blow. The RV subsequently self-destructed.

1987 September 16 An Ariane 3 launched (V19) from Kourou, French Guiana, carried the fourth European Communication Satellite, *ECS-4*, and the third Australian communication and broadcast satellite, *Aussat K3*, to orbit. With an on-orbit mass of 1,430 lb., *ECS-4* was positioned over 10° E, and services to Europe began November 1. The last of Australia's first-generation satellite buys, the 1,540-lb. *Aussat K3* was the first to be launched by Ariane. It was positioned above 164° E.

1987 September 23 At 00:22 UT *Progress 32* was launched to the *Mir* space station complex by an SL-4 from the Baikonur cosmodrome. Docking at the back of the *Kvant* module took place at 01:08 UT on September 26. *Mir* cosmonauts Yuri Romanenko and Alexandr Alexandrov began unloading the 4,410-lb. of stores, which included 695 lb. of food, camera film, oxygen bottles, new equipment to replace malfunctioning components, scientific instruments and mail for the crew. *Progress* also off-loaded 1,875 lb. of propellant. In a unique test, *Progress 32* undocked at 04:09 UT on November 10 and backed away 8,200 ft. before maneuvering back in for a second docking at 05:47 UT. The final undocking came at 19:25 UT on November 17, and *Progress 32* was de-orbited next day.

1987 September 25 The Los Alamos National Laboratory awarded a contract to build the facility which would house a Neutral Particle Beam Ground Test Accelerator (GTA). The GTA was needed by the Strategic Defense Initiative ballistic missile defense organization to demonstrate how neutral particle beams could be used to discriminate between targets and decoys in a space-based mode. If developed into an operational system, the neutral beam director would weigh about 80 tons and take a heavy-lift launch vehicle to place it in orbit. Elements of the concept would be tested in space on the BEAR experiment.

1987 September 29 The third Soviet biological mission carrying monkeys into space was flown under the designation *Cosmos 1887*. Launched by an SL-4 from Plesetsk, the 14,775-lb. satellite was a derivative of the Vostok manned spacecraft but used a solid propellant retro-rocket rather than the earlier liquid propellant system. Placed in an orbit of 133 x 237 mi., with an inclination of 62.8°, the satellite was recovered on October 12 after landing 2,000 mi. off target and just 25 mi. from the town of Mirny at the end of a flight lasting 13 days.

1987 October 5 The Strategic Defense Initiative organization announced contracts would be awarded to Martin Marietta, Lockheed and TRW Systems for a feasibility study on a space-based laser system known as Zenith Star. Key elements in the design of Zenith Star would be TRW's Alpha hydrogen fluoride laser and a large advanced mirror program (LAMP) device for focusing the beam to selected targets. LAMP was being developed by the Itek Corp. and was an adapted, segmented mirror made up from several separate segments which independently could be adjusted to correct for precise figure shaping of the beam.

1987 October 6 U.S. Air Force Secretary Edward C. Aldridge outlined plans for military satellite launchers. Because NASA projected a reduced payload capability for the Shuttle, and a reduced maximum launch rate of 14 flights per year when it returned to flight, the Air Force wanted to move away from this manned vehicle. It wanted to order an additional 25 Titan 4 launchers in addition to the 23 already approved by Congress, 8 of which would help compensate for the loss of Shuttle flights between this date and 1993. Martin Marietta could produce 6 Titan 4s per year, but the Air Force wanted to increase annual production capacity to 10.

1987 October 22 NASA announced selection of Martin Marietta, Rockwell International and United Technologies Corp., for contracts on the first of a two-phase study into Shuttle-C unmanned cargo-carrying launch vehicles. Systems definition studies were to focus on configuration details within optional second-phase study focusing on a selected configuration and operations concept. A major purpose of the study was to determine whether Shuttle-C would be cost-effective for launching elements of the Space Station.

1987 October 26　The Titan 34D returned to flight with the launch of a KH-11 digital imaging reconnaissance satellite from Vandenberg AFB, Calif. Ending an 18-month hiatus in Titan heavyweight launchers due to tests with suspect solid rocket motor design, the flight replenished badly stretched intelligence gathering capability made leaner through the loss of one KH-11 and one KH-9 in launch accidents during 1985 and 1986, respectively. A missile early warning satellite from the Defense Support Program was launched by another Titan 34D from Cape Canaveral, Fla., on November 29.

1987 October 30　NASA Administrator James Fletcher received the recommendations of the source evaluation board examining the four proposals for respective work packages on the Space Station. Fearing that U.S. budget cuts might compromise the letting of contracts, which NASA had hoped to complete during early November, Fletcher elected to postpone the final decision until after congressional action on the budget for fiscal year 1988. Cuts could seriously affect NASA's ability to work the four contracts.

1987 November 9　A test motor fired for 2 min. by Thiokol demonstrated the capacity of the new Shuttle solid rocket booster (SRB) segment joint to seal the motor from leaking gases in the event of a failure in the primary O-ring. With a deliberately gouged O-ring, the motor demonstrated that the backup O-ring would hold back the 950-lb./sq. in. pressure reached by the SRB at ignition. It was a leaking O-ring seal that destroyed the Shuttle *Challenger* on January 28, 1986.

1987 November 9–10　At a ministerial meeting of the 14 members of the European Space Agency, agreements were reached on the broad outline of future European space programs. With the exception of the United Kingdom, which refused to support any of the proposed major programs, and Canada, which had observer-status only, all countries agreed to help finance the Ariane 5 launcher, the Columbus space station module and free-flyer, and Hermes, the manned reusable space plane. France would fund 45% of Ariane 5 and Hermes, leaving Germany to lead the Columbus program with 38% of funds while France agreed to to pay 13.8%. Together France and Germany would fund 67% of Ariane 5, 51.8% of Columbus and 75% of Hermes.

1987 November 20　At 23:47 UT the *Progress 33* unmanned cargo tanker was launched by an SL-4 from the Baikonur cosmodrome. It docked to the back of the *Kvant* module at the *Mir* complex at 01:39 UT on November 23. Cosmonauts Yuri Romanenko and Alexandr Alexandrov unpacked equipment, including the Mariya plasma telescopes for astrophysical observations. Also offloaded was a furnace powered by solar energy through mirrors used for crystal growth studies. *Progress 33* undocked from the *Mir* complex on December 18 and was de-orbited the next day.

1987 November 21　An Ariane 2 (V20) launched from Kourou, French Guiana, carried the first West German direct broadcast TV satellite, *TV-SAT 1*, to orbit. Under an agreement signed in 1980, West Germany and France agreed to jointly build two direct broadcast satellites. The one built by

France would be known as *TDF-1*. *TV-SAT* was a three-axis-stabilized satellite based on the Eurosatellite Spacebus 300 design, also adopted for TDF and Sweden's Tele-X direct broadcast satellite. Comprising a box structure and top-mounted antennas, a total height of 21 ft. and a span of 63 ft. 1 in. across the two solar-array wings, *TV-SAT 1* was built to produce 3,215 W of electrical power. It had a launch weight of 4,579 lb. and an on-orbit mass of 2,260 lb. Five TV channels operating through the 18/12 GHz band were to have provided up to four simultaneous programs. In orbit, one solar wing failed to deploy and blocked the uplink antenna. Written off commercially, *TV-SAT 1* was used for engineering tests.

1987 November　The Strategic Defense Initiative antiballistic missile organization released preliminary design details of the Zenith Star space-based hydrogen fluoride laser, envisaging a structure 80 ft. in length, 15 ft. in diameter and weighing 100,000 lb. at launch. Martin Marietta and McDonnell Douglas were working on studies of a bootstrap-booster put together using existing rocket motors from Shuttle, Titan and Delta. With a liftoff thrust of around 10 million lb., the launcher could be built on a redundant launch pad and used only once to place the prototype Zenith Star in orbit. To comply with the 1972 ABM treaty, Zenith Star was not allowed to simulate a weapon.

1987 December 1　NASA announced the four industry teams selected to design and build the permanent manned Space Station. Managed by the Marshall Space Flight Center, Work Package 1 (pressure modules, logistics elements, resource nodes and environmental control equipment) was awarded to Boeing; managed by the Johnson Space Center, Work Package 2 (truss structure, mobile servicing transporter, air-locks, resource node outfitting, communications and tracking systems, guidance and navigation systems, propulsion system and thermal control) went to McDonnell Douglas; managed by Goddard Space Flight Center, Work Package 3 (free-flying polar orbiting platform) went to General Electric; managed by Lewis Research Center, Work Package 4 (electrical supply and control) went to Rocketdyne. Losing bidders for Work Packages 1 and 2 were Martin Marietta and Rockwell International, while General Electric and Rockwell International were the sole bidders for Work Packages 3 and 4, respectively. Letter contracts were issued December 23.

1987 December 8　The United States and the Soviet Union signed a historic treaty on the complete elimination of their short-range (310–620 mi.) and intermediate-range (620–3,420 mi.) missiles. Soviet short-range missiles included the Soviet SS-12 (726), SS-23 (200) and SSC-X-4 (84), a total 1,010 missiles each with one warhead. The only U.S. short-range missile was the Pershing IA, of which there were 178 missiles in storage, each with one warhead. Intermediate missiles included the Soviet SS-4 (65 + 105 in store), SS-5 (6 in store), and the SS-20 (405 + 245 in store) with a total 2,126 warheads. U.S. missiles in this class were the Pershing II (120 + 127 in store) and the ground-launched cruise missile, or GLCM (309 + 133 in store), a total 689 missiles and warheads.

1987 December 13　A Black Brant sounding rocket launched a space power experiment for the SDI ballistic missile defense program from Wallops Island, Va. Designated SPEAR-1, the mission involved a 10 min. ballistic flight during which 7 min. of useful data was obtained as two probes were charged to 44,000 V to observe their behavior in space. The test con-

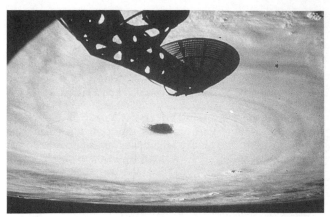

Some of the work conducted by Soviet cosmonauts in the Mir *space station involved Earth observation and measurement of storm centers like that shown here.*

firmed a theory that the vacuum of space could serve as an effective insulator and that great weight could be saved by not putting insulation on high-power laser and particle beam weapons. The payload weighed 800 lb. and reached a height of more than 250 mi.

1987 December 21   At 11:18 UT cosmonauts Vladimir G. Titov, Musa K. Manarov and Anatoly S. Levchenko were launched in their *Soyuz-TM 4* spacecraft by an SL-4 from the Baikonur cosmodrome. They rendezvoused with the *Mir* complex, in an orbit of 207.5 × 223 mi., on December 23 and docked to the back of the *Kvant* module at 12:51 UT. For almost eight days the cosmonauts conducted joint activities with the *Mir* crew, Yuri Romanenko and Alexandr Alexandrov. The latter pair returned to Earth in the *Soyuz TM-3* spacecraft on December 29, taking Levchenko with them and leaving Titov and Manarov on board. At 09:10 UT on December 30, they undocked *Soyuz-TM 4* from the *Kvant* port and redocked the spacecraft to the forward axial port.

1987 December 29   *Mir* cosmonauts Yuri Romanenko, Aleksandr Aleksandrov and Anatoly Levchenko returned to Earth in the *Soyuz-TM 3* spacecraft brought to the station July 22 at the end of a record-breaking endurance flight. *Soyuz-TM 3* had been in space for more than six months when it undocked from the axial port on the forward docking module at 05:55 UT with the complex in an orbit of 208 × 222.5 mi. The retrofire burn began at 08:23 UT and the spacecraft landed in the Soviet Union at 09:16 UT. Romanenko had been in space for 326 days 11 hr. 38 min. and was found to be in good health as a result of a strenuous physical exercise program that had him exercising 2½ hr. a day toward the end of the flight. Launched to the *Mir* complex on July 22, Alexandrov had been in space 160 days 7 hr. 17 min. Levchenko, launched aboard *Soyuz-TM 4* on December 21, had been in space for just 7 days 22 hr. 58 min.

1987   The Soviet Union began to deploy the world's first rail-mobile ICBM, the SS-24 Scalpel, known in the Soviet Union as the RS-22. Following five years of development, the solid propellant missile had emerged with a complement of 10 MIRV warheads each of 100 KT and an accuracy of 650 ft. across a range of 6,000 mi. Launched from presurveyed railcar sites, each missile was contained on a flat-bed with a length of 85 ft. and a width of 10 ft. Elevated by hydraulic ram to a vertical launch position, the missile could be fired within 12 min. By 1990 some 60 SS-24s had been deployed.

1988 January 1   Full-scale development of the Ariane 5 launcher began in Europe. The basic launcher would comprise a liquid hydrogen–liquid oxygen core stage (H155) containing 342,000 lb. (155 metric tons) of propellant consumed by one 252,000-lb. (vacuum) thrust HM60 engine with a specific impulse of 430 sec. The H155 would have a length of 100 ft., a diameter of 17 ft. 8 in. and a burn time of 10 min. At launch two P230 solid propellant boosters would provide an additional 2.43 million lb. of thrust. In this configuration Ariane 5 could lift 48,500 lb. to low Earth orbit. To put a single 15,000-lb. satellite, or two satellites with a combined weight of 13,000 lb., in a geosynchronous transfer path, Ariane 5 would carry an MMH-N_2O_4 second stage (L5) powered by one 4,400-lb. thrust motor.

1988 January 7   In an effort to encourage commercial space activity, senior officials in the Reagan administration forced NASA to make a commitment to lease the Industrial Space Facility (ISF) proposed by Space Industries. The ISF was intended as a precursor to the space station, not a threat to it as NASA believed, said Maxime Faget, chief executive of SII and originator of the concept. NASA said it had no use for the ISF, but the Department of Commerce believed that it would attract experimenters from the commercial sector. It wanted NASA to agree to a leasing term between 1991 and 1995, when the permanently manned space station would be available. NASA eventually resisted the move to lease the ISF.

Anticipating an upsurge of political will to define grand goals for U.S. space exploration, officials at the NASA Johnson Space Center and the Marshall Space Flight Center began a process of defining mission options. Under contracts with industry, Martin Marietta was studying the technology needs of a Mars base, STS Technologies was studying base operations through the year 2035 and Marshall was examining how the Boeing space station modules could be adapted as crew quarters for large manned vehicles traveling to Mars.

1988 January 11   The U.S. Air Force held a bidders conference for a launcher to carry the third generation Defense Satellite Communication System (DSCS-III) satellites into orbit. Defined as Medium Launch Vehicle-2 (MLV-2), the requirement stipulated a 5,800–6,200-lb. capability to geosynchronous transfer orbit. Originally planned for launch aboard the Shuttle, DSCS-III would be shifted to an expendable launch vehicle with a first flight in January 1991. The Air Force wanted to fly four DSCS-IIIs in 1991, and one each year between 1992 and 1997.

1988 January 12   NASA's Ames Research Center reported progress with what it considered the most advanced space suit ever built. Designated AX-5, the suit was made of aluminum for daily, routine EVA (extravehicular activity) work and comprised a solid metal shell operating at 14.7 PSI (sea-level) pressure. The suit was entered through the rear with the legs put in first, followed by the upper part of the body. The suit had a helmet 1 ft. 1 in. in diameter allowing a wide angle view. Covered by an ultrathin layer of pure gold to protect against corrosion, hydrazine propellant and other contaminants, the suit would protect the wearer from 400°F to −250°F. Budget cuts eventually forced NASA to abandon the AX-5 and rely on the Shuttle suit for Space Station–planned EVA operations.

1988 January 13–15 Officials from NASA field centers involved in defining a Mars rover/sample-return mission met to discuss preliminary reviews of studies performed by Martin Marietta, Lockheed and the FMC Corp. The Johnson Space Center and the Jet Propulsion Laboratory were working together to define technology options. Teamed with Honeywell, Lockheed wanted to employ the terrain-contour mapping radar from the cruise missile program in an orbiter so that accurate maps could be assembled before the descent vehicle went to the surface. Imaging data would then be fed into the landing vehicle so it would know where to go before it arrived.

1988 January 20 At 22:52 UT the 15,965-lb. *Progress 34* cargo tanker was launched by SL-4 from the Baikonur cosmodrome. It rendezvoused with the *Mir* complex, in an orbit of 207.5 × 221 mi., on January 23 and docked to the back of the *Kvant* module at 00:39 UT. Cosmonauts Vladimir Titov and Musa Manarov unpacked the more than 4,400 lb. of stores and, after a refueling operation had been completed, prepared for a space walk on February 26, after which the cargo tanker was used to replenish the docking module with air. *Progress 34* undocked from *Kvant* at 03:40 UT on March 4, and was de-orbited the same day.

1988 January 25 In his annual State of the Union Address, President Ronald Reagan outlined new goals and policy rationales aimed at restoring U.S. leadership in space and at establishing new objectives for NASA and the civilian space community. Reagan announced that a manned colony on the moon and manned expeditions to Mars should be the long-range goals of the U.S. space exploration program. NASA was to be allowed funds in fiscal year 1989 for the so-called Pathfinder program in which technologies crucial to these objectives could be developed. Only then, he said, could mission plans and hardware designs be laid out.

1988 January 26 General Electric (GE) signed a contract with Martin Marietta for the launch of 15 satellites on the company's Titan 3 launch vehicles. In an unprecedented commercial launch agreement, GE would pay $750 million–1.5 billion for the flights. The deal raised to 19 the total number of Titan launch options and included a cancellation fee if any of GE's satellite customers preferred to go on different launch vehicles. With the trend toward on-orbit delivery to the customer, launch arrangements fixed by the satellite builder suited operators and insurers.

1988 January In its budget submission to Congress for fiscal year 1989 the Department of Defense announced modified proposals for the Peacekeeper missile. Instead of persisting with requests for 50 missiles in addition to the 50 already deployed at Warren AFB, Wyo., the Air Force requested funds to move the existing missiles from their silos to a rail garrison basing mode. The Air Force still wanted a full complement of 100 missiles but preferred to have the mobile basing mode introduced as soon as possible.

1988 February 8 A Delta 3910 launch vehicle carried a complex set of test equipment into space on a sophisticated exercise for the Strategic Defense Initiative ballistic missile defense program. Launched from Cape Canaveral, Fla., the Delta second stage placed itself in a 149 × 219 mi., 28.5° orbit. Attached to the front end was a 12 ft. long sensor module containing seven instruments and fourteen subsatellites in two groups. At 1 hr. 9 min. into the mission the first of

six subsatellites was released, followed by the rest at 2 min. intervals. Each simulated a Soviet warhead, maneuvering bus or decoy, and the sensor module tracked these as though it were a battle station locking on for attack. At 2 hr. 33 min. the other five objects were deployed, and these too were scanned at varying distances as they drifted away. At 4 hr. a test device with a solid propellant motor was released, and it maneuvered as it simulated a MIRV bus. At 4 hr. 50 min. a 4 × 6 ft. object was ejected. It puffed different gases through vent ports so the sensors could "sniff" its presence. At 7 hr. 37 min. the sensors were directed to track a small Strypi rocket launched from Hawaii to measure tracking capabilities through different atmospheric layers. At 8 hr. 32 min. another satellite was ejected with a 1,700-lb. thrust motor that ignited for 16 sec. so the sensors could measure the plume. The second stage decayed on March 1.

1988 February 11 President Ronald Reagan announced a comprehensive new space policy which envisaged greater cooperation between NASA and other government agencies to cut waste and increase efficiency. It set broad goals without timelines for expanded human space exploration, but emphasized the need for greater reliance on commercial space ventures. It endorsed the NASA space station but wanted private sector involvement. The new directive also endorsed the early development of an advanced launch system for heavyweight payloads.

1988 February 16 Air Force Secretary Edward C. Aldridge asked the Air Force to put together a team to formulate a position on antisatellite (ASAT) systems. The Air Force had stated a preference for a ground-based ASAT rather than the one launched by F-15 aircraft, which the Department of Defense had canceled. Preliminary studies indicated much of the work on antiballistic missile technologies done by the Strategic Defense Initiative office was relevant to antisatellite applications.

1988 February 19 The first in a new generation of Japanese communication satellites, CS-3A was launched by H-I from Tanegashima to a geosynchronous transfer orbit from which it was placed in stationary orbit at 132° E. With a seven-year design life, *Sakura-3A* (Cherry Blossom) was a cylinder, 7 ft. in diameter and 10 ft. 8 in. tall, weighing 2,425 lb. at launch and 1,210 lb. on station. Each CS-3 generation satellite had 10 K-band and 3 C-band transponders with 50% redundancy and provided 6,000 simultaneous telephone circuits. They were the first satellites to employ highly efficient gallium arsenide solar cells as a primary power source. CS-3B was launched to 136° E on September 16.

NASA informed Shuttle-C study contractors Martin Marietta, Rockwell International and United Technologies Corp. that it would begin phase-two systems definition when the first phase ended March 19. Capable of lifting 100,000–150,000 lb. of payload, the Shuttle-C would have carried a cargo element with a length of 81 ft. and a diameter of 15 ft. Two Space Shuttle main engines in a boattail configuration would power a 100,000-lb. payload, while three would lift 170,000 lb. The contractors' estimated launch costs would average $2,000/lb. of payload versus $3,793/lb. for Delta II, $4,100/lb. for Titan 4 and $8,000 for a manned Shuttle.

1988 February 25 At a secret meeting in Brasilia, officials from China and Brazil reached agreement on the development of an Earth-imaging satellite to be launched in 1992 on a Chinese launch vehicle. Expected to weigh 3,000 lb., the

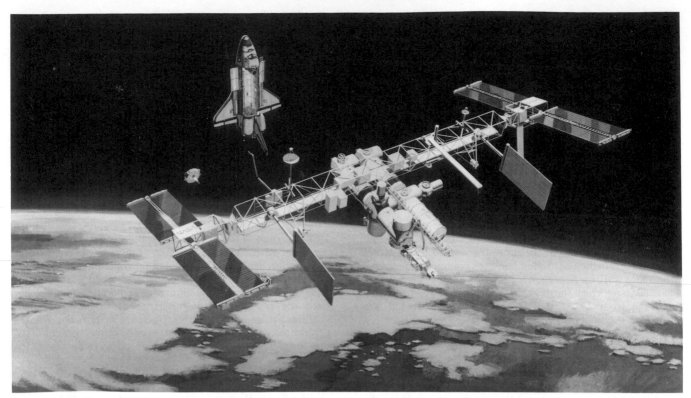

The development of NASA's single-keel space station progressed during 1988 with participation by designers from Canada, Europe and Japan.

satellite was to be designed for taking multispectral images with a resolution of 66 ft. from a 500-mi. orbit. China had been developing a multispectral-linear array scanner at a laboratory in Xian.

1988 February 26 Cosmonauts Vladimir Titov and Musa Manarov performed the first space walk of their long-duration stay aboard the *Mir* space station when they moved through a forward docking hatch at 09:30 UT. In activities that lasted 4 hr. 25 min. the two cosmonauts made their way to the solar array assembly mounted to the top of the *Mir* station by cosmonauts Yuri Romanenko and Alexandr Laveikin on June 12, 1987. Folding down the lower part of the concertina-shaped array, the cosmonauts replaced one of the two sections with an improved design. Before returning to the station, Titov and Manarov retrieved sample containers from the exterior wall.

1988 March 4 A group of Australian businessmen met with officials from Martin Marietta in Denver, Colo., to discuss the prospect of building a spaceport at Cape York in Queensland, Australia. Known as the Cape York Space Agency (CYSA), the group wanted to provide launch facilities for operators launching geosynchronous satellites for commercial customers. Located at 12° south latitude, a launcher fired from Cape York to geostationary orbit at the equator could lift heavier payloads than a similar one one from Cape Canaveral, situated at almost 29° north. This is because less of the launcher's energy would be needed to turn the trajectory to a due east orbit than it would from Cape Canaveral and so more of the velocity could be expended in gaining speed and not turning the angle of the trajectory. Plus, the Earth spins faster

at the equator and gives the launcher a better start. The Soviets, who were even more severely handicapped with their most southerly launch site (Baikonur) at 47° north, were also interested.

1988 March 7 The first Chinese domestic communication satellite, the 1,985-lb. *STTW-2* (Shiyong Tongbu Tongxin Weixing-2), was launched by CZ-3 from the Xichang site to a geostationary orbit above 88° E. One of the Dong Fang Hong 2 series, the satellite weighed 970 lb. at apogee motor burn-out and carried three channels for central TV. It was cylindrical in shape, with a height of 7 ft. 10 in. and a diameter of 6 ft. 6 in. A second satellite of the same type, *STTW-3*, was launched by CZ-3 to a position over 10° E on December 22. *STTW-4* was launched on February 4, 1990, to 98.5° E. Launched on December 28, 1991, *STTW-5* was dumped in an incorrect orbit of 127 × 21,804 mi. inclined 31.6° due to a malfunction in the third stage.

1988 March 11 An Ariane 3 (V21) launched from Kourou, French Guiana, carried the *Spacenet 3* and *Telecom 1C* communication satellites into orbit. Deployed first, GTE Corporation's *Spacenet 3* completed the initial three-satellite constellation of C-band and Ku-band services, leaving only one of two Ku-band GSTAR satellites yet to be launched. *Spacenet 3* was positioned at 87° W. The third of three data, TV and telephone relay satellites for France, *Telecom 1C* was located above 5° E to replace *Telecom 1B*, which had failed on January 15, 1988.

The governor of Florida was asked to study the feasibility of a commercial spaceport being built near Cape Canaveral and operated as a private facility leased to launch operators. A group of industrialists and businessmen believed they could build and run a commercial launch facility for less cost and with fewer people than NASA's facilities, which operators would have to lease from the government. Proponents calculated the revenue from just 12 private launches per year would contribute $540 million/year to the state's economy. The investment in infrastructure essential to these services was not forthcoming, however, and the low launch rate of Cape Canaveral traffic discouraged development.

1988 March 16   The U.S. Department of Defense formally authorized the Strategic Defense Initiative organization to proceed with plans for a space-based interceptor test in space for 1990, and cleared its compliance with the 1972 antiballistic missile treaty. The SDI organization believed the SBI could be based in space in large numbers to achieve with kinetic-kill what lasers and particle-beams were still too primitive to accomplish in the near term: relatively low-cost defenses. At a June 2 review, a Defense Acquisition Board advised the SDI organization that plans for space tests would have to be cut due to funding restrictions.

1988 March 17   In the first commercial launch for the Soviet Union, an SL-3 carried the 3,970–2,150-lb. Indian remote sensing satellite, *IRS-1*, from Baikonur cosmodrome to a 556 x 567 mi., 99° orbit. From a sun-synchronous orbit, the satellite was launched under the auspices of Glavkosmos, a semiautonomous organization set up during 1987 to market the Soviet space program. *IRS-1* was declared operational less than two months after launch and was used by India to obtain visible and near-infrared images of the Earth. It observed the Earth in four spectral bands with two camera systems at resolutions of 250 ft. and 115 ft., respectively.

1988 March 22   Japan's National Space Development Agency conducted the first development firing of the solid rocket booster (SRB) for the H-II heavy-lift launch vehicle. With a thrust of approximately 350,600 lb., the H-II SRB is the third most powerful solid propellant booster for a satellite launcher next to those used with the NASA Shuttle and the U.S. Air Force Titan. It has an overall length of 75 ft. 6 in., a diameter of 5 ft. 11 in. and consumes 130,000 lb. of propellant in a burn time of 1 min. 34 sec. Ignited on the pad, the boosters augment powered ascent with the liquid propellant first stage to a height of 22 mi. and a velocity of 4,900 ft./sec. SRB separation occurs 5 sec. after burnout.

1988 March 23   At 21:05 UT *Progress 35* was launched by SL-4 from the Baikonur cosmodrome. It docked to the back of the *Kvant* module at 22:22 UT on March 25. With the station in an orbit of 202.5 × 212 mi., cosmonauts Vladimir Titov and Musa Manarov unpacked the stores which included, in the 880 lb. of food, a very welcome supply of fresh fruit and vegetables. *Progress 35* remained attached to the *Mir* complex for six weeks. It undocked at 01:36 UT on May 5 and was de-orbited into the atmosphere the same day.

1988 March 31   In the first restructuring of Shuttle prices since the *Challenger* disaster on January 28, 1986, the Office of Management and Budget allowed NASA to charge only $110 million for a full orbiter payload bay. The White House interagency group on space had wanted to raise the price to $245 million. This higher figure would achieve full cost recovery, but market studies showed that charges in excess of $110 million would price it out of the market. NASA was

required to come up with prices for commercial customers wanting to launch research modules via the Shuttle.

1988 March 25   A four-stage Scout launch vehicle carried an international atmospheric research satellite, *San Marco D*, from the San Marco launch platform off the coast of Kenya to a 163.5 × 382 mi., 3° orbit. The Italian-built satellite carried instruments from Italy, West Germany and the United States. It comprised a sphere 3 ft. 4 in. in diameter, weighing 520 lb. and instrumented with five experiments designed to study equatorial airglow, neutral winds and neutral particle temperatures in the radiation belts, and electric fields and ion winds around the satellite. The project had been managed by NASA's Goddard Space Flight Center and the Aerospace Research Center at the University of Rome, Italy. It decayed December 6.

1988 March   NASA's Jet Propulsion Laboratory completed revision of its Comet Rendezvous/Asteroid Flyby (CRAF) mission following a delay in getting funds for a start on the project. Under the revised plan, a Mariner Mark II spacecraft would be launched in October 1994 for a rendezvous with the comet Wild 2 in February 2001. By June 1988 this plan had changed again to accommodate further delays. It now envisaged a launch in April 1995 toward the asteroid belt. From there the spacecraft would return for an Earth flyby in February 1997, gathering energy to fly past the asteroid Eunomia in October 1997 and going on to rendezvous with comet Wild 2 in July 2001. It would remain with that comet as it moved back toward the sun, outside Earth's orbit, dropping a penetrometer into the nucleus in June 2002 and concluding its mission in December 2003 as the comet passed perihelion. The penetrometer would relay scientific data via the spacecraft. In 1991 CRAF was canceled.

   Lowell Wood, one of the architects of the Strategic Defense Initiative antiballistic missile program, proposed a new method of countering Soviet ICBMs and their MIRVed warheads. Extending the analogy that directed-energy weapons were potentially "smart rocks," he suggested the use of highly sophisticated microminiaturized kinetic-kill vehicles, or "brilliant pebbles." Advocating several thousand placed in space awaiting targets from an aggressive enemy, Wood sought to satisfy SDI requirements with microelectronics rather than brute force.

1988 April 1   The U.S. Air Force issued a stop-work order to contractors on the SICBM Midgetman program. Budget restrictions forced a decision to cancel plans for deployment of the missile in hard mobile launcher vehicles capable of being dispersed to remote areas in time of crisis. Some money was made available to develop the missile itself and to conduct a limited number of flights, thus ensuring that if the technology was needed by a future president it would be available.

   Full definition of France's Hermes space plane got under way at prime contractors Aérospatiale and Dassault-Breguet. The design was to be completed by the end of 1990. Two vehicles would be built: The first would be used for aerodynamic tests beginning mid-1996, the second for unmanned orbital flights starting in June 1997. The no. 1 vehicle would fly its first manned spaceflight in April 1998. As now envisaged, Hermes would have a length of 50 ft. 10 in., a wingspan of 32 ft. 10 in., a payload capability of 6,600 lb. and be carried into orbit by Ariane 5. Changes brought about by lessons from the *Challenger* disaster included an escape module for the three-person crew, capable of being used to a speed of Mach 7 and a height of 31 mi.

James B. Odom replaced Andrew Stofan as associate administrator for Space Station. Odom had been at the Marshall Space Flight Center, where he was director of science and engineering. Between 1983 and 1986, Odom had been manager of the Space Telescope. Stofan retired for personal reasons and joined Martin Marietta on April 4.

1988 April 5–7   The second Symposium on Lunar Bases and Space Activities of the 21st Century was held in Houston, Tex. Sponsored by NASA, the American Institute of Aeronautics and Astronautics, the Lunar and Planetary Institute and other specialist organizations, there was a strong call for political action to mobilize a pro-active lobby in a campaign for the expansion of human exploration. For its part, NASA formed a Lunar Exploration Science Working Group in early 1989 to help develop strategies for the exploration of the moon and Mars.

1988 April 7   The U.S. Army officially formed a new Space Command to coordinate separate elements the Army could use to support ground forces. The new Space Command would employ more than 100 personnel, compared to only four space-dedicated officers in 1985, at headquarters located at Peterson AFB, Colorado Springs, Colo.—headquarters of the joint service U.S. Space Command and North American Aerospace Defense Command. Army Space Command would focus on harnessing space assets in support of ground forces and develop antiballistic missile systems should the president and Congress opt to deploy those forces. On July 1 Army Space Command took over the Defense Satellite Communication System ground facilities, previously operated by the Army's Information Systems Command.

1988 April 14   The first in a new series of Soviet materials processing and technology experiment satellites, *Photon 1*, was launched by SL-4 from Plesetsk to a 133 × 229 mi., 62.8° orbit. Based on the Vostok spacecraft, *Photon 1* weighed about 13,600 lb. and carried 1,540 lb. of instruments and equipment in the reentry module. About 400 W of electrical power was available, with up to 700 W available for 1 hr. 30 min. each day, and the vehicle had an on-orbit life of 15 days. *Photon 1* was recovered on April 28. Crystal production, the separation and purification of biological products and electrophoresis were conducted using adaptable instruments that could be flown on *Photon* or *Mir* stations.

1988 April 20   First in a new series of operational digital imaging satellites for remote sensing was launched by the Soviet Union under the designation *Cosmos 1939*. Placed in a 383 × 410 mi., 97.9° orbit by an SL-3 from Baikonur, the satellite was one of a series called Resurs-O and transmitted images to Earth on digital radio link instead of by film pods returned to Earth. The 4,450-lb. *Cosmos 1939* provided images that were marketed in the United States by Space Commerce Corp., but the 150-ft. resolution was poor compared to that of Landsat and SPOT.

1988 April 26   The United States submitted a proposal to the Soviet Union for greater cooperation in civilian space activities, outlining an exchange of scientific instruments carried on each others' vehicles. NASA wanted to fly ozone-mapping sensors on Soviet Meteor weather satellites and the Soviets were invited to put a receiver on the Mars Observer spacecraft. It was also suggested that the Soviets contribute life science experiments to a Shuttle/Spacelab mission (SLS-1) set for mid-1991, and that a more coordinated exploration of the

solar system be planned. The Soviets accepted the proposals three days later.

1988 May 1   Three U.S. aerospace companies were awarded nine-month study contracts on a successor to the Defense Satellite Communications System constellation of DSCS-III satellites. The companies were: TRW Systems, Hughes and General Electric. The satellites were conceived as supporting the existing duties of the DSCS series, and the requirements were sized around operational availability in the early part of the next century.

1988 May 9   Fourteen candidate names for the Space Station were submitted to NASA for short-listing by NASA Administrator James Fletcher: Aurora, Earth-Star, Freedom, Hercules, Independence, Jupiter, Landmark, Liberty, Minerva, Olympia, Pegasus, Pilgrim, Prospector, Starlight and Sky-base. When offered these names, President Ronald Reagan selected the name Freedom.

1988 May 13   At 04:30 UT *Progress 36* was launched to the Mir space station complex, then into an orbit of 205 × 222.5 mi., by an SL-4 from the Baikonur cosmodrome. It docked to the back of the *Kvant* module at 02:13 UT on May 15. Along with the usual provisions and stores, *Progress 36* carried equipment and instruments to support the visit of a French cosmonaut to be launched aboard *Soyuz-TM 7* on November 26. In preparation for the launch of *Soyuz-TM 5*, whose crew included the first Bulgarian cosmonaut, *Progress 36* boosted the orbiting complex to a height of 216 × 221 mi. It undocked at 11:12 UT on June 5 and de-orbited the same day.

Roald Sagdeyev, director of the Space Research Institute of the Soviet Academy of Sciences and chairman of the Committee of Soviet Scientists Against the Nuclear Threat, joined with a group of U.S. scientists in calling for a ban on nuclear reactors in Earth orbit. To date the United States had launched 22 satellites with radioisotope thermoelectric generators and one with a nuclear reactor. The Soviets had launched more than 30 reactor-powered ocean surveillance satellites carrying radar. The scientists said that an accident could contaminate large areas of Earth.

1988 May 18   The U.S. Air Force announced it had received authority from the Department of Defense to proceed with full-scale development of the rail-garrison basing mode for the 50 Peacekeeper missiles based in Minuteman silos. Westinghouse was contracted to build the missile launch car (MLC), each of which was required to carry two 97.5-ton missiles. Each of 25 trains would carry two locomotives, two MLCs, launch control car, two security cars and additional support cars. Warren AFB, Wyo., had been selected as the main operating base, with six other installations as operating locations: Barksdale AFB, La.; Dyess AFB, Tex.; Fairchild AFB, Wash.; Grand Forks AFB, N.Dak.; Little Rock AFB, Ark.; and Wurtsmith AFB, Mich. The trains would be housed in earthen igloos 800 ft. long and dispersed to 120,000 mi. of existing rail track in time of crisis. Analysis projected a 90% survivability against attack from 150 Soviet SS-18 ICBMs. The first train was scheduled to be deployed in 1992.

1988 May   The U.S. Air Force selected the General Dynamics Atlas-Centaur for the second medium launch vehicle (MLV-2) contract to put 10 Department of Defense communication satellites in orbit. The winning design was the Atlas 2 configuration with stretched Atlas and Centaur tanks capable of placing a payload weight of 6,100 lb. into geosynchronous transfer orbit. The decision by General Dynamics to build 18 Atlas-Centaurs for commercial marketing had been a bridge to the government contract.

NASA's Langley Research Center completed an in-house study into technological advances that could be applied to a fully reusable Shuttle successor, Shuttle II. In the proposed concept it would utilize a containerized cargo pod mounted atop the orbiter, with a 20,000–40,000 lb. low Earth orbit capability. New 670,000-lb. thrust Space Shuttle main engines, a more efficient thermal protection system and use of hardware developed for the U.S. Air Force's Advanced Launch System would cut launch costs and improve efficiency.

1988 June 7   At 14:03 UT Soviet cosmonauts Anatoly Y. Solovyov and Viktor P. Savinykh, together with Bulgarian cosmonaut Alexandr Alexandrov, were launched to the Mir station complex in the *Soyuz-TM 5* spacecraft by SL-4 from Baikonur cosmodrome. They docked to the back of the *Kvant* module at 15:57 UT on June 9, and joined Vladimir Titov and Musa Manarov inside *Mir*. The first attempt to put a Bulgarian cosmonaut aboard a Soviet space station had been unsuccessful when *Soyuz 33*, launched on April 10, 1979, failed to dock with *Salyut 6*. Over seven days of combined activity, 46 experiments were completed before the visiting crew undocked in the *Soyuz-TM 4* spacecraft at 06:18 UT on June 17, returning to Earth at an elapsed time of 9 days 20 hr. 10 min. The next day Titov and Manarov moved *Soyuz-TM 5* from the back of the *Kvant* module to the forward axial docking port.

U.S. officials testified to the multinational COCOM (Coordinating Committee on Export Control), substantiating objections to the use of Chinese launch vehicles for satellites built by U.S. companies. COCOM existed to control the transfer of technology to communist countries but U.S. officials also feared that Chinese launch vehicles could attract traffic away from U.S. launchers. Through the Great Wall Industry Corp., China had already signed up the Asia Satellite Telecommunication Co. for the launch of *Asiasat 1* on a CZ-3. Australia was considering a CZ-2E launch for its Aussat communications satellite.

1988 June 13   The board of governors of the International Telecommunications Satellite consortium, Intelsat, approved a recommendation from Director General Dean Burch that Ford Aerospace be awarded a development contract for the Intelsat 7 generation of global communication satellites. The Ford group included the French Alcatel company and Japan's Mitsubishi. The Intelsat 7 design was similar to the Ford Superbird satellites built for Japan. Failed contenders for the lucrative contract were a consortium of Aérospatiale, General Electric and Messerschmitt-Bölkow-Blohm, a group comprising Matra, British Aerospace, and TRW Systems, and Hughes Aircraft. Intelsat had plans to buy 11–13 Intelsat 7-series satellites.

1988 June 15   The first Ariane 4 launch vehicle (401) lifted off from the ELA-2 complex at Kourou, French Guiana, at 11:19:01 UT. With a liftoff weight of 919,732 lb., vehicle 401 carried a 6,100-lb. payload, comprising three satellites, to geostationary transfer orbit. Ariane 4 differed from Ariane 3 in having a longer (77 ft. 5 in.) first stage designated L220, 498,330 lb. of UDMH-N_2O_4 propellants and a burn time extended from 2 min. 15 sec. to 3 min. 25 sec. It also featured the option of two liquid propellant booster (LPB) assemblies and/or two solid propellant booster (SPB) units. For its first flight Ariane 4 had all four boosters and a liftoff thrust of 1.37

million lb. Ignition of the LPBs and the first-stage engines came at T−4 sec., with ignition of the SPBs and liftoff at T−0. The solids were jettisoned 1 min. 6 sec. later, the liquids 2 min. 25 sec. after liftoff and first stage shutdown at 3 min. 25 sec. Second- and third-stage flights occurred as they would on an Ariane 3 flight, with shutdown at 17 min. 39 sec. Enclosed within a bulbous shroud 43 ft. 11 in. long with a diameter of 13 ft. 1 in., the three satellites released in sequence (from the top down) were the 1,543-lb. *Meteosat 3* weather satellite placed at 0° longitude; the 331-lb. amateur radio satellite *Amsat III-C (Oscar 13)*, left in a 1,438 × 22,678 mi. orbit inclined 57.4°; and the 2,690-lb. *PAS-1 (Pan American Satellite 1)* placed at 45° W. A modified RCA Astro Series 3000 satellite, *PAS-1* comprises a rectangular structure 5 ft. 4 in. × 4 ft. 4 in. × 4 ft. 4 in., a 7 ft. 2 in. antenna and two solar cell wings. *PAS-1* had 24 C-band and Ku-band channels providing data and voice communications to small antennas in the United States, Latin America and Europe.

1988 June 16   Aussat, the Australian telecommunications entity, selected the Hughes HS 601 bus for its Aussat B series of satellites. This was the first sale of the HS 601, the first Hughes three-axis-stabilized communications satellite. HS 601 would provide almost 3,700 W of electrical power from solar arrays spanning 66 ft. A typical HS 601 would weigh 3,000 lb., although growth versions could weigh up to 10,000 lb. and provide 6,000 W of power. The series was specifically designed for direct broadcasting, where high-power traveling wave tubes were needed, and for a 15-year life. On July 29 the U.S. Navy selected the HS 601 for its UHF Follow-On satellite, UFO.

1988 June 30   Soviet cosmonauts Vladimir Titov and Musa Manarov performed a space walk from their orbiting *Mir* complex. Their objective was to replace a sensor detector on the back of the *Kvant* astrophysical laboratory. Contributed by a Dutch/British team, the instrument had not been designed for in-orbit replacement, and impeded by their life-support umbilicals, the crew were unable to complete the work by the end of the 5 hr. 10 min. space walk. Using a new autonomous back-pack that did away with the need for a backup umbilical, a second space walk on October 20 proved more successful and the detector was replaced in an activity lasting 4 hr. 12 min.

1988 July 1   Ten teams of U.S. arms reduction verification inspectors began visits to 133 sites in the Soviet Union where intermediate-range nuclear missiles were being destroyed. In an unprecedented agreement on the mutual destruction of an entire class of weapons, five teams of Soviet inspectors, each with ten personnel, began inspecting sites in the western United States on July 2. The inspection and verification process was to last 13 years, during which time all missiles in this class would be destroyed.

1988 July 4   Space Shuttle *Discovery* was rolled to Launch Complex 39B at the Kennedy Space Center, Fla., in preparation for a resumption of U.S. manned spaceflight more than two years after the *Challenger* disaster. *Discovery* was for launch on September 6 with a tracking and data relay satellite of a type that was being carried by *Challenger* when it blew up. From here on, mission designations would abandon the coded number/letter sequences, the last flight being 51-L, and revert to the STS number series, picking up at STS-26, the 26th Shuttle launch.

1988 July 5 *Okean 1*, the first in a series of operational Soviet oceanographic satellites, was launched by SL-14 from Plesetsk to a 393 × 412 mi., 82.5° orbit. The 3,500-lb. satellite carried an array of Earth observing instruments for monitoring ocean conditions in northern latitudes. It played a major part in maintaining a bank of data on ice conditions, water temperature, new ice fields and general weather data. *Okean 2* was launched on February 28, 1990, followed by *Okean 3* on June 4, 1991. The last two satellites each weighed 4,200 lb. and all three remain in space.

1988 July 6–14 At the International Seminar on Future Studies of Mars, held at the Space Research Institute of the Soviet Academy of Sciences, plans for the Soviet exploration of Mars were described to an international audience of scientists. L. A. Gorshkov, from Glavkosmos, the Soviet space-marketing organization, confirmed that the Soviet Union would send its next unmanned spacecraft to Mars in 1994 and not 1992 as earlier plans indicated. A roving vehicle was planned for the 1994 flight followed by a mission in 2000–2005 that would return samples of the Martian surface to Earth. Gorshkov said that plans envisaged a manned expedition to Mars in 2005–2010 using a nuclear-electric propulsion system.

1988 July 7 The Soviet Union launched the first of two 13,700-lb. Phobos spacecraft to Mars. The second was launched five days later, both sent on their way by SL-12 launchers. A completely new design from the Babakin Engineering Research Center characterized the Phobos spacecraft, equipped with a radical new propulsion system modularized for application to different mission objectives as required. The main unit would be jettisoned after placing Phobos spacecraft in orbit around Mars, leaving the secondary propulsion package to perform orbit maneuvers which, for this mission, included a "docking" with the moon Phobos by a special 220-lb. lander carried aboard each spacecraft. The Phobos mission to Mars was organized by the Soviet Union Academy of Sciences' Space Research Center and coordinated with participation by 13 other countries. All contact with *Phobos 1* was lost on August 28, 1988, before the spacecraft reached Mars, due to a human error in sending up a wrong command from Earth. *Phobos 1* passed Mars in January 1989.

1988 July 13 The second attempt to launch a satellite by four-stage ASLV eluded the Indian Space Research Organisation when a first stage failure at 2 min. 30 sec. brought the configuration tumbling into the Indian Ocean shortly after launch from Sriharikota. The ASLV launcher was carrying the 331-lb. *SRS-2* satellite instrumented with a special electro-optical scanner which had been developed under a joint program with West Germany.

1988 July 18 The Soviet unmanned cargo tanker *Progress 37* was launched by an SL-4 from the Baikonur cosmodrome to Earth orbit, from where it docked to the back of the *Kvant* module on the *Mir* space station July 20. In an orbit of 213 × 215.5 mi., the *Mir* cosmonauts Vladimir Titov and Musa Manarov unpacked supplies and provisions as well as supplementary experiments for activation aboard the space station. *Progress 37* was undocked from the back of the complex on August 12 and reentered the atmosphere to destruction on the same day.

1988 July 21 An Ariane 3 (V24) carried the fourth, and last, European Communication Satellite *(ECS-5)* and India's *Insat*

1C multipurpose satellite, from Kourou, French Guiana, to orbit. With a launch weight of 2,624 lb., *Insat 1C* was positioned over 93.5° E to provide communications and meteorological services via 12 C-band transponders and two S-band transponders. The Ford Aerospace satellite consisted of a box structure, 5 ft. 1 in. × 4 ft. 8 in. × 7 ft. 2 in., supporting a 123.79 sq. ft. solar array providing 1,200 W. On July 29 *Insat 1C* suffered a 50% power loss due to an electrical short circuit, but the satellite was able to provide limited service on half its transponders. *ECS-5* was positioned at 16° E, then moved to 13° E for operational use.

1988 July 29 The U.S. Navy issued a contract to Hughes Space and Communications for a UHF Follow-On (UFO) satellite based on the Hughes HS 601 bus. With a weight of 2,300 lb. on station, the UFO would have a solar array span of 60 ft. 6 in., producing 2,400 W electrical power, and 39 channels hardened against nuclear attack. UFO would have 70% more communications capability than the FltSatCom satellites it was designed to replace and would be built for a 14-year life versus 5 years for FltSatCom. The Navy planned to orbit 10 UFOs between 1992 and 1996.

1988 July A U.S. Defense Resource Board, chaired by Deputy Secretary of Defense William Taft IV, rejected a U.S. Air Force plea that funding for the Milstar advanced military communication satellites be spread across the four armed forces. Continuous delay and technical problems had resulted in cost overruns of this high-priority project which was to provide secure voice and data communications between the National Command Authority and nuclear forces. The giant cruciform satellite was designed with solar wings that span more than 50 ft. and a weight of around 8,000 lb. The Air Force planned to have six satellites in the constellation, with flights beginning mid-1993.

Concluding a four-year examination of the nation's space science program, the U.S. National Academy of Sciences recommended an ambitious plan for the future development of new programs. Showing little enthusiasm for the proposed Freedom space station, it specifically recommended the launch of five geosynchronous environment monitoring satellites, two to six polar orbiting platforms for low-altitude observation of regional surface and atmospheric changes, and specialized Explorer-type missions to observe specific aspects of the Earth-sun interface. Orbiters and deep-penetration probes into the atmospheres of the outer planets characterized planetary recommendations, with sample-return missions to Mercury, the moon, Mars, Venus, asteroids and comets.

The United States, the Soviet Union, Canada and France agreed to a 15-year pact on the SARSAT (search-and-rescue satellite system) network. To date, more than 1,150 lives had been saved by rescue forces alerted via distress signals relayed through transponders on selected satellites. The concept had been so successful that the International Maritime Organization now required all vessels over 300 tons displacement to carry emergency transmitters by 1993, and the U.S. Coast Guard issued a rule that all fishing boats carry them by August 1989.

The British government decided against further financial support for studies of the HOTOL reusable satellite launcher. Kenneth Clarke, outgoing minister for space (as well as mail, telecommunications and films) at the Department of Trade and Industry, said that the scale of funding required to get HOTOL built was "far too great for the United Kingdom to contemplate on a national basis." He was, however, hopeful of European participation, should that be forthcoming. As a result of this disinterest, HOTOL designers and propulsion expert Alan Bond took steps to garner international interest on HOTOL and its RB-545 engine.

1988 August 10 On Launch Complex 39B, the Shuttle *Discovery* completed a 21.8-sec. flight readiness firing (FRF) of the three Space Shuttle main engines, qualifying the vehicle for orbital flight. The test also served to give ground and launch crews a full dress rehearsal for a launch attempt. It had been 31 months since the launch control center at the Kennedy Space Center had sent a Shuttle into space, and the readiness firing helped tweak procedures in a real-time simulation culminating in ignition of the SSMEs. The crew were not on board *Discovery* for the FRF.

1988 August 17 Integrating Shuttle II studies conducted at a number of NASA field centers, the Office of Space Flight briefed managers on new concepts for a Shuttle successor. Called the Advanced Manned Launch System (AMLS), it was a multiphase program based on a buy-as-you-earn philosophy. In the first phase, a second-generation orbiter with canards and upswept wing tips would use the standard Space Shuttle main engines, external tanks and solid rocket boosters. When revenues warranted, the new orbiter would adopt second-generation liquid propellant boosters. For fully reusable operations, a new winged fly-back booster would carry the second-generation orbiter for a more efficient and cost-effective system.

1988 August 29 At 04:23 UT Soviet cosmonauts Vladimir Lyakhov and Valery Polyakov, and Afghan cosmonaut Abdul Ahad Mohmand were launched aboard the *Soyuz-TM 6* spacecraft by SL-4 from the Baikonur cosmodrome. They docked to the back of the *Kvant* module at 05:41 UT on August 31. Polyakov was a physician and remained with Titov and Manarov aboard the *Mir* complex on an extended stay. On September 6 Lyakhov and Mohmand made ready to return in *Soyuz-TM 5* and pulled away from *Mir*, leaving their spacecraft for Titov, Manarov and Polyakov. Detecting a fault in one of the attitude sensors, the reentry system shut down the retromotor after 1 min. of a 3 min. 50 sec. burn. Seven minutes later the motor restarted automatically, but was manually shut down by the crew 3 sec. later because it would have brought the spacecraft to Earth in China. Two orbits later they tried again, but the engine shut down seconds later because a computer program left in the spacecraft for the Bulgarian mission on June 7 gave incorrect instructions. In the cramped descent module, the crew had no food and no provision for waste disposal. The final attempt to get back came at 00:01 UT on September 7, and *Soyuz-TM 5* landed 49 min. later at an elapsed time of 8 days 20 hr. 27 min.

1988 August Further studies by NASA's Jet Propulsion Laboratory on the Comet Rendezvous/Asteroid Flyby (CRAF) mission finalized a flight plan that would target the comet Kopff as the primary objective. The Mariner Mark II would now be launched in August 1995 on a mission toward the asteroid belt, returning for an Earth flyby in July 1997 during which it would obtain the energy to move far out toward Jupiter, encountering the asteroid 449 Hamburga in January 1998. CRAF would rendezvous with Kopff in August 2000 at a distance of 124,000 mi., creeping up on the comet until it began orbiting it at a range of only 40 mi. In July 2001 CRAF would drop a penetrometer to the nucleus and relay data to Earth until the mission ended in March 2003 after the perihelion passage.

The U.S. Air Force selected three companies to carry out design work on an advanced launch system (ALS) for lifting payloads of 100,000–200,000 lb. into low Earth orbit. The winners in a competition for which six contractors placed bids, were Boeing Aerospace, General Dynamics' Space Systems Division and a Martin Marietta astronautics team. The work was to last 25 months, after which a final concept would be selected for full-scale development. In the competition, Boeing proposed a fly-back booster with six liquid propellant engines for launch and four air-breathing engines for atmospheric flight. General Dynamics proposed a side-by-side arrangement of first-stage and booster elements of identical size, but with more engines on the primary stage to which the bulbous payload module was attached. The Martin Marietta-McDonnell Douglas design envisaged a cryogenic core stage to which one or two winged fly-back boosters would be attached.

1988 September 2 A Titan 34D/Transtage launched from Cape Canaveral, Fla., failed to place its secret payload in geostationary orbit as planned. The launcher performed as scheduled up to the apogee motor burn, but the Transtage failed to reignite a second time to circularize the orbit. The payload, a Vortex electronic intelligence gathering satellite, was left in an orbit of approximately 100 × 25,660 mi. Previously known as Chalet, Vortex is flown in a four-satellite constellation. The first two in the series had been in orbit for almost 10 years.

1988 September 6 A new Chinese launch vehicle, the CZ-4 (Long March 4), operating from a new site, Tai Yuan, carried a new weather satellite, *Feng Yun 1,* to a sun-synchronous orbit of 545 × 559 mi. inclined 99.2°. The CZ-4 was a three-stage launcher powered by UDMH-N_2O_4 propellants, with a first-stage thrust of 661,500 lb. and a capacity to put 5,500 lb. in polar orbit. CZ-4 stood 137 ft. 10 in. tall with a launch weight of 549,000 lb. *Feng Yun 1* weighed 1,650 lb. with two very high-resolution scanning radiometers in visible and infrared for cloud, ice, snow, land and sea surveys. It measured 3 ft. 11 in. × 4 ft. 7 in. × 4 ft. 7 in. with solar arrays spanning 28 ft. 3 in. Attitude control problems caused *Feng Yun 1* to fail after 39 days.

A one-third scale model of Japan's H-II launch vehicle, the TR-1, was test fired for the first time from Tanegashima. Although the TR-1 simulated the exterior shape of the H-II, unlike the full-size H-II its main stage was a solid propellant motor delivering a thrust of 137,000 lb. for 51 sec. With two inert simulated strap-on boosters, the TR-1 had a length of 47 ft., a diameter of 3 ft. 7 in. and a gross weight of 26,000 lb. Rocket motor burnout occurred at a height of 19 mi., after which the TR-1 coasted to an apogee of 53 mi. from which it fell back and was recovered by ship two hours later. A second TR-1 flight was conducted on January 27, 1989, followed by a third flight on August 20, 1989. The flights were made to obtain aerodynamic data and confirm SRB (solid rocket booster) separation characteristics.

1988 September 8 An Ariane 3 (V25) launcher carried the *GSTAR 3* and *SBS-5* communication satellites from Kourou, French Guiana, to geostationary transfer orbits. Just 37 sec. into the *GSTAR 3* apogee motor burn the satellite veered off course due to improper mass balancing. It was placed in a 10,327 × 22,450 mi. orbit inclined 1.7°. Using on-board thrusters, *GSTAR 3* was very gradually shifted to a more usable orbit, but propellant consumption cut service life from eight to two years. Similar to its numerical predecessors, the Hughes HS 376 *SBS-5* was placed over 122° W.

1988 September 9 The Soviet unmanned cargo tanker *Progress 38* was launched by an SL-4 from the Baikonur cosmodrome to an Earth orbit rendezvous trajectory with the *Mir* complex. *Progress 38* docked to the back of the *Kvant* module on September 12, that port having been vacated by *Soyuz-TM 6* when the crew moved it to the forward axial port shortly after *Soyuz-TM 5* returned on September 7. *Progress 38* was several times used to modify the space station's orbit. In an orbit of 209 × 226 mi., *Mir* cosmonauts Vladimir Titov, Musa Manarov and Valery Polyakov unpacked the vehicle before it was undocked and returned to the atmosphere on November 23.

1988 September 19 Israel became the ninth country to build and launch its own satellite when a launch vehicle called Shavit (Comet) carried a 342-lb. satellite called *Offeq-1 (Horizon 1)* from a launch site in the Negev Desert to a 155 × 715 mi., 142.9° orbit. The launch was a qualification of the three-stage launcher, a converted Jericho missile, and essential satellite technologies. Solar cells provided 246 W of power and the payload consisted of S-band communications and an active thermal control system. It also carried sensors and equipment designed for a future surveillance satellite. The retrograde orbit was a consequence of launching in a northwesterly direction over the Mediterranean to avoid populated areas. *Offeq-1* decayed on January 14, 1989.

1988 September 29 After a 32-month hiatus in manned flight operations and a 1 hr. 38 min. hold for fuses to be replaced in a space suit, NASA returned to space with the orbiter *Discovery* from LC-39B at the Kennedy Space Center, Fla. Known as STS-26, the launch designation reverted to the system used prior to the 41-B flight. Launched at 15:37:00 UT, *Discovery* carried astronauts Frederick Hauck (CDR), Richard Covey (PLT), John Lounge (MS), David Hilmers (MS) and George Nelson (MS). The 46,478-lb. payload comprised the *TDRS-3* tracking and data relay satellite, plus its IUS boost stage, which was deployed at 6 hr. 13 min. elapsed time as well as a variety of scientific and technical investigations and experiments. The mission ended at Edwards AFB, Calif., 4 days 1 hr. 0 min. 11 sec. after liftoff. *TDRS-3* was eventually placed at 171° W.

1988 October 4 A U.S. Defense Acquisitions Board approved changes to Strategic Defense Initiative plans which would deploy space-based interceptor (SBI) kinetic-kill vehicles in the first phase. In a restructuring aimed at lowering costs, five teams would examine competing technologies and SBI options, including the idea Brilliant Pebbles, in which small, high-technology orbiting devices would be responsible for antimissile defense. Conceived at the Lawrence Livermore National Laboratory, it called for the orbiting of 100,000 tiny interceptors instrumented to act independent of control. The board concluded that the first phase of SBI deployment could take place in the late 1990s without the advanced launch system.

Preliminary NASA studies on a Mars Rover–Sample Return (MRSR) mission indicated that optimum launch time for budget and planetary alignments envisaged a typical launch in December 1998 with arrival at Mars in October 1999. After eight months of orbital and site surveys, the rover and sample-return elements would descend and land, leaving the orbiter to continue orbital science work. After retrieving samples, the ascent vehicle would lift off in December 2000 and return to Earth in August 2001. Two complete spacecraft

would be involved, typical weights being 4,900 lb. for one and 7,400 lb. for the other.

1988 October 17 An electrical fire broke out in a test battery and battery harness attached to the Magellan spacecraft at the Kennedy Space Center, Fla. Magellan was being prepared for a launch on Shuttle STS-30 currently planned for April 28, 1989. The fire broke out when a technician in the spacecraft assembly and encapsulation facility put a connector in the wrong socket and flames reached a height of 3 in. Soot and grime had to be cleaned from areas of the spacecraft itself.

1988 October 28 An Ariane 2 launch vehicle (V26) carried the 4,699-lb. French direct broadcast satellite *TDF 1* from Kourou, French Guiana, to a geosynchronous transfer orbit. The satellite fired its apogee motor to circularize the orbit at geostationary altitude over 19° W. Built by Aérospatiale as the French contribution to a Franco-German direct broadcast program announced in October 1979, *TDF 1* comprised a box structure and, on station, had a height of 20 ft. 4 in., a solar array wingspan of 63 ft. 4 in. and a weight of 2,900 lb. Six 230-W traveling-wave-tube amplifiers supported five simultaneous TV channels in the 18/12 GHz band. The solar arrays provided 4,300 W at beginning of life and 3,300 W after eight years, the design life. Due to the failure of one-half the German *TV-SAT 1* capacity, one of *TDF 1*'s five channels was provided to the Germans.

1988 October Late in the month representatives from all the NASA field centers met with headquarters staff to compile a strategy of Mars exploration culminating in a manned expedition. A "fast-track" concept envisaged a three-person Mars expedition departing an Earth-orbit assembly point early in the 21st century and using aerobraking for Mars orbit capture. The crew would explore the surface for 20 days and return to Earth after a round trip of 14 months. In parallel, lunar base build-up would proceed toward the use of industrial facilities for extraction of gaseous helium 3 for use on Earth in fusion reactors for electrical power.

1988 November 1 NASA's Goddard Space Flight Center issued a request for bids on design and development of Space Station *Freedom*'s flight telerobotic servicer (FTS). The FTS was to be capable of performing diverse tasks, including installing fixtures on the truss, changing orbital replacement units, relieving astronauts on EVAs (extravehicular activities), and so on. The FTS was expected to operate from *Freedom* but to be capable of being carried on the front of an orbital maneuvering vehicle for application in distant orbits.

1988 November 15 The first flight of the Soviet reusable shuttle, *Buran (Snowstorm)*, took place from Baikonur cosmodrome with a launch by an SL-17 at 03:00 UT (08:30 local time). The culmination of a 16-year effort that began when Valentin Glushko redirected the Soviet manned space effort from moon landings to shuttle development, this historic event lasted 3 hr. 25 min., ending when *Buran* touched down on a specially prepared landing strip 7.5 mi. from the launch pad. Carried into orbit by the massive SL-17 launcher, *Buran* had a length of 118 ft., a wingspan of 79 ft., a double delta (78°/45°) wing, a loaded weight of 231,400 lb., a payload capability of up to 66,100 lb. and a down-load capability of 44,100 lb. The cargo bay is exactly the same size as that of the NASA Shuttle (60 × 15 ft) and the pressurized volume in the crew area is 2,472 cu. ft. versus 2,325 cu. ft. for the Shuttle. Liftoff of the unmanned, 5.37-million-lb. shuttle came 8 sec.

Unlike NASA's Shuttle, the Soviet Union's Buran *reusable spacecraft was moved to its launch pad in the horizontal position on rails;* Buran *made its one and only flight on November 15, 1988.*

after ignition of the four SL-17 core engines and four strap-on boosters, producing a total thrust of 7.9 million lb., increasing to 8.9 million lb. at booster separation. The core stage continued to generate 1.8 million lb. of thrust until 8 min., when it shut down, separated, and began a gradual descent toward destructive reentry over the South Pacific Ocean, 40 min. after launch. About 45 min. after liftoff *Buran* fired its two 6,800-lb. thrust orbit insertion motors to achieve a 154 × 159 mi., 51.6° orbit. For this flight *Buran* was powered by large chemical batteries, but 30-kW fuel cells had been developed for manned space missions. Primary reaction control system thrusters have a thrust of 660 lb. At 2 hr. 20 min. these thrusters turned *Buran* around and its orbit thrusters fired for de-orbit. The shuttle turned round again and pitched up 40° for reentry at the end of the second orbit, where it was protected from burning up by 38,000 thermal insulating tiles. Auto-land systems brought *Buran* in on a 20° glideslope, and it touched down on the 15,000-ft. runway, 7.5 mi. from its launch pad, at 211 MPH. Three parachutes deployed 14 sec. later. Considerable thermal damage had been experienced, and in an area along the underwing near the leading edge, plasma entered and melted 7.8 in. of the alloy wing structure. In 1994 the Russian shuttle program was abandoned.

1988 November 17   Australia announced that it had selected China's CZ-2E launch vehicle to place its Aussat-B series communication satellites in orbit. Prime contractor Hughes Aircraft and the U.S. State Department had been involved in the negotiations because objections to technology transfer had been lodged with COCOM about the prospect of a Chinese launcher being used to orbit the *Asiasat 1* satellite built by Hughes. Hughes signed the launch contract on behalf of Aussat because schedules demanded that a decision be ratified. On December 17 the State Department withdrew all objections, on the basis that China would launch no more than nine satellites in the next six years.

President Ronald Reagan signed the NASA Authorization Bill for 1989 which included language to reconstitute the National Space Council, abolished by the Nixon administration more than 20 years earlier. Chaired by the vice president, it would become active February 1, 1989, and oversee national space policy and how that was being conducted. When formed, the National Space Council had been considered a valuable organ by all space-participating government agencies.

1988 November 18   A prototype space-based interceptor was tested for the first time at the National SBI Flight Test Facility, Edwards AFB, Calif. The 150-lb. SBI model was 4 ft. long and had a down-firing 335-lb. thrust rocket to counteract gravity, permitting it to hover as though in space. The object of the test was to demonstrate stability and the ability of sensors to orient the model using tiny thrusters. The test lasted only 2 sec. and was aborted when the sensors picked a false target and locked on to light coming through a nearby door! A second test on December 11 demonstrated on-board software calculations for target alignment.

1988 November 24–25   At a meeting in Paris, France, of the European Space Agency's Science Program Committee, members agreed to make the Cassini/Titan probe mission the next major science project for the ESA. In a collaborative agreement with NASA, the ESA would build the probe carried by a Mariner Mark II spacecraft built at the Jet Propulsion Laboratory. Launch on a Titan 4/Centaur was now scheduled to take place in April 1996, followed by a return to Earth and a gravity-assist flyby in June 1996. Cassini would encounter the asteroid 66 Maja in March 1997, fly by Jupiter in February 2000 and reach Saturn in October 2002. The ESA probe would be released into the atmosphere of Saturn's moon Titan later that year, and the orbiter would continue to send back information about Saturn until 2006. In 1989 the ESA named the entry probe *Huygens*, after the Dutch astronomer Christian Huygens.

1988 November 26   At 15:49 UT Soviet cosmonauts Aleksandr Volkov and Sergei Krikalov and French cosmonaut Jean-Loup Chrétien were launched aboard their *Soyuz-TM 7* spacecraft by an SL-4 from the Baikonur cosmodrome. Meeting up with the *Mir* space station, in an orbit of 202 × 217.5 mi., they docked to the back of the *Kvant* module at 17:16 UT on November 28 and joined cosmonauts Vladimir Titov, Musa Manarov and Valery Polyakov. For 23 days the six cosmonauts worked together on joint experiments, the longest period such a large team had been aboard a Soviet space station. When Shuttle STS-27 was launched on December 2, there was a total of eleven people in space aboard Soviet and U.S. space vehicles.

1988 December 1   A technician at the Kennedy Space Center, Fla., accidentally ruined the rocket nozzle of an inertial upper stage motor which was to have been launched aboard Shuttle *Discovery* for the STS-29 mission planned for February 18, 1989. The IUS would have been used to boost the TDRS-D tracking and data relay satellite into geostationary orbit. It was ruined when the technician fell against the exhaust cone, causing a 4 in. crack to appear. Rescheduled for February 23, STS-29 was delayed until March 13, 1989, when the turbopumps on all three orbiter main engines had to be replaced due to suspected faults.

1988 December 2   At 14:30:34 UT the Shuttle *Atlantis* was launched from LC-39B at the Kennedy Space Center, Fla., carrying astronauts Robert Gibson (CDR), Guy Gardner (PLT), Richard Mullane (MS), Jerry Ross (MS) and William Shepherd (MS), on mission STS-27. Postponed for 24 hr. due to poor weather, the flight was a classified Department of

Defense mission and carried the *Lacrosse* imaging radar satellite capable of detecting ground and air activity day or night, in good weather or bad, penetrating ground foliage and camouflage. Occupying almost the entire Shuttle payload bay, the Martin Marietta satellite carried sensors and antennas that gave it a total span of 150 ft. when fully deployed. Originally intended for a Shuttle launch from Vandenberg AFB, Calif., *Lacrosse* was released at about 7 hr. elapsed time and made its way by thruster power to a 258 × 272 mi., 57° orbit. Over the next several weeks that orbit was raised to 415 × 437 mi. *Atlantis* landed at Edwards AFB, Calif., 4 days 9 hr. 5 min. 37 sec. after liftoff.

1988 December 9   French cosmonaut Jean-Loup Chrétien and Soviet cosmonaut Alexandr Volkov left their comrades in the *Mir* space station for a 5 hr. 57 min. space walk. They attached a French materials experiment to the exterior of *Mir* to be retrieved later for analysis on Earth, and they erected a structure built by French aerospace manufacturer Aérospatiale. There were difficulties with the structure until, out of radio contact with Earth, Volkov gave it a hefty kick with his boot and it instantly deployed!

1988 December 11   The first commercial launch of an Ariane 4 configuration (44LP) on mission V27, with a liftoff weight of 919,679 lb., carried two communication satellites to geosynchronous transfer orbit. Ariane 4 lifted the 3,160-lb. *Skynet 4B*, a United Kingdom military communication satellite weighing 1,742 lb. on station and located at 1° W, and the 3,925-lb. *Astra 1A*, a direct-broadcasting satellite for Luxembourg, located at 19.2° W and with an on-station weight of 2,304 lb. *Skynet 4B* was the first of three scheduled for launch. Built by British Aerospace, it consisted of a box structure 6 ft. 11 in. × 6 ft. 3 in. × 4 ft. 7 in. with two solar array wings spanning 52 ft. 6 in. Military communications was facilitated in the 7–8 GHz band, and in the 250–260 MHz and 305–315 MHz bands with 40-W transmitters. *Skynet 4A* was launched to 6° E by Titan 3 on January 1, 1990, followed by *Skynet 4C* to 1° W by Ariane 4 on August 30, 1990.

1988 December 21   Soviet cosmonauts Vladimir Titov and Musa Manarov, and French cosmonaut Jean-Loup Chrétien, returned to Earth in the *Soyuz-TM 6* spacecraft. It had been delivered to the *Mir* after launch on November 26 and docked to the back of the *Kvant* module. Cosmonauts Aleksandr Volkov, Sergei Krikalov and Valery Polyakov remained aboard the orbiting complex. Titov and Manarov had set a new world spaceflight endurance record by spending more than one full year in space, returning to Earth at an elapsed time of 365 days 22 hr. 39 min. Chrétien's flight had lasted 24 days 18 hr. 8 min.

1988 December 25   The unmanned cargo tanker *Progress 39* was launched from Baikonur cosmodrome by an SL-4 launch vehicle. The 15,900-lb. spacecraft performed a rendezvous and docking with the *Mir* space station, in an orbit of 202.5 × 217.5 mi., two days later. *Progress 39* was employed to boost the orbit to higher altitude as well as to provide additional supplies, stores and experiment equipment for cosmonauts Alexandr Volkov, Sergei Krikalov and Valery Polyakov. *Progress 39* was undocked from the back of the *Kvant* module on February 7, 1989, and returned to destruction in the atmosphere on the same day.

1989 January 3   E. Ray Tanner replaced Thomas Moser as associate administrator for space station *Freedom*. Tanner

had been manager of the Space Station Projects Office at NASA's Marshall Space Flight Center. Moser had been appointed deputy associate administrator for Space Station effective December 1.

1989 January 6   In its fiscal year 1990 budget request, NASA included $30 million for a funded start on the Comet Rendezvous/Asteroid Flyby and Cassini missions. Each would use a Mariner Mark II spacecraft as the common bus, reducing development and operations costs. NASA envisaged a CRAF launch in August 1995, followed by Cassini in April 1996. Concern at the combined costs of these large spacecraft, and problems with NASA funding levels as large programs like the *Freedom* space station took an increasingly high percentage of the NASA budget, caused the CRAF mission to be dropped from the budget request for fiscal year 1993 when it was presented in January 1992.

1989 January 10   The first in a new series of Soviet geodetic satellites called *Etalon* was launched on an SL-12 from Baikonur cosmodrome to a 11,870 × 11,900 mi., 64.9° orbit. *Etalon* was launched along with two *Glonass* navigation satellites. The 3,120-lb. satellite took the form of a 4 ft. 3 in. sphere, its surface covered with 306 separate reflector arrays. Each array comprised a cubical corner reflector with metal faces to reflect laser beams transmitted from the ground. The great mass : volume ratio of *Etalon* ensured it would remain in stable orbit for range-finding data from the laser system.

1989 January 29   The Soviet spacecraft *Phobos 2* arrived at Mars and was successfully placed in an elliptical, 537 × 49,500 mi. orbit. The scientific instruments began surveying the planet on February 1, and NASA maintained a busy tracking schedule to assist the Soviets and provide accurate location data for a planned test of the 220-lb. lander that was scheduled to separate from the main spacecraft and "dock" with the moon Phobos. In preparation for this activity, the orbit of Phobos was circularized on February 12. The mission plan envisaged the *Phobos 2* spacecraft performing a slow flyby of Mars' moon at a distance of 160 ft. while releasing a lander. This was designed to land on the moon Phobos for about one year while a smaller "hopper" type mobile lander conducted separate experiments moving around on its surface. Before these activities could be carried out, on March 27 all contact with *Phobos 2* was irretrievably lost and the mission abandoned. Some scientists believed the spacecraft had been struck by meteorites, showers of which had been seen from Earth.

1989 January   Senior Congressional Armed Services Committee members urged President George Bush to examine a basing mode for land-based ballistic missiles. They presented an idea which foresaw varying numbers of missiles to be deployed in capsules hardened against nuclear attack and dispersed among several thousand mock shelters. This deceptive basing would have the advantage of being adapted to Peacekeeper, Midgetman or Minuteman III ICBMs. Successive U.S. political leaders had unsuccessfully wrestled with the concept of survivability for the land-based missile force, and it was felt that this idea would separate missiles from basing modes and thereby make it suitable for whatever missile was chosen. There was little enthusiasm for the concept.

1989 February 1   Lt. Gen. George L. Monahan replaced Lt. Gen. James Abrahamson as director of the Strategic Defense Initiative organization at the Department of Defense. Advocates of the antiballistic missile research program feared support for the program would slip without the energetic and deeply committed stance taken by the outgoing head. Opponents saw it as a move by the Bush administration to lower the visibility of this controversial research effort. Abrahamson sent a memo February 10 endorsing the Brilliant Pebbles concept for early operational deployment around the mid-1990s.

1989 February 10   The unmanned *Progress 40* cargo tanker was launched by SL-4 from the Baikonur cosmodrome. Weighing 15,985 lb., *Progress* docked to the *Mir* complex, in an orbit of 215.5 × 225.5 mi., two days later. In addition to supplies and stores, the cargo vehicle also brought a special, deployable structure which would automatically unfold from the docking tunnel when the vehicle separated from the back of the *Kvant* module to which it had docked. Undocking occurred on March 3, and cosmonauts Alexandr Volkov, Sergei Krikalov and Valery Polyakov took pictures of the structure as it unfolded. *Progress 40* conducted maneuvers for two days and then de-orbited.

1989 February 13   Florida Gov. Bob Martinez unveiled a plan seeking support from the State Legislature for commercial launch facilities at Cape Canaveral and an adjacent area. Called Spaceport Florida, the plan envisaged a sounding rocket launch complex and reactivation of launch pads 14, 15 and 16, first used for Atlas, Titan I, and Titan II and Pershing, respectively. Martinez claimed it would be cheaper to lease the sites from the government and refurbish them for commercial launch rights than to build new pads.

1989 February 14   The first U.S. Air Force Navstar Block II global positioning satellite was launched by the first Interim Delta II 6925 from Cape Canaveral, Fla. Delta 6925 differed from the 3920/PAM-D in having an 11 ft. 10 in. extension to the first stage, increasing height from 112 ft. to 125 ft. 11 in., with a burn time of 4 min. 25 sec. First in a series of Block II Navstars incorporating updated technology, the 3,670-lb. satellite (1,850 lb. on station) was placed in an orbit of 12,439 x 12,643 mi. inclined 55.1°. Each Navstar Block II had a length of 7 ft. 10 in., a diameter of 5 ft. 11 in. and a span of 17 ft. 4 in. across the solar cell wings. The Air Force planned to launch Navstar at an average of one every two or three months until a constellation of 21 operational satellites and 3 spares were in orbit. At this date the Air Force had 7 operational Block I Navstars in orbit. The 21st Navstar II was launched on June 26, 1993, the 13th in a series of Block IIA satellites.

1989 February 21   A Japanese Mu-3S II solid propellant launcher carried the 650-lb. satellite *Exos-D* to a 163 × 6,227 mi. orbit inclined 75.1°. The satellite was instrumented to observe and measure characteristics of the Aurora Borealis and the Aurora Australis, the "northern lights" and the "southern lights," respectively, caused by charged particles from the sun spiraling down the magnetic poles of the Earth. It remains in orbit.

1989 March 2   NASA's Marshall Space Flight Center announced the award of five contracts to three companies for technology development on new propulsion systems for the advanced launch system. The contracts went to Aerojet General, Rocketdyne and Pratt and Whitney. The work required the companies to produce turbomachinery for liquid oxygen, liquid hydrogen and liquid methane propellants. Despite its

early opposition to the ALS, preferring to project a Shuttle-C for heavy-lift cargos, NASA had been required by Congress to support what was now considered a joint Air Force/NASA venture, despite the fact that military payloads were the only justifiable reason for the heavy-lift concept.

1989 March 6   An Ariane 44LP (V29) with full complement of liquid and solid strap-on boosters lifted off from Kourou, French Guiana, carrying the 5,030-lb. Hughes *JCSAT 1* communication satellite for Japan and the first operational Meteosat weather satellite, *MOP-1*. Stowed for launch, *JCSAT 1* had a height of 11 ft. 10 in. and a diameter of 12 ft. With the telescoping lower solar array deployed, it had a height of 33 ft. and an on-orbit mass of 3,035 lb. First in the new series of HS 393 satellite designs, *JCSAT 1* carried 32 operating channels with 40 × 20-W power amplifiers serving in the 14/12 GHz band through a single 7 ft. 10 in. dish antenna feeding the four main Japanese islands and Okinawa. It was positioned over 150° E. Placed at 0° longitude, the 1,500-lb. Meteosat *(MOP-1)* satellite was almost identical to the previous three Meteosats sent into orbit.

1989 March 13   At 14:57:00 UT Shuttle *Discovery* was launched from LC-39B at the Kennedy Space Center, carrying astronauts Michael Coats (CDR), John Blaha (PLT), James Buchli (MS), Robert Springer (MS) and James Bagian (MS). Designated STS-29, the mission had been delayed 1 hr. 50 min. due to ground fog and upper winds. The 47,424-lb. payload included the *TDRS-D* tracking and data relay satellite, released at 21:10 UT and sent to geostationary orbit at 41° W by its inertial upper stage. Other experiments included an experiment to evaluate space station *Freedom* technologies. Early in the flight, one of the three fuel cells fluctuated and the crew switched off unnecessary equipment to conserve electrical energy. After ground tests simulated the malfunction the crew reactivated the cell on March 15 and it worked until the end of the mission. *Discovery* landed at Edwards AFB, Calif., at 4 days 23 hr. 38 min. 52 sec.

A huge blast of x rays and charged particles from the sun hit Earth, causing the collapse of the entire electrical power system across Canada's Quebec province. The blackout was total within 90 sec. and lasted nine hours. In the United States and other parts of Canada transformers burned out and a New Jersey nuclear electrical power plant had to be shut down for six weeks. From Washington, D.C., through New York and New England, a total collapse was only narrowly avoided. The loss of revenue caused U.S. and Canadian power companies to plan a solar-storm warning satellite which would allow engineers to reconfigure power lines in the event of a surge.

1989 March 13–14   The Council of the European Space Agency approved selection of the French Matra polar platform over a competing design from British Aerospace in the the United Kingdom. British Aerospace had done much of the basic definition and design of free-flying, man-tended space platforms that would form a component of the *Columbus* program. Designed for launch on Ariane 4 and Ariane 5, the platform would be electronically linked to the *Columbus* manned pressure module but operate from polar orbit rather than the 28.5° orbit of space station *Freedom* to which the manned module would be attached.

1989 March 15   The Intelsat board authorized the award of launch contracts to Arianespace and General Dynamics for its seventh-generation international telecommunications satellites. Ariane 4 would launch three Intelsat 7 satellites, and

General Dynamics would launch two on the newly proposed Atlas 2AS. Like the configuration of an Atlas 2A, the 2AS series would have four strap-on Castor II solid propellant motors attached in pairs either side of the sustainer engine housing. These would raise geosynchronous transfer orbit capability to 8,150 lb. Intelsat envisaged the first launch on an Ariane 4 in late 1992, followed by the two Atlas 2AS flights and one Ariane 4 in 1993 and one Ariane 4 in 1994.

1989 March 16 *Progress 41* was launched by an SL-4 from the Baikonur cosmodrome, bringing supplies and stores to cosmonauts Aleksandr Volkov, Sergei Krikalov and Valery Polyakov aboard the *Mir*. The unmanned cargo tanker docked to the back of the *Kvant* module two days later, in an orbit of 216 × 225.5 mi. *Progress 41* performed an orbital adjustment maneuver for the space shuttle on April 10 as the crew prepared to leave for Earth on April 26. It had been planned to have the crew replaced in April, but further delays to the set of modules planned for expanding the complex brought cancellation and a period of unmanned activity for *Mir*. *Progress 41* was undocked from *Kvant* on April 21 and was de-orbited on April 25.

1989 March 18–19 At a NASA management council meeting, administrators discussed technical changes to space station *Freedom* being considered as a result of budget pressures and doubts about whether the Shuttle could effectively launch the various elements necessary for on-orbit assembly. Under a revised plan, solar collector devices 50 ft. in diameter were proposed instead of solar cell arrays which had been planned for Phase I in the station build-up cycle.

1989 March 19 The last of 18 development flight tests in the Peacekeeper missile program was successfully completed. The missile was launched from Vandenberg AFB, Calif., carrying seven Mk.21 MIRVs across the Pacific Ocean to the Kwajalein Atoll in the Marshall Islands. Air Force planners expected to launch 20 flight development rounds, but this test accomplished all objectives. It was the first Peacekeeper to be controlled from an airborne launch control center.

1989 March 21 In congressional hearings into the Strategic Defense Initiative organization's projections for Phase 1 operational deployment, the director, Lt. Gen. George Monahan defended against criticism his $69 billion price tag for an initial defense capability. The General Accounting Office, in an independent analysis, said the true cost would swell to $120 billion if annual operating costs were included. Less than two years earlier, projected Phase 1 deployment cost ran as high as $145 billion, but this had been reworked when senior Department of Defense officials ordered cuts.

1989 March 24 The last Delta 3920 placed a military satellite in orbit for the Strategic Defense Initiative ballistic missile defense research program. Launched from Cape Canaveral, Fla., the Delta Star satellite was put in a circular, 311 mi., 47.7° orbit for the purpose of observing booster plume characteristics of up to 24 launches over six months for tests of sensors and target-tracking capabilities. The 1,538-lb. sensor module had a diameter of 7 ft. 2 in. and a height of 4 ft. 1 in., attached to a support bus built by McDonnell Douglas. Total spacecraft weight was 6,000 lb. in a package 18 ft. tall and 7 ft. 6 in. across. Two solar panels had a span of 16 ft. Sensors were unable to observe the second stage de-orbit burn as planned due to a slow sensor door. Delta Star reentered June 23, 1992.

1989 April 2 The last Ariane 2 (V30) carried the 4,697-lb. *Tele-X* direct broadcast satellite from Kourou, French Guiana, to geostationary transfer orbit from which it was later placed at 5° E on April 14. Built by an Aérospatiale/Eurosatellite consortium, the satellite had an on-orbit mass of 2,815 lb. and comprised a box structure, 7 ft. 10 in. long, 5 ft. 5 in. wide and 7 ft. 10 in. high, to which was attached an antenna tower that raised total height to 16 ft. 5 in. Solar array wings providing 3,600 W spanned 62 ft. 4 in. *Tele-X* provided two direct broadcast (18/12 GHz) and two business communication (14/12 GHZ) channels via 200-W transmitters.

1989 April 7 The Alpha hydrogen fluoride laser being built for the Zenith Star space-based laser (Sabir) experiment was fired for the first time in a 0.2-sec. test conducted at the TRW Systems facility near San Juan Capistrano, Calif. With a maximum power output of about 2 MW, achievable only after several seconds, the 2.7-micron Alpha laser was already being seen as a dinosaur in the Strategic Defense Initiative ballistic missile defenses research program. Heavy, difficult to operate and expensive to build, the Alpha laser would need a massive launcher to place it in space and was being outclassed by less exotic means of disabling missiles and warheads.

1989 April 26 A Soviet SL-4 launcher carried the 13,670-lb. *Photon 2* materials processing and biological products laboratory from Plesetsk to a 134 × 236 mi., 62.8° orbit. Among other instrumented payloads, the Vostok-derived satellite carried a French commercial microgravity experiment. The reentry module returned to Earth on May 11 after a flight lasting 15 days.

1989 April 27 Soviet *Mir* cosmonauts Alexandr Volkov, Sergei Krikalov and the physician Valery Polyakov returned to Earth in the *Soyuz-TM 7* spacecraft launched on November 26, 1988. Undocking at 23:28 UT on April 26, the spacecraft landed at 02:59 UT on April 27. Since his launch in *Soyuz-TM 6* on August 29, 1988, Polyakov had been in space for 240 days 22 hr. 36 min. Since the launch of Volkov and Krikalov aboard *Soyuz-TM 7* on November 26, 1988, their mission had lasted 151 days 11 hr. 10 min. The *Mir* had been permanently manned since February 7, 1987, for a duration of 808 days 4 hr. 31 min.

1989 April 28 Shuttle mission STS-30 with orbiter *Atlantis* was canceled at T − 31 sec. when a liquid hydrogen recirculation pump on the no. 1 main engine showed a problem and a vapor leak was detected in the 4-in. liquid hydrogen line between the orbiter and the external tank. *Atlantis* was to have carried the *Magellan* planetary spacecraft into orbit as mission STS-30. The launch was canceled due to cloud cover and high winds and was rescheduled for May 4. The Shuttle was eventually launched on May 4.

1989 April A panel commissioned by Congress to examine the plan for NASA to fly unmanned free-flying platforms in the run-up to operations recommended abandoning the idea. NASA had been asked by the Reagan administration to lease the Industrial Space Facility, a privately developed concept for research into the development of commercial activities. NASA had not wanted to spend money in this way and managed to convince the panel that expenditure of resources would not significantly enhance the commercial prospects of the station, which had been the intention.

NASA selected a team led by Lockheed and Aerojet to build the advanced solid rocket motor (ASRM) intended to replace the redesigned solid rocket booster for the Shuttle. NASA wanted to begin using the new motors in 1994, increasing Shuttle payload capability by 12,000 lb. to 65,000 lb. They would each have a length of 126 ft. and a diameter of 12 ft. 6 in., containing an additional 200,000 lb. of propellant.

1989 May 4   At 18:46:59 UT Shuttle *Atlantis* was launched from LC-39B at the Kennedy Space Center carrying the *Magellan* Venus radar mapper and astronauts David M. Walker (CDR), Ronald J. Grabe (PLT), Norman E. Thegard (MS), Mary L. Cleave (MS) and Mark C. Lee (MS). Designated STS-30, the mission's payload weighed 47,813 lb. From an Earth parking orbit of 184 mi., the Shuttle released *Magellan* and its attached inertial upper stage weighing 40,118 lb. at 22:02 UT. The first IUS burn began at 23:01 UT, after *Atlantis* had moved away, producing 42,000 lb. of thrust for 2 min. 40 sec., followed 2 min. 29 sec. later by the 1 min. 45 sec. second-stage burn producing 18,500 lb. of thrust. On course for Venus, *Magellan* separated from the IUS at 11:27 UT. Other experiments included a fluids and materials processing test. *Atlantis* landed at Edwards AFB, Calif., 4 days 56 min. 37 sec. after liftoff.

1989 May 11   The U.S. Air Force Small ICBM, Midgetman, made its first flight from Vandenberg AFB, Calif. The first stage performed as planned, but problems began to develop 1 min. 13 sec. into the flight when trouble with the second stage caused the missile to behave erratically. Range safety signals were not sent to destroy the missile until 2 min. 12 sec. so that telemetry would continue to send valuable data for as long as possible. At that time the third stage had fired. The Midgetman was intentionally blown up when it had traveled 1,300 mi. downrange, and after the nose cone shrouding the simulated Mk.21 warhead had been jettisoned.

1989 May 22   NASA's *Magellan* spacecraft completed its first course correction burn at 02:00 UT by firing four 100-lb. thrusters for 5.3 sec. to change velocity by 9.7 ft./sec. *Magellan* was 2.5 million mi. from Earth traveling at a relative speed of 8,500 ft./sec. with 155 million mi. to go to Venus. Weighing 7,593 lb., *Magellan* was built around a 10-sided bus 1 ft. 5 in. high and about 6 ft. across supporting the synthetic aperture radar (SAR) in a forward equipment module mounted on top, 3 ft. × 4 ft. × 5 ft. 4 in. tall. That supported the high gain antenna (12 ft. in diameter) through which the SAR propagated its 2.385 GHz signal which would strike the surface of the planet at angles of 19–52°. Two 8 ft. 2 in. square solar panels spanned 30 ft. 7 in. and would provide 1.2 kW of electrical energy at Venus. A propulsion module at the base of the spacecraft supported four equally spaced outriggers, each with a cluster of six thrusters: two 100-lb., one 5-lb. and three 0.2-lb. thrust. *Magellan* carried 293 lb. of monopropellant hydrazine for attitude control and course corrections. Attached to the center of the propulsion module, a 4,721-lb. solid propellant Star 48B rocket motor would put *Magellan* into orbit around Venus.

1989 May 24–26   NASA held a conference at the Marshall Space Flight Center for potential Shuttle-C users attended by nearly 400 contractors, government and military representatives. NASA had wanted to develop the unmanned cargo carrier for launching hardware. Other potential customers wanted it for a broader range of applications, including the Department of Defense, which was considering it for launching the 96,000-lb. *Zenith Star* prototype laser battle station. Increasingly, Shuttle-C was fouled by budget pres-

sures and uncertainty about America's fleet of expendable launch vehicles. The concept was abandoned during 1990.

1989 May 25   The first Soviet Earth-resource satellite officially designated as such was launched from Plesetsk when an SL-4 carried a 13,900-lb. Resurs F2 to a 158 × 170 mi., 82.3° orbit. Two types of Resurs satellite were to be employed, all derivatives of the Vostok series. Resurs F2 usually remained in orbit for about 30 days, while Resurs F1 types remained aloft for about 15 days. The first Resurs F2 returned to Earth on June 17 and the first Resurs F1 was launched on June 27, returning on July 11. Resurs satellites could carry three cameras: a KFA 100 spectrazonal camera with a resolution of 16 ft. and capable of taking 1,800 images; a panchromatic MK.4 with 26 ft. resolution on 2,500 frames; and a KATE 200 with 65-ft. resolution on 2,500 images. The second Resurs F2 mission was flown July 18–August 8, 1989, the third F2 was flown August 15–September 14 and the second F1 was flown September 6–22. Three to five Resurs missions were flown each year between May and September.

1989 June 5   An Ariane 44L launch vehicle (V31) carried two powerful communication satellites to geostationary transfer orbit. With a liftoff weight of 1,063,826 lb. this was the first flight of the most powerful variant of the Ariane 4 family. First to be released from the payload structure was the 5,495-lb. *Superbird A*, a Ford Aerospace satellite for Space Communications Corp. of Japan. Derived from the Intelsat V design, *Superbird A* had a main body 7 ft. 11 in. × 8 ft. 6 in. × 7 ft. 3 in., with solar arrays spanning 66 ft. 7 in. and providing 4,460 W. Primarily aimed at the business community from a position over 158° E, *Superbird A* had 19 Ku-band (35-W) and 10 Ka-band (29-W) transmitters. The second satellite, *DFS Kopernikus 1*, was positioned over 23.5° E after three apogee motor firings.

NASA's *Voyager 2* spacecraft began observing Neptune as it approached the planet for a close encounter on August 25. Now almost 12 years into its mission, 2.65 billion mi. from Earth and less than 70 million mi. from its next and last planetary rendezvous, *Voyager 2* began taking images of Neptune every 3 hr. 44 min. Assembled in a sequence that would provide a time-lapse movie of the planet's rotation, the images would continue to be transmitted until August 6. These showed changing patterns in the gaseous outer envelope of the planet of 31,000-mi. diameter. As with its earlier flyby of Uranus, *Voyager 2* would transmit data at 21.6 kilobits/sec. thanks to Earth station upgrades and the use of special receiving antennas on the ground.

1989 June 14   The first in a new generation of U.S. geostationary early warning satellites, the Defense Support Program Block 14 series was carried into space by the first Titan 4 from Cape Canaveral, Fla. In this configuration, the Titan 4 carried an inertial upper stage as third stage. The 5,200-lb. TRW Systems satellite comprised a lengthened cylinder with four square solar-cell paddles attached to the base providing 1,274 W of electrical energy. A 12-ft. infrared telescope built by Aerojet ElectroSystems collected light for a grid array of 6,000 detectors. With this atop the cylinder, the Block 14 series had a total height of 33 ft. and a span of 13 ft. 8 in. across the solar panels.

1989 July 1   Richard H. Truly formally took over the post of NASA administrator, the first astronaut to achieve the highest office in the U.S. civilian space agency. His appointment not only rewarded success as spaceman and manager but also honored a commitment made after the loss of *Challenger* on January 28, 1986, to put more astronauts in positions of authority within NASA. Truly presided over the return to flight of the Shuttle fleet and the inauguration of a replacement orbiter. In a radical restructuring of NASA, the Bush administration sought Truly's replacement and he stood down on April 1, 1992.

1989 July 4   A Soviet SL-8 launcher carried the *Nadezhda 1* civilian navigation satellite from Plesetsk to a 595.5 × 628 mi., 83° orbit. This was the first officially designated Soviet navigation satellite extracted from the Cosmos series, of which there had been a long line dedicated to this function. *Nadezhda* took the form of a cylinder, 6 ft. 6 in. in diameter and 6 ft. 11 in. long with a long boom protruding from the front and a weight of 1,820 lb. It belongs to the same generic family as the geodetic satellites beginning with *Cosmos 1312* and carries COSPAS/SARSAT search and rescue transponders. *Nadezhda 2* was launched on February 27, 1990, followed by *Nadezhda 3* on March 12, 1991.

1989 July 6   Major U.S. aerospace companies were given a secret briefing by Vice President Dan Quayle on a major new space goal President George Bush was to announce on July 20, the 20th anniversary of the first manned moon landing. Members of Congress were briefed seven days later. In his post as chairman of the National Space Council, Quayle had pressed hard for a moon base goal and a commitment to a manned Mars mission, but President Bush was reluctant to give the mandate a timetable or a specific directive. These two briefings were the first steps in a plan formulated by Quayle, who wanted to prepare the way for President Bush to make a formal commitment to these grand objectives in the State of the Union Address in early 1990.

1989 July 12   The last Ariane 3 launch vehicle (V32) carried the European experimental *Olympus* communications satellite to a geosynchronous transfer orbit. Developed at first as L-Sat, Olympus was built by a group led by British Aerospace. The satellite consisted of a box structure, 9 ft. 6 in. long, 8 ft. 11 in. wide and 18 ft. 1 in. high, with a launch mass of 5,722 lb. and an on-orbit weight of 3,197 lb. With a span of 84 ft. 3 in., two solar array wings produce 3,600 W of power. *Olympus 1* provided a 230-W direct broadcast channel for Italian TV and one for international programming throughout

In July 1989, NASA's Johnson Space Center produced a lunar habitat concept comprising an inflatable structure with a diameter of 52 feet and provision for 12 astronauts.

Europe, both at 12 GHz, with signals received on 17-in. dish antenna. A 12–14 GHz specialized service payload provided video conferencing, data distribution and frequency reuse experiments. A 20–30 GHz communications payload provided education, video conferencing and low-speed data handling.

1989 July 13 An Aries sounding rocket fired from the White Sands Missile Range, N.Mex., carried the first neutral particle beam device ever operated in space to an altitude of 125 mi. Launched on behalf of the Strategic Defense Initiative ballistic missile defense program, the BEAR (beam aboard rocket) test provided the first opportunity to study the performance of neutral beam propagation in space. Equipment formed the top 24 ft. of the sounding rocket. The accelerator was turned on about 1 min. 40 sec. after launch at a height of 87 mi. as the vehicle was ascending. The neutral particle beam was fired for about 4 min. with the payload maneuvering in a programmed sequence. The beam propagated about 650 ft. and measurements were made via instrument packages on board. BEAR was recovered by parachute.

1989 July 20 On the 20th anniversary of the first manned moon landing, President George Bush delivered a speech at the National Air and Space Museum, Washington, D.C., calling for a new initiative to put men back on the moon and send expeditions to Mars. He publicly called upon Dan Quayle to use the resources of the National Space Council to draw up a list of requirements for that objective. For his part, Quayle saw himself repeating history by assembling for the President the structure through which Bush could make the national commitment, just as Lyndon Johnson had defined a moon landing goal for President John F. Kennedy.

1989 July 29 A memorandum of understanding was signed between Technopribor of the Soviet Union and Space Commerce Corp. of the United States for the marketing of a satellite launcher called Start, to be based on the SS-20 medium range ballistic missile. With a length of 65 ft. 6 in., and a launch weight between 88,400 lb. and 99,450 lb., the three-stage solid propellant Start could place a 300-lb. satellite in a 310-mi. orbit. Technopribor was a newly formed organization charged with controlling SS-20 production facilities. Space Commerce Corp. specialized in marketing deals with Soviet space organizations.

1989 July Orbital Sciences Corp. won a U.S. Department of Defense contract for a standard small launch vehicle (SSLV) capable of placing relatively lightweight satellites in low Earth orbit. The Department of Defense wanted to speed the process of launching satellites by finding a launcher that could be readied for flight and launched at short notice by few people. The requirement was met by adapting Orbital Sciences' winged Pegasus launcher and the first stage of a Peacekeeper missile. Called Taurus, the launcher would be capable of putting a 3,000-lb. payload in low Earth orbit or placing 860 lb. in geosynchronous transfer orbit.

A study on second-generation Shuttle concepts completed by the Langley Research Center and the George Washington University Joint Institute for the Advancement of Flight Sciences was presented to NASA management. Optimizing existing technology, the two-stage, fully reusable Shuttle envisaged the use of three or four Space Shuttle main engines in a fly-back booster. Two such boosters would power the two-engine orbiter, which would have a payload capability of 37,000 lb., or 70,000 lb. in a configuration in which each booster had four SSMEs and the orbiter three. The study

envisaged a developed Shuttle II concept with an orbiter employing five advanced, 625,000-lb. thrust SSMEs with each of two boosters carrying three such engines. This vehicle would carry an 84,000-lb. payload.

1989 August 1 A kinetic-kill vehicle hover interceptor test (KHIT) was conducted for the first time with a 152-lb. model space-based interceptor built by Rockwell International. Tested at the U.S. Air Force Astronautics Laboratory, Calif., for the Strategic Defense Initiative ballistic missile defense research program, the test lasted 13.5 sec., during which a 325-lb. rocket motor compensated for gravity while hundreds of thruster firings took place to demonstrate rapid maneuvering and targeting capabilities. The vehicle lifted itself to a height of 30 ft. as infrared sensors picked up and tracked a simulated rocket firing at the far end of the test building.

1989 August 8 At 12:37:00 UT, following a 40 min. weather hold, Shuttle *Columbia* was launched on STS-28 into a 188 × 196 mi., 56.9° orbit from LC-39B at the Kennedy Space Center, Fla. Carrying astronauts Brewster Shaw (CDR), Richard Richards (PLT), David Leestma (MS), James C. Adamson (MS) and Mark N. Brown (MS), this mission carried two classified payloads for the Department of Defense. At approximately 7.5 hr., an uprated, high-technology version of the top secret KH-11 digital imaging reconnaissance satellite was deployed from a 2,357-lb. cradle in the payload bay; a second payload was deployed later. *Columbia* landed at Edwards AFB, Calif., at an elapsed time of 5 days 1 hr. 0 min. 56 sec. The 20,617-lb. reconnaissance satellite experienced an attitude control problem and was observed tumbling at a rate of 30 RPM. It was maneuvered several times in orbit and eight days later its altitude was raised to a 286-mi. orbit.

An Ariane 44LP launch vehicle carried the 4,729-lb. German *TV-SAT 2* direct broadcast satellite to a geosynchronous transfer path and the 2,514-lb. astronomical *Hipparcos* satellite to an orbit of 337 × 22,300 mi. inclined 6.9°. Positioned at 19° W, *TV-SAT 2* was identical to *TV-SAT 1*, which had to be abandoned after a solar array wing jammed, and would provide three separate program channels. *Hipparcos* was a European Space Agency astrometry satellite designed to measure the precise positions, parallaxes and motions of stars. The satellite comprised a central box structure with three solar array panels at 120° intervals providing 300 W of power. The apogee motor to put *Hipparcos* in geosynchronous orbit failed to fire, leaving the satellite in its highly elliptical orbit, but one from which astronomers still hoped to get 90% of expected data.

1989 August 15–16 At meetings in Washington, D.C., NASA officials explained to representatives from the European Space Agency that they might have to rescope the assembly sequence of space station *Freedom* once more. Budget pressures were forcing NASA to consider a two-stage station that would leave non-United States elements for launch in a later stage. The basic station would house four people and provide 38 kW of electrical power, 20 kW of which was for essential power demands and only 18 kW for experiments. Adding the European and Japanese modules and raising power to the basic design level of 75 kW would be accomplished around the year 2002. The rescoping had been performed by a team under Ray Hook at the Langley Research Center.

1989 August 22–24 At the second International Conference on Solar System Exploration, held at the Jet Propulsion

Laboratory, Calif., Soviet scientists said their plans for a manned mission to Mars in the year 2015 had been dropped due to the complexity of the mission. They did say, however, that plans were being developed for a flight to Venus with a penetrometer that could send back information on the chemical composition of the surface in 1998, and a mission to orbit Mercury, closest planet to the sun, in 2002–2003.

1989 August 23   The first of an improved series of unmanned cargo tankers for servicing the *Mir*, *Progress M-1* was launched at 03:10 UT by an SL-4 from Baikonur cosmodrome. The 15,990-lb. spacecraft pursued the *Mir*, in an orbit of 237 × 246.5 mi., and docked to the forward axial port at the station on August 25. Progress M–series transporters could remain docked to the space station for up to 108 days and perform independently for up to 30 days. Two solar panels were carried to power a new, autonomous rendezvous and docking system. Reserves of propellant usually wasted on de-orbit could be transferred to *Mir*'s tanks if the approach and docking had used minimal quantities. The cargo capacity had been increased, and future Progress M–series vehicles were to have provision for a reentry capsule carrying 331 lb. of cargo for safe return to Earth. *Progress M-1* undocked from *Mir* on December 1 and was de-orbited the same day.

1989 August 25   NASA's *Voyager 2* became the first spacecraft to fly past Neptune, at this date the most distant planet in the solar system, because at the low point of its orbit, Pluto moves within the orbit of Neptune. At 04:06 UT *Voyager 2* passed within 2.9 million mi. of Nereid, one of two moons known before the mission to orbit the planet. *Voyager 2* had already discovered four more small moons during the approach phase. At 07:10 UT the spacecraft crossed the ring plane, and 56 min. later made its closest approach to Neptune, swooping to within 3,044 mi. of the cloud tops. Earth was 2.75 billion mi. away, 4 hr. 6 min. 4 sec. at the speed of light. At 13:20 UT *Voyager 2* passed within 29,000 mi. of the moon Triton, which was confirmed by the spacecraft's imagery and occultation experiments to be 1,690 mi. in diameter. *Voyager 2* confirmed that Neptune has a complex magnetosphere, several distinct rings, at least eight moons and geysers 5 mi. tall erupting from the surface of its largest moon, Triton.

1989 August 27   The first satellite launch by a private U.S. company took place from Cape Canaveral, Fla., when McDonnell Douglas launched a Delta 4925 carrying the British direct broadcast satellite *Marcopolo 1* to geosynchronous transfer orbit. Based on a Hughes HS 376, *Marcopolo 1* had been built for British Satellite Broadcasting (BSB) and incorporated five 55-W Ku-band transponders. Any three channels could provide 110 W for direct broadcasting by linking any two transponders, enough for good reception at 12-in. dishes. The satellite was placed at 31° W and would be joined by *Marcopolo 2* launched on August 18, 1990.

1989 August   A variant of the Shuttle-C cargo lifter, Shuttle-Z, was discussed by NASA and industry engineers seeking a heavier payload capability than the basic Shuttle-derived vehicle (SDV). Shuttle-Z adopted filament-wound solid rocket boosters and a large bulbous payload module, 40 × 120 ft., attached to the side of the external tank. The Shuttle orbiter boattail with its three Space Shuttle main engines would be attached to the base of the module. Serving as a second stage, a single SSME, a 400,000-lb. propellant tank inside the module, would separate and propel the 300,000-lb. payload into orbit. With a total height of 231 ft., Shuttle-Z would have

required significant changes to fixtures at a Shuttle launch pad, which made the concept prohibitively expensive.

1989 September 4   The last of 15 U.S. Air Force Titan 34D heavyweight launch vehicles successfully carried two secret military payloads into orbit from Complex 40 at Cape Canaveral. It was also the last launch of a Transtage upper stage, 46 of which had flown on Titan III–class vehicles. Future heavy-lift tasks would be carried by the new generation Titan 4, with four versions on offer: Titan 401, with a Centaur cryogenic third stage capable of placing 10,000 lb. in geostationary orbit; Titan 402 with a solid inertial upper stage third stage capable of lifting 38,784 lb. to low Earth orbit, or 50,000 lb. with the Hercules solid rocket motor upgrade (SRMU) ; Titan 403/405 with a 66-ft. fairing, without third stage and capable of putting 32,160 lb. in polar orbit, or 41,400 lb. with the SRMU (Vandenberg launches were designated 403, Canaveral launches 405); and Titan 404, without third stage, but with a special payload and shroud configuration and a payload capacity of 29,800 lb., or 36,700 lb. with SRMU.

1989 September 5   Soviet cosmonauts Alexandr Viktorenko and Alexandr Serebrov were launched at 21:38 UT aboard the 15,765-lb. *Soyuz-TM 8* spacecraft atop an SL-4 from Baikonur cosmodrome. After a two-day chase through space, *Soyuz-TM 8* approached the *Mir* station, but the automated docking had to be aborted and a manual link-up with the docking port on the back of the *Kvant* module was achieved at 22:25 UT on September 7. With *Progress M-1* to unpack, the two cosmonauts ended a four-month hiatus in manned *Mir* activity, and prepared the station to receive *Kvant 2*, the first of the growth modules. Before that, the crew carried out minor repairs and fitted a new computer memory in the central command and control system.

An H-I launch vehicle carried Japan's fourth geostationary meteorological satellite (GMS-4) from Tanegashima to a geosynchronous transfer ellipse from which it was eventually placed at 140° E. Built for NASDA (National Space Development Agency) by Hughes Aircraft, GMS-4 was cylindrical with a diameter of 6 ft. 10 in. and a total height of 11 ft. 4 in. With a launch weight of 1,598 lb., and an on-station mass of 725 lb., GMS-4, also known as *Himawari-4* (Sunflower), incorporated visible and infrared spin scan radiometers, but with light emitting diodes instead of the tungsten lamp carried by its predecessor.

1989 September 7   In a revival of interest, the U.S. Department of Defense held a classified briefing for industry officials on antisatellite requirements. Led by the Army, scientists were developing a range of kinetic energy (impacting) weapons for use against Soviet and Chinese military satellites. Three conceptual designs encompassed land- and sea-based systems.

1989 September 14   A Soviet SL-4 carried the 13,900-lb. international biological laboratory *Cosmos 2044* from Plesetsk to a 129 × 166 mi., 82.3° orbit. Derived from the Vostok spacecraft, *Cosmos 2044* was a joint U.S.-Soviet mission to study the biological reaction of living things to the space environment. A total of 29 separate investigations were conducted on monkeys, rats, fish, insects and plants. The reentry module of *Cosmos 2044* returned to Earth on September 29 after a flight lasting 15 days.

1989 September 19   As it had requested, NASA received proposals from industry for a series of small Explorer-type satellites which it could fund at the rate of one per year. Small companies such as Globesat, DSI and Amsat were becoming increasingly interested in producing small, cheap, reliable satellites that would cost little to launch. Under the generalized term *lightsat* the concept also began to attract larger companies which saw the impending government budget cuts as a threat to the concept of big, expensive satellites. Lightsats became an increasingly attractive concept in this economic environment.

1989 September 25   The last Atlas-Centaur launched by NASA lifted off from pad 36B at Cape Canaveral, Fla., carrying the last of eight FltSatCom satellites for the U.S. Department of Defense. Future Atlas-Centaur flights would be managed by General Dynamics on missions contracted by the private sector. In more than 27 years of flight operations with Atlas-Centaur, NASA had launched 68 vehicles of this type, of which 7 had failed. The U.S. Air Force and NASA had launched 494 Atlas launchers, either with or without upper stages, in the slightly more than 32 years since the first Atlas lifted off from pad 14 at Cape Canaveral on June 11, 1957.

1989 September 28   A Soviet SL-14 carried *Intercosmos 24* from Plesetsk to a 314 × 1,548 mi., 82.6° orbit at the start of an international mission to investigate the propagation of electromagnetic waves in the ionosphere. Also known as *Aktivny* [Active] *1K* , *Intercosmos 24* released a Czechoslovakian subsatellite called *Magion-2* on October 3 to an independent orbit of 313 × 1,550 mi., at 82.5°. The primary satellite transmitted 10.5 kHz signals to *Magion-2* for exploration of the ionosphere. The Soviets invited the U.S. company TRW Systems to participate in the experiment, and the company issued 100 low-cost receiver kits to schools in the hope they could assemble the sets and receive transmissions. Technical problems on both satellites prevented that from happening.

1989 October 18   Postponed twice since its planned October 12 liftoff date, the Shuttle *Atlantis* was launched by NASA at 16:53:40 UT from LC-39B at the Kennedy Space Center, carrying astronauts Donald Williams (CDR), Michael J. McCulley (PLT), Ellen S. Baker (MS), Franklin Chang-Díaz (MS) and Shannon Lucid (MS). Mission STS-34 carried a 48,643-lb. payload including a 38,323-lb. *Galileo*/IUS combination, which was released from the payload bay on the sixth orbit. After 1 hr., with *Atlantis* 43 mi. behind and 21 mi. above *Galileo*, the intermediate upper stage fired its first stage for 2 min. 29 sec., followed 2 min. 19 sec. later by a 1 min. 45 sec. burn of the second stage. On course for Jupiter by way of Venus and two Earth flybys, the 5,991-lb. *Galileo* separated from the spent second stage at a mission time of 8 hr. 14 min. Other equipment carried by *Atlantis* included a solar backscatter ultraviolet experiment for measuring the height and distribution of ozone in the upper atmosphere, and growth hormone and polymer experiments. Atlantis landed at Edwards AFB, Calif., at an elapsed time of 4 days 23 hr. 39 min. 20 sec.

1989 October 27   An Ariane 44L launched from Kourou, French Guiana, carried the *Intelsat VI F2* international telecommunication satellite to a geostationary transfer ellipse. First in a new generation of Intelsat satellites built by Hughes Aircraft, the *F2* comprised a spin-stabilized cylinder with a diameter of 11 ft. 10 in. and a height of 17 ft. 5 in., weighing 9,295 lb. at launch—by far the heaviest communication satellite to date. Intelsat VI had a downward telescoping, cylindrical solar array which deployed in orbit to expose a second set of cells on the main drum. Together, they provided a minimum 2,252 W of electrical energy for 38 C-band and 10 Ku-band transponders capable of handling 120,000 telephone calls and three color TV programs. With antennas extended, the satellite had a height of 39 ft. *Intelsat VI F2* reached geostationary orbit at 24.5° E in six burns of its liquid propellant apogee motor, weighing 5,645 lb. on station. Four more Intelsat VI satellites were launched: *F3* by a Titan 3 on March 14, 1990, stranded in a 343 × 355 mi. orbit but rescued by Shuttle mission STS-49 and boosted to geostationary orbit on May 21, 1992; *F4* by a Titan 3 on June 23, 1990; *F5* by an Ariane 44L on August 14, 1991; *F6* by an Ariane 44L on October 29, 1991; and *F1* by an Ariane 44L on June 10, 1992.

1989 October 31   NASA Administrator Richard H. Truly briefed Congress on changes to the space station *Freedom* build-up plan. The first element would be launched in March 1995 with man-tended capability in April 1996 providing 37.5 kW of electrical power and a laboratory module capable of supporting limited experiments with the Shuttle attached. Permanent manned capability for a four-person crew would be attained in July 1997. The Japanese experiment module (JEM) would be launched in February 1998, followed by the European Space Agency's *Columbus* in July 1998. The full, eight-person capability station would be available in 1999 with 75 kW. Fresh studies showed that only 30 kW would be available for research, 45 kW being needed for station systems. The dual-keel truss assembly that had been a feature of earlier station rework designs was eliminated.

1989 October   Scientists from the Lawrence Livermore National Laboratory briefed representatives of the NASA Johnson Space Center on a radical and innovative approach to lunar colonies and manned Mars bases. Conceived by Lowell Wood, architect of the Strategic Defense Initiative's Brilliant Pebbles concept of ballistic missile defense, the Mars flight envisaged Earth orbit assembly between 1994 and 1996, followed by the flight to Mars and return to Earth in 1999. Using existing launchers and low-cost technology, Wood believed the moon and Mars missions could be done by the year 2000 for $10 billion as against NASA's projected cost of $400 billion for completion in 2025.

1989 November 2   President George Bush approved a new national U.S. space policy in which he placed goals and activities in the following order of priority: strengthening U.S. security; obtaining technical and economic benefits; encouraging private investment in space programs; promoting international cooperation; maintaining freedom of space; and expanding the human presence from Earth into the solar system. This was the first space policy document from the Bush administration but fell short of a more dynamic set of objectives sought by Vice President Dan Quayle.

1989 November 8   NASA announced changes to its organizational structure for control of space station *Freedom*, the Shuttle and related programs. A single organization would integrate activities and operations previously conducted by the Office of Space Flight (OSF) and the Office of Space Station in a new OSF headed by veteran astronaut William B. Lenoir.

1989 November 9   The U.S. Department of Defense held a briefing for industry on a planned competition for the design of an exo-atmospheric ground-based ballistic missile interceptor (GBI). This work was to be conducted in parallel with another Strategic Defense Initiative project called ERIS, for exo-atmospheric reentry interceptor subsystem. ERIS test launches were scheduled to begin in 1990 and include development tasks for the GBI, which was expected to move the technology closer to an operational antiballistic missile system.

1989 November 9-11   The *Galileo* spacecraft performed its first trajectory correction maneuver comprising a sequence of more than 5,500 one-second firings of 0.2-lb. thrusters to effect a 56 ft./sec. velocity change. Firings came at 19 sec. intervals as the spacecraft's spun section rotated at 3.15 RPM. *Galileo* consisted of spun and de-spun sections. The spun section comprised an octagonal bus 4 ft. 10 in. across and 1 ft. 6 in. high, to the top of which was attached the unfurlable high gain antenna with 16-ft. diameter which would transmit science data from Jupiter at a maximum rate of 134 kilobits/sec. through a 43-W transmitter. Two nuclear radioisotope thermoelectric generators (RTGs) were mounted on separate booms extending almost 17 ft. from the spacecraft centerline at 120° intervals, providing 570 W at launch and 486 W after 6 years. A deployable magnetometer boom extending 36 ft. from the spacecraft centerline was attached to the spun bus opposite the two RTG booms. Attached beneath the bus was a retropropulsion module built by Germany that supported two horizontally opposed booms with a span of 14 ft. Each boom supported a cluster of six 2.25-lb. thrusters. The module carried four tanks containing 2,055 lb. of MMH–nitrogen tetroxide propellants and a single 90-lb. thrust motor for up to 4,730 ft./sec. in orbital injection and orbit changes. Below the retropropulsion module was the de-spun section containing the atmospheric probe, scan platform of planetary science instruments and a relay antenna. The probe weighed 747 lb. and the flat-cone aeroshell had a diameter of 4 ft. 1 in. and a height of 2 ft. 10 in.

1989 November 13-15   The Smithsonian Institution's National Air and Space Museum held a seminar to encourage aerospace corporations to archive their valuable project and program histories. Attention was drawn to the fact that 50 years hence most records would have dissolved because they originated on acid-based paper. Most of the engineering records on the Saturn V program had already been destroyed. Most acknowledged the dilemma: Companies were unwilling to vacate production, research and administration space for archives when, frequently, the time staff required to get classified material sifted for dissemination was expensive.

1989 November 17   NASA Administrator Richard Truly presented the report of a 90-day study on the human exploration of the moon and Mars to National Space Council Chairman Dan Quayle. Called the *Human Exploration Initiative*, the study described an evolutionary path over a 30-year period beginning with space station *Freedom* in the 1990s, flowing to a permanent lunar outpost at the beginning of the next century and culminating in a permanent Mars base. The study defined transportation systems, spacecraft, resource needs and timescales.

1989 November 18   The last NASA-owned expendable launch vehicle was launched from Vandenberg AFB, Calif., carrying the first in a new and powerful series of astrophysical satellites. Called *COBE (Cosmic Background Explorer)*, the satellite was carried by a Delta 5920 to a 550.5 × 556 mi., 99° orbit, where it would measure diffuse infrared radiation using three instruments. *COBE* had been designed for a Shuttle launch, but after the *Challenger* disaster of January 28, 1986, it was scaled down for a Delta launch for which weight was cut

from 10,000 lb. to 5,000 lb. It had a diameter of 8 ft. which increased to 13 ft. on deployment of a protective shield. A communications antenna increased the length to 19 ft. About 1,000 W of power was provided by three 28 ft. solar arrays. Within months of beginning operations, *COBE* provided astonishing information which dramatically advanced cosmological theory.

1989 November 23   Delayed two days for installation of replacement electronics boxes on a solid rocket booster, Shuttle mission STS-33 began with the launch of *Discovery* at 00:23:30 UT from LC-39B at the Kennedy Space Center, Fla. On board were Frederick Gregory (CDR), John Blaha (PLT), Story Musgrave (MS), Kathryn Thornton (MS) and Manley "Sonny" Carter Jr. (MS). The 45,735-lb. payload comprised a signal-intelligence gathering (sigint) satellite and its associated inertial upper stage and tilt-table used to support it in the cargo bay. Operated by the National Security Agency, the satellite was identical to the one launched by Shuttle mission 51-C on January 24, 1985. Boosted to geostationary orbit, it had a length of 35 ft. 8 in. and a maximum diameter of 11 ft. 10 in. This classified Department of Defense mission was extended by one day due to high winds at the Edwards AFB, Calif., landing site. Touchdown came at 5 days 0 hr. 6 min. 46 sec.

1989 November 26   The long-awaited delivery of the first growth module to the Soviet *Mir* began when *Kvant 2* was launched atop an SL-13 from Baikonur cosmodrome at 13:01 UT. The 43,140-lb. module had a length of 45 ft., a maximum diameter of 14 ft. 3 in. and two 287 sq. ft. solar cell arrays providing 6.9 kW of electrical energy. *Kvant 2* comprised three defined interior areas: a service/cargo section where propulsion systems and freight were carried; a payload/equipment section where experiments and additional station systems were housed; and an airlock section providing better access to the exterior for space walks. (*Mir* permitted egress only through the narrow docking ports). Using automatic rendezvous and docking equipment, the link-up with *Mir* was delayed until 12:21 UT on December 6, when *Kvant 2* attached itself to the forward axial port. *Kvant 2* carried an East German remote sensing experiment called Ikar, a shower unit for the crew and a 485-lb. manned maneuvering unit for moving away from the complex on space walks. This device had a 197 ft. tether, 32 attitude thrusters, and a total delta-velocity of 98 ft./sec. Two days after docking, *Kvant 2* was moved by the Liappa docking arm to the upper radial docking port at the forward section of the *Mir* station. Shortly thereafter, *Soyuz-TM 8* was moved to the forward axial docking port.

1989 December 1   A massive, 7,700-lb. astrophysical observatory called *Granat* was launched by the Soviet Union by SL-12 from Baikonur to a 1,096 × 125,820 mi., 51.8° orbit. Derived from the Venera and VEGA spacecraft, *Granat* comprised a cylinder 8 ft. 3 in. in diameter and 13 ft. long, with two solar panels deployed from either side. Science instruments included a French Sigma gamma-radiation telescope, 11 ft. 6 in. × 4 ft., a spectroscope and an x-ray imaging telescope, all to observe high-energy emissions from galactic and extragalactic sources. The suite of instruments had been put together by teams from France, Bulgaria and Denmark. It remains in orbit.

1989 December 8   The U.S. Air Force received a directive from the Department of Defense ordering it to stop design work on the proposed advanced launch system heavy-lift vehicle. Budget cuts anticipated for fiscal year 1991, to be announced in early 1990, would not sustain the effort required to press ahead with designs as planned. Three contractors (Boeing, General Dynamics and Martin Marietta/McDonnell Douglas) conducting design studies would present interim reports to the Air Force in late January 1990, and at that point the Air Force would scale back the program to a technology research effort only.

1989 December 20   At 03:31 UT the *Progress M-2* cargo tanker was launched by SL-4 from the Baikonur cosmodrome. Two days later it docked to the back of the *Kvant 1* module at the rear of the *Mir* space station, in an orbit of 244 × 245.5 mi. Cosmonauts Alexandr Viktorenko and Alexandr Serebrov, busied with the recent arrival of the *Kvant 2* module, unpacked stores and provisions from *Progress M-2*. Before it was undocked at 02:33 UT on February 9, 1990, and de-orbited to destruction over the Pacific Ocean, the unmanned cargo tanker supported four space walks by replenishing air supplies in the airlock module on *Kvant 2*.

1989 December 22   *Galileo* conducted its second trajectory correction maneuver, shifting the Venus flyby distance to about 6,200 mi. above the cloud tops. Four pulse chains with the 0.2-lb. thrusters took more than 2 hr. 15 min. and altered the velocity of the spacecraft by 2.5 ft./sec. Five days later mission controllers at the Jet Propulsion Laboratory began checking the science payload in a four-day operation. *Galileo* carried a solid-state imaging instrument with an effective focal length of 1,500-mm composing pictures with 800 picture elements on each of 800 lines. A near-infrared mapping spectrometer would measure Jupiter's atmospheric composition, cloud structure and temperature profiles, a photopolarimeter radiometer would determine energy balances in the atmosphere, and an ultraviolet spectrometer would observe the atmosphere above the cloud tops. All three would conduct extensive surveys of the Jovian moons.

1989 December   The United States and the Soviet Union reached agreement on cooperation in the exchange of biomedical information during discussions at Kislovodsk in the Soviet Union. Twenty U.S. life-science specialists met 30 of their Soviet counterparts and worked out a plan whereby Soviet *Mir* data would be provided to U.S. scientists. For their part, U.S. biomedical specialists would provide data obtained on Shuttle flights. The prospect of NASA astronauts visiting *Mir*, and Soviet cosmonauts flying on the Shuttle, was also raised.

1989   During the year the U.S. Department of Defense Have Regio single-stage-to-orbit program was reduced to low-key technology research. The concept of a rocket-powered, horizontal-takeoff-and-landing space launcher was almost totally eclipsed by success in designs of the National Aero-Space Plane (NASP). However, the Strategic Defense Initiative antiballistic missile program funded a research effort into reusable satellite launchers with vertical launch and landing capability. Throwing out the notion of exotic and expensive new materials for high-speed aerospace vehicles like NASP, the SDI team concentrated on basic technology and low cost. The result was a concept from McDonnell Douglas called DC-X. Funds were provided for a DC-X technology demonstrator in August 1991.

1990 January 1   The first commercial Titan 3 launch vehicle carried two communication satellites to geosynchronous

transfer ellipse from Cape Canaveral, Fla. The satellites were stowed in tandem above the second stage, on a cluster of fixtures resting on a 773-lb., 3 ft. high aft payload adaptor. The lower satellite was positioned atop an aft extension module, 4 ft. tall weighing 475 lb. Above that was an 11 ft. tall forward extension module weighing 918 lb. It supported the 2,964-lb. fairing, 34 ft. tall and 13 ft. 1 in. in diameter, encapsulating the entire payload cluster. The upper satellite rested on a 467-lb. forward adapter, 5 ft. 6 in. tall. With a launch weight of 5,027 lb. and an on-station mass of 2,914 lb., the lower payload comprised the Japanese satellite *JCSAT-2* and was placed in geostationary orbit at 154° E. The upper satellite was the 3,226-lb. British military *Skynet 4A* satellite, placed in geostationary orbit at 6° E.

1990 January 7   A flurry of space walking activity for cosmonauts Alexandr Viktorenko and Alexandr Serebrov aboard the orbiting *Mir* complex began with a 2 hr. 56 min. excursion through the *Kvant 2* airlock; materials samples left outside on an earlier space walk were retrieved and brought inside, and two new star sensors installed on the exterior of the *Kvant 1* module, attached to the rear of *Mir*. Better control of the complex was necessary because of the asymmetric alignment of *Kvant 2* deployed at 90° to the main space station. A second space walk, lasting 2 hr. 54 min., took place on January 11, during which the cosmonauts retrieved more sample packages and left others in their place for retrieval during a later space walk. The third activity, lasting 3 hr. 2 min., occurred on January 26 when new closed-loop space suits were tested in preparation for flight evaluation of the manned maneuvering unit brought to *Mir* by *Kvant 2*.

1990 January 9   After a 24-hr. delay to avoid bad weather, Shuttle mission STS-32 began with the launch of *Columbia* at 12:35:00 UT from LC-39A at the Kennedy Space Center, Fla.

Returned to Earth by Shuttle Columbia *on January 20, 1990, after almost six years in space, the Long Duration Exposure Facility (LDEF) was moved to a transportation canister in the Orbiter Processing Facility.*

Originally scheduled for December 18, 1989, it had been delayed while modifications to pad A were completed. The crew comprised Daniel Brandenstein (CDR), James Wetherbee (PLT), Bonnie Dunbar (MS), Marsha Ivins (MS) and David Low (MS). Included in the 26,488-lb. payload was *Syncom IV-5 (Leasat 5)*, released at 13:18 UT on January 9. Sent to geosynchronous transfer ellipse by its Minuteman III solid propellant perigee motor and then circularized in geosynchronous orbit by successive burns of liquid propellant engines, *Syncom IV-5* was the last of its type, only one of the five having failed to reach its proper orbit and operate as planned. An important objective of STS-32 was the retrieval of the LDEF satellite left in orbit by mission 41-C, achieved at 15:16 UT on January 12. *Columbia* landed at Edwards AFB, at the end of the longest Shuttle mission to this date: 10 days 21 hr. 0 min. 37 sec.

1990 January 22   The first Ariane 40 launch vehicle configuration carried seven satellites to orbit from Kourou, French Guiana. Essentially an Ariane 4 without strap-ons, and with a liftoff thrust of 531,343 lb., the V35 launcher deployed France's 4,123-lb. *SPOT-2* Earth resources satellite to a 510 × 511.5 mi. orbit inclined 98.7°. The remaining six satellites were all part of the Oscar series of amateur radio relays. Built by the United Kingdom's University of Surrey, the 95-lb. *Uosat 3* and the 99-lb. *Uosat 4* (1 ft. 2 in. × 1 ft. 2 in. × 2 ft) had solar cells providing 30 W of power. *Uosat 4* failed at 30 hr. Four 22-lb. Microsat satellites (9 in. × 9 in. × 2 ft. 7 in. including extended boom) with cells producing 6 W provided

different amateur radio services and were named *Pacsat*, *Dove*, *Webersat* and *Lusat*. All six satellites were strung out in slightly different orbits at approximately 488 × 497 mi. inclined 98.7°.

1990 January 24   An ambitious Japanese science mission began with the launch of *Muses-A* from the Kagoshima Space Center. Carried to escape velocity by a three stage Mu-3S II, the 434-lb. *Muses-A* (Mu-launched space engineering satellite) was put in an elliptical orbit of the Earth which intersected the moon's gravitational sphere of influence. After launch *Muses-A* was renamed *Hiten*, after the musical Buddhist angel. On March 19, just 10,235 mi. from the moon, it released a 26.5-lb. orbiter named *Hagoromo*, after the veil worn by Hiten. *Hagoromo*, a 1 ft. 4 in. × 1 ft. 2 in. polyhedron, was decelerated by a small solid propellant motor to a 5,600 × 13,670 mi. orbit of the moon. At the point of separation all contact with *Hagoromo* was unexpectedly lost. Cylindrical in shape, 4 ft. 7 in. × 2 ft. 7 in. tall, *Hiten* passed the moon and continued on the simulated path of a future mission called Geotail.

1990 January 26   The first flight of a rail-launched endoatmospheric defense interceptor (HEDI) took place at White Sands, N.Mex., when a two-stage missile based on the Sprint was fired to a height of 6 mi. and a speed of 5,500 MPH. The high-velocity missile was traveling at more than 200 MPH when it left a vertical rail to which it had been attached for launch and the first stage fired for just 1.5 sec., 0.2 sec. less

In development for the U.S. Strategic Defense Initiative (SDI) missile defense program, the first HEDI kinetic-kill vehicle sits on a rail-mounted launcher at White Sands Missile Range, New Mexico.

than planned. HEDI was part of the KITE (kinetic-kill vehicle integrated technology experiments) program which sought to develop an antiballistic missile system as part of the Strategic Defense Initiative research effort. The test was conducted for the Army Strategic Defense Command and demonstrated a kinetic-kill (impact) warhead fired on a simulated ascent profile. It was not intended to impact anything and there was no target against which the HEDI was fired.

1990 January 30   Senior space officials from Europe and Japan testified before Congress on their disquiet over recent redesign and build-up changes to space station *Freedom*. Plans to defer the launch of *Columbus* and the Japanese Experiment Module had caused concern, exacerbated when NASA told the European Space Agency that it could allocate only 3 kW of electrical power during the first stage of assembly, just enough to "stay alive" as NASA put it. The Europeans felt it was not unreasonable to want sufficient power to run experiments and not just to turn the lights on. NASA offered to advance the JEM and *Columbus* launch dates by a few months but could offer little comfort on the power demand, which would remain low until additional power became available later in the build-up sequence.

1990 January   NASA announced the names of 23 new pilot (PLT) and mission specialist (MS) Shuttle astronauts constituting Group 13: Lt. Cdr. Daniel W. Bursch (USN); Leroy Chiao; Lt. Col. Michael R. U. Clifford (USA); Kenneth D. Cockrell; Maj. Ellen M. Collins (USAF); Maj. William G. Gregory (USAF); Maj. James D. Halsell Jr. (USAF); Bernard A. Harris Jr.; Maj. Susan J. Helms (USAF); Thomas D. Jones; Lt. Col. William S. McArthur Jr. (USA); James H. Newman; Ellen Ochoa; Maj. Charles J. Precourt (USAF); Maj. Richard A. Searfoss (USAF); Ronald M. Sega; Capt. Nancy J. Sherlock (USA); Donald A. Thomas; Janice E. Voss; Carl E. Walz; Maj. Terrence W. Wilcutt (USMC); Peter J. K. Wisoff; and David A. Wolff.

1990 February 1   The first Soviet manned maneuvering unit was tested in space during a 4 hr. 59 min. space walk from the *Mir* station. Cosmonaut Alexandr Serebrov evaluated the unit on the first of two check-out walks to be conducted by the *Mir* crew. Tethered to the *Kvant 2* and with a docking unit on the front, Serebrov backed away 16 ft. 6 in., then moved in again, repeating the maneuver twice before backing all the way out to a distance of up to 132 ft. Serebrov was attached by a tether connected to an electric motor, which was evaluated in its simulated role of rescuing a stranded crew member by reeling him in again. Alexandr Viktorenko monitored Serebrov from the open hatch but got his own turn on February 5 when he flew the unit during a space walk lasting 3 hr. 45 min. Backing out as far as 148 ft., he demonstrated rolls, yaw maneuvers, inverted "flight" and the winch tether.

1990 February 7   The second Marine Observation Satellite, *MOS-1B*, was launched by H-I from Tanegashima to a 564 × 565 mi. orbit inclined 99.2°. Carrying a payload similar to that of *MOS-1A*, the 1,632-lb. satellite carried three radiometers and a data collection system transponder for collecting information from buoys and ground sites. Along with *MOS-1B*, the H-I launcher also carried a Japanese technology satellite, called *Debut*, and the second Japanese amateur radio satellite, *JAS-1B* , to an orbit of 565 × 1,082 mi. inclined 99°. The 110-lb. *Debut* (deployable boom and umbrella test) satellite carried a 4 ft. 9 in. boom and 24-panel umbrella device to test the concept of a deployable aerobrake for future planetary

spacecraft. The 110-lb. *JAS-1B* was developed by the Japan Amateur Radio League.

1990 February 10   At 05:58:48 UT the *Galileo* spacecraft passed within 6,200 mi. of the cloud tops of Venus at 41° south latitude. Earth and Venus were about 30 million mi. apart at the time, when *Galileo* gained 4,990 MPH from the gravity-assist. Minor technical problems plagued the imaging instrument during the scheduled acquisition of about 80 pictures of Venus, an entirely incidental part of the mission. *Galileo* continued to move in a circular orbit in toward the sun until, 15 days later, it reached perihelion of 65 million mi. and began to swing out toward the orbit of Earth, which it was scheduled to encounter on December 8, 1990.

1990 February 11   Soviet cosmonauts Anatoly Solovyov and Alexandr Balandin were launched aboard their 15,700-lb. *Soyuz-TM 9* spacecraft atop an SL-4 from the Baikonur cosmodrome at 06:16 UT. As discovered later, about 2 min. 40 sec. after liftoff, when the boost protective shroud and launch escape system were jettisoned, three thermal blankets covering a broad portion of the descent module were torn loose. These would be observed by the departing *Mir* crew in *Soyuz-TM 8* on February 18, prompting a space walk on July 17 to carry out repairs. Meanwhile, *Soyuz-TM 9* docked to the back of the *Kvant* module at 06:38 UT on February 13, with the complex in an orbit of 253.5 × 237.5 mi. After five days of joint activities, cosmonauts Aleksandr Viktorenko and Aleksandr Serebrov returned home in *Soyuz-TM 8*, leaving Solovyov and Balandin on *Mir*.

1990 February 14   A Delta II launched from Cape Canaveral, Fla., launched two U.S. satellites for the Strategic Defense Initiative ballistic missile defense program. Called *LACE* (low-power atmospheric compensation experiment), the first was placed in a 320 × 335 mi., 43.1° orbit from which the 3,155-lb. satellite measured laser beams targeting it from Earth. It did this through visible, infrared and phased array sensors, but it also carried an ultraviolet plume detector to measure wavelengths best suited to tracking rocket exhaust. *LACE* was built by the Naval Research Laboratory. The second satellite, *RME* (relay mirror experiment), weighed 2,295 lb. and went into a 266 × 283 mi., 43.1° orbit. Built by Ball Space Systems Division, *RME* carried a mirror 2 ft. in diameter to reflect laser beams back to Earth. Both satellites remain in orbit. *LACE* remains in space but *RME* decayed on May 24, 1992.

Working to a series of commands generated at the Jet Propulsion Laboratory, *Voyager 1* took a sequence of 64 images looking back toward the sun from a distance of about 3.7 billion mi. Included in this historic "first" were all the planets of the solar system with the exception of Mercury, Mars and Pluto. The multiple exposures were made with the 1,500-mm wide-angle camera and three color filters to form individual images. Both *Voyager 1* and *Voyager 2* are departing the solar system upstream of the heliosphere, the magnetic and radiation environment of the sun. Both spacecraft are heading in the direction of the bow shock wave which they may cross at a distance of about 9.5 billion mi. The Voyagers are likely to survive and send back information from a maximum distance of 14 billion mi., which they will reach in the year 2015.

1990 February 19   Mir cosmonauts Aleksandr Viktorenko and Aleksandr Serebrov returned to Earth after a stay aboard *Mir* and a mission that lasted 166 days 6 hr. 58 min. Returning to Earth in the same *Soyuz-TM 8* spacecraft that had carried them into space on September 5, 1989, the cosmonauts landed back in the Soviet Union 1,056 mi. southeast of Moscow at 04:36 UT. Meanwhile, back aboard *Mir*, cosmonauts Anatoly Solovyov and Alexandr Balandin moved *Soyuz-TM 9* from the back of the *Kvant* module to the forward axial port. They would continue to operate the complex of modules for the next six months.

Engineers at the European Space Agency sent a signal to the *Giotto* spacecraft to reactivate it for electronic diagnosis. *Giotto* had conducted a close flyby of Halley's comet in early 1986 and, if possible, it was to be programmed for a flyby of the comet Grigg-Skjellerup via a slingshot maneuver past Earth. The signal went out at 13:00 UT and was answered at 15:06. In March engineers began a series of maneuvers to prepare it for the gravity-assist by aligning its trajectory for a perfect flyby of Earth on July 2.

1990 February 22   An Ariane 44L carrying two powerful communications satellites for Japan was destroyed 1 min. 40 sec. after it had been launched from Kourou, French Guiana, at 23:17:00 UT. Mission V36 was carrying the 5,512-lb. *Superbird B* direct broadcast satellite for Space Communications Corp. and the 2,756-lb. *BS-2X*. Built by General Electric (RCA) for Comsat Corp., *BS-2X* was bought back by the builder when the customer canceled plans for domestic U.S. direct broadcasting. With a deployed height of 6 ft. 5 in., a solar array span of 55 ft. 9 in. and an on-orbit mass of 1,610 lb., *BS-2X* carried three 200-W transmitters operating in the 14/12 GHz band. A V36 failure review board identified a blocked first stage water pipe as the cause of the accident.

1990 February 27   A team within NASA formed to look at potential space station problems forecast extensive EVA (extravehicular activity) to repair or service the manned facility. The team projected an estimated 2,284 hr. of EVA per year and a space walk every other day to carry out essential work. They said that, the way the station was designed at present, astronauts would have their hands full with general work just to keep it working. Space station managers projected just 400 hr. EVA per year, or one space walk every 11 days, for servicing. The study prompted support from the astronauts for a redesign of the station to alleviate this heavy work demand.

1990 February 28   At 7:50:22 UT Shuttle mission STS-36 was launched from LC-39A at the Kennedy Space Center, Fla., carrying astronauts John Creighton (CDR), John Casper (PLT), David Hilmers (MS), Richard Mullane (MS) and Pierre Thuot aboard *Atlantis*. The mission had been postponed three days from February 22 because the commander was ill and weather was bad, then for 24 hr. on February 25 when a computer failed and for 24 hr. on February 26 due to more bad weather. This classified Department of Defense mission accommodated a 33,470-lb. payload, part of which comprised the second Block 14 Defense Support Program geostationary early warning satellite and an inertial upper stage. *Atlantis* landed at Edwards AFB, Calif., at 4 days 10 hr. 18 min. 22 sec.

At 23:11 UT the *Progress M-3* cargo tanker was launched by an SL-4 from Baikonur cosmodrome to a rendezvous with the *Mir* space station, orbiting 250 × 235.5 mi. above the Earth. *Progress M-3* docked to the back of the *Kvant* module at 01:05 UT on March 3. *Mir* cosmonauts Anatoly Solovyov and Alexandr Balandin unpacked the orbital module and over the next several weeks installed replacement equipment for

several items aboard the *Mir* station. *Progress M-3* conducted a propulsion burn on April 23, raising the orbit of the complex, and on April 27, it undocked from the *Kvant 1* and was de-orbited over the Pacific Ocean.

1990 March 2   Commenting on NASA's human exploration initiative, the National Research Council suggested the space agency was too conservative and entrenched in traditional ways of implementing grand goals. Commenting on the conclusions of a 90-day study conducted by NASA into lunar colonies and Mars bases, the NRC said that it believed the application of innovative technologies and a radical approach would cut costs. Acting on the NRC's view, President George Bush ordered that at least two completely separate proposals be drawn up for new manned missions to the moon and Mars.

1990 March 13   A second course correction burn was performed by the Venus-bound *Magellan* spacecraft 107 million mi. from Earth. The thrusters fired to adjust speed by 2.3 ft./sec. A final course correction maneuver was conducted on July 25, 1990, when the spacecraft was more than 136 million mi. from Earth with 3 million mi. to go to planetary encounter on August 10. *Magellan*'s small 0.2-lb. thrusters were fired in pulse-mode for 5 min. 21 sec., changing velocity by 2.2 ft./sec. The thrusters had adjusted the velocity of the spacecraft to within 2.5 in./sec. of the desired value. At this date *Magellan* was moving at a heliocentric velocity of 86,200 MPH, moving around the sun on an inner track more than 20,000 MPH faster than the Earth.

1990 March 29   An experiment to demonstrate that a high-resolution imaging laser could discriminate between an inflatable decoy and a reentry warhead in space was conducted from Wallops Island, Va. Called *Firefly*, the test began with the launch by NASA of a Terrier-Malemute sounding rocket carrying a decoy simulator in the form of an inflatable cone 6 ft. long. Deployed on a ballistic trajectory at a height of 300 mi., the simulated decoy was successfully tracked by microwave radar and locked on to by an argon ion laser.

1990 April 3   Israel's second satellite, *Ofeq 2*, was carried by a Shavit launcher from the Yavne site in the Negev desert to a 130 × 980 mi., 143.2° orbit. The 353-lb. satellite took the form of an octagonal prism, 7 ft. 6 in. tall, 4 ft. across at the bottom and 2 ft. 3 in. across at the top. Solar cells attached to the body of the satellite provided 246 W of electrical power. Built by Israel Aircraft Industries, *Ofeq 2* was launched by the Israeli Space Agency. It decayed out of orbit on July 9.

1990 April 5   *Pegsat*, the world's first satellite launched from the air, was propelled from a converted Boeing B-52 after having first been carried into the air from Edwards AFB, Calif. Developed by Orbital Sciences Corp. and Hercules Aerospace Co., the three-stage solid propellant Pegasus had a length of 50 ft. and wings for aerodynamic stability and directional control in the atmosphere. Released at a height of 43,000 ft., Pegasus ignited for a 1 min. 13 sec. first-stage burn producing 109,575 lb. of thrust, followed by the second stage producing 28,000 lb. of thrust for 1 min. 12 sec. and a third-stage burn which put it and the payload in a 281 × 400 mi., 94.1° orbit. *Pegsat* remained attached to the third stage, but a U.S. Navy subsatellite was released to a 303 × 411 mi. orbit.

1990 April 7   In the first commercial flight for a Chinese launcher, a CZ-3 carried the communication satellite *Asiasat 1* into a geostationary transfer orbit from Xichang. The refurbished Hughes HS 376–series Westar VI retrieved by Shuttle mission STS-51, *Asiasat 1* weighed 3,180 lb. at launch and 1,345 lb. on station. Dimensions remained the same as the original Westar VI. Each of the 24 C-band transponders could handle 2,400 one-way voice circuits or one color TV channel. Stationed at 105° E, it was the first in a privately owned series planned by the Asia Satellite Telecommunications Co., with two beams covering the area from Korea to the Middle East and from Mongolia to Malaysia, respectively.

1990 April 9 Edward Teller, father of the hydrogen bomb and the physicist who convinced President Reagan to set up a Strategic Defense Initiative ballistic missile defense program, proposed a constellation of 1,000 satellites for Earth observation. In a concept dubbed Brilliant Eyes, delegates at a space symposium in Colorado Springs, Colo., heard Teller describe how unit costs could be cut dramatically if the Brilliant Pebbles SDI concept was adopted. Under that scheme, 4,000 highly instrumented microsats (satellites smaller than lightsats) would be used for missile defense. The production of an additional 1,000 microsats would, said Teller, make the concept very attractive. Each Brilliant Eye was projected to weigh only 110 lb. and orbit 124 mi. up, using lasers and radar to observe and monitor Earth and its environment.

U.S. Air Force Gen. Donald J. Kutyna asserted that the military should have control over its own fleet of expendable launchers and not rely on NASA and commercial organizations for availability of boosters for military satellites. Coupled with the lack of antisatellite weapons, the long lead-time required to get a launch vehicle up and ready for the launch of replacement satellites was an impediment to national security, he said.

1990 April 9–12 At a distance of 83 million mi., a complicated four-day trajectory correction maneuver was carried out by NASA's *Galileo* spacecraft, now on its way to an Earth flyby on December 8. For 6 hr. 30 min. each day the thrusters fired briefly once every 19 sec. to slow the spacecraft by about 20.3 ft./sec. Over the four days the thrusters pulsed almost 6,400 times and *Galileo*'s velocity was reduced by 82 ft./sec. This had the effect of moving the flight path of the spacecraft from an Earth flyby distance of 1.5 million mi. to about 300,000 mi. Another two-part maneuver on May 11–12 required 2,920 thruster pulses to change speed by 36 ft./sec. with *Galileo* about 96 million mi. from Earth. A third sequence of 288 pulses on July 17 tweaked the course by 3 ft./sec., 85 million mi. from Earth. The penultimate 1.7-ft./sec. fine tuning before Earth flyby occurred on October 9, with *Galileo* at a distance of 33 million mi. The final adjustment was made on November 13.

1990 April 11 A Soviet SL-4 launcher carried the 13,700-lb. materials processing laboratory *Photon 3* from Plesetsk to a 132 × 227 mi., 62.8° orbit. It was dedicated to a French microgravity research payload put together by Centre National d'Etudes Spatiales (CNES), the French national space agency. Called *Crocodile*, the payload contained materials and equipment for producing organic crystals. These were part of an experiment into the production of high-quality components for the electronics industry. The reentry module returned to Earth with *Crocodile* intact on April 27.

1990 April 13 More than five years after it had been returned to Earth at the end of Shuttle mission 51A, the Hughes HS 376 communication satellite *Palapa B-2R* was put back in orbit by

Heralding a new era of commercial operations by U.S. launch vehicle manufacturers, a Delta carries an Indonesian Palapa satellite toward orbit from Cape Canaveral, Florida, on April 13, 1990.

a Delta II launched from Cape Canaveral, Fla. The 2,650-lb. satellite fired its apogee motor, reducing weight to 1,435 lb., and was positioned in geostationary orbit over 107.7°. As *Palapa B2*, the satellite had first been launched on the STS 41B mission, February 3, 1984. Stranded in low Earth orbit by a PAM-D malfunction, it was recovered by Shuttle in November 1984. *Palapa B-2R* was the first communication satellite returned to Earth and relaunched.

1990 April 17 The U.S. Air Force Space Division awarded a contract to TRW Systems for 12 lightweight satellites, or lightsats, each of which would weigh under 1,000 lb. In the Space Test Experiments Platform (STEP) program, the satellites would have a common bus providing up to 300 W of electrical power, three-axis or gravity-gradient stabilization and telemetry. The Air Force wanted the satellites to carry payloads supporting general space technology experiments, tests on National Aero-Space Plane materials and missile research. The first was scheduled for launch in 1992.

1990 April 24 At 12:33:51 UT the Hubble space telescope was launched aboard the Shuttle *Discovery* from LC-39B at the Kennedy Space Center, Fla., on STS-31. The mission had been postponed from April 24, when a faulty valve in an auxiliary power unit halted the countdown at T − 4 min. The 28,673-lb. payload comprised the HST, several technology experiments and materials processing tests. The crew consisted of Loren Shriver (CDR), Charles Bolden (PLT), Steven

Hawley (MS), Bruce McCandless (MS), and Kathryn Sullivan (MS). Hubble was deployed when Hawley operated the remote manipulator arm to remove the bulky satellite from the cargo bay and leave it in orbit at 19:38 UT on April 25. *Discovery* landed at Edwards AFB, Calif., at 5 days 1 hr. 16 min. 6 sec.

The first commercial space firm to extend a public share offering began selling stock when trading opened on Orbital Sciences Corp. In less than one week, the sale raised $16.275 million for OSC and $14.973 million for shareholders, with shares at around $14. Half of its revenue came from the transfer orbit stage, and other promising developments included the air-launched Pegasus launcher, which accounted for 11% of its revenue, the Taurus ground-launched vehicle, and Cygnus, a more advanced launcher, as well as two sounding rockets.

1990 May 5 At 20:44 UT *Progress 42*, one of the original space station cargo tanker designs, was launched by an SL-4 from the Baikonur cosmodrome. It docked to the back of the *Kvant* module attached to the *Mir* space station complex at 22:45 UT on May 7 carrying 4,410 lb. of cargo. Cosmonauts Anatoly Solovyov and Alexandr Balandin emptied the orbital module and, in one of the less lengthy visits to *Mir*, *Progress 42* was undocked at 07:29 UT on May 27, to be de-orbited later the same day. This was the last of the original design of Progress cargo tanker, the remainder being M-series vehicles. On May 28 Solovyov and Balandin moved *Soyuz-TM 9* from the forward axial docking port to the back of the *Kvant 1* module.

1990 May 9 Two storable data and communications satellites were launched for the Department of Defense by a four-stage Scout from Vandenberg AFB, Calif. Called *Macsat* (multiple access communications) *1* and *2*, each 150-lb. satellite provided global store-and-forward capability from its 376 × 475 mi., 89.9° orbit. Both satellites remain in space.

1990 May 29 The annual round of Soviet Resurs remote sensing satellites got under way for 1990 with the launch of *Resurs F6* on an SL-4 from Plesetsk to a 158.5 × 166.5 mi., 82.3° orbit. *F6* and the fourth satellite, *F9*, launched on September 7, also carried microgravity experiments for Intospace, a German consortium. These two missions, and *F8* launched on August 16, carried the Priroda 4 camera system comprising two long-focus cameras, three multispectral cameras and a stellar camera for attitude reference. Launched on July 17, *Resurs F7* carried just multispectral and stellar cameras. The *F9* returned to Earth on September 21.

1990 May 30 After a postponement from May 16 to replace a coolant line in Shuttle orbiter *Columbia,* engineers at the Kennedy Space Center halted the countdown for STS-35 when high concentrations of hydrogen gas were detected in the orbiter's aft compartment. A tanking test on June 6 confirmed a leak in the 17-in. fuel line connecting the orbiter to the external tank, and the Shuttle was returned to the Vehicle Assembly Building on June 12. After repairs, it was rolled back to pad A on August 9 for a launch on September 1, though payload problems delayed that attempt. Subsequent attempts on September 6 and 18 were delayed when hydrogen was again detected in the aft compartment. Because of scheduling delays with other flights, *Columbia* was moved from pad A to pad B on October 8, making way for STS-38 four days later. Tropical storm Klaus caused a roll-back to the VAB again on October 9. Five days later it returned to pad A, and although hydrogen leaks were again

detected, they were within acceptable limits and STS-35 was launched December 2.

1990 May 31 At 14:33 UT a Soviet SL-13 launch vehicle lifted away from the Baikonur cosmodrome carrying the second *Mir* space station growth module called *Kristall*. With a weight of 43,300 lb., *Kristall* has the same general, cylindrical shape as the *Kvant 2* module launched December 6, 1989, having a length of 39 ft. and a diameter of 14 ft. 3 in. *Kristall* is a habitable materials processing laboratory comprising two main sections: an instrument-payload compartment containing a processing unit with four pieces of equipment, and a junction-docking compartment. The docking unit takes the place of the airlock in the design of *Kvant 2* and provides a node with two androgynous docking units to which the Shuttle Buran and a small x-ray telescope could be attached. *Kristall* docked to the forward axial port on the *Mir* space station at 14:47 UT on June 10, and was moved by the Liappa docking arm to the lower radial docking port opposite *Kvant 2* a day later. *Mir* cosmonauts Solovyov and Balandin entered *Kristall* on June 12 to check out the Krater-5 electric furnace, Optizon beam unit, Zona-02 and Zona-03 electric furnaces, the Marina and Glazar-2 telescopes and a biomedical evaluation unit.

1990 June 1 A Delta II launched from Cape Canaveral, Fla., carried the West German astrophysical satellite *Rosat* from Cape Canaveral, Fla., to a 348 × 359 mi., 53° orbit. Built by Dornier for the German Federal Ministry of Science and Technology, the 5,350-lb. satellite measured 15 ft. 5 in. × 7 ft. 3 in. × 14 ft. 1 in., and deployed three solar panels which formed a 14 ft. 9 in. octagon supplying 1,040 W. The primary payload comprised a 7 ft. 10 in. (focal length) x-ray telescope with 1.8 arc-sec. resolution. A consortium from the United Kingdom provided a 1 ft. 2 in. (focal length) extreme ultraviolet camera. By the end of the year, *Rosat* had identified 1,000 new ultraviolet radiation sources and revealed a structure to the universe linked with quasar galaxies.

1990 June 4 NASA rescoped the orbital maneuvering vehicle (OMV) program into a satellite servicer system that effectively killed the OMV and awarded Martin Marietta and TRW Systems a definition contract for the new telerobot. The servicer would integrate a flight telerobotic servicer that the Martin Marietta company was developing for space station *Freedom*, with the OMV bus that TRW had been designing under a previous contract. The servicer was defined as a vehicle capable of moving from orbit-to-orbit and conducting servicing operations on satellites, or remaining in the vicinity of the station to carry out external maintenance work.

1990 June 12 The last of India's Insat series of multipurpose weather and communications satellites, *Insat 1D*, was launched by Delta from Cape Canaveral, Fla., to a geostationary orbit at 83.1° E. Built by Ford Aerospace (now Loral) the satellite had a launch weight of 2,624 lb., an on-orbit weight of 1,550 lb. and consisted of a box-shaped bus measuring 5 ft. 1 in. × 4 ft. 7 in. × 7 ft. 2 in. Overall length was increased to 63 ft. 7 in. by a solar sail and a 124 sq. ft. solar panel which provided 1,200 W of power. It carried a radiometer for weather data, storm warnings and limited resource analysis, and 12 C-band transponders for telephone and data as well as two S-band transponders for direct-broadcast TV. A replacement for *Insat 1B*, *Insat 1D* had been damaged during the 1989 San Francisco Earthquake and delayed a year when a cable hoist broke and struck the C-band antenna.

Congress took a first look at NASA's manned lunar colonies and Mars bases plan and all but eliminated funds for the Office of Exploration to conduct essential innovative technological research. There was little support in Congress for the ambitious resurrection of U.S. manned space objectives, particularly when the space station program was fighting hard for survival. A bolder, more expensive initiative was viewed by the majority in Congress as both unwise and untimely.

1990 June 14 Engineers completed a vigorous set of mirror movements on the Hubble Space Telescope to correct apparent faults in the imaging system. Since activation, scientists had been concerned that the telescope's performance was far below specification. Tests revealed that faults in the main mirror had occurred during manufacture and test and that the problem could not be solved from the ground. By late June, scientists and engineers were beginning to plan for a Shuttle mission to carry astronauts to Hubble to make repairs.

1990 June 28 Aerojet, Pratt and Whitney, and Rocketdyne teamed to develop the propulsion system for the Air Force/ SDI/NASA Advanced Launch System. The first stage was being designed around a cluster of 580,000-lb. thrust liquid propellant motors in a launch system capable of lifting 50,000–200,000 lb. Boeing, General Dynamics and a Martin Marietta/McDonnell Douglas team were competing for overall vehicle design.

1990 June 29 Tests on Shuttle orbiter *Atlantis* on LC-39A confirmed a hydrogen leak in the 17-in. external tank–orbiter fuel connection. Rolled to pad A on June 18, 1990, *Atlantis* had been scheduled to fly the STS-38 mission in mid-July, but during the countdown engineers discovered a leak. Further tests were conducted on July 13 and July 25. *Atlantis* was rolled back to the Vehicle Assembly Building on August 9 but a hail storm began. It was moved to the Orbiter Processing Facility, where tile damage was repaired, and returned to the VAB on October 2 where a platform fell on it, causing further damage. It was returned to pad A on October 12, but the planned launch on November 9 was postponed for six days when the U.S. Air Force found problems with the classified satellite *Atlantis* was scheduled to launch.

1990 July 1 A panel chaired by NASA's Dr. William F. Fisher and Charles R. Price (informally dubbed the Fisher-Price group, after the U.S. toy manufacturer) raised serious doubts about the viability of NASA's new baseline design for space station *Freedom*. In a final report on EVA (extravehicular activities) needs, the group found that during mantended assembly, astronauts would have to spend 6,267 man-hours on external maintenance with only the Shuttle to support habitation on brief visits. Moreover, in the first 35 years of operation, astronauts would have to spend 3,276 hr. per year on routine EVA activities. NASA had hoped to keep annual EVA needs below 500 hours/year. In addition, where NASA had previously expected to use 45 kW of the 75 kW power for station needs, the systems were now projected to draw 60 kW, leaving only 15 kW for experimenters in all three modules. Weight growth was also a problem, the station design presently being 21% over its 512,000-lb. baseline limit.

1990 July 2 In the first application of the gravity-assist technique at an Earth encounter, the European Space Agency's *Giotto* comet hunter skimmed past the planet at a distance of 14,300 mi. When launched on July 2, 1985, *Giotto*

had been placed on a ⅚ solar trajectory where in exactly five Earth years (orbits of the sun) the spacecraft would perform six passes around the sun and return to exactly the same point in space where the Earth would be. Gaining energy from the Earth encounter, *Giotto* was now on a trajectory that would bring it to a very close encounter with the comet Grigg-Skjellerup on July 10, 1992.

1990 July 10 Henry F. Cooper replaced Lt. Gen. Monahan as director of the Strategic Defense Initiative ballistic missile defenses research program in the Department of Defense. The third SDI director, and the first civilian in that position, Dr. Cooper was a nationally recognized expert on strategic arms and nuclear weapons and had served two years as assistant director of the Arms Control and Disarmament Agency.

1990 July 11 An international gamma-ray observatory launched by the Soviet Union was carried from Baikonur by an SL-4 to a 245.5 × 247.5 mi., 51.6° orbit. Called *Gamma*, the 16,210-lb. satellite had been built upon the Progress cargo tanker employed by the *Mir* program and incorporated a giant instrument module, 26 ft. 3 in. long and 7 ft. 6 in. across. Two solar arrays spanned 40 ft. Carrying equipment supplied by France and Poland, *Gamma* incorporated a large gamma-ray telescope and two smaller telescopes, Disk and Pulsar Kh-2. *Gamma* decayed back into the atmosphere on February 28, 1992.

1990 July 16 The first flight of China's CZ-2E (Long March 2E) launcher placed the experimental Pakistani satellite *Badr-1* in a 125 × 611 mi. orbit inclined 28.4°. Similar to the CZ-2 with stretched first and second stages, the CZ-2E had four liquid-propellant strap-on boosters, each equipped with one YF-2 (first-stage) engine to produce a total liftoff thrust of 1,323,000 lb. The CZ-2E can lift 19,400 lb. to low Earth orbit or 9,920 lb. to geostationary transfer orbit. Launched from Xichang, the 115-lb. satellite had been built by the Pakistani space agency, Suparco, and carried a digital transmitter which sent signals to Earth until it reentered on December 8. A 16,200-lb. mass model of an Aussat satellite which CZ-2E was scheduled to launch in 1991 was retained atop the second stage, which decayed on October 9, 1990.

1990 July 17 At 13:06 UT *Mir* cosmonauts Anatoly Solovyov and Aleksandr Balandin began a space walk from *Kvant 2* to inspect damage to *Soyuz-TM 9* that had been caused when the aerodynamic shroud separated at launch on February 11. The crew had been unable to carry out the space walk inspection until the *Kristall* module, launched May 31, had brought special access ladders and work restraints to the station. On July 3 *Soyuz-TM 9* had been moved from the back of the *Kvant 1* module to the forward axial docking port. Making their way down the *Kvant 2* module, the cosmonauts were soon out of reach of the standard 82 ft. long tether and had to use mountaineering-style clamps to inch their way along hand rails. After 2 hr. they reached *Mir*'s forward docking unit and erected access ladders and restraints to work on the *Soyuz-TM 9* descent module. The cosmonauts found three blankets, each 6 ft. 6 in. square, were affected. Two were clamped back down and one was cut free. Back at *Kvant 2* they found they were unable to close the ingress/egress hatch and had to wait while controllers depressurized the main *Kvant 2* work compartment. With Solovyov and Balandin inside, the hatch between the work compartment and the airlock was closed and the work compartment was repressu-

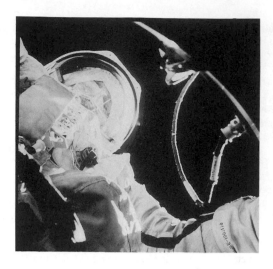

A Soviet cosmonaut takes a walk in space on one of many EVAs to perform experiments, develop engineering tasks or simply repair a failing component of the Mir *space station.*

rized, leaving the airlock module unpressurized and its hatch stuck partly open. The space walk had lasted 7 hr. 16 min.

1990 July 24　Ariane 4 mission V37 carried the French direct broadcast satellite *TDF 2* and the German telecommunications satellite *DFS Kopernikus 2* from Kourou, French Guiana, to geostationary transfer orbit. In the 44L configuration, Ariane 4 marked a resumption in launch operations following a five-month hiatus during which modifications were made to prevent a repeat of the February 22 accident. Identical to their numerical predecessors, *TDF 2* was positioned over 19° W, and *Kopernikus 2* over 28.5° E.

First in a new generation of space-based interceptor (SBI) for the Strategic Defense Initiative ballistic missile defenses program successfully performed a demonstration test in a building at Edwards AFB, Calif. The SBI model weighed only 40 lb., compared to 220 lb. for the earlier SBI model, with significant advances in systems miniaturization. In the advanced hover interceptor test (AHIT), the vehicle performed for 14 sec. during which it hovered and demonstrated repeated pulsed firing of its thrusters as it maneuvered to track a simulated target.

1990 July 25　The first commercial Atlas-Centaur launch vehicle carried a NASA/Department of Defense satellite designed to study the Earth's magnetic field from Cape Canaveral, Fla., to a 201 × 20,863 mi. orbit inclined 18.2°. Built by Ball Space Systems, the 3,760-lb. *Combined Release and Radiation Effects Satellite (CRRES)* carried barium, lithium, strontium and calcium in 24 canisters. When these elements were released as vapor, the sun's ultraviolet light would ionize the chemicals, creating color clouds by which scientists could study the interaction of the upper atmosphere with solar electric fields. The first chemical releases occurred on September 10 and 12. *CRRES* remains in space. On the day that General Dynamics launched its first commercial Atlas-Centaur, the company had 23 firm sales for Atlas-Centaur and

eight optional orders, and was committed to building 60 commercial launchers through 1997.

1990 July 26　At 11:15 UT *Mir* cosmonauts Anatoly Solovyov and Aleksandr Balandin began a 3 hr. 31 min. space walk to remove work ladders from the forward *Mir* docking unit which had been used July 17 to repair *Soyuz-TM 9*. With the equipment in place, the docking port would be fouled. Leaving a TV camera viewing the egress/ingress hatch on *Kvant 2*'s airlock module, which had refused to close on July 17 because of a buckled hinge, the crew stowed the work area access equipment aboard *Soyuz-TM 9* and by the time they had returned to the top of *Kvant 2*, ground controllers had a procedure for them to use in finally securing the hatch, which they accomplished at 14:46 UT, pending a full repair operation attempted on October 29

1990 August 1　At 09:32 UT Soviet cosmonauts Gennady Manakov and Gennady Strekalov were launched in the *Soyuz-TM 10* spacecraft by SL-4 from the Baikonur cosmodrome. They docked to the back of the *Kvant 1* module at the *Mir* station complex two days later and joined cosmonauts Anatoly Solovyov and Aleksandr Balandin, who were coming to the end of their six months aboard *Mir*. The four cosmonauts carried out joint activities for six days before the departure of Solovyov and Balandin. *Soyuz-TM 10* would be left at the *Kvant 1* docking port for the duration of its stay in space.

1990 August 8　The first six in a new series of Soviet second-generation military communication satellites was launched by an SL-14 from Plesetsk. Called Locyst, the system would comprise groups of six 500-lb. satellites placed in four orbital planes at 90° intervals around the globe. Designated *Cosmos 2090–2095*, they were each placed in slightly differing orbits closely around an average of 870 × 877 mi. with an inclination of 82.6°. A second set of six Locyst satellites *(Cosmos 2114–2119)* was launched by an SL-14 on December 22, followed by the third set *(Cosmos 2157–2162)* on September 28, 1991, and the fourth set *(Cosmos 2165–2170)* on November 12, 1991. Replacement sets consisted of: *Cosmos 2197–2202*, two of which were precursor demonstration satellites for a new system called Gonets, on July 13, 1992; *Cosmos 2211–2216* on October 20, 1992; *Cosmos 2245–2250* on May 11, 1993; and *Cosmos 2252–2257* on June 24, 1993.

1990 August 9　*Mir* cosmonauts Anatoly Solovyov and Aleksandr Balandin returned to Earth after a mission lasting 179 days 1 hr. 18 min. They departed in the *Soyuz-TM 9* spacecraft they brought with them to the *Mir* complex on February 11 and had recently positioned at the forward axial docking port. There was some unjustified concern about the undocking, retrofire and separation of the orbital module, descent module and instrument module because of damage to insulation on the spacecraft caused during launch. On *Mir*, routine operations were sustained by cosmonauts Gennady Manakov and Gennady Strekalov, in their second week of a four-month stay.

1990 August 10　At 16:32:32 UT the 15,232-lb. thrust solid propellant Star 48B retro-rocket on *Magellan* fired for 1 min. 24 sec., placing the spacecraft in a 183 × 5,253 mi. orbit of Venus inclined 85.5° to the equator and with a period of 3 hr. 16 min. At this date, Earth was 144 million mi. away. With the motor expended, *Magellan* weighed 3,160 lb. Early problems with the spacecraft delayed a start to synthetic aperture radar mapping of the surface of Venus, but on September 16 the first planned mapping program began. The first mapping cycle was measured as the time taken for Venus to rotate once beneath the polar orbiting spacecraft, thus allowing *Magellan* to radar-map almost all the surface. Venus rotates slowly once in 243 Earth days, so the first map cycle would end May 15, 1991. Data was sent to Earth at 268.8 kilobits/sec. Magellan was expected to return 3,070 gigabits of data, more than all previous NASA planetary missions combined.

1990 August 15 At 04:01 UT the unmanned *Progress M-4* cargo tanker was launched from Baikonur cosmodrome by an SL-4 launch vehicle. It docked to the forward axial docking port on the *Mir* space station, then in an orbit of 228.5 × 237 mi., at 05:26 UT two days later. After being unloaded by Gennady Manakov and Gennady Strekalov, *Progress M-4* was undocked from the forward axial port on September 17 and a cooperative exercise was performed with the *Mir* cosmonauts, who observed a plasma arc test conducted in the docking unit. *Progress M-4* was de-orbited to destruction in the atmosphere over the Pacific Ocean on September 20.

1990 August 18 Britain's second private direct broadcast satellite, *Marcopolo 2*, was launched by a Delta II from Cape Canaveral, Fla., and eventually positioned adjacent *Marcopolo 1* over 31° W. Kept in close proximity to each other, the two satellites put their identical footprints down on the same geographic area, allowing five linked channels of ten 55-W transponders to double-up and beam 110-W signals to small ground dish antennas.

1990 August 25 The first in a series of planned tests supporting development of the Brilliant Pebbles concept from the Strategic Defense Initiative ballistic missile defenses program failed to achieve its objectives. Launched from NASA's Wallops Island facility, the Black Brant rocket carried the SDI payload to a height of 124 mi. in a 13-min. flight, but telemetry was lost shortly after liftoff. The payload consisted of test components for the ultra-miniaturized Brilliant Pebbles interceptors some scientists said would be a more cost-effective defense against ballistic missiles.

1990 August 28 The first satellite in the third generation of Japanese broadcast satellite, *BS-3A* was launched by H-I to a transfer orbit from which it was moved to a geostationary position at 110° E. With a launch weight of 2,460 lb. and an on-station mass of 1,210 lb., *BS-3A* had three 120-W transponders operating at 14/12 GHz frequencies. Built by General Electric Astro-Space around the design of its Astro 3000 series, the satellite carried Japanese transponders, antennas and apogee kick motor. Box shaped, it measured 4 ft. 3 in. × 5 ft. 3 in. × 5 ft. 3 in. Delivering 1,443 W, the two solar-cell wings had a total span of 49 ft. 3 in. Deployment of the shaped parabolic antenna gave *BS-3A* a total height of 10 ft. 6 in.

1990 August 30 An Ariane 4 in the 44LP configuration (two liquid/two solid strap-ons) launched the United Kingdom's *Skynet 4C* military communication satellite and the *Eutelsat II F1* European communication satellite. Third of three satellites in the series, the 3,153-lb. *Skynet 4C* was positioned at 1° W. The first of six operational successors to the European Communication Satellite series, *Eutelsat II F1* was built by Aérospatiale and would provide telephone, data, telex, business services and TV distribution. With a deployed height of 8 ft. 10 in. and a body structure 4 ft. 3 in. × 4 ft. 3 in., the satellite had solar cells providing 3,000 W on wings with a span of 73

The first of six second-generation Eutelsat communication satellites launched by Ariane 44LP on August 30, 1990, significantly expanded regional telecommunications services for the European continent.

ft. 6 in. With an on-orbit weight of 2,062 lb., *Eutelsat II F1* provided services from its position at 13° E via 16 × 50 W transponders.

1990 September 3 The first completely successful Chinese sun-synchronous meteorological satellite, *Feng Yun 1-F2*, was launched by CZ-4 from Jiaquan to a 547 × 557 mi. orbit inclined 99°. It measured 4 ft. × 4 ft. 7 in. × 4 ft. 7 in., with a span of 28 ft. 3 in. across two solar arrays. The satellite had two scanning radiometers in five spectral bands from the visible to the near infrared, providing cloud and surface cover day and night. With an improved attitude control system, the 1,943-lb. satellite operated well. It was accompanied by two 9-lb. atmospheric research balloons inflated to a diameter of 8 ft. 3 in. on the third revolution, one to an orbit of 481.5 × 500 mi., which decayed on March 11, 1991, and the other to an orbit of 518 × 550.5 mi., which decayed on July 24, 1991. *Feng Yun 1-F2* remains in orbit.

1990 September 17 The Science Program Committee of the European Space Agency selected the experiments that would be carried to Saturn's moon Titan by the Huygens entry probe. Huygens was the European contribution to NASA's *Cassini* mission, scheduled for launch in April 1996. With a total weight of 424 lb., the Huygens probe would enter the atmosphere of Titan at about 20,000 ft./sec. and decelerate via atmospheric friction. Parachutes deployed at successively lower altitudes would reduce the descent rate at which instruments were lowered to the surface. Measurements in the atmosphere would be made during descent, and the probe would be designed to send back information to the *Cassini* orbiter for about two years.

1990 September 19 In closed session, aerospace leaders met in Washington, D.C., to critique space station *Freedom* plans and its newly estimated cost of $37 billion. NASA officials said they would have to redesign the station once more if its budget allocation of $2.451 billion for fiscal year 1991 was cut by as little as 8%. In fact, Congress cut the request by 22.5%. One option being studied by Marshall Space Flight Center was to cut the two U.S. pressure modules in half, making four. Whatever the outcome, NASA wanted to maintain a first-element launch date of March 1995. However, the build-up schedule was based on 10–12 Shuttle flights a year, of which 4–5 would be required for the station, while in reality NASA was still unable to demonstrate more than 6–8 missions per year.

1990 September 22 *Pioneer 10* reached a distance from the sun of more than 4.648 billion mi. (50 AU), equal to the greatest distance reached by the planet Pluto in its elliptical path around the solar system. At this date Pluto was inside the orbit of Neptune at a distance from the sun of less than 2.8 billion mi. Round-trip communication time at the speed of light was 13 hr. 47 min., and although transmitted at 8 W, the strength of the signal reaching Earth was less than 4 billionths of a trillionth of a watt. Electrical power was expected to drop below the transmission threshold in 1998, when *Pioneer 10* would be 6.7 billion mi. from the sun. *Pioneer 10* was traveling down the heliosphere in the opposite direction to the movement of the solar system in the galaxy. Within the next 862,000 years, the closest it would come to a known star will be in the year 34,582, 32,610 years after launch, when it passes within 3.26 light years of Ross 248.

1990 September 27 The unmanned cargo tanker *Progress M-5* was launched by an SL-4 from the Baikonur cosmodrome. Planned for some time, this variant of the Progress M–series incorporated a ballistic reentry capsule capable of returning to Earth intact 331 lb. of cargo. All previous Progress cargo tankers had been one-way supply freighters and were unable to bring experiment results or equipment back to Earth. *Progress M-5* docked to the forward axial port on the *Mir* station complex on September 29. After offloading supplies and provisions, cosmonauts Gennady Manakov and Gennady Strekalov closed up the orbital module, and *Progress M-5* separated on November 28 and the capsule returned to Earth at 11:04 UT that day carrying crystals produced in the Kristall laboratory at *Mir*.

1990 September NASA's Langley Research Center unveiled the mock-up of a space plane it wanted to develop as a ferry for carrying up to 10 people to and from space station *Freedom*. Conceived as a relatively low-cost transportation system, the HL-20 evolved from earlier U.S. lifting-body configurations. It had a length of 29 ft. 6 in., a wingspan across the upswept fins of 23 ft. 6 in. and in the flight configuration would weigh about 24,000 lb. Presented by William M. Piland, chief of Langley's space systems division, as an assured crew-return vehicle (ACRV) for docking at the space station for crew escape in emergency, it was also posed as a crew delivery vehicle which could be launched by a Titan 4–class launcher.

1990 October 5 A CZ-2C launcher carried the Chinese recoverable satellite *Fanhui Shi Weixing-1* from Jiuquan to a 124 × 183 mi. orbit inclined 54.9°. The beehive shaped, 5,735-lb., satellite had a length of 10 ft. 4 in. and a diameter of 7 ft. 3 in., including a 4,585-lb. recoverable component with a length of 4 ft. 11 in. and a diameter of 5 ft. 3 in. After eight days the capsule returned to Earth with its "cargo" of biological samples, including plants and animals. The main body of the satellite reentered on October 23.

1990 October 6 At 11:47:15 UT the Shuttle *Discovery* lifted off from LC-39B, Kennedy Space Center, Fla., at the start of STS-41, the mission to launch the 807-lb. ESA *Ulysses* solar polar spacecraft. The total payload weighed 49,999 lb. Astronauts Richard N. Richards (CDR), Robert D. Cabana (PLT), M. Shepherd (MS), Bruce E. Melnick (MS) and Thomas D. Akers (MS) controlled release of the 38,604-lb. inertial upper stage/PAM/Ulysses spacecraft combination from the cargo bay at an elapsed time of 6 hr. 1 min. The first stage of the solid propellant internal upper stage fired for 2 min. 30 sec. starting at 7 hr. 6 min., the second stage fired for 1 min. 45 sec. starting at 7 hr. 10 min., and the PAM-S stage fired for 1 min. 27 sec. beginning at 7 hr. 14 min. to increase velocity to 34,130 MPH. The spacecraft separated 9 min. later. *Ulysses* comprised a box-shaped structure supporting on top a parabolic high-gain antenna 5 ft. 6 in. in diameter with a nuclear radioisotope thermoelectric generator on a boom providing 285 W at the start of the mission and 255 W at the end. Folded for launch, *Ulysses* had a length of 10 ft. 6 in., a width of 10 ft. 9 in. and a height of 6 ft. 10 in. A magnetometer boom deployed to a length of 18 ft. 4 in. after separation, as did a 237-ft. dipole antenna and a 24 ft. 7 in. axial monopolar antenna. *Discovery* also carried several scientific experiments which were carried out before landing at Edwards AFB, Calif., at 4 days 2 hr. 10 min. 3 sec.

1990 October 12 An Ariane 44LP launch vehicle (V39) carried two spin-stabilized Hughes satellites from Kourou, French Guiana, to geostationary transfer orbit. The combined payload of 8,136 lb. comprised a 5,464-lb. Hughes HS 393 for Hughes Communications designated *SBS-6*, and a 2,672-lb. Hughes HS 376, also for Hughes Communications, designated *Galaxy 6*. *SBS-6* had an on-orbit weight of 3,338 lb. and an on-board power level of more than 3 kilowatts. With a 15-year life, it had 19 × 41-W transponders and 11 spare amplifiers providing backup to older satellites in the Hughes Communications network from its position at 99° W. *Galaxy 6* had an on-orbit mass of 1,561 lb., a power level of 900 W and 24 × 10-W transponders and 6 spare amplifiers. It was positioned at 91° W.

1990 October 15 The European Space Agency's *Ulysses* solar polar spacecraft began a series of tiny thruster firings to conduct a course correction maneuver. For a total of 26 hours in activity also conducted on October 16 and 18, the spacecraft changed velocity by approximately 325 ft./sec. On October 19 all nine science experiments were turned on. The 121.5-lb. of instruments included equipment for measuring solar wind plasma, solar wind ion composition, cosmic dust, magnetic fields, solar x rays and gamma-ray bursts, interstellar gas, low-energy ions and electrons, cosmic-ray particles, and radio and plasma waves. Data would be sent to Earth at up to 4.096 kilobits/sec. via a 20-W transmitter. A second course correction maneuver was performed on November 2, 1990, when the spacecraft's thrusters were fired for 1 hr. 43 min., followed by the third and last correction before Jovian encounter with a 8 min. 30 sec. firing on July 8, 1991.

1990 October 30 Mir cosmonauts Gennady Manakov and Gennady Strekalov performed a 3 hr. 45 min. space walk from the *Kvant 2* module docked to the space station complex and attempted to replace a buckled hinge on the outer door. They were unsuccessful in getting the procedure, developed in ground simulators, to work in space, and had to abandon the effort. They found that the hinge pin itself was bent and would have to be replaced by another space walk carried out by the replacement crew, Musa Manarov and Viktor Afanasyev, launched on December 2. Space-walking cosmonauts would continue to use the main experiment section of *Kvant 2* for depressurization.

A Delta II launch vehicle carried the first in a series of second-generation Inmarsat satellites from Cape Canaveral, Fla., to a geosynchronous transfer orbit. From there, *Inmarsat 2–1* was placed in a geostationary position above 64.5° E. Built by a consortium led by British Aerospace, the satellite was box-shaped, 8 ft. 2 in. tall, 4 ft. 10 in. long and 4 ft. 7 in. wide. With solar arrays spanning 50 ft. and providing 1,200 W of power, *Inmarsat 2–1* weighed 3,054 lb. at launch and 1,520 lb. on station. Via a communication payload provided by Hughes Aircraft, it could provide 250 ship-to-shore, aeronautical and mobile links. Three more satellites of the same series were launched: *F2* on March 8, 1991, to 15.5° E; *F3* on December 16, 1991, to 178° E; and *F4* on April 15, 1992, to 55° W.

1990 October   At a series of meetings between NASA and Soviet space officials in Moscow agreement was reached on the broad outline of a plan to launch a U.S. astronaut from Baikonur on a Soyuz spacecraft to the *Mir* space station, and for a Soviet cosmonaut to fly aboard the Shuttle. The two countries agreed to set up working parties to prepare for this activity, which was not expected to occur before the middle of the decade due to the need for extensive technical preparation and crew familiarization and training.

1990 November 2   Believing that the baseline design of space station *Freedom* was basically flawed and in need of significant reworking, the new Director of Space Station Richard M. Kohrs issued an 11-point directive calling for an agency-wide review of the international facility. In attempting to hold to a first-element launch in March 1995, NASA wanted to redesign the station for a maximum four people by reducing module size and eliminating the transverse truss assembly, minimize the impact on international partners, establish life sciences and materials science as the most urgent activities for early research, reduce the need for EVA (extravehicular activity) and reduce the required number of Shuttle assembly flights.

1990 November 13   NASA selected principal investigators and science teams for the *Cassini* mission to Saturn, including members from 14 countries supporting 62 separate investigations of the planet and its moon Titan. During four years of operations in orbit beginning in 2002, *Cassini* would make up to 30 close passes of many of Saturn's icy moons and use radar to construct a picture of Titan's surface beneath the dense obscuring clouds. A formal agreement between NASA and the European Space Agency for the Europeans to collaborate in the mission and build a probe for studying Titan's atmosphere was signed in December 1990.

1990 November 15   At 23:48:15 UT Shuttle *Atlantis* was launched from LC-39A at the Kennedy Space Center, Fla., carrying astronauts Richard Covey (CDR), Frank Culbertson (PLT), Charles Gemar (MS), Carl Meade (MS) and Robert Springer (MS) on the STS-38 mission. Dedicated to carrying Department of Defense payload AFP-658, this was the last Shuttle mission to operate under the veil of total secrecy that had prevailed for military missions since the launch of 51-C on January 24, 1985. Including STS-38, there had been seven missions dedicated to carrying classified military payloads, although the first (STS-4) had been a development flight. AFP-658 comprised a 22,000-lb. intelligence gathering satellite placed in an orbit from which it could gather data about

military movements in Iraq and Kuwait. STS-38 ended at Edwards AFB, Calif., at 4 days 21 hr. 54 min. 27 sec.

1990 November 20   An Ariane 4 (V40) in the 42P configuration with two solid propellant strap-ons and a liftoff weight of 701,657 lb., carried two U.S. communication satellites to geosynchronous transfer orbit. First to be released, the 2,578-lb. *Satcom C1* three-axis-stabilized satellite for General Electric's American Communications had a box structure (8 ft. 2 in. × 4 ft. 3 in. × 5 ft. 3 in.), a total height of 11 ft. 6 in. and solar arrays producing more than 1,100 W on wings with a span of 51 ft. 6 in. *Satcom C 1* had 24 × 9.5 W (6/4 GHz) transponders and provided a variety of services from 137° W. The 2,855-lb. *GSTAR 4* completed the Spacenet communications network, supplying services down 16 Ku-band transponders from 125° W.

1990 November 23   In a clear sign of the gathering momentum of political change and the impending breakup of the Soviet Union, the central authority of the Soviet Union launched a Gorizont communication satellite for the newly declared Russian Soviet Federal Republic. Sent to a geostationary position above 40° E, the 2,480-lb. *Gorizont 22* satellite was operated by the republic and not the Soviet Union. It was also the first Soviet satellite to be insured. *Gorizont 22* was launched from Baikonur on an SL-12.

1990 November 26   The first flight of the more powerful three-stage Delta II, the 7925 version, launched the 10th Navstar Block II global positioning satellite. With an improved Rocketdyne first-stage engine, the Delta 7925 had nine lightweight graphite epoxy Hercules Castor IVA solid propellant strap-on motors instead of the Thiokol Castor IV. These improvements increased geosynchronous transfer orbit payload capability from the 3,190-lb. with a Delta 6920 to 4,010 lb. Both versions had the lengthened second stage, increasing overall vehicle height from 116 ft. to 125 ft. 11 in. For low Earth orbit missions, only the first two stages would be used, the Delta 6920 lifting 8,780 lb., compared to 7,610 lb. for the Delta 3920, and 11,110 lb. for the 7925.

1990 December 2   At 06:49:01 UT the Shuttle *Columbia* was launched from LC-39B at the Kennedy Space Center, Fla., on the STS-35 mission carrying Vance Brand (CDR), Guy Gardner (PLT), Jeffrey Hoffman (MS), John Lounge (MS), Robert Parker (MS), Ronald Parise (PS) and Samuel Durrance (PS). Delayed 21 min. for clouds to clear, *Columbia* carried a 33,067-lb. payload comprising the first mission dedicated to astrophysics. It included ASTRO-1, an observatory with four telescopes for ultraviolet and x-ray astronomy measurements mounted on a Spacelab pallet. Several irritating systems problems plagued *Columbia* and the mission was cut short by a day due to impending bad weather at the Edwards AFB landing site, reached at 8 days 23 hr. 5 min. 7 sec.

Soviet cosmonauts Musa Manarov and Viktor Afanasyev, together with the Japanese cosmonaut-journalist Toyohiro Akiyama, were launched aboard their *Soyuz-TM 11* spacecraft by an SL-4 from the Baikonur cosmodrome at 08:13 UT. Including the seven astronauts aboard the Shuttle STS-35 mission and *Mir* cosmonauts Gennady Manakov and Gennady Strekalov, there were, for the first time, 12 people in space at the same time. *Soyuz-TM 11* docked with the forward axial port on the *Mir* space station at 09:57 UT on December 4. The first professional journalist to fly in space, Akiyama had been financed by the Japanese television station TBS, and it was said that costs to provide all the supporting broadcast equip-

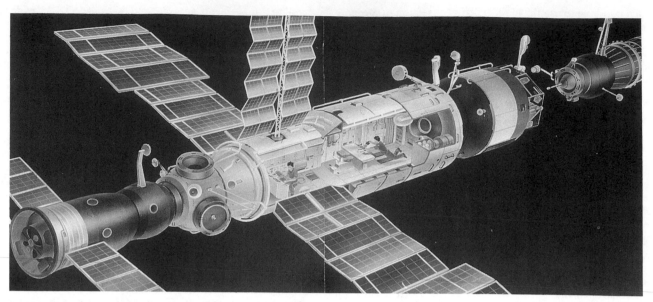

With a Kvant *module attached to the aft docking port, and a* Progress *tanker docked to* Kvant, *the* Mir *station carries a Soyuz spacecraft at the forward port and has room for four cosmonauts.*

ment and 150 technicians to cover the mission exceeded the more than $12 million paid to the Soviets for the flight. Manarov and Afanasyev were to replace Manakov and Strekalov aboard Mir for a stay of almost six months in space.

1990 December 8   *Galileo* made a close pass of the Earth, coming within 590 mi. of its surface at 20:35 UT where it gained 11,500 MPH. During the flyby period *Galileo* used its imaging instrument to take pictures of features on the moon under lighting conditions never seen before. As *Galileo* swooped around the Earth it imaged regions of Antarctica and South Africa. A minor course correction maneuver took place on December 19. The spacecraft continued moving around the sun inside Earth's orbit until, on January 11, 1991, it began to move out toward the asteroid belt, crossing the orbit of the Earth on February 15, 1991, in a wide arc that would return it to a second encounter with Earth in December 1992.

1990 December 10   At the end of a flight lasting 130 days 20 hr. 36 min., Soviet cosmonauts Gennady Manakov and Gennady Strekalov returned to Earth from the *Mir* space station complex, leaving Musa Manarov and Viktor Afanasyev in their place. Undocking from the back of the *Kvant 1* module at 02:48 UT, Manakov and Strekalov were accompanied by the Japanese journalist Toyohiro Akiyama, ending a 7-day 21 hr. 55 min. flight to *Mir*. All three cosmonauts returned to Earth in *Soyuz-TM 10*. *Mir* cosmonauts Manarov and Afanasyev transferred *Soyuz TM-11* from the back of the *Kvant 1* module to the forward axial docking port.

1991 January 8   *Mir* space station cosmonauts Viktor Afanasyev and Musa Manarov conducted a 5 hr. 18 min. space walk beginning at 17:03 UT. The objective was to dismantle the damaged hinge on the airlock door at the forward end of the *Kvant 2* module and to replace it with a new one. They were successful in this task, as they were in installing a metal truss structure to the outside of the *Mir* station in readiness for

a mobile crane assembly to be attached on a subsequent space walk. The crew also retrieved some sample canisters attached to the exterior.

The first of two new military communications satellites, *NATO 4A* was launched by a Delta II from Cape Canaveral, Fla., to a geosynchronous transfer path from where it was positioned in geostationary orbit over 18° W. Weighing 3,160 lb. at launch, the satellite was built by British Aerospace for the countries of the North Atlantic Treaty Organization and carried a payload supplied by Marconi Space Systems in the United Kingdom. The bus comprised a box-shaped structure, 6 ft. 11 in. high, 6 ft. 3 in. long and 4 ft. 7 in. wide. Solar arrays spanning 52 ft. 6 in. provided more than 1,200 W of electrical power. *NATO 4B* was launched by Delta II on December 8, 1993, to 263° E.

1991 January 14   At 14:50 UT the *Progress M-6* cargo tanker was launched by SL-4 from the Baikonur cosmodrome and placed initially in an orbit of 119.5 × 139 mi. The unmanned vehicle docked to the back of the *Kvant 1* module at the *Mir* complex at 16:35 UT two days later. *Progress M-6* had brought to the station a mobile telescopic arm which would be used later to move solar panels on the *Kristall* microgravity laboratory to a new location on the *Kvant 1* astrophysics laboratory. After using the *Progress* propulsion system to make an orbital adjustment to the station, the unmanned freighter departed from the aft docking port at 12:47 UT on March 15, and was de-orbited over the Pacific Ocean.

1991 January 15   Italy's first national communications satellite, *Italsat 1*, and the *Eutelsat II F2* European Communications Satellite, were launched aboard Ariane 44L flight V41 from Kourou, French Guiana, to geosynchronous transfer orbit. Built by Selenia Spazio, Rome, the 4,112-lb. *Italsat 1* comprised a box structure, 8 ft. 9 in. × 7 ft. 3 in. × 20 ft. high when deployed with two parabolic dish antennas 6 ft. 6 in. in diameter. The structure supported two solar array wings with a span of 68 ft. 11 in. *Italsat 1* provided six 30/20 GHz multibeam transponders, three 30/20 GHz global beam transponders and one 54/40 GHz propagation experiment. With a launch weight of 4,112 lb. and an on-orbit mass of 2,095 lb., it was so accurately placed at 13.2° E that the five-year design life was extended by two years. *Eutelsat II F2* was almost identical to its predecessor.

1991 January 18 One day after coalition forces began the air offensive prior to expelling Iraqi forces from Kuwait, Iraqi Scud missiles were fired against targets in Saudi Arabia. Between this date and February 25, three days before Iraqi forces surrendered, a total 81 Scuds were fired. Of that total, 43 were against Operation Desert Storm coalition forces in the Persian Gulf and 38 were against other targets, mostly urban areas in Israel. Coalition forces quickly adapted the Patriot surface-to-air missile as defense against Iraqi Scuds. However, while 70% of Scuds attacked by Patriots defending the Persian Gulf were destroyed, only 40% of Scuds targeting Israel were destroyed. The Iraqi missiles were modified, extended-range SS-1B Scud A and SS-1C Scud B types known as al-Hussein and al-Abbas, respectively.

1991 January 23 The first of two space walks by *Mir* cosmonauts Viktor Afanasyev and Musa Manarov to install equipment for moving solar panels from the *Kristall* laboratory to the *Kvant 1* module began at 10:59 UT. In an operation lasting 5 hr. 33 min. they moved a canister brought up by *Progress* freighter to the side of the *Mir* station and attached it to one of the fixtures used originally to secure the launch shroud. Compacted into a container 6 ft. 6 in. long, the telescopic arm could extend a maximum distance of 46 ft. Fabricated from composite materials, it weighed 99 lb. but could maneuver a maximum load of 1,543 lb. Each *Kristall* solar array had a mass of 1,100 lb. During the second space walk, which began at 09:00 UT on January 26, Afanasyev and Manarov spent 6 hr. 20 min. erecting a second telescopic boom to the *Kvant 1* module and attaching laser reflectors for new docking equipment.

1991 January 28 An ERIS interceptor launched from the U.S. Army missile test site at Kwajalein Atoll in the Pacific Ocean successfully intercepted a simulated warhead launched by a missile from Vandenberg AFB, Calif., 4,200 mi. away and destroyed it. ERIS (exo-atmospheric reentry interceptor subsystem) had evolved from the homing overlay tests of 1984 and was managed by the Army for the Strategic Defense Initiative ballistic missile defense program office. ERIS is 13 ft. 5 in. long with a diameter of 2 ft. 6 in. In the test a Minuteman I launched from VAFB carried a dummy reentry vehicle and several decoys. Launched by a two-stage Aries rocket, ERIS was launched in rain about 21 min. later. The reentry vehicle was distinguished from decoys released by Minuteman and impact was brought about at an altitude of 167 mi., 575 mi. from Kwajalein.

1991 January 29 In his State of the Union Address, President George Bush announced a redirection for the Strategic Defense Initiative ballistic missile defense research program. SDI would no longer pursue defense against at least 50% of a strategic nuclear arsenal involving a preemptive strike by a major superpower launching upwards of 10,000 nuclear warheads. It would now concentrate on giving the United States protection from missile attack by small countries in a rescoped endeavor known as global protection against limited strikes (GPALS). It was proposed that about 1,000 Brilliant Pebbles kinetic interceptors would be based in space for an estimated

Phase 1 deployment cost of $10 billion, a further $22 billion being spent on ground-based interceptors and space-based early warning and tracking systems known as Brilliant Eyes.

1991 January The Department of Defense refined a requirement for a successor to the Block 14 satellite early warning defense support program. Called the follow-on early warning system (FEWS), the concept had replaced an earlier specification called advanced warning system which evolved from a requirement issued by the Strategic Defense Initiative. TRW Systems and Aerojet General were teamed to build the 23 Block 14 DSP satellites which would be replaced by FEWS beginning with launches in 2003. A request for industrial proposals was issued in June 1991 and three parties competed: TRW with Grumman, Aerojet with Rockwell International, and Lockheed. The winning FEWS bidder was expected to build 12–18 satellites but the Clinton administration canceled FEWS in early 1994.

1991 February 7 Almost nine years after it had been launched by SL-13 from the Baikonur cosmodrome, *Salyut 7* descended into the atmosphere and burned up just before 04:00 UT. With the heavy *Cosmos 1686* module still attached to the aft port, the complex broke up over South America, showering areas in Argentina with debris. Although fragments came down within 12 mi. of Buenos Aires, the closest shave came when a large lump of hardware crashed into the garden of a house. To have sent *Salyut 7/Cosmos 1686* to a controlled destruction over an unpopulated region of the globe would have required 1,100 lb. of propellant, and *Cosmos 1686* had only 155 lb. remaining. The Soviet Shuttle *Buran* had been scheduled to retrieve *Salyut 7* and return it to Earth for extensive examination, but delays to that program prevented this.

1991 February 25 Two Defense Support Program early warning satellites detected the launch of an Iraqi al-Abbas missile that killed 28 U.S. soldiers during Operation Desert Storm. The No. 12 DSP was operating from above the Indian Ocean and had a "nose-on" view of the launch but failed to trigger an alert. The No. 13 DSP scanned the launch and alerted controllers at Darmstadt, Germany, and Colorado Springs, Colo., who reported this to air defense units in Dhahran, Saudi Arabia. Sirens were started 5 min. before the missile hit but no Patriots were fired because a missile tracking projection had not been made due to a combination of unusual circumstances. Admirably placed to track the al-Abbas, a third DSP had been off-line and was not integrated into the network.

1991 March 2 The *Astra 1B* direct broadcast satellite, second of three for Luxembourg, and the *MOP 2 (Meteosat 5)* weather satellite were launched by Ariane 44LP mission V42 from Kourou, French Guiana, to geosynchronous transfer orbit. The 5,777-lb. *Astra 1B* was placed at 19.2° E, which, with *Astra 1A*, doubled to 32 the number of program transponders for Société Européenne des Satellites. The 1,502-lb. *MOP 2* weather satellite joined four other Meteosats for the Eumetsat organization and was placed on standby at 4° E.

1991 March 7 Shuttle orbiter *Discovery* was rolled back to the Vehicle Assembly Building from LC-39A at the Kennedy Space Center, Fla., after cracks were found on external tank umbilical doors. *Discovery* had been taken to the pad for the STS-39 mission, but repairs were necessary that made it essential to return the Shuttle to the VAB. It was returned to

LC-39A on April 1 in readiness for a launch on April 23, but the flight was postponed when an electrical component on one of the three main engines gave a bad reading. STS-39 was launched on April 28.

1991 March 19 The 15,900-lb. *Progress M-7* cargo tanker was launched at 13:07 UT by an SL-4 from Baikonur cosmodrome. During a docking attempt at the back of the *Kvant 1* module (attached to the *Mir* space station) two days later, the spacecraft flew on past the complex and had to be brought back in for a second attempt on March 23. This also failed and tests of the automatic rendezvous and docking radar systems were conducted. Old Progress spacecraft used the Igla system, while Progress M and Soyuz TM spacecraft had the Kurs system. On March 26 *Mir* cosmonauts Viktor Afanasyev and Musa Manarov entered the *Soyuz TM-11* spacecraft, undocked and made a simulated Progress-type approach at the *Kvant 1* module. Discovering that one *Kvant* antenna was transmitting a false beam, they overrode the automatic system and docked manually at *Kvant 1*, leaving *Progress M-7* to dock, successfully, at the forward axial port at 12:03 UT on March 28. *Progress M-7* undocked at 23:00 UT on May 6, and was de-orbited the following day, but a recoverable capsule was lost.

1991 March 20 Moving away from Earth's orbit in the general direction of the asteroid belt, *Galileo* conducted a course correction maneuver by repeatedly pulsing its thrusters between 18:00 UT and 21:00 UT. This adjusted velocity by 7 ft./sec. and began to set up the flight path for a close look at the asteroid Gaspra, which the spacecraft was scheduled to pass on October 29. *Galileo* was now 36 million mi. from Earth. Further correction maneuvers in early July and on October 9 and 24 moved the flight path to a 1,000-mi. pass of Gaspra on October 29, 1992.

1991 March 21 NASA presented Congress with plans for the redesigned space station *Freedom*. The U.S. habitability and experiment pressure modules were reduced in size to 14 ft. 6 in. × 27 ft. A much shorter horizontal truss assembly would provide 18.75 kW of power (versus 37.5 kW) in man-tended mode and 56.25 kW (versus 75 kW) at permanent manned capability of which 30 kW would be dedicated to research needs. Simpler communications equipment would cut data transmission from 300 megabits/sec. to 50 megabits/sec. and reduce the number of computers from 22 to 7. Total length of the transverse truss assembly had been reduced from 493 ft. to 353 ft. With a first-element launch in January–March 1996, the station would be man-tended by April–June 1997 with the U.S. laboratory module, receive the Japanese Experiment Module in June 1998, Europe's *Columbus* module in September 1998 and the U.S. habitation module in June 1999, reaching full capability with four crew members by the end of 1999.

1991 March 30 As required by the U.S. House of Representatives Appropriations Conference Committee, the Department of Defense submitted a theater missile defense (TMD) report covering a range of antimissile systems for defense at battlefield and theater level. The Gulf War involving Iraq and the coalition forces brought together to liberate Kuwait had emphasized the importance of defense against missiles like the Scud. The report explained how the recently announced GPALS (global protection against limited strikes) and existing hardware could combine to provide protection against missiles at theater level.

1991 March 31 An SL-13 launched the massive, 40,900-lb. *Almaz 1* synthetic aperture radar remote-sensing satellite from Baikonur to a 182 × 189.5 mi., 72.7° orbit. This was the first time the Russians, or the Soviets before them, had referred to Almaz by name. It was in fact the sixth Almaz launched, including three as habitable space stations *(Salyuts 2, 3* and *5)* and two previous radar remote sensing satellites (a failed flight in 1985 and *Cosmos 1870*). The Chelomei bureau that designed the Almaz series was now known as the Mashinostroyeniye. *Almaz 1* measured 39 ft. 3 in. long × 13 ft. 7 in. in diameter and carried two 5 × 50 ft. synthetic aperture radar antennas, one on each side. Two 463 sq. ft. solar cell arrays either side of the forward module provided 2.5 kW of electrical energy, most of which was needed for the powerful radar. After tests and calibration, *Almaz 1* began transmitting in July, providing images of a 124 × 217 mi. area. *Almaz 1* decayed on October 17, 1992.

1991 April 4 The first of two fifth-generation communication satellites for Telesat Canada, *Anik E2* was launched by an Ariane 44P configuration from Kourou, French Guiana, to geosynchronous orbit on mission V43. With a liftoff weight of 782,334 lb. and four solid propellant strap-on boosters, *Anik E2* was placed on track for a geosynchronous slot at 107.3° W. Built by Spar Aerospace, incorporating a bus provided by GE Astro-Space (incorporating the former RCA), the two Anik E satellites had a body 9 ft. 4 in. × 7 ft. 6 in. × 7 ft. 10 in. high, solar arrays delivering 4 kW across a span of 70 ft. 6 in. Anik E had 24 × 12-W C-band transponders, plus six spares, and 16 x 50-W Ku-band transponders, with two spares. *Anik E2* failed to deploy its C-band antenna, but repeated efforts to shake it loose worked and the satellite completed deployment on October 2. Launched by an Ariane 44P on September 26 (V46), *Anik E1* went to 111.1° W.

1991 April 5 After a delay of 4 min. 44 sec. for low clouds, Shuttle orbiter *Atlantis* was launched on the STS-37 mission at 14:22:44 UT from LC-39B at the Kennedy Space Center, Fla. The crew consisted of Steven Nagel (CDR), Kenneth Cameron (PLT), Jay Apt (MS), Linda Godwin (MS) and Jerry Ross (MS). The 40,591-lb. payload included the Gamma Ray Observatory satellite, after the Hubble Space Telescope the second of NASA's four so-called great observatories, and at 34,445 lb. the heaviest payload deployed to Earth orbit by a Shuttle. After six failed attempts at deploying the GRO's high-gain dish antenna, astronauts Ross and Apt performed a 4 hr. 24 min. contingency EVA (extravehicular activity) and in 17 min. had it fixed. The GRO was deployed at 23:35 UT on April 7. The next day Ross and Apt performed a scheduled 6 hr. 11 min. EVA to test three alternate space carts which might be used to move equipment around on space station *Freedom*. Waved off for a day from a planned landing at Edwards AFB, Calif., by high crosswinds, the mission ended April 11 at an elapsed time of 5 days 23 hr. 32 min. 44 sec.

1991 April 11 At 20:50 UT computers aboard the *Galileo* spacecraft sent commands to unfurl the 16 ft. diameter high-gain antenna for the first time, but 50 sec. later the dual drive motors stalled and 8 minutes after that the motors were shut off by a safety feature. The antenna was needed to send high-data-rate transmissions from a flyby of the asteroid Gaspra and for all subsequent phases of the mission to Jupiter. Indications were that the antenna was stuck halfway to full deployment by a failure in one of the 18 umbrella-like ribs or by solar heat–caused differential expansion that locked up rib pins. A succession of tests failed to free the antenna.

1991 April 12   In celebrations of the 30th anniversary of the world's first human spaceflight, Soviet officials gathered with U.S. representatives at the Baikonur cosmodrome, where a mock-up of *Energia M*, a smaller version of the full-size Energia, was on a launch pad. The Soviets wanted to market the down-scaled Energia to foreign users. *Energia M* would be capable of placing 88,000 lb. in low Earth orbit rather than the 250,000 lb. of the full-size Energia. One of the latter was being prepared for an early 1992 flight, carrying the Soviet shuttle *Buran* to *Mir 1*. During the year, this plan was changed. The *Energia/Buran* flight was abandoned and a decision made to go for a *Mir 2*, perhaps in cooperation with the United States.

1991 April 13   A Delta 2 launched from Cape Canaveral, Fla., carried a 2,980-lb. communications satellite, *ASC-2*, to a geosynchronous transfer orbit. From there it was positioned in geostationary orbit at 101° W. The second of two satellites built for American Satellite Company/Contel, it was redesignated *Spacenet 4* in orbit because Contel and GTE Spacenet had absorbed ASC after the satellite had been ordered. GTE had bought RCA, and the launch of *ASC-2/Spacenet 4* completed a network of nine satellites which also included three other Spacenets, four G-Stars, and *ASC-1*. *Spacenet 4* was designed to provide eighteen C-band and six Ku-band transponders for TV and interactive data transmissions to small networks but three Ku-band transponders failed shortly after launch.

1991 April 18   An Atlas-Centaur carrying the Japanese BS-3H communication satellite failed to put its payload into orbit when contamination in one of the two Centaur engines prevented the stage from firing properly. It was launched from Complex 36 at Cape Canaveral, Fla., but the range safety officer gave the vehicle the self-destruct command at 6 min. 1 sec. and the upper stack fell back into the Atlantic Ocean. Coming at a time when General Dynamics was actively marketing the launcher, this failure was particularly unwelcome. It was the first time in more than five years that an Atlas-Centaur had failed.

The second, and the first successful, test flight of the small intercontinental ballistic missile (SICBM) took place from Vandenberg AFB, Calif. The 37,000-lb. missile flew 4,000 mi. across the Pacific to the Kwajalein Atoll in the Marshall Islands where reentry of the test vehicle was monitored 30 min. after launch.

1991 April 25   Beginning at 20:29 UT *Mir* cosmonauts Viktor Afanasyev and Musa Manarov conducted a space walk to examine the antenna for the Kurs radar approach system which, on an attempt to dock the *Progress M-7* cargo tanker on March 21, had failed. The cosmonauts discovered that a cosmonaut on a previous space walk had knocked off the dish antenna with his boot. They reported this to ground controllers, retrieved samples of materials left outside the station and conducted a small experiment in thermomechanical joints. The space walk ended at an elapsed time of 3 hr. 34 min.

Rockwell International rolled out the Shuttle orbiter built to replace *Challenger* destroyed on January 28, 1986. Designated OV-105 and named *Endeavour* after the ship used by British Captain James Cook on his first voyage of exploration in 1768–1771, the orbiter had a total dry weight of 172,876 lb. It incorporated an updated avionics system with new and more powerful computers which were being incorporated into the other three orbiters, improved nosewheel steering, updated mechanical systems and a drag parachute designed to reduce the orbiter's rollout distance. *Endeavour* also carried equip-

ment which could give it a 28-day orbital endurance capability if additional systems were fitted. It was delivered to the Kennedy Space Center, Fla., on May 7, 1991.

1991 April 28   After a 32-min. delay resolving minor technical problems, Shuttle mission STS-39 began with the launch of *Discovery* from LC-39A at the Kennedy Space Center, Fla., at 11:33:14 UT. On board were Michael Coats (CDR), Blaine Hammond (PLT), Guion Bluford (MS), Richard Hieb (MS), Gregory Harbaugh (MS), Donald McMonagle (MS) and Charles Veach (MS). Launched to a 56.9° orbit, the flight was an unclassified Department of Defense mission and the first Department of Defense Shuttle flight purely for research and development. The 26,324-lb. payload included AFP-675, a collection of sensors and detectors in support of work on ballistic missile defense, and Space Test Program-01 (STP-01) experiments. On May 1 at 08:18 UT *Discovery* released the 4,012-lb. *SPAS/Infrared Background Signature Survey (SPA-S/IBSS)* satellite to a 154 × 163 mi. orbit. Consisting of the twice-flown *SPAS* platform purchased from the German concern Messerschmitt-Bölkow-Blohm, *IBSS* was set free to observe from a distance Shuttle thruster and orbital maneuver system (OMS) burns with ultraviolet and cryogenically cooled infrared sensors. These provided data to characterize plume behavior in space, to ensure that any future Strategic Defense Initiative antiballistic missile system would target the motor creating the plume and not the plume itself. Two 180-lb. chemical release observation (CRO) satellites, *CRO-C* and *CRO-B*, were released from the Shuttle's cargo bay on May 2. These emitted gases that Soviet reentry vehicles could use to veil their presence. The 10,203-lb. AFP-675 pallet attached to the rear of the cargo bay sniffed and sensed the satellites through the gases to test discrimination. The *SPAS/IBSS* was then retrieved before *CRO-A* was released on May 3. After Shuttle orbit had been raised out of the way of the CRO satellites, the manipulator arm extended the IBSS sensors above the cargo bay for further tests. On May 6, just hours before reentry, a classified lightsat satellite was released into orbit. Because of bad weather at Edwards AFB, Calif., STS-39 was rerouted to a landing at Kennedy Space Center, 8 days 7 hr. 22 min. 23 sec. after launch.

1991 May 3   Lt. Gen. Thomas P. Stafford, USAF (Ret.) sent Vice President Dan Quayle the report of the Synthesis Group set up by President George Bush to plot a future course for U.S. space goals. NASA implemented an Outreach Program in support of the Synthesis Group to solicit views and recommendations from a wide cross-section of the aerospace community and concerned citizens. The final report proposed a permanent presence on the moon in 2004 and a manned landing on Mars in 2014. A range of scenarios were presented, all supporting extensive exploration, but with a strong emphasis on the urgent need for a long-range strategic plan.

1991 May 16   NASA's Gamma Ray Observatory (GRO), after the Hubble telescope the second of the so-called Great Observatories, began a 15-month-long sky survey after five weeks of tests and calibration. The GRO was officially renamed the Compton Observatory after Arthur Holly Compton (1892–1962), an American physicist whose pioneering work on high-energy radiation underpins much of modern physics. The Compton Observatory measured 15 ft. × 29 ft. 10 in., with a span of 70 ft. 3 in. across solar cell panels that provided 4,300 W, degrading to 2,000 W by the end of its projected life. It carried four instruments, with contributions from Germany, the Netherlands, the European Space Agency, as well as universities and research laboratories in the United States. Compton made major discoveries in high-energy astrophysics that significantly modified astronomers' interpretation of the universe.

NASA's *Magellan* began its second mapping cycle of Venus as the planet started a second rotation beneath the polar orbiting spacecraft. Stunning images returned by *Magellan* had unpeeled the obscuring atmosphere to reveal a surface dynamic and probably still active. The basic mission of *Magellan* was to map 70% of the surface thoroughly and to date the spacecraft had covered 83.7% of the planet. Emphasis in this second map cycle would go on those regions not covered during the first cycle, notably the south polar region. *Magellan* radar data had a resolution of about 400 ft., compared to 3,300 ft. for the Russian *Venera 15* and *16*. The second mapping cycle would end January 15, 1992, and scientists were debating what to do with *Magellan* after that. Some wanted to put *Magellan* in a circular orbit to improve resolution to 150 ft.

1991 May 18 At 12:50 UT Soviet cosmonauts Anatoly Artsebarski and Sergei Krikalov, accompanied by British cosmonaut Helen Sharman, were launched aboard the *Soyuz TM-12* spacecraft by an SL-4 from Baikonur cosmodrome. During the approach to the *Mir* space station on May 20, the long-distance approach system malfunctioned and the pilots had to dock manually, which was accomplished at the forward axial port at 14:31 UT that day. Joint experiments were conducted for five days, after which Sharman returned to Earth on May 26 with *Mir* cosmonauts Viktor Afanasyev and Musa Manarov, leaving Artsebarsky and Krikalov to man the *Mir* space station.

1991 May 21 The first of four Soviet Resurs remote sensing satellites flown during the 1991 agricultural season in the northern hemisphere was launched by an SL-4 from Plesetsk to a 141 × 143.5 mi., 82.3° orbit. Designated *Resurs F-10*, its reentry vehicle was recovered on June 20. *Resurs F-10* was one of the Resurs types equipped with the MK-4 system incorporating multispectral camera and stellar camera. Launched on June 28, *Resurs F-11* carried the *Priroda 4* system with two wide-field cameras, three multispectral cameras and one stellar reference camera. It was placed initially in a 157 × 166.5 mi., 82.3° orbit and was recovered on July 21. Of the same type, *Resurs F-12* was placed in a 163.5 × 177 mi., 82.3° orbit on July 23 and recovered on August 8. With *Priroda 4*, *F-13* was launched August 21 to a 140 × 143 mi., 82.3° orbit and was recovered on September 20.

1991 May 26 Soviet cosmonauts Viktor Afanasyev and Musa Manarov returned to Earth at the end of their almost six months in space aboard *Mir*, together with British cosmonaut Helen Sharman, who had visited them on a flight lasting 7 days 21 hr. 14 min. The three cosmonauts returned to Earth in the *Soyuz TM-11* spacecraft that had carried Afanasyev and Manarov into orbit on December 2, 1990. They undocked from the back of the *Kvant 1* module at 06:13 UT on May 26, and returned to Earth at 10:04 UT that day. Afanasyev and Manarov had been in space for 175 days 1 hr. 52 min., but this second long-duration flight for Manarov brought his total spaceflight time to 541 days 31 min. On May 23 *Mir* cosmonauts Artsebarsky and Krikalov transferred the *Soyuz TM-12* spacecraft from the forward axial port to the port at the back of the *Kvant 1* module.

1991 May 29 A Delta 2 launch vehicle carried the 2,945-lb. *Aurora 2* satellite from Cape Canaveral to a geosynchronous transfer path from where it was later positioned in geostationary orbit at 130° W. Built by General Electric Astro-Space (formerly RCA), it was a successor to the Alascom (Satcom) satellite which during its nine-year orbital life had been renamed *Aurora 1*. *Aurora 2* was box-shaped, 3 ft. 3 in. × 5 ft. 5 in. × 4 ft. 7 in., with two 3-panel solar arrays. Operating in the 6/4 GHz, C-band range, *Aurora 2* carried 32 transponders, of which 6 were redundant spares, providing 16 channels for customers in Alaska and 8 for U.S. businesses. With a 12-year life, *Aurora 2* began operations during July.

1991 May 30 At 08:04 UT the *Progress M-8* cargo tanker was launched by an SL-4 from Baikonur cosmodrome. It docked to the forward axial port at the *Mir* space station at 00:45 UT on June 1. Since discovering a damaged Kurs docking radar antenna when *Progress M-7* failed to dock at the *Kvant 1* port, ground controllers planned to have unmanned freighters use their automatic docking mechanisms with the viable radar at the forward port. *Progress M-8* would remain docked to Mir during an intensive period of space walking by cosmonauts Anatoly Artsebarsky and Sergei Krikalov between June 25 and July 27. It undocked at 22:17 UT on August 15, and deorbited that day. A balloon intended for deployment from *Progress M-8* burst on release.

1991 May Arianespace, the marketing organization for Europe's Ariane launcher, received a contract to launch the *Arabsat 1C* satellite. The Arabsat telecommunications organization had previously booked the satellite to fly on China's Long March but canceled the reservation when they realized how much extra work and cost there would be in getting the satellite to the site, checked out and installed aboard the rocket.

TRW Systems filed a request with the U.S. Federal Communications Commission for a license to develop a global mobile telephone network. The company wanted to put up a constellation of 12 satellites in 6,440-mi. orbits, with 4 satellites in each of three 55° inclination orbits at 120° intervals around the world. Called Odyssey, the system would carry voice, data and paging services and eventually have the capacity to support up to 6 million subscribers.

1991 June 5 Originally set for May 22 and delayed, first until June 1 and then to June 5, due to suspect electrical sensors, Shuttle mission STS-40 began at 13:24:51 UT from LC-39B at the Kennedy Space Center, Fla. The crew for *Columbia* included Bryan O'Connor (CDR), Sidney Gutierrez (PLT), James Bagian (MS), Tamara Jernigan (MS), Rhea Seddon (MS), F. Drew Gaffney (PS) and Millie Hughes-Fulford (PS). The 33,737-lb. payload comprised the Spacelab Life Sciences-1 package, a pressurized Spacelab module containing equipment for 18 life science investigations, 10 using humans as subjects, 7 using rodents and 1 using a jellyfish. The mission ended at Edwards AFB, Calif., at 9 days 2 hr. 14 min. 20 sec.

1991 June 22 Officials from the Department of Defense Strategic Defense Initiative antiballistic missile research program unveiled the Alpha chemical laser at a facility run by TRW Systems near San Juan Capistrano, Calif. The megawatt-range laser was being designed for use in space against ascending ballistic missiles and was considered the next step after the GPALS (global protection against limited strikes) program. The SDI office wanted to match Alpha to a mirror of 13-ft. diameter being developed by Martin Marietta for

Earth orbit tests in 1996–1998. Called StarLite, the 40,000–50,000-lb. device could be launched by a Titan 4. Starlite was essentially a slimmed down version of the massive Zenith Star, for which an entirely new heavyweight launch vehicle would have been needed.

1991 June 25 *Mir* cosmonauts Anatoly Artsebarski and Sergei Krikalov completed a 4 hr. 58 min. space walk that began at 21:11 UT on June 24. Working their way from the airlock module hatch at the top of the *Kvant 2* module to the back of the *Kvant 1* module attached to *Mir*, the cosmonauts replaced the broken antenna on the Kurs rendezvous and docking radar. They also put out a package to test a new concept of thermomechanical joints—materials that "remember" a preformed shape after being heated. A second space walk began at 19:02 on June 28 and involved the cosmonauts' being moved around the outside of the station by the telescopic arm that had been attached to *Mir* by Viktor Afanasyev and Musa Manarov on January 23 and deploying experiment packages delivered by *Progress M-8*. The space walk lasted 3 hr. 24 min.

1991 June 29 A four-stage Scout launched from Vandenberg AFB, Calif., carried the 188-lb. U.S. Air Force satellite *REX* to a 478 × 541 mi., 89.6° orbit. *REX* was an acronym for radiation experiment, a military satellite designed to test the response of communication equipment in a high-radiation environment. Such information was sought to perfect the development of survivable, secure military communications during a nuclear attack. *REX* remains in orbit.

1991 June Officials at the European Space Agency were briefed on a new version of the United Kingdom's proposed HOTOL reusable winged satellite launcher. Following studies between British Aerospace and a group of industries in the Soviet Union, the new HOTOL was projected to carry four RD-120 hydrogen-oxygen engines developed for the second stage of the Energia heavy-lift launcher. Carried into the air on the back of the giant Antonov An-225 transport aircraft, only one of which had ever been built, HOTOL would separate at 30,180 ft. and fire its engines, accelerating toward orbit.

1991 July 4 As a secondary payload to the eleventh Global Positioning System (GPS) Navstar 2–series satellite, a Delta 2 carried the experimental satellite *Losat-X* to Earth orbit. The GPS satellite was placed in a 12,475 × 12,610 mi., 55.3° orbit. *Losat-X*, an acronym for low altitude satellite experiment, was designed to measure drag effects and was placed in an orbit of 250 × 258.5 mi., with an inclination of 40°. The 165-lb. *Losat-X* decayed out of orbit on November 15.

1991 July 15 An intensive period of space walks from the *Mir* station began with cosmonauts Anatoly Artsebarsky and Viktor Krikalov working for 5 hr. 45 min. on building up a platform for a Sofora girder employing thermo-mechanical joints. Using the telescoping arm on *Mir* to move the platform to the side of the *Kvant 1* module, the two cosmonauts erected and installed the platform ready to receive the girder assembly. During a second, 5 hr. 28 min. space walk on July 19, the cosmonauts moved the Sofora assembly module and construction kit to the platform and erected it before getting back inside. On the third space walk, on July 23, the cosmonauts built 14 of the 20 modular sections of the Sofora girder in a period of 5 hr. 34 min. On the last Sofora space walk, on July 27, they completed the assembly in little more than 3 hr.

Sofora comprised a lattice-like tower 46 ft. tall and capable of being pivoted backwards from the vertical by 11° or down by 85° almost to the horizontal position back out along the *Kvant 1* module. As a final act, Artsebarsky mounted a Soviet flag to the top of Sofora. The 6 hr. 49 min. space walk brought to 31 hr. 58 min. the total time spent by Artsebarsky and Krikalov on activities outside *Mir*.

1991 July 17 With a liftoff weight of 533,303 lb., an Ariane 40 configuration carried the first European remote sensing satellite, *ERS-1*, and four small satellites to Earth orbit on mission V44. Placed in a 481 × 481.5 mi. orbit inclined 98.5°, the 5,256-lb. *ERS-1* was designed to continuously monitor ocean surface temperatures, wave and wind behavior, sea-state conditions, ice formation and migration, and atmospheric water vapor. Filling a gap left by the absence of a U.S. ocean resource satellite, the *ERS-1* had been built by Dornier from a French SPOT bus. The box-shaped structure (6 ft. 6 in. x 6 ft. 6 in. × 10 ft.) supported an articulated solar array panel 7 ft. 10 in. × 38 ft. 4 in. producing more than 2 kW. A 3 ft. 4 in. × 34 ft. synthetic aperture radar antenna dominated the lower section, producing a satellite height of 38 ft. 8 in. The four other satellites were: *Uosat 5*, a 110-lb. amateur radio satellite from the United Kingdom's University of Surrey; 48.5-lb. *Orbcomm-X*, built by Orbital Sciences Corp. in the United States as a mobile communications test device; 84-lb. German *Tubsat*, which provided digital data links for small portable transmitters; 57-lb. *SARA* radio-astronomy satellite from France's ESIEESPACE astronomy club. All four satellites were in similar orbits to *ERS-1* and remain in space.

The second air-launched flight of the Pegasus solid propellant satellite launcher carried seven Microsat satellites into orbit for the Department of Defense. At 17:33 UT about one hour after takeoff from Edwards AFB, Calif., the 41,950-lb. Pegasus and its payload was dropped from the B-52 mother plane. A stage separation sequence malfunctioned and the resulting orbit was lower than the planned 447 mi. Each 12-sided Microsat measured 7.5 in. × 1 ft. 7 in. and weighed 49 lb. Built by Defense Systems for the Defense Advanced Research Projects Agency, they were placed in a 223 × 284 mi. orbit with an inclination of 82°. *Microsats 1-7* decayed January 23–25, 1992.

1991 July 31 Two significant agreements between the United States and the Soviet Union were signed by President George Bush and Premier Mikhail Gorbachev in the Kremlin, Moscow. In one, major strategic arms reductions were agreed by both sides. Ending nine years of START (strategic arms reduction talks), it limited each side to a maximum of 6,000 nuclear weapons, with no more than 4,900 on ballistic missiles, and a combined total of 1,600 delivery vehicles (bombers, ICBMs and submarine-launched ballistic missiles). The Soviets agreed to a sublimit of 1,100 mobile ICBM warheads and a sublimit of 1,540 warheads on 154 heavyweight SS-18 ICBMs. The Soviets currently had 308 SS-18s, but the United States had no mobile or heavyweight ICBMs. (Actual inventories would be higher due to definition terms. For instance, one warhead would be attributed to each bomber and the first 180 Soviet bombers could have eight warheads apiece.) The net result would cut U.S. deployment from 12,081 strategic nuclear warheads to 10,360 by the end of the century, while Soviet numbers would fall from 10,821 to 8,040. In the other accord, the United States and the Soviet Union agreed on a program of joint manned spaceflights for the first time since the Apollo-Soyuz Test Project mission in July 1975. A Soviet cosmonaut would fly in the Shuttle during a mission planned for late 1993 and a U.S. astronaut would be carried in a Soyuz to *Mir* for a six-month stay.

1991 July   Two U.S. aerospace companies teamed up for design efforts on a proposed single-stage-to-orbit (SSTO) reusable launcher when Boeing and Rockwell joined forces on a combined bid. Funded by the Department of Defense Strategic Defense Initiative office, the SSTO project envisaged a winged launcher capable of carrying 10,000–20,000 lb. to low Earth orbit. McDonnell Douglas and General Dynamics were also working on the SSTO concept. Low-cost, relatively heavy-lift launchers were considered a key to getting space-based antimissile systems operational at affordable cost.

Vice President Dan Quayle told an audience at Vandenberg AFB, Calif., that emphasis must be placed on the National Launch System as a family of launchers to complement the Shuttle. Keen to exploit Shuttle technology, Martin Marietta proposed a launcher based on the external tank carrying two liquid propellant engines at the base and four jettisonable engines in a 1½-stage arrangement similar to the Atlas launcher. Suitably sized, the proposed vehicle could launch 50,000 lb. or 80,000 lb. payloads to low Earth orbit.

1991 August 2   Delayed three times since July 23 due to technical trouble, Shuttle mission STS-43 began at 15:01:59 UT with the launch of *Atlantis* from LC-39A at the Kennedy Space Center, Fla. The crew comprised John Blaha (CDR), Michael Baker (PLT), James Adamson (MS), David Low (MS) and Shannon Lucid (MS). The 49,355-lb. payload included the *TDRS-E* tracking and data relay satellite, released into orbit at 21:14 UT. Designated *TDRS-5* after deployment, it was eventually placed in geostationary orbit at 174° W, replacing *TDRS-3*, which was moved to 62° W. STS-43 also accommodated a range of technology experiments and tests of prototype equipment designs that might be used aboard space station *Freedom*. The mission ended at Kennedy Space Center, Fla., at 8 days 21 hr. 21 min. 25 sec.

1991 August 15   A Soviet SL-14 launched the *Meteor 3 F5* meteorological satellite from Plesetsk to an orbit of 736.4 × 747.5 mi., with an inclination of 82.6°. The 4,885-lb. satellite carried the total ozone mapping spectrometer (TOMS) provided by the United States. Identical to the instrument flown on *Nimbus 7*, TOMS provided high-resolution data on global ozone levels. The flight had been timed to capture data on ozone depletion during winter over the South Pole. Meteor carried Soviet radiometers and multispectral spectrometers for weather imagery and data collection. It remains in orbit.

1991 August 18   McDonnell Douglas received authorization from the Department of Defense Strategic Defense Initiative antiballistic missile team to proceed with demonstration tests of a radical new form of space transportation that might hold promise for major cuts in launch costs. Called DC-X, or Delta Clipper, the demonstrator would look like the conical nose fairing of a ballistic missile. It would be 39 ft. tall, 12 ft. across the base, weigh 41,630 lb. at liftoff, with power from four modified Pratt and Whitney RL10A-5 engines similar to those used in the cryogenic Centaur stage. Essentially a one-third scale version of the definitive Delta Clipper single-stage-to-orbit, the DC-X was designed to liftoff from the ground, rise to a maximum 17,000 ft., hover, move sideways under full control and descend on rocket power to a gentle touchdown.

1991 August 20   At 22:54 UT the *Progress M-9* cargo tanker was launched by an SL-4 from the Baikonur cosmodrome. It carried a 772-lb. recoverable capsule capable of containing up to 331 lb. of items for return to Earth. *Progress M-9* docked to the forward axial port of the *Mir* station at 00:54 UT on August 23. After being offloaded by Anatoly Artsebarsky and Sergei Krikalov, *Progress M-9* was undocked at 01:54 UT on September 30, and its recoverable capsule was picked up safely in Kazakhstan after landing at 08:18 UT.

1991 August 25   An H-I launch vehicle carried the Japanese direct broadcast satellite *BS-3B* from Tanegashima to a transfer path from where it was later placed in geostationary orbit at 110° E. *BS-3B* provided two 120-W transponders for the Nippon Hoso Kyokai (NHK) broadcasting corporation and one 120-W transponder for the Japan Satellite Broadcasting Co., all three being in the 14/12 GHz band. *BS-3B* had been built by Nippon Electric to a design based on the General Electric Satcom 3000 bus. It weighed 2,460 lb. at launch, 1,210 lb. on station, with 16,000 solar cells on fold-out arrays spanning 49 ft. 3 in. producing 1,100 W at the end of its intended life.

1991 August 29   A Soviet SL-3 launcher carried India's second remote sensing satellite, *IRS-1B*, from Baikonur to a 553 × 570 mi., 99.2° orbit. The 2,160-lb. satellite carried three linear imaging self-scanning sensors for terrain mapping, resource management, and a variety of applications in farming, mineral exploration, crop control and pest eradication. LISS-I imaged a swath 92 mi. across at a resolution of 238 ft. LISS-IIA and IIB worked together across a 90-mi. swath with a resolution of 119 ft. The box-shaped satellite carried 92 sq. ft. of solar cells producing 620 W at the end of its life. *IRS-1B* was the second Indian satellite launched by the Soviets on a commercial basis and replaced *IRS-1A*. It was to be joined by *IRS-1C*, launched by an SL-3 in 1994, and *IRS-1D*, which India planned to launch on its PSLV launcher in 1996–1997. The refurbished *IRS-1A* engineering model, redesignated *IRS-1E*, was destroyed when its PSLV launcher malfunctioned on September 20, 1993.

1991 August 30   A solid propellant Mu-3S II launch vehicle carried a 925-lb. Japanese solar observation satellite, *Solar-A*, from Kagoshima to a 343 × 486.5 mi. orbit inclined 31.3°. The first solar telescope to carry instruments for monitoring both hard and soft x rays, *Solar-A* incorporated equipment contributed by Japanese, British and U.S. researchers. Sponsored by Japan's Institute for Space and Astronautical Sciences, *Solar-A* had a soft x-ray telescope supplied by the NASA Marshall Space Flight Center, a hard x-ray telescope from Japan, a crystal spectrometer from Britain and a wideband spectrometer from Japan. *Solar-A* was rectangular with a width of 4 ft. 7 in. and a height of 6 ft. 6 in.

1991 September 12   After a delay of 14 min. to repair faulty communication links, Shuttle mission STS-48 began with the launch of *Discovery* at 23:11:04 UT from LC-39A at the Kennedy Space Center, Fla. On board were John Creighton (CDR), Kenneth Reightler (PLT), Mark Brown (MS), Charles Gemar (MS) and James Buchli (MS). The 21,599-lb. payload included the *Upper Atmosphere Research Satellite (UARS)*, deployed at 16:23 UT on September 15 by Mark Brown who operated the remote manipulator arm. *UARS* had been designed to conduct the most systematic and comprehensive satellite study of the Earth's stratosphere, mesosphere and lower thermosphere ever carried out. With a length of 32 ft.

Tests with a mobile erector-launcher for the U.S. Air Force Midgetman missile involved a variety of different vehicle designs for cross-country movement.

and a diameter of 15 ft., *UARS* weighed 15,000 lb. and carried 10 science instruments powered by 1.6 kW of electrical energy from six 5 ft. × 11 ft. solar arrays. Deployed to an orbit of 356 x 360 mi., at an inclination of 57°, the satellite was the first element in NASA's Mission to Planet Earth Program, providing a nearly continuous stream of data that map stratospheric temperatures, wind velocities and ozone concentrations. The Shuttle mission ended at Edwards AFB, Calif., at 5 days 8 hr. 27 min. 38 sec.

1991 September 27 President George Bush announced further cuts in the U.S. strategic nuclear arsenal above and beyond those identified by the START (strategic arms reduction talks) agreement signed on July 31. In a unilateral act, the United States would abandon plans for the rail garrison mobile basing mode for the Peacekeeper ICBM and retain the 50 currently deployed in silos at Warren AFB, Wyo., cancel the Midgetman SICBM and remove the 450 single-warhead Minuteman II ICBMs from strategic alert. Minuteman II was to be dismantled once START had been ratified. Other moves included destruction or storage of all tactical nuclear weapons presently deployed on land, at sea and in the air as well as removal of the 850 Lance missiles and 1,300 nuclear artillery shells still in the inventory.

1991 September Critical elements of the Soviet space program began to break up as the Soviet Union itself came under stress from political disruption. The Soviet Ministry of Defense lost control of missile-warning radar sites, the Soviet Academy of Sciences was dissolved and the Ministry for General Machine Building, largely responsible for ballistic missile production, was broken up. Lt. Gen. Vladimir Shatalov, head of cosmonaut training at Zvezdny Gorodok (Star City) was ousted and replaced by another ex-cosmonaut, Pyotr Klimuk. Newly declared sovereign republics laid claim to ministry organizations on their territory, and Russia's

design bureau and assembly plants began to forge independent marketing links.

1991 October 2 At 05:59 UT Soviet cosmonaut Aleksandr Volkov, Austrian cosmonaut Franz Viehboeck, and Kazakh cosmonaut Toktar Aubakirov were launched aboard the *Soyuz TM-13* spacecraft by an SL-4 from Baikonur cosmodrome. The Soyuz docked at the forward axial port on the *Mir* complex at 08:39 UT on October 4, and the crew joined Anatoly Artsebarsky and Sergei Krikalov shortly thereafter. Aubakirov was the first fee-paying cosmonaut participant from inside the Soviet Union, now in its last days as a union of separate states. He, Viehboeck and Artsebarsky returned to Earth on October 10.

1991 October 4 A Soviet SL-4 carried the 13,700-lb. *Photon 4* materials research satellite from Plesetsk to a 137 × 251 mi., 62.8° orbit. The payload comprised materials processing experiments utilizing equipment and samples provided by scientists in France and Germany, as well as experiments from Russia and several other Federation states. As with earlier satellites in this series, *Photon 4* was on a relatively short mission, and the reentry vehicle returned the experiments to Earth on October 20 after a 16-day mission.

1991 October 10 *Mir* cosmonaut Anatoly Artsebarski returned to Earth with Austrian cosmonaut Franz Viehboeck and Kazakh cosmonaut Toktar Aubakirov, leaving Aleksandr Volkov and Sergei Krikalov in charge of the *Mir* space station. Artsebarski's mission had lasted 144 days 15 hr. 22 min., while the other two had logged 7 days 22 hr. 13 min. The three cosmonauts returned in the *Soyuz TM-12* spacecraft which had been docked to the back of the *Kvant 1* module. On October 15 Volkov and Krikalov got inside the *Soyuz TM-13* spacecraft and transferred it to the back of the *Kvant 1* module, vacating the forward axial port.

At meetings held October 16–17, 1991, the first flight of the Hermes space plane was put back from 1998 to 2002 at the earliest. In this artist's impression, an Ariane 5 lifts the spaceplane to orbit.

1991 October 16–17   The management council of the European Space Agency, concerned at the burgeoning cost projections of the Hermes space plane, pushed back the first, unmanned, flight date from 1998 to 2002, and the first manned flight to 2003. Cost projections for full-scale development were now in the region of $9 billion, an increase of 40% on the original projection of $6.4 billion. The trimmed weight of Hermes had grown to 49,380 lb., just within the capabilities of the Ariane 5 launcher, but not until a growth version could lift the full 52,910-lb. Hermes would the space plane be capable of carrying its design payload of 6,000 lb.

1991 October 17   At 00:05 UT the unmanned *Progress M-10* cargo tanker was launched by an SL-4 from the Baikonur cosmodrome. During the approach for docking with the *Mir* station complex on October 19, the freighter ran into problems and the automatic docking was aborted at a separation distance of just 500 ft. Ground controllers worked around the failure, and *Progress M-10* successfully latched up on the forward axial port at 03:41 UT on October 21, remaining with the complex for three months. *Progress M-10* finally undocked at 07:14 UT on January 20, 1992, after having an earlier undocking postponed because of a problem with the control-moment gyroscopes installed in *Kvant 1* and *Kvant 2* and used for attitude control.

1991 October 25   The opportunity to bid for purchase and privatization of Aussat, the Australian telecommunications satellite organization, closed with two teams competing for ownership. One team led by Bell South of the United States and Cable and Wireless of the United Kingdom joined with some Australian companies to establish the government's requirement that 51% of shares would be in Australian hands within five years of purchase. The second team included Bell Atlantic and Ameritech from the United States, Hong Kong's Hutchison Whampoa and Telecom of France. Aussat had three operational HS 376 series satellites on station, each with 15 transponders.

1991 October 29   At 22:37 UT *Galileo* passed within 1,000 mi. of the asteroid Gaspra, a 6 × 7 × 11 mi. potato-shaped rock imaged by the spacecraft to a resolution of about 300 ft. In all, about 150 images were taken in the 4 hr. prior to closest approach. The images were stored on a tape recorder and most would be transmitted to Earth via the low-gain antenna because the 16 ft., umbrella-shaped high-gain antenna was stuck halfway to full deployment. This mesh antenna supported by 18 ribs was almost identical to those used on NASA's tracking and data relay satellites. Now more than 250 million mi. from Earth, it would take 80 hr. to transmit each image at the 40 bits/sec. data rate available through the low-gain antenna. One image was sent back early in November 1991 and another in May 1992.

1991 November 24   Delayed from November 19 for replacement of a guidance unit on an inertial upper stage in the payload bay, Shuttle mission STS-44 was launched at 23:44:00 UT from LC-39A at the Kennedy Space Center, Fla. On this first unclassified Department of Defense deployment mission, *Atlantis* carried astronauts Frederick Gregory (CDR), Terence Henricks (PLT), Mario Runco (MS), James Voss (MS), Story Musgrave (MS) and Thomas Hennen (PS). The 47,265-lb. payload included the 5,280-lb. *Defense Support Program-16 (DSP-16)* satellite and its IUS, deployed at 17:03 UT on November 25. The IUS stage fired *DSP-16* into geostationary orbit where it would monitor missile launches and clandestine nuclear testing for NORAD (North American Air Defense Command). *Atlantis* passed within 25 mi. of the Russian space station *Mir* on the second day in orbit. The crew also conducted the first U.S. manned military reconnaissance tests from space. The mission ended at Edwards AFB, Calif., 6 days 22 hr. 50 min. 44 sec. after liftoff, cut short by three days due to the failure of an inertial measurement unit on *Atlantis*.

1991 December 6   NASA signed a memorandum of understanding with the Agenzia Spaziale Italiana (Italian Space Agency) for design and development of two mini pressurized logistics modules (MPLM) for space station *Freedom*. They were to be capable of transporting user payloads and resupply items to the station and returning unnecessary items to the ground. Each cylindrical MPLM would have a length of 13 ft. 9 in. and weigh about 8,800 lb. The first such MPLM was scheduled for launch by Shuttle in May 1997 to assist with construction of the facility, with a second flown up in August 1997.

1991 December 7   General Dynamics launched the first of its new and uprated generation of Atlas-Centaur launch vehicles (AC-69) carrying the 4,132-lb. *Eutelsat II F3* communication satellite to a geosynchronous transfer path. Designated Atlas 2, the new launcher had a 6-ft. extension on the Atlas and a 3-ft. extension on the Centaur, with extra propellant for longer burn times. The MA-5 propulsion system was upgraded to the MA-5A with a liftoff thrust of 474,500 lb. Vehicle height was 156 ft., supporting a payload shroud 13 ft. 10 in. in diameter and carrying up to 5,900 lb. to geosynchronous transfer orbit.

1991 December 16   The first in a second generation of three French domestic commercial and military communication satellites, *Telecom 2A* and the *Inmarsat II F3* satellite, shared a ride on the Ariane 44L flight (mission V48). Built by Matra Marconi Space and Alcatel Espace, the 5,016-lb. *Telecom 2A* had a 6 ft. 3 in. × 8 ft. 6 in. body with a total height of 10 ft. 2 in. and a solar array span of 72 ft. 2 in. With more than 3,450 W electrical power, the satellite provided three 20-W and 40-W transponders for secure government communication channels at 14/12 GHz, 11 × 55-W Ku-band transponders and 10 × 11-W C-band transponders.

1991 December 18   A Soviet SL-14 Tsyklon launcher carried the science satellite *Intercosmos 25* and a Czechoslovak satellite called *Magion 3* from Plesetsk to Earth orbit. Part of the active plasma experiment (Apex) program, *Intercosmos 25* consisted of a cylinder, 5 ft. 11 in. long × 4 ft. 11 in. in diameter, with eight solar cell panels extended like petals from one end. The 2,200-lb. satellite carried instruments to measure streams of electrons and beams of plasma and their effects on the ionosphere and the magnetosphere. The 115-lb. *Magion 3* was deployed from *Intercosmos 25* on December 28 and measured electron and ion beams with its parent as the distance between them varied. Both satellites were placed in an orbit of 270 × 1,905 mi., at an inclination of 82.6° and both remain in space.

1991 December 28   China's *STTW-5* communications satellite was stranded in a useless orbit when the third-stage motor of the CZ-3 launcher shut down prematurely. The third stage operated successfully to place the assembly in an interim Earth parking orbit, but when the cryogenic third stage fired a second time to place the vehicle on a geosynchronous transfer path, the stage cut off at 2 min. 15 sec., much earlier than planned. The satellite's apogee kick motor was fired next day to place it in a 127 × 21,804 mi., 31.6° orbit where limited communications could be effected from its highly elliptical path.

1992 January 11   NASA's *Galileo* spacecraft reached the aphelion of its elliptical path around the sun and would gradually move back toward the Earth where, in December 1992, it would receive its final gravity boost, taking it once again toward the asteroid belt—this time on to Jupiter. At aphelion, *Galileo* was 211 million mi. from the sun and 302 million mi. from the Earth which, at this point in its orbit, was on the opposite side of the sun. Between January 13 and January 30 the spacecraft was at solar conjunction and telemetry was switched off.

1992 January 12–16   At a conference on space nuclear propulsion, the U.S. Air Force publicly unveiled designs for a high-performance nuclear rocket completed under a program run by the Strategic Defense Initiative antiballistic missile research program called Timberwind. Technical breakthroughs, said the Air Force, would permit development of a 75,000-lb. thrust particle bed reactor with hydrogen propellant heated to 5,000°F that produced a specific impulse of 1,000 sec. After five years of work, officials were convinced that they could produce a working rocket stage in eight more years, serving as upper stage for cislunar, manned Mars, or military Earth-orbit missions.

1992 January 14   Senior officers of the top command structure from the recently (on December 11, 1991) dissolved Soviet Union met in the Kremlin to review the distribution of forces among the separate sovereign states of the former Soviet Union. A new Interstate Space Council had been hurriedly set up to prevent fragmentation of space activities while military space operations would be controlled by the Joint Strategic Armed Force. Nevertheless, some measure of fragmentation was inevitable, and separate states laid claim to facilities on their territory. Kazakhstan drove a hard bargain for user-fees at the Baikonur cosmodrome and in 1993–1994 Russia drew

up plans to expand the facilities at Plesetsk with the ultimate aim of running down the launch facilities at Baikonur.

1992 January 20   In his annual State of the Union Address, President George Bush announced an outline plan for deep cuts in the strategic missile force if the Russians would agree to eliminate all MIRV'ed missiles. In exchange, the United States would eliminate all 50 Peacekeeper ICBMs, each with 10 warheads; download all 500 Minuteman III reentry vehicles from three to one warhead apiece; and reduce the submarine-launched ballistic missile warhead inventory from a maximum 3,456 allowed under START (Strategic Arms Reduction Talks) to 2,160 Trident missile warheads on 18 submarines. These moves would reduce to 500 the total number of ICBMs, each with one warhead. Under this proposal, both sides would reduce their nuclear forces to 4,700 warheads by the end of the century.

1992 January 22   After a 1-hr. delay for weather, Shuttle mission STS-42 began at 14:52:33 UT with the launch of *Discovery* from LC-39A at the Kennedy Space Center, Fla. Carrying U.S. astronauts Ronald Grabe (CDR), Stephen Oswald (LPT), William Readdy (MS), Norman Thagard (MS), David Hilmers (MS), Canadian astronaut Roberta Bondar (PS) and German astronaut Ulf Merbold (PS) for the European Space Agency, *Discovery* had a total payload of 32,394 lb. This International Microgravity Laboratory-1 (IML-1) mission utilized a Spacelab pressure module to explore in depth the effects of weightlessness on living organisms and materials processing. IML-1 included 72 million roundworms, 480 fruit flies and 3,942 insects. The flies were killed by a sterilizing solution in the containers. Divided into red and blue teams, the multinational crew conducted extensive science investigations and the seven-day mission was extended, terminating at Edwards AFB, Calif., 8 days 1 hr. 14 min. 44 sec. after liftoff.

1992 January 24   The *Magellan* spacecraft in orbit around Venus began a third mapping cycle, having already completed two during which it had covered over 95% of the surface in 3,880 orbits. A technical problem with one of the two downlink transmitters meant that data would from now on be sent at 115 kilobits/sec. instead of 268.8 kilobits/sec. Beginning January 26 scientists began a stereo mapping operation of Maxwell Montes, the highest mountain on Venus, and the western portion of Aphrodite Terra. The third mapping cycle ended in mid-September 1992.

1992 January 25   At 07:50 UT the *Progress M-11* cargo tanker was launched by an SL-4 from the Baikonur cosmodrome. The unmanned freighter docked to the forward axial docking port on the *Mir* complex at 09:31 UT two days later. During this activity, flight controllers on the ground threatened to go on strike to protest their pay. In the event, they stayed at their posts. *Mir* cosmonauts Alexandr Volkov and Sergei Krikalov unstowed equipment brought to the complex for carrying out equipment repairs, most notably to the control-moment gyroscopes used for maintaining attitude control. *Progress M-11* also brought up equipment cosmonauts would use during two planned international missions later during the year. It undocked on March 13 and was de-orbited the same day. On March 14 Volkov and Krikalov transferred *Soyuz TM-13* from the back of the *Kvant 1* module to the forward axial port.

1992 February 3 Results from the NASA *Upper Atmosphere Research Satellite (UARS)* showed exceptionally high levels of chlorine monoxide (C_1O) at high northern latitudes. Concentrations of C_1O erode the ozone layer. Particularly affected were skies at latitudes in excess of 50° N, and thinning ozone levels were noted over the European cities of London, Moscow and Amsterdam. NASA had already reported large holes over the South Pole.

1992 February 8 The European Space Agency's *Ulysses* spacecraft made its close pass of Jupiter at 12:02 UT, coming to within 235,000 mi. of the cloud tops. The gravitational influence of Jupiter turned the trajectory of the spacecraft so that it exited the environment of the planet at a heliocentric inclination of 80°. *Ulysses* would continue to move back across the solar system, passing across the sun's south pole between June and November 1994 at a distance of about 180 million mi. After moving north to cross the plane of the ecliptic at a distance of 110 million mi. in February 1995, *Ulysses* would pass across the sun's north polar region between June and September 1995 at a distance of 180 million mi.

1992 February 11 The first flight of an Atlas 2 launch vehicle in the U.S. Air Force MLV-2 (Medium Launch Vehicle Program-2) series took place from Cape Canaveral, Fla., when the fourth DSCS-III military communication satellite was carried to an interim geosynchronous transfer orbit of 107 × 22,180 mi. Before separating, the Centaur performed two burns, the second of which was the 300th space firing of the Aerojet RL-10 engine, which was on its 80th mission. This was also the first flight of the integrated apogee boost subsystem (IABS) designed specifically for DSCS-III and incorporating two 110-lb. thrust $MMH-N_2O_4$ motors, which raised the satellite to a geostationary orbit before separating. At launch the DSCS-III/IABS weighed 5,810 lb. Ten DSCS-III remained to be launched, the fifth of 14 being carried by Atlas 2 on July 3.

The last flight of Japan's H-I launch vehicle took place, carrying the first Japanese Earth Resource Satellite, *JERS-1*, from Tanegashima into a 352 × 354 mi. orbit inclined 97.7°. The 2,955-lb. satellite had been built by Mitsubishi Electric Co. as Japan's first Earth resource satellite. Based on a box-structure, 10 ft. 2 in. × 2 ft. 11 in. × 5 ft. 11 in., *JERS-1* carried a solar array panel 26 ft. 3 in. long and 11 ft. 2 in. wide and a synthetic aperture radar antenna 39 ft. long by 7 ft. 10 in. wide. From a sun-synchronous orbit, *JERS-1* was to make radar and optical observations, the latter using a multispectral scanner in eight visible and infrared bands.

1992 February 15 Supporting an effort to produce a miniature winged space plane, Japan's Institute of Space and Astronautical Sciences conducted an important test of a ⅟₇th scale model of a vehicle called HIMES. Carried to a height of 11.7 mi. by a 33 ft. helium balloon, the 6 ft. 6 in. model with a 5 ft. wingspan ignited a small booster rocket which accelerated it to a height of 41.5 mi. in two minutes. From there the 385-lb. model returned to the denser layers of the atmosphere and was recovered 250 mi. southeast of the Kagoshima launch site.

1992 February 20 *Mir* cosmonauts Aleksandr Volkov and Sergei Krikalov began a 4 hr. 12 min. space walk at 20:09 UT to dismantle equipment used in erecting the Sofora thermomechanical demonstration structure during July 1991. Volkov had to stay behind at the *Kvant 2* airlock hatch, however, when the heat exchanger to his suit failed. Unable to use the

telescopic "crane" that Volkov would have operated, Krikalov had to make his way by hand across to the *Kvant 1* module at the back of *Mir* where he carried out the work. Sample containers were retrieved and others deployed before the two cosmonauts got back inside.

1992 February 24–25 Rioting broke out at the Baikonur cosmodrome in Kazakhstan when hundred of soldiers protested at poor food and bad living conditions. Three people were killed, four barrack blocks were burned down and many cars stolen as conscripts went on the rampage. The violence came at the culmination of three weeks of strikes by workers at the launch site. Soviet cosmonauts Aleksandr Viktorenko and Aleksandr Kalery and German cosmonaut Klaus-Dietrich Flade, were due to arrive at Baikonur in early March for their flight aboard *Soyuz TM-14* to *Mir*.

1992 February 26 *Superbird B1*, a replacement for *Superbird B* lost in the Ariane V31 failure, shared a ride with *Arabsat 1C*, on an Ariane 44L configuration (mission V49) from Kourou, French Guiana, to a geosynchronous transfer orbit. The 5,645-lb. *Superbird B1* had a body size of 7 ft. 10 in. × 7 ft. 2 in. × 11 ft. 2 in., solar array wings with a 66 ft. 7 in. span providing 3,500 W of power, and an on-orbit mass of 3,378 lb. The payload comprised 23 × 50-W (14/12 GHz) transponders and 3 × 29-W (30/20 GHz) transponders providing telephone, TV broadcast and business data services from a position at 162° E. With an on-orbit mass of 1,731 lb., *Arabsat 1C* was placed at 31° E.

1992 February Officials at the Strategic Defense Initiative ballistic missile defense organization formed a plan to acquire Russian scientists and technologies that could make a useful contribution to the U.S. research effort under way at the Department of Defense. During his visit to the United States this month, Marshal Yevgeny Shaposhnikov, supreme commander of the Commonwealth of Independent States forces, agreed with this idea and indicated his enthusiasm for working with the SDI organization to achieve a global system of protection against the use of ballistic missiles by minor powers.

1992 March 6 The second and last ERIS (exo-atmospheric reentry vehicle interceptor subsystem) antiballistic missile tests was conducted over the Pacific Ocean but failed to result in an impact. A Minuteman missile launched from Vandenberg AFB, Calif., was tracked as it closed on the Kwajalein Atoll tests area and a modified Aries rocket was launched in pursuit. The Aries intercept vehicle made a near pass at an altitude of 180 mi., 500 mi. downrange from launch, but failed to properly discriminate the simulated warhead from a decoy because too little time had been allowed for calculations. Managers said that on an actual intercept in war the interceptor would have more time to analyze the separate objects.

1992 March 14 General Dynamics launched the 3,110-lb. *Galaxy 5* cable television relay satellite by an Atlas 1 from Cape Canaveral, Fla. A Hughes HS 376, *Galaxy 5* was positioned at 125° W, replacing *Westar V* which had been operating since June 1982. It carried 24 × 16-W C-band transponders, each of which could carry one analog TV channel or a simultaneous video, voice and data signal. With solar cells providing 1,100 W, *Galaxy 5* had a 12-year life. An attempt to launch *Galaxy 1-R* on August 22 failed when the Centaur stage on Atlas 1 did not ignite.

1992 March 17 At 10:54 UT Soviet cosmonauts Alexandr Viktorenko and Alexandr Kalery accompanied by German cosmonaut Klaus-Dietrich Flade, were launched in their *Soyuz TM-14* spacecraft by an SL-4 from the Baikonur cosmodrome. After two days they approached the orbiting *Mir* space station and docked to the back of the *Kvant 1* module at 12:33 UT on March 19. After Sigmund Jähn, Ulf Merbold, Reinhard Furrer and Ernst Messerschmid, Flade was the fifth German to fly in space. Joint activities on 14 German experiments were performed over the next few days before the resident *Mir* cosmonauts, Alexandr Volkov and Sergei Krikalov, returned to Earth with Flade, leaving Viktorenko and Kalery in space.

1992 March 24 Delayed a day to check on excessive hydrogen levels in the orbiter's aft compartment, Shuttle mission STS-45 began at 13:13:00 UT with the launch of *Atlantis* from LC-39A at the Kennedy Space Center, Fla. Aboard were astronauts Charles Bolden (CDR), Brian Duffy (PLT), Kathryn Sullivan (MS), Michael Foale (MS), David Leestma (MS), Dirk Frimout (PS) and Byron Lichtenberg (PS). The 20,371-lb. payload consisted primarily of the first Atmospheric Laboratory for Applications and Science (ATLAS-1) equipped with 12 instruments from the United States, France, Germany, Belgium, Switzerland, the Netherlands and Japan mounted on a Spacelab pallet. The payload also included the Shuttle solar backscatter ultraviolet (SSBUV) instrument, and investigations included atmospheric soundings, measurements of solar energy, solar radiation and atmospheric molecular composition. The mission was extended one day for additional science and ended at the Kennedy Space Center, 8 days 2 hr. 9 min. 28 sec. after liftoff.

1992 March 25 *Mir* cosmonauts Alexandr Volkov and Sergei Krikalov returned to Earth with the visiting German cosmonaut Klaus-Dietrich Flade in the *Soyuz TM-13* spacecraft, landing at 08:51 UT on the snow-covered steppes. Krikalov had been in space since May 18, 1991, logging 311 days 20 hr. 01 min. Launched on October 2, 1991, Volkov had been in space for 175 days 2 hr. 52 min. Flade returned after a mission lasting just 7 days 21 hr. 57 min. The *Soyuz TM-14* spacecraft that carried Flade to *Mir* had been left at the back of the *Kvant* module while his colleagues on that flight, Alexandr Viktorenko and Alexandr Kalery remained behind to manage the space station.

1992 March 27 The White House announced that the United States would buy a Topaz space nuclear reactor from Russia, as well as electric thrusters and several kilograms of plutonium 238. Although Topaz was the most advanced reactor of its kind, the electric thrusters were considered the most important buy. Known as Hall thrusters, they operated by ionizing xenon gas and accelerating the ions down a magnetic field to produce a high specific impulse. Several U.S. communication satellite builders wanted to apply ion thrusters to satellites to extend their operating life and reduce the quantities of propellant required. Now that satellite lifetimes had extended to more than 12 years, propellant mass was becoming a limiting factor.

1992 March NASA announced the names of 24 new pilot (PLT) and mission specialist (MS) astronaut candidates, constituting Group 14: Daniel T. Barry; Cmdr. Charles E. Brady Jr. (USN); Maurizio Cheli; Jean-François Clervoy; Capt. Catherine G. "Cady" Coleman (USAF); Marc Garneau (formerly a payload specialist); Michael L. Gernhardt; John M. Grunsfeld; Chris A. Hadfield; Capt. Scott J. "Doc" Horowitz

(USAF); Lt. Cmdr. Brent W. Jett Jr. (USN); Kevin R. Kregel; Lt. Cmdr. Wendy B. Lawrence (USN); Cmdr. J. M. Linenger (USN); Richard M. Linnehan; Lt. Cmdr. Michael E. Lopez-Alegria (USN); Scott E. Parazynski; Lt. Cmdr. Kent V. Rominger (USN); Cmdr. Winston E. Scott (USN); Steven L. Smith; Joseph R. "Joe" Tanner; Andrew S. W. Thomas; Koichi Wakata; and Mary Ellen Weber.

A National Research Council report on U.S. launch vehicle status came out strongly against development of a new heavyweight National Launch System (NLS). Citing availability of the Titan 4, with its 40,000-lb. lift capability, as an adequate launcher for the known launch requirements for the foreseeable future, the NRC advised development instead of a 20,000-lb. capability launcher to bolster flagging U.S. success in the commercial launch vehicle market. The Bush administration wanted a 135,000-lb. capability launcher first despite the fact that the only known payload of that size was classified Strategic Defense Initiative antiballistic missile cargo.

1992 April 1 Richard Truly stood down as NASA administrator at the behest of President Bush and the urging of National Space Council chief, Vice President Dan Quayle, and its executive secretary Mark J. Albrecht. The Bush administration wanted a completely new and innovative approach to program management and engineering in which the economics of the commercial world were deemed essential for survival in an increasingly competitive and demanding environment. Truly was replaced on this date by Daniel S. Goldin, who had been with TRW Systems since 1967. Before that, Goldin had been a research scientist at the NASA Lewis Research Center.

1992 April 3 Spacehab unveiled its first Shuttle middeck augmentation module at the new Spacehab processing facility at Cape Canaveral, Fla. Under contract to Spacehab, McDonnell Douglas designed the 9,286-lb. module to fit inside the forward section of the Shuttle orbiter's cargo bay. Spacehab would be attached to the airlock interface on the forward bulkhead by a short Spacelab tunnel adapter, providing 1,000 cu. ft. of additional work volume. The truncated cylindrical module had a length of 10 ft., a diameter of 13 ft. and 1,000 cu. ft. of additional working volume. Shuttle's middeck area contained 42 lockers, of which 35 were stuffed with crew supplies for a standard mission, leaving only 7 lockers for science experiments. Spacehab provided up to 100 additional lockers for experimenters.

1992 April 6–12 At the Moscow Aero and Industry Engine '92 exhibition, the NPO Energomesh, later Energia, company displayed a revolutionary rocket engine capable of producing a thrust of 900,000 lb. from a tripropellant oxygen-kerosene-hydrogen combination. Designated RD-701, the twin-nozzle engine consisted of two main combustion chambers with two preburners feeding fuel to each one. One pump would handle liquid oxygen and kerosene, delivered to the preburner creating an oxygen-rich gas pressurized to 10,220 lb./sq. in. The second pump would deliver liquid hydrogen, fed to the combustion chambers at ambient temperature. Energomesh envisaged the use of this high performance engine in a single-stage-to-orbit Shuttle, where it would operate in two modes: at first with a tripropellant combination producing 900,000 lb. of thrust, then in a bipropellant (liquid oxygen–liquid hydrogen) mode at 180,000 lb. of thrust.

1992 April 10 The first Delta 11 7925 carried the 3,670-lb. GPS 13 Navstar Block II navigation satellite from Cape Canaveral, Fla., to a 12,426 × 12,656 mi., 55.3° orbit. Delta

II differed from the Interim Delta II 6925 in having an improved expansion ratio nozzle extension on the RS-27 first-stage engine and nine Hercules strap-on boosters that produced a total thrust of 890,325 lb. for 1 min. 4 sec. Interim Delta II had nine Castor 4A strap-ons producing a thrust of 861,525 lb. for 56 sec. Delta II 7925 was used for successive GPS satellites, its first commercial launch being the *Palapa B4* on May 14.

1992 April 15　The first Ariane 44AL configuration (V50) carried the 5,016-lb. *Telecom 2B* and the 2,888-lb. *Inmarsat II F4* satellites into geosynchronous transfer orbit from Kourou, French Guiana. *Telecom 2B* was positioned at 3° E and *Inmarsat II F4* at 55° W. The Ariane 44AL was the standard Ariane 44L, with four liquid strap-ons, with an additional 242.5-lb. of payload capability by adding 12.6 in. to the length of the third, H10, stage and increasing burn duration by 28 sec. Actual payload gain was 441 lb. over the previously stated 9,261 lb. for this version, the most powerful Ariane 4, due to a reduction in the mass of the second stage and performance gain through flight experience.

1992 April 19　At 22:29 UT the *Progress M-12* cargo tanker was launched by an SL-4 from the Baikonur cosmodrome. It docked to the forward axial port on the *Mir* space station at 00:11 UT on April 22. The usual load of supplies and provisions was moved across to *Mir* and equipment in need of replacement was stowed on board for installation. On May 14 and 15, the *Progress* propulsion system was used to raise the orbit to 232.5 × 257 mi., and on June 26 the orbit had been changed again to 242.5 × 256.5 mi. *Progress M-12* undocked and was de-orbited on June 28.

1992 April 22　Thiokol fired the Castor 120 solid propellant motor it had developed from the first stage of the Peacekeeper ICBM. With a length of 30 ft. and a diameter of 7 ft. 9 in., the 120,000-lb. motor produced a thrust of 400,000 lb. for 1 min. 18 sec. Utilizing advanced graphite epoxy construction and graphite exhaust skirts, it was being developed as a suitable first or second stage for future satellite launchers.

1992 April 29　*Resurs F-14*, the first of three Resurs remote sensing satellites launched during the 1992 agricultural growing season, was carried by an SL-4 from Plesetsk to a 145 × 159 mi., 82° orbit. The reentry module was recovered, together with its film, on May 29. *Resurs F-15* was launched on June 23 to a 112 × 143 mi., 82.3° orbit and recovered on July 9. *F-16* was launched to a 138 × 145 mi., 82.5° orbit on August 19. It carried the first U.S. Department of Defense payload carried on a Russian spacecraft, comprising an experiment to measure beryllium atoms in low orbit. *F-16* deployed two 110-lb. satellites, *Pion 5* and *Pion 6*, to measure the effects of the upper atmosphere on reentry. *F-16*'s reentry module was recovered on September 4.

1992 May 7　The first flight of the orbiter *Endeavour* began at 23:40:00 UT from LC-39B at the Kennedy Space Center, Fla., with the launch of OV-105 on STS-49. Carrying astronauts Daniel Brandenstein (CDR), Kevin Chilton (PLT), Bruce Melnick (MS), Thomas Akers (MS), Richard Hieb (MS), Kathryn Thornton (MS) and Pierre Thuot (MS), the orbiter had a payload of 37,474 lb. The mission was to rendezvous with *Intelsat VI F-3*, which had been stranded in low orbit by a Titan launched on March 14, 1990, and attach to it a new perigee kick motor enabling controllers to send the satellite to *geostationary* orbit. Intelsat controllers fired

thrusters on the satellite to lower it from 345 mi. to a 223 × 228 mi. orbit for Shuttle rendezvous, completed May 10. The first of three EVAs (extravehicular activities) by Thuot and Hieb was to have completed work on *Intelsat VI F-3*, leaving two opportunities to rehearse the assembly of space station structures. In actuality, three EVAs were needed to attach the new motor to the satellite and one was conducted for the structures test. The first EVA conducted on May 10 by Thuot and Hieb lasted 3 hr. 43 min. while the second by the same pair on May 11 lasted 5 hr. 30 min., but the massive satellite tumbled and would not respond to attempts at capturing it. At 8 hr. 29 min., the third EVA was an all-time duration record and involved three astronauts (Hieb, Thuot and Akers) for the first time in a successful attempt to grab the satellite at 120° intervals and attach it to a capture bar. From there the manipulator arm maneuvered *Intelsat VI* down to its new perigee motor. All four EVAs totalled 25 hr. 27 min., a Shuttle record. The 32,000-lb. *Intelsat VI/PKM* was released at 16:53 UT on May 14, and fired its new, 45,000-lb. thrust Orbis 21S perigee kick motor 30 min. later, sending it toward geostationary orbit. STS-49 ended at Edwards AFB, Calif., at 8 days 21 hr. 17 min. 38 sec.

1992 May 14　The *Palapa B4* communications satellite was carried by Delta II from Cape Canaveral, Fla., to a geosynchronous transfer path and from there to a geostationary orbit over 118° E. A Hughes 376 series satellite, *Palapa B4* had a weight of 2,765 lb. at launch and 1,525 lb. after the expenditure of apogee kick motor propellant. It could provide service to Indonesia, Thailand, Malaysia, the Philippines and Papua New Guinea through 24 C-band transponders. Operated by PT Telkom, formerly Perumtel, *Palapa B4* was the fourth second-generation communication satellite launched for Indonesia.

1992 May 20　India's Augmented Satellite Launch Vehicle (ASLV) placed the 234-lb. satellite *SROSS C* into a 159 × 271 mi., 46° orbit from the Sriharikota launch site. Built by the Indian Space Research Organization, *SROSS C* (stretched Rohini satellite series) comprised a 2 ft. 9 in. wide octagon, 3 ft. 7 in. tall. Eight solar panels deployed radially from the base of the octagon provided 100 W of electrical power. The planned circular orbit was not achieved due to a malfunction in the fourth stage. Nevertheless, *SROSS C* did return some useful data from its gamma-ray instrument and the retarding potential analyzer, which probed Earth's ionosphere, before it decayed on July 14, well short of its planned two-year mission.

1992 June 7　A Delta 2 launched the 7,220-lb. *Extreme Ultraviolet Explorer (EUVE)* science satellite from Cape Canaveral, Fla., to a 320 × 327.5 mi., 28.4° orbit. Built for NASA's Goddard Space Flight Center by Fairchild Space, *EUVE* detected and measured distant sources of extreme ultraviolet radiation utilizing three instruments comprising three 414-lb. scanning telescopes to gather field views and one 712-lb. survey telescope for examining specific objects. An EUV object outside the Milky Way galaxy was observed for the first time by this satellite. Measuring 9 ft. 2 in. × 14 ft. 9 in., *EUVE* carried solar panels providing 1 kW of electrical energy. It remains in space.

1992 June 10　General Dynamics launched the first Atlas 2A from Cape Canaveral, Fla., carrying the *Intelsat K* satellite serving video, broadcast and business communication customers. Atlas 2A was basically the same as Atlas 2 but with a more powerful Pratt and Whitney RL-10A-4 engine in the Centaur stage delivering 20,800 lb. of thrust, instead of 16,500 lb. thrust for previous Centaurs, and a higher specific impulse of 449 sec. versus 444 sec. achieved through a new extendable nozzle. Atlas 2A could lift 6,710 lb. to geosynchronous transfer orbit. Built by General Electric Astro-Space, the 6,456-lb. *Intelsat K* measured 9 ft. 2 in. × 7 ft. 6 in. × 7 ft. 6 in., and had solar arrays producing 4,100 W of power. It carried 16 Ku-band channels with 19 separate uplink/downlink combinations serving Europe, the United States and parts of South America.

1992 June 16–17 U.S. President George Bush and Russia's first democratically elected president, Boris Yeltsin, concluded the outline of new arms reduction measures known as START II (Strategic Arms Reduction Talks II). They agreed to cut strategic nuclear weapons to 3,000–3,500 each, a reduction of 60% in the current arsenals and significantly below the Bush proposal of January 20. In the first phase, to be completed within seven years, each side would cut 4,250 to 3,800 warheads, so that neither side would have more than 1,200 multiwarhead ICBMs or 2,160 warheads on submarine-launched ballistic missiles. In the second phase, to be completed by 2003 at the latest, all MIRV'ed ICBMs were to be eliminated, with each side restricted to 1,750 SLBM warheads. For the first time, bomber warheads would be counted not as one per bomber but as the maximum nuclear bomb load each aircraft was designed to carry. By 2003 the United States would have 1,272 warheads on bombers, 1,728 on SLBMs (submarine-launched ballistic missiles) and a maximum 500 ICBMs.

1992 June 25 At 16:12:23 UT Shuttle mission STS-50 began with the launch of *Columbia* from LC-39A at the Kennedy Space Center, Fla. The crew comprised Richard Richards

During arms control discussions in the early 1990s, multiple-warhead missiles were outlawed. Here, a mockup of the front end of a U.S. Peacekeeper missile supports MIRV warheads. These were increasingly viewed as destabilizing because of their relative invulnerability when used as a first-strike weapon.

(CDR), Kenneth Bowersox (PLT), Bonnie Dunbar (MS), Carl Meade (MS), Ellen Baker (MS), Lawrence DeLucas (PS) and Eugene Trinh (PS). Aboard *Columbia* was a 33,329-lb. payload comprising the first United States Microgravity Laboratory (USML-1) consisting of a Spacelab pressure module and a wide range of materials processing and bioscience experiments. USML-1 accommodated 870 separate experiments for 31 investigations. This was the first extended-duration orbiter mission, for which modifications to *Columbia* provided additional consumables for added time in space. Delayed a day because of bad weather at Edwards AFB, Calif., *Columbia* finally landed at the Kennedy Space Center, Fla., at an elapsed time of 13 days 19 hr. 30 min. 4 sec., making this the longest U.S. manned spaceflight since the third Skylab visit ended in early 1974.

1992 June 30 At 16:43 UT the cargo tanker *Progress M-13* was launched by an SL-4 from the Baikonur cosmodrome. The unmanned freighter approached the forward axial port on the *Mir* space station on July 2, but a fault in the computer program for roll axis control aborted the docking attempt. Software command changes were successful and the ship docked with *Mir* at 16:55 UT on July 4. After cargo transfer, a propulsion burn put the complex in an orbit of 253 × 264 mi., and at 04:49 UT *Progress M-13* undocked and was de-orbited later the same day.

1992 July 3 A four-stage NASA Scout carried the 348-lb. *Sampex* science satellite from Vandenberg AFB, Calif., to a 317 × 422.5 mi., 81.7° orbit. *Sampex*, an acronym for solar anomalous magnetospheric particle explorer, was instrumented to measure particles from the sun captured by the Earth's magnetosphere in the Earth's polar axis. It was the first in a new series of small explorer (SMEX) missions aimed at flying small instruments on low-cost missions with reduced lead time. On September 4 NASA asked for proposals from the scientific community for additional SMEX missions. At this date two more were scheduled: the fast auroral snapshot explorer due for launch in 1994, and the submillimeter wave astronomy satellite, set for launch in 1995. *Sampex* remains in space.

1992 July 8 At 12:37 UT *Mir* cosmonauts Aleksandr Viktorenko and Aleksandr Kalery began a 2 hr. 3 min. space walk from the *Kvant 2* airlock module to replace two gyroscopes used for attitude control. At first, they opened up the wrong piece of thermal insulation blanket and used shears to cut their way into the appropriate part of the covering. After successfully hooking up the new control units, the cosmonauts tested a special pair of binoculars which had been designed to give a close-up view of distant parts of the complex. The *Progress M-13* cargo tanker was used to supplement repressurization air for the airlock module.

1992 July 9 An Ariane 44L configuration (V51) carried the 4,203-lb. *Insat 2A* and 4,139-lb. *Eutelsat II F4* satellites from Kourou, French Guiana, to geosynchronous transfer orbits. Positioned over 7° E, *Eutelsat II F4* was almost identical to earlier satellites in the series. *Insat 2A* was the first domestically produced multimission satellite. Built by the Indian Space Research Organization (ISRO), it was a follow-on to the first-generation series built by Ford Aerospace (now Loral). *Insat 2A* had a cube-shaped body, 6 ft. 3 in. × 5 ft. 7 in. × 5 ft. 3 in., expanding on orbit to 75 ft. 6 in. with solar panel and antenna boom extended. Solar cells produced 1,500 W to power 18 C-band transponders, two S-band transponders, a

UHF data relay transponder and a very high resolution radiometer for weather pictures from 74° E.

1992 July 10 At 15:30 UT the European Space Agency's *Giotto* spacecraft skimmed past the comet Grigg-Skjellerup at a distance of only 124 mi. *Giotto* entered the dust coma of the comet at a distance of 12,500 mi. with 8 of its original 11 experiments still working, seven years after launch and more than six years after the encounter with Halley's comet. Between 11,000 mi. and 9,000 mi. the spacecraft detected the bow shock wave of the comet as it ploughed through the solar wind, setting up a disturbance ahead of its passage across the inner solar system. Closing speed at encounter was 8.7 mi./sec., when *Giotto* was 133 million mi. from Earth and more than 93 million mi. from the sun. *Giotto* operations were officially terminated on July 23, 1992. On July 1, 1999, the spacecraft will pass 136,000 mi. above Earth with 8.8 lb. of hydrazine maneuvering propellant left. Engineers may attempt a close flyby of the planet.

1992 July 13–14 Following agreement by U.S. President George Bush and President Boris Yeltsin of Russia to participate in a Global Protection System against ballistic missile attack during their meeting on June 16–17, senior U.S. and Russian negotiators met in Moscow to outline respective concepts. The Strategic Defense Initiative announced by President Ronald Reagan on March 23, 1983, had matured through the period when the probability of global nuclear war had receded. Both major nuclear powers recognized the increasing danger of attack from smaller countries in possession of ballistic missiles and sought to find a common defense against that threat.

1992 July 24 A space mission to explore the tail of the Earth's geomagnetic field began with the launch of *Geotail* on a Delta 2 from Cape Canaveral, Fla. This was the first launch bought commercially by NASA. A joint venture between NASA and Japan's Institute of Space and Astronautical Sciences, *Geotail* was launched to a highly elliptical orbit of 25,700 × 316,000 mi., inclined 22.4°, with repeated lunar swingbys for gravity-assist maneuvers. Built by the Nippon Electric Co., *Geotail* was a 7 ft. 2 in. cylinder, 5 ft. 3 in. tall, with four 164-ft. booms supporting electric field and plasma sensors, and two 20-ft. magnetometer booms. Body-mounted solar cells produced 350 W. The first lunar swingby, on September 8, pushed apogee to 539,200 mi., but SUCCESSIVE lunar passes pushed apogee to a maximum 851,000 mi. by April 1994. After that, thrusters were to reduce apogee to 119,000 mi. for studies of the inner components of the geomagnetic tail.

1992 July 27 At 06:08 UT Soviet cosmonauts Anatoly Solovyov and Sergei Avdeyev, accompanied by the French cosmonaut Michel Tognini, were launched aboard the *Soyuz TM-15* spacecraft by SL-4 from the Baikonur cosmodrome. During the final approach to the forward axial port on the *Mir* space station, July 29, the Kurs radar failed and the crew had to manually dock their spacecraft, achieving a link-up at 07:41 UT. During joint activities with resident *Mir* cosmonauts Alexandr Viktorenko and Alexandr Kalery 10 French experiments were conducted by the five crew members. Meanwhile, back on Earth, the French signed an agreement with the Russians for four more 14-day joint flights, one each in 1993, 1996, 1998 and 2000.

Europe's Eureca *unmanned platform seen above the coast of Florida with Cape Canaveral jutting into the Atlantic Ocean, after deployment from* Atlantis *during the STS-46 Shuttle mission launched July 31, 1992.*

1992 July 31 At 13:56:48 UT the Shuttle STS-46 mission began with the launch of *Atlantis* from LC-39B at the Kennedy Space Center, Fla. The crew comprised Loren Shriver (CDR), Andrew Allen (PLT), Jeffrey Hoffman (MS), Franklin Chang-Díaz (MS), Marsha Ivins (MS), Swiss astronaut Claude Nicollier (PS) and Italian astronaut Franco Malerba, both of whom represented the European Space Agency. The total payload weighed 34,090 lb. and included the joint NASA/Agenzia Spaziale Italiana (the Italian space agency) tethered satellite system (TSS), and the *Eureca* free-flying platform built by ESA. *Eureca* was deployed at 07:06 UT on August 2, but failed to fire its propulsion system for the time required to place it in the planned orbit of 310 mi. Controllers issued fresh commands and *Eureca* was placed in a circular, 316 mi. orbit on August 7. The tethered satellite was reeled out at 21:19 UT on August 4, but to a maximum distance of only 860 ft. from the orbiter instead of the planned 12.5 mi. After numerous attempts to free the line, the TSS was reeled in and stowed by 23:00 UT on August 5. The mission ended back at the Kennedy Space Center, 7 days 23 hr. 15 min. 3 sec. after liftoff, with *Eureca* remaining in orbit until retrieval planned for June 1993.

1992 August 4–7 *Galileo* performed a computer-controlled sequence of 5,400 thruster pulses, the first in a series of trajectory correction maneuvers to precisely align the spacecraft for a close flyby of Earth on December 8, 1992. The flight path around Earth would also set up the post-encounter course for a subsequent flyby of asteroid Ida in August 1993. During these four days of thruster firings, when *Galileo* was 108 million mi. from Earth and 162 million mi. from the sun, the velocity of the spacecraft was changed by 69 ft./sec. A second, tweaking, correction was conducted on October 7.

1992 August 9 The first flight of the Chinese CZ-2D satellite launcher took place from Juiquan when it carried a 5,510-lb. science satellite called *Fanhui Shi Weixing 2-1 (FSW 2)* incorporating a recoverable pod. The satellite was a larger version of the *FSW 1*, the first flight of which was to take place on October 6, with a length of 15 ft. and a diameter of 7 ft. 3 in. It comprised three sections: equipment module, retromodule and reentry module. The satellite hosted a combination of remote sensing and crystal growth experiments and the pod was returned to Earth on August 25, the main body of *FSW 2* decaying on September 1.

1992 August 10 A joint NASA/CNES (Centre National d'Etudes Spatiales, the French national space agency) project, *Topex/Poseidon*, was launched along with two piggyback satellites by Ariane 44P from Kourou, French Guiana, to a 826.5 × 834 mi. orbit inclined 66°. The 5,296-lb. satellite took the form of a rectangular box structure, 13 ft. 7 in. × 4 ft. 9 in. x 7 ft. 2 in., with a cantilevered solar array panel extending 28 ft. 6 in. to one side providing 2,200 W. Built by Fairchild for NASA's Jet Propulsion Laboratory, Topex/Poseidon carried a microwave radiometer for measuring atmospheric water vapor (Topex) and one U.S. and one French (Poseidon) altimeter for ocean surface topography. Also orbited were the 110-lb. Korean communications test satellite *Kitsat-A*, and the 110-lb. French VHF test package, *S80/T*. Both satellites were placed close together in orbits of 810 × 823 mi.

Mir space station cosmonauts Aleksandr Viktorenko and Aleksandr Kalery and the visiting French cosmonaut Michel Tognini, returned to Earth at 01:05 UT in the *Soyuz TM-14* spacecraft launched on March 17. Viktorenko and Kalery had spent 145 days 14 hr. 10 min. on their mission, and Tognini had returned to Earth just 13 days 18 hr. 57 min. after launch. The cosmonauts left the *Soyuz TM-15* spacecraft that carried Tognini into space attached to *Mir*'s forward axial docking port. Tognini's colleagues on the TM-15 launch, Solovyov and Avdeyev, remained aboard the *Mir* space station.

1992 August 13 The first of two Australian communications satellites, *Optus B1* was carried by Chinese CZ-2E to a geostationary transfer ellipse from where it was boosted to geostationary orbit over 160° E. Built by Hughes, the satellite weighed 16,868 lb. at launch with perigee kick motor and 3,488 lb. on station. Originally called Aussat, the satellite was renamed after Optus Communications purchased Aussat earlier in the year. *Optus B* was a three-axis-stabilized satellite, with a height of 36 ft. 5 in. and a span of 67 ft. 3 in. across the solar cell arrays, which provide 3,200 W of electrical energy. It carried TV, telephone, data and facsimile transmissions for Australia and New Zealand through 15 Ku-band channels. An L-band channel provided direct broadcasting.

1992 August 15 At 22:19 UT the unmanned cargo tanker *Progress M-14* was launched by an SL-4 from the Baikonur cosmodrome. Carrying 5,510 lb. of supplies, including equipment for an intensive period of four space walks to begin on September 3, *Progress M-14* docked to the back of the *Kvant* module at 00:21 UT on August 18. The freighter remained docked to the station complex throughout September, replenishing the *Kvant 2* airlock module at each repressurization, and undocked at 16:46 UT on October 21. On its way through the atmosphere, a recoverable capsule was ejected and landed safely at 00:09 UT on October 22.

1992 August 17 NASA announced selection of 31 experiments from the public and private sectors for accommodation on the advanced communications technology satellite. Managed by NASA's Lewis Research Center, *ACTS* consisted of a box-shaped structure, 7 ft. × 6 ft. 8 in. × 6 ft. 4 in., supporting a transmitting reflector 11 ft. in diameter, a receiving reflector 7 ft. in diameter and a width of 29 ft. 11 in. across both reflectors. Some 134.5 sq. ft. of solar arrays on "wings" with a span of 47 ft. 1 in. provide 1,800 W of power for the bus and payload sections. *ACTS* would operate in the 30/20 (Ka) GHz band hosting a wide range of satellite communications technologies. Businesses, medical, educational, video teleconferencing, high-definition TV, land mobile, aviation and science users would help test the revolutionary systems evaluated on ACTS.

1992 August 22 An Atlas-Centaur launch vehicle carrying the Hughes *Galaxy 1-R* communication satellite was destroyed by the range safety officer at Cape Canaveral, Fla., when the Centaur stage failed to start properly. In an accident reminiscent of that which befell the Atlas-Centaur launched on April 18, 1991, igniters for the two RL-10A-3-3A engines operated as scheduled at 4 min. 47 sec. into the flight but failed to run for a planned 5 min. 20 sec. The AC-71 mission terminated 3 min. later with destruction of the tumbling vehicle and its 3,114-lb. payload, at a height of 102 mi. high and 1,050 mi. downrange. A review board judged that nitrogen gas ingested by one of the engines froze the start-up fuel pump. A replacement *Galaxy 1-R* was successfully launched by a Delta 2 7925 on February 19, 1994.

1992 August 27 A space cooperation agreement was signed by Lockheed Missiles and Space Co. and Energia Scientific and Industrial Corp., builders of the Soyuz TM spacecraft. Energia wanted the United States to buy Soyuz TM spacecraft to serve as the assured crew return vehicle (ACRV) on space station *Freedom*, docked as an emergency escape spacecraft in case of catastrophe on the orbiting facility. Both NASA and Lockheed expected to save money by using Soyuz as the baseline ACRV and to have it available three years earlier than a completely new vehicle. Lockheed was competing with Rockwell to build the station ACRV.

1992 August 31 A Delta 7925 launched from Cape Canaveral, Fla., carried the first of two replacement satellites for the General Electric Americom Satcom network to a geosynchronous transfer orbit. Called *Satcom C-4*, the 3,032-lb. satellite comprised a box structure 3 ft. 4 in. × 5 ft. 3 in. × 4 ft. 11 in. and solar panel wings with a span of 51 ft. 6 in. providing 1,500 W of power. Built by General Electric Astro-Space, the C-series carried 24 C-band transponders for video, voice and data primarily for the U.S. cable networks. *Satcom C-4* was positioned at 135° W as an on-orbit spare.

1992 September 3 The first of four space walks for *Mir* cosmonauts Anatoly Solovyov and Sergei Avdeyev began at 13:32 UT and lasted 3 hr. 56 min., during which they deployed equipment later used to fix a propulsion module to the top of the Sofora thermomechanical lattice tower. On a second space walk that began at 11:47 UT, September 7, the cosmonauts laid a cable along the 46 ft. long Sofora tower that would provide power to the propulsion module. Before completing their 5 hr. 8 min. space walk, the cosmonauts removed the Soviet flag from the top of the tower, placed there by Artsebarsky in July 1991 and probably one of the last such emblems of the old Soviet Union to be removed from Commonwealth of Independent States (CIS) government property! During the space walk and in response to worsening tension between CIS states, personnel at Ukrainian ground stations discontinued operations, reducing the time the cosmonauts were in contact with Earth. A third, 5 hr. 44 min., space walk began at 10:06 UT on September 11, in which the crew maneuvered the Sofora tower down across *Progress M-14* which had the propulsion block attached to the outside and attached it to its new position. The final, 3 hr. 33 min. space walk began at 07:49 UT on September 11. The cosmonauts fixed a Kurs radar system to the androgynous docking unit of the *Kristall* module which would enable the Shuttle *Buran* to dock at the port.

1992 September 10 An Ariane 44LP carried Spain's first domestic telecommunications satellite, *Hispasat 1A*, and the *Satcom C-3* U.S. communications satellite from Kourou, French Guiana, to a geosynchronous transfer orbit. The 4,838-lb. *Hispasat 1A* had been built by Matra Marconi Space and the main body (8 ft. 3 in. × 7 ft. 3 in. × 7 ft. 3 in) and 73 ft. 2 in. solar arrays, producing 4,000 W of electrical power. Located at 30° W, *Hispasat 1A* had 3 × 110-W Ku-band transponders providing direct broadcast to small ground antennas only 1 ft. 4 in. across, and 10 Ku-band transponders for telephone, facsimile, data and message traffic. One beam covered North and South America. On reaching orbit it was discovered that the 7 ft. 3 in. diameter direct broadcast antenna was misaligned, but the satellite was moved to compensate. *Satcom C-3*, identical to *C-4*, was placed over 131° W to replace *Satcom 1R*, which had been in service since April 1983.

1992 September 12 At 14:23:00 UT Shuttle mission STS-47 began with the launch of *Endeavour* from LC-39B at the Kennedy Space Center, Fla. The crew included U.S. astronauts Robert Gibson (CDR), Curtis Brown (PLT), Mark Lee (MS), Jan Davis (MS), Jay Apt (MS), Mae Jemison (MS) and Japanese Mamoru Mohri (PS). The 32,510-lb. payload comprised Spacelab J, a joint NASA/National Space Development Agency of Japan (NASDA) mission, incorporating 24 materials science and 20 life sciences experiments. The crew were divided into red and blue teams and worked around the clock performing tests and investigations. The mission ended at the Kennedy Space Center at 7 days 22 hr. 30 min. 23 sec.

1992 September 14 Maneuver thrusters aboard the Venus-orbiting *Magellan* radar mapper were fired to lower periapsis from 160 mi. to 113 mi. in preparation for a fourth mapping cycle. By this date the apoapsis had drifted upward in altitude and now stood at 5,296 mi. By carefully tracking the Doppler shift of a radio signal continuously beamed to Earth, a global map of the Venus gravity field could be determined. When *Magellan* passed over a dense region on Venus' interior, the spacecraft accelerated very slightly in its orbit, and the location of the denser region was plotted. The fourth mapping cycle was due to end on May 15, 1993. The Jet Propulsion Laboratory sought funds for a fifth and sixth cycle taking spacecraft operations up to December 1994.

1992 September 25 With a thrust of 2.8 million lb. from two solid propellant boosters of 10-ft. diameter a commercial Titan 3 launch vehicle lifted off from Launch Complex 40, Cape Canaveral, Fla., at 17:05 UT carrying the 5,672-lb. *Mars Observer* spacecraft. The 90 ft. 10 in. solid propellant boosters separated 8 sec. after the 548,000-lb. thrust Titan first stage ignited at an elapsed time of 1 min. 49 sec. Stage 2 ignited 4 min. 29 sec. into flight and shut down just over 3 min. 30 sec. later, at which point the assembly was in an interim 100 × 337 mi. orbit. *Mars Observer* and its transfer orbit stage separated from Titan about 7 min. later and, after coasting halfway round the world, the 24,000-lb. solid propellant TOS ignited

for a 2 min. 30 sec. burn producing a thrust of 58,600 lb., accelerating the spacecraft to escape velocity on a course for Mars. All telemetry was lost from the TOS at ignition, and controllers had to wait 40 min. until the spacecraft separated, turned itself on and reported that all was well.

1992 September Congress voted to eliminate all funds for what was now called the National Launch System (NLS), formerly the Advanced Launch System (ALS). Under joint Department of Defense/NASA development with an anticipated flight date of 2002, NLS had progressed to a proposed series of three vehicles: NLS-1 with a 135,000-lb. payload capability; NLS-2 with a 50,000-lb. payload capability; and NLS-3 with a 20,000-lb. capability. NLS-1 was regarded as defunct since NASA had no need for it and the changing geopolitical scene removed the urgency for heavy-lift payloads in the Strategic Defense Initiative antiballistic missile program. Critics of NLS-2/3 asserted that they duplicated existing launchers.

1992 October 6 A Chinese CZ-2C launcher carried the 4,630-lb. *Fanhui Shi Weixing 1–1 (FSW 1)* recoverable satellite to a 131 × 198 mi. orbit inclined 63°. The launcher also released the 570-lb. Swedish science satellite *Freja*, 7 ft. 3 in. in diameter, lifted to a 370 × 1,093 mi. orbit. *FSW 1* was a smaller version of the *FSW 2* launched on August 9, cone-shaped, with a length of 10 ft. 2 in. and a diameter of 7 ft. 3 in. It carried microgravity research instruments and remote sensing equipment, returning its recoverable capsule to Earth on October 13 followed by decay of the instrument module on October 31. *Freja* carried seven magnetometers and auroral imaging instruments from Sweden, the United States, Canada and Germany.

NASA and the Russian Space Agency signed a Human Space Flight Agreement whereby an experienced cosmonaut would fly aboard the Shuttle and a NASA astronaut would visit the *Mir* space station. The first flight would carry a cosmonaut on Shuttle mission STS-60 planned for November 1993. Cosmonauts Vladimir Titov and Sergei Krikalov were selected as candidates and began training at the Johnson Space Center. This was the first time a Russian had trained to fly in a U.S. spacecraft. This would be reciprocated during 1995 when a NASA astronaut was scheduled to fly to the Russian *Mir* space station in the Shuttle and remain on board for about three months. NASA would also transport two cosmonauts in the Shuttle to replace two cosmonauts who would already have served a term aboard *Mir*. In 1992 cosmonaut Krikalov was selected for STS-60, but the mission was postponed to February 3, 1994.

1992 October 10 About 2.8 million mi. from Earth, the *Mars Observer* spacecraft conducted its first trajectory correction maneuver by firing two 110-lb. bipropellant thrusters for 2 min. 13 sec., altering the velocity of the spacecraft by 164 ft./sec. Already, a combination of the gravity effects from Earth and the sun had slowed the spacecraft from its escape velocity of 37,500 MPH to a speed of 7,600 MPH. A second correction burn was planned for January 8, 1993. *Mars Observer* was scheduled to arrive at Mars on August 24, 1993, when two 110-lb. thrusters would fire to put the spacecraft in a 343 × 23,972 mi. orbit with periapsis over the south pole. Over the succeeding four months, maneuvers would lower the spacecraft to a 218 × 235 mi. sun-synchronous mapping orbit for observation of the Mars environment for 687 Earth days, a full Martian year.

1992 October 22 After a 1 hr. 53 min. wait for weather to clear at a contingency abort site, *Columbia* was launched at 17:09:40 UT from LC-39B at the Kennedy Space Center on STS-52. The crew consisted of James Wetherbee (CDR), Michael Baker (PLT), William Shepherd (MS), Tamara Jernigan (MS), Charles Veach (MS) and Steven MacLean (PS). The 26,892-lb. payload included the *Laser Geodynamic Satellite II (LAGEOS-II)* and the first U.S. Microgravity Payload (USMP-1). From a Shuttle orbit of 180 × 186 mi., at 28.4°, a 181-lb. Canadian target assembly satellite was deployed on the first day to test a space vision system for robot servicing systems. *LAGEOS-II* was deployed on October 23 and boosted to its planned orbit of 3,490 × 3,697 mi., at an inclination of 52.7° by an IRIS perigee motor produced by the Agenzia Spaziale Italiana (the Italian Space Agency) and a solid propellant apogee kick motor. Activated on the first day, USMP-1 included three experiments, mounted on two support structures in the orbiter payload bay, one of which was contributed by Centre National d'Etudes Spatiales, the French national space agency. The mission ended at the Kennedy Space Center, 9 days 20 hr. 56 min. 13 sec.

1992 October 27 At 17:20 UT *Progress M-15* was launched by SL-4 from Baikonur cosmodrome and docked to the back of the *Kvant 1* module at the *Mir* complex at 19:06 UT two days later. The freighter brought additional supplies and experiment equipment to the orbiting complex, unstowed by *Mir* cosmonauts Anatoly Solovyov and Sergei Avdeyev. *Progress M-15* pumped fuel and air to *Mir* on January 15, 1993, as the crew got ready to receive the *Soyuz TM-16* spacecraft launched January 24, 1993. *Progress M-15* would remain attached to the back of the *Mir* complex throughout the next crew change, the first time that had been accomplished. It would undock from the *Mir* complex on February 4, 1993.

The U.S. Defense Acquisition Board approved a plan by the U.S. Air Force to award Lockheed a contract for development and test of a Milstar 2–series advanced communication satellite. With 100 times the communications capacity of the Milstar 1–series satellites, the first of which had yet to be launched, Milstar 2 would carry a 48 megabit/sec. data-rate transmission system in addition to 0.5 megabits across 192 low data rate channels. The original Milstar 1 contracted with Lockheed had been renegotiated to just three satellites, of which the third would be converted to Milstar 2 configuration. The Air Force wanted Lockheed to build an additional four Milstar 2 satellites for a total of seven.

1992 October 28 An Ariane 42P carrying the 6,544-lb. *Galaxy VII* Hughes communication satellite used for the first time a procedure designed to increase the satellite's in-orbit life by optimizing the performance of launcher and payload. Called "perigee velocity augmentation" (PVA), it called for the third stage to lift the satellite to an apogee of only 17,150 mi. instead of the geosynchronous transfer apogee of 22,354 mi. Liquid propellant in the satellite's apogee motor was used to push the satellite to geosynchronous altitude and then circularize at 91° W. *Galaxy VII* was a Hughes HS 601 comprising a body structure 11 ft. 10 in. × 10 ft. 2 in. × 8 ft. 10 in., a deployed height of 24 ft. 7 in. and solar-cell arrays with a span of 86 ft. 11 in. producing 4,700 W on orbit. *Galaxy VII* replaced *SBS-4* and carried 24 × 18-W C-band and 24 × 50-W Ku-band transponders.

1992 October The Department of Defense placed orders with Orbital Sciences Corp. and EER Systems for the launch of small payloads from the Strategic Defense Initiative antiballistic missile research program. Based on a 400,000-lb. thrust Peacekeeper first stage and the three-stage Pegasus winged, air-dropped launcher, OSC's Taurus launcher would be used in 1995 to launch the *Clementine 2* satellite. EER Systems, which had absorbed Space Services, was given a launch contract for the Conestoga which comprised a combination of Thiokol Castor 4/4A motors in the first three stages and a Star 63F in the fourth stage.

1992 November 10 A Council of Ministers meeting of the European Space Agency approved far-reaching changes to future programs. Because of serious cost increases and changing priorities regarding the U.S. space station *Freedom*, the Council ordered that current work on the Hermes space plane would stop at the end of the year. It was to be followed by a three-year effort to redefine the program, seeking cooperation with Russia on a new manned space vehicle, further cooperation with the United States or adoption of an autonomous European program. In what were called *EuroMir-94* and *EuroMir-95*, ESA astronauts would twice visit the *Mir 1* space station aboard Soyuz ferry vehicles. Launched in September 1994, *EuroMir-94* was to last 30 days. In August 1995 a 135-day visit would begin as *EuroMir-95*. The four ESA astronaut candidates were Ulf Merbold, Christer Fugelsang Pedro Duque and Thomas Reiter.

1992 November 15 A Commonwealth of Independent States SL-4 launcher carried the 13,890-lb. *Resurs 500* satellite from Plesetsk to a 139 × 225 mi., 82.5° orbit. The flight was organized as an international marketing effort for CIS and Russian space technology, hardware and services. Instead of the 700-lb. of cameras and remote sensing equipment usually carried by Resurs satellites, *Resurs 500* (named for the 500th anniversary of Columbus' landing in America) carried letters from Boris Yeltsin, promotional literature, painted eggs, a miniature Statue of Liberty and other minor items. The reentry module returned to Earth on November 22, splashing down 200 mi. southwest of Seattle, Wash. Recovered by the Russian ship *Marshal Kylov*, the reentry module and its contents was handed over to city officials for the Museum of Flight.

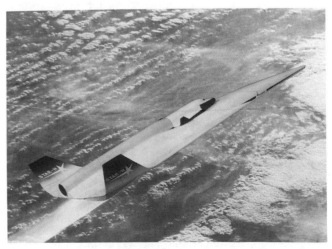

Free to consider radical new forms of space transportation after the cancellation of Hermes, the French manufacturer Dassault proposed this two-stage aerospace plane.

1992 November 16   Three U.S. aerospace companies submitted bids for the third U.S. Air Force Medium Launch Vehicle (MLV-3) requirement. McDonnell Douglas submitted its Delta, winner of the MLV-1 competition, General Dynamics proposed its Atlas-Centaur and Martin Marietta put up a version of the Titan II. The Air Force wanted to buy 36 launchers for the replacement Block 2 Navstar satellite. The MLV-1 order had already increased from 20 to 28 the number of launchers bought for the Block II Navstar. In an ironic twist, whereas once NASA was hailed as the fulcrum of space commercialization, the Air Force now provided vigorous incentives for launch vehicle manufacturers to sell their products.

1992 November 28   With a thrust of 3.2 million lb. from two solid propellant boosters, the third Titan 4 lifted off from Vandenberg AFB, Calif., at 21:34 UT carrying an advanced *KH-11* imaging reconnaissance satellite. This was the sixth launch of a Titan 4 and the third from Vandenberg on what the U.S. Air Force classified as a top secret mission.

1992 November 30   The Russian Flight Control Center published plans for the orbiting *Mir* complex. In addition to the scheduled flights for 1993, manned missions planned for 1994 included *Soyuz TM-19* on April 25 and *TM-20*, with an astronaut from the European Space Agency, on September 17. Missions planned for 1995 included *Soyuz TM-21* on March 1, the launch of NASA Shuttle *Atlantis* on June 3 for a docking at the *Kristall* module and the return of physician Valery Polyakov after more than a year in space, and *TM-22* on August 8 with another ESA astronaut. *Soyuz TM-23* would carry the French cosmonaut Claudie André-Deshays to *Mir* in January 1996. Plans for a much bigger *Mir 2* complex were approved by the Russian parliament in 1993, and called for the launch of the first elements in 1996 and permanent manning (for at least ten years) from 1997.

1992 November   General Electric Astro-Space successfully competed for a contract to build the *Landsat 7* remote sensing satellite. For the first time, the contract was awarded by a joint U.S. Air Force/NASA program office in the latest reshuffle to bolster the flagging U.S. remote sensing program. *Landsat 7* was to be equipped for cross-links with tracking data and relay satellites in geostationary orbit, massively increasing the downlink transmission time for individual user stations, which would hook up to the TDRS rather than the orbiting satellite itself. *Landsat 7* was tentatively scheduled for launch in 1998.

1992 December 1   An Ariane 42P, with two solid propellant strap-ons and the extended H10+ third stage, carried the 6,130-lb. *Superbird A1* communications satellite to geosynchronous transfer orbit on mission V55. From there it was lifted to position over 158° E, replacing *Superbird A*, lost in 1990 due to ground control errors. Identical to *Superbird B1* launched by Ariane on February 26, 1992, *Superbird A1* completed the two-satellite system for Japan's Space Communications Corp.

1992 December 2   After a wait of 1 hr. 25 min. for ice to clear from the external tank, Shuttle mission STS-53 began at 13:24:00 UT when *Discovery* was launched from LC-39A at the Kennedy Space Center, Fla. This was the last of 10 military Shuttle missions booked by the Department of Defense. Aboard the orbiter were David Walker (CDR), Robert Cabana (PLT), Guion Bluford (MS), James Voss (MS) and

Michael Clifford (MS). The 28,346-lb. payload included an advanced radar imaging reconnaissance satellite for the Department of Defense, known as *Department of Defense-1*, deployed to an orbit of 230 × 231 mi., with a 57° inclination, at 5 hr. 54 min. One experiment, the cryogenic heat pipe, was a joint Department of Defense/NASA test of technology that would make it easier to reject excess heat from infrared sensors. After adjusting the orbit to 201 × 202.5 mi., the crew released to space six spheres, between 6 in. and 2 in. in diameter, to calibrate the Haystack Radar at Tyngsboro, Mass., which was used to detect and track objects in space as small as 0.4 in. at a range of 710 mi. The mission had been expected to terminate at the Kennedy Space Center, Fla., but poor weather diverted the landing to Edwards AFB, Calif., touchdown occurring at an elapsed time of 7 days 7 hr. 19 min. 17 sec.

1992 December 8   NASA's *Galileo* spacecraft made its second flyby of the Earth-moon system and received a 17,100 MPH velocity boost that would take it on to Jupiter for orbital insertion on December 7, 1995. *Galileo* sped past the moon at a distance of 75,000 mi. at 03:58 UT and obtained images of the north polar region difficult to view from Earth. At 15:09:25 UT the spacecraft passed within 189 mi. of the South Atlantic Ocean over a position at 34° S × 6° W. This highly accurate trajectory was within 1 mi. of the planned flight path, conserving propellant for course correction maneuvers on the way to Jupiter and for orbital maneuvers in the Jovian system of moons when the spacecraft arrived and began 22 months of intensive observation after relaying data from the entry probe.

1992 December 20   The deadline arrived for completion of a high-level report on U.S. space policy. After four months of effort to define a new concept of space leadership, which the report said should be earned rather than proclaimed, recommendations centered on strong coordination between separate government departments with interests in space. It said that security barriers should come down and start from the basis that all programs are unclassified unless national security interests dictate otherwise.

1992 December 21   The second of two satellites launched on a Chinese CZ-2E for Optus Communications of Australia, *Optus B2*, exploded 45 sec. after launch from the Jiuquan site. Most of the satellite fell back to Earth but the rest continued into an orbit of 132 × 640 mi. inclined 28.1°. Concern that the explosion may have been attributed to the launcher was fiercely challenged by the Chinese and responsibility for the accident was finally judged to have been caused by the payload shroud breaking up at an altitude of 23,000 ft., just 45 sec. after liftoff. Destroyed by a 1,000 MPH head wind, the satellite was torn away from the forward section of the launcher.

1992 December 23   The State Department sanctioned an agreement signed between Lockheed Missiles and Space Co. and the Krunichev Enterprise near Moscow. Lockheed would market the Proton launcher throughout the world and propose Proton for the launch of a Soyuz TM as the assured crew return vehicle for NASA's space station *Freedom*. *Proton 1* could place 44,000 lb. in low Earth orbit or 9,000 lb. into a geosynchronous transfer path. Baikonur cosmodrome had four Proton pads and these, it was assumed, would continue to be available.

1992 December 29   A Commonwealth of Independent States SL-4 launched the biological science satellite *Cosmos 2229*, *Bion 10*, from Plesetsk to a 133.5 × 226 mi., 62.8° orbit. The 13,900-lb. satellite carried an automatic, programmable incubator supplied by the European Space Agency. Called Biobox, it had been developed jointly with the Russian Institute of Biomedical Problems and was used to carry out research on bone tissues. The Russians supplied two monkeys, as well as plants and insects for the Bion project, the eighth with U.S. participation, and the NASA Ames Research Center provided diagnostic equipment for measuring physiological, neurological and biological factors. Thermal control problems brought premature termination of the flight, and *Bion 10* was returned to Earth on January 10, 1993.

1993 January 3   U.S. President George Bush and President Boris Yeltsin of Russia signed the START II (Strategic Arms Reduction Talks II) agreement, making the deepest cuts in strategic nuclear weapons ever. It required the United States and Russia to cut strategic nuclear weapons to a maximum of 3,500 by the year 2003. All MIRV warheads on ICBMs would be outlawed and the United States would eliminate Peacekeeper and deactivate all but 500 Minuteman III ICBMs, which would be fitted with one warhead rather than the three carried at this time. Only 432 MIRVed submarine-launched ballistic missiles would remain, comprising 192 Trident Is on 8 submarines and 240 Trident IIs on 10 submarines. During 1993 the Russians decided that by 2005 their land-based missile force would consist of only 900 SS-25s, each with one warhead.

1993 January 13   Delayed about 7 min. waiting for high winds to abate, Shuttle mission STS-54 began with the launch of *Endeavour* at 13:59:30 UT from LC-39B at the Kennedy Space Center, Fla. The crew comprised astronauts John Casper (CDR), Don McMonagle (PLT), Mario Runco (MS), Gregory Harbaugh (MS) and Susan Helms (MS). The 49,069-lb. payload included the *TDRS-6* tracking and data relay satellite and its inertial upper stage, deployed at an elapsed time of 6 hr. and eventually placed in geostationary transfer orbit. Initially, *TDRS-6* was placed at 210° E, but later moved

During late 1993, military leaders decided to opt for the SS-25 as the single missile that would form the mainstay of Russia's strategic nuclear deterrent of the future.

to 189° E. Other experiments included a Hitchhiker experiment collecting data on x-radiation from deep space, and microgravity bioprocessing tests. Runco and Harbaugh performed an EVA (extravehicular activity) on January 17 lasting 4 hr. 28 min., testing new foot restraints and conducting experiments in how to move large objects around the payload bay. After the *Gemini IV* mission of June 1965 and the *Gemini XII* mission of December 1966, this was NASA's third generic EVA wholly designed around advancing space walk procedures. The mission ended at the Kennedy Space Center, Fla., 5 days 23 hr. 38 min. 17 sec. after liftoff.

1993 January 19   Results from observations of the Antarctic ozone hole by *Nimbus 7* and *Meteor 3* TOMS (total ozone mapping spectrometer) instruments indicated that in September 1992 the hole encompassed an area of 8.9 million sq. mi., almost as great as the area of the North American continent (which is 9.4 million sq. mi.). NASA was focusing measurements on man-made chlorine levels, now recognized as a primary cause of ozone depletion and hole formation.

1993 January 24   At 05:58 UT Commonwealth of Independent States cosmonauts Gennady Manakov and Alexandr Poleshchuk were launched aboard the *Soyuz TM-16* spacecraft on a crew exchange flight to the *Mir* space station. They were to make a historic first docking at the androgynous unit on the *Kristall* laboratory, docked to the lower radial port of the space station. This would be achieved with the Kurs approach and docking radar installed by Anatoly Solovyov and Sergei Avdeyev during a space walk in September 1992. In a unique docking at right-angles to the longitudinal axis of the orbiting complex, Manakov and Poleshchuk successfully locked on to the APAS-89 docking unit at 07:41 UT on January 26. For the first time, three Soyuz-type spacecraft were docked to *Mir*: *Soyuz TM-15* at the forward axial port, *Progress M-15* at the back of the *Kvant 1* module, and *Soyuz TM-16* at the *Kristall* module. It was the biggest assembly of separately launched hardware ever put together in space. In addition to the three Soyuz/Progress vehicles, the complex itself comprised *Mir*, *Kvant 1*, *Kvant 2*, and *Kristall*, for a total mass of more than 200,000 lb. (100 tons).

1993 January   NASA Administrator Daniel S. Goldin ordered a comprehensive evaluation of future space launch system needs and what role, if any, the Shuttle should play in the launcher mix. The agency was to focus on three options for study: utilizing the Shuttle as the primary NASA launcher through the year 2030, involving major investment in technology upgrades for the 1960s design/1970s technology vehicle; replacing the Shuttle in about 2005 with a new expendable launch system derived from NLS (National Launch System) studies; utilizing the Shuttle until early in the next century; and replacing it with a radical new single-stage-to-orbit system.

1993 February 1   Commonwealth of Independent States cosmonauts Anatoly Solovyov and Sergei Avdeyev returned to Earth, landing at 03:52 UT in the *Soyuz TM-15* spacecraft which had been docked to the forward axial port of the *Mir* space station for more than six months. Their mission had lasted 188 days 20 hr. 43 min. and their management of the orbiting complex, about to begin its eighth year in space, had been transferred to Gennady Manakov and Alexandr Poleshchuk, launched aboard *Soyuz TM-16* on January 24.

1993 February 2   President Bill Clinton's newly appointed Secretary of Defense Les Aspin ordered the Strategic Defense

Initiative antiballistic missile organization to peg expenditure at last year's level and warned that they should expect no increase thereafter. The new administration favored research into theater and battlefield defenses, but was unlikely to fund anything other than a limited defense against strategic weapons. There was now little political support for work on laser and particle-beam weapons, and the Brilliant Pebbles concept of space based interceptors was being cut to a technology research effort.

1993 February 4   At 00:45 UT *Progress M-15* undocked from the back of the *Kvant 1* module at the *Mir* complex and withdrew to a distance of approximately 525 ft. Keenly watched by Gennady Manakov and Alexandr Poleshchuk aboard the space station, *Progress* turned through 190° so that its propulsion system faced the rear of the space station. From the forward docking area of the orbital module, a special container called Znamya began to spin to a rate of 94 RPM, creating inverse centripetal force which caused a furled solar sail to unfold like the iris over a camera lens. The mirror was 65 ft. 7 in. in diameter and comprised an aluminized Kevlar sheet, 5 microns (^{5}/1000 of a millimeter) thick. At a distance of 755 ft. from *Mir*, the rate of rotation was purposely slowed to just 14 RPM. Six minutes after deployment, the mirror was jettisoned and the test, of a possible solar sail device, was over. *Progress M-15* was de-orbited over the Pacific Ocean three days later.

1993 February 8   The *Mars Observer* spacecraft performed its second trajectory correction maneuver (TCM-2) at 22:00 UT when two of the large, 110-lb. thrust maneuver jets were fired for 35 sec. in a 32 ft./sec. burn. The spacecraft was more than 25 million mi. from Mars traveling at about 18,000 MPH relative to the red planet. A third correction burn was conducted at 17:00 UT on March 18. The spacecraft fired four 5-lb. thrusters in a 1.5-ft./sec. velocity adjustment with *Mars Observer* 18 million mi. from the planet.

1993 February 9   Two small satellites were propelled into orbit by an air-launched Pegasus released from a B-52 mother plane at 14:30 UT. The 32-lb. *Capabilities Demonstration Satellite* was the first to be deployed, to an orbit of 451.5 × 491.5 mi., at 24.97° inclination. It was launched for Orbital Communications Corp., as a pathfinder satellite designated *OXP-1*. The second satellite, *SCD-1 (Satelite de Coleta de Dados-1)* was designed and built by Brazil's civilian space agency, the Instituto Nacional de Pesquisas Espaciais, as a relay satellite for data transmitted by environmental monitoring stations at remote locations on the ground. Weighing 253 lb., it was placed in an orbit of 450.5 × 491 mi., at 24.97°.

1993 February 16   The first firing of the solid propellant rocket motor in development for the European Ariane 5 heavy-lift launcher took place on a test stand at Kourou, French Guiana. Generating a thrust of more than 1 million lb., it was the most powerful rocket motor produced and fired in Europe to date. The solid propellant rocket booster, two of which would be attached to the Ariane 5 core stage, had a length of 102 ft. 3 in., a diameter of 9 ft. 10 in. and a launch weight of 581,000 lb., producing a unit thrust of 1.2 million lb. at ground level. The two boosters and the 252,000-lb. thrust cryogenic core stage together produce a liftoff thrust of 2.652 million lb.

1993 February 18   The incoming Clinton administration revealed its economic plan and announced changes to funding

Released by NASA in July 1993, this Option A concept for the revamped space station was put together in response to President Bill Clinton's edict that the space agency cut costs on the proposed facility.

and management of NASA's space station *Freedom*, calling for a complete redesign to cut costs. Projected costs for the baseline *Freedom* design had burgeoned to a staggering $31.3 billion, of which $25.1 billion would be needed over the next seven years just to get it built. Total running costs for 10 years thereafter were estimated at a further $31.6 billion, for a total $62.9 billion over 17 years, an average $3.7 billion each year. Supporters in Congress were horrified to learn that the *Freedom* budget for fiscal years 1993–1995 was already expected to overrun by $1.08 billion.

1993 February 20   A Mu-3 S2 launcher carried the 926-lb. Japanese Astro-D astronomy satellite from Uchinoura launch site to a 334 × 402 mi., 31.1° orbit. Named *Asuka*, the satellite was the product of a cooperative venture between Japan's Institute of Space and Aeronautical Sciences and NASA and combined imaging and spectroscopy in one satellite. A powerful function of the satellite was in its combination of conical-foil mirror technology of the broad band x-ray telescope with charged-couple detector devices being developed for NASA's Advanced X-Ray Astrophysics Facility (AXAF). The mission was specifically designed to pursue a broad range of questions in astrophysics.

1993 February 21   Weighing 15,990 lb., the Commonwealth of Independent States' cargo tanker *Progress M-16* was launched from Baikonur at 18:32 UT to an orbit of 118.5 × 158 mi. and from there to the *Mir* space station in an orbit of 241 × 243.5 mi., at 51.62° inclination. At 20:18 UT on February 23, it was automatically docked to the back of the *Kvant 1* module. Carrying 5,729 lb. of supplies for the space station, *Progress M-16* was unpacked and its contents moved to *Mir*. The cargo tanker was undocked from the aft axial port at 06:50 UT on March 26, and backed away to a distance of about 230 ft. from where it moved back and redocked at 07:07 UT. The final undocking took place at 04:21 UT on March 27, followed by a de-orbit burn over the Pacific Ocean the same day.

1993 February 23   Japan's National Space Development Agency (NASDA) successfully completed the first firing of the

H-II first stage in a test lasting 10 sec. at the Yoshinobu launch site, Tanegashima. The first stage had a length of 91 ft. 10 in., a diameter of 13 ft. 1 in. and contained 190,000 lb. of cryogenic hydrogen and oxygen for a total stage liftoff weight of 216,100 lb. Propellants were fed to the 190,000-lb. thrust LE-7 engine for a burn time of 5 min. 46 sec. at a specific impulse of 445 sec. Six test firings were completed by the end of the development program on July 8, the longest being a 5 min. 53 sec. firing on June 15. The H-II second stage has a length of 36 ft., a diameter of 13 ft. 1 in. and contains 37,500 lb. of liquid hydrogen and liquid oxygen for a total stage mass of 44,100 lb. A development of the LE-5 stage built for the H-I launcher, the H-II's LE-5A produces a thrust of 26,500 lb. at a specific impulse of 452 sec. for a total burn time of 10 min. 9 sec.

1993 March 4   For the first time, scientists were able to measure the effect on atmospheric temperatures of a volcanic eruption on Earth. Instruments on NASA's *Earth Radiation Budget Experiment Satellite (ERBS)* measured the effect of the Mount Pinatubo (Philippines) eruption in June 1991 and discovered the Earth's atmosphere had cooled by 1°F as a result of the gas and dust thrown out by the volcano. Global observation of the Earth's temperature began in 1976. This event was 60% higher than any recorded since that date. It proved that clouds and poisonous gases act as an insulator, reflecting solar heat back into space and cooling the Earth. Some scientists had thought the reverse to be the case, claiming that heat would be trapped and that temperatures would rise.

NASA selected the University of Texas' Southwestern Medical Center, Dallas, as its Specialized Center of Research and Training in integrated physiology. At this center, physiologists would study how different organ systems from various species react to spaceflight. One component would focus on cellular and molecular mechanisms and another would concentrate on how spaceflight affects skeletal muscle, bone and minerals, and the cardiovascular system. The center would help NASA coordinate life sciences research in preparation for extensive cooperative experiments with the Russians.

1993 March 9   NASA Administrator Daniel Goldin informed all field centers that President Bill Clinton had authorized a complete redesign of space station *Freedom*. On March 10 a hand-picked redesign team began work under former Apollo program manager, Dr. Joseph Shea, until he walked off the team after a stormy meeting on April 22. He was replaced by Dr. Bryan O'Connor. The 45-member station redesign team had until May 15 to produce an interim report, with a final report due June 1. Their mandate stipulated that the space station have an initial operational capability by 1997, complete development within a five-year budget plan, a 10-year, rather than a 15-year operating life, and that it retain options for Russian participation. On March 25 Vice President Albert Gore set up an Advisory Committee on the Redesign of the Space Station.

1993 March 22   After two postponements in the projected launch date (March 14 and 21), Shuttle mission STS-55 was set for launch when the flight was aborted at T − 3 sec. shortly after main engine ignition. This was the third launch pad abort (after missions 51-F and 41-D) in 56 Shuttle liftoff attempts since April 12, 1981. Incomplete ignition of the no. 3 main engine due to an internal oxygen preburner leak caused the premature shutdown. After replacing all three main engines

with the Shuttle sitting on LC-39A, NASA launched STS-56 from LC-39B and reset the STS-55 launch for April 24. That attempt was scrubbed early on launch day when three inertial measurement units gave faulty readings.

1993 March 25   The Russian Federation carried the *Start-1* satellite into space using the first solid propellant launcher employed by Russia, the Commonwealth of Independent States, or the Soviet Union. Derived from the SS-20/SS-25 family of ballistic missile, *Start-1* was a major step forward in the use of missiles for "civilian" satellite flights. The 573-lb. test satellite was carried from Plesetsk to an orbit of 425 × 603 mi., at 75.76° inclination. It was designed to serve as a precursor checkout model for a planned series of small communication satellites in the lightsat category increasingly favored by the United States.

An Atlas 1 failed to carry the first UHF Follow-On satellite, *UFO-1*, to a planned geosynchronous transfer path after launch from Cape Canaveral, Fla. *UFO-1* had been designed around the Hughes HS 601 satellite bus for the U.S. Navy as a replacement for the aging FltSatCom series. The 5,050-lb. satellite had been designed for an operating life of 14 years. Shortly after launch the Atlas booster engine performed erratically and the sustainer engine shut down early. Despite the Centaur's burning until its propellants were depleted, the stage and its payload were left stranded in an orbit of 133 × 21,934 mi. at an inclination of 27.76°. *UFO-2* was successfully launched by an Atlas 1 on September 3, 1993, to a geostationary position at 186° E.

1993 March 30   Accompanying a Navstar global positioning satellite on its launch from Cape Canaveral, Fla., a 57-lb. *SEDS* subsatellite was released from the second stage of the Delta 2 launcher in a controlled tether experiment as part of NASA's Flight Demonstration Program. *SEDS* (small expendable deployer system) comprised a 13 in. × 16 in. × 8 in. box attached to a canister containing a 12.4-mi. tether on a reel. From the interim transfer orbit of 115.5 × 12,700 mi., at an inclination of 34.86°, the Navstar satellite was released. At 04:12 UT, 1 hr. 3 min. after liftoff, *SEDS* was set free from the second-stage package at an altitude of 447.5 mi. The polyethylene fiber tether deployed to full extension, and measurements on motions and dynamic forces were obtained.

1993 March 31   The Russian cargo tanker *Progress M-17* was launched at 03:34 UT from Baikonur cosmodrome by an SL-4 launch vehicle to a 110.5 × 148 mi. orbit. From there it performed rendezvous maneuvers to link up with the orbiting *Mir* space station. It docked to the axial port on the *Kvant 1* module attached to the back of the space station at 05:16 UT on April 2, in an orbit of 244 × 245.5 mi. *Progress M-17* carried 5,740 lb. of supplies to the crew, Gennady Manakov and Alexandr Poleshchuk, aboard Mir since January 26. The cargo tanker remained docked to the aft axial port during the launch, rendezvous and docking of *Progress M-18*, launched from Baikonur on May 22. With the station in an orbit of 241 × 244 mi., *Progress M-17* undocked from *Mir* on August 11. It was expected to remain in space until about the end of 1994.

1993 April 8   After a two-day delay following a countdown scrub at T − 11 sec. due to faulty sensors, Shuttle mission STS-56 began with the launch of *Discovery* at 05:29:00 UT from LC-39B at Kennedy Space Center, Fla. Carrying astronauts Kenneth Cameron (CDR), Stephen Oswald (PLT), Kenneth Cockrell (MS), Michael Foale (MS) and Ellen Ochoa (MS), the primary payload comprised the second Atmospheric

Laboratory for Applications and Science (ATLAS-2). This comprised an assembly of sensors and detectors, mounted on a Spacelab pallet, designed to collect data on the interaction of solar radiation and the Earth's atmosphere. At 05:11 UT on April 11 the crew deployed the 2,842-lb. Shuttle Point Autonomous Research Tool for Astronomy-201 (SPARTAN-201), a free-flying platform observing the solar wind and the solar corona. It was retrieved at 07:20 UT on April 13. The mission lasted 9 days 6 hr. 8 min. 19 sec. and ended at Kennedy Space Center.

1993 April 9   McDonnell Douglas won the U.S. Air Force MLV-3 competition with its Delta 2 7925 launcher. Competing bids had come from General Dynamics, with a modified Atlas 2, and Martin Marietta, with a Titan 2S accommodating six Thiokol strap-on rockets. The Air Force wanted production options on 36 launchers through the year 2002, and deliveries would follow on from the MLV-1 contract for Delta 2 supporting the Navstar program. Current Navstar satellites weighed 4,010 lb., but the updated model for MLV-3 launch would weigh 4,480 lb. McDonnell Douglas was given a firm contract for 25 Deltas 2s, with the first mission as early as 1996.

1993 April 19   At 18:25 UT *Mir* cosmonauts Gennady Manakov and Alexandr Poleshchuk opened the exterior hatch on the *Kvant 2* astrophysics module attached to the *Mir* space station and prepared to conduct the 200th Soviet/Russian space walk. Poleshchuk moved down to the telescopic boom that would transport him to the *Kvant 2* work site in preparation for installing electric drive motors on the *Kristall* solar batteries that were to be moved to *Kvant*. The 5 hr. 25 min. space walk was a success, although the cosmonauts had difficulty in attaching the drive unit to the truss structure built previously on *Kvant*.

1993 April 25   An air-dropped Pegasus satellite launcher, released from a B-52 mother plane, carried the 240-lb. satellite *ALEXIS* to an approximate orbit of 460 × 519 mi., at 70° inclination. An acronym for Array of Low Energy X-ray Imaging Sensors, *ALEXIS* was to measure x rays in space for one year. After release from the B-52 at 13:56 UT, at an altitude of 43,600 ft., the three stages of the Pegasus launcher fired in sequence, but no telemetry was received from the satellite for several weeks. It is possible the communications antenna was inadvertently dislodged at third stage separation. In July telemetry was successfully picked up and the satellite declared operational.

1993 April 26   Following a two-day postponement when the April 24 attempt was scrubbed due to faulty instrument readings, Shuttle mission STS-55 began with the launch of *Columbia* at 14:50:00 UT from LC-39A, Kennedy Space Center, Fla. Carrying NASA astronauts Steven Nagel (CDR), Terence T. Henricks (PLT), Jerry Ross (MS and Payload Commander), Charles Precourt (MS) and Bernard Harris (MS), and German astronauts Ulrich Walter (PS) and Hans Wilhelm Schlegel (PS), the mission comprised a Spacelab flight managed by German technicians. Designated *Spacelab D-2*, it incorporated 88 experiments in materials and life sciences, technology applications, Earth observations, astronomy and atmospheric physics. Communications were lost for 80 min. on May 4 due to an errant ground signal. Extended by one day, the landing was diverted from Kennedy Space Center, Fla., to Edwards AFB, Calif., because of bad weather. Landing occurred at 9 days 23 hr. 39 min. 59 sec.,

bringing to more than one year (365 days 23 hr. 48 min) the total accumulated Shuttle mission time since the first flight on April 12, 1981.

1993 April   The International Maritime Satellite Organization (Inmarsat) completed negotiations for the launch of the fourth Inmarsat 3–series satellite on a Russian Proton launcher in 1995. This would mark the first occasion when the satellite of a Western organization had been launched by the Russians. The Inmarsat 3–series satellites were being built by a team headed by Martin Marietta Astro Space (formerly GE Astro-Space) .

1993 May 10–11   Scientists meeting in Wiesbaden, Germany, NASA, the European Space Agency, the Russian Space Research Institute and the space agencies of France, Italy and Germany set up an International Mars Exploration Working Group to plan a multilateral strategy for the exploration of Mars. Russian officials described their October 1994 Mars mission which would put a lander on the surface in 1995 and the 1996 launch of a Mars roving vehicle. NASA scientists explained a Mars environmental survey (Mesurs) series beginning with a 1996 Pathfinder flight to follow the *Mars Observer* mission then en route to Mars. The second phase of Mesurs would put 16 low-cost landers on the surface via four Delta 2 7925 launches.

1993 May 12   An Ariane 42L carried the 6,152-lb. *Astra 1C* satellite, and the 340-lb. amateur radio satellite *Arsene*, to orbit from Kourou, French Guiana, on mission V56. Built by Hughes Space and Communications, *Astra 1C* had a box structure 11 ft. 2 in. × 9 ft. 2 in., with solar cells producing 3,500 W on wings with a span of 68 ft. 11 in. With an on-orbit mass of 3,748 lb., *Astra 1C* carried 34 Ku-band channels fed through 18 transponders. Positioned at 12° E for checkout, the satellite was moved later to its operational slot over 19.2° E. Launched for Radio Amateur Club de France, the *Arsene* satellite took the form of a cube 3 ft. 2 in. × 3 ft. 7 in. covered in solar cells producing 42 W for three channels. It was placed in an orbit of 10,978 × 23,017 mi. inclined 1°.

1993 May 21   The first of four Resurs remote sensing satellites launched during 1993, *Resurs F-17* was carried by an SL-4 from Plesetsk to a 143 × 147 mi., 82.57° orbit. The 13,890-lb. satellite was a long-duration variant equipped with multispectral and sky-survey cameras. It was recovered on June 20. The second Resurs, *F-18*, was launched from Plesetsk on June 25 and carried to a 138.5 × 150 mi., 82.58° orbit. It was of the short-duration type and had topographic, spectrazonal and sky camera systems. *F-18* returned its reentry module to Earth on July 12. Ostensibly the first in a new series of Resurs-T remote sensing satellites, *Cosmos 2260* was launched by an SL-4 on July 22 to a 150 × 184.5 mi., 82.29° orbit. The reentry module was recovered on August 5. *Resurs F-19*, a short-duration type, was launched by an SL-4 on August 24 to a 139 × 145 mi., 82.6° orbit and recovered on September 10.

1993 May 22   At 06:42 UT a Russian SL-4 launch vehicle carried the unmanned cargo tanker *Progress M-18* into orbit from the Baikonur cosmodrome. At 08:25 UT on May 24 it docked to the forward axial port on the *Mir* space station, in an orbit of 242 × 243 mi., with *Progress M-17* still docked at the aft *Kvant 1* port. For the first time, two Progress cargo tankers were docked to the *Mir* station. *Mir* cosmonauts Gennady Manakov and Alexandr Poleshchuk unstowed equipment brought to the orbiting complex by the freighter.

Progress M-18 undocked from *Mir* on July 3 and de-orbited next day. During descent it jettisoned a recoverable capsule.

1993 May 25   The *Magellan* spacecraft began the first planetary aerobraking maneuver when it was lowered into the top of the Venus atmosphere to begin the process of circularizing its elliptical orbit. A 10-min. thruster burn starting at 17:31 UT put the spacecraft in a 5,262 × 107 mi. orbit of Venus in which it would graze the atmosphere, raising periapsis and lowering apoapsis. On August 5, after two days of orbit trim maneuvers, *Magellan* was in an orbit of 336 × 122 mi. For an expenditure of 70 lb. of propellant the atmosphere of Venus had slowed *Magellan* by 4,000 ft./sec. over more than 900 aerobraking passes, and as the spacecraft spiralled closer in, the orbital period had been cut from 3 hr. 14 min. to about 1 hr. 30 min. This greatly enhanced the ability of the spacecraft to conduct gravity anomaly tests over the polar regions.

1993 May 27   Hurting its image as a reliable satellite launcher, a Russian Proton (SL-12) failed in its attempt to carry a Gorizont communications satellite to geostationary orbit. Launched from Baikonur cosmodrome in Kazakhstan, the Proton's second stage malfunctioned and the upper elements fell back into the atmosphere and were destroyed. This was the 15th Proton failure in 180 flight attempts, a success rate of only 91.7%, compared to 96.6% for Ariane 4, its European competitor.

1993 May   Les Aspin, U.S. secretary of defense, announced that the Strategic Defense Initiative organization was to be renamed the Ballistic Missile Defense Organization (BMDO) to refocus efforts toward defense against battlefield and theater threats. Gone now was the commitment to provide defense against intercontinental strategic weapons with ground- and space-based weapons systems. Work on Brilliant Pebbles was to be scaled down in favor of developing ground-based missile defenses. The BMDO was to report to a lower level of authority at the Department of Defense. So ended a $35 billion research program that had begun March 23, 1983, the most expensive ever mounted by the Pentagon.

1993 June 7   NASA's Station redesign team presented its final report to the Advisory Committee set up by Vice President Albert Gore. On June 10 the committee submitted its report to President Bill Clinton describing three options. Option A resembled a scaled-down *Freedom* design with solar arrays stretching from a truss structure providing 57 kW, and 13 buildup flights; Option B encompassed three alternatives, all a close shadow of the *Freedom* design, all with 20–25 assembly flights (Options A and B incorporated two Soyuz TM as assured crew return vehicles); Option C was a radical new design based around a 64 ft. long core cylinder with seven decks launched on a Shuttle-Derived Vehicle, 56–61 kW of power and 12–18 assembly flights, depending on orbit inclination. Estimated development costs (10-year operations costs) for Options A, B and C, respectively, were $17 billion ($18.7 billion), $19.3 billion ($19 billion), and $15.2 billion ($14.4 billion).

1993 June 17   The Advisory Committee on the Redesign of the Space Station recommended Option A from a menu of several alternative configurations. On June 24 President Bill Clinton directed NASA to develop a Program Implementation Plan by September 7 which would define the way space station Alpha, as Option A was now called, would be built. On July 21 the international partners (Canada, Europe and Japan)

were invited on to the Space Station Transition Team, which had been formed to redirect the effort.

1993 June 18   *Mir* cosmonauts Gennady Manakov and Aleksandr Poleshchuk began a space walk from the *Kvant 2* astrophysics module at 17:25 UT to replace handles on the telescopic boom that broke off during the previous space walk on April 19. This arm was used to move the cosmonauts to the work station for the second part of the work carried out two months before. A second electric drive unit was installed on the *Kvant* module; and the crew completed their 4 hr. 33 min. walk as planned.

Orbital Sciences Corp. test launched their first Prospector sounding rocket from Cape Canaveral, Fla., in a mission called Joust 1. Comprising a single Thiokol Castor 4 solid propellant motor, the rocket veered off course 15 sec. after launch and was destroyed by a range safety officer 10 sec. later. Carrying a payload of 10 materials science and biotechnology experiments on a flight that should have produced 13 min. of weightlessness during ballistic free-fall, the rocket and its payload sank in the Atlantic Ocean. Orbital Sciences Corp. wanted to offer Prospector for commercial launches.

1993 June 21   After a 24-hr. waveoff at T − 5 min. due to bad weather on June 20, Shuttle mission STS-57 began with the launch of *Endeavour* at 13:07:22 UT from LC-39B, Kennedy Space Center, Fla., carrying astronauts Ronald Grabe (CDR), Brian Duffy (PLT), David Low (MS and Payload Commander), Nancy Sherlock (MS), Janice Voss (MS) and Peter Wisoff (MS). The payload comprised the Spacehab pressure module carrying two experiments in materials and life sciences, and waste water recycling equipment for space stations. The 9,424-lb. *Eureca* platform deployed by STS-46 on August 2, 1992, was captured and stowed in the cargo bay at 13:53 UT on June 24. A 5 hr. 50 min. contingency EVA (extravehicular activity) on June 25 was necessary to manually fold stuck antennas on *Eureca* which would have fouled the payload bay doors and prevented them from closing for reentry. The mission ended at the Kennedy Space Center, 9 days 23 hr. 44 min. 54 sec. after liftoff.

1993 June 23–24   At a council meeting of the European Space Agency, the Hermes space plane studies were redirected toward a crew and cargo transportation system based on nonwinged reentry vehicles. The vehicle was still to be defined around the Ariane 5 launcher but would no longer have the characteristics of a Shuttle-type vehicle. The three-year study on redirection of the Hermes program was scheduled to end in December 1995.

1993 June 25   An Ariane 42P launch vehicle (V57) carried the 6,588-lb. *Galaxy IV* satellite to a geosynchronous transfer orbit with an apogee of 12,962 × 22,285 mi. using the "perigee velocity augmentation" technique first exploited for the launch of *Galaxy VII* on October 28, 1992. Through a series of propulsion maneuvers, *Galaxy IV* was raised to geostationary orbit over 99° W. Based on a Hughes HS 601 bus, the satellite was virtually identical to *Galaxy VII* and carried the same communications payload. It weighed approximately 3,730 lb. on station.

A four-stage Scout launcher carried the 192-lb. *Radcal* radar calibration satellite from Vandenberg AFB, Calif., to a 472 × 552 mi., 89.6° orbit. Launched for the U.S. Air Force, *Radcal* would broadcast for a year, transmitting a calibration signal for Navstar global positioning system satellites and global radar stations. *Radcal* also carried a special device to evaluate the way solar cells could charge batteries on satellites more efficiently.

1993 June 26 NASA's Johnson Space Center hosted a tether experiment as a piggyback payload on a Delta 221 which carried the Navstar 21 global positioning system satellite from Cape Canaveral, Fla., to an elliptical transfer orbit of 116 × 12,666 mi., at 34.7° inclination. Designated Plasma Motor Generator (PMG), the experiment was a follow-on to the *SEDS* (small expendable deployer system) launch of March 30 and involved a small package deployed to a distance of 1,640 ft. on an electrodynamic tether. The experiment was designed to test the possibility of generating an electric current through the tether. A current of 0.3 amp was generated by the line as it passed through the Earth's magnetic field, acting like an armature in an electric motor.

1993 June Arianespace broke the U.S. monopoly on launching Indonesia's Palapa communication satellites when it successfully negotiated a contract for the launch of *Palapa C* in mid-1995 on an Ariane 4. The contract came two months after Satelindi (PT Satelit Palapa Indonesia), the former Perumtel, placed an order with Hughes Communications for two HS 601 series satellites, each of which would have a launch mass of 6,590 lb. and weigh 2,800 lb. on orbit after depletion of perigee kick motor propellant. Indonesia's PTT Telecom relinquished control of Palapa in January to Satelindi, which is 60% privately owned.

1993 July 1 At 14:33 UT the *Soyuz TM-17* spacecraft was launched by an SL-6 from the Baikonur cosmodrome carrying Russian cosmonauts Vasily Tsibliyev and Aleksandr Serebrov and French cosmonaut Jean-Pierre Haignere to the *Mir* space station. *TM-17* docked to *Mir*'s forward port at 16:24 UT on July 3, and the three cosmonauts joined Gennady Manakov and Alexandr Poleshchuk in the orbiting facility where they would remain for three weeks before the resident crew returned to Earth.

1993 July 7 The European Space Agency and NPO Energia signed an agreement covering visits by two ESA astronauts to the Russian *Mir* station in 1994 and 1995. NPO Energia was the "privatized" design bureau responsible under the former communist regime of developing the heavy-lift launcher and other Russian space projects. It was responsible for marketing key Russian space ventures. Under the agreement, an ESA astronaut would be carried to *Mir 1* by a Soyuz spacecraft and spend 30 days aboard the orbiting laboratory. The second flight, in 1995, was planned to last 135 days. Four ESA astronauts, Ulf Merbold, Pedro Duque, Christer Fugelsang and Thomas Reiter, began training for the mission at Star City, near Moscow, in August.

1993 July 15 NASA and the German space agency (DRA) signed an agreement covering the use of the German shuttle pallet satellite *(SPAS)* on four science missions, including the STS-51 flight. The SPAS platform was to carry dedicated experiments in atmospheric and climate physics, including infrared spectrometers and telescopes comanifested with NASA's *ATLAS-3* flight planned for launch in September 1994 on STS-66 with the Shuttle *Endeavour* Re-flights with the STS-51 and STS-66 *SPAS* missions were scheduled for 1995 and 1996, respectively. *SPAS* flights would prepare NASA for the big Mission to Planet Earth (MTPE) programs later in the decade.

1993 July 17 A technical hitch halted the countdown for Shuttle mission STS-51 and orbiter *Discovery* at the T − 20 min. point. A further launch attempt on July 24 was similarly halted because of a technical failure, at T − 19 sec. The launch was postponed to August 4, but managers decided not to fly until the Perseid meteor shower had passed Earth. These annual visitors are the trailing remains of a fragmented comet, Swift-Tuttle, traveling at 133,000 MPH in a fixed orbit around the sun. Scientists noted that when Earth passed through the trail in 1987, debris caused 150,000 shooting stars an hour. In space, the Hubble Space Telescope was turned end-on to the shower to minimize impact damage. At its third launch attempt on August 12 the countdown for STS-51 was again halted at T − 3 sec. because of faulty engine sensors, only the fourth time in Shuttle launch history that the engines have shut down after ignition.

1993 July 19 An Atlas 2 launched the third second-generation DSCS-IIIB military communications satellite from Cape Canaveral, Fla., to a geosynchronous transfer orbit. From there it was lifted to geostationary orbit. The fourth DSCS-IIIB series satellite was carried by Atlas 2 launched from Cape Canaveral on November 28, 1993.

1993 July 22 Russian cosmonauts Gennady Manakov and Alexandr Poleshchuk were returned to Earth in the *Soyuz TM-16* spacecraft along with French cosmonaut Jean-Pierre Haignere. Cosmonauts Vasily Tsibliyev and Aleksandr Serebrov remained aboard the *Mir* space station for a mission expected to last four months and include three separate space walks. As it turned out, their return to Earth was postponed from November 1993 to January 1994 to save money. *Soyuz TM-16* undocked from the *Kristall* module at the *Mir* complex and landed at 06:42 UT. Manakov and Poleshchuk had been in space for just under six months: 179 days 0 hr. 44 min.

An Ariane 44L with a liftoff weight of 1,059,566 lb. carried the 4,873-lb. Spanish communication satellite *Hispasat 1B* and the 4,258-lb. Indian multimission satellite *Insat 2B*, from Kourou, French Guiana, to geosynchronous transfer orbit on mission V58. Identical to its respective numerical predecessor, *Hispasat 1B* had an on-orbit mass of 2,933 lb. and was positioned at 30° W. With a mass on station of 2,617 lb., *Insat 2B* was placed over 83° E.

1993 July 26 The first photograph of Mars from NASA's *Mars Observer*, and the first Mars picture since the Viking missions of the 1970s, was taken from a distance of 3.6 million mi. The photograph was taken using the spacecraft's high-resolution, narrow-angle camera that provided a resolution of about 13.4 mi. per picture element (pixel). From Mars orbit the spacecraft would obtain low-resolution, wide-angle images showing objects down to 4.7 mi./pixel, or medium-resolution images with the same camera at 787 ft./pixel. A separate camera would provide very-high-resolution images at 4.6 ft./pixel over areas of special interest.

1993 August 2 A Titan 4 launcher carrying an advanced *KH-11* derivative imaging satellite failed to reach orbit after launch from Vandenberg AFB, Calif., at 19:59 UT. An explosion in one of the strap-on solid propellant boosters caused destruction of the vehicle 1 min. 41 sec. into the flight, 15 sec. before the core first stage was due to ignite and 26 sec. before the solids were due to burn out. The vehicle was 70 mi. down range at a height of 105 mi. John Pike of the Federation of American Scientists said the launch cost $200 million and the satellite cost $1.7 billion to build.

1993 August 9 The 3,775-lb. *NOAA-13* meteorological satellite placed in orbit by an Atlas E launch vehicle failed to operate as planned when a failure in the power distribution circuit prevented energy from the solar panels getting into the main electrical bus. As a contingency measure, to provide cover, *NOAA-9* was reactivated from a dormant posture. Launched from Vandenberg AFB, Calif., *NOAA-13* was placed in a 528 × 536 mi., 98.3° orbit to replace *NOAA-11*, launched on September 24, 1988. *NOAA-9* had been launched by Atlas E on December 12, 1984.

1993 August 10 At 22:23 UT a Russian SL-4 launcher carried the 13,900-lb. cargo tanker *Progress M-19* from Baikonur cosmodrome to the *Mir* complex. *Progress* docked to the rear *Kvant 1* port at midnight August 12–13. After Vasily Tsibliyev and Aleksandr Serebrov recovered the logistical supplies brought up by the unmanned vehicle, *Progress M-19* undocked from the *Kvant 1* port on October 12 and de-orbited, releasing a separate recovery capsule which returned to Earth early the next day.

1993 August 12 A computer-commanded abort in the countdown for launch on STS-51 at the Kennedy Space Center, Fla., shut down the three main engines on *Discovery* after ignition at the T − 3 sec. mark. The problem was found to be a faulty fuel-flow sensor. This was the fourth launch delay for mission STS-51 and forced managers to postpone to 1994 the planned November launch of *Discovery* on STS-60, scheduled to carry the first Russian cosmonaut in an American manned space vehicle. On August 20 Shuttle Program Operations Director Brewster Shaw ordered a thorough review of safety and launch procedures. Following a similar pad abort on March 22, this was the second time this year that the launch had been scrubbed after main engine ignition.

1993 August 17 NASA announced sweeping changes to the contractual base and management structure for the international space station, now known as the Alpha station. Instead of four separate work packages allocated to four separate industrial teams, the Johnson Space Center would be in charge of development and operations in a single contract to the Boeing Co. Existing work package contractors Grumman, McDonnell Douglas and Rocketdyne agreed to become subcontractors to Boeing.

1993 August 18 The McDonnell Douglas DC-X single-stage-to-orbit technology demonstrator successfully made its first flight from White Sands Missile Range, N.Mex. It lifted off at 22:43 UT (16:43 MDT), rose to a height of 150 ft., moved sideways 350 ft. under full control and descended for a landing on a prepared surface at an elapsed time of 60 sec. Powered by four throttleable Pratt and Whitney RL10A-5 hydrogen-oxygen engines, the DC-X had a maximum thrust of 54,400 lb. against a maximum liftoff weight of 41,630 lb.

1993 August 22 Communication with the *Mars Observer* spacecraft was lost at approximately 01:00 UT (18:00 hr. PDT). *Mars Observer* had an on-board command sequence that would put the spacecraft into orbit about Mars as planned on August 24, beginning with a Mars orbit insertion (MOI) burn at 20:40 UT, 1,072 mi. from the planet. Two 110-lb. thrusters would fire for 29 min. to place the spacecraft in an elliptical orbit with a periapsis of 308 mi. A complex and evolving sequence of maneuvers would modify the orbit for various science tasks. Despite repeated attempts to contact the *Mars Observer* nothing more was heard from the spacecraft,

and its fate is unknown. A review board subsequently decided a propellant leak causing an explosion was the most likely cause.

1993 August 28 The *Galileo* spacecraft made a close encounter with the asteroid Ida at 16:52 UT when it passed to within 1,500 mi. of the potato-shaped rock, more than 319 million mi. from Earth. *Galileo* was moving at a relative speed of approximately 28,000 MPH as it imaged the asteroid, which has a length of about 35 mi. and a mean diameter of about 18 mi. and rotates once every 4 hr. 35 min. Following a preliminary examination of images received on Earth, Ann Harch of the *Galileo* imaging team discovered that Ida has a small moon, 1 mi. in diameter, which appears to orbit the asteroid in a period of several days. This is the first known sighting of a moon orbiting an asteroid.

1993 August 31 An Italian satellite rode piggyback on an SL-14 Tsyklon launched from Plesetsk primarily assigned to carrying the 21st Meteor 2 communication satellite into orbit. The 4,400-lb. *Meteor 2–21* went into an orbit of 582.9 × 602 mi., at 82.5° inclination. The 66-lb. Italian satellite, named *TEMISAT* (Telesapzio Micro Satelite) was released to almost exactly the same orbit, but with perigee at 582.2 mi. *TEMISAT* was an Italian project utilizing a satellite bus provided by the Kayser-Threde company in Germany. *TEMISAT* was launched to demonstrate a commercial data relay service for business users, an effective marketing tool for new ventures.

1993 August Lockheed Missiles and Space Co. completed negotiations with Motorola for a system of 125 Earth-orbiting satellites as part of a global mobile telephone network called Iridium. The operational constellation would require only 66 satellites, but the name had been chosen when the company thought it would need 77, the atomic number of Iridium. Eleven 1,500-lb. satellites would be spaced across six orbital planes, each plane displaced 60° from the next. Each satellite would have 48 spot beams and use Ka-band for intersatellite links and ground commands. Voice and data would be handled through hand sets, each costing around $3,000 and each call to anywhere in the world costing $3/min. The system was expected to become operational in 1998 and to capture 1–2% of the 100 million worldwide user market by the year 2000. Motorola selected America's Delta 2, Russia's Proton and China's Long March launchers to put the 66 satellites in their 483-mi. orbits.

A high-level Russian working group submitted to NASA a report on how $2.5 billion worth of former Soviet space hardware could be sold to the United States for use on the space station, saving $7 billion on new equipment. The 27 members of the working group represented Russian aerospace industries including NPO Energia and the Krunichev State Space Scientific Production Center. Of the $7 billion savings, $3 billion would come from fewer Shuttle flights and the use of Russian expendable launchers. In the United States the Clinton administration preferred to invite full Russian partnership rather than proceed on a customer/supplier basis.

1993 September 1 NASA Administrator Daniel Goldin announced the members of a new team that was to examine the possibilities for a return to Mars in the wake of the *Mars Observer* failure. The follow-on mission study team was to examine low-cost options involving possible flights in 1994 or 1996. Over the next several months the team would argue the relative merits of a repeat *Mars Observer* mission or a dual mission using low-cost spacecraft splitting experiments between the two vehicles.

1993 September 7 The Alpha Station Program Implementation Plan was completed, defining the way the international space station would be built. NASA would assemble the station at an orbital inclination of 28.8° over 18 flights between 1998 and 2003. The implementation plan acknowledged that considerable political progress had been made in the previous two months to involve the Russians in Alpha station and concluded that this was advisable, although in that event the station would have to be assembled at 51.6° to permit access by Russian launchers, bringing a payload penalty to the NASA Shuttle. Cost and schedule data was submitted September 20.

1993 September 11 The McDonnell Douglas DC-X test rocket made its second flight from White Sands Missile Range, N.Mex. The 33,000-lb. vehicle lifted off at 18:12 UT, ascended to a height of 300 ft. and hovered, translated laterally for a distance of 350 ft. and descended to a touchdown, extending landing legs shortly before landing, at a mission time of 1 min. 5.8 sec. Within five minutes ground handling personnel were attending to the vehicle and pronounced it in fine order. A third flight was conducted on September 30, when the DC-X reached a height of 1,200 ft., an ascent rate of 120 ft./sec. and a descent rate of 75 ft./sec., twice the previous levels.

1993 September 12 The STS-51 mission began at 11:45:00 UT from Launch Complex 39B with the flight of Shuttle orbiter *Discovery* to an orbit of 169.5 × 192 mi., at 28.5° inclination. Carrying astronauts Capt. Frank Culbertson, USN (CDR), William Readdy (PLT), James H. Newman (MS-1), Cmdr. Daniel Bursch, USN (MS-2) and Maj. Carl Walz, USAF (MS-3), *Discovery* carried the 26,756-lb. Advanced Communications Technology Satellite–Transfer Orbit Stage which was to have been deployed at 7 hr. 58 min. mission time. At 6 hr. 35 min. the *ACTS*-TOS assembly was elevated 45° from the floor of the payload bay by a special cradle. The assembly was finally released at 9 hr. 28 min., but incorrect wiring nearly brought disaster when both primary and backup separation charges exploded, showering the cargo bay with debris. At 10 hr. 13 min. TOS fired for 1 min. 50 sec., producing 59,000-lb. thrust to place the 6,108-lb. *ACTS* to geosynchronous transfer orbit. A solid propellant perigee motor fired at apogee to place the satellite, now weighing 3,250 lb., on station at 100° W. The 7,070-lb. Orbiting and Retrievable Far and Extreme Ultraviolet Spectrometer-Shuttle Pallet Satellite *(ORFEUS-SPAS),* a free-flying platform, was released by the remote manipulator at 27 hr. 21 min. to conduct astronomical observations at very short wavelengths. Part of the *ASTRO-SPAS* (astronomy-shuttle pallet satellite) program, the platform was provided by the German space agency and would remain separated from *Discovery* by 35 mi. for more than five days. Astronauts Newman and Walz began a 7 hr. 5 min. 28 sec. EVA (extravehicular activity) at 08:32 UT on September 16 to evaluate hand tools and equipment designed for the first Hubble servicing mission, STS-61. After rendezvous with *ORFEUS-SPAS*, the platform was retrieved at 11:50 UT on September 18. After a 24-hr. delay due to bad weather, *Discovery* made the first night landing at Kennedy Space Center in the early hours of September 22 at a mission time of 9 days 20 hr. 11 min. 11 sec.

1993 September 18 The first flight model of the Ariane core stage was rolled out of the launcher integration building at Kourou, French Guiana. Flanked by two P230 solid propellant booster mock-ups, the stage was being prepared for a firing test with its HM60 Vulcain engine, scheduled for January 1994. The first of two Ariane 5 flight tests was scheduled for late 1995 from ELA-3, a specially built launch complex at Kourou. The stage would fire for 10 min., and on a geosynchronous transfer orbit delivery mission a special spin-up maneuver would ensure a ballistic flight path for the H155 with splashdown in the Atlantic Ocean 124 mi. from the launch site.

1993 September 20 India's Polar Satellite Launch Vehicle, a development of the augmented satellite launch vehicle, failed on its first attempt to place a satellite in orbit. Launched from the Sriharikota launch site, the 144 ft. 4 in. long PSLV was carrying the 1,865-lb. remote-sensing satellite *IRIS 1E*, the refurbished *IRIS 1A* engineering model. The core, six strapons and third stages use solid propellant, while the second and fourth stages use liquid propellant. The problem was traced to incorrect separation of second and third stages causing the third stage to be 48 mi. too low at ignition, making it impossible to reach orbit with the fourth stage. PSLV is capable of lifting a 2,200-lb. satellite to a 560 mi., sunsynchronous orbit.

1993 September 26 An Ariane 40 launch vehicle lifted off from Kourou, French Guiana, carrying the *SPOT-3* Earth resource satellite and six supplementary satellites for a total payload weight of 10,297 lb. The 4,205-lb. *SPOT-3* carried two high-resolution vidicon imaging instruments operating in three spectral bands with a resolution of 12 ft. and one panchromatic band with a resolution of 6 meters. Other satellites were: a 198-lb. spherical geodetic laser reflector called *Stella*; a 22-lb. Italian amateur radio satellite called *Itamsat*; a 23-lb. satellite called *Eyesat-A* evaluating a location identification system for mobile equipment; a 105-lb. Portuguese satellite for technology research called *Posat-1*; a 105-lb. British satellite called *Healthsat-1* for relaying medical information to hospitals in Africa; and the 105-lb. Korean telecommunications technology research satellite from Korea, *Kitsat-B*.

1993 October 4 Outward bound for Jupiter, the *Galileo* spacecraft began a four-day series of thruster firings to execute a trajectory correction maneuver that would put it precisely on course for the planet. In a sequence of 10,000 firings the spacecraft would change velocity by a mere 127 ft./sec., increasing speed and changing course slightly. *Galileo* was now 297 million mi. from the sun and 383 million mi. from Earth (because the Earth was on the opposite side of the sun with respect to *Galileo*) with 412 million mi. to go before reaching Jupiter on December 7, 1995.

1993 October 5 A Titan 2G launched from Vandenberg AFB, Calif., carried the 6,100-lb. *Landsat 6* remote sensing satellite to an initial, elliptical orbit with an apogee of 450 mi. From there, 13 min. 39 sec. after liftoff, a Thiokol Star 37 solid propellant kick motor was to have fired for 1 min. and circularized the orbit at 460 mi. From the moment the Titan 2 shut down, nothing more was heard from *Landsat 6*. Built by General Electric, now Martin Marietta, *Landsat 6* was to have been operated by *Eosat* and provided a near continuous flow of data from an enhanced thematic mapper (ETM) with a panchromatic resolution of 49 ft. *Eosat* expected to operate

Landsat 7, scheduled for launch in 1998, carrying a multispectral imaging instrument with a resolution of 16 ft.

1993 October 11 A Russian SL-4 carried the *Progress M-20* cargo tanker from Baikonur cosmodrome to the *Mir* complex in an orbit of 241 × 246.5 mi., at 51.6° inclination and from there to a docking at the rear of *Kvant 1*. While it was docked, Tsibliyev and Alexandr Serebrov retrieved supplies brought to the space station. *Progress M-20* also carried a crystallization experiment containing 12 specimens from the Boeing Company. *Progress M-20* undocked from *Kvant 1* on November 21 and reentered that day while its capsule carrying the U.S. experiment was recovered. The crew change aboard *Mir* that had been planned to commence with the launch of *Soyuz TM-18* on November 17 was postponed to January 8, 1994, to save money. That flight carried cosmonauts Valery Polyakov, Victor Afanasyev and Yuri Usachyov to *Mir* with Tsibliyev and Serebrov returning in *Progress TM-17* on January 14. Their mission had lasted 197 days. Polyakov returned to Earth in the early hours of March 22, 1995, aboard *Soyuz TM-21*. He had been in space for 437 days 18 hr. 1 min.

1993 October 12–14 At the Inmarsat International Conference on Mobile Satellite Communications, Paris, France, the international maritime satellite organization declared its interest in setting up a global network of mobile telephone subscribers. After Iridium and Motorola's Globalstar system, Inmarsat was the third organization to plan mobile communications nets. At the same meeting, the U.S. government indicated its willingness to ease regulations to allow Inmarsat to compete in the United States, denied so far because it was set up to handle only international maritime and air traffic. Analysts projected that by 2001, annual net revenues from global mobile systems would amount to $28 billion on 14 million handsets. Supporting launch contracts would net $5 billion.

1993 October 14 The launch of Shuttle STS-58 on the Spacelab Life Sciences-2 flight was aborted when a combination of weather and technical problems caused flight controllers to abandon the attempt. The flight had been scheduled for launch from LC-39B at 14:53 UT, but the launch was scrubbed just 31 sec. before resuming the countdown in a planned 10 min. hold at T − 9 min. A second attempt was made the following day, but similar problems plagued preparations and, adhering to standard procedure for technical reasons, the flight hardware had to be recycled for another attempt three days later.

1993 October 15 The Jet Propulsion Laboratory, Calif., issued a request for bids on a *Mars Observer* recovery mission whereby two small spacecraft would separately carry instruments duplicating those lost on the failed *Mars Observer* mission. The low-cost effort envisaged the two launches on Delta rockets in 1996 and one in 1998. NASA had already received approval to fly the first Pathfinder/Mesurs minirover mission in 1996, and the Russians still hoped to fly their Mars '94 roving vehicle flight. Through the International Mars Exploration Working Group, the Russians hoped to build on this and the Mars balloon mission, a Franco-Russian venture in 1998.

1993 October 18 At 15:53:10 UT Shuttle mission STS-58 began with the launch of *Columbia* from LC-39B at the Kennedy Space Center, Fla. With Col. John Blaha, USAF (CDR), Lt. Col. Richard Searfoss, USAF (PLT), Rhea Sed-

don, M.D. (MS-1), Lt. Col. William McArthur, U.S. Army (MS-2), David A. Wolff, MD (MS-3), Shannon Lucid (MS-4) and Martin Fettman (PS), *Columbia* carried a 3,588-lb. extended duration orbiter (EDO) kit of supplementary consumables for a planned 14-day flight during which extensive medical experiments would be conducted. The primary payload comprised a pressurized module for the Spacelab Life Sciences-2 set of 14 experiments concentrating on cardiovascular, regulatory, neurovestibular and musculoskeletal body systems. SLS-2 had a total weight of 21,840 lb. For the first time on a spaceflight, astronauts collected tissue and drew blood from themselves and rats carried on board. Also for the first time, a portable scientific work station was carried, enabling the pilot astronauts to maintain proficiency in orbiter handling and landing skills. Using software from the ground simulator, Blaha and Searfoss practiced landings as they would have done during training. The mission ended at Edwards AFB, Calif., at an elapsed time of 14 days 12 min. 34 sec.

The Council of the European Space Agency was presented with a significantly readjusted European space agenda for the next several years when proposals were given for a change in direction. The *Columbus* space station module was to be reduced in size to a loaded weight of 20,000 lb., and funding for development was cut from $3.1 billion to $2.2 billion. A new manned capsule was proposed for Ariane 5, replacing the canceled Hermes space plane. Called the crew transfer vehicle (CTV), the capsule would carry astronauts and cargo from Earth to *Columbus* and return for ballistic descent and recovery. ESA was also keen to develop a space tug, called the automated transfer vehicle (ATV).

1993 October 20 NASA announced that the Office of Space Flight, headed by Associate Administrator Jeremiah W. Pearson III, would assume responsibility for the Space Station Program. Integration of the Shuttle program, managed by the Office of Space Flight, and the space station was becoming increasingly important as the latter depended so heavily upon the former. Space station *Freedom* Director Richard Kohrs had retired on September 3.

1993 October 22 The first in a new generation of satellites for the International Telecommunications Satellite Organization, *Intelsat 701* was launched by Ariane 44LP from Kourou, French Guiana, at 05:46 UT. The launch vehicle placed the 8,030-lb. satellite in a geosynchronous transfer path of 124 × 22,332 mi., at 7° inclination. The on-board apogee motor put *Intelsat 701* in geostationary orbit at 121° E from which it could be checked by Space Systems/Loral, the manufacturer. The main body of the three-axis-stabilized satellite was 7 ft. 3 in. × 8 ft. × 13 ft. 9 in. high. With solar "wings" deployed, producing 4,000 W of power, it had a span of 71 ft. 6 in. and a weight of 4,850 lb. in geostationary orbit. *Intelsat 701* carried 26 × 30-W C-band and 10 × 50-W Ku-band transponders for service over the Pacific Ocean. Later, it was to be moved to its operational position at 174° E where it could serve the Asia-Pacific region. The current Intelsat manifest was for the launch of 13 more satellites by 1996.

Funds for the Strategic Defense Initiative antiballistic missile research program's DC-X single-stage-to-orbit technology demonstrator ran out, halting the first in a series of three test flights aimed at showing quick turnaround from one launch to another. The first of those flights had been scheduled for the following day, but the SDI office ran out of the $67 million permitted for the program. On January 31, 1994, NASA provided an extra $1 million to prepare the vehicle for a possible flight and preserve the McDonnell Douglas team intact for two months until a launcher study by the White House Office of Science and Technology Policy had reviewed DC-X.

1993 October 27   Lockheed Missiles and Space Co. delivered the first flight satellite in the Milstar 1 series to Cape Canaveral, Fla., for a Titan 4 flight scheduled for early 1994. Initial plans to buy seven Milstar 1 and Milstar 2 satellites had been changed by the U.S. Air Force in response to a "bottom-up" review of all U.S. military systems called for by the Clinton administration. The Air Force now envisaged six satellites, four of which would form an operational constellation in geostationary orbit. The 10,000-lb. Milstar 2 would carry a 4,000-lb. communication payload. The Air Force was formulating plans for a Milstar 3 which it would configure for post-Cold War requirements after the year 2000. The first Milstar 1 series satellite was successfully launched by Titan 4 on February 8 (February 7 local time).

1993 October 29   The first flight version of Japan's H-II cryogenic satellite launcher was delivered to the Tanegashima Space Center. With a height of 164 ft. and a gross liftoff weight of 582,100 lb., the H-II had the capacity to send 8,000 lb. to geosynchronous transfer orbit, or 2,200 lb. to the planets. The first flight of the H-II, the first expendable launcher to feature cryogenic liquid oxygen–liquid hydrogen first and second stages, occurred February 4, 1994. Launched from the new Yoshinobu launch complex, the most modern in the world, the H-II generated a liftoff thrust of 891,000 lb. from the two solid rocket boosters and the first stage LE-7 engine. The stage burnt out at 6 min. 1 sec. and was followed by stage separation 8 sec. later and a 7 min. 1 sec. second-stage burn beginning at 6 min. 15 sec. The first payload, the 1,910-lb. Orbital Reentry Experiment (OREX) was separated at 14 min. 10 sec. and completed one orbit of the Earth at 282 mi. before reentering the atmosphere for recovery. OREX took the form of a flat cone with a rounded nose, a diameter of 11 ft. 1 in. and a height of 4 ft. 9 in. It featured a decelerating aeroshell for testing thermal protection tiles such as might be applied to a Japanese space plane. At 25 min. 5 sec. the second stage reignited for 2 min. 49 sec., placing a 5,280-lb. vehicle evaluation payload (VEP) into a 279 × 22,320 mi. simulated geosynchronous transfer path. A second H-II flight took place on August 28, 1994, with the launch of ETS-6, but the satellite failed to put itself in the correct final orbit. The third H-II launch on March 18, 1995, carried the GMS-5 weather satellite and a test platform. For this launch the H-II was equipped with four solid propellant strap-on rockets.

1993 October   A group of U.S. bipartisan congressional staff members presented a report urging development of a new medium-sized launch vehicle. They cited the apparent inefficiency of U.S. launchers, all of which had their pedigree in the 1950s. Whereas Europe's Ariane 4, a product of 1970s technology, required a launch crew of 100 and only 10 days on the pad prior to flight, the U.S. Delta 2 required 300 people and 23 days, the Atlas-Centaur required 300 people and 55 days and the Titan 4 required more than 1,000 people and 100 days. The staffers recommended development of a new launcher which would cut these man-hour levels through modern technology.

1993 November 1   In an addendum to the Alpha Station Program Implementation Plan submitted September 7, NASA defined the integrated schedule for Russian participation around *Mir 2*, the core module of which was scheduled for launch in May 1997. In a sequence of 31 flights between that date and October 2001, the fully habitable station would be assembled at an orbital inclination of 51.6°. Phase I would consist of joint flights with Shuttle and *Mir 1*, comprising five U.S. visits to *Mir*, one *Mir 1* visit by an European Space Agency astronaut and one by a French cosmonaut, all between May 1995 and September 1997, and 10 Shuttle flights to *Mir* between June 1995 and September 1997. Phase II involved 10 Russian and American flights between May and December 1997 to build a new station to man-tended level. Phase III encompassed 21 Russian, American and European flights between January 1998 and October 2001, with the Japanese Experiment Module installed during October 1999 and the European *Columbus* module installed during April 2000. Phase III would mark permanent habitation.

1993 November 7   Heads of the space agencies involved in the U.S.-led *Freedom* space station program met in Montreal, Canada, and agreed to support a move to integrate *Freedom* and Russia's *Mir 2* station. However, Europe and Japan were concerned that by adding Russia as a partner and incorporating design and technology from the *Mir* program, they would lose out as primary partners devoid of the enormous backlog of manned flight experience possessed by the former Soviet Union. Government representatives of participating nations met on December 6 to ratify the intergovernmental agreement (IGA) covering space station participation.

1993 November 9   The Brazilian Ministry of Science and Technology and the China Great Wall Co. signed a contract covering the launch of two remote sensing satellites. The first was scheduled to fly in October 1996. Called *China-Brazil Earth Resource Satellites (CBERS)*, they grew out of an earlier agreement between the two countries for mutual development of Earth observation techniques. China's Academy of Space Technology was responsible for 70% of the program and for construction of the satellites.

1993 November 20   A Mexican communications satellite built by Hughes Space and Communications Co. and a European weather satellite were launched by Ariane 44LP from Kourou, French Guiana. Built by Hughes, the 6,120-lb. *Solidaridad 1* three-axis-stabilized communication satellite had a main body 9 ft. 3 in. × 7 ft. 3 in. × 9 ft. 10 in. tall when folded for launch and had a weight of 3,690 lb. in geostationary orbit. With a span of 69 ft., the solar arrays produced 3,370 W of power. It carried 18 C-band channels, 16 Ku-band channels and one L-band channel and would eventually take up operational position at 109° E from where transmitter beams would cover the Pacific Coast of South America, Central America and portions of the eastern states of the United States. The 1,550-lb. geosynchronous *Meteosat 6* weather satellite was positioned over the prime meridian at 0° long, where it would be used by the Eumetsat organization for hemispheric weather observations.

1993 November 29   In a 90-min. meeting at the White House congressional leaders agreed to support President Clinton's proposal that the United States integrate its space station with that of Russia. Early in 1994, however, cooler relations between the United States and Russia due both to spying charges and to the release by Russia's parliament of political activists that had tried to topple President Boris Yeltsin, stiffened opposition in Congress to full integration. Moreover, a new government in Canada indicated it wanted to withdraw completely from the international space station, and NASA redefined the station assembly schedule again after Congress imposed an annual spending cap of $2.1 billion. The first assembly sequence launch was now scheduled for late 1997, with full operational capability by 2002.

1993 December 2 After a one-day delay caused by bad weather, Shuttle mission STS-61 began from LC-39B at 09:27:00 UT with the launch of *Endeavour* and its crew: Col. Richard Covey, USAF (CDR), Cmdr. Kenneth Bowersox, USN (PLT), ESA astronaut Claude Nicollier, Jeffrey Hoffman, Story Musgrave and Lt. Col. Tom Akers, USAF The sole purpose of the mission was to service the Hubble Space Telescope, in a 360 × 367 mi. orbit, and correct anomalies in the main mirror. Insertion to a 35.7 × 352 mi., 28.5° orbit came with shutdown of the main engines at 8 min. 40 sec., followed by an Orbital Maneuvering System burn 34 min. later for a 345 × 357 mi. orbit, 6,800 mi. behind Hubble. The first maneuver to rendezvous with Hubble was at 5 hr. 24 min., raising perigee to 313 mi. Several more maneuvers brought *Endeavour* alongside it, and Nicollier used the manipulator arm to grapple the 24,500-lb. Hubble Space Telescope at 08:48 UT on December 4, drawing it down onto a 4,200-lb. support cradle in the payload bay 38 min. later. Starting 03:30 UT on December 5, Musgrave and Hoffman performed a 7 hr. 54 min. EVA (extravehicular activity) to change Hubble's rate-sensing gyroscopes and electronic control unit. The second EVA began at 03:29 UT on December 6, with Akers and Thornton on a 6 hr. 36 min. space walk. Because it was kinked and would not retract as planned, Akers detached one of Hubble's 700-lb. solar arrays and set it free and both astronauts replaced it with a new one. Successfully retracted, the other array was removed and stowed in the payload bay, and it too was replaced with a new one. At 03:43 UT on December 7 Musgrave and Hoffman began a 6 hr. 47 min. EVA, replacing the 613-lb. wide field/planetary camera and installing two new magnetometers. At 03:12 UT on December 8 Akers and Thornton began a 6 hr. 50 min. EVA, removing a blurred high-speed photometer and installing Costar, a 660-lb. set of corrective optics for the main mirror. At 02:27 UT on December 9 *Endeavour* conducted a 12-ft./sec. maneuver, putting the Shuttle and Hubble in a 368 × 369.5 mi. orbit, 1 hr. 3 min. before Musgrave and Hoffman began a 7 hr. 21 min. EVA. Solar array drive mechanisms were replaced and other instruments were fine-tuned before the two new solar arrays were deployed by 08:35 UT. The Hubble Space Telescope was lifted from the payload bay and restored to orbit at 10:27 UT on December 10. At touchdown at the Kennedy Space Center shortly after midnight, local time, December 13, the mission had lasted 10 days 19 hr. 58 min. 33 sec.

1993 December 6 NASA and the Russian space agency agreed to a program of up to 10 Shuttle flights to the orbiting *Mir 1* complex between 1995 and the end of the century, clearing the way for assembly of an International space station (ISS) combining elements of the U.S. Alpha station and Russia's *Mir 2*. To begin the sequence, Cosmonaut Vladimir G. Titov would be taken aboard the Shuttle for the 1995 flight of STS-63 and its rendezvous, and station-keeping, with *Mir 1*. Later in 1995 another Shuttle would dock with *Mir 1* for combined research activity and possible EVAs to replace solar array panels on the Russian station. As of this date, NASA and the Russian space agency anticipated construction of the ISS with 19 Shuttle flights and 12 Russian expendable launchers. Limited research could be conducted on the ISS as early as

1998, with full occupancy by 2002. In 1995 the number of shuttle visits to the *Mir 1* complex was cut from 10 to 7.

1993 December 11 The first of its kind, an auction of Soviet and Russian space hardware and artifacts was held at Sotheby's in New York. Bids generally exceeded expectations, an autographed speech given by Yuri Gagarin two days before his historic flight went for $110,000 versus the estimate of $20,000–30,000; an autographed draft of Gagarin's flight instructions went for $65,000 versus an estimated $7,000–10,000; and a record of Gagarin's flight signed by the cosmonaut brought $320,000 against an estimated $20,000–30,000. The first item of space mail delivered to Vladimir Shatalov aboard *Soyuz 5* fetched $110,000, and the first chess set used in space brought $30,000. Someone paid $60,000 for ownership of the *Luna 17* descent stage and Lunokhod rover on the moon, although the *Soyuz TM-10* spacecraft fetched only $1.5 million against an estimate of $5 million.

1993 December 16 The first Atlas 2AS lifted off from Cape Canaveral, Fla., at 00:38 UT with the 7,666-lb. *Telstar 401* television satellite. The new launcher carried four 97,520-lb. thrust Castor 4A solid propellant rockets attached in pairs between the lower skirt fairings of the liquid propellant booster motors. With a combined liftoff thrust of 616,000 lb. from the liquid propulsion main stage and two Castor 4As, the launcher ascended for 52 sec. before the solids burned out. Ignition of the remaining two Castor 4As came 7 sec. later, followed 5 sec. after that by the ejection of the first pair. The second pair were jettisoned at 1 min. 56 sec. into flight. The bulbous payload fairing with its 14-ft. diameter was jettisoned at 3 min. 35 sec., and the Atlas main stage separated from the Centaur at 4 min. 50 sec. A 5-min. Centaur firing established a 92 × 312 mi. parking orbit and a 14-min. coast, after which a second, 1 min. 38 sec. firing established a geosynchronous transfer path of 105 × 22,336 mi., at 26°. With an overall length of 156 ft. and a launch mass of 525,000 lb., the Atlas 2AS could put a maximum 19,000 lb. in low Earth orbit, 8,000 lb. in geosynchronous transfer orbit or send a 5,890-lb. spacecraft to the moon or the planets. Based on the RCA (now Martin Marietta Astro-Space) series 7000 bus, *Telstar 401* carried a 159.4 sq. ft. solar array producing 5 kW of power feeding 24 × 12-W C-band solid-state power amplifiers capable of operating as paired 24-W transponders, and 24 × 60-W Ku-band amplifiers pairable as 12 × 120-W transponders. The box-shaped, 8 ft. × 7 ft. × 10 ft. 4 in. satellite reached geostationary orbit at 24° W on December 24, from which it was moved later to 97° W to replace *Telstar 301*.

1993 December 18 An Ariane 44L carried the first American direct broadcast satellite, *DBS-1*, and the first satellite for Thailand, *Thaicom 1*, to a geosynchronous transfer path of 123 × 22,587 mi., at 3.97° inclination. From there *DBS-1* was positioned at 258° E, where it would be checked out before moving to its operational slot at 100.8°. Essentially a three-axis-stabilized Hughes HS 601 series satellite, *DBS-1* had a main body 7 ft. 6 in. × 7 ft. 6 in. × 8 ft. 6 in. high with two solar "wings" providing 4,000 W of power across a span of 86 ft. It weighed 6,310 lb. at launch, 3,800 lb. on station, and was to be operated by two U.S. companies: DirecTV and United States Satellite Broadcasting (USSB), a group led by Hughes Aircraft. It carried sixteen 120-W transponders for digital transmission, reconfigurable to eight 240-W transponders, that would serve cable TV and video rental users with up to 150 channels. Hughes expected to attract video rental users with a subscription service for customers with dish antennas 1 ft. 6 in. in diameter. A second satellite was to be launched in mid-1994 to 101.2° W. *Thaicom 1* was a Hughes 376 bus weighing 2,380 lb. at launch and operated by its owners, Shinawatra Satellite Public Co., from a slot at 78.5° E.

1993 December The White House asked industry and users for input by January 14, 1994, on U.S. launch vehicle development to support a report the Office of Science and Technology Policy was preparing. Martin Marietta was at this time in the process of buying General Dynamics' Space Division, thereby securing both Titan and Atlas launch vehicle lines. Concerned that Europe's Ariane program was securing well over 50% of commercial launch orders, the White House hoped to spur revitalization in the U.S. launcher market by offering innovative investment aids and cost-sharing schemes between government and industry.

GLOSSARY

ablative materials substances that char under the influence of kinetic heating generated through friction with the atmosphere, thereby preventing the buildup of excessive thermal energy on the reentry shield of a spacecraft

abort the act of terminating a launch or a spaceflight

accelerometer a device for measuring acceleration

actuator a device that produces mechanical action or motion, usually in response to a signal delivered from a manually operated or electronic sensor

aerobraking the practice of using the Earth's atmosphere as a brake to slow the speed of a space vehicle, ideally by deploying a convex shield to dissipate the heat generated through friction

aerospike engine an engine in which a ring-shaped combustion chamber discharges gases against the outer surface of a truncated cone and having the appearance of a short conventional exhaust nozzle with a plug in the center; also called plug-nozzle engine

alpha-particle scatterometer the name given to an instrument on the U.S unmanned Surveyor spacecraft designed to bombard the top few microns of the lunar surface with alpha particles from a radioactive source of curium 242. By measuring the changes to reflected alpha particles, scientists could conduct a crude chemical analysis of the lunar soil.

ap-, apo- the prefix, from the Greek for *away from*, *off*, indicating the high point of an elliptical orbit around a primary mass; most commonly applied to artificial satellites in Earth orbit (apogee), lunar orbit (apolune or aposelene) or solar orbit (aphelion)

apogee boost motor a rocket motor used to transfer a satellite from a geosynchronous transfer orbit to a geosynchronous or geostationary orbit

artificial gravity the application of inverse centripetal force to large vehicles in space to simulate the effect of gravity and alleviate some of the biophysical disadvantages of long space journeys; usually accomplished by rotating the vehicle around its center of mass at such a speed that acceleration at the extremities reaches 1 g

Atlantic Missile Range a firing range out across the Atlantic Ocean where rockets and missiles launched from Cape Canaveral, Fla., were tested in the 1950s and 1960s

attitude-control thrusters small rocket motors or similar reaction devices that control the attitude of a spacecraft or satellite in pitch, roll and yaw

attitude stabilization the stabilization of a satellite or spacecraft around all three axes (pitch, roll and yaw), usually by the use of attitude-control thrusters

bearing and power transfer assembly (BAPTA) the rotating interface between spinning and de-spun sections of a satellite that allows electrical power to flow across from one section to the other

bio-instrumentation instrumentation attached to an animal or a human for the purpose of obtaining biophysical data

bipropellant motor a motor in which two chemicals, usually hypergolic propellants, are mixed to initiate a chemical reaction producing gases that induce an action at the exhaust nozzle

bit the smallest unit of information in a computer or communications system which can have only one of two binary values, 1 or 0

boilerplate an engineering test vehicle, devoid of internal fittings, that simulates the size and mass of a spacecraft, satellite, rocket, missile or launch vehicle

bow wave (with reference to the Earth's magnetosphere) the turbulence caused by pressure from the solar wind disturbed by the belts of radiation and trapped particles that cause a wave to flow along the magnetosphere and form a long tail on the antisolar side

breadboard components or systems for a space vehicle laid out without regard to how they will be installed so that basic tests can be conducted to check operability or compatibility among components

Brilliant Eyes a concept whereby many low-cost, high-technology sensors in space could detect the launch of potentially offensive ballistic missiles; part of the SDI program

broadcast satellite a satellite, invariably in geostationary orbit, used for transmitting television or radio programs to consumers either directly, via a rooftop dish antenna, or to large antennas operated by a cable company and then to the consumer

building block standard rocket stages arranged according to mission requirements and stacked into a variety of optional configurations

bus (electrical) the voltage rating of a spacecraft's electrical power system, which in most cases is direct current

bus (satellite) the functional part of a satellite that provides electrical power, stabilization, attitude control and thermal control and to which the payload is attached

byte a unit of computer information usually comprising six or eight bits, two or four bytes being required to make a simple computer word, or quantum of information

canard a set of small wings or foreplanes attached for directional control

checkout the process of determining the operational status of a space vehicle, rocket or missile and its readiness to perform an assigned mission

circularization (of an orbit) a general term that describes the process of making an elliptical orbit circular; in fact every orbit is elliptical, but a circular orbit is one in which the low and the high points of the orbit are almost the same

clean room a space for the assembly of satellites and space vehicles that is kept free of aerosol or particulate contamination likely to damage delicate components or systems

cold-gas jet a reaction control thruster utilizing pressurized gas instead of combusted propellants to produce small amounts of thrust; useful for maneuvering devices inside large space structures or where contamination poses a problem

communication satellite a satellite, invariably in geostationary orbit, that provides a relay for television, telephone, telegraphic or facsimile transmissions between distant points on the surface of the Earth

control-moment gyroscopes spinning wheels with large mass whose rotational axes are normalized with the three primary axes (pitch, roll and yaw) of the space vehicle and through which attitude control is maintained by torquing the axis of rotation

cross range the distance by which an aerospace vehicle returning from space can fly to the left or right of its orbital groundtrack by using on-board propulsion or, in the case of the NASA shuttle, its wings for aerodynamic lift

cryogenics the science of very low temperatures; when applied to propellants, the liquefaction of a gas so as to reduce the volume and hence the size and weight of the tanks

Decree a U.S. military communication satellite program set up in May 1958 and aimed at providing services from a network of satellites in geostationary orbit

deep space generally considered to be space beyond geostationary orbit, or 22,300 miles above the surface of the Earth

deep-space network a set of ground stations operated by NASA for communication with interplanetary (deep) space probes

de-spin one component of a spin-stabilized space vehicle made to rotate in the opposite direction and at the same speed so that it appears to remain at rest, usually to permit instruments to point to a fixed direction in space

direct ascent the launch of a rocket or space vehicle from Earth directly to its ultimate objective, usually the moon or a planet, without first going to an interim orbit of the Earth

drop sonde the name given to entry probes proposed in June 1968 for a U.S. mission to Venus, later changed to multiprobe

drop test a test in which an object is allowed to fall under its own weight and without propulsion to evaluate its aerodynamic efficiency

Earth radiation budget the balance between the reflected energy from the sun and the energy emitted by activity within the Earth's environment; a key factor in environmental studies

Earth resource satellite a satellite that carries remote sensing instruments for obtaining data on the resources of the planet, including geophysical, soil, water, ice and crop conditions in addition to mineral and fossil fuel deposits

ecliptic the plane in which the Earth orbits the sun; also used as a reference standard for measuring the deviation (expressed in degrees of inclination) of the orbital plane of each of the other planets

electric propulsion a collective term for three applications of electric energy for propulsion, including electrothermal heating of a gaseous propellant, electrostatic ionization of a gaseous propellant, and electromagnetic generation of a gaseous plasma to produce thrust

electrophoresis the motion of charged particles under the influence of an electric field in a fluid: positive groups to the cathode and negative groups to the anode

equigravisphere the point between two bodies where the strength of respective gravity fields are of equal force

escape velocity the minimum velocity necessary for a space vehicle to escape the gravitational grip of a celestial body

expendable launch vehicle a launch vehicle designed for one flight, usually comprising several rocket stages, each jettisoned at burnout without provision for reuse

external tank the large tank containing liquid hydrogen and liquid oxygen from which the Shuttle orbiter gets propellant for its main engines

extravehicular activity (EVA) the name given by NASA to the act of space walking, involving humans outside of their space vehicle

fractional orbit bombardment system (FOBS) the designation given by the United States to a Soviet nuclear missile system designed to enter low Earth orbit for a brief period, with reentry prior to one complete circumnavigation of the Earth

free-return trajectory   a trajectory in which the spacecraft would loop around the moon and, deflected by lunar gravity, return to the vicinity of Earth without a major propulsion maneuver (used in connection with the Apollo program)

fuel cell   an electrical power source that generates energy from the combination of two fluids, usually hydrogen and oxygen, over a catalyst

gauleiter   a provincial governor in Germany under Hitler

geodetic satellite   a satellite designed to measure physical and geomagnetic properties of the Earth and used to establish an accurate grid, or geoid, for the precise location of surface features

geostationary orbit   a circular orbit, 22,300 miles above the surface of the Earth in the same plane as the Earth's equator, in which the orbital period of a satellite equals the rotation period of the Earth and the satellite appears to remain stationary with respect to a fixed point on the surface

geosynchronous orbit   an orbit in which the period of the satellite is equal to the rotation of the Earth but at an inclination different from that of the equator, the satellite describing a figure 8 as it periodically moves between northern and southern latitudes fixed by the orbital inclination at an altitude of 22,300 miles. For example, if the satellite is in a geosynchronous orbit of 12°, it will appear to loop between 12° north and 12° south.

geosynchronous transfer orbit   an elliptical trajectory in which apogee is at geosynchronous altitude (22,300 miles above the surface of the Earth), with perigee at some lower altitude to which the satellite or spacecraft has been delivered by the launch vehicle. It implies that the satellite or spacecraft will perform a maneuver at apogee to raise perigee to the same altitude, thereby achieving geosynchronous orbit. If, during this maneuver, the satellite is placed in the same plane as the Earth's equator, the satellite is then said to be in geostationary orbit.

go-around capability   the ability of an aerospace vehicle such as NASA's Shuttle to abort a landing and approach again for a second attempt using air-breathing engines

gravitational anomaly   a variation in the level of gravity across the surface of a planet caused by clumps of heavier or lighter surface material producing greater or less gravitational force than the mean of that planet

gravity gradient stabilization   a method by which a space vehicle in low Earth orbit is stabilized along its longitudinal axis, in which the minute difference in gravitational attraction between the two ends of the vehicle aligns it with the center of the Earth

hard landing   a landing in which a spacecraft survives an impact on the surface of another planet after an uncontrolled descent

heat sink   a material capable of removing thermal energy without charring or ablating and usually capable of being reused

heliosphere   the environment of the sun out to a distance of several billion miles, where interstellar radiation dominates

Hermes   a U.S. Army rocket research program started on November 15, 1944, and extended in 1947 to include ramjet missiles. Later, Hermes also included studies into satellite launchers. All work under the Hermes program came to an end in 1953.

hot launch   the ignition of a ballistic missile in its silo without first being raised to the surface. This requires provision for the escape of exhaust gases through tunnels or up the sides of the silo.

hypergolic propellants   oxidizer and fuel that ignite on contact, obviating the need for an ignition system to induce combustion

impact point   the precise location on the surface of a planet or moon where a space vehicle, rocket or missile makes contact and is destroyed

inert environment   the environment of a moon or a planet in which there is no biological activity

intravehicular activity   a phrase adopted by NASA to describe astronaut activity conducted inside a space vehicle

ion engine   a propulsion unit in which electrical energy is used to ionize a gaseous propellant and thrust is produced by an electrostatic field that accelerates the positive ions. This has the advantage of operating for very long periods, albeit at low thrust levels.

kick stage   a secondary propulsion system carried by a satellite or space vehicle to provide some final, or terminal, function, such as converting a geosynchronous transfer orbit to a geostationary orbit; alternatively, a purpose-built stage or tug to position satellites in their final orbits

kinetic kill   the destructive release of energy caused when one object (the interceptor) hits another (the target) at high speed; used in the Strategic Defense Initiative to signify the use of kinetic energy rather than a chemical or nuclear explosion to destroy an object in space or in the atmosphere

kinetic kill interceptor   a device designed to intercept a space vehicle and destroy it through impact alone

launch window   a period of time during which a satellite or space vehicle can be launched to achieve a preplanned objective

lift : drag   the ratio of lift to drag on an aerodynamic body

lifting body   a blended wing-body structure designed so that at least 50% of the lift derives from the design of the fuselage underbody functioning as wings

liquid propellant rocket motor   a rocket motor in which at least two chemicals, a fuel and an oxidizer, are stored in separate tanks and brought together in a combustion chamber, where they are ignited to produce thrust

lunar rays   bright lines that radiate from fresh craters on the surface of the moon and gradually dissipate with time

magnetic detector   an instrument or probe designed to detect and measure the intensity of magnetic fields

magnetosphere   the region of trapped radiation held by a planet's magnetic field lines; in the case of Earth, a protective sheath that screens the Earth's atmosphere from hard radiation

maneuvering thrusters   rocket motors or similar reaction devices by which a satellite or spacecraft is maneuvered in space, usually by changing its velocity and hence its trajectory

mare   (from the Latin for *sea*) the name given to dark circular basins on the moon; until the early 20th century thought to be filled with large expanses of water. Modern geology displaced this idea in favor of a theory that the "seas" were in fact lakes of lava, which was proven by the exploration of the lunar surface.

mass fraction   the fraction of the rocket's total mass that pertains to the structural mass without propellants

mass ratio   the ratio of the structural mass of a rocket to the total mass at launch. If a given rocket weighs 3 tons and carries a payload of 1 ton and a fuel load of 8 tons (total 12 tons), the mass ratio is obtained by dividing the takeoff mass by the remaining mass at burnout (4 tons), in this case 3:1.

monopropellant   a chemical that comprises a single propellant component, such as hydrazine made to decompose over a catalyst to form a gas

M-Stoff   German for methanol adopted by the rocket research teams 1933–1945

near space   generally considered to be space in the vicinity of the Earth out to geosynchronous orbit (22,300 miles), beyond which lies deep space

occultation blocking   the blockage of the line of sight, hence also of electromagnetic signals, between two objects when a third object passes between them, such as when the sun passes between the Earth and a spacecraft on the far side of the sun

orbit   the movement of a secondary object about a primary spherical body in such a manner that the curvature of the primary body causes its surface to fall away at approximately the same rate as the secondary falls toward the primary. In reality all orbits are elliptical, the degree of ellipticity dictated by the speed and direction of the secondary object.

orbit and attitude maneuver system (OAMS)   the propulsion system carried by the NASA shuttle orbiter for attitude and in-orbit maneuvers

parameter   an arbitrary constant that determines the specific form of a mathematical expression

passive communication   a communication satellite with no active components, such as metallized balloons from which radio signals are bounced between a ground transmitter and a ground receiver

payload   that component of a rocket or missile comprising the equipment or cargo

penetrometer   a needlelike device designed to pierce the surface of the moon or a planet and transmit signals from the instruments attached to it back to an orbiter or direct to Earth

peri-   the prefix, from the Greek for *around, near*, indicating the low point of an elliptical orbit around a primary mass. Most commonly applied to artificial satellites in Earth orbit (perigee), lunar orbit (perilune or periselene) or solar orbit (perihelion).

pixel   a picture element, usually the smallest binary-coded quantum of information used to constitute an image

plug-nozzle engine   *see* aerospike engine

Project Overcast   a U.S. Army intelligence operation set up on July 19, 1945, with the aim of obtaining the services of German scientists and rocket engineers for employment in the United States

Project Paperclip   the name by which Project Overcast was known from March 16, 1946

radiation detector   a device or instrument designed to locate and measure the intensity and type of radiation

radioisotope thermoelectric generator   an electrical power source for a space vehicle in which heat produced by radioactive decay is converted into electricity by thermocouples

rate gyroscope   a stabilization device which applies a mathematical operation on an input signal such as position error, angular deviation or control surface position. A rate gyro has one degree of freedom and senses movement that is translated into a mechanical command to correct an observed deviation.

reaction control system (RCS)   (for Apollo spacecraft) the attitude control thrusters used for stabilization

recoilless devices   devices such as rockets and missiles which release the reaction from an action in the form of exhaust gases. In this way they avoid the recoil action of, say, a gun.

return beam vidicon   an imaging system used on U.S. Landsat satellites whereby three cameras provide converging images in three portions of the visible and near-infrared spectrum. Images are stored on photosensitive surfaces and scanned by an internal electron beam to produce a video picture.

Reynolds number nondimensional (relative) parameter representing the ratio of the momentum forces to the viscous forces in the fluid flow

rocket bomber a rocket-propelled bomber equipped with wings for flight through the atmosphere but generally considered to fly through space on a ballistic, suborbital trajectory for much of its flight path

rollout (preflight activity) the transport of a launch vehicle to its launch pad; (Shuttle landing) the distance or time taken for the orbiter to come to a complete stop after touchdown

R-Stoff the name used by German rocket research engineers (1933–1945) for a mixture of 57% crude oxide M-xylidene and 43% tri-ethylamine (also known as Tonka)

scramjet a supersonic ramjet that obtains thrust from the ingestion of supersonic air through a duct in the nose, enabling the combustion of fuel

semihard coffin the term indicates the use of a steel casing to protect an Atlas ICBM in the horizontal position from which it could be raised to a vertical position for launch. The term "hard" indicates the degree of protection from nuclear attack. A semihard structure is one which will survive only superficial damage.

servo-actuated describes a device linked to a mechanism for converting a small mechanical force into a large mechanical force for control purposes

shaped charge the grain of a solid propellant cut so as to control the thrust and burn rate to a specified time profile

soil-mechanics surface sampler a device by which soil and small rock fragments can be obtained from the surface of another world for analysis on the landed spacecraft, or returned to Earth for examination

sol one day on Mars, which is equal to 24.6229 Earth hours

solar cell a device for converting solar energy into electrical energy, also known as a photovoltaic cell

solar power satellite the concept of a large orbiting structure carrying millions of individual solar cells for converting sunlight into electricity, which is then transmitted down to Earth as a microwave beam for reconversion to electricity on Earth

solar sailing the use of the solar wind to control the position and orientation of a spacecraft

solar wind charged particles and solar radiation which define the extent of the heliosphere

solid propellant rocket motor a rocket motor in which a combination of fuel and oxidizer is cast into a single plug of solid propellant which can be stored for long periods prior to use, although they generally produce less thrust per unit mass than liquid propellants

solid rocket boosters (SRBs) the solid propellant boosters specially designed to assist the Shuttle orbiter off the pad and toward space, together with a recovery system designed to lower each booster for splashdown and reuse

sounding rocket a single or multistage rocket fired into the upper atmosphere or near-Earth space on a ballistic trajectory, instrumented to record environmental phenomena before transmitting information to the ground or returning to Earth by parachute

space law law pertaining to agreements reached between parties concerning outer space or objects from Earth sent to outer space

Space Shuttle main engines the three engines carried by each NASA Shuttle orbiter as its primary propulsion into orbit

space tug defined by NASA during the 1970s as a propulsion unit capable of moving satellites and space vehicles from low Earth orbit to any other desired orbit up to geostationary orbit

specific impulse a measure of the thrust available as energy from the consumption of a specified amount of propellant in a given unit of time. Usually, the amount of thrust in pounds (or kilograms) generated by one pound (or one kilogram) of propellant in one second.

spectrophotometer an instrument used for measuring the photometric intensity of each color or wavelength in an optical spectrum

spin-stabilized the action of stabilizing a satellite or spacecraft by spinning it around its longitudinal axis

S-Stoff a rocket propellant comprising 90–98% nitric acid and 2–10% sulfuric acid under a name adopted by German rocket engineers working 1933–1945 (also known as Salbei)

stand-off bomb a nuclear bomb with propulsion attached, giving it a greater range than that of the bomber that carried it. The name is derived from the bomber standing-off some distance from the target to drop its bomb.

Statsionar a regional Soviet (Russian) communication satellite

steer a U.S. military communication satellite program set up in May 1958 and aimed at providing services for bombers of the Air Force Strategic Air Command over polar regions of the Earth

store-dump the means by which military surveillance satellites collect data or information while passing over a foreign country, store it on magnetic tape and transmit it to ground receivers over a designated part of the orbit, thus preventing others from listening in on what has been retrieved

stratigraphic describes the dating of geological events placed in chronological sequence according to superposition such that older events underlie more recent ones

synthetic aperture radar a radar in which a signal emits pulses continuously and at a precisely controlled frequency with all the echoes processed in such a way as to simulate an antenna as long as the flight path, thus improving resolution and enhancing definition

Tackle a U.S. military communication satellite program set up in May 1958 and aimed at providing advanced services for military operations over polar regions of the Earth

telemetry information or measurements transmitted by radio signal from a space vehicle to the surface of the Earth

tether a coupling or restraining line attaching a space-walking astronaut on EVA (extravehicular activity) to his space vehicle

thematic mapper (as applied to U.S. Landsat remote sensing satellites) a seven-band multispectral scanner measuring energy reflected from the surface of the Earth

thermionic conversion the conversion of heat into electrical energy, usually by liberated electrons leaving a heated cathode

thermocouple a loop comprising two different metals which, when heated by a thermal source, will cause a current to flow due to the unbalanced potentials

throw weight the weight of the warhead, or explosive charge, carried by a rocket or missile

thrust vector control the directional control of a rocket's thrust vector by controlling the pointing direction of the exhaust nozzle or deflecting the exhausted gases in some way

transient signals signals that are sent from a transmitter at sporadic and unpredictable intervals

traveling wave tube a device for the amplification of a radio signal utilizing an electron beam to greatly increase the energy of the signal

trimarese the configuration whereby two winged boosters sandwich a winged aerospace vehicle so that all the rocket motors are at the same level for simultaneous firing at liftoff

trim burn a small propulsive maneuver, usually involving an engine burn, to adjust or fine-tune the precise parameters of a trajectory or orbit

umbilical a line carrying electrical power or consumables

venturi a tube with a constriction so as to alter the flow of a gas passing through it

vernier motor a small propulsion system designed to control relatively small-scale velocity changes

warhead the part of a military missile carrying an explosive device, invariably forming the nose cone at the top, or front, of the missile

Z-Stoff the name adopted by German rocket engineers (1933–1945) for an aqueous solution of sodium or calcium permanganate

ABBREVIATIONS

ABM antiballistic missile

ALS advanced launch system

ALSEP Apollo lunar surface experiments package

AMU astronaut maneuvering unit

ASAT antisatellite

ASRM advanced solid rocket motor

ASSET aerothermodynamic/elastic structural systems environmental tests

ATS applications technology satellite

BEAR Beacon Experiments and Auroral Research

BMDO Ballistic Missile Defense Organization

BMEWS ballistic missile early warning system

BSE broadcasting satellite, experimental

CC&S central computer and sequencer

CDR commander (Space Shuttle missions and Apollo program)

CELV complementary expandable launch vehicle

CETEX contamination by extraterrestrial exploration

CLAW clustered atomic warhead

CM command module (Apollo program)

CMP command module pilot (Apollo program)

CNES Centre National d'Etudes Spatiales

COSPAR Committee on Space Research (of the ICSU)

CRAF Comet Rendezvous/Asteroid Flyby

CSE communications satellite, experimental

CSM command and service modules (Apollo program)

CTS communications technology satellite

DPS descent propulsion system

DSN Deep Space Network

EASEP early Apollo scientific experiments package

ECS European Communications Satellite

ELDO European Launcher Development Organization

elint electronic intelligence

ERIS exo-atmospheric reentry interceptor subsystem

ERTS Earth resources technology satellite

ESA European Space Agency

ESRO European Space Research Organization

ESSA Environmental Sciences Services Administration

ETM enhanced thematic mapper

EVA extravehicular activity

FCC flight control center

FDL flight dynamics laboratory

FOBS fractional orbit bombardment system

GALCIT Guggenheim Aeronautical Laboratory at the California Institute of Technology

GARP Global Atmospheric Research Project

GEOS geodynamics experimental observation satellite

GOES geostationary operational environmental satellite

GPS global positioning system

GRU Glavnoye Razvedyvatelnoe Upravleniye (Soviet military intelligence)

HATV high-altitude test vehicle

HCMM heat capacity mapping mission

HEAO   high-energy astronomy observatory

HEOS   highly eccentric observation satellite

HHMU   hand held maneuvering unit

HOTOL   horizontal takeoff and landing

IBSS   infrared background signature survey

ICBM   intercontinental ballistic missile

ICSU   International Council of Scientific Unions

ILRV   integrated launch and reentry vehicle

INMARSAT   International Maritime Satellite

INTASAT   Instituto Nacional de Técnica Aerospacial Satellite

INTELSAT   International Telecommunications Satellite (consortium)

IRBM   intermediate-range ballistic missile

IRS   infrared radiometer spectrometer

ISEE   International Sun-Earth Explorer

Isp   specific impulse

ITOS   improved Tiros operation system

IUE   International Ultraviolet Explorer

IUS   inertial upper stage (originally: interim upper stage)

KSC   Kennedy Space Center

LAGEOS   laser geodynamic satellite

LEM   lunar excursion module (Apollo program)

LK   lunar Korabl

LM   lunar module

LMP   lunar module pilot (Apollo program)

LOI   lunar orbit insertion

LOX   liquid oxygen

LRV   lunar roving vehicle

LTD   long-tank Delta

LTTAD   long-tank thrust-augmented Delta

LTTAT   long-tank thrust-augmented Thor

LTV   Ling-Temco-Vought

MARV   maneuverable reentry vehicle

MIDAS   missile defense alarm system

MIRV   multiple independently targeted reentry vehicle

MMH   monomethylhydrazine

MOS   Marine Observation Satellite

MRSR   Mars Rover–Sample Return

MRV   multiple reentry vehicle

MS   mission specialist (Space Shuttle missions)

MSBS   Mer-Sol Balistique Stratégique

MSC   Manned Spacecraft Center

MSFN   manned spaceflight network

N$_2$O$_4$ nitrogen tetroxide

NACA   National Advisory Committee for Aeronautics

NASA   National Aeronautics and Space Administration

NASDA   National Space Development Agency (Japan)

NERVA   nuclear engine for rocket vehicle application

NOAA   National Oceanic and Atmospheric Administration

OAO   orbiting astronomy observatory

OGO   orbiting geophysical observatory

OTS   operations technology satellite

PARD   Pilots Aircraft Research Division (NACA)

PILOT   piloted low-speed tests

PLT   pilot (Space Shuttle missions)

PRIME   precision recovery including maneuvering reentry

PS   payload specialist (Space Shuttle missions)

RAE   Royal Aircraft Establishment

RAF   Royal Air Force

RAND   Research and Development Corporation

RBV   return beam vidicon

RCA   Radio Corporation of America

RFNA   red fuming nitric acid

RIFT   reactor in-flight test

RP-1   rocket propellant-1

RTCC   real-time computer complex

RTG   radioisotope thermoelectric generator

RV   reentry vehicle

SALT   Strategic Arms Limitation Talks

SAMOS   space and missile observation system

SAMSO   Space and Missile Systems Organization (United States)

SBI   space-based interceptor

SDI   Strategic Defense Initiative

SDV   Shuttle-derived vehicle

SEPS   solar electric propulsion system

SERT   space electric rocket test

SIM   scientific instrument module

SLBM   submarine-launched ballistic missile

SM   service module (Apollo program)

SMEAT   Skylab medical experiments altitude test

SMS   synchronous meteorological satellite

SNECMA   Société Nationale d'Etude et de Construction de Moteurs d'Aviation

SPS   service propulsion system

SRB   solid rocket booster

SRMU   solid rocket motor upgrade

SSBN   submarines strategic ballistic nuclear

SSBS   Sol-Sol Balistique Stratégique

SSME   Space Shuttle main engine

SSUS   solid spinning upper stage

STAR   self-test and repair

START   spacecraft technology and advanced reentry tests

START   Strategic Arms Reduction Talks

STEM   stay-time extension module

TEI   trans-Earth injection

TLI   translunar injection

TOS   Tiros operational system

TRW   Thompson Ramo Woolridge

TVC   thrust vector control

TWT   traveling wave tube

UDMH   unsymmetrical dimethyl hydrazine

USA   United States Army

USAAF   United States Army Air Force

USAF   United States Air Force

USGS   United States Geological Survey

USN   United States Navy

UTC   United Technologies Corporation

UVS   ultraviolet spectrometer

NAME INDEX

How To Use: This index is to be used in conjunction with the chronologically ordered text. The year, month and day in every index entry serves as the page locator. For example: **Armstrong, Neil A. 1962 Mar 14** takes you directly to the *year* (1962), the *month* (March) and the *day* (14) for the citation on Neil Armstong. Asterisks * following the date indicate more than one text citation.

A

Abrahamson, James A.: MOL selection 1967 Jun 30*

Acton, Loren: STS 51-F 1985 Jul 29

Adams, Michael: NASA selection 1965 Nov 12*

Adamson, James C.: NASA selection 1984 May; STS-28 1989 Aug 8; STS-43 1991 Aug 2

Afanasyev, Viktor: Soyuz-TM11 1990 Oct 30/Dec 2*/Dec 10, 1991 Jan 8/Jan 23/Mar 19/Apr 25/May 18/May 26; Soyuz-TM18 1993 Oct 11

Akers, Thomas D.: NASA selection 1987 Aug*; STS-41 1990 Oct 6; STS-49 1992 May 7; STS-61 1993 Dec 2

Akiyama, ToyohIro: Soyuz-TM11 1990 Dec 2*; Soyuz-TM10 1990 Dec 10

Aksyonov, Vladimir Viktorovich; Soyuz 22 1976 Sep 15; Soyuz-T2 1980 Jun 5

Aldridge, Edward: first shuttle flight from Vandenberg 1985 Sep 11*

Aldrin Jr., Edwin E.: NASA selection 1963 Oct 18; GT-10 1966 Jan 25*; GT-9 1966 Mar 19*; GT-12 1966 Jun 17*/Nov 11; AS-505 1967 Nov 20; AS-504 1968 Jul 23; Apollo 8 1968 Aug 19; Apollo 11 1969 Jan 9/Jul 16*

Alexandrov, Alexander (Bulgarian): Soyuz-TM5 1988 Jun 7

Alexandrov, Alexander Pavlovich: Soviet selection 1978 Mar; Soyuz-T9 1983 Jun 27/Aug 14/Sep 26/Oct 20; 1984 Feb 8; Soyuz-TM3 1987 Jul 22/Aug 3/Sep 23/Nov 20/Dec 21/Dec 29

Allen, Andrew M.: NASA selection 1987 Aug*; STS-46 1992 Jul 31

Allen, Joseph P.: NASA selection 1967 Aug 4*; STS-5 1982 Nov 11; STS 51-A 1984 Nov 8

Al-Saud, Bin Salman: STS 51-B 1985 Jun 17

Anders, William A.: NASA selection 1963 Oct 18; GT-11 1966 Mar 19*; third Apollo 1966 Dec 22; AS-505 1967 Nov 20, 1968 Jul 23; Apollo 8 1968 Aug 19/Dec 21; Apollo 11 1969 Jan 9

Andre-Deshays, Claudie: Soyuz-TM23 1992 Nov 30

Anikeyev, Ivan Nikoleyevich: reported 1960 Mar 14

Apt, Jerome: NASA selection 1985 Jun; STS-37 1991 Apr 5; STS-47 1992 Sep 12

Armstrong, Neil A.: NASA selection 1962 Mar 14/Sep 17; assignments 1963 Jan 26; GT-5 1965 Feb 8; GT-8 1965 Sep 20; 1966 Mar 16; GT-11 1966 Mar 19*; AS-505 1967 Nov 20; Apollo 8 1968 Aug 19; Apollo 11 1969 Jan 9/Jul 16*

Artsebarsky, Anatoly: Soyuz-TM12 1991 May 18/May 30/Jun 25/Jul 15/Aug 20/Oct 2/Oct 10

Artyukhin, Yuri P.: Soviet selection 1963 Jan 11*; Almaz 1966 Sep 2; lunar landing 1967*; Soyuz 14 1974 Jul 3

Atkov, Oleg Yuryevich: Soyuz-T10 1984 Feb 8/Feb 21; Soyuz-T11 1984 Apr 3/May 18/May 28/Jul 17/Aug 14/Oct 2

Atwell, Alfred L.: NASA selection 1962 Oct 22*

Aubakirov, Toktar: Soyuz-TM13 1991 Oct 2; Soyuz-TM12 1991 Oct 10

Avdeyev, Sergei: Soyuz-TM15 1992 Jul 27/Aug 10/Sep 3/Oct 27, 1993 Jan 24/Feb 1

B

Bagian, James P.: NASA selection 1980 Jan; STS-29 1989 Mar 13; STS-40 1991 Jun 5

Baker, Ellen S.: NASA selection 1984 May; STS-34 1989 Oct 18; STS-50 1992 Jun 25

Baker, Michael A.: NASA selection 1985 Jun; STS-43 1991 Aug 2; STS-52 1992 Oct 22

Balandin, Alexander: Soyuz-TM9 1990 Feb 11/Feb 19/Feb 28*/May 5/May 31/Jul 17/Jul 26/Aug 1/Aug 9

Barry, Daniel T.: NASA selection 1992 Mar

Bartoe, John-David F.: STS 51-F 1985 Jul 29

Bassett II, Charles A.: NASA selection 1962 Oct 22*, 1963 Oct 18; GT-9 1965 Nov 8; killed 1966 Feb 28

Baudry, Patrick: association of European astronauts 1984 Oct; STS 51-E 1985 Mar 1; STS 51-G 1985 Mar 1/Jun 17

Bean, Alan L.: NASA selection 1963 Oct 18; GT-10 1966 Mar 19*; AS-504 1967 Nov 20; Apollo 12 1969 Apr 10/Nov 14; SL-3 1972 Jan 19; 1973 Jul 28; ASTP 1973 Jan 30

Belousov (Soviet cosmanaut): Almaz 1966 Sep 2

Belyayev, Pavel Ivanovich: reported 1960 Mar 14; Voskhod 2 1964 Mar*; 1965 Mar 18; Almaz 1966 Sep 2; lunar landing 1967*

Benefield, Tommie D. 1962 Oct 22*

Beregovoi, Georgi Timofeyevich: Voskhod 3 1965 Apr; Almaz 1966 Sep 2; lunar fly-by 1966 Dec 7*; Soyuz 2 1968 Oct 25

Berezovoy, Anatoly Nikolayevich: Soyuz 25 selection 1977 Feb; Soyuz-T5 1982 May 13/May 23/Jun 24/Jul 10/Jul 30/Aug 19/Sep 18/Oct 31; Soyuz-T7 1982 Dec 10

Blaha, John E.: NASA selection 1980 Jan; STS-29 1989 Mar 13; STS-33 1989 Nov 23; STS-43 1991 Aug 2; STS-58 1993 Oct 18

Bluford Jr., Guion S.: NASA selection 1978 Jan; STS-8 1983 Aug 30; STS 61-A 1985 Oct 30; STS-39 1991 Apr 28; STS-53 1992 Dec 2

Bobko, Karol J.: NASA selection 1966 Jun 17*; 1969 Aug 14; SMEAT test 1972 Jul 26*; ASTP 1973 Jan 30; STS-6 1983 Apr 4; STS 51-E 1985 Mar 1; STS 51-D 1985 Mar 1/Apr 12; STS 51-J 1985 Oct 3*

Bock Jr., Charles C.: NASA selection 1962 Apr 20*

Bolden, Charles F.: NASA selection 1980 Jan; STS 61-C 1986 Jan 12; STS-31 1990 Apr 24; STS-45 1992 Mar 24

Bondar, Roberta: STS-42 1992 Jan 20

Bondarenko, Valentin Vasilyevich: reported 1960 Mar 14; killed 1961 Mar 23*

Borman, Frank: NASA selection 1962 Sep 17*; assignments 1963 Jan 26; GT-4 1964 Jul 10; GT-7 1965 Jul 1/Dec 4; third Apollo 1966 Sep 29/Dec 22; refused entry to USSR 1967 Apr 23; AS-505 1967 Nov 20; 1968 Jul 23; Apollo 8 1968 Aug 19/Dec 21

Bowersox, Kenneth D.: NASA selection 1987 Aug*; STS-50 1992 Jun 25; STS-61 1993 Dec 2

Boyle, Tony: NASA selection 1984 Mar 15

Brady, Charles E.: NASA selection 1992 Mar

Brand, Vance D.: NASA selection 1966 April 4; Apollo 15 1970 Mar 26; SL-3/SL-4 1972 Jan 19; ASTP 1973 Jan 30; 1975 Jul 15; shuttle flight selection 1978 Mar 17; STS-5 1982 Nov 11; STS 41-B 1984 Feb 3; shuttle weaknesses 1985 Nov 5; STS-35 1990 Dec 2

Brandenstein, Daniel C.: NASA selection 1978 Jan; STS-8 1983 Aug 30; STS 51-G 1985 Jun 17; STS-32 1990 Jan 9; STS-49 1992 May 7

Bridges Jr., Roy D.: NASA selection 1980 Jan; STS 51-F 1985 Jul 29

Brown, Mark N.: NASA selection 1984 May; STS-28 1989 Aug 8; STS-48 1991 Sep 12

Brown Jr., Curtis L.: NASA selection 1987 Aug*; STS-47 1992 Sep 12

Buchli, James F.: NASA selection 1978 Jan; STS 51-C 1985 Jan 24; STS 61-A 1985 Oct 30; STS-29 1989 Mar 13; STS-48 1991 Sep 12

Bull, John S.: NASA selection 1966 Apr 4; AS-505 1967 Nov 20

Bursch, Daniel W.: NASA selection 1990 Jan; STS-51 1993 Sep 12

Bykovsky, Valery Fyodorovich: reported 1960 Mar 14; candidate for flight 1961 Mar 24; Vostok 5 1963 Jun 14; Soyuz 1966 Sep 2; circumlunar 1966 Sep 2; 1968 Nov; lunar fly-by 1966 Dec 7*; lunar landing 1967*; Soyuz 2 1967 Apr 23; Soyuz 22 1976 Sep 15; Soyuz 31 1978 Aug 26

C

Cabana, Robert D.: NASA selection 1985 Jun; STS-41 1990 Oct 6; STS-53 1992 Dec 2

Cameron, Kenneth D.: NASA selection 1984 May; STS-37 1991 Apr 5; STS-56 1993 Apr 8

Carpenter, Lt. Malcolm Scott: Soviet selection 1959 Apr 9; MA-6 1961 Nov 29*; MA-7 1962 Mar 1/May 24; assignments 1963 Jan 26

Carr, Gerald: NASA selection 1966 Apr 4; AS-505 1967 Nov 20; SL-4 1972 Jan 19; 1973 Nov 16

Carter Jr., Manley Lanier: NASA selection 1984 May; STS-33 1989 Nov 23

Casper, John H.: NASA selection 1984 May; STS-36 1990 Feb 28; STS-54 1993 Jan 13

Cenker, Robert: STS 61-C 1986 Jan 12

Cernan, Eugene A.: NASA selection 1963 Oct 18; GT-6 1965 Apr 5; Jul 22; Apollo 17 1965 Jun 29*; 1971 Aug 13; 1972 Dec 7;

GT-9 1965 Nov 8; 1966 Feb 28/May 17; GT-9A 1966 Jun 1/Jun 3; GT-12 1966 Jun 17*; Apollo 2 1966 Dec 22; Apollo 7 1967 May 9*; Apollo 10 1969 Mar 24/May 18; Apollo 14 1969 Aug 6

Chaffee, Roger B.: NASA selection 1963 Oct 18; AS-204 204 1965 Feb 25; 1966 Mar 19*/Dec 22; Apollo fire 1967 Jan 27*

Chang-Diaz, Franklin R.: NASA selection 1980 Jan; STS 61-C 1986 Jan 12; STS-34 1989 Oct 18; STS-46 1992 Jul 31

Chapman, Philip K.: NASA selection 1967 Aug 4*

Chekh, A.: TKS selection 1977 Feb

Cheli, Maurizio: NASA selection 1992 Mar

Chelomei, S.: TKS selection 1977 Feb

Chiao, Leroy: NASA selection 1990 Jan

Chilton, Kevin P.: NASA selection 1987 Aug*; STS-49 1992 May 7

Chretien, Jean-Loup: Soyuz-T6 1982 Jun 24; 1986 Mar 7; association of European astronauts 1984 Oct; Soyuz-TM7 1988 Nov 26/Dec 9; Soyuz-TM6 1988 Dec 21

Chuchin, S.: TKS selection 1977 Feb

Cleave, Mary L.: NASA selection 1980 Jan; STS 61-B 1985 Nov 27; STS-30 1989 May 4

Clervoy, Jean-Francois: NASA selection 1992 Mar

Clifford, Michael R. U.: NASA selection 1990 Jan; STS-53 1992 Dec 2

Coats, Michael L.: NASA selection 1978 Jan; STS 41-D 1984 Jun 25/Jun 26/Aug 30; STS-29 1989 Mar 13; STS-39 1991 Apr 28

Cockrell, Kenneth D.: NASA selection 1990 Jan; STS-56 1993 Apr 8

Coleman, Catherine G.: NASA selection 1992 Mar

Collins, Ellen M.: NASA selection 1990 Jan

Collins, Michael: NASA selection 1962 Oct 22*; 1963 Oct 18; GT-7 1965 Jul 1; GT-10 1966 Jan 25*/Jul 18/Jul 20*; Apollo 3 1966 Sep 29/Dec 22; AS-505 1967 Nov 20; 1968 Jul 23; AS-504 1968 Jul 23; Apollo 11 1969 Jan 9/Jul 16*

Conrad Jr., Charles: NASA selection 1962 Sep 17*; assignments 1963 Jan 26; GT-5 1965 Feb 8/Aug 21; GT-8 1965 Sep 20; GT-10 1966 Jan 25*; GT-11 1966 Mar 19*/Sep 12; Apollo 3 1966 Dec 22; AS-504 1967 Nov 20; Apollo 12 1969 Apr 10/Nov 14; SL-2 1972 Jan 19; 1973 May 25

Cooper, Leroy Gordon: NASA selection 1959 Apr 9; MA-8 1962 Jun 27; assignments 1963 Jan 26; MA-9 1963 May 15/Jun 6*; GT-5 1965 Feb 8/Aug 21; GT-12 1966 Jun 17*; refused entry to USSR 1967 Apr 23; Apollo 10 1969 Mar 24

Covey, Richard O.: NASA selection 1978 Jan; STS 51-I 1985 Aug 27; STS-26 1988 Sep 29; STS-38 1990 Nov 15; STS-61 1993 Dec 2

Creighton, John O.: NASA selection 1978 Jan; STS 51-G 1985 Jun 17; STS-36 1990 Feb 28; STS-48 1991 Sep 12

Crews, Albert H.: NASA selection 1962 Apr 20*/Sep 19; 1965 Nov 12*; 1969 Aug 14

SUBJECT INDEX

How To Use: This index is to be used in conjunction with the chronologically ordered text. The year, month and day in every index entry serves as the page locator. For example: **Apollo 13 1969 Feb 3** takes you directly to the *year* (1969), the *month* (February) and the *day* (3) for the citation on Apollo 13. Asterisks * following the date indicate more than one text citation.